Chemistry

experiment and theory

Bernice G. Segal

BARNARD COLLEGE
COLUMBIA UNIVERSITY

Chemistry
experiment and theory

John Wiley & Sons
NEW YORK · CHICHESTER · BRISBANE · TORONTO · SINGAPORE

Production supervised by Ruth Greif
Book and cover designed by Madelyn W. Lesure
Photo research by Linda Gutierrez
Illustrations by John Balbalis with the assistance of
    the Wiley Illustration Department
Manuscript edited by Jeanette Stiefel under the
    supervision of Deborah Herbert

Cover photograph by **Steve Feingold.**

Urea crystals $(NH_2)_2C{=}O$,
photographed by cross-polarized light,
at $10\times$ magnification, through a
Nikon microscope.

Library of Congress Cataloging in Publication Data:

Segal, Bernice G.
   Chemistry : experiment and theory.

   Includes index.
   1. Chemistry. I. Title.
QD31.2.S438 1985   540   84-25786
ISBN 0-471-80811-3

To the students and
faculty of Barnard College

preface

This book is intended for use as a text in the first-year general chemistry course that is offered in colleges and universities for students who plan to study other sciences in addition to chemistry, and who anticipate taking a year of organic chemistry following general chemistry. Although not all of these students will major in a science, they are, generally, planning to pursue a career that will require a knowledge of the fundamentals of chemistry.

A significant portion of such students find chemistry difficult. The majority of students who *do* have difficulty with chemistry report that they "understand the concepts but just can't do the problems." In writing this text, therefore, one of my principal goals has been to provide as helpful an explication as possible of the reasoning processes used in solving problems in chemistry. Both straightforward and more complex problems are worked out in detail. My purpose has been to make clear to students how you attack problems you do not know how to solve when you first read the question. Both in the text and in the Solutions Manual, particular attention has been given to explaining how to solve the more challenging problems that are encountered in what is frequently called the "science majors" general chemistry course. I have tried to help current students avoid the mistakes of the past by calling attention to common errors and misconceptions that I have seen often on students' examination papers.

Frequently students come to my office to ask for help with problem-solving. I usually begin by asking them to define the term or quantity they are asked to calculate in the problem. Very often they reply by giving me part of the definition, or by saying something about the topic. They are, however, unable to state a complete and exact definition. An extensive Glossary is provided at the end of this book to help students learn definitions. All new terms that appear in the text, and important terms that have been previously defined but have not been used for a while, are printed in color. A definition of any term that appears in color will be found in the Glossary. Students should refer to the Glossary frequently. It has been written to be of help when they are having trouble solving a problem.

Many capable students arrive at college without having learned all of the basic math skills necessary for successful problem-solving in chemistry. Appendix B covers those math techniques that are frequently used in chemistry, and contains both worked-out examples and exercises for the student to try. Appendix B is designed so that it should not be necessary to use a supplementary book on math techniques with this text. Material on significant figures, rounding, dimensional analysis, and computational techniques that are not routinely covered in most high schools is included in the Introduction.

At the end of each chapter there are three types of questions about the material covered in that chapter. The Exercises are keyed to specific sections and are designed to be straightforward. Working on the Exercises will enable students to find out if they have understood the basic concepts of a particular section. In addition, multiple choice questions are included at the end of every chapter. These vary in difficulty and are designed to cover the entire chapter, not just specific sections. In many universities today all (or a significant portion of) the examinations in general chemistry consist of multiple choice questions. The multiple choice questions at the end of each chapter are similar to the questions students may encounter on examinations and on national standardized tests such as the MCAT, DAT, and GRE. There are also questions at the end of each chapter labeled *Problems.* These are lengthier than the Exercises, and are integrative, that is, they bring together concepts from different parts of the chapter, and also from previous chapters where appropriate. Practice in drawing and using graphs is included in the Problems. The Problems are designed to be more challenging than the Exercises.

The broad topic of chemical equilibrium is introduced early in this book so that it can be covered in the first term of the one-year course. There are several reasons for this. Most of the experiments students perform in the laboratory portion of a typical general chemistry course involve an understanding of the nature of chemical equilibrium. If these experiments are done at the same time that students are learning the theory, the principles are reinforced and become more meaningful. Furthermore, the study of ionic equilibria provides an excellent way to integrate descriptive material, problem-solving techniques, and theory. The solution of problems involving equilibrium constants requires a thorough grasp of stoichiometry and reinforces the concepts covered early in almost every general chemistry course. In addition, it is easier to motivate the study of thermodynamics after an exposure to equilibrium. The concepts of thermodynamics are further removed from the student's experience and from observations made in the laboratory than are the principles of dynamic equilibrium. When we ask "What criteria should we use to predict whether a given reaction will proceed spontaneously?", and tie this question to the determination of equilibrium constants, the reasons for studying thermodynamics become clearer to the student.

In many general chemistry courses equilibrium is not covered until the second semester. There is no reason why an instructor using this text could not, following a discussion of stoichiometry, go directly to the material on atomic theory and chemical bonding and then return to the chapters on chemical equilibrium later.

I strongly believe that a student should complete a year of general chemistry not only having mastered the fundamental principles of chemistry but also having learned a great deal of descriptive chemistry. For that reason, descriptive chemistry is not contained in separate chapters at the end of the text, but is dispersed throughout, exposing the student to both theory and descriptive chemistry simultaneously. The relationship between the descriptive chemistry of the elements and the Aufbau principle is fascinating, and both topics are covered in Chapter 13. The descriptive chemistry necessary to be able to *predict* the products of many oxidation–reduction reactions is covered in Chapter 15, which also discusses the principles of oxidation and reduction and the techniques for balancing oxidation–reduction equations.

The chapter on organic chemistry, Chapter 23, deserves special mention. It is coauthored by Leroy G. Wade, Jr., of the Department of Chemistry of Colorado State University. Professor Wade has taught the large sophomore year course in organic chemistry for many years. He chose the topics to be covered in Chapter 23. Both of us believe that in the short time that can be allotted to organic chemistry in the first-year course, it is most helpful to emphasize the applications of those concepts already covered. Consequently, Chapter 23 focuses on structure and bonding.

Supplemental aids to accompany this book include a Study Guide and a Solutions Manual. The Study Guide was written by Peter S. Shenkin of Columbia University. Each chapter consists of three types of material. First, there is a review of the topics covered in the corresponding chapter of the text. Secondly, there are additional problems and review questions that are keyed to the sections within the chapter. Finally, there are amplifications of certain of the more difficult topics covered in the text. A different emphasis is offered on particular subjects, or more details are provided. Thus, although most of the Study Guide is designed to be of help to the student who is having difficulty, there is additional material to interest and challenge those students who have already mastered the text.

The Solutions Manual contains worked-out solutions to every exercise, multiple choice question, and problem in the text. If there are two ways to do a problem, they are both discussed. Comments about common student errors are included, and for the multiple choice questions the reasons for not selecting the incorrect options are provided.

In addition, figures and diagrams from the book are available on acetate transparencies for instructors who use an overhead projector to accompany their lectures.

Throughout the book and in the Solutions Manual my goal has been to be as helpful as possible to motivated, capable students who plan to pursue a career in any of the broad spectrum of scientific disciplines, and who need a strong foundation in the fundamentals of chemistry. I urge students who find the study of chemistry difficult to persevere. The successful meeting of a challenge is one of the great satisfactions of life. I hope the students who use this book will find that learning chemistry is not only thought-provoking and stimulating, but also rewarding, and above all, exciting.

New York
1984

Bernice G. Segal

acknowledgments

Many people reviewed and criticized the early drafts of the manuscript. I owe a great deal to them; I learned a lot from their comments and suggestions. I especially thank the following:

Edwin H. Abbott, Montana State University
Mario E. Baur, University of California at Los Angeles
John DeKorte, Northern Arizona State University
David Harris, University of California at Santa Barbara
Forrest C. Hentz, Jr., North Carolina State University
Nicholas K. Kildahl, Worcester Polytechnic Institute
Leslie Lessinger, Barnard College, Columbia University
Tobin J. Marks, Northwestern University
Louis H. Pignolet, University of Minnesota
Richard M. Stratt, Brown University
Eugene S. Stevens, S.U.N.Y. at Binghamton
Leroy G. Wade, Jr., Colorado State University
Lewis W. Walker, Goucher College
Albert Zabady, Montclair State College
Steven S. Zumdahl, University of Illinois at Urbana

I am indebted to the expert assistance of the following people at John Wiley & Sons: Dennis Sawicki, chemistry editor; Jeanette Stiefel, copy editor; John Balbalis, artist; Ruth Greif, production manager; Deborah Herbert, editing manager; Madelyn Lesure, designer, and Linda Guitierrez, photo researcher. To them, and the many others whose names I do not know, my deepest appreciation.

It means a great deal to me to be able to say "thank-you" here to Professor Leonard Nash of Harvard University, who taught me general chemistry many years ago. His zest and enthusiasm for chemistry, his entertaining and challenging lectures, and above all his ability to make a student reason rather than memorize, inspired me when I was a freshman, and have stayed with me through all the years since then.

I cannot close without a special note of thanks to my husband, Norman, and my children, Elizabeth and Daniel, who encouraged and cheered me on during the several years it has taken to write this book. I am particularly indebted to my son, Daniel, whose urgent importunings (I am tempted to say "nagging") were a significant part of my incentive to undertake this project. He selected and tested the computer hardware and software I used for word-processing and printing the manuscript. He wrote two sophisticated computer programs, specifically tailored to the needs of this project: one for compiling the glossary and one for writing the index, which saved me much time and effort. Writing and debugging the index program cost him several sleepless nights, and I want him to know how much I appreciate all the work he put in. My entire family has been a source of support, and I am very glad to be able to express my thanks to them in this way.

B.G.S.

contents (chapter titles)

detailed contents

introduction

Each year, I ask students to fill out an information sheet as they begin their course in General Chemistry. The response most frequently given to the question "What is your reason for taking this course?" is "Because it's required," followed by: for medical school, for biology, for geology, for dental school, for nursing, for engineering, for optometry, for art restoration, and so on. There certainly can be no doubt that a great many people who aren't chemists have to be familiar with the fundamentals of chemistry in order to do well in their chosen professions.

Why is a knowledge of chemistry useful for such a variety of fields? The dictionary defines chemistry as "the science of the composition, structure, properties, and reactions of matter, especially of atomic and molecular systems." Every material thing is a chemical or a mixture of chemicals: the foods we eat, the clothes we wear, the materials we use to build houses and bridges, the medicines we take, and the air we breathe. Anyone who needs to understand how various substances interact, or how they are changed when conditions change, will have to be familiar with some of the fundamental principles of chemistry.

The things that chemists actually do to investigate the properties and reactions of matter vary widely, and are constantly undergoing change as new methods are developed, new equipment is invented, and new techniques devised. The introduction of the high-speed computer, for example, made it possible to investigate a number of problems that could not previously be tackled. In the past twenty-five years, with the use of the computer, we have been able to elucidate the structure and function of many proteins and other macromolecules of biological importance. The substances chemists study vary widely, and are constantly changing. Because of this it has humorously been suggested that "chemistry is what chemists do."

Chemistry has traditionally been subdivided into four areas: organic, inorganic, analytical, and physical chemistry. Organic chemistry is the study of the compounds of a single element, carbon; inorganic chemistry is the study of compounds of all the other elements. These divisions have blurred somewhat in recent years and there are subtopics, such as organometallic chemistry, being investigated by both organic and inorganic chemists.

Analytical chemistry is the study of the methods used to determine the identity of the components of a mixture or a compound (qualitative analysis), and the relative amounts of each component (quantitative analysis). Much of the equipment developed by analytical chemists in past years is now routinely used by organic and inorganic chemists, as well as by physical chemists.

Physical chemistry is the study of the properties of matter and the development of theories that explain the observed properties. Physical chemists investigate both organic and inorganic substances, and chemists in all areas make use of the theories and techniques developed by physical chemists. Physical organic chemistry, for example, is one of the major subdivisions of organic chemistry.

Not only have the lines between the divisions within chemistry become less sharp during the past thirty years, but the lines dividing other scientific disciplines from chemistry have also blurred. Biochemistry and geochemistry are substantial fields, and new cross-disciplines such as bioinorganic chemistry, appear regularly.

Another way of differentiating the activities of chemists is to distinguish between experimental and theoretical chemists. The interplay between theorists and experimentalists has been extremely productive. When a theory is proposed to explain some observations that have already been made, it often stimulates new investigations. These investigations may or may not produce results predicted by the theory; if they do not, the theory has to be discarded and a new proposal put forth. Sometimes one person is both a theorist and an experimentalist, but more often chemists specialize in one or the other of these areas. No experimental chemist, however, can afford to be ignorant of the theories proposed to explain the experimental observations, and no theoretical chemist can afford to be ignorant of experimental techniques.

Some Fundamental Definitions

You will find, in the course of studying chemistry, that a great many new terms will be defined. The importance of learning the definition of each new term you encounter cannot be overemphasized. If you do not know the precise definition of a quantity whose value you are asked to determine, you will have difficulty with the reasoning required to solve the problem. To help you learn every definition perfectly, new terms are printed in color. Definitions of all terms printed in color are collected in a glossary at the end of the text.

We begin with definitions of terms you undoubtedly already are familiar with; terms so fundamental that it is almost impossible to grow up today without having heard them: element, compound, and mixture.

An **element** is a substance that cannot be separated into two or more substances different from itself by ordinary chemical means. There was a time when the phrase "by ordinary chemical means" would not have been included in this definition. Since 1938 scientists have learned how to break some of the heavier elements apart into lighter ones (this is the **fission** process used in the atom bomb). Research is also being carried out on the **fusion** of some of the lighter elements to form heavier ones. Fission and fusion are processes that are excluded by the phrase "ordinary chemical means."

There are 108 known elements at the present time. Each element is denoted by a one or two letter symbol. Some of the elements, such as silver (Ag), gold (Au), lead (Pb), and iron (Fe), have been known for a very long time. Others, such as astatine (At), berkelium (Bk), curium (Cm), and fermium (Fm), have been made only during the past fifty years. In many cases, you can tell by a glance at the symbols which elements have been known for hundreds of years: Their symbols are derived from the Latin names for the elements, and are unrelated to the English names. The Latin name for silver was *argentum,* for gold, *aurum,* for lead, *plumbum* (from which the

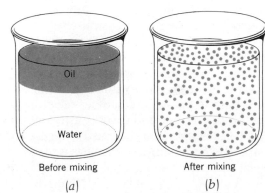

Before mixing   **After mixing**

(a)   *(b)*

Fig. I.1. Oil and water — a heterogeneous mixture. *(a)* Before mixing. *(b)* After mixing.

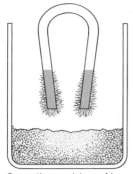

Separating a mixture of iron filings and sulfur by using a magnet to remove the iron.

Fig. I.2. A mixture of iron filings and powdered sulfur is a heterogeneous mixture. Pure iron is magnetic and insoluble in liquid carbon disulfide. Pure sulfur is nonmagnetic and soluble in carbon disulfide. In the mixture, both iron and sulfur retain their characteristic properties so that the iron filings can be removed from the mixture by using a magnet. Similarly, the sulfur can be separated from the iron by adding carbon disulfide, filtering, and then allowing the carbon disulfide to evaporate.

word plumber is derived), and for iron, *ferrum.* The names of the elements and their symbols are listed on the inside front cover of this book.

A **compound** contains two or more elements in a definite proportion by weight. There may be many different methods of preparing a compound, but the composition of a pure compound is independent of the method of preparation. A pure chemical compound always contains the same elements, united in the same proportions by weight, and with the same internal arrangement. Some familiar compounds are water (H_2O), which is 11.2% hydrogen and 88.8% oxygen by weight, and common table salt (NaCl), which is 39.3% sodium and 60.7% chlorine by weight. Compounds have well-defined properties that can be used to identify them.

A **mixture** differs from a compound in that its composition may vary widely. A mixture is composed of two or more substances, each of which retains its own characteristic properties. Mixtures do not have fixed or invariant properties. Many familiar materials are mixtures: sea water, milk, air, steel, marble, and concrete are examples. Mixtures are subdivided into two categories: **heterogeneous** and **homogeneous mixtures.** Homogeneous mixtures are uniform in their properties, and are called **solutions.** There are solutions of several gases (air is the most common example), of two or more liquids, of solids and of gases in a liquid, and of one solid in another.

Heterogeneous mixtures are not uniform in their properties throughout the sample. It is most likely that any rock you find is a heterogeneous mixture. A heterogeneous mixture can be so thoroughly mixed that it appears to be uniform, even when it is not truly homogeneous. (See Fig. I.1.) Until about forty years ago, the milk sold in the United States was not homogenized, and the fatty components (cream) rose to the top, while the watery components remained at the bottom. Today, milk appears to the eye to be a single substance because it has been mechanically homogenized.

Because each component of a mixture retains its own characteristic properties, the various components can be separated by physical methods. Consider, for example, a mixture of powdered calcium carbonate (the principal constituent of marble) and ordinary table salt (NaCl). This mixture can be separated by adding water. The salt dissolves in water, while calcium carbonate does not. After filtering to separate the solid calcium carbonate from the water solution, the water can be evaporated to obtain the pure salt. As a second example, consider a mixture of powdered sulfur and iron filings. A magnet can be used to remove the iron filings from the mixture (see Fig. I.2), or the sulfur can be dissolved in liquid carbon disulfide and separated from the iron by filtration. If the heterogeneous mixture of sulfur and iron filings is heated sufficiently, a chemical reaction occurs and the compound FeS, iron(II) sulfide, is formed. This compound is nonmagnetic and does not dissolve in carbon disulfide.

Units of Measurement

Communication between people living in different countries was much poorer several hundred years ago than it is today, and often different systems of measuring common quantities were developed. The two systems in common use today are the English system (in which distances are measured in feet and objects are weighed in pounds), and the **metric system** (in which distances are measured in meters or kilometers, and objects are weighed in grams or kilograms).

Scientists have long used the **metric,** rather than the English, system of units. During the past quarter century there has been a continuing effort to eliminate duplication and adopt a single set of units for scientific measurement. In 1960, **SI units** (Système Internationale d'Unités) were adopted by an international organization, the General Conference on Weights and Measures, and progress has been made in achieving the goal of a single unit for each of the fundamental quantities. Nevertheless, for historical reasons, for convenience, and because people find it difficult to give up units they are accustomed to, many quantities are still measured in more than one unit. In order to be able to read past as well as current scientific literature, you will need to be familiar with more than just the SI units. In this book we will emphasize SI units, but when we discuss a quantity that many chemists, biochemists, or biologists in the United States still measure in a unit different from the SI unit, we will give both values.

The seven fundamental SI units are listed in Table I.1. The units of all other quantities are derived from these. Appendix A is a summary of important units of measurement and their relationships to each other. In this chapter we will discuss the units of only a few quantities: mass, length, volume, density, and temperature. Units of other quantities will be discussed as we encounter them.

In the metric system, a series of prefixes is used to indicate decimal fractions or multiples of the various basic units. The same prefixes are used with all units of measurement. The most frequently used of these prefixes are listed in Table I.2.

The numbers that occur in scientific problems vary from extremely small numbers to extremely large ones, and to write these numbers most conveniently, **exponential notation** is employed. (Because scientists almost always use exponential notation, it is also called **scientific notation.**) In exponential notation a number is written as

$$\left(\begin{array}{c}\text{a number between}\\ \text{1 and 10}\end{array}\right) \times 10^n$$

where n is some integer, either positive or negative. For instance, 492 is 4.92×10^2, and 0.00492 is 4.92×10^{-3}. A summary of information about exponents is contained in Appendix B.

Table I.1. **The Basic SI Units**

Physical Quantity	Name of Unit
Mass	kilogram (kg)
Length	meter (m)
Time	second (s)
Electric current	ampere (A)
Temperature	kelvin (K)
Luminous intensity	candela (cd)
Amount of substance	mole (mol)

Table I.2. **Frequently Used Prefixes in the Metric System**

Prefix	Symbol	Factor	Example
pico	p	10^{-12}	1 picometer (pm) $= 1 \times 10^{-12}$ m
nano	n	10^{-9}	1 nanometer (nm) $= 1 \times 10^{-9}$ m
micro	μ	10^{-6}	1 microgram (μg) $= 1 \times 10^{-6}$ g
milli	m	10^{-3}	1 milligram (mg) $= 1 \times 10^{-3}$ g
centi	c	10^{-2}	1 centimeter (cm) $= 1 \times 10^{-2}$ m
deci	d	10^{-1}	1 decimeter (dm) $= 1 \times 10^{-1}$ m
kilo	k	10^{3}	1 kilometer (km) $= 1 \times 10^{3}$ m
mega	M	10^{6}	1 megameter (Mm) $= 1 \times 10^{6}$ m
giga	G	10^{9}	1 gigameter (Gm) $= 1 \times 10^{9}$ m

Mass and Weight

The fundamental unit of **mass** in the international system is the **kilogram** (kg). It is the mass of a specific cylinder, 90% platinum and 10% iridium, known as the international kilogram, and maintained in the archives of the International Bureau of Weights and Measures, near Paris. The kilogram is a convenient unit for measuring the mass of animals and people (1 kg is equal to 2.2046 lb), but it is too large for the samples of substances usually investigated in a chemistry lab. Chemists usually measure mass in grams (g) or milligrams (mg).

$$\text{one gram (1g) is } 1 \times 10^{-3} \text{ kilograms } (10^{-3} \text{ kg})$$
$$\text{one kilogram (1 kg) is 1000 grams } (10^3 \text{ g}) \tag{I-1}$$

$$\text{one milligram (1 mg) is } 10^{-3} \text{ g } \quad \text{or} \quad 10^{-6} \text{ kg} \tag{I-2}$$

A word is in order here on the distinction between **mass** and **weight.** Mass is a property measuring the quantity of matter in a body, and is independent of the body's location with respect to any other body of matter. Weight is a force, the magnitude of the gravitational attraction between the body and any large mass, such as the earth. If you were to go to the moon, your mass would remain constant, but your weight would change drastically, because the moon exerts a much smaller gravitational attraction than does the earth. Even here on earth, the force of gravity is not constant but varies slightly from place to place. The weight of a body, W, is directly proportional to its mass, m,

$$W = mg \tag{I-3}$$

but the proportionality factor, g (the **acceleration of gravity**), depends on the location where the weight is being measured.

Chemists measure the mass of a body by comparing its weight with the weight of a body whose mass is known. Since the weights are compared at the same location on earth, the masses are the same if the weights are the same. Thus the procedure used to determine the mass of an object is called *weighing.* As a result, it is a common practice to use the words weight and mass interchangeably, but it is important to understand the distinction between them.

The equipment used to determine the mass of an object is called a **balance.** Figure I.3 shows three balances widely used today. The single-pan analytical balance is by far the most commonly used balance in chemistry laboratories. Figure I.4 shows a double-pan analytical balance, only rarely used today.

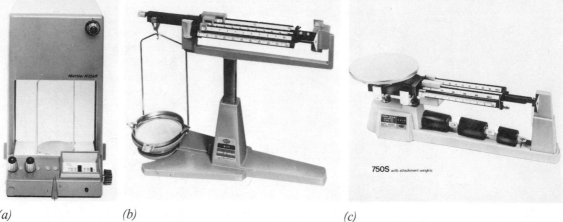

(a)   (b)   (c)

Fig. I.3. Three types of laboratory balances. *(a)* Single-pan analytical balance, Mettler Co. The mass of a sample can be determined to ±0.0001 g using an analytical balance. *(b)* A triple-beam, stirrup type balance, Ohaus Scale Corp. With this balance the mass of a sample can be determined to ±0.01 g. *(c)* A triple-beam platform balance, Ohaus Scale Corp. With this balance the mass of a sample can be determined to ±0.1 g.

Length

The SI unit of length is the **meter,** abbreviated m. Distances of experimental importance in a chemistry lab are rarely as long as a meter, and the units most commonly used to measure length are the centimeter (cm) and the millimeter (mm).

$$1 \text{ cm is } 10^{-2} \text{ m} \qquad 1 \text{ m is } 100 \text{ cm} \qquad \text{(I-4)}$$
$$1 \text{ mm is } 10^{-1} \text{ cm} \quad \text{or} \quad 10^{-3} \text{ m}$$
$$1 \text{ m is } 10^{3} \text{ mm} \qquad 1 \text{ cm is } 10 \text{ mm} \qquad \text{(I-5)}$$

Volume

Volume has the dimensions of (length)^3, and therefore the SI unit of volume is the cubic meter, m^3. The cubic meter, however, is too large for convenience, and the unit most commonly used by chemists is the **cubic centimeter** (cc or cm^3), which is also called a **milliliter** (mL). One **liter** (L) is exactly 1000 mL.

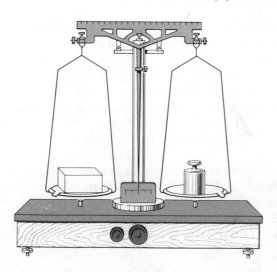

Fig. I.4. A traditional two-pan balance. Known masses are added to the right pan until the pointer is centered. The contents of each pan then have the same weight and therefore also possess the same mass.

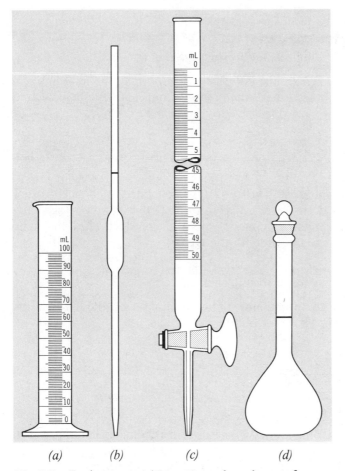

Fig. I.5. Equipment used to measure the volumes of liquids. *(a)* A graduated cylinder; least precise of the devices shown. The scale on a typical graduated cylinder can be read to ±0.2 mL. *(b)* A volumetric pipet; delivers a specified volume of liquid. *(c)* A buret; delivers any volume from 0 to 50 mL. With experience, a buret can be read to ±0.01 mL. *(d)* A volumetric flask; contains a specified volume of liquid.

Figure I.5 shows apparatus commonly used in a chemistry laboratory for measuring volumes of liquids.

Density

The **density** of an object is defined as the ratio of its mass to its volume:

$$\text{density} = \frac{\text{mass}}{\text{volume}} \tag{I-6}$$

As density must have units derived from those of mass and volume, the SI unit of density is kilograms per cubic meter (kg · m^{-3}). The densities of solids and liquids are usually reported in grams per cubic centimeter (g · cm^{-3}), or, equivalently, in grams per milliliter (g · mL^{-1}). Gas densities are reported in grams per liter (g · L^{-1}).

The density of a substance enables you to calculate the mass if you know the volume, and vice versa.

EXAMPLE I.1. Calculations involving density

The density of ethanol* is 0.794 g · mL⁻¹ at 25 °C. If a sample weighing 138 g is desired, what volume should be measured out?

Solution. This is a problem of practical importance because it is generally easier to measure volumes of liquids than to weigh out samples of a fixed amount. Equation (I-6) can be rearranged to solve for the volume:

$$\text{volume} = \frac{\text{mass}}{\text{density}}$$

Always use the units of quantities to help you figure out just what arithmetic process you should use. The units must be treated algebraically exactly as the numbers themselves are treated. If you multiply two numbers together, you must also multiply their units to obtain the units of the product. The units on both sides of any equation must be the same. In this case

$$\frac{\text{mass in g}}{\text{density in g} \cdot \text{mL}^{-1}} = \text{volume in mL}$$

The grams in the numerator and denominator on the left-hand side cancel, and $1/\text{mL}^{-1} = \text{mL}$. Thus the volume desired is

$$\frac{138 \text{ g}}{0.794 \text{ g} \cdot \text{mL}^{-1}} = 174 \text{ mL}$$

Temperature

Temperature is almost always measured in a chemistry laboratory using a thermometer graduated in degrees centigrade, or **Celsius.** The centigrade and Celsius scales are indistinguishable for practical purposes, but have different theoretical definitions, as will be discussed shortly. On the centigrade scale, the temperature of melting ice in water saturated with air at a pressure equal to one atmosphere† is defined as zero degrees exactly, and the temperature at which water boils at one atmosphere is defined as exactly one hundred degrees. The symbol used to designate temperature on the centigrade or Celsius scale is °C.

The SI temperature scale is the **Kelvin** or **absolute temperature scale,** named after a British physicist, William Thomson, Lord Kelvin (1824–1907). The zero point of the Kelvin scale is called **absolute zero,** because it is the lowest temperature possible, according to theory. Absolute zero is denoted 0 K. The degree sign (°) was formerly employed for the Kelvin scale, but SI units simply designate the unit of temperature as the kelvin (K). The second temperature used to fix the Kelvin scale is the **triple point of water,** the temperature at which air-free water freezes at the pressure of its own vapor. The triple point of water (which will be discussed in Section 5.10) is assigned the value 273.160 K. The freezing point of air-saturated water at 1 atm, 0 °C, is 273.15 K on the Kelvin scale.

The Celsius scale is *defined* relative to the Kelvin scale by the relation

$$\text{degrees Celsius (°C)} = \text{kelvins (K)} - 273.15 \tag{I-7}$$

---

* Ethanol or ethyl alcohol is the name for the alcohol that human beings drink. The word alcohol alone denotes a large number of chemically related compounds.

† Pressure units will be defined in Chapter 3. The pressure of the air at sea level is very close to 1 atm.

Although they are defined differently, the Celsius and centigrade scales are numerically identical. The boiling point of water at 1 atm is 373.15 K. It is customary to denote the Celsius temperature by the symbol t, and temperature on the absolute or Kelvin scale by the symbol T. While temperature is measured in the laboratory using thermometers that read in degrees Celsius, many fundamental laws of nature take a simple form only when temperature is given on the Kelvin scale. Accordingly, you should acquire the habit of converting experimental temperatures to kelvins before substituting them into theoretical equations.

On the **Fahrenheit** scale, the melting point of ice (the freezing point of water) is 32 °F, and the boiling point of water is 212 °F. A Fahrenheit degree is therefore smaller than a Celsius degree, as there are 180° between the freezing point and boiling point of water on the Fahrenheit scale, but only 100° between these points on the Celsius scale. The size of a degree on the Kelvin and Celsius scales is, in contrast, identical. The three temperature scales are compared in Fig. I.6.

The relation between the Celsius and Fahrenheit scales is

$$°C = \tfrac{100}{180}(°F - 32°) = \tfrac{5}{9}(°F - 32°) \tag{I-8}$$

or

$$°F = \tfrac{9}{5}(°C) + 32°$$

EXAMPLE I.2. The three temperature scales

The melting point of the element cesium, Cs, is 28.4 °C. What is the melting point of Cs on the Fahrenheit and Kelvin scales?

Solution

$$°F = \tfrac{9}{5}(°C) + 32 = \tfrac{9}{5}(28.4) + 32 = 51.1 + 32 = 83.1°F$$
$$K = °C + 273.15 = 28.4 + 273.15 = 301.55 \text{ K} = 301.6 \text{ K}$$

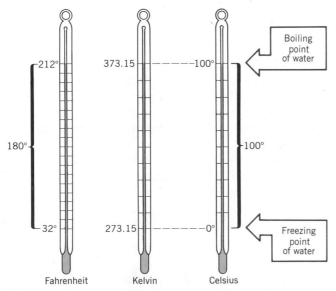

Fig. I.6. The relationships between the Celsius, Kelvin, and Fahrenheit temperature scales.

Physical and Chemical Properties

One of the jobs of a chemist is to be able to identify substances, whether pure or as components of a mixture. This is accomplished by being familiar with the properties of known substances. The **physical properties** of a substance are those that do not depend on its reaction with any other substance and that can be observed without changing the identity of the substance. Examples of physical properties are physical state (gas, liquid, or solid), color, melting point, boiling point, solubility in water and in other common solvents, and density.

The **chemical properties** of a substance are those that describe its reactions with other substances: Does it dissolve in acid, but not in water? Does it decolorize a deep purple solution of potassium permanganate? Does a solid substance (a **precipitate**) form when a solution of salt water is added?

There is no sharp dividing line between physical and chemical properties. Many substances dissolve in water because of a chemical reaction between water and the dissolving particles, yet solubility in water is usually classified as a physical property. Some solid compounds do not simply melt when heated, but decompose into other substances.

By examining a small number of both physical and chemical properties we can usually unambiguously identify a pure substance. Separating the components of a mixture in order to identify the ingredients may be a more challenging task.

Intensive and Extensive Properties

A property that does not depend on the amount of material present is an **intensive property.** Examples are color, density, and temperature. In contrast, properties that do depend on the amount of material, such as mass and volume, are called **extensive properties.** A burning twig and a bonfire may be at the same temperature, but the bonfire emits much more heat. The heat emitted is an extensive property, whereas the temperature is intensive.

Dimensional Analysis

It is frequently necessary in chemical problem solving to convert from one unit to another. To do this, conversion factors are employed. If you include the units of all quantities, and perform exactly the same algebraic operation on the units as you do on the numbers, you can often figure out how to solve a problem simply by insuring that the answer has the proper units. This type of reasoning, based on units alone, is called **dimensional analysis.**

Suppose we want to convert from pounds to grams. We know that 1 kg is 2.2046 lb, and also that 1 kg is 1000 g. We can therefore write two unit conversion factors, as follows:

$$\frac{2.2046 \text{ lb}}{1000 \text{ g}} = \frac{2.2046 \times 10^{-3} \text{ lb}}{1 \text{ g}}$$

and

$$\frac{1000 \text{ g}}{2.2046 \text{ lb}} = \frac{453.60 \text{ g}}{1 \text{ lb}}$$

If we want to convert a weight in pounds to a weight in grams we must multiply by the unit conversion factor that cancels the unit "lb" and gives the answer in g. For instance, if an object weighs 0.793 lb, its weight in grams is

$$(0.793 \text{ lb}) \left(\frac{1000 \text{ g}}{2.2046 \text{ lb}} \right) = 3.60 \times 10^2 \text{ g}$$

Any unit conversion factor is numerically equal to 1. Since 1000 cm³ = 1 L,

$$\frac{1000 \text{ cm}^3}{1 \text{ L}} = 1 \quad \text{and} \quad \frac{1 \text{ L}}{1000 \text{ cm}^3} = 1$$

EXAMPLE I.3. Unit conversion factors

Determine the unit conversion factor for converting a volume in liters to the SI unit of volume, cubic meters (m³).

Solution. A liter is 1000 cm³. We therefore need the conversion factor from cubic centimeters (cm³) to cubic meters (m³). We know the conversion factor from centimeters to meters: (100 cm)/(1 m). Simply cube this entire quantity to obtain the conversion factor from cubic centimeters to cubic meters.

$$\left(\frac{10^2 \text{ cm}}{1 \text{ m}} \right)^3 = \frac{10^6 \text{ cm}^3}{1 \text{ m}^3}$$

Multiplying two factors yields the desired conversion factor from liters to cubic meters:

$$\left(\frac{1 \text{ L}}{10^3 \text{ cm}^3} \right) \left(\frac{10^6 \text{ cm}^3}{1 \text{ m}^3} \right) = \frac{10^3 \text{ L}}{1 \text{ m}^3}$$

Thus to convert 2.54 L to cubic meters we obtain:

$$(2.54 \text{ L}) \left(\frac{1 \text{ m}^3}{10^3 \text{ L}} \right) = 2.54 \times 10^{-3} \text{ m}^3$$

No matter how many conversion factors are involved, if you write down every unit, and apply the conversion factors so that all units but the desired ones cancel out, you can always obtain the correct answer.

EXAMPLE I.4. Multiplying conversion factors

A certain synthetic procedure yields 4.85×10^{-2} g of product/s. How many kilograms will be produced in five days of continuous reaction?

Solution

$$\left(4.85 \times 10^{-2} \frac{\text{g}}{\text{s}} \right) \left(\frac{60 \text{ s}}{1 \text{ min}} \right) \left(\frac{60 \text{ min}}{1 \text{ h}} \right) \left(\frac{24 \text{ h}}{1 \text{ day}} \right) (5 \text{ days}) \left(\frac{1 \text{ kg}}{10^3 \text{ g}} \right) = 21.0 \text{ kg}$$

EXAMPLE I.5. Converting from English to SI units

A car is moving with a speed of 50.0 miles per hour (miles · h⁻¹). What is its speed in meters per second (m · s⁻¹)?

Solution. Use Appendix A for conversion factors:

$$\left(50.0 \frac{\text{miles}}{\text{h}} \right) \left(\frac{1 \text{ km}}{0.6214 \text{ miles}} \right) \left(\frac{10^3 \text{ m}}{1 \text{ km}} \right) \left(\frac{1 \text{ h}}{60 \text{ min}} \right) \left(\frac{1 \text{ min}}{60 \text{ s}} \right) = 22.4 \text{ m} \cdot \text{s}^{-1}$$

An alternative procedure is

$$\left(50.0\ \frac{\text{miles}}{\text{h}}\right)\left(\frac{5280\ \text{ft}}{1\ \text{mile}}\right)\left(\frac{1\ \text{yd}}{3\ \text{ft}}\right)\left(\frac{1\ \text{m}}{1.0936\ \text{yd}}\right)\left(\frac{1\ \text{h}}{60\ \text{min}}\right)\left(\frac{1\ \text{min}}{60\ \text{s}}\right) = 22.4\ \text{m}\cdot\text{s}^{-1}$$

Significant Figures

Every time we make a measurement we are limited by the equipment we use. When we write down a numerical value that is the result of an experimental measurement, therefore, that number has an **uncertainty** associated with it, a range of reliability that we call the **experimental error.** The word "error" should not be taken to mean that the experimenter has made a mistake; there is no way to avoid uncertainties of measurement, although we try to improve both equipment and techniques in order to make the experimental errors as small as possible.

A scientist indicates how reliable or how "good" a number is by including in it only those digits that are known with certainty, plus one additional digit that has uncertainty associated with it. For instance, on a triple-beam platform balance, the mass of an object can be measured to the nearest 0.1 g. If an object weighed on such a balance has a mass of 13.7 g, the 1 and 3 are known exactly, but the 7 is uncertain. If the value 13.7 g is written, with nothing specified about the magnitude of the uncertainty, it is usually understood that the limits of error are ± 1 in the last digit, so that the value 13.7 g would mean 13.7 ± 0.1 g. If the same object is now weighed on an analytical balance (Fig. I.3), and its mass is found to be 13.7164 g, the uncertainty is understood to be ± 0.0001 g. If the experimental error is anything other than ± 1 in the last digit, the magnitude of the uncertainty should be specifically stated.

When numbers are written in scientific notation, the number of digits in the number between 1 and 10 is equal to the number of **significant figures** in the quantity. Thus there are three significant figures in both 4.92×10^2 and 4.92×10^{-3}. This is a distinct advantage of exponential notation, because in certain cases the number of significant figures in a number written in ordinary decimal notation cannot be determined unambiguously.

Consider, for example, the three values 5.08×10^4, 5.080×10^4, and 5.0800×10^4. These values have three, four, and five significant figures, respectively. Writing 5.08×10^4 implies $(5.08 \pm 0.01) \times 10^4$, that is, the number is only known with certainty to be some value between 50,700 and 50,900. Writing 5.080×10^4 implies $(5.080 \pm 0.001) \times 10^4$, that is, the number is known to lie between 50,790 and 50,810. Similarly, 5.0800×10^4 tells us that the value is between 50,799 and 50,801. There is no way to express all of these possibilities in decimal notation. Simply writing 50,800 is ambiguous; either three, four, or five significant figures are possible for the number 50,800. In order to err on the side of caution, in the absence of specific knowledge, we would assume that 50,800 has no more than three significant figures.

There is no ambiguity about zeros to the right of the decimal point, however. The figure 14 implies 14 ± 1, that is, a number between 13 and 15, and has two significant figures. The number 14.00 implies 14.00 ± 0.01, a number between 13.99 and 14.01, and has four significant figures.

Zeros that serve only to tell us the power of 10 are not significant. Thus both 0.0738 and 0.00000738 have three significant figures, as the first value is 7.38×10^{-2}, while the second is 7.38×10^{-6}. If, as a matter of routine, you express every number in exponential notation, you will not be in doubt about the number of significant figures. The following exercise will give you practice in expressing numbers in exponential notation, and determining the number of significant figures.

EXAMPLE I.6.

Express each of the following numbers in scientific notation. How many significant figures are in each number? 409.10, 4091.00, 0.004091, 0.004091000, 308,000, 30,860, 0.00056030

Solution. For convenience, the answers are tabulated.

Number	Exponential Notation	Number of Significant Figures
409.10	4.0910×10^2	5
4091.00	4.09100×10^3	6
0.004091	4.091×10^{-3}	4
0.004091000	4.091000×10^{-3}	7
308,000	3.08×10^5	3
30,860	3.086×10^4	4
0.00056030	5.6030×10^{-4}	5

Rounding Numbers

Most calculators will give you answers to 10 digits, but it is unlikely that an experimental value will be known to more than 4 or 5 digits, and many experimental quantities are less well known than that. Most of the digits that will appear on your calculator display will therefore be meaningless, and you must round the value shown to the correct number of significant figures. The rules for rounding off are as follows:

1. Eliminate all digits that are not significant.
2. Examine the digits that you have discarded.

(a) If the first digit discarded (the one adjacent to the last significant figure retained) is larger than a 5, or if it is a 5 followed by other digits, at least one of which is not zero, increase the last retained digit by one. This is called **rounding up.**
Example: The number 1.863 rounded to two digits is 1.9, since the first discarded digit is a 6. Similarly, 1.8507 rounded to two digits is 1.9, since the first digit discarded is a 5 followed by two other digits, one of which is not zero.

(b) If the first digit discarded is less than 5, leave the last retained digit unchanged. This is called **rounding down.**
Example: The number 1.84987 rounded to two digits is 1.8, since the first discarded digit is a 4.

(c) If the first discarded digit is a 5 followed only by zeros, or by no other digits at all, round up if the last digit retained is odd, and round down if the last digit retained is even. In this case, the last digit retained will always be an even number.
Example: The number 1.8450 rounded to three figures is 1.84, whereas 1.8550 rounded to three figures is 1.86.

The rules for rounding are designed so that if a large number of roundings are carried out, you will round up just about as often as you round down, and rounding errors will tend to cancel out.

A word of caution: It is best to express numbers in exponential notation before rounding. Rounding *never* changes the power of 10! If you are asked to round 67,832 to four figures, for instance, the answer is 6.783×10^4. (Note that 67,832 does *not* become 6783 when you round!)

EXAMPLE I.7. Rounding numbers

(a) Round 8.21498 to (1) five figures, (2) three figures.

Solution. (1) 8.2150. The digit being discarded is an 8, and therefore the preceding digit must be increased by 1. (2) 8.21. The digits being discarded are 498. The first digit discarded is less than 5. Do *not* use a previously rounded version of a value and round twice in succession.

(b) Round 643.455 to (1) five figures, (2) three figures, (3) two figures.

Solution. (1) 643.46 or 6.4346×10^2. The figure being discarded is a 5 with nothing following it, and the last retained digit is odd. Therefore we round up. (2) 643 or 6.43×10^2, (3) 6.4×10^2

(c) Round 15.9994 to (1) five figures, (2) four figures.

Solution. (1) 15.999, (2) 16.00

(d) Round 59,648 to (1) four figures, (2) two figures.

Solution. (1) 5.965×10^4, (2) 6.0×10^4

Determining the Number of Significant Figures in a Calculated Value

Very often we measure several quantities in the laboratory and then perform calculations using the measured values to obtain a desired answer. The question then arises: How reliable is the calculated answer?

If the calculated value is obtained by doing *only* multiplication and/or division, we employ the following rule for a quick estimate of the number of significant figures in the answer:

In a product or quotient of experimental numbers, the final answer has only as many significant figures as the factor with the smallest number of significant figures. For instance, if you had to do the following calculation

$$\frac{(19.02)(323.16)(725.372)}{(5.4)(21.09)}$$

the answer should be expressed only to two significant figures, because the 5.4 in the denominator is only known to two figures. The fact that all the other factors have four or more significant figures does not improve the reliability of the answer.

You should get in the habit of deciding how many significant figures you are entitled to have in the answer to a problem before you pick up your calculator to evaluate the answer.

EXAMPLE I.8.

How many significant figures should be used for the answer to each of the following calculations?

(a) $\dfrac{(0.082056)(298.15)(0.379)}{(0.9480)}$

Solution. The factor with the least number of significant figures is 0.379, which has three significant figures. Therefore the answer should be expressed to three significant figures, as 9.78.

(b)
$$\frac{(0.46307)(0.0805)}{(63.54)(0.052)(2.809)}$$

Solution. The factor with the least number of significant figures is 0.052, which has two significant figures. Therefore the answer should be expressed to two significant figures, as 4.0×10^{-3}.

If a calculated value is obtained by doing only addition and/or subtraction, the result should be reported with the same number of decimal places as that of the term with the least number of decimal places. For example, the sum $45.6 + 0.368 + 83.51$ should not be reported to more than the first decimal place, as 45.6 is only known to the first decimal place. The answer is 129.5:

$$
\begin{array}{r}
45.6 \\
+ \quad 0.368 \\
\underline{83.51} \\
129.478 \rightarrow \text{round to } 129.5
\end{array}
$$

It is important to bear in mind that the number of significant figures in a sum or difference cannot be simply related to the number of significant figures in the terms added. The number of significant figures in a sum can be greater than the number of significant figures in any of the quantities added. Similarly, the number of significant figures in a difference can be smaller than the number of significant figures in any of the terms used in the calculation.

EXAMPLE I.9.

How many significant figures are there in the answers to the following calculations?

(a) $\qquad\qquad 0.0325 + 0.0812 + 0.0631$

Solution. The answer is 0.1768, valid to four significant figures, even though each term added has only three significant figures.

(b) $\qquad\qquad 37.596 - 36.802$

Solution. The answer is 0.794, valid to three significant figures, even though the two terms used in the calculation are each known to five significant figures.

When a calculation involves both addition (or subtraction) and multiplication (or division), do the addition first in order to determine the number of significant figures in the answer.

EXAMPLE I.10.

To how many significant figures should the answer to the following problems be expressed?

(a)
$$\frac{29.837 - 29.241}{32.064}$$

Solution. First do the subtraction in the numerator: $29.837 - 29.241 = 0.596$. Then the division problem is $(0.596)/(32.064)$, and the answer can only be expressed to three significant figures. Thus,

$$\frac{29.837 - 29.241}{32.064} = 1.86 \times 10^{-2}$$

(b)
$$\frac{752.1 + 26.3}{760.00}$$

Solution. Do the addition first. The division problem is then $(778.4)/(760.00)$, and the result should be given to four significant figures. The correct answer is therefore 1.024.

Exact versus Experimental Numbers

You should make a clear distinction between numbers that are known exactly, and experimentally determined quantities that have an uncertainty. If there are 9 people in a room, the integer 9 is exactly known, and there is no uncertainty in the value. Thus, although we write it as 9, it is quite clear that it is not a value with only one significant figure. An **exact number** has an infinite number of significant figures. A dozen eggs is defined as 12 eggs. The number 12 is known exactly and does *not* have only two significant figures. Conversion factors within the same system of units are exact. One kilogram is exactly 1000 g, and 1 mile is exactly 5280 ft.

EXAMPLE I.11.

To how many significant figures should we express the answer to the following question? If one steel ball has a mass of 4.364 g, what is the mass of five identical steel balls?

Solution. The product $5(4.364)$ should be reported to four significant figures, as 21.82 g. The number 5 is an exact value, and the factor with the least number of significant figures is the 4.364 g.

The foregoing rules enable us to obtain a quick estimate of the number of significant figures in a calculated value. For an accurate evaluation of the uncertainty in a number calculated using experimental quantities, one needs to know the **absolute** and the **relative uncertainty** of each quantity used in the calculation. These terms are defined and discussed in Appendix C.

Computational Techniques

After you have set up a problem, you should decide how many significant figures you are entitled to include in the final answer to the problem. To avoid rounding errors, carry one or perhaps two more significant figures than the number warranted by the data until the last step, and round only *once,* at the end, to the correct number of significant figures.

Suppose you have to add a set of numbers, as

$$
\begin{array}{r}
13.2 \\
22.38 \\
+\ \ 1.47832 \\
\underline{10.2609} \\
\end{array}
$$

You must first recognize that the answer can only be given to the first decimal place, because 13.2 is not known beyond the first decimal place. Keep each number known to two or more decimal places to the second place, and round the *final answer only* to the first place, as

$$
\begin{array}{r}
13.2 \\
22.38 \\
+\ \ 1.48 \\
\underline{10.26} \\
47.32 = 47.3 \\
\end{array}
$$

The correct answer to this addition problem is 47.3. If you round each value to the first decimal place *before* adding, the final answer will be too large, because it so happens that three values will be rounded up, as

$$
\begin{array}{r}
13.2 \\
22.4 \\
+\ \ 1.5 \\
\underline{10.3} \\
47.4 \\
\end{array}
$$

which is too large by one in the last significant digit.

The same procedure applies when you are multiplying or dividing: Avoid rounding to the correct number of significant figures until the last step, but do *not* carry a long string of figures that are not significant.

Often one calculates one or more intermediate quantities in the course of a long calculation. In such intermediate quantities retain one more significant figure than you are entitled to use, and round only the final result to the correct number of significant figures.

The following examples illustrate how several of the techniques discussed above are combined in a single problem.

EXAMPLE I.12.

The following items were placed into one container: 26 red beads (mass 4.27 g each), 58 blue beads (mass 3.91 g each), 28 yellow beads (mass 4.63 g each), 43 green beads (mass 4.79 g each), and 67 brown beads (mass 5.04 g each). What is the **percentage by weight** of the yellow beads in the contents of this container?

Solution. To answer this, we need to know both the total mass of all the beads, and the mass of the yellow beads. The mass of the yellow beads is $28(4.63 \text{ g}) = 129.64$ g, but that value is only good to three significant figures, as 4.63 is only known to three figures. If you were asked to report the mass of the yellow beads, the correct answer would be 1.30×10^2 g. However, since the answer we want is the weight percentage of yellow beads, we retain four significant figures.

The total mass of the contents of the container is

$$
\begin{aligned}
&26(4.27) + 58(3.91) + 28(4.63) + 43(4.79) + 67(5.04) \\
&=\ \ 111.0\ +\ 226.8\ +\ 129.6\ +\ 206.0\ +\ 337.7\ = 1011.1 \text{ g}
\end{aligned}
$$

One extra figure is retained in each of these intermediate values. The weight percentage of yellow beads is calculated as follows:

$$\text{percentage by weight of yellow beads} = \left(\frac{129.6}{1011.1}\right) \times 100 = 12.8\%$$

The final answer is valid to only three significant figures, as the numerator is only good to three figures.

EXAMPLE I.13.

A child's sandbox is 4.0-ft wide, 4.0-ft long, and 9.0-in. deep. If there are, on the average, 55 grains of sand/mm^3, how many grains of sand are in the sandbox when it is full? (See Appendix A for conversion factors.)

Solution. Since we know the number of grains of sand in a unit volume (1 mm^3), we must find the volume of the box.

$$\text{volume of box} = (4.0 \text{ ft})(4.0 \text{ ft})(9.0 \text{ in.})\left(\frac{1 \text{ ft}}{12 \text{ in.}}\right) = 12 \text{ ft}^3$$

We now have the volume of the sandbox in cubic feet, but we know the number of grains of sand per cubic millimeter. We must therefore calculate the conversion factor from cubic feet to cubic millimeters. We first obtain the conversion factor from feet to millimeters.

$$\left(\frac{12 \text{ in.}}{1 \text{ ft}}\right)\left(\frac{2.54 \text{ cm}}{1 \text{ in.}}\right)\left(\frac{10 \text{ mm}}{1 \text{ cm}}\right) = \frac{304.8 \text{ mm}}{1 \text{ ft}}$$

These conversion factors are exact numbers.

We must cube the conversion factor 304.8 mm/ft to obtain the conversion factor from cubic feet to cubic millimeters.

$$\left(\frac{304.8 \text{ mm}}{1 \text{ ft}}\right)^3 = \frac{2.832 \times 10^7 \text{ mm}^3}{1 \text{ ft}^3}$$

The number of grains of sand in the sandbox when it is full is therefore calculated to be

$$\left(\frac{55 \text{ grains}}{1 \text{ mm}^3}\right)\left(\frac{2.83 \times 10^7 \text{ mm}^3}{1 \text{ ft}^3}\right)(12 \text{ ft}^3) = 1.9 \times 10^{10} \text{ grains}$$

The answer is good to only two significant figures because both the volume of the sandbox and the 55 grains of sand/mm^3 are only known to two figures.

Problems

Use Appendix A for conversion factors needed.

I.1. State the number of significant figures in each of the following:
(a) 0.01000   (b) 2400   (c) 0.0000706   (d) 0.1023
(e) 60,200   (f) 0.004   (g) 208,842,000   (h) 0.003240

I.2. Round each of the following numbers to four, three, and two significant figures:
(a) 15.9994   (b) 1.00728   (c) 28,755   (d) 2,603,702   (e) 0.0020451

I.3. The density of chloroform, $CHCl_3$, is $1.4832\ g \cdot mL^{-1}$ at 20 °C. What volume should be measured out if 59.69 g are needed?

I.4. Express each of the following numbers in scientific notation, and state the number of significant figures in each:
(a) 0.0002008   (b) 20,772,000   (c) 0.010570   (d) 7030

I.5. The following calculations have been performed with an electronic calculator, and the calculator display is shown. Round the answers given to the correct number of significant figures, and report the answer in exponential notation.

(a) $\dfrac{(2.1)(0.0821)(295)}{4.32} = 11.77336806$

(b) $\dfrac{(0.00323)(107.87)}{1.023} = 0.340586608$

(c) $\dfrac{(0.928)(0.00520)}{(0.082056)(297.25)} = 0.000197842$

(d) $\dfrac{9.753 - 9.512}{15.9994} = 0.015063065$

(e) $\dfrac{26.923 - 26.062}{63.54} = 0.013550519$

I.6. What is the Kelvin temperature, expressed to the correct number of significant figures, corresponding to each of the following temperatures on the Celsius scale?
(a) 24 °C   (b) 25.0 °C   (c) 26.8 °C   (d) 24.35 °C   (e) 77 °C

I.7. (a) Round the following values to three significant figures:
    (1) 757.5   (2) 0.010185   (3) 0.0064048   (4) 30.97376
(b) Round the following values to two significant figures:
    (1) 435   (2) 19.9314   (3) 0.0078506   (4) 68,587

I.8. Obtain the factor for converting density in grams per liter to the SI unit of density, kilograms per cubic meter.

I.9. State whether each of the numbers in the following sentences is exact or has experimental uncertainty:
(a) The newspaper reported a crowd of 52,000 people at the concert in the park.
(b) There are 3 ft in a yard.
(c) Suzanne weighs 128 lb.
(d) At our local hospital, 1307 babies were born during 1983.
(e) One (short) ton weighs 2000 lb.
(f) I bought 18 oranges.
(g) I bought 4 lb of apples.

I.10. A sample of water has a mass of 234.9 g at 25.0 °C. The density of water at 25.0 °C, given in the *Handbook of Chemistry and Physics,* is $0.99707\ g \cdot mL^{-1}$. What is the volume of the water? Express the answer both in milliliters and in liters, to the correct number of significant figures.

I.11. In a certain industrial plant, 3.87 g of a product are synthesized each minute. How many pounds will be produced in a week of continuous reaction?

I.12. Calculate the mass of solution obtained when 1.464 g of sodium chloride (NaCl) and 3.74 g of glycine are added to 5.00×10^2 g of water.

I.13. Carry out the following operations, reporting the answers to the correct number of significant figures:
(a) $4.02 + 15.9 + 0.823 = ?$ (b) $1.00797 + 126.90 = ?$
(c) $213 - 1.579 = ?$ (d) $40.08 + 15.9994 = ?$
(e) $137.33 + 32.064 + 63.9976 = ?$ (f) $6.3 \times 10^4 + 1.28 = ?$

I.14. Sally Student jogs 2.0 miles in 18.0 min. What is Sally's speed in meters per second?

I.15. In a carton of a dozen eggs, the average mass of one egg is 56.49 g. What is the total mass of the dozen eggs? Express your answer to the correct number of significant figures.

I.16. In order to determine the volume of a certain flask, it is weighed when empty, and then weighed filled with distilled water. The temperature of the water used is measured, and the density of water at that temperature is obtained by consulting the *Handbook of Chemistry and Physics.* In a certain experiment, the following data were obtained:

$$\text{weight of flask full of water} = 45.0078 \text{ g}$$
$$\text{weight of empty flask} = 20.0324 \text{ g}$$
$$\text{temperature of water} = 26.0 \text{ °C}$$
$$\text{density of water at } 26.0 \text{ °C} = 0.99681 \text{ g} \cdot \text{mL}^{-1}$$

Calculate the volume of the flask.

I.17. The dimensions of a block of copper are 5.0 in. $\times$ 3.0 in. $\times$ 2.0 in. The density of copper (Cu) is $8.92 \text{ g} \cdot \text{cm}^{-3}$. How much does the block weigh (a) in grams? (b) in pounds?

I.18. Carry out the following operations, reporting the answers to the correct number of significant figures:
(a) $\dfrac{(0.8600)(0.0742)}{(0.08206)(297.4)}$ (b) $\dfrac{32.9764 - 32.1402}{112.41 + 32.064}$

I.19. If a bee flies with an average speed of $3.4 \text{ m} \cdot \text{s}^{-1}$, what is its average speed expressed in miles per hour?

I.20. The *Handbook of Chemistry and Physics* gives the normal boiling points of substances in degrees centigrade. The following values are listed for the boiling points of a few liquids:
(a) Ethanol, 78.5   (b) Cycloheptene, 115   (c) Octane, 125.66
Give the boiling points of these three liquids in kelvins.

I.21. Mercury is one of the densest liquids known. At room temperature its density is $13.594 \text{ g} \cdot \text{cm}^{-3}$. Some mercury is poured into a tube with a uniform diameter of 9.0 mm. The height of the column of mercury is 683 mm. What is the mass of mercury in the tube?

I.22. A pycnometer is a flask specially designed for measuring the density of liquids. A certain pycnometer is known to have a volume of 50.02 mL. The empty pycnometer is weighed. It is then filled with liquid and reweighed. In a certain experiment, the temperature of the liquid was 25.0 °C and the following data were obtained:

$$\text{mass of pycnometer full of liquid} = 62.8365 \text{ g}$$
$$\text{mass of empty pycnometer} = 18.4631 \text{ g}$$

What is the density of this liquid at 25.0 °C? Report the answer to the correct number of significant figures and specify the units of the density.

I.23. The population of Brooklyn, New York, in a certain year is listed as 2,602,000. If, in the following year, 142,320 babies are born, 204,190 deaths occur, and there is a net influx of 33,286 people moving into Brooklyn, what will the population be at the end of the year? Express the answer in exponential notation.

I.24. A solution of the simple sugar, fructose, in water, is 20.0% fructose by weight, and has a density of $1.082 \text{ g} \cdot \text{mL}^{-1}$. How many grams of fructose are in 45.00 mL of this solution?

I.25. A sample of hydrogen gas is collected over liquid water at 23 °C. To obtain the pressure of dry hydrogen, the vapor pressure of water is subtracted from the barometric pressure. The reading of the barometer is 758.3 mmHg. The *Handbook of Chemistry and Physics* lists the value of the vapor pressure of water at 23 °C as 21.068 mmHg. Calculate the pressure of dry hydrogen.

I.26. The normal boiling point of methane, CH_4, is −164 °C. What is its boiling point on the Fahrenheit and Kelvin scales?

I.27. A solution of ethanol and water is prepared by mixing 40.0 mL of ethanol with 62.5 mL of water at 20.0 °C. The densities of ethanol and water at this temperature are, respectively, 0.7893 and $0.99823 \text{ g} \cdot \text{mL}^{-1}$. What is the percentage by weight of ethanol in this mixture?

I.28. The freezing point of propane is −189.69 °C. What is its freezing point in degrees Fahrenheit and in kelvins?

I.29. What factor should be used for converting density in grams per milliliter to the SI unit of density, kilograms per cubic meter?

chapter 1 Atomic Structure and Stoichiometry

Amedeo Avogadro (1776–1856), an Italian physicist, was a professor of higher physics at Turin University. After Avogadro learned of the investigations of J. L. Gay-Lussac on the combining ratios of various gases, he proposed, in 1811, that equal volumes of gases at the same temperature and pressure should contain equal numbers of molecules. This hypothesis, now known as Avogadro's Law, distinguished between atoms and molecules, and is one of the basic concepts of modern chemistry. As a corollary to his hypothesis, he proposed the existence of polyatomic elements. Avogadro's work was disregarded for nearly fifty years, and was not accepted until after his death, when, in 1858, Stanislao Cannizzaro constructed a logical system of chemistry based on Avogadro's hypothesis.

If we take a portion of matter that appears to be homogeneous, such as a piece of copper, and divide it in two, each smaller piece has all the properties required to identify it as being the same substance as the original, that is, as copper. Can we keep on dividing it indefinitely into successively smaller pieces, or will we eventually find some unit of matter that cannot be further subdivided without losing the characteristics of the original? That question so intrigued some of the early Greek philosophers that it was discussed for many centuries. We now know, based on conclusive experimental evidence, that matter cannot be subdivided indefinitely. An **atom** of copper is the smallest portion of matter that can be considered to be copper; if we subdivide the atom, we will have particles that are no longer copper.

A great deal of experimental evidence, obtained during the nineteenth century and the early part of the twentieth century, served to convince scientists of the particulate nature of matter. The **atomic theory of matter,** universally accepted by scientists today, states that all material substances are composed of minute particles or atoms of a comparatively small number of substances, called **elements.** An **atom** is the smallest particle of an element that can combine chemically with other elements. Atoms are therefore the building blocks of all matter.

Much of the evidence that led to the acceptance of the atomic theory of matter consists of experiments elucidating the quantitative relationships between substances as they undergo chemical changes. The study of these quantitative relationships is known as **stoichiometry.**

section 1.1
Subatomic Particles and the Structure of Atoms

At one time it was believed that an atom could not be subdivided, but we now know that is not true. In Chapter 12 we will discuss some of the fascinating experiments that provided evidence about the structure of atoms and the nature of the **subatomic particles** of which atoms are composed. There are many subatomic particles, and new ones are still being discovered. Three fundamental subatomic particles are particularly important in chemistry: **electrons, protons,** and **neutrons.**

An **electron** is an extremely light particle with a mass of 9.1095×10^{-31} kg (9.1095×10^{-28} g) and a negative charge of -1.602×10^{-19} C.* The charge on the electron is used as the fundamental unit of electrical charge in chemistry. That is, we generally measure charge in multiples of the magnitude (the absolute value) of the charge on an electron. When we say a particle has a charge of $3-$, we mean that its charge is three times as large as the charge on an electron, with the same sign. A charge of $2+$ means a charge with magnitude twice that of the electron, and of opposite sign. Most chemists denote the electron by the symbol e^-. Nuclear and radiochemists use the symbol β^-, for historical reasons that will be discussed in Chapter 22.

A **proton** has a mass of 1.6726×10^{-27} kg (1.6726×10^{-24} g), about 1836 times the mass of an electron. The proton, denoted p^+ by physicists and nuclear chemists, and H^+ by other chemists, has a positive charge of exactly the same magnitude as that of the electron, $+1.602 \times 10^{-19}$ C.

A **neutron** is an uncharged particle with a mass very nearly (but not exactly) equal to the mass of a proton. It actually has a mass 1.00138 times that of a proton, or 1.67495×10^{-27} kg. It is useful to remember that

$$\text{mass of neutron} \cong \text{mass of proton} \cong 1836(\text{mass of electron}) \tag{1-1}$$

---

* A coulomb is the amount of charge transferred by a current of one ampere in one second. $1 \text{ C} = 1 \text{ A} \cdot \text{s}$. See Appendix A.

Fig. 1.1. A representation of the electronic charge cloud of the hydrogen atom at room temperature. The density of dots indicates the density of electronic charge. Electrons spend most of their time close to the nucleus, but there is no fixed boundary for the volume of the atom.

Each atom consists of a **nucleus** and a number of electrons. The nucleus contains all the protons and neutrons, and therefore nearly all the mass of the atom. The electrons move extremely rapidly about the nucleus, and the space they occupy as they move defines the volume of the atom. The volume of the nucleus is very small relative to the volume of the atom. To give you some idea of the relative sizes of the atom and its nucleus, consider the following: If the diameter of an atom were about the length of a football field, the nucleus would be the size of an orange seed situated at the center of the 50 yard line. All the rest of the space of the atom is filled by the very light, rapidly moving electrons.

The motions of the electrons cannot be traced in detail. Electrons do not travel in fixed paths or orbits as do, for instance, the planets about the sun. Sometimes, in advertisements about atomic energy or on the packaging of certain children's toys, one sees symbols meant to represent an atom consisting of several ellipses oriented in different directions, centered about a small dot. These symbols convey an erroneous impression. The electrons in an atom do not remain in an elliptical orbit. Negative charge density occupies the entire three-dimensional volume of the atom. The distribution of electronic charge surrounding the nucleus of an atom is referred to as an **electron cloud.** Figure 1.1 illustrates one way we represent the electron cloud of a hydrogen atom.

Atoms of different elements differ in size, but the diameters of all atoms are roughly 1 or 2 angstroms (Å). An **angstrom** is a unit of length equal to 1×10^{-10} m, 1×10^{-8} cm, or 1×10^{-1} nm. Although an angstrom is not an SI unit, it is still frequently used by chemists, and one should be familiar with it. Nuclear diameters are very much smaller than atomic diameters. A typical value for the diameter of a nucleus is about 1×10^{-14} m, or 1×10^{-4} Å.

All atoms are electrically neutral and therefore must contain equal numbers of electrons and protons. The number of protons in the nucleus of an atom is called its **atomic number,** and is usually denoted by Z. Each element is distinguished from all others by its atomic number. All atoms with the same atomic number are atoms of the same element. Thus the element with $Z = 1$ (one proton, one electron) is hydrogen, while the element with $Z = 2$ (two protons, two electrons) is helium, and so on.

We can explain many of the known chemical properties of an element if we know its atomic number. The electrons in a neutral atom determine the chemistry of that atom. Those electrons that, on the average, tend to stay farthest away from the nucleus are primarily responsible for its chemical behavior, and are called the outer or **valence electrons.** Atoms with the same number of valence electrons have similar chemical properties, and are spoken of as a family, or a group.

The nucleus plus the inner electrons constitute the **kernel** or the **core** of the atom. We often use the symbol of an element to stand for the core, and place dots around it to indicate the valence electrons. Thus sodium, atomic number 11, may be denoted Na · , which tells us that the nucleus plus 10 inner electrons constitute the core, and there is only a single valence electron. Aluminum, atomic number 13, has 3 valence electrons and 10 inner electrons, and is symbolized Al ⋮ , while oxygen, atomic number 8, has 2 inner electrons and 6 valence electrons, and is represented as ⋮Ö⋮. In Chapter 13 we will discuss how to determine the number of valence electrons in any given atom.

Table 1.1. **The Isotopes of Hydrogen**[a]

Name of Isotope	Symbol for Isotope	Number of Electrons	Number of Protons	Number of Neutrons
Hydrogen or protium	$_1^1\text{H}$	1	1	0
Deuterium	$_1^2\text{H}$ or D	1	1	1
Tritium	$_1^3\text{H}$ or T	1	1	2

[a] Hydrogen is the only element whose isotopes are given separate names and symbols.

section 1.2
Isotopes of an Element

Atoms that have the same number of protons but differ in the number of neutrons, are called **isotopes.** For instance, there are three isotopes of the element hydrogen, for which $Z = 1$, as described in Table 1.1.

Note that the symbol for an isotope is $_Z^A\text{Element}$, where A is the **mass number,** that is, the sum of the number of protons and the number of neutrons, and Z is the atomic number. In older chemical literature you may also see the symbol written as $_Z\text{Element}^A$, but that form should no longer be used, according to the most recent international convention.

Since each atomic number uniquely defines an element, it is unnecessary to include both the atomic number, Z, and the symbol for the element. Consequently, when writing the symbol for an isotope, the atomic number is sometimes omitted, particularly for the more common, lighter elements. Thus the hydrogen isotopes (see Table 1.1) deuterium and tritium may be denoted ^2H and ^3H, respectively.

Almost every element has one or more stable isotopes, and every element has a number of unstable, **radioactive isotopes,** that is, isotopes that spontaneously decay to become other elements. For example, oxygen, as found in nature, consists of 99.759% $_8^{16}\text{O}$, 0.037% $_8^{17}\text{O}$, and 0.204% $_8^{18}\text{O}$.* These three stable isotopes all have 8 electrons and 8 protons, but have 8, 9, and 10 neutrons, respectively. There are also several known radioactive isotopes of oxygen, namely ^{13}O, ^{14}O, ^{15}O, ^{19}O, and ^{20}O. In nature, most elements are found as a mixture of isotopes, with each isotope having a distinct natural abundance, as is the case for oxygen.

section 1.3
The Atomic Weight Scale

Because an atom is such a tiny particle, in the laboratory it is rarely possible to work with individual atoms, or a few atoms, or even a few hundred atoms. The very smallest quantity of material we can actually see or handle contains billions and billions of atoms. When we speak of the **atomic weight** of an element, therefore, we do not mean the mass of an individual atom. We refer rather to an average mass of an atom in a sample containing all the naturally occurring isotopes, in their natural abundances.

The atomic weight scale is a relative scale, not an absolute one. It compares the average mass of an atom of one element to the mass of a standard. The standard atom, chosen by an international committee in 1961, is the carbon-12 (^{12}C) isotope. To deal with masses as small as atoms we define a new unit of mass. This unit is called the **atomic mass unit** (amu) or the **dalton** (after the English chemist, John Dalton). By

* Natural abundances of atoms are given as atom percentages. Thus 0.204% of all O atoms in nature are atoms of ^{18}O.

definition, an atomic mass unit is one twelfth the mass of a single atom of ^{12}C. The atomic weight of the ^{12}C isotope is then *exactly* 12 amu. All other atomic weights are given relative to that value. The mass of a ^{13}C atom is given, on that scale, as 13.00335 amu. What this means is that

$$\frac{\text{mass of one } ^{13}C \text{ atom}}{\text{mass of one } ^{12}C \text{ atom}} = \frac{13.00335}{12} \tag{1-2}$$

The atomic weight of an element is the average mass of an atom in any naturally occurring sample, relative to the value 12 amu as the mass of a carbon-12 atom. Carbon itself has three naturally occurring isotopes: $^{12}_{6}C$ (98.892%), $^{13}_{6}C$ (1.108%), and $^{14}_{6}C$ (1×10^{-10} %). The atomic weight of carbon is therefore calculated to be

$$(0.98892)(12) + (0.01108)(13.00335) = 11.867 + 0.1441 = 12.011$$

John Dalton collects marsh gas.

John Dalton (1766–1844), English chemist and physicist, is considered the founder of the modern atomic theory of matter. Dalton was a teacher of mathematics and natural philosophy at the New College in Manchester from 1793 until 1799, when the college was moved to York. He then became a "public and private teacher of mathematics and chemistry." For fifty-seven years, beginning in 1787, Dalton kept a meteorological diary, in which he entered more than 200,000 observations. The idea of atomic structure arose in his mind as a result of his study of the physical properties of the air and other gases. Dalton proposed that his atomic theory explains three observations that had been made previously by others: 1. the discovery of the elements, 2. the law of conservation of mass, and 3. the law of definite proportions, which states that every pure compound contains fixed and definite proportions by weight of its constituent elements. In 1808 Dalton published "A New System of Chemical Philosophy" in which he proposed the following postulates:

1. Matter consists of indivisible atoms.
2. All the atoms of a given element have identical properties, including identical mass.
3. Different elements have atoms that differ in mass.
4. Atoms are indestructible and chemical reactions are merely a rearrangement of atoms.
5. The formation of a compound from its elements takes place through the formation of "compound atoms" containing a definite and small number of atoms of each element.

The amount of ^{14}C in nature is too small to make a significant difference in the atomic weight of carbon.

The value of the atomic weight of potassium is 39.102. Therefore

$$\frac{\text{average mass of an atom of K}}{\text{mass of one }^{12}C\text{ atom}} = \frac{39.102}{12} \tag{1-3}$$

This can be rearranged to obtain

$$\text{average mass of an atom of K} = \frac{39.102}{12}(\text{mass of one }^{12}C\text{ atom})$$
$$= 39.102 \text{ amu}$$

The atomic weight of any element is the average mass of an atom of that element, in atomic mass units.

Using the known masses and natural abundances of the isotopes of an element, we can calculate the atomic weight of that element. A typical calculation is illustrated in Example 1.1.

EXAMPLE 1.1. The atomic weight as an average weight of different isotopes

The element silicon, Si, makes up 25.7% of the earth's crust by weight, and is the second most abundant element, with oxygen being the first. Three isotopes of silicon occur in nature: $^{28}_{14}Si$ (92.21%), which has an atomic mass of 27.97693 amu; $^{29}_{14}Si$ (4.70%), with an atomic mass of 28.97649 amu; and $^{30}_{14}Si$ (3.09%), with an atomic mass of 29.97376 amu. Calculate the atomic weight of silicon.

Solution. In averaging the atomic masses of the three isotopes, each mass must be weighted according to the percentage of that isotope that occurs in nature. Thus the atomic weight of silicon is determined to be

$$
\begin{aligned}
& (0.9221)(27.97693) + (0.0470)(28.97649) + (0.0309)(29.97376) \\
= \quad & \quad\;\; 25.80 \quad\quad + \quad\quad 1.36 \quad\quad + \quad\quad 0.926 \quad\quad = 28.09
\end{aligned}
$$

The atomic weight of silicon is 28.09 amu.

Clearly the atomic weight of an element depends on the relative abundances of the isotopes of that element. Variations in the relative abundances of the isotopes of some elements in samples obtained from different sources cause the atomic weights of these elements to vary somewhat from place to place in the world. Fortunately, these variations are usually very small. If you examine the table of atomic weights on the inside front cover, you will see that while some atomic weights are given to six or seven significant figures, as, for example, fluorine (18.99840) or bismuth (208.9804) others are given to only four significant figures, such as boron (10.81) and lead (207.2). For boron and lead, the variations in the percentages of different isotopes found in terrestrial samples limit the precision of the atomic weight given.

section 1.4
Avogadro's Number and the Concept of a Mole

A **molecule** is a combination of atoms, so tightly bound together that the molecule behaves as, and can be recognized as, a single particle. Both atoms and molecules are so small that it is convenient to define a new unit called a **mole,** which contains a specific, very large number of atoms or molecules. This particular number is known as **Avogadro's number.** Avogadro's number, N_A, is defined as the number of atoms contained in exactly 12 g (or 0.012 kg) of the isotope ^{12}C.

Avogadro's number is tremendously large and has the value 6.022045×10^{23}. To impress upon you the enormity of Avogadro's number, we will write it out with all its zeros:

$$N_A = 602,204,500,000,000,000,000,000 \qquad (1\text{-}4)$$

although, of course, the zeros after the 5 are not significant figures.

If we assigned every single person in the world today (4×10^9 people) to count the atoms in a mole of ^{12}C, and each person worked 48 h/wk, 52 wk/yr (no vacations) and counted 1 atom/s, it would take 17 million years to get the job done!

The value of Avogadro's number must be determined experimentally. A great many experiments have been performed in order to obtain the value of N_A, because it is one of the fundamental constants upon which many calculations depend. Two different experimental methods for obtaining Avogadro's number will be described in problems 1.11 and 1.12 at the end of this chapter, but at least half a dozen different experimental methods have been used.

Avogadro's number is so large because the mass of one atom is so small. For instance, the mass of one atom of ^{12}C is

$$\frac{12.000 \text{ g/mol}}{6.0220 \times 10^{23} \text{ atoms/mol}} = 1.9927 \times 10^{-23} \text{ g/atom}$$

$$= 0.000000000000000000000019927 \text{ g/atom} \qquad (1\text{-}5)$$

With the knowledge of Avogadro's number, we can determine the magnitude of one atomic mass unit in grams.

$$\left(\frac{1 \text{ atom } ^{12}C}{12 \text{ amu}}\right)\left(\frac{12 \text{ g}}{N_A \text{ atoms } ^{12}C}\right) = \frac{1}{N_A} = 1.660565 \times 10^{-24} \text{ g/amu}$$

or

$$1 \text{ amu} = 1.660565 \times 10^{-27} \text{ kg} \qquad (1\text{-}6)$$

It clearly is more convenient to give atomic masses in atomic mass units rather than in grams or kilograms.

Equation (1-6) is used to calculate the mass of a single atom in the same units we use to weigh samples in the laboratory. Most often, however, there will be no need for the mass of an atom in grams, and the relative scale of atomic weights will suffice to answer the quantitative questions we encounter. This distinction is illustrated in Examples 1.2 and 1.3.

EXAMPLE 1.2. Atomic mass units and the atomic weight scale

What is the average mass of one magnesium (Mg) atom, in grams?

Solution. An atomic weight is the average mass of an atom of the specified element in a naturally occurring sample, expressed in atomic mass units. Thus the average mass of one magnesium atom is 24.305 amu. To see this in detail, we can write the ratio

$$\frac{\text{average mass of an atom of Mg}}{\text{mass of one } ^{12}C \text{ atom}} = \frac{24.305}{12}$$

Solving this equation for the average mass of one magnesium atom we obtain

$$\text{average mass of an atom of Mg} = \left(\frac{24.305}{12}\right) (\text{mass of one } ^{12}C \text{ atom})$$

$$= \left(\frac{24.305}{12}\right) (12 \text{ amu}) = 24.305 \text{ amu}$$

Equation (1-6) can now be used to obtain the average mass in grams.

$$\text{average mass of an atom of Mg} = (24.305 \text{ amu})(1.660565 \times 10^{-27} \text{ kg/amu})$$
$$= 4.0360 \times 10^{-26} \text{ kg} = 4.0360 \times 10^{-23} \text{ g}$$

The reason this is an average mass, and not the actual mass of an atom of magnesium, is that three isotopes of magnesium, ^{24}Mg, ^{25}Mg, and ^{26}Mg, are found in nature.

EXAMPLE 1.3. Relative masses of atoms

What is the ratio of the mass of one hundred billion atoms of magnesium to the mass of one hundred billion atoms of hydrogen?

Solution. Since atomic weights are relative masses,

$$\frac{\text{mass of one hundred billion Mg atoms}}{\text{mass of one hundred billion H atoms}} = \frac{24.305}{1.0079}$$

Since all atomic weights are expressed relative to the value 12 for ^{12}C, and since one mole is defined as exactly 12 g of ^{12}C, the mass of one mole of any element, in grams, is equal to its atomic weight. This is easy to understand if we actually write down the ratio; we can then see that the common factor of N_A cancels out.

$$\frac{\text{mass of } N_A \text{ atoms of K}}{\text{mass of } N_A \text{ atoms of } ^{12}\text{C}} = \frac{N_A(\text{average mass of one atom of K})}{N_A(\text{mass of one } ^{12}\text{C atom})} \qquad (1\text{-}7)$$
$$= \frac{39.102}{12}$$

Rearranging this equation to solve for the mass of N_A atoms of K we obtain

$$\text{mass of } N_A \text{ atoms of K} = \left(\frac{39.102}{12}\right)(\text{mass of } N_A \text{ atoms of } ^{12}\text{C})$$
$$= \left(\frac{39.102}{12}\right)(12 \text{ g}) = 39.102 \text{ g}$$

Thus 1 mol of potassium contains 6.022×10^{23} atoms of K and has a mass of 39.102 g.

> To sum up: One **mole** of any element contains 6.022×10^{23} atoms of that element, and has a mass, in grams, equal to the atomic weight of that element.

1 mol of H contains 6.022×10^{23} atoms and has a mass of 1.008 g.
1 mol of C contains 6.022×10^{23} atoms and has a mass of 12.011 g.
1 mol of U contains 6.022×10^{23} atoms and has a mass of 238.03 g.

Think of a mole as a fixed number, Avogadro's number, of particles, in much the same way that you think of a dozen as the number 12. One can have a dozen grapefruit or a dozen golf balls. A dozen grapefruit weighs more than a dozen golf balls. All they have in common is the number 12. Similarly, one can have a mole of hydrogen (which weighs 1.008 g), and a mole of uranium (which weighs 238.03 g). What they have in common is the number of atoms, Avogadro's number, 6.022×10^{23}. (See Fig. 1.2.)

Fig. 1.2. One mole of four different elements: — copper, iron, lead, and mercury. Each sample contains the same number of atoms.

section 1.5
Molecular Formulas and Molecular Weights for Discrete Molecules

A compound is a substance formed when atoms combine chemically. Almost everything in the world is a compound or a mixture of compounds. The great majority of compounds have a fixed and definite atomic composition; we call such compounds **stoichiometric.** All gaseous compounds are stoichiometric and are composed of discrete (individual) molecules or atoms. The **formula** of a substance specifies its atomic composition. Carbon dioxide has the formula CO_2. This means that one molecule of carbon dioxide contains 1 atom of carbon and 2 atoms of oxygen. Butane, the combustible gas frequently used in cigarette lighters, has the formula C_4H_{10}. One molecule of gaseous butane contains 4 atoms of carbon and 10 atoms of hydrogen.

Three common gases of industrial importance are combinations of the elements nitrogen and oxygen. Dinitrogen oxide (nitrous oxide*), N_2O, is known as **laughing gas,** and is used as an anesthetic, particularly by dentists. It is somewhat soluble in cream, and has been widely used to dispense whipped cream from aerosol cans. Dinitrogen oxide can be used for these purposes because it is quite unreactive. The formula N_2O tells us that each molecule of dinitrogen oxide contains two atoms of N and one atom of O. Nitrogen oxide (nitric oxide), NO, is a very reactive, colorless, gaseous compound. The formula NO tells us that one molecule of nitric oxide contains one atom of N and one atom of O. Nitrogen oxide combines readily with oxygen to give the deep reddish-brown gas nitrogen dioxide, NO_2. Most gases are colorless or pale in color; the red-brown color of NO_2 is quite distinctive and easily recognized. Nitrogen dioxide is produced in automobile exhausts and is one of the major air pollutants. It is found in the atmospheric smog above many urban areas today. One molecule of nitrogen dioxide contains one atom of N and two atoms of O.

Because of their exceedingly small size, it is impossible to handle single molecules of any substance. It is, therefore, more convenient to speak in terms of moles of butane, or moles of nitrous oxide. Each mole contains Avogadro's number of molecules, that is, 6.022×10^{23} molecules of C_4H_{10} or of N_2O, respectively. One mole of C_4H_{10} contains 4 mol of C atoms and 10 mol of H atoms, in just the same way that a dozen eggs contain a dozen egg yolks plus a dozen egg whites, or a dozen bicycles consist of a dozen handlebars, two dozen wheels, and two dozen pedals.

---

* Many common compounds have two names: A systematic name using rules of nomenclature adopted by the International Union of Pure and Applied Chemistry (IUPAC), and a common name, in use before the IUPAC rules were formulated. We have given both names, with the common name in parentheses.

> A mole of a substance that consists of discrete molecules of formula A_pB_q contains p moles of atom A and q moles of atom B.

The samples we work with in the laboratory do not usually consist of exactly 1 mol of a substance; we may have 0.047 mol or 2.38 mol, and so on. The formula provides the same information regardless of how many moles of substance we actually have, because it tells us about **molar ratios.** In any sample of the gas A_pB_q the following ratios apply:

$$\frac{\text{number of moles of A}}{\text{number of moles of B}} = \frac{p}{q} \tag{1-8}$$

$$\frac{\text{number of moles of A}}{\text{number of moles of } A_pB_q} = \frac{p}{1} = p \tag{1-9}$$

$$\frac{\text{number of moles of B}}{\text{number of moles of } A_pB_q} = \frac{q}{1} = q \tag{1-10}$$

These ratios should be treated just like any other algebraic equation. We may write Eq. (1-10), for example, as

$$\text{number of moles of B} = q(\text{number of moles of } A_pB_q) \tag{1-10a}$$

If we had a sample containing 0.0470 mol of butane, the formula C_4H_{10} tells us immediately that we have $4(0.0470) = 0.188$ mol of C and $10(0.0470) = 0.470$ mol of H.

The **molecular weight** of a substance that consists of discrete molecules is the mass of one molecule of that substance in atomic mass units, or the mass of one mole of that substance (Avogadro's number of molecules) in grams. The molecular weight is simply the sum of the masses of each atom contained in the molecule. Thus, the molecular weight of butane, C_4H_{10}, is $4(12.011) + 10(1.008) = 58.12$. This means that a molecule of butane has a mass of 58.12 amu, and a mole (6.022×10^{23} molecules) of butane has a mass of 58.12 g.

A molecular formula contains a wealth of information in a very compact form. Examples 1.4 and 1.5 illustrate the many quantities that can be calculated using only a molecular formula and a table of atomic weights.

EXAMPLE 1.4. Making use of a molecular formula

Dopamine is a neurotransmitter, a molecule that serves to transmit messages in the brain. The chemical formula of dopamine is $C_8H_{11}O_2N$.

(a) What is the molecular weight of dopamine?

Solution. We must look up the atomic weights of carbon, hydrogen, oxygen, and nitrogen in a table of atomic weights. The molecular weight of dopamine is then given by

$$8(12.011) + 11(1.0079) + 2(15.9994) + 14.0067 = 153.180$$

The units of a molecular weight are either atomic mass units or grams per mole.

(b) What is the percentage by weight of oxygen in dopamine?

Solution. The formula $C_8H_{11}O_2N$ tells us that there are 2 mol of oxygen per mole of dopamine. One mole of O atoms has a mass of 15.9994 g.

Hence,

$$\frac{\text{mass of oxygen in 1 mol of } C_8H_{11}O_2N}{\text{mass of 1 mol of } C_8H_{11}O_2N} = \frac{2(15.9994)}{153.180} = 0.2089$$

Thus dopamine is 20.89% oxygen, by weight.

(c) What is the mass of 6.91×10^{-3} mol of $C_8H_{11}O_2N$?

Solution. The mass of 1 mol of dopamine is 153.180 g. The mass of 6.91×10^{-3} mol is therefore

$$(6.91 \times 10^{-3} \text{ mol})(153.180 \text{ g} \cdot \text{mol}^{-1}) = 1.06 \text{ g}$$

Note that as the number 6.91×10^{-3} is only given to three significant figures (Introduction, pages 14–16), the answer can only be reported to three figures; 1.058 rounds to 1.06.

(d) How many moles of dopamine are in a sample weighing 0.547 g?

Solution. Since the mass of 1 mol of dopamine is 153.180 g, a 0.547-g sample is considerably less than a mole. Make use of units to figure out how to proceed. The units of molecular weight are grams per mole ($g \cdot mol^{-1}$). We must divide grams by grams per mole in order to obtain an answer in moles.

$$(0.547 \text{ g}) \left(\frac{1 \text{ mol}}{153.180 \text{ g}} \right) = \frac{0.547 \text{ g}}{153.180 \text{ g/mol}} = 3.57 \times 10^{-3} \text{ mol}$$

A 0.547-g sample of $C_8H_{11}O_2N$ contains 3.57×10^{-3} mol of $C_8H_{11}O_2N$.

(e) How many moles of carbon are in a sample of dopamine of mass 0.547 g?

Solution. The molar ratio of carbon to dopamine is

$$\frac{\text{number of moles of carbon}}{\text{number of moles of dopamine}} = \frac{8}{1}$$

Since in part **(d)** we determined that there are 3.57×10^{-3} mol of $C_8H_{11}O_2N$ in a 0.547-g sample,

$$\text{number of moles of carbon} = 8(\text{number of moles of dopamine})$$
$$= 8(3.57 \times 10^{-3} \text{ mol}) = 0.0286 \text{ mol}$$

(f) How many molecules are in a sample of 0.547 g of dopamine?

Solution. There are 6.022×10^{23} molecules/mol, and 3.57×10^{-3} mol of dopamine in a 0.547-g sample, so the number of molecules of dopamine is

$$\left(\frac{6.022 \times 10^{23} \text{ molecules}}{1 \text{ mol}} \right)(3.57 \times 10^{-3} \text{ mol}) = 21.5 \times 10^{20}$$

$$= 2.15 \times 10^{21} \text{ molecules}$$

(g) How many C atoms are in a sample of $C_8H_{11}O_2N$ of mass 0.547 g?

Solution

$$\text{number of C atoms} = 8 \text{ (number of dopamine molecules)}$$
$$= 8(2.15 \times 10^{21}) = 1.72 \times 10^{22} \text{ carbon atoms}$$

EXAMPLE 1.5. Molar ratios

The commonly used pain reliever, aspirin, has the molecular formula $C_9H_8O_4$. If a sample of aspirin contains 0.968 g of carbon, what is the mass of hydrogen in the sample? What is the mass of oxygen in the sample?

Solution. In aspirin, the molar ratio of carbon to hydrogen is $9:8$.

$$\frac{\text{No. moles C}}{\text{No. moles H}} = \frac{9}{8}$$

Rearranging this ratio to solve for the number of moles of hydrogen we obtain

$$\text{No. moles H} = \tfrac{8}{9}(\text{No. moles C})$$

We must therefore calculate the number of moles of carbon in 0.968 g of carbon. The mass of 1 mol of carbon is 12.011 g.

$$\text{No. moles C} = (0.968 \text{ g})\left(\frac{1 \text{ mol}}{12.011 \text{ g}}\right) = \frac{0.968 \text{ g}}{12.011 \text{ g/mol}} = 8.06 \times 10^{-2} \text{ mol}$$

$$\text{No. moles H} = \tfrac{8}{9}(8.06 \times 10^{-2}) = 7.16 \times 10^{-2} \text{ mol}$$

The mass of 1 mol of hydrogen is 1.008 g. Hence the mass of hydrogen in the sample is

$$(7.16 \times 10^{-2} \text{ mol})(1.008 \text{ g} \cdot \text{mol}^{-1}) = 7.22 \times 10^{-2} \text{ g}$$

The molar ratio of oxygen to carbon in aspirin is $4:9$, so that

$$\text{No. moles O} = \tfrac{4}{9}(\text{No. moles C})$$
$$= \tfrac{4}{9}(8.06 \times 10^{-2}) = 3.58 \times 10^{-2} \text{ mol}$$

The mass of 1 mol of oxygen is 16.00 g. Hence the mass of oxygen in the sample is

$$(16.00 \text{ g} \cdot \text{mol}^{-1})(3.58 \times 10^{-2} \text{ mol}) = 0.573 \text{ g}$$

A sample of aspirin that contains 0.968 g of carbon also contains 7.22×10^{-2} g of hydrogen and 0.573 g of oxygen.

section 1.6
The Significance of "Molecular" Formulas and "Molecular" Weights for Ionic Crystalline Solids

In the previous section we discussed compounds that consist of discrete molecules. All gaseous compounds, most liquids, and some solids consist of discrete molecules. There are a number of liquids and solids, however, that do *not* consist of discrete molecules. Consider, for example, ordinary table salt, sodium chloride (NaCl). Sodium chloride exists as **ionic crystals** at room temperature. An **ion** is a charged particle, formed when an atom or a group of atoms loses or gains one or more electrons. The building blocks of ionic crystalline solids, such as sodium chloride, are ions and not atoms.

Cations, positively charged ions, are formed when atoms lose one or more electrons. **Anions,** negatively charged ions, are formed when atoms gain one or more electrons. Table 1.2 describes six common ions and the atoms from which the ions are formed.

Table 1.2. **The Relationship Between Atoms and Monatomic Ions**[a]

Name	Number of Protons	Number of Electrons	Kernel	Number of Valence Electrons
Sodium atom, Na·	11	11	Nucleus plus 10 electrons	1
Sodium ion, Na$^+$	11	10	Nucleus plus 10 electrons	0
Potassium atom, K·	19	19	Nucleus plus 18 electrons	1
Potassium ion, K$^+$	19	18	Nucleus plus 18 electrons	0
Calcium atom, Ca:	20	20	Nucleus plus 18 electrons	2
Calcium ion, Ca^{2+}	20	18	Nucleus plus 18 electrons	0
Oxygen atom, :Ö:	8	8	Nucleus plus 2 electrons	6
Oxide ion, :Ö:$^{2-}$	8	10	Nucleus plus 2 electrons	8
Chlorine atom, :Ċl:	17	17	Nucleus plus 10 electrons	7
Chloride ion, :Ċl:$^-$	17	18	Nucleus plus 10 electrons	8
Sulfur atom, :Ṡ:	16	16	Nucleus plus 10 electrons	6
Sulfide ion, :S̈:$^{2-}$	16	18	Nucleus plus 10 electrons	8

[a] Cations are formed when atoms lose valence electrons. Anions are formed when atoms gain valence electrons.

An ionic crystalline compound is itself electrically neutral; although it contains both anions and cations, its net charge is zero. The ions that make up sodium chloride are the sodium ion, Na$^+$, a cation with one positive charge equal in magnitude but opposite in sign to the charge on an electron, and the chloride ion, Cl$^-$, an anion with one negative charge equal to the charge on an electron. Crystals of sodium chloride contain an equal number of Na$^+$ and Cl$^-$ ions. (See Fig. 1.3.) Each Na$^+$ ion is

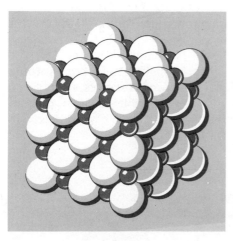

Fig. 1.3. Sodium chloride crystal. The larger spheres are the chloride (Cl$^-$) ions, and the smaller spheres are the sodium (Na$^+$) ions.

surrounded by six Cl^- ions, and each Cl^- is surrounded by six Na^+ ions, all equidistant. It is impossible to pick out any pair, that is, one Na^+ and one Cl^- ion, and to think of that pair as an identifiable unit, as a molecule. There is so such thing as a "molecule" of sodium chloride in the crystal.

Because crystals of sodium chloride are electrically neutral, the ratio of Na^+ ions to Cl^- ions must be $1:1$. To specify this ratio we write the formula of sodium chloride as NaCl.

The properties of ionic crystalline solids are quite different from the properties of solids consisting of discrete molecules. In Chapter 2 we will discuss the distinguishing characteristics of ionic and molecular solids. It is important to realize, however, that the *formula* does not tell you whether a substance is ionic or molecular. We must understand through previous familiarity that when we write NaCl it does not have the same meaning as when we write CO_2, or C_4H_{10}, or N_2O. The formulas C_4H_{10} and N_2O represent discrete, that is, individual molecules. The formula NaCl represents only the $1:1$ molar ratio of Na:Cl in the crystal.* The information the formula provides about **molar ratios** is the same whether the formula describes an ionic crystal or a discrete gaseous molecule. The formula C_4H_{10} tells us that the ratio of C atoms to H atoms in butane is $4:10$ (or $2:5$). The formula NaCl tells us that the ratio of Na atoms to Cl atoms in sodium chloride is $1:1$, and that the ratio of Na^+ cations to Cl^- anions is $1:1$.

The term "mole" comprehensively means Avogadro's number of particles, whatever the nature of the particles. Thus 1 mol of solid NaCl contains 1 mol of Na^+ ions and 1 mol of Cl^- ions.

Calcium phosphate, $Ca_3(PO_4)_2$, is another example of an ionic crystalline compound. Calcium phosphate, a white insoluble solid, is the principal ingredient of phosphate rock, which is the source in nature of the phosphates so widely used for fertilizers. The building blocks of calcium phosphate are calcium ions, Ca^{2+}, and phosphate ions, PO_4^{3-}. In order for the crystal to be electrically neutral, there must be three calcium ions for every two phosphate ions. Thus the formula $Ca_3(PO_4)_2$ means

$$\frac{\text{No. moles } Ca^{2+} \text{ ions}}{\text{No. moles } PO_4^{3-} \text{ ions}} = \frac{3}{2} \tag{1-11}$$

The formula $Ca_3(PO_4)_2$ also means that the ratio of Ca:P:O in this compound is $3:2:8$. We can interpret the formula correctly either by stating that 1 mol of $Ca_3(PO_4)_2$ contains 3 mol of Ca^{2+} ions and 2 mol of PO_4^{3-} ions, or by saying that 1 mol of $Ca_3(PO_4)_2$ contains 3 mol of Ca, 2 mol of P, and 8 mol of O.

Table 1.3 contrasts the information contained in the formula for each of two species: C_4H_{10}, a discrete gaseous molecule, and K_2CO_3, potassium carbonate, a white, soluble, ionic crystalline solid. The knowledge that C_4H_{10} consists of discrete molecules, while K_2CO_3 is an ionic crystalline solid, comes from a study of the physical and chemical properties of these two substances, and not from the formulas. Thus the information on lines (1) and (4) of Table 1.3 can be obtained only after we know whether a substance is molecular or ionic. We will discuss how to obtain this information in Chapter 2.

---

* Above 1413 °C, sodium chloride exists as a vapor. In the vapor phase discrete Na^+Cl^- units exist, and are termed "ion-pairs." The vapor phase of NaCl will be discussed in Chapter 14.

Table 1.3. Comparison of Information Contained in the Molecular Formula of a Discrete Molecule and an Ionic Crystalline Solid

C_4H_{10}, *Butane*	K_2CO_3, *Potassium Carbonate*
(1) 1 molecule of C_4H_{10} contains 4 atoms of C and 10 atoms of H	(1) No discrete molecules exist
(2) Molar ratio $C:H = 4:10$	(2) Molar ratio $K:C:O = 2:1:3$
(3) 1 mol C_4H_{10} contains 4 mol C plus 10 mol H	(3) 1 mol K_2CO_3 contains 2 mol K, 1 mol C, and 3 mol O
(4) No ions exist	(4) 1 mol K_2CO_3 contains 2 mol K^+ ions and 1 mol CO_3^{2-} ions

section 1.7
Distinction between the Terms Molecular Weight, Formula Weight, and Gram Atomic Weight

The mass of one mole of any substance that exists as discrete molecules is the molecular weight and is usually expressed in grams per mole ($g \cdot mol^{-1}$). (The SI unit is kilograms per mole.) Because the mass in grams of Avogadro's number of atoms (1 mol) of any element is equal to the atomic weight of that element, the mass of a mole of atoms is called its **gram atomic weight.** A **gram-atom** is simply a mole of atoms. Unless it is important, in a given context, to emphasize that atoms, rather than molecules, are being discussed, most chemists use the term mole for Avogadro's number of atoms, rather than the term gram-atom.

The weight of one mole of an ionic crystalline solid, or of any substance that does not exist as discrete molecules, is the **formula weight,** usually given in grams. Consider calcium oxide, CaO, a white, ionic crystalline solid, commercially called **lime.** The building blocks of CaO are calcium ions, Ca^{2+}, and oxide ions, O^{2-}. One mole of CaO contains 6.022×10^{23} Ca^{2+} ions and 6.022×10^{23} O^{2-} ions. Although a molecule of CaO does not exist in the solid, a mole of solid CaO is clearly defined.

The term **formula weight** (as opposed to molecular weight) is employed to emphasize the fact that the formula CaO does not imply the existence of a molecule of CaO. If we wish to distinguish between molecular species like N_2O and C_4H_{10}, and those compounds that do not exist as discrete molecules, such as NaCl and CaO, we make use of the term *molecular weight* for N_2O and C_4H_{10}, but *formula weight* for NaCl and CaO. In this text we will usually make this distinction, but there are chemists who use the term *molecular weight* for all compounds, and simply remember which substances do not actually consist of discrete molecules. For a substance that does exist as discrete molecules, the formula weight and the molecular weight are identical. (See Fig. 1.4.)

section 1.8
Empirical Formulas

Equation (1-8) describes the information about the molar ratio of two different atoms that are constituents of a particular compound:

$$\text{In compound } A_pB_q, \qquad \frac{\text{No. moles A}}{\text{No. moles B}} = \frac{p}{q} \qquad (1\text{-}8)$$

Fig. 1.4. The mass of one mole of sodium bicarbonate is its formula weight, because there are no molecules of "NaHCO$_3$," only a regular array of sodium (Na$^+$) ions and bicarbonate (HCO$_3^-$) ions. Similarly, the mass of one mole of sodium chloride is the formula weight of NaCl. The mass of one mole of water, however, is the molecular weight of H$_2$O, as discrete H$_2$O molecules exist.

Can we reason backwards from a knowledge of this ratio to obtain the formula of the compound? Suppose we know that in a certain gaseous compound containing only carbon and hydrogen

$$\frac{\text{No. moles C}}{\text{No. moles H}} = \frac{1}{3}$$

Is the formula of this compound CH$_3$, C$_2$H$_6$, C$_3$H$_9$, or C$_4$H$_{12}$? Clearly there is no way to tell which of an infinite number of possibilities is correct if all the information we have is in the form of the molar ratio, Eq. (1-8). The **simplest formula** is CH$_3$, but we cannot determine the correct molecular formula without additional information. All we can state with any certainty is that the formula is (CH$_3$)$_n$, where n is some integer. The simplest formula is sometimes called the **empirical formula,** although it is useful to consider the empirical formula to include the unknown integer, n. Thus we may say for this gas: The simplest formula is CH$_3$, and the empirical formula is (CH$_3$)$_n$. We can determine the molecular formula only after we have more information, specifically a knowledge of the molecular weight or the formula weight. If we know from some experimental evidence that the molecular weight of this gas is 30, we can reason as follows to determine the value of n:

The formula weight of the unit CH$_3$ is $12 + 3(1) = 15$.

The molecular weight of (CH$_3$)$_n$ is therefore $15n$.

If $15n = 30$, then $n = 2$. The molecular formula is therefore C$_2$H$_6$, and the exact molecular weight is 30.069 g · mol^{-1}.

As a second example, let us consider the simple sugar glucose, which contains the elements C, H, and O. If by experiment we determine that in glucose the molar ratios are C : H : O $= 1 : 2 : 1$, we have established that the simplest formula is CH$_2$O, and the empirical formula is (CH$_2$O)$_n$, where n is as yet unknown. A different experiment is required to provide the information that the molecular weight of glucose is 180. As the formula weight of CH$_2$O is 30, the molecular weight is $30n$, so that $n = 6$ in this case. The correct molecular formula is therefore C$_6$H$_{12}$O$_6$. It is preferable to write C$_6$H$_{12}$O$_6$ rather than (CH$_2$O)$_6$, because the latter would imply that six CH$_2$O units are bonded together, which is something about which we have no information from stoichiometric calculations. Other experimental evidence proves, in fact, that the structure of glucose is *not* described by the formula (CH$_2$O)$_6$.

Example 1.6 illustrates how to obtain the empirical formula of a compound from its elemental analysis, that is, from a knowledge of the percentage by weight of each element in the compound.

EXAMPLE 1.6. Simplest formula from elemental analysis

Esters are a class of compounds known for their pleasant fruit like odors and flavors. They are widely used in synthetic flavorings. Ethyl butyrate is an ester with an odor very much like that of pineapple. It is a colorless liquid at room temperature, and is known to contain only the elements C, H, and O. A sample of pure ethyl butyrate is analyzed and found to be 62.04% C and 10.41% H by weight. What is the empirical formula of ethyl butyrate?

Solution. To determine the empirical formula we need the molar ratio of C : H : O for ethyl butyrate. Imagine that we have a sample of ethyl butyrate that weighs exactly 100 g. (One may use *any* weight of ethyl butyrate, but it is simplest by far to begin with exactly 100 g because we are given the percent composition by weight, and multiplication by 100 is particularly easy.)

Our 100.00-g sample of ethyl butyrate contains 62.04 g of C and 10.41 g of H. The mass of oxygen is then obtained by difference.

$$\text{mass of O} = 100.00 \text{ g} - 62.04 \text{ g} - 10.41 \text{ g} = 27.55 \text{ g}$$

The number of moles of each element in the compound is calculated using its atomic weight:

$$\text{No. moles C} = \frac{62.04 \text{ g}}{12.011 \text{ g/mol}} = 5.165 \text{ mol}$$

$$\text{No. moles H} = \frac{10.41 \text{ g}}{1.008 \text{ g/mol}} = 10.33 \text{ mol}$$

$$\text{No. moles O} = \frac{27.55 \text{ g}}{16.00 \text{ g/mol}} = 1.722 \text{ mol}$$

To obtain the molar ratios, divide each of these by the smallest value, which in this case is 1.722. Hence,

$$\text{moles C} : \text{H} : \text{O} = \frac{5.165}{1.722} : \frac{10.33}{1.722} : \frac{1.722}{1.722} = 3.00 : 6.00 : 1.00$$

The simplest formula for ethyl butyrate is therefore C_3H_6O, and the empirical formula is $(C_3H_6O)_n$. It is not possible to determine the value of n from the information given. (The correct molecular formula for ethyl butyrate is $C_6H_{12}O_2$, so that n is 2.)

section 1.9
The Significance of Balanced Chemical Equations

When we speak of a **balanced chemical equation,** we mean an equation that describes a physical or chemical change, and is consistent with the requirement that *in any such process both mass and charge are conserved,* that is, they remain the same before and after the change has taken place.

Let us illustrate this with a specific example, the reaction between zinc metal and dilute hydrochloric acid, which produces hydrogen gas, H_2. The simplest laboratory method of preparing $H_2(g)$ involves the reaction of certain metals with aqueous acids, so that the reaction between zinc and hydrochloric acid is typical of a large number of chemical reactions. In order to write the equation correctly, one has to know that what we call hydrochloric acid is a solution of the colorless gas hydrogen chloride, $HCl(g)$, in water. The (g) written after the formula of a substance indicates that it is a gas; the symbol (s) indicates a solid, (ℓ) a liquid, and (aq) indicates a species that is in aqueous (water) solution.

If we dissolve $HCl(g)$ in water and examine the solution formed, we find that, as long as the solution is sufficiently dilute, no molecules of HCl can be observed in the solution. Three species are observed in aqueous hydrochloric acid: water molecules, **hydronium ions,** $H^+(aq)$,* and chloride ions, $Cl^-(aq)$. Thus we say that HCl ionizes in aqueous solution. Ionization in aqueous solution will be discussed in Chapter 7. The important thing to recognize here is that in order to write a correctly balanced equation we must first know the formulas of the substances we are mixing together. In this reaction, we are mixing solid zinc, $Zn(s)$, with aqueous hydrochloric acid, $H^+(aq)$ and $Cl^-(aq)$. Water is an essential part of this reaction, and its presence is indicated by the (aq) after the ions. Figure 1.5 is a photograph of the reaction between zinc and hydrochloric acid; we can readily observe that a gas is being produced. There are simple tests to prove that the gas is H_2. If the reaction is allowed to continue until it is complete, and if sufficient hydrochloric acid has been added, we will observe that all the zinc metal has dissolved. What remains is a clear, colorless solution, that contains zinc ions, $Zn^{2+}(aq)$, and chloride ions, $Cl^-(aq)$. Note that, in solution, the $Zn^{2+}(aq)$ ions and $Cl^-(aq)$ ions are uncombined; they move through the solution as separate entities.

If we simply write down on the left side of the arrow the formulas of all the species present when we start, and on the right side all the species present when the reaction is complete, we will have the following *unbalanced* equation:

$$Zn(s) + H^+(aq) + Cl^-(aq) \rightarrow Zn^{2+}(aq) + Cl^-(aq) + H_2(g) \qquad (1\text{-}12)$$

Note that in this reaction the only charged species are the ions that exist in aqueous solution.

Equation (1-12) is unbalanced with respect to hydrogen atoms, because there is only one H on the left-hand side {in $H^+(aq)$}, but there are two on the right-hand side in the molecule H_2. Hence we must multiply the $H^+(aq)$ by two. Since the H^+ came from HCl, and the formula HCl indicates a $1:1$ ratio of $H:Cl$, the solution of hydrochloric acid must contain H^+ and Cl^- in a $1:1$ ratio. For this reason, if we multiply H^+ by two we must also multiply Cl^- by two, on both sides of the equation. We now have a balanced total equation:

$$Zn(s) + 2H^+(aq) + 2Cl^-(aq) \rightarrow Zn^{2+}(aq) + 2Cl^-(aq) + H_2(g) \qquad (1\text{-}13)$$

Equation (1-13) tells us exactly what species were mixed together as reactants (on the left-hand side) and what species are present at the end of the reaction (on the right-hand side). It indicates the conservation of charge, as the net charge is the same on both sides of the equation, and it indicates the conservation of mass for each element.

If we look at Eq. (1-13) carefully, however, we can clearly see that the chloride ions take no part in this reaction. They are present in the hydrochloric acid and remain unchanged during the reaction, so they are present after the reaction is over as well. If

---

* In aqueous solution H^+ ions bond to one or more water molecules. The ion formed when one H^+ bonds to one H_2O is called a hydronium ion, H_3O^+. The hydronium ion is itself hydrated, that is, associated with other water molecules. (See Sections 7.1 and 7.4.) We use the symbol $H^+(aq)$ here as the simplest representation of the species formed by the bonding between H^+ and water.

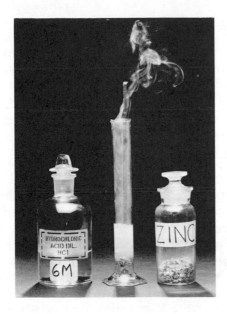

Fig. 1.5. The reaction between zinc metal and hydrochloric acid. The product of this reaction is hydrogen gas. The net ionic equation for the reaction is

$$Zn(s) + 2H^+(aq) \rightarrow Zn^{2+}(aq) + H_2(g)$$

we want to focus on the changes that occur during this chemical reaction, the Cl^- ions should not be included. We can properly write

$$Zn(s) + 2H^+(aq) \rightarrow Zn^{2+}(aq) + H_2(g) \qquad (1\text{-}14)$$

to describe all the changes taking place in this reaction.

Equation (1-14) is called a **net ionic equation.** Species that are present in solution, but take no part in the reaction that occurs, are omitted when one writes a net ionic equation. Ions that are omitted, like the Cl^- ions in Eq. (1-13), are called **spectator ions** or **bystander ions,** to indicate that they do not take part in the reaction.

Chemists almost always write net ionic equations rather than total equations. Note that both the left-hand side and the right-hand side of Eq. (1-14) have the same net charge of $+2$. ***All equations must be balanced with respect to charge as well as with respect to the number of moles of each kind of atom.*** The progression from Eq. (1-13) to Eq. (1-14) is entirely consistent with the rules of elementary algebra: A term that appears on both sides of an equation is canceled out. The rules of elementary algebra apply to chemical equations just as they do to algebraic equations.

Sometimes one may see the equation for the reaction between zinc and hydrochloric acid written as $Zn + 2HCl \rightarrow ZnCl_2 + H_2$. This is a balanced equation, but it does not describe the constituents of the solution, either before or after reaction. A dilute aqueous solution of HCl contains, in addition to water, only the ions $H^+(aq)$ and $Cl^-(aq)$. After the reaction, the solution contains zinc ions and chloride ions; it does not contain any "$ZnCl_2$." If we took the final solution of the products of this reaction, and allowed all the water to evaporate, we would obtain the white, crystalline, ionic solid whose formula is $ZnCl_2$ (zinc chloride), but in the aqueous solution there are separate Zn^{2+} and Cl^- ions. The equations we write should indicate the principal species actually present. For reactions in solution, we will use net ionic equations exclusively in this text. If pure gaseous HCl is passed over solid zinc, the reaction

$$Zn(s) + 2HCl(g) \rightarrow ZnCl_2(s) + H_2(g) \qquad (1\text{-}15)$$

occurs. In solution, the reaction that occurs is correctly represented by Eq. (1-14).

Significance of the Coefficients in a Balanced Equation

We may inquire precisely what information is conveyed by the numerical coefficients that appear in the balanced net ionic equation, Eq. (1-14). It is certainly correct to read this as "one atom of zinc reacts with two hydronium ions in aqueous solution to produce one zinc ion plus one molecule of hydrogen gas." That is not the most useful way of thinking about the equation, however, because we never deal with only one or two atoms, ions, or molecules.

Algebraically, an equation is still valid if we multiply each term by the same number. If we consider that multiplicative constant to be Avogadro's number, we can read the equation as "1 mol of solid zinc reacts with 2 mol of aqueous H^+ ions to form 1 mol of zinc ions plus 1 mol of H_2 gas." That is also correct, but it is still not the most useful way of thinking about the coefficients. It is too limiting. For the equation also means that "13 mol of solid zinc react with 26 mol of aqueous H^+ ions to form 13 mol of $Zn^{2+}(aq)$ ions and 13 mol of H_2 gas," or any other number of moles we might think of. *Remember: An equation tells you nothing about the actual amount of material used in any experiment.* You go to a cupboard containing chemicals and get out a bottle of granular zinc and pour some into a beaker. The amount you pour is totally unrelated to the equation you write down to describe what happens when you add hydrochloric acid to that zinc. The coefficients in the equation tell you only the **molar ratios** in which the species combine or are formed. The information contained in the net ionic Eq. (1-14) is a series of ratios:

$$\frac{\text{No. moles Zn used}}{\text{No. moles } H_2 \text{ formed}} = \frac{1}{1} \tag{1-16}$$

$$\frac{\text{No. moles } H^+ \text{ used}}{\text{No. moles } H_2 \text{ formed}} = \frac{2}{1} \tag{1-17}$$

or

$$\frac{\text{No. moles } Zn^{2+} \text{ formed}}{\text{No. moles } H^+ \text{ used}} = \frac{1}{2} \tag{1-18}$$

and so on. Remember that you can treat these ratios algebraically just like any other equation. Equation (1-18), for example, can be rearranged as

$$\text{No. moles } Zn^{2+} \text{ formed} = \tfrac{1}{2}(\text{No. moles } H^+ \text{ used}) \tag{1-18a}$$

Sometimes students ask if it is correct to write

$$\tfrac{1}{2}Zn(s) + H^+(aq) \rightarrow \tfrac{1}{2}Zn^{2+}(aq) + \tfrac{1}{2}H_2(g) \tag{1-19}$$

Obviously this makes no sense if you are thinking about the equation in terms of atoms or molecules; one cannot have one half of a zinc atom or one half of a molecule of H_2! It is certainly true that half a mole of solid zinc reacts with 1 mol of aqueous H^+ ions to form half a mole of Zn^{2+} ions and half a mole of H_2 gas. However, this information is conveyed by Eq. (1-14) as well. In principle, it makes no difference whether one writes the equation for the reaction as Eq. (1-14) or Eq. (1-19). The significant information is contained in the ratios of Eqs. (1-16), (1-17), and (1-18), and other ratios that can be written. In practice, we try to write the simplest equation, and most people would agree that Eq. (1-14) is simpler than Eq. (1-19), but Eq. (1-19) is not *wrong*. We will come across examples where two equations, differing only by a constant numerical factor, seem equally simple, and in that case it is a matter of taste which you choose to write. The reaction between aluminum metal and hydrochloric acid, for instance, may be written either as

$$Al(s) + 3H^+(aq) \rightarrow Al^{3+}(aq) + \tfrac{3}{2}H_2(g)$$

or as

$$2Al(s) + 6H^+(aq) \rightarrow 2Al^{3+}(aq) + 3H_2(g)$$

To summarize, consider a balanced equation of the form

$$2A + 3B \rightarrow 4C + D \tag{1-20}$$

What this equation means is that

$$\frac{\text{No. moles C formed}}{\text{No. moles B used}} = \frac{4}{3} \tag{1-21}$$

or

$$\frac{\text{No. moles C formed}}{\text{No. moles D formed}} = \frac{4}{1} \tag{1-22}$$

or

$$\frac{\text{No. moles B used}}{\text{No. moles A used}} = \frac{3}{2} \tag{1-23}$$

or

$$\frac{\text{No. moles A used}}{\text{No. moles C formed}} = \frac{2}{4} = \frac{1}{2} \tag{1-24}$$

and so on.

Once you are in the habit of thinking about chemical equations in terms of molar ratios, quantitative problems become straightforward. If we know, for instance, that 0.318 mol of D is formed in a given experiment utilizing reaction (1-20), then we know immediately that $3(0.318) = 0.954$ mol of B was used up, because

$$\text{No. moles B used} = 3(\text{No. moles D formed})$$

We also know that $4(0.318) = 1.272$ mole of C were formed, and that $2(0.318) = 0.636$ mol of A was used. The use of these molar ratios is illustrated in Examples 1.7 and 1.8.

EXAMPLE 1.7. Calculations using balanced equations

Aluminum metal reacts with hydrochloric acid to produce hydrogen gas. The balanced net ionic equation for the reaction is

$$Al(s) + 3H^+(aq) \rightarrow Al^{3+}(aq) + \tfrac{3}{2}H_2(g)$$

In a laboratory experiment, a student dissolved 3.85 g of solid aluminum in excess hydrochloric acid.

(a) How many moles of HCl were used up in this reaction?

Solution. The equation tells us the **molar ratio** in which the Al(s) and $H^+(aq)$ ions combine:

$$\frac{\text{No. moles Al used}}{\text{No. moles H}^+ \text{ used}} = \frac{1}{3}$$

which can be rearranged to read

$$\text{No. moles H}^+ \text{ used} = 3(\text{No. moles Al used})$$

We must therefore calculate the number of moles of Al that were used up in this reaction. We know that 3.85 g of Al were used up, so the mass in grams must be converted to moles. To do this we need the atomic weight of Al, which is 26.98 $g \cdot mol^{-1}$.

$$(3.85 \text{ g}) \left(\frac{1 \text{ mol}}{26.98 \text{ g}} \right) = \frac{3.85 \text{ g}}{26.98 \text{ g/mol}} = 0.1427 \text{ mol} = 0.143 \text{ mol}$$

The number of moles of H^+ used up was therefore

$$3(\text{No. moles Al used}) = 3(0.1427) = 0.428 \text{ mol } H^+$$

Note that in order to avoid a rounding error in the number of moles of $H^+(aq)$ used, we carry the fourth figure in the number of moles of Al, and round the answer to the correct number of significant figures (Introduction, pages 16–17) only once. You can perhaps see this more readily if we do not explicitly calculate the number of moles of aluminum used, and write simply

$$\text{No. moles } H^+(aq) \text{ used} = \frac{3(3.85 \text{ g})}{26.98 \text{ g/mol}} = 0.428 \text{ mol}$$

Since the molar ratio of $H^+(aq)$ ions in solution to HCl is $1:1$, 0.428 mol of HCl was used up in this reaction.

(b) How many grams of $H_2(g)$ were formed in this reaction?

Solution. The balanced equation tells us

$$\frac{\text{No. moles } H_2 \text{ formed}}{\text{No. moles } H^+ \text{ used}} = \frac{3/2}{3} = \frac{1}{2}$$

Thus,

$$\text{No. moles } H_2 \text{ formed} = \tfrac{1}{2}(0.428 \text{ mol}) = 0.214 \text{ mol}$$

Each mole of H_2 gas has a mass of $2(1.0079 \text{ g}) = 2.016 \text{ g}$. To obtain the mass of H_2 formed, multiply (mole)(gram/mole):

$$(0.214 \text{ mol})(2.016 \text{ g} \cdot mol^{-1}) = 0.431 \text{ g of } H_2$$

The reaction produces 0.214 mol or 0.431 g of H_2 gas.

EXAMPLE 1.8. Balancing chemical equations

Butane gas, C_4H_{10}, burns in oxygen to yield CO_2 and H_2O.

(a) Balance the equation for the combustion of butane.

Solution. Begin with the skeleton equation showing the formulas of the reactants and products:

$$C_4H_{10} + O_2 \rightarrow CO_2 + H_2O$$

Balance the C atoms. There are 4 C atoms on the left and only one on the right, so the CO_2 must be multiplied by four. Balance the H atoms. There are 10 H atoms on the left and only 2 on the right, so multiply the H_2O by five. The equation now appears as

$$C_4H_{10} + O_2 \rightarrow 4CO_2 + 5H_2O$$

Balance the O atoms. There are 13 O atoms on the right, 8 in the $4CO_2$, and 5 in the $5H_2O$. There are only 2 on the left. Multiply the O_2 by $\frac{13}{2}$. The correct balanced equation is

$$C_4H_{10}(g) + \tfrac{13}{2} O_2(g) \rightarrow 4CO_2(g) + 5H_2O(liq)$$

(b) How many grams of CO_2 are obtained by the complete combustion of 8.8347 g of C_4H_{10}?

Solution. The molecular weight of C_4H_{10} is 58.123 g/mol. Thus the number of moles of C_4H_{10} burned is

$$\frac{8.8347 \text{ g}}{58.123 \text{ g/mol}} = 0.15200 \text{ mol}$$

The coefficients of the equation tell us that

$$\text{No. moles } CO_2 \text{ formed} = 4(\text{No. moles } C_4H_{10} \text{ used up})$$

so that the number of moles of CO_2 formed = 4(0.15200) = 0.60800. The molecular weight of CO_2 is 44.010 g/mol, so the mass of CO_2 formed is

$$(0.60800 \text{ mol})(44.010 \text{ g/mol}) = 26.578 \text{ g}$$

(c) How many moles of O_2 are consumed in this combustion?

Solution
$$\frac{\text{moles } O_2 \text{ used}}{\text{moles } C_4H_{10} \text{ used}} = \frac{13}{2}$$

Thus the number of moles of O_2 consumed = $(\frac{13}{2})(0.15200) = 0.9880$ mol.

---

section 1.10
Stoichiometric Calculations

A great many of the questions that chemists want to answer require a knowledge of the quantitative relations that exist between the substances involved in chemical reactions. Quantitative calculations of chemical composition and reaction are referred to as **stoichiometry.** While the questions we seek to answer are diverse, the methods used to perform the calculations are basically the same and involve only the following three concepts.

Concept 1. The mole. One mole of any substance contains 6.022×10^{23} particles of that substance (atoms, molecules, or ions), and has a mass equal to the formula weight, in grams. Hence,

$$\text{No. moles} = \frac{\text{mass in grams}}{\text{formula weight in grams/mole}} \tag{1-25}$$

which can be rearranged as

$$\text{mass in grams} = (\text{No. moles})(\text{formula weight in grams/mole}) \tag{1-26}$$

In using either Eq. (1-25) or Eq. (1-26) it is important to remember that all three quantities refer to the *same* chemical substance. That is, to obtain the number of moles of a given substance, you must know the mass in grams of that substance, and the formula weight in grams per mole of the same substance.

A common student error, particularly during the stress of an exam, is to write one of these equations incorrectly, for example, to interchange the numerator and denominator in Eq. (1-25). If you will always check the *units* on both sides of any equation you write down, you will spot an error immediately, and will be able to correct the equation you have written before you begin to do the numerical calculations. Treat the units of each quantity algebraically in exactly the same manner you treat the quantities themselves. (Refer to the section on **dimensional analysis.** Introduction, pages 10–12.) On the right side of Eq. (1-25), for instance, the units are

$$g/(g \cdot mol^{-1}) \quad \text{or} \quad g(mol \cdot g^{-1}) = mol$$

the same units as on the left side. Note that if you had incorrectly inverted the fraction on the right, its units would be

$$(g \cdot mol^{-1})/g = mol^{-1}$$

If you are in the habit of *always* checking the units on both sides of an equation, you would say to yourself, "I have moles on the left side and reciprocal moles (mol^{-1}) on the right side—something is wrong and I'd better not go on until I recheck the equation I used."

Concept 2. The conservation of mass. In any chemical reaction mass is conserved, that is, the number of moles of atoms of any particular element is the same after the reaction as it is before the reaction. Atoms are neither created nor destroyed during a chemical reaction. The element may have combined with something new, but the total number of moles of that element does not change. Thus you may convert the carbon in sucrose ($C_{12}H_{22}O_{11}$, common sugar) to carbon dioxide by burning it in excess oxygen, but the number of moles of carbon in the sucrose before the combustion is identical to the number of moles of carbon in the carbon dioxide after the combustion. In stoichiometric calculations, it is essential to equate the number of moles of any particular element before and after reaction.

Concept 3. The distinction between an empirical formula and an exact molecular formula. A knowledge of the percentage by weight of each element in a compound allows us to calculate only an **empirical formula** for the compound, giving the molar ratios of the atoms combined in the compound. In order to determine the exact molecular formula from the empirical formula, the molecular weight of the compound must be known.

We will now apply these three concepts to a number of different problems. You may think of a way to solve some of these problems that is different from the solution provided. There are often several correct methods of solving a given problem. We try to use the simplest method, the one that involves the least amount of numerical calculation.

EXAMPLE 1.9. Significance of a chemical formula and definition of a mole

Acetone, C_3H_6O, is a colorless liquid at room temperature with a somewhat pungent odor. It is an excellent solvent for a wide variety of organic substances, and is used, for instance, in nail polish remover. A sample of acetone has a mass of 27.56 g.

(a) How many moles of acetone are in this sample?

Solution. Equation (1-25) relates the mass of a sample of a compound to the number of moles of that compound. The molecular weight of $C_3H_6O = 3(12.011) + 6(1.0079) + 15.9994 = 58.080 \, g \cdot mol^{-1}$. Note that the mass of the sample is given to four significant figures. (Introduction, pages 12–13.) That means the molecular weight used should have at least four figures. To avoid accumulating rounding errors in calculating a molecular weight to four significant figures, use five figures in each atomic weight.

Applying Eq. (1-25) to this sample of acetone we obtain:

$$\text{No. moles } C_3H_6O = (25.76 \text{ g})\left(\frac{1 \text{ mol}}{58.08 \text{ g}}\right) = \frac{27.56 \text{ g}}{58.08 \text{ g/mol}} = 0.4745 \text{ mol}$$

(b) How many moles of carbon are in this sample?

Solution. The formula C_3H_6O tells us that

$$\frac{\text{No. moles C}}{\text{No. moles } C_3H_6O} = \frac{3}{1} \quad \text{as in Eq. (1-9)}$$

Rearranging this equation to solve for the number of moles of carbon we obtain

$$\text{No. moles C} = 3(\text{No. moles } C_3H_6O) = 3(0.4745 \text{ mol})$$
$$= 1.424 \text{ mol C}$$

(c) If all the carbon in this sample is converted to carbon dioxide, CO_2, by burning the acetone in excess oxygen, how many grams of CO_2 will be produced?

Solution. The number of moles of CO_2 produced will be equal to the number of moles of C in the sample of acetone, because the molar ratio of $C:CO_2$ is $1:1$, and the number of moles of carbon is conserved. Hence, there are 1.424 mol of CO_2 produced by the combustion. To calculate the mass of this CO_2 we use Eq. (1-26). We need the molecular weight of CO_2, which is $12.011 + 2(15.9994) = 44.010 \text{ g} \cdot \text{mol}^{-1}$. Therefore,

$$\text{mass of } CO_2 = 3(0.4745 \text{ mol})(44.010 \text{ g} \cdot \text{mol}^{-1}) = 62.65 \text{ g}$$

Note particularly that we need 3(0.4745) rather than the rounded value, 1.424, for the number of moles of carbon dioxide. To avoid rounding errors in a desired answer, carry one additional figure in all intermediate values, and round only once, at the end of the calculation.

EXAMPLE 1.10. Empirical formula and molecular formula from chemical analysis of an organic compound

(a) A white crystalline solid is known to contain only C, H, and O. A sample of this compound weighing 0.4647 g is burned with excess oxygen, which completely converts all the carbon to gaseous CO_2 and all the hydrogen to liquid H_2O. These products are trapped separately and weighed. It is found that the mass of the CO_2 produced is 0.8635 g, and the mass of the H_2O is 0.1767 g. What is the empirical formula of the compound?

Solution. The way to approach this problem is to focus on the conservation of mass. From the mass of CO_2 produced we can obtain the number of moles of CO_2 produced, and therefore the number of moles of carbon in the product, which must be equal to the number of moles of carbon in the original sample of compound. The molecular weight of CO_2 is $44.01 \text{ g} \cdot \text{mol}^{-1}$, therefore

$$\text{No. moles } CO_2 = \frac{0.8635 \text{ g}}{44.01 \text{ g/mol}} = 1.962 \times 10^{-2} \text{ mol}$$

Since the molar ratio of $C:CO_2 = 1:1$, there must have been 1.962×10^{-2} mol of C in the original sample also. We can therefore calculate the mass of carbon in the original sample:

$$\text{mass of C in sample} = (1.962 \times 10^{-2} \text{ mol})(12.011 \text{ g} \cdot \text{mol}^{-1}) = 0.2357 \text{ g}$$

Now we apply the law of conservation of mass to the hydrogen atoms, in order to calculate the mass of H in the original sample. We will need the molecular weight of H_2O, which is $18.0152 \text{ g} \cdot \text{mol}^{-1}$.

$$\text{No. moles } H_2O = \frac{0.1767 \text{ g}}{18.015 \text{ g/mol}} = 9.808 \times 10^{-3} \text{ mol}$$

Since the molar ratio of $H : H_2O$ is $2 : 1$,

$$\text{No. moles H} = 2(\text{No. moles } H_2O) = 2(9.808 \times 10^{-3}) = 1.962 \times 10^{-2} \text{ mol}$$

By conservation of mass, there was 1.962×10^{-2} mol of H in the original sample of compound. The mass of H in the original sample is calculated as follows:

$$\text{mass of H} = (1.962 \times 10^{-2} \text{ mol})(1.0079 \text{ g} \cdot \text{mol}^{-1}) = 0.01977 \text{ g}$$

We are now able to calculate the mass of oxygen in the original sample. Since the mass of the sample was 0.4647 g,

$$\text{mass of sample} = 0.4647 \text{ g} = \text{mass of C} + \text{mass of H} + \text{mass of O}$$
$$= 0.2357 \text{ g} + 0.01977 \text{ g} + \text{mass of O}$$

Hence,

$$\text{mass of O} = 0.4647 \text{ g} - 0.2357 \text{ g} - 0.01977 \text{ g} = 0.2092 \text{ g}$$

From the mass of oxygen, we can obtain the number of moles of O in the sample.

$$\text{No. moles O} = \frac{0.2092 \text{ g}}{15.9994 \text{ g/mol}} = 1.308 \times 10^{-2} \text{ mol}$$

The molar ratios are therefore $C : H : O = 1.962 : 1.962 : 1.308$, since the factor of 10^{-2} in each term cancels out when you take the ratio. To express this as a ratio of simple integers, divide by the smallest number (1.308), and keep at least three significant figures.

$$\frac{1.962}{1.308} : \frac{1.962}{1.308} : \frac{1.308}{1.308} = 1.500 : 1.500 : 1.000$$

Since we do not have integers after dividing by the smallest number, multiply through by the smallest integer that will yield only integers in the ratios. In this case, multiply through by two.

$$1.500 : 1.500 : 1.000 = 3.000 : 3.000 : 2.000$$

and the simplest formula of this compound is $C_3H_3O_2$. The empirical formula is $(C_3H_3O_2)_n$, where n is some integer.

Note particularly that you do not round the number of moles of each element to obtain integers; the molar ratios must *be* integral within the limits of experimental error. Thus we did not round 1.962 to 2, nor did we round 1.308 to 1. Always divide first by the smallest number of moles.

(b) In a separate experiment the molecular weight of the compound was determined to be 140 ± 5. What is the correct molecular formula and the exact molecular weight of this compound?

Solution. The molecular weight given has a relatively large experimental uncertainty, and there is therefore no need to calculate the formula weight of $C_3H_3O_2$ to more than two figures. The formula weight of $C_3H_3O_2$ is 71, so the molecular weight of $(C_3H_3O_2)_n$ is $71n$. The only integer for which $71n$ is within the range betwen 135 and 145 is $n = 2$. Hence the correct molecular formula is $C_6H_6O_4$, and the exact molecular weight is $142.11 \text{ g} \cdot \text{mol}^{-1}$.

In Example 1.10(b) note that it was not necessary to know the molecular weight very precisely in order to obtain the correct molecular formula. Even with a large uncertainty in the experimentally determined molecular weight, it is possible to decide which value of n, in the empirical formula, is correct. The value of n must, of course, be an exact integer.

EXAMPLE 1.11. Determination of an atomic weight

Element X reacts with oxygen to produce a compound of formula X_2O_5. In an experiment it is found that 2.0769 g of pure X produce 3.7076 g of pure X_2O_5. If the atomic weight of oxygen is known to be 15.9994 g · mol^{-1}, calculate the atomic weight of X.

Solution. The mass of oxygen in the X_2O_5 produced in this experiment is

$$3.7076 \text{ g} - 2.0769 \text{ g} = 1.6307 \text{ g of O atoms}$$

Therefore the number of moles of oxygen in the X_2O_5 produced in this experiment is

$$(1.6307 \text{ g}) \left(\frac{1 \text{ mol}}{15.9994 \text{ g}} \right) = \frac{1.6307 \text{ g}}{15.9994 \text{ g/mol}} = 0.10192 \text{ mol}$$

The molar ratio of X : O in this compound is

$$\frac{\text{No. moles X}}{\text{No. moles O}} = \frac{2}{5}$$

This ratio can be rearranged to solve for the number of moles of X:

$$\text{No. moles X} = \tfrac{2}{5}(\text{No. moles O}) = \tfrac{2}{5}(0.10192 \text{ mol})$$
$$= 0.040769 \text{ mol X}$$

The sample of pure X used in this experiment weighed 2.0769 g. Therefore the product X_2O_5 also contained 2.0769 g of X, because mass is conserved; the amount of X before and after reaction must be the same. Since we know both the number of moles of X and the mass of X, we can calculate the atomic weight of X as follows:

$$\frac{2.0769 \text{ g of X}}{\text{atomic weight of X in g/mol}} = 0.040769 \text{ mol of X}$$

Rearranging this to solve for the atomic weight of X we obtain

$$\text{atomic weight of X} = \frac{2.0769 \text{ g}}{0.040769 \text{ mol}} = 50.943 \text{ g} \cdot \text{mol}^{-1}$$

The element is vanadium, atomic number 23.

EXAMPLE 1.12. The reaction of a mixture of substances to yield a single product

All the thallium metal is extracted from a 5.9018-g sample of a mixture of thallium(I) chloride, TlCl, and thallium(I) oxide, Tl_2O. The mass of thallium obtained is 5.3995 g. What was the percentage by weight of TlCl in the original sample?

Solution. Let

$$x = \text{mass of TlCl in the sample}$$

Then,

$$5.9018 - x = \text{mass of } Tl_2O \text{ in the sample}$$

The formula weights of TlCl and Tl_2O are, respectively, 239.82 and 424.74

$g \cdot mol^{-1}$. Thus the number of moles of each of the two compounds in the mixture is given by:

$$\text{No. moles TlCl} = \frac{x \text{ g}}{239.82 \text{ g/mol}}$$

$$\text{No. moles Tl}_2\text{O} = \frac{5.9018 - x}{424.74 \text{ g/mol}}$$

The atomic weight of thallium is 204.37 g/mol, so the number of moles of thallium extracted from the mixture is

$$\frac{5.3995 \text{ g}}{204.37 \text{ g/mol}} = 0.026420 \text{ mol Tl}$$

The molar ratio of Tl : TlCl is 1 : 1. The molar ratio of Tl : Tl$_2$O = 2 : 1. Thus the total number of moles of thallium in the mixture is

$$\frac{x}{239.82} + 2\left(\frac{5.9018 - x}{424.74}\right) = 0.026420 \text{ mol Tl}$$

Multiply by 424.74. We obtain $1.77108x + 2(5.9018 - x) = 11.22163$.
Collecting terms yields $0.5820 = 0.22892x$, so that

$$x = 2.5424 \text{ g} = 2.542 \text{ g of TlCl in the sample}$$

The percentage by weight of TlCl in the original sample was

$$\left(\frac{2.5424 \text{ g}}{5.9018 \text{ g}}\right) \times 100 = 43.08\%$$

section 1.11
The Limiting Reagent and the Yield of Product

Although substances always combine in the simple integral ratios specified by the equation for the reaction, we can mix them together in any arbitrary way we choose. Usually one or more species are in excess when we mix reactants. The maximum possible yield of product is then determined by the reactant that is not in excess, which is called the **limiting reagent**. The following example illustrates how to determine which reactant is the limiting reagent.

EXAMPLE 1.13. **The limiting reagent**

What is the maximum amount of the red-brown gas nitrogen dioxide, NO_2, that can be produced by mixing 3.823 g of NO(g) and 2.886 g of O_2(g)?

Solution. The balanced equation for the reaction is

$$2NO(g) + O_2(g) \rightarrow 2NO_2(g)$$

This equation tells us that NO and O_2 combine in a 2 : 1 molar ratio. They have not been mixed in that ratio, however. First calculate the number of moles of each reactant that have been mixed together. The molecular weights of NO and O_2 are 30.006 and 31.999 $g \cdot mol^{-1}$, respectively.

$$\text{No. moles NO} = \frac{3.823 \text{ g}}{30.006 \text{ g/mol}} = 0.1274 \text{ mol}$$

$$\text{No. moles O}_2 = \frac{2.886 \text{ g}}{31.999 \text{ g/mol}} = 0.09019 \text{ mol}$$

The 2 : 1 molar ratio of NO : O_2 tells us that for the reaction to completely consume 0.1274 mol of NO, only 0.06370 mol of O_2 is required. There is, therefore, excess O_2 present. The maximum amount of NO_2 that can be produced is determined by the amount of NO present, and not by the amount of O_2 present. The **limiting reagent is** therefore NO.

Since

$$\frac{\text{No. moles NO}_2 \text{ formed}}{\text{No. moles NO used}} = \frac{2}{2} = \frac{1}{1}$$

the maximum amount of NO_2 that can be produced in this reaction is 0.1274 mol. As the molecular weight of NO_2 is 46.0055 g · mol^{-1}, the maximum mass of NO_2 that can be produced is

$$(0.1274 \text{ mol})(46.0055 \text{ g} \cdot \text{mol}^{-1}) = 5.861\text{-g NO}_2$$

It is not at all uncommon for the actual yield of a product to be less than the theoretical maximum yield, particularly for reactions involving organic reagents. We therefore define the **percentage yield of product** as

$$\text{Percentage yield of product} = \frac{\text{actual yield}}{\text{theoretical maximum yield}} \times 100$$

Some of the reasons why the actual yield may be less than the maximum possible yield are the following:

1. Side reactions may occur, that is, other reactions in addition to the principal one may take place.
2. Procedures may be necessary to separate the product from the reaction mixture and obtain it in a pure state. Product may be lost during the separation and purification processes.
3. The reaction may not go to completion. As we will discuss in detail in later chapters, not all reactions proceed completely to the right.
4. There may have been impurities in one or more of the reactants.

The following example involves the determination of the limiting reagent and the calculation of the percentage yield of product.

EXAMPLE 1.14. Theoretical yield and actual yield

A 28.36-g sample of bright green crystalline nickel(II) chloride hexahydrate, $NiCl_2 \cdot 6H_2O$, was dissolved in water to yield an apple green solution. When 30.0 mL of ethylenediamine, $NH_2CH_2CH_2NH_2$, a colorless, musky-smelling liquid, was added to the solution while stirring, the solution turned purple. After the resulting solution was cooled in an ice bath, beautiful deep violet crystals precipitated out. The equation for the reaction that occurred is

$$Ni^{2+}(aq) + 3NH_2CH_2CH_2NH_2(\ell) + 2Cl^-(aq) + 2H_2O \rightarrow$$
$$Ni(NH_2CH_2CH_2NH_2)_3Cl_2 \cdot 2H_2O(s)$$

(a) If the density of ethylenediamine is 0.900 g · mL^{-1}, what is the theoretical maximum yield, in grams, of the solid product?

Solution. We must first find out how many moles of each of the reactants are present at the start of the reaction.

Crystals of nickel(II) chloride contain 6 mol of water per mol of $NiCl_2$ in the crystalline structure, and the compound is called a hexahydrate (hexa = six; hydrate = water). It is quite common for water molecules to be included in the crystal structures of inorganic compounds. The formula weight of $NiCl_2 \cdot 6H_2O$ is 237.70 $g \cdot mol^{-1}$. Hence,

$$\text{No. moles } NiCl_2 \cdot 6H_2O = \frac{28.36 \text{ g}}{237.70 \text{ g/mol}} = 0.1193 \text{ mol}$$

The apple green solution therefore contains 0.1193 mol of $Ni^{2+}(aq)$ ions and $2(0.1193 \text{ mol}) = 0.2386$ mol of $Cl^-(aq)$ ions.

The molecular weight of ethylenediamine is 60.11 $g \cdot mol^{-1}$. We are not given the mass of ethylenediamine added, however, but the volume, as this is the most convenient way of measuring liquid reagents. The **density** of any substance (Introduction, page 7) is defined as the mass per unit volume, and is the quantity needed if we want to find the mass given the volume, or vice versa. We can rearrange the definition

$$\text{density in } g \cdot mL^{-1} = \frac{\text{mass in g}}{\text{volume in mL}}$$

to obtain

$$\text{mass in g} = (\text{density in } g \cdot mL^{-1})(\text{volume in mL})$$

The mass of ethylenediamine used in this experiment was therefore

$$\text{mass of } NH_2CH_2CH_2NH_2 = (30.0 \text{ mL})(0.900 \text{ g} \cdot mL^{-1}) = 27.0 \text{ g}$$

We can now calculate the number of moles of ethylenediamine used:

$$\text{No. moles } NH_2CH_2CH_2NH_2 = \frac{27.0 \text{ g}}{60.11 \text{ g/mol}} = 0.449 \text{ mol}$$

The number of moles of $NH_2CH_2CH_2NH_2$ is only known to three significant figures because both the volume and the density of ethylenediamine are only given to three figures. The equation tells us that in this reaction

$$\frac{\text{No. moles } NH_2CH_2CH_2NH_2}{\text{No. moles } Ni^{2+}} = \frac{3}{1}$$

We began with 0.1193 mol of Ni^2, so if *all* the Ni^{2+} ions in the solution are used to form the precipitate, they would react with $3(0.1193 \text{ mol}) = 0.3579$ mol of ethylenediamine. There is, therefore, excess ethylenediamine in the solution. The Ni^{2+} (or the $NiCl_2 \cdot 6H_2O$) is the limiting reagent; the maximum possible yield of product is determined by the amount of Ni^{2+} in solution, and not by the amount of ethylenediamine. Since the molar ratio of Ni^{2+} to product is 1:1, the maximum possible amount of product is 0.1193 mol. The formula weight of the product is 345.98 $g \cdot mol^{-1}$, and hence the theoretical maximum yield of the violet crystals is

$$(0.1193 \text{ mol})(345.98 \text{ g} \cdot mol^{-1}) = 41.28 \text{ g}$$

(b) The deep violet crystalline product was separated from the solution by filtration, dried in air, and weighed. The mass of product collected was 32.48 g. What was the percentage yield of product in this reaction?

Solution

$$\text{percentage yield} = \frac{\text{actual mass of product}}{\text{theoretical mass of product}} \times 100$$

In this experiment, therefore,

$$\text{percentage yield} = \frac{32.48 \text{ g}}{41.28 \text{ g}} \times 100 = 78.68\%$$

A word of caution is in order before you attempt to solve the problems at the end of this chapter. Many students think that they should be able to read a problem and know right away how to go about solving it, that is, that they should see the way to the final answer immediately. Except for the simplest problems, that is not a correct assumption. You may often begin working a problem without a complete idea of how you will end up. You must learn to reason your way through a problem, and to ask yourself many intermediate questions as you proceed. Begin by writing the definitions and units of the terms involved in the question. Remember that there are often several correct methods of solving a given problem. You should never try to memorize the solution to one of the problems worked out in the text. It is the reasoning that is important, not any one method of approach.

Summary

An **atom** of any element consists of a positively charged **nucleus** (containing **protons** and **neutrons**) surrounded by negatively charged, rapidly moving **electrons.** All atoms are electrically neutral. When an atom loses or gains one or more electrons, charged species called **ions** are formed.

Elements can be classified by their **atomic number,** the number of protons in the nucleus. All atoms of any given element have the same atomic number. The sum of the number of protons and neutrons in the nucleus is the **mass number.** Atoms of the same element with different mass numbers are called **isotopes.** Most elements exist in nature as a mixture of several isotopes.

The **atomic weight scale** compares the average mass of an atom of an element to the mass of a ^{12}C atom. The atomic weight of a ^{12}C atom is defined to be exactly 12 **atomic mass units.** Exactly 12 g of ^{12}C is called one **mole** of ^{12}C. The number of atoms in 12 g of ^{12}C is **Avogadro's number,** $N_A = 6.022 \times 10^{23}$. One mole of any element has a mass, in grams, equal to the atomic weight of that element, and contains Avogadro's number of atoms of that element.

Atoms can combine to form **molecules.** The **formula** of a compound describes its atomic composition. Subscripts on the symbol for the element indicate the number of atoms of that element in one molecule of the compound. One mole of a compound that consists of discrete molecules of formula $A_p B_q$ contains p moles of atom A and q moles of atom B. The **molecular weight** of a compound is the sum of the atomic weights of all atoms in the compound. The mass of one mole (Avogadro's number of molecules) of the compound is equal to its molecular weight in grams.

Many inorganic solids do not consist of discrete molecules, but are **ionic crystals** in which the building blocks are **ions** and not atoms. All ionic crystalline solids contain both positively charged ions **(cations)** and negatively charged ions **(anions),** in ratios such that the entire crystal is electrically neutral. The formula of an ionic crystalline solid indicates the *ratio* of ions in the crystal, as no discrete molecule exists. The mass of one mole of an ionic crystalline solid is called its **formula weight.**

A **balanced chemical equation** describes a physical or chemical change and shows that in any such process both mass and charge are conserved. The coefficients in a balanced equation indicate the **molar ratios** in which the substances taking part in the reaction combine, or are formed. The equation provides no information about the actual amount of material used or formed in any given experiment.

Stoichiometric calculations are used to obtain quantitative relationships between substances as they undergo chemical changes. In solving stoichiometric problems, three basic ideas are employed: 1. the concept of a mole, 2. the conservation of mass, and 3. the distinction between an **empirical formula** and a molecular formula.

The maximum possible yield of a product of a reaction is determined by the number of moles of the **limiting reagent,** the reacting species that is not present in excess. Therefore, in determining the maximum possible amount of product formed, one should first calculate the number of moles of each reactant, and ascertain which reactant is the limiting reagent.

Exercises

Sections 1.1 and 1.2

1. What is the number of protons, neutrons, and electrons, in atoms of ^{39}K, ^{40}K, and ^{41}K?

2. State the number of protons, neutrons, and electrons in atoms of all the stable isotopes of mercury: ^{196}Hg, ^{198}Hg, ^{199}Hg, ^{200}Hg, ^{201}Hg, ^{202}Hg, and ^{204}Hg.

3. Atoms with the same mass number but different atomic number are called **isobars.** What is the number of protons, neutrons, and electrons in the isobars ^{19}F, ^{19}Ne, and ^{19}O?

4. The following elements all have two valence electrons: Be, Mg, Ca, Sr, and Ba. How many electrons are in the core of each of these atoms?

Section 1.3

5. There are only two naturally occurring isotopes of the element gallium: ^{69}Ga (60.4%) with an atomic mass of 68.9257, and ^{71}Ga (39.6%) with an atomic mass of 70.9249. Calculate the atomic weight of gallium.

Section 1.4

6. Calculate the average mass, in grams, of an atom of chlorine, Cl, in a naturally occurring sample. Why is this an average mass and not the actual mass of an atom of chlorine?

7. What is the mass, in kilograms, of a mole of copper?

8. What is the ratio of the average mass of an atom of gold, Au, to the average mass of an atom of silver, Ag?

Section 1.5

9. Sucrose (ordinary sugar) has the formula $C_{12}H_{22}O_{11}$. (a) What is the molecular weight of sucrose? (b) What is the percentage by weight of carbon in sucrose?

10. Citric acid is a solid substance with formula $C_6H_8O_7$. How many moles of carbon are there in a sample of citric acid with a mass of 4.892 g?

11. Cholesterol has the formula $C_{27}H_{46}O$. How many moles of cholesterol are in a sample of mass 13.97 g? How many moles of hydrogen are in this sample?

12. The compound commonly known as bicarbonate of soda is $NaHCO_3$, sodium hydrogen carbonate, or sodium bicarbonate. What is the mass of oxygen in 0.8561 g of $NaHCO_3$?

13. Potassium permanganate, $KMnO_4$, is a dark purple crystalline solid used as an antiseptic, and in deodorizers and dyes. How many oxygen atoms are in a sample of $KMnO_4$ of mass 7.238 g?

Section 1.6

14. The elements Li, Na, K, Rb, and Cs all have a single valence electron. Write the formulas of the cations formed when each of these atoms loses its valence electron.

15. What is the formula of sodium sulfate, an ionic crystalline solid composed of sodium ions, Na^+, and sulfate ions, SO_4^{2-}? Indicate your reasoning.

16. Lanthanum sulfate is a white ionic solid with nine *waters of crystallization* in its crystal structure. Its formula is $La_2(SO_4)_3 \cdot 9H_2O$. In 0.394 mol of $La_2(SO_4)_3 \cdot 9H_2O$, how many moles of SO_4^{2-} ions are there? How many moles of H_2O?

17. How many protons and electrons are in each of the following species? Mg, Mg^{2+}, Al, Al^{3+}, F, F^-, Rb, Rb^+, Br, Br^-, Se, and Se^{2-}.

Section 1.8

18. The empirical formula for the organic solvent benzene is $(CH)_n$. The molecular weight of benzene is $78.1 \text{ g} \cdot \text{mol}^{-1}$. What is the molecular formula of benzene?

19. *Para*-dichlorobenzene is a white solid that has been used as moth balls. The compound contains the elements carbon, hydrogen, and chlorine, in the molar ratios $C:H:Cl = 3:2:1$. What is the empirical formula of *para*-dichlorobenzene? If the molecular weight is $147.0 \text{ g} \cdot \text{mol}^{-1}$, what is the molecular formula?

20. Propane gas is used as a fuel. Propane is 81.71% carbon and 18.29% hydrogen by weight. What is the empirical formula of propane? If its molecular weight is 44 $\text{g} \cdot \text{mol}^{-1}$, what is the molecular formula of propane?

Section 1.9

21. Copper metal can be dissolved in warm dilute nitric acid. The net ionic equation for the reaction is

$$3Cu(s) + 8H^+(aq) + 2NO_3^-(aq) \rightarrow 2NO(g) + 3Cu^{2+}(aq) + 4H_2O$$

What is the value of each of the following molar ratios?

(a) $\dfrac{\text{No. moles NO(g) formed}}{\text{No. moles Cu(s) used}}$ (b) $\dfrac{\text{No. moles Cu(s) used}}{\text{No. moles H}^+\text{(aq) used}}$

22. A common laboratory method for producing oxygen gas is by heating potassium chlorate, $KClO_3$. The balanced equation for the reaction is

$$2KClO_3(s) \xrightarrow{\text{heat}} 2KCl(s) + 3O_2(g)$$

How many grams of $KClO_3$ must be decomposed to produce 1.758 g of O_2?

23. The net ionic equation for the reaction that occurs when a piece of zinc metal is inserted into a solution of silver nitrate, $AgNO_3$, in water is

$$Zn(s) + 2Ag^+(aq) \rightarrow Zn^{2+}(aq) + 2Ag(s)$$

How many moles of $AgNO_3$ would be required to react completely with 6.283 g of zinc metal? How many grams of solid silver would be produced by this reaction?

24. The net ionic equation for the reaction between potassium permanganate, $KMnO_4$, and potassium iodide, KI, in acidic aqueous solution is

$$2MnO_4^-(aq) + 10\ I^-(aq) + 16H^+(aq) \rightarrow 2Mn^{2+}(aq) + 8H_2O + 5\ I_2$$

If 0.8155 mol of iodine, I_2, are produced by this reaction, how many moles of permanganate ion, MnO_4^-, are used up? How many grams of the solid $KMnO_4$ were needed to produce the 0.8155 mol of I_2?

Section 1.10

25. If all the carbon in a 37.570-g sample of propane, C_3H_8, is converted to carbon dioxide by burning the propane in excess oxygen, how many grams of CO_2 will be produced?

26. A 30.2022-g sample of liquid bromine, Br_2, reacts completely with a piece of aluminum metal to form aluminum bromide, $AlBr_3(s)$. How many grams of $AlBr_3$ will be produced?

Section 1.11

27. Calcium oxide combines with carbon dioxide to produce calcium carbonate, $CaCO_3$, commonly known as limestone:

$$CaO(s) + CO_2(g) \rightarrow CaCO_3(s)$$

If 0.892 g of CaO and 0.739 g of CO_2 are combined, which is the limiting reagent? What is the maximum yield of $CaCO_3$, in grams?

28. The combustion of propane, C_3H_8, in oxygen proceeds as follows:

$$C_3H_8(g) + 5\ O_2(g) \rightarrow 3CO_2(g) + 4H_2O(\ell)$$

If 2.760 g of C_3H_8 are burned in 14.886 g of O_2, which is the limiting reagent? What is the maximum yield of CO_2, in grams?

29. The reaction between methanol and acetic acid is

$$\underset{\text{Methanol}}{CH_3OH(\ell)} + \underset{\text{Acetic acid}}{CH_3COOH(\ell)} \rightarrow \underset{\text{Methyl acetate}}{CH_3COOCH_3(\ell)} + H_2O(\ell)$$

If 18.36 g of CH_3COOCH_3 are obtained when 9.04 g of CH_3OH and 15.78 g of CH_3COOH are reacted, what is the percentage yield of product?

Multiple Choice Questions

1. Which of the following samples contains the largest number of atoms?
   (a) 1 g of Ni(s)   (b) 1 g of Ca(s)   (c) 1 g of $N_2(g)$ (d) 1 g of B(s)
   (e) 1 g of $O_2(g)$

2. Which of the following samples contains the smallest number of atoms?
   (a) 1 g of $CO_2(g)$ (b) 1 g of $C_8H_{18}(liq)$ (c) 1 g of $C_2H_6(g)$
   (d) 1 g of LiF(s)   (e) 1 g of $B_4H_{10}(g)$

3. The ratio of oxygen atoms to hydrogen atoms in $(NH_4)_2SO_4$ is
   (a) $1:1$ (b) $2:1$ (c) $1:2$ (d) $1:4$ (e) $4:1$

4. A sample of ammonium phosphate, $(NH_4)_3PO_4$, contains 3.18 mol of hydrogen atoms. The number of moles of oxygen atoms in the sample is
   (a) 0.265   (b) 0.795   (c) 1.06   (d) 3.18   (e) 4.00

5. A sample of copper sulfate pentahydrate, $CuSO_4 \cdot 5H_2O$, contains 3.782 g of Cu. How many grams of oxygen are in this sample?
   (a) 0.952 g   (b) 3.809 g   (c) 4.761 g   (d) 7.618 g   (e) 8.570 g

6. Element X reacts with oxygen to produce a pure sample of X_2O_3. In an experiment it is found that 1.0000 g of X produces 1.1596 g of X_2O_3. If the atomic weight of oxygen is known to be 15.9994 g $\cdot$ mol^{-1}, what is the atomic weight of X?
   (a) 20.70   (b) 66.85   (c) 100.2   (d) 150.4   (e) 168.9

7. An ore contains 1.34% of the mineral argentite, Ag_2S, by weight. How many grams of this ore would have to be processed in order to obtain 1.00 g of pure solid silver, Ag?
   (a) 74.6 g   (b) 85.7 g   (c) 107.9 g   (d) 134.0 g   (e) 171.4 g

8. When pentane, C_5H_{12}, is burned in excess oxygen, the products of the reaction are $CO_2(g)$ and $H_2O(\ell)$. In the balanced equation for this combustion

$$C_5H_{12}(g) + \underline{8}\,O_2(g) \rightarrow 5CO_2(g) + 6H_2O(\ell)$$

the coefficient of oxygen should be
   (a) 16   (b) 12   (c) 11   (d) 8   (e) 6

9. The complete combustion of ethane gas yields $CO_2(g)$ and $H_2O(\ell)$ as indicated by the equation below:

$$C_2H_6(g) + \underline{\tfrac{7}{2}}\,O_2(g) \rightarrow \underline{2}\,CO_2(g) + \underline{3}\,H_2O(\ell)$$

In this equation the ratio of the coefficients of CO_2 to H_2O is
   (a) 1:1   (b) 2:3   (c) 3:2   (d) 1:3   (e) 3:1

10. Ethanol, C_2H_5OH, is the substance commonly called "alcohol." The density of liquid ethanol is 0.7893 g $\cdot$ mL^{-1} at 20 °C. If 1.2 mol of ethanol are needed for a particular experiment, what volume of ethanol should be measured out?
    (a) 55 mL   (b) 58 mL   (c) 70 mL   (d) 79 mL   (e) 120 mL

11. What is the total number of atoms present in 25.0 mg of camphor, $C_{10}H_{16}O$?
    (a) 9.89×10^{19} (b) 6.02×10^{20} (c) 9.89×10^{20} (d) 2.57×10^{21}
    (e) 2.67×10^{21}

12. A compound having the empirical formula $(C_3H_4O)_n$ has a molecular weight of 170 ± 5. The molecular formula of this compound is
    (a) C_3H_4O (b) $C_6H_8O_2$ (c) $C_6H_{12}O_3$ (d) $C_9H_{12}O_3$ (e) $C_9H_{16}O_3$

13. Analysis of a compound yields the following percent composition by weight: 65.03% Ag, 15.68% Cr, 19.29% O. What is the simplest formula of this compound?
    (a) Ag_2CrO_4 (b) $Ag_2Cr_2O_7$ (c) $AgCrO_2$ (d) $AgCr_2O_3$ (e) $Ag_2Cr_2O_3$

14. Cortisone is a molecular substance containing 21 atoms of carbon per molecule. The weight percentage of carbon in cortisone is 69.98%. What is the molecular weight of cortisone?
    (a) 176.5   (b) 252.2   (c) 287.6   (d) 312.8   (e) 360.4

15. Methyl benzoate is prepared by the reaction between benzoic acid and methanol, according to the equation

$$\underset{\text{Benzoic acid}}{C_6H_5COOH} + \underset{\text{Methanol}}{CH_3OH} \rightarrow \underset{\text{Methyl benzoate}}{C_6H_5COOCH_3} + H_2O$$

In an experiment 24.4 g of benzoic acid were reacted with 70.0 mL of CH_3OH. The density of CH_3OH is 0.791 g $\cdot$ mL^{-1}. The methyl benzoate produced in this reaction had a mass of 21.6 g. What was the percentage yield of product?

(a) 91.7%   (b) 79.3%   (c) 71.5%   (d) 21.7%   (e) 9.17%

16. An atom that has the same number of neutrons as ^{85}Rb is

   (a) ^{85}Kr (b) ^{87}Rb (c) ^{85}Sr (d) ^{86}Sr (e) ^{86}Kr

17. Which of the following samples contains 2.0×10^{23} atoms?

   (a) 8.0-g O_2 (b) 3.0-g Be   (c) 8.0-g C   (d) 12.0-g He   (e) 19.0-g F_2

18. The number of electrons in the telluride ion, Te^{2-}, is

   (a) 50   (b) 51   (c) 52   (d) 53   (e) 54

19. Which of the following gases is red brown in color?

   (a) N_2O (b) NO_2 (c) NO   (d) O_2 (e) N_2

20. Which of the following substances is commonly called laughing gas?

   (a) N_2O (b) NO_2 (c) NO   (d) O_2 (e) N_2

21. The atomic weight of Cu is 63.546. There are only two naturally occurring isotopes of copper, ^{63}Cu and ^{65}Cu. The natural abundance of the ^{63}Cu isotope must be approximately

   (a) 10%   (b) 30%   (c) 50%   (d) 70%   (e) 90%

Problems

1.1. An unknown compound is known to contain the element phosphorus, P. It is desired to determine the percentage of P in this compound. A 0.7970-g sample of this compound was weighed out. By a series of chemical reactions, all of the P in this 0.7970-g sample was oxidized to phosphate, precipitated as magnesium ammonium phosphate ($MgNH_4PO_4$), and finally converted to magnesium pyrophosphate, $Mg_2P_2O_7$, a white insoluble solid. The $Mg_2P_2O_7$ was collected and found to weigh 1.0864 g.

   (a) How many moles of $Mg_2P_2O_7$ are in the 1.0864 g of $Mg_2P_2O_7$?
   (b) How many moles of P are in the 1.0864 g of $Mg_2P_2O_7$?
   (c) How many grams of P were in the original sample of the unknown compound?
   (d) What was the percent by weight of P in the unknown compound?

1.2. From the following isotopic masses and abundances, calculate the atomic weight of magnesium.

Mass Number	Natural Abundance	Mass
24	78.70%	23.98504
25	10.13%	24.98584
26	11.17%	25.98259

Be careful about the number of significant figures in the answer.

1.3. There are 20 amino acids that are the building blocks of proteins. The amino acid histidine is known to contain only carbon, hydrogen, oxygen, and nitrogen. The composition by weight of histidine is 46.45% C; 5.85% H; 27.08% N. What is the empirical formula of histidine?

1.4. Gadolinium is a member of the lanthanide series of metals. Gadolinium forms an oxide that is 86.76% Gd by weight. What is the empirical formula of this oxide?

1.5. A 2.9310-g sample of benzoic acid, C_6H_5COOH, is completely burned in excess oxygen. The equation for the reaction is

$$C_6H_5COOH(s) + \tfrac{15}{2} O_2(g) \rightarrow 7CO_2(g) + 3H_2O(\ell)$$

How many moles of $CO_2(g)$ are formed?

1.6. Amphetamine (also called dexedrine) is a solid with molecular formula $C_9H_{13}N$. Its density is 0.949 g · mL^{-1}. Calculate the following quantities:
(a) The number of moles of amphetamine in a 62.87-g sample.
(b) The total number of atoms in 18.04 g of amphetamine.
(c) The volume of 4.50 mol of amphetamine.
(d) The weight of amphetamine that contains exactly the same number of H atoms as there are in 117.1 g of water.

1.7. The net ionic equation for the reaction between ferrous ion, Fe^{2+}, and permanganate ion, MnO_4^-, in acid solution is

$$5Fe^{2+}(aq) + MnO_4^-(aq) + 8H^+(aq) \rightarrow 5Fe^{3+}(aq) + 4H_2O + Mn^{2+}(aq)$$

In a certain experiment in which the reaction above occurred, 1.360×10^{-3} mol of ferrous ion were used up.
(a) How many moles of Mn^{2+} were formed?
(b) How many moles of Fe^{3+} were formed?
(c) How many moles of MnO_4^- were used up?
(d) If the ferrous ion came from the light green crystalline solid ferrous sulfate heptahydrate, $FeSO_4 \cdot 7H_2O$, how many grams of $FeSO_4 \cdot 7H_2O$ were used up?

1.8. The element ytterbium forms a compound $YbBr_3$, ytterbium(III) bromide. To a solution of 1.3209 g of $YbBr_3$ in water, excess silver nitrate solution is added, precipitating all the bromide as AgBr, an insoluble pale yellow solid. The AgBr is collected, dried, and found to weigh 1.8027 g. Calculate the atomic weight of Yb from these data.

1.9. A certain solid is known to be a mixture of $Al_2(SO_4)_3$, $CoSO_4$, and $(NH_4)_2SO_4$. A sample of this mixture weighing 2.7022 g is dissolved in water. A solution of excess ammonia and ammonium chloride is added, and the solution is heated to boiling. White, gelatinous $Al(OH)_3$ precipitates out of the boiling solution. The precipitate is filtered, dried in air, and then heated in a crucible for 10 min to drive off all water and convert the $Al(OH)_3$ to Al_2O_3. When cool, the dried solid Al_2O_3 was found to weigh 0.3272 g.
(a) How many moles of Al_2O_3 were obtained in this reaction?
(b) How many moles of Al were in the original sample, assuming complete conversion to Al_2O_3?
(c) What was the percentage by weight of Al in the original sample?
(d) What was the percentage by weight of $Al_2(SO_4)_3$ in the original sample?

1.10. Show that the atomic weight of oxygen can be calculated correctly to five significant figures even if the exact isotopic masses of ^{17}O and ^{18}O are only known to two significant figures. Use the following data to calculate the atomic weight of oxygen:

Mass Number	Natural Abundance	Mass
16	99.759%	15.99491
17	0.037%	17
18	0.204%	18

1.11. This problem illustrates an early experimental determination of Avogadro's number. Radium is a naturally occurring radioactive element that decays by emitting alpha (α) particles. An α-particle is the nucleus of a helium atom. As each emitted α-particle travels through the air, it picks up two electrons and becomes a neutral He atom. Thus, He gas can be collected from the emissions of a sample of radium. The number of α-particles emitted per second can be obtained by counting the light flashes produced when the α-particles strike a zinc sulfide screen. In 1910, Ernest Rutherford and Thomas Royds determined Avogadro's number by collecting and measuring the amount of He gas emitted by a sample of radium for 1 yr. They also measured the number of α-particles emitted per second by an identical sample of radium. In an experiment reproducing the measurements of Rutherford and his co-workers, 22.0×10^{-3} mg of He gas were collected in 1 yr from a sample of radium. This sample was observed to emit 10.6×10^{10} α-particles per second.

Calculate (a) the mass of one He atom, in grams, and (b) Avogadro's number, using only these data and the atomic weight of helium.

As the experimental uncertainty in these data is fairly large, do not expect the answer you obtain to be correct to three figures.

1.12. This problem illustrates an experimental method used to determine Avogadro's number. It depends on a previous determination of the magnitude of the charge on an electron, 1.602×10^{-19} C.

Electrolysis is a process in which electrical current is passed through a solution containing ions. Electrolysis of a solution containing silver ions, $Ag^+(aq)$, causes solid silver to be deposited. The reaction that occurs is

$$Ag^+(aq) + e^- \rightarrow Ag(s)$$

To deposit 1 mol of silver requires the passage of 1 mol (Avogadro's number) of electrons.

In an electrolysis experiment, a steady current of 5.00×10^{-2} amperes (A) is passed through an aqueous solution of silver nitrate for 1.75 h. The solid silver deposited weighs 0.352 g.

(a) By definition, a coulomb is the amount of charge transferred when a current of one ampere passes for one second. How many coulombs of charge were passed through this solution during the electrolysis?

(b) How many moles of Ag(s) were deposited in this experiment?

(c) How many coulombs of charge would be required to deposit 1 mol of Ag? [This amount of charge is referred to as 1 $\mathscr{F}$ (Faraday).]

(d) Using your answer to part (c), and the charge on one electron, calculate Avogadro's number.

1.13. A green pigment called hemovanadin can be isolated from the sea squirt, a marine chordate resembling a tadpole. The hemovanadin molecule has a molecular weight of 2.4×10^5 daltons, and is 0.51% vanadium by weight. How many vanadium atoms are there per hemovanadin molecule?*

1.14. In a certain experiment, 5.8750 g of sodium oxide, Na_2O, were obtained from the reaction of 4.5980 g of sodium metal, Na(s), with excess O_2 gas. What is the percentage yield of Na_2O?

1.15. A solution of 2.63 g of triphenyl phosphine, $(C_6H_5)_3P$, dissolved in 20 mL of the organic solvent dichloromethane is added to a solution of 3.5 mL of liquid antimony

---

* Courtesy of F. C. Hentz, Jr., and G. G. Long.

pentachloride, $SbCl_5$, dissolved in 10 mL of dichloromethane. After cooling and adding ether, a precipitate of $[(C_6H_5)_3PCl][SbCl_6]$ is formed. The precipitate is collected, dried, and found to weigh 5.314 g. The equation for the reaction that occurred is

$$(C_6H_5)_3P + 2SbCl_5 \rightarrow [(C_6H_5)_3PCl][SbCl_6] + SbCl_3$$

(a) If the density of liquid $SbCl_5$ is 2.336 g · mL^{-1}, which of the two reactants is the limiting reagent?

(b) Calculate the percentage yield of product in this experiment.

1.16. A certain compound X contains only C, H, and N. Analysis shows that the compound contains 77.91% carbon by weight. When 4.8102 g of compound are burned in excess oxygen, all the hydrogen is converted to H_2O weighing 1.6862 g.

(a) What is the mass of each element in the 4.8102-g sample of X?

(b) What is the empirical formula of X?

(c) An experiment shows that, within a precision of 5%, the molecular weight of X is 155. What is the molecular formula of X? What is the exact molecular weight of X?

1.17. Propane gas, C_3H_8, burns in air to form CO_2 and H_2O.

(a) Write a balanced equation for the combustion of C_3H_8.

(b) How many grams of CO_2 will be for.ned when a mixture containing 3.907 g of C_3H_8 and 10.848 g of O_2 is ignited?

1.18. When iron is exposed to air, it rusts according to the following equation:

$$2Fe(s) + \tfrac{3}{2} O_2(g) \rightarrow Fe_2O_3(s)$$

If 2.2943 g of O_2 react completely with a sample of iron of mass 11.2811 g, what percentage of the iron rusts? How many grams of Fe_2O_3 are formed?

1.19. A mixture of NaI and KI weighing 3.9762 g was dissolved in water and treated with aqueous Ag^+ (for example, a solution of silver nitrate, $AgNO_3$). All of the iodide ion in the mixture was recovered as 5.8622 g of pure silver iodide, AgI, an insoluble, pale yellow solid. What was the percentage by weight of KI in the original sample?

chapter 2 Introduction to the Periodic Table and to Inorganic Nomenclature

Dmitri Ivanovich Mendeleev (1834–1907), a Russian chemist, was a Professor of Chemistry at the Technological Institute of St. Petersburg (now Leningrad) and at the University of St. Petersburg. In 1893 he became Director of the Bureau of Weights and Measures in Russia. Mendeleev arranged the elements according to their atomic weights and the periodicities of their properties. He predicted the existence of the elements gallium, scandium, and germanium, and correctly described many of their properties, several years before they were discovered. The accuracy of his predictions led to the rapid acceptance of his periodic arrangement of the elements.

Several million compounds are known, and new ones are being synthesized all the time. Some of them are molecular compounds like C_4H_{10} and N_2O, while others are ionic crystalline solids like CaO (composed of Ca^{2+} and O^{2-} ions in a 1 : 1 molar ratio), or K_2SO_4 (composed of K^+ and SO_4^{2-} ions in a 2 : 1 molar ratio). You may be wondering how you are to know whether a chemical formula represents a molecular compound or an ionic compound. Fortunately, there are some broad generalizations that can be made about the compound or compounds formed when various atoms combine chemically. We will discuss the principles needed to understand the nature of chemical bonding in a detailed and systematic manner in Chapters 12 through 14. In this chapter we will discuss information contained in the periodic table of the elements that clarifies and unifies a great deal of descriptive information about chemical substances.

section 2.1
The Sections of the Periodic Table

The principles that enable us to predict the formulas and properties of chemical compounds are concisely represented in the **periodic table of the elements,** which can be found on the inside front cover, as well as in Fig. 2.1. When the elements are arranged in order of increasing atomic number, Z, we find that there are regular recurrences of a great many physical and chemical properties. Properties that vary with atomic number in a regular, recurring pattern are called **periodic properties.**

An example of a periodic property is the volume of one mole of atoms in the solid state. A plot of the atomic volume per mole as a function of atomic number is shown in Fig. 2.2. The atomic volume is obtained from the density of the solid and the atomic weight:

$$\text{atomic volume in } cm^3 \cdot mol^{-1} = \frac{\text{atomic weight in } g \cdot mol^{-1}}{\text{density in } g \cdot cm^{-3}} \tag{2-1}$$

Most elements are solids at room temperature; for those that are gases or liquids, the convention is to use the density at the melting point to calculate the atomic volume per mole. An examination of Fig. 2.2 reveals the periodicity of atomic volumes.

The existence of a relationship between the atomic weights and the properties of the elements was proposed by the German chemist Johann Dobereiner in 1829 and by the English chemist John Newlands in 1865. Working independently, during the same general period of time, both Dmitri Mendeleev (1834–1907) in Russia and Lothar Meyer (1830–1895) in Germany arranged the elements in the order of their atomic weights and showed that they fall into groups in which similar chemical and physical properties are repeated at periodic intervals. Our modern periodic table is an extension of their pioneering work. Mendeleev predicted the existence and properties of six elements that had not yet been discovered on the basis of gaps he found in the recurring series of elements with similar chemical and physical properties. The subsequent discoveries of these elements quickly convinced other scientists that Mendeleev had succeeded in arranging the elements in the proper periods.

Elements with similar chemistry are said to constitute a **family** or a **group.** In the periodic table, elements in the same group are placed in a vertical column, in order of increasing atomic number, Z. The feature that all the elements in one family or group have in common is similar electronic structure; in particular they have the same number of **valence electrons,** those electrons that are, on the average, farthest from the nucleus.

The periodic table of the elements

IA	IIA	IIIB	IVB	VB	VIB	VIIB	VIIIB	VIIIB	VIIIB	IB	IIB	IIIA	IVA	VA	VIA	VIIA	Noble Gases
1 H 1.008																	2 He 4.003
3 Li 6.941	4 Be 9.012											5 B 10.81	6 C 12.011	7 N 14.007	8 O 15.999	9 F 18.998	10 Ne 20.179
11 Na 22.990	12 Mg 24.305											13 Al 26.982	14 Si 28.086	15 P 30.974	16 S 32.06	17 Cl 35.453	18 Ar 39.948
19 K 39.098	20 Ca 40.08	21 Sc 44.956	22 Ti 47.90	23 V 50.942	24 Cr 51.996	25 Mn 54.938	26 Fe 55.847	27 Co 58.933	28 Ni 58.70	29 Cu 63.546	30 Zn 65.38	31 Ga 69.72	32 Ge 72.59	33 As 74.922	34 Se 78.96	35 Br 79.904	36 Kr 83.80
37 Rb 85.468	38 Sr 87.62	39 Y 88.906	40 Zr 91.22	41 Nb 92.906	42 Mo 95.94	43 Tc (98)	44 Ru 101.07	45 Rh 102.905	46 Pd 106.4	47 Ag 107.868	48 Cd 112.41	49 In 114.82	50 Sn 118.69	51 Sb 121.75	52 Te 127.60	53 I 126.905	54 Xe 131.30
55 Cs 132.905	56 Ba 137.33	57* La 138.905	72 Hf 178.49	73 Ta 180.948	74 W 183.85	75 Re 186.2	76 Os 190.2	77 Ir 192.22	78 Pt 195.09	79 Au 196.966	80 Hg 200.59	81 Tl 204.37	82 Pb 207.19	83 Bi 208.98	84 Po (209)	85 At (210)	86 Rn (~222)
87 Fr (223)	88 Ra 226.02	89† Ac 227.028	104 Unq (261)	105 Unp (262)	106 Unh (263)												

The Active Metals — Transition Elements — The Nonmetals — Inner Transition Metals

*Lanthanides

58 Ce 140.12	59 Pr 140.907	60 Nd 144.24	61 Pm (145)	62 Sm 150.4	63 Eu 151.96	64 Gd 157.25	65 Tb 158.925	66 Dy 162.50	67 Ho 164.930	68 Er 167.26	69 Tm 168.934	70 Yb 173.04	71 Lu 174.96

†Actinides

90 Th 232.038	91 Pa 231.036	92 U 238.029	93 Np 237.048	94 Pu (244)	95 Am (243)	96 Cm (247)	97 Bk (247)	98 Cf (251)	99 Es (254)	100 Fm (257)	101 Md (258)	102 No (259)	103 Lr (260)

Fig. 2.1. The periodic table of the elements.

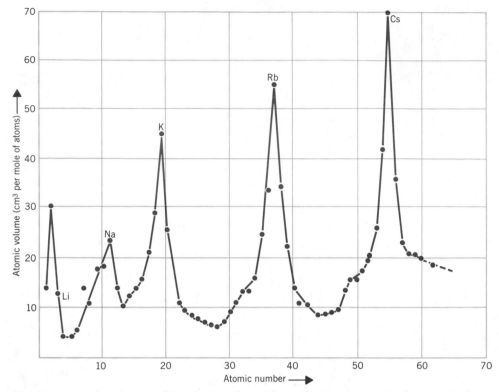

Fig. 2.2. Atomic volumes of the elements in cubic centimeters per mole of atoms, in the solid state, as a function of atomic number. Elements with $Z > 70$ have been omitted because for some of them density data is not available.

The horizontal rows of the periodic table are called **periods.** There is a very short period of 2 elements, hydrogen and helium, and then two short periods of 8 elements each (Li, atomic number 3, through Ne, atomic number 10, and then Na, atomic number 11, through Ar, atomic number 18). Following the two short periods, there are two long periods of 18 elements each, and one very long period of 32 elements. The seventh period would also consist of 32 elements, but many of them have not yet been synthesized. None of the elements of the seventh period has any stable isotopes; they are all radioactive, and decay to form elements of lower atomic number.

A great deal of information is contained in the periodic table. We will return to it often throughout the text. It is helpful to acquire a preliminary understanding of the periodic table by dividing it into six sections (see Fig. 2.1), and becoming familiar in some detail with four of these sections. The four sections we will discuss here are 1. the active metals; 2. the noble gases; 3. the nonmetals; and 4. the transition metals. The two sections we will not discuss in detail are the metallic elements of Groups IIIA, IVA, and VA, which, with the active metals of Groups IA and IIA, constitute the **representative metals,** and the **inner transition elements.** The inner transition elements consist of two rows of 14 elements each, the **lanthanides** and the **actinides.** The **lanthanides** (also called the **rare earths**) are elements 58 through 71, which follow lanthanum in the periodic table. The **actinides** are elements 90 through 103, which follow actinium in the periodic table.

section 2.2
The Metals

Metals are good conductors of heat and electricity. In the solid state, metals are malleable (they can be flattened out by rolling or beating with a hammer) and ductile (they can be drawn into a wire). Almost all the metals are solids at room temperature. The only metal with a melting point well below room temperature (−39 °C) is mercury, Hg, which is therefore a liquid at room temperature. Three metals have melting points just slightly above normal room temperature, and are liquid on a particularly warm day. They are cesium (melting point 28.4 °C), francium (melting point 27 °C), and gallium (melting point 29.8 °C). Gallium readily **supercools,** that is, remains liquid even at temperatures several degrees below its melting point.

Metals tend to transfer their valence electrons to other substances and become **cations** (positively charged ions) when they participate in chemical reactions. Energy is always required to remove one or more electrons from a neutral atom. The easier it is (that is, the less energy that is required) for a metal to lose its valence electrons, the more reactive the metal will be and the more metallic character it will have. Because the amount of energy required to remove the valence electrons from a metal atom varies considerably, the chemical reactivity of the metals also varies a good deal. Sodium, for instance, reacts rapidly and violently with water at room temperature:

$$2Na(s) + 2H_2O \rightarrow 2Na^+(aq) + 2OH^-(aq) + H_2(g) \qquad (2\text{-}2)$$

Nickel and copper, on the other hand, do not react with water.

Roughly 70% of all the known elements are metals. The metallic elements in the periodic table are shown in Fig. 2.3. A comparison of this figure with the entire periodic table, Fig. 2.1, shows that metals are found at the far left, and at the bottom of the periodic table. Metallic character increases as one goes to the left across the periodic table, and as one proceeds downward (increasing Z) within any group or family. Thus the most metallic elements are cesium (Cs, atomic number 55) and francium (Fr, atomic number 87).

The groups of the periodic table are identified by a Roman numeral followed either by the letter A or B. (See Fig. 2.1.) Groups IA through VIIA plus Group 0 (the noble gases) constitute the **representative elements.** The representative metals are therefore all the elements of Groups IA and IIA, plus the metallic (heavier) elements of Groups IIIA, IVA, and VA. The two families at the far left of the periodic table are called the **active metals** because they are extremely reactive chemically.

The Active Metals

At the extreme left side of the periodic table is Group IA, the **alkali metals:** lithium (Li), sodium (Na), potassium (K), rubidium (Rb), cesium (Cs), and francium (Fr). All the alkali metals have a single valence electron and form cations with only a single positive charge when they take part in compound formation. These cations are Li^+, Na^+, K^+, Rb^+, and Cs^+. (Francium is a radioactive element, rarely encountered.) The alkali metals are silvery in color, have a shiny luster, low melting points, and low densities. (See Table 2.1.) Lithium is the least dense of all metals, with a density roughly half that of water. The alkalis are soft metals that can easily be cut with a knife. Figure 2.4 is a photograph of a sample of potassium.

The alkali metals are chemically very reactive, that is, they combine readily with many other substances because they have a strong tendency to lose their single valence electron and become cations with a single positive charge. The sample of

Li	6.9
3	
Na	23.0
11	
K	39.1
19	
Rb	85.5
37	
Cs	132.9
55	
Fr	(223)
87	
Group IA	

					Transition Elements											
3 Li 6.941	4 Be 9.012															
11 Na 22.98977	12 Mg 24.305											13 Al 26.98154				
19 K 39.098	20 Ca 40.08	21 Sc 44.956	22 Ti 47.90	23 V 50.942	24 Cr 51.996	25 Mn 54.9380	26 Fe 55.847	27 Co 58.9332	28 Ni 58.70	29 Cu 63.546	30 Zn 65.38	31 Ga 69.72	32 Ge 72.59			
37 Rb 85.4678	38 Sr 87.62	39 Y 88.9059	40 Zr 91.22	41 Nb 92.9064	42 Mo 95.94	43 Tc (98)	44 Ru 101.07	45 Rh 102.905	46 Pd 106.4	47 Ag 107.868	48 Cd 112.41	49 In 114.82	50 Sn 118.69	51 Sb 121.75		
55 Cs 132.9054	56 Ba 137.33	57* La 138.905	72 Hf 178.49	73 Ta 180.9479	74 W 183.85	75 Re 186.207	76 Os 190.2	77 Ir 192.22	78 Pt 105.09	79 Au 196.966	80 Hg 200.59	81 Tl 204.37	82 Pb 207.2	83 Bi 208.9804	84 Po (209)	
87 Fr (223)	88 Ra 226.0254	89† Ac (227)	104 Unq (261)	105 Unp (262)	106 Unh (263)											

* Lanthanides

58 Ce 140.12	59 Pr 140.907	60 Nd 144.24	61 Pm (145)	62 Sm 150.4	63 Eu 151.96	64 Gd 157.25	65 Tb 158.925	66 Dy 162.50	67 Ho 164.930	68 Er 167.26	69 Tm 168.934	70 Yb 173.04	71 Lu 174.96

† Actinides

90 Th 232.038	91 Pa 231.036	92 U 238.029	93 Np 237.048	94 Pu (244)	95 Am (243)	96 Cm (247)	97 Bk (247)	98 Cf (251)	99 Es (254)	100 Fm (257)	101 Md (258)	102 No (259)	103 Lr (260)

Fig. 2.3. The metallic elements. Ge, Sb, and Po are metalloids.

Table 2.1. **Some Properties of the Alkali and Alkaline Earth Metals**

Element	Atomic Number	Melting[a] Point (°C)	Boiling[a] Point (°C)	Density (g · cm⁻³)
Li	3	181	1330 ± 10	0.53
Na	11	97.8	888 ± 5	0.97
K	19	63.6	760 ± 10	0.86
Rb	37	38.9	688 ± 8	1.53
Cs	55	28.5	680 ± 10	1.89
Be	4	1278 ± 5	2970	1.85
Mg	12	650 ± 1	1105 ± 15	1.74
Ca	20	845 ± 5	1480 ± 15	1.54
Sr	38	769 ± 1	1382 ± 2	2.6
Ba	56	715 ± 10	1639 ± 1	3.5

[a] The limits listed are ranges of reported values.

potassium shown in Fig. 2.4 is enclosed in a sealed glass tube to protect it from the air, because the alkali metals react rapidly with both H_2O and O_2. The reaction between an alkali metal and H_2O (Eq. 2-2) is often violent. So much heat is evolved that the hydrogen produced by the reaction can ignite and cause a fire. The reaction between any of the alkali metals and oxygen also evolves a large amount of heat. Lithium reacts directly with oxygen to form the simple oxide, Li_2O, an ionic crystalline solid composed of Li^+ ions and O^{2-} (oxide) ions:

$$2Li(s) + \tfrac{1}{2} O_2(g) \rightarrow Li_2O(s) \tag{2-3}$$

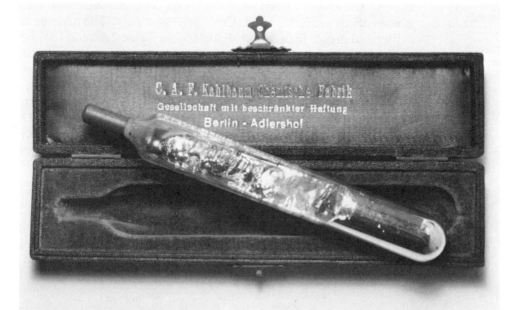

Fig. 2.4. Potassium metal. The sample is enclosed in a sealed, evacuated glass tube to protect the potassium from exposure to the air.

Sodium reacts with oxygen to yield a mixture of sodium oxide, $Na_2O(s)$, and sodium peroxide, $Na_2O_2(s)$, which contains the **peroxide ion,** O_2^{2-}

$$2Na(s) + O_2(g) \rightarrow Na_2O_2(s) \tag{2-4}$$

$$2Na(s) + \tfrac{1}{2}O_2(g) \rightarrow Na_2O(s) \tag{2-5}$$

The heavier alkali metals react with oxygen to form **superoxides** containing the superoxide ion, O_2^-

$$Rb(s) + O_2(g) \rightarrow RbO_2(s) \tag{2-6}$$

Be 4	9.0
Mg 12	24.3
Ca 20	40.1
Sr 38	87.6
Ba 56	137.3
Ra 88	(226)
Group IIA	

Because the reactions of the alkali metals with water and oxygen produce so much heat, they can cause a fire or even be explosive. It is therefore dangerous to expose a sample of an alkali metal to the air, which contains both oxygen and water vapor. Alkali metals are usually stored under kerosene or some other nonreactive organic liquid. Of course, one should never try to put out a fire involving one of the alkali metals by throwing water on it! If you are working with one of the alkali metals in the lab, it is a good idea to keep a bucket of sand handy for dousing accidental fires.

The second vertical column of the periodic table on the left-hand side consists of Group IIA, the **alkaline earth metals**: beryllium (Be), magnesium (Mg), calcium (Ca), strontium (Sr), and barium (Ba). The last member of the family, radium (Ra), is radioactive. Radium was discovered in 1898 by Marie and Pierre Curie. Marie Curie produced pure radium in 1911; it was the first radioactive element to be isolated. Radium is extracted from pitchblende, a uranium ore; about 7 tons of pitchblende are required to produce 1 g of pure radium. All radium isotopes decay spontaneously to other elements, emitting dangerous radiation. The final product of the nuclear disintegration is lead. Radium is rarely encountered in the chemistry laboratory, both because the radiation is a health hazard, and because there are only small amounts of radium available.

All of the alkaline earth metals have two valence electrons and form cations with two positive charges when they react with other substances to form compounds. These cations are Be^{2+}, Mg^{2+}, Ca^{2+}, Sr^{2+}, and Ba^{2+}. The alkaline earth metals are harder and denser than the alkali metals (see Table 2.1), and less chemically active, although they are still active metals. Because these metals react with so many substances, none of the alkaline earths is found in nature as the free metal.

Calcium is the most common member of the alkaline earth family, and is the fifth most abundant element on earth. Calcium is found in marble, limestone, and chalk, as well as in the shells of marine animals, as the carbonate, $CaCO_3$. Barium and strontium are found most frequently as the sulfates, $BaSO_4$ and $SrSO_4$, or the carbonates, $BaCO_3$ and $SrCO_3$. The alkaline earth metals combine directly with oxygen to yield the oxides. A typical reaction is

$$Ca(s) + \tfrac{1}{2}O_2(g) \rightarrow CaO(s) \tag{2-7}$$

Barium reacts to form the peroxide, BaO_2

$$Ba(s) + O_2(g) \rightarrow BaO_2(s) \tag{2-8}$$

With the exception of beryllium, the alkaline earth metals react with nitrogen gas to form **nitrides,** compounds containing the nitride ion, N^{3-}. A typical reaction is

$$3Sr(s) + N_2(g) \rightarrow Sr_3N_2(s) \tag{2-9}$$

Table 2.1 lists the melting points, boiling points, and densities of the alkali and alkaline earth metals.

Note that the alkaline earth metals, as a group, have higher melting and boiling points, and greater densities than the alkali metals. While there are irregularities, it is generally true that the melting points decrease, and the densities increase, as the atomic number increases in these two groups of metals. The alkali metals show most clearly the effects of increasing size and mass on physical properties of the elements within one family of metals.

section 2.3
The Noble Gases

The family of elements at the extreme right of the periodic table consisting of the gases helium (He), neon (Ne), argon (Ar), krypton (Kr), xenon (Xe), and radon (Rn), is known as the **rare gases** or the **noble gases.** For many years it was believed that these gases were chemically inert, that is, that they could not combine with any other element, and they were therefore also called the **inert gases.**

Although these gases are not all inert, they are not chemically active substances. The name of the first noble gas to be isolated, argon, comes from the Greek word *argos,* which means inactive, or not working. Argon constitutes 1% of our atmosphere by volume (1.3% by weight). It was first obtained by Lord Rayleigh and Sir William Ramsay, British scientists, by the fractional distillation of liquid air. Helium was the first of the noble gases to be detected. In 1868 strong lines due to an element unknown on earth were detected in the spectrum of the sun, and the element was named helium from the Greek word for sun, *helios.* In 1895 Ramsay isolated helium from uranium minerals, where it is produced by radioactive decay. Within five years after argon had been isolated, careful fractionation of air by Ramsay led to the discovery of neon, krypton, and xenon. The last member of this family, radon, is radioactive.

In 1962 Neil Bartlett synthesized the first compound of xenon, and several other xenon compounds were synthesized by others later that same year. A number of compounds of Xe and Kr have now been prepared, but none involving He, Ne, or Ar.

Xenon reacts directly with fluorine, F_2, to form three different crystalline xenon fluorides:

$Xe(g) + F_2(g) \rightarrow XeF_2(s)$	(2-10)
$XeF_2(s) + F_2(g) \rightarrow XeF_4(s)$	(2-11)
$XeF_4(s) + F_2(g) \rightarrow XeF_6(s)$	(2-12)

Other xenon compounds that have been made are $XeOF_2$, $XeOF_4$, $CsXeF_7$, and Cs_2XeF_8. Krypton difluoride, KrF_2, has also been prepared.

All the members of this group are monatomic gases, that is, they exist as discrete, individual atoms. All other elements that are gases at room temperature and normal atmospheric pressure are diatomic: H_2, O_2, N_2, F_2, and Cl_2. In fact, the noble gases are the only gases that are monatomic at temperatures close to room temperature and at normal atmospheric pressure.

2	He	4.0
10	Ne	20.2
18	Ar	39.9
36	Kr	83.8
54	Xe	131.3
86	Rn	(222)
	Group 0	

section 2.4
The Nonmetals

Nonmetals are poor conductors of heat and electricity, and may be either gases, liquids, or solids. Solid nonmetals are neither malleable nor ductile; they are either hard and brittle, or soft and crumbly. Solid sulfur, for instance, is easily powdered and in that form resembles pale yellow flour. Solid iodine consists of diatomic molecules,

I_2, and the shiny greyish-black crystals are easily broken. Phosphorus, which exists in several different solid forms, consists of P_4 molecules. Red phosphorus is an amorphous powder, and yellow (or white) phosphorus is wax-like. The P_4 molecules of white phosphorus are tetrahedral in shape, with each P atom bonded to three other P atoms situated at the corners of a tetrahedron. (See Fig. 2.5.) White phosphorus reacts so vigorously with oxygen that it ignites and burns in air, and is used in military incendiary devices. It is stored under water to protect it from the air. Red phosphorus, however, is stable in air.

The nonmetals occupy a triangular region on the right side of the periodic table. (See Fig. 2.6.) The essential characteristic of a nonmetallic element is its tendency to gain one or more electrons during chemical reactions, thereby becoming an **anion** (a negatively charged ion). The greater the tendency to accept electrons, the more nonmetallic the element. Elements become increasingly nonmetallic in character as one goes to the right across the periodic table, and upwards within a group. The most nonmetallic element is fluorine, F, atomic number 9. As one moves to the left, or goes down a family to the members of higher atomic number, the elements become more metallic. Indeed, boron (B), silicon (Si), arsenic (As), and tellurium (Te), are considered **metalloids** or **semimetals.**

The Halogens

9	F **19.0**
17	Cl **35.5**
35	Br **79.9**
53	I **126.9**
85	At **(210)**
Group VIIA	

There are two groups of nonmetals of great importance chemically. The first is the **halogen family,** the vertical column of elements on the far right of the periodic table, directly adjacent to the rare gases. The **halogens,** Group VIIA, consist of the elements fluorine (F), chlorine (Cl), bromine (Br), and iodine (I). The last member of the family, astatine (At), is radioactive. Astatine is a man-made element, and was synthesized in 1940 by Emilio Segré and his co-workers at the University of California. The total amount of astatine in the earth's crust is less than 1 oz, so it is unlikely you will ever come into contact with any!

The halogens normally exist as diatomic molecules. Fluorine, F_2, is a pale yellow corrosive gas that reacts with practically all other substances, both organic and inorganic. Chlorine, Cl_2, a pale greenish-yellow gas, is also highly reactive. Bromine, a reddish-brown liquid of high density, is the only liquid nonmetallic element at room temperature. The density of liquid bromine is about three times that of water. Bromine is volatile; its vapor is also red brown. Both liquid and vapor are diatomic, Br_2. Iodine, I_2, is a greyish-black, lustrous solid that vaporizes, yielding a violet gas. Figure 2.7 is a photograph of a sample of I_2. The melting and boiling points of the halogens are listed in Table 2.2.

The halogens form anions with single negative charges (mononegative ions) called **halide ions:** fluoride (F^-), chloride (Cl^-), bromide (Br^-), and iodide (I^-) ions. All the elements in the periodic table form halide compounds except for He, Ne, and Ar.

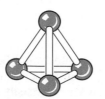

Fig. 2.5. The tetrahedral structure of white phosphorus, P_4.

5 B 10.81	6 C 12.011	7 N 14.0067	8 O 15.9994	9 F 18.99840
	14 Si 28.086	15 P 30.97376	16 S 32.06	17 Cl 35.453
		33 As 74.9216	34 Se 78.96	35 Br 79.904
			52 Te 127.60	53 I 126.9045
				85 At (210)

※ the only liquid nonmetallic element at rm. temp.

Fig. 2.6. The nonmetals. Boron (B), silicon (Si), arsenic (As), and tellurium (Te) are metalloids.

Some typical reactions of the halogens are the following:

$$2Na(s) + Cl_2(g) \rightarrow 2NaCl(s) \tag{2-13}$$

$$Br_2(\ell) + H_2O \rightarrow H^+(aq) + Br^-(aq) + HOBr(aq) \tag{2-14}$$

$$I_2(s) + H_2(g) \rightarrow 2HI(g) \tag{2-15}$$

$$Sr(s) + Br_2(\ell) \rightarrow SrBr_2(s) \tag{2-16}$$

The Chalcogens

The **oxygen family,** also called the **chalcogens,** contains three nonmetals, oxygen (O), sulfur (S), and selenium (Se). Oxygen is the most abundant element on earth. It is a principal component of water and of the rocks and minerals in the earth's crust. About 21% of the components of dry air consists of O_2 molecules. A second form of

Fig. 2.7. Crystals of I_2, a molecular solid. I_2 is volatile, low melting, and the crystals are easily broken.

Table 2.2. **Some Properties of the Halogens**

Element	Atomic Number	Color and Form	Melting Point (°C)	Boiling Point (°C)	Density[a]
F_2	9	Pale yellow gas	−220	−188.1	$1.7\ \text{g} \cdot \text{L}^{-1}$
Cl_2	17	Greenish-yellow gas	−101	−34.6	$3.2\ \text{g} \cdot \text{L}^{-1}$
Br_2	35	Reddish-brown liquid	−7.2	+58.8	$3.1\ \text{g} \cdot \text{cm}^{-3}$
I_2	53	Greyish-black shiny solid	+113.5	+184	$4.9\ \text{g} \cdot \text{cm}^{-3}$

[a] The densities of gases are reported at 0 °C and 1 atm. The densities of liquids and solids are reported at 20 °C.

16.0	
O	
8	
32.1	
S	
16	
79.0	
Se	
34	
127.6	
Te	
52	
(209)	
Po	
84	
Group VIA	

the element oxygen, called **ozone,** O_3, is present in very small amounts in the upper atmosphere. The molecules O_2 and O_3 are **allotropes,** different forms of the same element. Ozone absorbs ultraviolet radiation in the upper atmosphere, and is of vital importance in protecting the earth's surface from excessive exposure to ultraviolet light. On absorbing ultraviolet radiation, ozone decomposes to diatomic O_2:

$$2\ O_3(g) \xrightarrow{\text{UV light}} 3\ O_2(g) \tag{2-17}$$

With metallic elements, oxygen forms ionic compounds containing either the **oxide ion,** O^{2-}, the **peroxide ion,** O_2^{2-}, or the **superoxide ion,** O_2^{-}. With nonmetallic elements oxygen forms a wide variety of molecular compounds. Oxygen is an essential component of many organic compounds, notably fats, carbohydrates, and proteins.

There are two allotropic crystalline forms of sulfur, orthorhombic and monoclinic. (See Fig. 2.8.) Both forms consist of puckered rings of eight sulfur atoms, so that the molecular formula is S_8. Orthorhombic sulfur is the stable form at room temperature and atmospheric pressure. When heated to 112.8 °C, the orthorhombic crystals melt, forming a straw-colored liquid, that also consists of S_8 molecules. When this liquid cools, long needle-like crystals of monoclinic sulfur, the stable form above 96 °C, are formed. If the monoclinic crystals are allowed to cool to room temperature, they gradually change to the orthorhombic form.

Sulfur is a constituent of most animal and plant proteins. Both coal and petroleum contain sulfur, as a result of their biological origin. When sulfur-containing fuels are burned, sulfur dioxide is formed:

$$S(s) + O_2(g) \rightarrow SO_2(g) \tag{2-18}$$

Sulfur dioxide is one of the principal air pollutants, and along with nitrogen dioxide, NO_2, is the cause of *acid rain.* Sulfur dioxide reacts with water and oxygen in the air, in the presence of sunlight, to form sulfur trioxide, $SO_3(g)$, and aqueous droplets containing sulfuric acid, H_2SO_4. The overall reaction is

$$SO_2(g) + H_2O(g) + \tfrac{1}{2}O_2(g) \xrightarrow[\text{energy}]{\text{radiant}} H_2SO_4(aq) \tag{2-19}$$

Acid rain has killed fish and plant life in many freshwater lakes. Because acid reacts with calcium carbonate (limestone), marble statues and gargoyles on many ancient buildings are being slowly corroded. Acid rain is also corrosive to metals and some paints.

As the atomic number increases in this group of elements, metallic character also

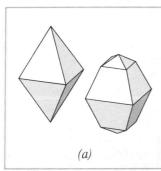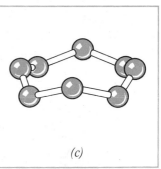

Fig. 2.8. Elemental sulfur.
(a) The shape of orthorhombic crystals, the stable form at room temperature. *(b)* The shape of monoclinic crystals, the stable form above 96 °C. *(c)* The molecular structure of S_8 molecules. *(d)* Large multicrystalline mass of rhombic sulfur. *(e)* Needlelike crystals of monoclinic sulfur.

(d)

(e)

increases, and the fourth member of the family, tellurium, is a **metalloid** or **semimetal.** Tellurium reacts both with metals and with nonmetals. There are compounds such as sodium telluride, Na_2Te, and hydrogen telluride, H_2Te, in which tellurium behaves like a nonmetal. (Hydrogen telluride is reputed to be the most foul smelling of all inorganic compounds.) There are also compounds such as tellurium dibromide, $TeBr_2$, and tellurium tetrachloride, $TeCl_4$, in which tellurium transfers some electron density to a halogen, and thus behaves in a way characteristic of metals.

When they react with metals, members of the oxygen family form dinegative ions: oxide (O^{2-}), sulfide (S^{2-}), selenide (Se^{2-}), and telluride (Te^{2-}) ions.

section 2.5
The Unique Position of Hydrogen

The first element in the periodic table, hydrogen, is unique. It does not belong to any group. Hydrogen, atomic number 1, has only a single valence electron. It can lose an electron to form the cation H^+, or it can gain an electron to form the anion H^-, **hydride ion,** (although this is less common).

Free elemental hydrogen is a diatomic gas, H_2. It is the lightest of all gases, and is colorless, odorless, and tasteless. Both its boiling point (-252.7 °C or 20.5 K) and melting point (-259 °C or 14 K) are very low. Only helium has lower boiling and melting points than hydrogen. Thus at very low temperatures, when all other substances are either liquids or solids, hydrogen and helium are still gases.

Hydrogen is very reactive and combines with most other elements to form compounds. It combines with oxygen, for example, to form water:

$$H_2(g) \ + \ \tfrac{1}{2} O_2(g) \ \rightarrow \ H_2O(\ell) \tag{2-20}$$

Because of its reactivity, the amount of free $H_2(g)$ in the atmosphere is very small, less than one part per million, by volume.

Hydrogen is the most abundant of all elements in the universe. On earth it is found chiefly in combination with oxygen in water, but it is also present combined with carbon in organic material such as petroleum, cellulose, and the sugars. Hydrogen is found in the sun and most stars, where it is dissociated into protons and electrons because of the extremely high temperatures. The composition of the sun is 73% hydrogen, 26% helium, and only 1% of all other elements. The planets Jupiter, Saturn, Uranus, and Neptune are composed mostly of hydrogen and helium.

Hydrogen combines with the halogens to form the hydrogen halides, HF, HCl, HBr, and HI. These compounds dissolve readily in water and their aqueous solutions are called, respectively, hydrofluoric acid, hydrochloric acid, hydrobromic acid, and hydroiodic acid. All these compounds are ionized in water (HF is only slightly ionized), but the free proton, H^+, does not exist in aqueous solution. It is bonded to one or more water molecules, and is usually symbolized as either H_3O^+ or $H^+(aq)$, and called the **hydronium ion.** A more detailed discussion of the hydronium ion will be found in Section 7.4.

Hydrogen also combines with the alkali and alkaline earth metals (with the exception of beryllium), to form ionic crystalline solids called hydrides, composed of the metal cations and the **hydride ion,** H^-. Examples are lithium hydride, LiH, sodium hydride, NaH, calcium hydride, CaH_2, and barium hydride, BaH_2. The hydrides can be prepared by heating the metal in the presence of hydrogen gas:

$$2Na(s) \ + \ H_2(g) \ \xrightarrow{\text{heat}} \ 2NaH(s) \tag{2-21}$$

section 2.6
The Transition Metals

The middle section of the periodic table contains the **transition metals.** There are three series of transition metals, each containing 10 elements. The first series contains the elements from scandium (Sc) through zinc (Zn), atomic numbers 21 through 30. The chemistry of the transition elements is much more complex than is the chemistry of the elements at either side of the periodic table. This is due to their electronic structure and will be discussed in detail in Chapters 13 and 20. You should, however, be aware of the fact that most of the transition metals can form more than one cation, in contrast to the alkali and alkaline earth metals. A few common examples of transition metal cations are

Cr^{2+}, chromium(II) ion Cr^{3+}, chromium(III) ion
Cu^+, copper(I) ion Cu^{2+}, copper(II) ion
Fe^{2+}, iron(II) ion Fe^{3+} iron(III) ion

section 2.7
Ionic Crystalline Solids

When an alkali or alkaline earth metal reacts with a nonmetallic element, such as a halogen or a chalcogen, the compound formed is an ionic crystalline solid. Direct reactions between an alkali or alkaline earth metal and a nonmetal are **exothermic,** that is, heat is produced as the reaction occurs. Formulas such as $NaBr$, K_2S, RbI, MgO, $BaCl_2$, and SrF_2, all represent ionic crystalline solids. The anions in ionic crystalline solids may also be polyatomic, such as NO_3^-, nitrate ion, SO_4^{2-}, sulfate ion, CO_3^{2-}, carbonate ion, and others listed in Table 2.4.

Ionic crystalline solids are hard compared to molecular solids, and have relatively high melting points. For instance, the melting point of NaCl is 801 °C, and Li_2O melts above 1700 °C, a temperature so high that it has not been possible to measure it more accurately.

Electrical conductivity is due to moving charged particles. Molten ionic compounds are good conductors of electricity, because the ions can move through the liquid melt. In contrast, liquid or molten molecular compounds do not conduct electricity. Neither ionic crystalline solids nor molecular solids are good conductors of electricity. Ionic crystalline solids are poor conductors of electricity because the ions are not free to move, as they are in the molten compound. The force of attraction between ions of opposite charge is very large, and at temperatures below the melting point of the solid, the ions are held at fixed positions in the crystal structure.

Aqueous solutions of soluble ionic compounds are good conductors of electricity because the separated ions move relatively independently through the solution. In general, aqueous solutions of molecular compounds do not conduct electricity, although some molecular compounds, such as HCl(g), react with water to form ions, producing a solution that does conduct electricity.

section 2.8
Molecular Compounds

When nonmetals combine with other nonmetals, the compounds produced are generally **molecular** rather than ionic. Many molecular compounds are gases at room temperature. Examples are N_2O, NO, NO_2, PF_3, PF_5, SO_2, SO_3, Cl_2, NF_3, Cl_2O, CH_4, CO, and CO_2. Note particularly that the same two atoms may combine to give several molecular compounds with different stoichiometries, that is, different molar ratios of the constituent atoms.

Crystals of molecular compounds are relatively soft and have low melting points compared with those of ionic crystalline solids. Examples of molecular compounds that are solids at room temperature are sucrose (ordinary sugar, $C_{12}H_{22}O_{11}$, melting point 185 °C), naphthalene (often used as moth balls, $C_{10}H_8$, melting point 80.6 °C),

Table 2.3. **Characteristics of Molecular and Ionic Solids**

Solids	Chemical Units at Lattice Sites	Some Properties	Examples
Molecular	Molecules or atoms	Soft, generally low melting, nonconductors of electricity both solid and molten	H_2O (ice) $C_6H_{12}O_6$ (glucose) CO_2 (dry ice) I_2 (iodine) $C_{10}H_{16}O$ (camphor)
Ionic crystalline	Cations and anions	Hard, high melting, brittle, nonconductors solid, but conductors when molten	$NaCl$, CaO, KI, $BaSO_4$, $MgBr_2$, KNO_3, Li_2O, CsF, Na_2S, $Ca_3(PO_4)_2$

and phosphorus oxybromide ($POBr_3$, melting point 56 °C). Compounds consisting only of the elements carbon and hydrogen (**hydrocarbons**) are molecular, as are most other organic compounds.*

All gaseous compounds are molecular. With a few exceptions, such as the red-brown nitrogen dioxide, NO_2, most gases are colorless. Solid and liquid molecular compounds have no metallic luster and are frequently white or colorless. Solid, liquid, and gaseous molecular compounds are poor conductors of electricity. Unless there is a reaction between the molecular compound and water that produces ions, aqueous solutions of molecular compounds also have very low electrical conductivity, in contrast to aqueous solutions of ionic crystalline solids. For example, a solution of sodium chloride, $NaCl$, in water is a good conductor of electricity, whereas a solution of sugar in water has a negligible electrical conductivity.

Table 2.3 summarizes the differences between molecular solids and ionic crystalline solids.

section 2.9
Some Simple Inorganic Nomenclature

The fascinating part of learning a foreign language is becoming familiar with another literature, history, and culture, and having conversations with others who speak that language. But in order to read the literature or have a conversation, you must first learn the vocabulary. A similar situation exists in the study of chemistry. One has to have a basic working vocabulary in order to be able to read about and understand the fundamental principles and the interesting applications.

An international committee has developed a system of rules for naming both inorganic and organic chemical compounds. A portion of the rules for naming the more common ions encountered in inorganic chemistry is contained in Table 2.4.

In naming ionic crystalline solids, the name of the cation is given first, followed by the name of the anion. There is no need for indicating the number of ions of each kind per formula unit, because that is determined by the charge on the ions and the requirement that the solid be electrically neutral. Examples 2.1 and 2.2 illustrate the way in which ionic crystalline solids are named.

---

* All compounds of carbon, with the exception of CO, CO_2, CS_2, the carbonates, and the cyanides, are termed organic. There are more than three million organic compounds.

Table 2.4. **Naming Common Inorganic Ions**

Cations

1. Monatomic ions of the metals that form only one cation (the alkali and alkaline earth metals) are named simply by adding the word ion to the name of the element:

Li^+, lithium ion Na^+, sodium ion K^+, potassium ion
Be^{2+}, beryllium ion Ca^{2+}, calcium ion Mg^{2+}, magnesium ion

2. Elements of the transition metals and the metals closer to the right-hand side of the periodic table often form more than one cation. For monatomic ions the internationally accepted, modern (Stock) system of nomenclature is to indicate the magnitude of the charge on the ion by a Roman numeral in parentheses directly following the name of the metal. In an older system of nomenclature, the name of the ion with the lower charge ends in -ous, while the name of the ion with the higher charge ends in -ic. These endings are added to the root of the Latin name of the element (from which the symbol for the element is obtained). In the examples below, the older name is given in parentheses.

Cr^{2+} chromium(II) (chromous) ion Cr^{3+} chromium(III) (chromic) ion
Co^{2+} cobalt(II) (cobaltous) ion Co^{3+} cobalt(III) (cobaltic) ion
Cu^+ copper(I) (cuprous) ion Cu^{2+} copper(II) (cupric) ion
Fe^{2+} iron(II) (ferrous) ion Fe^{3+} iron(III) (ferric) ion
Mn^{2+} manganese(II) (manganous) ion Mn^{3+} manganese(III) (manganic) ion
Hg_2^{2+} mercury(I) (mercurous) ion Hg^{2+} mercury (II) (mercuric) ion
Sn^{2+} tin(II) (stannous) ion Sn^{4+} tin(IV) (stannic) ion
Tl^+ thallium(I) (thallous) ion Tl^{3+} thallium(III) (thallic) ion

3. Most cations are monatomic, but there are three common polyatomic cations:

NH_4^+ ammonium ion Hg_2^{2+} mercurous ion H_3O^+ hydronium ion

The transition metal cations form a large number of complex ions by bonding to other species. Some examples of complex cations are $[Cu(NH_3)_4]^{2+}$, $[Fe(H_2O)_6]^{3+}$, and $[Ag(NH_3)_2]^+$. The rules of nomenclature for complex ions will be found in Section 20.6.

Anions

1. Monatomic anions are named by dropping the ending of the name of the element and adding -ide. Some common monatomic anions are

F^- fluoride ion Cl^- chloride ion Br^- bromide ion I^- iodide ion
O^{2-} oxide ion S^{2-} sulfide ion N^{3-} nitride ion P^{3-} phosphide ion

2. Polyatomic anions containing oxygen (**oxyanions**) have names that end in -ite or -ate. If there is only a single oxyanion for an element, its name ends in -ate. If there are two oxyanions, the one with less oxygen ends in -ite, the one with more oxygen ends in -ate. Examples are

NO_2^- nitrite ion SO_3^{2-} sulfite ion AsO_3^{3-} arsenite ion
NO_3^- nitrate ion SO_4^{2-} sulfate ion AsO_4^{3-} arsenate ion
CO_3^{2-} carbonate ion CrO_4^{2-} chromate ion PO_4^{3-} phosphate ion

In some cases there are more than two oxyanions of an element. The prefix *hypo-* is used to designate the anion with the least number of oxygen atoms, and the prefix *per-* is used to designate the anion with the largest number of oxygen atoms. The halogens, in particular, form more than two oxyanions. The names of the oxyanions of chlorine illustrate the rules of nomenclature:

ClO^-, hypochlorite ion ClO_2^-, chlorite ion
ClO_3^-, chlorate ion ClO_4^-, perchlorate ion

(continues)

Table 2.4. *(Continued)*

3. Many doubly or triply charged anions readily add an H^+ to form an ion of lower charge. These ions are named by prefixing the word hydrogen to the name of the anion. In the older nomenclature, the prefix *bi-* was added. Examples are

HCO_3^-, hydrogen carbonate ion (bicarbonate ion)

HSO_4^-, hydrogen sulfate ion (bisulfate ion)

HS^-, hydrogen sulfide ion (bisulfide ion)

HSO_3^-, hydrogen sulfite ion (bisulfite ion)

4. Other common polyatomic ions whose names one should be familiar with are the following:

OH^- hydroxide ion	O_2^{2-} peroxide ion	O_2^- superoxide ion
$C_2O_4^{2-}$ oxalate ion	I_3^- triiodide ion	$S_2O_3^{2-}$ thiosulfate ion[a]
CN^- cyanide ion	OCN^- cyanate ion	SCN^- thiocyanate ion[a]
$Cr_2O_7^{2-}$ dichromate ion	S_2^{2-} disulfide ion	MnO_4^- permanganate ion

[a] The prefix *thio-* indicates that a sulfur atom has replaced an oxygen atom. Thus thiosulfate is obtained by replacing an O atom in sulfate ion with an S atom.

EXAMPLE 2.1. Nomenclature of ionic crystalline solids

Name the following compounds: **(a)** Cu_2CO_3, **(b)** $CuCO_3$, **(c)** CrO, **(d)** Cr_2O_3, **(e)** $KHSO_4$, **(f)** $Fe(OH)_2$, **(g)** $Fe(OH)_3$, **(h)** $(NH_4)_2C_2O_4$.

Solution. The systematic (Stock) names are given, with the older names in parentheses.

(a) Cu_2CO_3 is copper(I) carbonate (cuprous carbonate).
(b) $CuCO_3$ is copper(II) carbonate (cupric carbonate).
(c) CrO is chromium(II) oxide (chromous oxide).
(d) Cr_2O_3 is chromium(III) oxide (chromic oxide).
(e) $KHSO_4$ is potassium hydrogen sulfate (potassium bisulfate).
(f) $Fe(OH)_2$ is iron(II) hydroxide (ferrous hydroxide).
(g) $Fe(OH)_3$ is iron(III) hydroxide (ferric hydroxide).
(h) $(NH_4)_2C_2O_4$ is ammonium oxalate.

EXAMPLE 2.2. Determining the formulas of ionic solids

What are the formulas of the following compounds? **(a)** Sodium carbonate, **(b)** iron(III) nitrate, **(c)** iron(II) sulfate, **(d)** strontium thiocyanate trihydrate, **(e)** barium thiosulfate, **(f)** aluminum oxide.

Solution. **(a)** Sodium ion has a charge of $+1$, and carbonate ion, CO_3^{2-}, has a charge of -2. In order for the solid to be electrically neutral, there must be 2 Na^+ ions for each CO_3^{2-} ion. The formula of sodium carbonate is therefore Na_2CO_3.
(b) Three nitrate ions, NO_3^-, must combine with one Fe^{3+} ion in order for the solid to be electrically neutral. The formula is therefore $Fe(NO_3)_3$.
(c) The molar ratio of $Fe^{2+}:SO_4^{2-}$ must be $1:1$, so the formula is $FeSO_4$.
(d) The molar ratio of $Sr^{2+}:SCN^-$ must be $1:2$ for electroneutrality. The formula of the trihydrate is therefore $Sr(SCN)_2 \cdot 3H_2O$.
(e) The molar ratio of $Ba^{2+}:S_2O_3^{2-}$ must be $1:1$, so the formula is BaS_2O_3.
(f) Aluminum is in Group IIIA, and the aluminum ion is Al^{3+}. The molar ratio of $Al^{3+}:O^{2-}$ must be $2:3$ for electroneutrality, and the formula is Al_2O_3.

For molecular compounds, it is necessary to indicate how many atoms of each kind are contained in one molecule, because the same two atoms may combine to give several compounds. The prefixes di-, tri-, tetra-, penta-, hexa-, and so on, are used to indicate the number of atoms of each kind in the molecule. Example 2.3 illustrates the way in which molecular compounds are named.

EXAMPLE 2.3. Nomenclature of molecular compounds

Give the systematic names for the following compounds: (a) N_2O, (b) NO_2, (c) N_2O_4, (d) N_2O_5, (e) PCl_3, (f) PCl_5, (g) S_2F_2, (h) SF_4, (i) S_2F_{10}, (j) IF_5, (k) IF_7.

Solution. Older (common) names are given in parentheses.
(a) N_2O is dinitrogen oxide (nitrous oxide, laughing gas).
(b) NO_2 is nitrogen dioxide.
(c) N_2O_4 is dinitrogen tetroxide (nitrogen tetroxide).
(d) N_2O_5 is dinitrogen pentoxide (nitrogen pentoxide).
(e) PCl_3 is phosphorus trichloride.
(f) PCl_5 is phosphorus pentachloride.
(g) S_2F_2 is disulfur difluoride.
(h) SF_4 is sulfur tetrafluoride.
(i) S_2F_{10} is disulfur decafluoride.
(j) IF_5 is iodine pentafluoride.
(k) IF_7 is iodine heptafluoride.

section 2.10
Nonstoichiometric Compounds

The majority of compounds are **stoichiometric,** that is, they have a fixed and definite atomic composition, and the molar ratios of the different atoms that make up the compound are ratios of small whole numbers (simple integral ratios). Stoichiometric compounds are also called **daltonides,** in honor of John Dalton, the English chemist and physicist whose pioneering work led to the acceptance of the atomic theory of matter. There are, however, compounds that are **nonstoichiometric,** whose composition may vary within a certain range. Such compounds are also called **berthollides,** after the French chemist Claude Louis Berthollet, who proposed that the composition of a compound depends on the manner of its preparation.

Many nonstoichiometric compounds are ionic crystalline solids. In growing ionic crystals of any size, enormous numbers of ions must be deposited at exactly the proper location in the crystal structure. Depending on the method of preparation, that is, on how quickly the crystal is formed, at what temperature and pressure, what else is present in the solution, what the relative concentrations of the different ions of which the crystal is composed may be, and so on, certain **defects** in the crystal may occur. A common defect is a **vacancy;** an ion is missing from its proper place in the crystal structure. Figure 2.9 illustrates a vacancy in the iron(II) oxide crystal.

One rule that is never violated is that the crystal as a whole must remain electrically neutral, whether it has defects or not. If there is a vacancy or any irregularity that results in an excess of one kind of ion, the crystal will gain or lose electrons in order to maintain electrical neutrality. As an example, let us consider copper(I) sulfide, Cu_2S, which is a nonstoichiometric compound. If a crystal of copper(I) sulfide were pure Cu_2S, the molar ratio of Cu : S would be 2 : 1. Frequently, however, there are some Cu^+ vacancies in the crystal. In order to preserve electrical neutrality, a Cu^{2+} ion

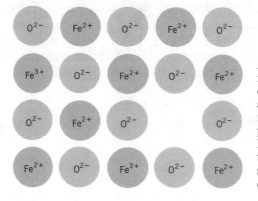

Fig. 2.9. A schematic representation of the nonstoichiometry of iron(II) oxide, an ionic crystalline solid. One Fe^{2+} ion is missing, and in order to maintain electrical neutrality for the solid, two Fe^{3+} ions are present in the crystal structure. The ratio of Fe : O in crystals of iron(II) oxide is therefore less than 1 : 1, and varies from sample to sample depending on the number of vacancies present.

[copper(II) or cupric ion] is formed at some other point in the crystal. For each Cu^+ vacancy, one Cu^+ ion at some other location loses an electron and becomes a Cu^{2+} ion. The overall ratio of Cu : S is therefore not a simple integral ratio. The molar ratio of Cu : S will lie somewhere between 2 : 1 and 1 : 1, and will vary with the number of Cu^+ vacancies in the crystal. Crystals with compositions anywhere from $Cu_{1.7}S$ to Cu_2S have been observed.

Many transition metal oxides, sulfides, and hydrides are nonstoichiometric, because transition metals are capable of forming more than one kind of cation. For instance, pure FeO [iron(II) oxide or ferrous oxide] has never been observed; there are always some Fe^{3+} ions as well as Fe^{2+} ions in the crystal. Iron(II) oxide has a range of compositions from about $Fe_{0.84}O$ to $Fe_{0.94}O$.

Summary

In the **periodic table,** the elements are arranged in order of increasing atomic number. Elements with similar physical and chemical properties are placed in a vertical column and constitute a **family** or **group.**

About 70% of all the known elements are **metals.** Metals are found at the left side and at the bottom of the periodic table. Metallic character is associated with the tendency of an element to transfer electrons to other substances. Metallic character increases as one goes to the left across the periodic table horizontally, or goes down a group to elements of increasing atomic number.

The two groups of elements (Group IA and IIA) at the extreme left side of the periodic table, the **alkali** and **alkaline earth metals,** are called **active metals** because of their great chemical reactivity. When an alkali metal takes part in a chemical reaction it invariably transfers its single valence electron to the other reactant, and becomes a cation with a single positive charge. The alkaline earth metals tend to transfer their two valence electrons and become cations of charge 2 +.

At the extreme right side of the periodic table is the family of **rare** or **noble gases.** These elements are almost completely unreactive chemically. No compounds of He, Ne, or Ar have been prepared as yet, but compounds of Xe, Kr, and Rn are known. The noble gases are the only elements that exist as monatomic gases at room temperature and normal atmospheric pressure.

The **nonmetals** are found in a triangular region at the upper right side of the periodic table. Nonmetals tend to accept electrons from other substances during chemical reactions. The chemical activity of a nonmetal increases as its tendency to gain electrons increases; nonmetallic character increases as one goes to the right across the periodic table horizontally, and upwards within a group to elements of lower atomic number.

The most chemically active family of nonmetals is the **halogens,** Group VIIA. The halogens form mononegative ions called **halide ions.** All elements in the periodic table, with the exception of He, Ne, and Ar, combine with the halogens to form halide compounds. Directly to the left of the halogens in the periodic table is the **oxygen family** of elements, Group VIA, which contains three nonmetals: oxygen, sulfur, and selenium. Members of the oxygen family tend to gain two electrons and become dinegative ions when they react with metallic elements.

Hydrogen is a unique element in the periodic system. It belongs to no family. It has only a single electron, as it is atomic number 1. Hydrogen can lose its electron to form the cation H^+, or it can gain an electron to form the **hydride ion,** H^-. The cation H^+ is a single proton, and does not exist free in aqueous solution. It bonds to one or more water molecules, and the resulting species is called a **hydronium ion,** H_3O^+ or $H^+(aq)$. Hydrogen is the most abundant element in the universe and is the principal component of most stars. Hydrogen is a reactive element and combines with most of the other elements to form compounds.

The middle section of the periodic table contains three series of **transition metals** with each series consisting of 10 elements. The chemistry of the transition metals is more complex than the chemistry of the elements at either side of the periodic table. Many of the transition elements form more than one cation. This is in contrast to the behavior of the alkali metals, which form only cations of charge $1+$, and the alkaline earths, which form only cations of charge $2+$.

The reaction between an alkali or alkaline earth metal and a halogen or other nonmetal is **exothermic,** and produces an ionic crystalline solid. Ionic crystals are hard and have high melting points compared to molecular solids. Molten ionic compounds are good conductors of electricity, and aqueous solutions of soluble ionic compounds also have high electrical conductivity.

Nonmetals combine with other nonmetals to produce **molecular compounds.** Many molecular compounds are gases at room temperature. Molecular solids are relatively soft and have low melting points compared to ionic crystalline solids. Solid, liquid, and gaseous molecular compounds are poor conductors of electricity.

Molecular compounds have a fixed and definite atomic composition, but ionic crystalline solids may be **nonstoichiometric,** that is, their composition may vary within a certain range.

Multiple Choice Questions

1. In the liquid state the poorest electrical conductor among the following substances is
   (a) $SrBr_2$ (b) NaI (c) KCN (d) CBr_4 (e) $Ba(NO_3)_2$

2. The correct formula for tin(II) sulfide is
   (a) TiS (b) $SrSO_3$ (c) SnS (d) TiS_2 (e) $SnSO_3$

3. The correct formula for silver perchlorate is
   (a) AgCl (b) $SiCl_4$ (c) Si_2Cl_2 (d) $AgClO_3$ (e) $AgClO_4$

4. An aqueous solution of which of the following compounds will not conduct electricity?
   (a) CH_3OH (b) $Ba(OH)_2$ (c) $SrBr_2$ (d) KOH (e) $Ca(NO_3)_2$

5. Which of the following compounds is not a gas at room temperature?

   (a) NF_3 (b) H_2S (c) SF_6 (d) SrF_2 (e) N_2O

6. The element with atomic number 117 has not yet been discovered or synthesized. It should be

   (a) a transition metal   (b) a member of the oxygen family
   (c) a representative metal   (d) a noble gas   (e) a halogen

7. The element dysprosium, Dy, forms an oxide with formula Dy_2O_3. The corresponding bromide of dysprosium has the formula

   (a) Dy_2Br_3 (b) $DyBr$ (c) $DyBr_2$ (d) $DyBr_3$ (e) Dy_3Br_2

8. In the compound sodium oxalate, the molar ratio of sodium to carbon atoms is

   (a) $1:1$ (b) $2:1$ (c) $1:2$ (d) $1:3$ (e) $2:3$

9. Two elements that have similar properties because they have the same number of valence electrons are

   (a) C and Cl   (b) Ca and Al   (c) O and Ar   (d) Si and S   (e) Se and Te

10. When cerium(III) carbonate pentahydrate is heated strongly it decomposes to yield carbon dioxide, cerium(III) oxide, and water vapor. How many moles of carbon dioxide can be obtained by completely decomposing 1.50 mol of cerium(III) carbonate pentahydrate?

    (a) 1.50   (b) 3.00   (c) 4.50   (d) 6.00   (e) 7.50

11. The name of the ion BrO_3^- is

    (a) bromide ion   (b) hypobromite ion   (c) bromate ion
    (d) perbromate ion   (e) bromite ion

12. The first ionization energy of an element is the energy required to remove one electron from a gaseous atom of that element, that is, it is the energy required for the reaction

$$X(g) \rightarrow X^+(g) + e^-$$

where X stands for any element. Which of the following elements would you expect to have the lowest first ionization energy?

    (a) Mg   (b) Rb   (c) Li   (d) Ca   (e) Be

13. Gallium and sulfur react directly to form gallium sulfide, Ga_2S_3. If we heat 2 mol of gallium with 2 mol of sulfur, what is the maximum number of moles of Ga_2S_3 that can be formed?

    (a) one third   (b) one half   (c) two thirds   (d) one   (e) two

14. From its position in the periodic table, at room temperature the element yttrium, Y, is expected to be

    (a) a liquid   (b) a high-melting metal   (c) a metalloid   (d) a molecular solid
    (e) a soft metal

15. The most metallic element in the fifth period is

    (a) Ag   (b) Rb   (c) Sn   (d) Pd   (e) Rh

16. The name of the SO_3^{2-} ion is

    (a) sulfite   (b) sulfate   (c) sulfide   (d) sulfurate   (e) bisulfide ion

Problems

2.1. What are the formulas of the compounds hydrogen forms with rubidium (Rb), sulfur (S), strontium (Sr), and iodine (I)? Classify these compounds as ionic or molecular.

2.2. Give both the systematic and the common name (if they differ) for each of the following compounds:
   (a) $Sr(NO_3)_2$
   (b) Ag_2CO_3
   (c) $CoCl_2$
   (d) $CoCl_3$
   (e) $AuCN$
   (f) $(NH_4)_3PO_4$
   (g) $KHSO_4$
   (h) Na_2S
   (i) $Ca(HSO_3)_2$
   (j) $CuBr$
   (k) $Fe_2(SO_4)_3$
   (l) $FeSO_4$
   (m) Na_2SO_3
   (n) $KMnO_4$
   (o) CuI_2

2.3. Consult the periodic table and arrange the following pure solid elements in order of increasing electrical conductivity at room temperature: Tl, Ca, S, Sn, and Si. Explain the reason for the order you chose.

2.4. Give the formula of each of the following compounds:
   (a) calcium oxalate
   (b) potassium hydrogen carbonate
   (c) barium nitrite
   (d) ammonium permanganate
   (e) lithium nitrate
   (f) aluminum perchlorate hexahydrate
   (g) sodium thiosulfate
   (h) rubidium peroxide
   (i) copper(I) oxide
   (j) cobalt(III) sulfate
   (k) iron(III) oxide
   (l) strontium hydroxide
   (m) chromium(II) fluoride
   (n) copper(II) selenate pentahydrate

2.5. Name the following compounds and classify them as either ionic or molecular: Rb_2S, H_2Se, NH_3, CCl_4, BaI_2, OF_2, BrF_5, SrS, and CO.

2.6. Write a balanced equation for the reaction between the following substances:
   (a) potassium metal with water
   (b) potassium metal with oxygen
   (c) strontium metal with water
   (d) strontium metal with oxygen

2.7. Give the molecular formulas and physical state (solid, liquid, or gas) of the following free elements at room temperature and normal atmospheric pressure: nitrogen, oxygen, fluorine, neon, phosphorus, sulfur, chlorine, hydrogen, argon, bromine, iodine, and krypton.

2.8. A group of scientists at the University of California, as well as a group in the Soviet Union, are trying to synthesize elements with atomic numbers above 106. (a) What will be the atomic number of the next rare gas? (b) What will be the atomic number of the element expected to have properties similar to copper, silver, and gold? (c) Which existing element will have properties most similar to those of element number 113? Explain your answers.

2.9. Arrange the following series of elements in order of increasing metallic character:
   (a) Si, Sn, C, Ge, Pb   (b) Si, Cl, Al, Na, P

2.10. Write a balanced equation for the reaction of
   (a) Chlorine gas with water.
   (b) Chlorine gas with hydrogen gas.
   (c) Chlorine gas with strontium metal.
   (d) Hydrogen gas with oxygen gas.

2.11. The reaction between 1 mol of gaseous dinitrogen trioxide and 1 mol of solid barium oxide yields 1 mol of a single product that is an ionic crystalline solid. What is the name and formula of that product?

2.12. Arrange the following series of elements in order of increasing nonmetallic character:
(a) As, P, Bi, Sb, N   (b) N, B, Li, F, O

2.13. The atomic volume is the volume occupied by one mole of atoms. Using the table of atomic weights and the densities listed in Table 2.1, calculate the atomic volumes of the alkali and alkaline earth metals. Tabulate your values, and after considering the results, answer the following questions:
(a) Generally, how does the atomic volume vary within a family of the periodic table?
(b) Are the atomic volumes of the alkaline earth metals smaller, larger, or the same as the atomic volumes of the adjacent alkali metals?

2.14. Magnesium metal reacts directly with oxygen gas to form magnesium oxide, with nitrogen gas to form magnesium nitride, with hydrogen gas to form magnesium hydride, and with chlorine gas to form magnesium chloride. Write balanced equations for each of these reactions.

2.15. Classify each of the following compounds as either ionic or molecular: C_2H_6, $Mg(NO_3)_2$, N_2O_5, CS_2, $SiBr_4$, $(NH_4)_2SO_4$, NCl_3, and Na_2O_2.

2.16. When solid ammonium nitrate is heated it decomposes, yielding only dinitrogen oxide (nitrous oxide) gas and water vapor. Write a balanced equation for this reaction.

2.17. Classify the following elements as either representative elements, transition metals, or inner transition metals: Rb, Cu, Al, Ag, Sr, P, U, Zn, Li, Fe, Sb, Gd, Mg, Mn, Th, Pb, Mo, Ce, Pt, and Hg.

chapter 3 The Gas Laws and Stoichiometry Involving Gases

Robert Boyle (1627–1691), an English natural philosopher and one of the founders of modern chemistry, spent most of his professional life in Oxford and London. He performed a series of experiments on the properties of air, and in 1660 enunciated the law that the volume of a gas varies inversely as the pressure, at constant temperature. Boyle was one of the founders of the "Royal Society of London for improving natural knowledge." Although he investigated many areas of the physical sciences, chemistry was his favorite study. He distinguished between mixtures and compounds, and demonstrated that a compound may have quite different properties from those of its constituent elements. He studied the chemistry of respiration and of combustion. Boyle was also interested in theology, and was learned in Latin, French, Hebrew, Greek, and Syriac.

The study of the properties of gases and of reactions between gases has played an important role in the history of chemistry, particularly in the development of atomic theory. Dalton's ideas about atomic theory were a result of his investigations of properties of the air and other gases. Observations made early in the nineteenth century about the volumes of gases that combined in chemical reactions led to conclusions about the molecular nature of matter and chemical stoichiometry. By studying the relationships between the temperature, pressure, volume, and number of moles of gas, we learn how the molecular weight of a gas can be determined. Furthermore, experimental observations about the properties of gases provide the foundation for the theory that describes the nature of molecules of gases, the kinetic–molecular theory, which will be discussed in Chapter 4.

section 3.1
Avogadro's Law and Its Use in Determining Molecular Formulas

In 1808 the French scientist Joseph Louis Gay-Lussac published a paper discussing the results of a series of experiments he and others had performed on the ratios of the volumes of gases that combine to form a compound. He described experiments that showed, for example, that reacting two volumes of hydrogen with one volume of oxygen produced two volumes of water vapor, when the volumes of all the gases were measured at the same temperature and pressure. The use of the word "volumes" means that the actual volume does not affect the ratio: 120 mL of hydrogen combine with 60 mL of oxygen to produce 120 mL of water vapor, or 4 L of hydrogen combine with 2 L of oxygen to produce 4 L of water vapor, and so on. The ratio of combining volumes of H_2 to O_2 is 2 : 1 when both gases are at the same temperature and pressure, regardless of what volumes are used. Gay-Lussac also found that one volume of oxygen combines with two volumes of carbon monoxide to produce two volumes of carbon dioxide, and that one volume of nitrogen combines with one volume of oxygen to produce two volumes of nitric oxide, $NO(g)$.

As a result of his investigations of a number of different reactions, Gay-Lussac formulated the **law of combining volumes:** The volumes of gases that react with one another, or are produced in a chemical reaction, are in the ratio of small integers, provided that all the gases are at the same temperature and pressure.

Avogadro's Law

Gay-Lussac apparently did not grasp the theoretical significance of his experimental results, but in 1811 the Italian physicist Amadeo Avogadro proposed that Gay-Lussac's law of combining volumes implied that *equal volumes of different gases at the same temperature and pressure contain equal numbers of molecules.* Avogadro's hypothesis was disregarded for nearly fifty years, and was not accepted until after his death. In 1858 Stanislao Cannizzaro extended Avogadro's hypothesis and used it as a basis for establishing molecular formulas.

Avogadro had proposed, in 1811, that in order to explain Gay-Lussac's law of combining volumes, some gaseous elements must consist of polyatomic molecules. Consider, for instance, the reaction between hydrogen and oxygen. We now know that both of these gases are diatomic molecules, H_2 and O_2, so that we write the equation for the reaction as

$$2H_2(g) + O_2(g) \rightarrow 2H_2O(g) \tag{3-1}$$

which we read to mean 2 molecules of H_2 combine with 1 molecule of O_2 to form

2 molecules of H_2O, or $2n$ molecules of H_2 combine with n molecules of O_2 to form $2n$ molecules of H_2O, where n is any integer. Gay-Lussac's work showed that the equation can also be interpreted to mean that 2 volumes of H_2 combine with 1 volume of O_2 to yield 2 volumes of H_2O, provided that all the gases are at the same temperature and pressure. Gay-Lussac's law of combining volumes therefore implies that the volume of a gas is directly proportional to the number of molecules in the sample, and that, if different gases are compared at the same temperature and pressure, equal volumes of the different gases must contain the same number of molecules.

Avogadro's hypothesis means that if you had in front of you three containers of identical volume, such as three 500-mL bulbs, and one was filled with NO_2, one with C_4H_{10}, and one with H_2 gas, all at the same temperature and pressure, then each flask would contain exactly the same number of molecules. This hypothesis of Avogadro's was of tremendous importance in developing atomic theory, because it means that we can "count" molecules not only by weighing substances, but also by measuring volumes of gases at the same temperature and pressure. The usefulness and validity of Avogadro's hypothesis have been thoroughly investigated since 1811, and it is now called **Avogadro's law.**

If Avogadro's law is correct, one mole of all gases should occupy exactly the same volume, provided all the gases are at the same temperature and pressure. Careful experimental observations show that the molar volumes of different gases at the same temperature and pressure are not *exactly* the same, but they differ only slightly.* It is customary to refer to 0 °C and 1-atm pressure as "standard conditions of temperature and pressure" (STP). One mole of any gas occupies a volume that is very close to 22.4 L at 0 °C and 1 atm. Values of the molar volumes at STP for a few common gases are O_2, 22.394 L/mol; H_2, 22.431 L/mol; CO_2, 22.264 L/mol; and NH_3, 22.076 L/mol.

Examples 3.1 and 3.2 illustrate how Avogadro's law can be used to determine the molecular weight and molecular formula of an unknown gaseous compound.

EXAMPLE 3.1. The molar volume of a gas at STP

A 2.640-g sample of an unknown gas occupies a volume of 705.6 mL at 0 °C and 1 atm (STP). What is the molecular weight of this gas?

Solution. Using Avogadro's law, we can find the number of moles of gas in the sample, since 1 mol of any gas occupies 22.4 L at STP. The volume of the gas should be expressed in liters.

$$\text{No. moles gas} = (0.7056 \text{ L})\left(\frac{1 \text{ mol}}{22.4 \text{ L}}\right) = 0.0315 \text{ mol}$$

The molecular weight is obtained by rearranging Eq. (1-25):

$$\text{molecular weight in g} \cdot \text{mol}^{-1} = \frac{\text{mass in grams}}{\text{No. moles}} = \frac{2.640 \text{ g}}{0.0315 \text{ mol}}$$

$$= 83.8 \text{ g} \cdot \text{mol}^{-1}$$

---

* At extremely high pressure (for instance, above 50 or 100 atm), the ratios of combining volumes of gases may vary significantly from simple integral ratios, and therefore Avogadro's law may be in error by a significant amount. The results of many experiments show that predictions based on Avogadro's law usually disagree with experimental observation by 1% or less, if the gases are at temperatures fairly close to, or higher than, room temperature, or at pressures close to, or lower than, 1 atm. The error made in assuming Avogadro's law is correct becomes large only at very high pressure or very low temperature.

EXAMPLE 3.2. Determination of the molecular formula of a gaseous compound using Avogadro's law

The elements boron and hydrogen combine to form a great many different compounds called boron hydrides or boranes. The following experiment was performed in order to determine the molecular formula of a gaseous compound known to contain only B and H. An evacuated gas bulb of volume 952 mL and mass 73.2684 g was filled with the gaseous borane at 0 °C and a pressure of exactly 1 atm. The mass of the gas-filled bulb was 75.5398 g. This sample of the compound was then burned in excess oxygen, which completely converted it to water, H_2O, and the solid white boron oxide, B_2O_3. The mass of the B_2O_3 was found to be 5.9316 g. What is the correct molecular formula of the borane?

Solution. To find the molecular formula, we need the molar ratio of boron atoms to hydrogen atoms. Let us first calculate the number of moles of B atoms in the sample of the unknown borane. This must be the same as the number of moles of B in the boron oxide, by the law of conservation of mass. The formula weight of B_2O_3 is 69.62 g/mol.

$$\text{No. moles } B_2O_3 = \frac{5.9316 \text{ g}}{69.62 \text{ g/mol}} = 8.520 \times 10^{-2} \text{ mol}$$

$$\text{No. moles } B = 2(\text{No. moles } B_2O_3) = 1.704 \times 10^{-1} \text{ mol}$$

We can now determine the mass of B in the sample:

$$\text{mass of } B = (1.704 \times 10^{-1} \text{ mol})(10.81 \text{ g} \cdot \text{mol}^{-1}) = 1.842 \text{ g of B}$$

We need to know the number of moles of hydrogen in the borane sample. We can calculate this if we know the mass of hydrogen in the sample. To determine the mass of H atoms, we need to know the mass of the sample itself, since

$$\text{mass of borane sample} = \text{mass of } B + \text{mass of } H$$

The mass of the sample is obtained by subtracting the mass of the evacuated bulb from the mass of the gas-filled bulb:

$$\text{mass of borane sample} = 75.5398 \text{ g} - 73.2684 \text{ g} = 2.2714 \text{ g}$$

Hence,

$$\text{mass of } H = 2.2714 \text{ g} - 1.842 \text{ g} = 0.429 \text{ g of H atoms}$$

The number of moles of H atoms is obtained using Eq. (1-25):

$$\text{No. moles } H = \frac{0.429 \text{ g}}{1.008 \text{ g/mol}} = 0.426 \text{ mol}$$

The molar ratio of B:H in the unknown borane is therefore

$$\frac{\text{moles B}}{\text{moles H}} = \frac{0.1704}{0.426} = \frac{1}{2.5} = \frac{2}{5}$$

and the simplest formula is B_2H_5. The empirical formula is $(B_2H_5)_n$.

We can determine the value of n in the empirical formula by utilizing Avogadro's law at 0 °C and 1-atm pressure. The molar volume of any gas at STP is 22.4 L. This 2.2714-g sample of the gaseous borane occupies a volume of 952 mL at STP. Therefore the number of moles of gas in this sample is

$$\text{No. moles borane} = \frac{0.952 \text{ L}}{22.4 \text{ L/mol}} = 4.25 \times 10^{-2} \text{ mol}$$

An approximately correct value for the molecular weight can now be obtained.

$$\text{molecular weight borane} = \frac{2.2714 \text{ g}}{4.25 \times 10^{-2} \text{ mol}} = 53.4 \text{ g} \cdot \text{mol}^{-1}$$

The formula weight of B_2H_5 is 26.66, and therefore the molecular weight of a gas with empirical formula $(B_2H_5)_n$ is $26.66n$, where n must be an integer. The only integral value for which $26.66n$ is close to 53.4 is $n = 2$. The correct molecular formula of this compound is B_4H_{10}.

Note particularly that the exact molecular weight of B_4H_{10} is 53.32 g · mol^{-1}. Using Avogadro's law we obtained 53.4 g · mol^{-1} as the molecular weight of the gas, which is very close to the true value, but is not exactly correct. As was pointed out in Example 1.10, it is not necessary to know the molecular weight exactly in order to determine the correct molecular formula from the empirical formula.

Avogadro's law is called a **limiting law,** because it becomes exact only in the limit of extremely low pressure. A correct statement of Avogadro's law is the following: *In the limit of extremely low pressure, that is, as the pressure approaches zero, and at a constant temperature, equal volumes of different gases contain equal numbers of molecules.* If we use Avogadro's law when we do calculations involving gases at pressures of a few atmospheres or less, and at temperatures close to room temperature or above, we will obtain answers that will be in error by a very small amount: No more than 1 or 2%, and usually considerably less. Very often experimental uncertainties, the limits of accuracy of the measurements we make in the laboratory, are of the order of magnitude of 1 or 2%, and hence Avogadro's law is perfectly satisfactory.

section 3.2
The Pressure of Gases: Definition, Measurement, and Units

Significant quantities of gases exist for about 80 km (50 miles) above the surface of the earth. These gases are subject to the earth's gravitational pull, and the density of gases decreases as the distance from the earth's surface increases. The gaseous atmosphere exerts a force on the earth and on everything on the earth's surface. **Pressure** is defined as the force exerted per unit area:

$$\text{pressure} = \text{force/area} \tag{3-2}$$

The pressure of the gases in the atmosphere is exerted on our bodies. We are used to living with this pressure, so we do not notice it. If we travel to localities with very high altitudes, such as Aspen, Colorado, or Mexico City, where the atmospheric pressure is less, we do notice the difference and may feel ill, particularly if we try to be active immediately after arrival. This phenomenon is known as altitude sickness and is due to cellular oxygen deprivation. The higher the altitude, the lower the atmospheric pressure. Because airplanes fly at high altitudes, airplane cabins are **pressurized.**

The instrument most frequently used to measure the pressure of gases in the atmosphere is a **barometer,** invented by Evangelista Torricelli in 1643. In constructing a barometer, a long glass tube, closed at one end, is completely filled with mercury. All air bubbles are carefully removed. The open end is covered so that no mercury can spill out and no air can get in, and the tube is inverted into a dish containing a pool of mercury. The cover is then removed. The open end of the tube is under the reservoir of mercury, but does not rest on the bottom of the dish.

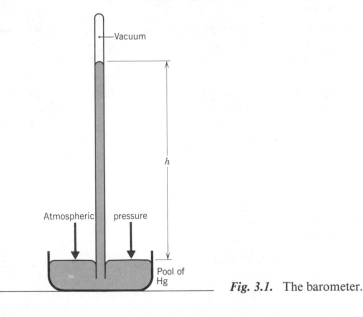

Fig. 3.1. The barometer.

Figure 3.1 shows a completed barometer, prepared in the laboratory as described. You will observe that the mercury in the tube drops down when the tube is inverted, leaving a vacuum at the top, a space empty of all gases.* Not all the mercury falls out of the tube, however; a column of mercury remains. The mass of mercury in the tube exerts a pressure that is exactly balanced by the pressure of the atmosphere on the surface of the pool of mercury in the dish, and this equality determines the height of the column of mercury above the surface of the pool.

$$\begin{array}{l}\text{pressure due to column} \\ \text{of mercury in the tube}\end{array} = \begin{array}{l}\text{pressure of the atmosphere} \\ \text{on the surface of the pool}\end{array} \qquad (3\text{-}3)$$

The exact height of the column of mercury in a barometer depends upon the atmospheric pressure, which varies with the weather, the temperature, and the height above sea level. Weather reports usually include the barometric pressure, and indicate whether it is rising or falling. A falling barometer means the atmospheric pressure is decreasing, and often indicates a storm is on the way. Normal variations in the height of the mercury column are not large. At sea level, the column height is usually between 750 and 770 mmHg (75.0 and 77.0 cmHg), or between 29 and 30 in.Hg.

Since we can measure pressure by measuring the height of a column of mercury, it is common to report the pressure in units of millimeters of mercury (mmHg). Another unit of pressure, very widely used, is the **atmosphere.** One standard atmosphere is, by definition, the pressure that supports a column of mercury exactly 760.00-mm high at 0 °C and standard gravity.† Thus

$$1 \text{ atm} = 760.00 \text{ mmHg} \qquad (3\text{-}4)$$

We have previously defined pressure as the force per unit area (Eq. 3-2). The SI unit of force (see Appendix A) is the newton (N), and the unit of area is the square meter (m²), so the SI unit of pressure is newtons per square meter ($N \cdot m^{-2}$). One newton per square meter is designated a **pascal** (Pa), to honor the French physicist, mathematician, and religious writer, Blaise Pascal (1623–1662). To convert a pressure expressed in atmospheres to SI units, we use the relation

* There is a very small amount of mercury vapor in this space.
† See Appendix D for the definition of standard gravity.

$$1 \text{ atm} = 1.0132 \times 10^5 \text{ N} \cdot \text{m}^{-2} = 1.0132 \times 10^5 \text{ Pa} \tag{3-5}$$

The derivation of this conversion factor is given in Appendix D.

There are two other pressure units in use that you should be familiar with. The **bar,** which is used by meteorologists and scientists who study systems at high pressures, is being used more frequently since the introduction of SI units. One bar is exactly 1×10^5 Pa. It is therefore just slightly less than 1 atm.

$$1 \text{ bar} = 1 \times 10^5 \text{ Pa} = 0.98692 \text{ atm}$$

One **torr,** named in honor of Torricelli, is the pressure that supports a column of mercury exactly 1-mm high. One torr is therefore $(1/760)$ of an atmosphere. The torr has been replaced as a pressure unit by the millimeter of mercury.

EXAMPLE 3.3. Interconversion of pressure units

A sample of gas contained in a bulb supports a column of mercury 680.2-mm high. What is the pressure of this gas in **(a)** atmospheres, **(b)** newtons per square meter or pascals, and **(c)** bars?

Solution. (a) Since 1 atm supports a column of mercury exactly 760.00-mm high,

$$(680.2 \text{ mmHg}) \left(\frac{1 \text{ atm}}{760.00 \text{ mmHg}} \right) = 0.8950 \text{ atm}$$

(b) To convert to pascals, use Eq. (3-5):

$$(0.8950 \text{ atm}) \left(\frac{1.0132 \times 10^5 \text{ N} \cdot \text{m}^{-2}}{1 \text{ atm}} \right) = 9.068 \times 10^4 \text{ N} \cdot \text{m}^{-2} = 9.068 \times 10^4 \text{ Pa}$$

(c)
$$\left(\frac{1 \text{ bar}}{10^5 \text{ Pa}} \right) (9.068 \times 10^4 \text{ Pa}) = 0.9068 \text{ bar}$$

You may be wondering why mercury is the liquid used in a barometer. Mercury is chosen because it is one of the densest liquids known; its density is $13.6 \text{ g} \cdot \text{cm}^{-3}$ at temperatures close to room temperature. The force of gravity pulls the mercury in the barometer tube down; the atmospheric pressure on the surface of the pool supports the column. The gravitational force, f, on the column of mercury is defined by Newton's second law:

$$f = mg \tag{3-6}$$

where m is the mass of mercury in the tube, and g is the acceleration due to gravity. The mass of mercury in the tube can be calculated from the relation

$$(\text{mass in g}) = (\text{density in g} \cdot \text{cm}^{-3})(\text{height in cm}) \left(\begin{array}{c} \text{cross-sectional} \\ \text{area in cm}^2 \end{array} \right) \tag{3-7}$$

where the volume of mercury in the barometer tube is obtained by multiplying the height of the column by the cross-sectional area of the tube. The pressure of the mercury column is given by

$$\begin{array}{c} \text{pressure of} \\ \text{mercury column} \end{array} = \frac{\left(\begin{array}{c} \text{mass of} \\ \text{mercury} \end{array} \right) \left(\begin{array}{c} \text{acceleration} \\ \text{of gravity} \end{array} \right)}{\left(\begin{array}{c} \text{cross-sectional} \\ \text{area of tube} \end{array} \right)} \tag{3-8}$$

$$= (\text{density})(\text{height of column}) \left(\begin{array}{c} \text{acceleration} \\ \text{of gravity} \end{array} \right) \tag{3-9}$$

Since the pressure of the column of liquid must equal atmospheric pressure, the denser the liquid, the less the height of the column.

EXAMPLE 3.4. Height of the column of liquid in a barometer

What is the height of a column of water that can be supported in a barometer tube by a pressure of exactly 1 atm at 25 °C? The density of water is 0.997 g · cm^{-3} at 25 °C.

Solution. One atmosphere supports a column of mercury 76.0-cm high; the density of mercury is 13.6 g · cm^{-3}. If we let h = height of the column of water supported by 1 atm, then using Eq. (3-9) we obtain

$$(13.6 \text{ g} \cdot \text{cm}^{-3})(76.0 \text{ cm}) \left(\frac{\text{acceleration}}{\text{of gravity}} \right) = (0.997 \text{ g} \cdot \text{cm}^{-3})(h) \left(\frac{\text{acceleration}}{\text{of gravity}} \right)$$

Each side of this equation is equal to 1 atm. The acceleration of gravity cancels out of the calculation, and we obtain

$$h = \frac{(13.6)(76.0)}{(0.997)} \text{ cm} = 1.037 \times 10^3 \text{ cm} = 10.4 \text{ m}$$

The conversion factor for converting meters to feet is given in Appendix A.

$$(10.37 \text{ m}) \left(\frac{3.28 \text{ ft}}{1 \text{ m}} \right) = 34.0 \text{ ft}$$

Clearly, water is not used as the liquid in a barometer that measures atmospheric pressures, because 34.0 ft is much too high to be practical.

To measure the pressure of a gas in the laboratory, we employ a **manometer,** illustrated in Fig. 3.2. The type of manometer shown has one closed arm and one open arm that can be connected to the container of gas whose pressure is to be measured. The space above the liquid (usually mercury) in the closed arm is evacuated, and the difference in height, h, between the liquid levels in the two arms measures the pressure of the gas.

section 3.3
Boyle's Law

The first scientist to investigate the relationship between the pressure and volume of a sample of gas was Robert Boyle, in 1662. Boyle used a J-tube, and trapped a fixed quantity of air in the closed end by pouring mercury in the tube, as shown in Fig. 3.3. The other end was open to the atmosphere. If the trapped air is at atmospheric pressure, the two arms of the tube have mercury filled to the same height, as shown in

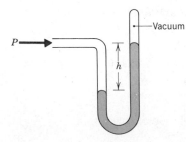

Fig. 3.2. The closed manometer. The space above the manometer fluid in the closed end is evacuated. The difference in height, h, between the liquid levels in the two arms measures the pressure of the gas entering the open arm. Usually mercury is used as the manometer fluid. It is assumed that the vapor pressure of the manometer fluid in the evacuated space is negligible. For mercury at room temperature this is a reasonable assumption.

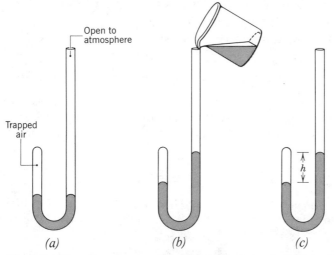

Fig. 3.3. The apparatus used by Robert Boyle to study the relationship between the pressure and volume of a fixed quantity of gas at constant temperature. *(a)* Trapped air at atmospheric pressure. *(b)* Pouring mercury into the open end to increase the pressure of the trapped air. *(c)* Trapped air at a pressure greater than atmospheric. The difference in height between the two levels, *h*, plus the atmospheric pressure is the pressure of the trapped air.

Fig. 3.3*(a)*. By pouring more mercury into the open end, the pressure of the trapped air is increased. There is then a difference in the heights of the mercury levels in the two arms, the quantity *h* shown in Fig. 3.3*(c)*. Some of Boyle's original data are presented in Table 3.1.

Boyle measured the volume of the trapped air in terms of the length of the column of air. Because the cross-sectional area of the tube is constant, the volume is directly proportional to the length. Atmospheric pressure was $29\frac{2}{16}$ in. of Hg when Boyle made these measurements. Then

$$\text{pressure of the trapped air} = \text{atmospheric pressure} + h$$
$$= 29\tfrac{2}{16} + \text{column 2} \tag{3-10}$$

Boyle observed that as the pressure of the gas increased, its volume decreased, that is, the pressure and volume were inversely proportional: $P = k/V$ or $PV = k$, at constant temperature for a fixed quantity of gas. Because of experimental uncertain-

Table 3.1. Robert Boyle's Original Data[a] (1662)

1 Length of Air Column (Arbitrary Units) Proportional to Volume	2 Difference in Hg Level in Inches	3 Difference Plus Atmospheric Pressure = Pressure, P	4 Product of Columns 1 and 3 = PV
12	0	$29\frac{2}{16}$	349
10	$6\frac{3}{16}$	$35\frac{5}{16}$	353
8	$15\frac{1}{16}$	$44\frac{3}{16}$	353
6	$29\frac{11}{16}$	$58\frac{13}{16}$	353
4	$58\frac{3}{16}$	$87\frac{4}{16}$	351
3	$88\frac{7}{16}$	$117\frac{9}{16}$	353

[a] All Boyle's measurements were made at room temperature, for a fixed quantity of gas trapped in the closed end of the J-tube.

ties in the third significant figure in Boyle's data for PV (column 4), all the values listed are the same, 3.5×10^2, within experimental error.

We can correctly summarize Boyle's results by stating what is now called **Boyle's law**:

$$PV = \text{a constant, at constant temperature, for a fixed quantity of gas} \qquad (3\text{-}11)$$

To be completely rigorous, all we can conclude from Boyle's measurements is that, within an experimental uncertainty of slightly less than ±1%, the product of the pressure and volume of a gas at constant temperature is a constant for a fixed quantity of gas.

EXAMPLE 3.5. Boyle's law

A sample of air occupies a volume of 450.0 mL at 20 °C and 1.00-atm pressure. What will the pressure of this air be if the sample is transferred to a 2.000-L bulb and kept at the same temperature?

Solution. Since the temperature is constant, and the amount of gas is fixed, Boyle's law states that $P_1 V_1 = P_2 V_2$, where the subscripts 1 and 2 indicate the initial and final states, respectively. If we solve this relation for P_2 we obtain

$$P_2 = \frac{P_1 V_1}{V_2} = \frac{(1.00 \text{ atm})(450 \text{ mL})}{2000 \text{ mL}} = 0.225 \text{ atm} = 171 \text{ mmHg}$$

A plot of P versus V for a sample of gas at a fixed temperature is called a **$P\text{-}V$ isotherm.** Typical plots are shown in Fig. 3.4. The shape of a plot of P versus V at constant temperature, if $PV = k$, where k is a constant, is one half of a hyperbola.

Another way of plotting Boyle's data, or any other data obtained to investigate the validity of Boyle's law, is to plot PV as a function of P at a fixed temperature, for a fixed amount of gas. Since Boyle's law says that PV is a constant at constant temperature for a given sample of gas, this means that for any value of P the product PV should be invariant (constant). Thus a plot of PV versus P is predicted to be a horizontal straight line, as shown in Fig. 3.5.

Boyle's equipment permitted him to make measurements that were uncertain in the third significant figure, and the range of pressures he covered was roughly from 1 to 4 atm. The "gas" he used was air, which is not a pure compound but a mixture of several gases, and all his measurements were carried out at room temperature. Since Boyle's time, equipment has been developed that makes it possible to measure the product of P and V to six significant figures, and to cover a much wider range of

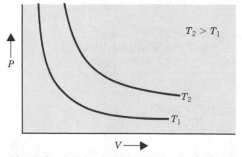

Fig. 3.4. $P\text{-}V$ isotherms. Each plot is drawn at a constant temperature, for a fixed amount of gas.

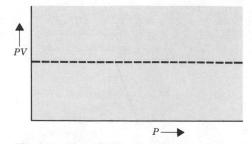

Fig. 3.5. Plot of PV versus P at constant temperature, for a gas that obeys Boyle's law.

Table 3.2. **PV as a Function of P for 1 mol of Gas at 0 °C**

CO₂		O₂	
P (atm)	*PV (L · atm)*	*P (atm)*	*PV (L · atm)*
1.00000	22.2643	1.00000	22.3939
0.66667	22.3148	0.75000	22.3987
0.50000	22.3397	0.50000	22.4045
0.33333	22.3654	0.25000	22.4096
0.25000	22.3775		
0.16667	22.3897		

pressures, from a small fraction of an atmosphere up to several hundred atmospheres. Measurements have also been made at a variety of temperatures. Some modern data on 1-mol samples of CO_2 and O_2 at 0 °C and low pressures (1 atm and below), are given in Table 3.2.

Examine the data in Table 3.2 carefully and ask yourself: Is Boyle's law correct? Is the product of the pressure and volume of a fixed sample of gas at constant temperature a *constant?* Clearly, it is not. From these data, which are valid to six significant figures, it is seen that as P decreases, the product PV increases slightly for both gases below 1 atm. For O_2, the four values of PV given are all 22.4 L · atm to three significant figures, but the measurements are reliable to six significant figures, and differ by amounts greater than experimental uncertainty. For 1 mol of CO_2 at 0 °C, the values of PV are not even constant to three significant figures.

In Fig. 3.6 a plot of PV versus P at constant temperature for one mole of each of these two gases is shown. It is clear that Boyle's law does not describe the behavior of these gases to six significant figures. Values calculated assuming Boyle's law is correct will not agree exactly with experimentally measured values, but the difference between the two will be small.

At pressures below a few atmospheres, and at temperatures close to room temperature or above, the error made using Boyle's law is usually 1% or less. Deviations from Boyle's law increase as the pressure increases or the temperature is lowered. We recall that the same is true for Avogadro's law. The cause of deviations from both laws is the same and will be discussed in Section 4.3. We can emphasize the similarity between these two laws by writing the expression for the volume of a gas as described by each law. Avogadro's law states that the volume, V, of a gas is directly proportional to the number of molecules, and therefore to the number of moles of gas, n, if the temperature, T, and the pressure, P, are constant. Boyle's law states that the volume of a gas is inversely proportional to its pressure, if the temperature and the number of moles of gas are constant. Algebraically, these laws may be written:

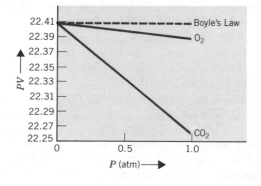

Fig. 3.6. Plot of PV in L · atm versus P in atm for 1 mol of two real gases at 0 °C, at pressures less than 1 atm. The plot for a gas that obeys Boyle's law is shown for comparison.

$$\text{Avogadro's law} \qquad V = k_1 n \qquad \text{if } T, P \text{ constant} \qquad (3\text{-}12)$$

$$\text{Boyle's law} \qquad V = k_2/P \qquad \text{if } T, n \text{ constant} \qquad (3\text{-}13)$$

where k_1 and k_2 are constants. Agreement between each of these laws and experimental measurements becomes increasingly better as the pressure decreases for a fixed temperature. Each is a **limiting law,** that is, each becomes exact as the pressure approaches zero as a limit.

A gas that obeys both Boyle's law and Avogadro's law is, by definition, an **ideal gas.** For real gases, deviations from ideal behavior are small at low pressures and temperatures close to room temperature or above. One mole of an ideal gas occupies 22.414 L at 0 °C and 1-atm pressure (STP). We see from the data in Table 3.2 that neither O_2 nor CO_2 is ideal, because the volume of 1 mol of O_2 is 22.3939 L, while the volume of 1 mol of CO_2 is 22.2643 L, at 0 °C and 1 atm. We say that O_2 is "more ideal" than CO_2, which means that we make a smaller error if we assume O_2 is ideal, than if we assume CO_2 is ideal.

Any calculation for a real gas that assumes the molar volume at 0 °C and 1 atm is 22.4 L will be in error by a very small amount, usually less than 1%. If we know the volume of a sample of gas at 0 °C and 1 atm, we can calculate the number of moles of gas in the sample by assuming that 1 mol occupies 22.4 L, as in Examples 3.1 and 3.2. The number of moles calculated in this way will be very close to the true value.

A word of caution is in order here. The molar volume of an ideal gas depends on the temperature and pressure. At a temperature different from 0 °C or a pressure different from 1 atm, the molar volume may be quite far from 22.4 L. We will shortly develop a general relation that holds at all temperatures and pressures. (Section 3.5.) A common student error is the use of 22.4 L as the molar volume when it is not appropriate, that is, when the temperature is not 0 °C, or the pressure is not 1 atm, or even when the substance in question is a liquid or a solid! You can avoid this by asking yourself: "Am I dealing with a gas at 0 °C and 1 atm?" every time you are about to use 22.4 L for a molar volume.

section 3.4
The Law of Charles and Gay-Lussac

Both Avogadro's and Boyle's law concern measurements made at a constant temperature. Towards the end of the eighteenth century two scientists, Jacques Alexandre Charles and Joseph Louis Gay-Lussac, independently investigated the way in which the volume of a gas varies with temperature. (Both men were interested in using hot gases for the inflation of balloons.) They observed that *the volume of any gas increases linearly with increasing temperature when the pressure and the quantity of gas (that is, the number of moles of gas) are held constant.* The algebraic equation that expresses the linear variation of volume, V, with temperature, t, is

$$V = mt + b \qquad (3\text{-}14)$$

Figure 3.7 shows a plot of V versus t for a fixed sample of gas at constant pressure. The value of m in Eq. (3-14) is the slope of the plot of V versus t, and b is the intercept on the V axis. (See Appendix B, Section 4.) In Eq. (3-14), t can represent temperature on any scale, but it is customary in scientific work to use the Celsius scale to measure temperature. (For a definition of the Celsius scale, refer to the Introduction, pages 8–9.) If we substitute $t = 0$ °C into Eq. (3-14), we see that $V = b$ when $t = 0$ °C. Thus b, the intercept on the V axis, is the value of the volume of the gas at 0 °C. To indicate this, we will change the symbol we use for the intercept, and set $b = V_0$. The symbol

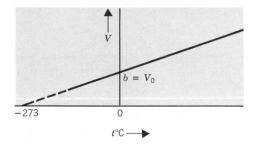

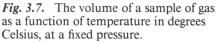

Fig. 3.7. The volume of a sample of gas as a function of temperature in degrees Celsius, at a fixed pressure.

V_0 indicates "the volume at 0 °C" more clearly than the letter b. With this substitution, Eq. (3-14) becomes

$$V = mt + V_0 \tag{3-15}$$

By factoring V_0 out of each term on the right, and rearranging the terms for convenience, we can write the expression for V as

$$V = V_0\left(1 + \frac{mt}{V_0}\right) \tag{3-16}$$

To simplify the form of Eq. (3-16), we define a new variable, α, as $\alpha = m/V_0$. Equation (3-16) then becomes

$$V = V_0(1 + \alpha t) \tag{3-17}$$

The remarkable fact about the experimental results of both Charles and Gay-Lussac was that the numerical value of α was the same for all the gases they studied, within experimental uncertainty. For their data, and later data as well, the value of α is

$$\alpha = 3.66 \times 10^{-3} = \tfrac{1}{273} \tag{3-18}$$

If we set $V = 0$ in Eq. (3-17) and solve for the temperature at which the volume of the gas would reach zero, we obtain $t = -1/\alpha = -273$ °C. Examine Fig. 3.7 carefully, and note that if the plot is extrapolated back to cross the temperature axis, the temperature at which V would be zero is -273 °C. Of course, a zero volume is physically meaningless, and there is no substance that is still gaseous at -273 °C, or even for a few degrees above that temperature at pressures close to an atmosphere. It is striking, however, that regardless of the gas used to make the measurements, a plot of volume as a function of temperature in degrees Celsius, when extrapolated back to cross the temperature axis, crosses at $t = -273$ °C.

When more accurate measurements of α were made in the late nineteenth and twentieth centuries, at a variety of different pressures, it was found that the value of α is not *exactly* the same for all gases, and varies somewhat with pressure. For different gases, the values of α are quite close to one another, but they are not identical when data accurate to five or six significant figures are available. However, if the measurements are made at lower and lower pressures, the different values of α for different gases become more nearly alike, and in the limit as the pressure approaches zero,* the value of α is the same for all gases.

$$\lim_{P \to 0} \alpha = \frac{1}{273.15} = \text{a universal constant, the same for all gases} \tag{3-19}$$

Suppose we measure the volume of a given quantity of gas at a very low constant pressure, at two different temperatures, t_1 and t_2. Then

---

* The symbol $\lim\limits_{P \to 0} \alpha$ is read as "the limiting value of α as the pressure P approaches zero."

$$V_1 = V_0(1 + \alpha t_1)$$

and

$$V_2 = V_0(1 + \alpha t_2)$$

by substitution into Eq. (3-17). The ratio V_1/V_2 is then given by

$$\frac{V_1}{V_2} = \frac{1 + \alpha t_1}{1 + \alpha t_2} = \frac{1 + (t_1/273.15)}{1 + (t_2/273.15)} = \frac{273.15 + t_1}{273.15 + t_2} \tag{3-20}$$

since the V_0 term cancels when V_1 is divided by V_2. The final term on the right is obtained by clearing fractions. A careful consideration of the term on the far right of Eq. (3-20) suggests that it would be convenient to define a new temperature scale by the relation

$$T = 273.15 + t \tag{3-21}$$

where T is called the **absolute temperature** or the **Kelvin temperature** (see Introduction, page 8), and t is temperature in degrees Celsius. In SI units, a degree on the Kelvin scale is called a **kelvin,** and is denoted K. It should be clear from Eq. (3-21) that the size of a kelvin and of a degree Celsius is exactly the same.

With temperature expressed on the absolute scale, Eq. (3-20) becomes simply

$$\frac{V_1}{V_2} = \frac{T_1}{T_2} \qquad provided \text{ the pressure and the quantity of gas are constant} \tag{3-22}$$

This can be rearranged to read

$$\frac{V_1}{T_1} = \frac{V_2}{T_2} \qquad \text{if } P \text{ and } n \text{ are constant} \tag{3-23}$$

which holds for any two temperatures. Equation (3-23) can only be true if

$$\frac{V}{T} = k_3 = \text{a constant at constant } P \text{ and } n \tag{3-24}$$

Equation (3-24) is known as **Charles' Law** or **Gay-Lussac's Law.**

Note that **absolute zero,** that is, zero degrees on the Kelvin scale, corresponds to $-273.15\ °C$. Absolute zero is the lowest possible temperature, because a lower temperature would correspond to a negative volume of gas, which is physically meaningless. Actually, no substances remain gaseous at temperatures close to 0 K and pressures close to 1 atm. At 1 atm all substances are either solids or liquids below 4 K. Helium remains a gas at lower temperatures than any other substance; the boiling point of He is 4.22 K. Other substances with extremely low boiling points are hydrogen (bp 20.28 K) and neon (bp 27.10 K).

section 3.5
The Ideal Gas Law

Avogadro's law, Boyle's law, and the Law of Charles and Gay-Lussac are all obeyed very closely, but not exactly, by real gases at temperatures close to or above room temperature and at pressures below a few atmospheres. All three are "limiting laws" in that they become more and more exact as the pressure decreases. These ABC's of gases can be written as

Avogadro's law	$V = k_1 n$	if T, P constant	(3-12)
Boyle's law	$V = k_2/P$	if T, n constant	(3-13)
Charles' law	$V = k_3 T$	if P, n constant	(3-24)

We see that the volume of a gas depends on the temperature, the pressure, and the quantity (that is, the number of moles) of the gas. The volume is a function of three variables. A single equation that describes the way the volume depends on all three variables is known as an **equation of state.** Consideration of Avogadro's, Boyle's, and Charles' laws should convince you that they can all be correct only if

$$V = (\text{constant})nT/P \tag{3-25}$$

For instance, if n and T are held fixed, Eq. (3-25) reduces to $V = (\text{constant})/P$, which is Boyle's law. We call the constant that appears in the equation of state the **universal gas constant** and denote it by the letter R. Thus Eq. (3-25) is written

$$V = RnT/P \quad \text{or} \quad PV = nRT \tag{3-26}$$

The form of Eq. (3-26) on the right is commonly called the **ideal gas law equation.**

No real gas obeys the ideal gas equation exactly. At pressures below a few atmospheres, and at temperatures close to room temperature or above, the errors we make if we assume the ideal gas law is correct are very small, usually less than 1%. We are often unconcerned about errors of this magnitude because they are as small as, or smaller than, experimental uncertainties; consequently we use the ideal gas equation extensively.

We can determine the numerical value of the gas constant, R, by considering how the product PV changes as P is decreased at constant temperature for a fixed sample of gas. We have already seen in Fig. 3.6 that for a real gas the plot of PV versus P at constant temperature does not obey Boyle's law (and therefore does not obey the ideal gas law). However, as the pressure is decreased, deviations of the value of PV for real gases from the value predicted by the ideal gas law become smaller and smaller, and in the limit of zero pressure, the ideal law becomes exact. Thus although $PV = nRT$ is an ideal law, the equation

$$\lim_{P \to 0} (PV) = nRT \tag{3-27}$$

is exact. The data given in Table 3.2 for CO_2 and O_2 are typical of the behavior observed for all gases. At pressures below an atmosphere, a plot of PV versus P for any real gas at constant T and a fixed number of moles, is a straight line, but not a horizontal line. By extending the experimental line back until it crosses the $P = 0$ axis, we can obtain the limiting value as $P \to 0$. The process of extending an experimental plot back to values that cannot be obtained experimentally is known as **extrapolation.**

The data for O_2 given in Table 3.2 are plotted in Fig. 3.8 to illustrate this procedure. Similar plots for a great many other gases lead to the result that for 1 mol of any gas at 0 °C,

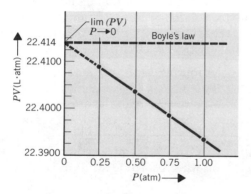

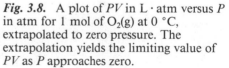

Fig. 3.8. A plot of PV in L · atm versus P in atm for 1 mol of $O_2(g)$ at 0 °C, extrapolated to zero pressure. The extrapolation yields the limiting value of PV as P approaches zero.

$$\lim_{P \to 0} (PV) = 22.414 \text{ L} \cdot \text{atm} \tag{3-28}$$

Thus

$$nRT = (1 \text{ mol})(R)(273.15 \text{ K}) = 22.414 \text{ L} \cdot \text{atm} \tag{3-29}$$

and therefore

$$R = \frac{22.414}{273.15} = 0.082057 \text{ L} \cdot \text{atm} \cdot \text{mol}^{-1}\text{K}^{-1} \tag{3-30}$$

The units of R given above are read as "liter atmospheres per mole per kelvin" and are the appropriate units for $R = PV/nT$ when P is expressed in atmospheres, V is given in liters, n is in moles, and T is in kelvins.

If we measure P or V in other units, the numerical value of R, and its units, will change. For instance, suppose we measure P in millimeters of mercury rather than in atmospheres. Then we must convert R so that it is expressed in $\text{L} \cdot \text{mmHg} \cdot \text{mol}^{-1}\text{K}^{-1}$. Be sure to write down the units of conversion factors and check to see that units cancel to give the desired result.

$$R = \left(0.082057 \frac{\text{L} \cdot \text{atm}}{\text{mol} \cdot \text{K}}\right)\left(760 \frac{\text{mmHg}}{\text{atm}}\right) = 62.363 \frac{\text{L} \cdot \text{mmHg}}{\text{mol} \cdot \text{K}} \tag{3-31}$$

If the volume is given in cubic centimeters (cm^3) rather than in liters, $R = 82.06$ $\text{cm}^3 \cdot \text{atm} \cdot \text{mol}^{-1}\text{K}^{-1}$, since there are 1000 cm^3 in a liter.

A great many practical problems can be solved using the ideal gas law, and the examples discussed below illustrate its usefulness. When we use the ideal gas law we should bear in mind that it is not precisely correct so that, unless the pressure is considerably less than 1 atm, there is no point in performing calculations to more than three significant figures.

In many problems involving gases, the initial conditions are described, then conditions are changed, and one has to calculate the value of some variable after the change. In such problems, the ideal gas law should be written for both the initial state and the final state. By rearranging the equations so that all the quantities that remain constant are on one side, the numerical calculation is simplified greatly. The value of R always cancels out of such calculations. Examples 3.6 through 3.9 illustrate this type of problem.

EXAMPLE 3.6. The gas volume thermometer; Charles' Law

The volume of a fixed quantity of gas can be used to measure temperature provided the pressure remains constant. A suitable apparatus is shown in Fig. 3.9. Such a gas volume "thermometer" is filled with a sample of N_2 gas at 1-atm pressure, and immersed in an ice-water bath at 0 °C. The volume of the gas is found to be 58.7 mL. The bulb containing the gas is then immersed in boiling liquid isopropyl alcohol (ordinary rubbing alcohol), and the volume of the same sample of gas is found to be 76.4 mL at a pressure of 1 atm. What is the boiling point of isopropyl alcohol? Give the answer both in kelvins and in degrees Celsius.

Solution. The ideal gas law involves four variables: pressure, volume, temperature, and the number of moles of gas. First identify the variables that remain constant during the process described. In this problem, both the number of moles of N_2 gas and the pressure remain constant. If we rearrange the ideal gas law so that all the constant terms are on one side of the equation, we have $nR/P = V/T$. The initial state (state 1) has the gas at a volume $V_1 = 58.7$ mL and a temperature $T_1 = 273.15$ K. (The temperature *must* be in kelvins when using the ideal gas law.) When the gas is

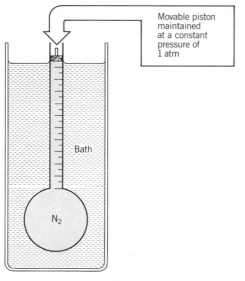

Movable piston
maintained
at a constant
pressure of
1 atm

Bath

N$_2$

Fig. 3.9. A gas volume thermometer.

immersed in boiling isopropyl alcohol (state 2), its volume $V_2 = 76.4$ mL, and its temperature, T_2, is to be determined. We therefore have

$$\frac{nR}{P} = \text{constant} = \frac{V_1}{T_1} = \frac{V_2}{T_2} \qquad \text{(Charles' Law)}$$

Rearranging this to solve for T_2, we obtain

$$T_2 = \left(\frac{V_2}{V_1}\right) T_1 = \frac{76.4}{58.7} (273.15 \text{ K}) = 355.5 \text{ K} \qquad \text{or} \qquad 82.4 \text{ °C}$$

The boiling point of isopropyl alcohol is 82.4 °C at 1 atm.

EXAMPLE 3.7. The effect of a simultaneous change of pressure and temperature

A sample of a gas is found to occupy a volume of 346 mL at 25 °C when the pressure is 751 mmHg. What volume will this sample occupy if the pressure is increased to 1.40 atm and the temperature is increased to 35 °C?

Solution. In this problem the only variable that remains constant is the number of moles of gas. We rearrange the ideal gas law to put all constant terms together: $nR = PV/T$. Since nR is a constant in this problem

$$\frac{P_1 V_1}{T_1} = \frac{P_2 V_2}{T_2}$$

Both P_1 and P_2 must be expressed in the same units. It makes no difference whether both are given in millimeters of mercury, or both are given in atmospheres. We will express both pressures in millimeters of mercury.

$P_1 = 751$ mmHg $\qquad P_2 = (1.40 \text{ atm})(760 \text{ mmHg/atm}) = 1064$ mmHg
$T_1 = 298.2$ K $\qquad\quad T_2 = 308.2$ K
$V_1 = 346$ mL $\qquad\quad V_2 = ?$

Solving the equation given above for V_2 we obtain

$$V_2 = \left(\frac{P_1}{P_2}\right)\left(\frac{T_2}{T_1}\right) V_1 = \frac{(751 \text{ mmHg})}{(1064 \text{ mmHg})} \frac{(308.2 \text{ K})}{(298.2 \text{ K})} (346 \text{ mL}) = 252 \text{ mL}$$

EXAMPLE 3.8. Change in the number of moles of gas with temperature, at constant volume and pressure

A bulb is filled with 20.0 g of CO_2 gas at 23.0 °C and 1 atm. The bulb is heated to 63.0 °C and the stopcock is opened. How much CO_2 escapes, if the external pressure is 1 atm when the stopcock is opened, and the bulb is kept at 63.0 °C? When the pressure inside the bulb reaches 1 atm, the stopcock is closed again.

Solution. In this problem we must focus on the gas inside the bulb. The quantities that are the same at the beginning and the end of the experiment are the volume of the bulb and the pressure inside the bulb. Heating the gas to 63.0 °C increases the pressure in the bulb above 1 atm, so that when the stopcock is opened, gas escapes from the bulb until the pressure in the bulb is again equal to the external pressure of 1 atm. To solve this problem, rearrange the ideal gas law to read

$$\frac{PV}{R} = \text{constant} = n_1 T_1 = n_2 T_2$$

where n_2 is the number of moles of gas remaining inside the bulb after the pressure has returned to 1 atm at 63.0 °C.

$$n_1 = \frac{20.0 \text{ g}}{44.01 \text{ g/mol}} = 0.454 \text{ mol} \qquad n_2 = ?$$
$$T_1 = 296.2 \text{ K} \qquad\qquad\qquad T_2 = 336.2 \text{ K}$$

Therefore,

$$(0.454 \text{ mol})(296.2 \text{ K}) = n_2(336.2 \text{ K})$$

and

$$n_2 = \frac{(0.454)(296.2)}{(336.2)} = 0.400 \text{ mol}$$

Since 0.400 mol remains in the bulb after the stopcock has been opened, 0.054 mol of CO_2 must have escaped. Therefore the mass of CO_2 that escaped is $(0.054 \text{ mol})(44.01 \text{ g} \cdot \text{mol}^{-1}) = 2.4 \text{ g}$.

EXAMPLE 3.9. More general use of the ideal gas law

An evacuated bulb of unknown volume is filled with a sample of H_2 gas at a temperature T. The pressure of the gas in the bulb is 756 mmHg. A portion of the H_2 gas is transferred to a different flask and found to occupy a volume of 40.0 mL at 1.00 atm and the same temperature, T. The pressure of the H_2 remaining in the original bulb drops to 625 mmHg at the same temperature, T. Assuming H_2 is an ideal gas, what is the volume of the bulb?

Solution. In this problem, the temperature and the total number of moles of H_2 both remain constant throughout, but the sample is divided into two portions. If you do not know the numerical value of a quantity, assign it a symbol. Then write the ideal gas equations using both symbols and numbers. After you look at the equations you have written, a method of solving them algebraically will be more apparent.

Let V = volume of the bulb, in milliliters

n = total number of moles of H_2

Then, for the initial condition, $756V = nRT$. There is no reason to convert the 756-mmHg pressure into atmospheres, because no numerical value of R has been selected. The only restriction on the unit of pressure used is that it must be consistent with the pressure unit used in the value of R.

We don't know how big a sample was transferred to the 40.0-mL flask. Let n_1 = the number of moles transferred to the 40.0-mL flask. Then, for the gas in that flask $(760)(40.0) = n_1 RT$. In this equation, we have written the 1-atm pressure as 760 mmHg, so that we use the same pressure unit in all the equations we write.

Let n_2 = the number of moles of H_2 remaining in the original bulb. Then, for the n_2 moles of H_2 after the transfer, $625 V = n_2 RT$.

Since the total number of moles of H_2, $n = n_1 + n_2$, then $nRT = n_1 RT + n_2 RT$. We have an expression for each term in this equation:

$$nRT = 756 V = (760)(40.0) + 625 V = n_1 RT + n_2 RT$$

Solving for V we obtain

$$(756 V - 625 V) = 131 V = (760)(40.0) \text{ mmHg} \cdot \text{mL}$$

Hence,

$$V = \frac{3.04 \times 10^4}{131} \text{ mL} = 232 \text{ mL}$$

Note that to calculate V we did not need to use any numerical value of R, or know the value of T.

In some problems involving gases, only a single set of conditions is described. In such problems a numerical value of R must be selected, and you must be careful about units. Example 3.10 is an illustration of this type of problem.

EXAMPLE 3.10. Use of the ideal gas law when a numerical value must be chosen for R

A sample of butane of mass 3.728 g is placed in an evacuated bulb of volume 489 mL at 25.0 °C. What is the pressure of the gas in the bulb?

Solution. We will need the number of moles of butane in order to use the ideal gas law. The molecular weight of butane, C_4H_{10}, is 58.12 g · mol⁻¹. Hence

$$\text{No. moles } C_4H_{10} = \frac{3.728 \text{ g}}{58.12 \text{ g/mol}} = 6.414 \times 10^{-2} \text{ mol}$$

We must decide what value of R to use. We will do the calculation with two different choices for R, to show that as long as the units of all quantities are consistent with the units of R, it makes no difference which value of R is used.

Choice 1. Use $R = 62.36 \text{ L} \cdot \text{mmHg} \cdot \text{mol}^{-1}\text{K}^{-1}$. Then the volume must be expressed in liters and the pressure calculated will be in millimeters of mercury.

$$P = \frac{nRT}{V} = \frac{(6.414 \times 10^{-2} \text{ mol})}{(0.489 \text{ L})} \left(62.36 \frac{\text{L} \cdot \text{mmHg}}{\text{mol} \cdot \text{K}} \right) (298.2 \text{ K}) = 2439 \text{ mmHg}$$

$$= 2.44 \times 10^3 \text{ mmHg} \quad \text{or} \quad (2.44 \times 10^3 \text{ mmHg}) \left(\frac{1 \text{ atm}}{760 \text{ mmHg}} \right) = 3.21 \text{ atm}$$

Choice 2. Use $R = 0.08206 \text{ L} \cdot \text{atm} \cdot \text{mol}^{-1}\text{K}^{-1}$. Then the volume must be in liters and the pressure calculated will be in atmospheres.

$$P = \frac{nRT}{V} = \frac{(6.414 \times 10^{-2} \text{ mol})}{(0.489 \text{ L})} \left(0.08206 \frac{\text{L} \cdot \text{atm}}{\text{mol} \cdot \text{K}} \right) (298.2 \text{ K}) = 3.21 \text{ atm}$$

A very useful form of the ideal gas law is obtained by specifically including the density of the gas in the equation. The density of any substance is the ratio of its mass to its volume. Gases have very low densities relative to solids and liquids, and gas densities are usually expressed in grams per liter $(g \cdot L^{-1})$, whereas the densities of solids and liquids are usually given in grams per cubic centimeter $(g \cdot cm^{-3})$. The definition of the density, δ, of a gas is

$$\delta = m/V \tag{3-32}$$

where $m =$ mass in grams, and $V =$ volume in liters.

The ideal gas equation involves n, the number of moles of gas. We know the relation between the mass in grams of any substance, and the number of moles of that substance. It is Eq. (1-25). Let $M =$ the molecular weight of the gas in grams per mole. Then the number of moles of gas, n, is $n = m/M$. If we substitute this expression for n into the ideal gas law we obtain

$$PV = \frac{mRT}{M} \tag{3-33}$$

which can be rearranged to yield

$$P = \left(\frac{m}{V}\right)\left(\frac{RT}{M}\right) = \frac{\delta RT}{M} \tag{3-34}$$

by introducing the definition of density, Eq. (3-32). Rearranging this equation to solve for the molecular weight of the gas, we obtain

$$M = \frac{\delta RT}{P} \tag{3-35}$$

Equation (3-35) enables us to calculate a value for the molecular weight, M, of a gas, correct to within the ideal gas approximation, if we know the gas density, δ, at a given T and P. The use of Eq. (3-35) is illustrated in Example 3.11.

EXAMPLE 3.11. **Use of the vapor density to calculate a molecular formula and molecular weight**

At room temperature, pyridine is a colorless liquid with an extremely unpleasant odor. Pyridine occurs naturally in tobacco and hemlock. It contains only the elements C, H, and N, and is 75.92% C and 6.37% H by weight. At 110 °C and a pressure of 630 mmHg, the density of gaseous pyridine has been measured as 2.12 $g \cdot L^{-1}$.

(a) Calculate the molecular formula of pyridine.

Solution. The percentage of N in pyridine is obtained by difference: $100.00 - 75.92 - 6.37 = 17.71\%$ N. Consider exactly 100.00 g of pyridine. This sample contains 75.92 g of C, 6.37 g of H, and 17.71 g of N.

$$\text{No. moles C} = \frac{75.92 \text{ g}}{12.011 \text{ g/mol}} = 6.321 \text{ mol of C}$$

$$\text{No. moles H} = \frac{6.37 \text{ g}}{1.008 \text{ g/mol}} = 6.32 \text{ mol of H}$$

$$\text{No. moles N} = \frac{17.71 \text{ g}}{14.007 \text{ g/mol}} = 1.264 \text{ mol of N}$$

The molar ratio $C:H:N = 6.321:6.32:1.264 = 5:5:1$. Thus the simplest formula of pyridine is C_5H_5N, and the empirical formula is $(C_5H_5N)_n$, where n is an integer. An approximate value of the molecular weight is obtained using Eq. (3-35):

$$M = \left(2.12\,\frac{g}{L}\right)\left(8.206 \times 10^{-2}\,\frac{L \cdot atm}{mol \cdot K}\right)\frac{(383.2\ K)}{(630/760\ atm)} = 80.4\ g/mol$$

The pressure of 630 mmHg was converted to atmospheres in order to be consistent with the unit of pressure in the value of R chosen. The formula weight of C_5H_5N is 79.10 $g \cdot mol^{-1}$. The only value of n for which $79.10n \sim 80.4$ is $n = 1$. Thus the molecular formula of pyridine is C_5H_5N, and the correct molecular weight is 79.10 g/mol.

(b) How great an error is made by using the ideal gas equation to calculate the molecular weight of pyridine from the gas density at this temperature and pressure?

Solution. The error that is made is given by

$$percentage\ error = \frac{(calculated\ value - true\ value)}{(true\ value)} \times 100 \tag{3-36}$$

The calculated value of the molecular weight using the ideal gas equation was 80.4 g/mol. The percentage error made is therefore

$$percentage\ error = \frac{(80.4 - 79.1)}{79.1} \times 100 = 1.6\%$$

The error of 1.6% is due to the fact that pyridine is not an ideal gas, particularly close to the temperature and pressure where it condenses to a liquid. The normal boiling point of pyridine is 115 °C at 1-atm pressure.

section 3.6
Dalton's Law of Partial Pressures

Until now we have been discussing a gas, that is, a single pure gaseous substance. All gases mix with one another, and mixtures of gases are quite common. Indeed, the air we breathe is a mixture of the gases N_2 (78.08%), O_2 (20.95%), Ar (0.934%), CO_2 (0.0314%), plus trace amounts of many other gases. The percentages given are **mole percents** of dry air. (Air also contains varying amounts of water vapor, depending on the weather and the location.) In a mixture of gases enclosed in a volume, V, each gas occupies the entire volume. The volume of a gas is the volume of the container, and each gas fills the entire container.

The ideal gas law applies to a mixture of gases as well as it does to a pure gas. Indeed, it will be recalled that Boyle's measurements were made with air as the gas. Suppose we have two different gases, which we may denote gas 1 and gas 2, and each is at the same temperature T and occupies a container of volume V. To the ideal gas approximation, $P_1 = n_1RT/V$, and $P_2 = n_2RT/V$. If we now put both gases into a single container of volume V at the same temperature T, each gas will exert the same pressure it would exert if it alone occupied the volume V, provided the gases mixed do not react with one another chemically.

In a mixture of gases, the pressure each component would exert if it alone occupied the entire volume is called the **partial pressure** of that component. Thus the partial pressure of gas 1 in the mixture is $P_1 = n_1RT/V$. If we sum the partial pressures of the two gases we obtain

$$P_1 + P_2 = n_1RT/V + n_2RT/V = (n_1 + n_2)(RT/V)$$
$$= n_{total}\,(RT/V) \tag{3-37}$$

But since the ideal gas law applies to the mixture of gases

$$P_{total} = n_{total}(RT/V) \qquad (3\text{-}38)$$

and we have therefore shown that for a mixture of two gases *the total pressure is the sum of the partial pressures.*

We can readily extend this principle to a mixture of any number of gases. To distinguish the various gases in a mixture, they are numbered, and their properties are denoted using the corresponding numerical subscript. A particular gas in a mixture is frequently called the "*i*th component," where the letter *i* represents any number. The partial pressure of the *i*th component in the mixture is given by $P_i = n_i(RT/V)$, and the sum of the partial pressures is denoted

$$\sum_i P_i = P_1 + P_2 + P_3 + \cdots \qquad (3\text{-}39)$$

where the Greek letter $\sum$ (a capital sigma) is used to indicate a summation. The symbol $\sum_i P_i$ is read as "the sum over all values of *i* of the partial pressures P_i." Thus

$$\sum_i P_i = \frac{RT}{V}(n_1 + n_2 + n_3 + \cdots) = \frac{RT}{V}\sum_i n_i \qquad (3\text{-}40)$$

or

$$\sum_i P_i = n_{total}(RT/V) = P_{total} \qquad (3\text{-}41)$$

Equation (3-41) is **Dalton's law of partial pressures:** The total pressure in a mixture of gases is the sum of the partial pressures of the components. This is an ideal law and applies exactly only if each gas and the mixture can be assumed to be ideal.

A gas prepared in the laboratory is often collected by the displacement of water, as illustrated in Fig. 3.10. The gas in the collection vessel then contains both molecules of H_2O and molecules of the prepared gas. The water molecules exert a pressure, called the vapor pressure of water, that depends *only* on the temperature of the liquid water. (The vapor pressure of liquids will be discussed in Section 5.6.) The total pressure of the gas in the collection vessel is given by Dalton's law as

$$P_{total} = P_{gas} + P_{H_2O}$$

If the level of the water is the same both inside and outside the collection vessel, as shown in Fig. 3.10, then the pressure inside the collection vessel must be the same as the outside pressure, namely, the atmospheric pressure. The atmospheric pressure is determined by reading a barometer in the laboratory. The vapor pressure of water is determined by consulting a reference work listing the vapor pressure of water as a

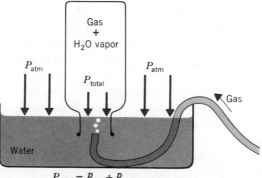

Fig. 3.10. Collection of a gas by displacement of water. The total pressure in the collection vessel is the sum of the pressure of the gas and the vapor pressure of water, according to Dalton's Law of partial pressures.

Table 3.3. **The Vapor Pressure of Water as a Function of Temperature between 20 and 30 °C**

Temperature (°C)	Vapor Pressure (mmHg)
20.00	17.535
21.00	18.650
22.00	19.827
23.00	21.068
24.00	22.377
25.00	23.756
26.00	25.209
27.00	26.739
28.00	28.349
29.00	30.043
30.00	31.824

function of temperature. Table 3.3 gives the vapor pressure of water at temperatures near room temperature. The partial pressure of the gas prepared is then obtained by difference,

$$P_{gas} = P_{total} - P_{H_2O}$$

A very useful form of Dalton's law can be obtained if we divide the expression for the partial pressure, P_i, of the ith component, $P_i = n_i(RT/V)$, by the expression for the total pressure, Eq. (3-38), $P_{total} = n_{total}(RT/V)$. In the division, the (RT/V) term cancels out (remember that each gas in the mixture is at the same temperature T and occupies the entire volume V), so that we obtain

$$\frac{P_i}{P_{total}} = \frac{n_i}{n_{total}} \tag{3-42}$$

The quantity n_i/n_{total} that appears on the right-hand side of Eq. (3-42) is extremely useful, and is so important that it is given a special name and symbol. It is called the **mole fraction,** and is usually denoted by X_i.

$$X_i = n_i/n_{total} \tag{3-43}$$

Thus, if we have a mixture of He, CO_2, and N_2, the mole fraction of N_2 is given by

$$X_{N_2} = \frac{n_{N_2}}{n_{He} + n_{CO_2} + n_{N_2}} = \frac{n_{N_2}}{n_{total}}$$

We can calculate the mole fraction of a component in a mixture of solids, liquids, or gases. Often, as we shall see later in Section 6.8, one has both a liquid mixture and a gas mixture present, and they have different compositions, so that it is useful to specify whether one is discussing a gas or a liquid mixture. Thus we may write Eq. (3-42) as

$$\frac{P_i}{P_{total}} = \frac{n_i}{n_{total}} = X_i^{(gas)} \quad \text{or} \quad P_i = X_i^{(gas)} P_{total} \tag{3-44}$$

where P_i is the partial pressure of the ith component and $X_i^{(gas)}$ is the mole fraction of the ith component of the gaseous mixture. Eq. (3-44) is another way of expressing Dalton's law of partial pressures.

For a two component system,

$$X_1 = \frac{n_1}{n_1 + n_2} \quad \text{and} \quad X_2 = \frac{n_2}{n_1 + n_2}$$

and therefore $X_1 + X_2 = 1$. By extending this argument to a many component system, where $X_i = n_i/n_{total}$, we obtain

$$\sum_i X_i = \sum_i \frac{n_i}{n_{total}} = \frac{n_{total}}{n_{total}} = 1$$

We see, therefore, that in any mixture the sum of the mole fractions of all components is 1:

$$\sum_i X_i = 1 \qquad (3\text{-}45)$$

which is a very useful relationship.

A number of stoichiometric calculations involving gases require knowledge of Dalton's law and the definition of mole fraction. Examples 3.12, 3.13, and 3.14 illustrate the techniques used in solving typical problems.

EXAMPLE 3.12. Use of Dalton's law

Benzoic acid is a white, crystalline solid, important as the starting point in the manufacture of a large group of chemicals used in medicine, food preservatives, dyes, and insect repellents. A 1.5388-g sample of benzoic acid, C_6H_5COOH, is completely burned with excess oxygen in a 6.00-L container. The equation for the combustion reaction is

$$C_6H_5COOH(s) + \tfrac{15}{2} O_2(g) \rightarrow 7CO_2(g) + 3H_2O(\ell)$$

When the reaction is complete, the container and its contents are maintained at a constant temperature of 25.0 °C.

(a) What is the partial pressure of CO_2 in the container after the reaction is complete? Assume that the volume of the liquid water formed is negligible compared to the volume of the container.

Solution. The partial pressure of CO_2 in the container is given by $P_{CO_2} = n_{CO_2}(RT/V)$. We must therefore calculate the number of moles of CO_2 formed in the combustion reaction. The equation for the combustion reaction tells us that the number of moles of CO_2 formed is seven times the number of moles of benzoic acid burned. The molecular weight of C_6H_5COOH is 122.12 g/mol. The number of moles of benzoic acid used in the combustion reaction is

$$\frac{1.5388 \text{ g}}{122.12 \text{ g/mol}} = 1.260 \times 10^{-2} \text{ mol of } C_6H_5COOH$$

Hence,

$$\text{No. moles } CO_2 \text{ formed} = 7(1.260 \times 10^{-2}) = 8.820 \times 10^{-2} \text{ mol}$$

Therefore,

$$P_{CO_2} = \frac{(8.820 \times 10^{-2} \text{ mol})}{(6.00 \text{ L})} \left(8.2057 \times 10^{-2} \frac{\text{L} \cdot \text{atm}}{\text{mol} \cdot \text{K}} \right) (298.2 \text{ K})$$
$$= 0.360 \text{ atm} = 273 \text{ mmHg}$$

(b) The total pressure in the container is 814 mmHg after the combustion. Calculate the partial pressure of the excess (unused) oxygen in the container. How many moles of O_2 remain in the container after the combustion is complete?

Solution. The contents of the container at the end of the reaction are CO_2, O_2, water vapor, and liquid water. There will always be water vapor in the gas phase in a closed container that has any liquid water in it. The vapor pressure of water depends only on the temperature. From Table 3.3 we obtain $P_{H_2O} = 23.8$ mmHg at 25 °C. Dalton's law, Eq. (3-41), applied to this mixture of three gases is

$$P_{total} = P_{CO_2} + P_{H_2O} + P_{O_2}$$
$$814 \text{ mmHg} = 273.3 \text{ mmHg} + 23.8 \text{ mmHg} + P_{O_2}$$

Therefore $P_{O_2} = 517$ mmHg. We can calculate the number of moles of excess oxygen in the container from the partial pressure of O_2 as follows:

$$n_{O_2} = \frac{P_{O_2} \cdot V}{RT} = \frac{(517 \text{ mmHg})(6.00 \text{ L})}{(62.36 \text{ L} \cdot \text{mmHg} \cdot \text{mol}^{-1}\text{K}^{-1})(298.2 \text{ K})} = 0.167 \text{ mol of } O_2$$

EXAMPLE 3.13. Stoichiometry involving gases

A closed bulb contains 0.0100 mol of inert helium gas, and a sample of solid white ammonium chloride, NH_4Cl. Assume that the volume of the solid NH_4Cl is negligible compared to the volume of the bulb. The pressure of the He is measured at 27 °C and is found to be 114 mmHg. The bulb is then heated to 327 °C. All the NH_4Cl decomposes according to the equation

$$NH_4Cl(s) \rightarrow NH_3(g) + HCl(g)$$

The final total pressure in the bulb after complete decomposition of the solid is 908 mmHg. Assume all the gases are ideal.

(a) What is the partial pressure of $HCl(g)$ in the bulb at 327 °C when reaction is complete?

Solution. The contents of the bulb at 327 °C are the three gases He, NH_3, and HCl. Using Dalton's law, the total pressure is

$$P_{total} = 908 \text{ mmHg} = P_{He} + P_{NH_3} + P_{HCl}$$

We can calculate the pressure of the He at 327 °C, because we know its pressure at 27 °C. The number of moles of He and the volume of the bulb remain constant, so that

$$P_2 = P_1(T_2/T_1) = \frac{(114 \text{ mmHg})(600 \text{ K})}{(300 \text{ K})} = 228 \text{ mmHg} = P_{He}$$

The equation for the reaction tells us that NH_3 and HCl are formed in a 1 : 1 molar ratio. Since the number of moles of NH_3 and HCl are equal, and they are at the same temperature, and occupy the same volume, both gases must be at the same pressure.

$$\text{let} \quad x = P_{NH_3} = P_{HCl}$$

Then Dalton's law for the total pressure becomes

$$908 \text{ mmHg} = 228 \text{ mmHg} + 2x \quad \text{and hence} \quad x = P_{HCl} = 340 \text{ mm}$$

(b) How many grams of $NH_4Cl(s)$ were in the bulb at 27 °C?

Solution. Since all the $NH_4Cl(s)$ was converted to $HCl(g)$ and $NH_3(g)$, if we calculate the number of moles of $HCl(g)$ formed, we will know the number of moles of solid NH_4Cl originally present. We do not know the volume of the bulb, so we cannot use the ideal gas law to calculate the number of moles of $HCl(g)$. We do know the number of moles of He present, and its temperature and pressure. We could calculate the volume of the bulb from the data we have about the helium, but it is not necessary to do so. Of the terms in the ideal gas law, V and T are the same for all the gases, so write the term V/RT, which is constant, for both He and HCl:

$$V/RT = n_{He}/P_{He} = n_{HCl}/P_{HCl}$$

Solve this for the number of moles of HCl:

$$n_{HCl} = n_{He}(P_{HCl}/P_{He})$$

Thus,

$$n_{HCl} = (0.0100 \text{ mol})(340 \text{ mmHg})/(228 \text{ mmHg}) = 0.0149 \text{ mol}$$

There was, therefore, 0.0149 mol of $NH_4Cl(s)$ in the bulb originally. The formula weight of NH_4Cl is 53.49 g/mol. Hence the mass of solid NH_4Cl in the bulb was $(0.0149 \text{ mol})(53.49 \text{ g/mol}) = 0.798$ g of NH_4Cl.

EXAMPLE 3.14. Stoichiometry in the combustion of a mixture of hydrocarbons. Use of the definition of mole fraction*

A gas is known to be a mixture of propane, C_3H_8, and methane, CH_4. (Natural gas, which we use for cooking and heating our homes, is a mixture of methane and other hydrocarbons, such as propane.) The gas is contained in a vessel of unknown volume, V, at a temperature, T, and exerts a pressure of 320 mmHg. The gas is burned in excess oxygen and all the carbon is recovered as CO_2. The CO_2 is collected in a container of volume V and found to have a pressure of 448 mmHg at the same temperature, T. Calculate the mole fraction of propane in the original mixture. Assume all gases are ideal.

Solution. This is a problem that should be attacked by using symbols for unknown quantities, writing down the ideal gas law for each mixture, and then looking at the equations you have written for a clue as to how to proceed. Begin by writing down the definition of the quantity you are asked to determine, the mole fraction of propane in the original mixture. To do that, some symbols must be defined. Let

$$n_1 = \text{No. moles } C_3H_8 \text{ in the mixture}$$
$$n_2 = \text{No. moles } CH_4 \text{ in the mixture}$$

Then the mole fraction of C_3H_8, X_1, is given by

$$X_1 = \frac{n_1}{n_1 + n_2}$$

Write the ideal gas law for the original mixture of the two gases:

$$P_{total} = 320 \text{ mmHg} = n_{total}(RT/V) = (n_1 + n_2)(RT/V)$$

We can also write down the equation for the pressure of the CO_2 collected: $P_{CO_2} = n_{CO_2}(RT/V)$. What is the relation between n_1, n_2, and the number of moles of CO_2 formed? The law of conservation of mass, applied to the carbon atoms, determines how many moles of CO_2 are formed in the combustion reaction. Each mole of C_3H_8, after reaction with O_2, yields 3 mol of CO_2. Since we started with n_1 mol of C_3H_8, the combustion produces $3n_1$ mol of CO_2. Similarly, each mole of CH_4, when burned in O_2, produces 1 mol of CO_2, so that burning n_2 mol of CH_4 yields n_2 mol of CO_2.

---

* This problem is more difficult than others in this chapter.

Hence the total number of moles of CO_2 produced by burning n_1 mol of C_3H_8 and n_2 mol of CH_4 is

$$n_{CO_2} = 3n_1 + n_2$$

(*Note:* It is not necessary to write down equations for the combustion of C_3H_8 and of CH_4. You can reason entirely from the conservation of C atoms. If you do try to write equations, you must write *separate* equations, one for the combustion of C_3H_8 and one for the combustion of CH_4.)

The equation for the pressure of CO_2 is, therefore,

$$P_{CO_2} = 448 \text{ mmHg} = n_{CO_2}(RT/V) = (3n_1 + n_2)(RT/V)$$

We now have two equations involving the unknowns n_1, n_2, T, and V:

$$320 \text{ mmHg} = (n_1 + n_2)(RT/V)$$

and

$$448 \text{ mmHg} = (3n_1 + n_2)(RT/V)$$

By dividing one of these equations by the other, the factor (RT/V) cancels out, and we obtain

$$\frac{320}{448} = \frac{n_1 + n_2}{3n_1 + n_2}$$

Reduce the fraction on the left: $\frac{320}{448} = \frac{5}{7}$. Our equation has therefore been simplified to $\frac{5}{7} = (n_1 + n_2)/(3n_1 + n_2)$. Multiplying this fraction out yields $15n_1 + 5n_2 = 7n_1 + 7n_2$, which reduces to $8n_1 = 2n_2$ or $4n_1 = n_2$. We can now substitute $4n_1$ for n_2 in the definition of the mole fraction, X_1:

$$X_1 = \frac{n_1}{(n_1 + n_2)} = \frac{n_1}{(n_1 + 4n_1)} = \frac{n_1}{5n_1} = \frac{1}{5}$$

The mole fraction of propane in the original mixture is $\frac{1}{5}$ or 0.200 (20.0 mol %).

Summary

The volume of a gas depends on its temperature, its pressure, and on the quantity of gas (that is, on the number of moles of gas). **Avogadro's law** states that the volume of a gas is directly proportional to the number of moles of gas, if the temperature and pressure are constant. The proportionality constant is the same for all gases at the same temperature and pressure, so that equal volumes of different gases at the same temperature and pressure contain equal numbers of moles of gas, and therefore equal numbers of molecules of gas. **Boyle's law** states that the volume of a gas is inversely proportional to its pressure, provided that the temperature and the number of moles of gas are constant. **The law of Charles and Gay-Lussac** states that the volume of a gas is directly proportional to its absolute temperature, provided that the pressure and number of moles of gas remain constant. These three gas laws are combined into a single equation, applicable to all gases, at all values of P, T, and V, and known as the **ideal gas law**, $PV = nRT$, where R is called the **universal gas constant**. Predictions of the properties of gases made by using the ideal gas law are in good agreement with experimental observation at pressures less than a few atmospheres, and at temperatures close to, or above, room temperature. Significant errors are made using the ideal gas law at very high pressures or low temperatures.

A great many practical problems involving gases can be solved using the ideal gas law. In particular, the molecular weight of a gas can be determined if the density of the gas is measured at a known temperature and pressure, because the ideal gas law can be rearranged to $M = \delta RT/P$.

The pressure of a gas is frequently measured by measuring the height of a column of mercury that the gas supports. The gases in the earth's atmosphere support a column of mercury that varies in height depending on the temperature, the location, and the weather, but is almost always between 750 and 770 mmHg at sea level. By definition, **one atmosphere** is the pressure that supports a column of mercury exactly 760-mm high at 0 °C and standard gravity. Pressure is defined as the force exerted per unit area, and the SI unit of pressure is the **pascal**, Pa, or newton per square meter $(N \cdot m^{-2})$.

In a mixture of gases, each gas exerts the same pressure it would exert if it alone occupied the volume of the container. The pressure of each component in a mixture of gases is called the **partial pressure** of that component. The sum of the partial pressures of all the gases in the mixture is the total pressure of the gas mixture, according to **Dalton's law of partial pressures**. A useful way of expressing Dalton's law is $P_i = X_i^{(gas)} P_{total}$, where P_i is the partial pressure of the ith component of the mixture, and $X_i^{(gas)}$ is its mole fraction.

Exercises

Section 3.1

1. A sample of gas of mass 2.929 g occupies a volume of 426 mL at 0 °C and 1.00-atm pressure. What is the molecular weight of the gas?

2. A sample of nitrogen gas, N_2, is in a container of volume V at 25 °C, and exerts a pressure of 640 mmHg. A sample of propane gas, C_3H_8, is in a container of the same volume V at 25 °C, and exerts a pressure of 320 mmHg. Is each of the following statements about these two samples of gas TRUE or FALSE? You must *explain* your answer.

    (a) The number of moles of $N_2(g)$ is the same as the number of moles of $C_3H_8(g)$.
    (b) There are Avogadro's number of molecules in each of the two samples.

3. When asked to calculate the number of moles of gas in a container of 175 mL at 25 °C and 1.00-atm pressure, a student submitted the following answer:

$$\text{No. moles gas} = \frac{0.175 \text{ L}}{22.4 \text{ L/mol}} = 7.81 \times 10^{-3} \text{ mol}$$

    The student's answer is incorrect. Explain the error the student has made.

4. What volume of O_2, measured at 24 °C and 752 mmHg, is needed to completely burn all the methane in a 3.00-L container at the same temperature and pressure? The combustion reaction is

$$CH_4(g) + 2 O_2 \rightarrow CO_2(g) + 2H_2O(\ell)$$

Section 3.2

5. A sample of gas contained in a bulb supports a column of mercury 43.7-cm high. What is the pressure of this gas in (a) atmospheres and (b) pascals?

6. Consider two samples of gas at the same pressure, P, and temperature, T. One gas is in a cubical container of volume 1000 cm³, the other is in a cubical container of volume 8000 cm³. What is the ratio of the force exerted on one wall of the larger container to the force exerted on one wall of the smaller container?

7. The element gallium, Ga (mp 29.8 °C), is sometimes used in a barometer at elevated temperatures, because it has a much lower vapor pressure than mercury at high

temperatures. If the density of liquid gallium is 6.09 g · cm^{-3} at 35 °C, how high a column of gallium can a pressure of 0.500 atm support at 35 °C? The density of mercury at 0 °C is 13.6 g · cm^{-3}.

Section 3.3

8. In the laboratory, it is often necessary to measure the pressure of a gas in a vacuum line, a system containing gases at very low pressures. The device used to measure very low pressures is called a McLeod gauge. In a McLeod gauge, a large volume of gas from the vacuum line is compressed to a much smaller volume, so that the pressure increases to a value large enough to be readily measurable. If 450 cm^3 of gas from a vacuum line are compressed to a volume of 0.074 cm^3 at constant temperature, and the pressure of the condensed gas is 395 mmHg, what is the pressure in the vacuum line? Give the answer both in millimeters of mercury and in atmospheres.

9. The following question appeared on a general chemistry exam: A balloon is filled with 0.0100 mol of He at 27 °C and 1.00-atm pressure. Another 0.0100 mol of He is added, at the same temperature, and the balloon's volume increases by a factor of 1.60. Calculate the final pressure in the balloon.

A student gave the following answer:

$$P_1 V_1 = P_2 V_2 \quad \text{and therefore} \quad (1.00)(V) = P_2(1.60\,V)$$

so that $P_2 = 1.00/1.60 = 0.625$ atm. The student's answer is incorrect. Explain the nature of the error made.

Section 3.4

10. A gas is confined in a graduated cylinder enclosed by a movable piston. When the temperature of the room is 24.4 °C, the volume of the gas at atmospheric pressure is found to be 39.4 mL. The cylinder is then immersed in a bath containing a mixture of liquid and solid benzene, maintained at the normal freezing point of benzene. The piston moves to maintain constant atmospheric pressure. The volume of the gas at the freezing point of benzene is 36.9 mL. Calculate the freezing point of benzene.

Section 3.5

11. A certain tank can safely withstand a maximum pressure of 40.0 atm. When filled with 36.4 g of N_2 gas, the pressure in the tank is found to be 703 mmHg. How many more grams of N_2 gas can safely be added to the tank at the same temperature?

12. The reaction

$$Al(s) + 3H^+(aq) \rightarrow Al^{3+}(aq) + \tfrac{3}{2}H_2(g)$$

is used to make H_2 gas in a laboratory experiment. The H_2 formed occupies a volume of 120.0 mL at 25.0 °C and 772 mmHg. Calculate the number of moles of Al that were used to generate this hydrogen.

13. At what temperature will 1.00 g of O_2 occupy 650.0 mL at a pressure of 762 mmHg?

14. What is the molar volume of an ideal gas at 25.0 °C and 1.00 atm?

15. A sample of a gaseous compound of mass 0.9539 g occupies a volume of 280.0 mL at 120.0 °C and a pressure of 742.1 mmHg. Calculate the molecular weight of this gas.

16. A 1.50-L container is filled with N_2 gas at 60.0 °C and 1.00 atm. (a) What is the density of the N_2 gas? (b) How many grams of N_2 must be removed from the container if the pressure is to remain at 1.00 atm when the temperature is raised to 100.0 °C?

17. A gas is placed in a storage tank at a pressure of 30.0 atm at 22.3 °C. As a safety device, there is a small metal plug in the tank made of a metal alloy that melts at 125 °C. If the tank is heated, what is the maximum pressure that will be attained in the tank before the plug will melt and release gas?

18. A gas occupies a certain volume, V, at 25.0 °C and a pressure of 2.00 atm. The gas is allowed to expand to a volume exactly three times the original volume. If it is desired to have the pressure of the gas be 1.00 atm in the expanded volume, what should the final temperature of the gas be?

19. When propane, C_3H_8, is burned with excess oxygen, the reaction is

$$C_3H_8(g) + 5\ O_2(g) \rightarrow 3CO_2(g) + 4H_2O(\ell)$$

If 160.0 mL of CO_2 at 22.0 °C and 750 mmHg are produced by this reaction, how many moles of O_2 were used up?

20. A balloon is filled with 2.40×10^{-2} mol of He at 23.0 °C and 750-mmHg pressure. An additional 1.20×10^{-2} mol of He is added at the same temperature, and the balloon's volume increases by a factor of 1.50. Calculate the final pressure in the balloon.

Section 3.6

21. A mixture of $N_2(g)$ and $O_2(g)$ is collected over water at 22.4 °C in a 900.0-mL container. The total pressure of the gas mixture is 758.3 mmHg. If the sample is known to contain 0.0162 mol of N_2, how many moles of O_2 are in the mixture? The vapor pressure of water is 20.316 mmHg at 22.4 °C.

22. A mixture of N_2 and CO_2 is 38.4% N_2 by weight.
    (a) What is the mole fraction of N_2 in this mixture?
    (b) If the total pressure of the gas mixture is 1.46 atm, what is the partial pressure of N_2 in this mixture?

23. Hydrogen is prepared in the laboratory by the reaction of Mg with dilute hydrochloric acid, according to the reaction

$$Mg(s) + 2H^+(aq) \rightarrow H_2(g) + Mg^{2+}(aq)$$

A 0.8995-g sample of Mg is reacted with dilute hydrochloric acid, and the H_2 gas is collected over water, so that both H_2O vapor and H_2 are in the gas collection tube. The temperature of the water and the gas is 24.8 °C, and the barometric pressure is 767.6 mmHg. The vapor pressure of water at 24.8 °C is 23.48 mmHg.
    (a) How many moles of H_2 will be formed, assuming complete reaction of all the Mg?
    (b) What is the partial pressure of H_2 in the collection tube?
    (c) What volume of $H_2(g)$ will be collected?

Multiple Choice Questions

1. A sample of gas occupies a volume of 430.0 mL at 28.2 °C and 754.2 mmHg. If the sample is cooled to 20.0 °C, what is the pressure of the gas in millimeters of mercury?
   (a) 534.9   (b) 733.7   (c) 760.0   (d) 775.3   (e) 842.3

2. A 0.128-g sample of ethane gas, C_2H_6, is sealed in a glass tube at 24.4 °C and a pressure of 765.3 mmHg. The tube can withstand a maximum pressure of 2.50 atm without bursting. To what maximum temperature may the tube safely be heated?
   (a) 60.6 °C   (b) 100.0 °C   (c) 333.8 °C   (d) 465.6 °C   (e) 738.8 °C

3. A sample of a pure gas has a density of 1.60 g · L^{-1} at 26.5 °C and 680.2 mmHg. Which of the following could the sample be?
   (a) CH_4 (b) C_2H_6 (c) CO_2 (d) Xe   (e) SF_6

4. The combustion of butane, C_4H_{10}, yields only $CO_2(g)$ and H_2O. A sample of butane gas at 23.8 °C and 753.4 mmHg occupies a volume of 1.85 L. How many liters of

$CO_2(g)$, measured at the same temperature and pressure, will be formed on complete combustion of this C_4H_{10}?

(a) 7.40 L   (b) 9.25 L   (c) 12.0 L   (d) 16.7 L   (e) 24.0 L

5. A 0.8763-g sample of sucrose, $C_{12}H_{22}O_{11}$, is completely burned with excess oxygen in a 3.40-L container that is maintained at a constant temperature of 25.00 °C. The equation for the combustion is

$$C_{12}H_{22}O_{11}(s) + 12 O_2(g) \rightarrow 12CO_2(g) + 11H_2O(\ell)$$

What is the partial pressure, in millimeters of mercury, of CO_2 in the container after the combustion?

(a) 0.221   (b) 7.24   (c) 14.0   (d) 86.4   (e) 168

6. At room temperature mercury has a density of 13.6 $g \cdot cm^{-3}$, while liquid bromoform, $CHBr_3$, has a density of 2.89 $g \cdot cm^{-3}$. How high a column of bromoform will be supported by a pressure that supports a column of mercury 200-mm high?

(a) 94.1 mm   (b) 272.0 mm   (c) 42.5 cm   (d) 94.1 cm   (e) 272.0 cm

7. A sample of gas at 35.0 °C and 1.00 atm occupies a volume of 3.75 L. At what temperature should the gas be kept, if it is desired to reduce the volume to 3.00 L, but keep the pressure constant?

(a) −26.6 °C   (b) 0.00 °C   (c) 3.98 °C   (d) 28.0 °C   (e) 112.0 °C

8. How many moles of helium gas occupy 22.4 L at 30.0 °C and 1.00 atm?

(a) 0.110   (b) 0.900   (c) 1.00   (d) 1.11   (e) 1.90

9. If a sample of H_2 gas, collected over water at 26.4 °C and 753.2 mmHg, occupies a volume of 94.6 mL, what volume will the same H_2 occupy when it is dried, and at a temperature of 20.0 °C and a pressure of 1.00 atm? The vapor pressure of water at 26.4 °C is 25.8 mmHg.

(a) 68.6 mL   (b) 72.1 mL   (c) 88.6 mL   (d) 91.8 mL   (e) 92.5 mL

10. What is the density of chlorine gas at 50.0 °C and 1.60 atm?

(a) 2.14 g/L   (b) 4.28 g/L   (c) 8.46 g/L   (d) 13.8 g/L   (e) 27.7 g/L

11. A 0.4153-g sample of gas is collected over water at 23.5 °C and 758.3 mmHg. The collected gas occupies 130 mL. The vapor pressure of water at 23.5 °C is 21.7 mmHg. The molecular weight of this gas, in grams per mole, is

(a) 70   (b) 80   (c) 90   (d) 100   (e) 110

12. At 27 °C and 1.00 atm, the density of a gaseous hydrocarbon is 1.22 $g \cdot L^{-1}$. The hydrocarbon is

(a) CH_4 (b) C_2H_4 (c) C_2H_6 (d) C_3H_8 (e) C_3H_6

13. A mixture of helium and argon contains 3 mol of He for every 2 mol of Ar. The partial pressure of argon is

(a) two thirds the total pressure   (b) one third the total pressure
(c) three fifths the total pressure   (d) one half the total pressure
(e) two fifths the total pressure

14. A 1.7093-g sample of Al metal is reacted with excess aqueous hydrochloric acid to produce hydrogen gas:

$$2Al(s) + 6H^+(aq) \rightarrow 3H_2(g) + 2Al^{3+}(aq)$$

What volume of $H_2(g)$ is collected at a temperature of 24.8 °C and a pressure of 738.5 mmHg?

(a) 88.45 mL   (b) 199.0 mL   (c) 1.063 L   (d) 2.258 L   (e) 2.391 L

15. The molar volume of helium is 51.4 L at
    (a) 100 °C and 1.00 atm   (b) 25 °C and 0.250 atm   (c) 0 °C and 0.500 atm
    (d) 300 °C and 1.00 atm   (e) 40 °C and 0.500 atm

16. A mixture of 50.0 mL of ammonia and 60.0 mL of oxygen gas reacts according to the equation below:

$$4NH_3(g) + 5 O_2(g) \rightarrow 4NO(g) + 6H_2O(g)$$

If all gases are at the same temperature and pressure, and the reaction continues until one of the gases is completely consumed, what volume of water vapor is produced?
    (a) 48.0 mL   (b) 60.0 mL   (c) 72.0 mL   (d) 75.0 mL   (e) 110 mL

17. What is the partial pressure of SO_2 if 100 g of O_2 are mixed with 100 g of SO_2, and the total pressure is 600 mmHg?
    (a) 500 mmHg   (b) 400 mmHg   (c) 300 mmHg   (d) 200 mmHg
    (e) 100 mmHg

18. A 0.20-mol sample of a hydrocarbon C_xH_y yields, after complete combustion with excess O_2 gas, 0.80 mol of CO_2 and 1.0 mol of H_2O. The molecular formula of C_xH_y is
    (a) C_4H_{10} (b) C_4H_8 (c) C_4H_5 (d) C_8H_{16} (e) C_8H_{10}

Problems

3.1. When 0.2553 g of a gaseous hydrocarbon was burned in excess oxygen, 0.8010 g of CO_2 gas was recovered.
(a) Assuming all of the carbon in the hydrocarbon was recovered as $CO_2(g)$, what is the empirical formula of the compound?
(b) An experimental measurement of the density of this gaseous hydrocarbon gave a value of $1.87 \text{ g} \cdot \text{L}^{-1}$ at 273.2 K and 1.00 atm. What is the molecular formula and exact molecular weight of this compound?

3.2. The following statements are made about two samples of gas, each in a 1.00-L container at the same temperature. One container is filled with ethane, $C_2H_6(g)$, at 340-mmHg pressure. The other contains hydrogen gas, H_2, at 680-mmHg pressure. Indicate whether each statement is TRUE or FALSE, and *explain* the answer you chose.
(a) The number of atoms in the ethane sample is one half the number of atoms in the hydrogen sample.
(b) The density of the ethane gas is one half the density of the hydrogen gas.

3.3. A good vacuum produced in common laboratory apparatus corresponds to 1.00×10^{-6}-mmHg pressure at 25 °C. Calculate the number of molecules per cubic centimeter at this pressure and temperature. What important fact does this calculation illustrate?

3.4. In a laboratory experiment, 0.3404 g of Mg was reacted with dilute hydrochloric acid, generating H_2 gas:

$$Mg(s) + 2H^+(aq) \rightarrow Mg^{2+}(aq) + H_2(g)$$

The hydrogen gas was collected over water at 26.4 °C. The barometric pressure was 764.8 mmHg. The vapor pressure of water at 26.4 °C is 25.8 mmHg. Calculate the volume of hydrogen gas collected.

3.5. A certain compound is known to contain only C, H, and N. Analysis shows that it is 65.42% carbon by weight. When 1.3704 g of the compound are burned in excess

oxygen, all of the hydrogen is converted to H_2O of mass 1.1206 g.
(a) What is the mass of each element in the 1.3704-g sample of this compound?
(b) What is the simplest formula of this compound?
(c) An experiment shows that, within $\pm 4\%$, the molecular weight of this compound is $110\ g \cdot mol^{-1}$. Calculate the molecular formula and the exact molecular weight of this compound.

3.6. A chloride of antimony is a solid at room temperature, but vaporizes when heated. When 2.359 g of this compound are vaporized in an evacuated 1.00-L bulb at 580 K, the pressure in the bulb is 376 mmHg.
(a) What is the molecular weight of this chloride of antimony, assuming the vapor is an ideal gas?
(b) The atomic weight of Sb is 121.75. What is the maximum number of atoms of Sb per molecule of this compound?
(c) What is the molecular formula of this chloride of antimony? Explain the reasoning you used to obtain the formula.
(d) What is the exact molecular weight of this compound?

3.7. The reaction

$$2H_2(g) + O_2(g) \rightarrow 2H_2O(g)$$

goes essentially to completion to the right. If 12.00 mL of H_2 gas and 20.00 mL of O_2 gas react completely, what is the volume of the resulting gas mixture (the H_2O vapor and any excess reactant gas)? All the volumes are measured at the same temperature and pressure.

3.8. Excess aqueous hydrochloric acid was added to a 0.415-g sample of an alloy of aluminum and copper. The aluminum dissolved according to the reaction

$$Al(s) + 3H^+(aq) \rightarrow Al^{3+}(aq) + \tfrac{3}{2}H_2(g)$$

but the copper did not dissolve or react; it remained as the pure metal. The gaseous H_2 produced was collected over water. The volume of the collected gas was 436 mL at 24.6 °C and 746.6-mmHg pressure. What is the weight percentage of Al in the alloy? The vapor pressure of H_2O at 24.6 °C is 23.2 mmHg.

3.9. Ethylamine is a gas with a pungent, disagreeable, fish-like odor (a characteristic of the class of compounds called amines). Ethylamine is known to contain only the elements C, H, and N. An elemental analysis shows that it is 53.28% C and 31.07% N by weight.
(a) Calculate the empirical formula of ethylamine.
(b) The density of gaseous ethylamine is found to be $1.47\ g \cdot L^{-1}$ at 77 °C and 714.4 mmHg. Calculate the correct molecular formula and the exact molecular weight of this gas.

3.10. *Rutherford's determination of Avogadro's number.* (See problem 1.11.) Rutherford and his co-workers found that radium, a radioactive element, disintegrates and emits α-particles, which are the nuclei of helium atoms, He^{2+}. Charged particles can be counted with a Geiger counter. A certain radium sample emits 2.23×10^{12} α-particles/min. Each α-particle picks up two electrons as it travels through the air, and becomes a neutral He atom:

$$He^{2+} + 2e^- \rightarrow He\ (g)$$

The He gas is collected for 1 yr. At the end of that time it is found to occupy 0.237 mL at 27 °C and 152.0 mmHg. Calculate Avogadro's number from the data given.

3.11. A gas mixture contains n_1 mol of hydrogen sulfide, H_2S, and n_2 mol of carbon disulfide, CS_2. The mixture is in a volume of V liters, at a temperature T, and a pressure of 330.0 mmHg. The mixture is burned in excess oxygen. All of the sulfur is converted to SO_2 gas, all of the carbon to CO_2 gas, and all of the hydrogen to liquid H_2O.

(a) Write an expression for the number of moles of (i) SO_2 formed and (ii) CO_2 formed, in terms of the initial number of moles of H_2S and CS_2, that is, in terms of n_1 and n_2. $u = n_1 + 2n_2$

(b) The SO_2 and CO_2 gases are collected together in a container of the same volume, V, at the same temperature, T, and the pressure is measured to be 440.0 mmHg. Assuming all the gases are ideal, write an expression for (i) the pressure of the original gas mixture and (ii) the pressure of the mixture of SO_2 and CO_2. Each expression should involve n_1 and n_2.

(c) Calculate the mole fraction of CS_2 in the original mixture.

3.12. You are grading papers. Each of the responses to the following questions contains one or more errors. Explain precisely the nature of the error or errors made. Correct the answers given.

(a) Question: At what temperature will 1.20 g of O_2 gas occupy 900.0 mL at 742.0-mmHg pressure?

Alex A's answer: $T = (742.0)(900.0)/(62.36)(1.20)$

(b) Question: A 2.08-g sample of a certain gaseous compound occupies a volume of 350.0 mL at 150 °C and a pressure of 756 mmHg. Calculate the molecular weight of this gas.

Bonny B's answer: $M = \dfrac{(756)(350)}{(2.08)(0.08206)(150)}$

(c) Question: State Avogadro's hypothesis.

Caspar C's answer: Equal volumes of different gases contain equal numbers of molecules.

(d) Question: A balloon is filled with 0.0100 mol of CO_2 at 25 °C and 1.00-atm pressure. Another 0.0200 mol of CO_2 are added at the same temperature, and the balloon's volume increases by a factor of 1.50. Calculate the final pressure in the balloon.

Dolly D's answer: $P_1V_1 = P_2V_2$. Therefore $(1.00)(V) = P_2(1.50V)$
and $P_2 = (1.00/1.50)\text{atm} = 0.667$ atm

(e) Question: If a certain vessel is filled with 0.0600 mol of N_2 gas at 27 °C and 1.00-atm pressure, how many moles will be required to fill it at 127 °C and 1.00-atm pressure?

Edgar E's answer: Let n = number of moles of N_2 required.

$$\text{Then } \frac{0.0600}{300} = \frac{n}{400}$$

(f) Question: When cyclohexane, C_6H_{12}, is burned with excess oxygen the reaction is

$$C_6H_{12}(\ell) + 9\,O_2(g) \rightarrow 6CO_2(g) + 6H_2O(\ell)$$

If 420 mL of $CO_2(g)$ at 30 °C and 745 mmHg are produced by this reaction, how many moles of O_2 were used up?

Fanny F's answer: No. moles $O_2 = \left(\dfrac{2}{3}\right)$ (No. moles CO_2) $= \left(\dfrac{2}{3}\right)\dfrac{PV}{RT}$

No. moles $O_2 = \dfrac{2(745)(0.420)}{3(0.08206)(30)}$ mol

3.13. A mixture of neon and hydrogen occupies a certain volume, V, at 20.0 °C and 1.80 atm. The mixture is allowed to expand to a volume exactly 2.50 times the original volume, and the temperature is then changed until the pressure is 1.20 atm. Consider the sample an ideal gas mixture.
(a) Calculate the final temperature of the gas in degrees Celsius and in kelvins. (b) The gas mixture is 78.5% neon by weight. What is the final partial pressure of neon?

3.14. A mixture of the gases ethane, C_2H_6, ethylene, C_2H_4, and helium, He, is contained in a bulb of unknown volume at 300 K and a total pressure of 680 mmHg. The mixture is burned with excess O_2. All of the carbon is converted to $CO_2(g)$ and all of the hydrogen is converted to liquid H_2O. Helium does not react with O_2. The $CO_2(g)$ is collected in a bulb of 250.0-mL volume, and found to exert a pressure of 561.3 mmHg at 300 K. The helium is also collected in a separate bulb of 250.0 mL and found to exert a pressure of 187.1 mmHg at 300 K.
(a) How many moles of He were in the original vessel?
(b) How many moles of CO_2 were produced when the mixture was burned?
(c) How many moles of hydrocarbon (that is, the total number of moles of C_2H_6 *plus* C_2H_4) were contained in the original vessel?
(d) What was the volume of the original vessel?
(e) What was the partial pressure of He in the original vessel?

3.15. A sample of butane gas, C_4H_{10}, of unknown mass, is contained in a vessel of unknown volume, V, at 24.8 °C and a pressure of 560.0 mmHg. To this vessel 8.6787 g of neon gas are added in such a way that no butane is lost from the vessel. The final total pressure in the vessel is 1420.0 mmHg at the same temperature. Calculate the volume of the vessel and the mass of butane.

3.16. When heated carefully in the presence of a very small amount of MnO_2 (a black powder that acts as a catalyst), potassium chlorate, $KClO_3$, decomposes to yield oxygen gas, according to the following reaction:

$$2KClO_3(s) \rightarrow 3O_2(g) + 2KCl(s)$$

(*Caution:* The decomposition of chlorates may become explosive if the reaction mixture is allowed to become too hot.)

A white solid is known to be a mixture of only $KClO_3$ and KCl. A sample of this mixture of mass 7.0950 g is heated in the presence of a small amount of MnO_2, and the O_2 produced is collected over water at a total pressure of 741.5 mmHg, and a temperature of 27.2 °C. The vapor pressure of H_2O at 27.2 °C is 27.1 mmHg. The O_2 is found to occupy a volume of 865.5 mL.
(a) How many moles of $KClO_3$ decomposed to produce this O_2 gas?
(b) Assuming all the $KClO_3$ in the sample decomposed, what was the percentage by weight of $KClO_3$ in the original mixture?

3.17. A sample of mass 7.8902 g is known to be a mixture of $CaCO_3$ and $NaHCO_3$. When excess dilute hydrochloric acid is added to this sample, carbon dioxide gas is produced:

$$CaCO_3(s) + 2H^+(aq) \rightarrow CO_2(g) + Ca^{2+}(aq) + H_2O(\ell)$$
$$NaHCO_3(s) + H^+(aq) \rightarrow CO_2(g) + Na^+(aq) + H_2O(\ell)$$

The carbon dioxide is collected in an evacuated 2.004-L flask at 24.5 °C. The pressure of CO_2 in the flask is found to be 785.70 mmHg. Calculate the percentage by weight of $NaHCO_3$ in the original mixture.

chapter 4 The Properties of Gases and the Kinetic – Molecular Theory

Ludwig Boltzmann (1844 – 1906), Austrian physicist, received his doctorate from the University of Vienna in 1866. He held professorships in mathematics, experimental physics, and theoretical physics in several universities in Austria and Germany. In addition to his extensive calculations in the kinetic theory of gases, he laid the foundations of the field of statistical mechanics, and explained the second law of thermodynamics on the basis of the atomic theory of matter. Boltzmann's work in statistical mechanics, now universally recognized, was strongly attacked by Wilhelm Ostwald (see Chapter 9) and others who did not believe in atoms. Ill and depressed, Boltzmann took his own life in 1906. On his tombstone in Vienna is engraved the fundamental law of statistical mechanics.

Chemists are interested in ascertaining what the laws of nature are, and how various properties of matter are related to one another. The only way to learn about these things is to perform experiments. In addition, chemists want to have a theory about, or a model of, the nature of molecules, so that it is possible to understand why the relationships we observe experimentally take the form they do. We are not content simply to know that at room temperature and above gases obey Boyle's law, Avogadro's law, and Charles' law as long as the pressure is below a few atmospheres. We want to be able to start with a theoretical model, that is, a description of the system we are investigating, and then *derive* the observed laws from this model.

The study of gases offers an excellent example of how theory and experiment are interrelated. Observations about the properties of matter lead to the formulation of a theory about the nature of matter; once the theory has been proposed, it suggests new experiments that can be performed to test the theory. Even if the theory turns out to be incorrect (which is not an infrequent occurrence), it serves a useful function because it stimulates new experimental observations, and we then learn more about the relationships between the various properties of matter. It is the combination of theory and experiment and their interplay that contributes to the growth of our knowledge.

section 4.1
The Postulates of the Kinetic Theory of Gases

A model of the nature of gases was developed during the latter half of the nineteenth century. This model is called the **kinetic–molecular theory** of gases, and was primarily the work of three outstanding scientists: Rudolf Clausius (1822–1888), James Clerk Maxwell (1831–1879), and Ludwig Boltzmann (1844–1906). A theory begins with a set of postulates, or assumptions, and the kinetic–molecular theory of gases consists of the following assumptions about the nature of gases:

Postulate 1. A pure gas is composed of a huge number of identical molecules.

Postulate 2. The diameter of a molecule is negligible compared to the average distance between two molecules in the gas phase. An equivalent statement is that the volume occupied by the molecules themselves is negligible compared with the total volume occupied by the gas, which is the volume of the container. Figure 4.1 illustrates what is meant by this postulate.

Postulate 3. The molecules of a gas are in rapid motion, and they move about their container in a random fashion. On the average, the number of molecules moving in any one direction is the same as the number moving in any other direction. The word "kinetic" is derived from the Greek *kinetikos,* which means "in motion," so that this is the postulate from which the theory derives its name.

Postulate 4. These identical molecules exert no force on one another; they neither attract nor repel one another.

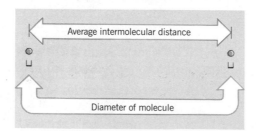

Fig. 4.1. Relative magnitudes of the diameter of a molecule and the average intermolecular distance in the gas phase. While the average intermolecular distance depends on the pressure and temperature, the model proposed is that at any pressure and temperature the average intermolecular distance is so large compared to the diameter of a molecule that we may consider the molecule as a point mass.

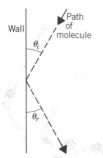

Fig. 4.2. Path of a molecule before and after elastic collision with a wall. For an elastic collision, the angle of incidence, θ_i, is equal to the angle of reflection, θ_r.

Postulate 5. The pressure that a gas exerts on the walls of its container is due to the collisions that the constantly moving molecules make with the walls. The walls of the container are continually being struck by the randomly moving molecules. The pressure the walls experience is the force of these collisions per unit area.

Postulate 6. All collisions that the molecules make with the walls of the container, or with one another, are elastic collisions. This means that there is no loss of energy on collision, and that the total amount of energy of the molecules remains constant, as long as the temperature is constant, despite billions upon billions of collisions. Another consequence of **elastic collisions** is that when a molecule strikes a wall, it rebounds along a path such that the angle of incidence equals the angle of reflection. This is illustrated in Fig. 4.2.

section 4.2
Correlations between the Observed Properties of Gases and the Postulates of the Kinetic–Molecular Theory

Gases have a number of properties that are strikingly different from the properties of liquids or solids. In trying to explain the observed properties in terms of the fundamental nature of gases, scientists were led to the postulates listed in the preceding section as a description of the behavior of molecules in a gas. Let us, therefore, examine some of the properties of gases, and consider the relationship between the postulates of the kinetic theory and the observed behavior of gases.

1. Gases expand rapidly into an evacuated container. Refer to Fig. 4.3. Almost immediately after opening a stopcock connecting an evacuated container to a container filled with a gas, the gas fills both vessels. This can only occur if the molecules of the gas are in rapid motion. The fact that the motion is random, that is, that all directions in space are equivalent, is shown by the fact that the expansion into the evacuated container occurs regardless of the orientation of the bulbs in space.

2. All gases diffuse into one another. If two bulbs, each containing a different gas, for example, N_2 and CH_4, are connected by a stopcock, and the stopcock is opened to both bulbs, after a certain amount of time both bulbs will be filled with a homogeneous mixture of both gases. This indicates that the molecules of a gas are in constant motion. **Diffusion** is defined as the intermingling of molecules as a result of their random motion. Diffusion is not instantaneous because so many collisions between molecules take place as the molecules move from one bulb to another. (See Fig. 4.4.)

3. A gas expands to fill all available space in its container. This is the defining property of a gas, the property that distinguishes gases from liquids and solids. The fact that a gas expands indefinitely if the volume of its container is increased indicates that there are negligible forces of attraction between the molecules of a gas. If there

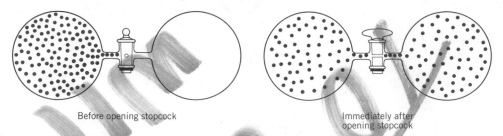

Before opening stopcock   Immediately after
                                                 opening stopcock

Fig. 4.3. The expansion of a gas into an evacuated container.

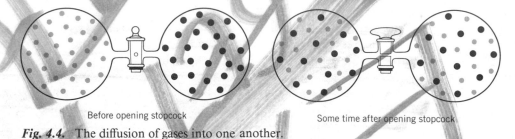

Before opening stopcock   Some time after opening stopcock

Fig. 4.4. The diffusion of gases into one another.

were cohesive forces holding the molecules together, the molecules would cluster in one part of the container instead of filling the entire container. If there are no forces of attraction to keep the molecules close to one another, the only limitations on the volume of a gas are the walls of its container. In contrast, the magnitude of the force of attraction between molecules in liquids and solids is relatively large. The volume of a liquid or solid does not change when it is placed in a larger container.

4. Gases are highly compressible compared to liquids or solids. If pressure is applied to a gas, its volume decreases significantly. Indeed, as we have seen in Chapter 3, the pressure and volume of a fixed quantity of a gas are inversely proportional at constant temperature. In contrast, very large pressures must be applied to liquids or solids to decrease their volume even a small amount. The compressibility of gases is evidence for the postulate that there are large distances, on the average, between gas molecules. Because there is so much empty space between gas molecules, it is easy to decrease the average distance between gas molecules by decreasing the volume of the container.

5. Gases have low densities compared to liquids and solids. Gas densities are about a thousand times smaller than the densities of liquids and solids. This again is evidence for the postulate of very large intermolecular distances compared to molecular diameters. The density, δ, is very small because $\delta = m/V$ [Eq. (3-32)], and V is large due to the large average intermolecular distances.

6. If a fixed sample of gas is left undisturbed at constant volume and temperature, the pressure of the gas remains constant indefinitely. This is evidence for the postulate of elastic collisions. If the molecules lost energy on colliding with the walls, the pressure would decrease.

EXAMPLE 4.1. Use of the postulates of the kinetic theory

At constant temperature and volume, the pressure of a gas is directly proportional to the number of moles of gas. Explain this observation in terms of a postulate of the kinetic theory of gases.

Solution. Postulate 5 states that the pressure of a gas is due to the collisions that the constantly moving molecules make with the walls of the container. The number of collisions the walls experience is directly proportional to the density of the gas, that is, to the number of molecules per milliliter. If the number of moles of gas in a given volume is doubled, the number of molecules per milliliter is doubled. The walls of the container will then be struck by molecules twice as often. Thus the pressure of a gas is directly proportional to the number of moles of gas, in a container of fixed volume at constant temperature.

section 4.3
Criticisms of Two of the Postulates of the Kinetic Theory of Gases

Any gas can be liquefied by compressing it and cooling it sufficiently. Without intermolecular forces, it would be impossible to condense a gas into a liquid. Thus postulate 4 cannot be entirely correct. The forces of attraction between gas molecules cannot be zero, but they must be very small (or else a gas would not expand to fill all available space in its container). At low gas densities (that is, low pressures at room temperature and above), it is a good approximation to treat a gas as if the attractive forces between molecules are negligibly small. The magnitude of the force of attraction between gas molecules must, however, increase when the density of the gas increases, that is, when the pressure increases and/or the temperature decreases, because when the gas density is increased sufficiently, liquefaction of the gas occurs.*

Thus, while postulate 4 is not entirely correct, it may be expected to be a very good approximation at low gas densities, but to be a poorer approximation at high pressures and low temperatures.

Furthermore, if we think about postulate 2, we realize that it cannot be entirely correct either. As we increase the pressure of a gas at constant temperature, we are decreasing the average intermolecular distance between gas molecules. Since the diameter of a gas molecule is finite (although very small), as the average distance between molecules decreases, there must be a point at which it is no longer a good approximation to consider the molecular diameter to be negligibly small compared to the average distance between molecules. Thus postulate 2 is also a very good approximation at low gas densities, but a poorer approximation at high pressures and/or low temperatures.

At low pressures and high temperatures the volume of a gas is large. Gas molecules are very far apart, and intermolecular distances are so large compared to molecular diameters that we may consider the molecules to be "point masses" of zero diameter, and we can treat the forces of attraction between gas molecules as if they are negligibly small. At high pressures and low temperatures, however, the gas molecules are pushed closer together. We must then recognize the finite, nonzero size of the molecules. The magnitude of the force of attraction between gas molecules is not negligible

---

* In fact, for many substances, the force of attraction between two molecules varies inversely as the seventh power of the distance between the molecules. The attractive potential energy of interaction varies inversely as the sixth power of the distance between the molecules. For such substances, the force of attraction can be written as k/r^7, where k is some constant, and r is the distance between the molecules. At low gas densities, r is large and the force of attraction is negligibly small. It is only at high pressures and/or low temperatures, when the distance, r, between two molecules becomes very small, that the magnitude of the force of attraction becomes significantly large.

at high pressures and low temperatures, and postulates 2 and 4 are not a correct description of the nature of a gas.

We recall that at high pressures and low temperatures the ideal gas law is a poor approximation to reality, but that it becomes a better approximation as the pressure decreases and the temperature increases. We may conclude from this that the model described by the postulates of the kinetic theory corresponds to an ideal gas rather than to a real one. We should not be surprised, therefore, to find that the laws derived from the kinetic theory of gases are the ideal gas laws.

EXAMPLE 4.2. Deviations from ideal gas behavior

(a) Calculate the volume of 3.50 mol of an ideal gas at 16.0 atm and 47.0 °C.

Solution

$$V_{ideal} = \frac{nRT}{P} = \frac{(3.50 \text{ mol})}{(16.0 \text{ atm})} \left(0.08206 \frac{\text{L} \cdot \text{atm}}{\text{mol} \cdot \text{K}}\right) (320.2 \text{ K})$$
$$= 5.75 \text{ L}$$

(b) At 47.0 °C and 16.0 atm, 3.50 mol of ammonia gas, NH_3, occupy a volume of 5.20 L. Calculate the percentage error made when the ideal gas law is used to determine the volume of 3.50 mol of ammonia at 47.0 °C and 16.0 atm.

Solution

$$\text{percentage error} = \left(\frac{V_{ideal} - V_{real}}{V_{real}}\right) \times 100$$
$$= \left(\frac{5.75 - 5.20}{5.20}\right) \times 100 = 10.5\%$$

(c) Is the percentage error made in using the ideal gas law to compute the volume of 3.50 mol of NH_3 at 25 °C and 1.00 atm greater than, less than, or the same as the error made at 47.0 °C and 16.0 atm? Explain your answer.

Solution. The error made using the ideal gas law depends on the average distance between molecules, which depends on the number of molecules per cubic centimeter. The larger the gas density, the more the molecules are crowded together, and the larger the error we will make, if we use the ideal gas law. For 3.50 mol of an ideal gas occupying a volume of 5.75 L or 5.75×10^3 cm^3,

$$\frac{\text{No. molecules}}{\text{cubic centimeter}} = \frac{(3.50)(6.022 \times 10^{23}) \text{ molecules}}{5.75 \times 10^3 \text{ cm}^3} = 3.67 \times 10^{20} \text{ molecules/cm}^3$$

For 3.50 mol of an ideal gas at 25.0 °C and 1.00 atm, the volume is

$$V = \frac{(3.50 \text{ mol})}{(1.00 \text{ atm})} \left(0.08206 \frac{\text{L} \cdot \text{atm}}{\text{mol} \cdot \text{K}}\right) (298.2 \text{ K}) = 85.6 \text{ L}$$

The number of molecules per cubic centimeter is, therefore,

$$\frac{(3.50)(6.022 \times 10^{23}) \text{ molecules}}{85.6 \times 10^3 \text{ cm}^3} = 2.46 \times 10^{19} \text{ molecules/cm}^3$$

Because the number of molecules per cubic centimeter is significantly less at 1.00 atm and 25 °C, the average distance between molecules is greater, and the percentage error made using the ideal gas law is smaller at 25 °C and 1.00 atm than it is at 47.0 °C and 16.0 atm.

section 4.4
An Outline of the Derivation of Boyle's Law from the Postulates of Kinetic Theory

We would like to start with the postulates of the kinetic theory and derive an equation for the pressure of a gas. The theory tells us that the pressure of a gas is due to the collisions the randomly moving molecules make with the walls of the container. Furthermore, the definition of pressure (force per unit area) tells us that what we need to calculate is the force of the collisions the molecules make on a wall. If we know the force of the collisions one molecule makes on a particular wall, we can divide by the area of the wall to obtain the pressure due to that one molecule. We then sum the contributions of each molecule in the container to obtain the total pressure due to all the molecules.

To calculate a force we must use Newton's second law of motion:

$$\text{force} = (\text{mass})(\text{acceleration}) \quad \text{or} \quad f = ma \tag{4-1}$$

Since we know the mass of a molecule of any specified gas, what we have to determine is the **acceleration.** Acceleration is defined as the change in the velocity per unit time (also called the time rate of change of velocity). Velocity is a **vector** quantity; it has both magnitude and direction. The magnitude of the velocity is called the **speed.** Any vector can be resolved into components along mutually perpendicular axes. Figure 4.5 displays three vectors, all of the same magnitude (length of the arrow) but with different directions, resolved into their components along the x and y axes. The symbol **u** indicates a vector having both magnitude and direction, whereas u represents the magnitude (length) of that vector, or the speed. For two dimensions, as is clear from Fig. 4.5, the Pythagorean theorem states that $u^2 = u_x^2 + u_y^2$. In three-dimensional space a vector can be resolved into components along the x, y, and z axes, and the Pythagorean theorem is

$$u^2 = u_x^2 + u_y^2 + u_z^2 \tag{4-2}$$

When a molecule strikes a wall, its velocity changes, because the direction of the velocity vector changes, although the magnitude of the velocity, the speed, remains constant. If we can calculate the total change in the velocity of a molecule in one second due to the collisions the molecule makes on the wall, we will be able to calculate the force due to the collisions, and therefore the pressure. We can express this as

$$\begin{pmatrix} \text{force of collisions of} \\ \text{one mole on a wall} \end{pmatrix} = \begin{pmatrix} \text{mass of} \\ \text{molecule} \end{pmatrix} \begin{pmatrix} \dfrac{\text{change in velocity}}{\text{unit time}} \end{pmatrix} \tag{4-3}$$

We must direct our attention, therefore, to calculating the change in velocity per unit time due to the collisions a molecule makes on a wall of the container.

Let us consider a cube of side L (Fig. 4.6), that has one gas molecule of mass m whizzing about in it randomly, as described by the postulates of kinetic theory. The velocity of the molecule will change only when the molecule strikes a wall of the container. Let's focus our attention on one wall of this container, a wall perpendicular to the y direction. (See Fig. 4.6.) Because collisions are elastic, when a molecule strikes that wall, the y component of the velocity will change direction, but not

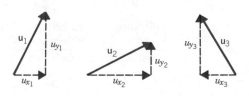

Fig. 4.5. Resolution of a vector into components along the x and y axes.

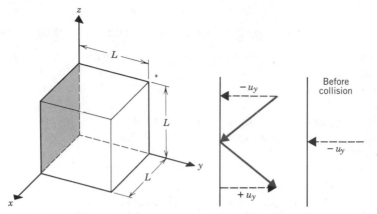

Fig. 4.6. Cubical box of side L. The shaded wall is perpendicular to the y axis.

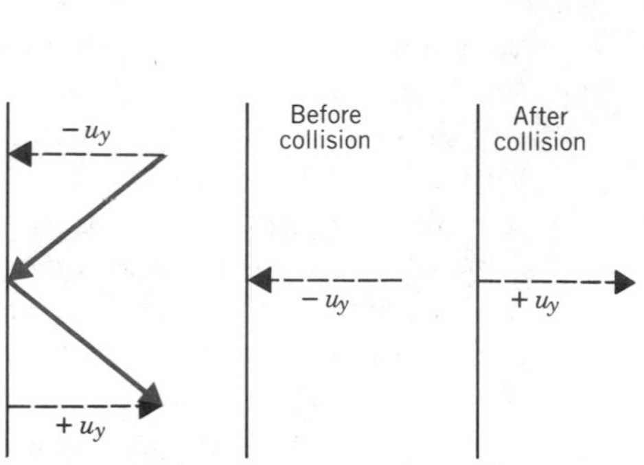

Fig. 4.7. The change in the velocity after an elastic collision.

magnitude. No other component of the velocity changes. If before collision the y component of velocity has the value $-u_y$, then after the collision it has the value $+u_y$. This is illustrated in Fig. 4.7. Thus we have

$$\frac{\text{change in velocity}}{\text{collision}} = u_y - (-u_y) = 2u_y \tag{4-4}$$

For instance, if the y component of velocity changes from -13 m/s to $+13$ m/s, the magnitude of the *change* in velocity is 26 m/s.

What we have calculated in Eq. (4-4) is the change in velocity per collision, and what we need in Eq. (4-3) is the change in velocity per unit time, that is, per second. If we know how many collisions there are per second we can use the relation

dimensional analysis
$$\frac{\text{change in velocity}}{\text{second}} = \left(\frac{\text{change in velocity}}{\text{collision}}\right)\left(\frac{\text{No. of collisions}}{\text{second}}\right) \tag{4-5}$$

to calculate the change in the velocity per unit time. Note that dimensional analysis (see Introduction) enables us to write down the correct relation.

If we rearrange the definition of velocity,

$$\text{velocity} = \frac{\text{distance}}{\text{time}}$$

to solve for the time,

$$\text{time} = \frac{\text{distance}}{\text{velocity}}$$

we can calculate how long, in seconds, a period of time elapses between successive collisions on a single wall. Since we are considering only the wall perpendicular to the y direction, and the only velocity component that is involved is the y component, we must consider distance in the y direction. Immediately after a collision on the wall perpendicular to the y direction, the molecule must travel across the cube and back, a distance $2L$ in the y direction, before it strikes the same wall again. Thus,

$$\begin{matrix}\text{time in seconds between collisions}\\ \text{on one wall perpendicular}\\ \text{to the } y \text{ direction}\end{matrix} = \frac{\left(\begin{matrix}\text{distance traveled between}\\ \text{successive collisions}\end{matrix}\right)}{\text{speed in the } y \text{ direction}} = \frac{2L}{u_y} \tag{4-6}$$

If you know that the time between successive collisions on one wall is $\frac{1}{9}$ of a second, how many collisions are there in one second? There are 9, which is obtained by inverting $\frac{1}{9}$. That is,

$$\frac{\text{number of collisions}}{\text{second}} = \frac{1}{\left(\begin{array}{c}\text{time in seconds between} \\ \text{successive collisions}\end{array}\right)} = \frac{u_y}{2L} \qquad (4\text{-}7)$$

Always check the units of quantities to make sure you have written the relation correctly. The units of u_y, a velocity, are meters per second (m · s^{-1}), and the units of $2L$, a distance, are meters (m). Thus the units of $u_y/2L$ are reciprocal seconds (s^{-1}), that is, $u_y/2L$ is the number of collisions *per* second.

We can now substitute Eqs. (4-7) and (4-4) into Eq. (4-5):

$$\frac{\text{change in velocity}}{\text{second}} = (2u_y)(u_y/2L) = u_y^2/L \qquad (4\text{-}8)$$

Substitution of Eq. (4-8) into Eq. (4-3) yields

$$\begin{array}{c}\text{force of collisions of a single molecule on} \\ \text{a wall perpendicular to the } y \text{ direction}\end{array} = \frac{mu_y^2}{L} \qquad (4\text{-}9)$$

The area of one wall is L^2. Using the definition of pressure, Eq. (3-2), we obtain

$$P = \frac{\text{force}}{\text{area}} = \frac{mu_y^2}{L \cdot L^2} = \frac{mu_y^2}{L^3} = \frac{mu_y^2}{V} \qquad (4\text{-}10)$$

since L^3 is the volume of the cube, V. At this point we can see that there is an inverse relationship between the pressure and the volume. Equation (4-10) is the pressure due to the collisions a single molecule makes on the wall perpendicular to the y direction. We must now consider the pressure due to many molecules.

Suppose we had two molecules inside the cube. Each would have its own velocity, and each would make a contribution to the pressure experienced by the wall, of the form given by Eq. (4-10). Thus,

$$\text{pressure due to two molecules} = \frac{mu_{y_1}^2}{V} + \frac{mu_{y_2}^2}{V} = \frac{m}{V}(u_{y_1}^2 + u_{y_2}^2) \qquad (4\text{-}11)$$

Now let us imagine that we have 1 mol of gas in the cube, so that there are N_A molecules. If we extended the idea expressed in Eq. (4-11), we would have on the right side the sum of N_A terms of the form $u_{y_i}^2$, one for each molecule in the gas. An equivalent procedure is to multiply the average value of u_y^2 by N_A, and write

$$\text{pressure due to one mole of gas} = \left(\frac{m}{V}\right)N_A\langle u_y^2 \rangle \qquad (4\text{-}12)$$

where the symbol $\langle u_y^2 \rangle$ represents the average value of the square of u_y.*

There is nothing to distinguish one direction in space from any other; indeed, it is a postulate of the kinetic theory that the motion of molecules is random, and that all directions in space are equivalent. Hence we may write

$$\langle u_x^2 \rangle = \langle u_y^2 \rangle = \langle u_z^2 \rangle \qquad (4\text{-}13)$$

as a mathematical statement of the postulate that the motion of gas molecules is

---

* There are two notations in use to represent an average value. The average value of x is denoted either $\bar{x}$ or $\langle x \rangle$. We will use the latter notation, because it has wider applications in other branches of physics and chemistry.

random, and that, on the average, one direction is the same as any other. By the Pythagorean theorem, Eq. (4-2), for each molecule, $u^2 = u_x^2 + u_y^2 + u_z^2$. The same relation holds for the average values. Thus,

$$\langle u^2 \rangle = \langle u_x^2 \rangle + \langle u_y^2 \rangle + \langle u_z^2 \rangle \tag{4-14}$$

Since the three terms on the right-hand side of Eq. (4-14) are equal, according to Eq. (4-13), we obtain

$$\langle u^2 \rangle = 3\langle u_y^2 \rangle \quad \text{or} \quad \langle u_y^2 \rangle = \tfrac{1}{3}\langle u^2 \rangle \tag{4-15}$$

Substituting this into Eq. (4-12) yields

$$P = \frac{\tfrac{1}{3}N_A m \langle u^2 \rangle}{V} \quad \text{or} \quad PV = \tfrac{1}{3}N_A m \langle u^2 \rangle \tag{4-16}$$

Equation (4-16) is equivalent to Boyle's law, but it is not immediately apparent that the term on the right-hand side is a constant at constant temperature.

To gain some insight into the significance of the right-hand side of Eq. (4-16), and to prove that it is a constant at constant temperature, we must introduce the definition of **kinetic energy.** Kinetic energy is the energy a body possesses due to the motion of that body (in contrast to **potential energy,** which is the energy a body possesses due to the position of the body relative to other bodies). The kind of motion we are discussing here is the motion of the entire molecule through space;* this is called **translational motion,** and the kinetic energy associated with that motion is called **translational kinetic energy.** (Other kinds of motion that molecules possess are vibrational and rotational motion, which will be discussed in Section 17.7.) If we consider one molecule in a sample of gas, the ith molecule, and it is moving through space with a speed u_i, then by definition

$$\text{translational kinetic energy} = \tfrac{1}{2}m u_i^2 \tag{4-17}$$

In a sample of gas that contains a large number of molecules, the **average value of the translational kinetic energy per molecule** is denoted $\langle \varepsilon_k \rangle$ and is given by

$$\begin{array}{l}\text{average value of translational}\\ \text{kinetic energy per molecule}\end{array} = \langle \varepsilon_k \rangle = \tfrac{1}{2}m \langle u^2 \rangle \tag{4-18}$$

We see, therefore, that $m \langle u^2 \rangle = 2\langle \varepsilon_k \rangle$, and if we substitute this into Eq. (4-16), we obtain

$$PV = \tfrac{2}{3}N_A \langle \varepsilon_k \rangle \tag{4-19}$$

The term $N_A \langle \varepsilon_k \rangle$ is the product of the average translational kinetic energy per molecule and the number of molecules per mole, which is the translational kinetic energy of one mole of gas, $\overline{E}_{\text{trans}}$.

$$\text{molar translational kinetic energy} = N_A \langle \varepsilon_k \rangle = \overline{E}_{\text{trans}} \tag{4-20}$$

Thus Eq. (4-19) becomes simply

$$PV = \tfrac{2}{3}\overline{E}_{\text{trans}} \tag{4-21}$$

A postulate of the kinetic theory is that all collisions are elastic, so that there is no loss of energy on collision. The translational kinetic energy of one mole of gas is therefore a constant; it does not change as long as the temperature is constant. Since

---

* The center-of-mass of a molecule changes position during translational motion; it does not move during pure rotational or vibrational motion. A spinning top that remains fixed at one point in space is rotating; when you walk across a room your motion is translational.

two thirds of any constant is also a constant, we see that the right-hand side of Eq. (4-21) is indeed a constant at constant temperature, and we have therefore arrived at Boyle's law, reasoning only from the postulates of kinetic theory.*

This discussion illustrates how theory enhances our understanding of the laws of nature. We know from experimental observation that for one mole of gas at a fixed temperature, PV is a constant (Boyle's law). In attempting to derive that relation from a set of postulates describing the nature of gas molecules, we have learned a new and very important fact about the "constant" in Boyle's law: The value of the constant is $\frac{2}{3}$ the total translational kinetic energy of the mole of gas.

section 4.5
The Kinetic Theory of Temperature

The connection between heat and motion, as evidenced by frictional phenomena, has been known for a very long time. We rub our hands together to warm them, and we know that it is possible to start a fire by rubbing two dry twigs together. (It isn't *easy,* but it is possible!) In order to understand the properties of gases that are temperature dependent, we must add one more postulate to the defining assumptions of the kinetic theory of gases:

Postulate 7. The absolute temperature of a sample of gas is directly proportional to the average translational kinetic energy of the gas molecules. Both the temperature and the pressure of a gas are due to the motion of the molecules. According to kinetic theory, the temperature of a gas is just a measure of the kinetic energy the molecules possess. When we raise the temperature of a sample of gas, the molecules move faster, on the average. Therefore, the average value of the speed of a molecule increases as the temperature increases, and the average translational kinetic energy, $\langle \varepsilon_k \rangle$, increases.

EXAMPLE 4.3. The kinetic theory of temperature

At constant volume, the pressure of a fixed sample of a gas is directly proportional to the absolute temperature. Explain this observation in terms of the postulates of the kinetic–molecular theory of gases.

Solution. Postulate 7 states that an increase in temperature means an increase in the average translational kinetic energy of the gas molecules. Since $\langle \varepsilon_k \rangle = \frac{1}{2} m \langle u^2 \rangle$, Eq. (4-18), an increase in temperature must be due to faster moving molecules, on the average. If the molecules are moving faster, they collide with the walls more frequently, and with a greater force per collision. Postulate 5 tells us that the pressure of a gas is due to the collisions the molecules make with the walls. In sum, an increase in temperature means an increase in the average speed of the gas molecules, which means more frequent collisions with the walls, as well as a greater force per collision. Both these factors result in a greater gas pressure, provided the volume of the container is constant.

We know experimentally that $PV = RT$ for one mole of gas at low pressures and high temperatures, the conditions for which postulates 2 and 4 of the kinetic–

---

* The argument given should not be considered a rigorous derivation of Boyle's law. We have neglected the effects of collisions between the molecules themselves, and introduced other simplifications.

molecular theory are valid approximations. In the preceding section we discussed the derivation of the relation

$$PV = \tfrac{2}{3}\overline{E}_{\text{trans}}$$

Eq. (4-21). These two expressions for PV must be equal to one another, and we can therefore obtain the relationship between the absolute temperature and the translational kinetic energy. Since $PV = \tfrac{2}{3}\overline{E}_{\text{trans}} = RT$,

$$\overline{E}_{\text{trans}} = \tfrac{3}{2}RT \qquad \text{per mole} \tag{4-22}$$

Equation (4-22) tells us that the molar translational kinetic energy is a function of temperature *only*. It is the same for all gases at the same temperature, and does not depend on the pressure or the molecular mass of the gas. Since $\overline{E}_{\text{trans}} = N_A \langle \varepsilon_k \rangle$, Eq. (4-20), we obtain by substitution into Eq. (4-22), $N_A \langle \varepsilon_k \rangle = \tfrac{3}{2}RT$, or

$$\langle \varepsilon_k \rangle = \frac{3}{2}\frac{RT}{N_A} = \frac{3}{2}kT \tag{4-23}$$

The value of R divided by Avogadro's number (that is, the gas constant per molecule) is an especially important fundamental constant. It is called **Boltzmann's constant,** and is denoted by the symbol k.

$$k = \frac{R}{N_A} = \text{Boltzmann's constant} \qquad \langle \varepsilon_k \rangle \tag{4-24}$$

Equation (4-23) states that the average translational kinetic energy per molecule in a sample of gas is a function *only* of the temperature. Thus it is the same for CO_2, for He, for N_2O, and for all gases at any pressure, as long as the temperature is constant.

You should be careful to distinguish between the average translational kinetic energy per molecule, the translational kinetic energy per mole, and the total translational kinetic energy for a sample of n moles of gas. Expressions for these three quantities are given in Table 4.1.

Note particularly that Boltzmann's constant is used when calculating the average kinetic energy of a single molecule, but that the gas constant, R, is used for molar quantities of gas. Example 4.4 in the following section illustrates the distinctions between these three quantities.

section 4.6
Energy Units for PV and RT

In Chapter 3, when we used the ideal gas law, we expressed PV in liter · atmospheres, or in liter · millimeters of mercury. Equation (4-21), however, tells us that $PV = \tfrac{2}{3}\overline{E}_{\text{trans}}$, and hence PV must be an energy, and can also be expressed in the units of energy. By definition, work, a form of energy, is the product of a force and the distance through which the force moves: energy = (force)(distance). It may not be immediately apparent that the quantity PV has units of energy, but if we insert the definitions of pressure (force/area) and volume (area × length), we see that

$$PV = \frac{\text{force}}{\text{area}}\,(\text{area})(\text{length}) = (\text{force})(\text{length}) \tag{4-25}$$

Table 4.2 summarizes the units that are in use to measure energy.

With the introduction of SI units, the international scientific community decided to eliminate the duplication of energy units and to use the joule exclusively. With the

Table 4.1. **Different Ways of Describing the Translational Kinetic Energy of a Sample of Gas**

average translational kinetic energy per molecule	$= \langle \varepsilon_k \rangle = \frac{3}{2}kT$
translational kinetic energy per mole	$= \overline{E}_{trans} = N_A \langle \varepsilon_k \rangle = \frac{3}{2}RT$
total translational kinetic energy for n moles	$= E_{trans} = \frac{3}{2}nRT$

passage of time, both the erg and the calorie should disappear as units. At the moment, however, there are many chemists and biologists who are accustomed to using calories to measure energy, and you will see publications that still employ this unit.

When it is important to consider the product PV or nRT as an energy, we should not use liter · atmospheres or liter · millimeters of mercury as units. We have already stated the relation between atmospheres and newtons per square meter: $1\ atm = 1.0132 \times 10^5\ N \cdot m^{-2}$, Eq. (3-5). We also know that 1 L is $10^3\ cm^3$ or $10^{-3}\ m^3$. Therefore,

$$1\ L \cdot atm = (1.0132 \times 10^5\ N \cdot m^{-2})(1 \times 10^{-3}\ m^3)$$
$$= 1.0132 \times 10^2\ N \cdot m = 101.32\ J \tag{4-26}$$

This enables us to express R in energy units per mole per kelvin:

$$R = \left(8.2057 \times 10^{-2} \frac{L \cdot atm}{mol \cdot K}\right)\left(1.0132 \times 10^2 \frac{J}{L \cdot atm}\right) = 8.314\ J \cdot mol^{-1}K^{-1}$$

Because of the various units that are in use to measure energy, the gas constant R can also be expressed in many different units. These are summarized in Table 4.3. For instance, if we want to express R in calories per mole per kelvin, we must use the conversion factor from joules to calories given in Table 4.2.

$$R = \left(8.314 \frac{J}{mol \cdot K}\right)\left(\frac{1\ cal}{4.1840\ J}\right)$$
$$= 1.987\ cal \cdot mol^{-1}K^{-1}$$

Table 4.2. Energy Units

Unit, with Symbol	Definition	Relation to Other Units
joule, J kilojoule, kJ	One joule is the energy expended in exerting a force of one newton a distance of one meter.	$1\ J = 1\ N \cdot m$ $1\ kJ = 10^3\ J$ $1\ J = 10^7\ erg$
erg[a]	One erg is the energy expended in exerting a force of one dyne through a distance of one centimeter.	$1\ erg = 1\ dyn \cdot cm$ $1\ erg = 10^{-7}\ J$
calorie, cal kilocalorie, kcal	*Original:* The amount of heat required to raise the temperature of one gram of water from 14.5 to 15.5 °C. *Modern:* One calorie is equal to 4.1840 joules.	$1\ cal = 4.1840\ J$ $1\ kcal = 10^3\ cal$

[a] An erg is a very tiny unit of energy. It has been estimated that an erg is the energy an average fly (mass 10 mg) has to exert to do one push-up (a distance of ~1 mm) against the force of gravity.

Table 4.3. Values of the Gas
Constant, R, in Several Units

8.3144 J $\cdot$ mol^{-1}K^{-1}
8.3144 $\times$ 10^7 erg $\cdot$ mol^{-1}K^{-1}
1.9872 cal $\cdot$ mol^{-1}K^{-1}
8.2057 $\times$ 10^{-2} L $\cdot$ atm $\cdot$ mol^{-1}K^{-1}
62.363 L $\cdot$ mmHg $\cdot$ mol^{-1}K^{-1}

We can also express Boltzmann's constant, $k = R/N_A$, in joules per molecule per kelvin.

$$k = \frac{8.3144 \text{ J} \cdot \text{mol}^{-1}\text{K}^{-1}}{6.022045 \times 10^{23} \text{ molecules} \cdot \text{mol}^{-1}} = 1.3807 \times 10^{-23} \frac{\text{J} \cdot \text{K}^{-1}}{\text{molecule}}$$

The following problem illustrates the use of various energy units and emphasizes the distinction between the average kinetic energy per molecule and the total kinetic energy.

EXAMPLE 4.4. Calculations of the translational kinetic energy of a gas

A glass bulb of volume 2.0 L contains 0.625 mol of helium gas at 25 °C. Helium is a monatomic gas; the only kind of kinetic energy monatomic gases possess is translational kinetic energy.

(a) What is the total kinetic energy of this sample of He gas, expressed both in joules and in calories?

Solution. The total translational kinetic energy for a sample of n moles of gas is $\frac{3}{2}nRT$. In this case, $n = 0.625$ mol and $T = 298.2$ K. Since monatomic gases possess only translational kinetic energy,

$$\text{total kinetic energy} = \tfrac{3}{2}\,(0.625 \text{ mol})\,(R)\,(298.2 \text{ K})$$

By choosing the proper value of R, the total kinetic energy can be given in any desired unit. Thus,

$$\text{total kinetic energy} = \frac{3}{2}\,(0.625 \text{ mol})\left(8.314\,\frac{\text{J}}{\text{mol} \cdot \text{K}}\right)(298.2 \text{ K})$$
$$= 2.32 \times 10^3 \text{ J} = 2.32 \text{ kJ}$$

Or,

$$\text{total kinetic energy} = \frac{3}{2}\,(0.625 \text{ mol})\left(1.987\,\frac{\text{cal}}{\text{mol} \cdot \text{K}}\right)(298.2 \text{ K})$$
$$= 555 \text{ cal}$$

(b) What is the average kinetic energy of an atom of He in this sample? Express the answer both in ergs and in joules.

Solution. The average translational kinetic energy per He atom is independent of the size of the sample of gas, and is

$$\langle \varepsilon_k \rangle = \tfrac{3}{2}kT = \tfrac{3}{2}(1.38066 \times 10^{-23} \text{ J} \cdot \text{atom}^{-1}\text{K}^{-1})\,(298.2 \text{ K})$$
$$= 6.175 \times 10^{-21} \text{ J} \cdot \text{atom}^{-1} = 6.175 \times 10^{-14} \text{ erg} \cdot \text{atom}^{-1}$$

(c) How much energy, expressed in joules, is required to heat this sample of gas from 25.0 to 125.0 °C?

Solution. The kinetic energy of the sample at 25 °C is $\frac{3}{2}nR(298.2\text{ K})$, whereas after heating it to 125 °C the kinetic energy is $\frac{3}{2}nR(398.2\text{ K})$. It is the difference between these two values that must be supplied in order to raise the temperature to 125 °C. Thus the amount of energy required is $\frac{3}{2}nR(398.2 - 298.2) = \frac{3}{2}nR(100.0)$.

$$\frac{\text{energy to heat sample}}{\text{from 25 to 125 °C}} = \frac{3}{2}(0.625\text{ mol})\left(8.3144\ \frac{\text{J}}{\text{mol}\cdot\text{K}}\right)(100.0\text{ K}) = 779\text{ J}$$

Note that the method of calculation shown above takes less time and less numerical calculation than if you calculate the kinetic energy at 125 °C and also at 25 °C, and then subtract. Write out expressions for the energies at the two different temperatures, using symbols, and *subtract before substituting any numerical values.* In general, the amount of energy required to raise a sample of n moles of a monatomic gas from T_1 to T_2 (temperature in kelvins) is $\frac{3}{2}nR(T_2 - T_1)$.

section 4.7
The Mean-Square Speed, the Root-Mean-Square Speed, and the Mean Speed

The translational kinetic energy per mole of gas (see Table 4.1) is given by

$$\overline{E}_{\text{trans}} = N_A\langle\varepsilon_k\rangle = \tfrac{3}{2}RT$$

If we substitute the definition of the average translational kinetic energy per molecule, $\langle\varepsilon_k\rangle = \frac{1}{2}m\langle u^2\rangle$, Eq. (4-18), into this relation, we obtain

$$\tfrac{1}{2}N_A m\langle u^2\rangle = \tfrac{3}{2}RT$$

or

$$N_A m\langle u^2\rangle = 3RT \tag{4-27}$$

What is $N_A m$? Let's write down the units of each term, multiply them, and then look at the units of the product.

$$\left(N_A\ \frac{\text{molecules}}{\text{mol}}\right)\left(m\ \frac{\text{mass}}{\text{molecule}}\right) = M\ \frac{\text{mass}}{\text{mol}} \tag{4-28}$$

The product $N_A m$ is the molecular weight of the gas, denoted M. From Eqs. (4-27) and (4-28) we obtain

$$\langle u^2\rangle = 3RT/M \tag{4-29}$$

We can also obtain a similar relation by equating the two expressions for the average translational kinetic energy per molecule, Eqs. (4-18) and (4-23):

$$\langle\varepsilon_k\rangle = \tfrac{1}{2}m\langle u^2\rangle = \tfrac{3}{2}kT$$

from which it follows that

$$\langle u^2\rangle = 3kT/m \tag{4-30}$$

It should be clear to you that Eqs. (4-29) and (4-30) are equivalent, since Eq. (4-29) can be converted to Eq. (4-30) by dividing both numerator and denominator by N_A: $R/N_A = k$ and $M/N_A = m$.

The quantity $\langle u^2\rangle$ is called the **mean-square speed.** Equation (4-29), which is used to calculate numerical values of $\langle u^2\rangle$, is very easy to remember, but it is important to be careful about units when you use it. The quantity $\langle u^2\rangle$ is the average value of the square of the speeds of all the molecules in the sample, and therefore must have units of velocity squared, that is, meters squared per second squared. The molecular weight, M, must be expressed in kilograms per mole (remember that the SI unit of mass is the kilogram), and R must be given in joules per mole per kelvin. Thus $3RT/M$ has units

$$\frac{(\text{J} \cdot \text{mol}^{-1}\text{K}^{-1})(\text{K})}{(\text{kg mol}^{-1})} \quad \text{or} \quad \text{J/kg}$$

Refer to Appendix A, and you will see that $1 \text{ J} = 1 \text{ kg} \cdot \text{m}^2\text{s}^{-2}$, so that $1 \text{ J/kg} = 1 \text{ m}^2\text{s}^{-2}$, as required for the square of a speed. Note that consistency of units is essential. Every single term on the right-hand side of Eq. (4-29) must be in SI units.

The **root-mean-square speed**, u_{rms}, is by definition the square root of the mean-square speed:

$$u_{\text{rms}} = \langle u^2 \rangle^{1/2} = \left(\frac{3RT}{M}\right)^{1/2} = \left(\frac{3kT}{m}\right)^{1/2} \tag{4-31}$$

u_{rms} is a speed, and therefore has units of velocity, m/s.

The root-mean-square speed is not the same as the **mean** or **average speed**, $\langle u \rangle$, and you should be sure that you understand the distinction between them. A simple exercise will illustrate the difference between $\langle u \rangle$ and u_{rms}.

EXAMPLE 4.5. The difference between $\langle u \rangle$ and u_{rms}

There are four flies in a jar, and at a given instant of time the speeds of the four flies are 3.00, 4.00, 7.00, and 10.00 m/s. Calculate $\langle u \rangle$, $\langle u^2 \rangle$, and u_{rms} for the flies at that instant.

Solution. The mean speed, $\langle u \rangle$, is obtained simply by adding the four speeds and dividing by four:

$$\langle u \rangle = \frac{3.00 + 4.00 + 7.00 + 10.00}{4} \text{ m/s} = 6.00 \text{ m/s}$$

The **mean-square speed** is obtained by squaring each speed first, and then averaging:

$$\langle u^2 \rangle = \frac{3^2 + 4^2 + 7^2 + 10^2}{4} = \frac{9 + 16 + 49 + 100}{4} = \frac{174}{4} = 43.5 \text{ m}^2/\text{s}^2$$

Hence

$$u_{\text{rms}} = \langle u^2 \rangle^{1/2} = (43.5)^{1/2} \text{ m/s} = 6.60 \text{ m/s}$$

Note that while u_{rms} and $\langle u \rangle$ are not identical, they are fairly close to one another. For any sample of a gas, u_{rms} is always greater than $\langle u \rangle$.

EXAMPLE 4.6. Calculation of u_{rms} for two different gases

The gas with the lowest molecular weight is hydrogen, H_2. A substance with a high molecular weight that is gaseous at fairly low temperatures is uranium hexafluoride, UF_6. Uranium is a heavy metal, and most compounds of uranium are solids at room temperature. Uranium hexafluoride, however, is a liquid at room temperature, and boils at 56.6 °C.*

Calculate the root-mean-square speeds of H_2 and UF_6 at 80.0 °C.

* Most high molecular weight compounds are solids at room temperature. For instance, hydrocarbons with molecular weights between 300 and 360, such as $C_{24}H_{50}$, $C_{25}H_{20}$, and $C_{26}H_{54}$, typically boil at several hundred degrees Celsius. For a compound with molecular weight 352 to boil at 56.6 °C is highly unusual. A number of transition metals form hexafluorides, all of which are volatile. For many of these metals, the hexafluoride is their most volatile compound. The metal atom is surrounded by six fluorines, and there are fairly weak interactions between neighboring molecules, which results in a high volatility.

Solution. We calculate u_{rms} using Eq. (4-31), $u_{rms} = (3RT/M)^{1/2}$. In SI units, the molecular weight, M, must be in kilograms per mole, and $R = 8.3144 \text{ J} \cdot \text{mol}^{-1}\text{K}^{-1}$. For H_2, $M = 2.016 \times 10^{-3}$ kg/mol. For UF_6, $M = 0.35202$ kg/mol. Thus, for H_2,

$$u_{rms} = \left[\frac{(3)(8.3144)(353.2)}{2.016 \times 10^{-3}}\right]^{1/2} = 2.090 \times 10^3 \text{ m/s} = 2.090 \text{ km/s}$$

For UF_6,

$$u_{rms} = \left[\frac{(3)(8.3144)(353.2)}{0.35202}\right]^{1/2} = 158.2 \text{ m/s} = 0.1582 \text{ km/s}$$

These speeds are quite large. Let's convert them to miles per hour, to see just how fast molecules move about, on the average. Conversion factors are given in Appendix A. For UF_6,

$$\left(0.1582\ \frac{\text{km}}{\text{s}}\right)\left(60\ \frac{\text{s}}{\text{min}}\right)\left(60\ \frac{\text{min}}{\text{h}}\right)\left(\frac{1 \text{ mile}}{1.6093 \text{ km}}\right) = 353.9 \text{ mile/h}$$

For H_2, u_{rms} is 4675 mile/h at 80 °C!

It is important to remember that at ordinary pressures a gas molecule does not move in a straight line for more than a very short period of time, because it is continually colliding with other molecules. In thinking about the calculations done in Example 4.6, note how much faster the lighter molecule, H_2, moves than does the heavier molecule, UF_6. Molecular speeds are inversely proportional to the square root of the molecular weight, so that at a given temperature, light molecules move faster than heavy ones, on the average. Molecular speeds are also directly proportional to the square root of the absolute temperature. *from equation*

Because the individual molecules in a sample of gas move at different speeds, we can only calculate *average* speeds when describing the sample as a whole. The root-mean-square speed, u_{rms}, and the mean speed, $\langle u \rangle$, are both average quantities. We need them both in order to discuss the various properties of gases, because some properties (such as the kinetic energy) depend on u_{rms} (or its square, $\langle u^2 \rangle$), while others (such as the viscosity) depend on $\langle u \rangle$. Both u_{rms} and $\langle u \rangle$ are proportional to $(T/M)^{1/2}$, but their proportionality constants differ.

section 4.8
Graham's Law of Effusion

If we fill a balloon with two different gases, He and CO_2 for instance, we find that the He escapes from the balloon (through the walls) much faster than the CO_2 does. If you have ever purchased a helium-filled balloon at a zoo or a circus, you will recall that immediately after the purchase the balloon will rise to the ceiling of a room if you let the string go (or disappear into the sky, if you are careless and let go out-of-doors), but the following morning you will find the balloon down on the floor. That is because the He atoms have effused through the tiny pores of the walls of the balloon. The passage of gas molecules through a very small orifice or through a porous membrane is called **effusion**. In effusion, the hole through which the molecules pass must be so small that molecules go through one at a time; molecules undergo no collisions in passing through the hole. Thomas Graham (1805–1869) studied the relative rates of effusion of different gases at the same temperature, and observed that

$$\frac{\text{rate of effusion for gas 1}}{\text{rate of effusion for gas 2}} = \left(\frac{M_2}{M_1}\right)^{1/2} \tag{4-32}$$

that is, the *relative* rates of effusion of two gases at the same temperature are inversely proportional to the square root of their molecular weights. This relation is known as **Graham's law.** What Graham observed experimentally we can now understand from our discussion of the kinetic theory of gases. From the expression derived for the root-mean-square speed of a gas, Eq. (4-31), we can obtain the *ratio* of the root-mean-square speeds of two different gases at the same temperature:

$$\frac{(u_{rms}) \text{ gas 1}}{(u_{rms}) \text{ gas 2}} = \frac{(3RT/M_1)^{1/2}}{(3RT/M_2)^{1/2}} = \left(\frac{M_2}{M_1}\right)^{1/2} \tag{4-33}$$

While we have not derived the expression for the mean speed in a sample of gas, it is possible to obtain that expression from kinetic theory, and it is found that the mean speed is also inversely proportional to the square root of the molecular weight and directly proportional to the square root of the absolute temperature.*

The ratio of rates of effusion of two different gases is the same as the ratio of their mean speeds, or of their root-mean-square speeds. Since lighter molecules move more quickly, on the average, than heavier molecules, they escape through tiny orifices more quickly as well. For example, if both He and CO_2 are at the same temperature,

$$\frac{\text{rate of effusion of He}}{\text{rate of effusion of } CO_2} = \left(\frac{M_{CO_2}}{M_{He}}\right)^{1/2} = \left(\frac{44}{4}\right)^{1/2} = 11^{1/2} = 3.3$$

Helium escapes through a porous membrane 3.3 times faster than does CO_2.

The time it takes for a sample of gas to effuse is inversely proportional to the rate of effusion, so that we can use the relative times of effusion to determine the molecular weight of an unknown gas, as in Example 4.7.

EXAMPLE 4.7. Relative times of effusion of different gases

A known volume of nitrogen gas at a fixed temperature and pressure effuses through a small orifice in 1 min and 24 s. It takes 1 min and 43 s for the same volume of an unknown gas to effuse at the same temperature and pressure. What is the approximate molecular weight of the unknown gas? If the empirical formula of the unknown gas is determined from its elemental analysis to be $(CH_2)_n$, what are the molecular formula and exact molecular weight of the gas?

Solution. Using Eq. (4-33) we obtain

$$\frac{\text{time for gas 2}}{\text{time for gas 1}} = \frac{\text{rate for gas 1}}{\text{rate for gas 2}} = \left(\frac{M_2}{M_1}\right)^{1/2}$$

The molecular weight of nitrogen is $28.013 \text{ g} \cdot \text{mol}^{-1}$. Expressing the times in minutes we obtain

$$\frac{1.72}{1.40} = \left(\frac{M_2}{28.013}\right)^{1/2}$$

---

* The mean speed is given by $\langle u \rangle = (8RT/\pi M)^{1/2}$. The rate of effusion is not equal to either the mean speed or the root-mean-square speed, but like each of them, it is proportional to $(T/M)^{1/2}$.

Squaring both sides yields

$$1.509 = \frac{M_2}{28.013}$$

and $M_2 = (28.013 \text{ g} \cdot \text{mol}^{-1})(1.509) = 42.3 \text{ g} \cdot \text{mol}^{-1}$. This is only approximately correct due to experimental uncertainties in measuring the times of gas flow. Since the formula weight of CH_2 is 14, the molecular weight of $(CH_2)_n$ is $14n$, and $14n \sim 42.3$. Therefore n is exactly 3, and the molecular formula is C_3H_6. The exact molecular weight is $42.080 \text{ g} \cdot \text{mol}^{-1}$.

The difference in rates of effusion of gases with different molecular weights was put to practical use during the Second World War, to separate two isotopes of uranium, ^{235}U and ^{238}U. The isotope needed to make the atomic bombs that the United States dropped on Hiroshima and Nagasaki was ^{235}U, but naturally occurring uranium is, of course, a mixture of isotopes. The natural abundances of the different isotopes of uranium are ^{238}U, 99.27%; ^{235}U, 0.72%; ^{234}U and several others, 0.01%. Clearly the separation of ^{235}U from the more abundant ^{238}U is a difficult task. Gaseous effusion was chosen as the most suitable method of separation. Since this requires a gaseous compound of uranium, the ore must first be converted to UF_6, the most volatile uranium compound. (See Example 4.6.)

At the Oak Ridge National Laboratories in Tennessee, a plant was built to separate $^{235}UF_6$ from $^{238}UF_6$. The gas emerging from a porous barrier is slightly richer in ^{235}U than the original mixture. The molecular weights are so close together, however, that the difference between the rates of effusion of $^{235}UF_6$ and $^{238}UF_6$ is quite small. Equation (4-33) tells us that

$$\frac{\text{rate of effusion of } ^{235}UF_6}{\text{rate of effusion of } ^{238}UF_6} = \left[\frac{238 + 6(19)}{235 + 6(19)}\right]^{1/2} = \left(\frac{352}{349}\right)^{1/2} = 1.0043$$

In order to achieve a practical separation of the isotopes, the gas mixture must pass through a very large number of porous barriers sequentially, so that a gas mixture slightly enriched in $^{235}UF_6$ after emerging from one barrier is passed through another barrier, and emerges still richer in $^{235}UF_6$, and is then passed through a third barrier, and so on.

In many problems comparing properties of different gases, or of the same gas under different conditions, you will find it useful to compute a *ratio* of terms, as was done in obtaining Eq. (4-33). To compute a ratio, do not calculate each term separately and then divide. Rather, you should write down the expression for each term, and carry out the division using symbols for numerical quantities. Often many terms will cancel, and this simplifies the arithmetic you have to do. Example 4.8 illustrates how you can save time by making use of a ratio when doing a numerical calculation.

EXAMPLE 4.8. Utilization of the ratio of the root-mean-square speeds of two different gases

The root-mean-square speed of O_2 at 0.0 °C is 4.61×10^2 m/s. What is the root-mean-square speed of SO_2, **(a)** at 0.0 °C and **(b)** at 100.0 °C?

Solution

(a) It is much easier to make use of the ratio, Eq. (4-33), than to use Eq. (4-31). The $3R$ factor cancels when you take the ratio. Since both gases are at the same temperature,

$$\frac{u_{rms} \text{ for } SO_2}{u_{rms} \text{ for } O_2} = \left(\frac{M_{O_2}}{M_{SO_2}}\right)^{1/2} = \left(\frac{32.00}{64.06}\right)^{1/2} = 0.7068$$

Thus,

$$u_{rms} \text{ for } SO_2 \text{ at } 273 \text{ K} = (0.7068)(u_{rms} \text{ for } O_2 \text{ at } 273 \text{ K})$$
$$= (0.7068)(4.61 \times 10^2) = 3.26 \times 10^2 \text{ m/s}$$

(b)
$$\frac{u_{rms} \text{ for } SO_2 \text{ at } 373.2 \text{ K}}{u_{rms} \text{ for } SO_2 \text{ at } 272.3 \text{ K}} = \left(\frac{373.2}{273.2}\right)^{1/2} = 1.169$$

Thus,

$$u_{rms} \text{ for } SO_2 \text{ at } 373.2 \text{ K} = (1.169)(u_{rms} \text{ for } SO_2 \text{ at } 273.2 \text{ K})$$
$$= (1.169)(3.26 \times 10^2 \text{ m/s}) = 3.81 \times 10^2 \text{ m/s}$$

section 4.9
The Distribution of Molecular Velocities

If the molecules of a gas are in motion, they cannot all have the same speed, nor will they tend to move in any one direction in preference to another. In their continual random motion, the molecules of a gas collide with one another many times, and these collisions provide the mechanism through which the velocities of the individual molecules constantly change.

Because a molecule is so tiny, we cannot see an individual molecule, not even with a powerful microscope. But we can imagine a creature so tiny that he can ride on a molecule; such a creature is named "Maxwell's demon." Picture this totally imaginary fellow riding on a gas molecule as it moves through space. He has a most uncomfortable ride. He is continually being bumped into by other molecules and having both his speed and direction changed (something like those rides in amusement parks where you drive an electric car and everyone else bumps into you and turns you about—except that in the case of gas molecules, the occurrence of collisions is completely random).

As a result of all these collisions, there exists a **distribution of velocities**; at any instant there are molecules with every speed from zero on up. Only a very small fraction of the molecules have either very low or very high speeds; the majority have a speed in the vicinity of the mean speed. It is of interest to know how the velocities are distributed when a **steady state** is attained. The term steady state does *not* mean that any given molecule will maintain its speed unchanged. On the contrary, the speed of any given molecule is constantly changing. At a steady state, however, the fraction of molecules that have speeds between any two values remains fixed. This means, for instance, that if one eighth of the molecules have speeds between 420 and 450 m/s right now, then an hour later, or a week later, one eighth of the molecules will still have speeds between 420 and 450 m/s (as long as the temperature remains constant), even though each individual molecule will have changed its speed and direction many times during the elapsed period. The **distribution** of speeds is unchanged, even though energy and velocity exchanges between the molecules are proceeding continually, as a result of impacts.

The velocity distribution function, $D(u)$, was derived by Maxwell and Boltzmann in 1860. It is *defined* as a function with the following property: When $D(u)$ is plotted as a function of u,

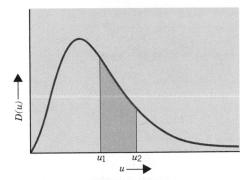

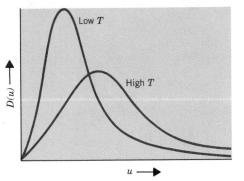

Fig. 4.8. The distribution of speeds of gas molecules at a given temperature. The shaded area is the area under the curve between the speeds u_1 and u_2.

Fig. 4.9. The distribution function for gas velocities at two different temperatures.

$$\frac{\text{the area under the curve}}{\text{between the values } u_1 \text{ and } u_2} = \frac{\text{the fraction of molecules with}}{\text{speeds between } u_1 \text{ and } u_2}$$

Figure 4.8 illustrates the important features of the distribution function, which are listed below.

1. The general shape. There is always a maximum value of the plot of $D(u)$ versus u. The value of $D(u)$ is zero at $u = 0$; it then rises as u increases, reaches a maximum, and falls off, more slowly than it rises, approaching zero asymptotically. The value of the speed at which $D(u)$ is a maximum is called the **most probable speed.** The most probable speed is slightly less than both the mean speed and the root-mean-square speed, but it is not very different from either of these values, which you recall are always fairly close to one another.*

2. Very few molecules have speeds close to zero, and very few molecules have extremely high speeds.

3. The distribution is temperature dependent. At low temperatures, the maximum is higher and the curve has little spread. As the temperature increases, the curve gets flatter and broader, and the most probable speed increases. Figure 4.9 displays the shape of the distribution function for the same gas at two different temperatures.

section 4.10
van der Waals' Equation

We have seen that the assumptions of the kinetic theory contain two postulates that are only approximately valid, which become poorer approximations to reality as the pressure increases. There have been many equations of state proposed in order to fit the experimental data for gases at high pressures and low temperatures better than the ideal gas law does. All of these equations lack the simplicity and universality of the ideal gas law, because the deviation from ideality of each gas is individual, and depends upon the size of the molecules and the force of attraction between them.

Of the many equations proposed, the best known is probably **van der Waals' equation,** proposed by the Dutch scientist Johannes van der Waals (1837–1923). In

* The most probable speed is $(2RT/M)^{1/2}$, the mean speed is $(8RT/\pi M)^{1/2}$, and the root-mean-square speed is $(3RT/M)^{1/2}$.

Table 4.4. Values of the van der Waals' Constants,
a and *b*, for Some Common Gases

Gas	$a\left(\dfrac{L^2 \cdot atm}{mol^2}\right)$	$b\left(\dfrac{L}{mol}\right)$
He	0.034	0.0237
H_2	0.244	0.0266
O_2	1.360	0.0318
N_2	1.390	0.0391
CH_4	2.253	0.0428
CO_2	3.592	0.0427
H_2O	5.464	0.0305
C_2H_6	5.489	0.0638
SO_2	6.714	0.0564

van der Waals' equation, a term of small numerical value is subtracted from the volume of the container to correct for the finite volume occupied by the molecules themselves. Furthermore, since the molecules actually are attracted to one another, when they collide they tend to stay together somewhat longer than if there were no forces of attraction between them, and this reduces the pressure experienced by the walls of the container, because it lengthens the time between collisions with the walls. Therefore a correction term is applied to take this into consideration, and the equation proposed by van der Waals to improve upon $P = nRT/V$ is

$$P = \frac{nRT}{(V - nb)} - \frac{n^2 a}{V^2} \tag{4-34}$$

In this equation, nb is a correction term subtracted from V, the volume of the container, to take care of the nonzero diameter of the gas molecules. The correction term $n^2 a/V^2$ is due to the nonzero force of attraction between molecules, which makes the measured pressure, P, slightly less than the "ideal pressure." Van der Waals' equation may be rearranged to the form

$$\left(P + \frac{n^2 a}{V^2}\right)(V - nb) = nRT \tag{4-35}$$

which shows the parallelism to $PV = nRT$.

The terms a and b in van der Waals' equation are **empirical constants.** That means they are determined by experiment (as opposed to being calculated theoretically). The values of a and b are different for different gases. A larger value of a means a larger force of attraction between gas molecules. A larger value of b means a larger molecular diameter. Table 4.4 lists the values of a and b for a few common gases. Examining these values we find that H_2 and He obey the ideal gas law more closely than any other gases, while CO_2, H_2O, C_2H_6, and SO_2 all display fairly large deviations from ideality.

The van der Waals' equation is not exactly correct, but for high pressures and low temperatures a smaller error is made using van der Waals' equation than using the ideal gas law. This is illustrated in Example 4.9.

EXAMPLE 4.9. Use of van der Waals' equation

(a) Calculate the pressure of 10.00 mol of ethane, C_2H_6, in a 4.86-L flask at 300.0 K, using (1) the ideal gas law and (2) van der Waals' equation.

Solution

$$P_{\text{ideal}} = \frac{nRT}{V} = \frac{(10.00)(0.08206)(300.0)}{4.86} = 50.7 \text{ atm}$$

To evaluate P using van der Waals' (v der W) equation, we obtain a and b for ethane from Table 4.4, and use Eq. (4-34).

$$P_{\text{v der W}} = \frac{(10.00)(0.08206)(300.0)}{4.86 - (10.00)(0.0638)} - \frac{(100.0)(5.489)}{(4.86)^2} = 58.3 - 23.2$$

$$= 35.1 \text{ atm}$$

(b) The observed pressure for 10.00 mol of C_2H_6 at 300.0 K in a volume of 4.86 L is 34.0 atm. Calculate the percentage error using (1) the ideal gas law and (2) van der Waals' equation.

Solution

(1) percentage error $= \left(\dfrac{P_{\text{ideal}} - P_{\text{real}}}{P_{\text{real}}} \right) \times 100$

   percentage error $= \left(\dfrac{50.7 - 34.0}{34.0} \right) \times 100 = 49.1\%$

(2) percentage error $= \left(\dfrac{P_{\text{v der W}} - P_{\text{real}}}{P_{\text{real}}} \right) \times 100 = \left(\dfrac{35.1 - 34.0}{34.0} \right) \times 100 = 3.24\%$

At this relatively high pressure, the van der Waals' equation is closer to reality than is the ideal gas law.

The error made using the ideal gas law is smaller than one might suppose, because the two causes of nonideal gas behavior, (1) finite molecular volumes and (2) attraction between gas molecules, act in opposite directions. Finite molecular diameters make it more difficult to compress a gas than if it were ideal. If we correct the ideal gas law only for the nonzero diameter of the molecules, we would obtain the equation

$$P(V - nb) = nRT \qquad \text{or} \qquad PV = nRT + nbP \qquad (4\text{-}36)$$

A plot of PV as a function of P at constant temperature, for this equation when $n = 1$ mol, is shown in Fig. 4.10. We see that instead of remaining constant, as predicted by the ideal gas law, PV continually increases as P increases, if we correct only for the finite diameters of the molecules.

If, on the other hand, we correct the ideal gas law only for the force of attraction between gas molecules, we obtain the equation

$$\left(P + \frac{n^2 a}{V^2} \right) V = nRT \qquad \text{or} \qquad PV = nRT - \frac{n^2 a}{V} \qquad (4\text{-}37)$$

A plot of PV versus P at constant temperature for Eq. (4-37) when $n = 1$ mol, is shown in Fig. 4.11. Because V decreases as P increases, the term $n^2 a/V$ increases as P increases. Since we subtract a larger term from nRT as P increases, PV decreases as P increases. The force of attraction between gas molecules makes it easier to compress a gas than if it were ideal.

Because the two correction terms act in opposite directions, a plot of PV versus P at constant temperature for van der Waals' equation lies closer to the ideal gas plot than do the plots of Figs. 4.10 and 4.11. Such a plot is shown in Fig. 4.12. While van der Waals' equation is not in exact agreement with experimental observation, the shape of a plot of PV versus P at constant temperature for van der Waals' equation agrees

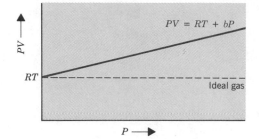

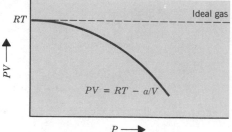

Fig. 4.10. A plot of PV versus P at constant temperature for one mole of a gas that obeys Eq. (4-36), which corrects the ideal gas law only for the finite diameters of gas molecules. If the molecules themselves occupy a significant fraction of the volume of the container, the product PV will increase as P increases.

Fig. 4.11. A plot of PV versus P at constant temperature for one mole of a gas that obeys Eq. (4-37), which corrects the ideal gas law only for the force of attraction between gas molecules. If the force of attraction between the gas molecules is significantly large, PV decreases as P increases.

very well with the actual shape of the plot for a real gas. At low pressures, PV decreases as P increases. Compare Fig. 3.6, in which PV for O_2 and CO_2 is plotted as a function of P for pressures less than 1 atm, with Fig. 4.12. The plots of Fig. 3.6 are just a very small portion of the plot of Fig. 4.12, the part close to zero pressure. The value of PV continues to decrease as pressure increases above 1 atm, but at some high pressure (often between 50 and 100 atm), the curve passes through a minimum and then begins to increase steadily as P increases.

Summary

The **kinetic–molecular theory of gases** consists of a set of postulates that describe the nature of gas molecules. Because gases expand rapidly into an evacuated container, and because all gases diffuse into one another, it is postulated that the molecules of a gas are in rapid, random motion. Because gases expand to fill all available space in their container, no matter how much the volume of the container is increased, it is postulated that there are no attractive forces to keep the molecules close to one another. Because gases are highly compressible and have such low densities compared to solids and liquids, it is postulated that the diameter of a molecule is negligible compared to the average intermolecular distance between gas molecules.

From the postulates of the kinetic–molecular theory and the laws of mechanics,

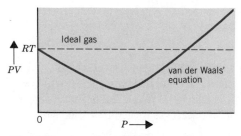

Fig. 4.12. A a plot of PV versus P at constant temperature for one mole of a gas that obeys van der Waals' equation. This is also the shape of such a plot for a real gas. The range of pressures is from zero to several hundred atmospheres. At lower pressures, the predominant cause of deviations from ideality is the nonzero force of attraction between the gas molecules. At higher pressures, the predominant cause of deviation from ideality is the finite diameter of the molecules.

we can derive the ideal gas laws. At high pressures and low temperatures, two postulates of the kinetic theory are not good descriptions of the nature of gas molecules. The magnitude of the force of attraction between gas molecules increases as the distance between the molecules decreases, and becomes significant when the pressure is larger than a few atmospheres at temperatures near room temperature. Furthermore, the finite molecular diameters are not negligible compared to the average intermolecular distance when the pressure is high or the temperature is low, that is, when the gas density is high. For these two reasons, therefore, real gases deviate from the ideal gas laws at high pressures and low temperatures.

When we derive Boyle's law from the postulates of kinetic theory, we learn the physical significance of the product PV, which is a constant at constant temperature. The product of the pressure and the volume of a gas is two thirds of the total **translational kinetic energy** of the gas. From this, and the ideal gas law, we can obtain the relation between the total translational kinetic energy and the absolute temperature, $E_{trans} = \frac{3}{2}nRT$. The **average translational kinetic energy per molecule**, $\langle \varepsilon_k \rangle$, is $\frac{3}{2}kT$, where k is **Boltzmann's constant**.

Each molecule in a sample of gas has a continually changing velocity, due to the many collisions between molecules and with the walls of the container. To describe motion in a gas sample, therefore, we need average quantities, and there are two speeds that are useful in calculating various properties of gases, the **root-mean-square speed,** and the **mean speed.** These two quantities are not the same, but they are close in value, and each is proportional to $(T/M)^{1/2}$, where T is the absolute temperature and M is the molecular weight of the gas. Thus lighter gases move more quickly, on the average, than heavier gases. Both the average speed and the root-mean-square speed increase as the temperature increases, proportional to the square root of the absolute temperature.

Graham's law states that the relative rates of effusion of two different gases at the same temperature are inversely proportional to the square root of their molecular weights. Gaseous effusion is used to separate the uranium isotopes, ^{235}U and ^{238}U. The most volatile uranium compound, UF_6, is used in this process.

While the velocity of any one molecule is continually changing, the fraction of molecules with speeds between any two specified values remains constant as long as the temperature is constant. This is known as the **steady state distribution of gas velocities,** and is described mathematically by the **Maxwell–Boltzmann distribution function.**

The **van der Waals' equation** is an equation of state that is designed to agree with experimental values of P, V, and T for a gas at high pressures and low temperatures, that is, at high gas densities. It contains two empirical constants, a and b. The larger the value of a, the larger the force of attraction between the gas molecules. The larger the value of b, the larger the diameter of the molecule. The force of attraction between the gas molecules makes it easier to compress a real gas than an ideal gas, while the finite molecular diameter makes it more difficult to compress a real gas than an ideal gas. Thus the two causes of nonideal behavior act in opposite directions.

Exercises

Sections 4.1 – 4.3

1. Would you expect two gases at the same pressure and temperature to mix by diffusion more rapidly at higher pressure or at lower pressure? Explain your answer in terms of the description of a gas given by the kinetic–molecular theory.

2. A student asks the following question: "For a sample of gas at fixed temperature, I can reduce the pressure almost to zero by making the volume larger and larger. Can I reduce the volume almost to zero by making the pressure larger and larger?" Explain why the answer to the student's question is "No."

3. The molar volume of CO_2 at 320 K is observed to be 2.52 L/mol when the pressure of the gas is 10.0 atm, and 0.54 L/mol when the pressure is 40.0 atm. Calculate the percentage error made using the ideal gas law to compute the molar volume of CO_2 at these two pressures.

4. Consider a fixed amount of a certain gas at temperature T. If its pressure is doubled from 0.10 to 0.20 atm, its final volume is one half the original value. If, however, its pressure is doubled from 40 to 80 atm, the final volume is larger than one half the original value. Account for this deviation from Boyle's law.

Section 4.4

5. What postulate of the kinetic–molecular theory justifies the following equation?
$$\langle u_x^2 \rangle = \langle u_y^2 \rangle = \langle u_z^2 \rangle$$

6. What postulate of the kinetic–molecular theory justifies the following statement? "The translational kinetic energy of one mole of any gas is a constant at constant temperature." Explain your answer.

7. A bug weighing 20.0 mg is flying with a speed of 5.0 m/s. What is the translational kinetic energy of the bug, in joules?

Sections 4.5 and 4.6

8. At what temperature will 0.80 mol of nitrogen gas, N_2, have the same total translational kinetic energy that 0.50 mol of methane, CH_4, has at 400.0 K?

9. At what temperature will the average translational kinetic energy per molecule for 0.75 mol of ethane gas, C_2H_6, be the same as the average translational kinetic energy per molecule for 2.4 mol of ammonia gas, NH_3, at 310.0 K?

10. A sample of 0.200 mol of CO_2 gas at 1.00-atm pressure occupies a volume of 5.40 L. What is the total translational kinetic energy of this sample of gas? What is the average translational kinetic energy per molecule in this sample?

Section 4.7

11. A child has caught five bugs and put them in a jar. At a given instant of time, the speeds of the bugs are, respectively, 1.00, 2.00, 3.00, 4.00, and 5.00 m/s. Calculate the mean speed, the mean-square speed, and the root-mean-square speed of the five bugs. Which is larger, the mean speed or the root-mean-square speed?

12. (a) Suppose there are four beads in a container, each with a different weight. Show that the total weight of the four beads is $4\langle w \rangle$, where $\langle w \rangle$ is the average weight of a bead.
(b) If there are N objects in a container, and the average mass of an object is $\langle m \rangle$, show that the total mass is $N\langle m \rangle$.

13. In which of the following two samples of gas is the root-mean-square speed greater? Justify your answer.
Sample A: 2.0 mol of neon at 25 °C in a 10.0-L container.
Sample B: 0.40 mol of $Cl_2(g)$ at 20 °C in a 2.5-L container.

Section 4.8

14. Methane gas, CH_4, and CO_2 gas are enclosed in the same container. A tiny pinhole is made in this container. Does the CH_4 effuse out of the container twice as fast as the CO_2? Show all work to justify your answer.

15. Two identical flasks contain samples of nitrogen gas. One flask is heated to 125 °C, while the other is kept at 25 °C. If each flask has a tiny pinhole, how many times faster will the N_2 effuse out of the hotter container?

Section 4.10

16. Which molecule has a larger molecular diameter, SO_2 or H_2O? Justify your answer.

17. One mole of $CO_2(g)$ at 320.0 K occupies a volume of 2.52 L at 10.00-atm pressure.

(a) Is the actual pressure less than, greater than, or the same as the ideal pressure?
(b) What is the pressure of 1 mol of CO_2 gas at 320.0 K in a volume of 2.52 L according to the van der Waals' equation? Which equation of state agrees better with experimental observation for CO_2 at this temperature and pressure, the ideal gas equation or van der Waals' equation?

Multiple Choice Questions

1. The average molecular speed is *greatest* in which of the following gas samples?
   (a) 1.0 mol of N_2 at 560 K   (b) 0.50 mol of Ne at 500 K
   (c) 0.20 mol of CO_2 at 440 K   (d) 2.0 mol of He at 140 K
   (e) 0.40 mol of O_2 at 480 K

2. A real gas most closely approaches the behavior of an ideal gas at
   (a) 15 atm and 200 K   (b) 1 atm and 273 K   (c) 0.50 atm and 500 K
   (d) 15 atm and 500 K   (e) 1 atm and 298 K

3. If both gases are at the same temperature, the rate of effusion of O_2 is very close to
   (a) 8 times that of He   (b) 4 times that of He   (c) 2.8 times that of He
   (d) 0.35 times that of He   (e) 0.125 times that of He.

4. When the pressure of a sample of gas is increased from 0.50 to 100 atm at constant temperature, its volume changes from 2.0 L to 13 mL. What could cause this deviation from Boyle's law?
   (a) The average molecular speed has increased.
   (b) The volume of the gas molecules is a significant fraction of the volume of the container at the higher pressure.
   (c) The force of attraction between the gas molecules is greater when the pressure is higher.
   (d) The molecules have dimerized at the higher pressure.
   (e) The collisions of the molecules on the walls of the container are no longer elastic at the higher pressure.

5. At what temperature will the rate of effusion of N_2 be 1.625 times the rate of effusion of SO_2 at 50.0 °C?
   (a) 87 K   (b) 111 K   (c) 230 K   (d) 373 K   (e) 431 K

6. If a sample of an ideal gas in a sealed container of fixed volume is heated from 10 to 40 °C, the value of which of the following quantities will remain constant?
   (a) The average force of a molecular collision with the walls of the container.
   (b) The pressure of the gas.
   (c) The root-mean-square speed of the molecules.
   (d) The total kinetic energy of the sample.
   (e) The density of the gas.

7. It takes 26 s for 10 mL of H_2 to effuse through a porous membrane. It takes 130 s for 10 mL of an unknown gas to effuse through the same membrane, when both gases are at the same temperature and pressure. What is the molecular weight of the unknown gas, in grams per mole?

   (a) 100   (b) 80   (c) 50   (d) 40   (e) 25

8. The root-mean-square speed of the molecules in a sample of gas is proportional to

   (a) the square of the absolute temperature; (b) the molecular weight of the gas; (c) the absolute temperature; (d) the reciprocal of the molecular weight of the gas; (e) the square root of the absolute temperature.

9. A sample of gas in a closed container of fixed volume is at 250 K and 400-mmHg pressure. If the gas is heated to 375 K, its pressure increases to 600 mmHg. By what factor will the average speed of the molecules increase?

   (a) 1.22   (b) 1.50   (c) 2.25   (d) 2.00   (e) 2.75

10. At 298 K, which of the following gases has the lowest average molecular speed?

    (a) CO_2 at 0.20 atm   (b) He at 0.40 atm   (c) Ne at 0.60 atm
    (d) CH_4 at 0.80 atm   (e) NO at 1.00 atm

11. The number of collisions of Ar atoms with the walls of their container per unit time,

    (a) Increases when the temperature decreases.
    (b) Remains the same when CO_2 is added to the container at constant temperature.
    (c) Increases when CO_2 is added to the container at constant temperature.
    (d) Decreases when the average kinetic energy per molecule increases.
    (e) Remains the same when the volume of the container is decreased at constant temperature.

12. At 47.0 °C and 16.0 atm, the molar volume of ammonia gas, NH_3, is about 10% less than the molar volume of an ideal gas. The reason the actual volume is *less* than the ideal volume is that

    (a) NH_3 decomposes to N_2 and H_2 at 47.0 °C.
    (b) The force of attraction between NH_3 molecules is significant at this temperature and pressure.
    (c) The volume occupied by the NH_3 molecules themselves is a significant fraction of the volume of the container at this pressure and temperature.
    (d) NH_3 molecules move more slowly than predicted by the kinetic theory at this pressure and temperature.
    (e) At 16.0 atm, the motion of NH_3 molecules is no longer random.

13. If the rate of effusion of ammonia, NH_3, is 3.32 times faster than that of an unknown gas when both gases are at 350 K, what is the molecular weight of the unknown gas?

    (a) 31.0   (b) 45.5   (c) 56.5   (d) 112   (e) 188

14. At what temperature will the molar kinetic energy of 0.30 mol of He be the same as the molar kinetic energy of 0.40 mol of Ar at 400 K?

    (a) 533 K   (b) 400 K   (c) 346 K   (d) 300 K   (e) 225 K

15. At what temperature will the total kinetic energy of 0.30 mol of He be the same as the total kinetic energy of 0.40 mol of Ar at 400 K?

    (a) 533 K   (b) 400 K   (c) 346 K   (d) 300 K   (e) 225 K

Problems

4.1. (a) What is the total kinetic energy of $\frac{1}{3}$ mol of Ar(g) at 400.0 K, in joules?

   (b) What is the average kinetic energy per atom in $\frac{1}{3}$ mol of Ar(g) at 400.0 K?

(c) What is the amount of energy, expressed both in calories and in joules, required to raise the temperature of $\frac{1}{3}$ mol of Ar from 400.0 to 520.0 K?

4.2. Calculate the root-mean-square speed of nitrogen gas, N_2, at 25.0 °C, in meters per second and in miles per hour.

4.3. The average speed of an O_2 molecule at 0.0 °C is 425 m/s. Using this value, calculate the average speed at 100.0 °C.

4.4. A balloon that is permeable to gases with molecular weights under 30 is filled with Ne and placed in a box containing pure He. Will the balloon expand or contract? Explain your answer.

4.5. Data for PV as a function of P for 1 mol of CO_2 at 0.0 °C is given in Table 3.2.
(a) Plot PV as a function of P on a scale sufficiently expanded so that the experimental variations in PV can be observed on the graph. Extend your plot back to $P = 0$. Use only millimeter ruled graph paper.
(b) From your plot, determine the value of RT at 0.0 °C.
(c) On the same sheet of graph paper, draw the plot of PV versus P for an ideal gas.
(d) Below $P = 1$ atm, the plot of PV versus P is linear and follows the equation: $PV = RT + BP$, where B is an empirical constant. Determine the value of B for CO_2 at 0 °C from the plot you drew.
(e) Calculate the value of PV at 0.900 atm for 1 mol of CO_2 at 0.0 °C, using the equation you obtained in part (d). What is the percentage error made using the ideal gas law to compute PV for 1 mol of CO_2 at 0.0 °C and 0.900 atm?

4.6. Is each of the following statements about 1-L samples of N_2 gas and C_4H_{10} gas at 25.0 °C and 1.00-atm pressure, TRUE or FALSE? *Explain* your answers.
(a) Both samples contain the same number of atoms.
(b) The root-mean-square speed of a molecule in the two samples is the same.
(c) The average translational kinetic energy per molecule is the same in the two samples.
(d) The density, in grams per liter, of the N_2 is less than the density of the C_4H_{10}.
(e) The total translational kinetic energy of the two samples is the same.

4.7. You are grading papers. Each of the responses to the following questions contains one or more errors. Explain precisely the nature of the error or errors made and correct them.
(a) Question: Calculate the total translational kinetic energy of 2.0 mol of O_2 gas at 27 °C. Express the answer in joules.

$$\text{Axel A's answer: } E_{trans} = \frac{(2.0)(3.0)(0.08206)(300)}{32.00} \text{ J}$$

(b) Question: State Graham's law.

Beula B's answer: The relative rates of effusion of two different gases are inversely proportional to their molecular weights.

(c) Question: A balloon is filled with a mixture of Ne(g) and He(g). Calculate the relative rates of effusion of the two gases from the balloon.

Cyril C's answer: The He effuses 5.0 times faster than the Ne.

(d) Question: Calculate the root-mean-square speed of 4.00 mol of CO_2 gas at 100.0 °C.

$$\text{Derry D's answer: } u_{rms} = \left[\frac{(3)(62.363)(100)}{44.01} \right]^{1/2}$$

4.8. The following statements are made about two samples of gas in 1.00-L cubical containers at the same temperature and pressure. One sample is He, and the other is C_3H_8 (propane). Is each of the following statements TRUE or FALSE? *Explain* your answers.

(a) The force experienced by one wall of the cube containing the C_3H_8 is greater than the force experienced by one wall of the cube containing the He.

(b) The total translational kinetic energy of the sample of C_3H_8 is greater than the total kinetic energy of the sample of He.

(c) A tiny pinhole is made in each container. The He effuses out of its container twice as fast as the C_3H_8.

4.9. The rate of effusion of an unknown gas, X, at 480 K is 1.60 times the rate of effusion of SO_2 gas at 300 K. Calculate the molecular weight of X.

4.10. Krypton gas is contained in a bulb of unknown volume at 298 K and a pressure of 200.0 mmHg. Methane gas, CH_4, is contained in a second bulb of unknown volume at 298 K and a pressure of 80.0 mmHg. When the two bulbs are connected, the total gas pressure after diffusion is complete is 160.0 mmHg. The sum of the volumes of the two bulbs is 3.00 L.

(a) Calculate the volume of the bulb that originally contained Kr. *Hint:* It is not necessary to substitute any numerical value for the gas constant, R.

(b) What is the total translational kinetic energy, in calories and joules, of the mixture of gases in the final 3.00-L volume?

(c) If the root-mean-square speed of the Kr in these bulbs is 298 m/s, what is the root-mean-square speed of the CH_4?

(d) What is the ratio of the average translational kinetic energy per atom of Kr to the average translational kinetic energy per molecule of CH_4?

4.11. In a volume of 1.40 L, 2.60 mol of CO_2 at 320.0 K are observed to exert a pressure of 40.0 atm.

(a) What is the pressure of 2.60 mol of an ideal gas in a volume of 1.40 L at 320.0 K?

(b) What is the pressure of 2.60 mol of CO_2 in a volume of 1.40 L at 320.0 K according to the van der Waals' equation?

(c) Calculate the percentage error made using each of these equations to calculate the pressure.

4.12. At what temperature will the root-mean-square speed of a sample of $N_2(g)$ be the same as that of a sample of $CH_4(g)$ at 100.0 °C?

4.13. The following data were obtained for methane, CH_4, at 0 °C.

P (atm)	1.00000	0.75000	0.50000	0.25000
δ (g/L)	0.71707	0.53745	0.35808	0.17893

(a) Starting with Eq. (3-27), derive the relation

$$M = RT \lim_{P \to 0} (\delta/P)$$

which is used to determine the exact molecular weight of a gas.

(b) For methane, plot (δ/P) as a function of P, and read on your plot the value of $\lim_{P \to 0} (\delta/P)$. Then calculate the exact molecular weight of CH_4, using the equation given in part (a).

(c) Calculate the approximate molecular weight that is obtained using the ideal gas law for the data at 1.0000 atm. By how many percent is this approximate molecular weight in error?

4.14. A mixture of hydrogen, H_2, and nitrogen, N_2, is prepared in which the number of collisions per second by molecules of each gas on the walls of the container is the same. Which gas has the greater number of moles in the container? Explain your answer.

4.15. Equimolar samples of boron trifluoride, BF_3, and ammonia, NH_3, are introduced into the opposite ends of a long glass tube that is kept horizontal. Both substances are gases, and diffuse down the tube. They react spontaneously to form an addition complex, F_3B-NH_3. Will the complex form at the center of the tube, on the BF_3 side of the center, or on the NH_3 side of the center? Explain your answer.

chapter 5 Intermolecular Forces, Condensed Phases, and Changes of Phase

Johannes Diderik van der Waals (1837–1923), Dutch physicist, was Professor of Physics at the University of Amsterdam from 1877 to 1907. He was awarded the Nobel Prize in Physics in 1910 for his research on the properties of gases and fluids. Van der Waals developed a kinetic theory of the fluid state that led to the conception of the continuity of the liquid and gaseous states. His equation of state for gases provided an explanation of the critical pressure, temperature, and volume, and agreed well with experimental observations on gaseous carbon dioxide, which displays relatively large deviations from ideal gas behavior.

In the preceding chapter we have seen that there are weak forces of attraction between molecules in the gas phase, and that these forces lead to deviations from the ideal gas law. More importantly, these **intermolecular forces** are responsible for the formation of the **condensed phases**: liquids and solids. Because all molecules exert some sort of attractive force on one another, it is possible to liquefy any gas by reducing the temperature or increasing the pressure (or doing both simultaneously). The origin of intermolecular forces is the electrical nature of matter, and the force of attraction between a negative charge and a positive charge, as well as the repulsive forces between like charges. Nonetheless, we can distinguish between several kinds of intermolecular forces, and consider their relative strengths. This enables us to understand why some substances have high boiling points and are liquid or solid at room temperature, and why other substances have low boiling points and are gases at room temperature. In general, the stronger the intermolecular forces, the higher the boiling point.

section 5.1
van der Waals Forces

Weak attractive forces between uncharged atoms or molecules are collectively referred to as **van der Waals forces**. These forces arise from the electrostatic attraction of the nuclei of one molecule for the electrons of a different molecule. The repulsions between the electrons of the two molecules and the nuclei of the two molecules counteract the electrostatic attractions, but there is always a small net attractive force. van der Waals forces are **short-range forces**, that is, they are only significant when the molecules are very close to one another. They are significant, of course, during collisions between two gas-phase molecules, and it is this attraction that leads to the correction term, $n^2 a/V^2$, in van der Waals' equation, Eq. (4-34). Let us consider two types of van der Waals forces: **dipole-dipole interactions** and **London** or **dispersion forces**.

Dipole-Dipole Interactions

Many molecules are **polar**, that is, one end of the molecule has a small net positive charge, while the other end has a small net negative charge. This is due to the fact that atoms differ in their ability to attract electrons, and, as a result, the center of the positive charge of the molecule does not coincide with the center of the negative charge. The ability of an atom in a molecule to pull electrons towards itself, and away from other atoms to which it is bonded, is called the **electronegativity** of that atom. (Electronegativity will be discussed in greater detail in Section 13.8.) In general, nonmetals are more electronegative than metals.

A very simple example is the gaseous molecule HCl. Chlorine is a nonmetal with a high electronegativity; it is more electronegative than hydrogen. The electron cloud of the HCl molecule is therefore pulled toward the Cl end of the molecule, which acquires a fractional net negative charge, while the H end has a fractional net positive charge. The charge on each end is much less than the charge on an electron or a proton; there are no ions in a gaseous HCl molecule.

The water molecule, H_2O, is a bent or V-shaped molecule, with the oxygen end having a small net negative charge, and the two hydrogen atoms having a small net positive charge. (See Fig. 5.1.) The center of the positive charges is halfway between the two H atoms, and the center of the negative charge is on the O atom. Thus the water molecule, like HCl, is polar. Figure 5.2 illustrates several polar molecules.

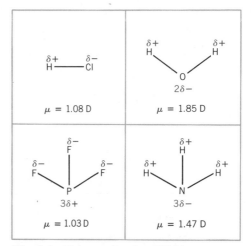

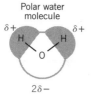

Fig. 5.1. Space-filling model of the H₂O molecule showing the separation of positive and negative charge. The magnitude of $\delta+$ and $\delta-$ is considerably less than 1.

Fig. 5.2. Polar molecules. Within any one molecule, the absolute magnitude of $\delta+$ is equal to the absolute magnitude of $\delta-$, so that each molecule is electrically neutral. The values of $\delta+$ and $\delta-$ differ from molecule to molecule. The dipole moment, μ, in debyes (see text) is given for each of these molecules.

Even if the individual bonds within a molecule are polar, the molecule as a whole is nonpolar if the center of all the negative charges and the center of all the positive charges coincide. Consider, for example, the molecule CCl_4, carbon tetrachloride. The carbon atom in carbon tetrachloride can be envisioned as being at the center of a regular tetrahedron, with the four chlorine atoms at the corners. (See Fig. 5.3.) Although each C—Cl bond is polar, with the Cl end having a net negative charge and the C end a net positive charge, the center of all the negative charges is at the center of the tetrahedron, that is, right at the carbon atom. Since there is no separation between the centers of positive and negative charge, the molecule is not polar. Another example is the planar molecule, BF_3, in which the three F atoms are situated at the corners of an equilateral triangle, with the B atom in the center. These molecules are depicted in Fig. 5.3.

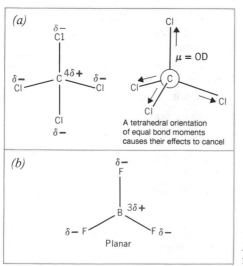

Fig. 5.3. Symmetric molecules with zero net dipole moment. (a) CCl_4 (b) BF_3.

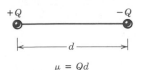

$$\mu = Qd$$

Fig. 5.4. Definition of the dipole moment, μ.

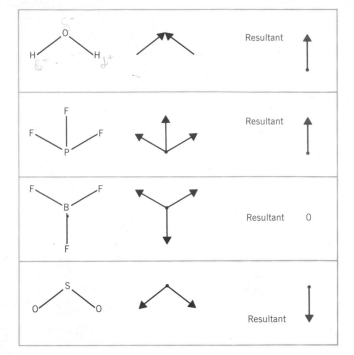

Fig. 5.5. Bond–dipole vector addition. The arrow points from the + to the − end of a polar bond. The resultant vector is the net dipole of the molecule.

The **dipole moment** of a polar molecule is defined as the product of the magnitude of the charge at each center and the distance separating the center of positive charge and the center of negative charge. If the magnitudes of the charges at the two centers are $+Q$ and $-Q$, and d is the distance between the two centers, the dipole moment, μ, is defined as Qd. (See Fig. 5.4.) The unit of dipole moment is the **debye** (D), named after Peter Debye (1884–1966), a Dutch physical chemist who emigrated to the United States at the beginning of the Second World War. A dipole consisting of charges $\pm e$, where e is the charge on the electron, separated by a distance of 0.10 nm (1 Å or 1×10^{-8} cm), has a moment of 4.80 D.*

The dipole moment of a molecule is the resultant of the individual bond dipoles. Each bond dipole can be considered a vector, represented by an arrow that points from the positive to the negative end. The net dipole moment of the molecule is then obtained by **bond-dipole vector addition.** A few examples are shown in Fig. 5.5.

The larger the dipole moment of a molecule, the greater the dipole-dipole attractive force between molecules. In the liquid state, although molecules are continually moving, they tend to align themselves so that, on the average, the intermolecular attractions are maximized. This means that, more often than not, the positive end of one molecule is close to the negative end of another. This is illustrated in Fig. 5.6.

London or Dispersion Forces

Gases of nonpolar molecules, such as N_2, CH_4, or Ar, can also be condensed. (Argon is, of course, monatomic, but the word "molecule" is used collectively here.) We must therefore inquire into the nature of the force of attraction between two nonpolar molecules.

---

* In electrostatic units, the charge on the electron is $e = 4.80 \times 10^{-10}$ esu. If the distance separating two charges of magnitude e is 1×10^{-8} cm, the dipole moment is 4.80×10^{-18} esu · cm. The unit 10^{-18} esu · cm is the debye (D).

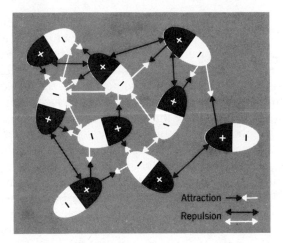

Fig. 5.6. Electrostatic interactions between dipoles.

Let us consider a noble gas atom such as Ar or He, which has a spherically symmetric electronic charge cloud, and is nonpolar. Because the electrons are constantly and continually moving, however, at any given instant the atom may be *temporarily* polar. At the next instant, the electrons have changed position and the polarity is different. Over time, the net dipole moment is zero, but the atom has a fluctuating, transitory dipole moment. This is illustrated for He in Fig. 5.7. When a second atom closely approaches one that has an instantaneous polarity, the mutual repulsion of the two electron charge clouds results in the momentary distortion or **polarization** of the electronic charge cloud of the second atom. Electrons in different atoms or molecules tend to *correlate* their motions so that they stay as far apart as possible and thereby reduce their mutual repulsion. A fluctuating dipole moment in one atom induces a temporary dipole moment in a nearby atom in such a way that there is a net attraction between the two rapidly fluctuating dipoles. This is illustrated in Fig. 5.8. The **induced dipole moment** of the second atom is transitory, and the

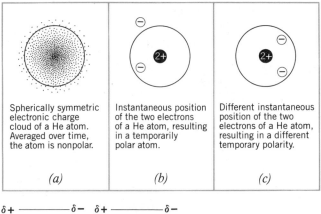

Spherically symmetric electronic charge cloud of a He atom. Averaged over time, the atom is nonpolar.

(a)

Instantaneous position of the two electrons of a He atom, resulting in a temporarily polar atom.

(b)

Different instantaneous position of the two electrons of a He atom, resulting in a different temporary polarity.

(c)

Fig. 5.7. The fluctuating dipole in a He atom, which averages to zero over time. The He atom is nonpolar; its dipole moment is zero.

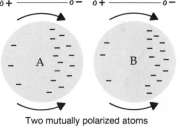

Two mutually polarized atoms

Fig. 5.8. The correlation of the electronic motion of two adjacent atoms, resulting in an attractive force between two atoms with temporary dipoles. A momentary distortion of the charge cloud of one atom induces a dipole moment in a nearby atom. Many such dipoles are formed and destroyed each second.

correlated motion of the two electronic charge clouds disappears when the atoms move apart. The attractive force between a pair of rapidly fluctuating dipoles whose electronic motions are correlated is called the **London dispersion force,** after Fritz London who first gave the correct quantitative explanation for it in 1930.

The strength of the dispersion force depends on how easily the electron cloud of the atom or molecule can be distorted or polarized. The further away the electrons are from the nuclei, the easier it is to polarize the electron cloud. For this reason larger molecules with greater numbers of electrons are usually more polarizable than smaller molecules. We find that the magnitude of the dispersion force generally increases as the number of electrons in the molecule increases. For example, the boiling points and melting points of the rare gases increase steadily as the atomic number, and therefore the number of electrons, increases. For a series of similar molecules, such as the hydrocarbons, boiling points also increase regularly as the molecular weights increase. The shape of a molecule affects its polarizability also, so that it is difficult to compare molecules that are not fairly similar.

Table 5.1 summarizes some data showing the effect of increasing London dispersion forces on the boiling points.

Polar molecules have higher melting points and boiling points than nonpolar compounds of similar nature and molecular weight, because in addition to the London dispersion force of attraction there is also the dipole-dipole attraction. In general, the larger the dipole moment, the higher the melting and boiling point. Some data illustrating this point are summarized in Table 5.2.

Be careful not to make predictions about relative boiling points on the basis of dipole moment alone. For instance, chloromethane, CH_3Cl, has a dipole moment of 1.87 D, while carbon tetrachloride, CCl_4, is symmetric and has a zero dipole moment. Yet CCl_4 is a liquid at room temperature (bp 76.5 °C), while CH_3Cl is a gas (bp −24.2 °C). The dispersion forces in CCl_4 (molecular weight 153.8) are stronger than the sum of the dispersion plus dipole-dipole forces in CH_3Cl (molecular weight 50.5).

Table 5.1. **The Effect of Increasing Dispersion Forces on the Boiling Point**

Rare Gas	Atomic Number	Boiling Point (°C)
He	2	−268.9
Ne	10	−246.0
Ar	18	−185.7
Kr	36	−152.3
Xe	54	−107
Rn	86	−61.8

Compound	Molecular Weight	Boiling Point (°C)
CH_4	16	−164
C_2H_6	30	−88.6
C_3H_8	44	−42.1
C_4H_{10}	58	−0.5
C_5H_{12}	72	+36

Table 5.2. The Effect of Dipole-Dipole Forces on Melting Points and Boiling Points

Compound	Molecular Weight	Dipole Moment (D)	Melting Point (°C)	Boiling Point (°C)
C_2H_6	30.1	0	−183.3	−88.6
CH_3F	34.0	1.85	−141.8	−78.4
CF_4	88.0	0	−184	−129
$CHClF_2$	86.5	1.42	−146	−40.8
CH_2Cl_2	84.9	1.60	−95	+40
SiH_4	32.1	0	−185	−111.8
PH_3	34.0	0.58	−133	−87.7
H_2S	34.1	0.97	−85.5	−60.7

section 5.2
The Hydrogen Bond

A **hydrogen bond** (or **bridge**) is formed when a bonded hydrogen atom in a molecule is situated between two strongly electronegative atoms, that is, atoms that have a strong tendency to attract electrons. Electronegative atoms are, of course, atoms of the nonmetals in the upper right-hand corner of the periodic table. (See Section 2.4.) The most strongly electronegative atoms are F, O, N, and Cl, and these are the only atoms for which hydrogen bonding is of significant strength.

A hydrogen bond may be thought of as a special type of dipole-dipole interaction. Molecules in which there is a polar O—H, F—H, or N—H bond will form hydrogen bonds by orienting themselves so that the hydrogen atom is located between two electronegative atoms. This interaction is represented as O—H $\cdots$ O, F—H $\cdots$ F, or O—H $\cdots$ N, for example. Hydrogen bonds are considerably stronger than ordinary dipole-dipole interactions or London dispersion forces, but they are less than one tenth as strong as typical chemical bonds between atoms *within* one molecule (**intramolecular forces**).

Hydrogen bonding results in abnormally high melting points and boiling points. Consider, for example, the series of similar compounds HF, HCl, HBr, and HI. London dispersion forces increase regularly from HF through HI, due to the increasing number of electrons, and we might therefore expect the boiling points to increase regularly also. In fact, the boiling point of HF is abnormally high, as can be seen by examining the data in Table 5.3, which is plotted in Fig. 5.9. In liquid HF very few

Table 5.3. The Effect of Hydrogen Bonding on the Melting Point and Boiling Point[a]

Compound	Melting Point (°C)	Boiling Point (°C)
HI	−50.8	−35.5
HBr	−86	−66.4
HCl	−111	−85
HF	−83	+19.4
H_2Te	−48	−1.8
H_2Se	−65.7	−41.3
H_2S	−85.5	−60.7
H_2O	0.0	+100.0

[a] The abnormally high melting and boiling points of H_2O and HF are due to extensive hydrogen bonding.

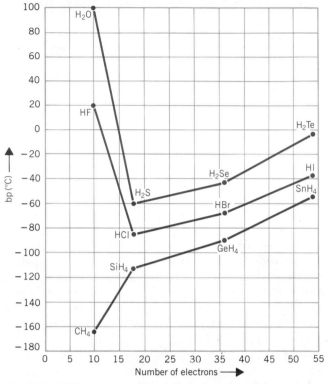

Fig. 5.9.   Boiling points of isoelectronic series of the hydrogen compounds of the halogens (Group VIIA) and the chalcogens (Group VIA), compared with those of the hydrogen compounds of Group IVA of the periodic table. The boiling points of HF and H_2O are seen to be unusually high, due to the extensive hydrogen bonding that exists in the liquid states of these molecules.

molecules exist as separate units. Long zig–zag chains are formed, due to extensive hydrogen bonding, as shown below:

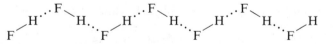

Because hydrogen bonds are much stronger than van der Waals forces, liquids that are hydrogen bonded are harder to vaporize, and therefore have higher boiling points, than nonhydrogen-bonded liquids.

Hydrogen bonds are strongest when all three atoms lie along a straight line as in HF. All O—H ··· O bridges in water are also linear, and this has a remarkable effect on the structure of ice. Each water molecule in ice is hydrogen bonded to four others in a three-dimensional array, as illustrated in Fig. 5.10. The structure of ice, shown in Fig. 5.11, is a very open structure, in which each water molecule is surrounded only by four nearest neighbors. As a result, ice has an unusually low density. When ice melts, some of these hydrogen bonds break, and water molecules enter some of the open spaces in the ice structure. Thus the density of ice is less than the density of liquid water. At 0 °C, the melting point, ice has a density of 0.917 g/mL, whereas liquid water has a density of 1.00 g/mL.

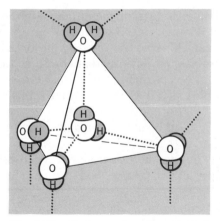

Fig. 5.10. Hydrogen bonding (dotted lines) between H_2O molecules in ice.

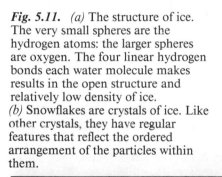

Fig. 5.11. *(a)* The structure of ice. The very small spheres are the hydrogen atoms: the larger spheres are oxygen. The four linear hydrogen bonds each water molecule makes results in the open structure and relatively low density of ice.
(b) Snowflakes are crystals of ice. Like other crystals, they have regular features that reflect the ordered arrangement of the particles within them.

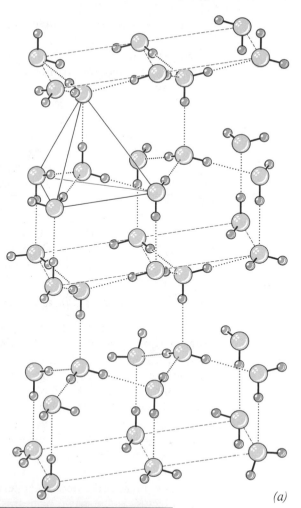

(a)

(b)

Because ice is less dense than liquid water, it floats on the surface of the liquid. You undoubtedly have seen chunks of ice floating on the surface of a river or stream during the winter, so it is familiar to you. This property of water, however, is highly unusual. The solid form of most other substances is denser than the liquid form. Most solids will sink, and not float on the surface of their liquids. Because ice floats on liquid water, the layer of ice on the surface of a river in the winter acts as a thermal insulator between the water below and the air above. Rivers do not freeze solid (as they would if the ice sank), and fish and other marine organisms are enabled to survive long periods of freezing weather.

Water has many unique properties due to the extensive hydrogen bonding in the solid and liquid phases. These properties will be discussed further in Sections 7.1 and 7.2.

A great many molecules of fundamental importance in biological systems are hydrogen bonded. In particular, hydrogen bonding has profound effects on the structure of proteins and of DNA (deoxyribonucleic acid), the basic genetic material.

section 5.3
The Intermolecular Potential Energy Function

Energy can be divided into two categories: kinetic and potential. **Potential energy** is the energy a body possesses because of its position relative to other bodies. For instance, a piece of chalk held high in the air has a greater potential energy than the same piece of chalk lying on the floor. Most of the energy changes in chemical reactions are potential energy changes. When new compounds are formed, the distances between atoms change, and this change in relative positions changes the potential energy of the system. By convention, the potential energy of particles separated by an infinite distance is zero. A force of attraction between two particles causes the potential energy to decrease; repulsion causes the potential energy to increase.

Let us consider the potential energy of interaction between two uncharged spherical molecules, such as two Ar atoms, or two CH_4 molecules. When the two particles are infinitely separated, that is, when the intermolecular distance, r, is very large, the potential energy is zero. As the molecules are brought closer together, they begin to attract one another, and the potential energy decreases, that is, it becomes negative. However, if the intermolecular distance, r, becomes very small, the electron clouds of the two molecules will come into contact with one another, and a very strong repulsive force will set in. The potential energy will then rise very steeply. As a result, the potential energy curve passes through a minimum value. A plot of the potential energy as a function of the distance, r, between the two particles is shown in Fig. 5.12. The difference between zero potential energy and the minimum potential energy is called the **depth of the potential well,** and can be denoted ε. The greater the force of attraction between two molecules, the larger ε will be. We therefore expect that molecules for which the van der Waals' empirical constant "a" is large will also have large values for ε, the depth of the potential well in the potential energy function. The magnitude of ε will be greater for polar molecules than for nonpolar molecules of approximately the same molecular weight.

section 5.4
Molecular Arrangements in the Solid, Liquid, and Gaseous States

The solid state is characterized both by orderly arrangements of atoms and by relatively small distances between the molecules or ions, so that solids generally have the

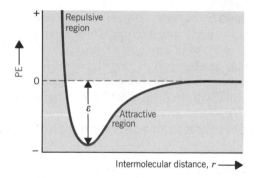

Fig. 5.12. The potential energy of interaction of two uncharged atoms or molecules, as a function of the distance, r, between them. The stronger the force of attraction between the two particles, the greater the depth of the potential well, ε.

highest densities of the three states of matter. In crystalline solids, the regular arrangement of atoms, molecules, or ions extends throughout the entire macroscopic crystal, except for minor flaws. The gaseous state, on the other extreme, is characterized by random motion of gas molecules, with relatively large distances between the molecules, with the result that gases have relatively low densities, as was discussed in Chapter 4. The liquid state is intermediate between the solid and the gaseous states, but is more similar to the solid state than to the gaseous. Molecules in a liquid are in incessant random motion, but the average distance between them is considerably smaller than in the gas phase. While there is local order in the liquid, there is no long-range order comparable to that which exists in a crystalline solid. Although the arrangement of molecules in the liquid state is irregular, there is contact between adjacent molecules. As a result, the densities of liquids are generally about 10% less than the densities of their corresponding solids, whereas gas densities are many times smaller. Figure 5.13 illustrates the differences between molecular arrangements in the solid and liquid states.

section 5.5
Vaporization, Fusion, and Sublimation

Vaporization

The conversion of a liquid to its vapor is called **vaporization.** Heat must be absorbed by the liquid for this process to occur. For instance, in order for 1 mol (18 g) of liquid water to be completely vaporized at 20 °C, 44.10 kJ (10.54 kcal) of heat energy must be absorbed. You have already observed this absorption of heat: Perspiration cools us

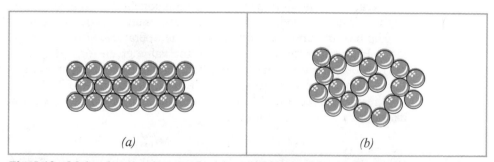

Fig. 5.13. Molecular arrangements in *(a)* a crystalline solid, and *(b)* a liquid. The spheres represent atoms, molecules, or ions.

because the heat required to vaporize the water comes from our bodies. Water has an unusually high heat of vaporization per *gram;* it requires more energy to vaporize one gram of water than to vaporize a gram of any other liquid. For example, 41.92 kJ (10.02 kcal) of heat energy must be absorbed to vaporize 1 mol of ethyl alcohol at 20 °C, but 1 mol of ethyl alcohol, C_2H_5OH, weighs 46.07 g, so that the energy required per gram is only $(41,920 \text{ J})/(46.07 \text{ g}) = 910 \text{ J/g}$, whereas for water it is $(44,100 \text{ J})/(18.015 \text{ g}) = 2448 \text{ J/g}$.

We might indicate the amount of heat required to vaporize a liquid by including it in the equation we write, as follows:

$$H_2O \text{ (liq)} + 44.1 \text{ kJ} \rightarrow H_2O \text{ (g)}$$

It is not customary, however, to write the amount of heat that is either given off or absorbed in any physical or chemical reaction as if it were a reactant or a product. We do not usually include energy terms in the chemical equations we write.

Most chemical reactions take place open to the atmosphere, and occur at constant pressure. We define the amount of heat absorbed by the system in any change that takes place at constant pressure as the **enthalpy change** of the system, and denote it **ΔH.**

If heat must be absorbed in order for a reaction to proceed, that is, if we would write the energy term as a reactant on the left-hand side of the equation (as is the case for any vaporization), ΔH is, by convention, a positive quantity, and the reaction is said to be **endothermic** (endo = into, therm = heat).

If heat is released (given off) as a reaction proceeds, that is, if we would write the energy term as a product on the right-hand side of the equation, ΔH is by convention a negative quantity and the reaction is said to be **exothermic** (exo = out of, therm = heat).

The amount of heat that is required to convert one mole of liquid into one mole of vapor at a given temperature and constant pressure is called the **heat of vaporization,** ΔH_{vap}, of that liquid. All vaporizations are endothermic reactions. We indicate this for water by writing, for instance,

$$H_2O \text{ (liq)} \rightarrow H_2O \text{ (g)} \qquad \Delta H_{vap}^{293} = +44.10 \text{ kJ/mol} \qquad (5\text{-}1)$$

which is read as "the heat of vaporization of water at constant pressure and 293 K (20 °C) is 44.10 kJ/mol."

It always requires heat to vaporize a liquid because of the greater magnitude of the force of attraction between the molecules in the liquid state as compared to the gaseous state. Energy must be supplied to overcome the force of attraction between molecules in the liquid, to pull them apart and increase the distance between the molecules. The energy supplied increases the potential energy of the molecules. The temperature of both liquid and gas remains constant during the vaporization process as there is no change in the average kinetic energy of the molecules.

The heat of vaporization varies with temperature, although the variation is not large. If no temperature is specified, the value given for ΔH_{vap} is the value at the normal boiling point, that is, the boiling point at 1-atm pressure. For water, the normal boiling point is 100 °C (373 K), and $\Delta H_{vap}^{373} = +40.7 \text{ kJ/mol}$.

The reverse reaction of vaporization is called **condensation.** All condensations are exothermic. For water, for example,

$$H_2O(g) \rightarrow H_2O(liq) \qquad \Delta H_{cond}^{293} = -44.10 \text{ kJ/mol} \qquad (5\text{-}2)$$

When the gas molecules are so close together that the attractive forces cause a liquid to form, the energy of the system decreases, and the extra energy is released in the

Table 5.4. Correlations between the Values of the Empirical Constant *a* in van der Waals' Equation, the Depth of the Potential Well, ε, in the Potential Energy Function, the Heat of Vaporization, and the Normal Boiling Point of Some Common Substances, Showing the Effects of Increasing Force of Attraction between Molecules

Substance	$a \left(\dfrac{L^2 \, atm}{mol^2} \right)$	$\varepsilon \left(\dfrac{kJ}{mol} \right)$	$\Delta H_{vap} \left(\dfrac{kJ}{mol} \right)$	T_{bp} (K)
He	0.034	0.084	0.084	4
H_2	0.244	0.305	0.904	20
N_2	1.390	0.757	5.56	77
O_2	1.360	0.941	6.82	90
CH_4	2.253	1.14	8.16	166

form of heat. The potential energy of a liquid is much lower than that of the gas at the same temperature.

If there is a relatively large force of attraction between the molecules of a given substance, we expect a relatively large heat of vaporization. We have already discussed the fact that a larger force of attraction between molecules results in larger deviations from the ideal gas law. In addition, the larger the forces of attraction between molecules, the greater the depth of the potential well in the potential energy plot, Fig. 5.12. These trends are illustrated by the data presented in Table 5.4.

Fusion

The conversion of a solid to its liquid is called **fusion,** or melting. Heat is always required for fusion; the amount of heat that must be absorbed in order to convert one mole of solid to one mole of liquid at the melting point at a constant pressure of one atmosphere, is called the **enthalpy of fusion,** or the **heat of fusion,** and is denoted ΔH_{fus}. All fusions are endothermic; therefore ΔH_{fus} is always a positive quantity. For water, ΔH_{fus} is 6.008 kJ/mol (1.436 kcal/mol). Note that for water, ΔH_{fus} is less than ΔH_{vap} by a considerable amount (6.01 kJ/mol compared to 40.7 kJ/mol). For all substances ΔH_{fus} is smaller than ΔH_{vap}. When a solid is converted to a liquid, the average distance between molecules is increased somewhat (as evidenced by the approximately 10% decrease in density), but the change is not nearly as great as when the liquid is converted to the gas. For all substances, evaporation produces a much greater separation between molecules, and therefore a greater change in the potential energy, than does fusion.

Figure 5.14 is a schematic diagram of the relative energies of the three states of matter for a typical substance, with the magnitudes of ΔH_{fus} and ΔH_{vap} indicated. The heat of vaporization is usually from 5 to 10 times larger than the heat of fusion.

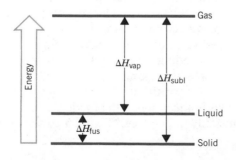

Fig. 5.14. Relative energies of the solid, liquid, and gaseous states of a typical substance, at constant pressure.

Sublimation

Sublimation is the process in which a solid is converted directly to the vapor, without passing through the liquid phase. While all solids sublime at sufficiently low pressure, only a few substances sublime at normal atmospheric pressures. The most common of these is "dry ice," which is really solid CO_2. The conversion of solid CO_2 to the gas is readily observable; this sublimation process has been used as a stage prop to make "smoke" in theatrical productions. The gaseous CO_2 subliming from a piece of solid CO_2 is colder than the air and causes water vapor in the air to condense. It is these droplets of water that appear to be smoke or haze. Other materials that sublime at 1 atm are camphor, naphthalene, and *para*-dichlorobenzene, all of which can be used as moth balls. The solid "disappears" in your closet or clothes drawer without any liquid being formed; you can smell the vapor strongly. As is apparent from Fig. 5.14, all sublimations are endothermic, and at any given temperature

$$\Delta H_{subl} = \Delta H_{fus} + \Delta H_{vap} \qquad (5\text{-}3)$$

Sublimation is put to practical use in the process known as freeze drying. Freeze-dried instant coffee is produced by freezing freshly brewed coffee and then placing the frozen material in a chamber that can be evacuated by removing the air until the pressure falls to 1 mmHg or less. At this low pressure ice sublimes, and the gaseous water molecules are removed as the evacuation continues. This method of removing the water in brewed coffee preserves the heat-sensitive molecules that give coffee its flavor. Many other foods are also freeze dried. The freeze-drying process not only preserves flavor, it also prevents spoilage, as the bacteria that might spoil the food cannot grow in the absence of water.

section 5.6
The Equilibrium between a Liquid and Its Vapor

You are certainly aware that if you leave a glass of water on the kitchen table one evening, when you look at it the next morning there will be much less water in the glass, and there may even be no water left at all. We say the water has *evaporated*, literally, become a vapor. All liquids evaporate if left in a container open to the atmosphere. Why does a liquid evaporate? The molecules of a liquid are in constant motion. Their kinetic energies range from very low to very high values, and are constantly changing as a result of collisions. There is a distribution of the speeds and of the kinetic energies of liquid molecules, just as there is a distribution of the speeds of gas molecules, as described in Section 4.9. Liquid molecules are held together by forces of attraction, but if, as a result of having a high enough kinetic energy, a molecule on the surface of a liquid moves a sufficiently large distance away from other liquid molecules, it will become a vapor molecule.

The force of attraction between molecules gets smaller very quickly as the distance between molecules increases. There is a certain minimum kinetic energy, which we may denote ε_{escape}, that is the smallest value of kinetic energy a liquid molecule must possess in order for it to overcome the binding forces that keep it in the liquid phase. If a molecule on the surface of a liquid possesses a kinetic energy greater than or equal to ε_{escape}, that molecule will be able to leave the liquid phase and become a vapor molecule. The *fraction* of molecules in the liquid with kinetic energy greater than or equal to ε_{escape} is a constant at constant temperature.* Figure 5.15 shows the distribu-

---

* It is proportional to the Boltzmann factor, $e^{-\varepsilon_{escape}/kT}$.

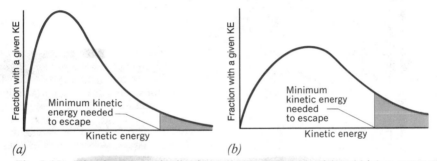

Fig. 5.15. Kinetic energy distribution at low temperature *(a)* and high temperature *(b)*. The same minimum kinetic energy is needed to escape at both temperatures, but the total fraction of molecules having at least this much energy (the shaded area) is larger at the higher temperature.

tion of kinetic energies for molecules of a typical liquid at two different temperatures. As a liquid evaporates, the size of the liquid phase decreases, but the fraction of liquid molecules that possesses sufficient kinetic energy to escape into the vapor phase remains constant, provided the temperature remains constant. If the container of liquid is open to the atmosphere, vapor molecules will move into the air above the container and will then be able to move away from the liquid surface. For this reason, evaporation will continue until no liquid is left.

The Equilibrium Vapor Pressure

Imagine a covered container, half-filled with a liquid. (See Fig. 5.16.) Molecules that leave the liquid and become vapor molecules must remain in a relatively small volume above the surface of the liquid. Since the gas molecules move about randomly, some gas molecules will strike the surface of the liquid, just as gas molecules collide with the walls of the container. Occasionally, a relatively slow-moving gas molecule strikes the surface of the liquid, is attracted by the molecules of the liquid, and joins the liquid phase. This process is called **condensation.** In a closed container, both evaporation and condensation occur simultaneously.

The number of molecules that leave the liquid (and enter the vapor phase) per unit time and per unit surface area is called the **rate of evaporation.** As long as the temperature is constant, the rate of evaporation is constant, because it depends only on the fraction of liquid molecules with kinetic energies greater than or equal to ε_{escape}. The **rate of condensation** (the number of molecules that leave the vapor phase per unit time) is proportional to the number of molecules per unit volume in the gas

Fig. 5.16. The equilibrium between liquid and vapor in a tightly covered container.

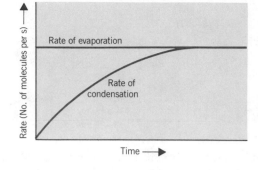

Fig. 5.17. Evaporation of a liquid in a closed container at constant temperature. The rate of evaporation is constant with time, whereas the rate of condensation increases with time until it becomes equal to the rate of evaporation. When the rate of evaporation and the rate of condensation are equal, the system is in a state of dynamic equilibrium.

phase. The number of vapor molecules striking a unit area on the surface of the liquid per unit time depends on the density of gas molecules, that is, on n/V. To the ideal gas approximation, $n/V = P/RT$, and since the temperature is constant, the rate of condensation is directly proportional to the pressure of the vapor.

Let us assume that time "zero" is the instant we cover the container, and that at that instant there are no molecules of the liquid species in the vapor phase. The initial rate of condensation is therefore zero. Since liquid is evaporating, but none is condensing, the number of molecules in the vapor phase begins to increase, and n/V also increases since V is constant. Thus the rate of condensation increases as time passes. If the rate of evaporation remains constant, but the rate of condensation increases continually from its initial value of zero, it is clear that eventually the vapor pressure will reach a value such that the rate of condensation is equal to the rate of evaporation. The time dependence of the rates of condensation and evaporation are depicted in Fig. 5.17.

Once the rates of evaporation and condensation are equal, both evaporation and condensation continue to occur, but since they go on at the same rate, the number of molecules entering and leaving the vapor phase is the same. The value of n/V in the gas phase, and therefore the value of the vapor pressure, P, will remain constant as long as the temperature is constant. The system has reached a **state of equilibrium.** It is a **dynamic equilibrium,** because both evaporation and condensation continue to occur, but the vapor pressure at equilibrium remains constant. The **equilibrium vapor pressure** is a function of temperature. For example, for water at 25 °C, the equilibrium vapor pressure is 23.76 mmHg, while at 40 °C it is 55.32 mmHg. For all liquids the equilibrium vapor pressure increases as the temperature increases.

The magnitude of the equilibrium vapor pressure is independent of the size or surface area of the liquid phase, as long as there is *some* liquid present. If there is an enclosed volume above the liquid, the vapor pressure of H_2O at 25 °C is 23.76 mmHg, regardless of whether we have a bathtub full of water or a teacup of water. Once equilibrium has been attained, the vapor pressure is constant at constant temperature. If we make a change in the volume of the vapor phase that momentarily changes the pressure of the vapor, the relative amounts of substance in the two phases will adjust in order to maintain constant vapor pressure.

Consider a system consisting of a liquid and its vapor contained in a cylinder enclosed by a piston that can move up or down, as shown in Fig. 5.18(a). Let us start from a state of equilibrium at constant temperature, T. Consider what happens if we suddenly move the piston upward. We have increased the volume of the gas phase. The pressure of the gas decreases, and since the rate of condensation depends directly on the pressure of the gas, the rate of condensation decreases as well. The system is no longer at equilibrium, because the rate of evaporation remains constant as long as the

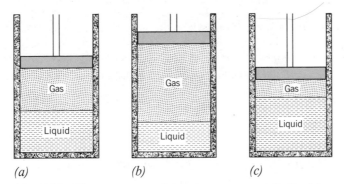

(a)   *(b)*   *(c)*

Fig. 5.18. The constancy of the vapor pressure of a liquid at constant temperature. *(a)* Original position of the piston, when the system is at equilibrium. *(b)* Equilibrium restored after piston is moved up. *(c)* Equilibrium restored after piston is moved down. See text.

temperature is constant. Thus immediately after the piston is pulled up, the rate of condensation is *less* than the rate of evaporation. Hence more molecules enter the gas phase than leave it, and as a result, the pressure of the vapor increases. After a short period of time, the vapor pressure will once again reach its equilibrium value. When equilibrium is restored, less of the substance will be in the liquid phase, and more will be in the gas phase, but the vapor pressure will be exactly what it was before. This is depicted in Fig. 5.18*(b)*.

Now consider what happens if, starting from our original equilibrium in Fig. 5.18*(a)*, we push the piston down and keep it there. Since we have decreased the volume of the gas, its pressure increases, and momentarily the system is not at equilibrium because the rate of condensation is greater than the rate of evaporation. More molecules leave the gas phase than enter it, and as a result the pressure decreases. After a short period of time, the vapor pressure will once again reach its equilibrium value. When equilibrium is restored, the liquid phase will be larger in size than it was originally, and the gas phase will be smaller in size than it was originally, but the vapor pressure will be exactly what it was before. Note that whatever change is made, the end result is only a change in the relative sizes of the liquid and vapor phases. Equilibrium is always restored, and the equilibrium vapor pressure remains the same as long as the temperature is constant.

EXAMPLE 5.1. The equilibrium vapor pressure

In an experiment using the apparatus shown in Fig. 5.18, the liquid phase consists of 24.00 g of water, and the vapor phase has a volume of 400.0 mL when the system is at equilibrium at 40.0 °C. The vapor pressure of water at 40.0 °C is 55.32 mmHg. If the piston is raised so that the volume of the gas is increased to 800.0 mL, how much water will be in the liquid phase when equilibrium is restored?

Solution. First calculate the number of moles of water in the gas phase when the system is at equilibrium at 40.0 °C and the gas volume is 400.0 mL. Assume the vapor is ideal at this low pressure.

$$n_{gas} = \frac{(55.32 \text{ mmHg}/760.0 \text{ mmHg} \cdot \text{atm}^{-1}) (0.4000 \text{ L})}{(8.2057 \times 10^{-2} \text{ L} \cdot \text{atm/mol} \cdot \text{K})(313.2 \text{ K})} = 1.133 \times 10^{-3} \text{ mol}$$

If the gas volume is doubled to 800.0 mL, when equilibrium is restored the number of moles of H_2O in the gas phase must be twice as large, as the pressure and temperature remain constant. Thus when the volume of the gas phase is 800.0 mL, $n_{gas} = 2.266 \times 10^{-3}$ mol. The increase in the number of moles of H_2O in the vapor phase is 1.133×10^{-3} mol, or 0.02041 g of H_2O, since the molecular weight of H_2O is 18.015 g/mol. Thus 0.02041 g of H_2O have left the liquid phase and entered the vapor phase. The number of grams of H_2O in the liquid phase is now 23.98 g.

section 5.7
The Nature of the Equilibrium State: The Tendency to Decrease the Potential Energy and to Increase Molecular Disorder

Two questions of great interest to chemists are these: Why does a reaction proceed in the way it does? What determines the direction of chemical reactions? We can try to obtain an understanding of the factors that drive reactions by thinking about processes that occur spontaneously.

There are a great many processes that proceed spontaneously from a state of high potential energy to a state of low potential energy: water runs downhill, weights fall down. We have already noted that for any given substance, the solid represents a state of lower potential energy than the liquid, and the liquid is a state of lower potential energy than the gas. (See Fig. 5.14.) We therefore conclude that the tendency to seek a state of lower potential energy cannot be the *only* driving force for reactions, or everything in the world would be solid, which is obviously not the case.

Let us think about another spontaneous process that we have already discussed in Section 4.2: The diffusion of gases. If we start with two different gases, each in a separate bulb, as in Fig. 5.19*(a)*, at the same temperature and pressure, and then open the stopcock between the bulbs, we will always find that the system spontaneously moves toward the more disordered state, in which both gases occupy both bulbs, as in Fig. 5.19*(b)*.

By considering a great many spontaneous processes, we observe that there is a tendency for systems to proceed to a state in which there is greater disorder or chaos on the molecular level, as well as a tendency to proceed to a state of lower potential energy. By comparing the gaseous, liquid, and solid states, we see that both potential energy and molecular disorder are lowest for the solid, and highest for the gaseous state. A summary is given in Table 5.5.

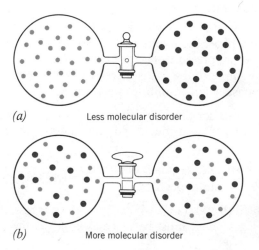

(a)   Less molecular disorder

(b)   More molecular disorder

Fig. 5.19. The spontaneous diffusion of gases at the same temperature and pressure. *(a)* Before opening stopcock. *(b)* After opening stopcock. The direction of spontaneous change is toward a state of greater molecular disorder.

Table 5.5. **Potential Energy and Molecular Disorder for the Solid, Liquid, and Gaseous States**

	Solid	*Liquid*	*Gas*
Potential energy	Least	Intermediate	Greatest
Molecular disorder	Least	Intermediate	Greatest

The tendency to achieve maximum molecular disorder favors the vapor phase, while the tendency to achieve the lowest potential energy favors the solid. Each substance, at a given temperature, finds a compromise between these two tendencies. The best compromise at any given temperature is the equilibrium state. Equilibrium is the condition of most favorable compromise between the natural tendencies to reach minimum potential energy and maximum molecular disorder. That compromise depends strongly on the temperature. For all substances, if the temperature is sufficiently high, the vapor phase is the stable state. We see therefore that as the temperature increases, the tendency to achieve maximum molecular disorder becomes increasingly dominant in determining the equilibrium state.

There is a property of every substance and of every system that is a measure of the amount of molecular disorder. That property is called the **entropy** and is denoted by the letter S. When there is any sort of reaction or change, the entropy may change. We define ΔS as the change in the entropy. For instance, if we consider the change liquid $\rightarrow$ vapor, we define ΔS_{vap} as the **entropy change of vaporization**:

$$\Delta S_{vap} = S_{gas} - S_{liq} \tag{5-4}$$

ΔS_{vap} is always positive because the gas is more disordered than the liquid, so that $S_{gas} > S_{liq}$.

Entropy will be discussed more fully in Chapter 18. What it is important to understand now is simply that one of the tendencies of all naturally occurring processes is to achieve maximum molecular disorder, that is, maximum entropy.

section 5.8
A Summary of Properties of the Equilibrium State

1. The direction of all spontaneous processes is towards a state of equilibrium.
2. Equilibrium in molecular systems is **dynamic**, that is, opposing reactions are occurring simultaneously at equal rates. In the case of the liquid $\rightleftharpoons$ vapor equilibrium, the opposing reactions are evaporation and condensation. We will discuss many examples of dynamic equilibria in the chapters that follow.
3. If a system is at equilibrium and something occurs to disturb it, a change will spontaneously occur to restore equilibrium. The direction of the change will be to oppose the disturbance, in order to return to a position of equilibrium. In Section 5.6 we have discussed how the vapor pressure of a liquid returns to its equilibrium value after a disturbance. (See Fig. 5.18.) Although we have only looked at a single example, the phenomenon is a general one, and the statement given is one form of **Le Chatelier's Principle**, which will be discussed in greater detail in Section 8.4.
4. The condition of equilibrium at a given temperature is a compromise between the tendency to minimize the energy and the tendency to maximize the entropy (the molecular disorder).

section 5.9
The Temperature Dependence of the Equilibrium Vapor Pressure and the Normal Boiling Point

For every liquid, the vapor pressure increases as the temperature increases. Data for two common liquids, water and chloroform, $CHCl_3$, are given in Table 5.6, and plotted in Fig. 5.20. Chloroform is a widely used organic solvent.

At any given temperature the vapor pressure of $CHCl_3$ is higher than that of H_2O. The force of attraction between $CHCl_3$ molecules is less than the force of attraction between water molecules, due to the extensive hydrogen bonding in liquid water, as discussed in Section 5.2. As a result, $CHCl_3$ is a more **volatile liquid** than H_2O. If you open a bottle of $CHCl_3$, within a few seconds you can detect its sweet musty odor in the air. Chloroform, like other halogenated carbon compounds such as CCl_4 and CH_2Cl_2, is toxic (it damages the liver). Breathing these compounds should be avoided.

A method of measuring the vapor pressure of a liquid is illustrated in Fig. 5.21.

The Normal Boiling Point

The vapor pressure (vp) of every liquid increases regularly as the temperature increases. There is, therefore, one temperature at which the equilibrium vapor pressure for a given liquid is 1 atm (760 mmHg). That temperature is, by definition, the **normal boiling point.** We can see from Table 5.6 that the normal boiling point of $CHCl_3$ is 61.2 °C, while for H_2O it is 100.0 °C. More volatile liquids boil at lower temperatures. The normal boiling point of each of these liquids is indicated on Fig. 5.20 as the intersection of the vapor pressure plot with the vp = 760-mmHg line.

As a liquid boils, bubbles of vapor form throughout the liquid. At the normal boiling point, the pressure of the gas in the bubbles is equal to the external pressure, and the bubbles rise to the surface and escape into the atmosphere. The formation of a bubble in the body of a liquid is a difficult process because it is necessary for many

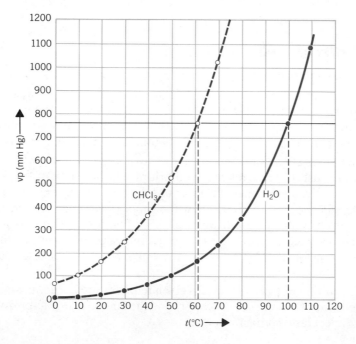

Fig. 5.20. The temperature dependence of the vapor pressure of $CHCl_3$ and H_2O.

Table 5.6. **The Vapor Pressures of H_2O and $CHCl_3$ as a Function of Temperature**

Temperature (°C)	$H_2O(liq) \rightleftharpoons H_2O(vap)$ Vapor Pressure (mmHg)	$CHCl_3(liq) \rightleftharpoons CHCl_3(vap)$ Vapor Pressure (mmHg)
0	4.58	61.0
10	9.21	100.5
20	17.54	159.6
30	31.82	246.0
40	55.32	366.4
50	92.51	526.0
61.2	157.8	760.0
70	233.7	1019
80	355.1	1403
100	760.0	2244
110	1074.6	· · ·

molecules with kinetic energies greater than or equal to ε_{escape} to be very close to one another. It may sometimes happen, therefore, that a liquid is heated to its boiling point, and yet no bubbles form. The continued addition of heat to such a liquid causes it to become **superheated,** that is, the temperature of the liquid will be higher than its normal boiling point.

A superheated liquid is in an unstable state and cannot persist for very long. If heat continues to be added, eventually a bubble will form, and when it does the vapor pressure in the bubble is greater than atmospheric pressure, and the bubble expands rapidly, sometimes explosively, spattering hot liquid. This occurrence is called **bumping.** When a liquid is to be heated to the boiling point, precautions should be taken to avoid bumping, both because it is dangerous to spatter hot liquids and also to prevent the loss of the substance being boiled. Small glass beads, porous pieces of ceramic material, and carborundum bits can all be used as **boiling chips,** to promote bubble formation and thereby prevent bumping.

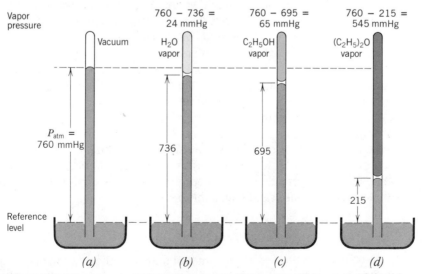

Fig. 5.21. Measurement of vapor pressure. When a small amount of liquid is introduced above the mercury in a barometer, the vapor pressure of the liquid forces the mercury down. *(a)* No liquid above the mercury. *(b)* H_2O. *(c)* Ethyl alcohol, C_2H_5OH. *(d)* Diethyl ether, $(C_2H_5)_2O$.

EXAMPLE 5.2. Le Chatelier's Principle and the temperature dependence of the equilibrium vapor pressure

Explain why the equilibrium vapor pressure increases with increasing temperature, in terms of Le Chatelier's Principle.

Solution. The conversion of liquid to vapor is an endothermic process and may be written as

$$\text{liquid} + \text{heat} \rightleftharpoons \text{gas}$$

If heat is added to a liquid that is in equilibrium with its vapor, the added heat constitutes a stress or a disturbance of the equilibrium state. A change will occur to oppose that disturbance. The stress of added heat is opposed by using up some of the heat to convert liquid into gas, and therefore the system moves to a state with higher vapor pressure. Equilibrium is restored, but at the higher temperature both the rate of evaporation and the rate of condensation are larger than at the lower temperature.

section 5.10
Phase Diagrams for Pure Substances

The Vapor Pressure of Solids

The vaporization of a solid is called sublimation. Each solid has an equilibrium vapor pressure at a given temperature. The vaporization of any solid is endothermic:

$$\text{solid} + \text{heat} \rightleftharpoons \text{vapor} \tag{5-5}$$

and therefore the vapor pressure of a solid increases with increasing temperature. The argument, in terms of Le Chatelier's principle, is the same as that given in Example 5.2. As the temperature is increased, eventually a temperature is attained at which the solid melts. If the temperature is increased above this melting point, there will be equilibrium between the liquid and vapor. The vapor pressure of the solid increases with increasing temperature faster than the vapor pressure of the liquid increases with increasing temperature, so the solid–vapor curve has a steeper slope than the liquid–vapor curve, and the two curves intersect.

The temperature at which the vapor pressure curve of the solid intersects the vapor pressure curve of the liquid is called the **triple point.** Figure 5.22 shows the temperature dependence of the vapor pressures of both solid and liquid for a typical substance. At the triple point, all three phases, solid, liquid, and gas, are in equilibrium. For water, the triple point occurs at a temperature of 0.0098 °C and the vapor pressure of both the solid and the liquid is 4.58 mmHg. For CO_2, the triple point occurs at −57 °C (216 K) and a pressure of 5.2 atm. At pressures below 5.2 atm only the solid and vapor phases are in equilibrium; it is for this reason that at normal atmospheric pressure solid CO_2 sublimes.

One-Component Phase Diagrams

A **one-component phase diagram** is a plot that gives the relation between pressure and temperature for the equilibria between the gaseous, liquid, and crystalline phases of a single pure substance. One more curve must be added to Fig. 5.22 to make it a phase diagram, the curve that gives the temperatures and pressures at which the solid and liquid phases are in equilibrium. The phase diagram for CO_2 is given in Fig. 5.23.

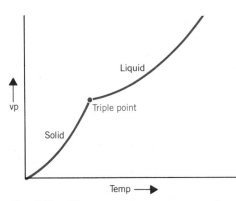

Fig. 5.22. The vapor pressure curves of both solid and liquid for a typical substance.

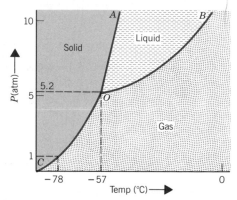

Fig. 5.23. Phase diagram for CO_2. Point O is the triple point, at which solid, liquid, and gas are in equilibrium.

Point O is the triple point. Curve OB is the vapor pressure curve for liquid CO_2. At any point along the curve OB, liquid CO_2 and gaseous CO_2 will be in equilibrium. Curve OA is the solid–liquid equilibrium curve. At any point along OA, the solid and liquid phases are in equilibrium. In the region between curves OA and OB, the liquid phase is the stable phase. The curve OC is the vapor pressure curve of the solid.

A phase diagram tells us what the stable phase is at any given temperature and pressure, and also enables us to predict phase changes that will occur as the pressure and/or temperature are changed. This is illustrated in Example 5.3.

EXAMPLE 5.3. Utilizing the phase diagram for CO_2

(a) What is the stable phase of CO_2 (1) At 4 atm and -75 °C? (2) At 2 atm and -10 °C?

Solution. (1) The point $P = 4.0$ atm and $t = -75$ °C lies in the region to the left of curve OC; in that region the solid phase is the stable phase. (2) The point $P = 2.0$ atm and $t = -10$ °C lies to the right of curve OC and below curve OB. In that region, the gas phase is the stable phase.

(b) Describe the phase change that occurs if CO_2 at 1.00-atm pressure is heated from -100 °C to room temperature.

Solution. At -100 °C and 1 atm, solid CO_2 is stable. If we keep the pressure constant at 1 atm and increase the temperature, we proceed along the horizontal dashed line at $P = 1$ atm. From -100 to -78 °C, the solid phase is stable. At -78 °C, the vapor forms, and solid CO_2 sublimes. Above -78 °C the gas phase is stable. At a pressure of 1 atm, no liquid CO_2 will form at any temperature.

The phase diagram of CO_2 is similar to that of the great majority of substances, in that the solid–liquid equilibrium curve, OA, has a positive slope. That means that as the temperature increases, the pressure at which the solid and liquid are in equilibrium also increases.

There are a very few substances for which the solid–liquid equilibrium curve has a negative slope. The most notable example is water. The phase diagram for water is shown in Fig. 5.24. The negative slope of the curve OA for water means that as the temperature increases, the pressure at which solid and liquid are in equilibrium

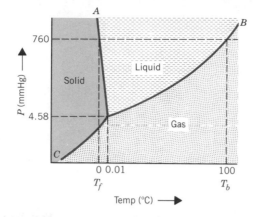

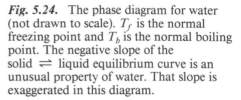

Fig. 5.24. The phase diagram for water (not drawn to scale). T_f is the normal freezing point and T_b is the normal boiling point. The negative slope of the solid $\rightleftharpoons$ liquid equilibrium curve is an unusual property of water. That slope is exaggerated in this diagram.

decreases. The unusual slope of this curve for water is due to the fact that ice has a lower density than liquid water, as discussed in Section 5.2. Increasing pressure on a system at equilibrium favors a shift to the denser phase. For most substances, the solid phase is denser than the liquid, and increasing pressure at constant temperature causes the liquid to be converted to solid, as is the case for CO_2. For water, because of the extensive hydrogen bonding of the solid phase (see Fig. 5.10), the liquid is denser than the solid. Increasing the pressure at constant temperature causes ice to be converted to liquid water.

EXAMPLE 5.4. Le Chatelier's Principle and the lowering of the melting point of ice with increasing pressure

The normal melting point of ice, at 1-atm pressure, is 0.0 °C. If the pressure is increased to 170 atm, the melting point decreases to −1.3 °C. Explain this phenomenon in terms of Le Chatelier's principle.

Solution. The density of liquid water is 1.00 g/cm³ at 0 °C, while the density of ice is 0.917 g/cm³. If we begin with a mixture of ice and liquid water in equilibrium at 1.00 atm, and then increase the pressure to 170 atm, the increase in pressure constitutes a stress, or a disturbance of the equilibrium. To relieve the stress of added pressure, the system shrinks in volume by shifting to the phase in which the molecules are more closely packed, that is, the phase with higher density. In this case, ice is converted to liquid water. At 170 atm and 0 °C only liquid water is stable. The conversion of ice to liquid water, however, is an endothermic process. If the ice and water are insulated so that heat does not enter from the surroundings, the heat required to melt the ice comes from the kinetic energy of the molecules, which therefore decreases. The molecules move more slowly, and when the system is at equilibrium at 170 atm, the temperature has dropped to −1.3 °C.

Critical Temperature and Pressure

We have previously stated that any gas can be liquefied by decreasing its temperature and increasing its pressure. Decreasing the temperature decreases the average kinetic energy of the molecules, and increasing the pressure decreases the average distance between molecules. When the molecules are close together, if their kinetic energy is lowered, they do not possess enough energy to overcome the forces of attraction between molecules, and the liquid forms. However, for each substance, there exists a temperature, called the **critical temperature,** above which the substance cannot be

Table 5.7. **The Critical Temperatures and Pressures of Some Common Substances**

Compound	Critical Temperature (°C)	Critical Pressure (atm)
Ammonia, NH_3	132.2	111.5
Argon, Ar	−122.5	48.0
Carbon dioxide, CO_2	31	72.8
Hydrogen, H_2	−239.9	12.8
Methane, CH_4	−82.1	45.8
Nitrogen, N_2	−147	33.5
Oxygen, O_2	−118.4	50.1
Sulfur dioxide, SO_2	157.8	77.7
Water, H_2O	374.1	218.3

liquefied, no matter how great the applied pressure. At temperatures above the critical temperature, the average kinetic energy of the gaseous molecules is sufficiently high to overcome the attractive forces between molecules, and a liquid will not form, no matter how much pressure is applied.

The minimum pressure required to liquefy a gas at its critical temperature is called the **critical pressure.** Table 5.7 lists the critical temperatures and pressures of some common substances.

Substances for which the intermolecular forces are large, such as the polar molecules NH_3, SO_2, and H_2O, have high critical temperatures. Substances for which the intermolecular forces of attraction are small, such as Ar, H_2, N_2, and O_2, have low critical temperatures.

The curve in the phase diagram that gives the temperatures and pressures at which the liquid and vapor phases are in equilibrium comes to an end at the critical temperature and pressure. Above that point, called the **critical point,** there is only a single fluid phase, and it is impossible to distinguish liquid and vapor. A phase diagram for a typical substance showing the critical point is shown in Fig. 5.25.

Summary

The **intermolecular forces** of attraction between molecules are responsible for the formation of the **condensed phases,** solid and liquid. The stronger the intermolecular forces, the higher the melting and boiling points of a substance.

Weak attractive forces between uncharged atoms or molecules are known as **van**

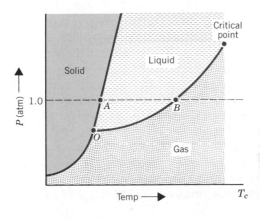

Fig. 5.25. Phase diagram of a typical substance. Point *O* is the triple point. Point *A* is the normal melting point of the solid, and point *B* is the normal boiling point of the liquid. The coordinates of the critical point are the critical pressure, P_c, and the critical temperature, T_c.

der Waals forces. The weakest type of van der Waals force is the **London** or **dispersion force.** The magnitude of the London force increases with an increase in the number of electrons, and therefore substances of higher molecular weight generally have higher melting and boiling points. Many molecules are **polar** and possess a permanent **dipole moment,** because the center of positive charge in the molecule does not coincide with the center of negative charge. If two substances have approximately the same molecular weight, and one is polar while the other is not, the polar substance will have the higher melting and boiling point.

Van der Waals forces are short range, that is, they do not become significantly large until the distance between molecules is very small. As two molecules come close together and the force of attraction increases, the potential energy of the two molecules decreases. If, however, the molecules get close enough so that their electron charge clouds begin to overlap, a strong repulsion sets in and the potential energy increases very steeply.

A very strong type of intermolecular force is the **hydrogen bond** that is formed when a hydrogen atom is situated between two highly electronegative atoms, such as O, F, N, or Cl. Hydrogen bonding results in unusually high melting and boiling points.

The conversion of a solid to its liquid is called **fusion** or **melting.** The conversion of a liquid to its vapor is **vaporization.** All fusions and vaporizations are endothermic. The amount of heat required to convert one mole of liquid to one mole of vapor, the **heat of vaporization,** is always considerably larger than the amount of heat required to convert one mole of solid to one mole of liquid, the **heat of fusion.** The transition from liquid to vapor results in a much greater increase in the average distance between molecules than does the transition from solid to liquid. At certain temperatures and pressures, a solid can be converted directly to the vapor without passing through the liquid phase. This process is called **sublimation.**

The **rate of evaporation** of a given liquid depends only on the temperature, and hence is a constant at constant temperature. The **rate of condensation** depends on the pressure of that substance in the gas phase. At a given temperature, there is only one value of the vapor pressure at which the rate of condensation is equal to the rate of evaporation, and that value is called the **equilibrium vapor pressure.**

The equilibrium vapor pressure is a constant at constant temperature. If we make a change in the volume of the vapor phase, the relative amounts of liquid and vapor will change to maintain constant vapor pressure. As the temperature increases, the equilibrium vapor pressure of every liquid increases. For each liquid there is one temperature, called the **normal boiling point,** at which the equilibrium vapor pressure is exactly 1 atm.

All processes, both chemical and physical, proceed spontaneously toward a state of equilibrium. There are two factors that cause reactions to proceed spontaneously: (a) the tendency to achieve minimum potential energy, and (b) the tendency to achieve maximum molecular disorder. The **entropy** is a measure of the amount of molecular disorder. All equilibria in molecular systems are **dynamic,** that is, opposing reactions occur simultaneously at the same rate. If a system is at equilibrium and a disturbance occurs, a change will spontaneously occur in a direction to oppose the disturbance and restore equilibrium. This description of what is observed for all dynamic equilibria is known as **Le Chatelier's Principle.**

A **phase diagram** for a pure substance is a plot that shows the relation between pressure and temperature for the equilibria between gaseous, liquid, and solid phases of that substance. The intersection of the vapor pressure curves of solid and liquid is known as the **triple point,** a unique pressure and temperature at which the solid, liquid, and gas phases of a single substance are all in equilibrium. The curve showing the temperatures and pressures at which liquid and vapor are in equilibrium ends at the **critical point**. At temperatures above the **critical temperature** it is impossible to liquefy a gas, no matter how great the applied pressure.

Exercises

Section 5.1

1. Account for the trends observed for the melting and boiling points of the halogens:

Halogen	*mp (°C)*	*bp (°C)*
Fluorine, F_2	−220	−188
Chlorine, Cl_2	−101	−35
Bromine, Br_2	−7	+59
Iodine, I_2	+113.5	+184

2. Account for the fact that silicon tetrafluoride, SiF_4, is nonpolar even though fluorine is much more electronegative than silicon.

3. Carbon tetrafluoride, CF_4, and phosphorus trifluoride, PF_3, have approximately the same molecular weight, $88 \text{ g} \cdot \text{mol}^{-1}$. The melting and boiling points of the two compounds are

Compound	*mp (°C)*	*bp (°C)*
CF_4	−184	−128
PF_3	−151.5	−101.5

Explain why CF_4 has a lower melting and boiling point than PF_3.

4. Silicon tetrachloride, $SiCl_4$, has a zero dipole moment, yet its boiling point (57.6 °C) is considerably higher than that of chlorosilane, SiH_3Cl (−30.4 °C) which has a dipole moment of 1.31 D. Explain why the polar molecule, SiH_3Cl, is lower boiling than the nonpolar $SiCl_4$.

5. Carbon dioxide, CO_2, has a zero dipole moment. Is the geometry of CO_2 "*V* shaped" (bent) or linear? Explain the reasons for your answer.

Section 5.2

6. A garden hose with water in it was left outdoors throughout the winter. In the spring the hose was found to have many cracks. Explain in detail why the hose was cracked.

7. Hydrogen bonding can occur in pure liquids of which of the following compounds? Explain your answers.

    (a) NH_3 (b) LiF   (c) HBr   (d) CH_3OH (e) LiH
    (f) F_2O (g) SCl_2 (h) CH_4 (i) CH_3CF_3 (j) CH_3COOH

8. Dimethyl ether and ethanol are isomers. They have the same numbers and kinds of atoms, and therefore identical molecular weights, $46.07 \text{ g} \cdot \text{mol}^{-1}$. The structures of these two molecules and their boiling points are shown below:

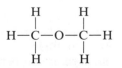

Dimethyl ether, bp −23 °C

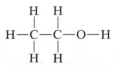

Ethanol, bp 78.5 °C

Explain why these two compounds have such different boiling points.

9. Arrange the following substances in order of increasing boiling points. Explain the reasons for the order you chose.

(a) F_2 (b) NaF   (c) HCl   (d) N_2 (e) Ne   (f) H_2O

Section 5.4

10. Account for the fact that liquids are virtually incompressible while gases are easily compressed.

11. At 25 °C and 1.0 atm, a certain substance has a density of $1.7 \, g \cdot L^{-1}$. At this temperature and pressure is the substance a solid, liquid, or a gas? Explain your answer.

Section 5.5

12. At 0 °C the heat of vaporization of liquid water is $44.86 \, kJ \cdot mol^{-1}$. Using this plus other data given in this section, calculate the heat of sublimation of ice at 0 °C.

13. Arrange the hydrogen halides (HF, HCl, HBr, and HI) in order of increasing heat of vaporization. Explain the reason for the order you chose.

14. Account for the fact that the heat of vaporization of H_2O decreases from 44.10 kJ/mol at 20 °C to 40.65 kJ/mol at 100 °C.

15. The heat of fusion of a solid is sometimes called the *latent* heat of fusion. The word latent (hidden) is used to emphasize that as heat is added to a solid at its melting point, the temperature remains constant. What happens to the heat that is added, since it does not result in an increase in temperature?

16. Explain why the heat of vaporization is usually from 5 to 10 times larger than the heat of fusion for the same substance.

17. The molar heats of vaporization and the boiling points of a number of common substances are given below:

Compound	ΔH_{vap} (kJ/mol)	bp (°C)
Methane, CH_4	8.907	−164
Ammonia, NH_3	23.6	−33
Water, H_2O	40.7	100
Hydrogen fluoride, HF	30.17	19
Hydrogen chloride, HCl	15.06	−85
Methanol, CH_3OH	39.23	65
Acetone, CH_3COCH_3	31.97	56.2
Formamide, $HCONH_2$	65.09	193
Benzene, C_6H_6	30.8	80.2

(a) For each of these substances, calculate the heat of vaporization per gram. Tabulate your results.

(b) Which of these substances cannot be hydrogen bonded in the liquid phase? Compare the heat of vaporization per gram of the hydrogen-bonded liquids with that of the nonhydrogen-bonded liquids.

18. The reverse of fusion is freezing. Is freezing an endothermic or an exothermic reaction? Explain.

Section 5.6

19. Naturally occurring iodine consists only of the single isotope ^{127}I, but a radioactive isotope, ^{131}I, is commercially available. Some methyl iodide, CH_3I, is prepared with radioactive iodine. Methyl iodide is a liquid at room temperature and normal atmospheric pressure.

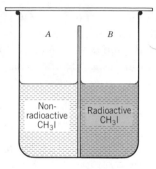

In an experiment using the apparatus shown above, one compartment *(A)* is half-filled with ordinary CH_3I, while the other compartment *(B)* is half-filled with radioactive CH_3I. The apparatus is tightly covered. The level of liquid in the two compartments remains constant with time. After many hours, the liquid in compartment *(A)* is removed, examined, and found to be radioactive. Explain in detail how the liquid in compartment *(A)* becomes radioactive even though the amount of liquid in each compartment remains constant. Discuss how this experiment proves that the equilibrium between liquid and vapor is dynamic and not static.

20. Two 500.0-mL bulbs are connected by a stopcock. One bulb is evacuated. The other bulb contains 100.0 mL of liquid chloroform in equilibrium with its vapor. Both bulbs are maintained at 20.0 °C. The vapor pressure of $CHCl_3$ at 20.0 °C is 159.6 mmHg. The density of liquid $CHCl_3$ is 1.4832 g/mL at 20.0 °C. The stopcock between the two bulbs is opened. What volume of liquid $CHCl_3$ remains in the bulb when equilibrium is restored? Neglect the volume decrease of the liquid in calculating the volume of the gas phase after the stopcock is opened.

Section 5.7

21. When the reaction

$$NH_4Cl(s) \rightarrow NH_3(g) + HCl(g)$$

occurs, does the entropy increase, decrease, or remain the same? Explain.

22. The entropy of fusion, ΔS_{fus}, is defined as the entropy change for the reaction

$$\text{solid} \rightarrow \text{liquid}$$

(a) Is ΔS_{fus} for all substances positive, negative, or is it impossible to predict unless the substance is known? Explain.

(b) For a given substance, which do you expect to be larger, ΔS_{fus} or ΔS_{vap}? Explain your answer.

Section 5.9

23. Using the plot in Fig. 5.20, predict at what temperature $CHCl_3$ will boil if the external pressure is reduced to 450 mmHg.

24. Some campers on the top of a mountain in the Rockies find that the temperature of boiling water at their campsite is only 94 °C. What is the atmospheric pressure at this altitude? Refer to Fig. 5.20.

25. Iron pentacarbonyl, $Fe(CO)_5$, and nickel tetracarbonyl, $Ni(CO)_4$, are both liquids at room temperature, but the vapor pressure of $Fe(CO)_5$ is slightly less than 40 mmHg, while that of $Ni(CO)_4$ is 400 mmHg. What factors can account for the difference in volatility of these two liquids?

26. If a sample of liquid acetone, CH_3COCH_3, is placed on your left forearm, while a sample of liquid water is placed on your right forearm, the acetone will feel considerably cooler, even though both liquids are at the same temperature as the room. Why does the arm on which the acetone is placed feel cooler?

Section 5.10

27. Using the phase diagram for CO_2, Fig. 5.23, describe the phase change(s) that will occur if

    (a) CO_2 at 8 atm and -60 °C is heated to 10 °C, keeping the pressure constant.
    (b) The pressure on a sample of CO_2 maintained at -20 °C is increased from 1 to 10 atm.
    (c) The pressure on a sample of CO_2 maintained at -60 °C is increased from 1 to 10 atm.

28. A certain compound has a normal melting point of 41 °C and a normal boiling point of 123 °C. The triple point of this substance occurs at 39 °C and 85 mmHg.

    (a) Sketch the phase diagram for this compound. Label the solid, liquid, and vapor regions. Label the axes.
    (b) Does the liquid phase of this compound have a density greater than, less than, or equal to the density of the solid phase? Explain your answer in terms of the phase diagram drawn in part (a).

29. Complete the following table by giving the stable phase of CO_2 at each of the specified temperatures and pressures. Refer to Fig. 5.23.

t (°C)	P (atm)	Stable Phase
-90	1.0	SOLID
-65	1.0	GAS
-65	10.0	SOLID
-35	2.0	GAS.
-35	8.0	LIQUID
-20	10.0	LIQUID

30. Pure sulfur has two different crystalline phases, rhombic and monoclinic. These phases differ in the geometric arrangements of the sulfur atoms in the crystal structure. The stable phase at room temperature and 1 atm is rhombic sulfur.
    A sketch of the phase diagram is shown below.

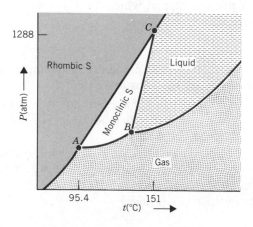

(a) The points A, B, and C are all triple points, that is, three different phases are in equilibrium at each of these points. Name the phases in equilibrium at points A, B, and C.

(b) Which crystalline phase is denser, monoclinic or rhombic? Explain your answer.

Multiple Choice Questions

1. On the basis of the relative strengths of intermolecular forces, one can predict that the order of decreasing boiling points of the following three substances is
   (a) $CH_3OH > CH_4 > H_2$ (b) $CH_3OH > H_2 > CH_4$ (c) $CH_4 > CH_3OH > H_2$
   (d) $CH_4 > H_2 > CH_3OH$ (e) $H_2 > CH_4 > CH_3OH$

2. A certain solid has a density of 4.0 g/cm^3. An educated guess of the value of the density of the liquid form of this substance would be
   (a) 4.0 g/cm^3 (b) 3.6 g/cm^3 (c) 1.8 g/cm^3 (d) 0.4 g/cm^3 (e) 4.0 g/L

3. Which of the following substances is most likely to exist as a gas at 25 °C and 1-atm pressure?
   (a) MgO   (b) $C_{10}H_{22}$ (c) B_2H_6 (d) AsI_3 (e) LiF

4. Which of the following oxides has the highest melting point?
   (a) MgO   (b) CO   (c) B_2O_3 (d) ScO_2 (e) N_2O_5

Questions 5–7 Refer to the Phase Diagram Shown Below:

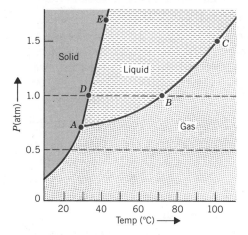

5. Which point is the normal boiling point?
   (a) A (b) B (c) C (d) D (e) E

6. If the temperature increases from 0 to 50 °C at a constant pressure of 0.5 atm, which of the following processes occurs?
   (a) sublimation   (b) freezing   (c) vaporization   (d) fusion   (e) condensation

7. If the temperature increases from 0 to 50 °C at a constant pressure of 1.0 atm, which of the following processes occurs?
   (a) sublimation   (b) fusion   (c) vaporization   (d) freezing   (e) condensation

8. Which of the following statements about the equilibrium vapor pressure of a pure liquid is TRUE?
   The equilibrium vapor pressure of a pure liquid

(a) Remains constant with increasing temperature.
(b) Decreases to half its original value if the volume of the gas phase is doubled.
(c) Increases to twice its original value if the volume of the liquid phase is doubled.
(d) Decreases to half its original value if the surface area of the liquid is reduced by one half.
(e) Is independent of the volume of the vapor phase.

9. Which of the following diatomic gases has a boiling point very close to the boiling point of the rare gas Ar?

(a) NO   (b) Cl_2 (c) F_2 (d) H_2 (e) HCl

10. The vapor pressure of a given liquid will decrease if

(a) The liquid is moved to a container in which its surface area is very much smaller.
(b) The volume of liquid in the container is decreased.
(c) The volume of the vapor phase is increased.
(d) The temperature is decreased.
(e) The number of moles of liquid is decreased.

11. Bromine, Br_2, boils at 58.8 °C, while iodine monochloride, ICl, boils at 97.4 °C. The principal reason ICl boils almost 40° higher than Br_2 is that

(a) The molecular weight of ICl is 162.4 while that of Br_2 is 159.8.
(b) ICl is an ionic compound, while Br_2 is molecular.
(c) London dispersion forces are stronger for ICl than for Br_2.
(d) ICl has a higher vapor pressure than Br_2 at room temperature.
(e) ICl is polar, while Br_2 is nonpolar.

12. One mole of ethanol, CH_3CH_2OH, is placed in a 250-mL beaker. Another mole of ethanol is placed in a 1.0-L beaker. Both beakers are covered and maintained at the same temperature. The equilibrium partial pressure of ethanol in the gas phase in the 250-mL beaker is 88 mmHg. In the 1.0-L beaker the equilibrium partial pressure, in millimeters of mercury, is

(a) 22   (b) 44   (c) 88   (d) 176   (e) 352

13. Which of the following substances has the highest boiling point?

(a) $CHCl_3$ (b) CH_2I_2 (c) CH_2Br_2 (d) CH_2Cl_2 (e) $CHICl_2$

14. Intermolecular forces in liquid <u>A</u> are considerably larger than intermolecular forces in liquid <u>B</u>. Which of the following properties is NOT expected to be larger for <u>A</u> than for <u>B</u>?

(a) The vapor pressure at 20 °C.
(b) The temperature at which the vapor pressure is 100 mmHg.
(c) The critical temperature.
(d) The heat of vaporization, ΔH_{vap}.
(e) The normal boiling point.

15. At a specified value of pressure and of temperature, which of the following gases will show the greatest deviation from the ideal gas law?

(a) N_2 (b) NH_3 (c) NO   (d) Ne   (e) NF_3

16. The normal boiling point of a liquid

(a) Is the temperature at which liquid and vapor are in equilibrium.
(b) Varies with the atmospheric pressure.
(c) Is the temperature at which the vapor pressure is 1 atm.
(d) Is the temperature at which the vapor pressure equals the external pressure.
(e) Is directly proportional to the molecular weight of the liquid.

17. The boiling point of NH_3 (-33 °C) is higher than the boiling point of PH_3 (-87.7 °C) because
    (a) The molecular weight of NH_3 is less than that of PH_3.
    (b) London dispersion forces are stronger for NH_3 than for PH_3.
    (c) The vapor pressure of NH_3 is greater than the vapor pressure of PH_3.
    (d) NH_3 is hydrogen bonded, but PH_3 is not.
    (e) NH_3 is polar, but PH_3 is not.

18. Which of the following substances is most likely to exist as a solid at room temperature?
    (a) NCl_3 (b) PF_3 (c) BrCl   (d) MgF_2 (e) SnH_4

19. A sample of liquid H_2O of mass 2.00 g is injected into an evacuated 10.0-L flask maintained at 30.0 °C, at which temperature the vapor pressure of H_2O is 31.824 mmHg. What percentage of the water will be vapor when the system comes to equilibrium? Assume water vapor behaves as an ideal gas. The volume occupied by the liquid water is negligible compared to the volume of the container.
    (a) 1.68%   (b) 15.2%   (c) 30.3%   (d) 50.0%   (e) 90.9%

20. What is the minimum volume the flask in question 19 should have, if no liquid water is to be present at equilibrium?
    (a) 6.5 L   (b) 22.4 L   (c) 32.0 L   (d) 66.0 L   (e) 118.8 L

chapter 6 *Properties of Dilute Solutions*

Francois Marie Raoult (1830–1901) was a French chemist who investigated the properties of solutions. From 1867 until his death in 1901, he was a professor of chemistry at the University of Grenoble. Raoult discovered that the freezing point depression of aqueous solutions is proportional to the solute mole fraction. His most valuable finding (now called Raoult's Law) was that the vapor pressure of the solvent in equilibrium with a solution is proportional to the mole fraction of the solvent in the solution.

A **solution** is a homogeneous mixture of two or more substances. Solids, liquids, and gases can all be dissolved in liquids to make solutions. Solutions of solids in other solids are possible also, but they are much less common. When a solution is prepared, the substance present in largest quantity is called the **solvent;** the other substances present in solution are called **solutes.** Sometimes it is very clear which is solvent and which is solute, as when a small amount of sugar is dissolved in water (sugar = solute, H_2O = solvent); other times it is essentially arbitrary, as when we mix roughly equal amounts of water and alcohol.

Most chemical reactions are carried out in solution because the movement of solute particles through the solution provides a mechanism for the reacting species to come into contact with one another. Most biochemical reactions in living organisms occur in aqueous solution. In this chapter and the one that follows we will study the properties of solutions, particularly aqueous solutions.

section 6.1
Dynamic Equilibrium in Saturated Solutions

Saturated and Unsaturated Solutions

Suppose we add 10.0 g of sucrose (ordinary sugar) to 100.0 mL of water in a beaker and stir. After several minutes, all the sugar will have dissolved, and the solution will be clear. No solid sucrose will be visible in the beaker. This solution is **unsaturated,** because it is possible to dissolve still more sugar in it. If we add another 10.0 g of sucrose, and stir again, all of that sugar will also dissolve, and the solution will still be unsaturated. If we continue to add sugar, with stirring, eventually we will not be able to dissolve any more, and there will be a layer of undissolved sugar on the bottom of the beaker, as shown in Fig. 6.1. The solution is now **saturated. A saturated solution** is, by definition, a solution that contains the maximum amount of solute that can be dissolved in a given quantity of solvent. A solution in equilibrium with an excess of undissolved solute is necessarily saturated. The concentration of solute in a saturated solution is a constant, that is, it remains fixed as time passes, as long as the temperature remains constant. If we add solid solute to a saturated solution that already has some excess solid on the bottom of the container, the size of the solid phase will increase, but the composition of the solution will not change.

The concentration of solute in an unsaturated solution may be any value above zero up to the concentration of the saturated solution. Once a saturated solution has been prepared, and there is solid solute material present, the concentration of solute in the solution cannot be increased as long as the temperature remains constant.

Equilibrium in Saturated Solutions

In Section 5.8 we summarized properties of the equilibrium state, and stated that

Fig. 6.1. A saturated solution. The concentration of sugar in the liquid phase remains constant as long as the temperature is constant.

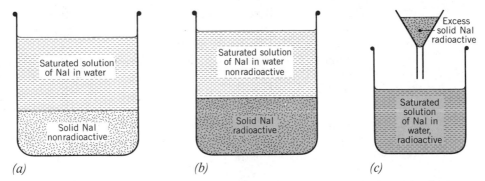

Fig. 6.2. Demonstration of dynamic equilibrium in a saturated solution. *(a)* Initial saturated solution, prepared using only nonradioactive NaI. *(b)* Contents *immediately* after adding some radioactive solid NaI to the already saturated solution. The amount of solid has increased. The concentration of dissolved material has not changed. *(c)* After several hours have passed, when the solution is filtered it is observed to be radioactive. Both the solid material and the solution are radioactive.

equilibrium in molecular systems is **dynamic,** that is, opposing reactions are occurring at equal rates. In order to understand the dynamic equilibrium in a saturated solution, let us try a simple experiment. We begin with a saturated solution of sodium iodide, NaI, with some excess solid white NaI on the bottom of the container, as depicted in Fig. 6.2*(a)*.

Iodine has only a single stable isotope, ^{127}I, but there are several radioactive isotopes of iodine. One of these, ^{131}I, is used quite extensively in medical research for studying, among other things, the function of the thyroid gland. Sodium iodide made with ^{131}I is commercially available. Suppose we add some radioactive solid NaI^{131} to the saturated solution of ordinary nonradioactive NaI. What will happen? The amount of solid material on the bottom of the container will, of course, increase. Now allow some time to pass, during which the temperature is kept constant and the container is not disturbed in any way. After a few hours filter the solution, separating the solid material from the liquid. Figure 6.2 illustrates the experimental procedure.

When the filtered solution is tested for radioactivity, it is always found to be radioactive. This proves that ^{131}I atoms originally present in the solid phase (the added solid NaI) appear, after some time, in an already saturated solution. It is clear that even though the concentration of solute does not change with time, solute particles are leaving the solid phase and entering the solution. The only way this can occur without a change in the concentration of the solute is for solute particles simultaneously to be leaving the solution and becoming part of the solid phase. Thus the equilibrium

$$\text{excess undissolved solid} \rightleftharpoons \text{saturated solution} \qquad (6\text{-}1)$$

is a dynamic one, with opposing reactions occurring simultaneously at equal rates. In this case, the opposing reactions are (1) the **dissolution** of solute into the solution, and (2) the **precipitation** of solute out of the solution.

section 6.2
Concentration Units for Solutions

In the previous section we stated that the concentration of a saturated solution remains constant as long as the temperature is constant, but the concentration of an unsaturated solution can be any value greater than zero up to the value in a saturated solution. The concentration of solute in a solution may be reported in a number of different ways. For example, we may give the number of grams of solute dissolved in a specified volume of liquid, such as "grams solute per 100 mL of solvent." That unit is frequently used in standard references, such as the *Handbook of Chemistry and Physics*. There are, however, three different units of concentration widely used by chemists when preparing solutions for experiments. These units are **molarity, molality,** and **mole fraction**. It is important to learn the definitions of these three units, to understand the differences between them, and to acquire computational skills for problems involving them.

The Mole Fraction

The **mole fraction,** X_i, has already been defined for a mixture of gases in Section 3.6, Eq. (3-43). That definition applies equally well to any component of a liquid solution. If n_i is the number of moles of the ith component of a solution,

$$X_i = \frac{n_i}{\sum\limits_i n_i} = \frac{n_i}{n_{total}} \tag{6-2}$$

If we sum this equation over all values of i we obtain

$$\sum_i X_i = \frac{\sum n_i}{n_{total}} = \frac{n_{total}}{n_{total}} = 1 \tag{6-3}$$

that is, the sum of the mole fractions of all the components of a solution is 1. If there are only two components, Eq. (6-3) is simply $X_1 + X_2 = 1$. If a solution of ethyl alcohol (EtOH) and water has $X_{EtOH} = \frac{1}{5}$, then $X_{H_2O} = \frac{4}{5}$. Mole fractions are numbers between 0 and 1:

$$0 \le X_i \le 1 \tag{6-4}$$

for all i. A **mole percentage** is merely a mole fraction expressed as a percentage. For example, if $X_{EtOH} = \frac{1}{5}$, the mole percentage of ethyl alcohol is 20%.

EXAMPLE 6.1. Definition of mole fraction and mole percentage

A solution is prepared by dissolving 16.6624 g of solid naphthalene, $C_{10}H_8$, in 200.0 mL of liquid benzene, C_6H_6, at 20 °C. The density of benzene at 20 °C is 0.87865 g · mL^{-1}. Calculate the mole fraction and mole percentage of benzene and naphthalene in this solution.

Solution. The molecular weight of $C_{10}H_8$ is 128.17 g · mol^{-1}. The number of moles of naphthalene in a sample of mass 16.6624 g is therefore

$$\frac{16.6624 \text{ g}}{128.17 \text{ g/mol}} = 0.13000 \text{ mol of naphthalene}$$

The mass of 200.0 mL of benzene is (200.0 mL)(0.87865 g/mL) = 175.7 g. The molecular weight of C_6H_6 is 78.113 g/mol, so that the number of moles of benzene in the 200.0 mL is

$$\frac{175.7 \text{ g}}{78.113 \text{ g/mol}} = 2.249 \text{ mol of benzene}$$

The mole fraction of naphthalene is

$$\frac{\text{No. mol } C_{10}H_8}{\text{No. mol } C_{10}H_8 + \text{No. mol } C_6H_6} = \frac{0.1300}{0.1300 + 2.249} = 0.05464$$

The mole percentage of naphthalene is 5.464%.

The mole fraction of benzene is calculated from the relation

$$X_{C_6H_6} = 1 - X_{C_{10}H_8} = 1 - 0.05464 = 0.94546$$

The mole percentage of benzene is therefore 94.546%.

The Molality

The **molality**, denoted m, is a quantity used to describe the concentration of a solute in a solution. It is defined as the number of moles of solute dissolved in one kilogram (1000 g) of solvent. If nothing else is specified, the solvent is assumed to be water.

A 1.00 m (molal) solution of NaCl contains 1 mol (58.44 g) of NaCl dissolved in exactly 1 kg of H_2O. Any portion of that solution is still 1.00 m. Thus, if we dissolve $\frac{1}{4}$ mol (14.61 g) of NaCl in 250.0 g of H_2O, the solution is also 1.00 m. The important factor is the *ratio* of moles of solute to kilograms of solvent.

Since the number of moles of solvent (solv) in 1000 g of solvent is $1000/M_{solv}$, where M_{solv} is the molecular weight of the solvent, the relation between the mole fraction of solute, X_{sol}, and the molality of the solute, m, for the case where there is only a single solute, is simply

$$X_{sol} = \frac{\text{mol solute}}{\text{mol solute} + \text{mol solvent}} = \frac{m}{m + (1000/M_{solv})} \tag{6-5}$$

If the solvent is H_2O, M_{solv} is 18.015 g/mol, and $1000/M_{solv}$ is 55.5. Example 6.2 illustrates the relationship between mole fraction and molality.

EXAMPLE 6.2. Calculation of the mole fraction and molality of a solution of a single solute

Glucose, $C_6H_{12}O_6$, is a simple sugar (a monosaccharide), which is a principal energy source for humans and other vertebrates. Other forms of sugar may be converted, in part, to glucose for transport in the body. In a complex series of reactions, glucose combines with the oxygen we breathe to produce, after many intermediate products, $CO_2(g)$ and H_2O, and the net reaction releases energy. A hospital patient being fed intravenously is usually receiving a solution of glucose in water, with added salts.

If 1.275 g of glucose are dissolved in 63.59 g of water, what is the molality of the resulting solution? What are the mole fractions of glucose and of water in this solution?

Solution. The molecular weight of glucose, $C_6H_{12}O_6$, is 180.16 g · mol⁻¹. Therefore the sample of 1.275 g of glucose contains 7.077×10^{-3} mol of glucose. The ratio of moles of glucose to kilograms of H_2O in this solution is

$$\frac{7.077 \times 10^{-3} \text{ mol } C_6H_{12}O_6}{63.59 \times 10^{-3} \text{ kg } H_2O} = \frac{0.1113 \text{ mol } C_6H_{12}O_6}{1 \text{ kg } H_2O} = 0.1113 \ m$$

Note that in solving this problem we use the molecular weight of glucose to five figures, and carry all intermediate calculations to five figures, so that we round only once, at the end, to the four significant figures called for by the data. (See Introduction, pages 14–16.)

The mole fraction of glucose can be calculated in either of two ways. We can use the original data as follows:

$$\text{No. mol } C_6H_{12}O_6 = 7.077 \times 10^{-3} \text{ mol}$$

$$\text{No. mol } H_2O = 63.59 \text{ g}/18.015 \text{ g} \cdot \text{mol}^{-1} = 3.530 \text{ mol}$$

$$X_{C_6H_{12}O_6} = \frac{7.077 \times 10^{-3}}{7.077 \times 10^{-3} + 3.530} = \frac{7.077 \times 10^{-3}}{3.537} = 2.001 \times 10^{-3}$$

$$X_{H_2O} = 1 - 0.002001 = 0.997999 = 0.9980$$

Alternatively, we can use the molality and Eq. (6-5):

$$X_{C_6H_{12}O_6} = \frac{0.1113}{0.1113 + 1000/18.015} = \frac{0.1113}{55.621} = 2.001 \times 10^{-3}$$

The Molarity

The unit most commonly used to give the concentration of the solute is the **molarity**, denoted M. The molarity is defined as the number of moles of solute in one liter of *solution*. Note particularly that the definition specifies a liter of the final solution, and *not* a liter of the solvent. Volume changes usually occur when solute and solvent are mixed. The volume of a solution is not exactly the same as the volume of the solvent, nor is it generally equal to the sum of the volumes of solute plus solvent. Because the intermolecular forces between solute and solvent molecules are not the same as the intermolecular forces between solute molecules, or between solvent molecules, the volume may either shrink or expand when solute and solvent are mixed.

To prepare a solution of a given molarity, we must use a **volumetric flask**. Volumetric flasks of various sizes are shown in Fig. 6.3. The method employed to prepare a solution of a specified molarity is described in Example 6.3.

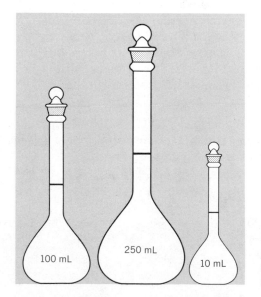

100 mL   250 mL   10 mL

Fig. 6.3. Volumetric flasks are designed to contain a specified volume of liquid at a specified temperature, usually 20 °C.

(a)   *(b)*   *(c)*   *(d)*

Fig. 6.4. Preparation of a solution of a specified molarity. *(a)* The solute is accurately weighed into a volumetric flask. *(b)* Distilled water is added. *(c)* The flask is stoppered and shaken to dissolve the solute. *(d)* Distilled water is added carefully to bring the volume up to the mark etched around the neck of the flask.

EXAMPLE 6.3. The preparation of a solution of a given molarity

What are the directions for preparing 250.00 mL of 0.1000 M sodium sulfate, Na_2SO_4?

Solution. The formula weight of Na_2SO_4 is 142.041 $g \cdot mol^{-1}$. The mass of 0.1000 mol of Na_2SO_4 is therefore 14.2041 g. The symbol 0.1000 M means 0.1000 mol/L of solution. However, we are not asked to prepare 1 L of solution, but only 250.0 mL, one quarter of a liter. Therefore we need $\frac{1}{4}$(14.2041 g) = 3.5510 g of Na_2SO_4.

The directions for preparing 250.00 mL of 0.1000 M Na_2SO_4 are as follows: Weigh out 3.5510 g of Na_2SO_4. Transfer this to a 250-mL volumetric flask. Add distilled water to fill most of the bulb in the flask, but not to reach the long neck of the flask. The initial amount of water added must be considerably less than the amount that will be needed to prepare the solution. Invert the flask several times and mix thoroughly until all the solute is dissolved. Add more water until the volume of solution is just below the mark (the ring etched on the long neck of the flask). Invert several times, mixing thoroughly. Now add water dropwise until the solution is only two or three drops short of the mark. Invert and mix thoroughly again. Add water one drop at a time, mixing thoroughly after each drop, until the volume of the solution is right up to the mark, that is, exactly 250.0 mL. This procedure is illustrated in Fig. 6.4.

The preparation of a Na_2SO_4 solution affords a good chance to observe the volume contracting when solute and solvent are mixed. If you place some solid Na_2SO_4 in a volumetric flask, add water right up to the mark, and *then* shake the solution thoroughly, you will see that the final volume of solution is substantially below the mark. In this case the volume of the solution is less than the sum of the volume of water plus the volume of the Na_2SO_4. A contraction occurs on dissolving Na_2SO_4 in water.

There are other solute plus solvent combinations for which the volume *expands* on mixing; it is for this reason that water is not added to the mark of a volumetric flask before the solution is thoroughly mixed and all solute is dissolved. Usually you will not know whether a volume expansion or contraction will take place, and therefore must follow the procedure described in Example 6.3.

The volume of a solution is temperature dependent, so that molarity is a temperature dependent concentration unit. If you prepare a solution at a certain temperature, and the temperature changes, the molarity will change. However, the volumes of liquids (unlike gases) do not change very much with small changes in temperature, and since room temperature in most laboratories does not usually vary by more than ±2 or 3°, the temperature dependence of the molarity does not present a problem except in the most precise work.

Molality and mole fraction concentration units are not defined in terms of any volume, so they do not change with temperature. This is an advantage of molality and mole fraction over molarity, and there are experiments in which concentrations are expressed as molalities to avoid using a temperature-dependent unit. However, because it is much easier to measure a precise volume of a liquid, using a volumetric flask, than to weigh out a precise number of grams of a liquid, molarity is the most convenient unit for ease of preparation, and is most widely used.

Another convenience of molarity as a unit is that moles per liter and millimoles per milliliter are identical. Thus the same numerical value of the concentration pertains whether we are dealing with large quantities of solution and prefer to cite the concentration in moles per liter, or we are dealing with small quantities of solution and find it more useful to give the concentration in millimoles per milliliter.

The two relations given below are particularly useful for numerical calculations involving molarities.

$$\text{No. mol solute} = \left(\text{concentration in } \frac{\text{mol}}{\text{L}}\right)(\text{volume in L}) \tag{6-6a}$$

$$\text{No. mmol solute} = \left(\text{concentration in } \frac{\text{mmol}}{\text{mL}}\right)(\text{volume in mL}) \tag{6-6b}$$

Often we have a stock solution that is moderately concentrated and want to prepare a more dilute solution for use in a particular experiment. Example 6.4 describes the method of determining just how we should carry out the dilution. The important thing to remember is that in any dilution only solvent is added, and therefore the number of moles (or millimoles) of solute remains constant.

EXAMPLE 6.4. Dilution of a more concentrated solution to produce a solution of specified concentration

What are the directions for preparing 50.00 mL of a 0.0400 M Na_2SO_4 solution from a stock solution of 0.1000 M Na_2SO_4?

Solution. The question can be rephrased as follows: To what volume, V, of the 0.1000 M solution should water be added so that the resulting solution has a concentration of 0.0400 M and a volume of 50.00 mL? The volume of the 0.1000 M solution must contain the same number of millimoles of Na_2SO_4 as the final solution, because adding water cannot change the number of millimoles of Na_2SO_4.

How many millimoles of Na_2SO_4 are in 50.00 mL of a 0.0400 M solution? We use Eq. (6-6b) to obtain

No. mmol Na_2SO_4 in the final solution $= (50.00 \text{ mL})\left(0.0400 \dfrac{\text{mmol}}{\text{mL}}\right) = 2.00 \text{ mmol}$

What volume, V, of 0.1000 M Na_2SO_4 contains 2.00 mmol of Na_2SO_4? Once again we make use of Eq. (6-6b).

$$(V \text{ mL})\left(0.1000 \dfrac{\text{mmol}}{\text{mL}}\right) = 2.00 \text{ mmol}$$

and therefore $V = 20.0$ mL.

Note that one can simply equate the number of millimoles of Na_2SO_4 in the final solution and in the volume, V, of the 0.1000 M solution:

$$(50.00 \text{ mL})\left(0.0400 \dfrac{\text{mmol}}{\text{mL}}\right) = (V \text{ mL})\left(0.1000 \dfrac{\text{mmol}}{\text{mL}}\right)$$

to obtain $V = 20.0$ mL.

The directions for preparing the dilute solution are as follows: Using a 20.0-mL **volumetric pipet** (see Fig. 6.5), transfer 20.0 mL of the stock 0.1000 M Na_2SO_4 solution to a 50-mL volumetric flask. Add distilled water to just below the mark, invert and mix thoroughly. Add water dropwise to the mark, mixing thoroughly.

In general, to determine the volume of a concentrated solution that should be used to prepare a specific volume, V_{dil}, of a dilute solution of concentration M_{dil}, we use the relation

$$(V_{conc})(M_{conc}) = (V_{dil})(M_{dil}) \tag{6-7}$$

Volume changes on diluting an already dilute solution, such as a 0.1000 M solution, are very small compared to the volume changes that occur when dissolving a solid solute in a pure liquid or when mixing two different pure liquids together. It is a very good approximation to assume that the volume of a mixture of two dilute solutions is the sum of the volumes of the two solutions.

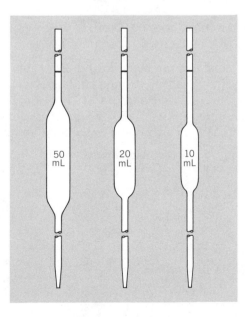

50 mL   20 mL   10 mL

Fig. 6.5. Volumetric pipets. A volumetric pipet (also called a transfer pipet) is used to deliver a specified volume of liquid at a specified temperature, usually 20 °C.

Formality versus Molarity

In the solutions of Na_2SO_4 that we have just been discussing, there are no molecules of Na_2SO_4. Sodium sulfate is an ionic crystalline solid, and when dissolved in water the Na^+ ions and the SO_4^{2-} ions separate from one another and move about in the solution as individual entities, although of course there is electrostatic attraction between ions of opposite charge, and repulsion between ions of like charge.

When we dissolve 0.1000 mol of Na_2SO_4 in enough water to make 1 L of solution, the solute particles are 0.2000 mol of Na^+ ions and 0.1000 mol of SO_4^{2-} ions. We use square brackets with the symbol of the solute species inside the brackets to indicate the concentration in moles per liter. Thus $[Na^+] = 0.2000$ M is read as "the concentration of sodium ion is 0.2000 mol/L." Rigorously, as there is no such species as Na_2SO_4 in this solution, we should not say that the solution is 0.1000 M in Na_2SO_4. To avoid this, we introduce the concept of **formality**. For solutions in which the solute particles are ions and not molecules, or are a combination of both ions and molecules, we describe the concentration as **formal** rather than **molar**, and use the symbol F rather than M. This is exactly the same distinction we made between formula weight and molecular weight in Section 1.7.

To emphasize the fact that no such species as Na_2SO_4 exists in aqueous solution, we should denote the solution prepared in Example 6.3 as 0.1000 F Na_2SO_4, and not 0.1000 M Na_2SO_4. Throughout this text we will distinguish between formality and molarity, but you should be aware that many working chemists do not bother to make this distinction, and use the term molarity to describe the concentration of solute particles whether the particles are molecules or ions. Thus the symbol M does not guarantee that the solute is actually a molecule, and you must have previous knowledge about the species that are ionic in solution. Some general principles to use in determining whether or not a compound is an ionic crystalline solid have been discussed in Section 2.7, and a further discussion of the nature of solutes will be given in Section 7.3.

Example 6.5 illustrates the method of calculating the concentrations of ionic species in solution when we dissolve ionic crystalline solids in water.

EXAMPLE 6.5. Concentrations of ions

If 40.00 mL of 0.200 F $NaNO_3$ is mixed with 60.00 mL of 0.100 F $Cu(NO_3)_2$, what is the concentration of each solute in the resulting solution?

Solution. Because two dilute solutions are being mixed, the volume of the final solution is 100.00 mL. The 0.200 F $NaNO_3$ solution contains Na^+ ions and NO_3^- ions, while the 0.100 F $Cu(NO_3)_2$ solution contains Cu^{2+} ions and NO_3^- ions. The final solution therefore contains three solutes: Na^+, Cu^{2+}, and NO_3^- ions. We need to know how many millimoles of each ion are present in the final solution.

$$\text{No. mmol } Na^+ \text{ in 40.00 mL 0.200 } F \text{ } NaNO_3 = (40.00 \text{ mL})\left(0.200 \frac{\text{mmol } NaNO_3}{\text{mL}}\right)\left(\frac{1 \text{ mmol } Na^+}{1 \text{ mmol } NaNO_3}\right)$$
$$= 8.00 \text{ mmol } Na^+$$

$$\text{No. mmol } NO_3^- \text{ in 40.00 mL 0.200 } F \text{ } NaNO_3 = (40.00 \text{ mL})\left(0.200 \frac{\text{mmol } NaNO_3}{\text{mL}}\right)\left(\frac{1 \text{ mmol } NO_3^-}{1 \text{ mmol } NaNO_3}\right)$$
$$= 8.00 \text{ mmol } NO_3^-$$

$$\text{No. mmol } Cu^{2+} \text{ in 60.00 mL 0.100 } F \text{ } Cu(NO_3)_2 = (60.00 \text{ mL})\left[0.100 \frac{\text{mmol } Cu(NO_3)_2}{\text{mL}}\right]\left[\frac{1 \text{ mmol } Cu^{2+}}{1 \text{ mmol } Cu(NO_3)_2}\right]$$
$$= 6.00 \text{ mmol } Cu^{2+}$$

$$\text{No. mmol NO}_3^- \text{ in } 60.00 \atop \text{mL } 0.100 \ F \text{ Cu(NO}_3)_2} = (60.00 \text{ mL}) \left[0.100 \ \frac{\text{mmol Cu(NO}_3)_2}{\text{mL}} \right] \left[\frac{2 \text{ mmol NO}_3^-}{1 \text{ mmol Cu(NO}_3)_2} \right]$$

$$= 12.00 \text{ mmol NO}_3^-$$

The total number of millimoles of NO_3^- in the solution is 20.00 mmol, 8.00 mmol from the $NaNO_3$, and 12.00 mmol from the $Cu(NO_3)_2$. Hence the concentration of each of these three ions in the final solution, which has a volume of 100 mL, is

$$[Na^+] = \frac{8.00 \text{ mmol}}{100 \text{ mL}} = 0.0800 \ M$$

$$[NO_3^-] = \frac{20.00 \text{ mmol}}{100 \text{ mL}} = 0.200 \ M$$

$$[Cu^{2+}] = \frac{6.00 \text{ mmol}}{100 \text{ mL}} = 0.0600 \ M$$

Weight Percentage

In order to calculate the molality or the mole fraction of solute if the molarity is known, it is necessary to know the density of the solution. Manufacturers of several commonly used solutions frequently report both the density and the percentage by weight of solute (the **weight percentage**) on the label. Example 6.6 illustrates how one calculates the molality, the molarity, and the mole fraction of solute from the weight percentage and the density of solution. This problem also shows how the three concentration units are related. For very dilute solutions, the density of the solution is very close to the density of the pure solvent. Since the density of water is $1.0 \text{ g} \cdot \text{cm}^{-3}$ at room temperature, the molality and molarity of *dilute* aqueous solutions differ very little. For concentrated solutions, the molarity and molality may be quite different, as in Example 6.6.

EXAMPLE 6.6. **Relation between weight percentage, molality, molarity, and mole fraction**

The label on a commercially available concentrated hydrochloric acid solution reads "37.4% HCl by weight, density 1.18 g/mL." Calculate the molality, formality, and mole fraction of HCl in this solution.

Solution. Since HCl is extensively dissociated into $H^+(aq)$ and $Cl^-(aq)$ ions, we should use the term formality rather than molarity. To calculate the formality we need to know the number of moles of HCl in 1 L of this solution. The mass of 1 L of this solution is obtained from the density, as follows:

$$\text{mass of 1 L of solution} = \left(1.18 \ \frac{\text{g}}{\text{mL}} \right) (1000 \text{ mL}) = 1180 \text{ g}$$

Of this 1180 g, only 37.4% is HCl, the rest is water. Thus the mass of HCl is $(0.374)(1180 \text{ g}) = 441 \text{ g}$. The formula weight of HCl is $36.46 \text{ g} \cdot \text{mol}^{-1}$, so that there are $441/36.46 = 12.1$ mol of HCl in a liter of this solution. The solution is therefore 12.1 F HCl.

The mass of water in this liter of solution is obtained as follows:

$$\text{mass of solution} - \text{mass of HCl} = 1180 \text{ g} - 441 \text{ g} = 739 \text{ g}$$

$$= \text{mass of } H_2O \text{ in 1 L}$$

To calculate the molality, we need the mass of water in kilograms, which is 0.739 kg. The molality of HCl is therefore 12.1-mol HCl per 0.739-kg water:

$$\frac{12.1 \text{ mol HCl}}{0.739 \text{ kg H}_2\text{O}} = \frac{16.4 \text{ mol HCl}}{1 \text{ kg H}_2\text{O}} = 16.4 \ m$$

Since the molecular weight of water is $18.015 \text{ g} \cdot \text{mol}^{-1}$, the number of moles of water in this liter of solution is

$$739 \text{ g}/18.015 \text{ g} \cdot \text{mol}^{-1} = 41.0 \text{ mol of H}_2\text{O}$$

The mole fraction of HCl is therefore

$$X_{\text{HCl}} = \frac{12.1}{12.1 + 41.0} = \frac{12.1}{53.1} = 0.228$$

Solutions sold commercially as concentrated hydrochloric acid vary in weight percentage of HCl from 36.5 to 38.0%. This means the formality ranges from 11.8 to 12.3 F. (Try to calculate these formalities yourself.) It is convenient to remember that concentrated HCl is roughly 12 F. (Most chemists would say "12 M.")

section 6.3
The Vapor Pressure of a Solution of a Nonvolatile Solute in a Volatile Solvent: Raoult's Law

In Sections 5.6 and 5.9 we discussed the equilibrium vapor pressure of a pure liquid. Suppose we dissolve a **nonvolatile solute** (that is, a solute with a negligible vapor pressure) in a **volatile liquid**. Examples of such solutions are sugar dissolved in water, or $NiSO_4$ dissolved in ethanol (ethyl alcohol). If we keep such a solution in a closed container, molecules of the solvent (the volatile liquid), will enter the gas phase, and the solution will have an equilibrium vapor pressure. Since the solute is nonvolatile, the vapor phase does not contain measurable amounts of solute molecules. The pressure of the solvent in the gas phase in equilibrium with the solution is called the **vapor pressure of the solution.**

If we measure the vapor pressure of many such solutions, the vapor pressure of the solution is always found to be lower than the vapor pressure of the pure solvent at the same temperature. Since we are dealing with a two-component system, let us denote the volatile solvent as component "1" and the nonvolatile solute as component "2." The symbol P_1 will be used to represent the vapor pressure of the solution, and P_1^0 will be used to represent the vapor pressure of the pure solvent. We can use the specific example of a solution of sucrose (a sugar) in water to summarize the observations that have been made about the vapor pressure of solutions of a nonvolatile solute in a volatile solvent.

solution of sugar in $\rightleftharpoons$ H_2O (vap) at the equilibrium
H_2O at temperature T vapor pressure P_1

pure liquid H_2O at $\rightleftharpoons$ H_2O (vap) at the equilibrium
temperature T vapor pressure P_1^0

For all such solutions P_1 is less than P_1^0, $(P_1 < P_1^0)$.

Raoult's Law

Francois Raoult (1830–1901), a French chemist, measured the vapor pressure of the solvent above solutions of nonvolatile solutes of varying concentrations, and ob-

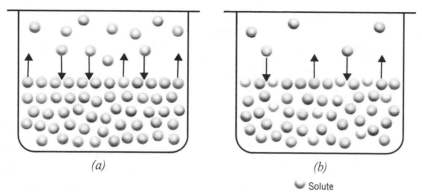

(a)   *(b)*

Solute

Fig. 6.6. Molecular view of Raoult's law. *(a)* Pure solvent *(b)* solution.

served that the higher the concentration of the solute, the lower the vapor pressure of the solvent. We can understand this phenomenon qualitatively. The surface of the solution has a smaller fraction of solvent molecules as the concentration of solute increases. (See Fig. 6.6.) Therefore the fraction of molecules at the surface that is volatile, and that has sufficient kinetic energy to overcome the binding forces of the liquid, decreases as more solute molecules are added.

For a given solvent in a specific closed container, adding a solute to the liquid decreases the rate of evaporation of the solvent, but does not affect the rate of condensation, because that depends only on the pressure of the solvent in the gas phase. The rate of evaporation is therefore equal to the rate of condensation at a lower vapor pressure than for the pure solvent. Just how much the vapor pressure is lowered depends strongly on the natures of the solute and solvent, and can be quite complex, but for *dilute** solutions, a very simple relationship is found. The solutions Raoult investigated were dilute, and within the precision of his measurements, his results are described by the equation

$$P_1 = P_1^0 X_1 = P_1^0 \left(\frac{n_1}{n_{\text{total}}} \right) \qquad (6\text{-}8)$$

which is known as **Raoult's law**. In this equation, X_1 is the mole fraction of the *solvent* in the solution, n_1 is the number of moles of solvent, and n_{total} is the number of moles of solute plus solvent. The symbols P_1 and P_1^0 have already been defined. Remember that the subscript "1" always refers to the solvent, and the subscript "2" to the solute. As the solute concentration increases, the mole fraction of solvent, X_1, decreases, and the vapor pressure of the solution decreases. Raoult's law states that the vapor pressure of a dilute solution of a nonvolatile solute is directly proportional to the mole fraction of solvent, and the proportionality constant is the vapor pressure of pure solvent. A plot of the vapor pressure versus the mole fraction of solvent for a solution that obeys Raoult's law is shown in Fig. 6.7.

Most real solutions deviate from Raoult's law, that is, Raoult's law is not an *exact* description of the behavior of most solutions. The more dilute the solution, the smaller the deviations from Raoult's law. A solution that obeys Raoult's law is called an **ideal solution**. Most solutions are not ideal, but there are a few ideal solutions. In general, if we use Raoult's law to predict the vapor pressure of a solution, we will be making an error, but it will be a small error provided the solution is dilute.

---

* There is no fixed concentration that distinguishes dilute from concentrated solutions. As a very rough guide, solutions less than 0.100 *M* are generally considered dilute.

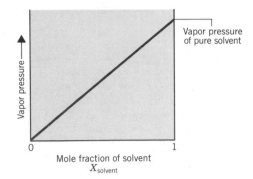

Fig. 6.7. A plot of the vapor pressure versus mole fraction of solvent for an ideal solution, that is, a solution that obeys Raoult's law.

We define the **vapor pressure lowering** as $P_1^0 - P_1 = \Delta P$. The vapor pressure lowering is the difference between the vapor pressure of the pure solvent and the vapor pressure of the solution, at the same temperature. If the solution is ideal, or sufficiently dilute so that Raoult's law is valid, the vapor pressure lowering is given by

$$\Delta P = P_1^0 - P_1 = P_1^0 - P_1^0 X_1 = P_1^0(1 - X_1) \tag{6-9a}$$

If there is only a single solute, then $X_1 + X_2 = 1$, and $1 - X_1 = X_2$. In that case, the vapor pressure lowering can be expressed as

$$\Delta P = P_1^0 X_2 = P_1^0 \left(\frac{n_2}{n_1 + n_2} \right) \tag{6-9b}$$

where X_2 is the mole fraction of solute, and n_2 is the number of moles of solute. All other terms are as defined for Eq. (6-8).

EXAMPLE 6.7. The use of Raoult's law

An aqueous solution of sucrose is prepared in which the molality of sucrose is 0.475 *m*.

(a) What is the vapor pressure of H_2O above this solution at 26.4 °C? The vapor pressure of pure H_2O at 26.4 °C is 25.812 mmHg.

(b) What is the magnitude of the vapor pressure lowering?

Solution

(a) We must calculate the mole fraction of H_2O in the solution. If there are 0.475 mol of sucrose, there are exactly 1000 g of H_2O, or 1000 g/18.015 g · mol^{-1} = 55.509 mol of H_2O. Therefore

$$X_{H_2O} = \frac{55.509}{55.509 + 0.475} = \frac{55.509}{55.984} = 0.99152$$

$$P_{H_2O} = (25.812 \text{ mmHg})(0.99152) = 25.593 \text{ mmHg}$$

(b) $\Delta P = 25.812 \text{ mmHg} - 25.593 \text{ mmHg} = 0.219 \text{ mmHg}$. Alternatively, we can calculate the vapor pressure lowering as follows:

$$X_{sucrose} = 1 - X_{H_2O} = 1 - 0.99152 = 0.00848$$
$$\Delta P = (0.00848)(25.812 \text{ mmHg}) = 0.219 \text{ mmHg}$$

If you know how much the vapor pressure is lowered when a given weight of a molecular solute is added to a known quantity of a pure liquid, you can use Eq. (6-9b) to calculate the molecular weight of the solute. However, vapor pressure lowerings are usually small (see Example 6.7) and subject to a large experimental uncertainty (that is, they are difficult to measure accurately), and therefore this is not, in general, a practical method for determining molecular weights.

section 6.4
The Elevation of the Boiling Point of Dilute Solutions of a Nonvolatile Solute

One consequence of the lowering of the vapor pressure of a solution relative to the vapor pressure of the pure solvent is that the boiling point of the solution will be higher than the boiling point of the pure solvent.

In Fig. 5.20 we plotted the vapor pressure of two pure liquids as a function of temperature. The general shape of the curve is the same for any liquid. If we dissolve a nonvolatile solute in some volatile liquid, the vapor pressure of the solution is lower than the vapor pressure of the pure liquid at each temperature. The result is that the vapor pressure curve for the solution is similar in shape to that for the pure liquid, but lies below it, as shown in Fig. 6.8.

We define the normal boiling point of the solution, T, as the temperature at which the vapor pressure of the solution is one atmosphere. This must necessarily be a higher temperature than the boiling point, T_0, of the pure solvent. (See Fig. 6.8.) The increase in the boiling point when a nonvolatile solute is dissolved in a liquid is called the **boiling point elevation,** ΔT_b, and is defined as

$$\Delta T_b = \text{bp solution} - \text{bp solvent} = T - T_0 \qquad (6\text{-}10)$$

The boiling point elevation is larger the more concentrated the solution, because the vapor pressure lowering is greater the more concentrated the solution. For dilute solutions, the boiling point elevation is directly proportional to the concentration, and it is customary to express this proportionality using molality as the concentration unit. We therefore have

$$\Delta T_b = K_b m \qquad (6\text{-}11)$$

The proportionality constant, K_b, is a property of the *solvent only,* and is called the **molal boiling point elevation constant,** or the **ebullioscopic constant.** Values of K_b for several common solvents are given in Table 6.1. Note the units of K_b. The product $K_b m$ must have units of temperature, and thus K_b is in (kilograms solvent)(kelvins)/ (moles solute) or kg · K/mol.

The value of K_b is equal to the boiling point elevation that would be observed for a 1 m solution of any nonvolatile solute, *if* the solution is ideal. Very few solutions are ideal at a concentration of 1 m.

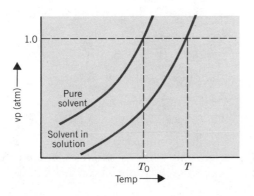

Fig. 6.8. The elevation of the boiling point of a solution of a nonvolatile solute. T_0 is the boiling point of the pure solvent. T is the boiling point of the solution. T is always higher than T_0.

Table 6.1. The Molal Boiling Point Elevation
Constants of Some Common Solvents

Substance	$K_b(kg \cdot K/mol)$	Normal bp ($°C$)
Acetone, C_3H_6O	1.71	56.2
Benzene, C_6H_6	2.53	80.2
Carbon tetrachloride, CCl_4	5.03	76.5
Chloroform, $CHCl_3$	3.63	61.2
Ethanol, C_2H_5OH	1.22	78.5
Water, H_2O	0.51	100.0

Note that the value of the boiling point elevation, ΔT_b, depends, for dilute solutions, only on the *concentration* of the solute particles and not on their nature. Properties that depend only on the concentration of the solute, and not on its nature, are known as **colligative properties.** Both the **boiling point elevation** and the **vapor pressure lowering** are colligative properties. Two other colligative properties, the freezing point depression and the osmotic pressure, will be discussed in Sections 6.5, 6.6, and 6.7.

The boiling point elevation can be used to determine the molecular weight of a solute. The method of calculation is described in Example 6.8.

EXAMPLE 6.8. Determination of the molecular weight of an unknown solute from the boiling point elevation

An unknown white molecular solid has been sent out for elemental analysis. From the weight percentages the empirical formula is determined to be $(C_5H_4)_n$. A 0.527-g sample of the compound is dissolved in 40.19 g of $CHCl_3$, and it is found that the solution boils 0.392° higher than does the pure $CHCl_3$ used as solvent. The unknown neither associates nor dissociates in $CHCl_3$ solution. Calculate the molecular formula and exact molecular weight of this compound.

Solution. We can easily calculate the molality of the solution from Eq. (6-11), since $\Delta T_b = 0.392°$ and K_b for $CHCl_3$ is 3.63 kg · K/mol (obtained from Table 6.1). Substitution into Eq. (6-11) yields

$$\Delta T_b = 0.392° = 3.63\, m \quad \text{and} \quad m = 0.108 \text{ molal}$$

Since we do not know the molecular weight of this compound, let us denote it by the symbol M. We have 0.527 g of compound, or (0.527/M) moles of compound. The molality is the ratio of moles of solute to kilograms of solvent. Hence

$$\frac{(0.527/M) \text{ mol solute}}{40.19 \times 10^{-3} \text{ kg } CHCl_3} = 0.108\, m$$

Solving for the molecular weight we obtain

$$M = \frac{527}{(40.19)(0.108)} = 121 \text{ g/mol}$$

The formula weight of C_5H_4 is 64.087, and the molecular weight of the compound is $64.087n$, where n is an integer. The only value of n that yields a molecular weight close to 121 is $n = 2$. The correct molecular formula is therefore $C_{10}H_8$, and the exact molecular weight is 128.17 g · mol^{-1}.

You may well ask: Why did the molecular weight obtained using the relation $\Delta T_b = K_b m$ in Example 6.8 come out to be 121, rather far from the true value of 128.17? There are two reasons. (1) There is a good deal of experimental uncertainty in measuring a difference in temperature of 0.392°. It is very difficult to measure temperature to a thousandth of a degree; room temperature can easily fluctuate by several thousandths of a degree. Since boiling point elevations are small, no more than a degree or two usually, they are subject to a large percentage uncertainty (percentage error). (2) The relation $\Delta T_b = K_b m$ is an ideal law. It is not exactly correct for real solutions, because it is derived from Raoult's law, which, as we have already discussed, does not accurately represent the behavior of many real solutions. Every time we use the relation $\Delta T_b = K_b m$ we can expect that we are making an error. Note, however, that even with a significant error in the value of the molecular weight, as in Example 6.8, it is possible to obtain the correct molecular formula and then the exact molecular weight. This has already been pointed out in Examples 1.10 and 3.2.

Suppose we wanted to make the error incurred in using the ideal equation, $\Delta T_b = K_b m$, small. Raoult's law becomes more exact as the solution becomes more dilute, so to reduce the error made in using Eq. (6-11) we should make the solution more dilute. But if we do that, the boiling point elevation becomes smaller, and such small temperature differences are exceedingly difficult to measure. Conversely, if we attempt to reduce the experimental uncertainty in measuring ΔT_b, we must increase the magnitude of ΔT_b, and this requires a more concentrated solution, which increases the error made in using Eq. (6-11). As a result, we cannot expect to obtain an exact molecular weight by measuring a boiling point elevation. If, however, we know the empirical formula, we will almost always be able to obtain values of the molecular weight close enough to the true value so that we can determine the correct molecular formula.

section 6.5
The Depression of the Freezing Point of Dilute Solutions of a Nonvolatile Solute

As has already been discussed in Section 5.10, a solid can also exist in equilibrium with its vapor in a closed container. The vapor pressure of a solid increases with increasing temperature, and it does so more rapidly than does the vapor pressure of the pure liquid. The intersection of the vapor pressure curve for the solid with the vapor pressure curve for the liquid is the **triple point,** at which all three phases are in equilibrium. (Refer to Section 5.10.) Because the vapor pressure curve for the solution is always below that of the pure liquid, the intersection of the solid and the solution vapor pressure curves will always occur at a lower temperature than the intersection of the solid and the pure liquid vapor pressure curves. This is shown in Fig. 6.9. It is clear that the triple point of the solution is at a *lower* temperature than the triple point of the pure liquid.

The curve showing the temperatures and pressures at which solid and liquid are in equilibrium must begin at the triple point. The curve for the solution is similar in shape to the curve for the pure liquid, but lies below it, at lower temperatures, as shown in Fig. 6.9.

The triple point is very close to, but not identical with, the normal freezing point. The normal freezing point is the temperature at which solid and liquid are in equilibrium in the presence of air, at an external pressure of one atmosphere. For water, for example, the freezing point is 0.0000 °C at 1 atm, while the triple point temperature is 0.0098 °C at 4.58-mmHg pressure.

The freezing point of a solution is defined as the temperature at which solid is in

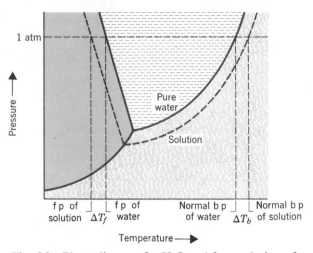

Fig. 6.9. Phase diagram for H_2O and for a solution of a nonvolatile solute in water (not drawn to scale). The freezing point of the solution is *lower* than the freezing point of pure water; the boiling point of the solution is higher than the boiling point of pure water.

equilibrium with the solution. It is usually (although not always) found that when a solution is cooled only the pure solvent precipitates, that is, the first solid material to appear is solvent only, and the solute particles remain in solution (which becomes more concentrated, since there is less solvent). Theoretically, the freezing point of a solution is the temperature at which the first solid particles begin to precipitate out of the solution as you cool the liquid. It is not uncommon, however, for solutions to become **supercooled,** that is, to be cooled to a temperature somewhat below the freezing point before any solid material appears.

At an external pressure of 1 atm, the temperature at which solid solvent is in equilibrium with the solution is *lower* than the temperature at which solid solvent is in equilibrium with the pure liquid. This is shown in Fig. 6.9 and is a consequence of the fact that the triple point of the solution is at a lower temperature than the triple point for the pure liquid. Within experimental uncertainty, the lowering of the normal freezing point is equal in magnitude to the lowering of the triple point. The difference between the freezing point (fp) of the pure solvent and of the solution is called the **freezing point depression,** ΔT_f, and is defined by the relation

$$\Delta T_f = \text{fp depression} = \text{fp pure solvent} - \text{fp solution} \qquad (6\text{-}12)$$

Note that both ΔT_f and ΔT_b are defined to be positive quantities. Compare Eqs. (6-10) and (6-12) carefully.

The **freezing point depression,** like the **boiling point elevation,** is a **colligative property.** For dilute solutions, the magnitude of the freezing point depression is directly proportional to the magnitude of the **vapor pressure lowering,** and therefore to the concentration of the solute. We express this proportionality in terms of the molality, m, by the relation

$$\Delta T_f = K_f m \qquad (6\text{-}13)$$

where K_f is a property of the *solvent only,* and is called the **molal freezing point depression constant,** or the **cryoscopic constant.**

Values of K_f for several substances are given in Table 6.2. Equation (6-13), like Eq. (6-11), is not exactly correct. It is a better approximation to reality the more dilute the solution. The value of K_f is equal to the freezing point depression that would be

Table 6.2. The Molal Freezing Point Depression Constants of Some Substances Commonly Used as the Solvent for Freezing Point Depression Measurements

Substance	K_f (kg · K/mol)	Normal fp (°C)
Acetic acid, CH_3COOH	3.90	16.6
Benzene, C_6H_6	5.1	5.5
Camphor, $C_{10}H_{16}O$	40.	174.
Cyclohexane, C_6H_{12}	19.2	6.5
Naphthalene, $C_{10}H_8$	6.8	80.2
Water, H_2O	1.86	0.0

observed for a 1 *m* solution of any nonvolatile solute *if* the solution is ideal. Very few solutions are ideal at a concentration of 1 *m*.

Freezing point depressions are larger than boiling point elevations for the same solution, because for any given solvent $K_f > K_b$. Compare the values for water and benzene in Tables 6.1 and 6.2. Because the freezing point depression is larger than the boiling point elevation, it is easier to measure and has a smaller percentage error, since the absolute error in reading the thermometer is the same for both a freezing point and a boiling point. Consequently, it is more convenient to obtain the molecular weight of an unknown solute by measuring the freezing point depression than by measuring the boiling point elevation. The method of calculation, however, is exactly the same as that described in Example 6.8.

The larger ΔT_f is, the smaller will be the experimental uncertainty in ΔT_f, and consequently, if the unknown is soluble in it, one tries to use a solvent such as camphor that has a very large K_f value. Of course that requires heating the camphor above its normal freezing point to prepare the solution.

The purpose of spreading common salt (NaCl), $CaCl_2$, or various other substances on roadways in the wintertime is to depress the freezing point of water and reduce the temperature at which ice forms. Another practical use of the freezing point depression is illustrated in Example 6.9.

EXAMPLE 6.9. The freezing point depression

Aqueous solutions of ethylene glycol, CH_2OHCH_2OH, are commonly used as antifreeze for automobile radiators. Pure ethylene glycol is a liquid with density 1.11 g · mL^{-1}.

(a) What should be the approximate molality of ethylene glycol in an aqueous solution, if the solution is to freeze at a temperature no higher than −15 °C?

Solution. Pure water freezes at 0.0 °C and has a molal freezing point depression constant, K_f, of 1.86 kg · K/mol. (See Table 6.2.) If the freezing point is lowered to −15 °C, $\Delta T_f = +15°$. Equation (6-13) is used to calculate the molality, as follows:

$$+15 = 1.86 \, m \quad \text{hence} \quad m = 8.1 \text{ molal}$$

(b) Explain why the value calculated in part **(a)** is only an approximation to the molality of ethylene glycol that lowers the freezing point of water to −15 °C.

Solution. Equation (6-13) is not exactly correct. It is a good approximation for dilute solutions, but 8.1 *m* is quite concentrated. To reduce the freezing point by as much as 15°, the concentration of solute must be quite high, and therefore a significant error is introduced by using Eq. (6-13).

(c) What volume of ethylene glycol should be added to 20.0 L of water (density 1.00 g/mL) to produce a solution with the molality calculated in part **(a)**?

Solution. Since the density of water is 1.00 g/mL, 20.0 L of water have a mass of 20.0×10^3 g, or 20.0 kg. The molality is the ratio of moles of solute to kilograms of solvent. Let x = the number of moles of ethylene glycol that must be added to 20.0 kg of water to produce a solution that is 8.1 m. Then

$$\frac{x \text{ mol } CH_2OHCH_2OH}{20.0 \text{ kg } H_2O} = \frac{8.1 \text{ mol } CH_2OHCH_2OH}{1.00 \text{ kg } H_2O}$$

Solving for x, we obtain $x = (20.0)(8.1) = 1.6 \times 10^2$ mol of ethylene glycol. The molecular weight of CH_2OHCH_2OH is 62.07 g/mol. Thus the mass of ethylene glycol needed to prepare this 8.1 m solution is $(62.07 \text{ g} \cdot \text{mol}^{-1})(1.6 \times 10^2 \text{ mol}) = 9.9 \times 10^3$ g. To find the volume required, we need the density of ethylene glycol.

$$\text{volume in mL} = \frac{\text{mass in g}}{\text{density in g/mL}} = \frac{9.9 \times 10^3 \text{ g}}{1.11 \text{ g/mL}} = 8.9 \times 10^3 \text{ mL}$$

Hence we must mix 8.9 L of ethylene glycol with 20.0 L of water to prepare a solution that is 8.1 m in ethylene glycol.

section 6.6
Freezing Point Depressions and Boiling Point Elevations for Solutions of Electrolytes

We have stressed the fact that both the freezing point depression and the boiling point elevation are colligative properties that depend on the *concentration* of solute particles and not, to a first approximation, on the nature of the solute particles. Freezing point depression data for several different aqueous solutions are given in Table 6.3.

An examination of the data in Table 6.3 shows that for 0.100 m solutions of glycerine, ethanol, dextrose, and sucrose, the freezing point depression is essentially the same, and the relation $\Delta T_f = 1.86 \, m$ is approximately correct for solutions of these four substances. The value of ΔT_f for 0.100 m solutions of HCl, KNO_3, or KCl, however, is almost (but not quite) twice as large as it is for 0.100 m solutions of dextrose, glycerine, ethanol, or sucrose. Note that the freezing point depression of 0.100 m solutions of 1 : 1 electrolytes is slightly less than the freezing point depression of 0.200 m solutions of nonelectrolytes. Similarly, the freezing point depression of

Table 6.3. **The Freezing Point Depression of Aqueous Solutions**[a]

Nonelectrolytes in H_2O			Electrolytes in H_2O		
Substance	*Concentration*	$\Delta T_f (K)$	*Substance*	*Concentration*	$\Delta T_f (K)$
Glycerine	0.100 m	0.187	(1 : 1) HCl	0.100 m	0.352
Ethanol	0.100 m	0.183	KNO_3	0.100 m	0.331
Dextrose	0.100 m	0.186	KCl	0.100 m	0.345
Sucrose	0.100 m	0.188	· · ·	· · ·	· · ·
Sucrose	0.200 m	0.376	(1 : 2) Na_2SO_4	0.100 m	0.434
Dextrose	0.200 m	0.372	and $CaCl_2$	0.100 m	0.494
Dextrose	0.300 m	0.558	(2 : 1) $NiCl_2$	0.100 m	0.538

[a] The molal freezing point depression constant, K_f, for water is 1.86.

0.100 m solutions of $1:2$ or $2:1$ electrolytes is a little less than the freezing point depression of 0.300 m solutions of nonelectrolytes.

We can understand the difference between the freezing point depression of 0.100 m solutions of HCl and of dextrose, because we know that there is no such species as "HCl" in dilute aqueous solutions; the solution actually contains 0.100 m $H^+(aq)$ ions plus 0.100 m $Cl^-(aq)$ ions. The total concentration of all solute particles, regardless of their nature, in 0.100 m HCl, is 0.200 m. Indeed, it was data such as this that first led scientists to propose that HCl and similar substances must dissociate into two particles in solution.

Substances that exist as ions in aqueous solution are called **electrolytes**, whereas substances that do not dissociate but exist as discrete molecules in aqueous solution are called **nonelectrolytes**. Ionic crystalline compounds are electrolytes, and there are molecules that dissociate in solution, such as HCl(g), that also are electrolytes. This will be discussed more fully in Sections 7.2, 7.3, and 7.4.

The freezing point depression and the boiling point elevation of solutions of electrolytes depend on the number of ions per *formula unit,* which we will denote by the symbol v. For HCl, KNO_3, and KCl, $v = 2$. Such electrolytes are called $1:1$ electrolytes, because the ratio of positive to negative ions is $1:1$. For Na_2SO_4, $v = 3$, because there are 2 Na^+ ions and one SO_4^{2-} ion per Na_2SO_4 unit. Thus a 0.100 m Na_2SO_4 solution has a total solute particle concentration of 0.300 m.

For solutions of electrolytes, if m is the molality of the *compound,* the expressions for the boiling point elevation and the freezing point depression should be written as

$$\Delta T_b = v K_b m \tag{6-14a}$$

and

$$\Delta T_f = v K_f m \tag{6-14b}$$

Deviations from these equations are larger than deviations from Eqs. (6-11) and (6-13) for solutions of nonelectrolytes. The reason for this is that the electrostatic force of attraction between positive and negative ions keeps oppositely charged ions closer together, on the average, than uncharged species would be, and prevents the ions from moving about in solution as freely as uncharged molecules do. These **interionic forces** result in the formation of **ion-pairs** (two ions of opposite charge moving togther as a unit for some space of time) or even larger aggregates of ions if the solution is more concentrated. Solutions of electrolytes deviate much more from ideal solution behavior than do solutions of nonelectrolytes, because of the interionic forces. Because the force of attraction between ions of opposite charge increases as the charge on the ions increases, deviations from ideal solution behavior are greater for solutions containing doubly and triply charged ions than for solutions containing only singly charged ions.

The experimental value of the ratio $\Delta T_f / K_f m$ is called the **van't Hoff mole number,** and is observed to be less than v, the actual number of moles of ions per formula unit. Consider the data in Table 6.4, for aqueous sodium chloride solutions.

Table 6.4. The van't Hoff Mole Number for Aqueous Solutions of Sodium Chloride, as a Function of the Molality of the NaCl[a]

Concentration of NaCl	ΔT_f (K)	$\Delta T_f / K_f m$
0.100 m	0.348	0.348/0.186 $= 1.87$
0.0100 m	0.0360	0.0360/0.0186 $= 1.94$
0.00500 m	0.0182	0.0182/0.0093 $= 1.96$

[a] $K_f = 1.86$ for water.

Note that the experimental value of $\Delta T_f / K_f m$ is not exactly 2 for any of the solutions in Table 6.4, but approaches 2 more closely as the solution becomes more dilute.

EXAMPLE 6.10. Freezing point depressions of solutions of electrolytes

Arrange the following solutions in order of decreasing freezing point.
(a) $0.050 \, F \, NaNO_3$ (b) $0.075 \, F \, CuSO_4$ (c) $0.060 \, F \, (NH_4)_2SO_4$
(d) $0.140 \, M$ sucrose, $C_{12}H_{22}O_{11}$ (e) $0.040 \, F \, BaCl_2$.

Solution. We will tabulate the calculations required to obtain the total concentration of all solute particles in each solution.

Substance	ν (ions/formula)	Solute Particle Concentrations	Total Concentration of All Solute Particles
$NaNO_3$	2	$[Na^+] = 0.050 \, M$ $[NO_3^-] = 0.050 \, M$	$0.100 \, M$
$CuSO_4$	2	$[Cu^{2+}] = 0.075 \, M$ $[SO_4^{2-}] = 0.075 \, M$	$0.150 \, M$
$(NH_4)_2SO_4$	3	$[NH_4^+] = 0.120 \, M$ $[SO_4^{2-}] = 0.060 \, M$	$0.180 \, M$
Sucrose	(nonelectrolyte)	$[sucrose] = 0.140 \, M$	$0.140 \, M$
$BaCl_2$	3	$[Ba^{2+}] = 0.040 \, M$ $[Cl^-] = 0.080 \, M$	$0.120 \, M$

The larger the total concentration of all solute particles, the lower the freezing point. Since we are only arranging the solutions in order of decreasing freezing point, the concentration unit is not important. The correct arrangement, starting with the solution with the highest freezing point, is as follows:

highest freezing point (a) $0.050 \, F \, NaNO_3$
next (e) $0.040 \, F \, BaCl_2$
next (d) $0.140 \, M$ sucrose
next (b) $0.075 \, F \, CuSO_4$
lowest freezing point (c) $0.060 \, F \, (NH_4)_2SO_4$

section 6.7
Osmotic Pressure

The phenomenon of osmotic pressure is a colligative property that requires the presence of a **semipermeable membrane.** A semipermeable membrane, also called a selectively permeable membrane, is a film with pores of a size such that small molecules can pass through the membrane, but larger molecules cannot. For aqueous solutions in particular, cell membranes and other biological membranes such as intestinal walls are permeable to water, but not to the larger solute particles.

Suppose there is pure water on one side of a semipermeable membrane, and an aqueous solution on the other side. We then observe that water flows through the membrane into the solution, decreasing the concentration of solute. Similarly, if there are solutions of different concentrations on either side of the membrane, water will flow from the less concentrated solution into the more concentrated solution.

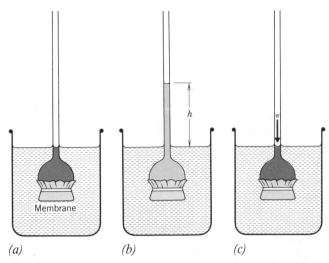

Fig. 6.10. A demonstration of osmosis and osmotic pressure. *(a)* The tube contains a solution and the beaker contains distilled water. *(b)* Water has flowed into the solution, diluting it. The solution rises to a height *h* above the surface of the water. *(c)* The pressure that must be applied to prevent the rise of the solution is the osmotic pressure, π.

The movement of water through a semipermeable membrane, from a solution of lesser solute concentration into one of greater solute concentration, is called **osmosis.** A simple apparatus for demonstrating osmosis is illustrated in Fig. 6.10.

The flow of solvent through the membrane is not affected by what is dissolved in the water, but only by the total concentration of solute particles, and it is therefore a colligative property. The only way to prevent the flow of solvent through the membrane is to apply a pressure to the solution. The amount of pressure that must be applied just to prevent the flow of solvent and thereby to keep the fluid level of the solution constant, is called the **osmotic pressure.**

Osmosis is an essential process for most living systems. Plant cells expand as water moves into them, and the pressure of the water causes the cell wall to expand. Osmosis is involved in the excretion of urine, and in the interchange of nutrients and waste products between tissue cells and their surroundings.

Consider a pure solvent and a solution of a nonvolatile solute, separated by a semipermeable membrane. Because the vapor pressure of the solvent is lower above the solution, solvent flows into the solution. The amount of pressure that must be exerted just to prevent this flow is the **osmotic pressure,** π. The relation between osmotic pressure and the concentration of solute can be very complex, but if the solution is dilute, the experimentally observed relation, known as the **van't Hoff equation,** is given by

$$\pi = cRT \tag{6-15}$$

where c is the solute concentration in moles per liter, R is the gas constant in $L \cdot atm \cdot mol^{-1}K^{-1}$, and π is the osmotic pressure in atmospheres. For solutions as dilute as 1×10^{-3} *M* or even 1×10^{-4} *M*, the osmotic pressure is readily measurable, and for this reason it is the colligative property most suitable for determining the molecular weights of biologically interesting molecules, which are usually species of very high molecular weight and fairly low solubility in water. Osmotic pressure measurements can be used to obtain the molecular weights of solutes whose molecular weights are as large as 30,000. For species with molecular weights much higher than 30,000, the concentration obtainable is generally too low to produce a measurable osmotic pressure. Example 6.11 illustrates the use of the van't Hoff equation for the determination of the molecular weight of a solute.

EXAMPLE 6.11. Molecular weight determination by measurement of the osmotic pressure

Insulin is a hormone that regulates carbohydrate metabolism by controlling the amount of glucose (a simple sugar) in the blood. Insulins from different animals are not identical, but they differ only slightly. A 2.0-g sample of insulin isolated from beef pancreas is dissolved in enough water to make 250.0 mL of solution, and the osmotic pressure of the solution is measured and found to be 26.1 mmHg at 30.0 °C. What is the molecular weight of this insulin?

Solution. Let M = molecular weight of insulin. The number of moles of insulin in this solution is therefore (2.0/M) mol, and the volume of the solution is 0.2500 L. Therefore the concentration is

$$\frac{(2.0/M) \text{ mol of insulin}}{0.2500 \text{ L}} = (8.0/M) \text{ mol/L}$$

In Eq. (6-15), if R is expressed in L · atm · mol^{-1}K^{-1}, the osmotic pressure must be given in atmospheres.

$$26.1 \text{ mmHg} = \frac{26.1}{760} \text{ atm}$$

Substituting in $\pi = cRT$ yields

$$\left(\frac{26.1}{760}\right) \text{atm} = \left(\frac{8.0}{M} \frac{\text{mol}}{\text{L}}\right)\left(0.08206 \frac{\text{L} \cdot \text{atm}}{\text{mol} \cdot \text{K}}\right)(303.2 \text{ K})$$

Solving this for M, we obtain

$$M = (8.0)(0.08206)(303.2)(760)/26.1 = 5.8 \times 10^3$$

The molecular weight of insulin is about 5800.

Because Eq. (6-15) is so similar to the ideal gas law, $P = (n/V)RT$, there have been attempts to explain the van't Hoff law for the osmotic pressure in terms of some sort of ideal gas behavior of the solute particles. Such interpretations are not really valid, as the osmotic pressure is not exerted by the solute particles. The osmotic pressure is a pressure that must be applied to a more concentrated solution to bring about equilibrium between the *solvent* in two solutions of different concentrations that are separated by a semipermeable membrane.

A method known as **reverse osmosis** can be used for desalinating water. We have defined the osmotic pressure as the pressure that must be applied just to prevent solvent flow from the less concentrated solution into the more concentrated solution. If a pressure greater than the osmotic pressure is applied, solvent is forced to flow from the more concentrated into the less concentrated solution. The average concentration of dissolved salts in sea water results in an osmotic pressure for sea water of roughly 22 atm. If sea water and fresh water are separated by a semipermeable membrane and a pressure in substantial excess of 22 atm is applied to the sea water, pure water is forced out of the sea water, enlarging the amount of fresh water on the other side of the membrane.

section 6.8
Solutions of Two Volatile Liquids

If two volatile liquids are mixed to form a solution, the gas phase above the solution will contain molecules of both species, and the vapor pressure of the solution is the sum of the vapor pressures of each component. If Raoult's law is obeyed by *both* liquids over the entire range of concentrations possible for such a mixture, the solution is said to be **ideal**. Most real solutions are not ideal, but there are some ideal solutions. If the two molecules are similar chemically and structurally, the solution is more likely to be ideal, because a solution of two liquids A and B will be ideal provided that the forces of attraction between molecules A and A, A and B, and B and B, are all the same.

A common example of an ideal solution is a mixture of benzene, C_6H_6, and toluene, $C_6H_5CH_3$. Benzene (B) and toluene (T) are both organic solvents, and are closely related structurally. We will write equations for the pressure of benzene and of toluene above solutions of the two components, but the equations will be general and will apply for any ideal solution of two volatile liquids. At 20 °C the vapor pressure of pure benzene, P_B^0, is 75 mmHg, while the vapor pressure of pure toluene, P_T^0, is 22 mmHg. Benzene, with a higher vapor pressure, is the more volatile of these two liquids. Since the solution is ideal, the pressure of each of the components is given by Raoult's law:

$$P_B = P_B^0 X_B^{liq} \qquad \text{and} \qquad P_T = P_T^0 X_T^{liq} \qquad\qquad (6\text{-}16)$$

where X_B^{liq} and X_T^{liq} are the mole fractions of benzene and toluene, respectively, in the liquid phase. Since both benzene and toluene are volatile, the vapor phase will be a mixture of both gases, and we can describe the composition of the vapor phase by giving the mole fraction of each component, X_B^{gas} and X_T^{gas}. Note that we need the superscripts "liq" and "gas" on the symbol for mole fraction, because both phases are mixtures of the two components, and the composition of the vapor phase is different from the composition of the liquid phase. Of course, for each phase, the sum of the mole fractions is unity, that is,

$$X_B^{gas} + X_T^{gas} = 1 \qquad \text{and} \qquad X_B^{liq} + X_T^{liq} = 1 \qquad\qquad (6\text{-}17)$$

Let us consider two different solutions of benzene and toluene. Solution 1 contains 20.0-mol % benzene and 80.0-mol % toluene in the liquid phase at 20 °C. For the vapor in equilibrium with this solution at 20 °C,

$$P_B = (75 \text{ mmHg})(0.200) = 15.0 \text{ mmHg}$$
$$P_T = (22 \text{ mmHg})(0.800) = 17.6 \text{ mmHg}$$
$$P_{total} = P_B + P_T = 32.6 \text{ mmHg}$$

Solution 2 contains 70.0 mol % benzene and 30.0-mol % toluene in the liquid phase at 20 °C. For the vapor in equilibrium with this solution at 20 °C,

$$P_B = (75 \text{ mmHg})(0.700) = 52.5 \text{ mmHg}$$
$$P_T = (22 \text{ mmHg})(0.300) = 6.6 \text{ mmHg}$$
$$P_{total} = P_B + P_T = 59.1 \text{ mmHg}$$

We can put all of this information (and more) on one diagram, as illustrated in Fig. 6.11. On the ordinate axis we plot vapor pressure in millimeters of mercury. The composition of the liquid phase is plotted as the abscissa. Since

$$X_B + X_T = 1$$

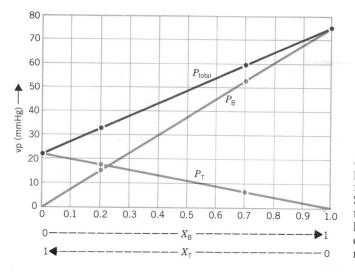

Fig. 6.11. Vapor pressure plots of benzene–toluene mixtures at 20 °C. Solutions of benzene and toluene are ideal. Raoult's law is valid for both components over the entire range of concentrations.

in each phase, when X_B runs from 0 to 1, X_T runs from 1 to 0, and the same scale can be read to give both values. We have four points for the plot of the vapor pressure of benzene, P_B, as a function of the composition of the solution. When $X_B^{liq} = 0$ (the solution is pure toluene, no benzene at all), clearly $P_B = 0$. Furthermore, when $X_B^{liq} = 1$ (the solution is pure benzene), then $P_B = P_B^0 = 75$ mmHg. If these points, plus the two points for solutions 1 and 2 are plotted, they lie on a straight line. (See Fig. 6.11.) The plot is linear because the equation $P_B = 75 X_B^{liq}$ is the equation of a straight line drawn through the two points ($X_B^{liq} = 0$, $P_B = 0$) and ($X_B^{liq} = 1$, $P_B = 75$). Similarly, the plot of the vapor pressure of toluene, P_T, is a straight line through the two points ($X_B^{liq} = 0$, $P_T = 22$) and ($X_B^{liq} = 1$, $P_T = 0$).

We also have four points for the plot of P_{total} versus X_B^{liq}, and we see that this plot is a straight line drawn through the two points ($X_B^{liq} = 0$, $P_{total} = 22$) and ($X_B^{liq} = 1$, $P_{total} = 75$). Thus we have, in Fig. 6.11, a vapor pressure diagram typical of that for any ideal **binary liquid solution** (that is, a solution of two volatile liquids) in which there are three separate straight line plots, for P_B, P_T, and P_{total}.

The composition of the vapor phase in equilibrium with a binary liquid solution will not be the same as the composition of the liquid. Let us calculate the composition of the vapor phase in equilibrium with solution 1, for which $X_B^{liq} = 0.200$. To do this, we must use **Dalton's law of partial pressures** in the form given as Eq. (3-44):

$$X_B^{gas} = P_B/P_{total} = 15.0/32.6 = 0.460 \quad \text{or} \quad 46.0 \text{ mol}\%$$
$$X_T^{gas} = P_T/P_{total} = 17.6/32.6 = 0.540 \quad \text{or} \quad 54.0 \text{ mol}\%$$

You can check arithmetic errors by remembering that $X_B^{gas} + X_T^{gas} = 1$. Note that X_B^{gas} (0.460) is larger than X_B^{liq} (0.200). This means that the vapor phase is richer in benzene (and consequently poorer in toluene) than is the liquid phase.

The pressure of benzene above the solution can be calculated both from the mole fraction of benzene in the liquid phase, using Raoult's law, and from the mole fraction of benzene in the gas phase, using Dalton's law. For a component, B, of an ideal solution of two volatile liquids

$$P_B = P_B^0 X_B^{liq} = P_{total} X_B^{gas} \tag{6-18}$$

Example 6.12 illustrates the use of these relations.

EXAMPLE 6.12. **Difference in composition of liquid and vapor phases for binary liquid mixtures**

Two liquids, $\underline{A}$ and $\underline{B}$, form ideal solutions. A solution of $\underline{A}$ and $\underline{B}$ is prepared at 30 °C in which the mole fraction of $\underline{A}$ is 0.256. In the vapor phase in equilibrium with this solution, the mole fraction of $\underline{A}$ is 0.318 and the total pressure is 673 mmHg.

(a) Calculate the partial pressures of $\underline{A}$ and $\underline{B}$ in the vapor phase.

Solution

$$P_{\underline{A}} = X_{\underline{A}}^{gas} P_{total} = (0.318)(673 \text{ mmHg}) = 214 \text{ mmHg}$$

$$P_{\underline{B}} = P_{total} - P_{\underline{A}} = 673 \text{ mmHg} - 214 \text{ mmHg} = 459 \text{ mmHg}$$

(b) Calculate the vapor pressures of pure liquid $\underline{A}$ and pure liquid $\underline{B}$ at 30 °C.

Solution

$$P_{\underline{A}} = P_{\underline{A}}^0 X_{\underline{A}}^{liq} = 214 \text{ mmHg} = P_{\underline{A}}^0 (0.256)$$

Therefore,

$$P_{\underline{A}}^0 = \frac{214 \text{ mmHg}}{0.256} = 836 \text{ mmHg} \qquad P_{\underline{B}}^0 = \frac{P_B}{1 - X_{\underline{A}}^{liq}} = \frac{459 \text{ mmHg}}{0.744} = 617 \text{ mmHg}$$

Fractional Distillation

Because the liquid and vapor phases have different compositions, it is possible to separate a binary liquid solution into its two pure components. The process for carrying out this separation is called **fractional distillation.**

The principle of fractional distillation can be understood by considering a mixture of benzene and toluene that we wish to separate into pure benzene and pure toluene. If we vaporize a small amount of the mixture, the vapor will be richer in benzene than the original liquid was. If we now draw off and then condense that vapor, we will obtain a small amount of a new liquid of the same composition as the vapor, that is, richer in benzene than the original liquid. A second vaporization of this new liquid will produce a vapor still richer in benzene. If we can arrange a very large number of successive vaporizations and condensations, we can eventually produce a vapor that is pure benzene.

The apparatus used for carrying out this succession of many vaporizations and condensations is called a **distillation column.** (See Fig. 6.12.) The **packing material** in the column, which may be steel wool, glass coils, or a twisted steel gauze, provides cooler surfaces to condense the vapor as it rises in the column. The flask at the bottom into which the mixture is placed (the pot) is heated to cause vaporization. As the vapor rises and reaches the cooler packing material, it condenses. Some of the liquid drops down into the pot again, but some remains in the column, and as warm vapor reaches it again, revaporizes and rises a little higher in the column before it recondenses. The progress of the vapor upward through the column takes a reasonably long period of time, and is readily observable.

The vapor that comes out of the top of the column (called the **distillate**) is pure benzene, and is condensed and led away from the material in the pot, into a receiving vessel. The liquid (called the **residue**) that remains in the pot at the end of the

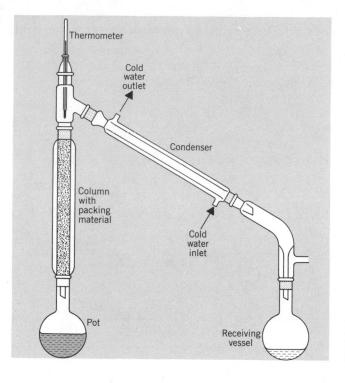

Fig. 6.12. The apparatus used to carry out a fractional distillation.

distillation, is toluene. In general, for ideal or nearly ideal binary liquid mixtures, the distillate is the more volatile of the two liquids, and the residue is the less volatile.

Nonideal Solutions

All the discussion so far has been about ideal solutions. Ideal solutions of A and B are obtained when the forces of attraction between A and A, B and B, and A and B are all alike. Thus a molecule of A is bound to the liquid phase just as strongly whether its neighboring molecules are A molecules or B molecules. In such a case, when you mix liquids A and B together, the volume of the solution (V_{soln}) is the sum of the volumes of the individual liquids, that is, $V_{soln} = V_A + V_B$, and there is no volume change on mixing. We generally express this by saying $\Delta V_{mix} = 0$, which is read as "the change in volume on mixing is zero." Furthermore, no heat is released when liquids A and B are mixed together, nor is any heat absorbed, if the solution is ideal. Thus the **heat of mixing,** ΔH_{mix}, is zero for ideal solutions. We have already noted, however, that only a few solutions are ideal. What changes do we observe when we deal with real solutions?

There are two possible types of deviations from ideal solution behavior. The first type occurs when the unlike molecules, A and B, are attracted to one another more strongly than are the like molecules A to A or B to B. A very strong force of attraction between A and B makes it harder to get A and B into the vapor phase than if the solution were ideal. Hence the vapor pressure of each component is less than the value predicted by Raoult's law.

$$P_A < P_A^0 X_A^{liq} \qquad \text{and} \qquad P_B < P_B^0 X_B^{liq}$$

This situation is called a **negative deviation from Raoult's law,** and the vapor pressure diagram for such a system is illustrated in Fig. 6.13. Note that the actual vapor

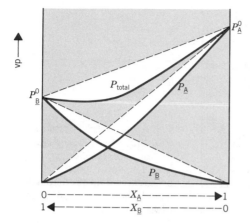

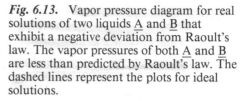

Fig. 6.13. Vapor pressure diagram for real solutions of two liquids $\underline{A}$ and $\underline{B}$ that exhibit a negative deviation from Raoult's law. The vapor pressures of both $\underline{A}$ and $\underline{B}$ are less than predicted by Raoult's law. The dashed lines represent the plots for ideal solutions.

pressure plot is always below the ideal plot (drawn as a dashed line) for each of the three curves, P_A, P_B, and P_{total}.

If there are negative deviations from Raoult's law, when liquids A and B are mixed the total volume of liquid will *decrease*, as A and B molecules are pulled toward one another by the strong forces of attraction. Thus an A molecule is pulled closer to a neighboring B molecule than it is to another A molecule, and the volume of the solution is less than the sum of the volumes of the individual pure liquids:

$$V_{soln} < V_A + V_B \quad \text{for negative deviations from Raoult's law} \quad (6\text{-}19)$$

Since the final volume of the solution is less than the original volume of the two separate liquids, there is a negative change of the volume on mixing, and $\Delta V_{mix} < 0$. Furthermore, a strong force of attraction between A and B leads to a decrease in potential energy as A and B molecules come close to one another (refer to Section 5.3), and as the system achieves a state of lower potential energy, heat is released. Thus, if A and B are more strongly attracted to one another than A to A or B to B, the solution will get warm when A and B are mixed. A release of heat corresponds to a negative value of ΔH. (See Section 5.5.) Therefore the heat of mixing is negative, $\Delta H_{mix} < 0$, when there are negative deviations from Raoult's law. Examples of binary liquid mixtures that have large negative deviations from Raoult's law are acetone–$CHCl_3$, and diethyl ether–HCl.

The second type of deviation from ideal solution behavior occurs when the unlike molecules A and B are less strongly attracted to one another than are the like molecules A to A or B to B. In this case, it is easier for A or B to escape from the solution and get into the vapor phase than if the solution were ideal. As a result, the vapor pressure of each component is greater than the value predicted by Raoult's law.

$$P_A > P_A^0 X_A^{liq} \quad \text{and} \quad P_B > P_B^0 X_B^{liq}$$

The total pressure is greater than predicted for an ideal solution also. This situation is an example of a **positive deviation from Raoult's law.** The vapor pressure diagram for such a system is illustrated in Fig. 6.14. Since A and B tend to stay farther apart from one another than A and A molecules or B and B molecules, the volume expands when the pure liquids are mixed, hence $\Delta V_{mix} > 0$. Furthermore, heat must be absorbed in order for A and B to mix and have the solution at the same temperature as the original pure liquids. The mixing of A and B is an endothermic reaction, and one can feel the solution getting cooler as A and B mix. Thus the heat of mixing is a positive quantity, $\Delta H_{mix} > 0$. Examples of binary liquid mixtures displaying positive deviations from Raoult's law are $CHCl_3$–ethyl alcohol, and CCl_4–methyl alcohol mixtures. Carbon

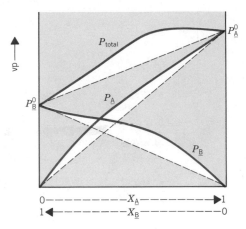

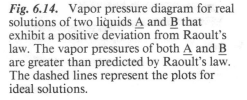

Fig. 6.14. Vapor pressure diagram for real solutions of two liquids $\underline{A}$ and $\underline{B}$ that exhibit a positive deviation from Raoult's law. The vapor pressures of both $\underline{A}$ and $\underline{B}$ are greater than predicted by Raoult's law. The dashed lines represent the plots for ideal solutions.

tetrachloride, CCl_4, is nonpolar, while methyl alcohol is polar and hydrogen bonded. Thus the forces of attraction between the unlike molecules CCl_4 and CH_3OH are considerably weaker than the forces of attraction between the like molecules.

section 6.9
Henry's Law

Gases dissolve in liquids, and the effect of the pressure of the gas above the solution on the solubility of the gas is described by **Henry's Law,** which states that the solubility of a gas, A, in a liquid is directly proportional to the equilibrium partial pressure of the gas above the solution:

$$[A] = k_H P_A \tag{6-20}$$

The value of the proportionality constant, k_H, depends on the nature of both the gas and the solvent.

Carbonated soft drinks and champagne contain dissolved CO_2 gas in equilibrium with pressures in excess of 1 atm. When the bottle is opened and the pressure falls to 1 atm (with a noticeable "pop"), the solubility of CO_2 decreases, and bubbles of CO_2 plus some water vapor escape from the solution. If the bottle is left open for any length of time, the champagne or soda becomes flat.

Another example of the operation of Henry's law occurs in deep-sea diving. The pressure of the ocean at great depths is several atmospheres, and in order to be able to exhale, the diver must breathe air at a pressure of several atmospheres. At the higher pressure, the solubility in the blood of the two major constituents of air, O_2 and N_2, is several times its normal value. The O_2 is metabolized, but the N_2 is not. When the diver is ready to return to sea level, and ascends to lower pressure, the solubility of N_2 decreases, and bubbles of nitrogen gas form in the blood. These escaping bubbles cause extreme pain and, if the diver ascends too rapidly, can result in death. This phenomenon is called "the bends" because the afflicted diver bends over in pain. To avoid the bends, divers are brought to the surface slowly. A chemical solution to the problem of the bends was proposed by Joel Hildebrand. Since helium is only about one half as soluble in blood as nitrogen, he suggested that divers should breathe a mixture of helium and oxygen. Divers air tanks now contain not ordinary air, but a mixture of He and O_2.

Summary

If a solution is unsaturated, it is possible to dissolve more solute in it than is presently dissolved, at the same temperature. A **saturated solution** contains the maximum amount of solute that can be dissolved at that temperature. Adding more solute to an already saturated solution results in a layer of excess solid at the bottom of the container. A **dynamic equilibrium** exists between the excess solid and the saturated solution. The two opposing reactions that occur at the same rate are (1) the dissolution of solute into the solution, and (2) the precipitation of solute out of the solution.

The three most widely used units to describe the concentration of a solute are the **mole fraction**, the **molality**, and the **molarity**. Definitions of these three units are summarized below:

$$X_{sol} = \frac{\text{moles solute}}{\text{total no. moles of all substances in solution}}$$

Molality, m = moles of solute per kilogram of solvent

Molarity, M or c, = moles of solute per liter of solution

Properties of solutions that depend only on the total solute concentration and not on the nature of the solute particles are known as **colligative properties**. There are four colligative properties: the **vapor pressure lowering**, the **boiling point elevation**, the **freezing point depression**, and the **osmotic pressure**.

The pressure of solvent in equilibrium with a dilute solution is always less than the equilibrium vapor pressure of the pure liquid. **Raoult's law** states that the vapor pressure of a dilute solution is directly proportional to the mole fraction of solvent, and the proportionality constant is the vapor pressure of the pure solvent. The vapor pressure *lowering* can then be shown to be directly proportional to the mole fraction of *solute*.

Because the vapor pressure of the solution is lower than that of the pure liquid, the solution must be raised to a higher temperature than the pure liquid before it will boil, that is, before its vapor pressure reaches 1 atm. The elevation of the boiling point is directly proportional to the concentration of the solute, usually expressed as a molality: $\Delta T_b = K_b m$. The proportionality constant, K_b, is a property of the solvent only.

The lowering of the vapor pressure also results in a depression of the freezing point, $\Delta T_f = K_f m$. The molal freezing point depression constant, K_f, is a property only of the solvent. Both the boiling point elevation and the freezing point depression can be used to determine the molality, and from that the molecular weight, of a solute whose molecular weight is not known.

For a given solution, the freezing point depression is larger than the boiling point elevation, and freezing point depressions can therefore be measured with a smaller percentage uncertainty. For this reason it is more common to determine a molecular weight by measuring the freezing point depression than by measuring the boiling point elevation.

For very large molecules, particularly macromolecules of biological interest, the solubility is usually so low that both the freezing point depression and the boiling point elevation are too small to be measured. The colligative property most suitable for determining the molecular weight of a large molecule is the **osmotic pressure**. **Osmosis** is the flow of solvent through a **semipermeable membrane** from a less concentrated solution into a more concentrated solution. The pressure that must be applied to the more concentrated solution just to prevent this solvent flow is the osmotic pressure. For very dilute solutions, the osmotic pressure is directly proportional to the molality or molarity of the solute, and the proportionality constant is RT.

A **binary liquid solution** of two volatile liquids is **ideal** if Raoult's law is obeyed by both components over the entire range of compositions. The vapor phase in equilibrium with such a mixture has a composition different from that of the liquid phase. Because of the difference in composition of the liquid and vapor phases, an ideal or nearly ideal binary liquid mixture can be separated into its two pure components by **fractional distillation.**

The solubility of a gas in a liquid is directly proportional to the partial pressure of the gas in equilibrium with the liquid solution. This relationship is known as **Henry's law.**

Exercises

Sections 6.1 and 6.2

1. How many grams of NaOH must be weighed out to prepare 250.0 mL of 0.100 F NaOH?

2. Calculate the molarity of (a) A solution containing 33.3295 g of glucose, $C_6H_{12}O_6$, per liter of solution. (b) A saturated solution of CO_2 in water that contains 0.145 g of CO_2 per 100 mL of solution at 25 °C.

3. Calculate the formality of (a) A solution containing 45.5880 g of ammonium sulfate per liter. (b) A solution containing 9.3065 g of sodium pyrophosphate, $Na_4P_2O_7$, per 250.0 mL of solution. (c) A saturated solution of silver chloride that contains 0.192 mg of AgCl per 100.0 mL of solution at 25 °C.

4. What is the formality of a K_2CrO_4 solution prepared by diluting 10.00 mL of a 0.500 F solution to 25.00 mL?

5. What are the molarities of the ions in each of the following solutions? (a) 0.100 F NH_4Cl (b) 0.200 F $BaCl_2$ (c) 0.180 F $La(NO_3)_3$.

6. What are the molarities of the ions in each of the following mixtures? Assume complete dissociation of the salts and additivity of the volumes.

   (a) A 50.00-mL portion of 0.100 F $CuSO_4$ is mixed with 200.0 mL of 0.040 F K_2SO_4.
   (b) An 80.00-mL portion of 0.200 F $Ba(NO_3)_2$ is mixed with 20.00 mL of 0.500 F NH_4NO_3.

7. What is the molality of ethanol, CH_3CH_2OH, in a solution of ethanol and water in which the mole fraction of ethanol is 0.0820?

8. A concentrated HCl solution is 12.0 F. Write directions for preparing 500.0 mL of 1.00 F HCl by diluting the concentrated acid.

9. A solution is prepared by dissolving 19.5392 g of solid benzoic acid, C_6H_5COOH, in 300.0 mL of benzene, C_6H_6, at 20 °C. The density of benzene at this temperature is 0.8787 g $\cdot$ mL^{-1}. Calculate the mole fraction and molality of benzoic acid in this solution.

10. A saturated solution of calcium hydroxide at 0 °C is 0.0250 F. Calculate the maximum number of grams of solid calcium hydroxide that can be dissolved in 100.0 mL of water at 0 °C, assuming the volume of solution is 100.0 mL.

11. You are given a liter of a clear solution and told that it is a saturated solution of NaCl in water. What procedure could you carry out to verify that the solution is indeed saturated?

12. Write a detailed set of directions for preparing 500.00 mL of 0.1000 F potassium dichromate, $K_2Cr_2O_7$.

13. What is the molarity of a solution of the simple sugar levulose, molecular weight 180.16, that is 5.0000% levulose by weight and has a density of 1.0199 $g \cdot cm^{-3}$?

Section 6.3

14. A 3.50-g sample of an unknown nonvolatile substance, Q, is dissolved in 109.359 g of benzene, C_6H_6, and is found to lower the vapor pressure of benzene from 200.0 to 196.4 mmHg.

    (a) What are the mole fractions of benzene and of Q in this solution?

    (b) What is the molecular weight of Q? Be careful about the number of significant figures used in reporting the molecular weight of Q.

15. Dextrose, $C_6H_{12}O_6$, is a simple sugar. An aqueous solution that is 5.500% dextrose by weight has a density of 1.0213 g/mL at 20 °C. The vapor pressure of pure H_2O at 20 °C is 17.535 mmHg. How much *lower* is the vapor pressure of this solution than the vapor pressure of pure water at 20 °C?

16. A solution is prepared by dissolving 34.09 g of C_6H_5OH in 100.0 g of H_2O at 25 °C. The measured vapor pressure of this solution is 21.9 mmHg. The vapor pressure of pure H_2O at 25 °C is 23.756 mmHg. Is this solution ideal? Show all work necessary to justify your answer.

Section 6.4

17. The boiling point of a solution of 2.933 g of naphthalene, $C_{10}H_8$, in 80.00 g of nitrobenzene, $C_6H_5NO_2$, is 1.50° higher than that of the pure nitrobenzene used to prepare the solution. What is the ebullioscopic constant for nitrobenzene?

18. A 2.076-g sample of an unknown solid is dissolved in 58.00 g of pure carbon tetra-chloride, CCl_4. The boiling point of this solution is found to be 78.2 °C. Calculate the molecular weight of the unknown compound using data from Table 6.1.

19. D-fructose is a simple sugar that occurs in fruits. It has a molecular weight of 180.16. An aqueous solution that is 4.000% D-fructose by weight has a density of 1.0158 $g \cdot mL^{-1}$ at 20 °C. What is the boiling point of this solution?

Section 6.5

20. What is the freezing point of the solution of Exercise 19?

21. Anthracene, $C_{14}H_{10}$, is a nonvolatile organic solid. What is the freezing point of a solution of 5.346 g of anthracene in 75.00 g of benzene, C_6H_6? Use data from Table 6.2.

22. A 2.736-g sample of an unknown organic liquid with empirical formula $(C_3H_6Cl)_n$ is dissolved in 93.47 g of cyclohexane, C_6H_{12}. The freezing point of this solution is 2.87 °C. The freezing point of the cyclohexane used to prepare the solution is 6.52 °C. Calculate the molecular formula and exact molecular weight of the un-known.

Section 6.6

23. What are the freezing and boiling points of the following solutions, assuming they are ideal?

    (a) 0.050 m $NaNO_3$ (b) 0.050 m $MgBr_2$ (c) 0.050 m $La(NO_3)_3$

24. Arrange the following solutions in order of increasing boiling point, starting with the solution with the lowest boiling point.

    (a) 0.040 F KBr   (b) 0.060 M ethanol   (c) 0.050 F $CaCl_2$
    (d) 0.060 F $LiNO_3$

Section 6.7

25. A semipermeable membrane separates a solution that is $0.012\ M$ in glucose from one that is $0.250\ M$ in glucose. On which of these solutions must pressure be applied to prevent a net flow of water through the membrane?

26. What is the osmotic pressure of a solution of 1.841 g of sucrose, $C_{12}H_{22}O_{11}$, in enough water to make 500.0 mL of solution at 25 °C? Express the answer both in atmospheres and in millimeters of mercury.

27. A saturated aqueous solution of a polypeptide contains 0.283 g/100.0 mL of solution at 35 °C. The osmotic pressure of this solution is measured and found to be 6.3 mmHg. Calculate the molecular weight of this polypeptide.

Section 6.8

28. At 30.0 °C the vapor pressure of pure benzene is 125 mmHg, while that of pure toluene is 39.0 mmHg. Solutions of benzene and toluene are ideal. A solution is prepared by mixing 0.300 mol of benzene with 0.300 mol of toluene. Calculate the partial pressures of benzene and toluene in the vapor phase in equilibrium with this solution. What is the vapor pressure of this solution?

29. The liquids octane, C_8H_{18}, and nonane, C_9H_{20}, form a very nearly ideal solution. At 66 °C, the vapor pressure of octane is 100.0 mmHg, while that of nonane is 40.0 mmHg. Assume solutions of these liquids are ideal. If 200.0 g of octane and 300.0 g of nonane are mixed together

    (a) What is the mole fraction of each component in the solution?

    (b) What is the partial pressure of each component in the vapor phase in equilibrium with this mixture at 66 °C?

    (c) What is the composition of the vapor phase in equilibrium with this mixture at 66 °C?

30. At 24 °C, the vapor pressure of pure liquid C is 328.0 mmHg, and the vapor pressure of pure liquid D is 174.6 mmHg. In the vapor in equilibrium with a binary solution of C and D at 24 °C in which the mole fraction of C is 0.048, the partial pressures of C and D, respectively, are 18.5 and 166.3 mmHg.

    (a) Do C and D obey Raoult's law? Show all calculations necessary to answer this question.

    (b) Does this system exhibit positive or negative deviations from Raoult's law, or is it ideal? Explain your answer.

Multiple Choice Questions

1. In an aqueous solution of strontium nitrate, the $[NO_3^-]$ is 0.080 M. This solution is correctly labeled

    (a) $0.080\ F\ Sr(NO_3)_2$ (b) $0.160\ F\ Sr(NO_3)_2$ (c) $0.080\ M\ Sr^{2+}$
    (d) $0.040\ F\ Sr(NO_3)_2$ (e) $0.080\ M\ NO_3^-$

2. Which of the following solutions has the lowest freezing point?

    (a) $0.10\ m$ sucrose   (b) $0.10\ m\ NiCl_2$ (c) $0.10\ m\ CuSO_4$
    (d) $0.10\ m\ NH_4NO_3$ (e) $0.20\ m$ glucose

3. An aqueous solution of acetone, CH_3COCH_3, is 10.00% acetone by weight. What is the mole percentage of acetone in this solution?

    (a) 3.332 mol%   (b) 5.000 mol%   (c) 10.00 mol%   (d) 11.11 mol%
    (e) 17.22 mol%

4. The density of an aqueous solution of acetone, CH_3COCH_3, that is 10.00% acetone by weight is $0.9867 \, g \cdot mL^{-1}$ at 20 °C. What is the molarity of acetone in this solution at 20 °C?

(a) $0.1722 \, M$ (b) $0.9867 \, M$ (c) $1.699 \, M$ (d) $3.332 \, M$ (e) $9.867 \, M$

5. A 10.00-g sample of $CuSO_4 \cdot 5H_2O$ was dissolved in sufficient water to make 500.0 mL of solution. What is the molarity of Cu^{2+} in this solution?

(a) $0.04005 \, M$ (b) $0.06266 \, M$ (c) $0.08010 \, M$ (d) $0.1253 \, M$ (e) $0.1574 \, M$

6. The freezing point of an aqueous solution of a nonelectrolyte is -0.14 °C. The molality of this solution is

(a) $1.86 \, m$ (b) $1.00 \, m$ (c) $0.15 \, m$ (d) $0.14 \, m$ (e) $0.075 \, m$

7. 1,2-Benzanthracene is a yellow–brown nonvolatile solid with molecular weight $228.29 \, g \cdot mol^{-1}$. An 18.2632-g sample of 1,2-benzanthracene is dissolved in 250.0 g of benzene, C_6H_6. What is the vapor pressure of this solution at 25 °C, in millimeters of mercury, if the vapor pressure of pure benzene is 93.4 mmHg at this temperature?

(a) 60.8   (b) 91.1   (c) 93.4   (d) 95.7   (e) 760

8. Assuming that all volumes are additive, how much water should be added to 25.00 mL of 6.00 F HNO_3 to prepare 0.500 F HNO_3?

(a) 350 mL   (b) 325 mL   (c) 300 mL   (d) 275 mL   (e) 250 mL

9. What is the $[NH_4^+]$ in 0.0520 F $(NH_4)_2SO_4$?

(a) $0.0260 \, M$ (b) $0.0520 \, M$ (c) $0.104 \, M$ (d) $0.520 \, M$ (e) $1.04 \, M$

10. An aqueous solution of ethanol, CH_3CH_2OH, that is 12.00% ethanol by weight, has a density of $0.9808 \, g \cdot mL^{-1}$ at 20 °C. What is the molality of ethanol in this solution?

(a) 0.05063   (b) 0.1200   (c) 2.555   (d) 2.960   (e) 12.00

11. The freezing point of a solution prepared by dissolving 20.5461 g of a nonvolatile nonelectrolyte with empirical formula $(C_3H_2)_n$ in 400.0 g of benzene is 4.33 °C. The benzene used to prepare the solution froze at 5.48 °C, using the same thermometer. The correct molecular formula of this compound is

(a) C_3H_2 (b) C_6H_4 (c) C_9H_6 (d) $C_{15}H_{10}$ (e) $C_{18}H_{12}$

12. The boiling point of a 1.00 m solution of $CaCl_2$ should be elevated by

(a) exactly 0.51°   (b) somewhat less than 1.02°   (c) exactly 1.02°
(d) somewhat less than 1.53°   (e) exactly 1.53°

13. What volume of a 0.0500 F $K_2Cr_2O_7$ solution should you pipet into a 500.0-mL volumetric flask in order to dilute the more concentrated solution to 0.0200 F?

(a) 250 mL   (b) 200 mL   (c) 150 mL   (d) 100 mL   (e) 50.0 mL

14. The liquids benzene, C_6H_6 (molecular weight = 78.11) and toluene, $C_6H_5CH_3$, (molecular weight = 92.14) form ideal solutions. At 35 °C the vapor pressure of benzene is 160.0 mmHg, while that of toluene is 50.0 mmHg. If 64.05 g of benzene and 106.26 g of toluene are poured into a large container which is then covered and maintained at 35 °C, what is the mole fraction of toluene *in the vapor phase* when the system comes to equilibrium?

(a) 0.305   (b) 0.584   (c) 0.624   (d) 0.695   (e) 0.762

15. How many grams of potassium permanganate, $KMnO_4$, are needed to prepare 250.00 mL of a 0.1000 F solution?

(a) 3.951 g   (b) 9.877 g   (c) 15.80 g   (d) 39.51 g   (e) 158.0 g

16. The concentration of a saturated solution of a certain polypeptide is $1.0 \times 10^{-3} \, M$ at 25 °C. The osmotic pressure of this solution, in millimeters of mercury, is

(a) 0.0245   (b) 0.760   (c) 18.6   (d) 24.5   (e) 156

17. A 20.00 mL portion of 0.100 F Ba(NO$_3$)$_2$ is mixed with 30.00 mL of 0.400 F NH$_4$NO$_3$. The [NO$_3^-$] in the resulting solution is
    (a) 0.250 M (b) 0.280 M (c) 0.320 M (d) 0.400 M (e) 0.500 M

18. Of the following measurements, the one most suitable for the determination of the molecular weight of oxyhemoglobin, a molecule with a molecular weight of many thousands, is
    (a) the vapor pressure lowering; (b) the elevation of the boiling point; (c) the depression of the freezing point; (d) the osmotic pressure; (e) any of the previous four, as they are all equally good.

19. A solution is prepared by dissolving 1.864 g of KCl and 8.293 g of K$_2$CO$_3$ in enough water to make the final volume 500.00 mL. What is the [K$^+$] in the solution?
    (a) 0.08500 M (b) 0.1200 M (c) 0.1450 M (d) 0.1700 M (e) 0.2900 M

20. At 35 °C, the vapor pressure of CS$_2$ is 512 mmHg, and of acetone, CH$_3$COCH$_3$, is 344 mmHg. A solution of CS$_2$ and acetone in which the mole fraction of CS$_2$ is 0.25 has a total vapor pressure of 600 mmHg. Which of the following statements about solutions of acetone and CS$_2$ is TRUE?
    (a) A mixture of 100.00 mL of acetone and 100.00 mL of CS$_2$ has a volume of 200.00 mL.
    (b) When acetone and CS$_2$ are mixed at 35 °C, heat must be absorbed in order to produce a solution at 35 °C.
    (c) When acetone and CS$_2$ are mixed at 35 °C, heat is released.
    (d) Raoult's law is obeyed by both CS$_2$ and acetone for the solution in which the mole fraction of CS$_2$ is 0.25.
    (e) A mixture of 100.00 mL of acetone and 100.00 mL of CS$_2$ will have a volume significantly less than 200.00 mL.

Problems

6.1. The liquids hexane, C$_6$H$_{14}$, and heptane, C$_7$H$_{16}$, form an ideal solution. At 49.6 °C, the vapor pressure of hexane is 400.0 mmHg, while that of heptane is 124 mmHg.
    (a) A solution is prepared by mixing equal weights of hexane and heptane. Calculate the mole fraction of hexane in this mixture.
    (b) What is the partial pressure of each component in the vapor phase in equilibrium with this solution at 49.6 °C?
    (c) If some of the vapor in equilibrium with this solution is condensed, what is the mole fraction of hexane in the new liquid?
    (d) If a mixture of hexane and heptane is distilled, which substance will be collected from the top of the distillation column? Explain your answer.

6.2. The chief constituent of lemon oil is a hydrocarbon, limonene, that is 88.16% C, 11.84% H. A solution of 8.362 g of limonene in 50.00 g of benzene, C$_6$H$_6$, boils at 83.28 °C. The benzene used to prepare the solution boils at 80.15 °C, using the same thermometer. Calculate the correct molecular formula and exact molecular weight of limonene. Explain why the molecular weight calculated using the boiling point elevation is not the exact molecular weight.

6.3. Phenol is a caustic, poisonous, white crystalline compound, also called carbolic acid. It is used in the manufacture of various plastics, disinfectants, and pharmaceuticals.
    (a) At 150.0 °C and 200.0 mmHg, phenol is a gas with a density of 0.713 g · L^{-1}. Calculate the molecular weight of phenol in the gaseous state.

(b) Phenol dissolves in a solvent called bromoform, which has a molal freezing point depression constant of $14.4 \text{ kg} \cdot \text{K} \cdot \text{mol}^{-1}$. A solution of 5.45 g of phenol in 100.0 g of bromoform has a freezing point 4.32° lower than the freezing point of the pure solvent. Calculate the molecular weight of phenol when it is dissolved in bromoform. What conclusion can you draw about the form of phenol when it is dissolved in bromoform?

6.4. What are the concentrations of all the ions in a solution that is prepared by mixing 30.00 mL of $0.120 \, F \, (NH_4)_2SO_4$, 70.00 mL of $0.200 \, F \, NH_4Cl$, and 100.00 mL of $0.080 \, F \, ZnCl_2$?

6.5. (a) You are hired as a lab technician for a summer job. The first day your employer hands you a jar of white, crystalline potassium iodate and says "Please prepare 2 L of $0.0800 \, F \, KIO_3$." Write a detailed set of instructions describing precisely how you would prepare this solution.
(b) Ten days later you need 100.0 mL of $0.0200 \, F \, KIO_3$. How would you dilute a portion of the stock solution you have already prepared to make this solution? Write precise instructions.
(c) Calculate how many millimeters of $0.0500 \, F \, FeSO_4$ are required to react completely with 20.00 mL of the $0.0200 \, F \, KIO_3$, according to the reaction

$$10Fe^{2+}(aq) + 12H^+(aq) + 2\,IO_3^-(aq) \rightarrow 10Fe^{3+}(aq) + I_2 + 6H_2O$$

6.6. Two liquids, $\underline{Y}$ and $\underline{Z}$, are mixed to form a solution at temperature T. At this temperature the vapor pressure of pure $\underline{Y}$ is 148.0 mmHg, while that of pure $\underline{Z}$ is 286.0 mmHg.
(a) Draw a diagram, like that below, and plot on it the partial vapor pressures of $\underline{Y}$ and $\underline{Z}$, and the total vapor pressure of solutions of $\underline{Y}$ and $\underline{Z}$, as a function of the composition of the solutions, assuming that solutions of $\underline{Y}$ and $\underline{Z}$ are ideal.

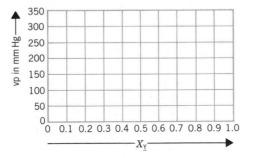

(b) A solution of $\underline{Y}$ and $\underline{Z}$ is prepared in which the mole fraction of $\underline{Y}$ is 0.250. The partial pressures of $\underline{Y}$ and $\underline{Z}$ in the vapor in equilibrium with this solution are 21.4 and 172.6 mmHg, respectively. Does this system exhibit negative or positive deviations from Raoult's law? Show all calculations necessary to prove your answer.
(c) Draw a second diagram like that in part (a), above, and sketch on it the approximate shapes of the actual vapor pressure diagrams of $\underline{Y}$ and $\underline{Z}$ and their binary solutions.
(d) When $\underline{Y}$ and $\underline{Z}$ are mixed, will the heat of mixing, ΔH_{mix}, be zero, greater than zero, or less than zero? Explain your answer.

6.7. An ester of camphoric acid contains only the elements C, H, and O. When analyzed, it is found to be 65.60% C and 9.44% H by weight. A solution of 0.785 g of this ester in 8.040 g of camphor is found to freeze 15.2° lower than the freezing point of pure camphor. Determine the exact molecular weight and formula of this ester. (The term **ester** is defined in Section 23.6.)

6.8. A saturated solution of lead iodide, PbI_2, contains 0.0701 g of PbI_2 per 100.0 mL of solution at 25 °C. Calculate the concentrations of the lead and iodide ions in a saturated solution at 25 °C.

6.9. At 35.2 °C the vapor pressure of pure acetone, $\underline{A}$, is 343.8 mmHg, and the vapor pressure of pure carbon disulfide, CS_2, is 512.3 mmHg. (a) In a binary solution of CS_2 and $\underline{A}$, in which the mole fraction of CS_2 is 0.062, the partial vapor pressures of CS_2 and $\underline{A}$ are, respectively, 110.7 and 331.0 mmHg. Does $\underline{A}$ obey Raoult's law? Show all calculations necessary to prove your answer.
(b) Calculate the mole fraction of CS_2 and of $\underline{A}$ in the vapor phase in equilibrium with the solution described in part (a).
(c) Is this solution ideal, or does it exhibit a positive or a negative deviation from Raoult's law? Explain.
(d) Is the force of attraction between an acetone molecule and a CS_2 molecule greater than, less than, or equal to that between two acetone molecules? Account for your answer in terms of properties of these molecules. (CS_2 is a linear molecule, $S{=}C{=}S$.)
(e) When 100.00 mL of $\underline{A}$ and 100.00 mL of CS_2 are mixed, will the volume of the solution be less than, greater than, or equal to 200.00 mL? Explain, in terms of your answer to part (d).

6.10. Thyroxine, a hormone that controls the rate of metabolism in the body, can be isolated from the thyroid gland. A solution containing 1.138 g of thyroxine in 25.00 g of benzene has an osmotic pressure of 1.24 atm at 20 °C. The density of benzene at this temperature is 0.8787 g · mL^{-1}. Calculate the molecular weight of thyroxine.

6.11. Solutions of isopropyl alcohol and propyl alcohol are ideal. At 40 °C, a solution that is $\frac{3}{4}$ isopropyl alcohol by weight has an equilibrium vapor pressure of 88.8 mmHg. A solution that is $\frac{1}{3}$ isopropyl alcohol by weight has an equilibrium vapor pressure of 68.3 mmHg. The two alcohols are isomers and have the same molecular weight. Calculate the vapor pressure of the pure alcohols at 40 °C.

6.12. Catechol, $C_6H_6O_2$, occurs naturally in many plants. A sample of 4.4044 g of catechol is dissolved in 200.0 g of benzene, C_6H_6, at 26.1 °C. The vapor pressure of pure benzene at this temperature is 100.0 mmHg.
(a) Calculate the molality of catechol in this solution.
(b) Calculate the mole fraction of catechol in this solution.
(c) What is the boiling point of this solution?
(d) What is the freezing point of this solution?
(e) What is the vapor pressure of this solution?

6.13. A compound with molecular formula C_6H_5ONa is a white solid, soluble in water. A 5.805-g sample of this solid is dissolved in 250.0 mL of water at 25 °C. The density of water at 25 °C is 0.997 g/mL. The freezing point of this solution is −0.72 °C. What conclusion can you draw about the form of this compound in aqueous solution?

6.14. Two liquids, $\underline{Y}$ and $\underline{Z}$, form ideal solutions at temperature T. When the total vapor pressure above a solution of $\underline{Y}$ and $\underline{Z}$ is 300.0 mmHg, the mole fraction of $\underline{Y}$ in the vapor phase is 0.650, and in the liquid phase is 0.300.
(a) Calculate the partial pressures of $\underline{Y}$ and of $\underline{Z}$ in the vapor phase.
(b) Calculate the vapor pressures of pure $\underline{Y}$ and of pure $\underline{Z}$ at temperature T.
(c) Draw a vapor pressure diagram for binary solutions of $\underline{Y}$ and $\underline{Z}$ at temperature T. (Refer to Fig. 6.11.)

6.15. The label on a commercially available hydriodic acid solution reads "47.3% HI by weight, density 1.50 g/mL." Calculate the formality, molality, and mole fraction of HI in this solution.

6.16. A solution containing 2.831 g of sulfur dissolved in 50.00 mL of CS_2 boils at a temperature 0.412° higher than the pure CS_2 used to make the solution. The density of CS_2 is 1.263 g/mL, and its ebullioscopic constant is 2.34 kg $\cdot$ K $\cdot$ mol^{-1}. Calculate the molecular formula and molecular weight of sulfur in this solvent.

6.17. Gaseous hydrogen chloride, HCl, is very soluble in water. Assume that HCl is an ideal gas.
(a) What volume of HCl(g), measured at 26.4 °C and 748 mmHg, is required to prepare 500.0 mL of 4.73 F hydrochloric acid?
(b) The density of 4.73 F HCl is 1.08 g/mL. Assuming that HCl is 100% dissociated into H^+ and Cl^- ions in this solution, what is the mole fraction of water in 4.73 F HCl?

6.18. From a measurement of the freezing point depression of benzene, the molecular weight of acetic acid in a benzene solution was determined to be 100, whereas its formula weight (CH_3COOH) is 60. Suggest an explanation for the observed value of the molecular weight.

6.19. Concentrated nitric acid is a solution of HNO_3 in water that is 69% HNO_3 by weight. The density of concentrated nitric acid is 1.41 g/mL at 20 °C. You want to prepare exactly 250.0 mL of 2.0 F HNO_3 by diluting a sample of the concentrated acid. What volume of the concentrated acid should you measure out for this preparation?

chapter 7 *Aqueous Solutions and Ionic Reactions*

Svante August Arrhenius (1859–1927), a Swedish chemist, was one of the founders of physical chemistry. His Ph.D. thesis at the University of Uppsala dealt with the conductivity of solutions of electrolytes. He proposed the then revolutionary idea that an aqueous solution of sodium chloride contained separate sodium ions and chloride ions. Arrhenius reported the following anecdote which describes the attitude about his theories when they were first proposed. "I came to my professor, Cleve, whom I admired very much, and I said 'I have a new theory of electrical conductivity as a cause of chemical reactions.' He said 'This is very interesting' and then he said 'Goodbye.' He explained to me later that he knew very well that there are so many different theories formed, and that they are almost all certain to be wrong, for after a short time they disappear." In 1903 Arrhenius was awarded the Nobel Prize in Chemistry "in recognition of the special services rendered by him to the development of chemistry by his electrolytic theory of association."

Water, which seems to us to be a common, ordinary material, is really a highly unusual substance with many unique properties, some of which have already been discussed in Sections 5.2 and 5.10. Water is certainly one of the best and most useful solvents that we have. While many substances do not dissolve in water (in particular, a large number of organic compounds), more compounds dissolve in water than in almost any other single liquid.* A great many ionic crystalline solids, such as NaCl, $Ba(NO_3)_2$, CaI_2, and $(NH_4)_2SO_4$, that contain no molecules but are composed of arrays of positive and negative ions, dissolve in water. Many molecular compounds also dissolve in water. Indeed, so many substances dissolve in water that we are led to ask: What are the properties of water that make it such a good solvent?

section 7.1
The Role of Water as a Solvent for Ionic Crystalline Solids

The Dipole Moment of Water and the Hydration of Ions

There are two properties of water that are primarily responsible for the fact that it is an excellent solvent for ionic crystalline solids. The first of these is that the water molecule is a **dipole;** it is a bent molecule with the oxygen end having a small net negative charge and the two H atoms having a small net positive charge. (See Fig. 7.1 and Section 5.1.) The entire molecule is, of course, electrically neutral, but there is a separation of positive and negative charges within the molecule, because oxygen is much more electronegative than hydrogen.

Because water is a polar molecule, there is a force of attraction between any ion and that end of the H_2O molecule that is of opposite sign. This is called an **ion-dipole force of attraction.** As a result, H_2O molecules tend to orient about any ion and to be loosely associated with it. We say that ions are **hydrated** or **solvated.** Figure 7.2 shows a typical orienting of water molecules about a positive ion and about a negative ion. Just how many water molecules are most closely associated with a given ion depends on the size of the ion and the magnitude of its charge. Water molecules and ions in solution are in constant motion, and the number of water molecules close to a given ion changes with time. There is an average number of water molecules most closely associated with a given ion, and this average is called the **hydration number.** The hydration number is large for small, highly charged cations. It is about 4 for Li^+ and 6 for Mg^{2+} ion. Anions generally have smaller hydration numbers than cations.

In Section 5.7 we discussed the two factors governing the direction in which a spontaneous process occurs: (1) the tendency to proceed to a state of lower energy, and (2) the tendency to proceed to a state of greater molecular disorder, that is, greater entropy. There is a regular, highly ordered arrangement of ions in an ionic crystalline solid, whereas ions move freely about in an aqueous solution. Thus dissolution of an ionic solid is favored by the increase in disorder of the solute particles, the ions. The change in energy that occurs when a solid dissolves at constant pressure is called the **heat of solution,** ΔH_{soln}. The heat of solution depends on the relative magnitudes of the forces of attraction between solute and solvent particles in solution, between solute particles in the solid, and between solvent molecules in the pure liquid.

To account for these several interactions we consider the dissolution process to be

---

* There is no "universal solvent," that is, a liquid that can dissolve everything, but liquid HF comes close. Most organic compounds that are insoluble in water dissolve in liquid HF. However, liquid HF erodes glass containers and is exceedingly dangerous because it burns the skin and destroys human tissues, so that it is not a generally useful solvent.

Fig. 7.1. Model and schematic diagram of the water molecule. The H—O—H bond angle is 104.5°. The small net positive charge on each H atom, denoted $\delta+$, is significantly less than the charge on a +1 cation. The H_2O molecule is electrically neutral, so there is a small net charge on the O atom of magnitude $2\delta-$.

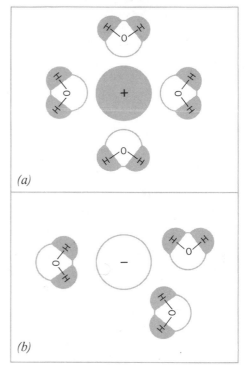

(a)

(b)

Fig. 7.2. The hydration of ions. *(a)* A hydrated positive ion. *(b)* A hydrated negative ion.

made up of several steps. For the solution of an ionic crystalline solid we first calculate the **crystal lattice energy,** which is defined as the amount of energy required to separate the ions in one mole of solid from their positions in the crystal structure to an infinite distance apart in the gaseous state. The lattice energy (which will be discussed more fully in Section 21.8) for a typical ionic solid with a singly charged cation and a singly charged anion is the amount of energy required for the process

$$MX(\text{crystal}) \rightarrow M^+(g) + X^-(g) \tag{7-1}$$

We then calculate the amount of energy released when these isolated gaseous ions are placed in water where they become hydrated due to the ion-dipole forces of attraction. The **hydrated ion,** that is, the ion surrounded by closely associated water molecules, is a configuration of lower potential energy than the separated ion and water molecules. Since the potential energy decreases as hydrated ions are formed, this step is exothermic. The hydration of both cation and anion can be represented as

$$M^+(g) + X^-(g) + xH_2O \rightarrow M^+(aq) + X^-(aq) \tag{7-2}$$

The overall dissolution process can therefore be thought of as the sum of the changes in Eqs. (7-1) and (7-2):

$$MX(\text{crystal}) \rightarrow M^+(g) + X^-(g)$$
$$\underline{M^+(g) + X^-(g) + xH_2O \rightarrow M^+(aq) + X^-(aq)}$$
$$MX(\text{crystal}) + xH_2O \rightarrow M^+(aq) + X^-(aq)$$

The heat of solution is the difference between the amount of energy required to separate the crystal into gaseous ions (the lattice energy) and the amount of energy released when the isolated ions are hydrated (the hydration energy). Both the lattice energy and the hydration energy generally have a magnitude of several hundred kilojoules, and the difference between them is relatively small, and may be either positive or negative.

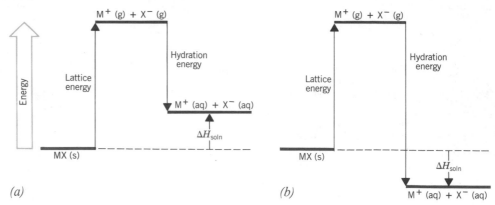

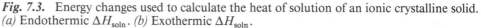

Fig. 7.3. Energy changes used to calculate the heat of solution of an ionic crystalline solid. *(a)* Endothermic ΔH_{soln}. *(b)* Exothermic ΔH_{soln}.

If ΔH_{soln} is positive, the dissolution process is endothermic and heat must be supplied in order to dissolve the crystal at constant temperature. Ionic crystalline solids for which dissolution is an endothermic process include KI and NH_4NO_3. These compounds dissolve readily in water even though the energy does not decrease when dissolution occurs, because of the increase in molecular disorder on dissolution. Figure 7.3*(a)* illustrates the relative magnitudes of the lattice and hydration energies that lead to an endothermic heat of solution. If ΔH_{soln} is positive, the dissolution process is exothermic and heat is released when the crystal dissolves. Ionic crystalline solids with negative heats of solution include LiBr, $CaCl_2$, and anhydrous Na_2SO_4. Figure 7.3*(b)* illustrates the relative magnitudes of the lattice and hydration energies that lead to an exothermic heat of solution.

The Dielectric Constant of Water

Water is not the only molecule that is a dipole; there are a great many polar molecules. Nor is the dipole moment of water particularly large, as can be seen from Table 7.1. Thus the fact that water has a dipole moment is not sufficient to explain why it is such a good solvent for ionic crystalline solids. A second factor is the unusually high **dielectric constant** of water.

In order to define the dielectric constant and to understand the significance of the high dielectric constant of water, we must be familiar with **Coulomb's Law.** Charles Augustin de Coulomb (1736–1806), a French physicist, investigated the force between two charged particles. For a system in which two particles with charges q_1 and q_2 are separated by a distance r that is large in comparison with the size of the particles, Coulomb found that the force varies directly as the product of the charges and inversely as the square of the distance between them. The force between charged particles also depends on the medium in which the charges are immersed. The force is much smaller if the charged particles are in water than it is if they are in air. Indeed, the force is inversely proportional to the dielectric constant of the medium in which the charged particles are immersed. Thus Coulomb's law may be written

$$F = \frac{kq_1q_2}{Dr^2} \tag{7-3}$$

where D is the dielectric constant of the medium. If the charges, q_1 and q_2, are of opposite sign, the force, F, is a force of attraction and is negative. If q_1 and q_2 are of the

Table 7.1. **Dipole Moments, in Debyes, of Selected Molecules**

Molecule	μ
Acetone, C_3H_6O	2.80
Ammonia, NH_3	1.47
Benzene, C_6H_6	0
Carbon disulfide, CS_2	0
Chloroform, $CHCl_3$	1.01
Ethanol, C_2H_5OH	1.69
Hydrogen fluoride, HF	1.82
Methanol, CH_3OH	1.70
Sulfur dioxide, SO_2	1.63
Water, H_2O	1.85

same sign, the force is one of repulsion and F is positive. The constant k is included because of the several systems of units in use; it has different numerical values depending on the units used for charge, distance, and force.*

The important point for us to focus on here is that the larger the dielectric constant of the medium, the smaller the force between charged particles, and water has an unusually large dielectric constant compared to other liquids. Table 7.2 gives the dielectric constants of several liquids at 25 °C. A glance at the table makes it clear that compared to the common organic solvents, water has an exceptionally large dielectric constant. Most organic solvents have dielectric constants between 2 and 10. The alcohols typically have dielectric constants between 20 and 35. Only a very few liquids have dielectric constants as large as, or larger than, that of water.

Since the force between charged particles is so much smaller in water than it is in air, for instance, it is much easier to move oppositely charged particles apart in water than in air. Indeed, the work required to separate a positive and a negative ion by a given distance is 1/78.54 as much in water as it is in air, at 25 °C. The dielectric constant of a liquid is defined as the *ratio* of the work required to separate two oppositely charged particles a given distance in a vacuum to the work required to separate them to that same distance when they are immersed in the liquid.

A similar proportion can be written for any two liquids because the work required to separate opposite charges a fixed distance is inversely proportional to the dielectric constant. Thus,

$$\frac{\text{work required in } H_2O}{\text{work required in } CH_3CH_2OH} = \frac{24.30}{78.54} = 30.9\% \tag{7-4}$$

This means that it is much easier to separate the ions of an ionic crystalline solid in water than it is in ethanol, and explains why water is a better solvent than ethanol for such solutes. Remember that ions are the components of ionic crystalline solids. The role of the H_2O as a solvent is to make it relatively easy to separate the ions already present in the solid, by greatly decreasing the electrostatic forces between them.

You may be wondering: Why is it that water has such an unusually large dielectric constant? It is important to distinguish between the dipole moment and the dielectric

---

* In SI units, q_1 and q_2 are expressed in Coulombs (C) and r in meters (m). The force is given in Newtons, and k then has the value 8.988×10^9 Jm/C^2. It is customary to express Coulomb's law as

$$F = q_1 q_2 / 4\pi\varepsilon_0 D r^2$$

where ε_0 is called the permittivity of a vacuum and has the value 8.854×10^{-12} $C^2J^{-1}m^{-1}$.

Table 7.2. **Dielectric Constants of Selected Liquids at 25 °C**

Medium	D
Air	1.0
Cyclohexane, C_6H_{12}	2.015
Carbon tetrachloride, CCl_4	2.228
Benzene, C_6H_6	2.274
Chloroform, $CHCl_3$	4.8
Acetone, CH_3COCH_3	20.7
Ethanol, CH_3CH_2OH	24.30
Methanol, CH_3OH	32.63
Water, H_2O	78.54
Anhydrous H_2SO_4	101.

constant. *The dipole moment is a property of an individual molecule.* The dipole moment of water is large, but not unusually so. For example, the dipole moments of H_2O and of ethanol are quite similar, being 1.85 and 1.69 D, respectively, whereas the dielectric constant of water (78.5) is very much larger than the dielectric constant of ethanol (24.3). *The dielectric constant is a property of the liquid as a whole,* that is, of the aggregate of a very great many water molecules. It is cooperation between H_2O molecules that accounts for the high dielectric constant of water, and hydrogen bonding (see Section 5.2) is the feature that makes this cooperation possible. Hydrogen bonding is responsible for the very high dielectric constant of water.

The structure of ice, in which each water molecule is hydrogen bonded to four other water molecules in a three-dimensional array through linear $O-H \cdots O$ bonds, has been discussed in Section 5.2, and is shown in Figs. 5.9 and 5.10. In liquid water, there is less order and fewer hydrogen bonds than in ice, because the water molecules are moving through the liquid, but there is still a good deal of hydrogen bonding. The liquid structure is probably best described as partially broken down ice. The higher the temperature of liquid water, the more hydrogen bonds are broken, and the less order that exists. As a result, the dielectric constant of water decreases as the temperature increases. For water, D is 88.0 at 0 °C, 78.5 at 25 °C, and 55.3 at 100 °C. Even at the boiling point of water, therefore, some hydrogen bonding still persists. Hydrogen bonding orients the individual water dipoles and produces an ordered arrangement of water molecules with a large separation of the centers of positive and negative charge in the hydrogen-bonded aggregates. It is this organization of many water molecules in the liquid that results in the high dielectric constant of water. The hydrogen bonds tie water molecules together into large aggregates with an effective dipole moment many times that of a single molecule.

Liquid alcohols are also hydrogen bonded, and this is the reason that the alcohols have larger dielectric constants than other organic liquids.

section 7.2
Nonelectrolytes

The ability of water to engage in hydrogen bonding also contributes to its being such a good solvent for many molecular solutes, such as the sugars (sucrose, glucose, lactose, etc.), ethanol, and acetone.

The interaction of water and acetone, which mix together in all proportions, is

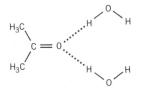

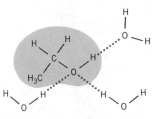

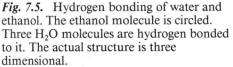

Fig. 7.4. Hydrogen bonding between acetone, $(CH_3)_2CO$, and water. Water and acetone are completely miscible in all proportions.

Fig. 7.5. Hydrogen bonding of water and ethanol. The ethanol molecule is circled. Three H_2O molecules are hydrogen bonded to it. The actual structure is three dimensional.

shown in Fig. 7.4. Acetone molecules cannot hydrogen bond to one another, so acetone itself is not a hydrogen-bonded liquid, but solutions of water and acetone are extensively hydrogen bonded.

A substance that dissolves in water without the formation of ions is called a **nonelectrolyte.** In general, molecular solutes that are polar molecules or that can form hydrogen bonds to water, will be soluble in water. Ethanol (ethyl alcohol), for example, dissolves readily in water and is hydrogen bonded to it, but no ionization takes place, so ethanol is a nonelectrolyte. The interaction of H_2O and ethanol, CH_3CH_2OH, is depicted in Fig. 7.5.

Organic molecules that are nonpolar, that is, that have zero dipole moment, such as CH_4 (methane) and CCl_4 (carbon tetrachloride), do not dissolve in water. Let us consider what happens when we mix CCl_4 and H_2O, and attempt to disperse CCl_4 molecules throughout water. The force of attraction between the polar water molecules is so much stronger than the very weak force of attraction between H_2O and CCl_4 that water molecules tend to stay close to one another, and to avoid having CCl_4 molecules as neighbors. The result is the separation of these liquids into two layers, with the denser CCl_4 forming the bottom layer. Water and carbon tetrachloride are said to be **immiscible.**

As a general rule, polar solute molecules will dissolve in polar solvents, such as water, while nonpolar solute molecules will dissolve in nonpolar solvents. This generalization is often stated as "like dissolves like."

section 7.3
Strong and Weak Electrolytes

The name **electrolyte** is given to a substance whose aqueous solutions contain ions, because ions are charged particles, which, when they move through a solution, conduct electricity. The ability of ions to carry electric charge through a solution can be demonstrated using the apparatus shown in Fig. 7.6.

Electrolytes can be classified as either strong or weak. **Strong electrolytes** are defined as those for which 100%, or very nearly 100%, of the solute particles in solution are ions, as long as the solution is dilute. **Weak electrolytes** are those for which substantially less than 100% of the solute particles are ions, and a significant fraction of molecules are present, even in dilute solution. This definition immediately raises problems: What is meant by "substantially less" than 100% or a "signifi-

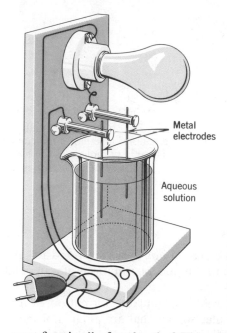

Fig. 7.6. Apparatus for demonstrating electrical conductivity in a solution. If an electrolyte is dissolved in the water, the light bulb will be "on," showing that current is flowing. If a nonelectrolyte is dissolved in the water, the light bulb will not shine.

cant fraction" of molecules? There is no distinct division between strong and weak electrolytes. All the solutes in nature cannot be placed into one or the other of just two categories. For the great majority of electrolytes it is usually quite clear whether the substance should be classified as strong or weak. There are compounds, however, that fall in between these two categories, and which we describe by using some phrase such as "moderately weak" or "moderately strong."

You will have to learn to recognize which electrolytes are strong and which are weak. Fortunately it is easy to do this because there are a few broad generalizations that can be made. To understand and utilize these generalizations, it is necessary to consider another way of classifying electrolytes. Electrolytes can be subdivided into one of three categories: **acids, bases,** and **salts.** Acids and bases can each be characterized by a set of common properties.

Acids

Acids have a sharp, sour taste, as in vinegar (acetic acid) or lemon juice (citric acid). Acids dissolve many metals with evolution of H_2 gas (see Section 1.9) and turn certain vegetable dyes a characteristic color. **Litmus,** for example, is a compound that can be extracted from some lichens; it turns red in an acid solution.

Bases

Bases, on the other hand, have a characteristic bitter taste; their solutions are slippery to the touch, and they turn litmus blue.

All the properties cited as being common to acids are properties of the hydrated proton, $H^+(aq)$. All the properties cited as being common to bases are properties of hydroxide ions, OH^-.

In 1884, the Swedish chemist Svante Arrhenius defined an acid as an electrolyte with cation H^+, and a base as an electrolyte with anion OH^-. With this definition, the formula for an acid must contain an H atom, usually written first to emphasize its acidic character. The six common strong acids are hydrochloric acid, HCl, hydrobromic acid, HBr, hydriodic acid, HI, nitric acid, HNO_3, perchloric acid, $HClO_4$,

Table 7.3. Summary of Information about Strong and Weak Electrolytes in Aqueous Solution

Rule	*Exceptions*
1. Most acids are weak electrolytes.	1. Common strong acids are HCl, HBr, HI, HNO_3, $HClO_4$, and H_2SO_4.
2. Most bases are weak electrolytes.	2. Common strong bases are LiOH, NaOH, KOH, RbOH, CsOH, $Ca(OH)_2$, $Sr(OH)_2$, and $Ba(OH)_2$.
3. Most salts are strong electrolytes.	3. Mercuric chloride, $HgCl_2$, and mercuric cyanide, $Hg(CN)_2$, are weak electrolytes. There are a number of moderately strong electrolytes, including the cadmium halides, $CdCl_2$, $CdBr_2$, and CdI_2, and lead acetate, $Pb(OAc)_2$.

and sulfuric acid, H_2SO_4. In the three acids listed last, the acidic hydrogens are bonded to oxygen atoms, although the molecular formula does not indicate this.

Using the Arrhenius definition, a base is an electrolyte whose anion is OH^-. Common strong bases include the alkali metal hydroxides, LiOH, NaOH, KOH, RbOH, CsOH, and the alkaline earth hydroxides, $Ca(OH)_2$, $Sr(OH)_2$, and $Ba(OH)_2$.

Salts

When an acid and a base react, water is formed, because

$$H^+(aq) + OH^-(aq) \rightarrow H_2O \qquad (7\text{-}5)$$

This reaction is called **neutralization.** In terms of the Arrhenius definition, the reaction of an acid and a base produces a **salt** plus H_2O. An ionic compound whose cation is any positive ion other than H^+ and whose anion is any negative ion other than OH^-, is a **salt.** The great majority of electrolytes are salts.

When electrolytes are divided into these three categories (acids, bases, and salts), we can summarize a great deal of information about which electrolytes are strong and which are weak. This information is contained in Table 7.3. Memorizing that table will make solving problems dealing with electrolytes very much easier for you.

The following example illustrates how to apply these rules and classify electrolytes as either strong or weak.

EXAMPLE 7.1. Recognizing strong and weak electrolytes

Classify the following electrolytes as acids, bases, or salts, and as either strong or weak electrolytes: $Ca(NO_3)_2$, HCN, NH_3, Li_2SO_4, CH_3COOH, ZnI_2, H_2SO_3, H_2S, $HClO_4$, NH_4MnO_4, HF, HNO_3, and $Ba(OH)_2$.

Solution

Compound	*Acid, Base, or Salt*	*Strong or Weak*	*Cation*	*Anion*	*Ref. in Table 7.3*
$Ca(NO_3)_2$	Salt	Strong	Ca^{2+}	NO_3^-	Rule 3
HCN	Acid	Weak	$H^+(aq)$	CN^-	Rule 1
NH_3	Base	Weak	NH_4^+	OH^-	Rule 2
Li_2SO_4	Salt	Strong	Li^+	SO_4^{2-}	Rule 3
CH_3COOH	Acid	Weak	$H^+(aq)$	CH_3COO^-	Rule 1
ZnI_2	Salt	Strong	Zn^{2+}	I^-	Rule 3
H_2SO_3	Acid	Weak	$H^+(aq)$	HSO_3^- and SO_3^{2-}	Rule 1

Compound	Acid, Base, or Salt	Strong or Weak	Cation	Anion	Ref. in Table 7.3
H_2S	Acid	Weak	$H^+(aq)$	HS^- and S^{2-}	Rule 1
$HClO_4$	Acid	Strong	$H^+(aq)$	ClO_4^-	Exception 1
NH_4MnO_4	Salt	Strong	NH_4^+	MnO_4^-	Rule 3
HF	Acid	Weak	$H^+(aq)$	F^-	Rule 1
HNO_3	Acid	Strong	$H^+(aq)$	NO_3^-	Exception 1
$Ba(OH)_2$	Base	Strong	Ba^{2+}	OH^-	Exception 2

We defined a strong electrolyte as one for which 100% of the solute particles are ions provided that the solution is dilute. The proviso that the solution must be dilute is very important. There is no sharp dividing line between a dilute solution and one that is concentrated, but in the context of this discussion, a solution with concentration less than 1 M would generally be considered dilute. In concentrated solutions of several substances classified as strong electrolytes, a substantial proportion of undissociated molecules may be present. Nitric acid, HNO_3, is a good example. For solutions less than 0.1 M all detectable solute particles are H_3O^+ and NO_3^- ions. For 6 M HNO_3, however, about 75% of the molecules are ionized, leaving 25% as undissociated HNO_3 molecules in solution. In 16 M HNO_3, only about 10% of the molecules are ionized.

section 7.4
Proton-Transfer Reactions: The Brønsted–Lowry Theory of Acids and Bases

Some molecular compounds, such as $HCl(g)$, $NH_3(g)$, $HCN(g)$, $CH_3COOH(\ell)$, and $HNO_3(\ell)$, that can hydrogen bond to water, are actually ionized or dissociated in aqueous solution. This occurs as a result of a **proton-transfer reaction,** in which the nucleus of the hydrogen atom in the hydrogen bond, originally bonded to one atom, moves along the hydrogen bond and becomes attached to a different atom. The ability of water to readily engage in proton-transfer reactions is another reason why it is such a good solvent.

Let us consider several examples of proton-transfer reactions in detail.

(a) The dissolution of gaseous HCl in H_2O. When HCl dissolves in water there is a strong attraction between the proton of the HCl and the oxygen of the H_2O, with the formation of a hydrogen bond. The small and mobile proton then shifts its position away from the Cl and toward the O atom, along the hydrogen bond. Only the proton, the nucleus of the hydrogen atom, shifts, so that a Cl^- ion is formed. The reaction can be described as follows:

$$Cl-H + O\genfrac{}{}{0pt}{}{\diagup H}{\diagdown H} \rightarrow \left\{ Cl-H\cdots O\genfrac{}{}{0pt}{}{\diagup H}{\diagdown H} \right\} \rightarrow Cl^- + \left(H-O\genfrac{}{}{0pt}{}{\diagup H}{\diagdown H} \right)^+ \tag{7-6a}$$

Hydrogen-bonded structure   Chloride ion   Hydronium ion

The equation for this reaction is normally written as

$$HCl(g) + H_2O \rightarrow H_3O^+(aq) + Cl^-(aq) \tag{7-6b}$$

In the reaction between $HCl(g)$ and H_2O, the water has accepted a proton from the HCl, forming H_3O^+, hydronium ion. Of course, each of the ions formed in this reaction, the chloride ion and the hydronium ion, is itself hydrated as pictured in Fig.

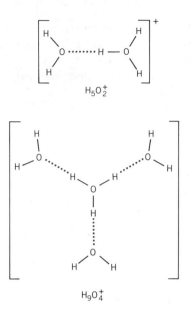

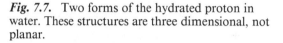

Fig. 7.7. Two forms of the hydrated proton in water. These structures are three dimensional, not planar.

7.2. Thus the H_3O^+ has additional water molecules closely associated with it, and there is experimental evidence for the existence of other species in the solid state, such as $H_5O_2^+$ and $H_9O_4^+$, shown in Fig. 7.7.

It is quite cumbersome to depict hydrated ions, and often we simply use the notation $Cl^-(aq)$ to indicate the fact that the chloride ion, for example, is hydrated. Similarly, the hydronium might be represented as $H^+(aq)$, and many chemists prefer to do so. There is much to be said for the argument that since the solution actually contains many species with varying numbers of H_2O molecules bonded with a proton, there is no adequate way to describe the hydronium ion with a single symbol, and therefore the simplest possible symbol, $H^+(aq)$, should be used. There are times, however, when we wish to emphasize the role of H_2O in proton-transfer reactions, and in those instances it is useful to denote the hydronium ion as H_3O^+.

You will recall, from Section 1.9, that a solution of the gas HCl in water is called hydrochloric acid. We have seen that when HCl dissolves in water, a proton-transfer reaction occurs in which HCl donates a proton to H_2O. In 1923 the Danish chemist J. N. Brønsted proposed that an **acid** be defined as a substance capable of donating or giving away a proton, and that a **base** be defined as a substance capable of accepting a proton. A similar concept was proposed by the British chemist T. M. Lowry, although he did not make the definitions of acid and base as explicit as Brønsted did. Accordingly, the following definitions are usually referred to as the **Brønsted–Lowry theory** of acids and bases:

> An acid is a proton donor.
> A base is a proton acceptor.

These definitions extend the concept of an acid and a base beyond that proposed by Arrhenius. Substances classified as acids and bases by the Arrhenius definition are also acids and bases using the Brønsted–Lowry definitions. In the reaction between HCl and H_2O, Eq. (7-6), HCl is an acid, and H_2O is a base.

(b) The reaction between acetic acid and water. At room temperature, pure acetic acid is a colorless liquid. It is readily soluble in water, and mixes with it in all proportions. Ordinary vinegar is a solution of acetic acid in water; 5% CH_3COOH by weight. The name "acetic" is derived from the Latin word for vinegar, *acetum*. The formula of acetic acid is CH_3COOH, but it is frequently abbreviated HOAc. Dissolving acetic acid in water involves the formation of a hydrogen-bonded structure, and we can write an equation similar to Eq. (7-6):

$$CH_3-\overset{\displaystyle O}{\overset{\|}{C}}-O-H + O\overset{H}{\underset{H}{\big\langle}} \rightleftharpoons \left\{ CH_3-\overset{\displaystyle O}{\overset{\|}{C}}-O-H\cdots O\overset{H}{\underset{H}{\big\langle}} \right\} \rightleftharpoons$$

Acetic acid   Hydrogen-bonded structure

$$CH_3-\overset{\displaystyle O}{\overset{\|}{C}}-O^- + H_3O^+ \quad (7\text{-}7a)$$

Acetate ion

The equation for this reaction is normally written as

$$CH_3COOH + H_2O \rightleftharpoons CH_3COO^- + H_3O^+ \qquad (7\text{-}7b)$$

or

$$HOAc + H_2O \rightleftharpoons OAc^- + H_3O^+ \qquad (7\text{-}7c)$$

There is one very important distinction between the proton transfers depicted in Eqs. (7-6) and (7-7). Proton transfer is virtually complete in the case of HCl in dilute solution. That means that essentially 100% of the HCl molecules dissolved in water react to form Cl^- and H_3O^+ ions. As long as the solution is dilute, there are no molecules of HCl detectable in aqueous solution.* Therefore HCl is a **strong acid**. In the reaction of acetic acid and water, however, the proton transfers proceed only to a slight extent. Most of the acetic acid in solution is present as undissociated molecules; only a small percentage exists as acetate and hydronium ions. Acetic acid is therefore **a weak acid.** It is quite easy to observe this difference between aqueous solutions of HCl and of CH_3COOH, because they differ greatly in their electrical conductivity. An aqueous solution of HCl is a much better conductor of electricity than an aqueous solution of CH_3COOH of the same concentration. Another experimental method that distinguishes between strong and weak electrolytes is the measurement of their colligative properties.

The freezing point depression and the boiling point elevation of weak electrolytes are significantly less than for strong electrolytes of the same concentration. The measurement of colligative properties of solutions of weak electrolytes can be used to determine the fraction ionized or the percentage of ionization of such solutes. Example 7.2 illustrates the method of calculation.

EXAMPLE 7.2. Calculation of the percentage of ionization of a weak acid from a freezing point depression

Nitrous acid, HONO or HNO_2, is a weak acid. The freezing point of an 0.0750 *m* aqueous solution of HONO is -0.150 °C. Calculate the percentage of ionization of 0.0750 *m* nitrous acid.

---

* Molecules of HCl have been detected in concentrated solutions. For example, in 10 *M* hydrochloric acid, about 0.3% of the acid is in molecular form.

Solution. The proton-transfer reaction between water and nitrous acid is

$$H_2O + \underset{\text{Nitrous acid}}{HONO} \rightleftharpoons H_3O^+ + \underset{\text{Nitrite ion}}{ONO^-}$$

Let

$$x = \text{molality of nitrite ion in this solution}$$

Then x is also the molality of the hydronium ion in this solution, because the molar ratio of H_3O^+ ions : ONO^- ions is 1 : 1. The molality of nitrous acid molecules in the solution is less than $0.0750\ m$ because some of the nitrous acid has ionized. Since

$$\frac{\text{No. mol } ONO^- \text{ formed}}{\text{No. mol } HONO \text{ ionized}} = \frac{1}{1}$$

the molality of the nitrous acid that has *not* ionized is $(0.0750 - x)$.

The total molality of all solute particles in solution is

$$m_{\text{total}} = \text{molality } HONO + \text{molality } H_3O^+ + \text{molality } ONO^-$$
$$= (0.0750 - x) + x + x = 0.0750 + x$$

We can calculate m_{total} from the freezing point depression. For H_2O, K_f is 1.86. (See Table 6.2.) Therefore

$$m_{\text{total}} = \frac{\Delta T_f}{K_f} = \frac{0.150}{1.86} = 0.0806$$

Hence $0.0750 + x = 0.0806$, and $x = 0.0056\ m$ or $5.6 \times 10^{-3}\ m$.

$$\frac{\text{percentage nitrous}}{\text{acid ionized}} = \frac{\text{molality of } H_3O^+}{\text{initial molality}} = \left(\frac{5.6 \times 10^{-3}}{7.50 \times 10^{-2}}\right) \times 100 = 7.5\% \text{ ionized}$$

Acidic Hydrogen Atoms

Although there are four hydrogen atoms in acetic acid, only one of them, the one bonded to an oxygen atom, is able to donate its proton. (See Fig. 7.8.) A hydrogen atom whose proton can be donated is said to be an **acidic hydrogen atom** (also called an **acidic proton**). Acetic acid has only a single acidic hydrogen atom and is therefore classified as a **monoprotic acid.** The other three H atoms in acetic acid are bonded to a carbon atom, and constitute a —CH_3 **(methyl)** group. The methyl group is very common in organic compounds. Methyl group hydrogen atoms are not acidic. Very frequently acidic protons are bonded to an oxygen atom. Because there is only a single acidic hydrogen atom in acetic acid, the one bonded to an O atom, we abbreviate acetic acid as HOAc. The acetate ion can then be denoted OAc⁻, showing the loss of a proton with the electron of the H atom left behind on the rest of the molecule.

In Eq. (7-7), as in Eq. (7-6), H_2O functions as a proton acceptor, that is, as a base. But H_2O can also serve as a proton donor, as we shall see in the following example.

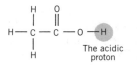

Fig. 7.8. The structure of acetic acid showing the single acidic proton. Hydrogen atoms bonded to carbon are generally not acidic.

(c) The dissolution of gaseous NH₃ in H₂O. Ammonia, NH_3, is a gas at room temperature. It has a very pungent odor, and is extremely soluble in water. A solution of NH_3 in water is commonly sold in supermarkets as a cleanser, labeled simply *ammonia.* The gas NH_3 also dissolves in water with the formation of a hydrogen bond, as shown below:

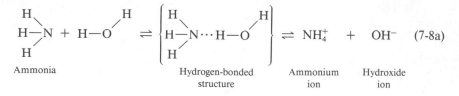

| Ammonia | Hydrogen-bonded structure | Ammonium ion | Hydroxide ion |

The equation for this reaction is normally written as

$$NH_3 + H_2O \rightleftharpoons NH_4^+ + OH^- \tag{7-8b}$$

In this reaction, a proton that was originally part of a water molecule is hydrogen bonded to both the N of NH_3 and the O atom of H_2O. The proton then shifts away from the O atom and towards the N atom, forming the ammonium ion, NH_4^+. The ion remaining after the H_2O molecule has lost a proton is OH^-, the hydroxide ion. An older name, ammonium hydroxide, is still in use by many chemical supply companies for solutions of NH_3 in water, but most chemists employ the terms ammonia or ammonia water. You may see the formula NH_4OH for "ammonium hydroxide," but it is not correct as no such molecule exists, and therefore that formula should not be used.

Note that in the proton-transfer reaction between NH_3 and H_2O, the water behaves as an acid, since it donates a proton to NH_3. As NH_3 is a proton acceptor, it is a base. The proton-transfer reaction between NH_3 and H_2O, like that between CH_3COOH and H_2O, proceeds only to a slight extent. Most of the ammonia in solution is present as molecules, NH_3; only a small percentage exists as NH_4^+ ions and OH^- ions. Thus NH_3, like CH_3COOH, is a weak electrolyte. We say that ammonia is a weak base.

We see, therefore, that molecular compounds capable of forming hydrogen bonds with water dissolve in water, and that ionization may result from a proton transfer after the formation of the hydrogen bond. This ionization should be distinguished from the dissolution of an ionic crystalline solid in water. In the latter case the ions are already present in the solid and are merely separated by the water. Water ionizes HCl; it does not ionize NaCl. In nonpolar solvents, such as benzene, HCl dissolves as a molecular solute with no dissociation into ions. A solution of HCl in benzene is a very poor conductor of electricity. Sodium chloride does not dissolve in benzene. Water is a good solvent for ionic solutes because of its high dielectric constant and its ability to form stable hydrated ions due to ion-dipole interactions.

section 7.5
A Summary of Information about the Solubilities of Electrolytes in Water

Whenever we write an equation for a reaction that occurs in aqueous solution, we want to represent as far as possible the actual species present. If a solid substance has only a very small solubility in water, it will be a precipitate at the bottom of the reaction vessel, and we represent it by writing the formula of the compound followed by the symbol (s) or the symbol ↓, which is used to indicate that the solid has

Table 7.4. Summary of Information about the Solubilities of Electrolytes in Aqueous Solution

Rule	Exceptions
1. Nitrates (NO_3^-), acetates (CH_3COO^-), chlorates (ClO_3^-), and perchlorates (ClO_4^-) are generally soluble.	1. No common exceptions. Silver acetate, AgOAc, is only slightly soluble.
2. Compounds of the alkali metals and of the ammonium ion, NH_4^+, are generally soluble.	2. No common exceptions. $KClO_4$ is only slightly soluble.
3. Chlorides (Cl^-), bromides (Br^-), and iodides (I^-) are generally soluble.	3. The halides of Ag^+, Pb^{2+}, Hg_2^{2+}, and Cu^+ are insoluble, except that $PbCl_2$ is slightly soluble. HgI_2 is insoluble. The oxychlorides of bismuth and antimony, BiOCl and SbOCl, are insoluble.
4. Sulfates (SO_4^{2-}) are generally soluble.	4. $PbSO_4$, $SrSO_4$, and $BaSO_4$ are insoluble. $CaSO_4$, Ag_2SO_4, and Hg_2SO_4 are slightly soluble.
5. Carbonates (CO_3^{2-}), sulfites (SO_3^{2-}), chromates (CrO_4^{2-}), and phosphates (PO_4^{3-}) are generally insoluble.	5. Carbonates, sulfites, chromates and phosphates of alkali metals and of NH_4^+ are soluble. (See Rule 2.) Li_3PO_4 is only slightly soluble.
6. Sulfides (S^{2-}) are generally insoluble.	6. Sulfides of the alkali metals and of NH_4^+ are soluble. (See Rule 2.) Sulfides of the alkaline earths, Cr_2S_3, and Al_2S_3 are decomposed by H_2O.
7. Hydroxides are generally insoluble.	7. Hydroxides of the alkali metals and of NH_4^+ are soluble. (See Rule 2.) Hydroxides of Ca^{2+}, Sr^{2+}, and Ba^{2+} are slightly soluble.

precipitated out of the solution as a result of the reaction that occurred. For instance,

$$Ag^+(aq) + Cl^-(aq) \rightarrow AgCl \downarrow \qquad (7\text{-}9)$$

states that silver ions and chloride ions react to precipitate solid silver chloride, a white, very slightly soluble solid. An ionic crystalline solid that has only a very small solubility in water is frequently said to be *insoluble.*

If a solid strong electrolyte dissolves in water, we represent it by writing the formulas of the ions that are actually the solute particles. To know how to write a particular substance in a reaction that takes place in aqueous solution, therefore, you must know which compounds are soluble and which are not. Table 7.4 summarizes information about the solubilities of most common electrolytes. If you have the information in Table 7.4 at your command, writing correct equations for reactions will be much easier for you.

Another generalization that can be made is that most acids are soluble in water, due to the proton-transfer reaction that takes place between the acid and water.

The following example illustrates how to utilize Table 7.4 to determine whether or not a given electrolyte is soluble in water.

EXAMPLE 7.3. Soluble and insoluble electrolytes

Classify the following electrolytes as either soluble or insoluble in water: KI, $Ca(NO_3)_2$, $BaSO_4$, $ZnBr_2$, $MgCO_3$, $(NH_4)_2SO_4$, $KMnO_4$, Hg_2Cl_2, MnS, PbI_2, $BaSO_3$, $Ca_3(PO_4)_2$, $K_2Cr_2O_7$, and $Al(OH)_3$.

Solution

Compound	Soluble or Insoluble	Ref. to Table 7.4
KI, potassium iodide	Soluble	Rules 2 and 3
Ca(NO$_3$)$_2$, calcium nitrate	Soluble	Rule 1
BaSO$_4$, barium sulfate	Insoluble	Exception 4
ZnBr$_2$, zinc bromide	Soluble	Rule 3
MgCO$_3$, magnesium carbonate	Insoluble	Rule 5
(NH$_4$)$_2$SO$_4$, ammonium sulfate	Soluble	Rules 2 and 4
KMnO$_4$, potassium permanganate	Soluble	Rule 2
Hg$_2$Cl$_2$, mercurous chloride	Insoluble	Exception 3
MnS, manganous sulfide	Insoluble	Rule 6
PbI$_2$, lead iodide	Insoluble	Exception 3
BaSO$_3$, barium sulfite	Insoluble	Rule 5
Ca$_3$(PO$_4$)$_2$, calcium phosphate	Insoluble	Rule 5
K$_2$Cr$_2$O$_7$, potassium dichromate	Soluble	Rule 2
Al(OH)$_3$, aluminum hydroxide	Insoluble	Rule 7

section 7.6
Writing Correctly Balanced Net Ionic Equations

There are only three rules to be followed in writing net ionic equations. These rules are

1. All soluble, strong electrolytes are written in ionic form.
2. All gases, insoluble solids, nonelectrolytes, and weak electrolytes are written in molecular form.
3. Any substance that is present but does not take part in the reaction is not written in the equation.

These are not arbitrary rules. We endeavor to represent, as far as possible, the species that predominate in the solution. When we dissolve a weak electrolyte in water, the solution contains molecules of that weak electrolyte plus the ions into which it dissociates. Because it is a weak electrolyte, the principal solute particles are the molecules; hence we use the molecular form to represent it. Similarly, since soluble strong electrolytes exist in aqueous solution as ions, we always write ions and not molecules for such substances.

The application of these three rules, plus the information contained in Tables 7.3 and 7.4, will enable you to write correctly balanced net ionic equations for proton-transfer reactions (acid–base reactions) and for simple **precipitation reactions.** A precipitation reaction is one in which an insoluble solid is formed when two or more soluble substances containing the ions of the insoluble solid are mixed together in aqueous solution. For instance, a solution of strontium nitrate contains strontium ions (Sr^{2+}) and nitrate ions (NO$_3^-$). A solution of potassium carbonate contains potassium ions (K$^+$) and carbonate ions (CO$_3^{2-}$). When a solution of strontium nitrate is mixed with a solution of potassium carbonate, the Sr^{2+} and the CO$_3^{2-}$ ions react to precipitate insoluble strontium carbonate, and the net ionic equation for the reaction is

$$Sr^{2+}(aq) + CO_3^{2-}(aq) \rightarrow SrCO_3 \downarrow \qquad (7\text{-}10)$$

The K$^+$ and the NO$_3^-$ ions take no part in any reaction. They are called **spectator or bystander ions,** and are not written in the equation.

Several examples are given below to illustrate how to utilize the information in Tables 7.3 and 7.4 to write correct net ionic equations. Important information about specific kinds of reactions is also described in the discussion of these equations.

EXAMPLE 7.4. Writing net ionic equations

Write correctly balanced net ionic equations for the reactions that occur when dilute aqueous solutions of the following reagents are mixed together:

(a) Ferric nitrate and sodium hydroxide.

Rule 1 of Table 7.4 states that all nitrates are soluble. Iron(III) nitrate (ferric nitrate) is a salt, and Rule 3 of Table 7.3 tells us that practically all salts are strong electrolytes. We conclude that iron(III) nitrate is a soluble, strong electrolyte, and should be written in ionic form. Sodium hydroxide is one of the common strong bases listed under the exceptions to Rule 2 in Table 7.3. Compounds of the alkali metals are generally soluble, hence sodium hydroxide is also a soluble, strong electrolyte and should be written in ionic form. You should therefore mentally record that what is being mixed together are aqueous solutions containing four different ions, namely Fe^{3+}, NO_3^-, Na^+, and OH^- ions.

Will any of these ions react with each other? *Ions will combine if an insoluble solid (a precipitate) is formed, or if a weak electrolyte is formed.* Rule 7 of Table 7.4 states that hydroxides are generally insoluble, and iron(III) hydroxide (ferric hydroxide) is not an exception to that rule. We therefore know that Fe^{3+} ions and OH^- ions will react to precipitate insoluble $Fe(OH)_3$. The formula for the precipitate is determined by the charges on each of the ions and the requirement that there be no net charge on the solid formed. The Na^+ and NO_3^- ions do not react with each other (nitrates are soluble and compounds of the alkali metals are soluble), so that the Na^+ and NO_3^- ions are spectator ions. The only reaction that occurs when these two aqueous solutions are mixed is the one between Fe^{3+} ions and OH^- ions. The net ionic equation is

$$Fe^{3+}(aq) + 3\,OH^-(aq) \rightarrow Fe(OH)_3\downarrow \qquad (7\text{-}11)$$

$Fe(OH)_3$ is a rust-colored, gelatinous precipitate. It is very fluffy and takes a fairly long time to settle.

(b) Barium hydroxide and sulfuric acid.

Barium hydroxide is a moderately soluble strong base. (See exceptions to Rule 2 of Table 7.3.) Therefore the species actually present in dilute solution are Ba^{2+} ions and OH^- ions. Sulfuric acid is a strong acid (Rule 1 of Table 7.3) and in dilute solution the principal species are $H^+(aq)$ and SO_4^{2-} ions.* Thus we must consider the possible reactions between any of the four ions, Ba^{2+}, OH^-, $H^+(aq)$, and SO_4^{2-}. Rule 4 of Table 7.4 tells us that while sulfates are generally soluble, $BaSO_4$ is a notable exception. Therefore $BaSO_4$ precipitates out when these solutions are mixed; it is a fine white

---

* A solution of sulfuric acid contains $H^+(aq)$, HSO_4^-, and SO_4^{2-} ions. While H_2SO_4 is a strong acid, HSO_4^-, the bisulfate ion, is only moderately strong. In dilute solutions there are significant concentrations of both SO_4^{2-} and HSO_4^- ions. The relative amounts of the two anions depend on the concentration (the formality) of the acid. If no concentration is mentioned, it is not possible to know whether the predominant anion is SO_4^{2-} or HSO_4^-. Therefore the equation for this reaction may be written as

$$Ba^{2+} + 2\,OH^- + H^+(aq) + HSO_4^- \rightarrow BaSO_4\downarrow + 2H_2O$$

as well as the form given as Eq. (7-12).

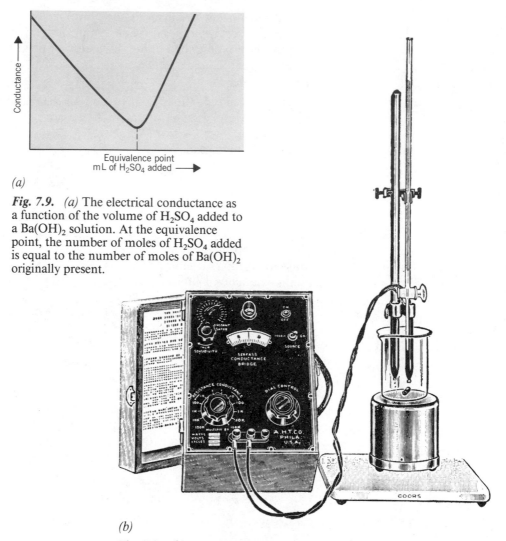

(a)

Fig. 7.9. (a) The electrical conductance as a function of the volume of H_2SO_4 added to a $Ba(OH)_2$ solution. At the equivalence point, the number of moles of H_2SO_4 added is equal to the number of moles of $Ba(OH)_2$ originally present.

(b)

Fig. 7.9. (b) Experimental set-up for a conductance titration.

crystalline solid. Since hydronium ions and hydroxide ions combine to form H_2O, all four of these ions are involved in reactions, and the net ionic equation is

$$Ba^{2+} + 2\,OH^- + 2H^+(aq) + SO_4^{2-} \rightarrow BaSO_4\downarrow + 2H_2O \qquad (7\text{-}12)$$

The white insoluble salt, $BaSO_4$, is widely used in medicine for diagnostic X-ray pictures. Barium sulfate, $BaSO_4$, is opaque to X-rays, and when a suspension of finely powered $BaSO_4$ in water is pumped into the intestinal tract, an X-ray picture of the area shows up ulcers or any abnormalities on the intestinal walls. The suspension of $BaSO_4$ in water can be swallowed, or administered as a "barium enema," depending on the area to be X-rayed.

The reaction between barium hydroxide and sulfuric acid can be followed by measuring the electrical conductance of the solution as H_2SO_4 is added in small increments to a solution of $Ba(OH)_2$. Such a procedure, called a **conductance titration,** demonstrates quite dramatically the significance of the net ionic equation. A solution of barium hydroxide contains Ba^{2+} ions and OH^- ions and is an excellent conductor of electricity. The initial conductance, before any sulfuric acid is added, is therefore quite high. As sulfuric acid is added, the conductance decreases because the ions being added combine with ions in solution to produce H_2O and solid $BaSO_4$, which precipitates out of solution. Note particularly that no ions appear on the right-hand side of the net ionic equation for the reaction, Eq. (7-12).

When the number of moles of H_2SO_4 added is exactly equal to the number of moles of $Ba(OH)_2$ originally present (the **equivalence point** of the titration), the conductance of the solution is very low. Indeed, if a light bulb is connected in series with the circuit, it burns brightly all during the titration until the equivalence point is reached. When the drop of H_2SO_4 solution is added that makes the number of moles of H_2SO_4 added exactly equal to the number of moles of $Ba(OH)_2$ originally present, the light goes out. The conductance of the solution has not dropped to zero, as shown by the plot in Fig. 7.9(a), but current flow has been reduced to a value insufficient to cause the filament in the bulb to glow. If more H_2SO_4 is then added, the light bulb immediately shines again, because there are no longer sufficient Ba^{2+} or OH^- ions in solution to react with the $H^+(aq)$, HSO_4^-, and SO_4^{2-} ions that are being added. Figure 7.9(b) shows the experimental apparatus used to measure the conductance as H_2SO_4 is added to a $Ba(OH)_2$ solution, and Fig. 7.9(a) shows how the conductance varies with the volume of added sulfuric acid solution. The **equivalence point** of the titration is attained when the conductance is at its minimum value.

(c) Methylamine and H_2O.

The word **amine** is derived from *ammonia* and indicates a compound in which one or more of the H atoms of NH_3 has been replaced by a group of atoms, usually an organic group. The word *methyl* is the name for the $—CH_3$ group, so that methylamine is CH_3NH_2, in which one H atom of NH_3 has been replaced by the $—CH_3$ group. **Amines**, like NH_3, are weak bases, and react with H_2O in a proton-transfer reaction analogous to Eq. (7-8). Thus the net ionic equation for the reaction between CH_3NH_2 and H_2O is

$$\underset{\text{Base}}{CH_3NH_2} + \underset{\text{Acid}}{H_2O} \rightleftharpoons \underset{\text{Acid}}{CH_3NH_3^+} + \underset{\text{Base}}{OH^-} \qquad (7\text{-}13)$$

The $CH_3NH_3^+$ ion is called the methyl ammonium ion. It is analogous to NH_4^+.

(d) Aluminum hydroxide and hydrochloric acid.

Rule 7 of Table 7.4 states that hydroxides are generally insoluble, and $Al(OH)_3$ is not an exception. Aluminum hydroxide, $Al(OH)_3$, is a white solid, with an extremely small solubility in water. It is therefore written in molecular form in any net ionic equation in which it appears. Insoluble hydroxides can, however, be dissolved in strong acids such as HCl, because of reactions such as the one shown below:

$$Al(OH)_3(s) + 3H^+(aq) \rightleftharpoons 3H_2O + Al^{3+}(aq) \qquad (7\text{-}14)$$

Hydronium ions combine with hydroxide ions from the $Al(OH)_3$ to form H_2O, and the Al^{3+} cations are then in the aqueous solution. Since chlorides are generally soluble, and aluminum chloride is not an exception, there is no reaction between Al^{3+} ions and Cl^- ions. The Cl^- ions from the hydrochloric acid are merely spectator ions, and do not appear in the net ionic equation. Of course, they are present in the solution at all times, and are required for electroneutrality. If the water is evaporated, crystals of $AlCl_3$ will be formed.

(e) Calcium carbonate and nitric acid.

Calcium carbonate, like most carbonates (Rule 5, Table 7.4), is only very slightly soluble in water. Calcium carbonate is a major constituent of the shells of clams, oysters, snails, and other marine animals. It is also the major component of marble, chalk, and limestone, which are, in part, the fossil remains of marine animals.

While carbonates generally are insoluble in water, they can be dissolved in strong acids such as nitric acid or hydrochloric acid. Hydronium ions combine with carbonate ions to form the weak acid H_2CO_3, carbonic acid. However, H_2CO_3 is unstable and decomposes to form H_2O and $CO_2(g)$. As an aid in remembering what occurs we can write the overall reaction in two steps, as

$$CaCO_3(s) + 2H^+(aq) \rightarrow Ca^{2+} + H_2CO_3 \qquad (7\text{-}15a)$$

$$H_2CO_3 \rightarrow H_2O + CO_2 \uparrow \qquad (7\text{-}15b)$$

but the correct net ionic equation shows just the final products of the reaction, and is

$$CaCO_3(s) + 2H^+(aq) \rightarrow Ca^{2+} + H_2O + CO_2 \uparrow \qquad (7\text{-}15)$$

Marble statues and structures such as the Parthenon in Athens, that have been outdoors for centuries, are now being eaten away, because of the reaction above, by rainfall that is acidic (**acid rain**) due to air pollutants such as SO_2 and NO_2.

(f) Sodium hydroxide and oxalic acid.

Oxalic acid, like most acids, is a weak acid, and is therefore written in molecular form. It is a **diprotic acid,** that is, it has two acidic protons. The molecular formula for oxalic acid is $H_2C_2O_4$, but the structural formula, which shows the bonding of the atoms, is

$$H-O-\overset{\overset{\displaystyle O}{\|}}{C}-\overset{\overset{\displaystyle O}{\|}}{C}-O-H$$

Note that the two acidic protons are bonded to oxygen atoms. The ion that remains when the two acidic protons are given away is called oxalate ion, and has a charge of -2. The names of most organic acids end in -ic, and the anion formed by removal of a proton or protons from the acid is named by substituting -ate for the -ic. Thus $H_2C_2O_4$ is oxalic acid and $C_2O_4^{2-}$ is oxalate ion. The net ionic equation for the proton-transfer reaction between oxalic acid and an excess of any strong base such as NaOH, is

$$\underset{\text{Acid}}{H_2C_2O_4} + \underset{\text{Base}}{2\,OH^-} \rightleftharpoons \underset{\text{Base}}{C_2O_4^{2-}} + \underset{\text{Acid}}{2H_2O} \qquad (7\text{-}16)$$

In the following examples, the correct net ionic equation is given without discussion. Use Tables 7.3 and 7.4 to deduce the equations.

EXAMPLE 7.5. Writing net ionic equations

Write correctly balanced net ionic equations for the reaction between the following reagents. Assume dilute aqueous solutions of all soluble substances. Indicate proton donors and acceptors in proton-transfer reactions.

(a) Silver nitrate and potassium sulfide.

$$2Ag^+(aq) + S^{2-}(aq) \rightleftharpoons Ag_2S \downarrow$$

(b) Acetic acid and potassium hydroxide.

$$\underset{\substack{\text{Acid}\\\text{H}^+ \text{ donor}}}{CH_3COOH} + \underset{\substack{\text{Base}\\\text{H}^+ \text{ acceptor}}}{OH^-} \rightleftharpoons \underset{\substack{\text{Base}\\\text{H}^+ \text{ acceptor}}}{CH_3COO^-} + \underset{\substack{\text{Acid}\\\text{H}^+ \text{ donor}}}{H_2O}$$

(c) Hydrogen sulfide gas and copper(II) chloride.

$$H_2S(g) + Cu^{2+}(aq) \rightleftharpoons CuS\downarrow + 2H^+(aq)$$

(d) Methylamine and perchloric acid.

$$\underset{\substack{\text{Base}\\ \text{H}^+\text{ acceptor}}}{CH_3NH_2} + \underset{\substack{\text{Acid}\\ \text{H}^+\text{ donor}}}{H_3O^+(aq)} \rightleftharpoons \underset{\substack{\text{Acid}\\ \text{H}^+\text{ donor}}}{CH_3NH_3^+} + \underset{\substack{\text{Base}\\ \text{H}^+\text{ acceptor}}}{H_2O}$$

(e) Sodium bromide and calcium nitrate.

No reaction occurs.

section 7.7
Stoichiometry of Ionic Reactions in Aqueous Solution

When aqueous solutions are mixed, it is often important to know the concentrations of the species in the resultant solution. Before any numerical calculations can be made, you must first determine whether or not a chemical reaction occurs when the solutions are mixed. If no reaction occurs, the computation proceeds as follows:

1. Calculate the number of moles (or millimoles) of each species in the solutions being mixed, using Eqs. (6-6):

$$\text{No. mol solute} = \left(\text{concentration in } \frac{\text{mol}}{\text{L}}\right)(\text{volume in L})$$

or

$$\text{No. mmol solute} = \left(\text{concentration in } \frac{\text{mmol}}{\text{mL}}\right)(\text{volume in mL})$$

2. Calculate the final volume of the mixture. When mixing dilute aqueous solutions, the final volume is simply the sum of the volumes of the solutions being mixed.
3. Calculate the concentration of each solute in the mixture by rearranging Eqs. (6-6) to read

$$\text{concentration of solute in mol/L} = \frac{(\text{No. of mol of solute})}{(\text{total volume in L})} \qquad (7\text{-}17a)$$

$$= \frac{(\text{No. of mmol of solute})}{(\text{total volume in mL})} \qquad (7\text{-}17b)$$

EXAMPLE 7.6. Solution stoichiometry when no reaction occurs

Calculate the concentrations of all species in the solution that results when 20.00 mL of 0.12 F NaCl, 40.00 mL of 0.15 F KNO$_3$, and 60.00 mL of 0.0800 F CaCl$_2$ are mixed.

Solution. The ions in the mixture are Na$^+$, K$^+$, Ca^{2+}, Cl$^-$, and NO$_3^-$. No precipitate is formed because nitrates and chlorides are generally soluble, as are salts of the alkali metals. No reaction occurs when these three solutions are mixed.

First we must determine the number of millimoles of each of these five ions in the resulting mixture:

$$\text{No. mmol Na}^+ = (20.00 \text{ mL})\left(0.120 \, \frac{\text{mmol}}{\text{mL}}\right) = 2.40 \text{ mmol}$$

$$\text{No. mmol Cl}^- = (20.00 \text{ mL})\left(0.120 \frac{\text{mmol}}{\text{mL}}\right) + (60.00 \text{ mL})(2)\left(0.0800 \frac{\text{mmol}}{\text{mL}}\right)$$

$$= \quad 2.40 \text{ mmol} \quad + \quad 9.60 \text{ mmol} \quad = 12.00 \text{ mmol}$$

There are two sources of chloride ions, both the NaCl and the $CaCl_2$. We must add the number of millimoles of Cl^- from each source. Because the molar ratio of Cl^- ions to $CaCl_2$ is $2:1$, the $[Cl^-]$ in 0.0800 F $CaCl_2$ is $2(0.0800 \text{ } M) = 0.160 \text{ } M$.

$$\text{No. mmol K}^+ = \text{No. mmol NO}_3^- = (40.00 \text{ mL})\left(0.150 \frac{\text{mmol}}{\text{mL}}\right) = 6.00 \text{ mmol}$$

$$\text{No. mmol Ca}^{2+} = (60.00 \text{ mL})\left(0.0800 \frac{\text{mmol}}{\text{mL}}\right) = 4.80 \text{ mmol}$$

The total volume of solution is 120.00 mL. Hence the final concentrations are

$$[\text{Na}^+] = \frac{2.40 \text{ mmol}}{120.00 \text{ mL}} = 0.0200 \text{ } M$$

$$[\text{Cl}^-] = \frac{12.00 \text{ mmol}}{120.00 \text{ mL}} = 0.1000 \text{ } M$$

$$[\text{K}^+] = [\text{NO}_3^-] = \frac{6.00 \text{ mmol}}{120.00 \text{ mL}} = 0.0500 \text{ } M$$

$$[\text{Ca}^{2+}] = \frac{4.80 \text{ mmol}}{120.00 \text{ mL}} = 0.0400 \text{ } M$$

Note that while the total number of moles (or millimoles) of each solute species is the same before and after mixing, the concentration of each species in the resulting mixture is less than that in the original solution because of dilution.

Suppose, however, that a reaction does occur when solutions are mixed. There are then two possibilities. The reaction may **go to completion,** that is, virtually 100% of one or more (but at least one) of the reacting species may be used up in the reaction. Such a reaction is said to be **quantitative.** Reactions in which a gas is formed (which then escapes into the atmosphere) are quantitative, as are most reactions in which an insoluble solid precipitates out of solution. The second possibility is that significantly less than 100% of each of the reacting species is used up in the reaction. The law of chemical equilibrium applies to such reactions, and that is the topic of the next chapter. We shall consider here only the stoichiometry of reactions that go to completion.

Example 7.7 illustrates a typical problem.

EXAMPLE 7.7. Precipitation reaction stoichiometry

A 25.00-mL portion of 0.112 F $Pb(NO_3)_2$ is mixed with a 45.00-mL portion of 0.0840 $F K_2CrO_4$, and a bright yellow precipitate of lead chromate is formed. Assuming the reaction has gone to completion, calculate the molarity of the principal ions present in solution after the reaction.

Solution. The net ionic equation for the reaction is

$$\text{Pb}^{2+}(\text{aq}) + \text{CrO}_4^{2-}(\text{aq}) \rightarrow \text{PbCrO}_4 \downarrow$$

Thus Pb^{2+} ions and CrO_4^{2-} ions react in a $1:1$ molar ratio.

$$\text{No. mmol Pb}^{2+} \text{ ions added} = (25.00 \text{ mL})\left(0.112 \frac{\text{mmol}}{\text{mL}}\right) = 2.80 \text{ mmol}$$

$$\text{No. mmol CrO}_4^{2-} \text{ added} = (45.00 \text{ mL})\left(0.0840 \frac{\text{mmol}}{\text{mL}}\right) = 3.78 \text{ mmol}$$

Consequently, there is an excess of CrO_4^{2-} ions. If virtually 100% of the Pb^{2+} ions are used up to form the $PbCrO_4$ precipitate, there will be $3.78 - 2.80 = 0.98$ mmol of CrO_4^{2-} in excess. It is not possible to precipitate exactly 100% of any ion, but it is possible to precipitate very close to that, say 99.999%. The volume of the final solution is 70.00 mL. The chromate ion concentration in the solution after reaction is therefore

$$[CrO_4^{2-}] = \frac{0.98 \text{ mmol}}{70.00 \text{ mL}} = 0.014 \ M$$

The NO_3^- and K^+ ions are spectator ions, and their concentrations are calculated just as in Example 7.6.

$$[NO_3^-] = \frac{(25.00 \text{ mL})(2)(0.112 \text{ mmol/mL})}{70.00 \text{ mL}} = 0.0800 \ M$$

$$[K^+] = \frac{(45.00 \text{ mL})(2)(0.0840 \text{ mmol/mL})}{70.00 \text{ mL}} = 0.108 \ M$$

Using Example 7.7 as a typical problem, we can summarize the procedure for calculating the concentrations of the species in solution when a reaction goes to completion as follows:

1. Write a balanced net ionic equation for the reaction.
2. Calculate the number of moles (or millimoles) of each species in the solutions being mixed.
3. Determine which of the species that react is completely used up, and which species, if any, is in excess.
4. Calculate the number of moles (or millimoles) of the species that is in excess.
5. Determine the volume of the final solution and the concentration of each of the species in solution, using Eq. (7-17).

Additional problems in solution stoichiometry when reactions occur that go to completion are illustrated in Examples 7.8 and 7.9.

EXAMPLE 7.8. Precipitation reaction stoichiometry

A 10.00-mL sample of $0.0930 \ F$ $AgNO_3$ is mixed with a 15.00-mL sample of $0.1200 \ F Na_2CO_3$, and a pale yellow precipitate of silver carbonate is formed. Assuming the reaction has gone to completion, calculate the mass of the silver carbonate precipitate and the concentrations of the principal ions remaining in solution.

Solution. The net ionic equation for the reaction is

$$2Ag^+(aq) + CO_3^{2-}(aq) \rightarrow Ag_2CO_3 \downarrow$$

Silver ions and carbonate ions react in a $2:1$ molar ratio.

$$\text{No. mmol Ag}^+ \text{ added} = (10.00 \text{ mL})\left(0.0930 \frac{\text{mmol}}{\text{mL}}\right) = 0.930 \text{ mmol}$$

$$\text{No. mmol } CO_3^{2-} \text{ added} = (15.00 \text{ mL})\left(0.1200 \frac{\text{mmol}}{\text{mL}}\right) = 1.800 \text{ mmol}$$

To precipitate the Ag^+ ions in solution completely, 0.465 mmol of CO_3^{2-} ions are needed, because

$$\text{mmol } CO_3^{2-} \text{ in precipitate} = (\tfrac{1}{2})(\text{mmol } Ag^+ \text{ in precipitate})$$

Thus there is excess CO_3^{2-} in the solution. The amount of the excess is $1.800 - 0.465 = 1.335$ mmol. The final volume of solution is 25.00 mL, so that

$$[CO_3^{2-}] = \frac{1.335 \text{ mmol}}{25.00 \text{ mL}} = 0.05340 \ M$$

Since

$$\text{No. mmol } Ag_2CO_3 \text{ formed} = (\tfrac{1}{2})(\text{No. mmol } Ag^+ \text{ used up})$$

there are 0.465 mmol, or 4.65×10^{-4} mol, of Ag_2CO_3 in the precipitate. The formula weight of Ag_2CO_3 is 275.75 g/mol, so that the mass of the precipitate is

$$(4.65 \times 10^{-4} \text{ mol})(275.75 \text{ g} \cdot \text{mol}^{-1}) = 0.128 \text{ g}$$

The Na^+ and NO_3^- ions are spectator ions.

$$[Na^+] = \frac{(15.0 \text{ mL})(2)(0.1200 \text{ mmol/mL})}{(25.00 \text{ mL})} = 0.1440 \ M$$

$$[NO_3^-] = \frac{(10.00 \text{ mL})(0.0930 \text{ mmol/mL})}{(25.00 \text{ mL})} = 0.0372 \ M$$

You can check your arithmetic by verifying the electroneutrality of the final solution. The total amount of negative charge must equal the total amount of positive charge. Since each CO_3^{2-} ion carries two negative charges, we must multiply the amount of CO_3^{2-} ions by two to obtain the negative charge on those ions. As the number of moles is directly proportional to the concentration of each ion in solution, the electroneutrality condition is expressed as

$$[NO_3^-] + 2[CO_3^{2-}] = [Na^+]$$
$$0.0372 + 2(0.0534) = 0.1440$$

Since $0.0372 + 0.1068$ does equal 0.1440, we have verified that the final solution is electrically neutral, as it must be.

In an **acid–base titration** a solution of known concentration of the acid, for example, is dispensed into the solution of the base from a **buret.** (See Fig. 7.10.) There are several methods of indicating when the number of moles of acid added is the exact amount required to react completely with the number of moles of base originally present (the **equivalence point**). Acid–base titrations will be discussed in detail in Chapter 10. The stoichiometry of such titrations is illustrated in Example 7.9.

EXAMPLE 7.9. Stoichiometry of titrations

Sodium tetraborate decahydrate, $Na_2B_4O_7 \cdot 10H_2O$, commonly called borax, is a white crystalline solid that reacts with strong acids as follows:

$$Na_2B_4O_7 \cdot 10H_2O(s) + 2H^+(aq) \rightarrow 4H_3BO_3 + 5H_2O + 2Na^+(aq)$$

The boric acid formed, H_3BO_3, is a weak acid, soluble in water. How many milliliters

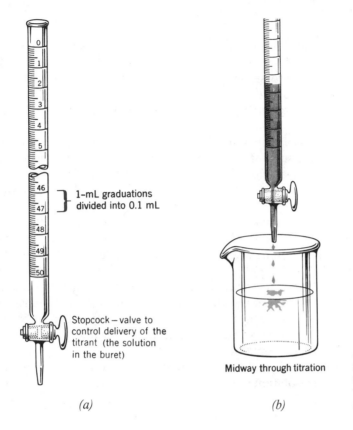

1-mL graduations
divided into 0.1 mL

Stopcock — valve to
control delivery of the
titrant (the solution
in the buret)

(a)

Midway through titration

(b)

Fig. 7.10. *(a)* A buret.
(b) Titration of a base with
an acid. The buret is filled
with a solution of the acid
of known concentration.

of 0.1082 F HCl are required to reach the equivalence point in a titration of a 0.6941-g sample of borax?

Solution. The equivalence point is attained when the number of moles of acid added is exactly enough to react completely with the borax, with no acid in excess. We must therefore determine the number of moles of borax (molecular weight $= 381.37$ g·mol^{-1}) in the sample.

$$\text{No. mol borax} = \frac{0.6941 \text{ g}}{381.37 \text{ g/mol}} = 1.820 \times 10^{-3} \text{ mol}$$

From the net ionic equation for the reaction, the following molar ratio is obtained:

$$\frac{\text{mol H}^+(\text{aq}) \text{ used}}{\text{mol Na}_2\text{B}_4\text{O}_7 \cdot 10\text{H}_2\text{O(s) used}} = \frac{2}{1}$$

We therefore need $2(1.820 \times 10^{-3}) = 3.640 \times 10^{-3}$ mol, or 3.640 mmol, of H$^+$(aq) to reach the equivalence point. The volume of 0.1082 F HCl required is determined by rearranging Eq. (7-17b) to read

$$\text{volume in mL} = \frac{\text{no. mmol}}{\text{concentration in mmol/mL}}$$

Hence,

$$\text{mL of 0.1082 } F \text{ HCl} = \frac{3.640 \text{ mmol}}{0.1082 \text{ mmol/mL}} = 33.64 \text{ mL}$$

Summary

Water is an excellent solvent for ionic solutes both because it has an unusually high **dielectric constant** and because it is a **polar molecule**. The large dielectric constant makes it much easier to separate ions in H_2O than in other liquids with lower dielectric constants. The force of attraction between ions and the dipolar H_2O molecules results in the formation of **hydrated ions** that are lower in energy than the separated ions and water molecules.

Many solutes dissolve in water because they form hydrogen bonds with H_2O molecules. In some cases, a **proton-transfer reaction** occurs, with the nucleus of the H atom moving along the hydrogen bond, to form ionic species. Water can function either as a **proton donor** (an acid), or as a **proton acceptor** (a base), in proton-transfer reactions.

Electrolytes can be classified either as **strong** (essentially 100% ions in dilute aqueous solution) or **weak** (significantly less than 100% ions in dilute aqueous solution). Rules for determining whether a given electrolyte is strong or weak have been summarized in Table 7.3.

When ionic solutions are mixed, two types of reactions that can occur are **precipitation reactions** and **proton-transfer reactions** (acid–base reactions). In order to determine whether or not a precipitation reaction occurs, one must be familiar with the common insoluble electrolytes. Rules summarizing information about the solubility of electrolytes are contained in Table 7.4.

Examples are given that illustrate the writing of net ionic equations and of solving problems involving solution stoichiometry in ionic reactions.

Exercises

Section 7.1

1. Compare the work required to separate a K^+ ion and a Cl^- ion in H_2O at 25 °C with that required to separate to an equal extent

   (a) A K^+ and a Cl^- ion in acetone at 25 °C.
   (b) A Mg^{2+} ion and a SO_4^{2-} ion in water at 25 °C.

2. Explain why the dielectric constant of acetone at 25 °C, 20.7, is less than that of methanol, 32.6, even though the dipole moment of acetone, 2.80 D, is greater than that of methanol, 1.70 D.

Section 7.2

3. Tetrabromomethane, CBr_4, is a molecular solid that is insoluble in water. Explain why CBr_4 does not dissolve in water.

Section 7.3

4. (a) For electrolytes, distinguish clearly between the terms **strong** and **concentrated**. Also distinguish between the terms **weak** and **dilute**.
   (b) A saturated solution of silver bromide is 7.1×10^{-7} M at 25 °C. Is silver bromide a strong or a weak electrolyte? Is a saturated solution of silver bromide concentrated or dilute?
   (c) A certain solution of NH_3 in water is 8.0 M. Is this solution concentrated or dilute? Is NH_3 a strong or a weak electrolyte?

5. Classify the following substances as acids, bases, salts, or nonelectrolytes. If the substance is an electrolyte, state whether it is weak or strong, and give the formula of the anion and cation that exist in aqueous solution.
(a) $NiBr_2$ (b) CH_3NH_2 (c) K_2SO_3 (d) CH_3OH (e) $Na_2C_2O_4$ (f) HNO_2
(g) $(NH_4)_2CO_3$ (h) H_3PO_4 (i) $Sr(OH)_2$ (j) $Co(NO_3)_2$ (k) $LiClO_4$
(l) C_6H_6 (m) $H_2C_2O_4$ (n) CH_3CH_2COOH (o) ZnI_2

6. Describe an experiment you could perform to determine whether phenol, C_6H_5OH, is an acid or a base in aqueous solution.

Section 7.4

7. Write the correctly balanced equation for the proton-transfer reaction between gaseous HBr and H_2O.

8. Propionic acid, CH_3CH_2COOH, has six hydrogen atoms. How many acidic hydrogen atoms does it have? Write the formula and circle any acidic hydrogen atoms.

9. Identify the proton donors and acceptors in the following proton-transfer reactions:

(a) $OH^- + NH_4^+ \rightleftharpoons NH_3 + H_2O$

(b) $H_2O + S^{2-} \rightleftharpoons HS^- + OH^-$

(c) $NH_3 + HSO_4^- \rightleftharpoons SO_4^{2-} + NH_4^+$

10. Describe an experiment you could perform to show that nitric acid, HNO_3, is a strong acid, but nitrous acid, HNO_2, is a weak acid.

11. Write the correctly balanced equation for the proton-transfer reaction between nitrous acid, HNO_2, and water. What is the name of the anion formed in this reaction?

12. Formic acid, HCOOH, is a weak acid. The freezing point of an 0.0500 m aqueous solution of formic acid is -0.098 °C. What is the percentage of formic acid that exists as ions in this solution?

Section 7.5

13. Classify the following electrolytes as either soluble or insoluble in water: $Cr(OH)_3$, $(NH_4)_3PO_4$, $NaHCO_3$, NiS, Hg_2Br_2, HNO_2, $SrSO_3$, $ZnCO_3$, $Al(NO_3)_3$, HI, $PbCrO_4$, AgI, $(NH_4)_2S$, $H_2C_2O_4$, and Ag_2CrO_4.

Section 7.6

14. Write a balanced net ionic equation for the reaction that occurs between each pair of reagents listed below. Assume dilute solutions of all soluble or moderately soluble reagents. If no reaction occurs, write NR.
(a) Calcium chloride and ammonium carbonate.
(b) Ammonium chloride and lithium hydroxide.
(c) Sodium chloride and ammonium carbonate.
(d) Zinc nitrate and ammonium sulfide.
(e) Barium hydroxide and hydrobromic acid.
(f) Silver nitrate and potassium chromate.
(g) Acetic acid and ammonia.
(h) Sodium sulfate and strontium hydroxide.
(i) Ammonium nitrate and magnesium chloride.
(j) Barium nitrate and potassium sulfite.
(k) Copper sulfate and H_2S.
(l) Nickel(II) chloride and sodium hydroxide.

15. Write balanced net ionic equations for the reactions between the following reagents. Assume dilute aqueous solution of all soluble and moderately soluble substances. Indicate proton donors and proton acceptors in proton-transfer reactions.

(a) Nitrous acid and sodium hydroxide.
(b) Calcium hydroxide and perchloric acid.
(c) Lead nitrate and sulfuric acid.
(d) Formic acid, HCOOH, and potassium cyanide.
(e) Sodium sulfide and iron(III) nitrate.
(f) Hydrogen sulfide and excess potassium hydroxide.

16. Write a balanced net ionic equation for each of the following reactions. Assume dilute aqueous solutions of all soluble substances. Water has been omitted as a reactant or product.

(a) Manganese(II) sulfide and hydrochloric acid $\rightarrow$ manganese(II) chloride and hydrogen sulfide gas.
(b) Silver chloride and sodium carbonate $\rightarrow$ silver carbonate and sodium chloride.
(c) Barium carbonate and hydrochloric acid $\rightarrow$ barium chloride and carbon dioxide gas.
(d) Sulfurous acid and strontium nitrate $\rightarrow$ strontium sulfite and nitric acid.
(e) Zinc hydroxide and acetic acid $\rightarrow$ zinc acetate.
(f) Iron(III) sulfate and barium hydroxide $\rightarrow$ barium sulfate and iron(III) hydroxide.

Section 7.7

17. Calculate the concentration of all ions in solution when 10.00 mL of 0.100 F KBr, 20.00 mL of 0.0850 F $Ba(NO_3)_2$, and 20.00 mL of 0.120 F KNO_3 are mixed.

18. A 40.00-mL sample of 0.0850 F KCl is mixed with a 20.00-mL sample of 0.110 F $AgNO_3$, and a white precipitate of silver chloride is formed. Calculate the molarity of the three principal ions present in solution after the reaction has gone to completion.

19. A 0.6506-g sample of calcium carbonate is completely dissolved in 80.00 mL of 0.2000 F HNO_3. All the CO_2 produced escapes into the air. What are the concentrations of the Ca^{2+}, H_3O^+, and NO_3^- ions in the resultant solution? Assume the volume remains constant. Write a balanced net ionic equation for the reaction that occurs.

20. What is the maximum number of grams of solid $Mg(OH)_2$ that can be dissolved in 100.00 mL of 0.100 F $HClO_4$?

21. If equal volumes of 0.0750 F $Ba(NO_3)_2$ and 0.1200 F K_2SO_4 are mixed, white barium sulfate precipitates. Assuming the reaction has gone to completion, what are the concentrations of the three principal ions remaining in solution?

Multiple Choice Questions

1. What is the maximum mass of $PbSO_4$ that can be precipitated by mixing 20.00 mL of 0.1000 F $Pb(NO_3)_2$ with 30.00 mL of 0.1000 F Na_2SO_4?
(a) 0.1516 g (b) 0.3033 g (c) 0.6065 g (d) 0.7591 g (e) 1.213 g

2. What is the $[Na^+]$ in a solution prepared by mixing 30.00 mL of 0.120 F NaCl with 70.00 mL of 0.150 F Na_2SO_4?
(a) 0.135 M (b) 0.141 M (c) 0.210 M (d) 0.246 M (e) 0.270 M

3. Which of the following is a weak electrolyte?
(a) CH_3COOH (b) CsOH (c) CH_3OH (d) $Ca(OH)_2$ (e) KOH

4. At 0 °C, 100.0 mL of a saturated solution of silver sulfate contains 0.57 g of dissolved silver sulfate. What is the silver ion concentration in this solution?

(a) 0.0057 M (b) 0.018 M (c) 0.028 M (d) 0.037 M (e) 0.056 M

5. Which of the following substances does not dissolve in water?

(a) $(NH_4)_2Cr_2O_7$ (b) C_6H_6 (c) CH_3CH_2OH (d) CH_3NH_2 (e) $H_2C_2O_4$

6. What volume of 0.1250 F HNO_3 is required to completely neutralize 25.00 mL of 0.1080 F $Ba(OH)_2$?

(a) 14.47 mL   (b) 21.60 mL   (c) 28.94 mL   (d) 43.20 mL   (e) 64.80 mL

7. In which of the following liquids would you expect the solubility of NaCl to be the smallest?

(a) HF   (b) CH_3OH (c) CH_3COCH_3 (acetone)   (d) H_2O (e) CCl_4

8. When 80.00 mL of 0.200 F HCl is mixed with 120.00 mL of 0.150 F KOH, the resultant solution is the same as a solution of

(a) 0.160 F KCl and 0.0200 F HCl   (b) 0.0800 F KCl
(c) 0.0800 F KCl and 0.0100 F KOH   (d) 0.160 F KCl and 0.0100 F KOH
(e) 0.0800 F KCl and 0.0100 F HCl

9. Equal volumes of 0.200 F $Ba(NO_3)_2$ and 0.120 F K_2CrO_4 are mixed. If the precipitation of yellow barium chromate is quantitative, what is the $[Ba^{2+}]$ in the final solution?

(a) 0.00 M (b) 0.0400 M (c) 0.0800 M (d) 0.100 M (e) 0.120 M

10. In an experiment, 40.00 mL of 0.100 F $Pb(NO_3)_2$ was mixed with 60.00 mL of 0.300 F HCl, and a white precipitate of $PbCl_2$ was formed. The precipitate was collected by filtration, dried, and found to weigh 1.068 g. What was the percentage yield of $PbCl_2$?

(a) 42.7%   (b) 73.2%   (c) 84.7%   (d) 96.0%   (e) 99.9%

11. A precipitate will be formed when an aqueous solution of hydrochloric acid is added to an aqueous solution of

(a) sodium nitrite   (b) barium nitrate   (c) mercury(I) nitrate
(d) zinc sulfate   (e) iron(II) sulfate

12. Which of the following statements is correct?

(a) It is impossible to prepare a 6.0 M solution of a weak electrolyte.
(b) One can prepare a 1.0 M solution of any strong electrolyte.
(c) All nonelectrolytes have low solubilities in water.
(d) Proton-transfer reactions between acids and bases always go to completion.
(e) A significant number of strong electrolytes have very low solubilities in water.

13. Equal volumes of 0.200 F HCl and 0.400 F KOH are mixed. The concentrations of the principal ions in the resulting solution are

(a) $[K^+] = 0.400\ M$; $[Cl^-] = 0.200\ M$; $[H^+(aq)] = 0.200\ M$
(b) $[K^+] = 0.200\ M$; $[Cl^-] = 0.100\ M$; $[OH^-] = 0.100\ M$
(c) $[K^+] = 0.100\ M$; $[Cl^-] = 0.100\ M$; $[OH^-] = 0.100\ M$
(d) $[K^+] = 0.200\ M$; $[Cl^-] = 0.100\ M$; $[H^+(aq)] = 0.100\ M$
(e) $[K^+] = 0.200\ M$; $[Cl^-] = 0.200\ M$

14. When solutions of iron(III) nitrate and sodium hydroxide are mixed, a red-brown gelatinous precipitate of iron(III) hydroxide is formed. Which of the following mixtures will produce the largest precipitate of iron(III) hydroxide?

(a) 10.00 mL of 0.100 F $Fe(NO_3)_3$ and 90.00 mL of 0.100 F NaOH

(b) 20.00 mL of 0.100 F Fe(NO$_3$)$_3$ and 80.00 mL of 0.100 F NaOH
(c) 25.00 mL of 0.100 F Fe(NO$_3$)$_3$ and 75.00 mL of 0.100 F NaOH
(d) 46.00 mL of 0.100 F Fe(NO$_3$)$_3$ and 54.00 mL of 0.100 F NaOH
(e) 64.00 mL of 0.100 F Fe(NO$_3$)$_3$ and 36.00 mL of 0.100 F NaOH

15. In a conductance titration of a 50.00-mL sample of a strontium hydroxide solution with 0.0452 F H$_2$SO$_4$, the minimum conductance was attained when 34.57 mL of acid had been added. The concentration of the original strontium hydroxide solution was

(a) 0.0313 F (b) 0.0625 F (c) 0.0654 F (d) 0.0904 F (e) 0.131F

16. What is the maximum number of millimoles of lanthanum oxalate, La$_2$(C$_2$O$_4$)$_3$, that can be precipitated by mixing 30.00 mL of 0.0860 F La(NO$_3$)$_3$ with 20.00 mL of 0.114 F Na$_2$C$_2$O$_4$?

(a) 0.760   (b) 1.29   (c) 1.52   (d) 2.28   (e) 2.58

17. An aqueous solution of barium chloride is added to an aqueous solution of each of the following reagents: sodium carbonate, silver nitrate, ammonium sulfate, and lead nitrate. In which cases will a precipitate form?

(a) Silver nitrate and lead nitrate.
(b) Silver nitrate and ammonium sulfate.
(c) Lead nitrate, ammonium sulfate, and sodium carbonate.
(d) Silver nitrate, lead nitrate, and ammonium sulfate.
(e) A precipitate forms with each of the four reagents.

18. Insoluble hydroxides can generally be dissolved in strong acids. Suppose you have solid samples of Pb(OH)$_2$, Al(OH)$_3$, Sr(OH)$_2$, and Zn(OH)$_2$. In which of the following strong acids could you dissolve all four solids to produce solely a clear solution?

(a) HCl   (b) HNO$_3$ (c) HBr   (d) H$_2$SO$_4$ (e) HI

Problems

7.1. Fill in the blanks in the chart below.

Substance	Formula	Soluble or Insoluble	Strong or Weak Electrolyte
Silver chloride	_____	_____	_____
Ammonia	_____	_____	_____
Barium sulfite	_____	_____	_____
Potassium permanganate	_____	_____	_____
Ammonium sulfide	_____	_____	_____
Mercurous iodide	_____	_____	_____
Lithium hydroxide	_____	_____	_____
Sodium acetate	_____	_____	_____
Hydrogen sulfide (hydrosulfuric acid)	_____	_____	_____
Silver chromate	_____	_____	_____
Oxalic acid	_____	_____	_____
Dimethylamine	_____	_____	_____
Iron(II) carbonate	_____	_____	_____
Perchloric acid	_____	_____	_____
Potassium chlorate	_____	_____	_____
Lead sulfate	_____	_____	_____
Hydrocyanic acid	_____	_____	_____

7.2. Insoluble hydroxides, carbonates, and sulfites can be dissolved in acidic solutions. Write balanced net ionic equations for dissolving
(a) Zinc hydroxide in acetic acid.
(b) Magnesium carbonate in nitric acid.
(c) Iron(III) hydroxide in perchloric acid.
(d) Barium sulfite in hydrochloric acid.
(e) Copper(II) carbonate in acetic acid.

7.3. Write balanced net ionic equations for the reactions between the following reagents. Assume dilute aqueous solutions of all soluble and moderately soluble substances. Indicate proton donors and proton acceptors in proton-transfer reactions.
(a) Lead hydroxide and hydriodic acid.
(b) Hydrogen sulfide and manganese(II) nitrate.
(c) Barium iodide and silver sulfate.
(d) Hydrocyanic acid and ammonia.
(e) Strontium carbonate and sulfuric acid.
(f) Magnesium carbonate and sulfuric acid.
(g) Sulfurous acid and zinc chloride.
(h) Barium acetate and sulfuric acid.
(i) Silver carbonate and hydrochloric acid.
(j) Ammonium carbonate and barium hydroxide.

7.4. A 30.00-mL sample of 0.0240 F $Hg_2(C_2H_3O_2)_2$, mercury(I) acetate, is mixed with 50.00 mL of 0.100 F HCl. Solid white mercury(I) chloride (also called calomel) precipitates. Assuming the reaction has gone to completion, calculate the mass of the precipitate and the $[Cl^-]$ remaining in the solution.

7.5. A 40.00-mL sample of an oxalic acid solution is treated with an excess of calcium chloride solution, and an excess of NH_3 is then added to the mixture. A white precipitate of $CaC_2O_4 \cdot H_2O$ is obtained, which is filtered, washed, and dried at 90 °C. At this temperature, the water of hydration is retained in the crystal structure. The precipitate is then found to weigh 0.8212 g.
(a) What was the concentration of oxalic acid in the original 40.00-mL sample?
(b) How many millimoles of ammonium ion were formed during the reaction?

7.6. A standard solution of 0.0200 F $KBrO_3$ is used to titrate 20.00 mL of a solution containing hydrazine, N_2H_4. The reaction that occurs is

$$2BrO_3^- + 3N_2H_4 \rightarrow 2Br^- + 3N_2(g) + 6H_2O$$

The equivalence point in the titration (when all the hydrazine has reacted) occurs when 42.80 mL of the potassium bromate solution have been used. What was the molarity of the hydrazine solution?

7.7. A 0.7075-g sample of a mixture of $CaCl_2$ and K_2SO_4 was analyzed by dissolving the sample in water and then completely precipitating the Ca^{2+} as CaC_2O_4, a white insoluble solid. The CaC_2O_4 was dissolved in sulfuric acid, and the resulting oxalic acid was titrated with a standard $KMnO_4$ solution. The titration required 40.05 mL of 0.04120 F $KMnO_4$ to reach the equivalence point. The equation for the titration reaction is

$$2MnO_4^- + 5H_2C_2O_4 + 6H^+(aq) \rightarrow 2Mn^{2+}(aq) + 10CO_2\uparrow + 8H_2O$$

(a) How many moles of $H_2C_2O_4$ were titrated?
(b) How many moles of $CaCl_2$ were in the original sample?
(c) What was the percentage by weight of $CaCl_2$ in the original sample?

7.8. In a chemistry lab, each student is to prepare 0.20 F HCl by diluting a stock 6.0 F HCl solution. The student is then to determine, to four significant figures, the concentration of the dilute acid he or she has prepared, by employing the following procedure:
(a) Dissolve 0.7000 g of pure solid HgO in excess KI solution. The reaction

$$HgO(s) + 4\,I^- + H_2O \rightarrow HgI_4^{2-} + 2\,OH^-$$

occurs quantitatively.
(b) Titrate the OH^- produced by this reaction against the prepared 0.20 F HCl solution.

   When Karla Kemist did this experiment, it required 31.73 mL of her dilute acid solution to reach the equivalence point in the titration. What was the concentration of the dilute HCl solution she had prepared?

7.9. A 1.3683-g sample of silver carbonate is treated with 50.00 mL of 0.3020 F HCl. Assuming the reaction that occurs has gone to completion, and the volume of solution is still 50.00 mL after the reaction, calculate
(a) The concentration of the two principal ions remaining in solution.
(b) The mass of silver chloride precipitated.
(c) The volume occupied by the CO_2 gas evolved, measured at 25.0 °C and 752 mmHg.

7.10. A 25.00-mL sample of 0.0812 F $Pb(NO_3)_2$ is mixed with 25.00 mL of 0.1024 F KIO_3, and white $Pb(IO_3)_2$ precipitates. Assuming the reaction has gone to completion, calculate the mass of the precipitate and the concentrations of the three principal ions remaining in solution.

7.11. A 25.00-mL sample of a solution of potassium iodate, KIO_3, is pipetted into a flask and treated with a strong acid and excess KI. The following reaction occurs:

$$IO_3^- + 5\,I^- + 6H^+(aq) \rightarrow 3\,I_2 + 3H_2O$$

The iodine formed in this reaction is titrated with a 0.1250 F $Na_2S_2O_3$ solution, according to the reaction

$$I_2 + 2S_2O_3^{2-} \rightarrow S_4O_6^{2-} + 2I^-$$

It requires 38.40 mL of the thiosulfate solution to reach the equivalence point in the titration.
(a) How many millimoles of I_2 were titrated?
(b) How many millimoles of IO_3^- were in the 25.00-mL sample?
(c) What was the formality of the potassium iodate solution?

7.12. A student, Arnie Analyst, is asked to determine the percentage of iron in a solid sample. Arnie weighs the sample and finds its mass is 3.6831 g. He dissolves the sample completely in acidic solution, converting all the iron to Fe^{2+} ions. The solution is then titrated with a standard 0.04000 F $K_2Cr_2O_7$ solution. The titration reaction is

$$14H^+(aq) + Cr_2O_7^{2-} + 6Fe^{2+} \rightarrow 2Cr^{3+} + 6Fe^{3+} + 7H_2O$$

It required 41.55 mL of the 0.04000 F $K_2Cr_2O_7$ to react completely with all the Fe^{2+} present, that is, to reach the equivalence point in the titration. What is the percentage by weight of iron in the original sample?

7.13. When water is added to solid $BiCl_3$, a reaction occurs that produces a white precipitate. Several possible reactions are

(a) $BiCl_3(s) + H_2O \rightarrow BiOCl\downarrow + 2H^+(aq) + 2Cl^-(aq)$

(b) $2BiCl_3(s) + 3H_2O \rightarrow Bi_2O_3\downarrow + 6H^+(aq) + 6Cl^-(aq)$

(c) $2BiCl_3(s) + H_2O \rightarrow Bi_2OCl_4\downarrow + 2H^+(aq) + 2Cl^-(aq)$

Devise an experimental procedure that would distinguish between these three possibilities. Start with a weighed sample of pure solid $BiCl_3$. Assume you have available a solution of sodium hydroxide whose concentration is known to be $0.2000\ F$. Describe the procedure and the calculations you would perform in detail.

7.14. The freezing point of an $0.0800\ m$ $NaHSO_4$ solution is $-0.345\ °C$. Calculate the percentage of HSO_4^- ions that transfer a proton to water to form SO_4^{2-} ions (the percentage HSO_4^- dissociated).

7.15. The mineral atacamite is a basic copper(II) chloride with formula $[CuCl_2 \cdot xCu(OH)_2]$. The mineral is insoluble in water, but reacts quantitatively with hydrochloric acid. In an experiment to determine the value of x, 45.05 mL of $0.5089\ F$ HCl were required to react completely with 1.6320 g of atacamite. What is the value of x?

chapter 8 Introduction to the Law of Chemical Equilibrium

Henri Le Chatelier (1850–1936), was a French inorganic chemist who received his degree from the Ecole des Mines in 1875 and became a professor there in 1877. In 1908 he was appointed professor at the University of Paris. Le Chatelier was an authority on metallurgy, on cements, glasses, fuels, and explosives. He is best known for the Principle he published in 1888 which states that if a system is in a state of equilibrium and one of the conditions is changed, the equilibrium will shift in such a way as to tend to restore the original condition.

When we mix two or more chemicals that react, if we wait a sufficiently long time (which may be a fraction of a second in some cases and many days in others), a condition of equilibrium will be achieved.* For reactions in solution, once equilibrium is attained, the concentrations of all substances present will remain fixed with time, as long as the temperature remains constant. For reactions involving gases, the partial pressure of each gas remains constant once equilibrium has been achieved, provided the temperature is constant. Although concentrations and partial pressures do not change with time, equilibrium in molecular systems is dynamic, with opposing reactions occurring simultaneously at equal rates. We have already considered two examples of dynamic equilibrium: The equilibrium between a liquid and its vapor in Section 5.6, and the equilibrium between a solid solute and a saturated solution of that solute, in Section 6.1. We are now ready to discuss many other kinds of equilibria, and to undertake a study of quantitative relationships in systems at equilibrium.

section 8.1
The Ideal Law of Chemical Equilibrium or the Law of Mass Action

A generalized equation to represent any reacting system that reaches a state of dynamic equilibrium may be written as

$$\alpha A + \beta B \rightleftharpoons \gamma C + \delta D \tag{8-1}$$

This equation states that there is a forward reaction in which the species A and B react in the molar ratio

$$\frac{\text{mol of A}}{\text{mol of B}} = \frac{\alpha}{\beta} \tag{8-2}$$

to form products C and D, and there is a reverse (backward) reaction in which species C and D react in the molar ratio

$$\frac{\text{mol of C}}{\text{mol of D}} = \frac{\gamma}{\delta} \tag{8-3}$$

to form the substances A and B.

Once this system has attained a state of equilibrium, the rate of the forward reaction is exactly equal to the rate of the backward reaction, with the result that, for a reaction in solution, the concentrations of the species A, B, C, and D do not change with time. Whether we start with large amounts of A and B, or with large amounts of C and D, or any mixture at all of A, B, C, and D, we find experimentally that when the system has achieved a state of equilibrium, the numerical value of the function $[C]^\gamma[D]^\delta/[A]^\alpha[B]^\beta$ is essentially the same, as long as the temperature does not change. We call the value of that function the **equilibrium constant** for the reaction, and denote it K_{eq}. The statement

$$\frac{[C]^\gamma[D]^\delta}{[A]^\alpha[B]^\beta} = K_{eq} = \text{a constant at constant temperature} \tag{8-4}$$

is known as the **law of mass action,** or the **ideal law of chemical equilibrium.** It is a convention, accepted by all chemists, always to write the ratio of concentrations as

* There are some reactions that take so long to reach equilibrium that for practical purposes equilibrium is not achieved.

shown in Eq. (8-4) for an equation of the form of Eq. (8-1). That is, *the concentrations of the products of the reaction (to the appropriate powers) are written in the numerator of the equilibrium constant, while the concentrations of the reactants are written in the denominator.*

Example 8.1 shows how to write the equilibrium constant expression for a reaction that takes place in solution.

EXAMPLE 8.1. Equilibrium constant expressions for reactions in solution

Write the equilibrium constant expression for each of the following reactions:

(a) The proton-transfer reaction between ammonia and acetic acid in dilute aqueous solution. The net ionic equation is

$$NH_3 + CH_3COOH \rightleftharpoons NH_4^+ + CH_3COO^-$$

Following the prescription of Eq. (8-4), the equilibrium constant expression is

$$K_{eq} = \frac{[NH_4^+][CH_3COO^-]}{[NH_3][CH_3COOH]}$$

The first power of each of these concentrations appears in the expression for the equilibrium constant because the numerical coefficient of each species in the correctly balanced net ionic equation is 1. Each term in square brackets has units of moles per liter. The numerator and denominator of this expression both have the same units, and the equilibrium constant itself, therefore, has no units (is dimensionless).

(b) The reaction between ferric ions and excess iodide ions in dilute aqueous solution. The correctly balanced net ionic equation for this reaction is

$$2Fe^{3+} + 3I^- \rightleftharpoons 2Fe^{2+} + I_3^-$$

Accordingly, the equilibrium constant expression is

$$K_{eq} = \frac{[Fe^{2+}]^2[I_3^-]}{[Fe^{3+}]^2[I^-]^3}$$

The numerator of this expression has units of $(mol/L)^3$ and the denominator has units of $(mol/L)^5$. The equilibrium constant itself therefore has units of $(mol/L)^{-2}$ or $(L/mol)^2$. These units are often abbreviated as M^{-2}.

The Equilibrium Constant for a Gas-Phase Reaction: K_p and K_c

Consider a reaction that occurs entirely in the gas phase, such as

$$H_2(g) + Cl_2(g) \rightleftharpoons 2HCl(g)$$

It is quite correct to write for the equilibrium constant

$$K_{eq} = \frac{[HCl]^2}{[H_2][Cl_2]}$$

but it is not customary to report the concentrations of gaseous species. We usually describe components in the gas phase in terms of their partial pressures. Since we recognize that

$$P_{HCl} = n_{HCl}(RT/V) = [HCl]RT$$

that is, the pressure of each gas in the mixture is directly proportional to its concentration in moles per liter at constant temperature, for reactions in the gas phase it is customary to write

$$K_{eq} = \frac{P^2_{HCl}}{P_{H_2}P_{Cl_2}} = K_p$$

The equilibrium constant using the concentrations of the species is denoted K_c, while the equilibrium constant using the pressures of the species is called K_p. In this particular example, since both the numerator and the denominator have exactly the same units, the equilibrium constant is dimensionless, and $K_c = K_p$.

It is not true in general, however, that $K_c = K_p$. The relationship between the numerical value of the equilibrium constant expressed as K_c and the value expressed as K_p must be determined by applying the relation $P_A = [A]RT$ to each gas present in the reaction mixture.

For the general gas-phase reaction

$$\alpha A(g) + \beta B(g) \rightleftharpoons \gamma C(g) + \delta D(g)$$

$$K_c = \frac{[C]^\gamma[D]^\delta}{[A]^\alpha[B]^\beta} \quad \text{and} \quad K_p = \frac{P_C^\gamma P_D^\delta}{P_A^\alpha P_B^\beta} \tag{8-5}$$

Substituting the relation between pressure and concentration for each gas, we obtain

$$K_p = \frac{[C]^\gamma(RT)^\gamma[D]^\delta(RT)^\delta}{[A]^\alpha(RT)^\alpha[B]^\beta(RT)^\beta} = \frac{[C]^\gamma[D]^\delta}{[A]^\alpha[B]^\beta}(RT)^{\gamma+\delta-\alpha-\beta} \tag{8-6}$$

We define

$$\Delta n_{gas} = \gamma + \delta - \alpha - \beta \tag{8-7}$$

so that

$$K_p = K_c(RT)^{\Delta n_{gas}} \tag{8-8}$$

Thus we can always determine K_p if we know K_c, or conversely, we can determine K_c if we know K_p. For reactions in which all species are gases, the equilibrium constant used for numerical calculations is almost always K_p.

EXAMPLE 8.2. K_p and K_c

Write expressions for both K_p and K_c for the reaction

$$NO_2(g) \rightleftharpoons NO(g) + \tfrac{1}{2}O_2(g)$$

Give the units of each of these equilibrium constants and the relation between them.

Solution

$$K_p = \frac{P_{NO}P_{O_2}^{1/2}}{P_{NO_2}} \text{ atm}^{1/2} \qquad K_c = \frac{[NO][O_2]^{1/2}}{[NO_2]} \left(\frac{mol}{L}\right)^{1/2}$$

For this reaction $\Delta n_{gas} = 1 + \tfrac{1}{2} - 1 = \tfrac{1}{2}$. Therefore,

$$K_p = K_c(RT)^{1/2}$$

Table 8.1 lists some data on the gas-phase reaction between H_2 and I_2 at 730.8 K. Whether one starts with pure HI(g), with equal pressures of H_2(g) and I_2(g), with excess H_2, or with excess I_2, the data show that the value of $K_p = P^2_{HI}/P_{H_2}P_{I_2}$ is constant, or very nearly constant.

Note that the values of K_p listed in Table 8.1 are not exactly the same to four

Table 8.1. The Constancy of K_p for Equilibrium at 730.8 K
between H_2, I_2, and HI in the Gas Phase[a]

Experiment Number	Partial Pressure (atm)			$K_p = P_{HI}^2/P_{H_2}P_{I_2}$
	I_2	H_2	HI	
1	0.03560	0.3368	0.7615	48.36
2	0.06343	0.2739	0.9262	49.39
3	0.09138	0.2304	1.0117	48.62
4	0.08592	0.08592	0.5996	48.70
5	0.10169	0.10169	0.7080	48.48
6	0.25265	0.25265	1.7651	48.81

[a] Data of A. Taylor and R. Crist, J. Am. Chem. Soc. **63**, 1377 (1941).

significant figures. That is why we call the law of mass action as stated in Eq. (8-4) an ideal law.

For reactions in solution the expression for K_c given in Eq. (8-4) is a constant at constant temperature only when the concentrations of all the substances involved in the equilibrium are very low. The numerical value of the function $[C]^\gamma[D]^\delta/[A]^\alpha[B]^\beta$ is not *exactly* the same for any and all mixtures containing A, B, C, and D, but the value of $[C]^\gamma[D]^\delta/[A]^\alpha[B]^\beta$ is almost constant, that is, it usually varies very little. Since the ideal law of chemical equilibrium is not exactly correct, we must remember that we make an error when we use it. However, it is a good approximation to the behavior of systems at equilibrium, and is very useful. We can learn a great deal by applying it, and we will make extensive use of it.

Reactions in Aqueous Solution in Which Water Is a Reactant or Product

Let us consider how to write the equilibrium constant for the proton-transfer reaction between acetic acid and water:

$$HOAc + H_2O \rightleftharpoons OAc^- + H_3O^+$$

A simple application of the law of mass action, Eq. (8-4), yields

$$K'_{eq} = \frac{[OAc^-][H_3O^+]}{[HOAc][H_2O]}$$

However, a special situation arises when reactions are carried out in aqueous solution and H_2O is also a reactant or a product. Since H_2O is the solvent, it is always present in large excess. For instance, suppose we prepared 1 L of a 0.100 M CH_3COOH solution. In that solution, the amount of water present is not exactly 1 L, but it is very close to it. Since the density of water at room temperature is 1.0 g/mL, there are very close to 1000 g of water present in that solution. The number of moles of H_2O present in a liter of 0.100 M HOAc is, therefore, very close to 1000/18 = 55.5 mol, and thus $[H_2O] \sim 55.5\ M$. Since only a very small fraction of all the CH_3COOH molecules in solution dissociates to yield acetate ions and hydronium ions at equilibrium, the number of moles of H_2O used up in this reaction is substantially less than 0.1. Hence the $[H_2O]$ remains ~55.5 M at all times. Indeed, the $[H_2O]$ is very close to 55.5 M in any dilute solution of acetic acid, or any other electrolyte. Since $[H_2O]$ does not change, we combine it with K'_{eq} and write

$$K_a(HOAc) = K'_{eq}[H_2O] \cong 55.5\ K'_{eq} = \frac{[OAc^-][H_3O^+]}{[HOAc]}$$

This particular equilibrium constant is called the **acidity constant** of acetic acid, and is denoted $K_a(HOAc)$. Other names for this equilibrium constant are the **ionization constant** of acetic acid, or the **dissociation constant** of acetic acid. The numerical value of the acidity constant of acetic acid has been carefully measured, and is 1.8×10^{-5} at 25 °C.

The argument we used to obtain the expression for the acidity constant of acetic acid is generally valid, and we therefore formulate the following rule: *In dilute aqueous solution water is present in such large excess that its concentration remains essentially constant during any reaction involving water. The numerical value of $[H_2O]$ is therefore always included in the value of the equilibrium constant. Thus the term $[H_2O]$ does not appear in any equilibrium constant expression for a reaction taking place in dilute solution.* If, however, the solvent is some liquid other than water, such as alcohol or acetone, and H_2O is one of the reacting species, then the concentration of water can change markedly during the reaction and $[H_2O]$ must be included in the expression for K_{eq}.

Equilibrium Constants for Heterogeneous Reactions

A reaction that occurs in more than a single phase is called a **heterogeneous reaction**, whereas a reaction that takes place in a single phase is a **homogeneous reaction**. Example 8.3 illustrates how to write the equilibrium constant expression for several typical heterogeneous reactions.

EXAMPLE 8.3. Equilibrium constants for heterogeneous reactions

Write the equilibrium constant expression for each of the following reactions:

(a) The reaction between solid zinc and a solution containing silver ions.

$$Zn(s) + 2Ag^+ = Zn^{2+} + 2Ag(s)$$

We must recognize that the solid Zn present at the beginning of this reaction is not in the aqueous solution; it is a pure phase, separate from the solution. Similarly, as soon as the silver is formed it precipitates out of the solution. The Zn(s) and the Ag(s) are pure solids of constant composition. The symbol [Zn] stands for the number of moles of zinc in a liter of pure solid zinc. That is a fixed number at a given temperature; it is easily calculated from the density of zinc in grams per milliliter. The essential point here is that both [Zn(s)] and [Ag(s)] are constant. Constant terms are not included in an equilibrium constant expression.

We formulate the general rule: *No term involving the concentration of a pure solid appears in an equilibrium constant expression.*

The equilibrium constant for the reaction between Zn(s) and Ag^+ ions is therefore

$$K_{eq} = \frac{[Zn^{2+}]}{[Ag^+]^2} \ (mol/L)^{-1}$$

It is important to remember that both solid zinc and solid silver must be present in the reaction vessel even though they do not appear in the equilibrium constant expression. For instance, if we simply mix together aqueous solutions of zinc nitrate and silver nitrate there is no reaction, and no relation between the concentrations of the zinc and silver ions.

(b) The equilibrium between any pure liquid and its vapor, as illustrated by $CCl_4(liq) = CCl_4(vap)$.

The concentration of pure liquid CCl_4 is a constant, and is not included in the equilibrium constant expression. The reasoning is the same as in the preceding example. The equilibrium constant is simply

$$K_p = P_{CCl_4}$$

The equilibrium constant is the vapor pressure of CCl_4, a constant at constant temperature, as has already been discussed in Section 5.9.

(c) The reaction between zinc and hydrochloric acid to produce hydrogen gas.

$$Zn(s) + 2H^+(aq) \rightleftharpoons Zn^{2+}(aq) + H_2(g)$$

We have already discussed balancing this equation in Section 1.9. The reaction involves a gas, a pure solid, and several species in aqueous solution. For heterogeneous reactions involving both the gas phase and a solution phase it is customary to use the pressure of gases rather than their concentrations, and so the equilibrium constant has the form:

$$K_{eq} = \frac{[Zn^{2+}]P_{H_2}}{[H^+]^2} \text{ (atm)(mol/L)}^{-1}$$

The numerical value of this equilibrium constant will depend on the unit used to report the pressure of H_2; the unit most frequently used is atmospheres.

It should be noted that in order for this system to attain equilibrium the H_2 gas must not be allowed to escape from the reaction vessel.

section 8.2
The Magnitude of the Equilibrium Constant and the Direction of Reaction

Let us consider two similar proton-transfer reactions:

$$HOAc + H_2O \rightleftharpoons OAc^- + H_3O^+ \tag{8-9}$$
$$\text{Acid} \qquad \text{Base} \qquad \text{Base} \qquad \text{Acid}$$

and
$$H_2S + H_2O \rightleftharpoons HS^- + H_3O^+ \tag{8-10}$$

Both H_2S and HOAc are weak acids. Hydrogen sulfide, H_2S, is a gas that dissolves in water. An aqueous solution of H_2S contains a small concentration of HS^- ions (hydrogen sulfide or bisulfide ions) and H_3O^+ ions as a result of the proton-transfer reaction, Eq. (8-10). Hydrogen sulfide is an extremely toxic gas with a foul odor. When eggs spoil, H_2S is produced, so the odor of H_2S is always described as the odor of rotten eggs. The unpleasant odor of H_2S is quite useful; its presence in the air is easily detected and one whiff will drive any one away!

The acidity constants for acetic acid and for H_2S are similar in form, but have different numerical values. At 25 °C, for example,

$$K_a(HOAc) = \frac{[OAc^-][H_3O^+]}{[HOAc]} = 1.8 \times 10^{-5} \tag{8-11}$$

$$K_a(H_2S) = \frac{[HS^-][H_3O^+]}{[H_2S]} = 1 \times 10^{-7} \tag{8-12}$$

We see that $K_a(HOAc) > K_a(H_2S)$. What do we learn from this? Since the numerator of the acidity constant expression contains the product of the ion concentrations, while the denominator contains the concentration of the molecular species, the fact that both of these numbers are small, that is, significantly less than 1, tells us that both of these acids are weak acids. Most of the HOAc or H_2S molecules dissolved in water are present at equilibrium as the undissociated molecules; only a small fraction exists as the ions. However, it is clear that the fraction present as ions is larger in the case of acetic acid than it is for H_2S. Thus while both HOAc and H_2S are weak acids, acetic acid is a stronger acid than H_2S because $K_a(HOAc)$ is larger (about 180 times larger) than $K_a(H_2S)$.

The Reaction Quotient, Q

There are two different, but related, questions that we might ask about any equilibrium reaction, $\alpha A + \beta B \rightleftharpoons \gamma C + \delta D$. One of these questions is "If we mix the species A, B, C, and D together in any specified amounts, will a reaction occur?" And if a reaction does occur, will it be *to the right,* that is, forming C and D and using up A and B, or *to the left,* that is, forming A and B and using up C and D?

To answer that question we define the **reaction quotient, Q.** The reaction quotient has the same form as the equilibrium constant, but is evaluated using any set of concentrations of the species involved in the reaction, and not necessarily equilibrium concentrations. Thus,

$$Q = \frac{[C]^\gamma[D]^\delta}{[A]^\alpha[B]^\beta} \qquad (8\text{-}13)$$

when the system is *not* at equilibrium. *A net reaction can occur only if a system is not at equilibrium.* When we mix reagents together we observe a change in the concentrations of several species *only* if the reaction mixture is not in a state of equilibrium. *All reactions occur to reach a state of equilibrium.* Hence if $Q \neq K_{eq}$, a reaction can occur. If $Q = K_{eq}$, the system is at equilibrium and we will not observe any net reaction taking place. The answer to the question of the direction of the reaction is obtained by comparing the value of the reaction quotient, Q, with the value of the equilibrium constant, K. If Q is less than K, reaction can occur to make Q bigger, that is, to produce more of the products C and D, and use up the reactants A and B. Thus the direction of the reaction will be to the right. If Q is greater than K, reaction can occur to make Q smaller, that is, to produce more A and B, and use up C and D. Thus the direction of the reaction will be to the left. The backward reaction can occur spontaneously. We sum up this information with the statements:

If $Q < K_{eq}$, a net forward reaction (to the right) can occur	(8-14a)
If $Q > K_{eq}$, a net backward reaction (to the left) can occur	(8-14b)
If $Q = K_{eq}$, the system is at equilibrium, and no net reaction will occur	(8-14c)

Think about the discussion presented until these three statements make sense to you, and you will always be able to deduce the correct consequences of a disparity between Q and K_{eq}. Example 8.4 illustrates the use of these relations.

EXAMPLE 8.4. Use of the reaction quotient to predict the direction of reaction

Ammonia is produced from the elements hydrogen and nitrogen by the gas-phase reaction $3H_2(g) + N_2(g) \rightleftharpoons 2NH_3(g)$. This is an extremely important reaction because it is the starting point in the production of many fertilizers. It is carried out at elevated temperatures because it is a very slow reaction at room temperature. For this reaction K_p is 6.5×10^{-3} atm^{-2} at 450 °C. Each of the following mixtures may or may not be at equlibrium at 450 °C at the instant of measurement. For each mixture, state whether or not it is in a state of equilibrium, and if it is not, indicate the direction in which reaction must proceed to achieve equilibrium.

(a) A mixture in which the partial pressures of the three gases are

$$P_{NH_3} = 60 \text{ atm} \qquad P_{H_2} = 150 \text{ atm} \qquad P_{N_2} = 80 \text{ atm}$$

Solution. At the instant of measuring these partial pressures, the reaction quotient, Q, for the reaction $3H_2(g) + N_2(g) \rightleftharpoons 2NH_3(g)$ has the value

$$Q = P_{NH_3}^2 / P_{N_2} P_{H_2}^3 = (60)^2 \text{atm}^2 / (80)(150)^3 \text{atm}^4$$
$$= 1.3 \times 10^{-5} \text{ atm}^{-2}$$

Since $Q < K_p = 6.5 \times 10^{-3}$ atm^{-2}, a reaction will occur to make Q bigger. The number of moles of NH_3 present, and therefore the partial pressure of NH_3, will increase, and some of the H_2 and N_2 present will be used up, so that the partial pressures of H_2 and N_2 will both decrease. Reaction will occur spontaneously to the right, and will continue until $Q = K_p$ and the system is at equilibrium.

(b) A mixture in which the partial pressures of the three gases are

$$P_{NH_3} = 60 \text{ atm} \qquad P_{H_2} - 2.0 \text{ atm} \qquad P_{N_2} = 5.0 \text{ atm}$$

Solution

$$Q = (60)^2 / (2)^3 (5) \text{ atm}^{-2} = 90 \text{ atm}^{-2}$$

Since $Q > K_p$, a reaction will occur to make Q smaller. Some of the NH_3 present will be used up, and more N_2 and H_2 will be formed. Reaction will occur spontaneously to the left, and will continue until $Q = K_p$ and the system is at equilibrium.

The Significance of the Magnitude of K_{eq}

The second question we might ask is this: Once the system reaches a state of equilibrium, will there be a lot of the products C and D relative to the amounts of the reactants A and B (a situation that is described by saying "the position of equilibrium is to the right"), or will there be a lot of A and B relative to the amounts of the products C and D? The answer to this question is obtained by comparing the magnitude of the equilibrium constant with the value 1. An examination of the form of the equilibrium constant, Eq. (8-4), makes it clear that if $K_{eq} \ll 1$, the concentrations of the products C and D are small relative to the concentrations of the reactants A and B. On the other hand, if $K_{eq} \gg 1$, the concentrations of the products C and D are large relative to the concentrations of the reactants A and B. We can summarize this information with the following three statements:

If $K_{eq} \ll 1$, the position of equilibrium is to the left	(8-15a)
If $K_{eq} \gg 1$, the position of equilibrium is to the right	(8-15b)
If $K_{eq} \sim 1$, there will be appreciable concentrations of both products and reactants present at equilibrium	(8-15c)

What is meant by the statement "K_{eq} is *approximately* 1" in Eq. (8-15c)? There are no hard and fast boundaries, but a reasonable range for the condition $K_{eq} \sim 1$ would be between 0.01 and 100.

The following two examples, about the same system at equilibrium, illustrate the concepts discussed in this section.

EXAMPLE 8.5. Heterogeneous equilibrium involving gases

At 275 °C, the equilibrium constant for the reaction

$$NH_4Cl(s) \rightleftharpoons NH_3(g) + HCl(g)$$

is 1.04×10^{-2} atm².

A 0.980-g sample of solid NH_4Cl is placed in a 1.000-L closed container and heated to 275 °C.

(a) Will there be any decomposition of the NH_4Cl when the system comes to equilibrium at 275 °C?

Solution. The reaction quotient expression for this equilibrium is

$$Q = P_{NH_3}P_{HCl}$$

Initially, $Q = 0$, as there is neither NH_3 nor HCl in the container. The equilibrium constant is 1.04×10^{-2} atm². Since $Q < K_p$, the reaction will occur spontaneously to the right. Some NH_3 and HCl will be formed, and there will be some decomposition of the NH_4Cl.

(b) At equilibrium at 275 °C what will the partial pressures of NH_3 and HCl in the container be?

Solution. NH_3 and HCl are formed in a 1:1 molar ratio in this reaction. Since the partial pressure of each gas is directly proportional to the number of moles of that gas, the partial pressures of NH_3 and HCl must be equal.

$$\text{Let } x = P_{NH_3} = P_{HCl} \text{ at equilibrium}$$

The equilibrium constant expression is

$$K_p = P_{NH_3}P_{HCl} = 1.04 \times 10^{-2} \text{ atm}^2$$

Hence
$$x^2 = 1.04 \times 10^{-2} \text{ atm}^2$$
$$x = 1.02 \times 10^{-1} \text{ atm}$$

When the system has come to equilibrium at 275 °C, the partial pressure of NH_3 and of HCl is 1.02×10^{-1} atm, or 77.5 mmHg.

(c) What will be the mass of solid NH_4Cl in the container at equilibrium at 275 °C?

Solution. We can calculate the number of moles of HCl or NH_3 in the container assuming the gases are ideal.

$$n_{NH_3} = n_{HCl} = \frac{(1.02 \times 10^{-1} \text{ atm})(1.000 \text{ L})}{\left(0.08206 \dfrac{\text{L} \cdot \text{atm}}{\text{mol} \cdot \text{K}}\right)(548.2 \text{ K})} = 2.268 \times 10^{-3} \text{ mol}$$

$$= 2.27 \times 10^{-3} \text{ mol}$$

Since

$$\frac{\text{No. mol } NH_4Cl \text{ decomposed}}{\text{No. mol } NH_3 \text{ formed}} = \frac{1}{1}$$

2.27×10^{-3} mol of solid NH_4Cl must have decomposed. The formula weight of NH_4Cl is 53.49 g · mol^{-1}, so that the mass of NH_4Cl that has decomposed is

$$(53.49 \text{ g} \cdot \text{mol}^{-1}) (2.268 \times 10^{-3} \text{ mol}) = 0.121 \text{ g}$$

As the mass of the original sample was 0.980 g, the mass of solid NH_4Cl remaining in the container at equilibrium is $0.980 - 0.121 = 0.859$ g.

EXAMPLE 8.6. Heterogeneous equilibrium involving gases

At 275 °C, the equilibrium constant for the reaction

$$NH_4Cl(s) \rightleftharpoons NH_3(g) + HCl(g)$$

is 1.04×10^{-2} atm^2.

The gases HCl and NH_3, both at 275 °C, are introduced into an evacuated 1.000-L flask. At the instant of mixing, the partial pressure of HCl in the flask is 0.800 atm and the partial pressure of NH_3 is 0.500 atm. The temperature of the flask is maintained constant at 275 °C.

(a) Show that the reaction between NH_3 and HCl to form solid NH_4Cl will take place when these gases are mixed.

Solution. At the instant of mixing, the reaction quotient is

$$Q = P_{NH_3}P_{HCl} = (0.500)(0.800) \text{ atm}^2 = 0.400 \text{ atm}^2$$

Since $K_p = 1.04 \times 10^{-2}$ atm^2, $Q > K_p$. Thus the backward reaction, the reaction to the left, will occur spontaneously, and NH_3 and HCl will combine to form solid NH_4Cl.

(b) Calculate the partial pressures of NH_3 and HCl at equilibrium at 275 °C.

Solution. At equilibrium, the partial pressure of NH_3 will be less than 0.500 atm, and the partial pressure of HCl will be less than 0.800 atm, because some NH_3 and some HCl will be used up to form $NH_4Cl(s)$. For every mole of NH_3 that is used up, a mole of HCl is also used up, since the two gases combine in a 1 : 1 molar ratio. Thus the decrease in the partial pressure of HCl is exactly the same as the decrease in the partial pressure of NH_3.

Let x = decrease in the partial pressure of each gas when system is at equilibrium

Then, at equilibrium,

$$P_{NH_3} = (0.500 - x) \text{ atm}$$
$$P_{HCl} = (0.800 - x) \text{ atm}$$

The equilibrium constant expression is therefore

$$(0.500 - x)(0.800 - x) = 1.04 \times 10^{-2} \text{ atm}^2$$

Multiplying this out and collecting terms, we obtain a quadratic equation in x:

$$x^2 - 1.300x + 0.3896 = 0$$

This can be solved using the quadratic formula (refer to Appendix B5)

$$x = \frac{1.300 \pm [1.690 - (4)(0.3896)]^{1/2}}{2}$$

which yields $x = 0.4686$ or 0.8314. The value 0.8314 must be rejected, as x must be smaller than 0.500, since both P_{NH_3} and P_{HCl} must be positive. With $x = 0.4686$, we obtain

$$P_{NH_3} = 0.500 - 0.4686 = 0.031 \text{ atm}$$
$$P_{HCl} = 0.800 - 0.4686 = 0.331 \text{ atm}$$

As the partial pressure of NH_3 has dropped from its initial value of 0.500 to an equilibrium value of 0.031 atm, we see that most of the NH_3 introduced into the flask has been used up to form NH_4Cl.

Consider the answers to Examples 8.5 and 8.6 carefully. Note that the individual equilibrium partial pressures of NH_3 and HCl are quite different in the two examples, but the value of the product $P_{NH_3}P_{HCl}$ at equilibrium is the same. That is what is meant by an equilibrium *constant*.

Note also that it is possible to have a system that is not at equilibrium. For instance, suppose we introduce $NH_3(g)$ and $HCl(g)$ into an evacuated flask at 275 °C so that at the instant of mixing the partial pressure of each gas is 0.020 atm. In that case, the reaction quotient is $4.0 \times 10^{-4} \text{ atm}^2$, and is significantly less than K_p for the reaction. To reach equilibrium there would have to be more NH_3 and HCl present. As there is no solid NH_4Cl to decompose, no reaction can occur. For a system to be at equilibrium, all the species in the reaction must be present in the reaction vessel. There is no term involving solid NH_4Cl in the expression for the equilibrium constant for this system, because the concentration of NH_4Cl in pure solid NH_4Cl is a constant, but there must be some solid NH_4Cl in the reaction vessel for the system to be at equilibrium.

section 8.3
Factors Affecting the Equilibrium Constant

The term *constant* means invariant, unchanging. The value of K_{eq} is independent of the concentration of any of the species in the reaction mixture as long as the temperature is constant. If one mixes A and B together in any proportion, reaction can occur to form C and D. The reaction will stop only when the reaction quotient, Q, is equal numerically to K_{eq} at the temperature specified. Similarly, if one mixes C and D together in any proportion, reaction can occur to form A and B, and the reaction will stop only when Q is numerically equal to K_{eq} at the given temperature. The individual values of [A], [B], [C], and [D] may vary, but the numerical value of the function $[C]^\gamma[D]^\delta/[A]^\alpha[B]^\beta$ will not change as long as the temperature remains constant. There are, however, two factors that will change the numerical value of K_{eq} for a given reaction. They are

1. *The temperature.* For any given reaction, the numerical value of K_{eq} will depend on the temperature. The K_{eq} will be different at 20 °C than at 50 or 0 °C. The *expression* for K_{eq} will be exactly the same, but the numerical value changes as the temperature changes. The temperature dependence of K_{eq} will be discussed in Chapter 18, Section 11.

2. *The solvent.* For reactions that take place in solution, the numerical value of K_{eq} will depend on the solvent. Generally we will be considering reactions in aqueous solution, but if the same reaction takes place in an ethanol–H_2O mixture, or in an acetone–H_2O mixture, the numerical value of K_{eq} will change.

section 8.4
Le Chatelier's Principle

We have already discussed the fact that if you disturb a system that is at equilibrium, a change will spontaneously occur to restore the system to a state of equilibrium. (See Sections 5.6 and 5.8.) A statement of the way a system at equilibrium responds to a disturbance was made by the French chemist Henri Le Chatelier in 1888. **Le Chatelier's Principle** says: If a system in a state of dynamic equilibrium is subjected to a disturbance (a stress) that changes any of the factors that determine the state of equilibrium, the equilibrium will be displaced in the direction that minimizes the effect of the disturbance.

Le Chatelier's Principle is extremely broad: It has a great many applications because there are many ways to disturb a system at equilibrium. We will consider several important examples of the operation of Le Chatelier's Principle under the headings: (1) concentration effects for reactions in solution, (2) pressure effects for gas-phase reactions, and (3) temperature effects.

1. The Effect on a System at Equilibrium of Changing the Concentration of One of the Reacting Species

Dissolving Solids That Are Insoluble in Water

We first consider the problem of dissolving insoluble solids. Table 7.4 summarized information about the solubilities of electrolytes in water. When we say that a substance is insoluble, we generally mean that its solubility *in water* is very small. Rule 7 of Table 7.4 states that most hydroxides are insoluble. Let us consider $Zn(OH)_2$ as a specific example. Zinc hydroxide, $Zn(OH)_2$, is a white, insoluble solid. Its solubility in water is extremely small, but the small amount that does dissolve is 100% in ionic form. The equilibrium

$$Zn(OH)_2(s) \rightleftharpoons Zn^{2+}(aq) + 2\,OH^-(aq) \qquad (8\text{-}16)$$

is established whenever $Zn(OH)_2$ precipitates out of an aqueous solution. The equilibrium constant* for reaction (8-16) is

$$K_{eq} = K_{sp} = [Zn^{2+}][OH^-]^2 = 5 \times 10^{-17} \qquad \text{at 25 °C} \qquad (8\text{-}17)$$

This particular type of equilibrium constant, the equilibrium constant for the dissolution of an insoluble electrolyte in water, is called a **solubility product**, and is denoted K_{sp}. Note that the denominator of the equilibrium constant expression is unity and *not* $[Zn(OH)_2]$, because $Zn(OH)_2$ is a pure solid, in a separate phase from the aqueous solution. [Refer to the rule formulated in discussing Example 8.3(a).]

Note also that the numerical value of this equilibrium constant is very much smaller than 1, which indicates that the position of equilibrium is far to the left [Eq. (8-15a)]. The very small magnitude of the solubility product of $Zn(OH)_2$ tells us that $Zn(OH)_2$ has a very small solubility in water; the concentrations of Zn^{2+} and OH^- ions in equilibrium with solid $Zn(OH)_2$ are very low. We call $Zn(OH)_2$ insoluble because its solubility in water is so small. Does this mean that there is no possible way to dissolve $Zn(OH)_2$? Not at all! We can utilize Le Chatelier's Principle, and reason that if we disturb the equilibrium by removing either Zn^{2+} ions or OH^- ions from solution, the equilibrium will shift to counter this change.

We can easily remove OH^- ions by adding any strong acid, because hydronium

---

* Equilibrium constants this small are difficult to measure and are often not reliable to better than one significant figure.

ions combine with hydroxide ions to form H_2O. If we add $H^+(aq)$ ions, we decrease the $[OH^-]$ in equilibrium with the solid $Zn(OH)_2$. The equilibrium (8-16) will then shift to the right to put more OH^- ions into solution, and some solid $Zn(OH)_2$ dissolves. We can write the equation for what occurs in two different ways. To indicate in detail the two reactions that are occurring we may write

$$Zn(OH)_2(s) \rightleftharpoons Zn^{2+}(aq) + 2\ OH^-(aq)$$
$$+$$
$$2H^+(aq)$$
$$\updownarrow$$
$$2H_2O$$

As we add more strong acid, we remove more OH^- ions from solution, and more solid $Zn(OH)_2$ dissolves to offset the disturbance caused by the added hydronium ions. If we add enough strong acid, all the $Zn(OH)_2$ will dissolve, and there will no longer be any equilibrium between solid $Zn(OH)_2$ and the aqueous solution.

While the scheme shown above is informative, it is also cumbersome, and the net ionic equation for the dissolution of $Zn(OH)_2$ in strong acid is written as

$$Zn(OH)_2(s) + 2H^+(aq) \rightleftharpoons Zn^{2+}(aq) + 2H_2O \tag{8-18}$$

Adding more strong acid drives this equilibrium to the right. The system offsets the addition of $H^+(aq)$ ions by reacting to use up some $H^+(aq)$. We see from this example that the addition of sufficient strong acid enables us to dissolve insoluble hydroxides.

Insoluble carbonates can also be dissolved by adding strong acid. We have already discussed dissolving $CaCO_3$ in nitric acid in Example 7.4(e). The dissolution of carbonates using $H^+(aq)$ is another example of the utilization of Le Chatelier's Principle. While $CaCO_3$ is called "insoluble," its solubility in water is not zero. The equilibrium

$$CaCO_3(s) \rightleftharpoons Ca^{2+}(aq) + CO_3^{2-}(aq) \tag{8-19}$$

is established whenever $CaCO_3$ is precipitated in aqueous solution, or when some solid $CaCO_3$ is stirred with water. We can remove CO_3^{2-} ions from a solution by adding $H^+(aq)$ ions, because hydronium ions and carbonate ions combine to form bicarbonate ions, HCO_3^-, and carbonic acid, H_2CO_3, both of which are weak acids. Carbonic acid, H_2CO_3, decomposes to H_2O and $CO_2(g)$. Since the $CO_2(g)$ escapes from the solution into the air, more and more CO_3^{2-} is removed from the solution, and the equilibrium (8-19) shifts to the right to counter this change. The net ionic equation for dissolving $CaCO_3$ in excess strong acid is

$$CaCO_3(s) + 2H^+(aq) \rightarrow Ca^{2+} + CO_2\uparrow + H_2O \tag{8-20}$$

Any insoluble carbonate can be dissolved by adding sufficient strong acid, because hydronium ions combine with carbonate ions. In general, if the anion of a salt combines with $H^+(aq)$ to form a weak acid, that salt will be more soluble in a strong acid than it is in water. Thus, carbonates, sulfites, sulfides, and phosphates are all more soluble in acid than they are in pure water. If the product of the reaction is a gas that escapes into the air in the laboratory, it is possible to dissolve the salt completely by adding excess strong acid. That is the case with sulfites as well as with carbonates, because the weak acid H_2SO_3 (sulfurous acid) decomposes to $SO_2(g)$ and H_2O.

EXAMPLE 8.7. Dissolving insoluble sulfites in strong acid

Write a balanced net ionic equation for dissolving barium sulfite in hydrochloric acid, and explain, using Le Chatelier's Principle, why barium sulfite dissolves in hydrochloric acid although it is insoluble in water.

Solution. The net ionic equation is

$$BaSO_3(s) + 2H^+(aq) \rightleftharpoons Ba^{2+}(aq) + H_2O + SO_2 \uparrow$$

When solid $BaSO_3$ is added to water very little dissolves, but the small amount that does dissolve is present in solution as Ba^{2+} ions and SO_3^{2-} ions. The equilibrium

$$BaSO_3(s) \rightleftharpoons Ba^{2+}(aq) + SO_3^{2-}(aq)$$

is established between the undissolved solid and the saturated solution. Adding $H^+(aq)$ ions to this system when it is at equilibrium disturbs the equilibrium by removing SO_3^{2-} ions from the solution. Since adding $H^+(aq)$ ions reduces the $[SO_3^{2-}]$, the equilibrium shifts to the right to counter this change, and more $BaSO_3$ dissolves. Because the final products include gaseous SO_2, which escapes into the air, solid $BaSO_3$ can be completely dissolved by adding an excess of hydrochloric acid.

Removing an Ion from Solution by Precipitation of an Insoluble Salt Containing That Ion

There are circumstances under which it is desirable to remove an ion from solution as completely as possible. Suppose, for instance, that we want to remove Pb^{2+} ions from solution. We can do this by adding a source of Cl^- ions, which will precipitate the Pb^{2+} ions as $PbCl_2$. The precipitate can then be separated from the remaining solution by filtration.

Lead chloride is one of the few chlorides that does not have a large solubility in water. (See Rule 3 of Table 7.4.) It is one of those salts that is considered *moderately soluble* or *slightly soluble*. The equilibrium constant for the reaction

$$PbCl_2(s) \rightleftharpoons Pb^{2+}(aq) + 2Cl^-(aq) \tag{8-21}$$

is

$$K_{eq} = K_{sp} = [Pb^{2+}][Cl^-]^2 = 1.6 \times 10^{-5} \, M^3 \quad \text{at 25 °C} \tag{8-22}$$

This equilibrium constant is a **solubility product,** just as that in Eq. (8-17).

Since $PbCl_2$ is moderately soluble, the $[Pb^{2+}]$ left in solution after the addition of chloride ions may not be insignificant, although it will not be large. How can we decrease the $[Pb^{2+}]$? We can utilize Le Chatelier's Principle, and add an excess of some soluble chloride such as HCl or NaCl, which will increase the $[Cl^-]$. Increasing the $[Cl^-]$ constitutes a stress or disturbance to the equilibrium, Eq. (8-21), which therefore shifts to offset the stress by using up some of the additional chloride ions. This can only be accomplished by a shift to the left, that is, by the precipitation of more $PbCl_2$. Since $[Pb^{2+}][Cl^-]^2$ is a constant at constant temperature, increasing the $[Cl^-]$ must result in a decrease in the $[Pb^{2+}]$, and we achieve our goal of more effectively removing Pb^{2+} ions from solution.

Using an excess of Cl^- ions to remove Pb^{2+} ions from solution is an example of the **common ion effect.** The solubility of an insoluble or slightly soluble electrolyte is less in a solution containing a soluble electrolyte that has an ion in common with the insoluble compound than it is in pure water. In the example discussed, we decreased the solubility of $PbCl_2$ by adding excess HCl or NaCl, or some other soluble chloride.

2. *Pressure Effects for Gas-Phase Reactions*

Consider the gas-phase decomposition of phosphorus pentachloride:

$$PCl_5(g) \rightleftharpoons PCl_3(g) + Cl_2(g) \tag{8-23}$$

Let us assume that at a given temperature we have an equilibrium mixture of these three gases, and the pressure of each gas has some specified value which we will denote as P'_{PCl_5}, P'_{Cl_2}, and P'_{PCl_3}. Then, since the system is at equilibrium at fixed temperature, the relation

$$K_p = P'_{Cl_2} P'_{PCl_3}/P'_{PCl_5}$$

is valid. Suppose we now disturb the system by decreasing the volume of the container by a factor of 2. This is easily accomplished if the gases are in a cylinder enclosed by a piston, and we push the piston down to halve the volume of the gaseous mixture. (See Fig.8.1.) At the instant we push the piston down, the pressure of each gas is doubled, and the system is no longer at equilibrium. What will the reaction quotient, Q, be at the instant the piston is pushed down?

$$Q = (2P'_{Cl_2})(2P'_{PCl_3})/(2P'_{PCl_5}) = 2K_p$$

Since $Q > K_p$, the system is not at equilibrium, and a reaction will occur. The reaction quotient, Q, must be decreased, and this can only be achieved by using up Cl_2 and PCl_3 and forming more PCl_5. Thus the system shifts to the left to restore equilibrium. When equilibrium is attained, the pressure of chlorine will be less than $2P'_{Cl_2}$, the pressure of PCl_3 will be less than $2P'_{PCl_3}$, and the pressure of PCl_5 will be greater than $2P'_{PCl_5}$. Each of these three pressures will be different from its original value, but the quotient $P_{Cl_2}P_{PCl_3}/P_{PCl_5}$ will have the same numerical value, K_p.

We can describe the shift that occurred in terms of Le Chatelier's Principle. The change that was made (decreasing the volume) had the effect of increasing the partial pressure of each gas, and therefore increasing the total pressure of the system. The system shifts to offset this stress, that is, to decrease the total pressure. It can accomplish this by shifting to the side with fewer moles of gas, since $P_{total} = n_{total}(RT/V)$. For the reaction considered, there are fewer moles of gas present when the equilibrium shifts to the left, since 2 mol of gas on the right combine to produce only a mole of PCl_5.

We formulate this as a general rule: *At constant temperature, increasing the total pressure on a gas-phase equilibrium due to an increase in the partial pressure of each gas involved in the equilibrium will cause a shift to the side with fewer moles of gas.*

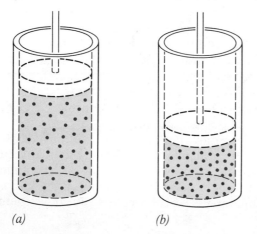

(a)   *(b)*

Fig. 8.1. Doubling the total pressure of a gas contained in a cylinder capped by a piston, by halving the volume of the gas.

It is important to remember that to interpret correctly the result of a change described in terms of an increase or decrease in the volume of the system, you must consider the effect of the disturbance on the total pressure of the system. That is because the "factors that determine the state of equilibrium" are the partial pressures of the gases. Note that Le Chatelier's Principle states it is a "stress that changes any of the factors that determine the state of equilibrium" that will be offset by a displacement of the equilibrium. Thus an increase in the volume of the system that results in a decrease in the total pressure will cause a shift to the side of the equilibrium with a greater number of moles of gas, in order to counter the decrease in the total pressure.

3. The Effect of a Change in Temperature

The numerical value of an equilibrium constant depends on the temperature. Let us consider the equilibrium

$$N_2O_4(g) \rightleftharpoons 2NO_2(g) \tag{8-24}$$

This is the dissociation of colorless dinitrogen tetroxide to produce the red-brown poisonous gas nitrogen dioxide. Dinitrogen tetroxide, N_2O_4, is a solid below $-11.2\ °C$, and boils at $21.2\ °C$, so that it is gaseous at $25\ °C$ and 1 atm. The equilibrium constant for this reaction is

$$K_p = P^2_{NO_2}/P_{N_2O_4} \tag{8-25}$$

Dissociation reactions involve the breaking of chemical bonds, and it always requires energy to break bonds. Dissociation reactions are therefore endothermic. For reaction (8-24), the heat absorbed at constant pressure, ΔH, is 58.02 kJ/mol at $25\ °C$.

In terms of Le Chatelier's Principle, the stress of raising the temperature by adding heat causes the reaction to shift in such a way as to offset the stress, that is, to shift to the right, because the forward direction uses up heat. We therefore expect that the numerical value of K_p will be larger at higher temperatures. This is indeed exactly what we observe. For this reaction K_p is 1.1×10^{-1} atm at $25\ °C$, but 1.3 atm at $60\ °C$. Thus K_p is more than 10 times larger at $60\ °C$ than it is at $25\ °C$. Increasing the temperature above $60\ °C$ will increase K_p still further. Raising the temperature drives this reaction to the right, toward the dissociation product, the red-brown gas, NO_2. We can readily observe this effect experimentally by measuring the increase in the intensity of the red-brown color as we heat a gaseous mixture of N_2O_4 and NO_2 above $25\ °C$.* (See Fig. 8.2.)

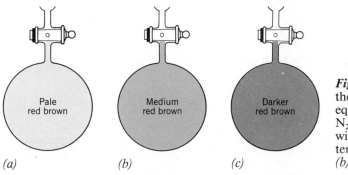

(a)   (b)   (c)

Fig. 8.2. The intensity of the red-brown color of an equilibrium mixture of N_2O_4 and NO_2 deepens with increasing temperature. *(a)* At $25\ °C$. *(b)* At $60\ °C$. *(c)* At $100\ °C$.

---

* A mixture of N_2O_4 and NO_2 should be handled with caution because it is highly toxic and reactive. All work with these gases should be carried out in a hood.

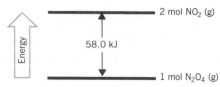

Fig. 8.3. Difference in energy between 1 mol of N_2O_4 and 2 mol of NO_2.

We state the general rule: *Raising the temperature of an endothermic reaction shifts the equilibrium to the right and thus increases the numerical value of the equilibrium constant. Raising the temperature of an exothermic reaction shifts the equilibrium to the left and therefore decreases the numerical value of the equilibrium constant.*

A consideration of the $N_2O_4(g) \rightleftharpoons 2NO_2(g)$ equilibrium serves as another illustration of the statement made previously in Sections 5.7 and 5.8 that the position of equilibrium at any temperature is a compromise between the tendency to maximize the entropy and the tendency to minimize the energy. One mole of N_2O_4 molecules is at a lower energy than 2 mol of NO_2 molecules. We know this because we must add energy to convert 1 mol of N_2O_4 to 2 mol of NO_2. (Refer to Fig. 8.3.) The tendency to minimize the energy therefore favors the N_2O_4, that is, it drives the equilibrium to the left. However, when two N atoms and four O atoms are bonded together as a single N_2O_4 molecule they must move as a unit, and there is a particular geometric arrangement (O_2N-NO_2) for the six atoms. When N_2O_4 dissociates to two NO_2 molecules, each NO_2 can move independently of the other. There is greater freedom of motion for the six atoms when they exist as two NO_2 molecules than when they are bonded together as N_2O_4. The tendency to achieve maximum entropy, that is, greater molecular disorder, favors the two NO_2 molecules and drives the equilibrium to the right. The actual position of equilibrium is a compromise between these two factors. At 25 °C, K_p is 1.1×10^{-1}, and since this is less than 1, the position of equilibrium is to the left. However, K_p is not very much less than 1 and significant amounts of both gases will be present in an equilibrium mixture. As the temperature is increased, the tendency to achieve maximum entropy becomes increasingly important, and the equilibrium shifts to the right. By 60 °C, K_p is just slightly greater than 1; it is 1.3 atm. As we raise the temperature we add thermal energy and drive the equilibrium further to the right.

EXAMPLE 8.8. **Le Chatelier's Principle and gas-phase equilibria**

Consider the exothermic formation of sulfur trioxide from sulfur dioxide and oxygen in the gas phase:

$$2SO_2(g) + O_2(g) \rightleftharpoons 2SO_3(g)$$

At 900 K, K_p for this reaction is 40.5 atm^{-1} and $\Delta H = -198$ kJ.

(a) Write the expression for the equilibrium constant for this reaction.

Solution

$$K_p = P^2_{SO_3}/P^2_{SO_2}P_{O_2}$$

(b) Will the equilibrium constant for this reaction at room temperature (~ 300 K) be greater than, less than, or equal to the equilibrium constant at 900 K? Explain your answer.

Solution. The equilibrium constant at 300 K will be greater than the equilibrium constant at 900 K. This is an exothermic reaction. If we decrease the temperature from 900 to 300 K, the equilibrium will be displaced to the right, releasing heat, to offset the stress of decreased temperature. If more SO_3 is produced and SO_2 and O_2 are used up, the equilibrium constant increases. In fact, there is a very large change in the numerical value of the equilibrium constant. At 25 °C (298 K), $K_p = 3.37 \times 10^{24}$ atm^{-1}.

(c) How will the mole fraction of SO_3 at equilibrium be affected by decreasing the volume in which the three gases are contained, keeping the temperature constant?

Solution. Decreasing the volume of the container will increase the partial pressure of each gas and therefore increase the total pressure. The system will shift to the side with fewer numbers of moles of gas, to minimize the effect of the disturbance. Decreasing the total number of moles of gas decreases the total pressure. Since there are two moles of gas on the right, but three on the left, the system will shift to the right. More SO_3 will be produced, and some SO_2 and O_2 will be used up. The mole fraction of SO_3 in the container will increase since the number of moles of SO_3 will increase and the total number of moles of gas will decrease.

(d) If, while keeping the temperature constant, more O_2 is pumped into a flask containing the three gases SO_3, SO_2, and O_2 in a state of equilibrium, will the number of moles of SO_2 increase, decrease, or remain the same? Explain.

Solution. The number of moles of SO_2 will decrease. The system will shift to use up some of the added O_2. Thus it will shift to the right. Some SO_2 will be used up, and more SO_3 will be formed.

(e) What is the effect on the equilibrium of adding 1 mol of He(g) to a flask containing SO_2, O_2, and SO_3 at equilibrium at constant temperature?

Solution. Adding He(g) has no effect at all. The partial pressures of SO_2, O_2, and SO_3 are unchanged by the addition of helium. The total pressure in the container increases, but as the partial pressures of the gases involved in the equilibrium are unaffected, the equilibrium does not shift.

EXAMPLE 8.9. Le Chatelier's Principle and solution equilibria

Consider the dissolution of the insoluble compound manganese(II) sulfide in water:

$$MnS(s) \rightleftharpoons Mn^{2}(aq) + S^{2-}(aq)$$

(a) Write the expression for the equilibrium constant for this reaction and specify its units.

Solution

$$K_{eq} = K_{sp} = [Mn^{2+}][S^{2-}](mol/L)^2 \quad \text{or} \quad M^2$$

(b) Will adding excess $Mn(NO_3)_2$ solution to a saturated solution of MnS in equilibrium with excess solid MnS increase, decrease, or leave unchanged the solubility of MnS?

Solution. Adding excess $Mn(NO_3)_2$ solution will decrease the solubility of MnS. A solution of $Mn(NO_3)_2$ is 100% Mn^{2+} ions and NO_3^- ions. When Mn^{2+} ions are added, the equilibrium shifts to the left to use up some of the added Mn^{2+} ions, and minimize the disturbance. Some Mn^{2+} ions and some S^{2-} ions react to precipitate more MnS. Since the amount of solid MnS precipitated has increased, the solubility of MnS decreased. This is an example of the **common ion effect**.

(c) Will adding excess HCl solution to a saturated solution of MnS in equilibrium with excess solid MnS increase, decrease, or leave unchanged the solubility of MnS? Explain.

Solution. Adding excess HCl will greatly increase the solubility of MnS. Indeed, it is possible to completely dissolve MnS in excess HCl. Hydronium ions combine with S^{2-} ions to form the weak acids HS^- and H_2S. If excess HCl is used, H_2S will be the principal product. Adding $H^+(aq)$ ions decreases the $[S^{2-}]$ in solution, and the equilibrium shifts to the right, producing more Mn^{2+} and S^{2-} ions, to offset the stress. The net ionic equation for dissolving MnS in HCl is

$$MnS(s) + 2H^+(aq) \rightleftharpoons Mn^{2+} + H_2S$$

section 8.5
Numerical Values of Equilibrium Constants for Reactions Written in the Reverse Direction and for Simultaneous Equilibria

The Effect of Reversing the Equilibrium Equation

Because all equilibria are dynamic it really does not matter which way we write the equation for the reaction. For the equilibrium between gaseous N_2O_4 and NO_2, for example, we may write either

$$N_2O_4(g) \rightleftharpoons 2NO_2(g) \qquad K_{p_1} = P^2_{NO_2}/P_{N_2O_4} \tag{1}$$

or

$$2NO_2(g) \rightleftharpoons N_2O_4(g) \qquad K_{p_2} = P_{N_2O_4}/P^2_{NO_2} \tag{2}$$

Written in form (1), the reaction is called the dissociation of N_2O_4; written in form (2), it is called the **dimerization** of NO_2. There is a simple relation between K_{p_1} and K_{p_2} which should be apparent on inspection of the two expressions, namely

$$K_{p_2} = K_{p_1}^{-1} = \frac{1}{1.13 \times 10^{-1} \text{ atm}} = 8.8 \text{ atm}^{-1} \qquad \text{at 25 °C}$$

The numerical value of the equilibrium constant depends on the way you write the equation. The expression for the equilibrium constant, and therefore its numerical value, must correspond to the way you have written the equation, but either way is perfectly correct.

EXAMPLE 8.10.   **Dependence of K_{eq} on the form of the equation**

Calculate K_p for the reaction

$$2SO_3(g) \rightleftharpoons 2SO_2(g) + O_2(g)$$

both at 298 K and at 900 K, using data given in Example 8.8.

Solution. For the reaction as written above

$$K_p = P^2_{SO_2}P_{O_2}/P^2_{SO_3}$$

This is just the inverse of the expression used in Example 8.8. Hence, written as above,

$$K_p \text{ at } 900 \text{ K} = \frac{1}{40.5 \text{ atm}^{-1}} = 2.47 \times 10^{-2} \text{ atm}$$

$$K_p \text{ at } 298 \text{ K} = \frac{1}{3.37 \times 10^{24} \text{ atm}^{-1}} = 2.97 \times 10^{-25} \text{ atm}$$

Even though it is not *wrong* to write a reaction in the reverse direction, it is sometimes much more common to write a certain type of reaction in one direction only. The proton-transfer reaction between a weak acid (represented generally by HA) and H_2O is usually written as

$$HA + H_2O \rightleftharpoons A^- + H_3O^+ \tag{8-26}$$

The equilibrium constant for the reaction above, called the **acidity constant** of the weak acid HA, is

$$K_a = \frac{[H_3O^+][A^-]}{[HA]} \tag{8-27}$$

It is simply not customary to write the reverse equation. Similarly, the equilibrium between an insoluble or slightly soluble solid and its saturated solution is customarily written with the solid on the left and the solution on the right-hand side.

The Effect of Multiplying an Equation by a Constant

When we write an equation, we always try to write the simplest coefficients possible, and usually it is quite clear what the "simplest" coefficients are. Occasionally, however, there is a choice. For example, the reaction for the formation of ammonia from the elements hydrogen and nitrogen may be written either as

$$N_2(g) + 3H_2(g) \rightleftharpoons 2NH_3(g) \qquad \text{for which} \qquad K_1 = P_{NH_3}^2/P_{N_2}P_{H_2}^3 \tag{1}$$

or as

$$\tfrac{1}{2}N_2(g) + \tfrac{3}{2}H_2(g) \rightleftharpoons NH_3(g) \qquad \text{for which} \qquad K_2 = P_{NH_3}/P_{N_2}^{1/2}P_{H_2}^{3/2} \tag{2}$$

An inspection of K_1 and K_2 will convince you that $K_1 = K_2^2$, or $K_2 = K_1^{1/2}$. Both ways of writing the equation are correct. The numerical value of the equilibrium constant depends on the way the equation is written. Both K_1 and K_2 are constants at constant temperature. It makes no difference which you use as long as there is a direct correspondence between the equation written and the equilibrium constant used.

It is possible, of course, both to reverse an equation and to multiply it by a constant. The following example illustrates how this affects the equilibrium constant.

EXAMPLE 8.11. Dependence of K_{eq} on the form of the equation

For the reaction

$$NO(g) + \tfrac{1}{2}Cl_2(g) \rightleftharpoons NOCl(g) \tag{I}$$

the equilibrium constant, K_{p_I}, is 3.7×10^3 atm$^{-1/2}$ at 25 °C.

Calculate $K_{p_{II}}$ for the reaction

$$2NOCl(g) \rightleftharpoons 2NO(g) + Cl_2(g) \tag{II}$$

Give the expression for $K_{p_{II}}$, its numerical value at 25 °C, and its units.

Solution

$$K_{p_1} = P_{NOCl}/P_{NO}P_{Cl_2}^{1/2}$$

and

$$K_{p_{II}} = P_{Cl_2}P_{NO}^2/P_{NOCl}^2$$

Equation (II) is obtained from Eq. (I) by multiplying the equation by two and reversing its direction. Inspection of the expressions for K_{p_1} and $K_{p_{II}}$ shows that

$$K_{p_{II}} = \frac{1}{K_{p_1}^2}$$

Hence

$$K_{p_{II}} = \frac{1}{(3.7 \times 10^3 \text{ atm}^{-1/2})^2} = 7.3 \times 10^{-8} \text{ atm}$$

Simultaneous Equilibria

Often several equilibria occur simultaneously in a given solution or mixture of gases. For instance, both the reactions

$$\text{(1)} \quad N_2O_4(g) \rightleftharpoons 2NO_2(g) \tag{8-28}$$

and

$$\text{(2)} \quad 2NO_2(g) \rightleftharpoons 2NO(g) + O_2(g) \tag{8-29}$$

occur in any mixture containing $N_2O_4(g)$ or $NO_2(g)$. At a given temperature, if you start with gaseous N_2O_4 (by heating pure solid N_2O_4), you will find all four species, N_2O_4, NO_2, NO, and O_2, present in the gas phase at equilibrium. Both the equilibrium constants

$$K_{p_1} = P_{NO_2}^2/P_{N_2O_4} \quad \text{and} \quad K_{p_2} = P_{NO}^2 P_{O_2}/P_{NO_2}^2$$

must be satisfied in the final equilibrium mixture.

It is also possible to write the equation for the decomposition of dinitrogen tetroxide all the way to nitrogen oxide (nitric oxide) and oxygen in a single step, by algebraically combining Eqs. (8-28) and (8-29) to yield

$$\text{(3)} \quad N_2O_4(g) \rightleftharpoons 2NO(g) + O_2(g) \tag{8-30}$$

The equilibrium constant for this reaction, K_{p_3}, is

$$K_{p_3} = P_{NO}^2 P_{O_2}/P_{N_2O_4}$$

Since the pressure of NO_2 that appears in both K_{p_1} and K_{p_2} is the same quantity (both constants are valid for the same mixture of gases, and there is only one partial pressure of NO_2 in that mixture), it is easy to see that

$$K_{p_1}K_{p_2} = \left(\frac{P_{NO_2}^2}{P_{N_2O_4}}\right)\left(\frac{P_{NO}^2 P_{O_2}}{P_{NO_2}^2}\right) = \frac{P_{NO}^2 P_{O_2}}{P_{N_2O_4}} = K_{p_3} \tag{8-31}$$

A typical example of simultaneous equilibria occurs in the aqueous solution of a diprotic acid, such as H_2S. The first stage of ionization is the proton-transfer reaction between H_2S and H_2O:

$$\text{(1)} \quad H_2S + H_2O \rightleftharpoons HS^- + H_3O^+ \tag{8-32}$$

The bisulfide (HS^-) ions formed also can donate their protons to water, and the second stage of ionization produces S^{2-} ions:

$$(2) \quad HS^- + H_2O \rightleftharpoons S^{2-} + H_3O^+ \tag{8-33}$$

The equilibrium constants for these two reactions are called either the first and second ionization constants of H_2S, or the acidity constants of H_2S and of HS^-.

$$K_1 = K_a(H_2S) = \frac{[HS^-][H_3O^+]}{[H_2S]} = 1 \times 10^{-7} \qquad \text{at } 25 \text{ °C} \tag{8-34}$$

$$K_2 = K_a(HS^-) = \frac{[S^{2-}][H_3O^+]}{[HS^-]} = 1.3 \times 10^{-13} \qquad \text{at } 25 \text{ °C} \tag{8-35}$$

The equation for the overall two-stage ionization of H_2S is obtained by summing Eqs. (8-32) and (8-33). That sum is

$$H_2S + 2H_2O \rightleftharpoons S^{2-} + 2H_3O^+ \tag{8-36}$$

The equilibrium constant for Eq. (8-36) is

$$K_{\text{overall}} = \frac{[S^{2-}][H_3O^+]^2}{[H_2S]} \tag{8-37}$$

Careful inspection shows that the product of $K_a(H_2S)$ and $K_a(HS^-)$ is identical with K_{overall}:

$$K_a(H_2S) \cdot K_a(HS^-) = \frac{[\cancel{HS^-}][H_3O^+]}{[H_2S]} \cdot \frac{[S^{2-}][H_3O^+]}{[\cancel{HS^-}]} = \frac{[H_3O^+]^2[S^{2-}]}{[H_2S]}$$
$$= (1 \times 10^{-7})(1.3 \times 10^{-13}) = 1.3 \times 10^{-20} = K_{\text{overall}}$$

Note that we can cancel $[HS^-]$ when we multiply $K_a(H_2S)$ and $K_a(HS^-)$ because all the species H_2S, HS^-, S^{2-}, and H_3O^+ are in the same aqueous solution, and the $[HS^-]$ in both equilibrium constants is the same quantity.

We have discussed two examples of a general phenomenon. If an overall reaction is the algebraic sum of two other reactions, denoted (1) and (2), then the equilibrium constant for the overall reaction is the product of the equilibrium constants of reactions (1) and (2):

$$K_{\text{overall}} = K_1 K_2 \tag{8-38}$$

Examples 8.12 and 8.13 illustrate further the relationships between equilibrium constants for reactions involved in simultaneous equilibria.

EXAMPLE 8.12. Equilibrium constants and simultaneous equilibria

The equilibrium constant for the reaction

$$2HCl(g) \rightleftharpoons H_2(g) + Cl_2(g) \tag{I}$$

is $K_I = 4.17 \times 10^{-34}$ at 25 °C.

The equilibrium constant for the reaction

$$I_2(g) + Cl_2(g) \rightleftharpoons 2 ICl(g) \tag{II}$$

is $K_{II} = 2.1 \times 10^5$ at 25 °C.

Calculate the equilibrium constant for the reaction

$$2HCl(g) + I_2(g) \rightleftharpoons 2 ICl(g) + H_2(g) \tag{III}$$

Solution. Equation (III) is simply the algebraic sum of Eqs. (I) and (II). The equilibrium constants for these three reactions are

$$K_I = \frac{P_{H_2}P_{Cl_2}}{P_{HCl}^2}; \qquad K_{II} = \frac{P_{ICl}^2}{P_{I_2}P_{Cl_2}}; \qquad \text{and} \qquad K_{III} = \frac{P_{H_2}P_{ICl}^2}{P_{HCl}^2 P_{I_2}}$$

We can see that

$$K_I K_{II} = \left(\frac{P_{H_2}P_{Cl_2}}{P_{HCl}^2}\right)\left(\frac{P_{ICl}^2}{P_{I_2}P_{Cl_2}}\right) = K_{III} = (4.17 \times 10^{-34})(2.1 \times 10^5)$$

$$K_{III} = 8.8 \times 10^{-29} \qquad \text{at } 25 \text{ °C}$$

EXAMPLE 8.13. Equilibrium constants and simultaneous equilibria

The equilibrium constant for the reaction

$$N_2(g) + \tfrac{1}{2}O_2(g) \rightleftharpoons N_2O(g) \tag{1}$$

is $K_1 = 7.1 \times 10^{-19}$ atm$^{-1/2}$ at 25 °C.

The equilibrium constant for the reaction

$$N_2(g) + O_2(g) \rightleftharpoons 2NO(g) \tag{2}$$

is $K_2 = 4.23 \times 10^{-31}$ at 25 °C.

Calculate the equilibrium constant at 25 °C for the reaction

$$N_2O(g) + \tfrac{1}{2}O_2(g) \rightleftharpoons 2NO(g) \tag{3}$$

Solution. To obtain Eq. (3) we must reverse Eq. (1) and then add it to Eq. (2). Let us call the reverse of Eq. (1) Eq. (4):

$$N_2O(g) \rightleftharpoons N_2(g) + \tfrac{1}{2}O_2(g) \tag{4}$$

Then

$$K_4 = \frac{P_{N_2}P_{O_2}^{1/2}}{P_{N_2O}} = (K_1)^{-1} = \frac{1}{7.1 \times 10^{-19}} = 1.4 \times 10^{18}$$

Adding Eqs. (2) and (4) yields Eq. (3). Thus K_3 is the product of K_2 and K_4:

$$K_3 = \frac{P_{NO}^2}{P_{N_2O}P_{O_2}^{1/2}} = \left(\frac{P_{NO}^2}{P_{N_2}P_{O_2}}\right)\left(\frac{P_{N_2}P_{O_2}^{1/2}}{P_{N_2O}}\right) = K_2 K_4 = K_2 K_1^{-1}$$

$$= (4.23 \times 10^{-31})(1.4 \times 10^{18}) = 6.0 \times 10^{-13} \qquad \text{at } 25 \text{ °C}$$

section 8.6
Some Typical Problems in Gas-Phase Equilibria

Since all the gases involved in a reaction are in the same container, with a constant volume, and at the same constant temperature, the pressure of each gas is directly proportional to the number of moles of that gas. Thus the decrease in pressure of a gas that is used up during a reaction is directly proportional to the number of moles used up, and the increase in pressure of a gas that is formed during a reaction is directly proportional to the number of moles formed. Determine the equilibrium pressure of each gas by considering the change from its initial pressure. The following examples illustrate the reasoning involved.

EXAMPLE 8.14. Determining K_p from pressure data

Some $SO_2(g)$ and $O_2(g)$ are mixed together in a flask at 1100 K in such a way that at the instant of mixing their partial pressures are, respectively, 1.00 and 0.500 atm. When the system comes to equilibrium at 1100 K, the total pressure in the flask is found to be 1.35 atm. Calculate K_p at 1100 K for the reaction

$$2SO_2(g) + O_2(g) \rightleftharpoons 2SO_3(g)$$

Solution. Since

$$\frac{\text{No. mol } SO_2 \text{ used up}}{\text{No. mol } SO_3 \text{ formed}} = \frac{1}{1}$$

the *decrease* in the partial pressure of SO_2 from its initial value will be equal to the partial pressure of SO_3 in the flask at equilibrium, since initially the pressure of SO_3 was zero. Remember that the partial pressure of a gas is directly proportional to the number of moles of that gas, at fixed volume and temperature. Hence,

$$\text{let } x = P_{SO_3} \text{ at equilibrium}$$

Then

$$(1.00 - x) = P_{SO_2} \text{ at equilibrium}$$

The decrease in the partial pressure of O_2 is only half as large as the decrease in the partial pressure of SO_2 because

$$\frac{\text{No. mol } O_2 \text{ used up}}{\text{No. mol } SO_2 \text{ used up}} = \frac{1}{2}$$

Therefore

$$(0.500 - x/2) = P_{O_2} \text{ at equilibrium}$$

By definition, $P_{\text{total}} = P_{SO_3} + P_{SO_2} + P_{O_2}$. Substituting into this equation yields

$$1.35 = x + (1.00 - x) + (0.500 - x/2)$$

Combining terms on the right-hand side we obtain

$$1.35 = 1.50 - x/2$$

so that

$$x = 3.00 - 2.70 = 0.30 \text{ atm}$$

Thus at equilibrium,

$$P_{SO_3} = x = 0.30 \text{ atm}$$
$$P_{SO_2} = (1.00 - x) = 0.70 \text{ atm}$$

and

$$P_{O_2} = (0.500 - x/2) = 0.500 - 0.15 = 0.35 \text{ atm}$$

The value of the equilibrium constant at 1100 K is therefore

$$K_p = P_{SO_3}^2 / P_{SO_2}^2 P_{O_2} = (0.30)^2/(0.70)^2(0.35) = 0.52 \text{ atm}^{-1}$$

EXAMPLE 8.15. Determining concentrations from K_c

At a given temperature the equilibrium constant of the reaction

$$H_2(g) + I_2(g) \rightleftharpoons 2HI(g)$$

is 6.0.

If 1.00 mol of H_2 and 1.00 mol of I_2 are introduced into a 1.00-L flask at this temperature, what will be the concentrations of all three species when the system comes to equilibrium?

Solution. The equilibrium constant for this reaction is dimensionless, so that both K_c and K_p are 6.0.

$$K_c = \frac{[HI]^2}{[H_2][I_2]} = 6.0 = K_p = \frac{P_{III}^2}{P_{H_2}P_{I_2}}$$

Let x equal the number of moles of I_2 used up when the system has come to equilibrium. Then, since H_2 and I_2 react in a 1 : 1 molar ratio, the number of moles of H_2 used up when the system comes to equilibrium is also x. The number of moles of each of these gases present at equilibrium is the initial number of moles *minus* the amount used up, that is, $1.00 - x$. Since

$$\frac{\text{No. mol HI formed}}{\text{No. mol } I_2 \text{ used up}} = \frac{2}{1}$$

the number of moles of HI formed when the system is at equilibrium is $2x$. The volume of the flask is 1.00 L. The concentration, in moles per liter, of each of these three species is, therefore,

$$[H_2] = [I_2] = (1.00 - x) \qquad [HI] = 2x \qquad \text{at equilibrium}$$

Substituting into the expression for K_c we obtain

$$K_c = 6.0 = \frac{(2x)^2}{(1.00 - x)^2}$$

To solve this equation, take the square root of both sides. We obtain

$$2.45 = \frac{2x}{(1.00 - x)}$$

Multiplying this out yields $2.45 - 2.45x = 2x$, which rearranges to

$$2.45 = 4.45x$$

so that

$$x = 0.55 \text{ mol/L}$$

At equilibrium, therefore,

$$[HI] = 2x = 1.10 \text{ mol/L}$$
$$[H_2] = [I_2] = 1.00 - x = 1.00 - 0.55 = 0.45 \text{ mol/L}$$

EXAMPLE 8.16. Determining the changes in the partial pressures of the reactants when a system at equilibrium is disturbed

Consider the equilibrium mixture of H_2, I_2, and HI of Experiment 4 of Table 8.1. If this equilibrium is disturbed by adding more HI so that the partial pressure of HI is suddenly increased to 1.0000 atm, what will the partial pressures of each of the gases be when the system returns to equilibrium? The equilibrium constant for the reaction $H_2 + I_2 \rightleftharpoons 2HI$ is 48.70 at 730.8 K.

Solution. At the instant the partial pressure of HI is increased to 1.0000 atm, the reaction quotient, Q, becomes

$$Q = \frac{(1.0000 \text{ atm})^2}{(0.08592 \text{ atm})(0.08592 \text{ atm})} = 135 > 48.7 = K_{eq}$$

Since $Q > K_{eq}$ the reaction must shift to the left to restore equilibrium. Let x equal the increase in the partial pressure of H_2 and I_2 that restores the system to equilibrium. Then the partial pressure of HI must decrease by an amount $2x$, because, for the reverse reaction,

$$\frac{\text{mol HI used up}}{\text{mol } H_2 \text{ formed}} = \frac{2}{1}$$

We can make a table to organize these calculations.

	$P_{I_2}(atm)$	$P_{H_2}(atm)$	$P_{HI}(atm)$
Instant of disturbance	0.08592	0.08592	1.0000
Change	x	x	$-2x$
New equilibrium	$(0.8592 + x)$	$(0.08592 + x)$	$(1.0000 - 2x)$

At the new position of equilibrium,

$$\frac{(1.0000 - 2x)^2}{(0.08592 + x)^2} = 48.70$$

Taking the square root of both sides of this equation yields

$$\frac{1.0000 - 2x}{0.08592 + x} = 6.979$$

so that $1.0000 - 2x = 0.5996 + 6.979x$. Solving this equation for x we obtain $0.4004 = 8.979x$, and $x = 0.04459$ atm. The partial pressure of both H_2 and I_2 when equilibrium is restored is $0.08592 + 0.04459 = 0.13051$ atm, and the partial pressure of HI is $1.0000 - 2(0.04459) = 0.9108$ atm. You can check your calculations by verifying that

$$\frac{(0.9108)^2}{(0.1305)^2} = 48.70$$

Summary

The **law of mass action** or the **ideal law of chemical equilibrium** states that for the general reaction $\alpha A + \beta B \rightleftharpoons \gamma C + \delta D$ there is a particular function that has the same numerical value regardless of how we mix the species A, B, C, and D together, as long as the system is at equilibrium at constant temperature. That function, called the **equilibrium constant,** is

$$K_{eq} = \frac{[C]^\gamma [D]^\delta}{[A]^\alpha [B]^\beta} = \text{a constant at constant temperature}$$

For reactions in the gas phase, the concentrations of the species are directly proportional to their partial pressures as long as the temperature is constant, and it is customary to use partial pressures, rather than concentrations, in the equilibrium constant expression.

For reactions in dilute aqueous solution, the concentration of water remains approximately 55.5 M throughout any reaction, even when water is a reactant or a product. For this reason, the term $[H_2O]$ does not appear in equilibrium constant expressions.

Each pure solid or pure liquid is in a phase by itself, and has a constant concentration at constant temperature. Therefore no term in the concentration of a pure solid or a pure liquid appears in an equilibrium constant expression.

The **reaction quotient,** Q, has the same form as the equilibrium constant, K_{eq}, but is evaluated using any given concentrations of the species involved in the reaction, and not necessarily equilibrium concentrations. To determine whether or not a reaction can occur when several substances are mixed together, evaluate Q at the instant of mixing. If the numerical value of Q is not the same as the value of the equilibrium constant, a reaction will occur. The direction of reaction is always such as to make Q equal to K_{eq}.

The magnitude of the equilibrium constant provides information about the relative concentrations of products and reactants. If the equilibrium constant is very large $(K_{eq} \gg 1)$, the position of equilibrium lies to the right, and at equilibrium there will be large concentrations of the products, relative to the concentrations of the reactants. If, on the other hand, the equilibrium constant is very small $(K_{eq} \ll 1)$, the position of equilibrium lies to the left. At equilibrium there will be small concentrations of the products, relative to the concentrations of the reactants.

The way in which a system at equilibrium responds to a change in any of the factors that determine the state of equilibrium is described by **Le Chatelier's Principle.** Some factors that determine the state of equilibrium are (1) the concentrations of the species taking part in the reaction, (2) the partial pressures of gaseous species taking part in the reaction, and (3) the temperature.

The numerical value of an equilibrium constant depends on the way the equation for the reaction is written. If the equation is reversed, the equilibrium constant is inverted. If the equation is multiplied by two, the equilibrium constant is squared. If the equation is multiplied by any numerical factor, the equilibrium constant is raised to that power.

When several reactions occur simultaneously in the same solution, or in the gas phase, the equilibrium constant for the overall reaction is related to the equilibrium constants for the individual steps.

Exercises

Section 8.1

1. Write the equilibrium constant expression for each of the following reactions:
   (a) $Ag_2CO_3(s) \rightleftharpoons 2Ag^+(aq) + CO_3^{2-}(aq)$
   (b) $H_2O_2(aq) \rightleftharpoons H_2O + \frac{1}{2}O_2(g)$
   (c) $NH_3(aq) + H_2O \rightleftharpoons NH_4^+(aq) + OH^-(aq)$
   (d) $HPO_4^{2-}(aq) + H_2O \rightleftharpoons H_3O^+(aq) + PO_4^{3-}(aq)$
   (e) $CHCl_3(liq) \rightleftharpoons CHCl_3(g)$
   (f) $CH_4(g) + 2O_2(g) \rightleftharpoons CO_2(g) + 2H_2O(liq)$
   (g) $Cu^{2+}(aq) + 4NH_3(aq) \rightleftharpoons Cu(NH_3)_4^{2+}(aq)$
   (h) $Al(s) + 3H^+(aq) \rightleftharpoons Al^{3+}(aq) + \frac{3}{2}H_2(g)$

(i) $3PbS(s) + 8H^+(aq) + 2NO_3^-(aq) \rightleftharpoons 3Pb^{2+}(aq) + 3S(s) + 2NO(g) + 4H_2O$
(j) $CaCO_3(s) \rightleftharpoons CaO(s) + CO_2(g)$
(k) $BaSO_4(s) + CO_3^{2-}(aq) \rightleftharpoons BaCO_3(s) + SO_4^{2-}(aq)$

2. Specify the units of the equilibrium constants for each of the following reactions:
   (a) $Fe(OH)_3(s) \rightleftharpoons Fe^{3+}(aq) + 3\ OH^-(aq)$
   (b) $PbI_2(s) \rightleftharpoons Pb^{2+}(aq) + 2\ I^-(aq)$
   (c) $HSO_3^-(aq) + H_2O \rightleftharpoons SO_3^{2-}(aq) + H_3O^+(aq)$
   (d) $N_2(g) + \frac{1}{2} O_2(g) \rightleftharpoons N_2O(g)$
   (e) $CH_3OH(liq) \rightleftharpoons CH_3OH(g)$
   (f) $N_2(g) + 2\ O_2(g) \rightleftharpoons 2NO_2(g)$
   (g) $Mg(s) + 2H^+(aq) \rightleftharpoons Mg^{2+}(aq) + H_2(g)$
   (h) $BaSO_3(s) \rightleftharpoons BaO(s) + SO_2(g)$

3. At 700 K, the equilibrium constant, K_p, for the reaction

$$2SO_3(g) \rightleftharpoons 2SO_2(g) + O_2(g)$$

is 1.80×10^{-5} atm. What is the numerical value, in moles per liter, of K_c for this reaction at 700 K?

4. For which of the following gas-phase reactions do K_p and K_c have the same numerical value? *Explain* your answers.
   (a) $CO(g) + \frac{1}{2} O_2(g) \rightleftharpoons CO_2(g)$
   (b) $C_2H_4(g) + H_2(g) \rightleftharpoons C_2H_6(g)$
   (c) $H_2O(g) + CO(g) \rightleftharpoons H_2(g) + CO_2(g)$
   (d) $N_2(g) + 3F_2(g) \rightleftharpoons 2NF_3(g)$
   (e) $F_2(g) + H_2(g) \rightleftharpoons 2HF(g)$

Section 8.2

5. The acidity constants of several weak acids at 25 °C are given below. Arrange these acids in order of decreasing acid strength, starting with the strongest of these acids.

Name	Formula	K_a
Acetic acid	CH_3COOH	1.8×10^{-5}
Benzoic acid	C_6H_5COOH	6.3×10^{-5}
Formic acid	$HCOOH$	1.8×10^{-4}
Hydrocyanic acid	HCN	4.0×10^{-10}
Hydrogen sulfate ion	HSO_4^-	1.2×10^{-2}
Nitrous acid	$HONO$	4.5×10^{-4}

6. The equilibrium constant for the reaction

$$H_2(g) + I_2(s) \rightleftharpoons 2HI(g)$$

is 0.35 at 25 °C. State whether or not each of the following mixtures is at equilibrium at 25 °C. If it is not at equilibrium, indicate the direction in which reaction will occur to achieve equilibrium.
   (a) A mixture in which the partial pressures of the two gases are $P_{H_2} = 0.10$ atm and $P_{HI} = 0.90$ atm, and there is solid I_2 in the container.
   (b) A mixture in which the partial pressures of the two gases are $P_{H_2} = 0.55$ atm and $P_{HI} = 0.44$ atm, and there is solid I_2 in the container.
   (c) A mixture in which the partial pressures of the two gases are $P_{H_2} = 2.5$ atm and $P_{HI} = 0.15$ atm, and there is solid I_2 in the container.

7. Ammonium hydrosulfide is a crystalline solid that decomposes when heated according to the reaction: $NH_4HS(s) \rightleftharpoons NH_3(g) + H_2S(g)$. Some solid NH_4HS is placed in an evacuated container at 25 °C. After a period of time long enough for equilibrium to be attained, the total pressure inside the container is found to be 0.659 atm. Some solid NH_4HS remains in the container. Give the expression for, the numerical value of, and the units of K_p for this reaction.

8. Gaseous NH_3 and HCl combine to form solid NH_4Cl as follows:

$$NH_3(g) + HCl(g) \rightleftharpoons NH_4Cl(s)$$

At 300 °C the equilibrium constant for this reaction is 17.8 atm^{-2}. Some gaseous NH_3 and HCl are introduced into an evacuated flask at 300 °C. Will any solid NH_4Cl be formed if the initial partial pressures are

(a) $P_{NH_3} = 684$ mmHg $P_{HCl} = 912$ mmHg?

(b) $P_{NH_3} = 30.4$ mmHg $P_{HCl} = 22.8$ mmHg?

9. Silver bromate, $AgBrO_3$, is a slightly soluble salt. The equilibrium constant for the reaction

$$AgBrO_3(s) \rightleftharpoons Ag^+(aq) + BrO_3^-(aq)$$

is 5.2×10^{-5} M^2 at 25 °C. Will any solid $AgBrO_3$ be formed if a solution of $AgNO_3$ and a solution of $KBrO_3$ are mixed so that at the instant of mixing

(a) $[Ag^+] = 1.0 \times 10^{-3}$ M and $[BrO_3^-] = 2.0 \times 10^{-3}$ M?

(b) $[Ag^+] = 1.0 \times 10^{-1}$ M and $[BrO_3^-] = 5.0 \times 10^{-1}$ M?

10. The equilibrium constant for the reaction

$$SO_2(g) + \tfrac{1}{2}O_2(g) \rightleftharpoons SO_3(g)$$

is 1.84×10^{12} atm$^{-1/2}$ at 25 °C. Some SO_2, O_2, and SO_3 gases are placed in a container so that at the instant of mixing at 25 °C the partial pressures of the three gases are $P_{SO_2} = 0.100$ atm; $P_{O_2} = 0.040$ atm; $P_{SO_3} = 3.00$ atm. When the system comes to equilibrium at 25 °C will the amount of SO_3 in the container be more than, less than, or the same as there was at the instant of mixing? Show all work necessary to prove your answer.

Sections 8.3 and 8.4

11. Ammonium bromide is a crystalline solid that decomposes when heated as follows:

$$NH_4Br(s) \rightleftharpoons NH_3(g) + HBr(g)$$

The decomposition is endothermic. Some solid NH_4Br is placed in an evacuated vessel at 400 °C, and the system comes to equilibrium.

(a) Suppose that extra $NH_3(g)$ is now injected into the vessel. Will the pressure of HBr(g) in the vessel increase, decrease, or remain the same? *Explain.*

(b) If the volume of the vessel is suddenly doubled, will the amount of solid NH_4Br in the vessel increase, decrease, or remain the same? *Explain.*

(c) Will K_p for this reaction at 25 °C be greater than, less than, or the same as the value of K_p at 400 °C? *Explain.*

12. Strontium sulfate, $SrSO_4$, and strontium sulfite, $SrSO_3$, are both white insoluble solids. Strontium sulfite can be dissolved by adding excess HCl solution, but strontium sulfate does not dissolve appreciably in excess HCl. Explain the difference between the solubility in HCl of these two salts.

13. Some solid pale yellow Ag_2CO_3 is shaken with 20 mL of water, and the following equilibrium is established:

$$Ag_2CO_3(s) \rightleftharpoons 2Ag^+(aq) + CO_3^{2-}(aq)$$

Describe the effect on this equilibrium of (a) adding another 20 mL of water; (b) adding some 0.5 F $AgNO_3$; (c) adding some 0.5 F HNO_3; (d) adding some 0.5 F Na_2CO_3; and (e) adding some solid Ag_2CO_3.

14. Consider the equilibrium $H_2(g) + Cl_2(g) \rightleftharpoons 2HCl(g)$. If the temperature and volume are kept constant and more Cl_2 is pumped into an equilibrium mixture of these three gases

(a) Will the equilibrium constant, K_p, increase, decrease, or remain the same? *Explain.*
(b) Will the amount of $HCl(g)$ increase, decrease, or remain the same? *Explain.*

15. The following solids, all insoluble in water, can be dissolved completely by adding excess HCl solution: $Cu(OH)_2$, $ZnCO_3$, $Ca_3(PO_4)_2$, and BaC_2O_4. Write a balanced net ionic equation for the dissolution of each of these solids in excess hydrochloric acid.

16. At 25 °C, ΔH for the reaction $2NO(g) + O_2(g) \rightleftharpoons 2NO_2(g)$ is -113 kJ. (a) Does the tendency for reactions to proceed in the direction that minimizes the energy favor the right- or the left-hand side of this reaction? *Explain.*
(b) Does the tendency for reactions to proceed in the direction of maximum entropy, that is, toward greater molecular disorder, favor the right- or the left-hand side of this reaction? *Explain.*
(c) If the temperature is raised from 25 to 500 °C will this equilibrium shift to the right or to the left? *Explain.*

17. The dissolution of lead nitrate in water is an endothermic reaction. Will the solubility of lead nitrate at 100 °C be greater than, less than, or the same as it is at 25 °C?

18. Calcium carbonate decomposes when it is heated as follows:

$$CaCO_3(g) \rightleftharpoons CaO(s) + CO_2(g)$$

Some solid $CaCO_3$ is placed in an evacuated vessel enclosed by a piston and heated so that a portion of it decomposes. If the piston is moved so that the volume of the vessel is doubled, while the temperature is held constant,

(a) Will the number of moles of CO_2 in the vessel increase, decrease, or remain the same? Explain your answer.
(b) Will the pressure of CO_2 in the vessel increase, decrease, or remain the same? Explain your answer.

19. Phosgene gas, $COCl_2$, dissociates as follows:

$$COCl_2(g) \rightleftharpoons CO(g) + Cl_2(g)$$

If a mixture of these three gases is compressed at constant temperature,

(a) Will the amount of CO in the mixture increase, decrease, or remain the same? Explain your answer.
(b) Will the equilibrium constant for the reaction increase, decrease, or remain the same? Explain your answer.
(c) Will the partial pressure of $COCl_2$ in the mixture increase, decrease, or remain the same? Explain your answer.

Section 8.5

20. The first and second ionization constants of oxalic acid, $H_2C_2O_4$, are, respectively, 5.9×10^{-2} and 6.4×10^{-5} at 25 °C. Calculate the equilibrium constant for the overall ionization constant of oxalic acid,

$$H_2C_2O_4 + 2H_2O \rightleftharpoons 2H_3O^+(aq) + C_2O_4^{2-}(aq)$$

21. At 25 °C the equilibrium constant for the proton-transfer reaction between formic acid and cyanide ion

$$HCOOH(aq) + CN^-(aq) \rightleftharpoons HCN(aq) + HCOO^-(aq)$$

is 4.5×10^5. Calculate the equilibrium constant for the proton-transfer reaction between formate ion and hydrocyanic acid,

$$HCOO^-(aq) + HCN(aq) \rightleftharpoons HCOOH(aq) + CN^-(aq)$$

22. The equilibrium constant for the reaction

$$2NO(g) + O_2(g) \rightleftharpoons 2NO_2(g)$$

is 1.62×10^{12} atm^{-1} at 25 °C. What is the expression for, the numerical value of, and the units of the equilibrium constant for the reaction

   (a) $NO(g) + \frac{1}{2}O_2(g) \rightleftharpoons NO_2(g)$?
   (b) $NO_2(g) \rightleftharpoons NO(g) + \frac{1}{2}O_2(g)$?

23. Equilibrium constants for the following two reactions are given at 25 °C:

$$P(s) + \tfrac{3}{2}Cl_2(g) \rightleftharpoons PCl_3(g) \qquad K_1 = 8.3 \times 10^{46} \tag{1}$$
$$PCl_3(g) + Cl_2(g) \rightleftharpoons PCl_5(g) \qquad K_2 = 3.4 \times 10^{6} \tag{2}$$

Calculate the equilibrium constant for the reaction that is the sum of Eqs. (1) and (2),

$$P(s) + \tfrac{5}{2}Cl_2(g) \rightleftharpoons PCl_5(g)$$

24. Show that the equilibrium constant for the reaction

$$Ag^+(aq) + 2NH_3(aq) \rightleftharpoons Ag(NH_3)_2^+(aq)$$

is the product of the equilibrium constants for the reactions below:

$$Ag^+(aq) + NH_3(aq) \rightleftharpoons Ag(NH_3)^+(aq) \tag{1}$$
$$Ag(NH_3)^+ + NH_3(aq) \rightleftharpoons Ag(NH_3)_2^+(aq) \tag{2}$$

Section 8.6

25. The gases Cl_2 and PCl_3 are mixed together in a flask at 250 °C. At the instant of mixing the partial pressures of these two gases are, respectively, 0.820 and 0.640 atm. When the system comes to equilibrium, the total pressure in the flask is found to be 1.295 atm. Calculate K_p at 250 °C for the reaction that occurred,

$$Cl_2(g) + PCl_3(g) \rightleftharpoons PCl_5(g)$$

26. A 0.600-mol sample of HBr(g) is introduced into an evacuated flask of volume 2.00 L, and kept at an elevated temperature, T. When the system comes to equilibrium, it is found that the flask contains 0.104 mol each of $H_2(g)$ and $Br_2(g)$. Calculate the numerical value of both K_p and K_c for the dissociation of HBr(g) at temperature T.

27. An evacuated flask was filled with 1.000 mol of $SO_3(g)$ at 24.8 °C and 1.000-atm pressure. The temperature was then raised to 900.0 K, and the total pressure in the flask was found to be 3.346 atm when the system reached equilibrium.

   (a) What would the pressure in the flask have been if no dissociation of SO_3 occurred?

(b) The dissociation of SO_3,

$$2SO_3(g) \rightleftharpoons 2SO_2(g) + O_2(g)$$

caused the pressure in the flask to be higher than the value calculated in part (a). Calculate the partial pressure of each of the gases in the equilibrium mixture at 900.0 K. Calculate also the numerical value of the equlibrium constant for the dissociation reaction at 900.0 K.

28. Sulfuryl chloride, SO_2Cl_2, is a highly reactive gaseous compound. When heated, it decomposes as follows:

$$SO_2Cl_2(g) \rightleftharpoons SO_2(g) + Cl_2(g)$$

A sample of 3.509 g of SO_2Cl_2 is placed in an evacuated 1.00-L bulb and the temperature is raised to 375 K.

(a) If no dissociation of the SO_2Cl_2 occurred, what would be the pressure in atmospheres in the bulb?

(b) When the system has come to equilibrium at 375 K, the total pressure in the bulb is found to be 1.43 atm. Calculate the partial pressures of SO_2, Cl_2, and SO_2Cl_2 at equilibrium at 375 K.

(c) Give the expression for the equilibrium constant, K_p, for the decomposition of SO_2Cl_2 and specify its units. Calculate the numerical value of K_p at 375 K.

Multiple Choice Questions

1. For which of the following reactions is the equilibrium constant called a solubility product?
   (a) $CaC_2O_4(s) + 2H^+(aq) \rightleftharpoons Ca^{2+}(aq) + H_2C_2O_4$
   (b) $La(OH)_3(s) \rightleftharpoons La^{3+}(aq) + 3\,OH^-(aq)$
   (c) $Ag^+(aq) + Cl^-(aq) \rightleftharpoons AgCl(s)$
   (d) $Cu(OH)_2(s) + 4NH_3 \rightleftharpoons Cu(NH_3)_4^{2+}(aq) + 2\,OH^-$
   (e) $Zn(OH)_2(s) + 2\,OH^- \rightleftharpoons Zn(OH)_4^{2-}$

2. For which of the following reactions are the numerical values of K_p and K_c the same?
   (a) $2NOCl(g) \rightleftharpoons 2NO(g) + Cl_2(g)$
   (b) $N_2 + 3H_2(g) \rightleftharpoons 2NH_3(g)$
   (c) $H_2(g) + Cl_2(g) \rightleftharpoons 2HCl(g)$
   (d) $H_2(g) + I_2(s) \rightleftharpoons 2HI(g)$
   (e) $COCl_2(g) \rightleftharpoons CO(g) + Cl_2(g)$

3. If the system $CaCO_3(s) \rightleftharpoons CaO(s) + CO_2(g)$ is at equilibrium at constant temperature, and the number of moles of CaO in the vessel is doubled,
   (a) The reaction quotient, Q, is doubled.
   (b) The reaction quotient, Q, is halved.
   (c) The number of moles of CO_2 present at equilibrium is halved.
   (d) The number of moles of $CaCO_3$ in the vessel increases.
   (e) The partial pressure of CO_2 in the vessel remains unchanged.

4. At a given temperature the equilibrium constant for the reaction

$$PCl_5(g) \rightleftharpoons PCl_3(g) + Cl_2(g)$$

is 2.4×10^{-3}.
   At the same temperature, the equilibrium constant for the reaction

$$PCl_3(g) + Cl_2(g) \rightleftharpoons PCl_5(g)$$

is
   (a) 2.4×10^{-3} (b) -2.4×10^{-3} (c) 4.8×10^{-2} (d) 4.2×10^2 (e) 2.3×10^5

5. Which of the following statements about the reaction quotient, Q, is FALSE?

(a) The reaction quotient, Q, and the equilibrium constant, K_{eq}, always have the same numerical value.

(b) The reaction quotient may sometimes be zero.

(c) The reaction quotient may be larger than the equilibrium constant.

(d) The reaction quotient may be smaller than the equilibrium constant.

(e) The numerical value of the reaction quotient changes with time as a reaction proceeds.

6. For which of the following systems at equilibrium at constant temperature will doubling the volume cause a shift to the right?

(a) $H_2(g) + Cl_2(g) \rightleftharpoons 2HCl(g)$

(b) $2CO(g) + O_2(g) \rightleftharpoons 2CO_2(g)$

(c) $N_2(g) + 3H_2(g) \rightleftharpoons 2NH_3(g)$

(d) $H_2 + CO_2(g) \rightleftharpoons H_2O(g) + CO(g)$

(e) $PCl_5(g) \rightleftharpoons PCl_3(g) + Cl_2(g)$

7. For the system $NH_4Cl(s) \rightleftharpoons NH_3(g) + HCl(g)$, if the concentration of NH_3 is doubled, the equilibrium constant will

(a) double   (b) increase, but by less than a factor of 2   (c) be halved

(d) remain the same   (e) decrease, but by less than a factor of 2

8. For which of the following reactions is the equilibrium constant called an acidity constant, K_a?

(a) $HCOOH + NH_3 \rightleftharpoons HCOO^-(aq) + NH_4^+(aq)$

(b) $H_3O^+(aq) + OH^-(aq) \rightleftharpoons 2H_2O$

(c) $HONO + H_2O \rightleftharpoons ONO^-(aq) + H_3O^+(aq)$

(d) $CH_3COOH + OH^-(aq) \rightleftharpoons CH_3COO^-(aq) + H_2O$

(e) $H_3O^+(aq) + CH_3NH_2 \rightleftharpoons CH_3NH_3^+(aq) + H_2O$

9. If the pressure on an equilibrium mixture of the three gases NO, Cl_2, and NOCl,

$$2NO(g) + Cl_2(g) \rightleftharpoons 2NOCl(g)$$

is suddenly decreased by doubling the volume of the container at constant temperature, when the system returns to equilibrium

(a) The concentration of NOCl will have increased.

(b) The value of the equilibrium constant K_c will have increased.

(c) The number of moles of Cl_2 will have increased.

(d) The number of moles of NOCl will have increased.

(e) The value of the equilibrium constant K_p will have increased.

10. For which of the following systems at equilibrium at constant temperature will decreasing the volume cause no shift at all?

(a) $H_2(g) + CO_2(g) \rightleftharpoons H_2O(g) + CO(g)$

(b) $2NO(g) + O_2(g) \rightleftharpoons 2NO_2(g)$

(c) $2NO_2(g) \rightleftharpoons N_2O_4(g)$

(d) $SO_3(g) \rightleftharpoons SO_2(g) + \frac{1}{2}O_2(g)$

(e) $N_2(g) + 3H_2(g) \rightleftharpoons 2NH_3(g)$

11. The decomposition of phosgene, $COCl_2 \rightleftharpoons CO(g) + Cl_2(g)$, is an endothermic process. Which of the following factors will cause the value of the equilibrium constant to increase?

(a) adding Cl_2 (b) adding $He(g)$ (c) decreasing the temperature

(d) decreasing the total pressure   (e) none of the above

12. At a given temperature, K_c for the reaction

$$H_2(g) + CO_2(g) \rightleftharpoons H_2O(g) + CO(g)$$

is 3.24. If 0.800 mol of both $H_2(g)$ and $CO_2(g)$ are placed in a 1.00-L container at this temperature, when the system comes to equilibrium the concentration of $CO(g)$ will be

(a) 1.60 M (b) 0.800 M (c) 0.611 M (d) 0.514 M (e) 0.247 M

13. For a reaction of the type $A(s) + 2B(g) \rightleftharpoons 2C(g)$, an equilibrium mixture consists of 3.0 mol of A, 0.80 mol of B, and 0.40 mol of C, in a 2.00-L flask. What is the value of K_c for this reaction?

(a) 5.0×10^{-1} (b) 3.3×10^{-1} (c) 2.5×10^{-1} (d) 1.7×10^{-1} (e) 8.3×10^{-2}

14. At a given temperature an equilibrium mixture of the reaction

$$2NO(g) + O_2(g) \rightleftharpoons 2NO_2(g)$$

contains 0.120 mol of NO_2, 0.080 mol of NO, and 0.640 mol of O_2 in a 4.00-L bulb. What is the value of K_c for this reaction at this temperature?

(a) 88   (b) 14   (c) 9.4   (d) 3.5   (e) 2.8

15. At a certain temperature the equilibrium constant, K_p, for the reaction $H_2(g) + I_2(g) \rightleftharpoons 2HI(g)$ is 9.0. At this temperature, an equilibrium mixture of these three gases contained 0.60 mol of HI and 0.40 mol of H_2 in a 2.00-L flask. How many moles of I_2 were in this equilibrium mixture?

(a) 0.40   (b) 0.17   (c) 0.10   (d) 0.085   (e) 0.050

Problems

8.1. If the equilibrium constant for reaction (1)

$$Ag^+(aq) + 2NH_3(aq) \rightleftharpoons Ag(NH_3)_2^+(aq) \tag{1}$$

is $K_1 = 1.8 \times 10^7$ at 25 °C, and for reaction (2)

$$Ag^+(aq) + Cl^-(aq) \rightleftharpoons AgCl(s) \tag{2}$$

is $K_2 = 5.6 \times 10^9$ at 25 °C, calculate the equilibrium constant for

$$AgCl(s) + 2NH_3(aq) \rightleftharpoons Ag(NH_3)_2^+(aq) + Cl^-(aq)$$

8.2. The gas carbon dioxide has a small solubility in water. The equation for the dissolution of CO_2 in water is simply

$$CO_2(g) \rightleftharpoons CO_2(aq)$$

(a) Write the expression for the equilibrium constant for this reaction. It is called the Henry's Law constant for CO_2.
(b) Explain, using Le Chatelier's Principle, why the solubility of CO_2 in water increases if the pressure of CO_2 above the solution is increased.

8.3. A 2.4156-g sample of PCl_5 was placed in an evacuated 2.000-L flask. The flask was then heated to 250.0 °C and maintained at that temperature. The PCl_5 completely vaporized, and the total pressure inside the flask was observed to be 358.7 mmHg. It is known that PCl_5 decomposes according to

$$PCl_5(g) \rightleftharpoons PCl_3(g) + Cl_2(g)$$

Calculate the partial pressure of each of these three gases in the equilibrium mixture at 250.0 °C. Calculate also the value of K_p for this reaction at 250.0 °C and specify its units.

8.4. For the reaction $N_2(g) + 3H_2(g) \rightleftharpoons 2NH_3(g)$ the equilibrium constant K_c at 500 °C is 6.00×10^{-2}. Calculate the numerical value of K_p for this reaction at 500 °C. Specify the units of both K_c and K_p.

8.5. The equilibrium constant for the reaction

$$NO(g) + \tfrac{1}{2} Cl_2(g) \rightleftharpoons NOCl(g)$$

is 3.7×10^3 atm$^{-1/2}$ at 25 °C. Some NO, Cl_2, and NOCl gases are placed in a container so that at the instant of mixing the partial pressures of these gases are 0.200 atm NO; 0.040 atm Cl_2; 1.00 atm NOCl.
(a) Give the expression for the reaction quotient, Q, and its numerical value at the instant of mixing.
(b) When the system comes to equilibrium at 25 °C, will the amount of NOCl in the container be more than, less than, or the same as there was at the instant of mixing? Explain your answer.
(c) If the *change* in the partial pressure of NO from 0.200 atm to its value at equilibrium is denoted by the symbol x, write an expression for the partial pressure of each of the three gases at equilibrium in terms of their initial values and x.

8.6. For the exothermic reaction

$$SO_2(g) + \tfrac{1}{2} O_2(g) \rightleftharpoons SO_3(g)$$

K_p is 1.84×10^{12} at 35 °C.
(a) Will K_p for this reaction at 675 K be greater than, less than, or equal to 1.84×10^{12}? Explain your answer.
(b) Some SO_2 and O_2 are mixed together in a flask in such a way that their partial pressures at the instant of mixing are 1.00 and 0.500 atm, respectively, at 675 K. The partial pressure of SO_3 at equilibrium at 675 K is 0.980 atm. Calculate K_p for the reaction above at 675 K.

8.7. Silver bromide, AgBr, silver carbonate, Ag_2CO_3, and silver phosphate, Ag_3PO_4 are all insoluble solids. Although Ag_2CO_3 and Ag_3PO_4 dissolve completely in excess dilute nitric acid, AgBr does not. Account for this difference in behavior.

8.8. A mixture of the three substances in the reaction

$$NH_4HS(s) \rightleftharpoons NH_3 + H_2S(g)$$

is at equilibrium at 200 °C in a container of volume V. The reaction is endothermic. For each of the following changes, state whether the partial pressure of NH_3 will increase, decrease, or remain the same, when equilibrium has been reestablished. Explain each answer.
(a) NH_3 is added.
(b) H_2S is added.
(c) NH_4HS is added.
(d) The temperature is increased.
(e) The volume of the container is doubled.

8.9. Consider the dissolution of the insoluble compound $Al(OH)_3$ in water:

$$Al(OH)_3(s) \rightleftharpoons Al^{3+}(aq) + 3\,OH^-(aq)$$

(a) Write an expression for the equilibrium constant for this reaction and specify its units.
(b) Will adding excess nitric acid to a saturated solution of $Al(OH)_3$ in equilibrium with excess solid $Al(OH)_3$ increase, decrease, or leave unchanged the solubility of $Al(OH)_3$? Explain your answer.

(c) Will adding excess $Al(NO_3)_3$ solution to a saturated solution of $Al(OH)_3$ in equilibrium with excess solid $Al(OH)_3$ increase, decrease, or leave unchanged the solubility of $Al(OH)_3$? Explain your answer.

8.10. For the endothermic reaction

$$N_2(g) + O_2(g) \rightleftharpoons 2NO(g) \qquad K_{c_1} = 4.23 \times 10^{-31} \qquad \text{at } 25 \text{ °C} \qquad \text{(I)}$$

(a) Calculate $K_{p_{II}}$ for the reaction

$$NO(g) \rightleftharpoons \tfrac{1}{2}N_2(g) + \tfrac{1}{2}O_2(g) \qquad \text{at } 25 \text{ °C} \qquad \text{(II)}$$

Give the expression for $K_{p_{II}}$, its numerical value, and its units.
(b) If 1.500 mol of N_2 and 1.500 mol of O_2 are placed in a 10.00-L bulb and maintained at a temperature of 1800 K, it is found that 1.60×10^{-2} mol of NO is present at equilibrium. Calculate K_c for reaction (I) at 1800 K.
(c) Does the tendency for reactions to proceed to a state of minimum energy drive reaction (I) to the right or to the left? Explain.
(d) Compare the values of K_c for reaction (I) at 298 K and at 1800 K, and explain the reason for the change with increase in temperature.

8.11. The decomposition of $H_2S(g)$, an endothermic reaction, proceeds as follows:

$$H_2S(g) \rightleftharpoons H_2(g) + S(s)$$

At 370 K, K_c for this reaction is 8×10^{-6}.
(a) What is the numerical value of K_p for this reaction at 370 K? Give the expressions for both K_c and K_p.
(b) A 1.00-L flask contains 3.00 mol of solid sulfur. If 0.200 mol of both $H_2S(g)$ and $H_2(g)$ are injected into this flask, and the temperature is maintained at 370 K, will the reaction proceed to the right or to the left to attain equilibrium? Explain briefly.
(c) Will the total amount of H_2S present at equilibrium increase, decrease, or remain the same, if the volume of the container is doubled while the temperature is maintained at 370 K?

8.12. At 150 °C the equilibrium constant for the reaction

$$I_2(g) + Br_2(g) \rightleftharpoons 2 IBr(g)$$

is 280. A quantity of IBr is placed in a sealed flask and maintained at 150 °C. When the system comes to equilibrium, the pressure of IBr is found to be 0.46 atm. What are the partial pressures of $I_2(g)$ and $Br_2(g)$ at equilibrium?

8.13. The decomposition of ammonium hydrosulfide

$$NH_4HS(s) \rightleftharpoons NH_3(g) + H_2S(g)$$

is an endothermic reaction. A 5.2589-g sample of solid NH_4HS is placed in an evacuated 3.000-L vessel at 25 °C. After a period of time long enough for equilibrium to be established, the total pressure inside the vessel is found to be 0.659 atm. Some solid NH_4HS remains in the flask.
(a) Give the expression for, the numerical value of, and the units of K_p for this reaction.
(b) What percentage of the solid placed in the flask has decomposed?
(c) Will the number of moles of NH_4HS in the vessel increase, decrease, or remain the same, if the volume of the vessel is halved? Explain your answer.
(d) Will the concentration of NH_4HS in moles per liter increase, decrease, or remain the same, if the volume of the vessel is halved? Explain your answer.

8.14. Gaseous NH_3 and HCl are injected into an evacuated 2.00-L bulb that is maintained at 300.0 °C. More NH_3 is added than HCl. White crystals of NH_4Cl are observed to form. When the system comes to equilibrium the total pressure inside the bulb is 1.086 atm. For the reaction

$$NH_4Cl(s) \rightleftharpoons NH_3(g) + HCl(g)$$

K_p is 5.67×10^{-2} at 300.0 °C.
(a) Calculate the partial pressure of each gas at equilibrium.
(b) If the partial pressure of HCl at the instant of mixing was 0.600 atm, how many grams of NH_4Cl are in the bulb at equilibrium?

8.15. The equilibrium constant for the reaction

$$BaF_2(s) + CO_3^{2-}(aq) \rightleftharpoons BaCO_3(s) + 2F^-(aq)$$

is 1.5×10^4 at 25 °C. Some solid BaF_2 is shaken with a solution that is 0.12 M in CO_3^{2-} and 0.24 M in F^-. Will any $BaCO_3$ form? Show calculations that prove your answer.

8.16. At 698.6 K, a mixture of the three gases H_2, I_2, and HI is at equilibrium when $P_{H_2} = P_{I_2} = 0.02745$ atm and $P_{HI} = 0.2024$ atm. If HI is added to the reaction vessel at constant temperature, so that P_{HI} is suddenly increased to 0.8000 atm, what will the pressure of each of these gases be when the system returns to equilibrium?

chapter 9 Acids, Bases, and Salts

Wilhelm Ostwald (1853 – 1932) was born in Riga, Latvia. From 1887 to 1906 he taught at the University of Leipzig, in Germany. He worked closely with Arrhenius and van't Hoff, and with them he established physical chemistry as a branch of chemistry. He won the 1909 Nobel Prize in Chemistry for his work in catalysis and chemical equilibrium. Ostwald's dilution law, which gives the relation between the fraction of a weak electrolyte present in solution in the form of free ions, the stoichiometric concentration of the weak electrolyte, and the dissociation constant of the weak electrolyte, was derived by Ostwald in 1888. Ostwald was probably the last great chemist who refused to believe in atoms. He believed all material phenomena could be explained solely by considering energy changes.

The great majority of reactions with which a chemist is concerned take place in solution, rather than in the solid or gaseous phases. We have already seen, in Chapter 7, that water is an exceptionally good solvent, both for ionic and molecular solutes. Because water is an essential component of almost all living systems, the chemistry of reactions occurring in aqueous solution is particularly important. Acids and bases are among our most common chemicals, and in this chapter we will begin to study acid–base reactions in aqueous solution. Many acid–base reactions do not go to completion, and we will make use of the law of chemical equilibrium, discussed in Chapter 8, to calculate the concentrations of the various species present in aqueous solutions of acids, bases, and salts.

section 9.1
The Ion Product of Water and the pH Scale

Self-Ionization Reactions

In liquid water, molecules of H_2O are hydrogen bonded to one another. The proton bridging two oxygen atoms can shift from one oxygen to the other, leaving the electron of the hydrogen atom behind. The result of this proton shift is the **self-ionization** of water, depicted by the equation

$$\begin{array}{c} H \\ \diagdown \\ O\cdots H\!-\!O \\ \diagup \\ H \end{array} \quad\rightleftharpoons\quad \underset{\substack{\text{Hydronium} \\ \text{ion}}}{H_3O^+} \;+\; \underset{\substack{\text{Hydroxide} \\ \text{ion}}}{OH^-} \tag{9-1a}$$

or more simply $2H_2O \rightleftharpoons H_3O^+ + OH^-$ \hfill (9-1b)

Self-ionization is not unique to water; it occurs in other hydrogen-bonded solvents also. For instance, in pure liquid ammonia the equilibrium

$$\begin{array}{cc} H & H \\ \diagdown & \diagdown \\ H\!-\!N\cdots H\!-\!N \\ \diagup & \diagup \\ H & H \end{array} \quad\rightleftharpoons\quad \underset{\substack{\text{Ammonium} \\ \text{ion}}}{NH_4^+} \;+\; \underset{\substack{\text{Amide} \\ \text{ion}}}{NH_2^-} \tag{9-2}$$

occurs. The **amide ion**, NH_2^-, is a strong base and does not exist in aqueous solution because it reacts with water to accept a proton and form NH_3 and OH^-:

$$NH_2^- + H_2O \rightarrow NH_3 + OH^- \tag{9-3}$$

This reaction proceeds essentially to completion in aqueous solution. In pure liquid ammonia, however, there is a very small concentration of both NH_4^+ and NH_2^- ions, due to the self-ionization reaction.

Ethanol is another hydrogen-bonded liquid capable of undergoing self-ionization:

$$2CH_3CH_2OH \rightleftharpoons CH_3CH_2OH_2^+ + CH_3CH_2O^- \tag{9-4}$$

Ethoxide ion, $CH_3CH_2O^-$, is also a strong base, which, like the amide ion NH_2^-, does not exist in aqueous solution. That is, the reaction

$$CH_3CH_2O^- + H_2O \rightarrow CH_3CH_2OH + OH^- \tag{9-5}$$

proceeds essentially to completion.

None of these self-ionization reactions proceeds to any substantial extent. The concentrations of H_3O^+ and OH^- ions in pure water are very small. At 25 °C, $[H_3O^+] = [OH^-] = 1.0 \times 10^{-7} M$. Self-ionizations are dynamic equilibria with the protons exchanging rapidly, via the hydrogen bonds. It has been estimated that the average lifetime of any single H_3O^+ ion in pure water is about 10^{-12} s. Of course, both the H_3O^+ ions and the OH^- ions are solvated, with roughly three waters of hydration per ion. Two forms of the hydrated proton in aqueous solution are depicted in Fig. 7.7. An H_3O^+ ion with three waters of hydration is $H_9O_4^+$.

The Ion Product of Water, K_w

Since the self-ionization reaction in water is a dynamic equilibrium, we can write an equilibrium constant for it. The equilibrium constant for reaction (9-1) is called the **ion product of water,** and is denoted K_w.

$$K_w = [H_3O^+][OH^-] \tag{9-6a}$$

Note that $[H_2O]$ does not appear in the denominator of this equilibrium constant, because in any dilute aqueous solution the $[H_2O]$ is essentially constant, at ~55.5 M, as discussed previously in Section 8.1.

As is the case for any equilibrium constant, the numerical value of K_w is temperature dependent. At 25 °C, $[H_3O^+] = [OH^-] = 1.0 \times 10^{-7} M$ in pure water, and hence

$$K_w = [H_3O^+][OH^-] = 1.0 \times 10^{-14} M^2 \quad \text{at 25 °C} \tag{9-7}$$

The units of K_w are (moles/liter)^2, abbreviated M^2.

The relation given in Eq. (9-7) is valid for any dilute aqueous solution at 25 °C, regardless of what else is present in the solution in addition to H_2O. Thus Eq. (9-7) is valid for a dilute hydrochloric acid solution, and for a dilute solution of CH_3COOH, or KOH, or NH_3, or any other solute. It is not valid, however, for a dilute solution in which the solvent is anything other than pure water. If we add alcohol to a solution, for instance, or acetone, then $[H_2O]$ will no longer be approximately 55.5 M, the dielectric constant of the solvent will change significantly, and Eq. (9-7) will no longer apply.

We can use the ion product of water to calculate the $[H_3O^+]$ and $[OH^-]$ in various dilute aqueous solutions. Example 9.1 illustrates a typical calculation.

EXAMPLE 9.1. Use of K_w to calculate either $[H_3O^+]$ or $[OH^-]$

Calculate both the $[H_3O^+]$ and $[OH^-]$ in a 0.400 F HCl solution at 25 °C.

Solution. Hydrochloric acid is a strong acid, completely ionized in dilute aqueous solution. (Refer to Table 7.3.) The self-ionization of water occurs to only a very slight extent in pure water, and the presence of the H_3O^+ from the HCl represses the self-ionization of water. This means that the equilibrium $2H_2O \rightleftharpoons H_3O^+ + OH^-$ is driven to the left, because the reaction between HCl and H_2O produces such a high $[H_3O^+]$ compared to that in pure water. This is another example of Le Chatelier's Principle. In pure water, the $[H_3O^+] = 1.0 \times 10^{-7} M$. In 0.400 F HCl, the total $[H_3O^+]$ comes from two sources: from the HCl and from the self-ionization of H_2O. The concentration of H_3O^+ due to the self-ionization reaction is very much smaller than $1 \times 10^{-7} M$, as reaction (9-1) is driven to the left by the H_3O^+ from the HCl.

Thus,

$$[H_3O^+] = \text{contribution from HCl} + \text{contribution from self-ionization of water}$$
$$= 0.400 + \ll 0.0000001$$
$$= 0.400$$

We can now calculate the $[OH^-]$ from the expression for K_w, Eq. (9-7).

$$[OH^-] = \frac{K_w}{[H_3O^+]} = \frac{1.0 \times 10^{-14}\,M^2}{0.400\,M} = 2.5 \times 10^{-14}\,M$$

Note that in pure water $[OH^-] = 1.0 \times 10^{-7}\,M$ at 25 °C, but it has been reduced to $2.5 \times 10^{-14}\,M$ in 0.400 F HCl. This illustrates quantitatively the operation of Le Chatelier's Principle, as discussed above.

In an aqueous solution of a strong acid, if the $[H_3O^+]$ from the strong acid is significantly larger than $10^{-7}\,M$, the contribution to the total $[H_3O^+]$ from the self-ionization of water is negligible compared to the contribution from the strong acid, as in Example 9.1. We can formulate the general rule: In an aqueous solution of a strong acid, if the concentration of strong acid is greater than or equal to $1 \times 10^{-6}\,M$, then virtually all the $H^+(aq)$ ions in the solution come from the strong acid, and we can neglect the contribution of the self-ionization of water to the total $[H^+(aq)]$.

The pH Scale

We can see from the foregoing discussion that the concentrations of hydronium and hydroxide ions in various dilute aqueous solutions cover a very wide range of values. The $[H_3O^+]$ can be 10 M or more in concentrated solution of strong acids, and can decrease to less than $1 \times 10^{-14}\,M$ in concentrated solutions of strong bases. The $[OH^-]$ in solution must cover an equally wide range of values, since $[H_3O^+][OH^-] = 1 \times 10^{-14}$ at 25 °C. Because the concentrations themselves vary so widely, it is convenient to introduce a new scale, called the **pH scale,** which is a measure of the hydronium ion concentration, but covers a much narrower range of values. The pH of a solution is defined as

$$pH = -\log_{10}[H^+(aq)] \tag{9-8}$$

For very accurate work this definition must be modified slightly to take into account the coulombic interactions between charged ions in solution (interionic forces), but the definition given is sufficient for most purposes. Neglecting to consider interionic forces introduces an uncertainty in the pH of approximately ±0.02 units, and therefore the pH calculated using Eq. (9-8) should not be reported to more than two decimal places. The most commonly used log tables give logarithms to four decimal places, and the display on your calculator when the "log" button is pushed may give many figures, but you should round the value shown to two decimal places.

The definition of base 10 logarithms and a discussion of some properties of logarithms can be found in Appendix B. It will be helpful to review this material before doing calculations involving pH.

In pure water at 25 °C, $[H_3O^+] = 1 \times 10^{-7}\,M$. Therefore, $\log_{10}[H_3O^+] = -7$, and pH = 7. We define a **neutral solution** as one in which $[H_3O^+] = [OH^-]$. Thus the pH of a neutral solution at 25 °C is 7. Table 9.1 summarizes the definition of acidic, basic, and neutral solutions and relates these definitions to the pH.

Table 9.1. **Comparison of Acidic, Neutral, and Basic Solutions**

Solution	*Definition*	*pH at 25 °C*
Acidic	$[H_3O^+] > [OH^-]$ $[H_3O^+] > 1 \times 10^{-7}\ M$ at 25 °C	<7
Neutral	$[H_3O^+] = [OH^-]$ $[H_3O^+] = 1 \times 10^{-7}\ M$ at 25 °C	7
Basic	$[H_3O^+] < [OH^-]$ $[H_3O^+] < 1 \times 10^{-7}\ M$ at 25 °C	>7

Examples 9.2 and 9.3 are typical calculations involving pH.

EXAMPLE 9.2. Calculation of the pH, given the $[H^+(aq)]$

Calculate the pH of **(a)** 2.00 *F* HCl, **(b)** 0.200 *F* HClO$_4$, and **(c)** 2.00×10^{-4} *F* HNO$_3$.

Solution. These three acids are all strong acids (refer to Table 7.3), 100% ions in dilute aqueous solution.

(a) $[H^+(aq)] = 2.00\ M$ in 2.00 *F* HCl. Hence, pH $= -\log 2.00 = -0.30$. Note that the pH is negative in this solution. The pH is negative in any solution for which the $[H^+(aq)]$ is greater than 1.0 *M*.

(b) Since perchloric acid is a strong acid, $[H^+(aq)] = 0.200\ M$ in this solution. Thus,

$$pH = -\log(2.00 \times 10^{-1}) = +0.70$$

(c) In 2.00×10^{-4} *F* HNO$_3$ $[H^+(aq)] = 2.00 \times 10^{-4}\ M$

Thus,

$$pH = -\log(2.00 \times 10^{-4}) = 3.70$$

Note that in each case the pH is given to the second decimal place only. Also note that the greater the $[H^+(aq)]$, the lower the pH.

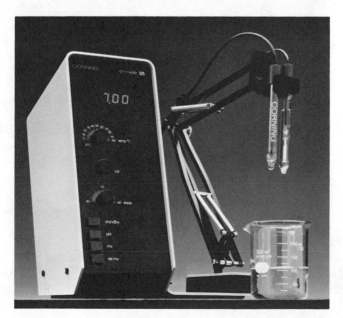

A pH meter is an instrument used to measure the pH of a solution. A glass electrode, sensitive to pH, and a reference electrode are immersed in the solution whose pH is to be measured. The meter is first calibrated by using a series of solutions of known pH. When the electrodes are immersed in the solution of unknown pH, the meter reads the pH directly.

EXAMPLE 9.3. Calculation of the [H$^+$] and [OH$^-$], given the pH

The pH of a sample of skim milk is found to be 6.6 at 25 °C. What are the concentrations of hydronium ion and hyroxide ion in this skim milk?

Solution. Equation (9-7) is applicable, because skim milk is a dilute aqueous solution. Since the pH = 6.6 = $-\log[H^+(aq)]$, then $\log[H^+(aq)] = -6.6$, and $[H^+(aq)]$ = antilog$(-6.6) = 10^{-6.6}$. If your calculator has a 10^x button, you will be able to obtain immediately

$$[H^+(aq)] = 10^{-6.6} = 2.5 \times 10^{-7}$$

Concentrations should be reported in standard scientific (exponential) notation, with integral powers of 10. It is also useful to remember that

$$10^{-x} = \frac{1}{10^x}$$

so that

$$10^{-6.6} = \frac{1}{10^{6.6}} = \frac{1}{4.0 \times 10^6} = 0.25 \times 10^{-6} = 2.5 \times 10^{-7}$$

We can calculate the [OH$^-$] in this skim milk as follows:

$$[OH^-] = \frac{K_w}{[H^+(aq)]} = \frac{1.0 \times 10^{-14}}{2.5 \times 10^{-7}} = 4.0 \times 10^{-8} \, M$$

If your calculator does not have a 10^x button, you will need to use a log table to obtain antilog$_{10}(-\text{pH}) = [H^+(aq)]$. Details of the method of obtaining the antilog of a negative number using a log table can be found in Appendix B.

The concept of a pH scale turns out to be so convenient that we find it useful to generalize, and to define pX as $-\log X$, where X can be any quantity. Thus we define pOH as $-\log_{10}[OH^-]$, pK_a as $-\log_{10}K_a$, pK_w as $-\log_{10}K_w$, and so on. With these definitions, it is easy to show that

$$\text{pH} + \text{pOH} = pK_w \tag{9-9a}$$

We begin with the definition of K_w,

$$[H_3O^+][OH^-] = K_w \tag{9-6a}$$

Using the equation for the logarithm of a product (Appendix B), take the logarithm of this equation to obtain

$$\log[H_3O^+] + \log[OH^-] = \log K_w \tag{9-6b}$$

Now multiply through by -1 and introduce the definitions of pH, pOH, and pK_w:

$$-\log[H_3O^+] - \log[OH^-] = -\log K_w$$

is therefore

$$\text{pH} + \text{pOH} = pK_w \tag{9-9a}$$

At 25 °C, $K_w = 1.0 \times 10^{-14}$, and therefore p$K_w = +14$. We obtain the convenient relation

$$\text{pH} + \text{pOH} = 14 \qquad \text{at 25 °C} \tag{9-9b}$$

For a neutral solution, $[H_3O^+] = [OH^-]$, and therefore pH = pOH. Thus Eq. (9-9a) becomes pH = $\frac{1}{2}$pK_w for a neutral solution, and at 25 °C we obtain pH = 7 for a neutral solution, from Eq. (9-9b).

Table 9.2. The Temperature Dependence
of the Ion Product of H_2O

Temperature (°C)	K_w (M^2)
0.0	0.12×10^{-14}
25.0	1.00×10^{-14}
60.0	9.6×10^{-14}

The ion product of water is temperature dependent, as is any equilibrium constant. Values of K_w as a function of temperature are given in Table 9.2. It is clear that a neutral solution has a pH of exactly 7 only at 25 °C, which is approximately room temperature.

It is important to remember that pH = 7 applies to a neutral solution *only* at 25 °C. The chart below provides a useful frame of reference to use when thinking about the pH of aqueous solutions at room temperature.

$[H_3O^+]$	10^{-1}	10^{-4}	10^{-7}	10^{-10}	$10^{-13}\ M$
pH	1	4	7	10	13
	←Acidic		Neutral at 25 °C		Basic→
pOH	13	10	7	4	1
$[OH^-]$	10^{-13}	10^{-10}	10^{-7}	10^{-4}	$10^{-1}\ M$

EXAMPLE 9.4. The pH of neutral solutions at temperatures other than 25 °C

Calculate the pH of a neutral solution at (a) 0 °C and (b) 60 °C.

Solution

(a) For a neutral solution, pH $= \frac{1}{2}pK_w$. From Table 9.2, we find $K_w = 1.2 \times 10^{-15}$, so that $pK_w = 14.92$, and pH $= 7.46$ for a neutral solution at 0 °C.

(b) At 60 °C, $K_w = 9.6 \times 10^{-14}$, so that $pK_w = 13.02$ and pH $= 6.51$ for a neutral solution.

EXAMPLE 9.5. The pH and pOH in solutions of a strong base

Calculate the $[OH^-]$, $[H_3O^+]$, pH, and pOH in 0.060 F $Ba(OH)_2$ at 25 °C.

Solution. Since $Ba(OH)_2$ is a strong base (refer to Table 7.3), $[OH^-] = 0.12\ M$ in 0.060 F $Ba(OH)_2$. We obtain the $[H^+(aq)]$ using Eq. (9-7).

$$[H^+(aq)] = \frac{1.0 \times 10^{-14}\ M}{0.12\ M} = 8.3 \times 10^{-14}\ M$$

$$pH = -\log(8.3 \times 10^{-14}) = 13.08$$

$$pOH = 14 - pH = 0.92$$

Of course, we could have obtained pOH directly from the $[OH^-]$:

$$pOH = -\log(0.12) = +0.92$$

section 9.2
Weak Acids and Bases

The Acidity Constant, K_a

The equilibrium constant for the proton-transfer reaction between a weak acid and water is called an **acidity constant** and denoted K_a. For the general weak monoprotic acid, HA, the proton-transfer reaction with water (refer to Section 7.4) is given by

$$HA + H_2O \rightleftharpoons H_3O^+ + A^- \qquad (9\text{-}10)$$

Acid   Base   Acid   Base

The general expression for K_a is therefore

$$K_a = \frac{[H_3O^+][A^-]}{[HA]} \qquad (9\text{-}11)$$

There are a great many weak acids, which may be either molecules or ions; a few of the most common are acetic acid, CH_3COOH; nitrous acid, HONO (often written HNO_2); benzoic acid, C_6H_5COOH; ammonium ion, NH_4^+; and hydrogen sulfide, H_2S. A selected list of K_a values is given in Table 9.3. A more extensive list can be found in Appendix E, Table E1.

It is important to remember that K_a is the equilibrium constant only for the reaction in which the weak acid donates a proton *to water;* that is, the base accepting the proton from the weak acid is H_2O.

The Basicity Constant, K_b

The equilibrium constant for the proton-transfer reaction between a weak base and water is called a **basicity constant,** and denoted K_b. For a general weak base, denoted simply B, the proton-transfer reaction with water (refer to Section 7.4) is given by

$$B + H_2O \rightleftharpoons BH^+ + OH^- \qquad (9\text{-}12a)$$

The general expression for K_b is therefore

$$K_b = \frac{[BH^+][OH^-]}{[B]} \qquad (9\text{-}13a)$$

If B is a neutral molecule, K_b is also called the **ionization constant of the base B.**

There are a great many weak bases, which may be either molecules or anions (or much more rarely, cations). A few of the most common weak bases are ammonia, NH_3; methylamine, CH_3NH_2; aniline, $C_6H_5NH_2$; acetate ion, CH_3COO^-; carbonate ion, CO_3^{2-}; and sulfide ion, S^{2-}. A selected list of K_b values is given in Table 9.4.

Table 9.3. **Acidity Constants for Some Weak Acids**

Weak Acid	Proton-Transfer Reaction	K_a
Acetic acid	$CH_3COOH + H_2O \rightleftharpoons CH_3COO^- + H_3O^+$	1.8×10^{-5}
Ammonium ion	$NH_4^+ + H_2O \rightleftharpoons NH_3 + H_3O^+$	5.7×10^{-10}
Formic acid	$HCOOH + H_2O \rightleftharpoons HCOO^- + H_3O^+$	1.8×10^{-4}
Hydrocyanic acid	$HCN + H_2O \rightleftharpoons CN^- + H_3O^+$	4.0×10^{-10}
Hydrogen sulfide	$H_2S + H_2O \rightleftharpoons HS^- + H_3O^+$	1.0×10^{-7}
Nitrous acid	$HONO + H_2O \rightleftharpoons ONO^- + H_3O^+$	4.5×10^{-4}

Table 9.4. **Basicity Constants for Some Weak Bases**

Weak Base	Proton-Transfer Reaction	K_b
Acetate ion	$CH_3COO^- + H_2O \rightleftharpoons CH_3COOH + OH^-$	5.7×10^{-10}
Ammonia	$NH_3 + H_2O \rightleftharpoons NH_4^+ + OH^-$	1.8×10^{-5}
Aniline	$C_6H_5NH_2 + H_2O \rightleftharpoons C_6H_5NH_3^+ + OH^-$	4×10^{-10}
Carbonate ion	$CO_3^{2-} + H_2O \rightleftharpoons HCO_3^- + OH^-$	2.1×10^{-4}
Cyanide ion	$CN^- + H_2O \rightleftharpoons HCN + OH^-$	2.5×10^{-5}
Methylamine	$CH_3NH_2 + H_2O \rightleftharpoons CH_3NH_3^+ + OH^-$	4.2×10^{-4}

Equations (9-12a) and (9-13a), applied to an anionic base like sulfide ion, are

$$\underset{\text{Base}}{S^{2-}} + \underset{\text{Acid}}{H_2O} \rightleftharpoons \underset{\text{Acid}}{HS^-} + \underset{\text{Base}}{OH^-} \qquad (9\text{-}12b)$$

and

$$K_b(S^{2-}) = \frac{[HS^-][OH^-]}{[S^{2-}]} \qquad (9\text{-}13b)$$

Remember that K_b is the equilibrium constant only for the reaction in which the weak base accepts a proton *from water*.

The Degree of Dissociation, α

For molecular weak acids and bases we frequently want to know: To what extent does this substance ionize, that is, what fraction of a base B is present in aqueous solution as BH^+ ions? Or, similarly, what fraction of an acid HA is present in aqueous solution as A^- ions?

For an aqueous solution of a weak acid we denote the **degree of dissociation** or the **fraction ionized** by the symbol α (alpha), and define it as

$$\alpha = \frac{[H_3O^+]}{C} = \frac{[A^-]}{C} \qquad (9\text{-}14)$$

where C is the stoichiometric concentration of the weak acid, that is, the formality of the acid, or the initial concentration of the weak acid.

For a strong acid like HCl for which the proton-transfer reaction (9-10) goes virtually to completion in dilute aqueous solution, $\alpha = 1$ or 100%. That is, 100% of HCl molecules dissolved in water are present as H_3O^+ ions and Cl^- ions, provided the solution is dilute. For a nonelectrolyte, no ions are present at all, and hence $\alpha = 0$. For a weak acid,

$$0 < \alpha < 1 \qquad (9\text{-}15)$$

Remember that the degree of dissociation of a weak acid, α, as defined by Eq. (9-14) applies only to a solution of HA in water; no other sources of H_3O^+ ions or A^- ions may be present.

Example 9.6 describes two methods for calculating the concentrations of the various species present in a solution of a weak acid in water, as well as the calculation of the degree of dissociation.

EXAMPLE 9.6. The ionization of an aqueous solution of a weak acid

Calculate the $[H_3O^+]$, $[CH_3COO^-]$, $[CH_3COOH]$, $[OH^-]$, pH, and the degree of dissociation, α, of a 0.100 F CH_3COOH solution at 25 °C.

Solution. There are two equivalent ways to solve this problem and we shall use both, to ensure that both methods are clear to you. For either method we begin by writing the proton-transfer reaction between acetic acid and water, and the expression for K_a:

$$CH_3COOH + H_2O \rightleftharpoons CH_3COO^- + H_3O^+ \qquad (9\text{-}16)$$

$$K_a = \frac{[H_3O^+][CH_3COO^-]}{[CH_3COOH]} = 1.8 \times 10^{-5} \qquad (9\text{-}17)$$

The numerical value of K_a for acetic acid at 25 °C is obtained from Table E1, Appendix E, or from Table 9.3. The stoichiometric concentration of acetic acid, the C in Eq. (9-14), is 0.100 M in this problem.

Method 1. We can calculate both the pH and α if we know the $[H_3O^+]$ in this solution, so we begin by saying

let $x = [H_3O^+]$ at equilibrium in 0.100 F CH_3COOH

It is extremely helpful to write down in detail the definition of any symbol introduced to solve a problem.

From Eq. (9-16) we see that in any solution prepared by mixing only acetic acid and water, the molar ratio of acetate ions to hydronium ions is 1 : 1, that is, the $[CH_3COO^-]$ must be equal to the $[H_3O^+]$.* We therefore have

$$[CH_3COO^-] = [H_3O^+] = x$$

The concentration of acetic acid molecules in this solution is slightly smaller than the stoichiometric concentration, 0.100 M, because although 0.100 mol of CH_3COOH molecules were dissolved per liter of solution, a fraction of those molecules exists as the ions, CH_3COO^- and H_3O^+, at equilibrium. The equation for the proton-transfer reaction, Eq. (9-16), tells us that

$$\frac{\text{moles of } CH_3COO^- \text{ formed}}{\text{moles of } CH_3COOH \text{ used up}} = \frac{1}{1}$$

and hence the concentration of CH_3COOH that is used up is also equal to x.

In solving equilibrium problems, it is helpful to write down the initial concentrations of all species, the change in concentration that occurs as the system comes to equilibrium, and then the final concentrations, at equilibrium. The following format may be useful:

	CH₃COOH	*CH₃COO⁻*	*H₃O⁺*
Initial concentration (M)	0.100	0	0
Change in concentration (M)	x	x	x
Equilibrium concentration (M)	$0.100 - x$	x	x

The equilibrium concentration of a substance that is used up during a reaction is its initial value *minus* the change in concentration. The equilibrium concentration of a

* This is true only if the contribution of the self-ionization of water to the $[H_3O^+]$ is negligible compared to the contribution due to reaction (9-16). Provided K_a is significantly larger than K_w, and the formality, C, of the weak acid is not too small, it is a valid approximation to neglect the self-ionization of water. A method of solving problems when this approximation is not valid can be found in Appendix K.

substance that is formed during a reaction is its initial value *plus* the change in concentration.

Since at equilibrium $[CH_3COOH] = (0.100 - x)$, the expression for the acidity constant of acetic acid is

$$K_a(HOAc) = \frac{[H_3O^+][CH_3COO^-]}{[CH_3COOH]} = \frac{x^2}{0.100 - x} = 1.8 \times 10^{-5}$$

This is a quadratic equation in one unknown and can be solved for x in a straightforward manner using the quadratic formula. (See Appendix B.) There is, however, a method of solving this equation that involves less numerical calculation than using the quadratic formula. Since we know that acetic acid, a weak acid, is only slightly ionized, we expect x, the $[H_3O^+]$, to be significantly smaller than $0.100\ M$. The arithmetic is very much easier if we introduce this assumption into our calculations.

Assume x is small compared to 0.100, so that $0.100 - x \sim 0.100$. Our equation then becomes

$$K_a(HOAc) = \frac{x^2}{0.100 - x} = \frac{x^2}{0.100} = 1.8 \times 10^{-5}$$

Thus

$$x^2 = 1.8 \times 10^{-6}$$

and

$$x = 1.3 \times 10^{-3}\ M$$

Now we must check the assumption we made. Every time an approximation is introduced in a calculation, you must check to see if the approximation is valid. Is 1.3×10^{-3} much less than 0.100? That depends, of course, on how many significant figures you are entitled to in your answer. Remember that we are using the ideal law of chemical equilibrium. Depending on the components of the solution, the ratio of concentrations used in the equilibrium constant expression, Eq. (8-4), may vary considerably. Errors as large as 10% are not uncommon if this ratio of concentrations is assumed to be a true constant at constant temperature. It is for this reason that acidity constants are given only to two significant figures. Consequently we will adopt the following rule when doing equilibrium constant problems: *A number, a, will be considered small relative to another number, n, provided that a is less than 10% of n.* Thus both $n - a \sim n$ and $n + a \sim n$, if $a < 0.10\ n$.

In the present calculation, we will consider $0.100 - x = 0.100$ provided that x, which we have calculated to be 1.3×10^{-3}, is less than 10% of 0.100, which is 1.0×10^{-2}. The relevant question is then: Is 1.3×10^{-3} less than 1.0×10^{-2}? The answer, of course, is yes, and therefore the approximation that was made is valid.

If the approximation introduced in order to solve for an unknown is *not* valid, the numerical value obtained will be inconsistent with the approximation made. The fact that 1.3×10^{-3} is less than 1.0×10^{-2} tells us that it is valid to substitute 0.100 for $0.100 - x$ in the equation for $K_a(HOAc)$. If you find you have made an invalid assumption, you can always solve the equation using the quadratic formula, or you can use an alternative approach that will be described later, in Example 9.8, when we have a problem that necessitates its use.

For the problem at hand, we have now shown that in $0.100\ F$ HOAc,

$$[H_3O^+] = [CH_3COO^-] = 1.3 \times 10^{-3}\ M$$

The concentration of acetic acid molecules is given by

$$[CH_3COOH] = 0.100 - 1.3 \times 10^{-3} = 0.100 - 0.001 = 0.099\ M$$

just slightly less than the stoichiometric concentration.

The degree of dissociation, α, is calculated to be

$$\alpha = \frac{[H_3O^+]}{C} = \frac{1.3 \times 10^{-3}}{0.100} = 1.3 \times 10^{-2} \quad \text{or} \quad 1.3\%$$

Thus 0.100 F CH$_3$COOH is only 1.3% dissociated at 25 °C.

The pH of the solution is readily obtained as

$$pH = -\log[H_3O^+] = -\log(1.3 \times 10^{-3}) = 3 - \log 1.3 = 2.89$$

Method 2. Rearranging the equation defining α

$$\alpha = \frac{[H_3O^+]}{C} = \frac{[OAc^-]}{C}$$

yields:

$$[H_3O^+] = [OAc^-] = \alpha C = 0.100\alpha$$

If α is the fraction dissociated, then $(1 - \alpha)$ is the fraction *not* dissociated, that is, the fraction that remains as undissociated CH$_3$COOH molecules. Thus,

$$1 - \alpha = \frac{[HOAc]}{C}$$

so that $[HOAc] = C(1 - \alpha) = 0.100(1 - \alpha)$. The expression for the acidity constant is therefore

$$K_a = \frac{[H_3O^+][OAc^-]}{[HOAc]} = \frac{(0.100\alpha)(0.100\alpha)}{0.100(1 - \alpha)} = \frac{0.100\alpha^2}{(1 - \alpha)} = 1.8 \times 10^{-5}$$

Multiplying both sides by 10 we obtain

$$\frac{\alpha^2}{1 - \alpha} = 1.8 \times 10^{-4}$$

Once again we have a quadratic equation to solve, and we can do this most simply by making an approximation based on our chemical knowledge. We know that $0 < \alpha < 1$ for any weak acid, and that for acetic acid α is quite small. Therefore we can make the approximation that $\alpha < 10\%$ of 1, so that $1 - \alpha \sim 1$. The equation to be solved is then simply

$$\alpha^2 = 1.8 \times 10^{-4} \quad \text{and thus} \quad \alpha = 1.34 \times 10^{-2} = 1.3 \times 10^{-2}$$

We must check our assumption. Is 1.3×10^{-2} less than 10% of 1? Yes, 1.3×10^{-2} is less than 0.1, and therefore $\alpha = 1.3 \times 10^{-2}$ is the correct answer. We then calculate the $[H_3O^+]$ from

$$[H_3O^+] = \alpha C = (1.3 \times 10^{-2})(0.100\ M) = 1.3 \times 10^{-3}\ M$$

The pH $= -\log(1.3 \times 10^{-3}) = 2.89$, just as in method 1.

Note that it makes absolutely no difference whether you solve first for the $[H_3O^+]$ (as in method 1) and calculate α afterward, or you solve first for α (as in method 2), and calculate the $[H_3O^+]$ afterward.

We still have to solve for the $[OH^-]$ in this acidic solution. We use the relation

$$[OH^-] = \frac{K_w}{[H_3O^+]} = \frac{1.0 \times 10^{-14}}{1.34 \times 10^{-3}} = 7.5 \times 10^{-12}\ M$$

Table 9.5. The Increase in the Degree of Dissociation of Acetic Acid with Dilution

Stoichiometric Concentration, C	$[H_3O^+]$ (M)	$\alpha = [H_3O^+]/C$	$[CH_3COOH]$ (M)	pH
1.000 F	4.2×10^{-3}	0.0042 or 0.42%	0.996	2.38
0.500 F	3.0×10^{-3}	0.0060 or 0.60%	0.497	2.52
0.100 F	1.3×10^{-3}	0.013 or 1.3%	0.099	2.89
0.050 F	9.5×10^{-4}	0.019 or 1.9%	0.049	3.02
0.010 F	4.2×10^{-4}	0.042 or 4.2%	0.0096	3.38

For an aqueous solution of any weak acid, HA, we may write $[H_3O^+] = [A^-] = \alpha C$, and $[HA] = (1 - \alpha)C$. Thus the expression for K_a can be written as

$$K_a(HA) = \frac{[H_3O^+][A^-]}{[HA]} = \frac{(\alpha C)(\alpha C)}{(1 - \alpha)C} = \frac{\alpha^2 C}{1 - \alpha} \qquad (9\text{-}18)$$

This particular expression for K_a is known as the **Ostwald Dilution Law**. It is merely another way of expressing the equilibrium constant. The German chemist Wilhelm Ostwald (1853–1932) was the first to point out that as the stoichiometric concentration of a weak acid is decreased, the degree of dissociation, α, increases. We can see this clearly from Eq. (9-18). Since $1 - \alpha \sim 1$ for most weak acids, Eq. (9-18) can be approximated as $K_a = \alpha^2 C$. Thus as C gets smaller, α must increase because K_a is a constant at constant temperature. Physically, the more dilute the solution, the greater the average distance between the ions, and the weaker the force of attraction between ions of opposite charge (refer to the discussion of Coulomb's Law in Section 7.1), so the easier it is for the molecules to be dissociated.

Table 9.5 shows the relation between the degree of dissociation of acetic acid and its stoichiometric concentration. In Example 9.6 we did the calculations that provide the entries in this table for 0.100 F HOAc. You can gain practice in this type of calculation by verifying the entries in Table 9.5 for a different stoichiometric concentration of acetic acid.

The entire discussion of weak acids can be applied to weak bases as well, provided that we replace Eq. (9-14) by the appropriate definition for the degree of dissociation of the weak base, B. The **degree or extent of dissociation of a weak base, B,** is also denoted α. It is defined as

$$\alpha = \frac{[OH^-]}{C} = \frac{[BH^+]}{C} \qquad (9\text{-}19a)$$

where C is the stoichiometric concentration (the formality) of the weak base. Thus in an aqueous solution of the base B,

$$[OH^-] = [BH^+] = \alpha C \qquad (9\text{-}19b)$$

and

$$[B] = (1 - \alpha)C \qquad (9\text{-}20)$$

so that

$$K_b = \frac{[BH^+][OH^-]}{[B]} = \frac{\alpha^2 C}{1 - \alpha} \qquad (9\text{-}21)$$

The Ostwald Dilution Law, therefore, applies to weak bases as well as to weak acids, that is, the degree of dissociation increases as the concentration of the weak base decreases.*

EXAMPLE 9.7. The ionization of an aqueous solution of a weak base

Calculate the $[OH^-]$, $[NH_4^+]$, $[NH_3]$, $[H_3O^+]$, and the pH of a 0.250 F NH_3 solution at 25 °C.

$$K_b(NH_3) = 1.8 \times 10^{-5} \quad \text{at 25 °C}$$

Solution. The proton-transfer reaction between NH_3 and H_2O is

$$NH_3 + H_2O \rightleftharpoons NH_4^+ + OH^- \tag{9-12c}$$

$$K_b(NH_3) = \frac{[NH_4^+][OH^-]}{[NH_3]} = 1.8 \times 10^{-5} \tag{9-13c}$$

Let $x = [NH_4^+]$ in 0.250 F NH_3 at equilibrium at 25 °C. From Eq. (9-12c) we have

$$[NH_4^+] = [OH^-] = x$$

The concentration of NH_3 *used up* in the proton-transfer reaction is also x, so that

$$[NH_3] = 0.250 - x$$

The equilibrium constant expression therefore becomes

$$K_b(NH_3) = \frac{[NH_4^+][OH^-]}{[NH_3]} = \frac{x^2}{0.250 - x} = 1.8 \times 10^{-5}$$

Assume x is small compared to 0.250, that is, assume $x < 10\%$ of 0.250. Then $0.250 - x \sim 0.250$, and the equilibrium constant expression becomes

$$\frac{x^2}{0.250 - x} = \frac{x^2}{0.250} = 1.8 \times 10^{-5}$$

or

$$x^2 = 0.45 \times 10^{-5} = 4.5 \times 10^{-6}$$

We obtain

$$x = 2.1 \times 10^{-3} \, M$$

We must now check the assumption we made. Is 2.1×10^{-3} less than 10% of 0.250? Yes, $2.1 \times 10^{-3} < 2.5 \times 10^{-2}$. Hence the assumption is valid, and

$$[NH_4^+] = [OH^-] = 2.1 \times 10^{-3} \, M$$

$$[NH_3] = 0.250 - 0.002 = 0.248 \, M$$

$$[H_3O^+] = \frac{K_w}{[OH^-]} = \frac{1.0 \times 10^{-14}}{2.1 \times 10^{-3}} = 4.8 \times 10^{-12} \, M$$

$$pH = -\log[H_3O^+] = 12 - \log 4.8 = 12 - 0.68 = 11.32$$

---

* In utilizing the Ostwald Dilution Law for calculations, the first step in computation is usually to assume that α is small compared to 1. This is justified only if $\alpha < 0.1$, or $\alpha^2 < 0.01$. Since

$$K/C = \alpha^2/(1 - \alpha)$$

where K can be either K_a or K_b, if $K/C > 0.01$ it will not be valid to assume that $1 - \alpha \sim 1$. In this case we must use either the quadratic formula or the method of successive approximations to solve for α. The method of successive approximations is described in Example 9.8.

EXAMPLE 9.8. Calculation of the degree of dissociation of a weak base using the method of successive approximations

Methylamine is a colorless gas at room temperature, with a strong unpleasant odor. It dissolves readily in water, with which it undergoes the proton-transfer reaction

$$CH_3NH_2 + H_2O \rightleftharpoons CH_3NH_3^+ + OH^- \tag{9-22}$$

At 25 °C, $K_b = 4.20 \times 10^{-4}$ for methylamine. Calculate the degree of dissociation of a 0.0200 F solution of methylamine at 25 °C.

Solution. Since all we are asked to find is the degree of dissociation, α, we will follow the procedure described in method 2 for Example 9.6. We begin by writing the equation that relates the degree of dissociation to the basicity constant, Eq. (9-21):

$$K_b(CH_3NH_2) = \frac{[CH_3NH_3^+][OH^-]}{[CH_3NH_2]} = \frac{\alpha^2 C}{(1-\alpha)} = 4.20 \times 10^{-4}$$

In this problem the stoichiometric concentration (the formality) of the base, C, is 0.0200. If we substitute the numerical value of C into the equilibrium constant expression, we can obtain

$$\frac{\alpha^2}{(1-\alpha)} = \frac{4.20 \times 10^{-4}}{2.00 \times 10^{-2}} = 2.10 \times 10^{-2}$$

First Approximation: To solve this equation for α assume that α is less than 10% of 1, so that $1 - \alpha \sim 1$. The equation then becomes

$$\alpha^2 = 2.10 \times 10^{-2} \quad \text{and} \quad \alpha = 1.45 \times 10^{-1}$$

Since 1.45×10^{-1} (0.145) is *not* less than 10% of 1 (0.100), the first approximation is not valid. We therefore make a second approximation that is better than the first.

Second Approximation: Assume $1 - \alpha = 1 - 0.145 = 0.855$. Note that we use the answer obtained from the first approximation to get a better value for $(1 - \alpha)$. We are no longer assuming that α is small compared to 1; we are assuming that 0.855 is a better approximation to the quantity $(1 - \alpha)$ than 1 is. We therefore obtain

$$\frac{\alpha^2}{(1-\alpha)} = \frac{\alpha^2}{0.855} = 2.10 \times 10^{-2}$$

Then $\alpha^2 = (0.855)(2.10 \times 10^{-2}) = 1.80 \times 10^{-2}$, and $\alpha = 1.34 \times 10^{-1}$.

Third Approximation: Assume $1 - \alpha = 1 - 0.134 = 0.866$. We use the result obtained from the second approximation to obtain a still better value for $(1 - \alpha)$. Hence

$$\frac{\alpha^2}{(1-\alpha)} = \frac{\alpha^2}{0.866} = 2.10 \times 10^{-2} \quad \text{and} \quad \alpha^2 = (0.866)(2.10 \times 10^{-2})$$

so that $\alpha^2 = 1.82 \times 10^{-2}$, and $\alpha = 1.35 \times 10^{-1} = 0.135$.

Fourth Approximation: $1 - \alpha = 1 - 0.135 = 0.865$. Then

$$\alpha^2 = (0.865)(2.10 \times 10^{-2}) = 1.82 \times 10^{-2} \quad \text{and} \quad \alpha = 0.135 \quad \text{or} \quad 13.5\%$$

The results of the third and fourth approximations for α are exactly the same to three significant figures. When two successive approximations give identical results to the correct number of significant figures, the answer obtained is the correct answer. The same answer would be obtained if the quadratic formula were employed, which you may want to verify for yourself.

Note that the first approximation gave $\alpha = 0.145$, whereas the correct answer is 0.135. Thus the first answer was not so terribly far off; the two values differ only by 1 in the second significant figure. One way that you can check that you have obtained the correct answer is to substitute back into the original equation:

$$K_b = \alpha^2 C/(1 - \alpha) = (0.135)^2 (2.00 \times 10^{-2})/0.865 = 4.2 \times 10^{-4}$$

In Example 9.8 we calculated that 0.0200 F methylamine is 13.5% dissociated at 25 °C. A more concentrated solution is dissociated to a smaller extent. The degree of dissociation increases as the solution becomes more dilute. For any weak base, B, as [B] decreases, [OH$^-$] decreases and [BH$^+$] decreases, but the fraction ionized, α, increases.

section 9.3
Conjugate Acids and Bases and Their Relative Strengths

An acid and a base related by the expression

$$\text{base} + \text{H}^+ = \text{acid} \tag{9-23}$$

are called a **conjugate acid–base pair.** Thus NH_3 and NH_4^+ constitute a conjugate pair, because $NH_3 + H^+ = NH_4^+$, and we may refer to them either by saying "NH_3 is a base and NH_4^+ ion is its conjugate acid" or by saying "NH_4^+ is an acid and NH_3 is its conjugate base." Acetic acid and acetate ion also constitute a conjugate pair, since

$$\underset{\text{Base}}{CH_3COO^-} + H^+ = \underset{\text{Acid}}{CH_3COOH}$$

For the general case, the weak acid HA has the weak base A$^-$ as its conjugate base. The acidity constant of HA is defined by Eq. (9-11), and is $K_a(\text{HA}) = [\text{H}_3\text{O}^+][\text{A}^-]/[\text{HA}]$. The basicity constant of the anion A$^-$ is the equilibrium constant for the proton-transfer reaction $\text{A}^- + \text{H}_2\text{O} \rightleftharpoons \text{HA} + \text{OH}^-$, and is defined as [see Eq. (9-13a)]

$$K_b(\text{A}^-) = \frac{[\text{HA}][\text{OH}^-]}{[\text{A}^-]}$$

If we multiply $K_a(\text{HA})$ by $K_b(\text{A}^-)$ we obtain

$$K_a(\text{HA}) \cdot K_b(\text{A}^-) = \frac{[\text{H}_3\text{O}^+][\text{A}^-]}{[\text{HA}]} \cdot \frac{[\text{HA}][\text{OH}^-]}{[\text{A}^-]} = [\text{H}_3\text{O}^+][\text{OH}^-] = K_w \tag{9-24a}$$

For any conjugate acid–base pair it is always true that

$$K_a(\text{weak acid}) \cdot K_b(\text{conjugate weak base}) = K_w \tag{9-24b}$$

In Fig. 9.1, the definitions of K_a and K_b for two typical acid–base conjugate pairs, NH_4^+/NH_3 and CH_3COOH/CH_3COO^-, are written in such a way as to emphasize the similarities between them.

Because of the relationship between K_a for any weak acid and K_b for its conjugate weak base, it is not necessary to tabulate both acidity and basicity constants. Many references tabulate only acidity constants of weak acids, as is done in Table E1, Appendix E. A desired basicity constant must then be calculated using Eq. (9-24b). Typical calculations are illustrated in Example 9.9.

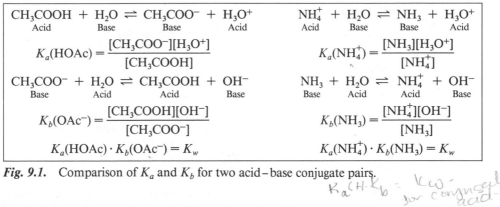

Fig. 9.1. Comparison of K_a and K_b for two acid–base conjugate pairs.

$K_a \cdot K_b = K_w$ for conjugate base acid pair.

EXAMPLE 9.9. Calculation of basicity constants

Calculate K_b at 25 °C for **(a)** sulfide ion, S^{2-}, and **(b)** monohydrogen phosphate ion, HPO_4^{2-}.

Solution

(a) The conjugate acid of S^{2-} ion is HS^- ion, the bisulfide ion or hydrogen sulfide ion. The acidity constant at 25 °C, $K_a(HS^-) = 1.3 \times 10^{-13}$, is given in Table E1, Appendix E. Using Eq. (9-24b) we obtain

$$K_b(S^{2-}) = \frac{K_w}{K_a(HS^-)} = \frac{1.0 \times 10^{-14}}{1.3 \times 10^{-13}} = 7.7 \times 10^{-2}$$

(b) The conjugate acid of HPO_4^{2-} is $H_2PO_4^-$, the dihydrogen phosphate ion. In Table E1 we find $K_a(H_2PO_4^-) = 6.2 \times 10^{-8}$. Thus

$$K_b(HPO_4^{2-}) = \frac{K_w}{K_a(H_2PO_4^-)} = \frac{1.0 \times 10^{-14}}{6.2 \times 10^{-8}} = 1.6 \times 10^{-7}$$

There are a great many weak acids, but although they are all called "weak" they may differ widely in their relative strengths. We can compare the relative strengths of many weak acids by tabulating them in order of the numerical value of their acidity constants. This has been done in Table 9.6 for several common weak acids.

Bisulfate ion, HSO_4^- (also called hydrogen sulfate ion), is the strongest acid of those listed in Table 9.6. Indeed, weak acids with acidity constants as large as 10^{-2} are sometimes called "moderately strong acids." The strengths of the acids in Table 9.6 decrease in the order given, that is, as K_a decreases. Thus methylammonium ion is the weakest acid of those listed.

Table 9.6. Relative Strengths of Weak Acids

Name of Acid	Proton-Transfer Reaction	K_a at 25 °C
Bisulfate ion	$HSO_4^- + H_2O \rightleftharpoons SO_4^{2-} + H_3O^+$	1.2×10^{-2}
Acetic acid	$CH_3COOH + H_2O \rightleftharpoons CH_3COO^- + H_3O^+$	1.8×10^{-5}
Hydroxylammonium ion	$^+H_3NOH + H_2O \rightleftharpoons H_2NOH + H_3O^+$	8.2×10^{-7}
Hydrogen sulfide	$H_2S + H_2O \rightleftharpoons HS^- + H_3O^+$	1.0×10^{-7}
Ammonium ion	$NH_4^+ + H_2O \rightleftharpoons NH_3 + H_3O^+$	5.7×10^{-10}
Hydrocyanic acid	$HCN + H_2O \rightleftharpoons CN^- + H_3O^+$	4.0×10^{-10}
Bicarbonate ion	$HCO_3^- + H_2O \rightleftharpoons CO_3^{2-} + H_3O^+$	4.7×10^{-11}
Methylammonium ion	$CH_3NH_3^+ + H_2O \rightleftharpoons CH_3NH_2 + H_3O^+$	2.4×10^{-11}

Table 9.7. **Relative Strengths of Weak Bases**

Name of Base	Proton-Transfer Reaction	K_b at 25 °C
Methylamine	$CH_3NH_2 + H_2O \rightleftharpoons CH_3NH_3^+ + OH^-$	4.2×10^{-4}
Carbonate ion	$CO_3^{2-} + H_2O \rightleftharpoons HCO_3^- + OH^-$	2.1×10^{-4}
Cyanide ion	$CN^- + H_2O \rightleftharpoons HCN + OH^-$	2.5×10^{-5}
Ammonia	$NH_3 + H_2O \rightleftharpoons NH_4^+ + OH^-$	1.8×10^{-5}
Bisulfide ion	$HS^- + H_2O \rightleftharpoons H_2S + OH^-$	1.0×10^{-7}
Hydroxylamine	$H_2NOH + H_2O \rightleftharpoons {}^+H_3NOH + OH^-$	1.2×10^{-8}
Acetate ion	$CH_3COO^- + H_2O \rightleftharpoons CH_3COOH + OH^-$	5.6×10^{-10}
Sulfate ion	$SO_4^{2-} + H_2O \rightleftharpoons HSO_4^- + OH^-$	8.3×10^{-13}

Equation (9-24b) tells us that as K_a decreases, K_b for the conjugate base must increase, since K_w is a constant. It should therefore be clear that the order of the strengths of the conjugate bases is just the reverse of the order of the strengths of the weak acids. Thus methylamine is the strongest base of those listed, and sulfate ion is the weakest. The relative strengths of these weak bases are given in Table 9.7. The values of K_b given in Table 9.7 have been obtained by using Eq. (9-24b).

We see that the weaker an acid, the stronger its conjugate base. It is important to remember that we are discussing relative (that is, comparative) strengths. Ammonia is a weak base; its conjugate acid, ammonium ion, is a weak acid. Acetate ion is a weak base; its conjugate acid, acetic acid, is a weak acid. Since $K_b(OAc^-) < K_b(NH_3)$, acetate ion is a weaker base than ammonia, and therefore acetic acid is a stronger acid than ammonium ion.

If a solution contains both a strong acid and a weak acid, the principal source of hydronium ions will be the strong acid, and the ionization of the weak acid will be repressed. This is illustrated in Example 9.10.

EXAMPLE 9.10. Calculating the pH of a solution containing both a strong and a weak acid

A solution is prepared by dissolving 0.120 mol of acetic acid and 0.100 mol of HCl in enough water to make exactly 1 L of solution. Calculate the pH of this solution and the acetate ion concentration at equilibrium.

Solution. The following equilibrium is established in this solution:

$$CH_3COOH + H_2O \rightleftharpoons CH_3COO^- + H_3O^+$$

Let us consider the initial concentrations of these species, and the changes that occur on reaction.

	CH₃COOH	CH₃COO⁻	H₃O⁺
Initial concentration (M)	0.120	0	0.100
Change in concentration (M)	x	x	x
Equilibrium concentration (M)	$0.120 - x$	x	$0.100 + x$

There are two sources of H_3O^+ in this solution. The strong acid, HCl, is 100% ionized. In addition to the 0.100 M H_3O^+ from the strong acid, there is a contribution to the $[H_3O^+]$ from the dissociation of the acetic acid. The $[CH_3COOH]$ at equilibrium is slightly less than the initial 0.120 M because some of it has ionized.

The equilibrium constant expression for the dissociation of acetic acid is

$$K_a(HOAc) = 1.8 \times 10^{-5} = \frac{[H_3O^+][OAc^-]}{[HOAc]} = \frac{(0.100 + x)(x)}{(0.120 - x)}$$

Because we know that acetic acid ionizes only slightly, and that its ionization is repressed by the extra H_3O^+ ions from the HCl, we expect x to be considerably smaller than 0.100 M. Assume x is less than 10% of 0.100. It is then necessarily also less than 10% of 0.120. Thus $0.100 + x \sim 0.100$, and $0.120 - x \sim 0.120$. The equilibrium constant expression becomes

$$K_a(\text{HOAc}) = 1.8 \times 10^{-5} = \frac{(0.100)(x)}{0.120}$$

and

$$x = (1.2)(1.8 \times 10^{-5}) = 2.2 \times 10^{-5} \, M$$

Is 2.2×10^{-5} less than 10% of 0.100? Yes, $2.2 \times 10^{-5} < 1.0 \times 10^{-2}$. Hence the approximation made is valid.

$$[\text{CH}_3\text{COO}^-] = 2.2 \times 10^{-5} \, M$$

$$[H_3O^+] = 0.100 + 2.2 \times 10^{-5} = 0.100 \, M \quad \text{and} \quad \text{pH} = 1.00$$

The contribution to the total $[H_3O^+]$ from the ionization of acetic acid is negligible compared to the contribution from the strong acid. The pH of this solution is the same as that of 0.100 F HCl.

section 9.4
Weak Polyprotic Acids

There are several weak **diprotic acids,** that is, acids that have two acidic protons. Some of the more common examples are hydrogen sulfide (H_2S), oxalic acid ($H_2C_2O_4$), carbonic acid (H_2CO_3 or $H_2O + CO_2$), sulfurous acid (H_2SO_3 or $H_2O + SO_2$), and tartaric acid ($H_2C_4H_4O_6$). For weak diprotic acids there are two stages of ionization. The general form of a diprotic acid may be denoted H_2A, and the two stages of ionization are given as:

1st stage $\quad H_2A + H_2O \rightleftharpoons HA^- + H_3O^+ \quad K_1 = K_a(H_2A) \quad$ (9-25)

2nd stage $\quad HA^- + H_2O \rightleftharpoons A^{2-} + H_3O^+ \quad K_2 = K_a(HA^-) \quad$ (9-26)

For any diprotic acid, $K_2 < K_1$. This is always true because it is more difficult to remove a proton from a negatively charged species (HA^-) than from an uncharged molecule (H_2A). Sometimes the difference between K_1 and K_2 is very large, as is the case with H_2S, where $K_1 = K_a(H_2S) = 1 \times 10^{-7}$, and $K_2 = K_a(HS^-) = 1 \times 10^{-13}$.[*] For some other diprotic acids K_1 and K_2 do not differ greatly, although K_1 is always larger than K_2. For example, K_1 for tartaric acid is 1.1×10^{-3} and K_2 is 4.3×10^{-5}.

By combining Eqs. (9-25) and (9-26) we can write the two stages of ionization as one overall reaction:

$$H_2A + 2H_2O \rightleftharpoons A^{2-} + 2H_3O^+ \quad (9-27)$$

The equilibrium constant for the overall reaction can be shown to be the product of K_1 and K_2, as follows:

$$K_{\text{overall}} = \frac{[A^{2-}][H_3O^+]^2}{[H_2A]} = \frac{[HA^-][H_3O^+]}{[H_2A]} \cdot \frac{[H_3O^+][A^{2-}]}{[HA^-]}$$

$$= K_1 \cdot K_2 \quad (9-28)$$

---

[*] The value of $K_a(HS^-)$ is difficult to measure and one can find values from 10^{-12} to 10^{-15} in various reference sources. The 64th edition of the *Handbook of Chemistry and Physics* lists 1.1×10^{-12} at 18 °C, while Lange's Handbook, 12th edition, gives 1.3×10^{-13} at 25 °C.

An aqueous solution of a diprotic acid contains the species H_2A, HA^-, A^{2-}, H_3O^+, and OH^- as well as H_2O. Since both the first and second stages of ionization of H_2A produce H_3O^+ ions, the total concentration of H_3O^+ ions in solution is determined by the requirement that the solution must be electrically neutral, that is, there must be one H_3O^+ ion for each A^- in solution, and two H_3O^+ ions for each A^{2-} in solution. The electroneutrality condition is therefore*

$$[H_3O^+] = [A^-] + 2[A^{2-}] \tag{9-29}$$

Note that it is *not* correct simply to look at the stoichiometry of either Eq. (9-25), Eq. (9-26), or Eq. (9-27) *separately* to write an expression for $[H_3O^+]$. However, for a solution prepared by mixing only the species H_2A and H_2O, as long as the concentration of H_2A is not too low, the contribution of the second stage of ionization to the $[H_3O^+]$ is usually negligible relative to the contribution of the first stage of ionization, and only Eq. (9-25) need be considered in calculating the pH.

The overall equilibrium constant is useful if the concentrations of two of the three species (H_2A, H_3O^+, and A^{2-}) are known. Let us consider how Eq. (9-28) is used in aqueous solutions of H_2S that have been acidified with HCl. The overall reaction for the two stages of ionization of H_2S is

$$H_2S + 2H_2O \rightleftharpoons S^{2-} + 2H_3O^+ \tag{9-30}$$

and

$$K_{overall} = \frac{[H_3O^+]^2[S^{2-}]}{[H_2S]} = K_a(H_2S) \cdot K_a(HS^-) \tag{9-31}$$

$$= 1 \times 10^{-20}$$

When H_2S is used as a reagent in aqueous solution, experimental conditions are usually such that the solution remains saturated with H_2S at all times. The concentration of H_2S in a saturated aqueous solution open to the atmosphere at room temperature is $0.10\ M$. If a strong acid is added to the solution, the pH of the solution is determined by the concentration of the strong acid alone. We can then use Eq. (9-31) to determine the $[S^{2-}]$ in equilibrium with a saturated aqueous solution of H_2S at any specified pH. A typical calculation is illustrated in Example 9.11.

EXAMPLE 9.11. Use of the overall equilibrium constant for a weak diprotic acid

A solution is acidified with HCl so that the $[H_3O^+] = 0.30\ M$, and is then saturated with H_2S gas at 25 °C, so that $[H_2S] = 0.10\ M$. Calculate the $[S^{2-}]$ in this solution.

Solution. It is important to remember that the hydronium ion concentration produced by the two stages of ionization of H_2S is very small; both H_2S and HS^- are extremely weak acids, with acidity constants very much less than 1. Thus the $[H_3O^+]$ in this solution is essentially identical to that provided by the strong acid, HCl. All the terms in Eq. (9-31) are therefore known except for the $[S^{2-}]$, and we have

$$K_{overall} = 1 \times 10^{-20} = \frac{[H_3O^+]^2\,[S^{2-}]}{[H_2S]} = \frac{(0.30)^2\,[S^{2-}]}{0.10}$$

which yields

$$[S^{2-}] = \frac{1 \times 10^{-21}}{9 \times 10^{-2}} = 1 \times 10^{-20}\ M$$

---

* We are assuming here that the contribution to the total $[H_3O^+]$ from the self-ionization of H_2O is negligible relative to the contribution from the first and second stages of ionization of H_2A. This is almost always a good approximation. It fails only if K_1 is not much larger than K_w, or if the concentration of H_2A is very low.

Note that in any saturated solution of H_2S at 25 °C, since $[H_2S] = 0.10\ M$,

$$[H_3O^+]^2[S^{2-}] = 1 \times 10^{-21} \tag{9-32}$$

This relation is sometimes referred to as the **ion product of H_2S.** It tells us that as the pH increases, that is, as the solution becomes more basic, the $[S^{2-}]$ in equilibrium with a saturated solution of H_2S increases. The calculation of Example 9.11 shows that the $[S^{2-}]$ in a saturated solution of H_2S acidified with a strong acid is extremely small.

A more extensive discussion of methods used to calculate the concentrations of species present in solutions of diprotic acids and their salts will be found in the following chapter, Section 10.5.

In addition to the diprotic acids there are a number of triprotic acids, of which the most common is orthophosphoric acid, H_3PO_4.* For a triprotic acid there are three acidic protons, three stages of ionization, and hence three acidity constants. For all triprotic acids, $K_1 > K_2 > K_3$. For phosphoric acid, the three stages of ionization and their equilibrium constants are

$$H_3PO_4 + H_2O \rightleftharpoons H_2PO_4^- + H_3O^+ \tag{9-33}$$

$$K_1 = \frac{[H_3O^+][H_2PO_4^-]}{[H_3PO_4]} = 7.5 \times 10^{-3}$$

$$H_2PO_4^- + H_2O \rightleftharpoons HPO_4^{2-} + H_3O^+ \tag{9-34}$$

$$K_2 = \frac{[HPO_4^{2-}][H_3O^+]}{[H_2PO_4^-]} = 6.2 \times 10^{-8}$$

$$HPO_4^{2-} + H_2O \rightleftharpoons PO_4^{3-} + H_3O^+ \tag{9-35}$$

$$K_3 = \frac{[PO_4^{3-}][H_3O^+]}{[HPO_4^{2-}]} = 3.6 \times 10^{-13}$$

An aqueous solution of a triprotic acid, H_3A, or of any of its salts, contains the species H_3A, H_2A^-, HA^{2-}, and A^{3-}, in addition to H_3O^+ and OH^-. The three equilibrium constants, K_1, K_2, and K_3, must all be satisfied simultaneously. However, the calculation of the concentrations of the species present in solutions of many triprotic acids can be greatly simplified by making suitable approximations, because usually only two of the successive species are present in appreciable concentration in any solution.

section 9.5
Acidic, Basic, and Neutral Salts

Arrhenius classified electrolytes (neutral molecules or ionic crystalline solids) into three categories: **acids, bases,** and **salts.** (Refer to Section 7.3.) In Arrhenius' classification system, the acids are neutral molecules such as HCl, H_2S, CH_3COOH (HOAc), or HNO_3, whose cation is H^+(aq) in aqueous solution. The bases are the uncharged crystalline solids such as $NaOH$, KOH, and $Ba(OH)_2$, whose anion is OH^-, or neutral molecules like NH_3 and CH_3NH_2, which form OH^- on being dissolved in water. **Salts** are electrolytes whose cation is not H^+ and whose anion is not OH^-.

Practically all salts are strong electrolytes, so that aqueous solutions of salts are solutions of two (or more) different ions. Using the **Brønsted–Lowry** definitions of

---

* There are three phosphoric acids: orthophosphoric acid, H_3PO_4; metaphosphoric acid, HPO_3; and pyrophosphoric acid, $H_4P_2O_7$. Orthophosphoric acid is the most common of the three, and is what is usually meant by the term "phosphoric acid."

acids and bases (Section 7.4) many ions are themselves either acids or bases, and we will therefore find salts that are acidic and salts that are basic, as well as salts that are neutral.

Acidic Salts

An **acidic salt** is a salt whose aqueous solution has a pH below 7 at 25 °C. Ammonium chloride, NH_4Cl, and ammonium nitrate, NH_4NO_3, are common examples of acidic salts. A solution of ammonium chloride contains NH_4^+ ions and Cl^- ions. Considering the **Brønsted–Lowry** definitions of acid and base we ask: Is either of these ions a proton donor? Is either a proton acceptor? The ammonium ion is a proton donor and transfers a proton to water:

$$NH_4^+ + H_2O \rightleftharpoons NH_3 + H_3O^+ \tag{9-36}$$

This reaction increases the $[H_3O^+]$ in solution. Chloride ions have virtually no tendency to accept protons in dilute aqueous solution because HCl is a strong acid, 100% ionized.

Therefore we say that a solution of NH_4Cl is acidic because it contains NH_4^+ ions, and NH_4^+ is a weak acid. The equilibrium constant for reaction (9-36) is very small, 5.7×10^{-10}, so that only a small fraction of NH_4^+ ions transfer protons to water, but the reaction makes the hydronium ion concentration in solution larger than the hydroxide ion concentration, and a solution of NH_4Cl is acidic.

The salt NH_4Cl is the product of the reaction between ammonia, a weak base, and HCl, a strong acid. The pure species react as follows:

$$NH_3(g) + HCl(g) \rightleftharpoons NH_4Cl(s) \tag{9-37}$$

The net ionic equation for the reaction between an aqueous solution of ammonia and of hydrochloric acid is

$$NH_3 + H_3O^+ \rightleftharpoons NH_4^+ + H_2O \tag{9-38}$$

Thus this reaction produces an aqueous solution of NH_4^+ and Cl^- ions, identical to a solution of the salt NH_4Cl, provided HCl and NH_3 are mixed in a 1 : 1 molar ratio.

Reaction between a weak base and a strong acid will always produce an acidic salt, that is, a salt whose cation is a weak acid.

EXAMPLE 9.12. Calculation of the pH of a solution of an acidic salt

Calculate the $[NH_4^+]$, $[Br^-]$, $[NH_3]$, $[H_3O^+]$, $[OH^-]$, and pH of a 0.300 F NH_4Br solution.

Solution. Ammonium bromide, a strong electrolyte, is 100% ions in aqueous solution. Bromide ions do not react with water. The Br^- ion has virtually no tendency to combine with H^+, because HBr is a strong acid. Thus $[Br^-] = 0.300$ M in this solution. Ammonium ion is an acid, and donates a proton to water, as in Eq. (9-36).

Let $x = [NH_3] = [H_3O^+]$ at equilibrium in 0.300 F NH_4Br. It is useful to choose as the unknown the concentration of a species present in very small concentration, as this makes it easier to make approximations that greatly simplify the mathematical solution of the equilibrium constant equation. The concentrations of ammonia and hydronium ions are equal because these two substances are formed in a 1 : 1 molar ratio by reaction (9-36). Let us tabulate the initial concentrations and the changes in concentration of the three species involved in the equilibrium of Eq. (9-36).

	NH_4^+	NH_3	H_3O^+
Initial concentration (M)	0.300	0	0
Change in concentration (M)	x	x	x
Equilibrium concentration (M)	$0.300 - x$	x	x

Remember that the concentration of NH_4^+ ions present at equilibrium is the initial concentration *minus* the concentration that is used up in reaction (9-36).

The equilibrium constant for Eq. (9-36) is the acidity constant of NH_4^+. The numerical value of $K_a(NH_4^+)$ is given in Table 9.6, and we therefore write

$$K_a(NH_4^+) = 5.7 \times 10^{-10} = \frac{[NH_3][H_3O^+]}{[NH_4^+]} = \frac{x^2}{0.300 - x}$$

Assume that x is small compared to 0.300, that is, that x is less than 10% of 0.300. Then

$$\frac{x^2}{0.300} = 5.7 \times 10^{-10}$$

and $x^2 = 1.7 \times 10^{-10}$, so that $x = 1.3 \times 10^{-5} M$.

We must now check our assumption. Is x less than 10% of 0.300? Yes, 1.3×10^{-5} is less than 3.00×10^{-2}. Hence

$$[NH_3] = [H_3O^+] = 1.3 \times 10^{-5} M$$
$$[NH_4^+] = 0.300 - 1.3 \times 10^{-5} = 0.300 M$$

We obtain the $[OH^-]$ from the relation

$$[OH^-] = \frac{K_w}{[H_3O^+]} = \frac{1.0 \times 10^{-14}}{1.3 \times 10^{-5}} = 7.7 \times 10^{-10} M$$

The pH $= -\log[H_3O^+] = 5 - \log 1.3 = 4.89$.

Note that the principal species in 0.300 F NH_4Br are, of course, NH_4^+ ions and Br^- ions. Both of these ions have concentrations of 0.300 M. The concentrations of NH_3 and H_3O^+ are small compared to the concentrations of the NH_4^+ and Br^- ions; nevertheless the solution is distinctly acidic, with pH $= 4.89$.

Basic Salts

A **basic salt** is a salt whose aqueous solution has a pH greater than 7 at 25 °C. There are a great many basic salts, since there are many anions that are weak bases. (See Table 9.7.) Remember that a base is a proton acceptor, and many anions combine with H^+. All anions except for the anions of strong acids accept a proton from water in aqueous solution. Since most acids are weak electrolytes (see Table 7.3), the great majority of anions are bases. Some common basic salts are Na_2CO_3, K_2S, $KOAc$, Na_3PO_4, Na_2SO_3, and $Ca(NO_2)_2$.

Reaction between a weak acid and a strong base will always produce a basic salt, that is, a salt whose anion is a weak base. For instance, passing gaseous H_2S into a solution of sodium hydroxide produces a solution containing Na^+ ions and S^{2-} ions, that is, a solution of the salt sodium sulfide. The net ionic equation for that reaction is

$$H_2S + 2\,OH^- \rightleftharpoons S^{2-} + 2H_2O \tag{9-39}$$

The calculation of the pH of a solution of a basic salt is essentially the same as the calculation of the pH of a solution of an acidic salt, except that we use K_b for the weak anion base where (as in Example 9.12) we used K_a for the weak cation acid.

EXAMPLE 9.13. Calculation of the pH of a solution of a basic salt

Calculate the $[Na^+]$, $[OAc^-]$, $[HOAc]$, $[H_3O^+]$, $[OH^-]$, and pH of a solution of 0.045 F NaOAc at 25 °C.

Solution. Sodium ions and acetate ions are the principal species in an aqueous solution of sodium acetate. Sodium ions do not react with water, so that $[Na^+] = 0.045\ M$ in 0.045 F NaOAc.

Acetate ion is a weak base, and accepts a proton from water:

$$OAc^- + H_2O \rightleftharpoons HOAc + OH^-$$

Since $K_b(OAc^-) = 5.6 \times 10^{-10}$ (Table 9.7), the proton-transfer reaction between OAc^- and H_2O does not proceed very far to the right, and we expect the concentration of HOAc molecules and OH^- ions in this solution to be low.

Let $x = [HOAc] = [OH^-]$ at equilibrium in 0.045 F NaOAc at 25 °C. Then

$$[OAc^-] = 0.045 - x$$

because one acetate ion is used up to form one HOAc molecule and one OH^- ion. Substitution of these expressions into the basicity constant of acetate ion yields

$$K_b(OAc^-) = 5.6 \times 10^{-10} = \frac{[HOAc][OH^-]}{[OAc^-]} = \frac{x^2}{0.045 - x}$$

Assume x is small compared to 0.045 M, that is, $0.045 - x \sim 0.045$. The equilibrium constant equation then becomes

$$5.6 \times 10^{-10} = \frac{x^2}{0.045} \quad \text{so that} \quad x^2 = (0.045)(5.6 \times 10^{-10}) = 25 \times 10^{-12}$$
$$\text{and} \quad x = 5.0 \times 10^{-6}\ M$$

We must now check the approximation made. Is x less than 10% of 0.045? Yes, 5.0×10^{-6} is less than 4.5×10^{-3}. Hence

$$[HOAc] = [OH^-] = 5.0 \times 10^{-6}\ M$$
$$[OAc^-] = 0.045 - 5.0 \times 10^{-6} = 0.045\ M$$

We obtain the $[H_3O^+]$ from the relation

$$[H_3O^+] = \frac{K_w}{[OH^-]} = \frac{1.0 \times 10^{-14}}{5.0 \times 10^{-6}} = 2.0 \times 10^{-9}\ M$$

Hence,

$$pH = -\log(2.0 \times 10^{-9}) = 9 - \log 2.0 = 8.70$$

Neutral Salts

If neither the cation nor the anion of a salt has any tendency to either accept or donate a proton, the salt is a neutral salt. This means that the anion must be the anion of a monoprotic strong acid, most commonly Cl^-, Br^-, I^-, NO_3^-, or ClO_4^-. These anions have virtually no tendency to combine with H^+ in dilute aqueous solution. The cation of a neutral salt is the cation of a strong base, that is, an alkali metal cation or an alkaline earth cation. Examples of neutral salts are $NaNO_3$, KCl, $CsBr$, BaI_2, $Ca(NO_3)_2$, and $KClO_4$.

The reaction between a strong base and a strong acid produces a neutral salt. If we mix aqueous solutions of NaOH and HCl containing equal numbers of moles of each of these substances, the resultant solution contains H_2O, Na^+ ions, and Cl^- ions, and is in every way identical with an aqueous solution of the neutral salt, NaCl.

EXAMPLE 9.14. The relative acidity of various solutions

Arrange the following solutions in order of increasing pH: **(a)** 0.200 F NH_4NO_3, **(b)** 0.200 F KNO_2, **(c)** 0.200 F KNO_3, **(d)** 0.200 F $HClO_4$, **(e)** 0.200 F Na_2CO_3, **(f)** 0.200 F HNO_2.

Solution. It is helpful first to classify each of these solutions as either acidic, basic, or neutral. There are three acidic solutions:

(a) 0.200 F NH_4NO_3	NH_4^+ ion is a weak acid
(d) 0.200 F $HClO_4$	perchloric acid is a strong acid
(f) 0.200 F HNO_2	nitrous acid is a weak acid

There are two basic solutions:

(b) 0.200 F KNO_2	NO_2^- ion is a weak base
(e) 0.200 F Na_2CO_3	CO_3^{2-} ion is a weak base

The remaining solution, **(c)** 0.200 F KNO_3, is neutral. Neither K^+ ions nor NO_3^- ions either accept or donate protons to water. Another way of recognizing that KNO_3 is a neutral salt is to note that KNO_3 is produced by the reaction of a strong base (KOH) and a strong acid (HNO_3).

To decide on the relative acidities of the two weak acids, we must look up their K_a values in Table E1, Appendix E. Nitrous acid is a stronger acid than ammonium ion, since $K_a(HNO_2) = 4.5 \times 10^{-4}$ is significantly larger than $K_a(NH_4^+) = 5.7 \times 10^{-10}$.

To decide on the relative basicities of the two weak bases, we must calculate their K_b values from the K_a values of their conjugate weak acids, using Eq. (9-24b). The conjugate acid of NO_2^- ion is nitrous acid, HNO_2. Thus

$$K_b(NO_2^-) = \frac{K_w}{K_a(HNO_2)} = \frac{1.0 \times 10^{-14}}{4.5 \times 10^{-4}} = 2.2 \times 10^{-11}$$

The conjugate acid of CO_3^{2-} is HCO_3^-, bicarbonate ion.

$$K_b(CO_3^{2-}) = \frac{K_w}{K_a(HCO_3^-)} = \frac{1.0 \times 10^{-14}}{4.7 \times 10^{-11}} = 2.1 \times 10^{-4}$$

Thus carbonate ion is a considerably stronger base than nitrite ion.

We can now arrange these six solutions in order of increasing pH, beginning with the most acidic, the strong acid, $HClO_4$.

lowest pH (most acidic)	**(d)** 0.200 F $HClO_4$	strong acid
	(f) 0.200 F HNO_2	$K_a(HNO_2) = 4.5 \times 10^{-4}$
	(a) 0.200 F NH_4NO_3	$K_a(NH_4^+) = 5.7 \times 10^{-10}$
	(c) 0.200 F KNO_3	neutral
	(b) 0.200 F KNO_2	$K_b(NO_2^-) = 2.2 \times 10^{-11}$
highest pH (most basic)	**(e)** 0.200 F Na_2CO_3	$K_b(CO_3^{2-}) = 2.1 \times 10^{-4}$

Salts of Both a Weak Acid and a Weak Base

Consider the reaction between acetic acid and ammonia:

$$CH_3COOH + NH_3 \rightleftharpoons CH_3COO^- + NH_4^+ \qquad (9\text{-}40)$$

If equimolar amounts of CH_3COOH and NH_3 are reacted in aqueous solution, a solution of the salt ammonium acetate is produced. The cation, NH_4^+, is a weak acid, and the anion, CH_3COO^-, is a weak base. Other examples of salts of both a weak acid and a weak base are ammonium cyanide, NH_4CN; methylammonium sulfide, $(CH_3NH_3)_2S$; and ammonium oxalate, $(NH_4)_2C_2O_4$. In general, the cation of such salts is a weak acid and the anion is a weak base.

A salt of a weak base and a weak acid is basic if $K_b(\text{anion}) > K_a(\text{cation})$, and acidic if $K_a(\text{cation}) > K_b(\text{anion})$. For instance, a solution of ammonium cyanide is basic because $K_b(CN^-) > K_a(NH_4^+)$. From Tables 9.6 and 9.7 we obtain $K_b(CN^-) = 2.5 \times 10^{-5}$ and $K_a(NH_4^+) = 5.7 \times 10^{-10}$. Similarly, solutions of methylammonium sulfide are basic because $K_b(S^{2-}) = 7.7 \times 10^{-2}$ (as calculated in Example 9.9), and $K_a(CH_3NH_3^+) = 2.4 \times 10^{-11}$. In summary, for salts that are formed by the reaction between a weak acid and a weak base,

> If $K_a(\text{cation}) > K_b(\text{anion})$ the solution is acidic.
> If $K_a(\text{cation}) = K_b(\text{anion})$ the solution is neutral.
> If $K_a(\text{cation}) < K_b(\text{anion})$ the solution is basic.

Ampholytes

A substance that can act either as an acid or a base is called an **ampholyte,** and such behavior is called **amphiprotic** behavior. We have already considered the amphiprotic behavior of water, but there are many other species that have this property as well. Both bisulfide ion (HS^-) and bicarbonate ion (HCO_3^-) are ampholytes. For instance, bisulfide ion acts as an acid in the reaction

$$HS^- + H_2O \rightleftharpoons S^{2-} + H_3O^+ \qquad K_a(HS^-) = 1.3 \times 10^{-13} \qquad (9\text{-}41)$$

and as a base in the reaction

$$HS^- + H_2O \rightleftharpoons H_2S + OH^- \qquad K_b(HS^-) = 1 \times 10^{-7} \qquad (9\text{-}42)$$

The basicity constant for HS^- is obtained from the acidity constant of its conjugate acid, H_2S, using Eq. (9-24b):

$$K_b(HS^-) = K_w/K_a(H_2S) = 1.0 \times 10^{-14}/1 \times 10^{-7} = 1 \times 10^{-7}$$

Since HS^- is an ampholyte, will solutions of potassium hydrogen sulfide, KHS, be acidic or basic? To answer that question we need only examine the relative magnitudes of $K_a(HS^-)$ and $K_b(HS^-)$. Since $K_b(HS^-)$ is ten million times larger than $K_a(HS^-)$, the basicity reaction proceeds to a much greater extent than does the acidity reaction. Solutions of KHS or NaHS are therefore basic.

The anion HA^- formed when any diprotic acid, H_2A, loses a proton is an ampholyte. Such a species can accept a proton to form H_2A or can donate a proton to form A^{2-}. In order to determine whether aqueous solutions of salts of these amphiprotic anions are acidic or basic, we must compare the relative magnitudes of K_b and K_a, for the *same species, HA^-.*

Are solutions of sodium binoxalate, $NaHC_2O_4$, acidic or basic? We use Table E1, Appendix E, for the K_a values needed.

$$K_a(HC_2O_4^-) = 6.4 \times 10^{-5}$$
$$K_b(HC_2O_4^-) = K_w/K_a(H_2C_2O_4) = 1.7 \times 10^{-13}$$

Since $K_a(HC_2O_4^-)$ is greater than $K_b(HC_2O_4^-)$, solutions of sodium binoxalate are acidic.

In order to find out whether solutions of sodium bicarbonate, $NaHCO_3$, are acidic or basic, we must look up the acidity constant of HCO_3^- and calculate its basicity constant. What is the conjugate acid of HCO_3^- ion? When HCO_3^- accepts a proton from water the molecule H_2CO_3 (carbonic acid) is formed, but this molecule is unstable and is largely decomposed into CO_2 plus H_2O. What we call carbonic acid is really an aqueous solution of carbon dioxide (ordinary soda water or seltzer!). Thus the conjugate acid of HCO_3^- is carbonic acid or aqueous CO_2. We combine the two equations

$$HCO_3^- + H_2O \rightleftharpoons H_2CO_3 + OH^- \tag{9-43}$$

and

$$H_2CO_3 \rightleftharpoons CO_2 + H_2O \tag{9-44}$$

to give the overall net ionic equation

$$HCO_3^- \rightleftharpoons CO_2 + OH^- \tag{9-45}$$

as the basicity reaction for bicarbonate ion. Therefore,

$$K_b(HCO_3^-) = \frac{[CO_2][OH^-]}{[HCO_3^-]} = \frac{[CO_2][OH^-][H_3O^+]}{[HCO_3^-][H_3O^+]} = \frac{K_w}{K_a(H_2O + CO_2)}$$
$$= \frac{1.0 \times 10^{-14}}{4.3 \times 10^{-7}} = 2.3 \times 10^{-8}$$

From Table E1, Appendix E, we obtain $K_a(HCO_3^-) = 4.7 \times 10^{-11}$. Since $K_b(HCO_3^-)$ is larger than $K_a(HCO_3^-)$, solutions of $NaHCO_3$ are basic. Sodium bicarbonate is commonly called bicarbonate of soda and is also known as baking soda. It is used both as a mild antacid for upset stomachs, and as a source of CO_2 in baking, where the CO_2 gas causes the dough to rise.

Acidic and Basic Oxides

Oxygen combines with almost every other element, and the acidity or basicity of oxides merits special consideration. Oxides of the active metals (Groups IA and IIA) are ionic crystalline salts. Crystalline Na_2O, sodium oxide, is composed of Na^+ ions and O^{2-} ions. Metallic oxides are basic, although oxide ions do not exist in aqueous solution. When Na_2O is dissolved in water the reaction

$$Na_2O(s) + H_2O \rightarrow 2Na^+ + 2\,OH^- \tag{9-46}$$

occurs. Thus the solution is identical with a solution of sodium hydroxide. Similar reactions occur for other alkali metal oxides and the alkaline earth oxides.

Oxides of the nonmetals, on the other hand, consist of discrete molecules, many of which are gases at room temperature. Such nonmetallic oxides are acidic, and the more electronegative the element, the more acidic its oxide will be. Some nonmetallic oxides are **acidic anhydrides**, that is, they react with water to give an acidic solution. Examples are the gases CO_2, SO_3, SO_2, N_2O_3, and the volatile white solid N_2O_5. Dinitrogen pentoxide dissolves in water to form nitric acid:

$$N_2O_5 + H_2O \rightarrow 2H^+(aq) + 2NO_3^-(aq) \tag{9-47}$$

Dinitrogen trioxide, an unstable red-brown gas, reacts with water to give nitrous acid, HONO (also written as HNO_2), which is a weak acid. The reaction is

$$N_2O_3(g) + H_2O \rightleftharpoons 2HONO \tag{9-48}$$

Sulfur trioxide, SO_3, is the anhydride of the strong acid H_2SO_4:

$$SO_3(g) + H_2O \rightleftharpoons 2H^+(aq) + SO_4^{2-}(aq) \tag{9-49}$$

while sulfur dioxide, SO_2, is the anhydride of sulfurous acid, a weak acid:

$$SO_2(g) + H_2O \rightleftharpoons H_2SO_3 \tag{9-50}$$

The anhydride of orthophosphoric acid is P_4O_{10}:*

$$P_4O_{10}(s) + 6H_2O \rightleftharpoons 4H_3PO_4 \tag{9-51}$$

The acidity of an oxide depends on one other property of the element combined with oxygen, in addition to its electronegativity. This property is the oxidation state, which will be discussed in Chapter 15. We will therefore return to a discussion of the acidity and basicity of oxides when we discuss the oxidation states of various elements in Chapter 15.

EXAMPLE 9.15. Classifying salts as acidic, basic, or neutral

Classify the following salts as acidic, basic, or neutral. Explain your reasoning. (a) $(NH_4)_2CO_3$, (b) $KClO_4$, (c) $KHSO_4$, (d) BaO, (e) $NaHSO_3$, (f) $CH_3NH_3NO_3$.

Solution

(a) Ammonium carbonate is a basic salt. Its cation, NH_4^+, is a weak acid; $K_a(NH_4^+) = 5.7 \times 10^{-10}$. Its anion, CO_3^{2-}, is a weak base.

$$K_b(CO_3^{2-}) = \frac{K_w}{K_a(HCO_3^-)} = \frac{1.0 \times 10^{-14}}{4.7 \times 10^{-11}} = 2.1 \times 10^{-4}$$

Since $K_b(CO_3^{2-}) > K_a(NH_4^+)$, solutions of ammonium carbonate are basic.

(b) Potassium perchlorate, $KClO_4$, is a neutral salt. Neither K^+ nor ClO_4^- has any tendency to donate or accept a proton. Reaction of the strong base KOH with the strong acid $HClO_4$ produces $KClO_4$.

(c) Potassium hydrogen sulfate (also called potassium bisulfate) is an acidic salt. The anion, HSO_4^-, is a moderately strong acid with $K_a = 1.2 \times 10^{-2}$. The cation K^+ is neither an acid nor a base.

(d) Barium oxide, BaO, is a basic salt. Oxides of the alkaline earth metals are basic. When dissolved in water, the reaction

$$BaO(s) + H_2O \rightleftharpoons Ba^{2+} + 2\,OH^-$$

occurs.

(e) The anion HSO_3^- is an ampholyte.

$$K_a(HSO_3^-) = 6.2 \times 10^{-8}$$
$$K_b(HSO_3^-) = K_w/K_a(H_2SO_3) = 1.0 \times 10^{-14}/1.2 \times 10^{-2} = 8.3 \times 10^{-13}$$

Since $K_a(HSO_3^-) > K_b(HSO_3^-)$, solutions of $NaHSO_3$ (sodium bisulfite or sodium hydrogen sulfite) are acidic.

(f) Methylammonium nitrate, $CH_3NH_3NO_3$, is an acidic salt. Its cation, $CH_3NH_3^+$, is a weak acid. Its anion, NO_3^-, has virtually no tendency to accept a proton in dilute aqueous solution.

---

*For a long time, the compound P_4O_{10} was considered to be P_2O_5, and was called phosphorus pentoxide. That common name is still used.

section 9.6
The Leveling Effect of Water on the Strengths of Strong Acids and Bases

In aqueous solution all strong acids donate a proton to water and are essentially 100% ionized to produce a solution containing hydronium ions plus the anions of the strong acid. For this reason we cannot distinguish among the strengths of the strong acids $HClO_4$, H_2SO_4, HCl, HBr, HI, and HNO_3, if we deal only with aqueous solutions. Aqueous solutions of all these strong acids contain only the same acid, namely hydronium ion. Water is said to have a **leveling effect** on the strengths of all very strong acids; their acidities are reduced to the level of the hydronium ion. Hydronium ion is the strongest acid that can exist in any significant concentration in aqueous solution.

Any acid with an acidity constant greater than 1 is a strong acid. By investigating the reactions of the strong acids in solvents less basic than water, such as any of the alcohols, we can obtain information about their relative strengths. These studies show that perchloric acid, $HClO_4$, is the strongest of all common acids. Figure 9.2 summarizes the information obtained about the relative strengths of the common strong acids.

Just as the hydronium ion is the strongest acid that can exist in aqueous solution, hydroxide ion is the strongest base that can exist in aqueous solution. Bases stronger than OH^- accept a proton from water to produce OH^-. We have already discussed three bases stronger than OH^- that cannot exist in aqueous solution, namely oxide ion, O^{2-}; amide ion, NH_2^-; and ethoxide ion, $C_2H_5O^-$.

Sodium amide, $NaNH_2$, is a crystalline ionic compound prepared by reacting sodium with pure liquid ammonia. If sodium amide is added to water there is a vigorous reaction that produces gaseous ammonia and a basic solution of sodium hydroxide with NH_3 dissolved in it. The equation for this reaction is Eq. (9-3). We have already discussed the fact that salts containing the oxide ion, such as Na_2O, K_2O, and BaO react with water to form OH^-, as in Eq. (9-46).

Another example of a base stronger than OH^- that reacts with H_2O to form OH^- is the hydride ion, H^-. Hydrides such as NaH, KH, and CaH_2 react with water to form $H_2(g)$ and OH^-:

$$CaH_2(s) + 2H_2O \rightarrow Ca^{2+} + 2H_2(g) + 2 OH^- \qquad (9\text{-}52)$$

These reactions are exothermic and can be explosive. Caution should always be exercised in handling metallic hydrides.

The relative strengths of bases stronger than hydroxide ion can be determined by studying reactions in nonaqueous solutions, using solvents more basic than water, such as liquid ammonia. Figure 9.3 summarizes the relative strengths of some bases

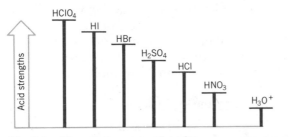

Fig. 9.2. The relative strengths of the common strong acids and the leveling effect of water on them. An aqueous solution of any of these strong acids contains only hydronium ions and the anions of the strong acid. All strong acids are leveled to the strength of the hydronium ion in aqueous solution.

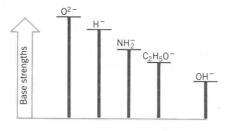

Fig. 9.3. The relative strengths of bases stronger than hydroxide ion and the leveling effect of water on them. All bases stronger than hydroxide ion are leveled to the strength of the OH^- ion in aqueous solution.

stronger than OH^-. All bases stronger than OH^- react with water to form OH^- ion. We describe this by saying that all bases stronger than hydroxide ion are leveled by water to the strength of the OH^- ion in aqueous solution.

A knowledge of the relative strengths of acids stronger than hydronium ion and of bases stronger than hydroxide ion is important in many reactions of organic chemistry that require a strong acid or base as a catalyst and must be run in nonaqueous solution.

Summary

Any aqueous solution contains both H_3O^+ and OH^- ions due to the **self-ionization** of water. The equilibrium constant for the proton-transfer reaction between two H_2O molecules is called the **ion product** of water, and is denoted K_w. The numerical value $K_w = [H_3O^+][OH^-] = 1.0 \times 10^{-14}$ at 25 °C.

Depending on what other substances are dissolved in water, the $[H_3O^+]$ and $[OH^-]$ in aqueous solutions may be as large as 10 or 15 M, or as small as 10^{-14} or 10^{-15} M. For convenience in discussing this wide range of values, the **pH scale** is utilized. The pH of a solution is defined as pH $= -\log[H^+(aq)]$. This concept is so useful it has been generalized to define pX as $-\log$ X.

The equilibrium constant for the proton-transfer reaction between a weak acid and water is called an **acidity constant,** and is denoted K_a. If the weak acid is a neutral molecule, its acidity constant is also called the **ionization constant** or the **dissociation constant** of the weak acid.

The equilibrium constant for the proton-transfer reaction between a weak base and water is called a **basicity constant** and is denoted K_b. If the weak base is a neutral molecule, its basicity constant is also called the **ionization constant** or the **dissociation constant** of the weak base.

For an aqueous solution of a weak acid with formality C, the **degree of dissociation** or the **fraction ionized,** α, is defined as $\alpha = [H_3O^+]/C$. For an aqueous solution of a weak base with formality C, the degree of dissociation or the fraction ionized, α, is defined as $\alpha = [OH^-]/C$. For both weak acids and weak bases $0 < \alpha < 1$.

The expression for K_a of a weak molecular acid or K_b of a weak molecular base in terms of the degree of dissociation, α, and the formality of the weak electrolyte, C, is known as the **Ostwald Dilution Law.** Consideration of the expression $K = \alpha^2 C/(1 - \alpha)$ leads to the conclusion that the more dilute the solution, the greater the degree of dissociation.

A base and an acid related by the expression

$$\text{base} + H^+ = \text{acid}$$

are called a **conjugate acid–base pair.** For any such pair, the relation

$$K_a(\text{weak acid}) \cdot K_b(\text{conjugate weak base}) = K_w$$

is valid. For this reason, it is customary to tabulate only acidity constants, and to calculate basicity constants from K_w and K_a of the conjugate acid. Since K_w is a constant, the weaker an acid, the stronger its conjugate base. The weak acids can be arranged in order of their relative strengths simply by listing them in order of decreasing acidity constant values. The larger the value of K_a, the stronger the acid. Similarly, the larger the value of K_b, the stronger the base.

Acids that have more than one proton that can be donated to a base are called **polyprotic acids.** These can be subdivided into **diprotic acids,** with two acidic protons, **triprotic acids,** with three acidic protons, and so on. For all diprotic acids the equilibrium constant, K_2, for the second stage of ionization is smaller than the equilibrium constant, K_1, for the first stage of ionization. The equilibrium constant for the overall two-step ionization is the product of K_1 and K_2: $K_{overall} = K_1 K_2$.

A salt is an electrolyte with a cation that is not H^+ and an anion that is not OH^-. Salts may be either acidic, basic, or neutral. A salt formed when a strong acid and a strong base react is a neutral salt. A salt formed when a strong acid and a weak base react is an acidic salt. A salt formed when a weak acid and a strong base react is a basic salt.

An **ampholyte** is a substance that can act both as an acid and as a base. Any anion of general form HA^- that can accept a proton to form the weak diprotic acid H_2A, or donate a proton to form the weak base A^{2-}, is an ampholyte. In order to determine whether aqueous solutions containing ampholytes are acidic, basic, or neutral, K_a and K_b for the ampholyte must be compared.

Oxides of metals are basic salts. The oxide anion is a stronger base than hydroxide ion, and therefore oxide ions cannot exist in aqueous solution. Oxide ions react with water to form OH^-. Dissolving an alkali metal oxide or an alkaline earth oxide in water produces a solution of the metallic cation and OH^-.

Oxides of the nonmetals are not salts, but molecular compounds, and many are gases at room temperature. Nonmetallic oxides are acidic, and in general the more electronegative an element, the more acidic its oxide will be.

No acid stronger than H_3O^+ can exist in water, because such a substance reacts with H_2O to produce H_3O^+ ions. No base stronger than OH^- can exist in water, because such a substance reacts with H_2O to produce OH^-. This phenomenon is known as the **leveling effect** of water on strong acids and bases.

Exercises

Section 9.1

1. Calculate the $[H_3O^+]$ and $[OH^-]$ at 25 °C in
   (a) 0.250 F $HClO_4$ (b) 0.160 F KOH
2. Calculate the pH at 25 °C of the following solutions:
   (a) 1.40 F HBr   (b) 0.30 F HNO_3 (c) 4.58×10^{-3} F $HClO_4$
3. Calculate the pH at 25 °C of the following solutions:
   (a) 1.25 F NaOH   (b) 0.37 F KOH   (c) 6.10×10^{-4} F NaOH
   (d) 0.045 F $Ba(OH)_2$
4. Calculate the pH and pOH of the following solutions at 25 °C:
   (a) 1.12 F HCl   (b) 4.74×10^{-3} F KOH   (c) 6.25×10^{-1} F NaOH
5. At 25 °C, the pH of a solution of lemon juice is 2.32. What are the $[H_3O^+]$ and $[OH^-]$ in this solution?

6. Calculate the concentrations of hydronium and hydroxide ions in a solution of washing soda ($Na_2CO_3 \cdot 10H_2O$) at 25 °C that has a pH $= 11.54$.

7. Determine the $[H^+(aq)]$ and pOH values of solutions at 25 °C with pH
   (a) 4.7   (b) 11.22   (c) -0.30

Section 9.2

8. Write the correctly balanced net ionic equation for the reaction whose equilibrium constant is
   (a) $K_a(C_6H_5COOH) = 6.3 \times 10^{-5}$ (b) $K_b(CH_3NH_2) = 3.7 \times 10^{-4}$
   (c) $K_a(H_2C_2O_4) = 5.4 \times 10^{-2}$ (d) $K_a(HSO_3^-) = 2.8 \times 10^{-7}$
   (e) $K_b(OCl^-) = 9.1 \times 10^{-7}$

9. Calculate the pH of a 0.250 F CH_3COOH solution at 25 °C.

10. Calculate the $[H_3O^+]$ and $[OH^-]$ of 0.370 F NH_3 at 25 °C.

11. Calculate the pH of 0.45 F CH_3NH_2 at 25 °C.

12. Calculate the $[H_3O^+]$ and $[OH^-]$ of 0.520 F HCN at 25 °C.

13. Calculate the degree of dissociation of 0.038 F C_6H_5COOH at 25 °C.

Section 9.3

Use Table E1, Appendix E, for K_a values.

14. State the formula and name of the conjugate base of each of the following acids:
    (a) H_3O^+ (b) HSO_4^- (c) NH_4^+ (d) HF   (e) H_3PO_4

15. State the formula and name of the conjugate acid of each of the following bases:
    (a) OH^- (b) HPO_4^{2-} (c) CH_3NH_2 (d) CO_3^{2-} (e) HS^-

16. Calculate the basicity constant of each of the following bases:
    (a) CN^- (b) CH_3COO^- (c) CH_3NH_2 (d) PO_4^{3-}

17. Arrange the following acids in order of decreasing acid strength:
    (a) benzoic acid   (b) formic acid   (c) hydriodic acid   (d) oxalic acid
    (e) hydrofluoric acid

18. Arrange the following bases in order of decreasing basicity:
    (a) S^{2-} (b) F^- (c) CH_3COO^- (d) CN^- (e) NH_3

Section 9.4

19. Calculate the equilibrium constant for the overall two stages of ionization of
    (a) oxalic acid, $H_2C_2O_4$ (b) sulfurous acid, H_2SO_3

20. A solution is acidified with HCl so that its pH is 1.0, and is saturated with H_2S at 25 °C. What is the sulfide ion concentration in this solution?

21. Write correctly balanced net ionic equations for the three stages of ionization of citric acid, $H_3C_6H_5O_7$.

Section 9.5

22. Classify the following salts as acidic, basic, or neutral:
    (a) NH_4Cl (b) KCN   (c) Na_2S (d) $NaNO_3$ (e) $NaNO_2$ (f) CH_3NH_3Br
    (g) LiCl   (h) Na_2CO_3 (i) KI   (j) NH_4NO_3

23. Classify the following ions as acids, bases, or ampholytes:
    (a) $C_2O_4^{2-}$ (b) NH_4^+ (c) PO_4^{3-} (d) HSO_3^- (e) $H_2PO_4^-$ (f) F^-
    (g) $CH_3NH_3^+$ (h) NO_2^-

24. Classify the following oxides as acidic or basic:
    (a) SrO   (b) SeO_3 (c) K_2O (d) As_2O_5

25. Arrange the following 0.1 F solutions in order of increasing pH:
    (a) NaCN   (b) HBr   (c) NaOAc   (d) KBr   (e) NH_4Br

26. Are aqueous solutions of $NaHSO_3$ acidic, basic, or neutral? Show all reasoning necessary to prove your answer.

Section 9.6

27. Explain what is meant by the "leveling effect of water." Write balanced chemical equations to show how this applies to perchloric acid, $HClO_4$, and sodium amide, $NaNH_2$.

Multiple Choice Questions

1. Which of the following species is a stronger acid than formic acid in aqueous solution?
   (a) CH_3COOH (b) NH_4^+ (c) H_2SO_3 (d) HCN   (e) HPO_4^{2-}

2. All of the following species can exist in aqueous solution EXCEPT
   (a) NH_4^+ (b) NO_3^- (c) NO_2^- (d) NH_3 (e) NH_2^-

3. The pH of a solution of 0.10 F CH_3COOH increases when which of the following substances is added?
   (a) $NaHSO_4$ (b) $HClO_4$ (c) NH_4NO_3 (d) K_2CO_3 (e) $H_2C_2O_4$

4. A solution has pH = 10.82. The $[OH^-]$ in this solution is
   (a) 1.5×10^{-11} (b) 6.6×10^{-10} (c) 1.0×10^{-7}
   (d) 1.5×10^{-5} (e) 6.6×10^{-4}

5. All of the following are acid–base conjugate pairs EXCEPT
   (a) $HONO, NO_2^-$ (b) H_3O^+, OH^- (c) $CH_3NH_3^+, CH_3NH_2$
   (d) HS^-, S^{2-} (e) $C_6H_5COOH, C_6H_5COO^-$

6. Which of the following species is an ampholyte?
   (a) CH_3COO^- (b) H_2O (c) NH_4^+ (d) C_6H_5COOH (e) S^{2-}

7. The pH of a 0.050 F HA solution is 5.35. What is K_a for HA?
   (a) 2×10^{-11} (b) 4×10^{-10} (c) 4.5×10^{-6} (d) 8.9×10^{-5}
   (e) 5.0×10^{-2}

8. The correctly balanced net ionic equation for the reaction that occurs when an aqueous solution of formic acid is mixed with an aqueous solution of potassium hydroxide is
   (a) $HCOOH + H_2O \rightleftharpoons HCOO^- + H_3O^+$
   (b) $H_3O^+ + OH^- \rightleftharpoons 2H_2O$
   (c) $HCOOH + KOH \rightleftharpoons H_2O + HCOOK$
   (d) $HCOOH + OH^- \rightleftharpoons H_2O + HCOO^-$
   (e) $HCOO^- + K^+ \rightleftharpoons HCOOK$

9. The symbol $K_b(HS^-)$ is the equilibrium constant for the reaction:
   (a) $HS^- + OH^- \rightleftharpoons S^{2-} + H_2O$
   (b) $HS^- + H_2O \rightleftharpoons H_2S + OH^-$
   (c) $HS^- + H_2O \rightleftharpoons H_3O^+ + S^{2-}$
   (d) $HS^- + H_3O^+ \rightleftharpoons H_2S + H_2O$
   (e) $HS^- + CH_3COOH \rightleftharpoons H_2S + CH_3COO^-$

10. The symbol $K_a(HS^-)$ is the equilibrium constant for the reaction:
   (a) $HS^- + OH^- \rightleftharpoons S^{2-} + H_2O$
   (b) $HS^- + H_2O \rightleftharpoons H_2S + OH^-$
   (c) $HS^- + H_2O \rightleftharpoons H_3O^+ + S^{2-}$
   (d) $HS^- + H_3O^+ \rightleftharpoons H_2S + H_2O$
   (e) $HS^- + CH_3COOH \rightleftharpoons H_2S + CH_3COO^-$

11. Which of the following solutions has the highest pH?
   (a) $2.0\ F\ HClO_4$ (b) $0.20\ F\ CH_3COOH$ (c) $0.020\ F\ HCl$ (d) $2.0\ F\ NaCl$
   (e) $0.200\ F\ HCOOH$

12. Which of the following solutions has the lowest pH?
   (a) $2.0\ F\ HClO_4$ (b) $0.20\ F\ CH_3COOH$ (c) $0.020\ F\ HCl$ (d) $2.0\ F\ NaCl$
   (e) $0.200\ F\ HCOOH$

13. When K_2O is added to water, the solution is basic because it contains a significant concentration of
   (a) K^+ (b) K_2O (c) O^{2-} (d) O_2^{2-} (e) OH^-

14. The correctly balanced net ionic equation for the reaction that occurs when a solution of acetic acid is mixed with a solution of sodium carbonate is
   (a) $CH_3COOH + CO_3^{2-} \rightleftharpoons CH_3COO^- + HCO_3^-$
   (b) $H_3O^+ + CO_3^{2-} \rightleftharpoons HCO_3^- + H_2O$
   (c) $CH_3COOH + Na_2CO_3 \rightleftharpoons CH_3COO^- + NaHCO_3 + Na^+$
   (d) $CO_3^{2-} + H_2O \rightleftharpoons HCO_3^- + OH^-$
   (e) $CH_3COOH + OH^- \rightleftharpoons CH_3COO^- + H_2O$

15. A solution has $[OH^-] = 4.8 \times 10^{-3}$. Its pH is
   (a) 11.7   (b) 8.4   (c) 4.8   (d) 3.7   (e) 2.3

16. The pH of $0.20\ F\ HCOOH$ is between
   (a) 1.5 and 2.0   (b) 2.0 and 2.5   (c) 2.5 and 3.0   (d) 3.0 and 3.5
   (e) 3.5 and 4.0

17. A solution that is $0.500\ F\ HCl$ is saturated with H_2S at 25 °C. Its pH is
   (a) 0.30   (b) 0.50   (c) 1.5   (d) 2.3   (e) 4.0

18. Which of the following solutions has the highest pH?
   (a) $0.10\ F\ KNO_3$ (b) $0.10\ F\ NaCN$ (c) $0.10\ F\ NH_4Cl$ (d) $0.10\ F\ NaOAc$
   (e) $0.10\ F\ CH_3NH_2$

19. What is the degree of dissociation of $0.048\ F\ C_6H_5COOH$?
   (a) 6.3%   (b) 4.8%   (c) 3.6%   (d) 2.5%   (e) 1.0%

20. What is the pH of a $0.080\ F\ CH_3COOH$ solution?
   (a) 0.80   (b) 1.10   (c) 2.92   (d) 4.76   (e) 5.26

21. Which of the following solutions is the most basic? All concentrations are $0.20\ F$.
   (a) NaI (b) NaF (c) $NaCN$ (d) $NaNO_3$ (e) $NaOAc$

22. What is the pH of a solution prepared by dissolving 0.100 mol of $NaOH$ and 0.100 mol of NH_3 in enough water to make 1 L?
   (a) 14.0   (b) 13.0   (c) 11.1   (d) 7.0   (e) 2.9

23. The pK_a of nitrous acid is
   (a) 4.65   (b) 4.35   (c) 4.00   (d) 3.65   (e) 3.35

24. The conjugate base of CH_3OH is
   (a) OH^- (b) CH_3O^- (c) O^{2-} (d) CH_4 (e) H_2O

25. The equilibrium constant for the reaction

$$HONO + CN^- \rightleftharpoons HCN + ONO^-$$

is $1.1 \times 10^{+6}$. From the magnitude of this constant one can conclude that
(a) CN^- is a stronger base than ONO^-.
(b) HCN is a stronger acid than HONO.
(c) The conjugate base of HONO is ONO^-.
(d) The conjugate acid of CN^- is HCN.
(e) HONO will react with F^- to form HF.

Problems

9.1. Calculate the $[C_6H_5COOH]$, $[C_6H_5COO^-]$, $[H_3O^+]$, $[OH^-]$, and pH of $0.020\ F$ C_6H_5COOH (benzoic acid) at 25 °C.

9.2. Calculate the concentrations of acetic acid, acetate ion, hydronium ion, and hydroxide ion in a solution that is prepared by dissolving 8.203 g of sodium acetate in enough water to make 500.0 mL of solution at 25 °C.

9.3. A $1.00\ F$ solution of nitrous acid, HONO, is 2.26% ionized at 25 °C. (a) Calculate the nitrite ion concentration and the pH of this solution. (b) Calculate K_a for nitrous acid at 25 °C. (c) Calculate the degree of dissociation of $0.250\ F$ HONO.

9.4. Calculate the $[H_3O^+]$, $[HOAc]$, $[OH^-]$, $[OAc^-]$, and pH of a solution that is prepared by dissolving 0.080 mol of HCl and 0.150 mol of CH_3COOH in enough water to make 1.00 L of solution. At the temperature for which the calculation is to be made the ionization constant of acetic acid is 1.78×10^{-5}.

9.5. To an aqueous solution containing 0.124 mol of NH_4NO_3, there is added 0.300 mol of solid NaOH pellets. When the pellets have all dissolved and the solution is sufficiently stirred to be homogeneous, the volume of the solution is found to be 1.000 L at 25 °C.
(a) Write a correctly balanced net ionic equation for the reaction that occurs in this solution.
(b) List the five chemical species present in this solution at equilibrium that have concentrations greater than $0.10\ M$. Arrange them in decreasing order of concentration.
(c) The basicity constant, K_b, for NH_3 is 1.8×10^{-5} at 25 °C. Calculate the $[NH_3]$, $[NH_4^+]$, and pH of this solution.

9.6. (a) A $0.10\ F$ solution of a certain base has a pH of 10.74. What is the basicity constant, K_b, of this base?
(b) A different solution of this base at the same temperature has a pH $= 10.36$. What is the formality of this solution?

9.7. Pivalic acid is a monoprotic weak acid. A $0.100\ F$ solution of pivalic acid has a pH $= 3.0$. What is the pH of $0.100\ F$ sodium pivalate at the same temperature?

9.8. (a) The weak monoprotic acid HA is 3.2% dissociated in $0.086\ F$ solution. What is the acidity constant, K_a, of HA?
(b) A certain solution of HA has a pH $= 2.48$. What is the formality of this solution?

9.9. Aniline, $C_6H_5NH_2$, is an industrially important amine used in the making of dyes. It is a weak base with $K_b = 3.8 \times 10^{-10}$ at 25 °C. At this temperature, aniline is soluble in water to the extent of 3.9 g/100.0 mL of solution. What is the pH of a saturated aqueous solution of aniline at 25 °C?

9.10. The hypobromite ion, BrO^-, is the conjugate base of the weak acid HOBr, hypobromous acid. The pH of a 0.100 F NaOBr (sodium hypobromite) solution is 10.85 at 25 °C.
(a) Write the correctly balanced net ionic equation for the reaction that occurs when NaOBr is dissolved in water, and the equilibrium constant expression for the reaction.
(b) Calculate the value of the basicity constant of OBr^- at 25 °C.
(c) Calculate the value of K_a for hypobromous acid at 25 °C.

9.11. Calculate the pH of a mixture of acetic acid and sodium acetate if the ratio $[CH_3COO^-]/[CH_3COOH]$ is
(a) $4:1$ (b) $2:1$ (c) $1:3$ (d) $1:6$

9.12. Equimolar quantities of each of the following substances is dissolved in enough water to make 1.00 L of solution. Arrange the solutions so prepared in order of increasing pH. Explain the reasons for your answers.
(a) $KHSO_4$ (b) $NaHCO_3$ (c) $HClO_4$ (d) Na_2O (e) CH_3NH_3Br

9.13. Classify aqueous solutions of the following substances as acidic, basic, or neutral. Explain the reasons for your answers.
(a) $(NH_4)_2S$ (b) $NaHSO_3$ (c) $Ba(OAc)_2$ (d) CaI_2 (e) $(CH_3NH_3)_2CO_3$

9.14. Trimethylamine, $(CH_3)_3N$, is a weak base that ionizes in aqueous solution due to the proton-transfer reaction:

$$(CH_3)_3N + H_2O \rightleftharpoons (CH_3)_3NH^+ + OH^-$$

A 0.120 F solution of $(CH_3)_3N$ is 2.29% ionized at 25 °C.
(a) Calculate the $[OH^-]$, $[(CH_3)_3NH^+]$, $[(CH_3)_3N]$, $[H_3O^+]$, and the pH of a 0.120 F solution of $(CH_3)_3N$ at 25 °C.
(b) Calculate K_b for $(CH_3)_3N$ at 25 °C.
(c) Calculate the degree of dissociation, α, of a 0.096 F solution of trimethylamine. Does the degree of dissociation increase, decrease, or remain unchanged as the concentration of $(CH_3)_3N$ decreases? Explain the reasons for your answer.

9.15. Write a correctly balanced net ionic equation for the reaction that occurs when solutions of the following reagents are mixed:
(a) Acetic acid and potassium hydroxide.
(b) Acetic acid and ammonia.
(c) Perchloric acid and ammonia.
(d) Nitric acid and sodium carbonate.
(e) Ammonium chloride and potassium cyanide.
(f) Ammonium nitrate and sodium hydroxide.
(g) Sodium sulfide and formic acid.
(h) Sodium acetate and hydrochloric acid.

9.16. Hypochlorous acid, HOCl, is a weak acid extensively used as a bleaching agent. Its acidity constant, K_a, is 3.2×10^{-8}.
(a) Calculate the pH and the degree of dissociation of a 0.0650 F HOCl solution.
(b) Write the expression for the basicity constant, K_b, for the conjugate base of HOCl, and calculate its numerical value.
(c) Calculate the pH of a 0.040 F NaOCl solution.
(d) Household bleach is made by dissolving chlorine in H_2O:

$$Cl_2(g) + 2H_2O \rightleftharpoons H_3O^+ + Cl^- + HOCl$$

Calculate the pH of such a solution if $[HOCl] = 0.0650\ M$.

9.17. Calculate the $[H_3O^+]$, $[HCOO^-]$, $[HCOOH]$, $[OH^-]$, and pH of 0.082 F HCOOH. What is the degree of dissociation of 0.082 F HCOOH?

9.18. A solution is prepared by dissolving 0.060 mol of NaOH and 0.045 mol of NH_3 in enough water to make 500.0 mL of solution. Calculate the $[H_3O^+]$, $[OH^-]$, $[NH_3]$, $[NH_4^+]$, and pH of this solution at 25 °C.

9.19. Calculate the $[H_3O^+]$, $[SO_4^{2-}]$, $[HSO_4^-]$, $[OH^-]$, and pH of 0.100 F H_2SO_4.

chapter 10 Buffer Solutions and Acid–Base Titrations

Johannes Nicolaus Brønsted (1879–1947), a Danish physical chemist, is best known for his theory of acids and bases. Brønsted received degrees in both chemical engineering and chemistry from the University of Copenhagen. He was an authority on the catalytic properties and strengths of acids and bases, and wrote texts on both inorganic and physical chemistry. In 1929 he was a visiting professor at Yale. In 1947 he was elected to the Danish parliament, and he died in December of that year, in Copenhagen.

There are many chemical reactions that are affected by a change in pH. Sometimes the yield of a product varies if the pH is changed; sometimes even the nature of the product changes if the pH is changed. Biochemical systems, in particular, are usually quite sensitive to changes in pH. For instance, the pH of human blood must remain within a very narrow range around the value 7.4 in order for a person to be healthy. Any considerable increase in the hydronium ion concentration causes increased and violent breathing. A person becomes ill if his blood pH falls below 7.3; this condition is known as acidosis. An increase in pH above 7.5 (alkalosis) is similarly a sign of disease.

Metabolic processes in animals and plants function well only if the fluids within the organism maintain an approximately constant pH. All organisms have naturally occurring chemical systems that serve to keep the pH of their vital fluids constant, even when extra acid or base is introduced. Similarly, in the laboratory we can prepare solutions that resist a change in pH. Such a solution is called a **buffer**. In ordinary terms, a buffer is anything that serves to deaden a shock or to bear the brunt of opposing forces. In chemistry, a **buffer** is a solution that maintains an approximately constant pH when small additions of either acid or base are made.

section 10.1
Buffer Solutions

Composition of Buffer Solutions

A solution containing substantial amounts of both a weak acid and its conjugate weak base (or of both a weak base and its conjugate weak acid) is a buffer solution. For instance, if we put acetic acid and sodium acetate into the same solution, substantial amounts of a weak acid, CH_3COOH, and its conjugate weak base, CH_3COO^-, will both be present, and the mixture will be a buffer. Another possibility is a mixture of NH_4Cl and NH_3. That solution contains both the weak acid NH_4^+ and its conjugate weak base NH_3 and is therefore a buffer. There are, of course, a great many possible buffer solutions because there are so many weak acids and bases.

Different buffer solutions will, in general, have different pH values. An acetic acid/acetate ($HOAc/OAc^-$) buffer can keep the pH constant at 4.8, while an ammonium ion/ammonia (NH_4^+/NH_3) buffer can maintain the pH at 9.3.

Mechanism of Buffer Action

In order to understand how a buffer solution serves to keep the pH approximately constant, we must consider the equilibrium between a weak acid and its conjugate weak base. We will use a specific example, namely the bicarbonate/carbonate (HCO_3^-/CO_3^{2-}) buffer, prepared by dissolving both $NaHCO_3$ and Na_2CO_3 in water. Buffers containing sodium bicarbonate (commonly called bicarbonate of soda) are produced commercially for use in the relief of acid indigestion. The equilibrium involved in the buffering action is

$$HCO_3^- + H_2O \rightleftharpoons CO_3^{2-} + H_3O^+ \tag{10-1}$$

The acidity constant for bicarbonate ion is

$$K_a(HCO_3^-) = \frac{[CO_3^{2-}][H_3O^+]}{[HCO_3^-]} \tag{10-2a}$$

which can be readily rearranged to solve for the $[H_3O^+]$. We obtain

$$[H_3O^+] = \frac{K_a[HCO_3^-]}{[CO_3^{2-}]} \tag{10-2b}$$

Since K_a is a constant, Eq. (10-2b) tells us that it is the *ratio* of bicarbonate ion concentration to the concentration of carbonate ion that determines the $[H_3O^+]$, that is, the acidity of the solution. We can control the $[H_3O^+]$, and therefore the pH, by adjusting the $[HCO_3^-]/[CO_3^{2-}]$ ratio.

Since we are interested in the pH, we take the base 10 logarithm of Eq. (10-2b) and obtain

$$\log[H_3O^+] = \log K_a + \log \frac{[HCO_3^-]}{[CO_3^{2-}]} \tag{10-3}$$

If we multiply this equation through by -1 and use the properties of logarithms (refer to Appendix B), we obtain

$$pH = pK_a + \log \frac{[CO_3^{2-}]}{[HCO_3^-]} \tag{10-4}$$

where $pK_a = -\log K_a$.

Equation (10-4) tells us that the pH of a solution containing substantial amounts of both HCO_3^- and CO_3^{2-} ions will remain approximately constant provided that the $[CO_3^{2-}]/[HCO_3^-]$ ratio remains approximately constant.

If we add H_3O^+ ions (any strong acid) to a solution containing both CO_3^{2-} and HCO_3^-, the CO_3^{2-} ions will combine with the H_3O^+ ions to form more HCO_3^-, according to the reaction

$$H_3O^+ + CO_3^{2-} \rightleftharpoons HCO_3^- + H_2O \tag{10-5}$$

This will change the $[CO_3^{2-}]/[HCO_3^-]$ ratio somewhat, but provided we had relatively large amounts of both carbonate ion and bicarbonate ion present originally, and we add only a relatively small amount of H_3O^+, the ratio will not change very much.

Similarly, if we add OH^- ions (any strong base) to a solution containing both CO_3^{2-} and HCO_3^-, the OH^- will combine with the HCO_3^- to form more CO_3^{2-} according to the reaction

$$OH^- + HCO_3^- \rightleftharpoons H_2O + CO_3^{2-} \tag{10-6}$$

This will increase the $[CO_3^{2-}]$ a little and decrease the $[HCO_3^-]$, but provided that we had relatively large amounts of both CO_3^{2-} and HCO_3^- present originally, and add only a relatively small amount of OH^-, the ratio will not change greatly.

For the general case of a buffer containing substantial amounts of both a weak acid, HA, and its conjugate weak base, A^-, Eq. (10-4) becomes

$$pH = pK_a + \log \frac{[A^-]}{[HA]} = pK_a + \log \frac{[\text{conjugate base}]}{[\text{weak acid}]} \tag{10-7}$$

Although this equation does not really need a special name, biochemists in particular frequently refer to it as the **Henderson–Hasselbalch equation.** It states the important fact that the pH of a buffer solution is controlled by the [conjugate base]/[weak acid] ratio.

We find that in order for a buffer solution to be really efficient at maintaining constant pH, two requirements must be met:

1. The number of moles of both the weak acid and its conjugate weak base must be relatively large.
2. The conjugate base to weak acid ratio must be close to $1:1$.

If the [conjugate base]/[weak acid] ratio is larger than 10 or smaller than 0.1, the solution will not be particularly effective at keeping the pH constant. The buffer efficiency is largest when the [conjugate base]/[weak acid] ratio is 1:1.

In thinking about the first of these requirements, you will want to know the meaning of the term "relatively large." The word relative compares the amounts of the weak acid and conjugate weak base in the buffer to the acid or base that will be added to the solution. No solution can keep the pH approximately constant if you add larger amounts of either acid or base than are present in the original buffer. A buffer solution maintains constant pH when *small* amounts of either acid or base are added. The buffer contains a weak acid that can react with OH^- as well as a weak base that can react with H_3O^+. The amounts of weak acid and weak base in the buffer must be significantly larger than the amounts of H_3O^+ or OH^- that will be added, or the pH cannot remain approximately constant.

Making the [conjugate base]/[weak acid] ratio reasonably close to 1:1 is a good way to ensure that relatively large amounts of both species are present.

EXAMPLE 10.1. Calculation of the pH of a buffer solution

A buffer solution is made by mixing 0.100 mol of acetic acid and 0.400 mol of sodium acetate per liter of solution. Calculate the pH of this buffer.

Solution. The equilibrium involved in this buffer solution is

$$HOAc + H_2O \rightleftharpoons OAc^- + H_3O^+$$

let $x = [H_3O^+]$ in this solution

then $[OAc^-] = 0.400 + x$ and $[HOAc] = 0.100 - x$

Note particularly that the $[H_3O^+]$ and the $[OAc^-]$ are *not* equal in this solution. The great bulk of the acetate ions present in the solution come from the sodium acetate, which is a strong electrolyte. The acetate ion from the sodium acetate represses the ionization of the acetic acid, so that the $[H_3O^+]$ in this solution is quite small relative to the $[OAc^-]$. Thus we expect x to be small relative to both 0.100 and 0.400. If we make this approximation, the equilibrium constant expression for HOAc is

$$K_a = \frac{[H_3O^+][OAc^-]}{[HOAc]} = \frac{(x)(0.400)}{(0.100)} = 1.8 \times 10^{-5}$$

Solving this equation for x we obtain $x = (1.8 \times 10^{-5})/4 = 4.5 \times 10^{-6}$. We must now check our assumption. Is 4.5×10^{-6} less than 10% of 0.100? Since it is, our assumption is valid and

$$[H_3O^+] = 4.5 \times 10^{-6}\ M$$

in this buffer. The pH is therefore $6 - \log 4.5 = 5.35$. This is an effective buffer solution because the [OAc⁻]/[HOAc] ratio is 0.400/0.100 = 4. The value of 4 lies between the acceptable limits of 10 and 0.10.

Note that we could have done this problem in a different, but equivalent way, by using Eq. (10-7). Because there must be substantial amounts of both a weak acid and its conjugate weak base in any buffer solution, the assumption we made above that the $[H_3O^+]$ is negligible compared to the concentrations of both acetic acid, HOAc, and its conjugate weak base, OAc^-, will always be valid, so that we can write immediately

$$pH = pK_a + \log(0.400/0.100) = pK_a + \log 4.00$$

Since K_a of HOAc is 1.8×10^{-5}, pH = $4.74_5 + 0.602 = 5.35$.

EXAMPLE 10.2. **The effect of diluting a buffer with water**

If 100.0 mL of water are added to 100.0 mL of a buffer that is 0.200 F NH_4NO_3 and 0.200 F NH_3, what effect does this dilution have on the pH of the buffer?

Solution. The pH of this buffer is given by Eq. (10-7) as

$$pH = pK_a(NH_4^+) + \log \frac{[NH_3]}{[NH_4^+]}$$

Since $K_a(NH_4^+) = 5.7 \times 10^{-10}$ (Table E1, Appendix E), $pK_a(NH_4^+) = 9.24$, so that pH $= 9.24 + \log([NH_3]/[NH_4^+])$.

Originally, both $[NH_4^+]$ and $[NH_3]$ are 0.200 M. The dilution with water decreases both these concentrations to 0.100 M, but the $[NH_3]/[NH_4^+]$ *ratio* remains 1 : 1. Thus the pH remains constant on dilution.

The conclusion of Example 10.2 is generally valid. Because dilution with water decreases both the concentration of the weak acid and its conjugate base by the same factor, their ratio remains constant, and therefore the pH of a buffer remains constant when water is added.

A buffer containing a weak acid, HA, and its conjugate base, A⁻, is able to maintain an approximately constant pH when either strong acid or strong base is added because the weak acid combines with OH⁻ ions to form more A⁻ ions

$$HA + OH^- \rightleftharpoons A^- + H_2O \tag{10-8a}$$

and A⁻ combines with H_3O^+ to form more HA molecules:

$$A^- + H_3O^+ \rightleftharpoons HA + H_2O \tag{10-9a}$$

Both these reactions proceed essentially to completion. Equation (10-8a) for acetic acid is, for example,

$$CH_3COOH + OH^- \rightleftharpoons CH_3COO^- + H_2O \tag{10-8b}$$

The equilibrium constant for this reaction is very large compared to 1:

$$K_{eq} = \frac{[CH_3COO^-]}{[OH^-][CH_3COOH]} = \frac{[CH_3COO^-][H_3O^+]}{[CH_3COOH]} \cdot \frac{1}{[H_3O^+][OH^-]} \tag{10-10}$$

$$K_{eq} = \frac{K_a(HOAc)}{K_w} = \frac{1.8 \times 10^{-5}}{1.0 \times 10^{-14}} = 1.8 \times 10^{+9} \gg 1$$

Because it has such a large equilibrium constant, reaction (10-8b) proceeds essentially to completion, all the way to the right. (Refer to Eq. 8.15.)

Equation (10-9a) for acetate ion is

$$CH_3COO^- + H_3O^+ \rightleftharpoons CH_3COOH + H_2O \tag{10-9b}$$

The equilibrium constant for this reaction is

$$K_{eq} = \frac{[CH_3COOH]}{[CH_3COO^-][H_3O^+]} = \frac{1}{K_a(HOAc)} = \frac{1}{1.8 \times 10^{-5}} = 5.6 \times 10^4 \gg 1 \tag{10-11}$$

Thus reaction (10-9b) also goes virtually to completion.

A specific numerical example will illustrate how effective a buffer solution is in resisting a change in pH.

EXAMPLE 10.3. Calculating the change in pH of a buffer solution on addition of a small amount of strong base

Calculate the change in pH that results when 0.50 mL of 0.200 F NaOH is added to 10.00 mL of a buffer solution containing acetic acid and potassium acetate, both at a concentration of 1.00 F.

Solution. To determine the change in the pH we must first know the pH of the solution before adding the 0.50 mL of 0.200 F NaOH. The equilibrium involved in this buffer is

$$HOAc + H_2O \rightleftharpoons H_3O^+ + OAc^-$$

let $x = [H_3O^+]$ in this solution

then $\qquad [OAc^-] = 1.00 + x \qquad$ and $\qquad [HOAc] = 1.00 - x$

If we assume that x is negligible compared to 1.00 because the acetate ions from the potassium acetate repress the ionization of acetic acid, the equilibrium constant expression becomes

$$K_a = \frac{[H_3O^+][OAc^-]}{[HOAc]} = \frac{[H_3O^+](1.00)}{(1.00)} = 1.8 \times 10^{-5}$$

Since 1.8×10^{-5} is negligible compared to 1.00, in this buffer the $[H_3O^+] = 1.8 \times 10^{-5}\ M$, and pH = p$K_a$ = 4.74.

In order to calculate how much the pH changes when we add 0.50 mL of 0.200 F NaOH, we must know how many millimoles of HOAc and OAc$^-$ we had present before the base is added.

$$\text{number of millimoles of HOAc originally present} = (10.00\ mL)(1.00\ M) = 10.00\ mmol$$

$$\text{number of millimoles of OAc}^-\text{ originally present} = (10.00\ mL)(1.00\ M) = 10.00\ mmol$$

How many millimoles of OH$^-$ are being added to this solution?

$$\text{No. mmol OH}^-\text{ added} = (0.50\ mL)(0.200\ M) = 0.10\ mmol$$

The OH$^-$ added to the solution reacts with HOAc present in the buffer as follows:

$$HOAc + OH^- \rightleftharpoons OAc^- + H_2O$$

This reaction proceeds essentially to completion (see Eq. 10-10), so adding the OH$^-$ serves to use up some of the HOAc and to increase the amount of OAc$^-$ in the solution.

Since OH$^-$ and HOAc combine in a 1:1 molar ratio, 0.10 mmol of OH$^-$ combines with 0.10 mmol of HOAc and forms 0.10 mmol of OAc$^-$. Thus, after the NaOH has been added, there are $10.00 - 0.10 = 9.90$ mmol of HOAc and $10.00 + 0.10 = 10.10$ mmol of OAc$^-$ present in the solution. These numbers should give you a feeling for what is meant by the requirement that the number of millimoles of weak acid and weak base in a buffer must be relatively large in order for the buffer to be effective at maintaining constant pH. The 10.00 mmol of both HOAc and OAc$^-$ originally present is large relative to the 0.10 mmol of OH$^-$ added to the solution.

The volume of the solution is 10.50 mL after the NaOH has been added, so that in the resultant solution

$$[HOAc] = (9.90\ mmol)/(10.50\ mL) = 0.943\ M$$

and

$$[OAc^-] = (10.10 \text{ mmol})/(10.50 \text{ mL}) = 0.962 \ M$$

Again we can assume that the $[H_3O^+]$ is small compared to the $[HOAc]$ and $[OAc^-]$, so that

$$[H_3O^+] = \frac{K_a[HOAc]}{[OAc^-]} = \frac{(1.8 \times 10^{-5})(0.943)}{(0.962)} = 1.7_6 \times 10^{-5}$$

Therefore $pH = 5 - \log 1.76 = 4.75$. Note that the pH has changed by only 0.01 unit after adding 0.50 mL of 0.200 F NaOH!

If you need to be convinced of just how effective the buffer solution of Example 10.3 is, calculate the change in pH when you add 0.50 mL of 0.200 F NaOH to 10.00 mL of water. You will find that the pH changes by 4.98 units, from 7.00 to 11.98.

pH Range of Efficient Buffer Action

The pH range over which a given buffer system (HA/A^-) will be effective at maintaining constant pH is determined by the acidity constant of the weak acid. This can be seen by considering Eq. (10-7). The buffer is most efficient at maintaining constant pH when the [conjugate base]/[weak acid] ratio is 1, and for that ratio, $pH = pK_a$, since $\log 1 = 0$.

If the [conjugate base]/[weak acid] ratio is 10, the pH calculated using Eq. (10-7) is $pK_a + 1$, since $\log 10 = 1$. Similarly, if the [conjugate base]/[weak acid] ratio is 0.1, the pH is $pK_a - 1$, since $\log 0.1 = -1$. Thus the range of pH over which we can expect a given conjugate pair (HA/A^-) to be an effective buffer is about two pH units, from $pK_a - 1$ to $pK_a + 1$.

Specifically, an acetic acid/acetate buffer will be useful for the pH range from 3.76 to 5.76, since $K_a(HOAc) = 1.75 \times 10^{-5}$ and $pK_a = 4.76$. On the other hand, the HCO_3^-/CO_3^{2-} buffer is useful between pH 9.3 and 11.3, since $K_a(HCO_3^-) = 4.7 \times 10^{-11}$ and $pK_a = 10.3$.

EXAMPLE 10.4. Preparation of a buffer of specified pH

For a biology project you are growing certain microorganisms that can only survive in a medium at $pH = 8.54$. You decide to use an NH_3/NH_4NO_3 buffer. What must be the molar ratio of NH_3 to NH_4NO_3 in your buffer?

Solution. The weak acid in this buffer is NH_4^+, for which K_a is 5.7×10^{-10}. Its pK_a is therefore 9.24, and substitution into Eq. (10-7) yields

$$pH = pK_a + \log([NH_3]/[NH_4^+]) = 9.24 + \log([NH_3]/[NH_4^+])$$

Setting the $pH = 8.54$, we obtain

$$8.54 = 9.24 + \log([NH_3]/[NH_4^+])$$

or

$$\log([NH_3]/[NH_4^+]) = -0.70$$

Hence the desired molar ratio is $[NH_3]/[NH_4^+] = 0.20$.

handwritten margin notes:
must be in excess
① weak acid + strong base
② strong acid + weak base in excess

Methods of Preparing Buffer Solutions

There is more than one way to prepare a buffer solution containing a weak acid and its conjugate weak base. If you wanted to prepare an acetic acid/acetate buffer, for instance, you could simply mix acetic acid with sodium acetate or potassium acetate, but you could also mix excess acetic acid with NaOH. Hydroxide ions react with acetic acid to form acetate ions, according to the reaction

$$CH_3COOH + OH^- \rightleftharpoons CH_3COO^- + H_2O \qquad (10\text{-}8b)$$

If the number of moles of acetic acid exceeds the number of moles of OH^- added, the final solution will contain both acetic acid and acetate ions and be identical with a mixture of acetic acid and sodium acetate. Thus mixing a weak acid with a strong base is one way to prepare a buffer solution, provided the weak acid is in excess.

Similarly, a buffer can be prepared by mixing a strong acid with a weak base, provided the weak base is in excess. Thus we can prepare an NH_4^+/NH_3 buffer either by adding NH_4Cl to an aqueous NH_3 solution, or by mixing HCl with excess NH_3. The hydronium ions in a solution of any strong acid react with NH_3 to form NH_4^+ ions:

$$NH_3 + H_3O^+ \rightleftharpoons NH_4^+ + H_2O \qquad (10\text{-}12)$$

and reaction (10-12) proceeds essentially to completion, as its equilibrium constant is $1/K_a(NH_4^+) = 1.8 \times 10^9$, which is large compared to 1. [See Eq. (8-15b).]

We have already discussed two ways to prepare an HOAc/OAc$^-$ buffer: (1) by mixing HOAc and NaOAc, and (2) by mixing NaOH with excess HOAc. A third possible method is to mix any strong acid with excess sodium acetate, which provides the required excess of the weak base, CH_3COO^- ions. A mixture of HCl and sodium acetate constitutes a buffer solution if the number of moles of sodium acetate is larger than the number of moles of HCl, since it contains both acetic acid and acetate ions. The reaction between any strong acid and acetate ion

$$H_3O^+ + CH_3COO^- \rightleftharpoons CH_3COOH + H_2O \qquad (10\text{-}9b)$$

proceeds essentially to completion. (See Eq. 10-11.)

The following three problems illustrate calculations involving buffers prepared by mixing a strong acid with a weak base or a strong base with a weak acid.

EXAMPLE 10.5. Preparation of a buffer solution by mixing a weak acid with a strong base

Calculate the pH of a buffer prepared by mixing 40.00 mL of 1.00 F propionic acid with 60.00 mL of 0.100 F NaOH. The acidity constant of propionic acid is 1.3×10^{-5}.

Solution. Propionic acid is an acid closely related to acetic acid, with very similar properties. Note that K_a for propionic acid is very close in value to K_a for acetic acid. The formula for propionic acid is CH_3CH_2COOH, but we will denote it HOPr, using an abbreviation analogous to the HOAc used for acetic acid. In aqueous solution, propionic acid is in equilibrium with hydronium ion and propionate ion, the anion formed when the acid donates a proton to some base. The equilibrium reaction is

$$HOPr + H_2O \rightleftharpoons H_3O^+ + OPr^-$$

When OH$^-$ ions are added to a solution containing HOPr, the reaction

$$OH^- + HOPr \rightleftharpoons H_2O + OPr^-$$

proceeds essentially to completion. Let us calculate how many millimoles of HOPr and how many millimoles of OH^- were mixed together to prepare this buffer.

$$\text{No. mmol HOPr} = (40.00 \text{ mL})(1.00 \text{ mmol/mL}) = 40.0 \text{ mmol}$$

$$\text{No. mmol OH}^- = (60.00 \text{ mL})(0.100 \text{ mmol/mL}) = 6.00 \text{ mmol}$$

Note that there is excess HOPr. The number of millimoles of HOPr is greater than the number of millimoles of OH$^-$ so that when OH$^-$ and HOPr react there will be HOPr remaining after the reaction.

Because the reaction between OH$^-$ and HOPr proceeds virtually to completion, 6.00 mmol of OH$^-$ react with 6.00 mmol of HOPr, forming 6.00 mmol of OPr$^-$ (propionate ion) and leaving $40.0 - 6.00 = 34.0$ mmol of HOPr in excess. The total volume of solution is 100.00 mL. The solution is identical to a mixture of 34.0 mmol of propionic acid and 6.00 mmol of sodium propionate in enough water to make 100.00 mL of solution.

$$\text{let } x = [\text{H}_3\text{O}^+] \text{ in this solution at equilibrium}$$

Then

$$[\text{OPr}^-] = \frac{(6.00 \text{ mmol})}{(100.00 \text{ mL})} + x = 0.0600 + x = 0.0600 \ M$$

$$[\text{HOPr}] = \frac{(34.0 \text{ mmol})}{(100.00 \text{ mL})} - x = 0.340 - x = 0.340 \ M$$

The propionate ion concentration is slightly greater than $0.0600 \ M$ because every time a hydronium ion is formed from the proton-transfer reaction between HOPr and H$_2$O, a propionate ion is also formed. We have assumed that the [H$_3$O$^+$] in this solution is small relative to either the [HOPr] or the [OPr$^-$] because the propionate ions from the sodium propionate repress the ionization of HOPr.

The equilibrium constant equation for propionic acid is

$$K_a = \frac{[\text{H}_3\text{O}^+][\text{OPr}^-]}{[\text{HOPr}]} = \frac{(x)(0.0600)}{(0.340)} = 1.3 \times 10^{-5}$$

If we solve this equation for x we obtain

$$x = \tfrac{34}{6}(1.3 \times 10^{-5}) = 7.4 \times 10^{-5}$$

Since 7.4×10^{-5} is small compared to 0.060, our assumption is justified. In this buffer the [H$_3$O$^+$] $= 7.4 \times 10^{-5} \ M$ and the pH $= 4.13$. We can show that this is indeed an effective buffer solution by calculating the [conjugate base]/[weak acid] ratio.

$$\frac{[\text{OPr}^-]}{[\text{HOPr}]} = \frac{0.0600}{0.340} = 0.176$$

which is within the limits of 10 and 0.1.

EXAMPLE 10.6. Preparation of a buffer solution by mixing a weak base with a strong acid

A 2.461-g sample of solid sodium acetate is dissolved in enough water to make 50.00 mL of solution. To the sodium acetate solution 100.00 mL of 0.120 F HCl is added.

(a) What is the pH of the resulting solution?

Solution. The reaction that occurs when sodium acetate solution and HCl are mixed is described by Eq. (10-9b). We must therefore find out how many millimoles of acetate ion and how many millimoles of hydronium ion we are mixing together. The gram formula weight of sodium acetate, CH_3COONa, is 82.03 g/mol. Thus we have added

$$\frac{2.461 \text{ g}}{82.03 \text{ g/mol}} = 3.000 \times 10^{-2} \text{ mol} = 30.000 \text{ mmol}$$

of sodium acetate. The solution therefore contains 30.00 mmol of acetate ions.

The number of millimoles of H_3O^+ added is

$$(100.00 \text{ mL})(0.120 \text{ mmol/mL}) = 12.0 \text{ mmol}$$

Note that there is excess weak base.

According to Eq. (10-9b), 12.0 mmol of H_3O^+ react with 12.0 mmol of CH_3COO^- to form 12.0 mmol of CH_3COOH. That reaction goes essentially to completion, and the number of millimoles of acetate ion in excess is $30.00 - 12.0 = 18.0$ mmol.

The solution resulting after mixing the two reagents contains 12.0 mmol of acetic acid and 18.0 mmol of acetate ion in a total volume of 150.0 mL. There is no significant contraction or expansion of volume on mixing two dilute aqueous solutions. The equilibrium

$$HOAc + H_2O \rightleftharpoons OAc^- + H_3O^+$$

is of course established in this solution.

$$\text{let } x = [H_3O^+] \text{ at equilibrium}$$

Then

$$[OAc^-] = \frac{18.0 \text{ mmol}}{150.0 \text{ mL}} + x = 0.120 + x \approx 0.120 \ M$$

and

$$[HOAc] = \frac{12.0 \text{ mmol}}{150.0 \text{ mL}} - x = 0.0800 - x \approx 0.0800 \ M$$

We have made the assumption that x is small compared to 0.0800 because we expect the $[H_3O^+]$ in this solution to be small relative to the $[HOAc]$ and $[OAc^-]$. The equilibrium constant expression is therefore

$$K_a(HOAc) = \frac{[H_3O^+][OAc^-]}{[HOAc]} = \frac{(x)(0.120)}{(0.080)} = 1.8 \times 10^{-5}$$

Solving this equation for x we obtain

$$x = \tfrac{8}{12}(1.8 \times 10^{-5}) = \tfrac{2}{3}(1.8 \times 10^{-5}) = 1.2 \times 10^{-5}$$

Since 1.2×10^{-5} is small compared to 0.080, our assumption is valid. Thus, $[H_3O^+] = 1.2 \times 10^{-5} \ M$, and pH = 4.92.

This is an effective buffer solution because the ratio

$$\frac{[OAc^-]}{[HOAc]} = \frac{0.120}{0.080} = 1.5$$

(b) A subsequent addition of 5.00 mL of 0.120 F HCl is made to the buffer solution of part **(a)**. What is the pH after this addition?

Solution. If we add 5.00 mL of 0.120 F HCl to the buffer, we are adding

$$(5.00 \text{ mL}) (0.120 \text{ mmol/mL}) = 0.600 \text{ mmol of } H_3O^+$$

This 0.600 mmol of H_3O^+ reacts with 0.600 mmol of OAc^- forming 0.600 mmol of HOAc. Thus after the addition, the solution will contain $12.0 + 0.600 = 12.6$ mmol of HOAc, and $18.0 - 0.600 = 17.4$ mmol of OAc^-, in a total volume of 155.0 mL.

We may assume that the $[H_3O^+]$ is small compared to the [HOAc] and $[OAc^-]$, so that

$$[OAc^-] = (17.4 \text{ mmol}/155.0 \text{ mL}) = 0.112 \ M$$

and

$$[HOAc] = (12.6 \text{ mmol}/155.0 \text{ mL}) = 0.0813 \ M$$

The $[OAc^-]/[HOAc]$ ratio has changed from $0.120/0.0800 = 1.50$ before the addition of the 5.00 mL of HCl, to $0.112/0.0813 = 1.38$ after the HCl was added. The $[H_3O^+]$ is now

$$[H_3O^+] = \frac{(1.8 \times 10^{-5})(0.0813)}{(0.112)} = 1.3 \times 10^{-5} \ M$$

and the pH is therefore 4.89.

The pH changed from 4.92 to 4.89 when 5.00 mL of 0.120 F HCl were added to the buffer. A pH change of only 0.03 units upon the addition of a moderately large amount of a strong acid is a very small pH change, and shows how effective the buffer solution is at maintaining the pH approximately constant.

EXAMPLE 10.7. Directions for preparing a buffer by mixing a weak acid and a strong base

What volume of 2.00 F NaOH must be added to 100.00 mL of 2.50 F HCOOH to prepare a buffer with pH = 4.00?

Solution. From Table E1, Appendix E, we find the acidity constant of formic acid, HCOOH, is 1.8×10^{-4}. The pK_a of formic acid is $-\log(1.8 \times 10^{-4}) = 3.74$. Using Eq. (10-7) we find, therefore, that

$$pH = 4.00 = 3.74 + \log \frac{[HCOO^-]}{[HCOOH]}$$

so $\qquad \log \dfrac{[HCOO^-]}{[HCOOH]} = 0.26 \qquad$ and $\qquad \dfrac{[HCOO^-]}{[HCOOH]} = 10^{0.26} = 1.8$

We see, therefore, that in order to make a buffer with pH = 4.00, we must make the $[HCOO^-]/[HCOOH]$ ratio equal 1.8.

The net ionic equation for the reaction that occurs when HCOOH and NaOH solutions are mixed is

$$HCOOH + OH^- \rightleftharpoons HCOO^- + H_2O$$

Let V be the volume, in milliliters, of 2.00 F NaOH that must be added to 100.00 mL of 2.50 F HCOOH to make the $[HCOO^-]/[HCOOH]$ ratio = 1.8. A sample of 2.00 F NaOH of volume V mL contains $(2.00V)$ mmol of OH^-. Since HCOOH and OH^- react in a 1:1 molar ratio, $2.00V$ mmol of OH^- combine with $2.00V$ mmol of HCOOH, and produce $2.00V$ mmol of $HCOO^-$ ions. We began with 100.00 mL of 2.50 F HCOOH, or 250 mmol of HCOOH. Since $2.00V$ mmol of HCOOH are used up in the reaction with OH^-, the amount of excess HCOOH is $(250 - 2.00V)$ mmol. Therefore,

$$\frac{\text{No. mmol } HCOO^- \text{ formed}}{\text{No. mmol } HCOOH \text{ remaining}} = \frac{2.00V \text{ mmol}}{(250 - 2.00V) \text{ mmol}}$$

This is precisely the ratio that must be 1.8 if the buffer is to have a pH = 4.00. Thus,

$$\frac{2.00V}{(250 - 2.00V)} = 1.8 \quad \text{or} \quad 2.00V = 450 - 3.6V$$

Thus,

$$5.6V = 450 \quad \text{and} \quad V = 80.4 \text{ mL}$$

If 80.4 mL of 2.00 F NaOH is added to 100.0 mL of 2.50 F HCOOH the solution will contain 160.8 mmol of $HCOO^-$ ions and 89.2 mmol of HCOOH (250 − 160.8 = 89.2). The [$HCOO^-$]/[HCOOH] ratio of this solution is 1.8, and the pH is 4.00. There will be 180.4 mL of this buffer. If you want a specified larger volume, for example 250 mL, you need only add water to bring the volume to 250 mL. Adding water decreases the concentrations of $HCOO^-$ and HCOOH by the same factor, but does not change the [$HCOO^-$]/[HCOOH] ratio, so the pH does not change. (See Example 10.2.)

section 10.2
Titration of a Strong Acid versus a Strong Base

A practical problem that arises in the laboratory is the determination of the concentration of acid or base in a solution. The method used to solve this problem is called an **acid–base titration.**

Suppose we have a solution we know to be acidic. What we do not know is the precise concentration of acid in the solution. To determine this, we measure out a known volume of the acidic solution using a **volumetric pipet,** and place it in a clean flask. We must have available a solution of a strong base, such as NaOH, with known concentration. The sodium hydroxide solution is placed in a **buret,** which allows us to add measured amounts of the base gradually to our unknown acid solution. Figure 10.1 shows an experimental set-up for performing an acid–base titration.

The titration consists of adding the NaOH solution of known concentration (called the **titrant**) to the acid solution of unknown concentration until an exactly equivalent amount of base has been added, that is, until the number of moles (or millimoles) of OH^- added is just enough to react completely with all the acid originally present. This point in the titration is called the **equivalence point.** We must have some way of knowing when that point is reached, and an **indicator** is usually added to signal the equivalence point. We will discuss **acid–base indicators** in the following section.

pH Changes During a Strong Acid versus Strong Base Titration

In most titrations, the concentration of either acid or base is not known and the purpose of the titration is to determine it. In order to understand how we can tell when the equivalence point has been reached, we will calculate how the pH of a specific acidic solution changes as base is added during a titration. In the titration of any strong acid with a strong base, the reaction that occurs as the titrant is added is simply

$$H^+(aq) + OH^-(aq) \rightleftharpoons H_2O \tag{10-13}$$

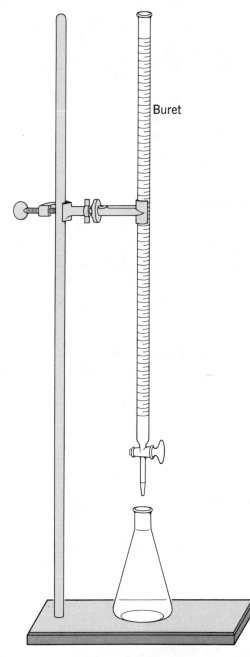

Buret

Fig. 10.1. Apparatus used to perform a titration.

We choose as our example the titration of 25.00 mL of 0.200 F HNO$_3$ with 0.200 F NaOH.

1. ***At the beginning of the titration.*** Before any base has been added, the [H$_3$O$^+$] = 0.200 M, the original concentration of the strong acid. The pH, therefore, is −log(0.200) = 0.70.

2. ***After 10.00 mL of base have been added.*** Because H$^+$ and OH$^-$ react in a 1 : 1 molar ratio, according to Eq. (10-13), we must calculate how many millimoles of H$_3$O$^+$ we started with, and how many millimoles of OH$^-$ have been added.

No. mmol H$^+$ at start = (25.00 mL) (0.200 mmol/mL) = 5.00 mmol

No. mmol OH$^-$ added = (10.00 mL) (0.200 mmol/mL) = 2.00 mmol

The 2.00 mmol of added OH^- combine with 2.00 mmol of H^+ to form 2.00 mmol of H_2O.

$$\left(\begin{array}{c}\text{No. mmol } H_3O^+ \\ \text{remaining}\end{array}\right) = \left(\begin{array}{c}\text{No. mmol } H_3O^+ \\ \text{at start}\end{array}\right) - \left(\begin{array}{c}\text{No. mmol } H_3O^+ \\ \text{combined with } OH^-\end{array}\right) = 5.00 - 2.00$$

There are, therefore, 3.00 mmol of H_3O^+ that are untitrated at this point in the titration. The volume of the solution is now 35.00 mL, as 10.00 mL of solution have been added to the original 25.00 mL. Thus,

$$[H_3O^+] = (3.00 \text{ mmol})/(35.00 \text{ mL}) = 8.57 \times 10^{-2} \ M$$
$$pH = -\log(8.57 \times 10^{-2}) = 1.07$$

3. **After 20.00 mL of base have been added.**

$$\text{No. mmol } OH^- \text{ added} = (20.00 \text{ mL}) (0.200 \text{ mmol/mL}) = 4.00 \text{ mmol}$$
$$\text{No. mmol } H_3O^+ \text{ remaining} = 5.00 - 4.00 = 1.00 \text{ mmol}$$
$$\text{volume of solution} = 25.00 \text{ mL} + 20.00 \text{ mL} = 45.00 \text{ mL}$$
$$[H_3O^+] = (1.00 \text{ mmol})/(45.00 \text{ mL}) = 2.22 \times 10^{-2} \ M$$
$$pH = -\log(2.22 \times 10^{-2}) = 1.65$$

4. **After 24.00 mL of base have been added.**

$$\text{No. mmol } OH^- \text{ added} = (24.00 \text{ mL}) (0.200 \text{ mmol/mL}) = 4.80 \text{ mmol}$$
$$\text{No. mmol } H_3O^+ \text{ remaining} = 5.00 - 4.80 = 0.20 \text{ mmol}$$
$$\text{volume of solution} = 25.00 \text{ mL} + 24.00 \text{ mL} = 49.00 \text{ mL}$$
$$[H_3O^+] = (0.20 \text{ mmol})/(49.00 \text{ mL}) = 4.08 \times 10^{-3} \ M$$
$$pH = -\log(4.08 \times 10^{-3}) = 2.39$$

5. **At the equivalence point.** When 25.00 mL of base have been added, 5.00 mmol of OH^- have been added, and therefore all the H_3O^+ originally present has been titrated. The concentration of hydronium ions in the solution is not zero, however, just as it is not zero in pure water. We can obtain the $[H_3O^+]$ from the relation

$$[H_3O^+] = [OH^-] \text{ at the equivalence point} \qquad (10\text{-}14)$$

Hence

$$K_w = 1.0 \times 10^{-14} = [H_3O^+][OH^-] = [H_3O^+]^2$$

and

$$[H_3O^+] = 1.0 \times 10^{-7} \ M$$

We conclude that *the pH is 7.00 at the equivalence point in the titration of any strong acid with any strong base.*

The solution at the equivalence point in this titration is identical with a solution of $0.100 \ F \ NaNO_3$. We have added 5.00 mmol of Na^+ ions and there were originally 5.00 mmol of NO_3^- ions present. The final volume of solution is 50.00 mL at the equivalence point, so

$$[Na^+] = [NO_3^-] = (5.00 \text{ mmol})/(50.00 \text{ mL}) = 0.100 \ M$$

Before the equivalence point we assumed all the hydronium ion in solution came from the untitrated acid, and we neglected the hydronium ion present due to the self-ionization of water. This is an excellent assumption because the self-ionization of water proceeds to such a small extent, and is repressed in the presence of any $H^+(aq)$ from the strong acid. At the equivalence point, however, there is no untitrated acid left, and the self-ionization of water is the only source of H^+(aq) in the solution.

Note particularly that the pH jumps from 2.39 to 7.00 on the addition of the last 1.00 mL of base before the equivalence point, whereas the addition of the first 24.00 mL of base causes a much smaller increase in pH, from 0.70 to 2.39.

What happens to the pH once the equivalence point is passed? If we add more than 25.00 mL of base, we are essentially adding OH^- to a given volume of water. Any OH^- added in excess of the 5.00 mmol needed to react with the 5.00 mol of H_3O^+ originally present remains in the solution unreacted.

6. **After 26.00 mL of base have been added.**

$$\text{No. mmol } OH^- \text{ added} = (26.00 \text{ mL})(0.200 \text{ mmol/mL}) = 5.20 \text{ mmol}$$
$$\text{No. mmol } OH^- \text{ in excess} = 5.20 - 5.00 = 0.20 \text{ mmol}$$
$$\text{volume of solution} = 25.00 \text{ mL} + 26.00 \text{ mL} = 51.00 \text{ mL}$$
$$[OH^-] = (0.20 \text{ mmol})/(51.00 \text{ mL}) = 3.92 \times 10^{-3} M$$
$$pOH = -\log(3.92 \times 10^{-3}) = 2.41 \quad \text{and} \quad pH = 14.00 - 2.41 = 11.59$$

7. **After 30.00 mL of base have been added.**

$$\text{No. mmol } OH^- \text{ added} = (30.00 \text{ mL})(0.200 \text{ mmol/mL}) = 6.00 \text{ mmol}$$
$$\text{No. mmol } OH^- \text{ in excess} = 6.00 - 5.00 = 1.00 \text{ mmol}$$
$$\text{volume of solution} - 25.00 \text{ mL} + 30.00 \text{ mL} = 55.00 \text{ mL}$$
$$[OH^-] = (1.00 \text{ mmol})/(55.00 \text{ mL}) = 1.82 \times 10^{-2} M$$
$$pOH = -\log(1.82 \times 10^{-2}) = 1.74 \quad \text{and} \quad pH = 14.00 - 1.74 = 12.26$$

We see that the pH jumps up quite markedly after the equivalence point, but as more base is added beyond the equivalence point, the pH increases only gradually again. A plot of pH versus the volume of added base is known as a **titration curve**. The titration curve for the titration of 25.00 mL of 0.200 F HNO_3 with 0.200 F NaOH is shown in Fig. 10.2.

The striking feature of this titration curve is the very sharp jump in pH at the equivalence point, compared to the gradual increase in pH both before and after the equivalence point. It is this rapid increase in pH at the equivalence point that makes it possible to detect the equivalence point and enables us to determine the volume of titrant required to reach the equivalence point.

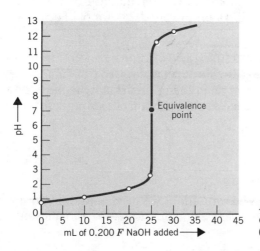

Fig. 10.2. Titration curve for the titration of 25.00 mL of 0.200 F HNO_3 versus 0.200 F NaOH.

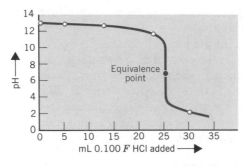

Fig. 10.3. Titration curve for the titration of 25.00 mL of 0.100 F NaOH versus 0.100 F HCl.

EXAMPLE 10.8. Titration stoichiometry

A 25.00-mL portion of an aqueous solution of HBr of unknown concentration is titrated with 0.1500 F NaOH. The equivalence point is reached when 32.80 mL of base has been added. Calculate the concentration of the hydrobromic acid.

Solution. Let C_A = molarity of the acid solution. Then,

$$\text{No. mmol acid} = (25.00 \text{ mL}) (C_A \text{ mmol/mL})$$

At the equivalence point, the number of millimoles of added base is exactly equal to the number of millimoles of acid originally present. Hence,

$$\text{No. mmol OH}^- \text{ added} = (32.80 \text{ mL}) (0.1500 \, M) = (25.00 \text{ mL}) (C_A)$$

Thus,

$$C_A = \frac{(32.80)(0.1500)}{(25.00)} = 0.1968 \, M$$

EXAMPLE 10.9. Titration of a strong acid versus a strong base

In order to determine the strength of a commercial antacid, one tablet was dissolved in 50.00 mL of 1.00 F HCl. The resulting solution, which was still acidic, was titrated with a standardized 0.500 F NaOH solution. The volume of the NaOH solution required to reach the equivalence point of the titration was 43.72 mL. Calculate the number of moles of acid that had been neutralized by the antacid tablet.

Solution. The number of millimoles of OH$^-$ added during the titration was (43.72 mL) (0.500 mmol/mL) = 21.86 mmol. Therefore 21.86 mmol of H_3O^+ reacted with the added OH$^-$ during the titration. As the tablet was dissolved in a solution containing 50.00 mmol of H_3O^+, the amount of acid neutralized by the tablet was 50.00 − 21.86 = 28.14 mmol, or 0.02814 mol of acid.

If the titrant is a solution of a strong acid of known concentration that is added to a solution of a strong base, the titration curve is inverted from that of Fig. 10.2, but the essential characteristic of a very rapid change in pH close to the equivalence point is the same. A typical titration curve for the titration of a strong base versus a strong acid is shown in Fig. 10.3.

section 10.3
Acid–Base Indicators

When performing an acid–base titration, we need some way to observe the very sharp rise in pH at the equivalence point. An **acid–base indicator** is a substance, usually a weak organic acid, that changes color within a fairly narrow range of pH values because the acid and its conjugate base are two different colors. A typical indicator can be symbolized HIn (In stands for indicator), to represent the fact that it is a weak acid. In aqueous solution, the equilibrium

$$HIn + H_2O \rightleftharpoons H_3O^+ + In^- \qquad (10\text{-}15)$$

is established. If the acid, HIn, has one color while its conjugate base, In$^-$, has a different color, the species may function as an indicator. To be a useful indicator, it is also necessary for the colors to be sufficiently intense so that even one or two drops colors the solution. There are a great many acid–base indicators. Bromthymol blue, for instance, is a species for which the acid form is yellow and the basic form is blue. Methyl red is an indicator for which the acid form is red and the basic form is yellow.

Of course, any aqueous solution to which indicator has been added will necessarily contain both HIn and In$^-$. What then determines the color we see? It is the relative amounts of the two forms. Although different people respond to colors somewhat differently, and there is no fixed ratio for everyone, as a generalization we find that if

$$[HIn]/[In^-] > 10 \qquad \text{we see the color of the acidic form} \qquad (10\text{-}16)$$

On the other hand, if

$$[HIn]/[In^-] < 1/10 \qquad \text{we see the color of the basic form} \qquad (10\text{-}17)$$

For values of the ratio [HIn]/[In$^-$] less than 10 but larger than 1/10, we see the color of a mixture of the two forms. The intermediate color for bromthymol blue, for instance, is green, while for methyl red it is orange.

The pH and the acidity constant, K_{In}, of the particular weak acid indicator determine the value of the ratio [HIn]/[In$^-$]. If we rearrange the equilibrium constant expression

$$K_{In} = \frac{[In^-][H_3O^+]}{[HIn]} \qquad (10\text{-}18a)$$

to solve for the desired ratio, we obtain

$$\frac{[HIn]}{[In^-]} = \frac{[H_3O^+]}{K_{In}} \qquad (10\text{-}18b)$$

Equation (10-18b) tells us that when a drop or two of a solution of a given indicator is placed in the titration flask, the color we see is determined by the [H$_3$O$^+$] in the flask, that is, by the pH of the solution being titrated. We must use only a very small amount of indicator so that the addition of the indicator does not affect the pH of the solution, and for that reason it is necessary for the colors of the indicator to be intense.

What must the pH of the solution be in order for [HIn]/[In$^-$] to be 10, that is, in order for the color we see to be the acidic color of the indicator? If [HIn]/[In$^-$] = 10, Eq. (10-18b) yields [H$_3$O$^+$] = 10 K_{In}. Taking logarithms of both sides, and multiplying through by -1, we obtain

$$pH = pK_{In} - \log(10) = pK_{In} - 1 \qquad \text{if} \quad [HIn]/[In^-] = 10 \qquad (10\text{-}19)$$

where $pK_{In} = -\log K_{In}$. If the pH is less than $pK_{In} - 1$, the ratio [HIn]/[In$^-$] will be greater than 10. Thus we will see the acidic color of the indicator provided that $pH \le pK_{In} - 1$.

If [HIn]/[In$^-$] = 1/10, Eq. (10-18b) yields [H$_3$O$^+$] = $K_{In}/10$, and therefore

$$pH = pK_{In} - \log(1/10) = pK_{In} + 1 \qquad \text{if} \quad [HIn]/[In^-] = 1/10 \qquad (10\text{-}20)$$

Table 10.1. **Summary of Relations Used to Determine the Color of a Solution Containing an Acid – Base Indicator**

If pH $\leq pK_{In} - 1$	we see the color of the acidic form.
If pH $= pK_{In}$	we see the intermediate color, a mixture of equal amounts of the acidic and basic forms.
If pH $\geq pK_{In} + 1$	we see the color of the basic form.

We see, therefore, that as long as pH $\geq pK_{In} + 1$, a solution containing the indicator will have the color of the basic form of the indicator.

When $[HIn] = [In^-]$, Eq. (10-18a) states that $[H_3O^+] = K_{In}$ and therefore pH $= pK_{In}$. At this pH we will see the intermediate color of the indicator. A summary of these relations is given in Table 10.1. Of course, different human beings differ in their sensitivity to varying shades of color, and the colors of some indicators are more intense than others. When solutions of varying pH containing an indicator are viewed, the pH range at which most people observe color changes will not be exactly two pH units (from $pK_{In} - 1$ to $pK_{In} + 1$) but it will be close to that.

Table 10.2 lists some common acid – base indicators and the pH range over which most people observe a change in color. Note that for phenolphthalein and thymolphthalein the acidic form is colorless. The pH range for an observed color change for thymolphthalein is smaller than for most other indicators.

Choosing an Indicator for a Titration

Which of these indicators should we choose for the titration of 0.200 *F* HNO$_3$ versus 0.200 *F* NaOH? We want an indicator that will change color with the addition of one drop of base, right at the equivalence point. Since there are about 20 drops/mL, a drop is roughly 0.05 mL. Any indicator that will change color as the volume of base added increases from 24.95 to 25.05 mL will be a satisfactory indicator. Let us see what the pH is one drop before, and one drop after, the equivalence point.

1. *One drop before the equivalence point.* When 24.95 mL of base have been added

$$\text{No. mmol OH}^- \text{ added} = (24.95 \text{ mL}) (0.200 \text{ mmol/mL}) = 4.99 \text{ mmol}$$

$$\text{No. mmol H}_3\text{O}^+ \text{ untitrated} = 5.00 - 4.99 = 0.01 \text{ mmol}$$

$$\text{volume of solution} = 25.00 \text{ ml} + 24.95 \text{ mL} = 49.95 \text{ mL}$$

$$[H_3O^+] = (0.01 \text{ mmol})/(49.95 \text{ mL}) = 2.0 \times 10^{-4} M \quad \text{and} \quad \text{pH} = 3.70$$

2. *One drop past the equivalence point.* When 25.05 mL of base have been added

$$\text{No. mmol OH}^- \text{ added} = (25.05 \text{ mL}) (0.200 \text{ mmol/mL}) = 5.01 \text{ mmol}$$

$$\text{No. mmol OH}^- \text{ in excess} = 5.01 - 5.00 = 0.01 \text{ mmol}$$

$$\text{volume of solution} = 25.00 \text{ mL} + 25.05 \text{ mL} = 50.05 \text{ mL}$$

$$[OH^-] = (0.01 \text{ mmol})/(50.05 \text{ mL}) = 2.0 \times 10^{-4} M$$

$$[H_3O^+] = K_w/[OH^-] = 5.0 \times 10^{-11} M \quad \text{and} \quad \text{pH} = 10.3$$

Any indicator that changes color at a pH between 3.7 and 10.3 will signal the equivalence point within one drop of the correct volume. We see, therefore, that many indicators will be perfectly satisfactory. Bromcresol green, methyl red, chlorophenol red, bromthymol blue, phenol red, and phenolphthalein can all be used. It is

Table 10.2 **The pH Range for Color Change of Some Common Acid–Base Indicators**

Indicator	Color of Acidic Form	Color of Basic Form	Range of Color Change, pH Units	pK_{In}
Methyl orange	Red	Yellow	3.1–4.4	3.7
Bromphenol blue	Yellow	Purple	3.0–4.6	3.8
Bromcresol green	Yellow	Blue	3.8–5.4	4.7
Methyl red	Red	Yellow	4.2–6.1	5.0
Chlorophenol red	Yellow	Red	5.2–6.8	6.2
Bromthymol blue	Yellow	Blue	6.0–7.6	7.1
Phenol red	Yellow	Red	6.4–8.2	7.8
Phenolphthalein	Colorless	Pink	8.0–9.8	9.7
Thymolphthalein	Colorless	Blue	9.4–10.6	10.0

because there is such a very rapid rise in pH very close to the equivalence point that we have such a wide choice of indicators.

If both the solution being titrated and the titrant are considerably more dilute than 0.200 F, the pH change at the equivalence point becomes less steep and greater care must be taken with the selection of an indicator.

Let us suppose, for instance, that we were titrating 25.00 mL of 0.0100 F HCl versus 0.0100 F NaOH. The number of millimoles of H_3O^+ being titrated is (25.00 mL)(0.0100 mmol/mL) = 0.250 mmol.

1. ***One drop before the equivalence point.***

$$\text{No. mmol OH}^- \text{ added} = (24.95 \text{ mL})(0.0100 \text{ mmol/mL}) = 0.2495 \text{ mmol}$$
$$\text{No. mmol H}_3\text{O}^+ \text{ untitrated} = 0.2500 - 0.2495 = 0.0005 \text{ mmol}$$
$$\text{volume of solution} = 25.00 \text{ mL} + 24.95 \text{ mL} = 49.95 \text{ mL}$$
$$[\text{H}_3\text{O}^+] = (5 \times 10^{-4} \text{ mmol})/(49.95 \text{ mL}) = 1 \times 10^{-5} \ M \quad \text{and} \quad \text{pH} = 5$$

2. ***One drop after the equivalence point.***

$$\text{No. mmol OH}^- \text{ added} = (25.05 \text{ mL})(0.0100 \text{ mmol/mL}) = 0.2505 \text{ mmol}$$
$$\text{No. mmol OH}^- \text{ in excess} = 0.2505 - 0.2500 = 0.0005 \text{ mmol}$$
$$\text{volume of solution} = 25.00 \text{ mL} + 25.05 \text{ mL} = 50.05 \text{ mL}$$
$$[\text{OH}^-] = (5 \times 10^{-4} \text{ mmol})/(50.05 \text{ mL}) = 1 \times 10^{-5} \ M$$
$$\text{therefore the } [\text{H}_3\text{O}^+] = 1 \times 10^{-9} \ M \quad \text{and} \quad \text{pH} = 9$$

For this titration, only an indicator that changes color between pH = 5 and pH = 9 will be satisfactory. Of those listed in Table 10.2, the only possibilities are bromthymol blue, chlorophenol red, and phenol red.

Figure 10.4 displays graphically the conclusion we can draw from these calculations: The more dilute the acid being titrated, the less sharply the pH rises close to the equivalence point, and the more careful we must be in selecting the proper indicator.

There is another method available for detecting the rapid change in pH very close to the equivalence point, and that is to use a piece of equipment called a **pH meter.** A pH meter is an instrument that measures the electrical potential difference between two electrodes, one of which (a glass electrode) is sensitive to the pH of the solution in which it is immersed. With a pH meter, you measure the pH of the solution after each addition of titrant, and in effect, plot the titration curve as you carry out the titration.

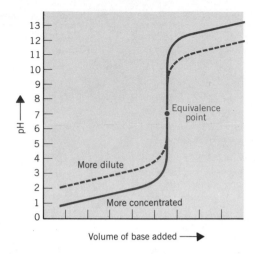

Fig. 10.4. Titration curves for a dilute and a concentrated solution of the same strong acid versus a strong base.

section 10.4
Titration of a Weak Monoprotic Acid versus a Strong Base

A solution of a weak acid, HA, in water has a considerably lower $[H_3O^+]$ than a solution of a strong acid of the same concentration, because most of the acid is present as undissociated HA molecules. The reaction that occurs as we add strong base as a titrant to a solution of a weak acid is

$$HA + OH^- \rightleftharpoons A^- + H_2O \tag{10-21}$$

In order to plot a titration curve for the titration of a weak acid versus a strong base, we will calculate how the pH changes as base is added during a typical titration, the titration of 50.00 mL of 0.100 F CH_3COOH with 0.100 F NaOH.

1. *At the beginning of the titration.* Before any base has been added, we simply have a solution of 0.100 F HOAc.

$$[H_3O^+] = [OAc^-] \quad \text{and} \quad [HOAc] = 0.100 - [H_3O^+]$$

The equilibrium constant equation, before the titration begins, is

$$K_a = \frac{[H_3O^+][OAc^-]}{[HOAc]} = \frac{[H_3O^+]^2}{0.100 - [H_3O^+]} = 1.8 \times 10^{-5} \tag{10-22}$$

Calculation of the hydronium ion concentration and pH of this solution was carried out in Example 9.6, and the pH was found to be 2.89.

As we add OH^- during a titration of a weak acid, the titration reaction (10-21) goes essentially to completion. Thus the number of millimoles of A^- (anion) formed is equal to the number of millimoles of OH^- added, and is also equal to the number of millimoles of HA titrated.

No. mmol HA untitrated = No. mmol HA at start − No. mmol HA titrated

The $[H_3O^+]$ in the solution at any point in the titration *prior* to the equivalence point is given by

$$[H_3O^+] = K_a([HOAc]/[OAc^-]) \tag{10-23}$$

and therefore we always need to know the [HOAc]/[OAc⁻] ratio. Since

$$[HOAc] = \frac{\text{No. mmol HOAc untitrated}}{\text{total volume of solution}}$$

and

$$[OAc^-] = \frac{\text{No. mmol OAc}^- \text{ formed}}{\text{total volume of solution}}$$

when we calculate the [HOAc]/[OAc⁻] ratio, the total volume of solution always cancels out.

2. *After 10.00 mL of base have been added.*

No. mmol HOAc at start = (50.00 mL)(0.100 mmol/mL) = 5.00 mmol

No. mmol OH⁻ added = (10.00 mL)(0.100 mmol/mL) = 1.00 mmol

No. mmol OAc⁻ formed = 1.00 mmol = No. mmol HOAc titrated

No. mmol HOAc untitrated = 5.00 − 1.00 = 4.00 mmol

$$\frac{[HOAc]}{[OAc^-]} = \frac{\text{No. mmol HOAc untitrated}}{\text{No. mmol OAc}^- \text{ formed}} = \frac{4.00}{1.00} = 4.00$$

Using Eq. (10-23),

$$[H_3O^+] = (1.8 \times 10^{-5})(4.00) = 7.2 \times 10^{-5} \ M$$

and
$$pH = -\log(7.2 \times 10^{-5}) = 4.14$$

In the same way we can calculate the pH when 20.00, 25.00, and 40.00 mL of base have been added to the acetic acid solution. The results of these calculations are tabulated in Table 10.3. You should perform these calculations yourself to insure that you can obtain the values listed.

Notice that during this part of the titration (as from 10.00 mL base to 40.00 mL base are added), the solution contains substantial amounts of both acetate ions and undissociated acetic acid, and is therefore a buffer solution. The pH does not change very much as the volume of base added increases from 10.00 to 40.00 mL. The increase in pH as the volume of base increases from 0.00 to 10.00 mL is as large as the increase caused by the addition of the next 30 mL of base.

3. *At the equivalence point.* When 50.00 mL of base have been added, the number of moles of OH⁻ added is exactly equal to the number of moles of HOAc originally present. Since OH⁻ and HOAc react in a 1:1 molar ratio, that is the equivalence point of the titration. As the titration reaction goes virtually to

Table 10.3 Titration of 50.00 mL of 0.100 *F* HOAc with 0.100 *F* NaOH

mL Base Added	Total Volume	$[H_3O^+]$	$[OH^-]$	$pH = -\log[H_3O^+]$
0.00	50.00	1.3×10^{-3}	7.5×10^{-12}	2.89
10.00	60.00	7.2×10^{-5}	1.4×10^{-10}	4.14
20.00	70.00	2.7×10^{-5}	3.7×10^{-10}	4.67
25.00	75.00	1.8×10^{-5}	5.6×10^{-10}	4.75
40.00	90.00	4.5×10^{-6}	2.2×10^{-9}	5.35
50.00	100.00	1.9×10^{-9}	5.3×10^{-6}	8.72
60.00	110.00	1.1×10^{-12}	9.1×10^{-3}	11.96

completion, the solution at the equivalence point contains 5.00 mmol of acetate ion and 5.00 mmol of Na^+ ion, in a total volume of 100.00 mL. The solution at the equivalence point is therefore identical to a 0.0500 F NaOAc solution. Since acetate ion is a weak base, a solution of sodium acetate is basic.

Before the equivalence point the titration reaction (10-21) goes essentially to completion to the right. At the equivalence point, when there is no longer any untitrated weak acid present, the reverse reaction

$$A^- + H_2O \rightleftharpoons HA + OH^- \qquad (10\text{-}24)$$

determines the pH of the solution. You should recognize that Eq. (10-24) is the basicity reaction of the weak base, A^-.

At the equivalence point

$$[HA] = [OH^-] \qquad (10\text{-}25)$$

and the basicity constant of the anion A^-

$$K_b(A^-) = \frac{K_w}{K_a(HA)} = \frac{[HA][OH^-]}{[A^-]} \qquad (10\text{-}26)$$

is used to determine the $[OH^-]$ in the solution.

For the titration we are considering, at the equivalence point

$$[OH^-] = [HOAc] \quad \text{and} \quad [OAc^-] = 0.0500 - [OH^-]$$

The acetate ion concentration is somewhat less than 0.0500 M because of the proton-transfer reaction between acetate ion and water, Eq. (10-24). If we let $x = [OH^-] = [HOAc]$, the equilibrium constant expression, Eq. (10-26), for this titration is

$$K_b(OAc^-) = \frac{1.0 \times 10^{-14}}{1.8 \times 10^{-5}} = 5.6 \times 10^{-10} = \frac{x^2}{0.0500 - x}$$

Assume that x is small compared to 0.0500, so that $0.0500 - x$ is approximately 0.0500. Our equation then becomes

$$x^2 = (5.6 \times 10^{-10})(5.00 \times 10^{-2}) = 28 \times 10^{-12}$$

and $\qquad\qquad x = 5.3 \times 10^{-6}$

Since 5.3×10^{-6} is less than 10% of 0.0500, our assumption is valid, and the $[OH^-]$ at the equivalence point of this titration is 5.3×10^{-6} M. We can calculate the $[H_3O^+]$ as follows:

$$[H_3O^+] = \frac{K_w}{[OH^-]} = \frac{1.0 \times 10^{-14}}{5.3 \times 10^{-6}} = 1.9 \times 10^{-9} \ M$$

and therefore $\qquad\qquad pH = 9 - \log(1.9) = 8.72$

Note particularly that the pH at the equivalence point of this titration is greater than 7. *The solution at the equivalence point in the titration of any monoprotic weak acid with a strong base is always basic.* The weaker the acid, the more basic the solution will be at the equivalence point, because the weaker the acid, the stronger its conjugate base. The solution at the equivalence point in the titration of the weak acid HA versus NaOH is simply a solution of the basic salt NaA in water.

Once the equivalence point has been passed, the $[OH^-]$ in the solution can be calculated in exactly the same way as for the titration of a strong acid versus a strong base.

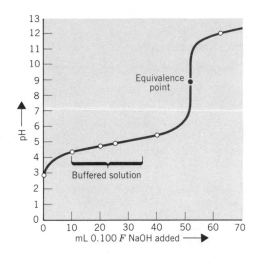

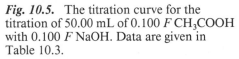

Fig. 10.5. The titration curve for the titration of 50.00 mL of 0.100 F CH$_3$COOH with 0.100 F NaOH. Data are given in Table 10.3.

4. *When 60.00 mL of base have been added.*

$$\text{No. mmol OH}^- \text{ added} = (60.00 \text{ mL})(0.100 \text{ mmol/mL}) = 6.00 \text{ mmol}$$

$$\text{No. mmol OH}^- \text{ in excess} = 6.00 - 5.00 = 1.00 \text{ mmol}$$

$$\text{volume of solution} = 50.00 \text{ mL} + 60.00 \text{ mL} = 110.00 \text{ mL}$$

$$[\text{OH}^-] = (1.00 \text{ mmol})/(110.00 \text{ mL}) = 9.09 \times 10^{-3} \, M$$

$$\text{pOH} = -\log(9.09 \times 10^{-3}) = 2.04 \quad \text{and} \quad \text{pH} = 14.00 - 2.04 = 11.96$$

One must be particularly careful in selecting the proper indicator for the titration of a weak acid versus a strong base. The data of Table 10.3 have been used to plot the titration curve for the titration of 50.00 mL of 0.100 F CH$_3$COOH with 0.100 F NaOH in Fig. 10.5. An examination of the pH rise at the equivalence point reveals that it is not as steep as in the titration of a strong acid with a strong base. It is best to select an indicator that changes color right at the equivalence point. Phenolphthalein is a suitable indicator for the titration of acetic acid with a strong base as it changes color between pH = 8 and pH = 9.8. (See Table 10.2.)

Compare Fig. 10.2, the titration curve for the titration of a strong monoprotic acid versus NaOH, with Fig. 10.5, the titration curve for the titration of a weak monoprotic acid versus NaOH. Note the following three points of difference:

1. There is an initial rise in pH at the very beginning of the titration of a weak acid with a strong base that is not present at the beginning of the titration of a strong acid with a strong base.
2. For equal concentrations, the rise in pH at the equivalence point is less steep for the titration of a weak acid versus a strong base than it is for the titration of a strong acid versus a strong base.
3. The pH at the equivalence point is exactly 7.0 for the titration of a strong acid with a strong base, but is greater than 7.0 for the titration of a weak monoprotic acid with a strong base.

From the foregoing discussion you should be able to show that the pH at the equivalence point in the titration of a weak base with a strong acid is *less* than 7.0. The following example illustrates the method of calculating the pH at the equivalence point in such a titration.

EXAMPLE 10.10. Titration of a weak base with a strong acid

A 25.00-mL portion of a solution of the weak base methylamine, CH_3NH_2, is titrated with 0.1009 F HCl, and the equivalence point is reached when 25.57 mL of the acid have been added.

(a) What is the concentration of the methylamine?

Solution. The concentration of methylamine is obtained by equating the number of millimoles of base originally present to the number of millimoles of acid needed to reach the equivalence point:

$$(25.57 \text{ mL})(0.1009 \text{ mmol/mL}) = (25.00 \text{ mL})(C_B)$$

and hence

$$C_B = \frac{(25.57)(0.1009)}{(25.00)} = 0.1032 \text{ } M$$

(b) What is the pH at the equivalence point in this titration? What indicator should be used?

Solution. The number of millimoles of methylamine titrated was

$$(25.00 \text{ mL})(0.1032 \text{ mmol/mL}) = 2.580 \text{ mmol}$$

The titration reaction is

$$CH_3NH_2 + H_3O^+ \rightleftharpoons CH_3NH_3^+ + H_2O \qquad (10\text{-}27)$$

At the equivalence point, the titration reaction has gone to completion and there are 2.580 mmol of methylammonium ion, $CH_3NH_3^+$, in solution. The total volume of the solution at the equivalence point is 50.57 mL. The solution is therefore identical with a solution of methylammonium chloride of concentration

$$\frac{(2.580 \text{ mmol})}{(50.57 \text{ mL})} = 0.05102 \text{ } M$$

The acidity constant of methylammonium ion (Table E1, Appendix E) is 2.40×10^{-11}. A solution of methylammonium chloride is acidic due to the proton-transfer reaction between $CH_3NH_3^+$ and water:

$$CH_3NH_3^+ + H_2O \rightleftharpoons CH_3NH_2 + H_3O^+ \qquad (10\text{-}28)$$

Note that Eq. (10-28) is the reverse of the titration reaction.

$$[CH_3NH_2] = [H_3O^+] \text{ at the equivalence point}$$
$$\text{let } x = [H_3O^+] \text{ at the equivalence point}$$

Then

$$[CH_3NH_3^+] = 0.05102 - x$$

The equilibrium constant equation is

$$K_a(CH_3NH_3^+) = \frac{[CH_3NH_2][H_3O^+]}{[CH_3NH_3^+]} = \frac{x^2}{0.05102 - x} = 2.40 \times 10^{-11}$$

Since methylammonium ion is such a weak acid, we expect the $[H_3O^+]$ to be small compared to 0.05102. If we assume $0.05102 - x$ is approximately 0.05102, the equilibrium constant equation becomes

$$x^2 = (2.40 \times 10^{-11})(5.102 \times 10^{-2}) = 1.22 \times 10^{-12}$$

and hence

$$x = 1.11 \times 10^{-6}$$

Since 1.11×10^{-6} is small compared to 0.05102, the assumption we made is valid, and therefore at the equivalence point in this titration the $[H_3O^+] = 1.11 \times 10^{-6} \ M$, and the pH = 5.96. The best indicator to use, of those listed in Table 10.2, is chlorophenol red. Note that the pH at the equivalence point is less than 7.

There is one particular point in the titration of a monoprotic weak acid with a strong base that merits special attention, and that is the "halfway point," when the number of moles of added base is exactly half the number of moles of weak acid originally present. At this point, half the acid has been titrated, and $[HA] = [A^-]$, so that $[H_3O^+] = K_a$. You can see this in Table 10.3. When 25.00 mL of base have been added in the titration of 50.00 mL of 0.100 F HOAc with 0.100 F NaOH, the $[H_3O^+] = 1.8 \times 10^{-5} = K_a(\text{HOAc})$.

section 10.5
Calculations Involving Weak Diprotic Acids and Their Salts

When a diprotic acid, H_2A, is dissolved in water there will be in solution the three species: H_2A, HA^-, and A^{2-}. We want to calculate the pH of a solution of a diprotic acid of specified concentration, and the concentrations of all the species present at equilibrium.

As an important example we will consider the case of carbonic acid, which is simply a solution of the gas CO_2. Any water left standing open to the atmosphere will become slightly acidic due to atmospheric CO_2, which dissolves in the water, and reacts with it to a slight extent. Carbon dioxide in the air is in equilibrium with dissolved CO_2 in the water:

$$CO_2(g) \rightleftharpoons CO_2(aq) \tag{10-29}$$

and the equilibrium constant for this reaction is called a **distribution constant**, because it describes how CO_2 is distributed between the solution phase and the gas phase.

$$K_{\text{dist}} = \frac{[CO_2]}{P_{CO_2}} = 0.0337 \ M/\text{atm} \tag{10-30a}$$

If we rearrange this to solve for the pressure of CO_2 we obtain

$$P_{CO_2} = (1/K_{\text{dist}})[CO_2] = K'[CO_2] = 29.7[CO_2] \tag{10-30b}$$

Equation (10-30b) is a form of **Henry's Law,** which can be stated either as "the partial pressure of a gas in equilibrium with a solution is directly proportional to the concentration of the gas in the solution" or as "the concentration of a gas in a solution is directly proportional to the partial pressure of the gas in equilibrium with the solution." (Refer to Section 6.9.)

When CO_2 dissolves in water, the reaction

$$CO_2 + H_2O \rightleftharpoons H_2CO_3 \tag{10-31}$$

occurs to a slight extent. The bulk of the material in solution, however, consists of CO_2 molecules, so that the first stage of ionization of carbonic acid

$$H_2CO_3 + H_2O \rightleftharpoons HCO_3^- + H_3O^+ \tag{10-32a}$$

is best represented as

$$CO_2 + 2H_2O \rightleftharpoons HCO_3^- + H_3O^+ \qquad (10\text{-}32b)$$

The first ionization constant of carbonic acid ($CO_2 + H_2O$) is

$$K_1 = \frac{[HCO_3^-][H_3O^+]}{[CO_2]} = 4.3 \times 10^{-7} \qquad \text{at } 25\ ^\circ C \qquad (10\text{-}33)$$

The second stage of ionization

$$HCO_3^- + H_2O \rightleftharpoons CO_3^{2-} + H_3O^+ \qquad (10\text{-}34)$$

has an acidity constant

$$K_2 = \frac{[CO_3^{2-}][H_3O^+]}{[HCO_3^-]} = 4.7 \times 10^{-11} \qquad \text{at } 25\ ^\circ C \qquad (10\text{-}35)$$

Note that K_1 is close to ten thousand times larger than K_2.

Composition of an Aqueous Solution of CO₂

A solution of CO_2 in water contains HCO_3^-, CO_3^{2-}, H_3O^+, and OH^- ions, as well as CO_2, H_2CO_3, and H_2O molecules. Note that there is only one kind of positive ion in solution, H_3O^+, but there are three kinds of negative ions. The solution must, of course, be electrically neutral. A relation that equates the total amount of negative charge in solution to the total amount of positive charge is called an **electroneutrality equation** or a **charge balance equation.** For a solution of CO_2 in water, the electroneutrality equation is

$$[H_3O^+] = [HCO_3^-] + [OH^-] + 2[CO_3^{2-}] \qquad (10\text{-}36)$$

There is a 2 multiplying the $[CO_3^{2-}]$ in this equation because there must be two H_3O^+ ions present in the solution for each CO_3^{2-} ion.

All the carbonate ions in the solution are a result of the second stage of ionization of carbonic acid. Since K_2 is so much smaller than K_1, the $[CO_3^{2-}]$ is very small compared to the $[HCO_3^-]$. It is generally true for any diprotic acid for which K_1 is at least 500 times larger than K_2, that we can neglect the second stage of ionization relative to the first in a solution containing nothing else but the diprotic acid and water.

The OH^- ions in the solution are the result of the self-ionization of water, which occurs to only a small extent for two reasons: (1) K_w is very small, only 1.0×10^{-14}, and (2) the presence of the H_3O^+ from the first stage of ionization of carbonic acid represses the self-ionization of water. As a result, the $[OH^-]$ is also small compared to the $[HCO_3^-]$. If we neglect both the second stage of ionization of carbonic acid and the self-ionization of water relative to the first stage of ionization of carbonic acid (since both K_w and K_2 are very small compared to K_1), Eq. (10-36) becomes simply

$$[H_3O^+] = [HCO_3^-] \quad \text{in an aqueous solution of } CO_2 \qquad (10\text{-}37)$$

We can then treat the problem of finding the pH of a solution of carbonic acid just as if the carbonic acid were monoprotic. Details of the calculation are described in Example 10.11.

EXAMPLE 10.11. Calculation of the pH of water in equilibrium with the CO₂ of the air

The average pressure of CO_2 in the air is 3.0×10^{-4} atm. Calculate the pH of water in equilibrium with the air at 25 °C, and the concentration of all species present in this solution.

Solution. The concentration of CO_2 in water in equilibrium with air is obtained from Eq. (10-30a):

$$[CO_2] = 0.0337 P_{CO_2} = (3.37 \times 10^{-2} \ M/atm)(3.0 \times 10^{-4} \ atm) = 1.0 \times 10^{-5} \ M$$

Since K_1 is almost ten thousand times larger than K_2, we can ignore the second stage of ionization of carbonic acid and write

$$[H_3O^+] = [HCO_3^-]$$

The equation for the first acidity constant of carbonic acid, Eq. (10-33), therefore becomes simply

$$K_1 = 4.3 \times 10^{-7} = \frac{[HCO_3^-][H_3O^+]}{[CO_2]} = \frac{[H_3O^+]^2}{1.0 \times 10^{-5}}$$

and thus

$$[H_3O^+]^2 = 4.3 \times 10^{-12} \quad \text{and} \quad [H_3O^+] = 2.1 \times 10^{-6} \ M$$

Note that in this example $[CO_2]$ is a constant, $1 \times 10^{-5} \ M$, because the CO_2 in solution is in equilibrium with the air which has a constant partial pressure of CO_2, 3.0×10^{-4} atm.

We can verify the assumptions we made when we neglected the second stage of ionization by calculating the $[CO_3^{2-}]$ in this solution, using K_2.

$$K_2 = \frac{[H_3O^+][CO_3^{2-}]}{[HCO_3^-]} = 4.7 \times 10^{-11}$$

If $[H_3O^+] = [HCO_3^-]$, we obtain $[CO_3^{2-}] = K_2 = 4.7 \times 10^{-11}$. Since 4.7×10^{-11} is very small compared to 2.1×10^{-6}, it is correct to say that the $[CO_3^{2-}]$ is negligible compared to the $[HCO_3^-]$.

We can check to see whether our neglect of the self-ionization of water was justifiable by calculating the $[OH^-]$.

$$[OH^-] = \frac{K_w}{[H_3O^+]} = \frac{1.0 \times 10^{-14}}{2.1 \times 10^{-6}} = 4.8 \times 10^{-9} \ M$$

Since $[OH^-] \ll [HCO_3^-]$, and also $2[CO_3^{2-}] \ll [HCO_3^-]$, it is valid to set the right-hand side of Eq. (10-36) equal to just the $[HCO_3^-]$.

We can summarize the concentrations of all species in this solution as follows:

$$[CO_2] = 1.0 \times 10^{-5} \ M$$
$$[H_3O^+] = [HCO_3^-] = 2.1 \times 10^{-6} \ M$$
$$[CO_3^{2-}] = 4.7 \times 10^{-11} \ M$$
$$[OH^-] = 4.8 \times 10^{-9} \ M$$

The pH is $6 - \log(2.1) = 5.7$. Thus pure water exposed to the air has a pH of 5.7.

Titration of a Weak Diprotic Acid

The Solution at the First Equivalence Point

When we titrate a diprotic acid there will be two equivalence points. The first equivalence point occurs when the number of moles of OH^- added is equal to the number of moles of H_2A originally present. The titration reaction up to the first equivalence point is

$$H_2A + OH^- \rightleftharpoons HA^- + H_2O \qquad (10-38)$$

If NaOH is used as the titrant, the solution at the first equivalence point is exactly the same as a solution of the salt NaHA. The anion HA^- is an ampholyte (refer to Section 9.5), and can act either as an acid

$$HA^- + H_2O \rightleftharpoons A^{2-} + H_3O^+ \tag{10-39}$$

or as a base

$$HA^- + H_2O \rightleftharpoons H_2A + OH^- \tag{10-40}$$

The acidity constant of HA^-, $K_a(HA^-)$, is the second ionization constant of H_2A, and has been denoted K_2. The basicity constant of HA^- is given by

$$K_b(HA^-) = \frac{[H_2A][OH^-]}{[HA^-]} = \frac{[H_2A]}{[HA^-][H_3O^+]} \cdot [H_3O^+][OH^-] = \frac{K_w}{K_1(H_2A)} \tag{10-41}$$

A solution of the salt NaHA will be acidic if $K_a(HA^-) > K_b(HA^-)$; it will be basic if $K_b(HA^-) > K_a(HA^-)$. Both reactions (10-39) and (10-40) occur when NaHA is dissolved in water.

The pH of Solutions of Salts of the Type NaHA

Consider a C molar aqueous solution of the salt NaHA. The principal species in solution, in addition to H_2O, are Na^+ and HA^- ions, but there will also be H_2A molecules and A^{2-}, H_3O^+, and OH^- ions. The electroneutrality or charge balance equation for this solution is

$$[H_3O^+] + [Na^+] = [HA^-] + [OH^-] + 2[A^{2-}] \tag{10-42a}$$

The relation that expresses the conservation of mass of the species denoted A is called the **material balance** or **mass balance equation**, and is

$$C = [H_2A] + [HA^-] + [A^{2-}] \tag{10-43}$$

There is no reaction involving Na^+ ions, so $[Na^+] = C$. Substituting C for $[Na^+]$ into the electroneutrality equation we obtain

$$[H_3O^+] + C = [HA^-] + [OH^-] + 2[A^{2-}] \tag{10-42b}$$

Substitution of Eq. (10-43) into Eq. (10-42b) yields

$$[H_3O^+] + [H_2A] = [OH^-] + [A^{2-}] \tag{10-44}$$

Provided that C is sufficiently large, it is usually a good approximation that

$$[H_2A] \gg [H_3O^+] \quad \text{and} \quad [A^{2-}] \gg [OH^-]$$

so that Eq. (10-44) can be simplified to

$$[H_2A] \cong [A^{2-}] \tag{10-45}$$

Equation (10-45) is the first approximation to make in calculating the pH of solutions of salts of the type NaHA, in which the ampholyte HA^- is a principal species. Another way to consider the approximation stated in Eq. (10-45) is to combine Eqs. (10-39) and (10-40) into one overall reaction that can be thought of as the principal reaction in a solution of NaHA:

$$HA^- + H_2O \rightleftharpoons A^{2-} + H_3O^+ \tag{10-39}$$
$$\underline{HA^- + H_2O \rightleftharpoons H_2A + OH^-} \tag{10-40}$$
$$2HA^- \rightleftharpoons H_2A + A^{2-} \tag{10-46}$$

In obtaining Eq. (10-46) we have summed Eqs. (10-39) and (10-40) and on the right-hand side converted $H_3O^+ + OH^-$ to $2H_2O$.

If we consider Eq. (10-46) we may conclude that in a solution of NaHA

$$[H_2A] \cong [A^{2-}] \tag{10-45}$$

This is, as we have already noted, the approximation usually made, but you should realize that it is based on the assumption that Eqs. (10-39) and (10-40) occur to *roughly* the same extent. This is usually a good approximation, but it is not always valid and should be checked by substituting the values obtained using this approximation into Eq. (10-44).

If we make the approximation of Eq. (10-45), it is easy to obtain a value for the $[H_3O^+]$ from the overall two stages of ionization of H_2A. [Refer to Section 9.4, Eqs. (9-27) and (9-28).] The overall equilibrium constant is

$$K_{overall} = K_1 K_2 = \frac{[H_3O^+]^2[A^{2-}]}{[H_2A]}$$

If $[A^{2-}] = [H_2A]$,

$$[H_3O^+]^2 = (K_1 K_2) \tag{10-47a}$$

and

$$[H_3O^+] = (K_1 K_2)^{1/2} \tag{10-47b}$$

Taking the logarithm of both sides of Eq. (10-47b) we obtain

$$pH = \tfrac{1}{2}(pK_1 + pK_2) \text{ for a solution of NaHA} \tag{10-47c}$$

Note that according to Eq. (10-47c), the pH of any solution of NaHA is the same, regardless of the concentration of the salt. The pH is independent of the concentration of the HA^- ion if the approximation made in Eq. (10-45) is valid. The following circumstances will result in significant errors in Eqs. (10-45) and (10-47a): (1) if K_1 is greater than 10^{-2}, (2) if K_2 is less than 10^{-13}, or (3) if C is low and of the same order of magnitude as K_1.

The following example illustrates the use of Eq. (10-47a).

EXAMPLE 10.12. Calculation of the concentrations of all species present in a solution of a salt of the type NaHA

Calculate the concentration of all species present in a 0.100 F NaHCO$_3$ solution at 25 °C.

Solution. The bicarbonate ion, HCO_3^- is an ampholyte. We have already discussed the fact that solutions of NaHCO$_3$ are basic in Section 9.5. The acidity constant for HCO_3^-, which is the second ionization constant of carbonic acid, is 4.7×10^{-11}. When HCO_3^- functions as a base, its conjugate acid is carbonic acid, or aqueous CO_2. The basicity reaction for bicarbonate ion is Eq. (9-45), and

$$K_b = \frac{K_w}{K_1(CO_2 + H_2O)} = \frac{1.0 \times 10^{-14}}{4.3 \times 10^{-7}} = 2.3 \times 10^{-8}$$

Since $K_b(HCO_3^-) > K_a(HCO_3^-)$, a solution of NaHCO$_3$ is basic.

The $[H_3O^+]$ in a solution of NaHCO$_3$ can be obtained using Eq. (10-47a).

$$[H_3O^+]^2 = K_1 K_2 = (4.3 \times 10^{-7})(4.7 \times 10^{-11}) = 20 \times 10^{-18}$$

so that

$$[H_3O^+] = 4.5 \times 10^{-9} \ M \quad \text{and} \quad pH = 8.35$$

The initial concentration of HCO_3^- is 0.100 M, but at equilibrium the $[HCO_3^-]$ is somewhat less than 0.100 M, because both CO_3^{2-} and CO_2 have been formed as HCO_3^-

reacts with water, as an acid and as a base. Conservation of mass requires that

$$0.100 = [HCO_3^-] + [CO_3^{2-}] + [CO_2]$$

The equation above is the **material balance equation.**

The sum of the acidity and basicity reactions for HCO_3^- is

$$2HCO_3^- \rightleftharpoons CO_3^{2-} + CO_2 + H_2O \qquad (10\text{-}48)$$

This is the principal reaction in a solution of $NaHCO_3$. It is Eq. (10-46) applied to HCO_3^-, but the species H_2CO_3 is written as $H_2O + CO_2$. According to this overall reaction, $[CO_2] = [CO_3^{2-}]$, and we have already used that approximation when we determined the pH using Eq. (10-47a). If we introduce that relation into the material balance equation, we obtain

$$0.100 = [HCO_3^-] + 2[CO_3^{2-}] = [HCO_3^-] + 2[CO_2]$$

This can be rearranged to solve for the $[HCO_3^-]$:

$$[HCO_3^-] = 0.100 - 2[CO_3^{2-}] = 0.100 - 2[CO_2]$$

The equilibrium constant for the principal reaction, Eq. (10-48), is

$$K_{prin} = \frac{[CO_2][CO_3^{2-}]}{[HCO_3^-]^2} = \frac{[CO_2]}{[H_3O^+][HCO_3^-]} \cdot \frac{[CO_3^{2-}][H_3O^+]}{[HCO_3^-]} = \frac{K_2}{K_1}$$
$$= (4.7 \times 10^{-11})/(4.3 \times 10^{-7}) = 1.1 \times 10^{-4}$$

Let $x = [CO_2] = [CO_3^{2-}]$ in 0.100 F $NaHCO_3$. Then

$$[HCO_3^-] = 0.100 - 2x$$

and

$$K_{prin} = 1.1 \times 10^{-4} = \frac{x^2}{(0.100 - 2x)^2}$$

Taking the square root of both sides of this equation we obtain

$$1.0_5 \times 10^{-2} = \frac{x}{0.100 - 2x}$$

which can be multiplied out and solved for x. We find

$$x = [CO_3^{2-}] = [CO_2] = 1.0 \times 10^{-3} M$$

and therefore $[HCO_3^-] = 0.100 - 2(0.001) = 0.098\ M$.

We had already found that $[H_3O^+] = 4.5 \times 10^{-9}\ M$, so that

$$[OH^-] = \frac{K_w}{[H_3O^+]} = \frac{1.0 \times 10^{-14}}{4.5 \times 10^{-9}} = 2.2 \times 10^{-6}\ M$$

Since both $[H_3O^+]$ and $[OH^-]$ are very much less than $1.0 \times 10^{-3}\ M$, Eq. (10-45) is a valid approximation for this solution.

In the titration of a diprotic acid, H_2A, with NaOH, the solution at the first equivalence point is a solution of the salt NaHA. Since HA^- is an ampholyte, such a solution may be either acidic or basic, depending on the relative magnitudes of $K_a(HA^-)$ and $K_b(HA^-)$. In order to select an indicator suitable for detecting the first equivalence point of such a titration, Eq. (10-47c) is used to determine the pH at that equivalence point.

The Solution at the Second Equivalence Point

The second equivalence point in the titration of a diprotic acid occurs when the number of moles of OH^- added is twice the number of moles of H_2A originally present. The overall reaction to reach the second equivalence point is

$$H_2A + 2\,OH^- \rightleftharpoons A^{2-} + 2H_2O \qquad (10\text{-}49)$$

and the reaction that occurs between the first and the second equivalence points is

$$HA^- + OH^- \rightleftharpoons A^{2-} + H_2O \qquad (10\text{-}50)$$

If the titrant is NaOH, the solution at the second equivalence point is therefore identical to a solution of the salt Na_2A. A solution of Na_2A is, of course, basic, because the dinegative anion A^{2-} is a base. The basicity reaction

$$A^{2-} + H_2O \rightleftharpoons HA^- + OH^- \qquad (10\text{-}51)$$

has the equilibrium constant

$$K_b\,(A^{2-}) = \frac{[HA^-][OH^-]}{[A^{2-}]} = \frac{[HA^-][OH^-][H_3O^+]}{[A^{2-}][H_3O^+]} = \frac{K_w}{K_2(H_2A)} \qquad (10\text{-}52)$$

The basicity reaction of A^{2-}, Eq. (10-51), is the principal reaction in a solution of Na_2A. Provided that $K_b(A^{2-})$ is at least 500 times larger than $K_b(HA^-)$, it is valid to neglect the basicity reaction of HA^-, and to set $[HA^-] = [OH^-]$ in a solution of Na_2A.

EXAMPLE 10.13. Calculation of the concentrations of all the species present in a solution of a salt of the type Na_2A

Calculate the concentrations of all species present in a 0.0500 F Na_2CO_3 solution.

Solution. The carbonate ion is a base, and the principal reaction in a solution of Na_2CO_3 is

$$CO_3^{2-} + H_2O \rightleftharpoons HCO_3^- + OH^-$$

The basicity constant of carbonate ion is

$$K_b(CO_3^{2-}) = \frac{[HCO_3^-][OH^-]}{[CO_3^{2-}]} = \frac{K_w}{K_a(HCO_3^-)} = \frac{1.0 \times 10^{-14}}{4.7 \times 10^{-11}} = 2.1 \times 10^{-4}$$

The basicity constant for HCO_3^- was found to be 2.3×10^{-8} in Example 10.12. This value is so much smaller than 2.1×10^{-4} that we can neglect the basicity of HCO_3^- compared to the basicity of CO_3^{2-}, and assume that $[HCO_3^-] = [OH^-]$ in a solution of Na_2CO_3.

As the solution is 0.0500 F Na_2CO_3, the mass balance equation, which indicates the conservation of mass, is

$$0.0500 = [CO_3^{2-}] + [HCO_3^-] + [CO_2]$$

The $[CO_2]$, however, is very much smaller than either the $[CO_3^{2-}]$ or the $[HCO_3^-]$, because it is formed by the basicity reaction of HCO_3^-, and $K_b(HCO_3^-)$ is so much smaller than $K_b(CO_3^{2-})$. Hence we can neglect the $[CO_2]$ in the mass balance equation, and rearrange it to solve for the $[CO_3^{2-}]$, as follows:

$$[CO_3^{2-}] = 0.0500 - [HCO_3^-]$$

$$\text{let } y = [HCO_3^-] = [OH^-] \quad \text{in } 0.0500\ F\ Na_2CO_3$$

then
$$[CO_3^{2-}] = 0.0500 - y$$

and the equilibrium constant equation is

$$K_b(CO_3^{2-}) = 2.1 \times 10^{-4} = \frac{y^2}{0.0500 - y}$$

To solve this equation, assume that $0.0500 - y = 0.0500$. Then

$$y^2 = (5.00 \times 10^{-2})(2.1 \times 10^{-4}) = 10.5 \times 10^{-6}$$

and
$$y = 3.2_4 \times 10^{-3}$$

Since 3.2×10^{-3} is not very much smaller than 5.00×10^{-2} (it is 6.4% of 0.0500) we might make a second approximation and set

$$0.0500 - y = 0.0500 - 0.0032 = 0.0468$$

The equilibrium constant equation then becomes

$$2.1 \times 10^{-4} = y^2/(4.68 \times 10^{-2}) \quad \text{or} \quad y^2 = 9.83 \times 10^{-6}$$

and therefore

$$y = 3.1 \times 10^{-3}$$

If we make a third approximation and set

$$0.0500 - y = 0.0500 - 0.0031 = 0.0469$$

we still obtain $y = 3.1 \times 10^{-3}$ on solving the equilibrium constant expression. Since two successive approximations yield the same value, this is the correct answer. A summary of our results is

$$[OH^-] = [HCO_3^-] = 3.1 \times 10^{-3} \, M$$
$$[CO_3^{2-}] = 0.0500 - 0.0031 = 0.0469 \, M$$
$$[H_3O^+] = K_w/[OH^-] = 3.2 \times 10^{-12} \, M \quad \text{and} \quad pH = 12 - \log(3.2) = 11.49$$

Note that this solution is distinctly basic. Carbonate ion is a moderately strong base, with $K_b = 2.1 \times 10^{-4}$.

We can also calculate the $[CO_2]$, using K_1 for carbonic acid.

$$K_1 = 4.3 \times 10^{-7} = \frac{[H_3O^+][HCO_3^-]}{[CO_2]} = \frac{(3.2 \times 10^{-12})(3.1 \times 10^{-3})}{[CO_2]}$$

and therefore

$$[CO_2] = 2.3 \times 10^{-8} \, M$$

Since 2.3×10^{-8} is negligible compared to the $[CO_3^{2-}]$ and the $[HCO_3^-]$, it is valid to neglect the $[CO_2]$ in the material balance equation.

If we titrate a solution of carbonic acid with NaOH, at the first equivalence point we have a solution of $NaHCO_3$ and at the second equivalence point we have a solution of Na_2CO_3. The methods of calculation described in Examples 10.12 and 10.13, therefore, enable us to determine the pH at the two equivalence points.

A typical titration curve for the titration of carbonic acid with NaOH is shown in Fig. 10.6. The two equivalence points are observable and well separated, because the two acidity constants for carbonic acid differ widely, by a factor of nearly 10,000.

For diprotic acids for which K_1 and K_2 are not widely separated (as an example, for succinic acid for which $K_1/K_2 = 27$), the two stages of ionization overlap and the titration curve does not display two separate, relatively steep rises in pH.

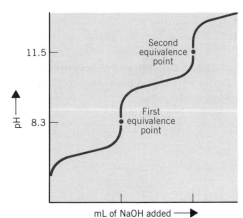

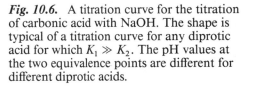

Fig. 10.6. A titration curve for the titration of carbonic acid with NaOH. The shape is typical of a titration curve for any diprotic acid for which $K_1 \gg K_2$. The pH values at the two equivalence points are different for different diprotic acids.

Summary

A **buffer solution** is a solution that maintains an approximately constant pH when small amounts of either a strong acid or a strong base are added.

A solution that contains both a weak acid and its conjugate weak base is a buffer, provided that (1) the number of moles of both the weak acid and its conjugate weak base is large compared to the number of moles of strong acid or base that may be added, and (2) the [weak acid]/[conjugate base] ratio is not larger than 10 nor smaller than 1/10. The buffering action will be most efficient if the ratio is 1 : 1.

A buffer in which the [weak acid]/[conjugate base] ratio is 1 : 1 has a pH equal to the pK_a of the weak acid. The range of pH values of efficient buffers for a given weak acid/conjugate base system is from $pK_a - 1$ to $pK_a + 1$.

A buffer solution containing the weak acid HA and its conjugate base A^- can be prepared in three different ways: (1) by mixing HA and a soluble salt such as NaA or KA, (2) by mixing a strong base such as NaOH with excess HA, and (3) by mixing a strong acid such as HCl with excess NaA.

To determine the concentration of acid in a solution, we **titrate** it with a strong base that has been **standardized,** that is, the concentration of the base has been accurately determined. The standardized base is added to the acid solution from a **buret.** When the number of millimoles of OH^- added is exactly equal to the number of millimoles of acid originally present, the **equivalence point** of the titration has been reached. Because the pH rises sharply at the equivalence point, the volume of base required to reach the equivalence point can be determined precisely.

An **acid–base indicator** is a substance that changes color within a fairly narrow range of pH values, usually two pH units or less. If the color change takes place close to the pH at the equivalence point of a titration, the indicator can be used to signal the end of the titration.

For the titration of a strong acid versus a strong base, the pH at the equivalence point is exactly 7.00. The more concentrated the acid and base, the steeper is the rise in pH right at the equivalence point. If the concentrations of acid and base are significantly smaller than $0.100\ F$, the pH rise at the equivalence point is not so steep, and only an indicator that changes color quite close to pH $= 7$ can be used.

For the titration of a monoprotic weak acid versus a strong base, the pH is always greater than 7.00 at the equivalence point. An indicator, such as phenolphthalein, that changes color only when the solution is distinctly basic must be used.

In the titration of a weak diprotic acid, H_2A, with a strong base, there are two equivalence points. If NaOH is used as the **titrant,** the solution at the first equivalence point is identical to a solution of the salt NaHA, while the solution at the second equivalence point is identical to a solution of Na_2A. Because HA^- is an ampholyte, the pH at the first equivalence point can be either less than or greater than 7, depending on the relative magnitudes of $K_a(HA^-)$ and $K_b(HA^-)$. The pH at the second equivalence point is always greater than 7.

Exercises

Section 10.1

1. A buffer solution is made by mixing 0.250 mol of NH_4Cl and 0.150 mol of NH_3 with enough water to make 1.00 L of solution. Calculate the pH of this buffer.

2. A buffer solution is made by dissolving 6.106 g of benzoic acid, C_6H_5COOH, and 14.411 g of sodium benzoate, C_6H_5COONa, in an alcohol/water mixture to make 500.00 mL of solution. (Benzoic acid is only slightly soluble in water at 25 °C.) Calculate the pH of this buffer.

3. To prepare an acetic acid/acetate buffer with pH $= 4.32$, what molar ratio of acetic acid to sodium acetate must you use?

4. Calculate the pH of a buffer prepared by mixing 20.00 mL of 0.500 F KOH with 80.00 mL of 0.500 F HOAc.

5. A 0.100 F CH_3COOH solution contains acetate ions as a result of the proton-transfer reaction between CH_3COOH and H_2O. Since this solution contains both a weak acid (HOAc) and its conjugate weak base (OAc^-), why is it not a buffer?

6. Calculate the pH of a buffer prepared by mixing 40.00 mL of 0.400 F NaOH with 80.00 mL of 0.500 F HCOOH, formic acid.

7. Calculate the pH of a buffer prepared by mixing 60.00 mL of 0.400 F sodium formate with 40.00 mL of 0.360 F HCl.

8. Calculate the pH of a buffer prepared by mixing 80.00 mL of 1.00 F NH_3 with 40.00 mL of 1.00 F HCl.

9. Calculate the pH of a buffer prepared by mixing 50.00 mL of 0.200 F NH_4NO_3 with 25.00 mL of 0.200 F NaOH.

10. Explain why it is not possible to make a buffer with pH $= 6.5$ using a mixture of NH_3 and NH_4Cl.

Section 10.2

11. Exactly 50.00 mL of an HBr solution of unknown concentration is titrated with 0.100 F NaOH. It requires 42.50 mL of the base to reach the equivalence point of the titration. What was the $[H_3O^+]$ in the HBr solution?

12. A 25.00-mL sample of a sulfuric acid (H_2SO_4) solution of unknown concentration is titrated with 0.200 F NaOH. The equivalence point is reached when 34.75 mL of base have been added. What was the formality of the sulfuric acid solution?

13. A 20.00-mL sample of 0.100 F HCl is titrated with 0.100 F NaOH. What is the pH of the solution when 19.50 mL of base have been added? What is the pH when 20.50 mL of base have been added?

14. A 25.00-mL sample of a sodium hydroxide solution of unknown concentration is titrated with 0.1000 F HCl. It requires 31.83 mL of the acid to reach the equivalence point of the titration. What is the concentration of the sodium hydroxide solution?

15. A student in a general chemistry lab is titrating a strong acid of unknown concentration with 0.100 F NaOH. By accident, the student overtitrates, that is, more base is added than is required to reach the equivalence point. To remedy this error the student adds exactly 5.00 mL of 0.100 F HCl to the titration vessel, and continues titrating. The equivalence point after the addition occurs when 46.72 mL of base have been added. What was the concentration of the student's original acid solution?

Section 10.3

16. Thymol blue is an acid–base indicator. The acidic form of thymol blue is yellow, the basic form is blue. If the acidity constant for thymol blue is 1.6×10^{-9}, what color will a solution containing several drops of thymol blue be at (a) pH $= 6.5$, (b) pH $= 7.5$, (c) pH $= 8.5$, and (d) pH $= 9.5$?

17. Which of the indicators listed in Table 10.2 is suitable for the titration of (a) 50.00 mL of 0.500 F HCl versus 0.500 F NaOH, (b) 50.00 mL of 0.050 F HCl versus 0.050 F NaOH, and (c) 50.00 mL of 0.0050 F HCl versus 0.0050 F NaOH?

Section 10.4

18. Write a correctly balanced net ionic equation for the reaction that occurs during the titration of

  (a) Benzoic acid with potassium hydroxide.
  (b) Ammonia with hydrochloric acid.
  (c) Formic acid with sodium hydroxide.
  (d) Aniline, $C_6H_5NH_2$, with perchloric acid.

19. Calculate the pH at the equivalence point in the titration of 50.00 mL of 0.0200 F benzoic acid with 0.0200 F NaOH.

20. Calculate the pH at the equivalence point in the titration of 25.00 mL of 0.200 F formic acid with 0.200 F NaOH.

21. A 20.00-mL sample of 0.500 F acetic acid is titrated with 0.500 F NaOH. Calculate the pH of the solution (a) before the titration begins, (b) when 10.00 mL of base have been added, (c) when 19.80 mL of base have been added, (d) at the equivalence point, and (e) when 20.20 mL of base have been added.

22. Which of the indicators in Table 10.2 would you use for the titration of 40.00 mL of 0.100 F HNO_2 versus 0.100 F NaOH?

23. Which of the indicators in Table 10.2 would you use for the titration of 50.00 mL of 0.100 F aniline, $C_6H_5NH_2$, with 0.100 F HCl?

24. A 20.00-mL sample of a solution of an unknown weak monoprotic acid is titrated with 0.1022 F NaOH. The pH when 14.82 mL of titrant have been added is measured and found to be 4.87. It requires 29.64 mL of base to reach the equivalence point.

   What is the acidity constant of this acid? What was the concentration of the original solution?

Section 10.5

25. Write a correctly balanced net ionic equation for the reaction that occurs during the titration of oxalic acid with potassium hydroxide,

  (a) Up to the first equivalence point.
  (b) Between the first and second equivalence points.
  (c) From the beginning of the titration up to the second equivalence point.

26. An aqueous solution that is saturated with CO_2 at 25 °C contains 0.145 g of CO_2/100 mL of solution. Calculate
   (a) The concentration of CO_2 in a saturated solution at 25 °C.
   (b) The pressure of CO_2 in the gas phase in equilibrium with a saturated solution at 25 °C.
   (c) The pH of a saturated solution of CO_2 at 25 °C.

27. Phthalic acid is a diprotic acid for which $pK_1 = 3.10$, and $pK_2 = 5.40$. What is the pH of a 0.100 F solution of potassium hydrogen phthalate (also called potassium bi-phthalate)?

28. Ascorbic acid (vitamin C) is a diprotic acid for which K_1 is 6.76×10^{-5}, and K_2 is 2.69×10^{-12}. If 50.00 mL of a 0.0600 F solution of ascorbic acid is titrated with 0.100 F NaOH,
   (a) What is the pH at the first equivalence point? Which of the indicators listed in Table 10.2 would you use to signal the first equivalence point in this titration?
   (b) How many milliliters of base are required to reach the first equivalence point?
   (c) How many milliliters of base are required to reach the second equivalence point?

29. The principal reaction that occurs in a solution of sodium binoxalate, $NaHC_2O_4$, is
$$2HC_2O_4^- \rightleftharpoons H_2C_2O_4 + C_2O_4^{2-}$$
   Calculate the equilibrium constant for this reaction.

30. List all the substances present in a 0.100 F Na_2S solution. Write the electroneutrality and material balance equations for this solution.

31. List all the substances present in a 0.100 F NaHS solution. Write the electroneutrality and material balance equations for this solution.

32. What is the pH of a 0.100 F $Na_2C_2O_4$ solution?

33. Calculate the concentrations of all species present in a 0.100 F K_2CO_3 solution.

34. Calculate the concentrations of all species present in a 0.0800 F $KHCO_3$ solution.

Multiple Choice Questions

1. To prepare a buffer with pH close to 3.4, you could use a mixture of
   (a) NH_4NO_3 and NH_3 (b) HOCl and NaOCl   (c) HOAc and NaOAc
   (d) HNO_2 and $NaNO_2$ (e) $NaHCO_3$ and Na_2CO_3

2. To prepare a buffer with pH close to 9.0, you could use a mixture of
   (a) NH_4NO_3 and NH_3 (b) HOCl and NaOCl   (c) HOAc and NaOAc
   (d) HNO_2 and $NaNO_2$ (e) $NaHCO_3$ and Na_2CO_3

3. In titrating 0.100 F HCOOH (formic acid) with 0.100 F NaOH, the solution at the equivalence point is
   (a) 0.100 F HCOONa   (b) H_2O (c) 0.0500 F HCOOH
   (d) 0.0500 F NaOH   (e) 0.0500 F HCOONa

4. A buffer that is a mixture of acetic acid and potassium acetate has a pH = 5.24. The $[OAc^-]/[HOAc]$ ratio in this buffer is
   (a) $1:1$ (b) $3:1$ (c) $5:1$ (d) $1:3$ (e) $1:5$

5. A 40.00-mL sample of benzoic acid of unknown concentration was titrated with 0.1025 F KOH. The equivalence point was reached when 29.38 mL of base had been added. The concentration of benzoate ion at the equivalence point is
   (a) 0.1025 M (b) 0.0753 M (c) 0.0591 M (d) 0.0434 M (e) 0.0376 M

6. The net ionic equation for the reaction that occurs when a solution of nitrous acid is titrated with potassium hydroxide is
   (a) $HNO_2 + KOH \rightleftharpoons KNO_2 + H_2O$
   (b) $H_3O^+ + OH^- \rightleftharpoons 2H_2O$
   (c) $NO_2^- + H_2O \rightleftharpoons HNO_2 + OH^-$
   (d) $HNO_2 + H_2O \rightleftharpoons NO_2^- + H_3O^+$
   (c) $HNO_2 + OH^- \rightleftharpoons NO_2^- + H_2O$

7. A 50.00-mL sample of $0.0100\ F\ Ba(OH)_2$ is titrated with $0.0100\ F\ HCl$. The solution at the equivalence point is
   (a) $3.33 \times 10^{-3}\ F\ BaCl_2$ (b) H_2O (c) $5.00 \times 10^{-3}\ F\ BaCl_2$
   (d) $2.50 \times 10^{-3}\ F\ BaCl_2$ (e) $1.00 \times 10^{-2}\ F\ BaCl_2$

8. A flask contains 100.00 mL of $0.100\ F\ HOAc$. To prepare a buffer with pH $= 4.75$, which of the following samples of barium acetate solution should be added to the flask?
   (a) 50.00 mL of $0.400\ F\ Ba(OAc)_2$ (b) 25.00 mL of $0.200\ F\ Ba(OAc)_2$
   (c) 50.00 mL of $0.200\ F\ Ba(OAc)_2$ (d) 100.00 mL of $0.100\ F\ Ba(OAc)_2$
   (e) 200.00 mL of $0.100\ F\ Ba(OAc)_2$

9. A 50.00-mL sample of acetic acid was titrated with $0.1200\ F\ KOH$, and 38.62 mL of base were required to reach the equivalence point. What was the pH of the titration mixture when 19.31 mL of base had been added?
   (a) 2.9   (b) 3.5   (c) 4.7   (d) 5.7   (e) 7.0

10. Which of the mixtures listed is a buffer with pH close to 9?
   (a) 50.00 mL of $0.100\ F\ NH_4Cl$ and 50.00 mL of $0.100\ F\ NaOH$
   (b) 50.00 mL of $0.100\ F\ NH_4Cl$ and 25.00 mL of $0.100\ F\ NaOH$
   (c) 50.00 mL of $0.100\ F\ NH_4Cl$ and 50.00 mL of $0.100\ F\ HCl$
   (d) 50.00 mL of $0.100\ F\ NH_4Cl$ and 25.00 mL of $0.100\ F\ HCl$
   (e) 25.00 mL of $0.100\ F\ NH_4Cl$ and 50.00 mL of $0.100\ F\ NaOH$

11. A solution prepared by dissolving 0.0100 mol of benzoic acid and 0.0200 mol of sodium benzoate in 500.0 mL of water has a pH between
   (a) 2 and 3   (b) 3 and 4   (c) 4 and 5   (d) 5 and 6   (e) 6 and 7

12. If 50.00 mL of $0.200\ F\ KOH$ is added to 40.00 mL of $0.500\ F\ HCOOH$, the pH of the resulting solution is
   (a) 7.00   (b) 5.50   (c) 4.00   (d) 3.75   (e) 3.45

13. Methylamine, CH_3NH_2, is a weak base with $K_b = 4.2 \times 10^{-4}$ at 25 °C. A 40.00-mL sample of an aqueous solution of methylamine is titrated with $0.150\ F\ HCl$, and the equivalence point is reached when 39.26 mL of the acid have been added. What is the pH at the equivalence point of this titration?
   (a) 8.35   (b) 7.00   (c) 6.87   (d) 6.37   (e) 5.87

14. When 75.00 mL of $0.100\ F\ HNO_3$ have been added to 45.00 mL of $0.100\ F\ NaOH$, the pH of the resulting solution is
   (a) 1.0   (b) 1.2   (c) 1.4   (d) 1.6   (e) 1.8

15. Carbon dioxide is dissolved in water under pressure until $[CO_2] = 0.050\ M$. The pH of this carbonated water is
   (a) 2.8   (b) 3.8   (c) 4.8   (d) 5.8   (e) 6.8

16. Which of the following mixtures will be a buffer solution when dissolved in 500.00 mL of water?

(a) 0.200 mol of aniline and 0.200 mol of HCl
(b) 0.200 mol of aniline and 0.400 mol of NaOH
(c) 0.200 mol of NaCl and 0.100 mol of HCl
(d) 0.200 mol of NaCl and 0.100 mol of NaOH
(e) 0.200 mol of aniline and 0.100 mol of HCl

17. A 25.00-mL sample of a monoprotic weak acid is titrated with 0.1016 F NaOH. The equivalence point is reached when 28.54 mL of base have been added. If K_a of this acid is 4.0×10^{-6}, a suitable indicator to use to detect the equivalence point is

(a) phenolphthalein   (b) thymolphthalein   (c) bromcresol green
(d) methyl red   (e) bromthymol blue

18. A weak base, B, has basicity constant $K_b = 2 \times 10^{-5}$. The pH of any solution in which $[B] = [BH^+]$ is

(a) 4.7   (b) 7.0   (c) 9.3   (d) 9.7   (e) 10.3

19. The net ionic equation for the reaction that occurs when barium hydroxide is titrated with sulfuric acid is

(a) $Ba^{2+} + 2\,OH^- + 2H^+ + SO_4^{2-} \rightleftharpoons BaSO_4 \downarrow + 2H_2O$
(b) $Ba(OH)_2 + 2H^+(aq) \rightleftharpoons 2H_2O + Ba^{2+}$
(c) $Ba^{2+} + SO_4^{2-} \rightleftharpoons BaSO_4 \downarrow$
(d) $Ba(OH)_2 + H_2SO_4 \rightleftharpoons BaSO_4 \downarrow + 2H_2O$
(e) $H_3O^+ + OH^- \rightleftharpoons 2H_2O$

20. A 20.00-mL sample of a 0.200 F solution of the weak diprotic acid H_2X is titrated with 0.250 F KOH. The solution at the *second* equivalence point is

(a) 0.100 F KHX   (b) 0.111 F KHX   (c) 0.100 F K_2X (d) 0.0667 F K_2X
(e) 0.0769 F K_2X

21. A solution saturated with H_2S at 25 °C and 1-atm pressure is 0.10 M in H_2S. What is the $[HS^-]$ in this solution?

(a) $1.0 \times 10^{-1}\ M$ (b) $1.0 \times 10^{-2}\ M$ (c) $1.0 \times 10^{-4}\ M$ (d) $1.0 \times 10^{-6}\ M$
(e) $1.0 \times 10^{-7}\ M$

22. What is the pH of 0.10 M H_2S at 25 °C?

(a) 1.0   (b) 2.0   (c) 4.0   (d) 6.0   (e) 7.0

23. What is the $[S^{2-}]$ in 0.10 M H_2S at 25 °C?

(a) $1 \times 10^{-20}\ M$ (b) $1 \times 10^{-13}\ M$ (c) $1 \times 10^{-8}\ M$ (d) $1 \times 10^{-7}\ M$
(e) $1 \times 10^{-4}\ M$

24. What is the $[H_3O^+]$ in a solution made by mixing 60.0 mL of 1.00 F sodium acetate with 40.0 mL of 0.50 F HCl?

(a) $5.4 \times 10^{-5}\ M$ (b) $3.6 \times 10^{-5}\ M$ (c) $1.8 \times 10^{-5}\ M$ (d) $9.0 \times 10^{-6}\ M$
(e) $6.0 \times 10^{-6}\ M$

25. It is necessary to maintain a solution at pH = 3.2. Which of the following pairs could be used to prepare a buffer with this pH?

(a) NH_4^+/NH_3 (b) $HCOOH/HCOO^-$ (c) HCO_3^-/CO_3^{2-} (d) HS^-/S^{2-}
(e) CH_3COOH/CH_3COO^-

   The following options are to be used for answers to questions 26–31. All solutions are at 25 °C.

(a) A solution with a pH less than 7 that is not a buffer.
(b) A buffer solution with a pH between 4 and 7.

(c) A solution with a pH of 7.

(d) A buffer solution with a pH between 7 and 10.

(e) A solution with a pH greater than 7 that is not a buffer.

Which of the statements listed above best describes each of the solutions in questions 26–31?

26. A mixture of 1 mol of KCl and 1 mol of HCl in enough water to make a liter of solution.

27. A mixture of 1 mol of Na_2CO_3 and 1 mol of CH_3COONa in enough water to make a liter of solution.

28. A mixture of 0.50 mol of HOAc and 1 mol of NaOAc in enough water to make a liter of solution.

29. A mixture of 0.50 mol of NaOH and 0.10 mol NH_3 in enough water to make a liter of solution.

30. A mixture of 40.00 mL of 1.00 F HCl with 40.00 mL of 1.00 F KOH.

31. A mixture of 50.00 mL of 1.00 F HCl with 25.00 mL of 1.00 F NaOAc.

Problems

10.1. Potassium hydrogen phthalate (also called potassium biphthalate) is a white crystalline solid with molecular weight 204.23. It is used to standardize solutions of bases. A sample of pure potassium hydrogen phthalate weighing 0.8097 g was dissolved in water and titrated with a sodium hydroxide solution. To reach the equivalence point in this titration required 40.25 mL of base. The same NaOH solution was then used to determine the molecular weight of an unknown weak monoprotic acid, HX. A sample of 1.8694 g of HX was dissolved in distilled water and the volume brought to exactly 100.00 mL in a volumetric flask. A 25.00-mL portion of this solution was titrated versus the standardized NaOH, and 28.02 mL of the base were required to reach the equivalence point.

(a) Calculate the concentration of the NaOH solution.

(b) Calculate the molecular weight of the weak acid HX.

10.2. Calculate the concentrations of all the species present at the equivalence point in the titration of 50.00 mL of 0.0450 F C_5H_5N (pyridine) with 0.0650 F HCl. Pyridinium ion, $C_5H_5NH^+$, is a weak acid with $pK_a = 5.17$.

10.3. A solution of glycine hydrochloride contains the chloride ion and the glycinium ion, ^+H_3N—CH_2—COOH. The glycinium ion, which may be abbreviated GH_2^+, is a diprotic acid. The two stages of ionization are

$$^+H_3NCH_2COOH + H_2O \rightleftharpoons {}^+H_3NCH_2COO^- + H_3O^+ \qquad K_1 = 4.47 \times 10^{-3} \quad (1)$$

$$^+H_3NCH_2COO^- + H_2O \rightleftharpoons H_2NCH_2COO^- + H_3O^+ \qquad K_2 = 1.66 \times 10^{-10} \quad (2)$$

Calculate the pH of a 0.0500 F solution of glycine hydrochloride.

10.4. To an aqueous solution containing 0.250 mol of NH_4Cl there is added 0.100 mol of solid NaOH pellets. The solution is stirred until the pellets have dissolved and it is homogeneous. Water is added until the volume is exactly 500.00 mL.

(a) Calculate the concentrations of all species in solution at equilibrium at 25 °C.

(b) Is this solution an effective buffer? Show all calculations required to prove your answer.

10.5. Complete the following table for the titration of 50.00 mL of 0.100 F propionic acid, CH_3CH_2COOH, with 0.100 F NaOH. For propionic acid, K_a is 1.3×10^{-5} at 25 °C. When the table is complete, plot pH versus milliliters of NaOH added, using millimeter ruled graph paper. Which of the indicators in Table 10.2 is a suitable indicator for this titration?

mL NaOH Added	Total Volume (mL)	$[H_3O^+]$	$[OH^-]$	pH
0.00				
5.00				
10.00				
20.00				
25.00				
35.00				
45.00				
49.00				
50.00				
51.00				
60.00				

10.6. You have available a large quantity of a 6.00 F solution of a strong monoprotic acid such as HCl or HNO_3, 1.00 mol of NH_3, and water. Write a precise set of directions for preparing 1.00 L of a buffer solution with pH = 9, using the listed substances. The entire mole of NH_3 is to be used.

10.7. Complete the following table for the titration of 40.00 mL of 0.200 F aniline, $C_6H_5NH_2$, with 0.200 F HCl. The basicity constant, K_b, for aniline is 4.2×10^{-10}. When the table is complete, plot pH versus milliliters of HCl added, using millimeter ruled graph paper. Which of the indicators in Table 10.2 is suitable for this titration?

mL HCl Added	Total Volume (mL)	$[H_3O^+]$	$[OH^-]$	pH
0.00				
5.00				
10.00				
20.00				
25.00				
30.00				
35.00				
38.00				
40.00				
42.00				
50.00				
60.00				

10.8. (a) A solution is prepared by dissolving 2.6250 g of solid sodium acetate in 250.00 mL of 0.0800 F HCl. Assume there is no volume change after the solid has been added. (That is not a bad approximation, but it is not exactly correct.) What is the pH of this buffer solution?

(b) A subsequent addition of 10.00 mL of 0.0800 F HCl is made to the solution described in (a). What is the new pH? By how many units did the pH change on addition of 10.00 mL of 0.0800 F HCl?

(c) By how many units does the pH change if 10.00 mL of 0.0800 F HCl are added to 250.00 mL of water? What important fact does this problem illustrate?

10.9. What volume of 1.00 F KOH must be added to 50.00 mL of 1.00 F HOAc to make a solution with pH = 5.00?

10.10. Into 500.0 mL of 0.500 F HCl is placed 0.750 mol of solid sodium formate. Assume that after the solid has dissolved and the solution is homogeneous, the volume is still 500.00 mL.
(a) Calculate the concentrations of formate ion, formic acid, hydronium ion, and hydroxide ion in this solution at 25 °C.
(b) Is this solution an effective buffer? Show calculations that prove your answer.

10.11. A 100.00-mL sample of 0.300 F NH$_4$NO$_3$ is mixed with 50.00 mL of 0.240 F NaOH.
(a) Calculate the $[NH_4^+]$, $[Cl^-]$, $[Na^+]$, $[OH^-]$, $[NH_3]$, $[H_3O^+]$, and pH of the resulting solution at equilibrium.
(b) Is the mixture an effective buffer? Explain your answer.

10.12. Write a balanced net ionic equation for each of the reactions between the following pairs of reagents. Assume dilute solutions of all soluble substances.
(a) Methylamine and formic acid.
(b) Ammonium sulfate and barium hydroxide.
(c) Solid silver carbonate and hydrochloric acid.
(d) Ammonium nitrate and sodium hydroxide.
(e) Sodium bicarbonate and hydrochloric acid.
(f) Sodium bicarbonate and sodium hydroxide.

10.13. A solution of an unknown monoprotic weak acid was titrated with 0.100 F NaOH. The equivalence point was reached when 37.48 mL of base had been added. From a second buret, exactly 18.74 mL of 0.100 F HCl were added to the titration solution. The pH was then measured with a pH meter and found to be 4.13. Calculate the acidity constant of the weak acid.

10.14. Calculate the pH of a buffer solution prepared by mixing
(a) 60.00 mL of 0.100 F NaOAc with 80.00 mL of 0.100 F HOAc
(b) 30.00 mL of 0.100 F NaOH with 70.00 mL of 0.100 F HOAc

10.15. In a laboratory practical exam, the students were asked to prepare a buffer having a pH anywhere between 4.6 and 5.0.

Addie A. mixed 50.00 mL 0.100 F NH$_4$Cl with 50.00 mL 0.100 F HCl.
Barney B. mixed 50.00 mL 0.100 F NaOAc with 50.00 mL 0.100 F HOAc.
Cecile C. mixed 25.00 mL 0.100 F NaOH with 50.00 mL 0.100 F HOAc.
Dawn D. mixed 80.00 mL 0.100 F HCOOH with 20.00 mL 0.100 F HCOONa.
Errol E. mixed 50.00 mL 0.100 F NH$_4$Cl with 50.00 mL 0.100 F NH$_3$.
Faye F. mixed 80.00 mL 0.100 F NaOAc with 40.00 mL 0.100 F HOAc.

Which of these students prepared the desired buffer? (More than one has used a correct procedure.) Explain what is wrong with each of the incorrect procedures listed above.

10.16. Ethylamine, $C_2H_5NH_2$, is a weak base that ionizes in aqueous solution as follows:

$$C_2H_5NH_2 + H_2O \rightleftharpoons C_2H_5NH_3^+ + OH^-$$

(a) A 0.150 F solution of ethylamine is 6.33% ionized at 25 °C. Calculate K_b for ethylamine.

(b) A buffer solution is prepared by dissolving 0.120 mol $C_2H_5NH_3Cl$ and 0.090 mol $C_2H_5NH_2$ in enough water to make 1.00 L of solution. Calculate the pH of this buffer.

10.17. Calculate the equilibrium constant for the reaction that occurs during each of the following titrations.
(a) 50.00 mL of 0.200 F HOAc versus 0.200 F KOH
(b) 50.00 mL of 0.100 F NH_3 versus 0.100 F HCl

10.18. An employer is interviewing three candidates for a job as a lab technician and asks each to describe how to prepare a buffer solution with a pH between 8.8 and 9.0.

Pansy P. says she would mix HOAc and KOAc solutions.
Quincy Q. says he would mix $NaHCO_3$ and Na_2CO_3 solutions.
Riva R. says she would mix NH_3 and NH_4NO_3 solutions.
Which of these three has proposed an appropriate procedure?

Explain your answer, and explain what is wrong with the two erroneous procedures.

10.19. A 60.00-mL sample of 0.100 F N_2H_4 is titrated with 0.150 F HCl. Hydrazine, N_2H_4, is a weak base with $K_b = 3.0 \times 10^{-6}$. The titration reaction is

$$N_2H_4 + H_3O^+ \rightleftharpoons N_2H_5^+ + H_2O$$

(a) Calculate the pH when 20.00 mL of 0.150 F HCl have been added.
(b) Calculate the pH at the equivalence point in this titration.
(c) What indicator would you use if you were performing this titration? Explain your answer.

chapter 11 Equilibria Involving Slightly Soluble Electrolytes

Arthur A. Noyes (1866–1936), a U.S. chemist, received his B.S. and M.S. degrees from the Massachusetts Institute of Technology (M.I.T.), and obtained his Ph.D. under Wilhelm Ostwald at Leipzig in 1890. He returned to M.I.T. to teach both organic and physical chemistry, and was the director of the research laboratory of physical chemistry. In 1915 he became director of the Gates Chemical Laboratory at the California Institute of Technology. For several years he was a professor at both institutions. He also served as president of the American Chemical Society. His research publications were in theoretical, analytical, and organic chemistry, as well as on educational subjects. Much of his work dealt with the properties of solutions of electrolytes. In 1892 he wrote a textbook on Qualitative Analysis. In later editions of this text he attempted to make clear how the laws of chemical equilibrium, and especially the principles relating to the solubility of slightly soluble electrolytes and to the ionization of weak electrolytes, were applied in constructing the analytical procedures. This book influenced several generations of chemists, was used all over the world, and was still in use during the 1950's. The tenth edition, revised and rewritten by Ernest H. Swift, was published in 1942.

Although a great many ionic crystalline solids dissolve in water, there are a significant number that have such low solubilities in water that we classify them as insoluble or slightly soluble. It is important to remember that the word *insoluble* does not imply that the substance has zero solubility; it means only that the solubility is very small, so small that we cannot, with our eyes, observe that any has dissolved if we put some in water and stir. Some chemists prefer to use only the phrases *sparingly soluble* or *very slightly soluble* to emphasize that the solubility is not zero, but the term insoluble solid is commonly used. As a very rough rule of thumb, if the solubility of a compound is 1×10^{-5} mol/L or less, the substance is usually described as insoluble. Substances with solubility larger than this, but less than about 2×10^{-2} mol/L are frequently called slightly soluble or moderately soluble.

When two solutions containing electrolytes are mixed and a reaction occurs that produces an insoluble solid, that solid precipitates out of the solution. This technique can be used to test for the presence of a particular ion in solution. The branch of chemistry called **inorganic qualitative analysis** is a series of procedures designed to identify the ions present in an unknown compound or mixture of compounds. Many of the tests involve the formation of an insoluble solid by precipitation. Often it is then necessary to dissolve the solid to perform further tests. In this chapter we will apply the principles of ionic equilibria to reactions involving insoluble or slightly soluble solids.

section 11.1
The Solubility Product

In writing the equation for the equilibrium between an insoluble or slightly soluble electrolyte and its saturated aqueous solution it is conventional to write the insoluble solid on the left-hand side of the equation and the ions that exist in the aqueous solution on the right-hand side. A typical example is

$$CaCO_3(s) \rightleftharpoons Ca^{2+}(aq) + CO_3^{2-}(aq) \tag{11-1}$$

Thus the forward reaction is the dissolution of the insoluble solid and the backward reaction is the precipitation of that solid.

The equilibrium constant for the dissolution of an insoluble or slightly soluble electrolyte is called a **solubility product.** Table 11.1 gives equations for several typical dissolution reactions with equilibrium constants that are solubility products, denoted K_{sp}. The expressions for and numerical values of their solubility products are also listed. A more extensive table of solubility products is given in Appendix E, Table E2.

The numerical value of a solubility product must be determined experimentally and is therefore subject to normal experimental uncertainties. The solubility product

Table 11.1. **Some Solubility Product Expressions for Various Types of Slightly Soluble Electrolytes**

Equilibrium Equation	*Solubility Product at 25 °C*
$AgCl(s) \rightleftharpoons Ag^+ + Cl^-$	$[Ag^+][Cl^-] = 1.8 \times 10^{-10}\ M^2$
$Ag_2CrO_4(s) \rightleftharpoons 2Ag^+ + CrO_4^{2-}$	$[Ag^+]^2[CrO_4^{2-}] = 9 \times 10^{-12}\ M^3$
$Al(OH)_3(s) \rightleftharpoons Al^{3+} + 3\ OH^-$	$[Al^{3+}][OH^-]^3 = 1.9 \times 10^{-33}\ M^4$
$Bi_2S_3(s) \rightleftharpoons 2Bi^{3+} + 3S^{2-}$	$[Bi^{3+}]^2[S^{2-}]^3 = 2 \times 10^{-72}\ M^5$
$MgNH_4PO_4(s) \rightleftharpoons Mg^{2+} + NH_4^+ + PO_4^{3-}$	$[Mg^{2+}][NH_4^+][PO_4^{3-}] = 2.5 \times 10^{-13}\ M^3$

of an electrolyte with very small solubility may be difficult to measure, and you will undoubtedly find some variation in the values given if you consult several different reference sources. Do not be surprised if the values vary by as much as a factor of 10 or more.

The solubility product principle states that a solubility product is a constant at constant temperature in a saturated solution of a slightly soluble electrolyte. Thus it is simply a specific example of the law of mass action, Eq. (8-14), just as a solubility product is a specific type of equilibrium constant. The solubility product principle is an ideal law, valid in dilute solutions of electrolytes.*

section 11.2
The Relationship between the Molar Solubility and the Solubility Product

The number of moles of an insoluble solid that dissolves per liter of solution is called the molar solubility of that solid. The molar solubility is not a fixed number, but changes as the constituents of the solution in equilibrium with the solid are changed. If no other constituents are mentioned, the term molar solubility refers to the number of moles of solid that dissolve in enough water to make exactly 1 L of solution. The simple term "solubility" is somewhat ambiguous as it may refer to the molar solubility, in moles of solute per liter of solution, or to the gram solubility, usually given in grams of solute per 100 mL of solution. Most Handbooks cite the solubility in units of grams per 100 mL of solution, whereas for theoretical calculations the molar solubility is used.

It is important to distinguish clearly between the terms solubility and solubility product. The solubility product is an equilibrium constant and does *not* change as the constituents of the solution in equilibrium with the solid are changed. The solubility product remains constant as long as the temperature is constant. The solubility, on the other hand, varies if other electrolytes are added to a saturated aqueous solution of the insoluble solid. The solubility and the solubility product are different properties, but they are related to one another. We can always determine one if we know the other. A number of typical problems illustrating the relationship between these two quantities are given below.

EXAMPLE 11.1. Calculation of the solubility product of a slightly soluble electrolyte from its solubility in water

The solubility of CaF_2 in water at 25 °C is 1.7×10^{-3} grams per 100 mL. Calculate the solubility product of CaF_2 at 25 °C.

Solution. In 1 L of solution the number of grams of CaF_2 that can be dissolved is 10 times the number that can be dissolved in 100 mL, and is thus 1.7×10^{-2} g. The

---

* The solubility product principle is exact if quantities called the activities of the ions are used in place of the ionic concentrations in the solubility product expression. Activities are directly proportional to the concentrations, but the proportionality constant (the activity coefficient) changes as the constituents of the solution change. If ionic concentrations are high, and particularly if highly charged ions are present, substantial errors may be introduced by using concentrations in the solubility product expression. Errors as large as 10 or 20% are common. We will assume the constancy at constant temperature of any solubility product as defined in Table 11.1, because the results calculated with this assumption are perfectly adequate for most purposes.

gram formula weight of CaF_2 is 78.08, so the number of moles of CaF_2 that can be dissolved per liter is

$$\frac{1.7 \times 10^{-2} \text{ g/L}}{78.08 \text{ g/mol}} = 2.18 \times 10^{-4} = 2.2 \times 10^{-4} \text{ mol/L}$$

The equation for the dissolution of CaF_2 in water is

$$CaF_2(s) \rightleftharpoons Ca^{2+}(aq) + 2F^-(aq)$$

Thus,

$$[Ca^{2+}] = 2.2 \times 10^{-4} \, M \quad \text{and} \quad [F^-] = 2(2.18 \times 10^{-4}) = 4.36 \times 10^{-4} \, M$$
$$K_{sp}(CaF_2) = [Ca^{2+}][F^-]^2 = (2.18 \times 10^{-4})(4.36 \times 10^{-4})^2 = 4.1 \times 10^{-11}$$

EXAMPLE 11.2. Calculation of the solubility of either a uni-univalent (1:1)* or a bi-bivalent (2:2) electrolyte in pure water

The value of K_{sp} for $BaSO_4$ at 25 °C is 1.1×10^{-10}. Calculate the molar solubility of $BaSO_4$ in water, and its solubility in grams per 100 mL, at 25 °C.

Solution. Begin by writing the equation for the reaction for which the equilibrium constant is the K_{sp} of $BaSO_4$:

$$BaSO_4(s) \rightleftharpoons Ba^{2+}(aq) + SO_4^{2-}(aq)$$

Remember that by convention the insoluble solid is written on the left-hand side of the equation. Barium sulfate is a strong electrolyte. Although its solubility is very low, 100% of the small amount that does dissolve is present in solution as the hydrated ions.

Let x equal the molar solubility of $BaSO_4$ at 25 °C. When x moles of $BaSO_4$ dissolve in enough water to make 1 L of solution, there are x moles of Ba^{2+} ions and x moles of SO_4^{2-} ions in that solution. Hence,

$$[Ba^{2+}] = [SO_4^{2-}] = x$$
$$K_{sp}(BaSO_4) = [Ba^{2+}][SO_4^{2-}] = x^2 = 1.1 \times 10^{-10} \, M^2$$
$$x = 1.0 \times 10^{-5} \, M$$

A saturated aqueous solution of $BaSO_4$ at 25 °C is $1.0 \times 10^{-5} \, M$.

To convert the molar solubility to the gram solubility, we need to know the gram formula weight of $BaSO_4$, which is 233.40. Hence the number of grams of $BaSO_4$ that dissolve in 1 L of solution at 25 °C is

$$(233.40 \text{ g/mol})(1.0 \times 10^{-5} \text{ mol/L}) = 2.3 \times 10^{-3} \text{ g/L}$$

The solubility in grams per 100 mL of solution is therefore 2.3×10^{-4}, or 0.00023 g.

The answers obtained in Example 11.2 will give you an indication of what is meant by the term insoluble solid. The solubility of $BaSO_4$ is not zero. However, the maximum amount that can be dissolved in enough water to make 100 mL of solution (~100 mL of H_2O) is 2.3×10^{-4} g, and this is so small that we call $BaSO_4$ insoluble.

EXAMPLE 11.3. Calculation of the solubility of a (1:2) or (2:1) electrolyte in pure water

The solubility product of lead iodate, $Pb(IO_3)_2$, is 2.6×10^{-13} at 25 °C. Calculate the molar solubility of $Pb(IO_3)_2$ in H_2O at 25 °C.

---

* Electrolytes can be classified into types by giving the magnitudes of the charges on cation and anion. Thus AgCl is a 1:1 electrolyte, $BaSO_4$ a 2:2 electrolyte, Ag_2CrO_4 a 1:2 electrolyte, and so on.

Solution. The reaction for which the equilibrium constant is the solubility product of lead iodate is

$$Pb(IO_3)_2(s) \rightleftharpoons Pb^{2+}(aq) + 2\,IO_3^-(aq)$$

The molar ratios needed in order to express the concentrations of the ions in terms of the molar solubility are

$$\frac{\text{moles Pb}^{2+}\text{ in solution}}{\text{moles Pb(IO}_3)_2\text{ dissolved}} = \frac{1}{1}$$

and

$$\frac{\text{moles IO}_3^-\text{ in solution}}{\text{moles Pb(IO}_3)_2\text{ dissolved}} = \frac{2}{1}$$

let s = the molar solubility of $Pb(IO_3)_2$ in water at 25 °C

then $[Pb^{2+}] = s$ and $[IO_3^-] = 2s$

For $Pb(IO_3)_2$,

$$K_{sp} = [Pb^{2+}][IO_3^-]^2 = (s)(2s)^2 = 2.6 \times 10^{-13}$$

We therefore obtain

$$4s^3 = 2.6 \times 10^{-13}$$

To determine the power of 10 of the answer in your head, make the power of 10 divisible by three before taking the cube root. We therefore write

$$4s^3 = 2.6 \times 10^{-13} = 260 \times 10^{-15}$$
$$s^3 = 65 \times 10^{-15} \quad \text{and} \quad s = 4.0 \times 10^{-5}\,M$$

We are only entitled to two significant figures in the answer, since K_{sp} is only given to two figures. Thus 4.0×10^{-5} mol of solid $Pb(IO_3)_2$ dissolves per liter of solution. In a saturated solution of $Pb(IO_3)_2$ in water, $[Pb^{2+}] = 4.0 \times 10^{-5}\,M$ and $[IO_3^-] = 8.0 \times 10^{-5}\,M$.

A word of caution is in order here. The values calculated for the molar solubility using the methods described in Examples 11.2 and 11.3 are correct only if no other reactions involving the ions of the electrolyte occur in the aqueous solution. In particular, we must always consider the fact that many anions are bases and engage in proton-transfer reactions with water. Such a reaction increases the solubility of the insoluble electrolyte. Consider the sulfides, for instance. Most sulfides are insoluble. Sulfide ion, however, is a base and reacts with water as follows:

$$S^{2-} + H_2O \rightleftharpoons HS^- + OH^- \tag{11-2}$$

The hydrolysis of sulfide ion increases the solubility of any insoluble sulfide above the value that would be calculated using the relations obtained in Examples 11.2 or 11.3. Consider silver sulfide, Ag_2S, a black, extremely insoluble salt, as a specific example. If we consider only the reaction

$$Ag_2S(s) \rightleftharpoons 2Ag^+(aq) + S^{2-}(aq) \tag{11-3}$$

we would obtain, following the procedure described in Example 11.3, $4s^3 = K_{sp}(Ag_2S) = 7 \times 10^{-50}$. The value of s obtained by solving this equation is smaller than the actual molar solubility of Ag_2S. The hydrolysis of sulfide ion, Eq. (11-2), serves to remove sulfide ions from the solution. By Le Chatelier's Principle this drives reaction (11-3) to the right, dissolving more Ag_2S. Even with the hydrolysis of the sulfide ion, the solubility of Ag_2S is very small, but it is larger than one would calculate ignoring reaction (11-2).

In Section 11.4 we will consider the calculation of the molar solubility of an insoluble sulfide including the effect of the proton-transfer reaction between S^{2-} and H_2O. (See Example 11.13.)

EXAMPLE 11.4. Calculation of the solubility of a (1:3) or (3:1) slightly soluble electrolyte in pure water

The solubility product of white, insoluble $La(OH)_3$ at 25 °C is 1×10^{-19}. Calculate the molar solubility of $La(OH)_3$ at 25 °C.

Solution

$$La(OH)_3(s) \rightleftharpoons La^{3+}(aq) + 3OH^-(aq)$$

let z = the molar solubility of $La(OH)_3$ in water at 25 °C

$$\text{then } [La^{3+}] = z \quad \text{and} \quad [OH^-] = 3z$$

For $La(OH)_3$,

$$K_{sp} = [La^{3+}][OH^-]^3 = (z)(3z)^3 = 27z^4 = 1 \times 10^{-19}$$

Therefore

$$z^4 = 3.7 \times 10^{-21}$$

and

$$z = 7.8 \times 10^{-6} \, M$$

Extract the fourth root using your calculator by taking the square root twice. Since 7.8×10^{-6} moles of $La(OH)_3$ dissolve per liter of solution, in a saturated solution of $La(OH)_3$ in water

$$[La^{3+}] = 7.8 \times 10^{-6} \, M \quad \text{and} \quad [OH^-] = 3(7.8 \times 10^{-6} \, M) = 2.3 \times 10^{-5} \, M$$

Until now all the problems we have considered have dealt with the solubility of an insoluble electrolyte in pure water. The solubility is considerably decreased by the presence in the solution of a soluble electrolyte having an ion in common with the insoluble electrolyte. We have already discussed this **common ion effect** qualitatively, in terms of Le Chatelier's Principle, in Section 8.4, but we can now examine the magnitude of this effect quantitatively. Consider the equilibrium between solid lead sulfate and its saturated aqueous solution:

$$PbSO_4(s) \rightleftharpoons Pb^{2+} + SO_4^{2-} \tag{11-4}$$

for which the equilibrium constant is

$$K_{sp}(PbSO_4) = 1.8 \times 10^{-8} = [Pb^{2+}][SO_4^{2-}] \tag{11-5}$$

In a saturated solution of $PbSO_4$ in water,

$$[Pb^{2+}] = [SO_4^{2-}] = (1.8 \times 10^{-8})^{1/2} = 1.3 \times 10^{-4} \, M$$

If we increase the $[Pb^{2+}]$ by adding another source of Pb^{2+} ions, such as a solution of $Pb(NO_3)_2$, the $[SO_4^{2-}]$ must decrease because the product $[Pb^{2+}][SO_4^{2-}]$ remains constant at $1.8 \times 10^{-8} \, M^2$. As the only source of SO_4^{2-} ions in the solution is the dissolution of $PbSO_4$, a decrease in the $[SO_4^{2-}]$ means a decrease in the solubility of $PbSO_4$.

The molar solubility of $PbSO_4$ in solutions of K_2SO_4 as a function of the concentration of K_2SO_4 is plotted in Fig. 11.1. The higher the concentration of the common ion, SO_4^{2-}, the greater the decrease in the solubility of $PbSO_4$ relative to its solubility

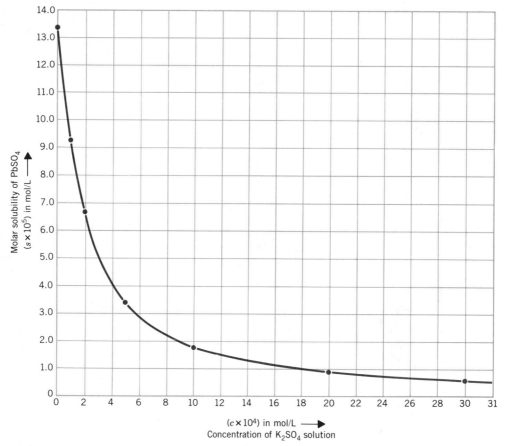

Fig. 11.1. The common ion effect. The molar solubility, s, of $PbSO_4$ in solutions containing K_2SO_4 is plotted as a function of the concentration, c, in mol/L, of the K_2SO_4. The relationship between s and c is $s(c + s) = 1.8 \times 10^{-8} = K_{sp}(PbSO_4)$. Thus as c increases, s decreases.

in pure water. A similar plot is obtained for the solubility of $PbSO_4$ in solutions of $Pb(NO_3)_2$ as a function of the molarity of $Pb(NO_3)_2$.

The following problems illustrate various types of calculations, but should also give you a feeling for the magnitude of the decrease in solubility when a common ion is present in the solution.

EXAMPLE 11.5. Calculation of the solubility of a slightly soluble (1:1) electrolyte in a solution containing a common ion

The solubility product of AgCl at 25 °C is 1.8×10^{-10}. Calculate the molar solubility of AgCl in 0.040 F NaCl at 25 °C.

Solution. The equation for the dissolution of AgCl is

$$AgCl(s) \rightleftharpoons Ag^+(aq) + Cl^-(aq)$$

In this solution there are two sources of chloride ion: The soluble NaCl is completely dissociated into Na^+ ions and Cl^- ions, and the small amount of AgCl that dissolves also provides Cl^- ions. The Cl^- ion is common to both the insoluble electrolyte, AgCl, and the soluble salt, NaCl. There is, however, only one source of Ag^+ ions, namely the dissolution of solid AgCl.

let s = the molar solubility of AgCl in 0.040 F NaCl at 25 °C

then $[Ag^+] = s$ and $[Cl^-] = 0.040 + s$

$$K_{sp}(AgCl) = [Ag^+][Cl^-] = (s)(0.040 + s) = 1.8 \times 10^{-10}$$

The equation we must solve for s is a quadratic equation; multiplying the two terms on the left-hand side yields $s^2 + 0.040s$. It is, of course, possible to solve this equation using the quadratic formula, but there is a simpler method that we can employ because we know something about this solution, and therefore about the value of the unknown, s. The solubility of AgCl in water is small, so small that we call AgCl insoluble. The addition of the NaCl to the solution decreases the solubility to a value less than the solubility in pure water. Thus we expect s to be a very small number, and in particular we expect it to be considerably smaller than 0.040. Hence we can assume that s is small compared to 0.040 and that $0.040 + s = 0.040$.

The solubility product equation therefore simplifies to

$$[Ag^+][Cl^-] = (s)(0.040) = 1.8 \times 10^{-10}$$

$$s = \left(\frac{1.8}{4}\right) \times 10^{-8} = 4.5 \times 10^{-9} \ M$$

We must, of course, check the assumption we have made. Is $0.040 + s \cong 0.040$? Yes, $0.040 + s = 0.040 + 0.0000000045 = 0.040$. The value of s is too small to be significantly added to 0.040, so that our assumption is certainly justified. Only 4.5×10^{-9} mol of AgCl dissolve per liter of 0.040 F NaCl.

In these solubility product problems, just as in the problems dealing with the dissociation of weak acids or bases discussed in Chapter 9, we will use the rule that one number is small compared to another provided that it is less than 10% of the larger number.

From the results of Example 11.5, we can see how much less the solubility of AgCl is in 0.040 F NaCl than it is in pure water. If we let x = the molar solubility of AgCl in water, then $x^2 = 1.8 \times 10^{-10}$, and $x = 1.3 \times 10^{-5} \ M$. Thus the molar solubility of AgCl decreases from 1.3×10^{-5} to $4.5 \times 10^{-9} \ M$ when the solution changes from pure water to 0.040 F NaCl. This is a decrease by a factor of about three thousand. You can see that the addition of a common ion can have a very large effect!

Note also that while the molar solubility of AgCl is less in any solution containing a common ion than it is in pure water, the solubility product is always the same as long as the temperature is constant.

EXAMPLE 11.6. Calculation of the solubility of a slightly soluble (1:2) or (2:1) electrolyte in a solution containing excess of a common ion

The solubility product of mercurous bromide is 1.3×10^{-21} at 25 °C. Calculate the molar solubility of mercurous bromide in 0.034 F KBr at 25 °C.

Solution. The equilibrium we must consider is

$$Hg_2Br_2(s) \rightleftharpoons Hg_2^{2+}(aq) + 2Br^-(aq)$$

Remember that mercurous ion is a diatomic cation.

There are two sources of bromide ion in this solution, the soluble KBr and the insoluble Hg_2Br_2. There is only one source of mercurous ions, namely the very small amount of Hg_2Br_2 that dissolves.

let y = the molar solubility of Hg_2Br_2 in 0.034 F KBr at 25 °C

then $[Hg_2^{2+}] = y$ and $[Br^-] = 0.034 + 2y$

The equilibrium constant expression is, therefore,

$$K_{sp}(Hg_2Br_2) = [Hg_2^{2+}][Br^-]^2 = (y)(0.034 + 2y)^2 = 1.3 \times 10^{-21}$$

If treated as a straightforward algebra problem, this is a cubic equation in y. However, we can make a simplifying assumption based on our knowledge of mercurous bromide. The extremely small value of K_{sp} tells us that the solubility of Hg_2Br_2 in water is small. Its solubility in this solution, which contains Br^- ions due to the presence of the KBr, will be significantly smaller than its solubility in water. We therefore expect y to be a very small number. Let us assume that $2y$ is small compared to 0.034, so that $0.034 + 2y = 0.034$. Note that the sum involves $2y$ and therefore our assumption must be about the relative magnitudes of the two terms in the sum, namely $2y$ and 0.034. With this assumption, the solubility product expression simplifies to:

$$y(0.034)^2 = 1.3 \times 10^{-21}$$
$$y = \frac{1.3 \times 10^{-21}}{1.16 \times 10^{-3}} = 1.1 \times 10^{-18}$$

We must now check the assumption we made. Is $2y = 2.2 \times 10^{-18}$ less than 10% of 0.034? Yes, 2.2×10^{-18} is very much smaller than 3.4×10^{-3}, and therefore our assumption is justified. The molar solubility of mercurous bromide in 0.034 F KBr is exceedingly small, only 1.1×10^{-18} M. You should be able to prove that the solubility of Hg_2Br_2 in 0.034 F KBr is about 30 million times smaller than it is in pure water.

There are problems in which the assumption we make does not turn out to be valid, and we must either tackle an equation of power higher than 1, or use the method of successive approximations. (Refer to Example 9.8.) The following example deals with such a calculation.

EXAMPLE 11.7. **Use of the method of successive approximations in a calculation of the solubility of a slightly soluble electrolyte in the presence of a common ion**

The solubility product of calcium sulfate is 2.4×10^{-5} at 25 °C. Calculate the molar solubility of $CaSO_4$ in 0.0080 F $CaCl_2$ at 25 °C.

Solution

$$CaSO_4(s) \rightleftharpoons Ca^{2+}(aq) + SO_4^{2-}(aq)$$

let s = the molar solubility of $CaSO_4$ in 8.0×10^{-3} F $CaCl_2$ at 25 °C

then $[Ca^{2+}] = 8.0 \times 10^{-3} + s$ and $[SO_4^{2-}] = s$

$$K_{sp}(CaSO_4) = [Ca^{2+}][SO_4^{2-}] = (8.0 \times 10^{-3} + s)(s) = 2.4 \times 10^{-5}$$

First Approximation: Assume

$$8.0 \times 10^{-3} + s = 8.0 \times 10^{-3}$$

The solubility product equation then simplifies to:

$$(8.0 \times 10^{-3})s = 2.4 \times 10^{-5}$$ and hence $s = 3.0 \times 10^{-3}$

We must now check the assumption we made. Is 3.0×10^{-3} less than 10% of 8.0×10^{-3}? No, 10% of 8.0×10^{-3} is 8.0×10^{-4} and 3×10^{-3} is larger than

8.0×10^{-4}. Therefore our assumption is not valid. We can, however, make a second approximation, which is better than the first one. We use the result of the first approximation to estimate the value of the sum $8.0 \times 10^{-3} + s$.

Second Approximation: Assume

$$8.0 \times 10^{-3} + s = 8.0 \times 10^{-3} + 3.0 \times 10^{-3}$$
$$= 1.1 \times 10^{-2}$$

The solubility product equation now becomes:

$$[Ca^{2+}][SO_4^{2-}] = (1.1 \times 10^{-2})(s) = 2.4 \times 10^{-5}$$

and hence

$$s = 2.2 \times 10^{-3}$$

In the method of successive approximations we must continue making approximations until two approximations in succession give the same answer. Thus we now utilize the result of the second approximation to make a still better approximation to the sum $8.0 \times 10^{-3} + s$.

Third Approximation: Assume

$$8.0 \times 10^{-3} + s = 8.0 \times 10^{-3} + 2.2 \times 10^{-3}$$
$$= 1.02 \times 10^{-2}$$

The solubility product equation is now

$$(1.02 \times 10^{-2})s = 2.4 \times 10^{-5}$$
$$\text{so that} \quad s = 2.35 \times 10^{-3}$$

Fourth Approximation: Assume

$$8.0 \times 10^{-3} + s = 8.0 \times 10^{-3} + 2.35 \times 10^{-3}$$
$$= 1.04 \times 10^{-2}$$

Then

$$(1.04 \times 10^{-2})s = 2.4 \times 10^{-5} \quad \text{and} \quad s = 2.31 \times 10^{-3}$$

The successive approximations are converging on the correct answer, which is therefore between 2.35×10^{-3} and 2.31×10^{-3} and consequently must be 2.3×10^{-3} to two figures. Note that the first approximation gave too large a result, the second too small, and so on. The first approximation yielded 3.0×10^{-3} M, which is at least the correct order of magnitude, although it is in error by more than 10%. The molar solubility of $CaSO_4$ in 0.0080 F $CaCl_2$ is 2.3×10^{-3} mol/L.

section 11.3
Rules for Determining Whether a Solution Is Unsaturated, Saturated, or Supersaturated

The general relations between the reaction quotient, Q, and the equilibrium constant that we have already discussed [Eqs. (8-14a)–(8-14c)] apply also to the equilibria between insoluble or slightly soluble electrolytes and aqueous solutions of these electrolytes. The reaction quotient, Q, has exactly the same form as the solubility product expression. The following rules are merely a specific example of Eqs. (8-14a)–(8-14c).

If $Q < K_{sp}$ the solution is unsaturated. No precipitation will occur. The reaction must proceed to the right to reach equilibrium. Any solid present will dissolve, putting more ions into solution. If there is no solid present, no reaction will occur. (11-6a)

If $Q > K_{sp}$ the solution is supersaturated. A precipitate will form. The reaction will proceed to the left, to the side with the insoluble solid, in order to reach equilibrium. The solid phase will increase in size until $Q = K_{sp}$ and equilibrium is established. (11-6b)

If $Q = K_{sp}$ the solution is saturated, and is therefore at equilibrium. No net reaction will occur. (11-6c)

As a practical matter we note that while in theory a precipitate will form if Q is even slightly larger than K_{sp}, in fact it is usually necessary for Q to be considerably larger than K_{sp} before any visible precipitation begins.

The following problem illustrates how these rules are applied.

EXAMPLE 11.8. **Determining whether or not a precipitate will form when solutions are mixed**

If 70.0 mL of 0.050 F $Ba(NO_3)_2$ are mixed with 30.0 mL of 0.020 F NaF, will any BaF_2 precipitate? K_{sp} of BaF_2 is 1.7×10^{-6} at 25 °C.

Solution. To answer a question like this, we must imagine the "instant of mixing," an instant before any reaction that may happen has had a chance to take place. We then calculate Q at the instant of mixing, and compare it with the equilibrium constant. Bear in mind that when two solutions are mixed, dilution always occurs. Because $NaNO_3$ is a soluble, strong electrolyte, the only reaction that can possibly occur when these two solutions are mixed is the formation of insoluble BaF_2. Let us calculate both the $[Ba^{2+}]$ and $[F^-]$ at the instant of mixing.

No. mmol Ba^{2+} added = (70.0 mL)(0.050 mmol/mL) = 3.50 mmol

total volume after mixing = 100.0 mL

Hence at the instant of mixing

$$[Ba^{2+}] = \frac{\text{mmol } Ba^{2+} \text{ added}}{\text{total volume}} = \frac{(70.0)(0.050)}{(100.0)} = 0.035 \ M$$

Note that we can consider the ratio of the initial volume of the solution containing Ba^{2+} ions to the final volume of the mixed solution a *dilution factor* and obtain the concentration at the instant of mixing by multiplying the initial concentration by the dilution factor. Similarly, the $[F^-]$ at the instant of mixing is given by

$$[F^-] = \frac{(\text{mmol } F^- \text{ added})}{(\text{total volume})} = \frac{(30.0)(0.020)}{(100.0)} = 0.0060 \ M$$

We are considering the reaction

$$BaF_2(s) \rightleftharpoons Ba^{2+}(aq) + 2F^-(aq) \tag{11-7a}$$

This is the reaction whose equilibrium constant is the K_{sp} of BaF_2. The reaction quotient for reaction (11-7a) is

$$Q_{7a} = [Ba^{2+}][F^-]^2 = (3.5 \times 10^{-2})(6.0 \times 10^{-3})^2 = 1.3 \times 10^{-6}$$

Since $Q_{7a} < K_{sp}$ the reaction does *not* proceed to the left. The solution is unsaturated and no BaF_2 will form.

In answering a question such as that posed in Example 11.8, students often like to begin by writing the precipitation reaction as the forward reaction, that is, by writing

$$Ba^{2+}(aq) + 2F^-(aq) \rightleftharpoons BaF_2(s) \tag{11-7b}$$

There is nothing wrong with this, but you must then remember that for this reaction

$$Q_{7b} = \frac{1}{[Ba^{2+}][F^-]^2} \quad \text{and} \quad K_{eq} = \frac{1}{K_{sp} \text{ of } BaF_2}$$

To ascertain whether reaction (11-7b) proceeds to the right or to the left, you must compare Q_{7b} with the reciprocal of the solubility product of BaF_2. You will then find that the precipitation reaction does not proceed to the right. Remember that the equation you write down, the expression for the reaction quotient, and the expression for the equilibrium constant, must all be consistent with one another. If you are consistent it does not matter in which direction you write any equilibrium, but most chemists find it convenient to stick to the convention of always putting the slightly soluble solid on the left-hand side of the equation and using Eqs. (11-6a)–(11-6c) to determine whether or not a precipitate will form.

When a precipitate does form, we will want to calculate the concentrations of the ions of the insoluble solid still in solution after precipitation is complete and the system is at equilibrium. The following example illustrates a calculation of this kind.

EXAMPLE 11.9. Calculation of the concentrations of the ions remaining in solution after precipitation

Calculate the $[Zn^{2+}]$ and $[C_2O_4^{2-}]$ remaining in solution after 15.00 mL of 0.120 F $Zn(NO_3)_2$ are mixed with 10.00 mL of 0.100 F $Na_2C_2O_4$. The K_{sp} of ZnC_2O_4 is 2.5×10^{-9}.

Solution. First let us calculate the $[Zn^{2+}]$ and $[C_2O_4^{2-}]$ at the instant of mixing.

$$\text{No. mmol } Zn^{2+} \text{ added} = (15.00 \text{ mL})(0.120 \text{ mmol/mL}) = 1.80 \text{ mmol}$$
$$\text{No. mmol } C_2O_4^{2-} \text{ added} = (10.00 \text{ mL})(0.100 \text{ mmol/mL}) = 1.00 \text{ mmol}$$
$$\text{total volume after mixing} = 25.00 \text{ mL}$$

At the instant of mixing, therefore,

$$[Zn^{2+}] = \frac{1.80 \text{ mmol}}{25.00 \text{ mL}} = 0.072 \ M$$

$$[C_2O_4^{2-}] = \frac{1.00 \text{ mmol}}{25.00 \text{ mL}} = 0.040 \ M$$

The value of the reaction quotient at the instant of mixing is

$$Q = [Zn^{2+}][C_2O_4^{2-}] = (7.2 \times 10^{-2})(4.0 \times 10^{-2}) = 2.9 \times 10^{-3}$$

Since $Q > K_{sp}(ZnC_2O_4)$, ZnC_2O_4 will precipitate. The precipitation reaction is

$$Zn^{2+}(aq) + C_2O_4^{2-}(aq) \rightleftharpoons ZnC_2O_4 \downarrow$$

Because Zn^{2+} and $C_2O_4^{2-}$ combine in a $1:1$ molar ratio, the reaction mixture has excess Zn^{2+} that cannot be precipitated, as we have added 1.80 mmol of Zn^{2+} but only 1.00 mmol of $C_2O_4^{2-}$. Virtually all of the $C_2O_4^{2-}$ will be in the precipitate. First let us calculate the concentration of Zn^{2+} that is in excess.

$$\text{No. mmol excess } Zn^{2+} = 1.80 - 1.00 = 0.80 \text{ mmol}$$

$$\text{excess } [Zn^{2+}] = \frac{0.80 \text{ mmol}}{25.00 \text{ mL}} = 0.032 \ M$$

Let $x = [C_2O_4^{2-}]$ after the precipitation is complete. We know this will be a very small value, although it will not be zero. For every $C_2O_4^{2-}$ ion that is not in the precipitate but remains in solution, there is also a Zn^{2+} ion in solution. So the total $[Zn^{2+}]$ consists of the excess plus an amount equal to the $[C_2O_4^{2-}]$. Thus,

$$[Zn^{2+}] = 0.032 + x$$

When equilibrium is attained,

$$[Zn^{2+}][C_2O_4^{2-}] = 2.5 \times 10^{-9} = K_{sp}$$

thus

$$(0.032 + x)(x) = 2.5 \times 10^{-9}$$

Assume x is small compared to 0.032. Then $0.032 + x = 0.032$, and the equation becomes simply $0.032x = 2.5 \times 10^{-9}$. We obtain

$$x = \frac{2.5 \times 10^{-9}}{3.2 \times 10^{-2}} = 7.8 \times 10^{-8} \, M$$

We must check our assumption. Is 7.8×10^{-8} small compared to 0.032? Yes, 7.8×10^{-8} is certainly much less than 10% of 3.2×10^{-2}. Hence,

$$[C_2O_4^{2-}] = 7.8 \times 10^{-8} \, M$$
$$[Zn^{2+}] = 3.2 \times 10^{-2} + 7.8 \times 10^{-8} = 3.2 \times 10^{-2} \, M$$

Note that when a precipitate forms after we mix two solutions, if one of the ions is in excess, the calculation is the same as determining the solubility of the insoluble solid in the presence of a common ion.

section 11.4
Equilibria Involving Both Slightly Soluble Electrolytes and Weak Acids or Bases

The equilibrium constants for certain specific reactions are tabulated, and we can find, for example, the values of acidity constants and of solubility products in many standard reference works. There are, however, many reactions for which the equilibrium constants are not available in tables, and we must learn how to calculate these equilibrium constants from the tabulated ones.

A typical problem involves the dissolution of an insoluble or slightly soluble electrolyte in acidic solution. Many substances that are insoluble in water dissolve readily in a strongly acidic solution, that is, a solution with a high $[H_3O^+]$. In particular, electrolytes with anions that are bases will be more soluble in acidic solution than in pure water.

If the anion of the insoluble electrolyte is a moderately strong base, such as CN^-, S^{2-}, or CO_3^{2-}, adding strong acid will significantly increase the solubility over its value in pure water. To answer a specific question about whether or not a given insoluble electrolyte will dissolve in an acidic solution, we must be able to calculate the equilibrium constant for the reaction involved. Several examples will make the technique utilized clear.

EXAMPLE 11.10. Dissolving insoluble sulfides in strong acid

Is it possible to dissolve MnS in excess HCl?

Solution. The equation for the dissolution of manganous sulfide in HCl is

$$MnS(s) + 2H^+(aq) \rightleftharpoons Mn^{2+} + H_2S \qquad (11\text{-}8)$$

Manganous sulfide is a pale pinkish-beige insoluble solid. The solubility product of MnS is 5×10^{-15} at 25 °C, so that its solubility in water is extremely small. To find out whether it is possible to dissolve much MnS in strong acid, we must evaluate the equilibrium constant of Eq. (11-8):

$$K_{eq} = \frac{[Mn^{2+}][H_2S]}{[H^+(aq)]^2} \tag{11-9a}$$

The equilibrium constant for any reaction involving an insoluble electrolyte depends, in some way, on the K_{sp} of that insoluble solid. Therefore this equilibrium constant must depend on $K_{sp}(MnS)$. We introduce the $[S^{2-}]$ into the expression for K_{eq} by multiplying both the numerator and denominator of Eq. (11-9a) by $[S^{2-}]$.

$$K_{eq} = \frac{[Mn^{2+}][H_2S]}{[H^+(aq)]^2} \cdot \frac{[S^{2-}]}{[S^{2-}]} = [Mn^{2+}][S^{2-}] \cdot \frac{[H_2S]}{[H^+]^2 [S^{2-}]} \tag{11-9b}$$

The terms in Eq. (11-9b) by which the solubility product of MnS is multiplied should be familiar. [Refer to Eq. (9-30).] The equilibrium constant for the overall two stages of ionization of H_2S is

$$K_{overall} = K_a(H_2S) \cdot K_a(HS^-) = \frac{[H^+(aq)]^2 [S^{2-}]}{[H_2S]}$$

The reciprocal of this expression appears in the K_{eq} we are trying to evaluate, Eq. (11-9b). Thus we have

$$K_{eq} = K_{sp}(MnS) \cdot \frac{1}{K_a(H_2S) \cdot K_a(HS^-)} = \frac{5 \times 10^{-15}}{(1 \times 10^{-7})(1.3 \times 10^{-13})} = 4 \times 10^{+5} \tag{11-9c}$$

Since the value of the equilibrium constant for Eq. (11-8) is much greater than 1, the position of equilibrium for reaction (11-8) is to the right. This tells us that it is indeed possible to dissolve manganous sulfide by adding a strong acid such as HCl.

EXAMPLE 11.11. Determination of the equilibrium constant for the dissolution of an insoluble electrolyte in acidic solution

How many grams of $BaSO_4$ dissolve in 1.00 L of 2.0 F HCl?

Solution. The first thing to do in answering a question like this is to think about the reason why $BaSO_4$ is more soluble in HCl than in pure water. You should refer to Section 8.4 and the discussion about making use of Le Chatelier's Principle to dissolve insoluble solids. When $BaSO_4$ dissolves in water, the two ions present in solution are Ba^{2+} and SO_4^{2-} ions. If we decrease the $[SO_4^{2-}]$ by adding H_3O^+ ions to form HSO_4^- ions:

$$H_3O^+ + SO_4^{2-} \rightleftharpoons HSO_4^- + H_2O \tag{11-10}$$

we drive the dissolution of $BaSO_4$

$$BaSO_4(s) \rightleftharpoons Ba^{2+} + SO_4^{2-} \tag{11-11}$$

to the right. Thus the overall reaction for the dissolution of $BaSO_4$ in strong acid is the sum of reactions (11-10) and (11-11), namely

$$BaSO_4(s) + H_3O^+ \rightleftharpoons Ba^{2+} + HSO_4^- + H_2O \tag{11-12}$$

To find out whether reaction (11-12) proceeds very far to the right, we must evaluate its equilibrium constant.

$$K_{eq} = \frac{[Ba^{2+}][HSO_4^-]}{[H_3O^+]} \tag{11-13a}$$

No table of equilibrium constants contains the value of K_{eq} desired. We can, however, readily calculate it from the solubility product of $BaSO_4$ and the acidity constant of HSO_4^-, both of which are available in standard reference works. (See Appendix E, Tables E1 and E2.)

$$K_{sp}(BaSO_4) = [Ba^{2+}][SO_4^{2-}] = 1.1 \times 10^{-10}$$

$$K_a(HSO_4^-) = \frac{[H_3O^+][SO_4^{2-}]}{[HSO_4^-]} = 1.2 \times 10^{-2}$$

One way to approach this problem is to utilize the fact that Eq. (11-12) is the sum of Eqs. (11-10) and (11-11). The equilibrium constant for reaction (11-10) is

$$K = \frac{[HSO_4^-]}{[H_3O^+][SO_4^{2-}]} = \frac{1}{K_a(HSO_4^-)}$$

The equilibrium constant for reaction (11-11) is the K_{sp} of $BaSO_4$. In Section 8.5 we discussed the following statement: If an overall reaction is the algebraic sum of two other equations, denoted (1) and (2), then the equilibrium constant for the overall reaction is the product of the equilibrium constants of reactions (1) and (2). Applying that method here we have

$$K_{eq} = \frac{[Ba^{2+}][HSO_4^-]}{[H_3O^+]} = \frac{[HSO_4^-]}{[H_3O^+][SO_4^{2-}]} \cdot [Ba^{2+}][SO_4^{2-}]$$

$$= \frac{1}{K_a(HSO_4^-)} \cdot K_{sp}(BaSO_4)$$

$$= \frac{1.1 \times 10^{-10}}{1.2 \times 10^{-2}} = 9.2 \times 10^{-9}$$

A second approach to this problem is to remember that if a reaction involves an insoluble electrolyte, the equilibrium constant for the reaction depends in some way on K_{sp} for that electrolyte. Similarly, if a reaction involves a weak acid, the equilibrium constant for the reaction depends in some way on K_a for that weak acid. Thus the equilibrium constant for reaction (11-12) depends on both the solubility product of $BaSO_4$ and the acidity constant of HSO_4^-.

Carefully examine the desired expression, Eq. (11-13a), and ask yourself: Since K_{eq} must depend in some way on the solubility product of $BaSO_4$, is any term present in the K_{sp} of $BaSO_4$ missing in this K_{eq}? Clearly the $[SO_4^{2-}]$ that appears in $K_{sp}(BaSO_4)$ is not present in this K_{eq}. Of course there actually is a $[SO_4^{2-}]$ in the solution. To introduce the $[SO_4^{2-}]$, multiply both the numerator and the denominator of the expression for K_{eq} by the $[SO_4^{2-}]$. That is equivalent to multiplying K_{eq} by 1, and cannot change the value of the equilibrium constant. We then have

$$K_{eq} = \frac{[Ba^{2+}][HSO_4^-]}{[H_3O^+]} \cdot \frac{[SO_4^{2-}]}{[SO_4^{2-}]} \tag{11-13b}$$

Now either mentally or actually rearrange the terms so that the $K_{sp}(BaSO_4)$ expression is grouped together, and you can examine the remaining terms. When that is done you should see that

$$K_{eq} = [Ba^{2+}][SO_4^{2-}] \cdot \frac{[HSO_4^-]}{[H_3O^+][SO_4^{2-}]} = \frac{K_{sp}(BaSO_4)}{K_a(HSO_4^-)} \tag{11-13c}$$

With either approach we find the value of the equilibrium constant to be $(1.1 \times 10^{-10})/(1.2 \times 10^{-2}) = 9.2 \times 10^{-9}$.

If we think about the magnitude of this equilibrium constant, we see that it is very small compared to 1. Using Eq. (8-15a), we note that the position of equilibrium is to the left if $K_{eq} \ll 1$. This tells us that not much $BaSO_4$ is going to dissolve in HCl or any other strong acid.

In order to calculate how many grams of $BaSO_4$ dissolve in 1 L of 2.0 F HCl, we must consider the molar ratios specified in Eq. (11-12). If x moles of solid $BaSO_4$ dissolve per liter, then

$$[Ba^{2+}] = [HSO_4^-] = x$$

The $[H_3O^+]$ was originally 2.0 M, but x mol/L are used up in forming HSO_4^-. Thus $[H_3O^+] = 2.0 - x$. The equilibrium constant expression then becomes

$$K_{eq} = 9.2 \times 10^{-9} = \frac{[Ba^{2+}][HSO_4^-]}{[H_3O^+]} = \frac{x^2}{2-x}$$

Assume x is small compared to 2. Then $2 - x = 2$, and

$$x^2 = 2(9.2 \times 10^{-9}) = 18.4 \times 10^{-9} = 1.84 \times 10^{-8}$$
$$x = 1.36 \times 10^{-4} = 1.4 \times 10^{-4} \, M$$

We must check our assumption. Is 1.4×10^{-4} less than 10% of 2.0? Yes, it is considerably less than 0.20, so our assumption is justified. The gram formula weight of $BaSO_4$ is 233.40, so the number of grams of $BaSO_4$ that dissolve in 1 L of 2.0 F HCl is

$$(1.36 \times 10^{-4} \text{ mol/L})(233.40 \text{ g/mol}) = 3.2 \times 10^{-2} \text{ g/L}$$

Thus 0.032 g of $BaSO_4$ dissolve per liter of 2.0 F HCl.

We can compare the solubility of $BaSO_4$ in 2.0 F HCl with its solubility in pure water, by comparing the answers to Examples 11.2 and 11.11. In pure water the solubility of $BaSO_4$ is 0.0023 g/L, while in 2.0 F HCl it is more than 10 times larger, 0.032 g/L. Although $BaSO_4$ is more than 10 times as soluble in 2.0 F HCl than it is in water, its solubility is very small in both solvents. Because 0.032 g is such a small amount, we conclude that we cannot dissolve much $BaSO_4$ in strong acid. Barium sulfate is considered to be only sparingly soluble in both water and strong acid.

The reason that $BaSO_4$ is not very soluble in strong acid is that HSO_4^- is a moderately strong acid, with $K_a = 1.2 \times 10^{-2}$. Thus its conjugate base, SO_4^{2-}, is an extremely weak base.

$$K_b(SO_4^{2-}) = \frac{K_w}{K_a(HSO_4^-)} = \frac{1.0 \times 10^{-14}}{1.2 \times 10^{-2}} = 8.3 \times 10^{-13}$$

and the reaction between H_3O^+ and SO_4^{2-}, Eq. (11-10), does not proceed very far to the right. In general, sulfates that are sparingly soluble in water will only be slightly more soluble in strong acid.

By thinking about the two calculations done in Examples 11.10 and 11.11, we gain some understanding of the ideas that enable us to predict whether or not a given insoluble electrolyte will dissolve in strong acid. Consider the two weak acids involved in these two questions. Hydrogen sulfate ion, HSO_4^-, has an acidity constant $K_a = 1.2 \times 10^{-2}$. It is one of those acids that does not fall clearly into either the strong or weak category, but has to be called moderately weak or moderately strong. It is certainly a much stronger acid than H_2S or HS^-. We see that $BaSO_4$ does not dissolve

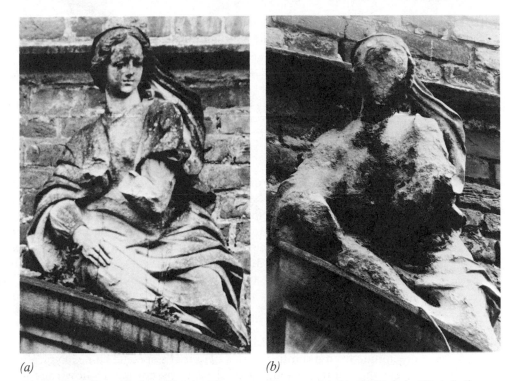

(a) *(b)*

The erosion of marble by acid rain. The statue shown, erected in 1702, is in Westphalia, Germany, in the industrial Rhine–Ruhr region. The photograph in *(a)* shows the statue in 1908, and in *(b)* shows the same statue in 1969, almost completely destroyed by acidic air pollutants. A major component of marble is calcium carbonate, $CaCO_3$. While $CaCO_3$ is insoluble in water, it dissolves readily in acid. The anion, CO_3^{2-}, is a base and reacts with H_3O^+ to form carbonic acid, H_2CO_3, a weak acid that decomposes to H_2O and CO_2. Air polluted with SO_2 and NO_2 from industrial smoke stacks and automobile exhausts becomes acidic due to the formation of sulfuric and nitric acids, and as the acid falls on marble and other stones containing carbonates, the marble dissolves.

in a strong acid, but MnS does. *The weaker the acid formed when the anion of the insoluble electrolyte reacts with hydronium ion, the more likely it is that the electrolyte will dissolve in strong acid.* The formation of the weak acid will remove anions from the solution and drive the dissolution reaction to the right.

Examine the two equilibrium constant expressions, Eqs. (11-9c) and (11-13c). Note that in both cases the acidity constant of the weak acid (or acids) formed when $H^+(aq)$ is added to the insoluble electrolyte appears in the denominator of the expression for K_{eq}. Thus the smaller K_a is, the more it will contribute to making the overall equilibrium constant large. The smaller the K_{sp} of the insoluble solid, on the other hand, the less likely it will be to dissolve.

Because H_2S is such a weak acid, sulfides are certainly much more soluble in strong acid than they are in water. Some sulfides, however, have such very small solubility products that their solubility even in concentrated HCl is very small. Let us consider CuS, for instance. Cupric sulfide is an exceedingly insoluble black salt, with a solubility product of 8.7×10^{-36} at 25 °C. To determine whether it will dissolve in a strong acid we need the equilibrium constant for the reaction

$$CuS(s) + 2H^+(aq) \rightleftharpoons Cu^{2+} + H_2S \tag{11-14}$$

which is

$$K_{eq} = \frac{[Cu^{2+}][H_2S]}{[H^+]^2} \tag{11-15}$$

The calculation is identical to that performed for MnS in Example 11.10, and you should be able to show that

$$K_{eq} = \frac{K_{sp}(\text{CuS})}{K_a(\text{H}_2\text{S}) \cdot K_a(\text{HS}^-)} = \frac{8.7 \times 10^{-36}}{(1.0 \times 10^{-7})(1.3 \times 10^{-13})} = 6.7 \times 10^{-16}$$

This equilibrium constant is so small that it is not possible to dissolve any significant quantity of CuS in concentrated HCl. The difference between MnS and CuS is in the magnitude of their solubility products. The K_{sp} of CuS is so very tiny that even the formation of the weak acid H_2S cannot drive the dissolution reaction far enough to the right.

Insoluble sulfides can be divided into two groups: Those that are insoluble in HCl, like CuS, and therefore can be precipitated in a hydrochloric acid solution, and those, like MnS, that will not precipitate in a hydrochloric acid solution. This is used as the basis for separating sulfides in that branch of chemistry known as **qualitative analysis,** a method of identifying the ions present in a given solution. The cations whose sulfides precipitate in a 0.3 M HCl solution saturated with H_2S are grouped together. The ions Cu^{2+}, Cd^{2+}, Hg^{2+}, and Bi^{3+} are in this group, among others. Cations whose sulfides do not precipitate in 0.3 M HCl, but will precipitate in a solution made alkaline with ammonia are in a different group that includes Mn^{2+}, Zn^{2+}, Ni^{2+}, and Fe^{2+}.

Instead of asking "Will MnS dissolve in HCl?" we can inquire whether MnS will precipitate from a solution with a pH set at a specific value. Typically in the lab we adjust the acidity to some desired value by adding a known amount of an HCl solution, and then pass H_2S gas through the solution. The solution is kept saturated with H_2S by continuously generating H_2S in it, and the concentration of H_2S remains constant at 0.1 M at 25 °C.

EXAMPLE 11.12. Determining whether an insoluble sulfide will precipitate in a solution of specified acidity

Will any MnS precipitate from a solution that is 0.20 M in Mn^{2+} ion if it is acidified with HCl so that the $[H_3O^+] = 0.040$ M, and is saturated with H_2S so that $[H_2S] = 0.10$ M?

There are two methods of approaching this problem and we will consider both of them.

Method 1. First let us calculate the $[S^{2-}]$ in the solution at the instant of mixing, that is, before any MnS can precipitate. Example 9.11 describes the way to calculate the $[S^{2-}]$ and should be referred to if the following relations are not obvious to you. We make use of the overall equilibrium constant for the two stages of ionization of H_2S:

$$\frac{[H_3O^+]^2 [S^{2-}]}{[H_2S]} = K_a(\text{H}_2\text{S}) \cdot K_a(\text{HS}^-) = 1.3 \times 10^{-20}$$

Simply substitute $[H_3O^+] = 0.040$ and $[H_2S] = 0.10$ into this equation and solve for the $[S^{2-}]$. We obtain

$$[S^{2-}] = \frac{(1.3 \times 10^{-20})(0.10)}{(0.040)^2} = \frac{1.3 \times 10^{-21}}{1.6 \times 10^{-3}} = 8.1 \times 10^{-19}$$

We then calculate the reaction quotient, Q, at the instant of mixing for the reaction

$$\text{MnS(s)} \rightleftharpoons \text{Mn}^{2+}\text{(aq)} + \text{S}^{2-}\text{(aq)}$$

and compare it with the K_{sp} of MnS, which is 5×10^{-15}.

$$Q = [\text{Mn}^{2+}][\text{S}^{2-}] = (0.20)(8.1 \times 10^{-19}) = 1.6 \times 10^{-19}$$

Since the value of Q is much smaller than $K_{sp}(\text{MnS})$, the solution is undersaturated with respect to Mn^{2+} and S^{2-} ions, and no MnS will precipitate.

Method 2. Consider the overall reaction for the precipitation of MnS in this solution:

$$\text{Mn}^{2+}(aq) + \text{H}_2\text{S} \rightleftharpoons \text{MnS}(s) + 2\text{H}^+(aq)$$

To find out whether this reaction proceeds to the right with the concentrations specified, we must compare Q for this reaction with the equilibrium constant for this reaction. First we evaluate the equilibrium constant, K_{eq}.

$$K_{eq} = \frac{[\text{H}^+]^2}{[\text{Mn}^{2+}][\text{H}_2\text{S}]} = \frac{[\text{H}^+]^2\,[\text{S}^{2-}]}{[\text{H}_2\text{S}]} \cdot \frac{1}{[\text{Mn}^{2+}][\text{S}^{2-}]} = \frac{K_a(\text{H}_2\text{S}) \cdot K_a(\text{HS}^-)}{K_{sp}(\text{MnS})}$$

$$= \frac{1.3 \times 10^{-20}}{5 \times 10^{-15}} = 2.6 \times 10^{-6}$$

At the instant of mixing,

$$Q = \frac{[\text{H}^+]^2}{[\text{Mn}^{2+}][\text{H}_2\text{S}]} = \frac{(0.040)^2}{(0.20)(0.10)} = 8.0 \times 10^{-2}$$

Since Q is larger than K_{eq}, the only possible reaction is the one that makes Q smaller, that is, the backward reaction. If there were any solid MnS present, it would dissolve. No reaction occurs because there is nothing that can happen to decrease Q. Certainly the reaction will not proceed to the right, and therefore no MnS will be formed.

Both methods of approach to this problem of course lead to the same conclusion: No MnS will precipitate from a solution that is saturated with H_2S and has $[\text{H}_3\text{O}^+] = 0.040\ M$ and $[\text{Mn}^{2+}] = 0.20\ M$.

In Section 11.2 we discussed the fact that anions that are weak bases react with water, and that this reaction increases the molar solubility of salts with basic anions above the values calculated by the methods described in Examples 11.2, 11.3, and 11.4. The following problem illustrates how to include the proton-transfer reaction between sulfide ion and water in calculating the molar solubility of a very slightly soluble sulfide.

EXAMPLE 11.13. Calculation of the molar solubility of a salt with an anion that is a weak base

The solubility product of MnS is 5×10^{-15} at $25\ °\text{C}$. Calculate the molar solubility of MnS in water at $25\ °\text{C}$, including the effect of the proton-transfer reaction between S^{2-} and H_2O.

Solution. The overall reaction that occurs on dissolving MnS in water is

$$\text{MnS}(s) + \text{H}_2\text{O} \rightleftharpoons \text{Mn}^{2+} + \text{HS}^- + \text{OH}^-$$

which is the sum of Eq. (11-2)

$$\text{S}^{2-} + \text{H}_2\text{O} \rightleftharpoons \text{HS}^- + \text{OH}^-$$

and of

$$\text{MnS}(s) \rightleftharpoons \text{Mn}^{2+} + \text{S}^{2-} \tag{11-16}$$

The equilibrium constant for Eq. (11-2) is the basicity constant of sulfide ion:

$$K_b(\text{S}^{2-}) = \frac{[\text{HS}^-][\text{OH}^-]}{[\text{S}^{2-}]} = \frac{K_w}{K_a(\text{HS}^-)}$$

$$= \frac{1.0 \times 10^{-14}}{1.3 \times 10^{-13}} = 7.7 \times 10^{-2}$$

The equilibrium constant for Eq. (11-16) is the solubility product of MnS:

$$K_{sp}(MnS) = [Mn^{2+}][S^{2-}] = 5 \times 10^{-15}$$

The equilibrium constant for the sum of Eqs. (11-2) and (11-16) is the product of these two equilibrium constants:

$$K_{eq} = K_{sp}(MnS) \cdot K_b(S^{2-}) = (7.7 \times 10^{-2})(5 \times 10^{-15}) = 3.8 \times 10^{-16}$$

$$\text{let } x = \text{molar solubility of MnS in water}$$

$$\text{then } x = [Mn^{2+}] = [HS^-] = [OH^-]$$

according to the overall reaction. Hence

$$x^3 = 3.8 \times 10^{-16} \quad \text{and} \quad x = 7.2 \times 10^{-6} \, M$$

If we use only the method of Example 11.2, we would calculate for the molar solubility $x^2 = 5 \times 10^{-15}$ and $x = 7.1 \times 10^{-8} \, M$. The true molar solubility is about 100 times larger than $7.1 \times 10^{-8} \, M$.

section 11.5
Analytical Separations Based on a Difference in Solubility Products

Ions of elements in the same family in the periodic table with the same charge have very similar chemistry. Consider, for example, the ions of the alkaline earth family: Mg^{2+}, Ca^{2+}, Sr^{2+}, and Ba^{2+}. All four cations form insoluble white carbonates when some soluble carbonate such as sodium carbonate is added to a solution containing these cations. However, while the solubility products of $BaCO_3$, $SrCO_3$, and $CaCO_3$ are very close in value, the solubility product of $MgCO_3$ is considerably larger than the others. Numerical values of the solubility products of the alkaline earth carbonates are given in Table 11.2. We can make use of the difference between the K_{sp} of $MgCO_3$ and that of the other alkaline earth carbonates to separate Mg^{2+} from any or all of the other alkaline earth cations. This is accomplished by controlling the $[CO_3^{2-}]$ carefully to prevent $MgCO_3$ from precipitating when $BaCO_3$, $SrCO_3$, and $CaCO_3$ are precipitated.

Suppose we have a solution that contains both Mg^{2+} and Ca^{2+}. It might be a mixture of any two soluble salts of these cations, such as $Mg(NO_3)_2$ and $CaCl_2$. If we wish to separate the Mg^{2+} from the Ca^{2+} we can attempt to precipitate $CaCO_3$ without precipitating any $MgCO_3$ at all. Filtration of the precipitate will then leave a solution containing Mg^{2+} ions, while the Ca^{2+} is contained in the precipitate, which has been physically separated from the solution.

There are two requirements for a successful **analytical separation:** (1) complete precipitation of the ion we want to remove from solution, and (2) no precipitation of the ion remaining in solution. What is meant by complete precipitation? Since no solid electrolyte has zero solubility in water, it is impossible to precipitate 100% of any ionic species. Thus complete precipitation does not mean 100% of the ions present

Table 11.2. **Solubility Products of Alkaline Earth Carbonates at 25 °C**

$MgCO_3$	4×10^{-5}
$CaCO_3$	4.8×10^{-9}
$SrCO_3$	9.4×10^{-10}
$BaCO_3$	8.1×10^{-9}

are incorporated into the precipitate, but it does mean that the ionic concentration is reduced to a very small value. We are immediately faced with the question: What is meant by a very small value? There is no hard and fast answer to this. It is usually a practical matter; the amount left must be small enough not to interfere with whatever reactions we want to carry out with the remaining solution. We can, however, formulate a reasonable rule of thumb: If the concentration of a given ion is reduced to 0.1% of its original value (or less), we will consider that precipitation is complete. That means 99.9% of the ions originally present in solution are in the precipitate at the end of the separation and only 0.1% remain in the solution. A specific problem will make the concepts involved in the calculation clear.

EXAMPLE 11.14. The analytical separation of Mg^{2+} and Ca^{2+}

A given solution is 0.080 M in Ca^{2+} and 0.060 M in Mg^{2+} ions.

(a) Is it possible to precipitate 99.9% of the Ca^{2+} as $CaCO_3$ without precipitating any $MgCO_3$?

Solution. We first calculate how large a concentration of carbonate ion is required to precipitate 99.9% of the Ca^{2+} ions present. *Remember:* If the ions are not part of the precipitate, they are still in solution! Since 99.9% of all the Ca^{2+} is in the precipitate, 0.1% remains in solution. We began with 0.080 M Ca^{2+}. Hence the $[Ca^{2+}]$ remaining in solution after the precipitation of the $CaCO_3$ is 0.1% of the 0.080 M, or

$$[Ca^{2+}] = (1 \times 10^{-3})(8.0 \times 10^{-2}) = 8.0 \times 10^{-5} \, M$$

What must the $[CO_3^{2-}]$ in solution be, in order for the precipitate of $CaCO_3$ to be in equilibrium with a solution with $[Ca^{2+}] = 8 \times 10^{-5} \, M$? Since there is equilibrium between the precipitate and the solution

$$[Ca^{2+}][CO_3^{2-}] = K_{sp}(CaCO_3) = 4.8 \times 10^{-9}$$

and hence

$$[CO_3^{2-}] = \frac{4.8 \times 10^{-9}}{8.0 \times 10^{-5}} = 0.60 \times 10^{-4} = 6.0 \times 10^{-5}$$

Thus the $[CO_3^{2-}]$ must be maintained at $6.0 \times 10^{-5} \, M$ in order to precipitate 99.9% of the Ca^{2+} originally present in solution.

To determine whether or not any Mg^{2+} will precipitate at this $[CO_3^{2-}]$, we evaluate Q for the equilibrium

$$MgCO_3(s) \rightleftharpoons Mg^{2+}(aq) + CO_3^{2-}(aq)$$
$$Q = [Mg^{2+}][CO_3^{2-}] = (0.060)(6.0 \times 10^{-5}) = 3.6 \times 10^{-6}$$

Since Q is smaller than the K_{sp} of $MgCO_3$, which is 4×10^{-5}, no $MgCO_3$ will precipitate. It is possible, therefore, to precipitate 99.9% of the Ca^{2+} in this solution as $CaCO_3$ and not precipitate any of the Mg^{2+}.

(b) What is the upper limit we must set on the $[CO_3^{2-}]$ in this solution in order to prevent any $MgCO_3$ from precipitating?

Solution. No $MgCO_3$ will precipitate until Q for $MgCO_3$ is equal to the K_{sp} of $MgCO_3$. The $[CO_3^{2-}]$ at which the very first speck of $MgCO_3$ will precipitate is, therefore,

$$[CO_3^{2-}] = \frac{K_{sp}(MgCO_3)}{[Mg^{2+}]} = \frac{4 \times 10^{-5}}{6 \times 10^{-2}} = 6.7 \times 10^{-4} \, M = 7 \times 10^{-4} \, M$$

The upper limit on the $[CO_3^{2-}]$ we can allow in this solution is 6.7×10^{-4} M. As long as the $[CO_3^{2-}]$ does not exceed that value, no $MgCO_3$ will precipitate.

These calculations tell us that an analytical separation of Mg^{2+} and Ca^{2+} is possible if there is a way to control the $[CO_3^{2-}]$ in the solution so that it is at least 6.0×10^{-5} M but never exceeds 6.7×10^{-4} M.

(c) If we keep the $[CO_3^{2-}]$ at this upper limit, what is the $[Ca^{2+}]$ in solution at the end of the separation?

Solution. If we maintain the $[CO_3^{2-}]$ at 6.7×10^{-4} M, the $[Ca^{2+}]$ in solution at equilibrium is

$$[Ca^{2+}] = \frac{K_{sp}(CaCO_3)}{[CO_3^{2-}]} = \frac{4.8 \times 10^{-9}}{6.7 \times 10^{-4}} = 7.2 \times 10^{-6} \; M$$

Thus the fraction of the Ca^{2+} ion originally present that is left unprecipitated is $(7.2 \times 10^{-6})/(8.0 \times 10^{-2})$, or 0.009%.

Summary

The equilibrium constant for the dissolution of an insoluble or slightly soluble electrolyte is called a **solubility product**. Like any equilibrium constant, a solubility product is a constant at constant temperature.

The **molar solubility** of an insoluble electrolyte is the number of moles of solid that dissolve per liter of solution. The molar solubility is not a constant, but changes as the constituents of the solution in equilibrium with the solid are changed. The molar solubility and the solubility product are always related to one another, but the relationship varies with the type of electrolyte ($1:1$, $1:2$, $1:3$, and so on) and with the other constituents of the solution.

When the aqueous solution contains a soluble electrolyte having an ion in common with the insoluble solid, the molar solubility of the insoluble solid is decreased by a significant amount. This is known as the **common ion effect**, and has been discussed qualitatively in Chapter 8. Methods of calculating the molar solubility of an insoluble electrolyte in a solution containing a common ion have been described in this chapter.

To determine whether a given insoluble electrolyte will precipitate when two solutions containing the ions of that insoluble solid are mixed, we must compare the reaction quotient, Q, for the dissolution reaction at the instant of mixing, with the solubility product, K_{sp}, for the reaction. Only if $Q > K_{sp}$ will any precipitation occur.

Insoluble electrolytes with anions that are bases will be more soluble in acidic solution than in pure water. When an insoluble electrolyte with an anion that is a base dissolves in acid, the weak acid conjugate to the anion base is formed. The weaker the acid formed, the more likely it is that the electrolyte will dissolve in strong acid, and the more its solubility will be increased above the solubility in pure water.

Much use is made of a difference between the solubility products of two insoluble solids having an ion in common, in carrying out an **analytical separation** of the ions that are not common to the two solids. The branch of chemistry known as **inorganic qualitative analysis** is a series of laboratory procedures for identifying the ions present in a given solution. The ions are divided into groups with similar chemistry, and then analytical separations are carried out in order to identify each ion in the group.

Exercises

Section 11.1

1. Write the solubility product expression for each of the following insoluble electrolytes: $AgBrO_3$, $Zn(OH)_2$, $Cu(IO_3)_2$, $SrCrO_4$, $AgSCN$, CaC_2O_4.

2. Write correctly balanced net ionic equations for the dissolution of each of the following slightly soluble electrolytes in water: Hg_2SO_4, $PbCO_3$, $AgIO_3$, $Cr(OH)_3$, $La(IO_3)_3$.

3. Is each of the following statements about the solubility product of an insoluble electrolyte TRUE or FALSE? If the statement is false, explain what is incorrect, and make an appropriate correction.

   (a) The solubility product of an insoluble electrolyte has a fixed numerical value.
   (b) The solubility product of an insoluble electrolyte is always a number smaller than 1.
   (c) A solubility product is an equilibrium constant.
   (d) The units of any solubility product are M^2, that is, $(mol/L)^2$.

Section 11.2

The K_{sp} values needed for these exercises can be found in Table E2 of the Appendix.

4. Calculate the molar solubility in water of each of the following insoluble substances, from its K_{sp} value: (a) silver bromide (b) barium iodate (c) strontium sulfate (d) lead iodide.

5. Calculate the solubility of (a) $AgIO_3$ and (b) $PbBr_2$ in grams per 100 mL in pure water, from the K_{sp} value of each of these salts.

6. Write the expression that relates the molar solubility in water, s, to the K_{sp} for each of the following insoluble electrolytes: (a) $PbSO_4$ (b) $Ce(IO_3)_3$ (c) $Zn(OH)_2$ (d) $AgBrO_3$ (e) Hg_2I_2.

7. Write the expression that relates y, the molar solubility in 0.15 F NaI, to the K_{sp} of each of the following insoluble iodides: (a) silver iodide (b) lead iodide (c) mercurous iodide.

8. Write the expression that relates z, the molar solubility in 0.086 F $Pb(NO_3)_2$, to the K_{sp} of each of the following electrolytes: (a) lead chloride (b) lead sulfate (c) lead oxalate (d) lead hydroxide.

9. Calculate the molar solubility of silver iodide in 0.20 F KI.

10. *The Handbook of Chemistry and Physics* gives the solubility of $Cd(OH)_2$ as 0.00026 grams per 100 mL at 25 °C. Calculate the solubility product of $Cd(OH)_2$ at 25 °C from this value.

11. The solubility of strontium sulfate at 0 °C is reported to be 0.0113 grams per 100 mL of solution. Calculate the solubility product of strontium sulfate at 0 °C from this value.

12. Calculate the molar solubility of barium iodate in 0.125 F KIO_3.

13. Calculate the molar solubility of $AgBr$ in pure water and in 0.12 F KBr. Explain the reason for the difference between these two values.

14. Calculate the molar solubility of $Mn(OH)_2$ in pure water and in 0.10 F $Mn(NO_3)_2$. Account for the difference between these two values.

Section 11.3

15. Write the correctly balanced net ionic equation for the reaction that occurs when aqueous solutions of the following reagents are mixed:

    (a) Lead nitrate and potassium chromate.
    (b) Sodium sulfide and silver nitrate.
    (c) Barium chloride and sodium oxalate.
    (d) Zinc iodide and ammonium sulfide.
    (e) Sodium phosphate and calcium chloride.
    (f) Cupric chloride and potassium iodate.

16. If 50.00 mL of 0.10 F AgNO$_3$ are mixed with 50.00 mL of 0.10 F KSCN (potassium thiocyanate), will any AgSCN precipitate? Show all calculations necessary to prove your answer. The solubility product of silver thiocyanate is 1×10^{-12}.

17. Will any silver acetate precipitate when 30.00 mL of 0.075 F AgNO$_3$ and 20.00 mL of 0.050 F NaOAc are mixed? Show all calculations necessary to prove your answer.

18. Will any barium iodate precipitate when 20.00 mL of 0.18 F KIO$_3$ and 40.00 mL of 0.12 F Ba(NO$_3$)$_2$ are mixed? Show all calculations necessary to prove your answer.

19. Will any lead chloride precipitate when 80.00 mL of 0.052 F Pb(NO$_3$)$_2$ and 20.00 mL of 0.035 F LiCl are mixed? Show all calculations necessary to prove your answer.

20. Calculate the [Ca^{2+}] and [CO$_3^{2-}$] remaining in solution after 40.00 mL of 0.100 F CaCl$_2$ and 10.00 mL of 0.200 F (NH$_4$)$_2$CO$_3$ are mixed, and the solution has come to equilibrium.

Section 11.4

21. Write correctly balanced net ionic equations for dissolving each of the following insoluble or slightly soluble electrolytes in nitric acid: (a) silver acetate (b) magnesium hydroxide (c) calcium oxalate (d) ferrous sulfide (e) lead carbonate (f) mercurous cyanide.

22. How many grams of solid SrSO$_4$ will dissolve in 100.0 mL of 1.20 F HNO$_3$?

23. Determine the numerical value of the equilibrium constant for the following reactions:
    (a) ZnS(s) + 2H$^+$(aq) $\rightleftharpoons$ H$_2$S + Zn^{2+}(aq)
    (b) AgCN(s) + H$^+$(aq) $\rightleftharpoons$ Ag$^+$(aq) + HCN
    (c) BaCO$_3$(s) + H$^+$(aq) $\rightleftharpoons$ Ba^{2+}(aq) + HCO$_3^-$(aq)
    (d) Tl$_2$S(s) + 2H$^+$(aq) $\rightleftharpoons$ 2Tl$^+$(aq) + H$_2$S
    (e) PbSO$_4$(s) + H$^+$(aq) $\rightleftharpoons$ Pb^{2+}(aq) + HSO$_4^-$(aq)

24. Determine the numerical value of the equilibrium constant for each of the following reactions:
    (a) Cd^{2+}(aq) + H$_2$S $\rightleftharpoons$ CdS(s) + 2H$^+$(aq)
    (b) Ag$^+$(aq) + HOAc $\rightleftharpoons$ AgOAc(s) + H$^+$(aq)
    (c) PbSO$_4$(s) + S^{2-} $\rightleftharpoons$ PbS(s) + SO$_4^{2-}$

25. Will any CdS precipitate from a solution that is 0.10 M in Cd^{2+} if it is acidified with HCl so that the [H$_3$O$^+$] = 1.0 M, and is saturated with H$_2$S so that [H$_2$S] = 0.10 M?

Section 11.5

26. A dilute solution of Na$_2$SO$_4$ is added dropwise to an aqueous solution that is 0.10 M in both Ba^{2+} and Ca^{2+}.
    (a) What is the minimum [SO$_4^{2-}$] necessary to initiate precipitation of BaSO$_4$?
    (b) What must the [SO$_4^{2-}$] be in order to precipitate 99.9% of all the Ba^{2+} originally present?

(c) Will any $CaSO_4$ be precipitated when 99.9% of the Ba^{2+} has been precipitated? Show all calculations necessary to prove your answer.

27. A solution is 0.10 M in Pb^{2+} and 0.12 M in Zn^{2+}. (a) Is it possible to precipitate 99.9% of the Pb^{2+} as PbS without precipitating any ZnS? (b) What is the upper limit that must be set on the $[S^-]$ in order to prevent any ZnS from precipitating?

Multiple Choice Questions

1. The solubility product expression for Hg_2Cl_2 is
   (a) $[Hg^+]^2[Cl^-]^2$ (b) $[Hg_2^{2+}][Cl^-]^2$ (c) $[Hg^+][Cl^-]$ (d) $[Hg_2^{2+}] + [2Cl^-]$
   (e) $[Hg_2^{2+}][2Cl^-]^2$

2. The anion of the insoluble salt MX is the conjugate base of the weak acid HX. The relationship between the equilibrium constant for the reaction

$$MX(s) + H^+(aq) \rightleftharpoons M^+(aq) + HX$$

   the solubility product of MX, and the acidity constant of HX, is that K_{eq} equals
   (a) $K_{sp}(MX)/K_a(HX)$ (b) $K_a(HX)/K_{sp}(MX)$ (c) $K_a(HX) + K_{sp}(MX)$
   (d) $K_{sp}(MX) - K_a(HX)$ (e) $\{K_{sp}(MX)\}^2/K_a(HX)$

3. The molar solubility, s, of $Mn(OH)_2$ in water in terms of its K_{sp} is
   (a) $s = (K_{sp})^{1/2}$ (b) $s = (K_{sp})^{1/3}$ (c) $s - (K_{sp}/4)^{1/3}$ (d) $s = (K_{sp}/6)^{1/3}$
   (e) $s = (K_{sp}/27)^{1/4}$

4. If 50.0 mL of 0.050 F $BaCl_2$ is mixed with 50.0 mL of 0.10 F K_2SO_4,
   (a) Potassium chloride will precipitate.
   (b) The $[Ba^{2+}]$ in the resulting solution will be 0.025 M.
   (c) No reaction will occur.
   (d) Barium sulfate will precipitate.
   (e) The $[K^+]$ in the resulting solution will be 0.050 M.

5. Which of the following insoluble electrolytes has the largest molar solubility in water?
   (a) $SrCO_3$ (b) MnS   (c) $PbSO_4$ (d) FeS   (e) AgCl

6. The maximum concentration of Ba^{2+} that can exist in a solution in which the sulfate ion concentration is 5.5×10^{-4} M is
   (a) 2.0×10^{-7} M (b) 5.5×10^{-4} M (c) 2.0×10^{-5} M
   (d) 6.0×10^{-14} M (e) 1.1×10^{-10} M

7. The relationship between the K_{sp} of AgBr and the molar solubility, z, of AgBr in 0.20 F KBr is that K_{sp} equals
   (a) z^2 (b) $z/0.20$ (c) $z^{1/2}$ (d) $4z^3$ (e) $0.20 z$

8. Which of the following insoluble substances has the largest molar solubility in water?
   (a) AgI   (b) PbI_2 (c) $SrSO_4$ (d) $Zn(OH)_2$ (e) MnS

9. When 80.0 mL of 0.120 F $Ca(NO_3)_2$ are mixed with 20.0 mL of 0.100 F K_2CO_3, in the resulting solution the
   (a) $[Ca^{2+}] = 0.096$ M (b) $[K^+] = 0.040$ M (c) $[NO_3^-] = 0.096$ M
   (d) $[CO_3^{2-}] = 0.100$ M (e) $[CO_3^{2-}] = 0.020$ M

10. At the temperature at which the K_{sp} of $PbSO_4$ is 1.7×10^{-8}, the molar solubility of $PbSO_4$ in 0.10 F K_2SO_4 is
    (a) 1.7×10^{-7} M (b) 1.7×10^{-8} M (c) 1.3×10^{-4} M (d) 1.7×10^{-9} M
    (e) 1.3×10^{-8} M

11. At the temperature at which the K_{sp} of $PbSO_4$ is 1.7×10^{-8}, the molar solubility of $PbSO_4$ in pure water is
    (a) $1.7 \times 10^{-7} M$ (b) $1.7 \times 10^{-8} M$ (c) $1.3 \times 10^{-4} M$ (d) $1.7 \times 10^{-9} M$
    (e) $1.3 \times 10^{-8} M$

12. In which of the solutions below is $CaCO_3$ the most soluble?
    (a) water   (b) $0.20 F CaCl_2$ (c) $0.20 F HCl$ (d) $0.20 F Na_2CO_3$
    (c) $0.20 F NaCl$

13. When 60.00 mL of $0.010 F Ca(NO_3)_2$ are mixed with 40.00 mL of $0.025 F K_2CrO_4$,
    (a) $CaCrO_4$ will precipitate.
    (b) The $[NO_3^-]$ in the resulting solution will be $6.0 \times 10^{-3} M$.
    (c) No reaction will occur.
    (d) The $[CrO_4^{2-}]$ in the resulting solution will be $0.025 M$.
    (e) KNO_3 will precipitate.

14. How many moles of SrF_2 will dissolve in 1 L of water if $K_{sp}(SrF_2) = 7.9 \times 10^{-10}$?
    (a) 2.8×10^{-5} (b) 7.9×10^{-10} (c) 5.8×10^{-4} (d) 5.8×10^{-5}
    (e) 9.2×10^{-4}

15. How many moles of SrF_2 will dissolve in 1 L of $0.10 F NaF$ if $K_{sp}(SrF_2) = 7.9 \times 10^{-10}$?
    (a) 2.8×10^{-5} (b) 7.9×10^{-8} (c) 7.9×10^{-9} (d) 2.0×10^{-8}
    (e) 4.0×10^{-9}

16. How many moles of SrF_2 will dissolve in 1 L of $0.10 F Sr(NO_3)_2$ if $K_{sp}(SrF_2) = 7.9 \times 10^{-10}$?
    (a) 2.8×10^{-5} (b) 7.9×10^{-8} (c) 7.9×10^{-9} (d) 4.0×10^{-9}
    (e) 4.4×10^{-5}

17. If $K_{sp}(PbSO_4) = 1.8 \times 10^{-8}$ and $K_a(HSO_4^-) = 1.0 \times 10^{-2}$, the equilibrium constant for the reaction

$$PbSO_4(s) + H^+(aq) \rightleftharpoons HSO_4^- + Pb^{2+}$$

    is   (a) 1.8×10^{-6} (b) 1.8×10^{-10} (c) 2.8×10^{-10} (d) 1.0×10^{-2}
    (e) 3.2×10^{-14}

18. Which of the following insoluble electrolytes will NOT be significantly more soluble in $1.0 F HNO_3$ than it is in pure water?
    (a) FeS (b) $SrCO_3$ (c) $AgCN$ (d) $BaSO_3$ (e) AgI

19. At the temperature at which the molar solubility of $PbBr_2$ in water is $2.3 \times 10^{-2} M$, what is the K_{sp} of $PbBr_2$?
    (a) 5.3×10^{-4} (b) 1.2×10^{-5} (c) 2.4×10^{-5} (d) 2.3×10^{-2}
    (e) 4.9×10^{-5}

20. The relationship between the molar solubility, y, of CaF_2 in $0.080 F NaF$ and the K_{sp} of CaF_2 is that y equals
    (a) $K_{sp}/0.080$ (b) $K_{sp}/(0.16)^2$ (c) $(K_{sp}/4)^{1/3}$ (d) $(K_{sp})^{1/2}$ (e) $K_{sp}/(0.080)^2$

21. At a certain temperature, a saturated solution of $Zn(OH)_2$ has a pH of 8.30. The value of K_{sp} for $Zn(OH)_2$ at this temperature is
    (a) 8.0×10^{-18} (b) 4.0×10^{-18} (c) 1.6×10^{-17} (d) 4.0×10^{-12}
    (e) 2.0×10^{-6}

22. A solution is $0.120 M$ in Pb^{2+}. If $K_{sp}(PbCrO_4) = 1.8 \times 10^{-14}$, in order to precipitate 99.9% of all the Pb^{2+} present, the $[CrO_4^{2-}]$ must be maintained at

(a) $1.8 \times 10^{-17} M$ (b) $1.8 \times 10^{-14} M$ (c) $1.5 \times 10^{-13} M$
(d) $1.5 \times 10^{-10} M$ (e) $1.3 \times 10^{-7} M$

23. The relationship between the molar solubility, z, of MgF_2 in $0.10\ F\ Mg(NO_3)_2$ and the K_{sp} of MgF_2 is that z equals
(a) $(K_{sp}/0.10)^{1/2}$ (b) $(K_{sp}/0.40)^{1/2}$ (c) $(K_{sp}/4)^{1/2}$ (d) $(K_{sp})^{1/2}$ (e) $(K_{sp}/4)^{1/3}$

24. When 20.00 mL of $0.100\ F\ Pb(NO_3)_2$ and 30.00 mL of $0.150\ F\ Na_2C_2O_4$ are mixed, PbC_2O_4 precipitates. The $[C_2O_4^{2-}]$ in the resulting solution is
(a) $0.0450\ M$ (b) $0.0500\ M$ (c) $0.0900\ M$ (d) $0.100\ M$ (e) $0.150\ M$

25. When 60.00 mL of $0.100\ F\ Ca(NO_3)_2$ and 40.00 mL of $0.125\ F\ Na_2CO_3$ are mixed, $CaCO_3$ precipitates. If $K_{sp}(CaCO_3) = 5 \times 10^{-9}$, the $[CO_3^{2-}]$ in the resulting solution is
(a) $0.10\ M$ (b) $0.050\ M$ (c) $5 \times 10^{-7}\ M$ (d) $5 \times 10^{-8}\ M$ (e) $8 \times 10^{-8}\ M$

26. If the K_{sp} of $Cu(OH)_2$ is 2×10^{-19} and the K_{sp} of CuS is 8.7×10^{-36}, the equilibrium constant for the reaction

$$Cu(OH)_2(s) + S^{2-} \rightleftharpoons CuS(s) + 2\,OH^-$$

is (a) 2×10^{-55} (b) 2×10^{16} (c) 4×10^{17} (d) 2×10^{18} (e) 4×10^{-17}

27. Which of the following insoluble solids is appreciably more soluble in $0.50\ M$ HCl than it is in water?
(a) $PbBr_2$ (b) AgI (c) $Ba(IO_3)_2$ (d) Hg_2Cl_2 (e) $ZnCO_3$

Problems

The K_{sp} and K_a values needed for solving these problems can be found in Tables E1 and E2, Appendix E.

11.1. The solubility of ceric(IV) iodate, $Ce(IO_3)_4$, is reported as 0.015 grams per 100 mL at 20 °C in the *Handbook of Chemistry and Physics*. Calculate the solubility product of $Ce(IO_3)_4$ at 20 °C.

11.2. Calculate the molar solubility of lead iodide in (a) pure water, (b) $0.10\ F\ Pb(NO_3)_2$, and (c) $0.10\ F$ KI. Account for the differences between these values.

11.3. The solubility product of magnesium hydroxide, $Mg(OH)_2$, is 1.5×10^{-11} at 25 °C.
(a) What is the molar solubility of $Mg(OH)_2$ in water at 25 °C?
(b) What is the $[OH^-]$ in the saturated solution?
(c) What is the pH of this solution?

11.4. Show all numerical calculations necessary to determine which member of each pair is more soluble in water.
(a) CaF_2 versus CaC_2O_4 (b) $Cu(OH)_2$ versus $La(OH)_3$
What important fact do these calculations illustrate?

11.5. Determine the numerical value of the equilibrium constant for each of the following reactions.
(a) $Hg_2^{2+}(aq) + 2HCN \rightleftharpoons Hg_2(CN)_2 \downarrow + 2H^+(aq)$
(b) $PbF_2(s) + 2\,IO_3^- \rightleftharpoons Pb(IO_3)_2(s) + 2F^-$
(c) $BaC_2O_4(s) + 2H^+(aq) \rightleftharpoons H_2C_2O_4 + Ba^{2+}$
(d) $2Ag^+(aq) + H_2S \rightleftharpoons Ag_2S \downarrow + 2H^+(aq)$
(e) $Ni(OH)_2(s) + 2H^+(aq) \rightleftharpoons Ni^{2+}(aq) + 2H_2O$
(f) $PbCl_2(s) + H_2S \rightleftharpoons PbS(s) + 2Cl^-(aq) + 2H^+(aq)$

11.6. A solution is 0.050 M in Mn^{2+}, Fe^{3+}, and La^{3+}. Ammonia gas is bubbled into the solution until the pH is 7.8. Assume there is no change in volume. If the pH is maintained at 7.8, which, if any, of the insoluble hydroxides of these three cations will precipitate? Show all work necessary to prove your answer.

11.7. (a) Write a correctly balanced net ionic equation for the reaction that occurs between each of the pairs of reagents listed below.
(b) Obtain the value of the equilibrium constant for the reaction you have written by combining constants in Tables E1 and E2.
(c) State whether or not the reaction will proceed to an appreciable extent to the right as you have written the equation, if all substances in solution are at 1.0 M concentration.

    1. Bismuth(III) nitrate and H_2S.
    2. Lead iodide and potassium sulfide.
    3. Nitric acid and cadmium hydroxide.
    4. Sodium hydroxide and nickel(II) chloride.
    5. Potassium bisulfate and sodium acetate.

11.8. The solubility of anhydrous barium oxalate in water at 25 °C is 9.4×10^{-3} grams per 100 mL of solution. The solubility product of barium oxalate at 25 °C is 1.1×10^{-7}. Calculate the solubility of barium oxalate in water in grams per 100 mL from its K_{sp} and explain why the value you have calculated does not agree with the observed value.

11.9. Calculate the molar solubility of $CaSO_4$ in (a) pure water, and (b) 2.500 F HCl.

11.10. If lead chloride is shaken with a solution that is 0.0010 M in Cl^- and 0.20 M in SO_4^{2-} will any $PbSO_4$ form? Calculate K_{eq} for $PbCl_2(s) + SO_4^{2-} \rightleftharpoons PbSO_4(s) + 2Cl^-$, and then indicate the reasoning used to answer the question.

11.11. The solubility of magnesium fluoride has been reported as 7.6×10^{-3} grams per 100 mL at 25 °C. Calculate the solubility product of magnesium fluoride from this value, and compare it with the K_{sp} tabulated in Appendix E. Account for the difference between your calculated K_{sp} and that in Table E2.

11.12. If AgCl is shaken with a solution that is 3.0×10^{-3} M in Cl^- and 0.45 M in CO_3^{2-}, will any AgCl be converted to Ag_2CO_3? Show all calculations necessary to prove your answer.

11.13. Calculate the concentration of each ionic species in solution after 40.00 mL of 0.162 F $AgNO_3$ and 20.00 mL of 0.144 F K_2CrO_4 are mixed, and the system has come to equilibrium.

11.14. Calculate the concentration of each ionic species in solution after 50.00 mL of 0.096 F KIO_3 and 30.00 mL of 0.112 F $Pb(NO_3)_2$ are mixed, and the system has come to equilibrium. K_{sp} of $Pb(IO_3)_2$ is 2.6×10^{-13}.

11.15. A white solid is known to be one of the following compounds: $BaSO_4$, Na_2CO_3, ZnS, NH_4NO_3, Hg_2Cl_2, $SrCO_3$, KI, or $PbSO_4$.
(a) The unknown is insoluble in water. Which of the possibilities is eliminated on the basis of this observation?
(b) The unknown dissolves readily in dilute HCl with the evolution of a colorless, odorless gas. Identify the unknown, giving the reasons for your choice.
(c) Write a correctly balanced net ionic equation for the reaction between the solid unknown and dilute HCl.

11.16. The solubility of Ag_2SO_4 in water at 0 °C is reported to be 0.568 grams per 100 mL of solution.

(a) Give the expression for, and the units of, the solubility product of Ag_2SO_4, and calculate its numerical value at 0 °C.

(b) Some solid Ag_2SO_4 is added to 0.300 F K_2SO_4 at 0 °C. Some, but not all, of the Ag_2SO_4 dissolves. What is the $[Ag^+]$ in the solution when the system has come to equilibrium? What is the molar solubility of Ag_2SO_4 in 0.300 F K_2SO_4 at 0 °C?

(c) Account for the difference between the molar solubility of Ag_2SO_4 in water and in 0.300 F K_2SO_4 in terms of Le Chatelier's Principle.

11.17. A solution is 0.100 M in both Ni^{2+} and Fe^{3+}. We would like to determine if it is possible to perform an analytical separation of these ions by controlling the pH of the solution.

(a) What must the pH of the solution be in order to precipitate 99.99% of the Fe^{3+} as $Fe(OH)_3$?

(b) Will any $Ni(OH)_2$ precipitate at the pH calculated in part (a)?

(c) At what pH will $Ni(OH)_2$ first begin to precipitate? Is an analytical separation of these two ions possible?

11.18. A solution of a leaf extract of the plant *Nyassa sylvatica* is found to be 0.100 M in Co^{2+} and 3.00×10^{-4} M in Fe^{2+}. In order to carry out a biochemical assay of the leaves of this plant, it is necessary to remove the Co^{2+} from this solution without removing any Fe^{2+}. This is done by controlling the pH so that CoS can be precipitated without precipitating FeS. The solution is kept saturated with H_2S, so that $[H_2S] = 0.10$ M.

$$K_{sp}(\text{CoS}) = 8 \times 10^{-23} \qquad K_{sp}(\text{FeS}) = 5 \times 10^{-18}$$

(a) What $[S^{2-}]$ is required to precipitate the first trace of FeS?

(b) What percentage of the original Co^{2+} is left unprecipitated if the $[S^{2-}]$ is maintained at the value you calculated in part (a)?

(c) What should the pH of the solution be in order to keep the $[S^{2-}]$ at the value calculated in part (a)?

11.19. Silver chloride, AgCl, is an insoluble white solid, and Ag_2CrO_4 is an insoluble dark red solid. Suppose that we have a solution in which $[Cl^-] = 0.036$ M and $[CrO_4^{2-}] = 0.012$ M. A dilute solution of $AgNO_3$ is added dropwise to this solution, and volume changes are so small that they can be neglected.

(a) What $[Ag^+]$ is required to precipitate 99.9% of the Cl^- originally present?

(b) Will any Ag_2CrO_4 precipitate at this $[Ag^+]$? Show calculations necessary to prove your answer.

(c) At what $[Ag^+]$ will Ag_2CrO_4 first begin to precipitate?

11.20. The basicity constant, K_b, for methylamine, CH_3NH_2, is 3.7×10^{-4}. If 2.1085 g of $Mn(NO_3)_2 \cdot 4H_2O$ is added to 100.0 mL of a solution containing 0.050 mol of CH_3NH_2 and 0.075 mol of CH_3NH_3Cl (methylammonium chloride), and the solution is stirred until equilibrium is attained, will any $Mn(OH)_2$ precipitate? Show the calculations that prove your answer. K_{sp} of $Mn(OH)_2 = 4.5 \times 10^{-14}$.

11.21. (a) Discuss the effect of adding ammonium nitrate to the equilibrium existing in aqueous ammonia.

(b) A buffer solution is 1.20 M in NH_4^+ and 0.150 M in NH_3. What is the maximum $[Al^{3+}]$ and the maximum $[Mg^{2+}]$ that can exist in this buffer?

(c) What use can you make of this information to devise an analytical separation of Al^{3+} and Mg^{2+}?

11.22. A 50.00-mL sample of 0.100 F $La(NO_3)_3$ is mixed with 50.00 mL of an NH_4^+/NH_3 buffer that is 0.200 M in NH_4^+ and 0.400 M in NH_3. What percentage of the original La^{3+} has been precipitated as $La(OH)_3$ when the mixture comes to equilibrium? The K_{sp} of $La(OH)_3$ is 1×10^{-19}.

11.23. Enough $Ni(NO_3)_2$ and $Cr(NO_3)_3$ are dissolved in 90.00 mL of 2.00 F HOAc to make $[Ni^{2+}] = 0.120$ M and $[Cr^{3+}] = 0.092$ M. If 10.00 mL of 6.00 F NaOH are added to this solution, will any $Ni(OH)_2$ precipitate? Will any $Cr(OH)_3$ precipitate?

11.24. A solution is prepared by dissolving 0.100 mol $Mn(NO_3)_2$, 0.100 mol $Cu(NO_3)_2$, 0.10 mol NaOAc, and 1.00 mol HOAc in enough water to make the volume 1.00 L. A 20.00-mL portion of this solution is saturated with H_2S until $[H_2S] = 0.100$ M. Will any CuS precipitate? Will any MnS precipitate? Show calculations that prove your answer.

11.25. Magnesium hydroxide is more soluble in a solution of NH_4Cl than it is in pure water due to the reaction

$$Mg(OH)_2(s) + 2NH_4^+ \rightleftharpoons Mg^{2+} + 2NH_3 + 2H_2O$$

(a) Express K_{eq} for this reaction in terms of $K_b(NH_3)$ and $K_{sp}\{Mg(OH)_2\}$.
(b) In an aqueous solution of NH_4Cl that is saturated with $Mg(OH)_2$, the pH = 9.07 and $[NH_4^+] = 0.200$ M.

   (i) If $K_b(NH_3) = 1.8 \times 10^{-5}$, calculate $[NH_3]$ in this solution.
   (ii) Calculate K_{eq} for the reaction given above.
   (iii) Calculate the solubility product of $Mg(OH)_2$ from the K_{eq} just calculated and the expression you wrote in part (a).

chapter 12 Atomic Structure, Atomic Spectra, and the Introduction of the Quantum Concept

Niels Henrik David Bohr (1885–1962), Danish physicist, received his doctorate at the University of Copenhagen in 1911, and immediately thereafter went to the Cavendish laboratories of Cambridge University to study under J. J. Thomson. The following year he worked in Manchester, England, under Ernest Rutherford. In 1913 he successfully applied the concept of energy quanta to explain the spectrum of atomic hydrogen. Bohr was awarded the Nobel Prize in 1922 for his work on the theory of atomic spectra. During the Second World War he was a key member of the team of scientists at the Los Alamos Laboratories in New Mexico that developed the atomic bomb. Bohr received the Atoms for Peace award in 1957.

In Chapter 1 we described the nature of the atom as it is understood today, without providing the experimental evidence for the description given. The series of experiments that led to our present understanding of atomic structure also provides a basis for understanding the periodic classification of the elements proposed by Dmitri I. Mendeleev and Lothar Meyer, and the nature of the chemical bond. A number of experiments carried out during the last half of the nineteenth century and the first two decades of the twentieth century could not be explained by the laws of "classical physics," that is, the laws governing mechanics, optics, electricity, and magnetism that had been used so successfully prior to 1900 to explain a great many diverse phenomena. An entirely new theoretical approach was needed to understand the properties of atoms and subatomic particles. These new ideas, the **quantum theory** of matter, were developed during the first thirty years of the twentieth century, and it is not an exaggeration to say that the concepts and methods of quantum theory revolutionized physics and chemistry.

In the following three chapters we will discuss the application of quantum theory to problems of atomic and molecular structure and chemical bonding. We begin by discussing the important and exciting series of experiments performed during the last decade of the nineteenth century and the early years of the twentieth century that served to elucidate properties of the electron and to provide us with a picture of the nuclear atom.

section 12.1
The Experimental Basis for Modern Concepts of the Atom
Cathode Rays

During the last half of the nineteenth century a number of physicists investigated the electrical conductivity of gases at low pressures. Gases do not ordinarily conduct electricity (they are **insulators** or nonconductors), but when subjected to high voltages (several thousand volts) at low pressures (less than 0.01 atm), electrical conduction does take place. An emission of light accompanies the conduction of electricity.

The apparatus used to study these phenomena is shown in Fig. 12.1, and is known as a **Crookes tube** (after Sir William Crookes, an English physicist). Two metal plates (electrodes) are sealed into a glass tube that can be evacuated and then filled with a gas to any desired pressure. One of these plates, the **anode,** is charged positively and the other, the **cathode,** is charged negatively. The space between the plates is then subjected to an electrical potential difference, measured in volts. The larger the magnitude of the charge on the plates, the larger the voltage. For low voltages nothing occurs, but when the voltage reaches a value between 5000 and 10,000 V, a glow discharge is observed. You have seen this yourself, because it is the glow of neon signs. Different colors are observed when different gases fill the tube. If neon itself is used, the glow is reddish orange.

Many experiments were performed in order to understand what causes the glow discharge. In one of them, a light weight pinwheel was placed in the space between cathode and anode and it was observed that the pinwheel begins to rotate when the gas begins to conduct electricity and the glow discharge ensues. This showed that particles travel through the Crookes tube, striking the pinwheel. It was also observed that any solid object placed between the cathode and anode casts a shadow. The location of the shadow proved that the direction of the particles traveling through the tube is always from the cathode to the anode. If the pressure of the gas in the tube is

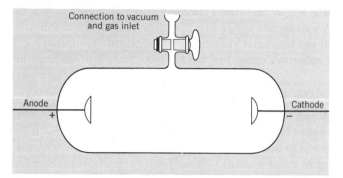

Fig. 12.1. Crookes tube. When the applied voltage reaches a value close to 10,000 V, the tube begins to glow.

very low, about 1×10^{-4} atm, the glass tube itself begins to glow or fluoresce.

We now understand that when the difference in electrical potential between cathode and anode reaches a value close to 10,000 V, the cathode emits a stream of electrons. These electrons travel across the tube and collide with the gas molecules, causing them to dissociate into electrons and positive ions. The collisions between electrons and gas molecules result in the glow discharge observed in neon signs. When electrons and positive ions strike the glass walls of the tube, they cause it to glow or fluoresce. At the time of these experiments, however, the existence of the electron had not yet been demonstrated. The experiments proved that the fluorescence observed is the result of the bombardment of the glass by rays of particles that emanate from the cathode or negative electrode, and are therefore called **cathode rays.** Cathode rays travel in straight lines until they collide with a gas molecule or strike the anode or the glass walls of the tube. If a metal plate with only a tiny slit in it is placed in front of the cathode, the cathode ray is confined to a thin beam. By 1890 it had been shown that this beam can be deflected by both an electric field and a magnetic field. (See Fig. 12.2.) The similar deflection by a magnetic field of a wire that carries current led to the hypothesis that the cathode rays consist of charged particles.

The Cathode-Ray Experiments of J. J. Thomson

Joseph J. Thomson (1856–1940) was a British physicist who made a careful and thorough study of cathode rays. His experiments provided the first evidence for the

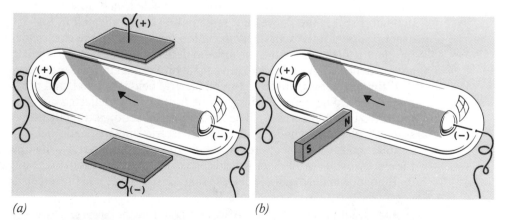

(a)   *(b)*

Fig. 12.2. The deflection of cathode rays by *(a)* an electric field and *(b)* a magnetic field.

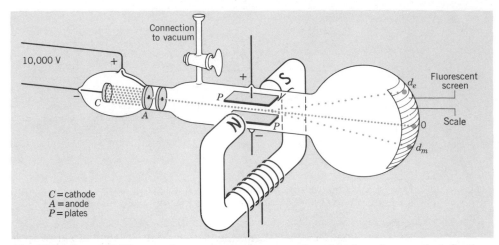

Fig. 12.3. Joseph J. Thomson's experiment demonstrating electric and magnetic deflection of a cathode ray. The deflection caused by an electric field applied to the plates, P, results in a bright spot on the screen at d_e. The undeflected beam strikes the screen at 0. The deflection caused by an applied magnetic field perpendicular to the paper results in a bright spot at d_m. Thomson used the cathode-ray tube to measure the charge-to-mass (e/m) ratio of the electron.

existence of electrons. The apparatus used by Thomson is illustrated schematically in Fig. 12.3. An evacuated tube contains a cathode, C, and an anode, A, near one end. The anode is pierced by a small hole at the center. Near the middle of the tube a pair of charged plates, labeled P, are arranged to establish an electric field in the vertical direction. In the same region there is a magnet. The two poles of the magnet produce a magnetic field in a direction perpendicular to the plane of the paper. The right-hand end of the tube contains a fluorescent screen, provided by a coating of zinc sulfide (ZnS) on the inner face of the tube. Zinc sulfide is a white solid that emits visible light when electrons strike it.

The difference in electrical potential between the anode and the cathode causes the cathode to emit electrons. Most of these electrons fall on the anode, but a number pass through the hole in the center of the anode, providing a narrow beam of electrons (the **cathode ray**) that travels through the tube. In the absence of any applied electric or magnetic field, these electrons travel in a straight line and hit the fluorescent screen at the point marked 0 in Fig. 12.3, causing a bright spot there. If only an electric field is applied, by charging the plates P, the beam is deflected upward toward the positive plate and strikes the fluorescent screen at a point indicated in Fig. 12.3 by d_e. The exact position of the spot depends on the magnitude of the electric field. The larger the electric field, the greater the upward deflection. The fact that the cathode rays are always deflected toward the positive plate proves that the rays are negatively charged.

If the magnet is turned on, but no electric field is applied, and the polarity of the magnetic field is properly arranged, the beam can be deflected downward to a point indicated on Fig. 12.3 as d_m. The downward deflection is directly proportional to the strength of the magnetic field.* If the electric and magnetic fields are turned on simultaneously and their strengths are adjusted appropriately, the upward force on the electrons due to the electric field can be made exactly equal to the downward force on them due to the magnetic field, so that the bright spot on the screen again occurs at

---

* Moving charged particles create a magnetic field. Thus the cathode ray itself produces a magnetic field that interacts with the applied magnetic field and this interaction causes the deflection of the ray.

the undeflected position, 0. By measuring the values of the electric and magnetic fields that result in no net deflection of the beam, the velocity of the electrons can be calculated. The observed velocity is very high, many thousand kilometers per second. The exact numerical value depends on experimental conditions, in particular on the potential difference between the cathode and the anode. From the equality of the potential and kinetic energies, the *ratio* of charge to mass, e/m, of the particles comprising the cathode ray can be determined.

An important result of Thomson's experiment is that the e/m ratio does not depend on the gas inside the tube or the metal used for the cathode or anode. The fact that the e/m ratio is the same whatever gas is present in the tube proves that the cathode ray does not consist of gaseous ions, for if it did, e/m would depend on the nature of the gas.

The value of e/m measured by Thomson for the particles of the cathode rays was more than one thousand times larger than the largest value of e/m observed for any particle prior to 1897. The very large value of e/m for cathode rays, plus the constancy of e/m regardless of the gas in the tube, led Thomson to conclude that the rays consist of a universal fragment of matter found in all atoms, and he adopted the name that had previously been proposed by G. J. Stoney, **electrons.** The numerical value of e/m for electrons obtained by the most recent accurate experiments is

$$e/m = 1.7588 \times 10^8 \text{ C/g*} \tag{12-1}$$

Millikan's Oil Drop Experiment

The very large value of e/m could result either from a large value of the electronic charge, or from a small value of the mass of the electron. An independent experiment that measures either e or m separately was therefore necessary, and in 1910 Robert A. Millikan determined the charge on the electron in a classic and famous experiment.

Robert A. Millikan (1868 – 1953) was an American physicist who used the apparatus illustrated schematically in Fig. 12.4 to measure the charge on an electron. Using an atomizer, microscopic spherical drops of oil are introduced into the space above two charged plates. Oil is used because it does not noticeably evaporate. Some of these drops fall through a hole in the upper plate and enter the observation

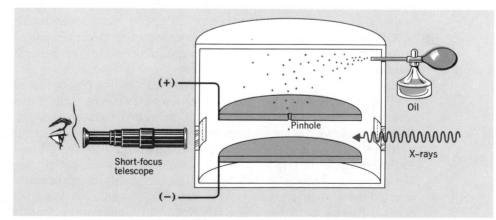

Fig. 12.4. Schematic diagram of Millikan's oil drop experiment to measure the charge on an electron.

* For a definition of the coulomb, C, the SI unit of charge, see Appendix A.

chamber, the space between the two charged plates. Oil drops in the observation chamber become charged by colliding with gaseous ions produced by the action of X-rays on the air in the chamber. Gravity causes the drops to fall, but they are slowed by friction, due to the viscosity of the air. The downward velocity of a falling drop when the plates are not charged is measured by observing the drop through a short-focus telescope. When a potential difference is applied to the plates so that the upper plate is positively charged, some drops move upward and some move downward even faster, as there are both positively and negatively charged drops. With the field turned on, a drop moving upward is selected for observation, and its upward velocity is measured. The charge on the drop can be calculated from the values of its upward and downward velocities, and from the magnitude of the potential difference, the known acceleration of gravity, the density of the oil, and the air viscosity.

Millikan's experiment provided independent evidence of the particulate nature of electricity. He found that different drops had charges of different magnitudes, and he listed all the values of charge observed in a very large number of measurements. All the charges were integral multiples of the value 1.60×10^{-19} coulomb (C), that is, 3.20×10^{-19} C, 4.80×10^{-19} C, and so on. Millikan therefore concluded that the fundamental unit of charge, the charge on the electron, is 1.60×10^{-19} C. Recent measurements of the electronic charge yield

$$e = 1.6022 \times 10^{-19} \text{ C} \tag{12-2}$$

which can be combined with the e/m ratio of Eq. (12-1) to obtain a value for the mass of the electron.

$$m = \frac{1.6022 \times 10^{-19} \text{ C}}{1.7588 \times 10^8 \text{ C/g}} = 9.1096 \times 10^{-28} \text{ g} = 9.1096 \times 10^{-31} \text{ kg} \tag{12-3}$$

Rutherford's Nuclear Atom

By 1910 it was well established that atoms are electrical in nature and that electrons are a universal fragment of all atoms, with a negative charge and a mass very much smaller than the mass of any atom. Atoms were known to be electrically neutral, but the structure of the atom was not known. The first suggestion about atomic structure came from Lord Kelvin in 1902; he proposed a model in which diffuse positive electricity was spread homogeneously throughout the atom, assumed to be spherical in shape. The exact nature of the positive electricity was not described but it seems to have been considered to be a rather viscous material, much like jello. The electrons were supposed to be embedded in this sphere of positive charge, like raisins in a pudding.

Lord Kelvin's proposal was investigated the following year by J. J. Thomson, who calculated the stability of such an arrangement for various numbers of electrons. As a result of this work, the model came to be called the "Thomson atom," and is generally described as having been proposed by Thomson. At the time this model was proposed, no experimental evidence existed that could either confirm or disprove it.

Ernest Rutherford (1871–1937), a New Zealand born physicist who studied under Thomson and became a Professor at the University of Manchester in England, realized that the correctness of this model could be tested by using alpha (α) particles emitted by radioactive atoms. Rutherford had previously proved that α rays are positively charged helium ions by collecting the emitted α-particles and showing that the gas accumulated in the collection chamber was identical with helium. He then reasoned that if a stream of α-particles is directed at a piece of thin metal foil, the particles incident on the foil will either pass right through or be deflected by the atoms

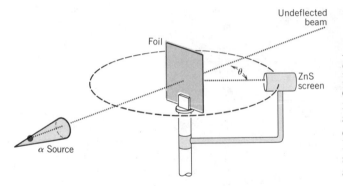

Fig. 12.5. The experiment of Rutherford, Geiger, and Marsden: The scattering of α-particles by a thin metal foil. By analyzing the angles by which the α-particles were deflected, Rutherford elucidated the structure of the nuclear atom.

comprising the foil. Information about the structure of the atoms of the metal foil can be obtained by analyzing the scattering of the α-particles.

Rutherford did not do the experimental work himself. It was carried out by Hans Geiger, a German physicist working in the laboratories at Manchester, and a young undergraduate, Ernest Marsden. It was Rutherford's suggestion, however, that prompted the experiment, which is illustrated in Fig. 12.5. An intense source of α rays was provided by radon gas confined in a conical tube that had a mica window at one end, through which the α-particles passed easily. The beam of α-particles was shot at thin foils of heavy metals (gold, platinum, and silver). The scattered α-particles were detected by using a ZnS screen, which produces a flash of light when hit by an α-particle.

Most of the α-particles were observed to pass straight through the foil undeflected; some were scattered through small angles. This was what was expected on the basis of the jello model proposed by Kelvin and Thomson. But what was totally unexpected was that some α-particles were observed on the same side of the foil as the source; they were deflected by angles greater than 90°. Geiger and Marsden observed a few α-particles deflected by very large angles, close to 180°. Scattering by any angle greater than 90° is totally inconsistent with the Kelvin–Thomson model. Rutherford's own account expresses the astonishment the observation of α-particles scattered by more than 90° caused: "I remember . . . Geiger coming to me in great excitement and saying, 'We have been able to get some of the α-particles coming backwards.' . . . It was quite the most incredible event that has ever happened to me in my life. It was almost as incredible as if you fired a 15-in. shell at a piece of tissue paper and it came back and hit you."

In 1911 Rutherford published a paper in which he presented a theoretical description of the structure of the atom that accounts for the scattering results observed by Geiger and Marsden. This paper describes the nuclear atom as we picture it today. Rutherford found that the only way to explain the observations was to assume that the mass and positive charge are contained in a particle (called the nucleus) whose dimensions are very small compared to the size of the atom itself. The very light electrons occupy most of the space of the atom. When α-particles pass through a metal foil most of them do not come close to the massive, positively charged nuclei, but encounter only the very light electrons which do not deflect the α-particles at all. On rare occasions, however, an α-particle approaches very closely to one of the nuclei in the foil and then the coulombic repulsion of the two positive charges causes the lighter α-particle to swerve sharply, or even to reverse its course. Figure 12.6 illustrates these situations.

This description of the atom, which is universally accepted today, seemed very surprising and unlikely to the scientists of 1911. Why should such a structure be

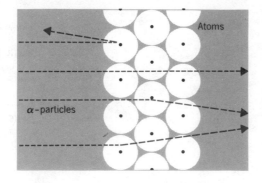

Fig. 12.6. Passage of α-particles through a gold foil. Most α-particles pass through undeflected, but a close approach to a gold nucleus causes a large deflection of the lighter α-particle, due to repulsion between the positively charged α-particles and the positively charged gold nuclei.

stable? Positive and negative charges attract one another—what keeps the negative electrons at some distance from the positive nucleus? Why aren't the electrons drawn into the nucleus as a result of the coulombic (electrostatic) attraction? Indeed, the laws of classical electromagnetic theory (the laws that physicists had deduced prior to 1900 by studying the electrical and magnetic properties of systems in the laboratory) predict unequivocally that an atom with such a structure could not exist.

Because the laws of electrostatics predict that stationary electrons would certainly be pulled into the nucleus by coulombic attraction, it was proposed that the electrons revolve in orbits about the nucleus, similar to the way the planets travel about the sun. But the planetary analogy does not work, because planets are uncharged, whereas electrons are negatively charged. In classical electromagnetic theory, a negatively charged particle traveling in an orbit about a stationary, positively charged nucleus will radiate energy. As it emits energy, its energy decreases and its radius of revolution decreases, that is, the distance separating the orbiting particle from the positive charge will continuously decrease. According to classical physics, then, if an electron at any distance from a nucleus were moving in an orbit around the nucleus, it would radiate energy, spiral in toward the nucleus, and very quickly merge with it. We would see a flash of white light, and the atom would no longer exist. Nothing in classical physics can explain the existence of a stable atom with the structure elucidated by the scattering experiments of Rutherford, Geiger, and Marsden.

section 12.2
Atomic Emission Spectra

The stability of the nuclear atom was not the only phenomenon that could not be explained by the laws of classical mechanics. There were a number of others, some of which had been perplexing scientists for many years. One of the most important of these was the radiation emitted by atoms, **atomic emission spectra.**

Ordinarily, atoms do not emit any radiation, but there are ways of causing radiation to be emitted. The simplest of these is to heat the atoms, and this is the basis of the **flame tests** used in qualitative analysis. A clean nichrome wire placed in a flame imparts no color to the flame, although the wire itself glows red-hot. If the wire is dipped into a solution of NaCl in HCl and then inserted into the flame, a persistent yellow-orange color is observed. If, instead of NaCl, the solution consists of $CaCl_2$ in HCl the flame is red orange, whereas if the solution is $BaCl_2$ in HCl, the color is yellow green. The color of the flame is so distinctive that it can be used to identify the metal ion of the salt. Table 12.1 gives the common flame colors of those metals that can be readily distinguished by a flame test.

Table 12.1. **The Colors Imparted to a Flame by Introducing Salts of these Elements Moistened with 12 *F* HCl**

Element	Color
Na	Strong yellow orange
K	Weak violet (appears red when viewed through cobalt blue glass)
Ca	Strong red orange
Sr	Deep red
Ba	Yellow green
Cu	Green
Sb	Bluish white
Pb	Pale blue

Electromagnetic Radiation and Atomic Spectra

Visible light is just a portion of the radiation emitted by atoms. Radiant energy is **electromagnetic radiation.** Experimental observations of the properties of light can be understood if light is described as being produced by the oscillating motion of an electric charge. An oscillating electric charge produces an oscillating electric field and an oscillating magnetic field, both of which are propagated through space and are called **electromagnetic radiation.** The electric field and the magnetic field are propagated as waves, with the two fields perpendicular to one another.

An instantaneous picture of a wave is shown in Fig. 12.7. If the wave is assumed to represent, for instance, the electric field, the maximum magnitude of the field has the value A, and is called the **amplitude** of the wave. The magnitude of the electric field oscillates between $+A$ and $-A$. The oscillation from $+A$ to $-A$ and back to $+A$ is repeated continuously; one complete repeating unit is called a **cycle**. The length of a cycle is known as the **wavelength** of the radiation and is denoted by λ.

Not only visible light, but also infrared, ultraviolet, microwaves, radio waves, and X-rays are forms of electromagnetic radiation. All electromagnetic radiation is propagated through space with the same speed, which is denoted by c, and called the **speed of light.** In SI units the speed of light in a vacuum has the value

$$c = 2.997925 \times 10^8 \text{ m/s} \tag{12-4}$$

A **monochromatic** (single frequency) wave can be described by specifying its wavelength, λ, or its **frequency,** ν. The frequency is defined as the number of cycles that pass a given point in space per second. Since the speed of light is c meters per second, and the wavelength is λ meters per cycle, the frequency is given by

$$\frac{c \text{ meters per second}}{\lambda \text{ meters per cycle}} = \nu \text{ cycles per second} \tag{12-5}$$

It is customary to omit the word cycles in stating the units of frequency; they are given

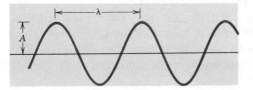

Fig. 12.7. Instantaneous picture of a monochromatic wave. A complete repeating unit is called a cycle.

as s^{-1}, or reciprocal seconds. Another term used for the unit of frequency is the hertz (Hz), in honor of Rudolf Hertz, who discovered electromagnetic radiation outside the visible range in 1896. The hertz is not an SI unit.

Since c is the same for all types of electromagnetic radiation, while λ and ν vary, it is convenient to express the inverse relationship between the wavelength and the frequency as

$$\lambda \nu = c \tag{12-6}$$

An alternative way to describe a monochromatic wave is to specify its **wavenumber,** $\tilde{\nu}$, which is defined as

$$\tilde{\nu} = 1/\lambda = \nu/c \tag{12-7}$$

The wavenumber is the reciprocal of the wavelength, and its SI units are reciprocal meters, m^{-1}.

It is important to be careful about units when using Eqs. (12-6) and (12-7). For most electromagnetic radiation, wavelengths are too small to be conveniently expressed in meters, and it is customary to report them in nanometers (nm), where $1 \text{ nm} = 1 \times 10^{-9}$ m. Another unit that is frequently used to express wavelengths, particularly of visible and ultraviolet light, is the **angstrom,** which is not an SI unit. The angstrom is denoted Å and is defined as

$$1 \text{ Å} = 1 \times 10^{-8} \text{ cm} = 1 \times 10^{-10} \text{ m} = 1 \times 10^{-1} \text{ nm} \tag{12-8}$$

EXAMPLE 12.1. Relationships between wavelength, frequency, and wavenumber for electromagnetic radiation

Electromagnetic radiation with a wavelength of 700 nm is visible red light. **(a)** What is the frequency and wavenumber of this radiation? **(b)** What is the wavelength expressed in angstroms? In cm?

Solution. Since the speed of light is given in meters per second, we must convert the wavelength to meters before using Eq. (12-6).

$$\lambda = 700 \text{ nm} = 700 \times 10^{-9} \text{ m} = 7.00 \times 10^{-7} \text{ m}$$

$$\nu = \frac{c}{\lambda} = \frac{2.998 \times 10^8 \text{ m/s}}{7.00 \times 10^{-7} \text{ m}} = 0.428 \times 10^{15} = 4.28 \times 10^{14} \text{ s}^{-1}$$

$$\tilde{\nu} = 1/\lambda = 1/(7.00 \times 10^{-7} \text{ m}) = 1.43 \times 10^6 \text{ m}^{-1}$$

To convert the wavelength to angstroms, use the factor given in Eq. (12-8):

$$700 \text{ nm} = (700 \text{ nm}) \left(\frac{1 \text{ Å}}{1 \times 10^{-1} \text{ nm}} \right) = 7000 \text{ Å}$$

To convert from meters to centimeters, multiply by 100:

$$\lambda = 7.00 \times 10^{-7} \text{ m} = 7.00 \times 10^{-5} \text{ cm}$$

The wavenumber can therefore also be expressed as $1.43 \times 10^4 \text{ cm}^{-1}$.

What we see during a flame test is the portion of the **spectrum** (the emitted radiation) in the visible region. Radiation may also be emitted that cannot be detected by the human eye. One way to determine the frequencies of the emitted radiation is to pass it through a prism, as shown in Fig. 12.8. Waves of different frequency are **refracted** or bent by different amounts when they pass through the prism.

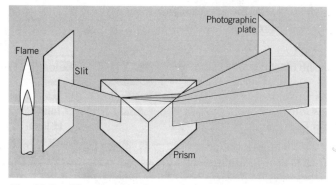

Fig. 12.8. Observation of an emission spectrum. The emitted radiation, after passing through a slit, enters the prism where waves of different wavelength are separated by refraction. A photographic plate serves to detect the various wavelengths.

The truly remarkable feature of atomic emission spectra is that each atom has its own distinctive spectrum. Atomic spectra are **discrete,** that is, certain specific frequencies are emitted by each atom, and other frequencies are never emitted. The spectrum of each type of atom is so distinctive that it is used to identify the presence of that element in analytical chemistry. Figure 12.9 shows the flame spectra of the common alkali and alkaline earth metals.

section 12.3
The Emission Spectrum of Atomic Hydrogen

Another way of producing atomic spectra is to pass a spark (an electrical discharge) through a sample containing the atom to be investigated. In this way one can obtain

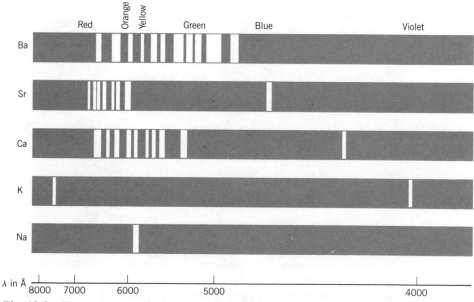

Fig. 12.9. Flame spectra of some alkali and alkaline earth metals.

Table 12.2. **Visible Region of the Spectrum of Atomic Hydrogen**

Intensity	Color	λ, Wavelength (nm)	ν, Frequency (s^{-1})
Strongest	Red	656.279	4.5681×10^{14}
Next	Blue green	486.133	6.1669×10^{14}
Next	Blue violet	434.047	6.9069×10^{14}
Weakest	Violet	410.174	7.3089×10^{14}

the spectrum of the simplest of all atoms, the hydrogen atom. When an electric discharge is passed through a bulb containing H_2 gas at low pressure, it dissociates the H_2 molecules into H atoms, and the bulb glows red. If the emitted radiation is passed through a slit, dispersed by a prism, and then detected by means of a photographic plate, it is observed to consist, in the visible region, of only four very distinct wavelengths. Specific details of the visible region of the spectrum of atomic H are given in Table 12.2.

The emitted frequencies are called "lines" since images of the slit appear at different positions on the photographic plate after the frequencies have been separated by the prism. In addition to the four lines in the visible region, there are many more in the near ultraviolet region. Figure 12.10 illustrates the procedure used to obtain the emission spectrum of atomic hydrogen.

It is helpful, in discussing spectra, to be familiar with the values of wavelength, frequency, and wavenumber that comprise the different regions of the electromagnetic spectrum. This information is contained in Table 12.3. You should use this table as a reference, but it is a good idea to remember that the principal regions of the spectrum are the microwave, infrared (IR), visible (VIS), ultraviolet (UV), and X-ray, in order of decreasing wavelength, and that the wavelengths of visible light are, roughly, those between 400 and 800 nm. The regions of the electromagnetic spectrum are illustrated in Fig. 12.11.

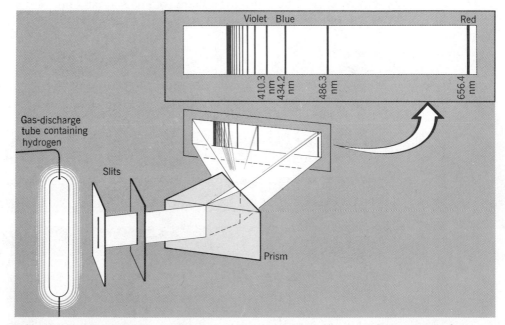

Fig. 12.10. The Balmer series of the emission spectrum of atomic hydrogen. The four lines whose wavelengths are given occur in the visible region, the remaining lines are in the near ultraviolet.

Table 12.3. Regions of the Electromagnetic Spectrum

Region of Spectrum	Wavelength (nm) λ	Frequency (s^{-1}) ν	Wavenumber (cm^{-1}) $\tilde{\nu}$
Microwave	2×10^8	1.5×10^9	5×10^{-2}
Far infrared	1×10^6	3×10^{11}	10
Near infrared	2×10^4	1.5×10^{13}	5×10^2
Visible	780	3.8×10^{14}	1.3×10^4
Near ultraviolet	380	7.9×10^{14}	2.6×10^4
Far ultraviolet	200	1.5×10^{15}	5×10^4
X-rays	10	3×10^{16}	10^6
	0.01	3×10^{19}	10^9

Note that the emission spectrum of atomic hydrogen in the visible and near ultraviolet region shown in Fig. 12.10 consists of a discrete set of frequencies, unique to H atoms. The prediction of classical mechanics about the radiation emitted by any atom is that it should consist of all possible frequencies, that is, the spectrum should be continuous. Nothing in classical mechanics is able to explain the existence of *discrete* atomic spectra.

The Balmer Formula

The wavelengths of the four lines of the H spectrum in the visible region were measured by the Swedish spectroscopist Anders J. Ångström during the second half of the nineteenth century. Johann Jacob Balmer (1825–1898), a Swiss physicist, studied the values obtained and noticed that as the wavelength decreased, the lines were closer and closer together. Balmer tried to obtain a simple formula that would represent all four of these wavelengths, and was, indeed, able to do so. The **Balmer formula,** published in 1885, is

$$\tilde{\nu} = 1/\lambda = 109{,}678 \left(\frac{1}{2^2} - \frac{1}{n^2} \right) \text{cm}^{-1} \tag{12-9}$$

where $n = 3$, 4, 5, and 6 for the four lines in the visible region. The value 109,678 cm^{-1} is an empirical constant, obtained by using the experimental wavelengths. If you substitute $n = 3$ into this equation and solve for λ, you will obtain

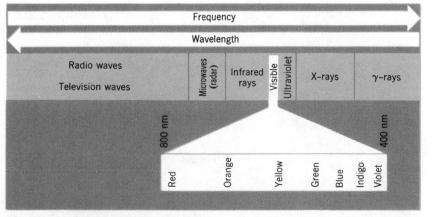

Fig. 12.11. The electromagnetic spectrum.

$\lambda = 656.28$ nm, the red line of the H spectrum. If you substitute $n = 4$ you will obtain $\lambda = 486.13$ nm, the blue-green line, and so on. Several lines in the near ultraviolet region had already been measured prior to 1885 and their wavelengths were found to fit Balmer's formula by setting $n = 7, 8, 9, \ldots$. All the wavelengths calculated by setting n in Eq. (12-9) at integral values greater than 2 are now called the **Balmer series** of the spectrum of atomic hydrogen.

EXAMPLE 12.2. The Balmer series in the spectrum of H atoms

Calculate the wavelength, frequency, and wavenumber of the fifth line in the Balmer series. In what region of the spectrum does this line occur?

Solution. The fifth line in the Balmer series is the line for which $n = 7$. The wavenumber, $\tilde{\nu}$, is given by Eq. (12-9) as

$$\tilde{\nu} = 1/\lambda = 109{,}678 \left(\frac{1}{2^2} - \frac{1}{7^2} \right) = 109{,}678 \left(\frac{1}{4} - \frac{1}{49} \right) = 25{,}181.2 \text{ cm}^{-1}$$

The wavelength, λ, is therefore $(1/25{,}181.2)$ cm, or

$$\lambda = 3.97122 \times 10^{-5} \text{ cm} = 3.97122 \times 10^{-7} \text{ m} = 397.122 \text{ nm}$$

The frequency, ν, is given by $\nu = c/\lambda$, so that

$$\nu = \frac{2.997925 \times 10^8 \text{ m/s}}{3.97122 \times 10^{-7} \text{ m}} = 7.54913 \times 10^{14} \text{ s}^{-1}$$

By referring to Table 12.3 you can see that this line is just on the borderline between the visible and the near ultraviolet region.

The Rydberg Constant

In 1906 the physicist T. Lyman observed additional lines in the spectrum of atomic hydrogen in the far ultraviolet region. The wavelengths of the lines discovered by Lyman fit the formula

$$\tilde{\nu} = 1/\lambda = 109{,}678 \left(\frac{1}{1^2} - \frac{1}{n^2} \right) \text{ cm}^{-1} \tag{12-10}$$

where $n = 2, 3, 4, \ldots, \infty$.

It is clear that the number $109{,}678$ cm^{-1} is an important constant and it has been named the **Rydberg constant** in honor of the Swedish spectroscopist, J. R. Rydberg. It is symbolized $\mathcal{R}$. (Do not confuse the Rydberg constant with the universal gas constant, R. There is no connection between them. We will always use a script $\mathcal{R}$ to denote the Rydberg constant.)

Additional lines in the spectrum of atomic hydrogen have been observed in the near and far infrared region. Every line that has been observed in every region of the spectrum fits the general formula

$$\tilde{\nu} = \frac{1}{\lambda} = \mathcal{R} \left(\frac{1}{n_L^2} - \frac{1}{n_H^2} \right) \text{ cm}^{-1} \tag{12-11}$$

where n_L is an integer (the lower value integer) and n_H is an integer with value higher than n_L.

The set of lines for a given value of n_L constitutes a series of lines in the spectrum of atomic hydrogen. Each series is named after its discoverer. Within each series there is

Table 12.4. **Principal Series in the Spectrum of Atomic Hydrogen**

Series	Spectral Region	n_L	n_H
Lyman	Far UV	1	2, 3, 4, . . . , ∞
Balmer	VIS and near UV	2	3, 4, 5, . . . , ∞
Paschen	IR	3	4, 5, 6, . . . , ∞
Brackett	IR	4	5, 6, 7, . . . , ∞
Pfund	IR	5	6, 7, 8, . . . , ∞

a striking pattern: As the wavelength decreases the spacing between adjacent lines decreases and the intensity also decreases. Each series converges to a limit: As $n_H \rightarrow \infty$, $1/\lambda \rightarrow \mathcal{R}/n_L^2$. Figure 12.12 shows the Balmer series of lines, which converge to the wavelength $\lambda = 364.70$ nm, in the near ultraviolet region. The principal series in the spectrum of atomic hydrogen are given in Table 12.4.

EXAMPLE 12.3. The spectrum of atomic hydrogen

What is the wavelength and frequency of the limiting line of the Lyman series in the spectrum of atomic hydrogen? In what region of the spectrum does this limiting line occur?

Solution. For the Lyman series, $n_L = 1$. As n_H gets larger and larger and approaches infinity ($n_H \rightarrow \infty$), $1/n_H^2$ gets smaller and smaller and approaches zero ($1/n_H^2 \rightarrow 0$), so that $1/\lambda \rightarrow \mathcal{R}/1^2 = \mathcal{R} = 109{,}678$ cm^{-1}. Therefore, for the limiting line

$$\lambda = (1/109{,}678) \text{ cm} = 9.1176 \times 10^{-6} \text{ cm} = 9.1176 \times 10^{-8} \text{ m}$$

$$\lambda = 91.176 \times 10^{-9} \text{ m} = 91.176 \text{ nm}$$

$$v = \frac{c}{\lambda} = \frac{2.997925 \times 10^8 \text{ m/s}}{9.1176 \times 10^{-8} \text{ m}} = 3.2881 \times 10^{15} \text{ s}^{-1}$$

According to Table 12.3, a line with a wavelength of 91.176 nm falls in the middle of the far ultraviolet region of the spectrum.

section 12.4
The Bohr Theory of the Hydrogen Atom and the Introduction of the Quantum Concept

Although the Balmer formula, Eq. (12-11), was well known, it could not be explained or derived from the laws of classical physics. The great Danish physicist, Niels Bohr (1885 – 1962), was the first to derive Eq. (12-11) from a set of fundamental hypothe-

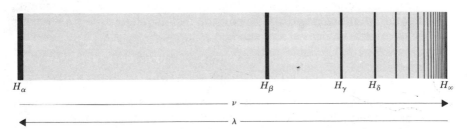

Fig. 12.12. The Balmer series of spectral lines of atomic hydrogen. The most intense line, H_α, is the red line with wavelength 656.3 nm. The intensity of the lines decreases as the wavelength decreases. The line H_∞ is the series limit in the near ultraviolet region, with wavelength 364.7 nm.

ses. Since the experimental relation could not be explained using only the assumptions of classical physics, Bohr began with a new and quite different set of assumptions. Bohr's paper, published in 1913, drew upon theoretical work of Max Planck (1858–1947) and Albert Einstein (1879–1955) concerning very different experimental problems, for which they also had been forced to discard some of the fundamental concepts of classical physics. To begin his derivation, Bohr chose a set of postulates that were quite radical; some of his assumptions are no longer considered to be correct. The only justification for Bohr's postulates was that by using them he could derive Balmer's empirical formula. We will list here only those assumptions of Bohr's that are still considered to be correct today.

1. *Classical physics does not apply to particles of atomic or subatomic dimensions.* Although Newtonian mechanics and electromagnetic theory correctly predict the behavior of systems with masses large enough to be measured directly in our laboratories, they are invalid when applied to atoms or subatomic particles.

2. *The energy of an electron in a hydrogen atom is quantized,* that is, it is limited to a discrete set of values. There is a lowest allowed value for the energy of an electron in a hydrogen atom. That energy is called the **ground state energy** and may be denoted E_1. The next allowed energy may be denoted E_2. Energies between E_1 and E_2 are forbidden. Each allowed value of the energy constitutes an allowed state of the electron: We speak of the set of **stationary states** of the hydrogen atom. The term "stationary" refers to the fact that in such a state the energy of the electron is fixed and remains constant with time. *When the electron is in a stationary state the atom is stable and does not radiate energy.* All states of the H atom above the ground state are referred to as **excited states**.

3. *A hydrogen atom radiates energy only when the electron "jumps" from one allowed stationary state to another.* When a transition between two allowed stationary states occurs, light is emitted with a frequency, v, given by

$$\Delta E = hv \qquad (12\text{-}12)$$

where ΔE is the difference in energy between the two allowed stationary states and h has the value 6.6262×10^{-34} J $\cdot$ s and is called **Planck's constant.** Similarly, radiation can be absorbed by H atoms only if the frequency of the radiation satisfies Eq. (12-12).

Equation (12-12) is an extremely important relation. It is valid for all types of spectra, atomic and molecular, and is known as the **Bohr frequency condition.** Since frequency is always given in units of reciprocal seconds, using Planck's constant in joule $\cdot$ seconds will yield the energy difference ΔE in joules. While the joule is the SI unit of energy, other units have been used and continue to be used by spectroscopists reporting energy differences. Two other commonly used units are the electron volt (eV) and the calorie (cal). Conversion factors relating these energy units to the joule are

$$1 \text{ eV} = 1.6022 \times 10^{-19} \text{ J} \qquad (12\text{-}13a)$$
$$1 \text{ cal} = 4.1840 \text{ J} \qquad (12\text{-}13b)$$

It is customary to use electron volts when discussing a single electron or a single atom, and to use joules or calories when discussing molar quantities.

Starting with the postulates listed above and a few others that are no longer considered to be correct, Bohr applied certain concepts of classical mechanics and derived the following expression for the allowed energies of the electron in the hydrogen atom:

$$E_n = -\frac{2\pi^2 m e^4}{h^2 n^2} = -\frac{K}{n^2} \qquad (12\text{-}14)$$

where $n = 1, 2, 3, \ldots, \infty$, m is the mass of the electron, e is the charge on the electron, h is Planck's constant, and K is a constant equal to $2\pi^2 m e^4/h^2$. The value of n is restricted to positive integers greater than zero; n is called the **quantum number**. Bohr's derivation of this equation is given in Appendix H.

The numerical value of K depends, of course, on the units used for mass, charge, and energy. The most commonly used values are $K = 313.5$ kcal/mol, 1312 kJ/mol, and 13.60 eV/atom. The lowest value of energy the electron in a hydrogen atom can have is obtained by setting $n = 1$ in Eq. (12-14).

$$E_1 = \text{ground state energy of the H atom} = -\frac{2\pi^2 m e^4}{h^2} = -13.60 \text{ eV} \qquad (12\text{-}15)$$

The next allowed energy (the first excited state) is

$$E_2 = -\frac{2\pi^2 m e^4}{2^2 h^2} = -\frac{13.60}{4} = -3.40 \text{ eV} \qquad (12\text{-}16)$$

The second excited state has energy

$$E_3 = -\frac{2\pi^2 m e^4}{3^2 h^2} = -\frac{13.60}{9} = -1.51 \text{ eV} \qquad (12\text{-}17)$$

and so on. It is clear that as n gets larger and larger and approaches infinity ($n \rightarrow \infty$), the energy of the electron gets less negative and approaches zero ($E_n \rightarrow 0$). Zero energy corresponds to infinite separation of the electron from the nucleus. Since there is attraction between the negative electron and the positive nucleus, the closer the electron comes to the nucleus the more negative its energy becomes, that is, the lower its energy becomes. The ground state, the state of lowest energy, is the state in which the electron is closest to the nucleus. When the electron is separated from the nucleus by an infinite distance, its energy is zero.

EXAMPLE 12.4. The Bohr frequency condition

How much energy is required to move one electron from the ground state of the H atom to the first excited state? How much energy is required to cause 1 mol of H atoms to undergo the transition from the ground state to the first excited state? What is the frequency of the radiation that must be absorbed to cause this transition to occur?

Solution. To move the electron of a hydrogen atom from the ground state, where it has energy E_1, to the first excited state, where it has energy E_2, we must supply the difference in energy, that is, $E_2 - E_1$. Using Eqs. (12-15) and (12-16) we obtain

$$\text{energy required for transition from ground state to first excited state} = E_2 - E_1 = -3.40 \text{ eV} - (-13.60 \text{ eV})$$

$$= 10.20 \text{ eV/atom}$$

To express this in joules, use the conversion factor of Eq. (12-13a):

$$10.20 \text{ eV} = (10.20 \text{ eV})(1.6022 \times 10^{-19} \text{ J/eV}) = 1.634 \times 10^{-18} \text{ J}$$

To obtain the energy required to cause a mole of H atoms to undergo this transition, multiply the energy needed per atom by Avogadro's number:

$$\left(1.634 \times 10^{-18} \frac{\text{J}}{\text{atom}}\right)\left(6.022 \times 10^{23} \frac{\text{atom}}{\text{mol}}\right) = 9.840 \times 10^5 \frac{\text{J}}{\text{mol}} = 984.0 \frac{\text{kJ}}{\text{mol}}$$

We could also have calculated this directly by using K in Eq. (12-14) in kilojoules per mole.

$$E_2 - E_1 = \frac{-K}{2^2} - \left(\frac{-K}{1^2}\right) = \frac{-K}{4} + K = \frac{3K}{4} = \frac{3}{4}(1312 \text{ kJ/mol}) = 984.0 \frac{\text{kJ}}{\text{mol}}$$

In using the Bohr frequency condition, Eq. (12-12), to calculate the frequency of the radiation that causes this transition, it is most convenient to give the energy difference in joules.

$$v = \frac{\Delta E}{h} = \frac{1.634 \times 10^{-18} \text{ J}}{6.6262 \times 10^{-34} \text{ J} \cdot \text{s}} = 2.466 \times 10^{15} \text{ s}^{-1}$$

This is the frequency of the first and most intense line of the Lyman series. By referring to Table 12.3 we see that this transition is in the far ultraviolet.

By using the Bohr frequency condition, postulate 3, we can obtain the frequencies predicted by Bohr for the spectrum of atomic hydrogen. At room temperature essentially all hydrogen atoms are in the ground electronic state. The electron has energy E_1 and no radiation is emitted, in accord with postulate 2. A spark excites the electron to one of the allowed higher energy states. The electron then spontaneously drops from the higher energy state to a lower energy state and emits radiation of frequency given by the Bohr frequency condition, Eq. (12-12).

Bohr's expression for the allowed energies of the hydrogen atom, Eq. (12-14), enables us to prepare an **energy level diagram** (Fig. 12.13) that contains, in one figure, the explanation for every line in the spectrum of atomic hydrogen. We plot energy along the ordinate (vertical) axis, and simply draw a horizontal line to represent the allowed energies. Vertical arrows indicate transitions between the allowed states. The length of the arrow is a measure of the difference in energy, ΔE, between the states and is therefore also a measure of the frequency emitted during the transition, according to Eq. (12-12). The Lyman series of lines is obtained when the lower energy state is that for which $n_L = 1$. When $n_L = 2$, the Balmer series of lines is obtained, and so on.

The frequency emitted during a transition from a higher energy state with quantum number n_H to a lower energy state with quantum number n_L is obtained by combining the Bohr frequency condition with Eq. (12-14):

$$E_{n_H} - E_{n_L} = hv = \frac{2\pi^2 me^4}{h^2}\left(\frac{1}{n_L^2} - \frac{1}{n_H^2}\right)$$

so that

$$v = \frac{2\pi^2 me^4}{h^3}\left(\frac{1}{n_L^2} - \frac{1}{n_H^2}\right) \tag{12-18}$$

The wavenumber, $\tilde{v}$, is $v/c = 1/\lambda$. Dividing Eq. (12-18) by c we obtain

$$\tilde{v} = 1/\lambda = \frac{2\pi^2 me^4}{ch^3}\left(\frac{1}{n_L^2} - \frac{1}{n_H^2}\right) \tag{12-19}$$

If Bohr's expression for the allowed energies, Eq. (12-14) is correct, then Eq. (12-19) must be identical with Eq. (12-11), the Balmer formula. The two equations are identical only if the Rydberg constant, $\mathcal{R}$, is given by

$$\mathcal{R} = \frac{2\pi^2 me^4}{ch^3} \tag{12-20}$$

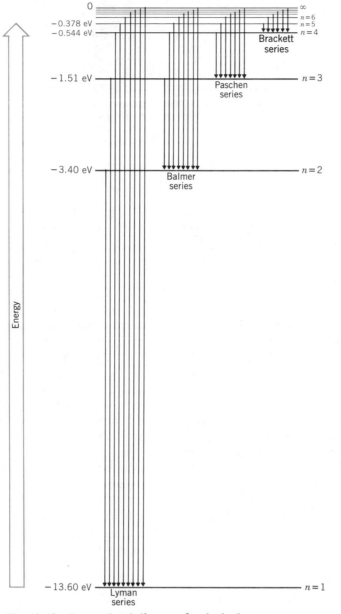

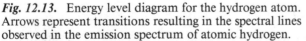

Fig. 12.13. Energy level diagram for the hydrogen atom. Arrows represent transitions resulting in the spectral lines observed in the emission spectrum of atomic hydrogen.

It was a great triumph for Bohr that substitution of the values of the fundamental constants into Eq. (12-20) yields 109,737 cm^{-1}, in very close agreement with the experimental value of the Rydberg constant, 109,678 cm^{-1}.

Every line in the spectrum of atomic hydrogen is accounted for by Eq. (12-19). Other useful forms of this equation are

$$v = \frac{c}{\lambda} = c\mathcal{R} \left(\frac{1}{n_L^2} - \frac{1}{n_H^2} \right)$$

(12-21a)

and

$$\Delta E = h\nu = hc\mathcal{R} \left(\frac{1}{n_L^2} - \frac{1}{n_H^2} \right)$$ (12-21b)

Using Eq. (12-21b) we can calculate the amount of energy that must be expended to remove an electron from its lowest energy state, $n_L = 1$, to an infinite distance from the nucleus, $n_H = \infty$. This is called the **ionization energy** (IE) of the hydrogen atom. If we set $n_L = 1$ and $n_H = \infty$ in Eq. (12-21b), we obtain:

$$\Delta E = \text{IE} = hc\mathcal{R} = 13.60 \text{ eV/atom} = 1312 \text{ kJ/mol}$$ (12-22)

The ionization energy is the amount of energy required to carry out the reaction

$$\text{H(g)} \rightarrow \text{H}^+\text{(g)} + e^-$$ (12-23)

The experimental value of the ionization energy of the hydrogen atom is precisely the value given by Eq. (12-22), which is another piece of evidence for the validity of Bohr's results.

Spectral Intensities

A comment is in order about the observed intensities of the spectral lines. You will remember that within each series the intensity decreases as the frequency increases (wavelength decreases), that is, as n_H increases. We are, of course, observing the transitions of an enormous number of hydrogen atoms. The observed intensity of a transition from $n_H \rightarrow n_L$ depends upon the number of hydrogen atoms in the state with quantum number n_H, which is called the population of the energy level with energy E_n. When an electric discharge is passed through a tube containing H atoms, some of the atoms are excited to the $n = 2$ state, some to the $n = 3$ state, and so on. Because it takes more energy to get the electron into the higher energy states, the population of a state decreases as n increases. Thus there are more hydrogen atoms with $n = 3$ than there are with $n = 4$, and so the intensity of the first Balmer line ($n_H = 3 \rightarrow n_L = 2$) is greater than the intensity of the second Balmer line ($n_H = 4 \rightarrow n_L = 2$). Bohr's theory made no attempt to explain the observed intensities, and one of the dissatisfactions with Bohr's theory that arose during the decade following his paper in 1913 was the absence of any way to explain the difference in intensities of the lines in atomic spectra.

section 12.5
The Dual Nature of Matter

Until 1900, most scientists believed that there was a clear distinction between matter and energy. Matter was considered to consist of a collection of particles (that is, to be **particulate** or **corpuscular** in nature), and was governed by a set of natural laws. Energy, when manifesting itself as radiation, was considered to be a collection of waves traveling through space at a constant speed, 2.997925×10^8 m/s, and was governed by an entirely different set of laws. Properties of waves such as refraction, interference, and diffraction had been carefully investigated; these appeared to be very different from the properties of matter. Radiant energy was considered to be continuous, and any combination of wavelength (or frequency) and energy was assumed to be possible.

Particles are, by their very nature, **quantized.** You can have 1 atom of copper, or 2, or 17, or 300, but you cannot have $1\frac{1}{2}$ or 99.37. Energy, prior to 1900, was not considered to consist of particles. It was thought to be noncorpuscular in nature, and therefore continuous. It was this distinction between matter and energy that had been abandoned by Planck in 1900, by Einstein in 1905, and again by Bohr in 1913. Each had found that in order to explain a particular experimental observation (and the three scientists were studying widely different phenomena), they had been forced to describe light (electromagnetic radiation) as consisting of small bundles or packets of energy of amount hv. A single packet of energy is called a **photon,** or a **quantum of energy.**

$$1 \text{ photon} = hv = 1 \text{ quantum of energy} \qquad (12\text{-}24)$$

When the concept of quantized, rather than continuous, energy is introduced, energy is being described as having some of the attributes of particles.

The Bohr postulate that was most radical was the concept of quantized energy. Bohr found that in order to account for the discrete line spectrum of hydrogen atoms he had to adopt Planck's proposal that all combinations of frequency and energy are *not* possible. On the contrary, the only energies possible are multiples of the basic unit: hv, $2hv$, $3hv$, $4hv$, and so on.

During the first quarter of the twentieth century an increasing number of puzzling phenomena were explained by assuming that light consists of discrete photons of energy hv. Physicists began to speak of the "dual nature" of light. When light travels through space it has wave properties, but when it exchanges energy with matter, light has corpuscular properies, that is, it behaves like a stream of particles.

The French physicist Louis de Broglie proposed in 1924 that not only light but *all* matter has a dual nature and possesses both wave and corpuscular properties. He reasoned that there should be symmetry in nature: If a radiant corpuscle — that is, a photon — has a frequency and a wavelength and therefore has wave properties, why should not a material particle also have wave properties?

A particle of mass m traveling with velocity v has a **linear momentum, p,** defined as $p = mv$. If a photon of light is corpuscular, it must also have momentum. By combining relations from Einstein's theory of relativity with the definition of a quantum of energy, $E = hv = hc/\lambda$, de Broglie derived the expression

$$p = h/\lambda \qquad (12\text{-}25)$$

for the momentum of a photon of light of wavelength λ.

Equation (12-25), known as the **de Broglie relation,** relates the wave property, λ, to the corpuscular property, p. In complete symmetry, de Broglie then proposed that a particle that has a momentum p has a wavelength associated with it given by the very same relation, that is $\lambda = h/p$. Since $p = mv$, the faster a particle moves, the greater its momentum, and the shorter is the wavelength that is associated with it. While all electromagnetic radiation travels with the same speed, c, species classically considered to be particles travel with a smaller speed, denoted v. The wavelength associated with a particle of mass m, traveling with speed v, can be obtained by rearranging the de Broglie relation $p = mv = h/\lambda$, to solve for the wavelength:

$$\lambda = h/mv \qquad (12\text{-}26)$$

In particular, de Broglie predicted the wavelength that should be associated with a beam of electrons accelerating through a given potential difference. A typical calculation of this quantity is carried out in Example 12.5.

Prince Louis Victor de Broglie, a French theoretical physicist born in 1892, received the Nobel Prize in 1929 for his work on the wave nature of the electron. He proposed that all matter has a dual nature, both wave and corpuscular, and worked out the formulas establishing the parallelism between the motion of a particle of matter and the propagation of the wave associated with that particle. His theoretical work laid the foundations for the development of the electron microscope and the technique of electron diffraction.

EXAMPLE 12.5. Wave properties of electrons

A beam of electrons is accelerated through a potential difference of 8000 V. Calculate the energy (in joules), the speed, the linear momentum, and the wavelength of these electrons.

Solution. An electron volt is, by definition, the energy an electron possesses when it is accelerated through a potential difference of 1 V. An electron accelerated through a potential difference of 8000 V therefore has 8000 eV of energy. This value can be converted to joules by using the conversion factor given in Eq. (12-13a).

$$\text{Energy} = (8000 \text{ eV})(1.6022 \times 10^{-19} \text{ J/eV}) = 1.2818 \times 10^{-15} \text{ J}$$

The definition of the kinetic energy of a particle of mass m traveling with speed v is

$$\text{kinetic energy} = E = \tfrac{1}{2}mv^2$$

This can be rearranged to solve for the speed. We obtain

$$v^2 = 2E/m$$

Using the known mass of the electron, given in Eq. (12-3), we can calculate the speed of an electron with a kinetic energy of 1.2818×10^{-15} J. We must be careful to use only SI units for all quantities. Mass must be expressed in kilograms, and the speed will be given in meters per second. We obtain

$$v^2 = \frac{2(1.2818 \times 10^{-15}) \text{ J}}{9.1096 \times 10^{-31} \text{ kg}} = 2.814 \times 10^{15} \text{ (m/s)}^2$$

Taking the square root we find that $v = 5.305 \times 10^7$ m/s, the speed of the electrons. The linear momentum, p, is given by

$$p = mv = (9.1096 \times 10^{-31} \text{ kg})(5.305 \times 10^7 \text{ m/s}) = 4.833 \times 10^{-23} \text{ kg} \cdot \text{m} \cdot \text{s}^{-1}$$

The de Broglie relation (12-26) can now be utilized to obtain the wavelength:

$$\lambda = \frac{h}{mv} = \frac{h}{p} = \frac{6.6262 \times 10^{-34} \text{ J} \cdot \text{s}}{4.833 \times 10^{-23} \text{ kg} \cdot \text{m} \cdot \text{s}^{-1}}$$

$$\lambda = 1.371 \times 10^{-11} \text{ m} = 0.01371 \text{ nm} = 0.1371 \text{ Å}$$

In checking units in this equation remember that $1 \text{ J} = 1 \text{ kg} \cdot \text{m}^2 \cdot \text{s}^{-2}$. By referring to Table 12.3 we see that a wavelength of 0.01371 nm is in the X-ray region.

Electron Diffraction

When visible light is passed through an optical grating that has a spacing between rulings close to (but slightly larger than) the wavelength of the light, the light is scattered and a **diffraction pattern** is obtained. (See Fig. 12.14.) For a wave to be diffracted, the spacings of the diffraction grating must be of the same order of magnitude as the wavelength of the light. X-rays have extremely short wavelength, but the spacing between rows of atoms in crystals is of the order of the wavelength of X-rays, so that a crystal serves as a natural "grating" for the diffraction of X-rays. By 1924, when de Broglie presented his theory, X-ray diffraction by crystals had already been demonstrated by Max von Laue and by William Bragg. De Broglie therefore proposed that a beam of electrons accelerated through a suitable potential difference would have a wavelength in the X-ray region and should also be diffracted by crystals. In de Broglie's own words:

> The lengths of the electron wave being thus of the same order as of X-rays, we may fairly expect to be able to obtain a scattering of this wave by crystals, in complete analogy to the Laue phenomenon in which, in a natural crystal like rock salt, the atoms of the substances comprising the crystal are arranged at regular intervals of the order of one Angstrom, and thus act as scattering centers for the waves.

When de Broglie first published his wave theory of matter, there was no experimental evidence to support his bold hypothesis. Within three years, however, two different experiments had been performed that demonstrated the diffraction of a beam of electrons. Clinton J. Davisson, assisted by L. H. Germer, working in the Bell Telephone Laboratories in New York, observed the diffraction of electrons when a beam of electrons was directed at a nickel crystal. Working independently, G. P. Thomson (the son of J. J. Thomson) and A. Reid also observed electron diffraction, shortly after Davisson and Germer. Thomson and Davisson shared the Nobel Prize

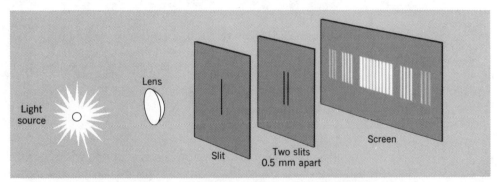

Fig. 12.14.   The production of a diffraction pattern using visible light.

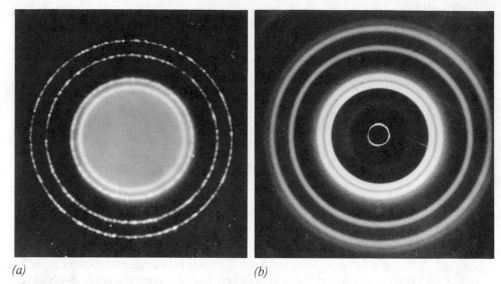

(a)   *(b)*

Fig. 12.15. Diffraction of waves by Al foil. *(a)* X-rays of wavelength 0.71 Å. *(b)* Electrons of wavelength 0.50 Å. The similarity of these patterns provides evidence for de Broglie's hypothesis that particles have wave properties.

in 1937 for their experimental demonstration of the de Broglie relationship: De Broglie had received the Nobel Prize in 1929. Diffraction patterns produced by scattering electrons from crystals are very similar to those produced by scattering X-rays from crystals, as can be seen in Fig. 12.15.

Since the original experiments of Thomson and Reid and of Davisson and Germer, many other experiments have demonstrated the interference of electron waves. All such experiments have shown that the wavelength of electron waves is exactly that predicted by the de Broglie relation, Eq. (12-26). Electron diffraction is now used as an experimental method for determining the distance between atoms in crystals.

Scientists today are no longer troubled by the seeming distinction between waves and particles. It is accepted that all matter has both wave and corpuscular properties. Why is it that we do not observe the wave nature of objects of the size that we deal with in the laboratory—a basketball or a marble? Because the wavelength is inversely proportional to the mass of the object [Eq. (12-26)], and when m is large, λ is so small that we cannot detect the wave nature of the particle.

EXAMPLE 12.6. Wave properties of macroscopic objects

What is the wavelength associated with a 5 oz (142 g) baseball moving with a speed of 30.0 m/s ($\sim$100 ft/s)?

Solution. Using the de Broglie relation, Eq. (12-26), we obtain:

$$\lambda = \frac{h}{mv} = \frac{6.6262 \times 10^{-34} \text{ J} \cdot \text{s}}{(0.142 \text{ kg})(30.0 \text{ m/s})} = 1.56 \times 10^{-34} \frac{\text{J} \cdot \text{s}^2}{\text{kg} \cdot \text{m}} = 1.56 \times 10^{-34} \text{ m}$$

$$= 1.56 \times 10^{-25} \text{ nm} = 1.56 \times 10^{-24} \text{ Å}$$

A wavelength so small is unobservable.

For substances of large mass, therefore, classical or Newtonian mechanics is perfectly satisfactory. Remember that Planck's constant is an exceedingly small number, and λ is directly proportional to h. It is only when we deal with particles of very small mass, that is, with atomic or subatomic particles, that λ becomes large enough to be observed. Both the wave nature and the corpuscular nature of such small particles can be manifested, and Newtonian mechanics is unable to account for their properties.

section 12.6
From Bohr's Quantum Theory to Wave Mechanics

Bohr's model of the hydrogen atom was so successful in explaining the atomic spectrum of hydrogen that attempts were immediately made to explain the spectra of other atoms and ions. As long as the system consists only of a single electron and a nucleus, for instance ions such as He^+ and Li^{2+} (which are called **H-like ions**), Bohr's method gives correct results, that is, results in agreement with experimental observations.

For an H-like ion with nuclear charge $+Ze$ (atomic number Z), the Bohr theory (see Appendix H) predicts that the allowed energies are given by

$$E_n = -\frac{2\pi^2 m e^4 Z^2}{h^2 n^2} \tag{12-27}$$

where $n = 1, 2, 3, \ldots, \infty$. The difference in energy between two states with quantum numbers n_L and n_H is therefore:

$$\Delta E = \frac{2\pi^2 m e^4 Z^2}{h^2} \left(\frac{1}{n_L^2} - \frac{1}{n_H^2}\right) = hc\mathcal{R}Z^2 \left(\frac{1}{n_L^2} - \frac{1}{n_H^2}\right) \tag{12-28}$$

Accordingly, the frequency and wavenumber of the radiation emitted during the transition $n_H \rightarrow n_L$ are given by

$$v = c\mathcal{R}Z^2 \left(\frac{1}{n_L^2} - \frac{1}{n_H^2}\right) \tag{12-29a}$$

and

$$\frac{1}{\lambda} = \tilde{v} = \frac{v}{c} = \mathcal{R}Z^2 \left(\frac{1}{n_L^2} - \frac{1}{n_H^2}\right) \tag{12-29b}$$

Calculations using these equations are illustrated in Example 12.7.

EXAMPLE 12.7. Spectra and ionization energy of H-like ions

(a) Calculate the energy of the photon emitted when the electron in a Be^{3+} ion drops from the $n = 2$ level to the ground state, in electron volts and in joules. How much energy is emitted if a mole of Be^{3+} ions undergoes this transition? In what region of the spectrum will the radiation emitted during this transition be observed?

Solution. The atomic number of Be is 4. The Be^{3+} ion has only a single electron and is therefore an H-like ion. Since $hc\mathcal{R}$ is 13.60 eV, Eq. (12-28) yields

$$\Delta E = (13.60)4^2 \left(\frac{1}{1^2} - \frac{1}{2^2}\right) = \frac{3}{4}(13.60)(16) = (12)(13.60) = 163.2 \text{ eV}$$

The photon emitted during this transition therefore has 163.2 eV of energy. In joules this is

$$(163.2 \text{ eV})(1.6022 \times 10^{-19} \text{ J/eV}) = 2.614 \times 10^{-17} \text{ J}$$

This is the energy of a single photon. If a mole of ions undergoes this transition the energy emitted is

$$(2.614 \times 10^{-17} \text{ J/ion})(6.022 \times 10^{23} \text{ ions/mol}) = 1.574 \times 10^7 \text{ J/mol}$$

or 1.574×10^4 kJ/mol, which is a very large amount of energy!

The wavelength of the radiation can be calculated using Eq. (12-29b):

$$1/\lambda = (109{,}678 \text{ cm}^{-1})(4^2)(\tfrac{3}{4}) = 12(109{,}678) \text{ cm}^{-1} = 1.316 \times 10^6 \text{ cm}^{-1}$$

Therefore

$$\lambda = (1/1.316) \times 10^{-6} \text{ cm} = 7.598 \times 10^{-7} \text{ cm} = 7.598 \times 10^{-9} \text{ m} = 7.598 \text{ nm}$$

This radiation is in the X-ray region, close to the borderline with the far ultraviolet.

(b) Calculate the energy required to ionize the electron in a Be^{3+} ion.

Solution. The amount of energy required to remove the electron from a single Be^{3+} ion is the difference in energy between the ground state ($n_L = 1$) and the state with $n_H = \infty$. Thus,

$$\text{IE} = hc\mathcal{R}Z^2 = (13.60 \text{ eV})(4^2) = 217.6 \text{ eV}$$

In other units the ionization energy is

$$\left(217.6 \, \frac{\text{eV}}{\text{ion}}\right)\left(1.6022 \times 10^{-19} \, \frac{\text{J}}{\text{eV}}\right)\left(6.022 \times 10^{23} \, \frac{\text{ions}}{\text{mol}}\right) = 2.099 \times 10^7 \, \frac{\text{J}}{\text{mol}}$$

Note that it requires 2.099×10^7 J to ionize a mole of Be^{3+} ions, compared with only 1.312×10^6 J to ionize a mole of H atoms. The reason so much more energy is needed to ionize Be^{3+} ions is that the nucleus of the Be^{3+} ion has four times as much positive charge as the nucleus of the H atom, and therefore attracts the single electron much more strongly.

For atoms containing many electrons, however, Bohr's method was not able to predict results in agreement with experimental observation. The Bohr theory was a mixture of classical Newtonian mechanics and nonclassical electrodynamics. The motion of the electron in a stationary state was assumed to obey the laws of Newtonian dynamics, but the transition of the electron from one stationary state to another was described by a new system of electrodynamics, which allowed the electron to emit or absorb radiation only in discrete packets (quanta) rather than continuously. As attempts were made to apply the Bohr model to a great many atomic processes, it became increasingly evident that Bohr's theory suffered from basic inconsistencies and in some systems gave incorrect results.

Between 1913 and 1926 many attempts were made to modify Bohr's approach, but none was completely successful. The year 1926, however, marked a dramatic change. In that year three men, all working independently, introduced a totally new approach to the theoretical description of atomic systems. These three men were Erwin Schrödinger (1887–1961), a Viennese physicist, Werner Heisenberg (1901–1976), a German physicist only twenty five years old when his work was published, and Paul A. M. Dirac, an English mathematician who was only twenty four years old in 1926. Each of these three presented his theoretical work in a different mathematical formalism, and it took some time before the equivalence of their basic concepts was fully appreciated. Their theory is called **quantum mechanics** or **wave mechanics**

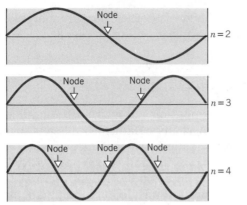

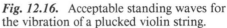

Fig. 12.16. Acceptable standing waves for the vibration of a plucked violin string.

and it has, quite literally, revolutionized the study of both physics and chemistry. Of the three mathematical approaches, Schrödinger's is the most familiar because it involves solving a differential equation. The study of differential equations is a standard part of calculus; anyone planning to pursue research in physics or chemistry today must become familiar with methods of solving differential equations.

The equations governing wave motion had been developed by mathematicians long before 1926. All waves — sound waves, ocean waves, electromagnetic waves — obey a differential equation of a specific form. A familiar problem that mathematicians had solved was the equation of motion of a plucked violin string. Because the ends of the string are tied down and cannot move, only certain wave forms are possible. This is illustrated in Fig. 12.16. If the length of the string is *L,* acceptable modes of vibration are called **standing waves** and are limited to waves with wavelength $\lambda = 2L/n$, where $n = 1, 2, 3, \ldots$. Note that in this problem a "quantum number," that is, a number restricted to a discrete set of values, is introduced naturally into the mathematical solution by the requirement that the string is tied down at both ends. This requirement is called a boundary condition.

Schrödinger began with de Broglie's concept of the dual nature of matter. If matter has wave properties, he reasoned, then the wave associated with matter must be described by a **wave equation,** a differential equation of wave motion. In essence, Schrödinger combined the de Broglie relation for the wavelength of a particle with the well-known differential equation for standing waves and obtained an equation now called the **Schrödinger wave equation.** Schrödinger's equation is widely used today to describe the behavior of atomic and molecular systems. The only limitation to its utility is the difficulty of finding solutions to the equation for a given system. The use of computers in recent years has greatly expanded our ability to utilize quantum mechanics to study more complex atomic and molecular systems.

The mathematics of the Schrödinger equation is beyond the scope of this text, but we shall outline the information contained in solutions of the Schrödinger equation and then proceed, in the following chapter, to describe specific results for hydrogen and other atoms.

The solutions to the Schrödinger equation for a particular system (an electron in an atom, for instance) are a set of mathematical functions called **wave functions** or **eigenfunctions** (eigen is the German word meaning "characteristic"). A **wave function** is usually denoted by the symbol ψ (the Greek letter *psi*). Every allowed state of a system has a wave function: When we know the wave function for a given state of a particular system we know as much information about that state as it is possible to know. The information we can obtain about a state of a given system is more limited than the information classical mechanics provides for systems that obey Newtonian dynamics. We can determine the energy of a particle in a stationary state exactly, but we cannot simultaneously know both the position and the velocity of the particle when it is in the specified state.

The Heisenberg Uncertainty Principle

The inability to determine both the position and the velocity simultaneously is not due to any theoretical flaw in wave mechanics: Rather it is inherent in the dual nature of matter. In its most rigorous mathematical form the statement that we are unable to determine simultaneously both the position and velocity of a particle is called the **Uncertainty Principle,*** and was first proposed by Heisenberg in 1927.

We can understand the Uncertainty Principle by thinking about experimental methods that are used to determine the position of a particle. Of course, if a particle is large, we can touch it with a ruler or some other measuring device without changing its position significantly. But what do we use to determine the position of a very small particle? When we use a microscope, we shine a beam of light on the particle. Light consists of a stream of photons of energy $h\nu$. The higher the frequency of the light used (that is, the shorter the wavelength), the higher the energy of the photons impinging on the particle.

It is a well-known property of optical systems that the resolution of a microscope or similar measuring device is of the same order of magnitude as the wavelength of the light. That means that if a particle is extremely small, in order to locate it accurately we must use radiation of very small wavelength, that is, high energy radiation. When high energy radiation strikes the particle, the position of the particle changes. The very act of measuring the location of the particle disturbs it, and it moves away with an uncertain momentum.

What information about the position of a particle in a given state can we obtain if we know the wave function for that state? We can obtain a **probability distribution,** the probability of locating the particle at any given point (x,y,z) in space. There will, in general, be many regions of space in which the probability of locating the particle is finite. The probability that the particle is within a small volume element around one position (x_1, y_1, z_1) may be 4.7%, while the probability that it is within an identical volume element around a different point (x_2, y_2, z_2) may be 2.9%, and so on. We cannot know exactly where the particle is, but we can determine those regions in space where there is a high probability of locating it, and those regions in which there is only a low probability of locating it.

Since we are interested primarily in the behavior of electrons in atoms and molecules, let us describe the physical significance of the wave function specifically for an electron in an atom. The wave function, $\psi(x,y,z)$, has some finite value (which may be zero) at every point in space. The *square* of the wave function, $|\psi(x,y,z)|^2$, is the relative **probability density** of locating the particle at the position (x,y,z).†

---

* The mathematical statement of the Uncertainty Principle is that the product of the uncertainty in the position, Δx, and in the momentum, Δp, must be greater than or equal to Planck's constant divided by 4π:

$$(\Delta x)(\Delta p) > h/4\pi$$

† You may be puzzled about the use of the absolute value sign in the symbol for the probability density, $|\psi|^2$. For real functions there is no need for this sign. In some cases, however, wave functions turn out to be complex, that is, to involve the imaginary number $i = \sqrt{-1}$. Since a probability must be real, we need a symbol to indicate how one squares a complex number to obtain a real number. The symbol $|\psi|^2$ indicates that process. For real functions, the probability density is simply ψ^2.

The physical significance of the probability distribution can, perhaps, be made clearer to you by the following two analogies:

1. Imagine that an electron is like a tiny firefly, moving rapidly and emitting light. (Electrons, of course, do *not* emit light and move much faster than real fireflies.) If we try to take a snapshot of this tiny moving light for a small fraction of a second, we will see nothing at all when we develop the film. But suppose we took a time exposure, for many minutes. What will we see when we develop the film? Regions where the electron spent a lot of time will be bright; regions where the electron spent very little time will be dark. There will be no sharp boundaries; we will see a fuzzy picture, a diffuse cloud-like brightness with different intensities in different regions of space. The intensity of brightness at any point will be proportional to $|\psi|^2$ at that point. We describe this by calling $|\psi|^2$ the **electron density** or the **charge cloud density**. Because electrons in atoms move so rapidly, experimental observations about electrons are consistent with the picture of a charge cloud density.

2. Imagine that you had some way of measuring the position of an electron in an atom, and you made the measurement again and again, billions of times. Each time you measure the position, you get a different answer! There are some regions of space in which you find the electron very often, others in which you find it only rarely, and some in which it is never found. If you had a set of coordinate axes in three dimensions and you could place a dot at the position at which you found the electron after each measurement, by the time you had plotted many billions of dots you would have a plot of $|\psi|^2$. The probability of finding the electron in any small region of space determines how often you will locate it there if you make an extremely large number of measurements. The density of dots is proportional to $|\psi|^2$, the probability distribution function.

Wave mechanics has now been applied to a great many problems involving atoms and molecules. In all cases where it has been possible to solve the mathematical equations appropriate to the given system, the theory has been able to predict results in agreement with experimental observation. Scientists were very quickly convinced of the fundamental validity of wave mechanics. Heisenberg received the Nobel Prize in 1932; Schrödinger and Dirac shared it the following year. The use of wave mechanics to chemists, however, has been limited by the difficulty of solving the Schrödinger equation for large atoms or molecules. Great advances have been made since the introduction of the high speed computer. We also rely heavily on the use of approximations that simplify the mathematics and give us very useful, but only approximately correct, wave functions. Dirac summed up the situation in 1929 in the following words, which are still valid today:

> The underlying physical laws necessary for the mathematical theory of a large part of physics and the whole of chemistry are thus completely known, and the difficulty is only that the exact application of these laws leads to equations much too complicated to be soluble.

Summary

When the potential difference between two separated, charged metal plates is several thousand volts, the negatively charged plate, or **cathode,** begins to emit streams of particles, called **cathode rays.** Joseph J. Thomson showed that the particles of cathode rays are negatively charged, are common to all matter, and have a very large charge to mass ratio (e/m). Thomson named the cathode-ray particles **electrons.**

The magnitude of the charge on the electron was measured by Robert A. Millikan in 1910, in his famous **oil drop experiment.** The value of the electronic charge, combined with the e/m ratio for electrons that had been measured by Thomson, enables us to calculate the mass of the electron. The electronic mass is very small, about 1800 times smaller than the mass of the lightest atom, H.

By directing a stream of α-particles at a gold foil and observing how the α-particles are scattered by the atoms of the foil, Ernest Rutherford (1911) elucidated the structure of the atom. Almost all the mass of the atom, and all the positive charge, is contained in a very tiny volume, called the nucleus; the light electrons occupy most of the atomic volume.

The stability of an atom with negatively charged electrons remaining some distance away from a positively charged nucleus cannot be explained by **classical electromagnetic theory.** Classical mechanics also fails to explain the existence of **discrete atomic spectra.** When atoms are heated or subjected to an electrical discharge (a spark), they emit **electromagnetic radiation.** Each element has its own distinctive **spectrum,** and emits radiation of characteristic frequencies, while other frequencies are not emitted.

The spectrum of the simplest of all atoms, the hydrogen atom, is obtained by passing an electrical discharge through a tube containing H_2 gas at low pressure. The spark dissociates the H_2 molecules into atoms, and the tube glows red. When the emitted radiation is passed through a slit, dispersed by a prism, and then detected using a photographic plate, only four frequencies in the visible region, and a much larger number in the ultraviolet are observed.

The wavelengths of all the radiation emitted by H atoms in the visible and near ultraviolet region are given by the **Balmer formula:**

$$\tilde{\nu} = \frac{1}{\lambda} = \mathcal{R}\left(\frac{1}{2^2} - \frac{1}{n^2}\right)$$

where $\mathcal{R}$ is the **Rydberg constant** (109,678 cm^{-1}), and n is a **quantum number,** ($n = 3, 4, 5, \ldots, \infty$). The **Balmer formula** was published in 1885, but could not be derived using the laws of classical mechanics.

In 1913 Niels Bohr proposed a new set of fundamental postulates for the behavior of atomic and subatomic particles. In order to derive the Balmer formula, Bohr was forced to abandon one of the basic concepts of classical mechanics: The concept that the energy of electromagnetic radiation is continuous. Instead Bohr adopted the concept previously proposed by Planck and Einstein that light consists of a stream of **photons** or **quanta of energy** of amount $h\nu$, where ν is the frequency of the radiation, and h is Planck's constant.

Bohr postulated that the energy of a hydrogen atom is quantized, and that there is a discrete set of values of allowed energies for the atom. The state of lowest energy is called the **ground state.** Allowed states of higher energy are called **excited states.** A hydrogen atom radiates energy only when it makes a transition from one allowed energy state to another allowed state of lower energy. When such a transition occurs, radiation is emitted of frequency ν, given by the **Bohr frequency condition,** $\Delta E = h\nu$, where ΔE is the difference in energy between the two allowed states. Starting with these bold new postulates, Bohr was able to derive an expression for the allowed energies of the H atom, and from that expression to obtain the Balmer formula.

A figure in which energy is plotted along the ordinate axis and a horizontal line is drawn to represent an allowed energy, is called an **energy level diagram.** Vertical arrows drawn from one allowed energy level to another represent the allowed transitions. Every frequency that has been observed in the spectrum of atomic hydrogen can be predicted using the Bohr frequency condition and the values of the allowed energies calculated using Bohr's formula.

Each allowed energy of the H atom is associated with a **quantum number,** n, ($n = 1, 2, 3, \ldots, \infty$). When $n = 1$, the electron of the H atom is in its lowest energy state and is closer to the nucleus than in any other state. When $n = \infty$, the electron is an infinite distance from the nucleus, and the atom has been ionized. The amount of energy required to ionize the H atom (the **ionization energy**) is correctly predicted by calculating the difference in energy between the state for which $n = \infty$ and the ground state.

Before 1900 scientists made a distinction between matter (which was considered to be a collection of discrete particles), and energy, which was considered to be continuous. With the introduction of the concept of a **photon,** energy has some of the attributes of particles. In 1924 Louis de Broglie proposed that, just as there are particle properties associated with light waves, so there are wave properties associated with particles. He derived the expression $p = mv = h/\lambda$ to relate the wave properties (λ) to the corpuscular properties (p), for a particle of mass m traveling with a speed v. Ordinarily we do not observe the wave properties of matter because Planck's constant is so small and m is so large that λ is too small to be observed. For atomic and subatomic particles, however, m is very small and wave properties become observable. Louis de Broglie proposed that a beam of electrons, accelerated through a potential difference of a few thousand volts, should have associated with it a wavelength in the X-ray region. Such a wave can be detected by observing the **diffraction pattern** when the beam is directed at a crystal. Three years after de Broglie's proposal, two different research teams observed the diffraction of a beam of electrons.

While the Bohr theory proved very useful in explaining the spectrum of atomic hydrogen and **H-like ions,** it proved inadequate to explain the spectra of atoms with more than one electron. In 1926, a new form of mechanics, **wave** or **quantum mechanics,** was introduced independently by three scientists, Erwin Schrödinger, Werner Heisenberg, and Paul A. M. Dirac, to describe the behavior of atomic and subatomic systems. The information we can obtain about a system in one of its allowed states is contained in a function, ψ, called the **wave function,** which is a solution of **Schrödinger's wave equation.** A plot of the square of the wave function is a **probability distribution** for the position of the particle. Because it is impossible to determine simultaneously both the position and the velocity (or momentum) of a particle in an allowed energy state (**Heisenberg's Uncertainty Principle**), the probability distribution is the maximum amount of information we can obtain about the position of the particle.

Exercises

Section 12.1

1. The ratio of charge to mass (e/m) for an electron is 1836 times larger than the e/m value for a proton. Calculate the mass of a proton.

2. What observations did Thomson make that led him to conclude that electrons are a universal fragment of all matter?

3. Why was Rutherford astonished when he discovered that some α-particles, directed at a heavy metal foil, are deflected by angles of more than 90°?

Section 12.2

4. Calculate the frequency and wavenumber of radiation that has a wavelength of 20,000 Å. Express the wavelength also in nanometers and centimeters. In what region of the electromagnetic spectrum does this radiation occur?

5. Calculate the wavelength and wavenumber of radiation of frequency 1.8×10^{15} s^{-1}. In what region of the electromagnetic spectrum does this radiation occur?

6. What is the function of the prism in an emission spectrometer?

Sections 12.3 and 12.4

7. What are the values of n_H and n_L for the third line in the Brackett series in the spectrum of atomic hydrogen?

8. Calculate the wavelength, frequency, and wavenumber of the principal (most intense) line in the Paschen series of the spectrum of H atoms. In what region of the electromagnetic spectrum does this line occur?

9. The Balmer series of lines in the spectrum of H atoms converges to a limit as the frequency increases. What is the frequency of this limiting line? In what region of the spectrum does it occur?

10. If a beam of white light containing every frequency from 4.2×10^{14} to 7.5×10^{14} s^{-1} is shined on a tube containing H atoms, which frequencies will be absorbed?

11. How much energy is required to cause 1 mol of H atoms to undergo the transition from the ground state to the third excited state? What is the frequency of the radiation that must be absorbed to cause this transition to occur?

12. An electron in a hydrogen atom makes a transition and loses 2.856 eV of energy. What is the frequency and wavelength of the emitted radiation? In what region of the electromagnetic spectrum does this radiation occur?

13. If the electron in a hydrogen atom were in the energy level with $n = 3$, how much energy would be required to ionize the atom?

14. Consider two monochromatic rays of light, one orange ($\lambda = 6.2 \times 10^{-5}$ cm), the other violet ($\lambda = 4.2 \times 10^{-5}$ cm). Which ray has higher energy?

Section 12.5

15. If a beam of electrons has associated with it a wave of wavelength 0.100 nm, what is the speed of the electrons?

16. What is the kinetic energy of an electron moving with a speed of 4.0×10^6 m/s?

17. What is the wavelength of a beam of electrons accelerated through a potential difference of 1000.0 V? In what region of the electromagnetic spectrum is this wavelength?

18. Calculate the wavelength of a photon of visible light with a frequency of 6.6×10^{14} s^{-1}. What is the energy of this photon? Specify units.

Section 12.6

19. The ground state energy of the electron in a hydrogen atom is -13.60 eV. What is the ground state energy of (a) a He$^+$ ion and (b) a Li^{2+} ion?

20. Calculate the energy in electron volts of the electron in a He$^+$ ion in the state with $n = 3$. Compare this with the energy of the electron in an H atom in the state with $n = 3$. Which is in a lower energy state? Account for this difference in energy.

21. Suppose we wanted to determine the position of an electron moving with a speed of 1.0×10^6 m/s by shining violet light of wavelength 400 nm on it. Compare the energy of one photon of this light with the energy of the electron to be located. What conclusion can you draw from this comparison?

22. Calculate the energy, in electron volts, of the photon emitted when the electron in a Li^{2+} ion drops from the $n = 2$ level to the ground state. What is the wavelength of the emitted radiation? In what region of the electromagnetic spectrum does this radiation occur?

Multiple Choice Questions

1. Which of the following wave properties is proportional to energy for electromagnetic radiation?
   (a) velocity   (b) wavenumber   (c) wavelength   (d) amplitude
   (e) time for one cycle to pass a given point in space.

2. Which of the following best describes the emission spectrum of atomic hydrogen?
   (a) A discrete series of lines of equal intensity and equally spaced with respect to wavelength.
   (b) A series of only four lines.
   (c) A continuous emission of radiation of all frequencies.
   (d) Several discrete series of lines with both intensity and spacings between lines decreasing as the wavenumber increases within each series.
   (e) A discrete series of lines with both intensity and spacings between lines decreasing as the wavelength increases.

3. Atomic emission spectra provide a measurement of
   (a) The average distance from the nucleus for each electron in the atom.
   (b) The number of protons in the nucleus of the atom.
   (c) The difference in energy between pairs of states of the atom.
   (d) The ionization energy of the atom.
   (e) The absolute value of all allowed energies for the atom.

4. The ionization energy of gaseous Na atoms is 495.80 kJ/mol. The lowest possible frequency of light that can ionize a Na atom is
   (a) $4.76 \times 10^{14} \text{ s}^{-1}$ (b) $7.50 \times 10^{14} \text{ s}^{-1}$ (c) $1.24 \times 10^{15} \text{ s}^{-1}$
   (d) $3.15 \times 10^{15} \text{ s}^{-1}$ (e) $5.82 \times 10^{15} \text{ s}^{-1}$

5. Joseph J. Thomson's cathode-ray tube experiments demonstrated that
   (a) α-Particles are the nuclei of He atoms.
   (b) The ratio of charge-to-mass for the particles of the cathode rays varies as different gases are placed in the tube.
   (c) The mass of an atom is essentially all contained in its very small nucleus.
   (d) Cathode rays are streams of negatively charged ions.
   (e) The charge-to-mass ratio of electrons is about 1800 times larger than the charge-to-mass ratio for a proton.

6. If the electron in a hydrogen atom drops from the $n = 6$ to the $n = 4$ level, the radiation emitted is in which series of lines in the spectrum of atomic hydrogen?
   (a) Lyman   (b) Balmer   (c) Paschen   (d) Brackett   (e) Pfund

7. In the spectrum of atomic hydrogen, a line is emitted with wavenumber 1.028×10^{-2} nm. In what region of the electromagnetic spectrum does this line occur?
   (a) far UV   (b) near UV   (c) VIS   (d) near IR   (e) far IR

8. Which of the following statements about Millikan's oil drop experiment is TRUE?
   (a) When the electric field is turned on, all the oil drops move toward the positively charged plate.

(b) The charge on each oil drop is the electronic charge.

(c) In the absence of the electric field, the speed with which the drop falls depends only upon the acceleration of gravity.

(d) Oil drops, rather than water drops, were used because oil is easier to see.

(e) Some oil drops become positively charged and some become negatively charged after colliding with gaseous ions.

9. Because of the observation that some α-particles directed at a gold foil are scattered backwards at angles larger than $90°$, Rutherford was able to conclude that

(a) All atoms are electrically neutral.

(b) The positively charged parts of atoms move with extremely large velocities.

(c) An electron has a very small mass.

(d) Negatively charged electrons are a part of all matter.

(e) The positively charged parts of atoms occupy only a very small fraction of the volume of the atom.

10. To move one electron in a hydrogen atom from the ground state to the second excited state 12.084 eV are needed. How much energy is needed to cause 1 mol of H atoms to undergo this transition?

(a) 728 kJ   (b) 984 kJ   (c) 1036 kJ   (d) 1166 kJ   (e) 1312 kJ

11. One of the lines in the spectrum of atomic hydrogen has wavenumber 5331.6 cm^{-1}. What is the frequency of this line?

(a) 5.623×10^6 s^{-1} (b) 1.876×10^{11} s^{-1} (c) 1.598×10^{12} s^{-1}

(d) 1.598×10^{14} s^{-1} (e) 5.623×10^{14} s^{-1}

12. The amount of energy required to remove the electron from a Li^{2+} ion in its ground state is how many times greater than the amount of energy needed to remove the electron from an H atom in its ground state?

(a) 2   (b) 3   (c) 4   (d) 6   (e) 9

13. A typical golf ball weighs 40.0 g. If it is moving with a velocity of 20.0 m/s, its de Broglie wavelength is

(a) 1.66×10^{-34} nm   (b) 8.28×10^{-32} nm   (c) 8.28×10^{-25} nm

(d) 1.66×10^{-24} nm   (e) 3.31×10^{-21} nm

14. What is the energy of a photon of infrared light with wavelength 4.0×10^3 nm?

(a) 5.0×10^{-20} J   (b) 7.5×10^{-20} J   (c) 4.0×10^{-16} J   (d) 2.5×10^{-14} J

(e) 7.5×10^{-13} J

15. The most intense line in the Brackett series of the spectrum of atomic hydrogen is the transition

(a) $n_H = \infty \rightarrow n_L = 1$ (b) $n_H = 8 \rightarrow n_L = 4$ (c) $n_H = \infty \rightarrow n_L = 4$

(d) $n_H = 4 \rightarrow n_L = 3$ (e) $n_H = 5 \rightarrow n_L = 4$

Problems

12.1. How many times larger is the spacing between the energy levels with $n = 3$ and $n = 4$ than the spacing between the energy levels with $n = 8$ and $n = 9$, for the hydrogen atom?

12.2. One of the lines in the spectrum of atomic hydrogen has a wavenumber of 97,491 cm^{-1}. To what series of the spectrum does this line belong? What transition (between what two quantum numbers) gives rise to the emission of this line?

12.3. One of the erroneous conclusions of the Bohr theory of the H atom was that the electron travels in circular orbits of fixed radius. In the ground state the electron was calculated by Bohr to be in an orbit 0.529 Å from the nucleus, moving with a speed of 2.188×10^6 m/s. Explain why this conclusion is considered to be incorrect today.

12.4. (a) The velocity of electrons striking the inner face of a television picture tube is 5.9×10^7 m/s. What is the de Broglie wavelength of such electrons? (b) A child throws a marble weighing 6.0 g with a velocity of 2.0 m/s. What is the de Broglie wavelength of the marble? (c) What important fact do you learn by comparing the answers to parts (a) and (b)?

12.5. Just as you cannot use an ordinary ruler to measure a distance of 1×10^{-8} cm, you cannot use visible light ($\lambda \sim 6 \times 10^2$ nm) to locate a particle that is only 0.2 nm in diameter, the size of a typical atom. How many times larger is the energy of a photon with $\lambda = 0.20$ nm than the kinetic energy of a typical gaseous hydrogen atom that is moving with a speed of 1.5×10^3 m/s? What important fact does this calculation illustrate?

12.6. The ground state energy of the electron in a He$^+$ ion is -54.38 eV. (a) Using only this value (no fundamental constants are needed), calculate the energy in electron volts of the state of the He$^+$ ion for which (1) $n = 5$, and (2) $n = 2$. (b) Calculate ΔE for the transition $n_H = 5 \rightarrow n_L = 2$ in electron volts and in joules. (c) Calculate both the wavelength and the frequency of the light that is emitted during the transition $n_H = 5 \rightarrow n_L = 2$. In what region of the electromagnetic spectrum does this line occur? (Note that this corresponds to the third line of the "Balmer series" of the He$^+$ ion.)

12.7. *The photoelectric effect.* When light strikes a metal surface, electrons may be emitted. The remarkable thing about this **photoelectric effect,** is that no matter how intense the light shined on the surface, if the frequency of the light is below a specific minimum value (called the threshold frequency) no electrons are emitted. Thus for a potassium surface, when a very intense beam of red light is used, no electrons are emitted, but electrons are emitted when even a weak beam of violet light is used. The greater the frequency of the light used (above the threshold frequency) the greater the speed and therefore the kinetic energy of the emitted electrons. The photoelectric effect was first successfully explained by Einstein in 1905. He proposed that there was a minimum amount of energy necessary to eject an electron from a metal surface. This minimum energy (called the work function of the metal) is a characteristic property of the metal. If a photon of the light striking the metal has an energy less than the work function of the metal, no electrons are emitted. If a photon of the light used has an energy greater than the work function of the metal, the kinetic energy of the emitted electron equals the difference in energy between the energy of the photon and the work function of the metal. (a) If yellow light of wavelength 589.0 nm is shined on a potassium surface, the emitted electrons have a kinetic energy of 5.77×10^{-20} J per electron. Calculate the value of the work function and the threshold frequency for potassium. (b) If ultraviolet light of wavelength 253.7 nm is shined on a potassium surface, what will the speed of the emitted electrons be?

12.8. Imagine a hypothetical atom that has only four allowed energy levels: A ground state with energy E_1 and the excited states E_2, E_3, E_4. When heated, this atom is observed to emit radiation of six different frequencies (in $\sec^{-1}$):

$$v_1 = 4.0 \times 10^{14} \qquad v_2 = 6.0 \times 10^{14} \qquad v_3 = 1.0 \times 10^{15}$$
$$v_4 = 2.0 \times 10^{15} \qquad v_5 = 2.6 \times 10^{15} \qquad v_6 = 3.0 \times 10^{15}$$

Draw an energy level diagram for this atom and use labeled arrows to assign the observed frequencies to the various transitions. The transition from $E_2 \rightarrow E_1$ gives rise to the frequency v_4. Make a corresponding statement about each of the other frequencies, correlated to your diagram. Indicate the reasoning used in assigning the frequencies to the stated transitions.

12.9. The energy level diagram for an H-like ion is given below. (a) Identify the ion by determining its atomic number, Z. (b) On the diagram draw an arrow to indicate the transition that gives rise to the *second* most intense line in the "Lyman series" for this ion. (c) Calculate the frequency and wavenumber of the radiation emitted when the transition you drew occurs. In what region of the electromagnetic spectrum does this line occur? (d) Using only the energy level diagram, how much energy is required to remove the electron from its nucleus? Write an equation for the reaction that occurs when the electron is removed.

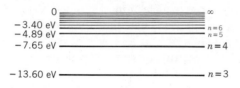

0	∞
-3.40 eV	$n=6$
-4.89 eV	$n=5$
-7.65 eV	$n=4$
-13.60 eV	$n=3$
-30.59 eV	$n=2$
-122.36 eV	$n=1$

chapter 13 The Electronic Structure of Atoms, the Periodic Table, and Periodic Properties

Erwin Schrödinger (1887–1961), an Austrian physicist, was born in Vienna and educated at the University of Vienna. He was a professor of physics at universities in Germany, Poland, and Switzerland. From 1940 to 1956 he was a professor at the Institute for Advanced Studies in Dublin, Ireland. Schrödinger's work was in mathematical physics, particularly the physics of the atom. In 1933 he and Paul A. M. Dirac were awarded the Nobel Prize for the development of wave mechanics. Schrödinger's wave equation, proposed in 1926, is the fundamental equation of quantum chemistry. The allowed energy levels of the electron in a hydrogen atom can be obtained by solving the Schrödinger equation for the hydrogen atom.

The previous chapter ended with a quotation of Dirac's about the difficulty of obtaining exact solutions to the wave equation. For a system consisting of a single electron and a nucleus, that is, for a hydrogen atom or an H-like ion, the Schrödinger equation can be solved exactly. The wave functions describing the electron in a hydrogen atom are particularly important. They have been used to predict properties of the hydrogen atom, and the results agree in every detail with experimental observation. The solutions to the wave equation for the hydrogen atom and H-like ions thus provided an important test of quantum mechanics and served to convince scientists of its validity. Even more importantly, the information derived from studying the wave functions for the hydrogen atom is used to describe and predict the behavior of electrons in many-electron atoms and molecules. In order to understand the periodicity of atomic properties and the nature of the chemical bond, we must first be familiar with the solutions to the Schrödinger equation for the electron in a hydrogen atom, the simplest of all atomic systems.

A complete theoretical treatment of the hydrogen atom, using a wave equation that incorporates Einstein's theory of relativity, yields a set of wave functions, each described by four quantum numbers: n, ℓ, m_ℓ, and m_s. These four quantum numbers arise naturally from the mathematics of solving the Schrödinger equation, just as a quantum number arises naturally when solving for the standing waves of a vibrating string that is tied down at both ends. (See Section 12.6.) No postulate, such as Bohr's, that energy is quantized is necessary. The quantization of energy is inherent in the wave properties of matter.

The solutions of the nonrelativistic Schrödinger equation for the hydrogen atom are a set of wave functions described by three quantum numbers: n, ℓ, and m_ℓ. These wave functions are called **atomic orbitals** (abbreviated AO's). We begin our study of the electronic structure of atoms with a discussion of the atomic orbitals of hydrogen.

section 13.1
Atomic Orbitals and the Quantum Numbers n, ℓ, and m_ℓ

Each wave function that is an exact solution to the nonrelativistic Schrödinger equation for the hydrogen atom is defined uniquely by stating the values of three quantum numbers. Restrictions on these quantum numbers and their relation to properties of the hydrogen atom and the atomic orbitals can be summarized as follows:

1. *The principal quantum number, n.* The **principal quantum number** may assume the values $n = 1,2,3,\ldots$, that is, any positive integer greater than zero. The value of n determines the overall size of the orbital and the energy of the electron when it is in the state associated with that particular orbital. For the hydrogen atom and H-like ions, the value of n alone determines the energy of the electron. The expression obtained using quantum mechanics for the energy of the electron in a state with quantum number n is exactly the same as the expression that Bohr had obtained, namely

$$E_n = -\frac{2\pi^2 me^4 Z^2}{n^2 h^2} = -\frac{K}{n^2} \tag{13-1}$$

where Z is the atomic number, e is the electronic charge, and m is the mass of the electron. Thus Bohr's result for the allowed energies of an electron in a hydrogen atom (or in an H-like ion) is correct even though two of his postulates are incorrect. [See Appendix H for Bohr's derivation of Eq. (13-1).]

The larger the value of n, the greater the energy of the electron and the greater the overall size of the orbital. The statement that the size of the orbital is larger means that there is a higher probability of finding the electron at greater distances from the nucleus.

2. ***The azimuthal or angular quantum number, ℓ.*** The value of ℓ may be 0, 1, 2, ... , $(n-1)$ for a given value of n. Note particularly that possible values of ℓ depend on the value of n.

If $n = 1$, ℓ can only be 0.

If $n = 2$, ℓ can be 0 or 1.

If $n = 3$, ℓ can be 0, 1, or 2, and so on.

The value of ℓ determines the shape of the atomic orbital. For each atomic orbital, we can separate the size of the orbital, that is, the average distance of the electron from the nucleus (which is determined by n) from the shape of the orbital, that is, its angular variation (which is determined by ℓ).

The value of ℓ determines the **angular momentum** of the electron; the larger the value of ℓ, the greater the angular momentum. An electron with angular momentum has kinetic energy corresponding to its angular motion. The amount of kinetic energy associated with the angular motion depends on ℓ, but this is limited by the total allowed energy, which is determined by the value of n. The value of ℓ therefore depends on the value of n.

The atomic orbitals that are solutions to the Schrödinger equation for an electron in a **many-electron atom** can still be defined by the same three quantum numbers. However, as we shall discuss more fully later, the energy of an electron in a many-electron atom is determined by both n and ℓ. This is in contrast to the situation for the hydrogen atom, for which the energy is determined only by n, as long as the atom is not placed in an electric or magnetic field. Equation (13-1) does not apply to any other atom than hydrogen. For atoms with more than one electron, there is no simple relationship between the energy and the two quantum numbers n and ℓ, but orbital energies depend on both n and ℓ. For the same value of n, the energy increases as ℓ increases.

For historical reasons, letters are assigned to different numerical values of ℓ. This occurred because spectroscopists studied atomic spectra long before the introduction of wave mechanics and they used symbols that afterwards were shown to be related to the quantum numbers that arose on solving the Schrödinger equation. The letters assigned, respectively, to the ℓ values 0, 1, 2, and 3 are s, p, d, and f. These are the first letters of the words sharp, principal, diffuse, and fundamental, which spectroscopists had employed to describe the lines observed in atomic spectra before the advent of wave mechanics. For larger n and ℓ, the symbols simply continue in alphabetical order:

ℓ	0	1	2	3	4	5
Symbol	s	p	d	f	g	h

Values of ℓ beyond 3 are not used for electrons in the ground state (the state of lowest energy) of any atom.

3. ***The magnetic quantum number, m_ℓ.*** Possible values of m_ℓ depend on the value of ℓ. The value of m_ℓ may be 0, ± 1, ± 2, ... , $\pm \ell$ for a given value of ℓ. Thus

If $\ell = 0$, m_ℓ can only be 0.

If $\ell = 1$, m_ℓ can be 1, 0, or -1.

If $\ell = 2$, m_ℓ can be 2, 1, 0, -1, or -2,   and so on.

Table 13.1. **Electronic Quantum Numbers and Atomic Orbitals**

Shell	n	ℓ	Orbital	m_ℓ	Degeneracy
K	1	0	$1s$	0	1
L	2	0	$2s$	0	1
	2	1	$2p$	1, 0, −1	3
M	3	0	$3s$	0	1
	3	1	$3p$	1, 0, −1	3
	3	2	$3d$	2, 1, 0, −1, −2	5
N	4	0	$4s$	0	1
	4	1	$4p$	1, 0, −1	3
	4	2	$4d$	2, 1, 0, −1, −2	5
	4	3	$4f$	3, 2, 1, 0, −1, −2, −3	7

For a given ℓ, therefore, m_ℓ can be ℓ, $\ell - 1, \ldots, 0, \ldots, -\ell$, and there are $(2\ell + 1)$ values of m_ℓ for each value of ℓ.

When a current flows through a loop of wire, a magnetic field is produced. Indeed, this is how magnetic fields are generated in the laboratory and for commercial use. A great many loops of wire are wound around a core, and a magnetic field is produced when current is passed through the wire. Such a device is called an **electromagnet**. A moving electron with a nonzero angular momentum also produces a magnetic field. The magnitude of this field is determined by the value of ℓ. The direction or orientation of the magnetic field is determined by the value of m_ℓ. In the absence of an external magnetic or electric field, states with different values of m_ℓ but the same values of n and ℓ have the same energy. Atomic orbitals with the same energy are said to be **degenerate**. Therefore there are $(2\ell + 1)$ degenerate orbitals for each value of ℓ. The number of orbitals having the same energy is called the **degeneracy** of the energy level.

An atomic orbital is given a symbol that lists the numerical value of n and the letter symbol that designates the value of ℓ. The orbital for which $n = 1$ and $\ell = 0$ is called the $1s$ orbital; that for which $n = 2$ and $\ell = 1$ is called the $2p$ orbital, and so on. In addition, orbitals are grouped into **shells** depending on the value of n. The K shell is the shell for which $n = 1$; for the L shell $n = 2$, and so on in alphabetical order. The shells are divided into **subshells** with different values of ℓ. All of the foregoing information about the atomic orbitals, their nomenclature, and the quantum numbers is summarized in Table 13.1.

EXAMPLE 13.1. Nomenclature of atomic orbitals

State whether or not each of the following symbols is an acceptable designation for an atomic orbital. Explain what is wrong with the unacceptable symbols. **(a)** $2d$, **(b)** $6g$, **(c)** $7s$, and **(d)** $5h$.

Solution

(a) The symbol $2d$ is unacceptable. For a d electron $\ell = 2$. The value of ℓ cannot exceed $n - 1$. Thus if $n = 2$, the largest value of ℓ allowed is 1. The only possible orbitals for $n = 2$ are the $2s$ and $2p$.

(b) The symbol $6g$ is acceptable. It stands for the nine degenerate orbitals with $n = 6$ and $\ell = 4$.

(c) The symbol $7s$ is acceptable. It stands for the atomic orbital with $n = 7$ and $\ell = 0$.

(d) The symbol $5h$ is unacceptable. For an h electron $\ell = 5$. If $n = 5$, the largest possible value of ℓ is 4. The only possible orbitals for $n = 5$ are the $5s$, $5p$, $5d$, $5f$, and $5g$.

Ways of Representing Atomic Orbitals

All s orbitals are spherical in shape. Figure 13.1 depicts several ways in which the $1s$ atomic orbital is represented. In Fig. 13.1*(a)* the electron density is represented by the density of dots (the stippling); we can of course draw only a cross section of the charge cloud, which is spherically symmetric. Since it takes time to draw so many dots, it is common to draw a contour to represent the boundary within which any specified large fraction of the electron density is located. For instance, we can draw the 95% contour, which for the $1s$ atomic orbital of hydrogen is a sphere of radius 1.6 Å (0.16 nm) centered about the nucleus. That means that only 5% of the total charge of the electron is outside a sphere of radius 0.16 nm from the nucleus. The radius of the 99% boundary contour is 2.2 Å, of the 90% contour is 1.4 Å, and of the 60% contour is 0.85 Å. We can therefore draw a series of contours, as in Fig. 13.1*(b)* or a single boundary surface, as in Fig. 13.1*(c)*.

Note that in Fig. 13.1*(a)* the electron density is greatest right at the nucleus and falls off as the distance from the nucleus increases. This can be seen clearly in Fig. 13.2, which is a plot of ψ^2 as a function of the distance, r, from the nucleus for the $1s$ atomic orbital. The probability of finding the electron in a small volume element around the point (x, y, z) at a distance r from the nucleus is proportional to ψ^2. Therefore ψ^2 is the probability density *per unit volume*. Note that on the plot in Fig. 13.2 the distance from the nucleus is measured in atomic units.

$$\text{1 atomic unit of length} = a_0 = 0.529 \text{ Å} = 0.0529 \text{ nm} \tag{13-2}$$

In the original Bohr theory, the electrons in the hydrogen atom were pictured as being restricted to moving on a set of allowed circular orbits around the nucleus. The radius of the first Bohr orbit, $a_0 = 0.529$ Å, is now used as the atomic unit of length.

The wave function for the $1s$ atomic orbital is

$$\psi_{1s} = ke^{-Zr/a_0} \tag{13-3}$$

where Z is the charge on the nucleus, a_0 is the radius of the first Bohr orbit, k is a constant, and r is the distance of the electron from the nucleus. Therefore the probability of finding the electron at a distance r from the nucleus decreases exponentially with increasing r, as shown in Fig. 13.2.

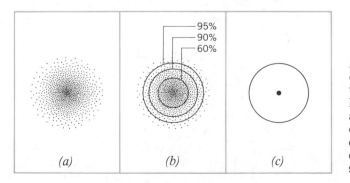

Fig. 13.1. Representation of the wave function for the ground state of the hydrogen atom, the $1s$ atomic orbital. *(a)* Charge cloud diagram. *(b)* Charge cloud with three boundary contours. *(c)* Boundary surface of the 95% contour.

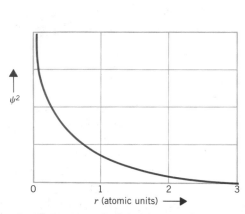

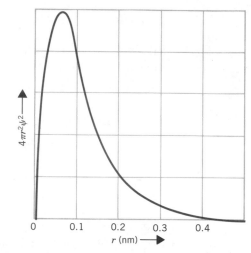

Fig. 13.2. Plot of ψ^2 for the $1s$ AO as a function of distance form the nucleus in atomic units.

Fig. 13.3. The radial probability distribution for the $1s$ AO of hydrogen. The value of r at which this function is a maximum is 0.0529 nm.

Often we need to know the probability that the electron is at a given distance from the nucleus, without any regard to direction. This is the probability of finding the electron in an infinitesimally thin spherical shell with radius r around the nucleus, and is called the **radial probability distribution.** It is obtained by averaging ψ^2 over all angular variables, and is expressed as $4\pi r^2 \psi^2(r)$. The radial probability distribution for a $1s$ atomic orbital is shown in Fig. 13.3. The value of r at the maximum radial probability is 0.529 Å (0.0529 nm), 1 atomic unit of length.

A new feature, called a **node,** appears when we make similar plots for the $2s$ atomic orbital. A **node** is a point or region in space where the wave function has zero amplitude. Both a plot of ψ^2 as a function of the distance of the electron from the nucleus and a charge cloud diagram for the $2s$ atomic orbital are depicted in Fig. 13.4. Note that when r is 1.06 Å (2 atomic units), the value of ψ^2 is zero. This means that if we imagine a series of concentric spherical shells surrounding the nucleus, there is a high probability of locating the electron at distances closer to the nucleus than 1.06 Å and also a high probability of finding the electron in a shell with radius somewhat greater than 1.06 Å, but zero probability of locating it exactly 1.06 Å from the nucleus. The nodal sphere is clearly seen in the charge cloud diagram. The $2s$ atomic orbital has a single nodal sphere, the $3s$ orbital has two nodal spheres, the $4s$ orbital has three nodal spheres, and so on.

The most important features of the s orbitals are the spherical boundary contour and the increase in overall size as n increases. Table 13.2 and Fig. 13.5 show the increase in size, with increasing value of n, for the $1s$, $2s$, and $3s$ atomic orbitals.

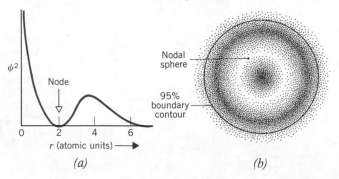

Fig. 13.4. The 2s atomic orbital of hydrogen. *(a)* Plot of ψ^2 versus r in atomic units. *(b)* Charge cloud.

Table 13.2. **The Increase in Size of the *s* Atomic Orbitals with Increase in Principal Quantum Number, *n***

Atomic Orbital	Radius of the 99% Boundary Contour
1s	0.22 nm
2s	0.65 nm
3s	1.2 nm

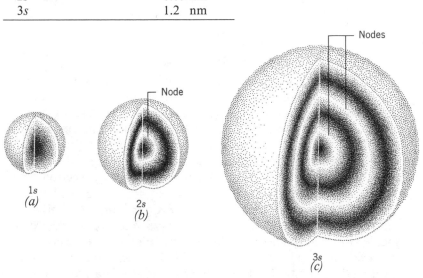

Fig. 13.5. The electron density distribution in the 1s, 2s, and 3s atomic orbitals. The overall size of the orbital increases with increasing *n*. The number of nodal spheres for an *ns* orbital is (*n* − 1).

The shape of an atomic orbital for which $\ell = 1$ (a *p* orbital) is very different from that of the *s* orbitals. The three *p* orbitals ($m_\ell = 0, \pm 1$) are dumbbell shaped (see Fig. 13.6) and consist of two lobes of electron density with a nodal plane of zero density separating the two lobes.

Each of the three *p* orbitals has the same shape, but they are oriented differently in space. One of the orbitals, denoted p_x, has maximum electron density along the *x* axis; its nodal plane is the *yz* plane. Similarly, the p_y atomic orbital has maximum electron density along the *y* axis and the *xz* plane is its nodal plane. Figure 13.7 shows all three *p* orbitals on the same coordinate axes.

For all values of *n*, the *p* orbitals have the same shape, but the overall size increases as *n* increases. The directionality of the *p* orbitals, that is, the fact that there is a

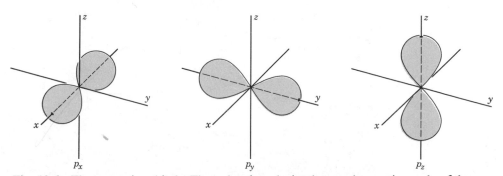

Fig. 13.6. The *p* atomic orbitals. These drawings depict the angular portion only of the electron density distribution. There is a zero probability of locating a *p* electron at the nucleus.

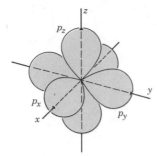

Fig. 13.7. The three p orbitals of a given p subshell point in directions that are mutually perpendicular. By imagining that they point along a set of x, y, and z axes, we can label them p_x, p_y, and p_z.

particular direction along which each p orbital has maximum electron density, plays a large role in determining the geometry of molecules, as we shall see in Chapter 14.

An important point that should be stressed is that the charge cloud of a *single* electron in, for instance, the $2p_x$ atomic orbital consists of *two* lobes of electron density. That means there is a high probability of locating the electron in a $2p_x$ atomic orbital at values of x greater than zero, and a high probability of locating it at values of x less than zero, but no probability at all of locating it anywhere in the yz plane along which $x = 0$. Thus there is no probability of finding a p electron right at the nucleus.

There are five d atomic orbitals ($\ell = 2$, $m_\ell = 0, \pm1, \pm2$, for $n \geq 3$), and the shapes of these five orbitals are shown in Fig. 13.8. Four of these orbitals have the same shape but differ in orientation in space. They have four lobes of electron density and two nodal planes. One of these four-lobed d orbitals has maximum electron density along the x and y axes. It is denoted $d_{x^2-y^2}$. For the other three of these four-lobed orbitals, the directions along which the electron density is a maximum lie at 45° to the coordinate axes. These three are denoted d_{xy}, d_{xz}, and d_{yz}. The d_{xy} atomic orbital, for example, has maximum electron density along the lines that bisect the x and y axes. The fifth d orbital, denoted d_{z^2}, has two lobes of electron density directed along the z axis, and a ring of electron density in the xy plane.

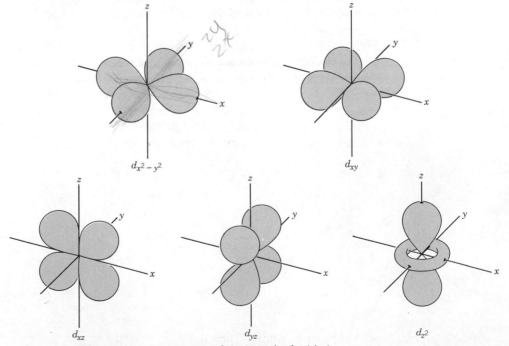

Fig. 13.8. The angular dependence of the atomic d orbitals.

Because of the way in which the p and d orbitals are oriented in space, electrons occupying different atomic orbitals are as far apart from one another as possible. These orientations therefore serve to minimize electron–electron repulsion in many-electron atoms.

section 13.2
The Spin Quantum Number and the Pauli Exclusion Principle

The three quantum numbers n, ℓ, and m_ℓ arise quite naturally from the mathematics of solving the Schrödinger equation for the hydrogen atom. For many-electron atoms the Schrödinger equation cannot be solved exactly, but approximate solutions can be obtained and it can be shown that the same three quantum numbers apply to electrons in all the atoms of the periodic table. Atomic spectra of many-electron atoms can be interpreted using these quantum numbers and their relationships to the properties of the orbitals. All features of atomic spectra, however, cannot be accounted for using only n, ℓ, and m_ℓ. In particular, when an atom is placed in an external magnetic field, states with different values of m_ℓ but the same values of n and ℓ are no longer equal in energy (the degeneracy is removed by the external field) and additional lines appear in the spectrum. This is known as the **Zeeman Effect,** after the Dutch physicist Pieter Zeeman (1865–1943), who first observed the splitting of spectral lines in a magnetic field. The additional fine structure observed (see Fig. 13.9) when an external magnetic field is applied cannot be accounted for if only the quantum numbers n, ℓ, and m_ℓ are considered.

Two young Dutch physicists, Samuel Goudsmit (1902–1979) and George E. Uhlenbeck (b. 1900) suggested in 1925 that a fourth quantum number is necessary in order to explain all features of atomic spectra. The spectra demonstrate that an electron in an atom possesses an **intrinsic magnetic moment,** that is, a magnetism not associated with the electron's orbital angular momentum. Goudsmit and Uhlenbeck pictured the electron as spinning about its own axis as well as moving about the nucleus with orbital angular momentum. This is in direct analogy to the planetary system; the planets revolve about the sun in elliptical orbits and simultaneously spin about an internal axis. Since an electron is a charged particle, the spin angular momentum and the orbital angular momentum each give rise to a magnetic moment. To explain the multiplicity of the fine structure of the lines in atomic spectra,

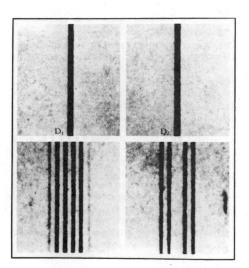

Fig. 13.9. The Zeeman effect. Above, the yellow-orange lines (*D*-lines) in the spectrum of atomic sodium in the absence of a magnetic field. Below, the same lines are split into multiplets when the sodium vapor is placed in an external magnetic field.

Goudsmit and Uhlenbeck found it necessary to assume that the spin (or intrinsic angular momentum) is quantized and that there are only two possible values for the intrinsic magnetic moment of an electron. While it is not correct to picture an electron as physically spinning, it is now well-established that an electron does possess an intrinsic angular momentum, independent of its orbital angular momentum, and that a fourth quantum number, called the **spin quantum number**, m_s, must be introduced to describe completely the state of an electron in an atom. There are only two possible values of m_s, $+\frac{1}{2}$ and $-\frac{1}{2}$.

While this fourth quantum number does not appear when the Schrödinger equation is solved, it had already been suggested by Wolfgang Pauli (1900–1958) that the theory of relativity requires four quantum numbers. Three-dimensional space requires three quantum numbers; in the theory of relativity there is a fourth dimension, time, that requires a fourth quantum number. Dirac included the principles of relativity in the fundamental postulates of his formulation of quantum mechanics, and in the Dirac method of obtaining the wave functions for the hydrogen atom the four quantum numbers n, ℓ, m_ℓ, and m_s do indeed arise naturally from the mathematics. An electron for which $m_s = +\frac{1}{2}$ is said to be in **spin state α,** and is depicted by an upward arrow, ↑. An electron for which $m_s = -\frac{1}{2}$ is said to be in **spin state β** and is depicted by a downward arrow, ↓. In order to describe fully the state of an electron in an atom you must specify both its atomic orbital and its spin state, that is, you must specify the value of all four quantum numbers: n, ℓ, m_ℓ, and m_s.

The result of the quantization of the electron spin is that the magnetic moment associated with the spin has two and only two orientations relative to the direction of an external magnetic field. These two directions are usually described as **spin up** (↑) and **spin down** (↓). In the absence of an external magnetic field, these two states of opposite spin have the same energy, but when a magnetic field is applied, they differ in energy by an amount proportional to the strength of the applied field.

A fundamental postulate about the quantum numbers of the electrons in many-electron atoms was proposed by Wolfgang Pauli, one of the giants of theoretical physics of the twentieth century. This postulate, now called the **Pauli Exclusion Principle,** states "In a many-electron atom, no two electrons can have identical values of all four quantum numbers." Each electron in the atom has a unique set of the four quantum numbers n, ℓ, m_ℓ, and m_s. Since there are only two possible values for m_s ($+\frac{1}{2}$ and $-\frac{1}{2}$), the Exclusion Principle states that *no more than two electrons can occupy the same atomic orbital.* Two electrons occupying the same orbital must be in different spin states, and their spins are said to be **paired.**

An analogy has been drawn between the familiar statement about our three dimensional world "No two bodies can occupy exactly the same space at the same time" with the Pauli Exclusion Principle. The Exclusion Principle should be understood as one of the fundamental postulates of the laws of nature; it is an experimental fact and has no explanation based on other principles.

EXAMPLE 13.2. Quantum numbers of the electrons in a set of degenerate atomic orbitals

Give the set of four quantum numbers for each of the electrons in the $3d$ subshell when it is fully occupied.

Solution. For any d orbital, $\ell = 2$ and there are five degenerate d orbitals for a given value of n. Each orbital can hold 2 electrons, so a total of 10 electrons is necessary to fill a d subshell, that is to fill all five d orbitals. For the $3d$ atomic orbitals the quantum numbers of the 10 electrons are

n	ℓ	m_ℓ	m_s	n	ℓ	m	m_s
3	2	0	$\frac{1}{2}$	3	2	0	$-\frac{1}{2}$
3	2	1	$\frac{1}{2}$	3	2	1	$-\frac{1}{2}$
3	2	−1	$\frac{1}{2}$	3	2	−1	$-\frac{1}{2}$
3	2	2	$\frac{1}{2}$	3	2	2	$-\frac{1}{2}$
3	2	−2	$\frac{1}{2}$	3	2	−2	$-\frac{1}{2}$

Because of the magnetic moment associated with the electron spin and with the orbital angular momentum of the electron, all substances have magnetic properties. Those substances in which all electron spins are paired are slightly repelled by an external magnetic field and are said to be **diamagnetic.** Those substances in which one or more electrons have unpaired spins are relatively strongly attracted to a magnetic field, and are said to be **paramagnetic.** The greater the number of unpaired electrons, the greater the paramagnetism per mole of substance.

section 13.3
The Aufbau Process and the Periodic Table

In addition to the Pauli Exclusion Principle one other basic concept is needed before we can begin to explain the electronic configurations of all the atoms in the periodic table. This is the principle that in the ground (lowest energy) state of a many-electron atom, each electron will occupy the lowest energy state available to it, consistent with the Exclusion Principle. The order of the energies of the different atomic orbitals of a given atom is therefore of utmost importance in understanding the electronic configuration of that atom.

It has already been stated, in Section 13.1, that for all other atoms except hydrogen, in the absence of an external magnetic or electric field, the energy of an electron depends on both n and ℓ. This is due to electron–electron repulsions within a many-electron atom, and to the fact that an electron some distance from the nucleus is **screened** or **shielded** from the full magnitude of the nuclear charge by other electrons that are closer to the nucleus.

The energy of an electron in a many-electron atom generally increases as n increases, but for a given n, the lower value of ℓ the lower the energy. This is due to an effect called **penetration,** the fraction of time an electron spends close to the nucleus. Because an electron close to the nucleus experiences a large electrostatic attraction, the greater the penetration, the lower the energy of the electron. For a given value of n, an s electron penetrates to the nucleus more than a p electron, which penetrates more than a d electron, and so on. Figure 13.10 depicts the radial distribution function of the $3s$, $3p$, and $3d$ atomic orbitals. Note that there are two small peaks very close to the nucleus for the $3s$ orbital, one small peak for the $3p$ orbital, and none for the $3d$ orbital. The physical significance of these small peaks close to the nucleus is a greater penetration to the nucleus. The different degrees of penetration to the nucleus of atomic orbitals with the same value of n but different values of ℓ lead to energies in the order $s < p < d < f < g < h$. Note that for the $n = 3$ wave functions (see Fig. 13.10), the value of r for which the radial distribution function is a maximum decreases in the order $s > p > d$. Nevertheless, the order of energies of these subshells is $s < p < d$. This indicates the importance of penetration in determining the overall energy of an electron for a given value of n.

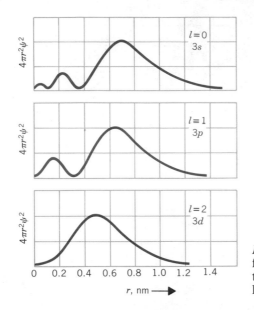

Fig. 13.10. The radial distribution function for the 3s, 3p, and 3d AO's. Penetration to the nucleus is shown by the small peaks at low values of r.

The Building-Up (Aufbau) of the Periodic Table

Let us apply these principles to the elucidation of the electronic configurations of the elements, starting with hydrogen, atomic number 1. The single electron in hydrogen occupies the atomic orbital of lowest energy, the $1s$. The value of m_s can be either $+\frac{1}{2}$ or $-\frac{1}{2}$; in a sample containing a large number of gaseous H atoms half will have $m_s = +\frac{1}{2}$ and half will have $m_s = -\frac{1}{2}$. There is no way to distinguish between these two states in the absence of an external magnetic field.

The next atom is He ($Z = 2$) with two electrons. These electrons both occupy the $1s$ atomic orbital, but their spins must be paired, as required by the Pauli Principle. We represent this as either

$$\underset{1s}{\uparrow \downarrow} \qquad \text{or} \qquad \underset{1s}{\textcircled{$\uparrow \downarrow$}}$$

The electronic configuration of a helium atom is written as $1s^2$, which is read "one s two" and signifies that there are two electrons in the $1s$ atomic orbital with opposite spins. Note that the $1s$ orbital, and therefore the K shell, is filled at the atom helium.

The third atom in the periodic table is Li ($Z = 3$), with three electrons. Since three electrons cannot occupy the same atomic orbital, one electron must go into the atomic orbital that is higher in energy than the $1s$, but lower than all other atomic orbitals, namely the $2s$ orbital. The electronic configuration of Li is denoted $1s^2 2s$. (Some people prefer to write $1s^2 2s^1$. If no superscript is written, a "1" is understood.) Half of the $2s$ electrons in a collection of a large number of gaseous Li atoms have $m_s = \frac{1}{2}$; the other half have $m_s = -\frac{1}{2}$.

After Li comes Be ($Z = 4$), which has the electronic configuration $1s^2 2s^2$. As the $1s$ and $2s$ atomic orbitals are now filled, the next element, B ($Z = 5$), must have one electron in a still higher energy atomic orbital, one of the three $2p$ orbitals, all of which are equal in energy in the absence of a magnetic field. It does not matter which of the three $2p$ orbitals is occupied. Indeed, all of the six possible combinations of m_ℓ and m_s are found in a sample containing a great many isolated boron atoms, and there are equal numbers of each of the six possibilities. We represent this schematically with the diagram shown in Fig. 13.11. It makes no difference whether the electron in the $2p$ orbital is represented as $\uparrow$ or $\downarrow$.

When we come to the next element, carbon ($Z = 6$), we have a choice that cannot be made on the basis of any principle previously discussed. Does the sixth electron go

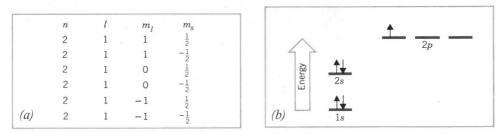

	n	l	m_l	m_s
	2	1	1	$\frac{1}{2}$
	2	1	1	$-\frac{1}{2}$
	2	1	0	$\frac{1}{2}$
	2	1	0	$-\frac{1}{2}$
	2	1	-1	$\frac{1}{2}$
(a)	2	1	-1	$-\frac{1}{2}$

Fig. 13.11. *(a)* Possible sets of quantum numbers for a $2p$ electron. *(b)* Energy level diagram and electronic configuration of boron, $1s^2 2s^2 2p$.

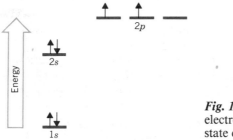

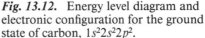

Fig. 13.12. Energy level diagram and electronic configuration for the ground state of carbon, $1s^2 2s^2 2p^2$.

into the same $2p$ atomic orbital as the fifth, with paired spin, or does it occupy a different $2p$ atomic orbital? And if it is in a different $2p$ orbital is the spin quantum number the same as that of the fifth electron, or different? The answer to these questions is given by **Hund's Rule.** Hund's Rule is obtained by correlating the lines observed in atomic spectra with the energies of the orbitals. It is therefore an empirical rule, deduced from experimental data. It states that the total energy of a many-electron atom with more than one electron occupying a set of degenerate orbitals is lowest if, as far as possible, electrons occupy different atomic orbitals and have **parallel spin,** that is, the same value of m_s.

Two electrons in different atomic orbitals are, on the average, farther apart than two electrons of paired spin in the same orbital, so that occupying different atomic orbitals of the same energy is a way to minimize **electron–electron repulsion.** Thus if vacant orbitals of equal energy are available, such as the three $2p$ atomic orbitals, electrons occupy different orbitals, and have parallel spin. In accordance with Hund's Rule, the ground state electronic configuration of carbon, $1s^2 2s^2 2p^2$, is shown schematically in Fig. 13.12. In a sample of a large number of free carbon atoms, some will have configuration $1s^2 2s^2 2p_x 2p_y$, others $1s^2 2s^2 2p_x 2p_z$, and others $1s^2 2s^2 2p_y p_z$. Some will have both electrons with spins $m_s = \frac{1}{2}$ and others will have both electrons with spin $m_s = -\frac{1}{2}$. All of these possibilities are understood to be comprised by the notation $1s^2 2s^2 2p^2$.

We have now established all the concepts necessary to construct the entire periodic table. This process is known as the **Aufbau** process (Aufbau means *building-up* or *construction* in German). We build up the periodic table by using three fundamental principles:

1. The Pauli Exclusion Principle. No two electrons in the same atom can have all four quantum numbers identical.

2. The atomic orbitals are filled in the order of increasing energy.

3. Hund's Rule. If more than one atomic orbital of the same energy is available, electrons will occupy different atomic orbitals with parallel spin, as far as possible, in the configuration of lowest energy.

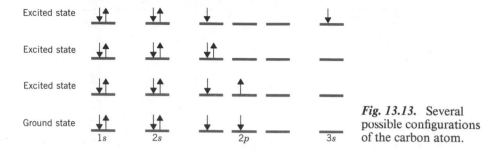

Fig. 13.13. Several possible configurations of the carbon atom.

It should be stressed that Hund's Rule does not forbid any configuration not already forbidden by the Pauli Exclusion Principle. Hund's Rule simply tells us which of the possible configurations is lowest in energy. Other configurations will be excited states, higher in energy than the ground state. Configurations for the ground state and a few excited states of carbon are depicted in Fig. 13.13.

Using the principles stated above, the electronic configurations of the first 18 elements in the periodic table have been written in Table 13.3. Study this table carefully. Note that the K shell is filled at He ($Z = 2$), and the L shell is filled at Ne ($Z = 10$). For the elements with $10 < Z < 18$, it is customary to indicate the electronic configuration $1s^2 2s^2 2p^6$ as (Ne)^{10}, or simply (Ne).

You should be able to write the electronic configuration of any of the first 18 elements without referring to Table 13.3, and it is a great convenience to have memorized the names of these 18 elements in order of atomic number. Examine the positions of these 18 elements in the periodic table, Fig. 13.14, while referring to the electronic configurations given in Table 13.3. Note that the first row of the table contains just 2 elements (filling the $1s$ AO), while the second and third rows contain 8 elements each (filling the $2s$ and $2p$ or the $3s$ and $3p$ AO's, respectively).

For these 18 elements, the Aufbau order of filling atomic orbitals is $1s$, $2s$, $2p$, $3s$, and then $3p$.

Table 13.3. Ground Electronic Configurations of the First 18 Elements of the Periodic Table

Z	K Shell	L Shell				M Shell				Electronic Configuration
1 H	↑									$1s$
2 He	↑↓									$1s^2$
3 Li	↑↓	↑								$1s^2 2s$
4 Be	↑↓	↑↓								$1s^2 2s^2$
5 B	↑↓	↑↓	↑							$1s^2 2s^2 2p$
6 C	↑↓	↑↓	↑	↑						$1s^2 2s^2 2p^2$
7 N	↑↓	↑↓	↑	↑	↑					$1s^2 2s^2 2p^3$
8 O	↑↓	↑↓	↑↓	↑	↑					$1s^2 2s^2 2p^4$
9 F	↑↓	↑↓	↑↓	↑↓	↑					$1s^2 2s^2 2p^5$
10 Ne	↑↓	↑↓	↑↓	↑↓	↑↓					$1s^2 2s^2 2p^6$
11 Na	↑↓	↑↓	↑↓	↑↓	↑↓	↑				(Ne)$3s$
12 Mg	↑↓	↑↓	↑↓	↑↓	↑↓	↑↓				(Ne)$3s^2$
13 Al	↑↓	↑↓	↑↓	↑↓	↑↓	↑↓	↑			(Ne)$3s^2 3p$
14 Si	↑↓	↑↓	↑↓	↑↓	↑↓	↑↓	↑	↑		(Ne)$3s^2 3p^2$
15 P	↑↓	↑↓	↑↓	↑↓	↑↓	↑↓	↑	↑	↑	(Ne)$3s^2 3p^3$
16 S	↑↓	↑↓	↑↓	↑↓	↑↓	↑↓	↑↓	↑	↑	(Ne)$3s^2 3p^4$
17 Cl	↑↓	↑↓	↑↓	↑↓	↑↓	↑↓	↑↓	↑↓	↑	(Ne)$3s^2 3p^5$
18 Ar	↑↓	↑↓	↑↓	↑↓	↑↓	↑↓	↑↓	↑↓	↑↓	(Ne)$3s^2 3p^6$
	$1s$	$2s$	$2p$			$3s$	$3p$			$3d$

	IA	IIA		IIIA	IVA	VA	VIA	VIIA	0
$n = 1$	H 1								He 2
$n = 2$	Li 3	Be 4		B 5	C 6	N 7	O 8	F 9	Ne 10
$n = 3$	Na 11	Mg 12		Al 13	Si 14	P 15	S 16	Cl 17	Ar 18

Fig. 13.14. The first 18 elements in the periodic table. The value of the principal quantum number, n, for the outermost or valence electrons increases by one for each horizontal row (period) of the periodic table.

As we examine the electronic configurations of the 18 elements listed in Table 13.3 we can begin to understand the relation between the chemistry of the atoms and their electronic configurations. Note that for all of these 18 elements, it is s or p subshells that are being filled as the atomic number increases. These elements, and those directly beneath them in the vertical columns of the periodic table comprise seven families of **representative elements**, Groups IA through VIIA, and the **rare** or **noble gases**, Group 0 (sometimes denoted Group VIII). Each one of these eight families has a chemistry quite distinct from the others, and that chemistry is determined by the electronic configuration of the outermost or **valence electrons** of the atoms in that family.

EXAMPLE 13.3. Possible electronic configurations of nitrogen

Given below are several electronic configurations that may be correct for the nitrogen atom. Indicate whether each of these configurations represents the ground state, or an excited state, or is an impossible (forbidden) configuration for nitrogen.

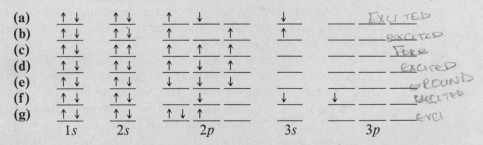

Solution. (a), (b), (d), (f), and (g) are all configurations of different excited states of N. The ground state electronic configuration of N is (e). Configuration (c) is forbidden because it violates the Pauli Principle. The two electrons in the $2s$ orbital have parallel spin.

section 13.4
Group 0, the Noble Gases: The Relationship between Their Chemistry and Their Electronic Configurations

The family of elements known as the **rare** or **noble gases** consists of He, Ne, Ar, Kr, Xe, and Rn. Experimental observation tells us that the rare gases are the most stable

Table 13.4. **Electronic Configurations of the Noble Gases**

Z	Rare Gas	Electronic Configuration
2	He	$1s^2$
10	Ne	$1s^2 2s^2 2p^6$
18	Ar	$(Ne)^{10}3s^2 3p^6$
36	Kr	$(Ar)^{18}3d^{10}4s^2 4p^6$
54	Xe	$(Kr)^{36}4d^{10}5s^2 5p^6$
86	Rn	$(Xe)^{54}4f^{14}5d^{10}6s^2 6p^6$

of all atoms and have very little tendency to combine with other substances. At room temperature, all members of this family are monatomic gases. Indeed, they are the only elements that exist as uncombined gaseous atoms at room temperature and 1-atm pressure. Until 1962, not a single compound involving any rare gas was known, and it was common to call them **inert gases.**

Three of these elements, He, Ne, and Ar are included among the 18 listed in Table 13.3. What do we notice when we examine the electronic configurations of He, Ne, and Ar? All of them have completely filled subshells. The K shell is filled at He; the L shell is filled at Ne; and at Ar the $3s$ and $3p$ subshells are both filled. Indeed if we examine the electronic configurations of the entire family of rare gases we see that in all of them (except for He) there are eight outer electrons that completely fill s and p subshells. After argon, the next rare gas, krypton ($Z = 36$) has a filled M shell and then eight electrons fill the $4s$ and $4p$ subshells. Electronic configurations of all the rare gases are given in Table 13.4. Except for He, the outer electronic configuration of all the rare gases is $(ns)^2(np)^6$ and all electrons in any of the rare gases are in completely filled subshells. The rare gases are chemically similar because their electronic configurations are similar, and our knowledge of the chemical inertness of these 6 elements tells us that these particular electronic configurations are especially stable. The rare gas configurations are of special importance because reactions of many other atoms can be understood in terms of a tendency to attain the same stable electronic configuration as one of the rare gases.

While the electronic configurations of the rare gases, with their completely filled subshells, are exceptionally stable, these gases are not completely inert as was thought at one time. No compounds of He, Ne, or Ar have been made, but several Xe compounds and a smaller number of Kr compounds are known. (A few Rn compounds have been prepared, but Rn is rarely used because it is radioactive.) The reason why it is easier to form compounds of Kr and Xe than of He, Ne, or Ar is that the outer electrons of Kr and Xe are so much further from the nucleus and are therefore less tightly held. Remember that coulombic attraction varies inversely with the square of the distance between the charged particles.

The first rare gas compound, $XePtF_6$, was synthesized by Neil Bartlett in 1962. Since then, XeF_2, XeF_4, XeF_6, XeO_3, XeO_4, $XeOF_2$, XeO_2F_2, $XeOF_4$, KrF_2, and $XeCl_2$ have all been synthesized. Note that more compounds of Xe have been synthesized than of Kr, because the outer electrons of Xe are less tightly held than are the outer electrons of Kr. Note also that in all the compounds of Kr and Xe, the rare gas has combined with a highly electronegative element, one with a strong tendency to pull electron density toward itself and leave the Xe or Kr with a net positive charge. The simple fluorides $XeF_2(s)$, $XeF_4(s)$, and $XeF_6(s)$ are stable at room temperature, but the simple oxides $XeO_3(s)$ and $XeO_4(g)$ decompose explosively to form $Xe(g)$ and $O_2(g)$.

section 13.5
The Periods of the Periodic Table and the Electronic
Configurations of the First Series of Transition Metals

A horizontal row of the periodic table that terminates in one of the rare gases is called a **period.** The first period is very short, consists of just two elements, H and He, and corresponds to the filling of the K shell. The second period contains eight elements and corresponds to filling the L shell. The third period contains another eight elements, because only the $3s$ and $3p$ subshells of the M shell are filled.

In order to continue the Aufbau process for elements with $Z > 18$, we must remember that the electronic configuration of a many-electron atom is the one that minimizes the total energy of *all* the electrons. Atomic orbitals are **one-electron wave functions,** that is, they are derived from the solutions to the Schrödinger equation for the hydrogen atom. For a many-electron atom an orbital is a subunit of the wave function for all the electrons. The repulsions between electrons must be included in a calculation of the total energy of all the electrons.

After the $3p$ orbital, which is filled at argon, the next two atomic orbitals are the $3d$ and $4s$. These two orbitals are very close in energy, particularly for elements 19 through 29. There is no permanent value of the energy of any atomic orbital. The energy of each orbital depends on the nuclear charge and on the repulsion between the electron in the given orbital and all the other electrons. The energy of an orbital decreases as the nuclear charge increases. Thus the $1s$ orbital of K is far below the $1s$ orbital of H in energy. The effect of **electron–electron repulsion** on the energy of an orbital depends not only on the number of other electrons but also on the degree of penetration to the nucleus of the given atomic orbital and of all the other electrons. For these reasons we cannot ask simply "Which orbital is lower in energy, the $3d$ or the $4s$?" but must determine the configuration that is lowest in energy for all the electrons.

The electronic configuration of K ($Z = 19$) is $(Ar)^{18}4s$. This configuration is lower in energy than $(Ar)^{18}3d$ because the $4s$ electron distribution penetrates to the nucleus more than the $3d$ does. The configuration $(Ar)^{18}3d$ is an excited state of K; the ground state (lowest energy) configuration is $(Ar)^{18}4s$, so that K is an alkali metal with chemistry similar to that of Na. Similarly, the electronic configuration of Ca ($Z = 20$) is $(Ar)^{18}4s^2$.

At Sc ($Z = 21$) the $3d$ level begins to be filled and the next 10 elements are a series of **transition elements** in which the inner $3d$ subshell is being filled. Electronic configurations of the **first transition series** (the elements scandium to zinc, $Z = 21$ to 30) are given in Table 13.5.

Table 13.5. Electronic Configurations of the First Series
of Transition Elements and Their Dipositive Ions

Z	*Atom*	*Configuration*	*Ion*	*Configuration*
21	Sc	$(Ar)^{18}3d^14s^2$	Sc^{2+}	$(Ar)^{18}3d^1$
22	Ti	$(Ar)^{18}3d^24s^2$	Ti^{2+}	$(Ar)^{18}3d^2$
23	V	$(Ar)^{18}3d^34s^2$	V^{2+}	$(Ar)^{18}3d^3$
24	Cr	$(Ar)^{18}3d^54s^1$	Cr^{2+}	$(Ar)^{18}3d^4$
25	Mn	$(Ar)^{18}3d^54s^2$	Mn^{2+}	$(Ar)^{18}3d^5$
26	Fe	$(Ar)^{18}3d^64s^2$	Fe^{2+}	$(Ar)^{18}3d^6$
27	Co	$(Ar)^{18}3d^74s^2$	Co^{2+}	$(Ar)^{18}3d^7$
28	Ni	$(Ar)^{18}3d^84s^2$	Ni^{2+}	$(Ar)^{18}3d^8$
29	Cu	$(Ar)^{18}3d^{10}4s^1$	Cu^{2+}	$(Ar)^{18}3d^9$
30	Zn	$(Ar)^{18}3d^{10}4s^2$	Zn^{2+}	$(Ar)^{18}3d^{10}$

The Half-Full Effect

There are several important points to notice about these electronic configurations. There are irregularities in the order of filling at both Cr and Cu. For Cr the ground state configuration is $3d^5 4s$, or schematically ↑___ ↑___ ↑___ ↑___ ↑___   ↑___ , rather than $3d^4 4s^2$, ↑___ ↑___ ↑___ ↑___ ___   ↑↓ (which is an excited state of chromium), because the energy of the Cr atom is lower when the six valence electrons are in different atomic orbitals (as far away from each other as possible) with parallel spin. An extra stabilization is achieved when all electrons have parallel spin. This effect is referred to as the *half-full effect*. A set of degenerate orbitals that is half-full, with all spins parallel, is an especially stable configuration. For Cu, the configuration $3d^{10} 4s$, with the $3d$ orbitals completely filled and the $4s$ orbital half-full, is lower in energy than $3d^9 4s^2$.

Transition Metal Ions

All the metals in the first transition series tend to lose two electrons to become dipositive ions, and the remaining electrons are always $3d$ electrons, because for these ions, the $4s$ orbital is higher in energy than the $3d$. When an atom is ionized, the electrons highest in energy are the most easily removed. Note that none of these ions (see Table 13.5) has the same configuration as any neutral atom. These ions do not have rare gas electronic configurations, and most of them are colored. In aqueous solution, Cu^{2+} is light blue, Ni^{2+} is bright apple green, Mn^{2+} is pale pink, and Co^{2+} is a rosy pink, for example. The Zn^{2+} ion, which has completely filled subshells, is colorless.

Because it is not possible for the transition metals to achieve a rare gas configuration by a simple loss or gain of one or two electrons, most of these elements can form more than one type of cation. You are already familiar with Fe^{2+} and Fe^{3+}, Cr^{2+} and Cr^{3+} ions, among others. (Refer to Table 2.3.)

The Long Periods of the Periodic Table

The fourth period of the periodic table contains 18 elements, the 10 transition metals (as the $3d$ orbitals are filled), and 8 elements corresponding to the filling of the $4s$ and $4p$ orbitals. Note that the order of filling for neutral atoms is first $4s$, then $3d$, then $4p$ (with some irregularities in the transition metals). The fifth period of the table also contains 18 elements as the $5s$, $4d$, and $5p$ subshells are filled, in that order. These two periods are shown in Fig. 13.15.

The energies of the $5s$ and $4d$ orbitals are very close, similar to the situation for the $4s$ and $3d$ orbitals. The **second series of transition metals,** elements 39 through 48 (yttrium to cadmium), are the 10 elements in which the $4d$ orbitals are being filled. Examine Table 13.8, which gives the electronic configurations of all the elements, and notice the irregularities in the orderly filling up of the $4d$ orbitals. Because inner d

$n = 4$	K 19	Ca 20	Sc 21	Ti 22	V 23	Cr 24	Mn 25	Fe 26	Co 27	Ni 28	Cu 29	Zn 30	Ga 31	Ge 32	As 33	Se 34	Br 35	Kr 36
$n = 5$	Rb 37	Sr 38	Y 39	Zr 40	Nb 41	Mo 42	Tc 43	Ru 44	Rh 45	Pd 46	Ag 47	Cd 48	In 49	Sn 50	Sb 51	Te 52	I 53	Xe 54

Fig. 13.15. The fourth and fifth periods of the periodic table. The first and second series of transition metals are shaded in color.

Table 13.6. **The Periods of the Periodic Table**

Period	Number of Elements	Orbitals Filled
1	2	$1s$
2	8	$2s,2p$
3	8	$3s,3p$
4	18	$4s,3d,4p$
5	18	$5s,4d,5p$
6	32	$6s,4f,5d,6p$
7	32	$7s,5f,6d,7p$

orbitals are being filled, the transition metals are more similar to one another than are the representative metals, Groups IA and IIA.

The sixth period begins with the filling of the $6s$ orbitals, and then the seven $4f$ orbitals are filled, which results in a series of 14 elements called the **lanthanides,** after the first element in the series, lanthanum ($Z = 57$). The lanthanides are also called the **rare earths.** They are very similar chemically, more so than the transition metals, because their outer electronic configurations are virtually identical. The rare earths are always found together in nature, and separating them is a challenging analytical procedure because of their chemical similarity. After the 14 lanthanides there is a third series of 10 transition metals, the elements 71 through 80 (lutetium to mercury), in which the $5d$ orbitals are filled. Another 6 elements, corresponding to filling the $6p$ subshell, completes the sixth period, which contains 32 elements.

The seventh period is a repetition of the sixth. A summary of the orbitals filled in each period is shown in Table 13.6. First the $7s$ orbital is filled and then there is another series of 14 elements that corresponds to filling the $5f$ orbitals. These 14 elements are called the **actinides,** as actinium ($Z = 89$) is the first member of the series. All the actinides are radioactive and those with $Z > 92$ are man-made and do not occur naturally on earth. The configurations of the actinides are somewhat irregular and involve both the $5f$ and $6d$ orbitals as these two subshells are very close in energy. All 28 elements in which inner f orbitals are being filled (that is, the lanthanides and the actinides) are referred to collectively as **inner transition elements.**

Six elements ($Z = 103$ to 107, plus 109) of the fourth series of transition elements have been made. It is customary for the discoverer of an element to have the privilege of naming that element. There is an unresolved dispute about whether elements 104 and 105 were made first by a research group in Berkeley, California, or by a group in Dubna, USSR. The Berkeley group proposed the name Rutherfordium for element 104, while the Soviet group proposed the name Kurchatovium. Until the dispute can be resolved, the International Union of Pure and Applied Chemistry has assigned names reflecting their atomic numbers to elements beyond 103. The names and symbols of the five elements beyond lawrencium (Lr), atomic number 103, that have been made at this time are given in Table 13.7.

Table 13.7. **Nomenclature of Elements with Z above 103**

Z	Name	Symbol
104	Unnilquadium	Unq
105	Unnilpentium	Unp
106	Unnilhexium	Unh
107	Unnilseptium	Uns
109	Unnilennium	Une

Table 13.8. **The Electronic Configurations of Gaseous Atoms**

Z	Element	Configuration	Z	Element	Configuration
1	H	$1s$	53	I	$(Kr)4d^{10}5s^25p^5$
2	He	$1s^2$	54	Xe	$(Kr)4d^{10}5s^25p^6$
3	Li	$(He)2s$	55	Cs	$(Xe)6s$
4	Be	$(He)2s^2$	56	Ba	$(Xe)6s^2$
5	B	$(He)2s^22p$	57	La	$(Xe)5d6s^2$
6	C	$(He)2s^22p^2$	58	Ce	$(Xe)4f5d6s^2$
7	N	$(He)2s^22p^3$	59	Pr	$(Xe)4f^36s^2$
8	O	$(He)2s^22p^4$	60	Nd	$(Xe)4f^46s^2$
9	F	$(He)2s^22p^5$	61	Pm	$(Xe)4f^56s^2$
10	Ne	$(He)2s^22p^6$	62	Sm	$(Xe)4f^66s^2$
11	Na	$(Ne)3s$	63	Eu	$(Xe)4f^76s^2$
12	Mg	$(Ne)3s^2$	64	Gd	$(Xe)4f^75d6s^2$
13	Al	$(Ne)3s^23p$	65	Tb	$(Xe)4f^96s^2$
14	Si	$(Ne)3s^23p^2$	66	Dy	$(Xe)4f^{10}6s^2$
15	P	$(Ne)3s^23p^3$	67	Ho	$(Xe)4f^{11}6s^2$
16	S	$(Ne)3s^23p^4$	68	Er	$(Xe)4f^{12}6s^2$
17	Cl	$(Ne)3s^23p^5$	69	Tm	$(Xe)4f^{13}6s^2$
18	Ar	$(Ne)3s^23p^6$	70	Yb	$(Xe)4f^{14}6s^2$
19	K	$(Ar)4s$	71	Lu	$(Xe)4f^{14}5d6s^2$
20	Ca	$(Ar)4s^2$	72	Hf	$(Xe)4f^{14}5d^26s^2$
21	Sc	$(Ar)3d4s^2$	73	Ta	$(Xe)4f^{14}5d^36s^2$
22	Ti	$(Ar)3d^24s^2$	74	W	$(Xe)4f^{14}5d^46s^2$
23	V	$(Ar)3d^34s^2$	75	Re	$(Xe)4f^{14}5d^56s^2$
24	Cr	$(Ar)3d^54s$	76	Os	$(Xe)4f^{14}5d^66s^2$
25	Mn	$(Ar)3d^54s^2$	77	Ir	$(Xe)4f^{14}5d^76s^2$
26	Fe	$(Ar)3d^64s^2$	78	Pt	$(Xe)4f^{14}5d^96s$
27	Co	$(Ar)3d^74s^2$	79	Au	$(Xe)4f^{14}5d^{10}6s$
28	Ni	$(Ar)3d^84s^2$	80	Hg	$(Xe)4f^{14}5d^{10}6s^2$
29	Cu	$(Ar)3d^{10}4s$	81	Tl	$(Xe)4f^{14}5d^{10}6s^26p$
30	Zn	$(Ar)3d^{10}4s^2$	82	Pb	$(Xe)4f^{14}5d^{10}6s^26p^2$
31	Ga	$(Ar)3d^{10}4s^24p$	83	Bi	$(Xe)4f^{14}5d^{10}6s^26p^3$
32	Ge	$(Ar)3d^{10}4s^24p^2$	84	Po	$(Xe)4f^{14}5d^{10}6s^26p^4$
33	As	$(Ar)3d^{10}4s^24p^3$	85	At	$(Xe)4f^{14}5d^{10}6s^26p^5$
34	Se	$(Ar)3d^{10}4s^24p^4$	86	Rn	$(Xe)4f^{14}5d^{10}6s^26p^6$
35	Br	$(Ar)3d^{10}4s^24p^5$	87	Fr	$(Rn)7s$
36	Kr	$(Ar)3d^{10}4s^24p^6$	88	Ra	$(Rn)7s^2$
37	Rb	$(Kr)5s$	89	Ac	$(Rn)6d7s^2$
38	Sr	$(Kr)5s^2$	90	Th	$(Rn)6d^27s^2$
39	Y	$(Kr)4d5s^2$	91	Pa	$(Rn)5f^26d7s^2$
40	Zr	$(Kr)4d^25s^2$	92	U	$(Rn)5f^36d7s^2$
41	Nb	$(Kr)4d^45s$	93	Np	$(Rn)5f^46d7s^2$
42	Mo	$(Kr)4d^55s$	94	Pu	$(Rn)5f^67s^2$
43	Tc	$(Kr)4d^55s^2$	95	Am	$(Rn)5f^77s^2$
44	Ru	$(Kr)4d^75s$	96	Cm	$(Rn)5f^76d7s^2$
45	Rh	$(Kr)4d^85s$	97	Bk	$(Rn)5f^97s^2$
46	Pd	$(Kr)4d^{10}$	98	Cf	$(Rn)5f^{10}7s^2$
47	Ag	$(Kr)4d^{10}5s$	99	Es	$(Rn)5f^{11}7s^2$
48	Cd	$(Kr)4d^{10}5s^2$	100	Fm	$(Rn)5f^{12}7s^2$
49	In	$(Kr)4d^{10}5s^25p$	101	Md	$(Rn)5f^{13}7s^2$
50	Sn	$(Kr)4d^{10}5s^25p^2$	102	No	$(Rn)5f^{14}7s^2$
51	Sb	$(Kr)4d^{10}5s^25p^3$	103	Lr	$(Rn)5f^{14}6d7s^2$
52	Te	$(Kr)4d^{10}5s^25p^4$	104	Unq	

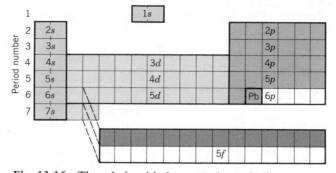

Fig. 13.16. The relationship between the periodic table and the order of filling the atomic orbitals in the Aufbau process. Shaded areas depict the subshells that are filled to obtain the electronic configuration of lead, atomic number 82.

Figure 13.16 depicts the relationship between the Aufbau order of the filling of atomic orbitals and the periodic table. The seventh period will end at $Z = 118$, with the next noble gas.

Table 13.8 lists the ground electronic configurations of all the elements.

section 13.6
A Periodic Property: The Ionization Energy

It always requires energy to remove an electron from a neutral atom, that is, to produce a singly positive ion. The amount of energy required to remove one electron varies greatly from atom to atom, however. If we examine the experimental values of this **ionization energy**, we learn a great deal about the stabilities of different electronic configurations.

We must first define the ionization energy (IE) carefully. It is the energy that must be expended to remove an electron from a single, *isolated* gas phase atom, that is, to carry out the reaction

$$X(g) \rightarrow X^+(g) + e^- \tag{13-4}$$

where X represents any element. We must start with a gaseous atom so that there are no chemical bonds being broken. The definition given is actually the first ionization energy, because we can, of course, remove a second electron, and a third, and so on (from atoms with many electrons). The **second ionization energy** is the energy required for the reaction

$$X^+(g) \rightarrow X^{2+}(g) + e^- \tag{13-5}$$

Ionization energies are usually reported either in electron volts per atom or kilojoules per mole. The relation between these units is obtained as follows [refer to Eq. (12-13a)]:

$$\left(\frac{1 \text{ eV}}{\text{atom}}\right)\left(1.6022 \times 10^{-19} \frac{\text{J}}{\text{eV}}\right)\left(6.0220 \times 10^{23} \frac{\text{atom}}{\text{mol}}\right)$$
$$= 9.6485 \times 10^4 \frac{\text{J}}{\text{mol}} = 96.485 \frac{\text{kJ}}{\text{mol}} \tag{13-6}$$

Table 13.9 lists the first ionization energies of the elements of the second period from Li to Ne. As we look down this list of values we are immediately struck by the fact that *the first ionization energy generally increases as we go across the period*. This

Table 13.9. First Ionization Energies of Elements of the Second Period

Z	Atom	Electronic Configuration	First IE	
			eV/atom	kJ/mol
3	Li	$1s^2 2s$	5.39	5.20×10^2
4	Be	$1s^2 2s^2$	9.32	8.99×10^2
5	B	$1s^2 2s^2 2p$	8.30	8.00×10^2
6	C	$1s^2 2s^2 2p^2$	11.3	1.09×10^3
7	N	$1s^2 2s^2 2p^3$	14.5	1.40×10^3
8	O	$1s^2 2s^2 2p^4$	13.6	1.31×10^3
9	F	$1s^2 2s^2 2p^5$	17.4	1.68×10^3
10	Ne	$1s^2 2s^2 2p^6$	21.6	2.08×10^3

occurs because the nuclear charge is increasing, but the orbitals being filled are all $n = 2$ orbitals, which are approximately the same distance from the nucleus. Thus as Z increases, the outer electrons are held more and more tightly and it requires more energy to remove one electron. A valence electron is screened or shielded from the full magnitude of the nuclear charge by all the inner electrons that are closer to the nucleus. Thus the **effective nuclear charge** a valence electron experiences is less than the actual nuclear charge. But from B ($Z = 5$) to Ne ($Z = 10$) the electron being removed on ionization is a $2p$ electron, and the core of inner electrons, $1s^2 2s^2$, provides approximately the same amount of shielding. Therefore as Z increases across the row the effective nuclear charge experienced by a $2p$ electron increases and the ionization energy also increases.

The alkali metal, Li, has the lowest first ionization energy of the elements of this period, as it is relatively easy to remove the single s electron and form an ion with rare gas configuration. The noble gas, Ne, has the highest ionization energy of the elements of this period because it is difficult to remove an electron from this very stable configuration. There are, however, two exceptions to the general trend of increasing ionization energy with increasing atomic number. These stand out very clearly if we plot the ionization energy as a function of the atomic number, Z, as in Fig. 13.17. The ionization energy of Be is higher than that of the succeeding element, B, and the ionization energy of N is higher than that of O which follows it. We can understand both of these observations in terms of the electronic configurations of the atoms. The electron being removed from Be is a $2s$ electron, while that being removed from B is a $2p$ electron. A $2p$ electron does not penetrate to the nucleus as much as a $2s$ electron

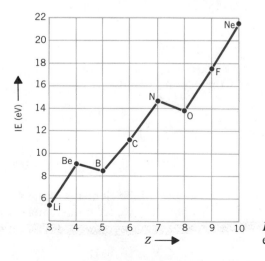

Fig. 13.17. First ionization energies of the elements of the second period.

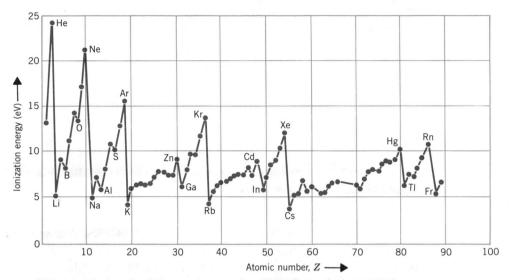

Fig. 13.18. First ionization energies as a function of atomic number, Z.

does. The $2p$ electron of B is more shielded from its nucleus by the inner electrons than is the $2s$ electron of Be. Consequently it is easier to remove the $2p$ electron from B than to remove the $2s$ electron of Be.

We can also explain why the ionization energy of O is lower than that of N in terms of the electronic configurations of these two atoms. In order to decrease electron–electron repulsion, electrons stay as far apart as possible. Thus the three $2p$ electrons in N occupy different atomic orbitals and have parallel spin (Hund's rule). In O, two of the four $2p$ electrons must occupy the same orbital with paired spins. These two electrons repel one another more than $2p$ electrons in different atomic orbitals, and it is easier to remove the fourth $2p$ electron from O than it is to remove one of the three $2p$ electrons from N, even though the nuclear charge is larger for O than it is for N.

The same trend that is observed for the second period is repeated for all the periods of the table. A plot of the first ionization energy as a function of atomic number for the entire periodic table is given in Fig. 13.18. The high ionization energies of the rare gases that terminate each period are clearly seen. Another striking feature of this plot is the decrease in ionization energy after each transition series. The first ionization energy of Zn ($Z = 30$) is 9.39 eV/atom, while that of Ga ($Z = 31$) is 6.00 eV/atom. Remember that for Zn, as for all the first transition series, the electron lost is a $4s$ electron; for Ga a $4p$ electron is lost on ionization. The electronic configurations of Zn, Cd, and Hg (the elements that end the three series of transition metals) consist of completely filled subshells and are $(n-1)d^{10}ns^2$. Consequently it requires a large amount of energy to remove one electron from these atoms. Within a family of the periodic table, the ionization energy decreases as the atomic number increases, because the electron being ionized, the valence electron, is further from the nucleus.

section 13.7
A Periodic Property: The Electron Affinity

Almost all atoms can accept one more electron than is present in the neutral atom and become a uninegative ion. For many atoms this process is accompanied by a release of energy, that is, the energy of the anion is less than the energy of the neutral

Table 13.10. Electron Affinities (eV/atom)
of Groups VA, VIA, and VIIA

N 0.0	O 1.47	F 3.45
P 0.77	S 2.08	Cl 3.61
As 0.80	Se 2.02	Br 3.36
Sb 1.05	Te 1.97	I 3.06
Bi 1.05	Po[a] (1.8)	At[a] (2.8)

[a] The values for Po and At are estimated.

atom. The electron affinity (abbreviated EA) is defined as the energy released when the following reaction occurs:

$$X(g) + e^- \rightarrow X^-(g) \qquad (13\text{-}7)$$

To be precise, what we have just defined is the *first* electron affinity, which is a positive quantity for many, but not all, atoms. The usual convention in discussing energy changes is to give the energy *absorbed* when a reaction takes place. As the electron affinity is the energy released when reaction (13-7) takes place, it is defined in contravention to the customary procedure.

Second and third electron affinities can also be defined. To add a second or third electron to an ion that already has a negative charge always requires an input of energy, and therefore second and third electron affinities are negative.

It is exceedingly difficult to measure electron affinities experimentally. Because of the large experimental uncertainty you may see different values in different references. The most accurately known values are those for the halogens. Many electron affinities have not been measured, and the values reported are calculated from theory. Electron affinity is also a periodic property. Values of the electron affinity for the elements of Groups VA, VIA, and VIIA are given in Table 13.10. As a general trend we note that the electron affinity increases as we go across a period from left to right. With some notable exceptions, the electron affinity decreases as we go down a family, that is, as the electron being added is further from the nucleus. The electron affinities of N, O, and F are, however, significantly less than the electron affinities of the elements directly beneath each of them. Why, for instance, is the electron affinity of fluorine less than that of chlorine? There have been many discussions of this point, but there seems to be general agreement that the $n = 2$ shell is so much smaller than the $n = 3$ shell that adding an additional electron to the $n = 2$ shell causes substantial electronic repulsion and thus reduces the electron affinity.

For atoms that do not readily accept an electron the experimental value of the electron affinity is subject to a large uncertainty. Interpretation of the values reported is difficult and has caused some controversy.

section 13.8
A Periodic Property: The Electronegativity

The concept of **electronegativity** was introduced by Linus Pauling in 1932. **Electronegativity** is a measure of the ability of a bonded atom to attract the electrons in the bond from the other atom or atoms to which it is bonded. An atom with a high electronegativity is able to pull electrons toward itself and away from an atom with a lower electronegativity. In these qualitative terms it is easy to understand the concept of electronegativity, but it is more difficult to obtain a quantitative value for the electronegativity of an atom. The principal reason for this is that the electronegativity is a property of the *bonded* atom (not the isolated gaseous atom) and therefore is not a constant for a particular atom but varies somewhat as the nature of the bonding differs in different molecules. When Pauling introduced the concept of electronegativity he devised a scale based on the energy required to break a bond between two atoms. Several other scales have since been suggested. The proposal that many consider to be the most sound theoretically is that of R. S. Mulliken, who defined electronegativity as the average of the first ionization energy and the electron affinity:

$$x_M = \text{Mulliken's electronegativity} = \tfrac{1}{2}(\text{IE} + \text{EA}) \qquad (13\text{-}8)$$

All that is needed is a relative scale of values and when Mulliken's values are multiplied by a proportionality constant to put them on the same scale as Pauling's, agreement is, for the most part, really very good. It is Pauling's scale that is most often used, and a set of values is given in Fig. 13.19.

Values of the electronegativities are very helpful for understanding certain properties of chemical bonds, but they should be used to make quantitative predictions cautiously. General features of the electronegativity scale with which you should be familiar are the following:

1. Fluorine is the most electronegative element and its electronegativity is set at 4.0. The least electronegative element is cesium, with an electronegativity of 0.7.

2. Generally, electronegativities increase as we go across a period and decrease as we go down a family in the table. The electronegativities of all the elements in a series of transition metals are very similar.

3. Very roughly, an electronegativity value of 2.1 divides the metals and the nonmetals. Most metals have an electronegativity value less than 2; most nonmetals have a value greater than 2. Elements classified as **metalloids** have an electronegativity very close to 2. The alkali metals have the lowest electronegativity values, decreasing from 1.0 for Li to 0.7 for Cs and Fr. Hydrogen has an electronegativity value in the middle of the scale, 2.1.

1 H 2.1																	2 He —
3 Li 1.0	4 Be 1.5											5 B 2.0	6 C 2.5	7 N 3.0	8 O 3.5	9 F 4.0	10 Ne —
11 Na 0.9	12 Mg 1.2											13 Al 1.5	14 Si 1.8	15 P 2.1	16 S 2.5	17 Cl 3.0	18 Ar —
19 K 0.8	20 Ca 1.0	21 Sc 1.3	22 Ti 1.5	23 V 1.6	24 Cr 1.6	25 Mn 1.5	26 Fe 1.8	27 Co 1.8	28 Ni 1.8	29 Cu 1.9	30 Zn 1.6	31 Ga 1.6	32 Ge 1.8	33 As 2.0	34 Se 2.4	35 Br 2.8	36 Kr —
37 Rb 0.8	38 Sr 1.0	39 Y 1.2	40 Zr 1.4	41 Nb 1.6	42 Mo 1.8	43 Tc 1.9	44 Ru 2.2	45 Rh 2.2	46 Pd 2.2	47 Ag 1.9	48 Cd 1.7	49 In 1.7	50 Sn 1.8	51 Sb 1.9	52 Te 2.1	53 I 2.5	54 Xe —
55 Cs 0.7	56 Ba 0.9	57–71 La–Lu 1.1–1.2	72 Hf 1.3	73 Ta 1.5	74 W 1.7	75 Re 1.9	76 Os 2.2	77 Ir 2.2	78 Pt 2.2	79 Au 2.4	80 Hg 1.9	81 Tl 1.8	82 Pb 1.8	83 Bi 1.9	84 Po 2.0	85 At 2.2	86 Rn —

Fig. 13.19. Electronegativities of the atoms: Pauling's scale.

section 13.9
Periodicity in Bonding: Ionic versus Covalent Bonding Across the Short Periods

We have seen that the electronic configurations of the rare gases are particularly stable. The chemistry of the alkali and alkaline earth metals, which will be discussed in the following section, can be understood in terms of a tendency to lose one or two electrons, respectively, in order to form an ion with the electronic configuration of one of the rare gases. It always requires energy to remove an electron from a neutral atom; losing or gaining more than two electrons to form an ion with rare gas electronic configuration requires a large amount of energy. In 1916 the American chemist Gilbert N. Lewis (1875–1946) proposed that atoms can achieve a rare gas electronic configuration not only by losing or gaining electrons, but also by *sharing* them. Since the rare gases (except for He) have eight valence electrons with configuration $(ns)^2(np)^6$, Lewis' proposal is known as the octet rule. This rule states that, except for hydrogen, an atom combines with other atoms to form bonds in order to have eight electrons in its valence shell. Hydrogen shares electrons with other atoms to achieve the helium electronic configuration.

Lewis Dot Formulas

To represent the valence electrons of atoms, Lewis introduced the **dot symbol.** In such a symbol, dots represent valence electrons and four pairs of dots surrounding the symbol of the element represent a rare gas configuration. Lewis dot symbols for elements 11 through 18 (Na to Ar) are shown below:

$$Na\cdot \quad Mg\colon \quad Al\colon \quad \colon\overset{\cdot}{Si}\cdot \quad \colon\overset{\cdot}{P}\cdot \quad \colon\overset{\cdot\cdot}{S}\colon \quad \colon\overset{\cdot\cdot}{Cl}\colon \quad \colon\overset{\cdot\cdot}{Ar}\colon$$

Note that Hund's rule is utilized in writing dot symbols. The three $3p$ electrons of phosphorus, for instance, are in different atomic orbitals, and P is shown with three unpaired electrons plus one pair of electrons, those in the $3s$ orbital.

How can a phosphorus atom achieve a rare gas configuration? A phosphorus atom needs three electrons and a chlorine atom needs one electron to achieve the argon configuration. If a phosphorus atom shares its three electrons with three chlorine atoms to form phosphorus trichloride with the electronic structure

$$\colon\overset{\cdot\cdot}{Cl}\colon\overset{\cdot\cdot}{P}\colon\overset{\cdot\cdot}{Cl}\colon \qquad or \qquad \colon\overset{\cdot\cdot}{Cl}-\overset{\cdot\cdot}{P}-\overset{\cdot\cdot}{Cl}\colon$$
$$\colon\overset{\cdot\cdot}{Cl}\colon \qquad\qquad\qquad \colon\overset{\cdot\cdot}{Cl}\colon$$

all four of these atoms will have an octet of electrons in their valence ($n = 3$) shell. A pair of electrons, shared by two atoms, constitutes a single covalent bond. It is customary to use a straight line to represent a shared pair of electrons, as in the structure on the right, above. The unshared pairs of electrons on the P and Cl atoms are called lone pairs; they are spin-paired electrons in atomic orbitals not used in bonding.

EXAMPLE 13.4. Lewis dot structures for simple compounds that obey the octet rule

Draw the Lewis dot structures for H_2O, SCl_2, and NH_3.

Solution

$$H\colon\overset{\cdot\cdot}{O}\colon \qquad \colon\overset{\cdot\cdot}{Cl}\colon\overset{\cdot\cdot}{S}\colon \qquad H\colon\overset{\cdot\cdot}{N}\colon H$$
$$H \qquad\qquad \colon\overset{\cdot\cdot}{Cl}\colon \qquad\qquad H$$

Note that there is one lone pair of electrons on the N atom, two lone pairs on O and S, and three lone pairs on the Cl atoms.

Validity of the Octet Rule

Although in most compounds shared pairs are formed so that each atom is surrounded by 8 electrons in its valence shell, there are some exceptions to the octet rule. There are a small number of **electron deficient** compounds of the elements with $Z < 10$, particularly of boron. An electron deficient compound is one in which an atom has less than 8 electrons in its outer shell. For elements with $Z > 12$ there are a number of compounds that have more than 8 electrons about a given atom, usually 10 or 12. This occurs when $n \geq 3$ for the outer electrons, because it is then possible to involve the d orbitals in bonding. Specific examples of exceptions to the octet rule will be considered when we discuss the chemistry of the elements of Groups IIIA, IVA, and VA, in the following chapters.

Ionic bonding and pure covalent bonding (a pair of electrons *equally* shared by two atoms) represent two extreme situations. In chemical bonds between two different atoms, the shared pair of electrons is shared unequally. When two atoms joined by a bond differ in electronegativity the more electronegative atom will have a greater share of the electron cloud that constitutes the bond between them. This type of bonding is called **covalent with partial ionic character** or **polar covalent**. If the sharing is very far from equal, the bonding is called **ionic with partial covalent character.**

Electronegativity and Bond Type

When the electronegativity of two bonded atoms differs greatly, the bond is **ionic.** A convenient rule of thumb is that if the absolute value of the difference in electronegativity between two atoms A and B, $|x_A - x_B|$, is greater than or equal to 1.8, the bond between them has more than 50% ionic character. If $|x_A - x_B|$ is less than 1.0, the bond has less than about 20% ionic character and is classified as a covalent bond. A knowledge of the electronegativity difference between two atoms enables us to predict whether the bond is largely ionic or largely covalent, and also to predict the bond polarity. We predict, for example, that the dipole moment (see Section 5.1) of BrF is greater than the dipole moment of ICl, because $x_F - x_{Br} = 1.2$, whereas $x_{Cl} - x_I = 0.5$.

In general, we find ionic bonding only between an active metal (a member of Groups IA or IIA) and one of the most electronegative elements, such as a halogen or oxygen. Even when the metal is from Group IA or IIA, bonding is not always ionic. The smaller the metal atom, the closer its valence electrons are to the nucleus, and the more tightly those electrons are held. Beryllium, in particular, forms many covalent and polar covalent bonds. We can observe this by comparing the melting points of the alkaline earth chlorides, which are listed in Table 13.11. Ionic crystalline solids generally have high melting points; the melting point of $BeCl_2$ is significantly lower than that of any of the other alkaline earth chlorides. The bonding in $BeCl_2$ will be discussed in detail in the following chapter.

Table 13.11. The Melting Points of the Alkaline Earth Chlorides

Compound	mp (°C)
$BeCl_2$	405
$MgCl_2$	708
$CaCl_2$	772
$SrCl_2$	873
$BaCl_2$	963

Table 13.12. **Compounds of Cl with the Elements from Na to Cl**

Name	Formula and Physical State at Room Temperature		Bonding[a]
Sodium chloride	NaCl (crystal)	mp 801 °C	Ionic
Magnesium chloride	$MgCl_2$ (crystal)	mp 708 °C	Ionic
Aluminum chloride	$AlCl_3$ (crystal)	sublimes at 178 °C	Ionic with partial covalent character
	Al_2Cl_6 (liquid and gas) 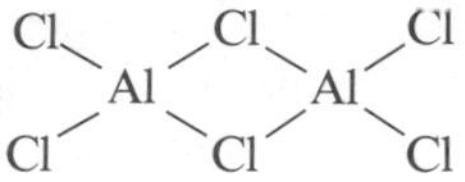		Covalent with partial ionic character

Tetrachlorosilane or silicon tetrachloride	$SiCl_4$ (liq) 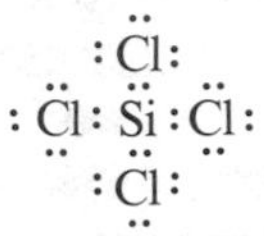	bp 57.6 °C	Polar covalent

Phosphorus pentachloride	PCl_5 (solid) 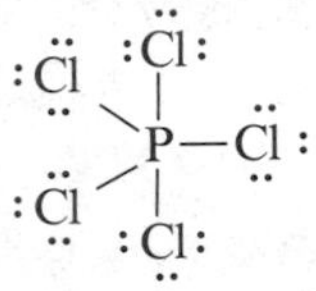	decomposes at 166.8 °C (gas-phase structure shown)	Covalent with partial ionic character. Note that there are 10 electrons around P; the octet rule is not obeyed as the $3d$ orbitals are utilized

Phosphorus trichloride	PCl_3 (liq)	bp 76 °C	Covalent with some ionic character
Sulfur dichloride	SCl_2 (liq)	decomposes at 59 °C	Covalent
Sulfur monochloride	S_2Cl_2 (liq) 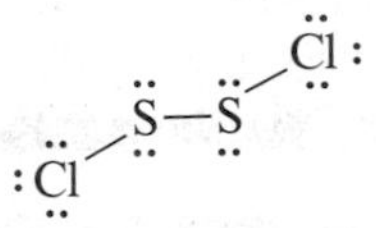	bp 135.6 °C	Covalent

Sulfur tetrachloride	SCl_4 (liq)	decomposes at −15 °C	Covalent; there are 10 electrons around S
Chlorine	Cl_2(g)	:Cl̈—C̈l:	Covalent

[a] Bonds with less than 10% ionic character are listed as covalent.

Bonding in compounds of the transition metals is ionic with partial covalent character, or polar covalent. Titanium, for instance, forms three different chlorides: $TiCl_2$, $TiCl_3$, and $TiCl_4$. Both $TiCl_2$ and $TiCl_3$ are crystalline at room temperature, but they are not as high melting as typical ionic solids; indeed they decompose when heated. Titanium tetrachloride is a yellow liquid that boils at 136.4 °C. The bonding in Ti(IV) compounds is essentially covalent. The chemistry of the first series of transition metals will be discussed further in Chapters 15 and 20.

Atoms in Groups IIIA, IVA, and VA tend to form compounds by sharing electrons rather than by forming ions. The bonds in these compounds are polar covalent. Trends in bonding as we go horizontally across the periodic table can be illustrated by considering the chlorides of the elements from sodium to chlorine. This information is summarized in Table 13.12. The bonds between chlorine and elements of Group IA and IIA are ionic, while bonds between chlorine and the elements of Groups IIIA,

IVA, and VA are partially ionic, partially covalent, with the covalent character increasing as we go across the table from left to right.

section 13.10
The Relationship between the Chemistry and the Electronic Configurations of the Alkali and Alkaline Earth Metals

All Group IA elements have a single s electron in their outer shell. Hydrogen is unique because it has only a single electron; its chemistry is distinctive and much of it has already been discussed in preceding chapters.

The Alkali Metals

All the other Group IA elements have one more electron than one of the rare gases and are known as the **alkali metals**. Electronic configurations of all the alkali metals are given in Table 13.13.

The alkali metals are extremely reactive substances. They all readily lose one electron, the outermost s electron, to form a singly positive ion that has the same electronic configuration as one of the rare gases. The relatively simple chemistry of the alkali metals is determined by the ease with which their single outermost electron can be lost in order to achieve a stable rare gas electronic configuration. Examine Fig. 13.18 and note that, as a group, the alkali metals have the lowest first ionization energies of all the elements in the periodic table.

The alkali metals react with nonmetallic elements to form ionic salts. These reactions are exothermic, frequently are violent, and can be explosive. The alkali metals must be protected from exposure to air because they react with O_2, H_2O, and CO_2. Typical reactions are

$$2Na(s) + O_2(g) \rightarrow Na_2O_2(s) \tag{13-9}$$

$$2Na(s) + 2H_2O(g) \rightarrow 2NaOH(s) + H_2(g) \tag{13-10}$$

$$Na(s) + H_2O(g) + CO_2(g) \rightarrow NaHCO_3(s) + \tfrac{1}{2}H_2(g) \tag{13-11}$$

The product of reaction (13-9) is sodium peroxide, Na_2O_2, a cream-colored solid that is the active ingredient of several commercial bleaching powders. Sodium bicarbonate, $NaHCO_3$, produced in reaction (13-11), is baking soda. It is used to make breads and cakes light and fluffy because when it is heated CO_2 is evolved:

$$2NaHCO_3(s) \xrightarrow{\text{heat}} Na_2CO_3(s) + H_2O(g) + CO_2(g) \tag{13-12}$$

Sodium carbonate, Na_2CO_3, is widely used in the manufacture of soaps, detergents, and glass.

Table 13.13. Electronic Configurations
of the Alkali Metals and Their Ions

Z	*Alkali Metal*	*Electronic Configuration*	*Ion*	*Electronic Configuration*
3	Li	$(He)^2 2s$	Li^+	$(He)^2$
11	Na	$(Ne)^{10} 3s$	Na^+	$(Ne)^{10}$
19	K	$(Ar)^{18} 4s$	K^+	$(Ar)^{18}$
37	Rb	$(Kr)^{36} 5s$	Rb^+	$(Kr)^{36}$
55	Cs	$(Xe)^{54} 6s$	Cs^+	$(Xe)^{54}$
87	Fr	$(Rn)^{86} 7s$	Fr^+	$(Rn)^{86}$

The alkali metals also react with liquid water:

$$2Na(s) + 2H_2O(liq) \rightarrow 2Na^+(aq) + 2\ OH^-(aq) + H_2(g) \qquad (13\text{-}13)$$

and this reaction is so exothermic that the heat generated often ignites the $H_2(g)$ produced. Even a small piece of alkali metal can cause a fire if reacted with liquid water, and in some cases the mixture of H_2 and air can explode. Caution must always be exercised when handling any of the alkali metals. A common practice is to store them under kerosene or some other inert liquid to protect the surface of the metal from exposure to air or moisture.

Alkali metal salts are white unless the anion happens to be colored. These salts are almost all soluble in water, and solutions containing alkali metal cations are colorless unless the anion in the solution happens to be a colored species. The visible colors of ions result from the absorption of light by the electrons of the ions. Ions with stable electronic configurations, and that means ions having the same electronic configuration as one of the rare gases, do not absorb light in the visible region and therefore are colorless.

Salts of all the alkali metals except francium are widely distributed in the earth's crust. Sodium and potassium salts are particularly abundant. Francium is radioactive and there is probably less than an ounce of francium at any time in the crust of the earth.

Group IIA, the Alkaline Earth Metals

All Group IIA elements have two electrons more than one of the rare gases and their outer electron configuration is $(ns)^2$. This family of elements is called the alkaline earth metals and their electronic configurations are given in Table 13.14. The chemistry of the alkaline earth metals results from their tendency to lose their two outer s electrons in order to achieve a rare gas electronic configuration when they combine with more electronegative elements. All the alkaline earth metals form only dipositive cations and react with nonmetallic elements to form ionic salts.

The alkaline earths are significantly less reactive than the alkali metals because it requires much more energy to pull two electrons away from the attraction of the nuclear charge than to remove a single electron. The first ionization energy of each alkaline earth is larger than the first ionization energy of the preceding alkali metal. The alkaline earth metals are also denser, harder, and less shiny than the alkali metals and they have higher melting and boiling points. (See Table 2.1.)

Table 13.14. Electronic Configurations of the Alkaline Earth Metals and Their Ions

Z	Alkaline Earth	Electronic Configuration	Ion	Electronic Configuration
4	Be	$(He)^2 2s^2$	Be^{2+}	$(He)^2$
12	Mg	$(Ne)^{10} 3s^2$	Mg^{2+}	$(Ne)^{10}$
20	Ca	$(Ar)^{18} 4s^2$	Ca^{2+}	$(Ar)^{18}$
38	Sr	$(Kr)^{36} 5s^2$	Sr^{2+}	$(Kr)^{36}$
56	Ba	$(Xe)^{54} 6s^2$	Ba^{2+}	$(Xe)^{54}$
88	Ra	$(Rn)^{86} 7s^2$	Ra^{2+}	$(Rn)^{86}$

All the alkaline earth metals react with water. The reaction of Sr is typical:

$$Sr(s) + 2H_2O(liq) \rightarrow Sr^{2+}(aq) + 2\,OH^-(aq) + H_2(g) \qquad (13\text{-}14)$$

These reactions are much less violent than the corresponding reaction of an alkali metal; indeed for beryllium and magnesium the reaction with water is slow. Alkaline earth metals become more reactive as the outer electrons are further removed from the nucleus. Thus barium is the most reactive member of the family, excluding radium, which is radioactive and used only for special purposes. Beryllium is less electropositive (that is, less likely to lose its two outer electrons) than the other alkaline earths because of its very small size. Its chemistry is quite different from all the others because it requires significantly more energy to remove the two outer electrons of Be than to remove the two outer electrons of the other alkaline earths. Many of the compounds of beryllium are not ionic.

Since the alkaline earth ions have rare gas configurations, solutions of these ions are colorless, and their salts are white (unless the anion happens to be colored). Salts of the alkaline earth metals are not as soluble as those of the alkali metals; fluorides, carbonates, sulfates, chromates, and oxalates of the alkaline earths are insoluble.

Hydroxides of the alkaline earths are only slightly soluble, and the solubility increases regularly from $Mg(OH)_2$ to $Ba(OH)_2$. Magnesium hydroxide, $Mg(OH)_2$, is only very slightly soluble; a saturated solution of $Ca(OH)_2$ at room temperature is 0.02 F, and a saturated solution of $Ba(OH)_2$ is 0.2 F.

More than 3% of the earth's crust is calcium, which is the fifth most abundant element in it. Calcium is never found uncombined and occurs widely in nature as $CaCO_3$. Calcium carbonate is the chief constituent of limestone, marble, chalk, coral reefs, and the shells of clams, oysters, and other mollusks. The Dolomite mountains are principally a carbonate of both magnesium and calcium, $MgCa(CO_3)_2$, and the white cliffs of Dover are chiefly limestone.

Fig. 13.20. The Dolomite mountains of Italy. The mineral dolomite is $CaMg(CO_3)_2$.

When $CaCO_3$ is heated to high temperatures it decomposes to form carbon dioxide and calcium oxide, commonly called **lime** or **quicklime**.

$$CaCO_3 \xrightarrow{\text{heat}} CO_2(g) + CaO(s) \tag{13-15}$$

Quicklime reacts readily with water to form the strong base $Ca(OH)_2$,

$$CaO(s) + H_2O(liq) \rightarrow Ca(OH)_2(s) \tag{13-16}$$

and this reaction is so exothermic that some of the water is converted to steam by the heat generated. The common name for $Ca(OH)_2$ is **slaked lime** (since it is formed when CaO has "slaked its thirst" for water). Because $CaCO_3$ is so abundant in nature, $Ca(OH)_2$ is the cheapest strong base, and is widely used in industry. When mixed with sand, calcium hydroxide hardens as mortar and cement, by absorbing CO_2 from the air:

$$Ca(OH)_2 + CO_2(g) \rightarrow CaCO_3 + H_2O \tag{13-17}$$

Strontium and calcium are very similar chemically. A radioactive isotope of strontium, ^{90}Sr, is a major product of the fission of uranium and plutonium. If tests of nuclear weapons are carried out in the atmosphere, ^{90}Sr is dispersed around the earth by the wind. If ^{90}Sr is ingested by a cow, it gets into her milk. The calcium and strontium in that milk then become incorporated into the bones and teeth of people who drink it. Calcium phosphate is an essential constituent of bones and teeth. The radiation emitted by ^{90}Sr damages bone marrow where red blood cells are made. Since the half-life of ^{90}Sr is 28 years, atmospheric testing of nuclear weapons poses a serious danger.

section 13.11
The Halogens: The Relationship between Their Chemistry and Their Electronic Configurations

The elements whose atomic numbers are one less than the rare gases constitute Group VIIA of the periodic table. They are called the **halogen family**: F, Cl, Br, I, and At. Atoms of these elements have seven valence electrons and their outer electronic configurations are $(ns)^2(np)^5$. All of these elements have a tendency to add one electron and thus attain a rare gas electronic configuration. The resulting uninegative ions are called **halide ions**. Electronic configurations of the halogens and their corresponding halide ions are listed in Table 13.15.

The halogens are extremely reactive and are not found uncombined in nature. Astatine is radioactive; its longest-lived isotope, ^{210}At, has a half-life of only 8.3 h, so that only minute quantities of astatine occur naturally. All the other halogens exist as diatomic molecules in their elemental form.

All halogen atoms have a single unpaired electron. If two Cl atoms, for example, share their unpaired electrons

$$:\ddot{C}l\cdot + \cdot\ddot{C}l: \rightarrow :\ddot{C}l:\ddot{C}l:$$

the product is the diatomic molecule, Cl_2, in which each Cl atom has, effectively, the electronic configuration of argon. The diatomic molecules are more stable, that is, lower in energy, than two separate halogen atoms and so the pure elements always occur as diatomic molecules. These diatomic molecules are themselves extremely reactive.

Fluorine, F_2, is a pale yellow gas that reacts with practically all organic and inorganic substances. It is therefore very difficult to handle, is corrosive, dangerous,

Table 13.15. **The Electronic Configurations of the Halogens and the Halide Ions**

Z	Halogen	Electronic Configuration	Halide Ion	Electronic Configuration
9	F	$(He)^2 2s^2 2p^5$	F^-	$(Ne)^{10}$
17	Cl	$(Ne)^{10} 3s^2 3p^5$	Cl^-	$(Ar)^{18}$
35	Br	$(Ar)^{18} 3d^{10} 4s^2 4p^5$	Br^-	$(Kr)^{36}$
53	I	$(Kr)^{36} 4d^{10} 5s^2 5p^5$	I^-	$(Xe)^{54}$
85	At	$(Xe)^{54} 4f^{14} 5d^{10} 6s^2 6p^5$	At^-	$(Rn)^{86}$

and must be used with extreme caution. Fluorine is the most electronegative of all the chemically reactive elements and has a great tendency to pull electrons away from other substances to form the fluoride ion. We have already mentioned that the ability of fluorine to attract the outer electrons of Kr, Xe, and Rn has led to the formation of rare gas fluorides.

Chlorine, a greenish-yellow gas, also reacts with nearly all other elements. While chlorine is not as electronegative as fluorine, it is still one of the most electronegative elements in the periodic table. Chlorine reacts explosively with metallic sodium to form sodium chloride, ordinary table salt, an ionic compound:

$$Na(s) + \tfrac{1}{2} Cl_2(g) \rightarrow Na^+ Cl^-(s) \qquad (13\text{-}18)$$

The bonding between the sodium and chloride ions is ionic; there are no shared pairs of electrons in the crystal. Each ion has a rare gas configuration. If we wish to use Lewis dot formulas to represent these ions we would write Na^+ and $:\!\overset{..}{\underset{..}{Cl}}\!:^-$ to indicate that $Na\cdot$ has lost its valence electron to become sodium ion, and $:\!\overset{..}{\underset{..}{Cl}}\!:$ has gained one electron to become chloride ion. More commonly, however, we do not bother to indicate a complete octet and write simply Cl^-.

Bromine, Br_2, is a very dense reddish-brown liquid that vaporizes readily to form a red-brown gas with a strong, disagreeable odor. (The name bromine comes from the Greek word for stench, *bromos*.) Although Br_2 is not as reactive as Cl_2 or F_2, it is still a highly reactive substance and combines with a great many elements and compounds. It must be handled with extreme care; it produces painful sores if it comes in contact with the skin and the vapor is extremely irritating to the eyes and throat.

As we go down the column of the halogen family, the outer electrons are further away from the attraction of the nucleus and both the electronegativity and the reactivity of the halogen decreases. Iodine, I_2, is the least active of the halogens, but it does form compounds with a great many elements. Because the outer electrons of an iodine atom are quite far from the nucleus, iodine exhibits some metallic-like properties and can acquire a net positive charge in a chemical bond. Iodine compounds are very important in medicine and in organic chemistry. The hormone L-thyroxine, which is produced by the thyroid gland, is 65% iodine by weight. A deficiency of iodine in the diet results in an enlargement of the thyroid gland, a condition known as goiter. To ensure a sufficient supply of iodine in our food, common table salt is "iodized," that is, a very small percentage of KI is added to the NaCl.

Since each halogen has a single unpaired electron, diatomic interhalogen compounds such as ICl, IBr, BrCl, BrF, and ClF are also formed. Lewis dot formulas showing the formation of a typical interhalogen compound are shown below:

$$:\!\overset{..}{\underset{..}{I}}\!\cdot \; + \; \cdot\overset{..}{\underset{..}{Cl}}\!: \; \rightarrow \; :\!\overset{..}{\underset{..}{I}}\!-\!\overset{..}{\underset{..}{Cl}}\!: \qquad (13\text{-}19)$$

The bond in ICl is a covalent bond with partial ionic character because the electronegativity of Cl is significantly larger than that of I. The pair of electrons shared to make the bond between I and Cl spends more time close to the Cl than to the I. Note that the compound is written ICl, indicating that I is the more metallic of the two atoms. Iodine monochloride, ICl, has a **dipole moment**; the I atom has a small net positive charge and the Cl atom has a small net negative charge.

The chemistry of the halogens is quite diverse because of their ability to form both ionic and covalent bonds and all gradations in between. All the elements in the periodic table with the exception of He, Ne, and Ar form halides. We will discuss halogen chemistry further in Chapters 14 and 15.

section 13.12
The Chalcogens (Oxygen Family): The Relationship between Their Chemistry and Their Electronic Configurations

The elements in the vertical column headed by oxygen all have outer electronic configuration $(ns)^2(np)^4$ and tend to gain two electrons to achieve a rare gas electronic configuration. The electronegativity of these elements decreases quite sharply as the atomic number increases, that is, as the size of the atom increases. Oxygen is highly electronegative and is an active nonmetal. Much of the chemistry of oxygen is unique and differs from that of the rest of the family. Sulfur and selenium are both nonmetals, but tellurium and polonium are **metalloids**, that is, elements that possess both metallic and nonmetallic properties. Electronic configurations of the chalcogens and their doubly negative ions are given in Table 13.16. Polonium is radioactive and is present in nature in very low abundance.

Ionic compounds are formed when these elements combine with the most electropositive of the elements, the alkali and alkaline earth metals. Thus the bonding in CaO, Rb_2S, or K_2Se is essentially ionic, but with less electropositive elements and with other nonmetals a great many compounds are formed in which the bonding is polar covalent or covalent. The chemistry of the members of the oxygen family will be discussed further in Chapters 14 and 15.

Isoelectronic Ions

As we discussed the chemistry of Groups IA, IIA, VIA, and VIIA, we have seen that these elements tend to form ions that have a rare gas configuration when they react. There are, therefore, a number of different ions that are **isoelectronic**, that is, ions that have the same electronic configuration. The nitride ion, N^{3-}, oxide ion, O^{2-}, fluoride ion, F^-, sodium ion, Na^+, magnesium ion, Mg^{2+}, and the aluminum ion, Al^{3+}, all have 10 electrons, and the same electronic configuration as the noble gas neon. For isoelectronic ions, the larger the nuclear charge, the stronger the attraction for the same number of electrons, and the smaller the ion. Thus, of the ions listed above, the ionic radius decreases regularly as the atomic number increases from 7 to 13. The

Table 13.16. Electronic Configurations of the Oxygen Family of Elements and Their Doubly Negative Ions

Z	Chalcogen	Electronic Configuration	Ion	Electronic Configuration
8	O	$(He)^2 2s^2 2p^4$	O^{2-}	$(Ne)^{10}$
16	S	$(Ne)^{10} 3s^2 3p^4$	S^{2-}	$(Ar)^{18}$
34	Se	$(Ar)^{18} 3d^{10} 4s^2 4p^4$	Se^{2-}	$(Kr)^{36}$
52	Te	$(Kr)^{36} 4d^{10} 5s^2 5p^4$	Te^{2-}	$(Xe)^{54}$
84	Po	$(Xe)^{54} 4f^{14} 5d^{10} 6s^2 6p^4$	Po^{2-}	$(Rn)^{86}$

anions are considerably larger than the cations, because the nuclear charge is less than 10 for N^{3-}, O^{2-}, and F^-, and there are 10 electrons, whereas for the cations the nuclear charge is greater than 10. The radii of the ions isoelectronic with neon are listed below, in angstrom units.

Ion	N^{3-}	O^{2-}	F^-	Na^+	Mg^{2+}	Al^{3+}
Radius (Å)	1.71	1.40	1.36	0.95	0.65	0.50

Summary

Each electron in an atom has a unique set of four quantum numbers: n, ℓ, m_ℓ, and m_s. The first three of these, n, ℓ, and m_ℓ, define an **atomic orbital**. The **principal quantum number**, n, may assume the values 1, 2, 3, ..., ∞. The larger the value of n, the larger the probability of finding the electron at greater distances from the nucleus, and the higher the energy of the electron. The **azimuthal quantum number**, ℓ, may assume the values 0, 1, ..., $(n-1)$ for a given value of n. The value of ℓ determines the shape or angular distribution of the electron cloud. If $\ell = 0$, the electron cloud is spherical and the orbital is called an s orbital. If $\ell = 1$, the electron cloud has two lobes, is shaped like a dumbbell, and is called a p orbital. For all atoms other than hydrogen, the energy of an atomic orbital depends on both n and ℓ. For a given n, the lower the value of ℓ, the lower the energy.

The *magnetic quantum number, m_ℓ,* can be 0, ± 1, ± 2, ..., $\pm \ell$, for a given value of ℓ. In the absence of an external magnetic or electric field, states with different values of m_ℓ, but the same value of n and ℓ are equal in energy and are said to be **degenerate**. There are $(2\ell + 1)$ values of m_ℓ for each value of ℓ, so that there are $(2\ell + 1)$ degenerate atomic orbitals for a given n and ℓ.

There are only two possible values of the **spin quantum number**, m_s, $+\frac{1}{2}$ and $-\frac{1}{2}$. Since each electron in the atom must have a unique set of the four quantum numbers, each atomic orbital can hold two electrons, one with $m_s = \frac{1}{2}$, and one with $m_s = -\frac{1}{2}$. Two electrons in the same orbital with opposite spins are said to be **paired**.

The electronic configurations of all the elements can be elucidated by applying three fundamental principles. This building-up of the periodic table one element at a time is known as the **Aufbau** process. The three principles are

1. **The Pauli Exclusion Principle:** No two electrons in one atom can have all four quantum numbers identical.
2. The atomic orbitals are filled so that the total energy of all the electrons is minimized. This means filling the orbitals in the order of increasing energy, taking into consideration electron–electron repulsions, which affect the energies of the orbitals.
3. **Hund's rule:** If more than one atomic orbital of the same energy is available, electrons will occupy different atomic orbitals with **parallel spin**, in the configuration of lowest energy.

The elements in which s and p orbitals are being filled are called **representative elements**. There are seven families of representative elements (Groups IA–VIIA) plus the **rare** or **noble gases** (Group 0), which have completely filled s and p subshells. The relationship between the chemistry of the representative elements and the electronic configurations of their **outermost** or **valence electrons** is described in this chapter.

The **octet rule**, proposed by G. N. Lewis, states that except for hydrogen, an atom will combine with other atoms to form bonds in order to have eight electrons in its valence shell, that is, to achieve the electronic configuration of one of the rare gases. Atoms can achieve a rare gas electronic configuration either by losing or gaining electrons, or by sharing them. The bond between a metal and a nonmetal in which the metal has lost one or more electrons and the nonmetal has gained one or more electrons, is an **ionic bond**. The bond between two atoms that are sharing a pair of electrons is a **covalent bond**. Most bonds are neither pure ionic nor pure covalent, but involve unequal sharing of a pair of electrons and are described as **polar covalent** or **covalent with partial ionic character**. A **Lewis dot symbol** or **formula** shows the electronic structure of atoms, molecules, or ions by using dots to represent valence electrons. Four pairs of dots surrounding the symbol of an element represent a rare gas configuration.

A horizontal row of the periodic table that terminates in one of the rare gases is called a **period**. There are seven periods of the elements that are now known. The first period consists of just 2 elements, hydrogen and helium, and corresponds to complete filling of the K shell, the $1s$ atomic orbital. The second and third periods consist of 8 elements each, and correspond to filling the L shell and the $3s$ and $3p$ orbitals, respectively. The fourth and fifth periods consist of 18 elements each. Each of these periods contains 10 **transition elements** (corresponding to filling the $3d$ and $4d$ orbitals, respectively), plus 8 representative elements. The sixth period consists of 32 elements: 14 **lanthanides**, 10 transition elements, and 8 representative elements. The seventh period begins with francium, element number 87, and would consist of 32 elements ending with $Z = 118$, but elements with atomic numbers above 109 have not yet been made. All elements of the seventh period are radioactive, and those for which $Z > 92$ are man-made.

Properties of atoms that depend on the electronic configuration of the atom are **periodic properties,** that is, they are properties that vary in a more or less regular way as we go across a period of the periodic table. Three important periodic properties are discussed in this chapter: the ionization energy, the electron affinity, and the electronegativity.

The **first ionization energy** (IE) is the energy required to remove one electron from a neutral, gaseous atom and produce a gaseous unipositive ion. All ionization energies are positive, and in general, the first ionization energy increases as we go across a period. There are some notable irregularities that can be explained with reference to the electronic configurations of the atoms. First ionization energies also decrease regularly as we go down a family of representative elements, and the electron being removed is further away from the nucleus.

The **electron affinity** (EA) is the amount of energy released when an electron is added to a neutral gaseous atom to form a gaseous uninegative ion. Most electron affinities are positive, but some are negative. A negative electron affinity means that energy must be absorbed when an electron is added. As a general trend, electron affinities increase as we go across a period from left to right, and decrease as we go down a family of representative elements and the electron being added is further away from the nucleus. Electron affinities are more difficult to measure than ionization energies.

The **electronegativity** is a measure of the ability of a bonded atom to pull electrons in the bond toward itself and away from the other atoms to which it is bonded. Because the electronegativity is a property of a bonded atom, it is not a constant for a given atom, but varies somewhat with bond type. Pauling's scale of electronegativities is most frequently used. The most electronegative of all atoms is F; its electronegativity is set at 4.0. Metals have electronegativities less than 2. The least electronegative elements are Cs and Fr, with electronegativity values of 0.7. Those elements classified as **metalloids** have electronegativities very close to 2. In general, electronegativities increase as we go across a period from left to right, and decrease as we go down a family of representative elements. The electronegativities of a series of transition metals are very similar and the electronegativities of all 14 lanthanides are virtually identical, with value 1.1.

Exercises

Section 13.1

1. State whether or not each of the following sets of quantum numbers describes an allowed state of an electron in an atom. Explain what is wrong with any set of values that does not describe an allowed state.
   (a) $n = 4$, $\ell = 4$, $m_\ell = 4$ (b) $n = 4$, $\ell = 3$, $m_\ell = 2$ (c) $n = 4$, $\ell = 3$, $m_\ell = -3$
   (d) $n = 4$, $\ell = 1$, $m_\ell = -2$ (e) $n = 4$, $\ell = 0$, $m_\ell = 0$ (f) $n = 4$, $\ell = -2$, $m_\ell = 0$

2. State whether or not each of the following symbols is an acceptable designation for an atomic orbital. Explain what is wrong with the unacceptable symbols. (a) $4g$ (b) $5g$ (c) $3f$ (d) $6d$

3. How many degenerate atomic orbitals are there that can be designated (a) $6p$ (b) $5d$ (c) $6f$? Give the values of n, ℓ, and m_ℓ for all the degenerate orbitals with each designation.

4. Calculate the total number of orbitals in the K, L, M, N, and O shells. Make a table of your results with the following headings:

   Shell   n   Total Number of Orbitals

   After considering your table, state the equation that relates the value of n and the total number of orbitals in each shell.

5. How many nodal spheres does a $5s$ electron charge cloud have?

6. What is the probability of finding a $4d$ electron right at the nucleus?

Section 13.2

7. Give explanations for your answers to the following questions. What is the maximum number of electrons that can be accommodated in (a) all the $5g$ orbitals, (b) all the $6f$ orbitals, and (c) all the $7s$ orbitals?

8. What is the maximum number of electrons that can be accommodated in the N shell, that is, in all the orbitals with $n = 4$?

Section 13.3

9. Write the complete ground state electronic configurations of the first three members of the nitrogen family: N ($Z = 7$), P ($Z = 15$), and As ($Z = 33$), showing all unpaired electrons. Underline the parts responsible for their similarity in chemical behavior.

10. Given below are several electronic configurations that may be correct for the oxygen atom. Indicate whether each of these represents the ground state, or an excited state, or is an impossible configuration for oxygen.

(a) ↑↓	↑↓	↓ ↑ ↓	↓	
(b) ↑↓	↑↓	↓ ↓ ↓↑	___	
(c) ↑↓	↑↓	↓ ↓ ↓↓	___	
(d) ↑↓	↑↓	↓ ↓ ↓	↓	
(e) ↑↓	↑↓	↓↑ ↑↓ ___	___	
(f) ↓↑	↑↓	↑ ↑↓↑	___	
1s	2s	2p	3s	

11. Draw an energy level diagram similar to that of Fig. 13.12 that shows the electronic configuration of (a) the ground state of O, (b) the ground state of F, and (c) the lowest excited state of Ne.

Section 13.5

12. Give the electronic configurations of the ground state of each of the following ions: (a) Sc^{3+} (b) Ti^{3+} (c) V^{3+} (d) Cr^{3+}

13. How many unpaired electrons are there in the ground state of each of the following ions? (a) Ca^{2+} (b) Ti^{2+} (c) Mn^{2+} (d) Cu^{2+} (e) Zn^{2+}

14. How many unpaired electrons are there in the ground state of each of the following gaseous atoms? (a) Ca   (b) Co   (c) Zn   (d) Ga

15. Iron forms two ions, Fe^{2+} and Fe^{3+}. Give the electronic configuration of each of these ions. Which would you expect to be more chemically reactive, Fe^{2+} or Fe^{3+}? Explain your answer.

16. Element number 117 has not yet been made. If it is ever made, to what group of the periodic table will it belong?

17. On the basis of the electronic configurations of the atoms, predict formulas for the following compounds: (a) potassium nitride   (b) aluminum sulfide   (c) strontium iodide   (d) titanium oxide   (e) zinc bromide

Section 13.6

18. The first ionization energies of the third period elements are

Element	Na	Mg	Al	Si	P	S	Cl	Ar
IE (eV)	5.14	7.64	6.00	8.15	11.0	10.36	13.01	15.80

(a) Plot the ionization energies of these elements versus Z, the atomic number. (b) As a general trend, the IE increases as Z increases within one period. Explain the reason for this general trend and account in detail for the irregularities in your plot.

19. Explain why the first ionization energy of In (5.78 eV) is much lower than that of Cd (8.99 eV), even though the nuclear charge of In is larger than that of Cd.

Section 13.7

20. Describe and account for the general trend of electron affinities across one period, using the data below. Also explain the low value for the electron affinity of P.

	Si (Z = 14)	P (Z = 15)	S (Z = 16)	Cl (Z = 17)
EA (eV/atom)	1.39	0.77	2.08	3.61

Section 13.8

21. Arrange the following atoms in order of *decreasing* electronegativity: As, F, S, Y, and Zn. Explain the reason for the order you chose.

22. Arrange each of the following groups of molecules in order of *increasing* dipole moment.

(a) HCl, HF, HI, HBr
(b) BrCl, ClF, NO, HCl, LiI

Section 13.9

23. Classify the bonds in each of the following compounds as either mainly ionic, mainly covalent, or polar covalent. Explain your answers. (a) NCl_3 (b) YCl_3 (c) CI_4 (d) BaO   (e) $BeBr_2$ (f) FeS

24. Arrange the following compounds in order of *increasing* ionic character in the bonds: $AsCl_3$, $BrCl$, $CaCl_2$, $GaCl_3$, $GeCl_4$, KCl

25. Draw Lewis dot formulas for the following compounds, which obey the octet rule:
(a) HF (b) AsI_3 (c) Cl_2O

26. Explain why the chemistry of beryllium differs significantly from that of all the other members of the alkaline earth family.

Sections 13.10, 13.11, and 13.12

27. The species Ar, K^+, Ca^{2+}, S^{2-}, and Cl^- are **isoelectronic**, that is, they all have the same electronic configuration. Arrange these five species in order of increasing radius, and explain the reason for the order you have chosen.

28. To what group of the periodic table does an element belong if the electronic configuration of its valence electrons is $(ns)^2(np)^2$?

29. What neutral atom has the same electronic configuration that iodine has in CsI?

30. Without looking at any table, write the ground state electronic configurations for the following atoms and indicate the number of unpaired electrons for each of these configurations.
(a) Si $(Z = 14)$ (b) Cl $(Z = 17)$ (c) Sr $(Z = 38)$ (d) Cs $(Z = 55)$

31. (a) Write a balanced net ionic equation for the reaction between $K(s)$ and H_2O.
(b) Which is more exothermic, the reaction between K and H_2O, or the reaction between Li and H_2O? Explain your answer in terms of the electronic configurations of Li and K.

32. Write balanced equations for the reactions of Cl_2 with (a) $K(s)$ and (b) $Sr(s)$. Which of these two reactions is more exothermic? Explain your answer in terms of the electronic configurations of K and Sr.

33. Magnesium combines with nitrogen gas to form solid magnesium nitride, a compound in which the bonding can be described as ionic with partial covalent character. Write a balanced equation for this reaction and give the electronic configurations of the ions in magnesium nitride.

Multiple Choice Questions

1. The following elements are in the fourth period of the periodic table.

$$Ca, V, Co, Zn, As$$

Of those listed the ones that have unpaired electrons in the ground state electronic configuration are
(a) Ca, V, and Co (b) V, Co, and Zn (c) Ca, Zn, and As
(d) V, Co, and As (e) Zn and As

2. Which of the series of elements listed below have most nearly the same atomic radius?
(a) Ne, Ar, Kr, Xe (b) Mg, Ca, Sr, Ba (c) B, C, N, O (d) Ga, Ge, As, Se
(e) Cr, Mn, Fe, Co

3. An element, X, has the electronic configuration $1s^2 2s^2 2p^6 3s^2 3p^3$. The formula of the most probable compound this element will form with calcium, Ca, is
(a) CaX (b) Ca_2X (c) CaX_2 (d) Ca_2X_3 (e) Ca_3X_2

4. Which of the following atoms has the largest number of unpaired electrons in its ground state configuration?
(a) Ag (b) Cd (c) Sn (d) Mo (e) Co

5. The first ionization energy of argon, Ar, is less than that of neon, Ne. An explanation of this fact is that

(a) The effective nuclear charge experienced by a valence electron in Ar is much larger than in Ne.

(b) The effective nuclear charge experienced by a valence electron in Ar is much smaller than in Ne.

(c) The atomic radius of Ar is larger than that of Ne.

(d) The atomic radius of Ar is smaller than that of Ne.

(e) The number of protons in the Ar nucleus is larger than the number of protons in the Ne nucleus.

6. The ground state electronic configuration of Fe^{3+} is

(a) $(Ar)^{18}3d^34s^2$ (b) $(Ar)^{18}3d^64s^2$ (c) $(Ar)^{18}3d^5$ (d) $(Ar)^{18}3d^44s$

(e) $(Ar)^{18}3d^6$

7. Which of the following sets of the four quantum numbers n, ℓ, m_ℓ, and m_s describes one of the outermost electrons in a ground state strontium atom?

(a) $5, 1, 1, \frac{1}{2}$ (b) $5, 0, 0, -\frac{1}{2}$ (c) $5, 0, 1, \frac{1}{2}$ (d) $5, 1, 0, \frac{1}{2}$ (e) $5, 2, 1, -\frac{1}{2}$

8. Of the following, the atom with the largest first ionization energy is the one with electronic configuration

(a) $(Ne)3s^23p^2$ (b) $(Ne)3s^23p^3$ (c) $(Ne)3s^23p^4$ (d) $1s^22s^22p^3$

(e) $(Ar)3d^{10}4s^24p^3$

9. Which of the following has the highest percentage of ionic character in its bonding?

(a) LiI   (b) $MgCl_2$ (c) CsF   (d) CsI   (e) $AlCl_3$

10. The Pauli Exclusion Principle requires that

(a) If the position of an electron is known exactly, its velocity cannot be simultaneously known precisely.

(b) Electrons in degenerate atomic orbitals have parallel spin.

(c) A particle of mass m moving with velocity v has a wavelength $\lambda = h/mv$.

(d) The velocity of all electromagnetic radiation equals the speed of light.

(e) No two electrons in the same atom can have the same set of four quantum numbers n, ℓ, m_ℓ, and m_s.

11. A representative element that is a metal is

(a) Zn   (b) Fe   (c) Sr   (d) Se   (e) Cr

12. The correct order for *decreasing* atomic radius of the five atoms H, Br, I, He, Cl is

(a) Br > I > He > Cl > H   (b) I > Br > Cl > He > H

(c) Cl > Br > I > H > He   (d) I > Br > Cl > H > He

(e) I > Br > He > Cl > H

13. Which of the following elements can form a 3+ ion with noble gas electronic configuration?

(a) Sr   (b) Y   (c) Ru   (d) In   (e) Sb

14. Which of the following ions has the smallest radius?

(a) Be^{2+} (b) Li^+ (c) N^{3-} (d) O^{2-} (e) F^-

15. Which of the following ions has the largest radius?

(a) Be^{2+} (b) Li^+ (c) N^{3-} (d) O^{2-} (e) F^-

16. In which of the following sets do all three compounds have bonds that are mainly ionic?

(a) $NaCl$, NCl_3, CCl_4 (b) CsBr, $BaBr_2$, SrO   (c) CsF, BF_3, NH_3

(d) Al_2O_3, CaO, SO_2 (e) RbI, ICl, HCl

17. Which of the ions listed below are isoelectronic with krypton?

$$Ag^+, Br^-, Cd^{2+}, Sc^{3+}, Se^{2-}, Sr^{2+}, Ti^{2+}, Zn^{2+}$$

(a) $Cd^{2+}, Sr^{2+}, Zn^{2+}$ (b) Br^-, Se^{2-}, Sr^{2+} (c) $Sc^{3+}, Se^{2-}, Zn^{2+}$
(d) Ag^+, Br^-, Cd^{2+} (e) $Cd^{2+}, Sc^{3+}, Sr^{2+}$

18.
$$I. \; n = 3, \ell = 2, m_\ell = -2$$
$$II. \; n = 3, \ell = 1, m_\ell = 0$$
$$III. \; n = 3, \ell = 0, m_\ell = -1$$
$$IV. \; n = 3, \ell = 2, m_\ell = 0$$
$$V. \; n = 3, \ell = 3, m_\ell = -2$$

Of the quantum state designations listed above, which does not describe an allowed state for an electron in an atom?

(a) I and V   (b) II and III   (c) III and IV   (d) I and IV   (e) III and V

19. Which of the following salts is not soluble in water?

(a) K_2CO_3 (b) Rb_2SO_4 (c) $BaCO_3$ (d) $SrCl_2$ (e) $NaIO_3$

20. The element with $Z = 110$ has not yet been made. It should be

(a) a halogen   (b) an actinide   (c) a chalcogen   (d) a rare gas
(e) a transition metal

21. The atom with lowest atomic number and a completely filled $3d$ subshell in the ground state is

(a) Cu   (b) Ar   (c) Cr   (d) Zn   (e) Kr

22. An ion with five unpaired electrons in its ground state is

(a) Cr^{3+} (b) Fe^{3+} (c) Mn^{3+} (d) Ni^{2+} (e) Cu^{2+}

23. The atom with lowest atomic number that has a ground state electronic configuration of $(n - 1)d^6ns^2$ is in the

(a) second period   (b) third period   (c) fourth period   (d) fifth period
(e) sixth period

24. When arranged in order of increasing atomic number, the elements exhibit periodicity for all of the following properties EXCEPT

(a) atomic radii   (b) atomic weights   (c) ionization energy
(d) boiling point   (e) electronegativity

25. The order of increasing ionization energy for the atoms neon, nitrogen, phosphorus, and sodium is

(a) $Na < P < N < Ne$ (b) $N < Ne < Na < P$ (c) $N < Na < Ne < P$
(d) $Na < N < P < Ne$ (e) $N < Na < P < Ne$

26. Which of the equations given below is the one whose ΔH value is the first ionization energy of Ba?

(a) $Ba(s) \rightarrow Ba^+(g) + e^-$
(b) $Ba(g) \rightarrow Ba^+(g) + e^-$
(c) $Ba(g) \rightarrow Ba^{2+}(g) + 2e^-$
(d) $Ba(s) + e^- \rightarrow Ba^-(g)$
(e) $Ba(g) + e^- \rightarrow Ba^-(g)$

Problems

13.1. What is the theoretical maximum number of electrons that can be accommodated in the K, L, M, N, and O shells? Tabulate your results. (Refer to Exercise 4.) State the equation that relates the value of n and the maximum possible number of electrons that can occupy each shell.

13.2. From the list below select (a) the ions that are isoelectronic with Ar and (b) the ions that do not have rare gas electronic configuration. Ag^+, Br^-, Ca^{2+}, Cd^{2+}, Fe^{3+}, I^-, Rb^+, S^{2-}, Sc^{3+}, Se^{2-}, Ti^{2+}, Zn^{2+}

13.3. How many nodal planes does the (a) d_{xz} and (b) $d_{x^2-y^2}$ atomic orbital have? Define the nodal planes of each of these orbitals in terms of the Cartesian coordinate system.

13.4. If CO_2 is bubbled into a saturated aqueous solution of calcium hydroxide, a fine white precipitate forms. Write a balanced net ionic equation for the reaction that occurs.

13.5. The energy required to remove one valence electron from a Sr atom is 5.69 eV; removal of a second electron requires almost twice as much energy, 10.98 eV. In comparison, although the first ionization energy of Rb is 4.18 eV, removal of a second electron requires almost seven times as much energy, 27.36 eV. Explain these facts on the basis of the electronic configurations of Sr and Rb.

13.6. (a) Explain why As ($Z = 33$) has a higher first ionization energy than Se ($Z = 34$), even though generally the first IE increases as Z increases across a period. (b) Explain why the first ionization energy of He (24.58 eV) is almost twice as large as the ionization energy of H (13.595 eV).

13.7. Examine the electronic configurations of elements 61 through 65 and account for the irregularity in the electronic configuration of gadolinium, element number 64.

13.8. The ionic radii of a series of ions are given below:

Ions	As^{3-}	Se^{2-}	Br^-	Rb^+	Sr^{2+}	Y^{3+}
Radius (Å)	2.22	1.98	1.95	1.48	1.13	0.93

Account for this trend in the ionic radii.

13.9. From the list below select (a) the ions that have rare gas configuration, (b) the ions that do not have rare gas configuration but have completely filled subshells, and (c) the ions that are colored (that is, those that absorb light in the visible region).

$$Ag^+, \quad Ba^{2+}, \quad Cd^{2+}, \quad Cr^{3+}, \quad Cu^{2+}, \quad Fe^{3+}, \quad I^-,$$
$$Mn^{2+}, \quad Ni^{2+}, \quad Pb^{2+}, \quad S^{2-}, \quad Ti^{3+}, \quad Y^{3+}, \quad Zn^{2+}.$$

13.10. For some elements the electron affinity has not been measured, but a value has been calculated using quantum mechanics. The calculated EA for Li is 0.54 eV, for Be it is -0.6 eV. Explain why these two values are both small, but differ in sign.

13.11. Given below are several electronic configurations that may be correct for the valence electrons of Cr. Indicate whether each of these represents the ground state, or an excited state, or is an impossible configuration for Cr. Note that the inner 18 electrons are in the argon configuration and are not shown.

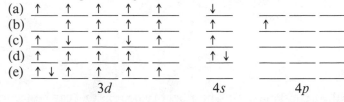

13.12. Give explanations for your answers to the following questions.

(a) Of the three ions listed, which has the largest ionic radius and which the smallest? Sc^{3+}, Cl^-, Ca^{2+}

(b) Of the three atoms listed, which has the largest first ionization energy and which the smallest? Al, Ar, Mg

(c) Of the three atoms listed, which has the largest *second* ionization energy and which the smallest? Na, Mg, Al

13.13. Indium ($Z = 49$) forms three chlorides and three sulfides.

InCl	mp 225±1 °C	In$_2$S mp 653±5 °C

$$
\begin{array}{llll}
InCl & mp\ 225\pm1\ °C & In_2S & mp\ 653\pm5\ °C \\
InCl_2 & mp\ 235\ °C & InS & mp\ 692\pm5\ °C \\
InCl_3 & mp\ 586\ °C & In_2S_3 & mp\ 1050\ °C
\end{array}
$$

Account for the formation of three compounds with each of these nonmetals by considering the electronic configuration of In. Explain the trend in the observed melting points.

13.14. The first ionization energies of the elements of Group VA are

Element	N	P	As	Sb	Bi
IE (eV/atom)	14.5	11.0	9.80	8.64	7.29
Atomic number Z	7	15	33	51	83

Plot the first ionization energy versus Z for these five elements. Explain the decrease in first IE with increasing Z. Explain why the decrease in the IE from N to P is much steeper than the decrease in IE with increasing Z for the rest of the plot.

13.15. Which of the following compounds would you expect to be colored (not white)? Explain your answers. $AlCl_3$, SrO, $Cu(NO_3)_2$, Tl_2S, $RaBr_2$, Sc_2O_3, HgS, $CdCl_2$, RhF_3, $PdBr_2$

13.16. The *second* ionization energy, $(IE)_2$, of the elements of the second period are given below:

Element	Li	Be	B	C	N	O	F	Ne
$(IE)_2$ (eV)	75.6	18.2	25.2	24.4	29.6	35.2	35.0	41.1

(a) Explain precisely

1. Why the $(IE)_2$ for Li is so much higher than for all other elements of this period.
2. Why the general trend from Be to Ne is of increasing $(IE)_2$.
3. Why the $(IE)_2$ of C is slightly less than the $(IE)_2$ of B.
4. Why the $(IE)_2$ of F is just about the same or very slightly less than that of O.

(b) Would you predict that the $(IE)_2$ of Na is (1) greater than 75.6 eV, (2) between 41.1 and 75.6 eV, or (3) less than 41.1 eV? Explain your answer.

chapter 14 The Nature of the Chemical Bond

Gilbert Newton Lewis (1875–1946) was born in Weymouth, Massachusetts and received his B.A., M.A., and Ph.D. degrees from Harvard. From 1905 to 1911 he did research and taught at the Massachusetts Institute of Technology. In 1912 he became Professor of Physical Chemistry at the University of California at Berkeley, where he remained until his death. Lewis' theoretical work on the significance of the electron pair in molecular bonding laid the foundation for a revised theory of chemical valence. In 1923 he published "Valence and the Structure of Atoms and Molecules" which greatly influenced the training of chemists in the twentieth century. In the same year he coauthored, with Merle Randall, "Thermodynamics and the Free Energy of Chemical Substances" which was the outstanding text on thermodynamics for more than a generation.

Table 14.1. The Nature and Properties of Interatomic Bonds

Bond Type	Nature of the Atoms Forming the Bond	Nature of the Bond	Properties Associated with the Bond	Examples, Comments
Ionic bond	Formed between a metallic element and a nonmetallic element whose electronegativities differ widely. While there is no hard and fast rule, usually the difference in electronegativity is > 1.8.	There is an actual transfer of an electron or electrons from the metal to the nonmetal. Electrostatic attraction holds the ions together.	Compounds with ionic bonds are crystalline solids with high mp and bp. When dissolved in a solvent, melted, or in the gas phase, the ions conduct electricity.	SrS, K_2O, NaCl, KH, BaO, CsF, and $Ca(NO_3)_2$. No individual molecules exist.
Polar covalent or covalent with partial ionic character	Formed between elements whose electronegativities differ somewhat, but not very widely. An approximate rule of thumb is that the electronegativity difference is < 1.8 but > 0.3.	There is an unequal sharing of electron pairs. The *bond* has a negative end and a positive end, e.g., $$H\!:\!\ddot{\underset{\cdot\cdot}{Cl}}\!:$$ The Cl end is more negative than the H end,	Compounds with polar covalent bonds have lower mp and bp than ionic compounds, but higher than nonpolar compounds of about the same molecular weight. The pure compounds do not conduct electricity, whether solid, liquid, or gas. Where the electronegativity difference is small (<0.5), bonds are almost pure covalent and properties reflect this.	HBr, H_2O, CH_3CH_2OH, NH_3, CO_2, and $CHCl_3$. Most bonds are of this type. While the individual *bonds* in a polyatomic molecule may be polar, the entire molecule may be nonpolar due to symmetry, e.g., CO_2 and CCl_4.
Pure covalent or nonpolar covalent	Formed between nonmetallic elements with the same electronegativities, or electronegativities that differ by a few tenths at most.	There is an equal sharing of electron pairs. The bond is described either by using a dot formula $$:\!\ddot{\underset{\cdot\cdot}{F}}\!:\!\ddot{\underset{\cdot\cdot}{F}}\!:$$ or as F—F.	Many compounds with pure covalent bonds are gases at room temperature. Covalent solids are generally poor conductors of heat and electricity. The liquid and gas phases do not conduct electricity.	H_2, Cl_2, N_2, P_4, O_2, F_2, NCl_3, and PH_3. Individual molecules exist.

Why do isolated atoms combine to form molecules or ions? Because the aggregates have a lower energy than the separate atoms. Particles combine to form a more stable polyatomic species. When the aggregate formed is lower in energy than the separate particles by about 40 kJ/mol ($\sim$10 kcal/mol) or more, we say that a chemical bond has been formed. Chemical bonds vary from ionic to pure covalent, with every possible gradation in between. The great majority of bonds are neither pure covalent nor ionic, but fall into the category of covalent bonds with partial ionic character. Table 14.1 summarizes information about interatomic bonds that has been discussed previously.

Note the distinction that is made between **intermolecular forces** and **interatomic bonds**. Intermolecular forces are weaker than chemical bonds; they are the forces of attraction between molecules that keep them close together in the liquid and solid states, and have been discussed in Chapter 5. Even hydrogen bonds, which are much stronger than van der Waals' forces, are only about one tenth as strong as most covalent bonds between atoms.

Theories of chemical bonding help us to understand both the geometry of molecules and their chemical reactivity. In this chapter we will discuss several theoretical approaches that serve to elucidate the nature of the chemical bond.

section 14.1
Ionic versus Covalent Bonding

Ionic Bonding

Ionic bonds produce aggregates that are lower in energy than isolated atoms because of the coulombic or electrostatic attraction between particles of opposite charge. Let us consider the case of sodium chloride as an example. We can easily calculate the difference in energy for the following gas-phase process

$$Na(g) + Cl(g) \rightarrow Na^+(g) + Cl^-(g) \tag{14-1}$$

because it is the difference between the first ionization energy of sodium and the electron affinity of chlorine.

$$Na(g) \rightarrow Na^+(g) + e^- \qquad \Delta H_a = IE(Na) = 496 \text{ kJ} \tag{14-1a}$$

$$Cl(g) + e^- \rightarrow Cl^-(g) \qquad \Delta H_b = -EA(Cl) = -348 \text{ kJ} \tag{14-1b}$$

so that ΔH for the overall process, the sum of these two reactions, is $496 - 348 = +148$ kJ per mole of Na and Cl atoms.

When thinking about the calculation above, remember that the electron affinity of Cl is defined as the energy *released* when reaction (14-1b) occurs, whereas ΔH is defined as the energy absorbed when a reaction occurs. Thus ΔH_b is the negative of the electron affinity of Cl. Because ΔH for the overall reaction (14-1) is positive, the reaction is endothermic. This means that a mole of isolated gaseous sodium atoms plus a mole of isolated gaseous chlorine atoms is 148 kJ lower in energy than a mole of isolated gaseous sodium ions plus a mole of isolated gaseous chloride ions. If we examine the vapor phase when gaseous sodium atoms and chlorine atoms are mixed at high temperature, however, we find that it contains neither isolated atoms nor ions, but consists of ion-pairs and aggregates of ion-pairs (dimers, trimers, and so on) in proportions that depend on the temperature. An ion-pair monomer consists of a sodium ion and a chloride ion with an internuclear separation of 2.51 Å (0.251 nm). When a sodium ion and a chloride ion come close together, energy is released due to the coulombic attraction between the opposite charges.

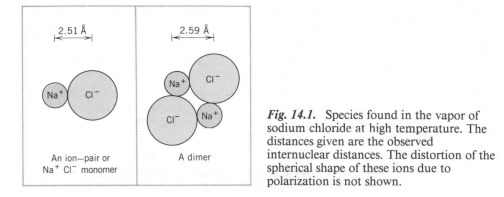

Fig. 14.1. Species found in the vapor of sodium chloride at high temperature. The distances given are the observed internuclear distances. The distortion of the spherical shape of these ions due to polarization is not shown.

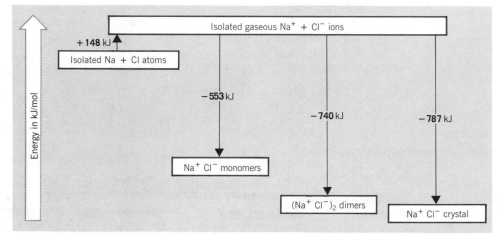

Fig. 14.2. Relative energies of various combinations of sodium and chlorine per mole of each atom.

Coulomb's law was discussed in Chapter 7, and the electrostatic *force* between two charged particles is given by Eq. (7-3). The electrostatic potential energy is defined as

$$\text{coulombic energy} = kq_1q_2/Dr \qquad (14\text{-}2)$$

where q_1 and q_2 are the charges, r is the distance between them, D is the dielectric constant of the medium between the charges, and k is a numerical constant whose value depends on the units used for q and r. For a sodium ion and a chloride ion in a vacuum (for which the dielectric constant, D, is 1.0), the coulombic energy is

$$-(8.988 \times 10^9 \text{ J} \cdot \text{m/C}^2) \frac{(1.602 \times 10^{-19} \text{ C})^2}{(2.51 \times 10^{-10} \text{ m})} = -9.19 \times 10^{-19} \text{ J}$$

as the magnitude of the charge on each of these ions is the electronic charge, 1.602×10^{-19} C. The coulombic energy is negative because the ions are oppositely charged. Energy is released as separate sodium ions and chloride ions come together until the distance separating them is only 2.51 Å, or 2.51×10^{-10} m. The formation of a mole of such ion-pair monomers releases

$$(9.19 \times 10^{-19} \text{ J/ion-pair})(6.022 \times 10^{23} \text{ ion-pair/mol})(10^{-3} \text{ kJ/J}) = 553 \text{ kJ/mol}$$

Thus a mole of ion-pair monomers is 553 kJ lower in energy than the separate gaseous ions. Further association to dimers lowers the energy even more. Figure 14.1 illustrates sodium chloride monomers and dimers in the vapor phase. At 750 °C, for instance, the vapor phase is roughly 65% monomers and 35% dimers.

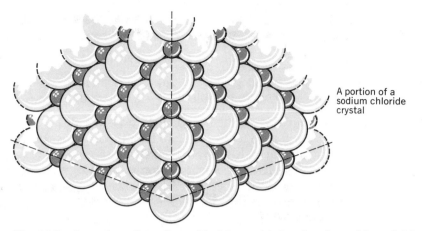

A portion of a sodium chloride crystal

Fig. 14.3. A portion of a sodium chloride crystal showing the packing of chloride ions (large spheres) and sodium ions (small spheres).

If the vapor is cooled, larger and larger aggregates are formed. At room temperature, the ionic crystal is the stable structure. Figure 14.2 diagrams the relative energies of various configurations and Fig. 14.3 shows the structure of crystalline sodium chloride.

Covalent bonds also result in a molecule lower in energy than the isolated atoms, but since in a pure covalent bond there is equal sharing of electrons and a symmetric charge distribution, a simple Coulomb's law calculation cannot explain the energy lowering as it does for ionic bonding. In order to understand many properties of molecules, such as their geometries and the strengths of the chemical bonds that hold the molecule together, we must investigate the nature of the covalent bond.

section 14.2
Lewis Structures

In the previous chapter we discussed briefly the proposal of G. N. Lewis that atoms can share electrons in order to achieve the electronic configuration of one of the rare gases. A shared pair of electrons is called a **single covalent bond**.

Multiple Bonds

For a number of compounds, an acceptable Lewis structure can be drawn only if we picture atoms as sharing more than one pair of electrons. Consider CO_2 as an example. Carbon has 4 valence electrons [configuration $(He)2s^22p^2$] and each oxygen has 6 valence electrons [configuration $(He)2s^22p^4$]. There are, therefore, a total of 16 electrons to form octets about each of the three atoms, and the only way this can be done is by having each atom share two pairs of electrons, as represented by

$$:\ddot{O}::C::\ddot{O}: \quad \text{or} \quad :\ddot{O}=C=\ddot{O}:$$

When atoms share two pairs of electrons we say there is a **double bond** between them. (If one is interested only in indicating the nature of the bonding, lone pairs may be omitted. The structure of CO_2 is then written simply as O=C=O.)

In order for each nitrogen atom in N_2 to attain a noble gas configuration using the

10 valence electrons from the two N atoms, three pairs of electrons must be shared between them, which constitutes a triple bond. The Lewis structure for N_2 is

$$:N:::N: \quad \text{or} \quad :N\equiv N:$$

How can you tell whether or not multiple bonds are used when drawing a Lewis structure? There are a few simple rules that will help you to figure this out.

1. Add up the number of valence shell electrons for all the atoms in the compound. You will need to use Table 13.8 for those atoms whose configurations you are not familiar with, but you should know the number of valence electrons for all the representative elements. If you are drawing a Lewis structure for an anion, add one more electron for each negative charge on the anion. If you are drawing the Lewis structure for a cation, subtract one electron for each positive charge on the cation.

2. Draw a single bond between each pair of atoms that is connected by a bond. You will need to have information about which atoms are bonded together before you can do this. A formula such as C_3H_6O is not sufficient. Some general guidelines are the following: In small molecules there is usually one central atom to which other atoms are bonded. Generally, the least electronegative atom is the central atom, except that hydrogen is never the central atom. Oxygen atoms do not bond to other oxygen atoms, except in peroxides (which contain the O_2^{2-} ion) and in superoxides (which contain the O_2^- ion). Molecules or ions with two central atoms, such as C_2H_4, N_2O_5, and $Cr_2O_7^{2-}$, are usually symmetrical. In forming octets, remember the dot symbols for the representative elements given in Section 13.9. The halogens need only one electron to achieve rare gas electronic configuration, and are therefore usually single bonded to the central atom, while the Group VIA elements need two electrons to achieve the electronic configuration of a rare gas and are frequently double bonded to the central atom.

3. After you have drawn single bonds, use the remaining electrons to put unshared (lone) pairs on each atom to satisfy the octet rule. When you have used up all the valence electrons in this way, if every atom has a complete octet, no multiple bonds are necessary. However, if you cannot satisfy the octet rule using only single bonds and lone pairs, use double or triple bonds, which will necessarily reduce the number of unshared electron pairs.

The following examples illustrate the use of these rules.

EXAMPLE 14.1. Lewis structures

Draw Lewis structures for the following substances: **(a)** methanol, CH_3OH, **(b)** chlorate ion, ClO_3^-, **(c)** N_2O (which has a nitrogen atom as the central atom), **(d)** phosgene, $COCl_2$ (both O and Cl are bonded to C), and **(e)** CO.

Solution

(a) The total number of valence electrons for CH_3OH is 14 (4 from C, 6 from O, and 1 from each of the four H atoms). The structure is

$$
\begin{array}{c}
H \\
| \\
H-C-\overset{..}{O}-H \\
| \\
H
\end{array}
$$

(b) There are 26 valence electrons for the ClO_3^- ion (7 from Cl, 6 from each of three O atoms, and 1 from the single negative charge on the ion). The Lewis structure is

$$\left(\begin{array}{c} \ddot{:O}: \\ | \\ :\ddot{O}-Cl-\ddot{O}: \\ \ddot{\;}\;\;\;\;\;\;\ddot{\;} \end{array} \right)^{-}$$

(c) There are 16 valence electrons for N_2O. The Lewis structure is $:\ddot{N}=N=\ddot{O}:$ Note that this is isoelectronic with CO_2. A second possible Lewis structure for N_2O is

$$:N:::N:\ddot{O}: \quad \text{or} \quad :N\equiv N-\ddot{O}:$$

(d) In $COCl_2$ there are 24 valence electrons. The central atom is C, the least electronegative of these three atoms. The Lewis structure is

$$\begin{array}{c} :O: \\ \| \\ :\ddot{Cl}-C-\ddot{Cl}: \\ \ddot{\;}\;\;\;\;\;\;\ddot{\;} \end{array}$$

(e) There are 10 valence electrons in CO, and the structure is isoelectronic with that of N_2, $:C\equiv O:$.

Other than satisfying the octet rule, is there any evidence for the existence of double and triple bonds? As a general rule, the larger the number of bonds we must draw between two atoms, the shorter the **bond length** (the distance between the nuclei of bonded atoms) and the larger the **bond energy** (the amount of energy that must be expended to break the bond). The data given in Table 14.2 on the carbon–oxygen bond distances and bond energies in three different compounds illustrate this point.

It is not difficult to understand why multiple bonds are shorter and stronger than single bonds. Nuclei of two different atoms repel one another as they are both positively charged. But each nucleus is attracted to the electron charge cloud of the electrons shared between them. The more electron density between the two nuclei, the greater this attraction, and the closer the nuclei can approach each other.

Formal Charges

The Lewis structure for CO, $:C:::O:$, raises an interesting problem. In assigning charges on atoms in molecules, it is customary to assume that each shared pair is equally shared by the atoms. Thus the oxygen atom in CO is assigned three of the electrons in the triple bond, plus the lone pair, for a total of five valence electrons. But an isolated oxygen atom has six valence electrons, so that in the structure $:C:::O:$, oxygen has a **formal charge** of $+1$. Similarly, the carbon atom is assigned five valence electrons in the structure $:C:::O:$, whereas an isolated carbon atom has only four valence electrons, so that carbon is assigned a formal charge of -1. To indicate the formal charges, we write the Lewis structure as $:\overset{\ominus}{C}\equiv\overset{\oplus}{O}:$.

Table 14.2. The Effect of Multiple Bonds on Bond Length and Bond Energy

Molecule	H_3C-O-H	$O=C=O$	$C\equiv O$
Type of carbon–oxygen bond	Single	Double	Triple
Carbon–oxygen bond length (Å)	1.43	1.16	1.13
Carbon–oxygen bond energy (kJ/mol)	358	803	1070

What is the physical significance of these formal charges? The assumption that each shared pair of electrons is shared *equally* by the two atoms is erroneous, so that the formal charges do *not* represent actual charges. You must not interpret the formal charges in CO to mean that CO is ionic or that there is anything like a full $+1$ charge on oxygen or a -1 charge on carbon. The electrons in the bond are *not* equally shared, but spend more time around oxygen, the more electronegative of the two atoms. Evidence for this is that the observed dipole moment of CO is very small, 0.1 D. The polarity of the dipole, however, is $:\overset{\ominus}{C}\equiv\overset{\oplus}{O}:$, as predicted by the formal charges.

The formal charge on the atoms in the Lewis structure for most molecules is zero. If at all possible we try to avoid Lewis structures with large formal charges on atoms, particularly on adjacent atoms. A Lewis structure that involves large formal charges in a covalently bonded molecule is probably not an adequate representation of the bonding in the molecule, but should be considered as a convenient symbol only. The sum of the formal charges on all the atoms must be zero for a molecule and equal to the charge on any ionic species.

EXAMPLE 14.2. Formal charge calculations

Calculate the formal charges on the atoms in the Lewis structure of **(a)** $COCl_2$, **(b)** CO_2, **(c)** N_2O, and **(d)** CrO_4^{2-}.

Solution

(a) The structure of $COCl_2$ is

$$:\underset{\cdot\cdot}{\overset{\cdot\cdot}{Cl}}-\overset{\overset{\textstyle :O:}{\|}}{C}-\underset{\cdot\cdot}{\overset{\cdot\cdot}{Cl}}:$$

There are three lone pairs on each Cl atom, and one of the electrons in the pair shared with C is assigned to each Cl, for a total of 7 valence electrons, just as in an isolated Cl atom. Hence the formal charge on Cl is 0. The C atom has four shared pairs, and is therefore assigned 4 electrons, just as in an isolated C atom, so its formal charge is also 0. Similarly, oxygen has a 0 formal charge with two lone pairs and two shared pairs.

(b) The formal charge on each atom in CO_2, $:\overset{\cdot\cdot}{O}=C=\overset{\cdot\cdot}{O}:$,   is 0.

(c) The formal charges in the two possible structures for N_2O are indicated by

$$:\overset{\ominus}{\underset{\cdot\cdot}{N}}=\overset{\oplus}{N}=O: \text{and}   :N\equiv\overset{\oplus}{N}-\overset{\cdot\cdot}{\underset{\cdot\cdot}{O}}\overset{\ominus}{}$$

(d) The atomic number of Cr is 24. Eighteen of its electrons are in closed shells (the argon configuration), so Cr has 6 valence electrons. Each of the four O atoms has 6 valence electrons. There are 2 additional electrons because the charge on the ion is -2, making a total of 32 valence electrons for the CrO_4^{2-} ion. The Lewis structure, showing the formal charges on each atom is

Note that the sum of the formal charges is -2, the charge on the CrO_4^{2-} ion.

The Concept of Resonance

There are many molecules and ions for which it is not possible to draw a single Lewis structure that is an adequate representation of the bonding, but it is possible to think of more than one structure that satisifies the octet rule.

Consider the gas SO_2 as an example. From spectroscopic and X-ray data the two sulfur–oxygen bonds in SO_2 are known to be equal in length and identical in every way. The SO_2 molecule has a dipole moment of 1.60 D, so the molecule must be bent, like H_2O. If SO_2 were linear, although each bond is polar, the entire molecule would be nonpolar. This is illustrated in Fig. 14.4. There are 18 valence electrons for SO_2, as each sulfur and oxygen has 6 valence electrons. A Lewis structure that satisfies the octet rule is

But the structure shown is clearly not an adequate representation of SO_2, because the two S—O bonds are not identical. A double bond is significantly shorter than a single bond. Because we know that both bonds in SO_2 are of equal length, this structure must be ruled out. The alternate structure,

however, is equally as likely as the first one. Neither is a good representation of SO_2, but we speak of the actual molecule as a **resonance hybrid** of the two structures, and indicate this with a double-headed arrow, as

$$\overset{\ominus}{:}\overset{..}{\underset{..}{O}}-\overset{\oplus}{\underset{..}{S}}=\overset{..}{O}: \leftrightarrow :\overset{..}{O}=\overset{\oplus}{\underset{..}{S}}-\overset{..}{\underset{..}{O}}:\overset{\ominus}{}$$

(with the formal charges indicated). Because the S—O bond lengths in SO_2 are

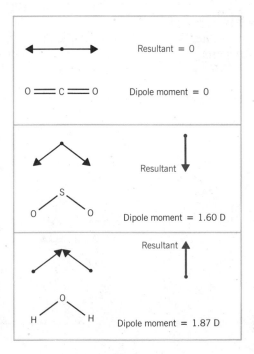

Fig. 14.4. The relation between molecular geometry and the dipole moment. The arrow points from plus to minus in a polar bond.

identical, and are in between the S—O single bond length and the S=O double bond length, we say that the two structures shown make equal contributions to the actual structure of SO_2.

It must be emphasized that the necessity for introducing the concept of resonance is due solely to our inability to represent the bonding with a single drawing using the rules set forth for Lewis structures. Neither of the two resonance structures for SO_2 has any reality. It has been said that depicting a molecule as a resonance hybrid is analogous to considering a rhinoceros as a cross between a dragon and a unicorn. The dragon and the unicorn are mythical or fictitious, just as the resonance structures we use are fictitious. A rhinoceros is real, and there is only one real structure for SO_2; it is just not possible to depict SO_2 with a single Lewis structure.

The term **resonance** is perhaps an unfortunate one, because in its common usage it is associated with vibration or oscillation. Nothing of that sort is involved when we write resonance structures. A Lewis structure is a *model* used to represent the bonding in molecules. For some molecules the model is not adequate, and resonance structures were introduced to compensate for that inadequacy.

As another example of the need for resonance structures, consider the carbonate ion, CO_3^{2-}. All three carbon–oxygen bonds are of equal length, and we represent this as a resonance hybrid of three structures:

$$
\begin{array}{ccc}
\overset{\displaystyle :\!O:}{\underset{}{\overset{\|}{}}} & & \overset{\displaystyle :\!\overset{..}{O}:^{\ominus}}{\underset{}{|}} \\
^{\ominus}:\!\overset{..}{\underset{..}{O}}\!-\!C\!-\!\overset{..}{\underset{..}{O}}\!:^{\ominus} & \leftrightarrow & ^{\ominus}:\!\overset{..}{\underset{..}{O}}\!-\!C\!=\!\overset{..}{O}: & \leftrightarrow & :\!O\!=\!C\!-\!\overset{..}{\underset{..}{O}}\!:^{\ominus}
\end{array}
$$

Note that the charge of 2– on the ion is equally shared by the three oxygen atoms. The nitrate anion, NO_3^-, is isoelectronic with CO_3^{2-} (there are 24 valence electrons for each of these ions), and also must be depicted as a resonance hybrid of three structures.

EXAMPLE 14.3. Using resonance hybrids to depict bonding

Draw appropriate resonance structures for SO_3. Spectroscopic data indicate that all three S—O bonds are identical, and that the bond length is shorter than a typical S—O single bond, but longer than a typical S=O double bond.

Solution. There are 24 valence electrons in SO_3, and we can draw three structures similar to those used for CO_3^{2-} or NO_3^-

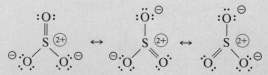

These three structures place a formal charge of +2 on S. Sulfur is certainly less electronegative than oxygen, but we can avoid a formal charge as large as +2 by including structures with more than 8 electrons about S, such as

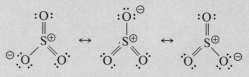

Exceptions to the Octet Rule

While the Lewis octet rule is both a useful and a simple model, there are a substantial number of molecules and ions for which it is inadequate. These exceptions fall into three categories: (1) species with an odd number of valence electrons, (2) electron deficient compounds, and (3) species with more than eight electrons about an atom.

Species with an Odd Number of Valence Electrons

Most molecules and ions have an even number of electrons, but there are several with an odd number of electrons. Some examples are NO, NO_2, and ClO_2. There is no way an odd number of electrons can be used to make only complete octets. Consider NO, with 11 valence electrons. The best Lewis structure we can draw is $:\ddot{N}=\ddot{O}:$, which gives the more electronegative oxygen atom a full octet, but has only 7 electrons around N, one of which is necessarily unpaired.

Unpaired Electrons and Paramagnetism

Substances that have unpaired electrons differ in their magnetic behavior from substances in which all spins are paired. Most molecules and ions have no unpaired spins and are **diamagnetic**, that is, they are very slightly repelled by a magnetic field. Species with unpaired electrons, however, are **paramagnetic**. They become magnetized by, and attracted to, a magnetic field. They lose their magnetism when removed from the field. The molecules NO, NO_2, and ClO_2 are all paramagnetic.

Electron Deficient Compounds

Boron, more than any other element in the periodic table, forms electron deficient compounds, in which the number of electron pairs available for bonding is less than the number needed to have a complete octet around the B atom. Boron has electronic configuration $1s^2 2s^2 2p$ and therefore has three valence electrons. Because the valence electrons are relatively close to the nucleus, the amount of energy required to lose three electrons to form a $+3$ cation with the same electronic configuration as He is very large. There are no ionic compounds of boron. If boron uses its three valence electrons to form three covalent bonds, as in the boron trihalides, BF_3, BCl_3, BBr_3, and BI_3, there are only six electrons, not eight, about the boron atom.

Consider boron trifluoride, BF_3, a colorless, pungent gas at room temperature. There are 24 valence electrons for BF_3, 3 from B and 7 from each of the three F atoms. The Lewis structure with B—F single bonds and no formal charge on any atom is

$$:\ddot{F}-B-\ddot{F}:$$
$$\underset{:\ddot{F}:}{\overset{|}{}}$$

which leaves boron with only 6 electrons around it.

The experimental value of the B—F bond length in BF_3 is 1.30 Å, somewhat shorter than expected for a single B—F bond. Several explanations for the shortening of the B—F bonds have been proposed. One suggestion is that the resonance structures shown below make contributions to the actual structure of the molecule:

$$:\ddot{F}-\overset{:F:^{\oplus}}{\underset{\ominus}{\overset{\|}{B}}}-\ddot{F}: \quad \leftrightarrow \quad {}^{\oplus}:\ddot{F}=\overset{:\ddot{F}:}{\underset{\ominus}{B}}-\ddot{F}: \quad \leftrightarrow \quad :\ddot{F}-\overset{:\ddot{F}:}{\underset{\ominus}{B}}=\ddot{F}:^{\oplus}$$

These structures involve back bonding from F to B, that is, electrons leaving the more electronegative fluorine atom to enter the vacant p orbital on B. Another suggestion is that because of the large electronegativity difference between F and B, $x_F - x_B = 2$, the electrons in the bond are pulled toward fluorine and the bond is both strengthened and shortened by having partial ionic character.

Boron trifluoride is very reactive, and combines with substances that have a lone pair of electrons, such as NH_3 and H_2O, to form **adducts** or **addition compounds.** The reaction with NH_3 to form the **adduct** BF_3NH_3, can be represented as

$$
\begin{array}{ccc}
:\!\ddot{F}: & H & :\!\ddot{F}:\ \ H \\
| & | & |\ \ \ | \\
:\!\ddot{F}\!-\!B\ \ +\ \ :\!N\!-\!H\ \ \rightarrow\ \ :\!\ddot{F}\!-\!B\!-\!N\!-\!H \\
| & | & |\ \ \ | \\
:\!\ddot{F}: & H & :\!\ddot{F}:\ \ H
\end{array}
\qquad (14\text{-}3)
$$

The bond between B and N is like any other polar covalent bond; the two atoms share a pair of electrons, but the sharing is unequal. A substance like BF_3 that can accept a pair of electrons is called a **Lewis acid.** A substance that has a lone pair of electrons it can donate, like NH_3, is called a **Lewis base.** These definitions extend the concept of acid–base reactions to include reactions such as that of Eq. (14-3). With the Brønsted–Lowry definitions of acid and base, the reaction between BF_3 and NH_3 is not considered an acid–base reaction. Boron trifluoride is one of the strongest Lewis acids known, and is widely used in organic chemistry to function as an acid in nonaqueous solvents. Lewis acids and bases will be discussed further in Chapter 20.

Expanded Valence Shells

There are many exceptions to the octet rule in which there are more than eight electrons about an atom. Such structures are possible only when the principal quantum number of the valence electrons of the central atom is three or greater and d electrons are involved in the bonding.

Examples of molecules with expanded valence shells are PCl_5 (10 electrons about P) and SF_6 (12 electrons about S). No elements in the second period form compounds with more than 8 electrons in the valence shell. While PF_5, PCl_5, AsF_5, and $SbCl_5$ are all known, there are no nitrogen analogs. Expanded valence shells are more common when the central atom is large, and is bonded to small, highly electronegative atoms such as O, F, and Cl.

EXAMPLE 14.4. Resonance structures for the CrO_4^{2-} ion

Experimental evidence indicates that the bond between Cr and O in the CrO_4^{2-} ion is in between a single and a double bond. Draw resonance structures that represent the actual structure of CrO_4^{2-}.

Solution. The Lewis structure with four Cr—O single bonds is shown in Example 14.2. Resonance structures with double bonds between Cr and O necessitate having 10 electrons around Cr. Since d orbitals can be used by Cr for bonding, an expanded valence shell is possible. The required resonance structures are

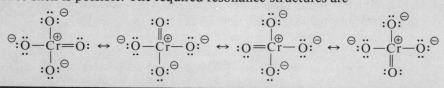

EXAMPLE 14.5. The dot structure for a central atom with an expanded valence shell

Draw an electron dot structure for XeF_2.

Solution. Xenon has 8 electrons in its valence shell and each F atom has 7, so there are a total of 22 valence electrons for XeF_2. There can only be an octet of electrons about each F atom, which uses 16 electrons, so there must be 6 more electrons about Xe. These must be lone pairs, giving Xe an expanded valence shell with 10 electrons. The structure is

$$:\ddot{F}—\ddot{X}\ddot{e}—\ddot{F}:$$

section 14.3
Valence-Shell Electron-Pair Repulsion (VSEPR) Theory

The Lewis structure of a molecule or ion conveys no information about the geometry of that species. Consider methane, CH_4, for example. The Lewis structure

$$H—\overset{\displaystyle H}{\underset{\displaystyle H}{\overset{|}{\underset{|}{C}}}}—H$$

does not indicate how the five atoms are arranged in space.

A useful and simple theory that enables us to predict correctly the geometries of most (but not all) substances, is entitled the **valence-shell electron-pair repulsion (VSEPR) theory.** This approach was first proposed by N. V. Sidgwick and H. M. Powell in 1940 and was expanded considerably by R. J. Gillespie and R. S. Nyholm in 1957, so that it is usually attributed to the latter two.

The basic idea is that the geometry of a molecule will be the structure that minimizes the repulsion between the pairs of electrons around an atom. If all the pairs of electrons in a molecule are shared, that is, are **bonded pairs,** there is only one particular geometry that keeps the electron pairs as far apart as possible and thereby minimizes the repulsion between them. Figure 14.5 shows the ideal geometry that minimizes repulsion for several different numbers of electron pairs about a central atom. This is the observed geometry if there are no lone pairs of electrons in the molecule.

The Lewis structure for methane involves four bonded pairs of electrons about C in CH_4. The VSEPR theory therefore predicts that the molecule is tetrahedral, and this has been experimentally verified. (The tetrahedral geometry of CH_4 was known before VSEPR theory was developed.) As illustrated in Fig. 14.6, methane is a symmetric tetrahedral molecule, and the H—C—H bond angle is $109°28'$.

When there are lone pairs of electrons on a central atom, the geometry is not the ideal structure shown in Fig. 14.5 because lone pairs exert a larger repulsive effect than do bonded pairs. This effect can be seen by considering the isoelectronic series of molecules CH_4, NH_3, and H_2O. Each of these molecules has eight valence electrons that are used in forming bonds. There are four bonds in CH_4, so all eight electrons are in bonded pairs, and the observed H—C—H bond angles are exactly $109.5°$. There are only three bonds in NH_3, so that six of the eight electrons are in three bonded pairs, and the remaining two electrons are a lone pair on N. The observed H—N—H bond angles are $107.3°$. In H_2O there are two lone pairs and two bonded pairs around O, and the observed H—O—H bond angle is $104.5°$. This is illustrated in Fig. 14.7.

According to VSEPR theory, the reason for this decrease in bond angle in the series CH_4, NH_3, H_2O, is that a lone pair of electrons takes up more space than a bonded

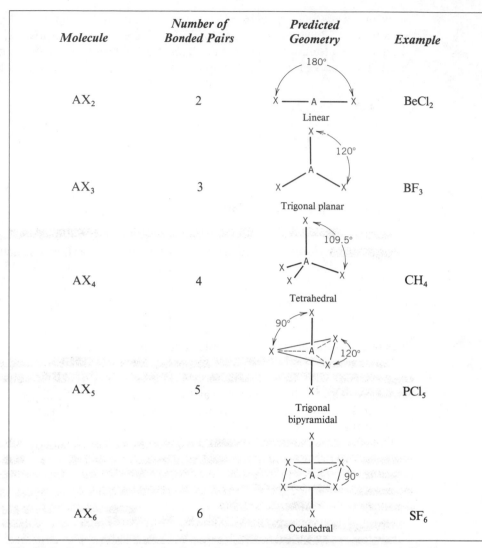

Molecule	Number of Bonded Pairs	Predicted Geometry	Example
AX₂	2	Linear	BeCl₂
AX₃	3	Trigonal planar	BF₃
AX₄	4	Tetrahedral	CH₄
AX₅	5	Trigonal bipyramidal	PCl₅
AX₆	6	Octahedral	SF₆

Fig. 14.5. Predicted geometries for molecules or ions with no lone pairs of electrons, as a function of the number of bonded pairs. For trigonal bipyramidal geometry, the three atoms at the corners of the equilateral triangle are said to be at *equatorial* positions, and the two atoms along the perpendicular to that triangle are at *axial* positions. The atoms at the axial positions are further from the central atom than are the atoms at the equatorial positions.

pair does. Lone pair electrons are not attracted to two nuclei, but only to one. The charge cloud distribution of a lone pair is closer to the central atom, and more spread out sideways, than the charge cloud distribution of a bonded pair of electrons. (You might think of the charge cloud of a lone pair as shorter and plumper than that of a bonded pair.) As a result, lone pair electrons exert a larger repulsive effect than bonded electrons do, and the bonded pairs move closer to one another, and away from the lone pairs, to minimize repulsion. This effect is illustrated in Fig. 14.8, which compares the geometry of NH_3 with that of the ammonium ion, NH_4^+.

We can predict the actual geometry of a molecule or ion with lone pairs by starting from the ideal geometry shown in Fig. 14.5 and moving bonded pairs away from lone pairs, with a consequent decrease in bond angle.

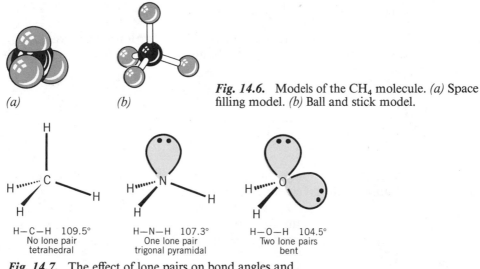

Fig. 14.6. Models of the CH_4 molecule. *(a)* Space filling model. *(b)* Ball and stick model.

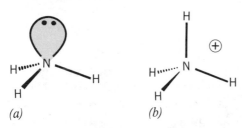

H—C—H 109.5°
No lone pair
tetrahedral

H—N—H 107.3°
One lone pair
trigonal pyramidal

H—O—H 104.5°
Two lone pairs
bent

Fig. 14.7. The effect of lone pairs on bond angles and molecular geometry.

(a)

(b)

Fig. 14.8. The effect of a lone pair on bond angle. *(a)* In NH_3 the H—N—H bond angle is 107.3°, and there is a lone pair on the N atom. *(b)* In NH_4^+ the H—N—H bond angle is 109.5°.

First count the total number of electron pairs, both bonded pairs and lone pairs. Start with the geometry expected if all the electron pairs were bonded pairs, and then consider the effect of the extra repulsion of the lone pairs.

An additional point of importance is the effect of a double or a triple bond. It is a very good approximation to assume that a multiple bond affects the molecular geometry as if it were a single bond. Thus in counting the total number of electron pairs to determine the ideal geometry, count multiple bonds as one pair.

Using VSEPR theory we could predict, for example, that CO_2 is linear but SO_2 is bent. (We know this is true because CO_2 has a zero dipole moment, while SO_2 has a dipole moment of 1.60 D.) There are two double bonds in CO_2 and no lone pairs on the central atom, C. It is therefore classified as an AX_2 molecule, with two bonded pairs and no lone pairs and is predicted to be linear. On the other hand, there is a lone pair of electrons on S in SO_2 (see the resonance structures in the preceding section), and so the total number of electron pairs is three (two bonded pairs and one lone pair). The bond angle predicted for three bonded electron pairs is 120°. A compound of the type $:AX_2$, with two bonded pairs and one lone pair is denoted AX_2E in VSEPR notation, where E is used to represent a lone pair of electrons. The X—A—X bond angle in an AX_2E molecule is predicted to be somewhat less than 120°. The observed bond angle in SO_2 is 119.5°. The shape of SO_2 is usually described as bent, angular, or *V*-shaped.

EXAMPLE 14.6. Predicting molecular geometry using VSEPR theory; molecules without multiple bonds

Predict the geometry of **(a)** SO_4^{2-} ion, **(b)** PCl_3, and **(c)** SCl_2.

Solution

(a) There are 32 valence electrons for the sulfate ion, 6 from S and each of the four O atoms, plus 2 from the charge on the anion. The Lewis structure that satisfies the octet rule,

has four bonded pairs and no lone pairs about S. It is therefore predicted to be tetrahedral, in agreement with observation.

Note that the CrO_4^{2-} ion, discussed in Examples 14.2(**d**) and 14.4, also has 32 valence electrons, and the same Lewis structure as SO_4^{2-}. The chromate ion is also tetrahedral, as predicted by VSEPR theory. The structure of CrO_4^{2-} ion is shown in Fig. 14.9.

(b) The Lewis structure for PCl_3 is

$$:\ddot{C}l-\underset{\displaystyle :\ddot{C}l:}{\overset{\displaystyle :\ddot{C}l:}{P}}-\ddot{C}l:$$

so that there are three bonded pairs and one lone pair about P. The idealized geometry for four bonded electron pairs is tetrahedral; VSEPR theory predicts that the bond angle should be less than 109.5°, and the geometry should be **trigonal pyramidal**, like that of ammonia. The observed bond angle is 100°, and the geometry is indeed trigonal pyramidal.

(c) The Lewis structure for SCl_2 is $:\ddot{C}l-\ddot{S}-\ddot{C}l:$, with two bonded pairs and two lone pairs about S. We therefore predict that the molecule is bent, and the bond angle should be less than 109.5°. The observed bond angle is 102°.

Let us now consider the geometry of formaldehyde, $H_2C=O$, a molecule with a double bond. Remember that in counting electron pairs a double bond is counted as one pair. We therefore count three bonded pairs about C and no lone pairs, and the ideal geometry is trigonal planar with 120° bond angles. This is an excellent approximation to the geometry of formaldehyde, which is planar, with bond angles very close

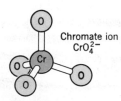

Fig. 14.9. The tetrahedral geometry of the CrO_4^{2-} ion. There are four bonded pairs and no lone pairs around the Cr atom.

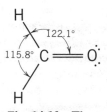

Fig. 14.10. The geometry of formaldehyde. The greater repulsion of the double bond pushes the single bonds closer together. The H—C—H bond angle is therefore a little less than 120°.

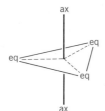

Fig. 14.11. Axial (ax) and equatorial (eq) positions in a trigonal bipyramid.

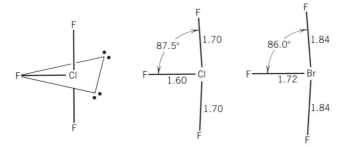

Fig. 14.12. The *T*-shaped geometry of AX_3E_2 molecules. Observed geometries of ClF_3 and BrF_3. Bond lengths in angstroms, Å.

to 120°. Because the electron cloud of a multiple bond takes up more room than does the electron cloud of a single bond, the bond angles in $H_2C=O$ deviate slightly from 120°. As a result of the increased repulsion of the double bond, the C—H bonds in formaldehyde move closer together and away from the C=O double bond. The observed geometry is shown in Fig. 14.10.

When there are five bonded pairs of electrons around a central atom, the geometry is trigonal bipyramidal. In this structure, while three of the X—A—X bond angles are 120°, six are only 90°. The five X atoms are not all equivalent. Three are in equatorial positions and two are in axial positions, as shown in Fig. 14.11. Repulsions between the electron pairs in bonds only 90° apart are greater than repulsions between electron pairs in bonds 120° apart. Each axial bond has three electron pairs 90° away, while each equatorial bond has only two electron pairs 90° away. Thus the axial bonds experience greater repulsion than the equatorial bonds. As a result, atoms at the equatorial positions are closer to the central atom than atoms at the axial positions. In PCl_5, for example, the equatorial P—Cl distances are 2.04 Å, whereas the axial P—Cl distances are 2.19 Å.

If the central atom has five electron pairs around it, but either one, two, or three of those pairs are lone pairs, the lone pairs will occupy the equatorial positions, where there are fewer 90° interactions. An AX_3E_2 molecule is predicted to be *T* shaped, as shown in Fig. 14.12. The one equatorial bond is stronger and shorter than the two axial bonds. Examples of AX_3E_2 molecules are ClF_3 and BrF_3, and details of their geometry are shown in Fig. 14.12. The F—Cl—F and F—Br—F bond angles are less than 90° as the bonded pairs move away from the lone pairs to minimize repulsion.

A summary of geometries predicted by VSEPR theory for molecules in which the central atom has lone pairs of electrons is shown in Fig. 14.13.

Molecule	Number of Electron Pairs	Number of Lone Pairs	Predicted Geometry	Example
AX_2E	3	1	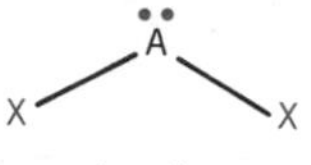 Angular	SO_2, O_3
AX_3E	4	1	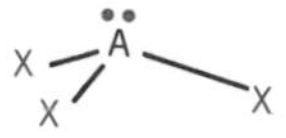 Trigonal pyramidal	NH_3, PCl_3
AX_2E_2	4	2	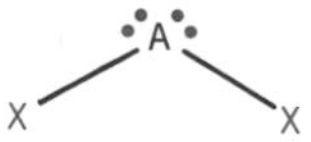 Angular	H_2O, SCl_2
AX_4E	5	1	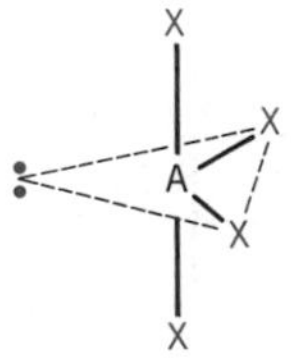 Sawhorse	SF_4
AX_3E_2	5	2	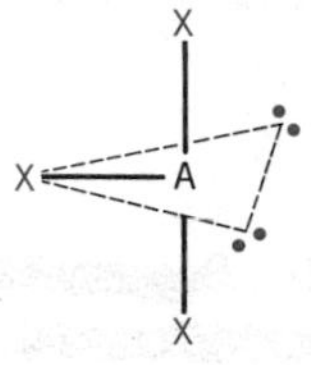 T shaped	ClF_3
AX_2E_3	5	3	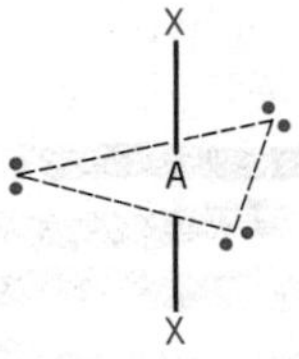 Linear	XeF_2, I_3^-
AX_5E	6	1	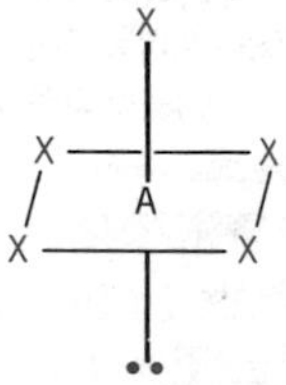 Square pyramidal	IF_5
AX_4E_2	6	2	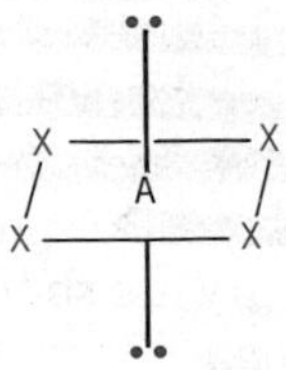 Square planar	IF_4^-

Fig. 14.13. Geometries predicted by VSEPR theory when there are lone pairs of electrons on the central atom.

EXAMPLE 14.7. Predicting molecular geometry using VSEPR theory

Predict the geometry of **(a)** ethylene, C_2H_4, **(b)** I_3^-, and **(c)** ICl_4^-.

Solution

(a) There are 12 valence electrons for ethylene, and the Lewis structure is $H_2C=CH_2$. Since the double bond is counted as one bonded pair, there are three bonded pairs about C and the ideal geometry is trigonal planar, with 120° bond angles. Because the double bond takes up more room than a single bond, we expect the H—C—H bond angles to be slightly less than 120°, and this is indeed what is observed. The actual geometry of ethylene is shown below.

(b) There are 22 valence electrons for the triiodide ion, I_3^-. The Lewis structure is

$$:\ddot{I}-\ddot{I}-\ddot{I}:$$

so that the central I atom has two bonded pairs and three lone pairs. (Recall the Lewis structure for XeF_2 in Example 14.5.) We expect the three lone pairs to occupy the equatorial positions of the trigonal bipyramidal geometry. The three iodine atoms should therefore lie on a straight line, and the ion is indeed observed to be linear.

(c) Each halogen atom has 7 valence electrons, and 1 electron has been added to make the anion, so that ICl_4^- has 36 valence electrons. Putting complete octets about each atom uses only 32 electrons. The remaining 4 are in two lone pairs on the I atom, which therefore has a total of 6 electron pairs around it, four bonded and two lone pairs. The geometry will be derived from distortion of an octahedral structure. There are two possibilities to consider:

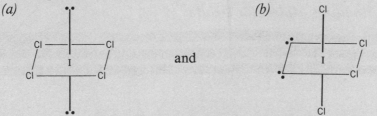

Since lone pairs repel one another more than bonded pairs do, structure *(a)*, which has the lone pairs 180° apart is preferred over structure *(b)*, which has the lone pairs only 90° apart. The observed **square planar** geometry of ICl_4^- is indeed that of structure *(a)*.

There are known examples of structures with seven, eight, and nine electron pairs around a central atom, but they are much less common than those shown in Figs. 14.5 and 14.13, and we will not consider them.

We have seen that by drawing Lewis structures and using VSEPR theory we can understand many features of the bonding and geometry of molecular species. These models, however, do not enable us to calculate bond energies or to explain molecular spectra. The most useful approach to a broader theoretical understanding of the covalent bond is called **molecular orbital (MO) theory**, and we shall now turn our attention to the molecular orbital description of the simplest of all covalently bonded molecules, H_2.

section 14.4
Molecular Orbitals of H$_2$

If two H atoms, each having a single $1s$ electron, approach one another so closely that their electron clouds begin to overlap, each electron feels the attractive pull of both nuclei. Provided that the two electrons have opposite spin, they will be pulled more and more into the region between the two nuclei. The two electron clouds overlap until they merge and each electron experiences the attraction of each nucleus equally. When each electron spends the same amount of time about each nucleus, an H$_2$ molecule has been formed. Each electron is no longer in an atomic orbital centered about a single nucleus; it is in a **molecular orbital,** centered about both nuclei. Figure 14.14 depicts the formation of this **bonding molecular orbital** of H$_2$ as a result of the overlap of two $1s$ atomic orbitals.

Sigma (σ)-Molecular Orbitals

The molecular orbital shown in Fig. 14.14 is called a **$\sigma 1s$ molecular orbital.** The symbol σ (sigma) signifies that the shape of the electron cloud in the molecular orbital has cylindrical symmetry. This means that if you imagine rotating the electron cloud about an axis connecting the two nuclei, the three-dimensional distribution of electron density will be the same whatever the angle of rotation. A sausage or salami has a shape similar to that of the $\sigma 1s$ MO; imagine passing a stick through the long axis of a sausage and rotating the stick. If an observer closes his eyes and you rotate the stick by any angle, when the observer opens his eyes he will not be able to tell whether or not you have rotated the sausage. Any molecular orbital possessing complete symmetry with respect to rotation about the internuclear axis is called a σ-type MO. The notation $1s$ following the σ indicates that this particular molecular orbital results from the overlap of two $1s$ atomic orbitals.

The Bonding Molecular Orbital

The $\sigma 1s$ MO is a **bonding molecular orbital** because when an electron occupies this molecular orbital its energy is lower than when it is in the $1s$ atomic orbital. In H$_2$, two electrons of opposite spin occupy the $\sigma 1s$ molecular orbital. They are shared

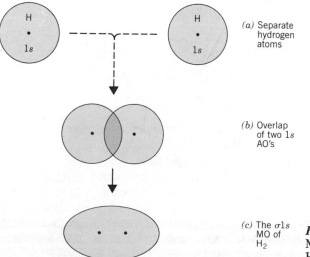

(a) Separate hydrogen atoms

(b) Overlap of two $1s$ AO's

(c) The $\sigma 1s$ MO of H$_2$

Fig. 14.14. The $\sigma 1s$ bonding MO of H$_2$, formed when two H $1s$ atomic orbitals overlap.

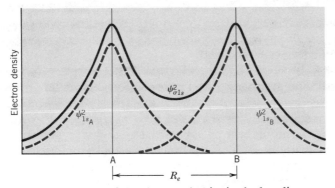

Fig. 14.15. Plots of the electron density in the bonding molecular orbital $\sigma1s$ (solid black curve) and in the two atomic orbitals, $1s_A$ and $1s_B$ (dashed red curve).

equally by the two nuclei and constitute a covalent bond. Note that the bulk of the electron density is in the region between the two nuclei. Figure 14.15 compares the electron density in the H_2 molecular orbital with the electron density in the two atomic orbitals.

The accumulation of the negative charge cloud in the region between the two nuclei provides the "glue" that keeps the molecule together. The two nuclei, having like charges, repel one another, but both are attracted to the electron charge cloud located between them. Thus, in a covalent bond, just as in an ionic bond, it is electrostatic attraction that keeps the molecule together. An ion-pair, such as an Na^+Cl^- monomer, features electrical asymmetry whereas there is complete electrical symmetry in a covalently bonded homonuclear diatomic molecule (a diatomic molecule in which both atoms are identical) such as H_2. The force of attraction that leads to a lowering of the energy for the bonded structure relative to the nonbonded atoms, however, is coulombic attraction in both cases.

The Potential Energy Curve for H_2

We can depict the energy changes that occur as two H atoms are brought close together and the two electrons occupy the $\sigma1s$ bonding molecular orbital by means of a potential energy diagram, shown in Fig. 14.16. We plot the energy of the two atoms as a function of the distance, R, between them. At infinite separation the potential energy is zero *by definition;* there is no interaction between particles an infinite distance apart. As the atoms come closer together and each electron feels the attrac-

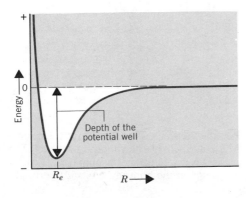

Fig. 14.16. The potential energy curve for two H atoms as a function of the distance, R, between them. The equilibrium internuclear distance, R_e, is 0.74 Å.

tion of both nuclei, the energy of the system decreases, that is, there is an increase in the stability of the system. The energy continues to decrease as the atoms get closer, but then reaches a minimum value, corresponding to the most stable configuration of the two nuclei plus the two electrons.

The distance between the nuclei at the minimum value of the energy is R_e, the equilibrium internuclear distance, which is called the **bond length**. If one tries to push the atoms together more closely than the equilibrium internuclear distance R_e, the strong repulsion of the positively charged nuclei sets in and the energy begins to rise very steeply. The two atoms behave very much as if they were connected by a spring. If they are pulled apart, the force of attraction of the nuclei for the electron cloud pulls them back to the distance R_e. If they are pushed together, the internuclear repulsion pushes them back to the distance R_e.

The difference in energy between the minimum value and zero (which corresponds to isolated gaseous H atoms) is a measure of the bond energy. The deeper the potential well, the stronger the bond, and the greater the amount of energy that is required to separate the molecule into its constituent atoms.

Molecular Orbitals as Linear Combinations of Atomic Orbitals

We have described the $\sigma 1s$ molecular orbital in terms of the electron charge cloud distribution in the H_2 molecule, but it also has a mathematical definition in quantum mechanics. A $1s$ atomic orbital is a mathematical function, the solution to the Schrödinger equation for the ground state of the hydrogen atom. A molecular orbital can be formulated as a **linear combination of atomic orbitals** (LCAO).

The $\sigma 1s$ molecular orbital of H_2 can be written as the sum of the two $1s$ atomic orbitals for the two H atoms that combine to make the molecule. If we designate the two H atoms as H_A and H_B, and the $1s$ atomic orbital as ψ_{1s}, then we can express the $\sigma 1s$ MO wavefunction as

$$\psi_{\sigma 1s}(H_2) = C[\psi_{1s}(H_A) + \psi_{1s}(H_B)] \tag{14-4}$$

where C is a numerical constant.

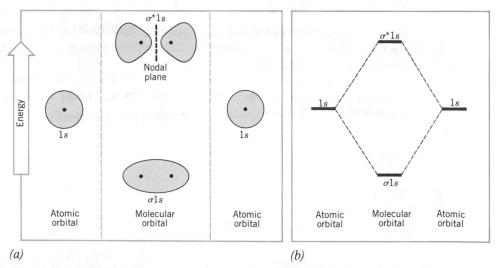

(a)   *(b)*

Fig. 14.17. *(a)* Energy diagram showing the shapes and the relative energies of the $1s$ atomic orbitals, the $\sigma 1s$ bonding MO, and the $\sigma^* 1s$ antibonding MO. *(b)* The energy level diagram. Note that the antibonding MO is raised in energy above the $1s$ atomic orbitals by an amount equal to the lowering in energy of the bonding molecular orbital.

An important rule of the mathematics used in quantum mechanics states that two atomic orbitals can be combined to produce two molecular orbitals, or more generally n atomic orbitals can be combined to produce n molecular orbitals. Thus there is another molecular orbital that can be formed from the two $1s$ atomic orbitals, $\psi_{1s}(H_A)$ and $\psi_{1s}(H_B)$, namely, the *difference* between the two atomic orbitals. This second MO is designated $\sigma*1s$ (read as sigma star $1s$), and is expressed as

$$\psi_{\sigma*1s}(H_2) = C[\psi_{1s}(H_A) - \psi_{1s}(H_B)] \tag{14-5}$$

The $\sigma*1s$ molecular orbital is higher in energy than the $1s$ atomic orbitals which combine to form it, and it is therefore called an **antibonding** orbital. An asterisk always denotes an antibonding molecular orbital. For the antibonding orbital there is very little electron density in the region between the two nuclei and therefore the nuclear repulsion pushes the nuclei apart. Figure 14.17 shows the shapes of the charge clouds and the relative energies of the $1s$ atomic orbitals and the two molecular orbitals that can be formed from them. The antibonding $\sigma*1s$ MO is higher in energy than the $1s$ atomic orbitals by an amount equal in magnitude to the lowering in energy of the bonding $\sigma1s$ MO below the $1s$ atomic orbitals.

Species with Valence Electrons in the $\sigma1s$ and $\sigma*1s$ MO's

The Pauli Exclusion Principle applies to molecular orbitals as well as to atomic orbitals. This means that each molecular orbital can accommodate two electrons of opposite spin, just as each atomic orbital can. Let us now consider all the different molecular species we can imagine using only the two molecular orbitals $\sigma1s$ and $\sigma*1s$. In filling molecular orbitals we use the same three principles we used in the Aufbau process, that is, in building up the periodic table. (See Section 13.3.)

The species H_2^+, called the **hydrogen-molecule ion**, is not formed in ordinary chemical reactions, but is formed when an electrical discharge passes through a sample of H_2 gas. It consists of two nuclei and a single electron, which occupies the $\sigma1s$ bonding molecular orbital. The hydrogen-molecule ion, H_2^+, is a stable species with a bond distance of 1.06 Å (0.106 nm) between the two nuclei, and a bond dissociation energy of 269 kJ/mol. Note that a single electron binds the two nuclei together and that it requires 269 kJ of energy to separate a mole of H_2^+ ions into the constituent H atoms and protons. The electronic configuration of H_2^+ is written as $(\sigma1s)^1$ or just $(\sigma1s)$.

For the H_2 molecule there are two electrons with opposite spin in the $\sigma1s$ molecular orbital. We denote the electronic configuration as $(\sigma1s)^2$. The additional electron increases the electron density between the two nuclei and pulls them closer together. The bond distance in H_2 is only 0.74 Å (0.074 nm), significantly smaller than the bond distance in H_2^+. Similarly, the bond is stronger; it requires 435 kJ/mol (104 kcal/mol) to dissociate H_2 molecules into H atoms.

The Species H_2^- and He_2^+

If we try to add a third electron to make the species H_2^-, it cannot go into the $\sigma1s$ bonding MO because that would violate the Pauli Exclusion Principle. The third electron must occupy the antibonding molecular orbital, and the electronic configuration of H_2^- has to be $(\sigma1s)^2(\sigma*1s)^1$. Another molecular species with three electrons and the same electronic configuration is the **helium-molecule ion**, He_2^+. As long as the number of electrons in bonding molecular orbitals is greater than the number in antibonding molecular orbitals, the molecular species should be stable, because there

Table 14.3. **Summary of Information about Homonuclear Diatomic Molecular Species Using the $\sigma1s$ and $\sigma*1s$ Molecular Orbitals**[a]

$\sigma*1s$	—	—	↑	↑ ↓
$\sigma1s$	↑	↑ ↓	↑ ↓	↑ ↓
Species	H_2^+	H_2	He_2^+	He_2
Electronic configuration	$(\sigma1s)$	$(\sigma1s)^2$	$(\sigma1s)^2(\sigma*1s)$	$(\sigma1s)^2(\sigma*1s)^2$
Number of bonding electrons	1	2	2	2
Number of antibonding electrons	0	0	1	2
Bond order	$\frac{1}{2}$	1	$\frac{1}{2}$	0
Bond energy (kJ/mol)	269	435	238	None
Bond length (Å)	1.06	0.74	1.08	None

[a] The energy of the $\sigma1s$ MO of He_2^+ is lower than the energy of the $\sigma1s$ MO of H_2.

will be a net decrease in energy when it is formed. In fact, the species H_2^- is not known, but He_2^+ has been observed. While H_2^- and He_2^+ are isoelectronic, the energies of the molecular orbitals of the two species are not the same. The $1s$ atomic orbital of helium is lower in energy than the $1s$ atomic orbital of hydrogen because the nuclear charge is twice as large in He. Similarly, the $\sigma1s$ molecular orbital of He_2^+ is lower in energy than the $\sigma1s$ molecular orbital of H_2.

If, however, we imagine putting four electrons into these two molecular orbitals, to form either He_2 or H_2^{2-}, the electronic configuration would have to be $(\sigma1s)^2(\sigma*1s)^2$, with the same number of electrons in both the bonding and antibonding molecular orbitals. Such species do not exist. We formulate the general rule: *In order for a molecular species to be stable, the number of electrons in bonding molecular orbitals must be greater than the number of electrons in antibonding molecular orbitals.* A stable molecule can exist only if there is an excess of bonding electrons.

Table 14.3 summarizes the information we have just described about molecular species with valence electrons in the $\sigma1s$ and $\sigma*1s$ MO's. Since a single covalent bond consists of a shared pair of electrons, we define the **bond order** as one half the excess of bonding electrons over antibonding electrons:

bond order $= \frac{1}{2}$(number bonding electrons − number antibonding electrons)

A stable species can exist only if the bond order is greater than 0. The greater the bond order, the stronger the bond and the shorter the bond length. Bond orders can be either integral or fractional. The bond order of both H_2^+ and He_2^+ is $\frac{1}{2}$, and these are stable species. Note also that a *single* electron can form a bond, as in the case of H_2^+.

section 14.5
Homonuclear Diatomic Molecules of the Second Period

When two different atoms of the same element combine, a **homonuclear diatomic molecule** is formed. Homonuclear diatomic molecules of the second period, from Li_2 to F_2, utilize molecular orbitals that are linear combinations of the $2s$ and $2p$ atomic orbitals, because these are the orbitals of the valence electrons of the atoms of the second period.

The σ2s and σ*2s Molecular Orbitals

Two $2s$ atomic orbitals can be combined to give a bonding MO called $\sigma2s$ and an antibonding MO called $\sigma*2s$, in exactly the same way that two $1s$ atomic orbitals are combined to form $\sigma1s$ and $\sigma*1s$. The elements of the second row that have only $2s$ electrons as valence electrons are Li $(1s^22s^1)$ and Be $(1s^22s^2)$. The diatomic molecule Li_2 is a known species, found only at high temperatures in the vapor of metallic lithium. Each lithium atom has a single unpaired $2s$ electron so that there are two electrons to put into molecular orbitals, and both will occupy the $\sigma2s$ bonding MO. We therefore expect Li_2 to be a stable molecule, as indeed it is. Because the $2s$ orbitals are much larger than the $1s$, the bond distance in Li_2, which is 2.67 Å, is considerably longer than the bond distance in H_2 (0.74 Å). As the two Li nuclei are farther apart than the two H nuclei in H_2, the electron cloud is spread out over a greater volume in space and the bond is weaker. It requires only 105 kJ/mol to dissociate Li_2 into two Li atoms, whereas 435 kJ/mol are required to dissociate H_2. The electronic configuration of Li_2 is depicted in Fig. 14.18. It is written as $KK(\sigma2s)^2$, where the symbol KK indicates that the K $(n = 1)$ shells of both lithium atoms are filled.

Since each Be atom has two $2s$ electrons, there are four electrons to put into molecular orbitals for the hypothetical Be_2. Two would then have to occupy the $\sigma*2s$ MO, and this would be a configuration with zero net bonding electrons. In accord with our general rule, the species Be_2 does not exist. Beryllium is a metallic solid at room temperature. It melts at 1280° and its vapor is monatomic.

π-Molecular Orbitals

There are two possible ways that two p atomic orbitals on different atoms can overlap, as illustrated in Fig. 14.19.

Mathematically, we can form two molecular orbitals from a linear combination of the two p atomic orbitals directed along the internuclear axis (head-on overlap). One is the sum of the two p atomic orbitals, and the other is the difference. Just as for the case of combining s atomic orbitals, we obtain a bonding and an antibonding molecular orbital. A head-on overlap of p atomic orbitals on two different atoms produces σ-type molecular orbitals, completely symmetric with respect to rotation about the internuclear axis. The bonding molecular orbital is denoted a σp MO, and the antibonding molecular orbital is denoted a $\sigma*p$ MO.

A *sideways* overlap of p atomic orbitals on two different atoms produces a new type of MO, with a nodal plane containing the internuclear axis. It is called a π-MO. While a π-MO does not have complete symmetry with respect to rotation about the internuclear axis, rotation by 180° about the internuclear axis results in a charge

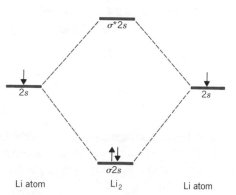

Fig. 14.18. Energy level diagram for the molecular orbitals of Li_2.

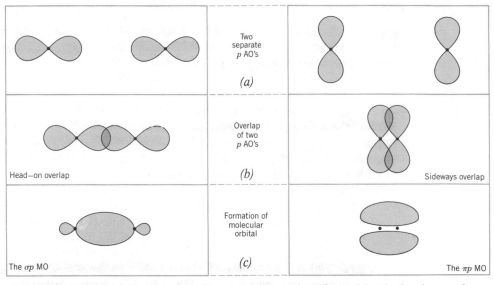

Fig. 14.19. Comparison of the bonding molecular orbitals formed by the head-on and sideways overlap of two p atomic orbitals.

distribution identical to the original. The combination of two p atomic orbitals directed along an axis perpendicular to the internuclear axis (sideways overlap) produces a bonding πp MO and an antibonding π^*p MO.

A bond formed when two electrons with paired spin occupy a σ-type bonding molecular orbital is called a **σ bond**. Similarly, a bond formed when two electrons with paired spin occupy a π-type bonding molecular orbital is called a **π bond**. It is an experimental fact that π bonds are not as strong as σ bonds, although the difference is not large.

In order to discuss the molecular orbitals formed on overlapping the $2p$ atomic orbitals of two different atoms, we must decide on the symbol to be used for the direction of the internuclear axis. The letter used to designate this direction is totally arbitrary; let us call it the z axis. Then the $2p_z$ atomic orbitals on each atom will combine to form a $\sigma 2p$ bonding molecular orbital and a σ^*2p antibonding molecular orbital, because there will be a head-on overlap of the p atomic orbitals directed along the internuclear axis.

The $2p_x$ atomic orbitals on each of the two atoms combine to form a $\pi 2p_x$ and a π^*2p_x molecular orbital, and the $2p_y$ atomic orbitals combine to form a $\pi 2p_y$ and a π^*2p_y molecular orbital. There is no way to distinguish the x and y directions; they are both perpendicular to the internuclear axis, which is the only physically distinguishable direction in space. As a result the $\pi 2p_x$ and $\pi 2p_y$ molecular orbitals are degenerate, that is, equal in energy. The π^*2p_x and π^*2p_y molecular orbitals are also degenerate.

Before we can apply the Aufbau principle and discuss the electronic configurations of the homonuclear diatomic molecules from B_2 to F_2, we must know the order of the energies of the various molecular orbitals. For B_2, C_2, and N_2, the ordering of the levels is given in Fig. 14.20, which also shows schematic diagrams of the shapes of the molecular orbitals.

For B_2, C_2, and N_2, the $\pi 2p$ bonding MO's are somewhat lower in energy than the $\sigma 2p$ bonding MO's. This ordering of the levels has been established experimentally by investigating the magnetic and spectroscopic properties of these molecules, as de-

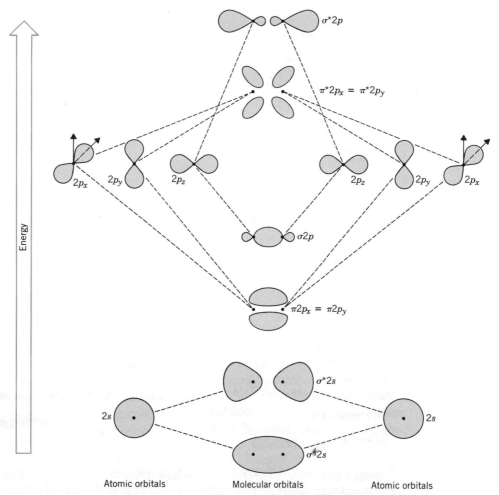

Fig. 14.20. Molecular orbitals formed from the $2s$ and $2p$ atomic orbitals. The order of the energy levels is correct for B_2, C_2, and N_2.

scribed below for B_2. The order of the energies of the $\pi 2p$ and $\sigma 2p$ MO's changes for O_2 and F_2. With increased nuclear charge the $\sigma 2p$ MO becomes lower in energy than the $\pi 2p$ MO. Energy level diagrams for these two cases are given in Fig. 14.21. With the use of these diagrams we can discuss the bonding in the five diatomic molecules B_2, C_2, N_2, O_2, and F_2.

Bonding in B_2, C_2, and N_2

While boron is a solid at room temperature, with the boron atoms linked in a three-dimensional network, the vapor phase of boron at temperatures above its boiling point (2550 °C) contains B_2 molecules. The electronic configuration of boron is $1s^2 2s^2 2p^1$ so that each B has three valence electrons for a total of six electrons to occupy molecular orbitals in B_2. The first four fill the $\sigma 2s$ and $\sigma^* 2s$ MO's, and there are two electrons to occupy the degenerate $\pi 2p$ MO's. In accord with Hund's rule, each occupies a different orbital and the two electrons have parallel spin. Species with unpaired electrons are paramagnetic, and B_2 is indeed observed to be paramagnetic. This observation proves conclusively that the $\pi 2p$ energy level is below the $\sigma 2p$ for B_2,

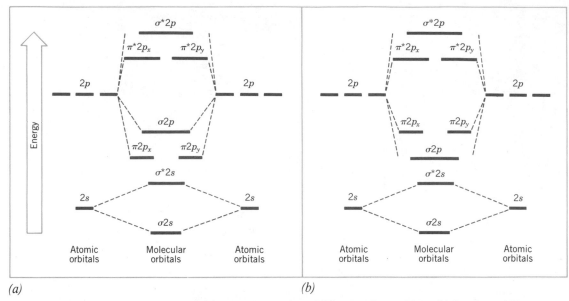

(a)   *(b)*

Fig. 14.21. Energies of molecular orbitals *(a)* for B_2, C_2, and N_2, *(b)* for O_2 and F_2.

because if the $\sigma 2p$ energy level were lower, both electrons would occupy the $\sigma 2p$ molecular orbital and there would be no unpaired spins.

The electronic configuration of B_2 is shown in Fig. 14.22, and is written as $KK(\sigma 2s)^2(\sigma^*2s)^2(\pi 2p_x)^1(\pi 2p_y)^1$. The number of electrons in bonding MO's is four, while the number in antibonding MO's is two, so that the net number of electrons in bonding MO's is $4 - 2 = 2$, and the bond order is 1. We say that a single covalent bond binds the B_2 molecule. Note that the net bond is a π bond, because the $\sigma 2s$ and σ^*2s MO's are both filled. There is no σ bond in B_2.

Because the nuclear charge of B is greater than that of Li, the bonding molecular orbitals of B_2 are lower in energy than the bonding molecular orbitals of Li_2, and are pulled in closer to the nuclei. As a result, the bond length in B_2, 1.59 Å, is shorter than that in Li_2, 2.67 Å, and the bond energy, 289 kJ/mol, is greater than that of Li_2, 105 kJ/mol.

The bonding in C_2, a species found at high temperature in the vapor of carbon, is described in Example 14.8.

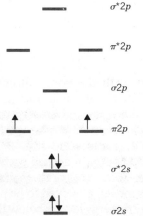

Fig. 14.22. The electronic configuration of B_2. Since there are two unpaired electrons, B_2 is paramagnetic.

EXAMPLE 14.8. Bonding in homonuclear diatomic molecules of the second period

Give the electronic configuration of the C_2 molecule and predict whether or not it is paramagnetic. What is the bond order? Do you expect the bond in C_2 to be stronger or weaker than that in B_2?

Solution. The electronic configuration of C is $1s^2 2s^2 2p^2$ so that each C has four valence electrons and there are eight electrons to occupy MO's in C_2. The energy level diagram for C_2 is shown below. The $\pi 2p$ MO's are filled and there are no unpaired electrons, so C_2 should be diamagnetic, which is in agreement with observation. There are six electrons in bonding MO's, and two in antibonding MO's, so that the bond order is $\frac{1}{2}(6-2) = 2$. Note that the net bonding electrons are in π-MO's. A bond order of 2 corresponds to a double bond, and we therefore expect the bond in C_2 to be considerably stronger than the single bond in B_2. Again, this agrees with observation. The bond energy of C_2 is $630 \text{ kJ} \cdot \text{mol}^{-1}$, somewhat more than twice the bond energy of B_2. The bond length in C_2, 1.24 Å, is shorter than the bond length in B_2, 1.59 Å. A double bond is shorter than a single bond. The electronic configuration of C_2 is $KK(\sigma 2s)^2(\sigma^* 2s)^2(\pi 2p)^4$.

	⎯⎯	$\sigma^* 2p$
⎯⎯	⎯⎯	$\pi^* 2p$
	⎯⎯	$\sigma 2p$
↑↓	↑↓	$\pi 2p$
	↑↓	$\sigma^* 2s$
	↑↓	$\sigma 2s$

It should be noted that with only eight valence electrons it is not possible to draw a Lewis dot structure for C_2 that includes an octet of electrons about each carbon atom. Because the bond order is 2, we show a double bond between the two C atoms, and the remaining four electrons are drawn as two lone pairs on each carbon. The lone pairs correspond, at least formally, to the $\sigma 2s$ and $\sigma^* 2s$ electrons, which confer no net bonding on the molecule. Thus the dot structure we draw for C_2 is :C=C:. Because C_2 is an electron deficient molecule with only six electrons around each carbon atom, it exists only at high temperatures. At lower temperatures carbon exists either as diamond, in which each C is covalently bonded to four other C atoms, or as graphite, in which each C is covalently bonded to three other C atoms. The structures of diamond and graphite will be discussed in Sections 14.7 and 14.8, respectively.

The diatomic molecule N_2 is the major constituent of air and is, of course, a stable molecule over a very wide range of temperatures. The electronic configuration of the N atom is $1s^2 2s^2 2p^3$. There are 10 electrons to occupy molecular orbitals in N_2, 5 valence electrons from each N atom. The electronic configuration, depicted in Fig. 14.23, is written as $KK(\sigma 2s)^2(\sigma^* 2s)^2(\pi 2p)^4(\sigma 2p)^2$.

As there are eight electrons in bonding MO's and two in antibonding MO's, the bond order is 3:

$$\text{bond order} = \frac{1}{2}(8-2) = 3$$

There is a triple bond in N_2, and the Lewis structure, :N≡N:, satisfies the octet rule. The MO energy level diagram tells us that the triple bond of the Lewis structure

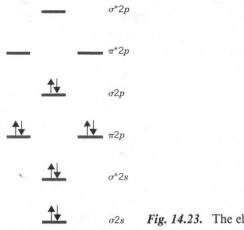

Fig. 14.23. The electronic configuration of N_2.

corresponds to one σ bond and two π bonds. The triple bond draws the two nuclei close together. The N_2 molecule has the shortest bond distance, 1.10 Å, and the largest bond energy, 946 kJ · mol^{-1}, of any homonuclear diatomic molecule of the second period.

Bonding in O_2

The next diatomic molecule, O_2, is also a stable species. Each O atom has 6 valence electrons so that there are 12 electrons to put into molecular orbitals. The electronic configuration, depicted in Fig. 14.24, is

$$KK(\sigma2s)^2(\sigma^*2s)^2(\sigma2p)^2(\pi2p)^4(\pi^*2p_x)^1(\pi^*2p_y)^1$$

Note that the order of the energies of the $\sigma2p$ and $\pi2p$ levels of O_2 is the reverse of the order for N_2. The striking thing about the MO description of the electronic configuration of O_2 is that it predicts that O_2 is paramagnetic, as there are 2 unpaired electrons in the π^*2p MO's. The paramagnetism of O_2 had been observed prior to the development of MO theory, and it was difficult to explain, because the Lewis structure $\ddot{\text{O}}{=}\ddot{\text{O}}$ has no unpaired electrons. There is no Lewis structure that is a proper representation of the O_2 molecule. One of the reasons why MO theory was widely

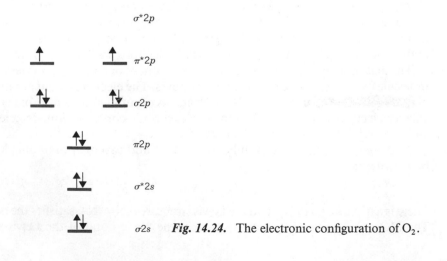

Fig. 14.24. The electronic configuration of O_2.

Table 14.4. **Properties of the Homonuclear Diatomic Molecules of the Second Period of the Periodic Table**

Molecule	Bond Order	Bond Length (Å)	Bond Energy (kJ/mol)	Magnetic Behavior
Li_2	1	2.67	105	Diamagnetic
Be_2	0	⋯	⋯	⋯
B_2	1	1.59	289	Paramagnetic
C_2	2	1.24	630	Diamagnetic
N_2	3	1.10	946	Diamagnetic
O_2	2	1.21	493	Paramagnetic
F_2	1	1.42	158	Diamagnetic
Ne_2	0	⋯	⋯	⋯

accepted in a relatively short period of time is because it can so simply explain the observed paramagnetism of O_2.

There are eight electrons in bonding molecular orbitals and four in antibonding molecular orbitals in O_2, so that the bond order is 2, $\frac{1}{2}(8 - 4) = 2$. Thus there is a double bond in the O_2 molecule, as indicated by the Lewis structure. The bond length, 1.21 Å, is longer than that of N_2, and the bond energy, 493 kJ · mol^{-1}, is less than that of N_2, which has a triple bond.

A summary of information about homonuclear diatomic molecules of the second period is given in Table 14.4. You should be able to explain the observed bond length and bond energy of F_2, and also to account for the absence of a stable Ne_2 species.

section 14.6
Heteronuclear Diatomic Molecules of the Second Period

One of the mathematical requirements for combining atomic orbitals of two different atoms to make molecular orbitals is that the two atomic orbitals must be fairly close in energy. It is for this reason that only the valence electrons of the two different atoms combine.

Consider, for example, the diatomic molecule lithium hydride, LiH, which is known to exist in the gas phase. (At room temperature LiH is an ionic crystalline solid. It melts at 680 °C. Individual LiH molecules exist only in the vapor at high temperatures.) Because of the greater nuclear charge on the Li nucleus, the $1s$ atomic orbital of Li is much lower in energy than the $1s$ atomic orbital of H, and there is no effective interaction between these two atomic orbitals. The $2s$ atomic orbital of Li, however, is screened from the Li nucleus by the two inner $1s$ electrons, and as a result the energies of the $2s$ atomic orbital of Li and the $1s$ atomic orbital of H are not very different, and they can be combined to form a molecular orbital. The $1s$ atomic orbital of H is somewhat lower in energy than the $2s$ atomic orbital of Li. When these two atomic orbitals combine, a bonding and an antibonding molecular orbital are formed, called simply σs and $\sigma^* s$. The σs bonding orbital is lower in energy than either the $1s$ AO of H or the $2s$ AO of Li, but it is closer in energy to the H $1s$ orbital. This is in accord with the fact that hydrogen is more electronegative than Li.

The energy level diagram for the molecular orbitals of LiH is shown in Fig. 14.25. An electron in the bonding σs MO spends more time close to H than it does close to Li. In the antibonding orbital, on the other hand, an electron spends more time close to Li than to H. Since there are two valence electrons for LiH, the σs MO is filled, and the molecule is stable. The bond is partly covalent, partly ionic, with the Li end

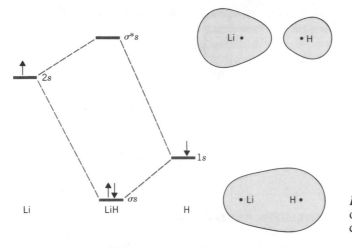

Fig. 14.25. Molecular orbital energy level diagram for LiH.

positively charged and the H end negatively charged. The observed dipole moment of LiH is 5.88 D (debyes). If the molecule consisted of a Li⁺ ion and an H⁻ ion separated by 1.60 Å, which is the observed bond distance in LiH, the dipole moment would be 7.68 D.* One simple and qualitatively useful definition of the **percent ionic character** is the ratio of the actual dipole moment to the hypothetical "pure ionic" moment, which is 5.88/7.68 or 77% ionic character for LiH. While the number 77% should not be considered to be exactly correct, it is certainly true that the bond in LiH is aptly described as an ionic bond with partial covalent character.

For a **heteronuclear diatomic molecule** or ion of the type AB, such as NO, CO, or CN⁻, the molecular orbital energy level diagram is similar to that for homonuclear diatomic molecules (Figs. 14.20 and 14.21), except that the energy levels of the more electronegative atom are lower than those of the more electropositive atom. In the bonding molecular orbitals, therefore, the electronic charge distribution is skewed toward the more electronegative atom, while in the antibonding molecular orbitals, it is skewed toward the less electronegative atom. An energy level diagram for a heteronuclear diatomic molecule of the type AB is shown in Fig. 14.26.

The diatomic molecule NO is of interest because it contains an odd number of valence electrons and is therefore paramagnetic. There are 11 electrons to occupy molecular orbitals, 5 from the N atom and 6 from the O atom. The electronic configuration of NO is $KK(\sigma s)^2(\sigma^* s)^2(\pi p)^4(\sigma p)^2(\pi^* p)^1$. With 8 electrons in bonding MO's and 3 in antibonding MO's, the bond order is $\frac{1}{2}(8 - 3)$ or $\frac{5}{2}$. Indeed, both the bond energy and bond length of NO are intermediate between those of O_2 (bond order = 2) and N_2 (bond order = 3). A bond order of $\frac{5}{2}$ cannot be represented by a Lewis structure. The best Lewis structure, $:\dot{N}=\ddot{O}:$, places the unpaired electron on N. In the MO description the unpaired electron is in the antibonding $\pi^* p$ MO, so that it does spend more time around the less electronegative N atom than it spends around the O atom.

Both the molecule CO and the ion CN⁻ have 10 valence electrons to occupy molecular orbitals. They are therefore isoelectronic with N_2. The electronic configuration of each of these species is $KK(\sigma s)^2(\sigma^* s)^2(\pi p)^4(\sigma p)^2$. It is interesting to compare the molecular orbital description of CO with the Lewis structure $:C\overset{\ominus}{\equiv}\overset{\oplus}{O}:$, which puts a formal charge of −1 on the C atom and +1 on the O atom. Since oxygen is

---

* A single positive and negative charge separated by 1.00 Å has a dipole moment of 4.80 D. Since the dipole moment is defined as the product of the charge and the distance separating the centers of positive and negative charge, the hypothetical dipole moment of a "pure ionic" species is given by (charge) × (bond distance) × (4.80) D.

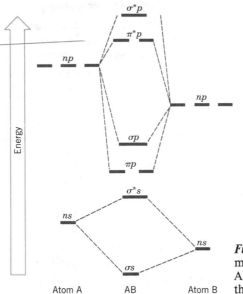

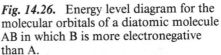

Fig. 14.26. Energy level diagram for the molecular orbitals of a diatomic molecule AB in which B is more electronegative than A.

more electronegative than carbon, the oxygen atomic orbitals contribute more to the bonding MO's than do the carbon atomic orbitals, whereas the carbon atomic orbitals contribute more to the antibonding MO's. As there are 8 electrons in bonding MO's and only 2 in antibonding MO's, the electron cloud is skewed toward oxygen. Thus the electrons in the bond are not equally shared by the two atoms, as is assumed in making the formal charge calculation. The net result is a CO molecule with a very small dipole moment, 0.1 D. The polarity of the dipole is, however, $\overset{\ominus}{C}\equiv\overset{\oplus}{O}$.

There is a triple bond in CO, just as in N_2, and the bond distance is just slightly larger than that of N_2, 1.13 Å as compared to 1.10 Å. As is expected for a triple bond, the bond energy of CO is very large, 1073 kJ $\cdot$ mol^{-1}.

section 14.7
Hybrid Atomic Orbitals and the Molecular Geometry of Polyatomic Molecules

The covalent bond in H_2 serves as a model for understanding all covalent bonds. When two atoms, each with an unpaired electron, come close enough together so that atomic orbitals begin to overlap, the two electrons are pulled into the region between the two nuclei. The greater the amount of overlap of the atomic orbitals in the region between the two nuclei, the stronger the bond between the two atoms will be. Charles Coulson, the British theoretical chemist, has termed this **"the criterion of maximum overlap."**

When we consider atoms other than hydrogen, we realize that the atomic orbitals of the valence electrons used for bonding are sometimes p or d orbitals that have quite different angular distributions than the s orbitals. Because of the specific angular distributions, the overlap of atomic orbitals will be concentrated in certain directions in space. The bond formed will be a directed covalent bond, localized in the portion of space in which the orbital is concentrated. This phenomenon determines the stereochemistry or three dimensional geometry of molecules, the existence of bond directions and bond angles.

While VSEPR theory enables us to predict molecular geometries, it does not explain the nature of the bonding between atoms in polyatomic molecules. It simply starts with the number of bonded pairs and lone pairs of electrons in the Lewis structure of a molecule, and uses that number to deduce the geometry. In the molecular orbital theory of bonding, we must describe the atomic orbitals that overlap to form a bonding molecular orbital with an electronic charge cloud distribution localized largely in the region between two nuclei, thereby forming a bond between them. It is more difficult to apply molecular orbital theory to polyatomic molecules than to diatomic molecules, however.

Let us consider the chemistry of beryllium, which we discussed briefly in Sections 13.9 and 13.10. The electronic configuration of Be is $1s^2 2s^2$. Although beryllium is an alkaline earth element, the amount of energy required to ionize both of its $2s$ electrons is quite large, and therefore beryllium is less electropositive than the other alkaline earths. The bonds in beryllium compounds, even those with the most electronegative elements, have at least partial covalent character. In compounds with the formula BeX_2, the $X—Be—X$ system is linear. Specific examples of linear compounds in which beryllium forms two polar covalent bonds are gaseous $BeCl_2$, $BeBr_2$, BeI_2, and BeH_2. Unlike hydrogen or any of the halogens, the electronic configuration of an isolated beryllium atom, $1s^2 2s^2$, has no unpaired electrons. What atomic orbital of Be is used to overlap with a $3p$ atomic orbital of chlorine, for example, to form the molecular orbital in $BeCl_2$?

A model that is used to account for the observed geometry of linear BeX_2 compounds and of other covalently bonded polyatomic molecules as well, is the concept of **hybrid atomic orbitals.** A hybrid atomic orbital is a combination of valence atomic orbitals *of the same atom.* Atomic orbitals on the same atom that are fairly close in energy can be combined to form hybrid atomic orbitals that can be used for molecule formation in place of the simple atomic orbitals. The number of hybrid atomic orbitals that can be formed is always identical to the number of simple atomic orbitals being combined.

The $2s$ and $2p$ atomic orbitals of Be are fairly close in energy. Mathematically we can combine the $2s$ and one of the $2p$ atomic orbitals of Be to produce two new hybrid atomic orbitals, which are called *sp* **hybrids.** If we arbitrarily call the internuclear axis of the linear $Cl—Be—Cl$ molecule the z axis, then we must combine the $2s$ atomic orbital with the $2p_z$ atomic orbital of Be, in order to form two hybrids that are directed in space along the z axis so that maximum overlap with an atomic orbital of Cl can occur along the z axis.

When we represent atomic orbitals with diagrams we usually draw the probability distribution or electron charge cloud, that is, we draw $|\psi|^2$. (Refer to Section 13.1.) In forming hybrid atomic orbitals we combine the atomic orbitals themselves, and not their squares. In Fig. 14.27 both the p_z wave function and its square are illustrated. The sign of the wave function is positive in one lobe and negative in the other. The

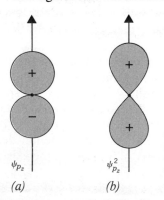

ψ_{p_z}

(a)

$\psi_{p_z}^2$

(b)

Fig. 14.27. Comparison of *(a)* the wave function ψ_{p_z} with *(b)* the square of the wave function, $\psi_{p_z}^2$. The sign of the wave function is plus in one lobe, minus in the other lobe. The electron density, which is proportional to the square of the wave function is, of course, positive everywhere.

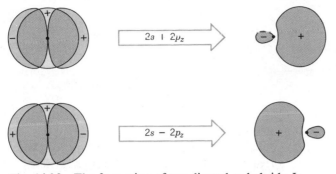

Fig. 14.28. The formation of two digonal *sp* hybrids. In general, the combination of the *ns* and an *np* atomic orbital of the same atom produces two *sp* hybrids with their directions of maximum extent 180° apart.

square of the wave function is, of course, positive everywhere, as it must be to be a probability distribution. (Refer to Section 12.6.) We can form two hybrid atomic orbitals from the 2*s* and $2p_z$ wave functions, $\psi_{2s} + \psi_{2p_z}$ and $\psi_{2s} - \psi_{2p_z}$. The formation of these two hybrids is shown in Fig. 14.28. The maximum extent of one of the two *sp* hybrids is in the $+z$ direction; the maximum extent of the other *sp* hybrid is in the $-z$ direction, so that the two *sp* hybrids (also called **digonal hybrids**) are at 180° to one another. The *sp* hybrids extend further along the *z* axis than either the *s* or the p_z orbitals.

In explaining the observed geometry of linear BeX_2 molecules by the use of *sp* hybrid atomic orbitals, we imagine a hypothetical valence state in which the two 2*s* electrons of a beryllium atom occupy the two *sp* hybrids with parallel spin. Because the 2*p* orbitals are higher in energy than the 2*s* orbital, a combination of 2*s* and 2*p* orbitals is necessarily higher in energy than the 2*s* orbital. This is illustrated in Fig. 14.29.

The energy required to **promote** the two 2*s* electrons to the higher energy *sp* hybridized state is more than compensated for by the formation of two strong covalent bonds when the *sp* hybrids overlap with atomic orbitals of other atoms.

Let us consider $BeCl_2$ as an example. Beryllium chloride is a white solid that sublimes readily. The vapor phase at high temperatures consists of monomeric linear $BeCl_2$ molecules. In these monomers beryllium forms two polar covalent bonds with Cl. The electronic configuration of Cl is $(Ne)3s^2 3p_x^2 3p_y^2 3p_z^1$. Since we have chosen the *z* axis to be the bond direction, the unpaired electron of Cl is in the $3p_z$ atomic

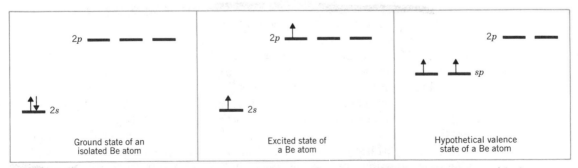

Fig. 14.29. Energy level diagram for beryllium comparing the ground state with the valence state used to form two covalent bonds. In the valence state there are two unpaired electrons in the two *sp* hybrid atomic orbitals. Note that there are always four atomic orbitals. The 2*s* and one of the 2*p* orbitals have been replaced by two *sp* hybrids.

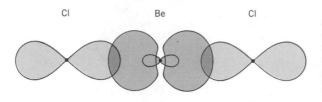

Cl   Be   Cl

Fig. 14.30. Schematic diagram of the overlap of two Cl $3p$ atomic orbitals with the Be sp hybrid orbitals to form linear $BeCl_2$.

orbital. A σ-type molecular orbital is formed by the overlap of the $3p_z$ atomic orbital of Cl with one of the sp hybrid atomic orbitals of Be. In order to have maximum overlap between the $3p_z$ atomic orbital of Cl and the sp hybrids of Be, both the Be nucleus and the Cl nucleus must be on the z axis. A second bond is formed by the overlap of the $3p_z$ atomic orbital of a second Cl atom with the other sp hybrid. Because the two hybrids are at 180° to one another, the molecule is linear. A schematic diagram of the overlap of atomic orbitals in monomeric $BeCl_2$ is depicted in Fig. 14.30. The two electrons shared by Be and Cl are more attracted to Cl, with its greater electronegativity, so that each bond is polar covalent. The entire $BeCl_2$ molecule is, however, nonpolar as the two bond dipoles sum to zero.

In the preceding discussion we have utilized the formation of two digonal sp hybrid atomic orbitals by Be in order to account for the observed linear geometry of BeX_2 compounds. Many other types of hybrid orbitals are used to explain other molecular geometries. Before we discuss any other specific molecules, let's summarize the information we have about hybrid atomic orbitals.

1. Hybridization is the process of mathematically combining or mixing orbitals on a single atom. Thus hybrid orbitals are *atomic orbitals*.
2. Only atomic orbitals fairly close in energy can be combined to form hybrid orbitals.
3. The number of hybrid atomic orbitals formed is always exactly equal to the number of simple atomic orbitals combined.
4. When atomic orbitals are combined to form hybrids, only the hybrids and any simple orbitals not utilized in hybrid formation are available for electrons to occupy. Thus an isolated Be atom has four atomic orbitals: one $2s$ and three $2p$. In its sp hybridized state Be still has four atomic orbitals: two sp hybrids and the two $2p$ atomic orbitals not used in hybrid formation.

Let us now consider bonding in BF_3, a colorless, pungent gas at room temperature (bp = −101 °C). Boron trifluoride is a planar molecule with three equivalent B—F bonds. The electronic configuration of isolated boron atoms is $1s^22s^22p$. In order to form three equivalent bonds in BF_3, boron must use three equivalent atomic orbitals to overlap atomic orbitals on fluorine and form molecular orbitals. Let us call the plane of the BF_3 molecule the xy plane. The valence atomic orbitals of boron with extent in the xy plane are the $2s$, $2p_x$, and $2p_y$. By hybridizing these three atomic orbitals, three equivalent sp^2 **hybrid orbitals** (Fig. 14.31) are formed. Each sp^2 hybrid orbital has one-third s and two-thirds p character. The three hybrids lie in the xy plane at 120° angles to one another, and extend further along their direction of maximum extent than do either the $2s$ or $2p$ atomic orbitals. By using the sp^2 hybrids rather than the simple unhybridized atomic orbitals, greater overlap with the atomic orbitals of other atoms is possible, and stronger covalent bonds can be formed. The geometry of BF_3 and the other boron trihalides is **trigonal planar,** and sp^2 hybrids are also called **trigonal hybrids.**

Let us now consider the use of hybrid atomic orbitals to account for the tetrahedral

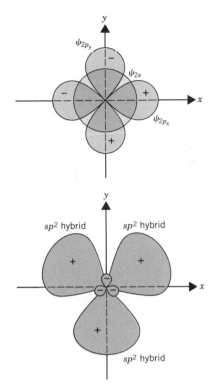

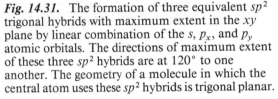

Fig. 14.31. The formation of three equivalent sp^2 trigonal hybrids with maximum extent in the xy plane by linear combination of the s, p_x, and p_y atomic orbitals. The directions of maximum extent of these three sp^2 hybrids are at 120° to one another. The geometry of a molecule in which the central atom uses these sp^2 hybrids is trigonal planar.

geometry of methane and a very large number of other compounds formed by carbon. The electronic configuration of an isolated gaseous carbon atom is $(He)2s^2 2p_x{}^1 2p_y{}^1$. With this configuration in mind we might at first predict the formation of compounds such as $:CH_2$ or $:CR_2$, where R can be any atom or group of atoms able to share an electron and form a single covalent bond. With this simple approach we would say that since an isolated carbon atom has two unpaired electrons it can form two covalent bonds. Note, however, that such a structure is unsatisfactory in that it does not result in an octet of electrons around the carbon atom. In fact, we do not find any stable molecules of the form $:CR_2$ in existence. Carbon forms millions of different compounds, and in all but a very small number (notably $:C\equiv O:$) carbon is **tetravalent**, that is, carbon forms four bonds to other atoms. In methane, CH_4, and carbon tetrachloride, CCl_4, carbon forms four equivalent bonds, and these molecules have tetrahedral geometry. There are many compounds, such as chloroform, $CHCl_3$, and methyl iodide, CH_3I, in which carbon forms four bonds that are not all equivalent, and the geometry is very close to tetrahedral, but the bond angles are not exactly 109.5°, the tetrahedral bond angle.

Since the $2s$ and the three $2p$ atomic orbitals of carbon are not very far apart in energy, they can be combined to form four hybrid atomic orbitals. In 1931 Linus Pauling showed that there are four mathematical combinations of the $2s$ and $2p$ atomic orbitals that are directed in space toward the corners of a regular tetrahedron. A simple way of depicting the four tetrahedral directions is to draw the lines connecting the center of a cube to alternate corners, as in Fig. 14.32. The tetrahedral directions make angles of 109°28′ to one another. The four tetrahedral hybrid orbitals are denoted sp^3 **hybrids.** The symbol sp^3 means that the hybrid is formed by combining one s and three p atomic orbitals, and the hybrid has one-quarter s character and three-quarters p character. The electron density distribution in any of the four tetrahedral sp^3 hybrids has the same shape, shown in Fig. 14.33. The four sp^3 hybrids are

Linus Pauling, a U. S. chemist who profoundly influenced the development of chemistry during the twentieth century, was born in 1901. He received his Ph.D. from the California Institute of Technology, and was a member of the faculty there from 1922 until his retirement.

This chapter takes its title from his book, *The Nature of the Chemical Bond and the Structure of Molecules and Crystals,* first published in 1939, with the third edition published in 1960. Pauling introduced the concepts of directed valence and the hybridization of atomic orbitals. Much of his research was on the effects of resonance on molecular structure and stability, and he investigated the connection between the magnetic properties and electronic structure of transition metal complexes. With E. B. Wilson he wrote a text, *Introduction to Quantum Mechanics,* that was used by a generation of chemists.

Pauling and his students elucidated the arrangements of polypeptide chains in proteins, particularly the alpha helix. His research group also discovered the abnormality in the molecular structure of hemoglobin associated with sickle cell anemia, a genetic disease.

Pauling received the Nobel Prize in Chemistry in 1954 and the Nobel peace prize in 1962. In recent years he has headed research investigations of cancer and the role of vitamin C in maintaining health.

completely equivalent, but each has its direction of maximum extent along a different axis.

We can therefore imagine a higher energy valence state of the carbon atom in which the four $n = 2$ electrons are each in one of the four hybrid tetrahedral orbitals, as shown in Fig. 14.34.

For maximum overlap of the $1s$ atomic orbitals of four H atoms with the four tetrahedral sp^3 hybrid orbitals of carbon, each of the four H atoms bonded to carbon in CH_4 must be located on one of the tetrahedral axes. Because the sp^3 hybrids are concentrated along the tetrahedral axes, the amount of overlap is large. The energy

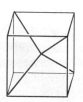

Fig. 14.32. The tetrahedral directions. These are the axes along which the four sp^3 hybrid orbitals have maximum extent.

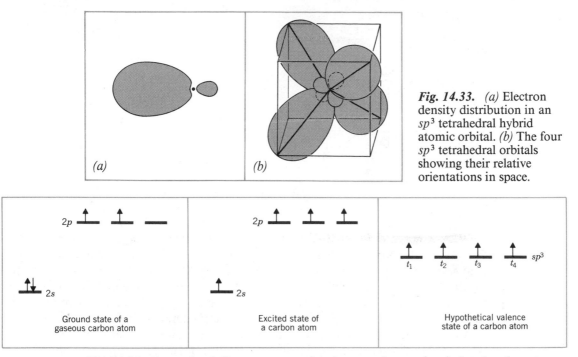

Fig. 14.33. *(a)* Electron density distribution in an sp^3 tetrahedral hybrid atomic orbital. *(b)* The four sp^3 tetrahedral orbitals showing their relative orientations in space.

Fig. 14.34. Energy level diagrams comparing the ground state of an isolated carbon atom with the valence state of a carbon atom that is ready to form four bonds using the four sp^3 tetrahedral hybrid atomic orbitals, which are denoted t_1, t_2, t_3, and t_4.

required to promote the four electrons of carbon to the sp^3 hybridized valence state is more than compensated for by the formation of four strong bonds.

Each of the four equivalent bonds in CH_4 involves a pair of electrons in a σ-type molecular orbital formed by the overlap of a $1s$ atomic orbital of hydrogen with an sp^3 hybrid atomic orbital of carbon. We know from spectroscopic evidence that methane is a symmetric tetrahedral molecule with H—C—H bond angles of exactly $109°28'$. By postulating the formation of hybrid orbitals for carbon and applying the criterion of maximum overlap we can account for the observed geometry of CH_4. When the four atoms bonded to carbon are not identical, small distortions from exact tetrahedral geometry occur. For example, in methyl chloride, CH_3Cl, the H—C—H bond angles are $110.5°$ and the H—C—Cl bond angles are $108.5°$, rather than the exact tetrahedral angle of $109.5°$. Nevertheless, we speak of the CH_3Cl molecule as being tetrahedral.

Elemental carbon exists in two different solid forms, diamond and graphite. The structure of diamond is shown in Fig. 14.35. Each carbon atom is bonded tetrahedrally to four other carbon atoms. Each carbon–carbon single bond in diamond is formed by the overlap of an sp^3 hybrid atomic orbital of one C atom with an sp^3 hybrid atomic orbital of another C atom. Because all the C atoms in the diamond lattice are interconnected by a three-dimensional network of strong carbon–carbon single bonds, diamond is a very hard and stable material. Indeed, diamond is the hardest substance known, and diamonds that are not of gem quality are used industrially for cutting materials that cannot be cut by ordinary saws. Diamond is exceptionally inert chemically, and has the highest melting point (above 3550 °C) of any element. The structure of graphite will be discussed in the following section, and is shown in Fig. 14.42.

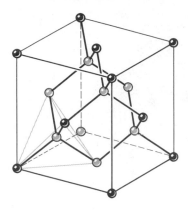

Fig. 14.35. The structure of diamond.

Other atoms in addition to carbon utilize sp^3 hybrid atomic orbitals to form bonds. Let us consider NH_3, in which the H—N—H bond angles are known to be 107.3°. (See Figs. 14.7 and 14.8.) If we assume that the N atom uses sp^3 hybrid orbitals to overlap the $1s$ atomic orbitals of H to form the N—H bonds in NH_3, we would predict that the bond angle should be 109.5°, as in CH_4. However, there are only three N—H bonds in NH_3, and the fourth tetrahedral orbital would have to be occupied by the lone pair of electrons on the N atom. We have already depicted the charge cloud of a lone pair of electrons as spreading out more laterally than the charge cloud of a bonded pair, because the lone pair electrons are attracted only to a single nucleus, not to two nuclei. This lateral spread repels the electrons in the bonded pairs, which move closer together, reducing the bond angle below the tetrahedral 109.5°. This is described, in terms of hybrid orbitals, by saying that the hybrid occupied by the lone pair electrons has slightly more s character (and consequently slightly less p character) than a pure sp^3 hybrid, and the hybrids used in bonding have slightly more p character and slightly less s character than a pure sp^3 hybrid. Depicting the orbitals used by N in NH_3 as very close to sp^3 hybrid atomic orbitals serves to explain the fairly large dipole moment of NH_3, 1.5 D. Each of the three N—H bonds is polar, with the N end negative and the H end positive, but there is also a moment due to the lone pair electrons, with the N end positive and the lone pair electrons negative. The molecular dipole moment, 1.5 D, can be regarded as the vector sum of the three N—H bond moments and the lone pair moment, with a large resultant, as shown in Fig. 14.36.

It is useful to contrast VSEPR theory with the hybrid atomic orbital model for explaining observed molecular geometries. The two models have different purposes. We use VSEPR theory to predict what the molecular geometry will be if we know how many bonded pairs and lone pairs are around the central atom. On the other hand, once we know the geometry of a given molecule, we use the concept of hybrid atomic orbitals to describe the nature of the bonds formed by the central atom. It is not generally possible to predict the hybrids that are used by a central atom; we construct hybrids that can account for the observed geometry.

We have described in detail hybrids that are used to explain a linear geometry, as in $BeCl_2$, hybrids that are used to account for a trigonal planar geometry, as in BF_3, and

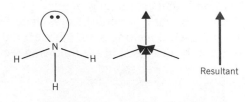

Resultant

Fig. 14.36. The molecular dipole moment of NH_3 as the sum of the three N—H bond dipole moments and the lone pair moment. The resultant molecular dipole moment is 1.5 D.

Table 14.5. Geometries of Frequently Used Hybrid Atomic Orbitals

Hybrid Orbitals	Number of Bonds	Molecular Geometry (Bond Angles)	Specific Example	AO's Combined
sp	2	Linear (180°)	$BeCl_2$	$2s$, one $2p$
sp^2	3	Trigonal planar (120°)	BF_3	$2s$, two $2p$
sp^3	4	Tetrahedral (109.5°)	CH_4	$2s$, all $2p$
dsp^2	4	Square planar (90°)	$Ni(CN)_4^{2-}$	$3d$, $4s$, two $4p$
dsp^3	5	Trigonal bipyramidal (120° and 90°)	PCl_5	$3s$, all $3p$, one $3d$
sp^3d^2	6	Octahedral (90°)	SF_6	$3s$, all $3p$, two $3d$
d^2sp^3	6	Octahedral (90°)	FeF_6^{3-}	two $3d$, $4s$, all $4p$

hybrids that are used to account for a tetrahedral geometry, as in CH_4. Table 14.5 lists several of the more frequently used hybrid orbitals and the geometries to which they correspond. Figure 14.37 illustrates the directional properties of hybrid atomic orbitals.

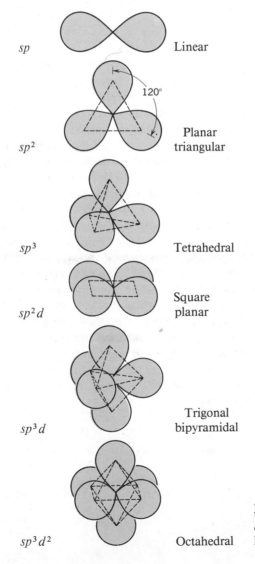

sp — Linear

sp^2 — Planar triangular — 120°

sp^3 — Tetrahedral

sp^2d — Square planar

sp^3d — Trigonal bipyramidal

sp^3d^2 — Octahedral

Fig. 14.37. The directional properties of the most commonly used hybrid atomic orbitals. The minor lobes of these orbitals have been omitted for the sake of clarity.

As can be seen by examining Table 14.5, hybrid orbitals can be formed using the d atomic orbitals as well as the s and p. In SF_6, for example, there are six equivalent S—F bonds, with 90° F—S—F bond angles. The electronic configuration of isolated sulfur atoms is $(Ne)^{10}(3s)^2(3p)^4$. In order to form six bonds, with 12 electrons around sulfur, an expanded valence shell using the $3d$ atomic orbitals is required. Since the F—S—F bond angles are all 90°, the S—F bonds are directed along the x, y, and z coordinate axes. Accordingly, the atomic orbitals combined to form the hybrids are those with maximum extent along the coordinate axes, namely the $3s$, $3p_x$, $3p_y$, $3p_z$, $3d_{z^2}$, and $3d_{x^2-y^2}$. These sp^3d^2 hybrids point towards the corners of a regular octahedron, and account for the octahedral geometry of SF_6.

For elements of the first transition series of metals, the valence electrons are in $3d$ and $4s$ atomic orbitals, which are very close in energy. (Refer to Section 13.5.) Cations of these metals form a large number of octahedral complex ions. The chemistry of these complexes will be discussed in Chapter 20. The hybrid atomic orbitals used by the metal in forming six equivalent bonds directed to the corners of a regular octahedron are the $3d_{x^2-y^2}$, $3d_{z^2}$, $4s$, $4p_x$, $4p_y$, and $4p_z$.

section 14.8
Multiple Bonds

In the previous section we discussed compounds in which a carbon atom uses four tetrahedral sp^3 hybrid orbitals to form bonds to four other atoms, as in CCl_4, CH_3Cl, or CH_2I_2. There are also a large number of compounds in which a carbon atom is bonded to only three other atoms. The simplest of these is ethylene, C_2H_4.

Ethylene is a planar molecule in which each carbon atom is bonded to two hydrogen atoms and to the other carbon atom, as in Fig. 14.38. For a carbon atom to bond to three other atoms, three hybrid atomic orbitals are constructed by combining the $2s$ and two of the $2p$ atomic orbitals. If the hybrids are a combination of the $2s$, $2p_x$, and $2p_y$ atomic orbitals, three sp^2 **(trigonal) hybrids** that lie in the xy plane and make angles of 120° to one another are formed. The fourth valence electron of the C atom occupies the $2p_z$ atomic orbital that is not used in constructing the sp^2 hybrids. The valence state of a carbon atom using sp^2 hybrid orbitals is depicted in Fig. 14.39.

In ethylene each carbon atom makes three σ bonds. Two of these bonds are the result of overlapping an sp^2 hybrid AO of C with the $1s$ AO of H. The third is the result of overlapping an sp^2 hybrid AO on each carbon. The p_z atomic orbital on each carbon atom is occupied by a single electron, and these orbitals overlap sideways to form a π molecular orbital, which is then occupied by the two electrons, forming a second bond between the two carbon atoms. We therefore say that there is a **double bond** between the two carbon atoms in C_2H_4, one σ bond and one π bond. In order to have maximum overlap of the p_z orbitals on the two C atoms, all six atoms of C_2H_4 must lie in one plane. The σ-bond framework of C_2H_4 is illustrated in Fig. 14.40. The Lewis formula for C_2H_4 is either

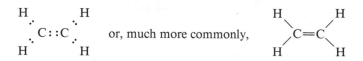

Note that each carbon atom is tetravalent (it forms four bonds to other atoms), and that the octet rule is satisfied.

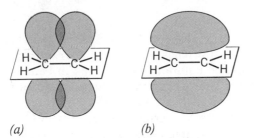

(a)   (b)

Fig. 14.38. The structure of ethylene, C_2H_4. The six atoms lie in a plane. *(a)* Overlapping of the p orbitals on each C atom perpendicular to the plane of the six atoms. *(b)* The π molecular orbital with density above and below the plane of the six atoms.

Ground state of
a carbon atom

Hypothetical
valence state

Fig. 14.39. Energy level diagram depicting the valence state of a C atom ready to form three σ bonds using sp^2 hybrid AO's.

Because a π bond is somewhat weaker than a σ bond, a double bond is not twice as strong as a single bond, but it is certainly considerably stronger than a single bond. A double bond is also shorter than a single bond. The C—C single bond length, as in the diamond structure, is 1.54 Å. The C═C double bond length, as in ethylene, is 1.34 Å. You may also remember, from the discussion of ethylene in Example 14.7, that the double bond takes up somewhat more room than a single bond, and as a result of the increased repulsion of electron pairs in other bonds, the bond angles in ethylene are not exactly 120°. Each H—C—H angle is 117° and each H—C—C angle is 121.5°.

A further discussion of bonding in carbon compounds with double and triple bonds will be found in Chapter 23, Section 23.1.

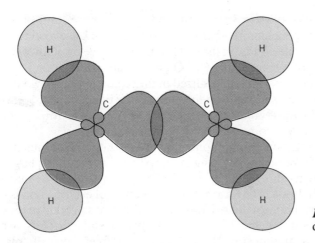

Fig. 14.40. The σ-bond skeleton of ethylene, C_2H_4.

Delocalized Orbitals

Until fairly recently one of the most widely used organic solvents was benzene, C_6H_6. (Its use has decreased recently because it has been found to be **carcinogenic**, that is, to cause cancer.) For many years the nature of the bonding in benzene remained a puzzle. The six carbon atoms in benzene are in a planar hexagonal ring, with one hydrogen attached to each carbon. All carbon–carbon bonds are equal, and the carbon–carbon bond distance is 1.39 Å, intermediate between the C—C single bond distance of 1.54 Å and the C=C double bond distance of 1.34 Å. The structure proposed in 1865 by Friedrich August Kekulé, a German chemist,

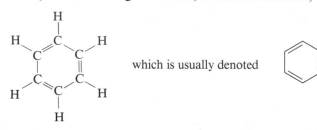

which is usually denoted

is clearly incorrect, because such a structure would contain alternate single and double bonds, and all six carbon–carbon bonds would not be equivalent. We therefore speak of the actual structure as a resonance hybrid of two Kekulé structures:

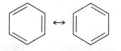

Since each carbon atom in benzene is bonded to three other atoms (two C, one H), with 120° bond angles, the σ-bond framework must involve sp^2 hybridization of the carbon orbitals. This leaves one electron in a p orbital perpendicular to the plane of the ring, and these p orbitals can overlap sideways *on both sides*. The result is a complete **delocalization** of the six p electrons into a ring of π electron density both above and below the plane of the molecule. This is illustrated in Fig. 14.41. It is now customary to indicate the delocalized ring of π electrons by drawing a circle in the center of the hexagon that represents the σ-bond framework of the carbon atoms, as

The bond order of the C—C bond in benzene is $\frac{3}{2}$ (1 σ bond plus $\frac{1}{2}$ a π bond). Fractional bond orders occur in other species represented by several resonance structures involving multiple bonds. In the CO_3^{2-} ion, with three resonance structures each of which has one double bond, the C—O bond order is $1 + \frac{1}{3}$, or $\frac{4}{3}$.

While electrons in σ bonds are localized in the region between two bonded atoms, there are many examples in addition to benzene of π bonds that are delocalized, that is, that extend over more than two bonded atoms. Another important example is graphite, one of the two common forms of elemental carbon.

In graphite each carbon atom is bonded to three others, with equal bond distances of 1.415 Å and 120° bond angles, lying in one plane. The result is a large planar sheet of hexagonal rings, as shown in Fig. 14.42. This geometry tells us that each carbon atom is using sp^2 hybrids. The fourth valence electron of each carbon is in a p orbital perpendicular to the plane of the sp^2 framework. Sideways overlap of the p orbitals on each carbon results in a delocalized π orbital that extends virtually over the entire plane. The electrons that occupy the π bonds move freely over the plane, with the

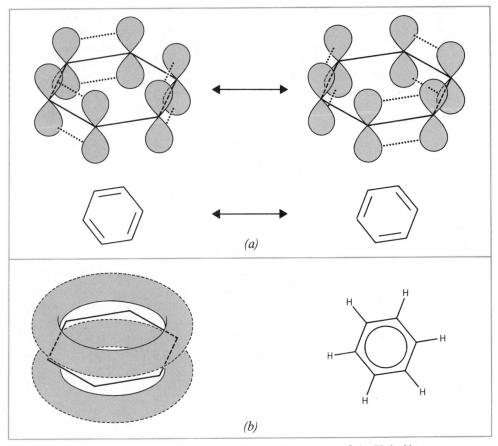

Fig. 14.41. Representations of benzene, C_6H_6. *(a)* Resonance of the Kekulé structures. *(b)* Actual structure.

result that graphite is a good conductor of heat and electricity, but only in directions along the plane of the carbon atoms. Because graphite is a good conductor of electricity, and is a relatively cheap material, a rod of graphite is used as the central electrode in ordinary dry cell (flashlight) batteries.

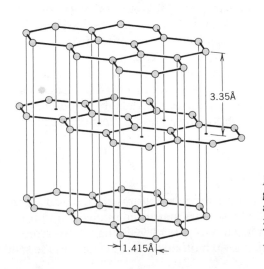

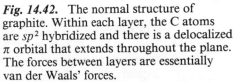

Fig. 14.42. The normal structure of graphite. Within each layer, the C atoms are sp^2 hybridized and there is a delocalized π orbital that extends throughout the plane. The forces between layers are essentially van der Waals' forces.

The planar sheets of hexagonal rings of carbon atoms in graphite are stacked loosely on top of one another, with 3.35 Å between layers. There are actually two different stacking patterns (two crystalline modifications), but in neither case do all the carbon atoms of one layer lie directly over those in an adjacent layer. The 3.35 Å distance between layers is too large for a chemical bond, and the forces between layers are essentially van der Waals' forces. As a result the layers slip easily over one another, and graphite is therefore a useful lubricant. In appearance, graphite is a very dark grey with a shiny luster. It is easily separated into sheets, and feels slippery. It is soft, and leaves a mark if pressed on paper. For this reason it is used as the "lead" in so-called "lead pencils." When one writes with a lead pencil, it is layers of graphite that slide off the pencil and onto the paper.

section 14.9
Hydrogen Bridge Bonds; the Boranes

Boron forms a large number of compounds with hydrogen, called **boranes** or **boron hydrides,** of which the first few are B_2H_6, B_4H_{10}, B_5H_9, B_5H_{11}, B_6H_{10}, and B_6H_{12}. The great variety of stoichiometries and the difficulty with drawing Lewis structures indicate that the bonding in the boranes is unusual. Let us consider diborane, B_2H_6, the simplest of these compounds. Diborane is a colorless, highly reactive, air sensitive gas. It bursts into flame spontaneously if exposed to the air. The reaction with O_2 is

$$B_2H_6(g) + 3 O_2 \rightarrow B_2O_3(s) + 3H_2O \tag{14-6}$$

Diborane also reacts rapidly with water:

$$B_2H_6(g) + 6H_2O \rightarrow 2B(OH)_3(s) + 6H_2(g) \tag{14-7}$$

The product of this reaction, $B(OH)_3$, is boric acid.

Each B atom has three valence electrons and each H atom only one, so that there are 12 valence electrons for B_2H_6. The experimentally determined structure is shown in Fig. 14.43(a). Clearly there are not enough electrons for 8 electron-pair bonds. The measured B—H bond distance in the B—H—B bridges is longer (1.33 Å) than the terminal B—H distance (1.19 Å). Furthermore, in an ordinary electron pair bond, hydrogen can only bond to a single atom. The B—H—B bonds are called **bridge bonds** or **three-center bonds.** They involve a single pair of electrons bonding three atoms, one H and two B atoms. The 1s orbital of the bridging hydrogen overlaps sp^3 hybrids on the boron atoms on each side. The bonding is illustrated in Fig. 14.43(b) and (c). The four terminal H atoms and the two B atoms lie in one plane. The bridging H atoms lie on a line perpendicular to that plane.

While hydrogen bridge bonds are found in compounds with elements other than boron, boron is unusual because of the large number of electron deficient compounds it forms, in which the number of electron pairs available for bonding is less than the number of bonds required for the structure, so that three-center bonds are utilized.

Summary

The formation of a chemical bond between two atoms produces an aggregate which is lower in energy than the separate particles that are combined. It requires significantly more energy to break a chemical or **interatomic bond** than to overcome the **intermolecular forces** of attraction that keep molecules close together in the solid and liquid states.

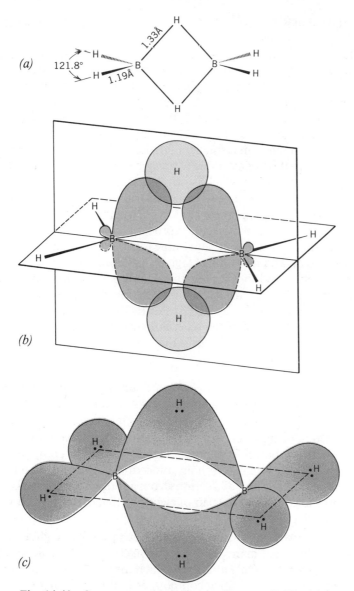

Fig. 14.43. Structure and bonding in diborane, B_2H_6. *(a)* Molecular geometry. Wedge bonds point out, in front of the plane of the paper. Dashed bonds point in, behind the plane of the paper. *(b)* Atomic orbitals overlapping to form the molecule. The B atom is sp^3 hybridized. *(c)* Schematic diagram of the bonding orbitals.

Ions of opposite charge are bound together in ionic crystalline solids by coulombic forces of attraction. The amount the energy is lowered when an ion-pair monomer is formed from isolated gaseous atoms can be calculated using Coulomb's law, the ionization energy of the metal, and the electron affinity of the nonmetal.

A single covalent bond results when two atoms share a pair of electrons. In most (but not all) compounds, each atom is surrounded by an octet of electrons in its valence shell and therefore has the same electronic configuration as one of the rare gases. Sometimes atoms share two or three pairs of electrons in order to achieve the electronic configuration of a rare gas. In these cases we say that a **double** or **triple bond,** respectively, has been formed. Examples of Lewis structures for molecules with double and triple bonds have been discussed in this chapter. In general, the larger the number of bonds (that is, the more electron pairs that are shared), the shorter the **bond length** and the larger the **bond energy.**

When a Lewis structure for a molecule or ion is drawn, each atom has a **formal charge** that is calculated in the following way. The sum of all electrons in lone pairs plus one half the electrons of each pair shared by an atom is assigned to that atom. If this sum is equal to the number of valence electrons in an isolated atom of that element, the formal charge on the atom is 0. A negative formal charge results if this sum is larger than the number of valence electrons in an isolated atom, and a positive formal charge results if this sum is smaller than the number of valence electrons in an isolated atom. This method of assigning formal charges assumes that a shared pair is equally shared by both atoms, which is not necessarily true, so that the calculated formal charges should not be considered necessarily to represent the actual charges on the atoms in the molecule. However, a Lewis structure that involves large formal charges in a covalently bonded molecule is probably not an adequate representation of the bonding in the molecule. The sum of the formal charges on all the atoms must be 0 for a molecule and equal to the charge on any ionic species.

There are molecules and ions for which it is not possible to draw a single Lewis structure that correctly represents the bonding, but for which it is possible to draw two or more structures that satisfy the octet rule. In this case we say that the actual structure is a **resonance hybrid** of the several incorrect structures.

There are three categories of substances that are exceptions to the octet rule. They are

1. Species with an odd number of electrons. Substances with unpaired spins are **paramagnetic,** that is, they are attracted to a magnetic field.

2. Electron deficient compounds. Compounds in which an atom has fewer than eight electrons in its valence shell are formed principally by elements with $Z < 10$. Of these, boron forms more electron deficient compounds than any other element. A substance with an unoccupied atomic orbital in its valence shell that can accept a *pair* of electrons from another substance is called a **Lewis acid.** A substance with a lone pair of electrons that can be shared with another atom is called a **Lewis base.**

3. Species with **expanded valence shells,** that is, more than eight electrons in the outer shell. There are many more exceptions to the octet rule in this category than in the other two. Only atoms for which the principal quantum number of the valence shell is $n \geq 3$ form compounds that require an expanded valence shell, and such structures are more likely when the central atom is large and is bonded to small, highly electronegative atoms.

A very useful theory for predicting the molecular geometry of a molecule or ion from its Lewis structure is the **valence-shell electron-pair repulsion (VSEPR) theory.** This theory asserts that the geometry of any substance will be the structure that minimizes the repulsion between pairs of electrons in the valence shell. A distinction must be made between **bonded pairs** and **lone pairs.** If all the electron pairs in the valence shell are shared with other atoms, that is, are **bonded pairs,** then there is only one geometry that keeps the electron pairs as far apart as possible. The geometry expected for a molecule or ion with a given number of bonded pairs around a central atom is listed in Fig. 14.5. There are no exceptions to this prediction of VSEPR theory.

Lone pairs spread out laterally more than bonded pairs do, and this increased repulsion causes bonded pairs to move closer together and away from the lone pairs.

To predict the geometry of a molecule or ion from its Lewis structure, count the total number of bonded pairs plus lone pairs around the central atom. Multiple bonds are counted as one bonded pair. Start with the geometry expected for all bonded pairs, and then move the bonded pairs away from the lone pairs, decreasing the bond angle. If a multiple bond is present, single bonds move away from the multiple bond slightly, causing small deviations from the ideal geometry shown in Fig. 14.5. A summary of the geometries predicted by VSEPR theory for molecules with lone pairs on the central atom is shown in Fig. 14.13.

An electron in an atomic orbital is centered about a single nucleus. An electron in a **molecular orbital** is spread out over two or more nuclei. A **localized molecular orbital** is centered about two nuclei. Mathematically a molecular orbital is a linear combination of atomic orbitals of two different atoms. When two atoms approach one another so closely that their electron clouds begin to overlap, a molecular orbital that is a combination of the overlapping valence atomic orbitals can be formed.

Two atomic orbitals can be combined to yield two molecular orbitals, a **bonding** and an **antibonding molecular orbital.** In the bonding MO most of the electron density is located in the region between the two nuclei. Both nuclei are attracted to this electron density, which constitutes the bond that keeps the nuclei close together. An electron in a bonding molecular orbital has a lower energy than an electron in either of the atomic orbitals that combined to form the MO. In the antibonding MO, only a small fraction of the electron density is in the region between the two nuclei. Antibonding MO's are denoted by the use of an asterisk. An electron in an antibonding MO has a higher energy than an electron in either of the atomic orbitals that combined to form the MO.

Molecular orbitals are classified according to symmetry type. The combination of an s atomic orbital with any other always results in a sigma (σ)-molecular orbital. A σ-MO has complete symmetry with respect to rotation by any angle about the internuclear axis. Molecular orbitals with one nodal plane that includes the internuclear axis are called pi (π)-MO's. Overlap of p atomic orbitals that are directed along an axis perpendicular to the internuclear axis (sideways overlap) results in π-type molecular orbitals. Appropriate overlap of d orbitals can also result in π-MO's.

The bond order is defined as one half the excess of electrons in bonding MO's over electrons in antibonding MO's. Molecular species are stable only if the bond order is greater than 0. The larger the bond order, the shorter and stronger the bond. Species with a bond order of 1 are said to be single bonded; those with a bond order of 2 are double bonded, and so on. Molecular orbital descriptions of both **homonuclear** and **heteronuclear diatomic molecules** of the second period are discussed in this chapter.

Valence atomic orbitals on the *same* atom that are fairly close in energy can be combined to form **hybrid atomic orbitals.** The number of hybrid orbitals that can be formed is always the same as the number of simple atomic orbitals being combined. Hybrid atomic orbitals are used to overlap atomic orbitals of other atoms to form molecular orbitals.

The combination of the s and one of the p valence atomic orbitals produces two sp (digonal) hybrids that are directed at 180° to one another. Polyatomic molecules with linear geometry can be accounted for by assuming the central atom is using sp hybrids.

The combination of the s and two p valence atomic orbitals produces three sp^2 (trigonal) hybrids that lie in a plane at 120° angles to one another, resulting in trigonal planar geometry.

The combination of the s and all three p valence atomic orbitals produces four sp^3 (tetrahedral) hybrid orbitals that are directed to the corners of a regular tetrahedron, with 109°28′ angles between them. Hybrid orbitals that can account for structures in which the central atom forms more than four bonds involve the *d* atomic orbitals as well as *s* and *p* orbitals.

If there are more than two atoms with unpaired electrons in *p* orbitals that can overlap sideways, a **delocalized π orbital** is formed. Electrons in delocalized orbitals move freely around all the atoms involved.

An unusual type of bonding is found in the **boranes,** compounds of boron and hydrogen. These are electron deficient compounds, with fewer electron pairs available for bonding than the number of bonds that are necessary to form the structure. These compounds utilize **three-center bonds** in which a single pair of electrons bonds three atoms, one H and two B atoms.

Exercises

Section 14.1

1. The internuclear distance for a Li^+Br^- ion-pair monomer has been reported to be 2.24 Å. (a) How much energy is released when one Li^+ ion and one Br^- ion, initially separated by an infinite distance, are brought together to form an ion-pair monomer? (b) How much energy is released when 1 mol of ion-pair monomers are formed from isolated gaseous Li^+ and Br^- ions?

2. What is the difference in energy between a mole of isolated gaseous Li atoms plus a mole of isolated gaseous Br atoms and a mole of isolated gaseous Li^+ ions plus a mole of isolated gaseous Br^- ions? Specify which is higher in energy and the magnitude of the difference. The ionization energy of Li(g) is 520 kJ/mol and the electron affinity of Br(g) is 324 kJ/mol.

Sections 14.2 and 14.3

3. Draw Lewis structures for (a) borohydride ion, BH_4^- (b) cyanide ion, CN^- (c) formic acid, HCOOH (d) H_2S (e) hypochlorite ion, ClO^-.

4. Draw resonance structures for the following species (a) nitrite ion, NO_2^- (b) acetate ion, CH_3COO^- (c) thiocyanate ion, SCN^- (d) ozone, O_3.

5. Calculate the formal charge on each atom in (a) sulfate ion, SO_4^{2-} (b) BeH_2 (c) chlorite ion, ClO_2^- (d) BF_3.

6. The following molecules are exceptions to the octet rule. Draw the most satisfactory Lewis structure for each. (a) NO_2 (b) AsF_5 (c) SF_6.

7. The following oxides of bromine (all of them unstable except at very low temperatures) have been reported: Br_2O, Br_3O_8, BrO_2, and Br_2O_7. Which of these oxides do you expect to be paramagnetic?

8. Predict the molecular geometry of (a) phosphate ion, PO_4^{3-} (b) OF_2 (c) NCl_3 (d) $SbCl_5$ (e) nitrate ion, NO_3^-.

9. Explain why BF_3 is a nonpolar molecule, even though F is considerably more electronegative than B.

10. Draw the structure of 1,1-dichloroethylene, $Cl_2C=CH_2$. What are the predicted bond angles?

11. Explain why the bond angle (a) in HOCl is 103° and not 109°28′, (b) in ClNO is 116° and not 120°, and (c) in NF_3 is 102° and not 109°28′.

12. Explain why in AsF_5 two of the As—F bond distances are 1.711 Å while the other three are 1.656 Å. Draw the structure and indicate which bonds are longer.

13. Explain why the Cl—C—O bond angle in Cl_2CO is 124° and not 120°.

14. Use VSEPR theory to predict that IF_4^- is square planar and not tetrahedral.

15. (a) Predict whether each of the following triatomic molecules is linear or bent: O_3, CS_2, NO_2, HCN, H_2S. Explain your answers. (b) Predict whether each of the molecules in part (a) is polar or nonpolar. Explain your answers.

Sections 14.4 and 14.5

16. Using an energy level diagram, depict the electronic configuration of F_2. Explain why the bond length in F_2, 1.42 Å, is longer than the bond lengths in C_2, N_2, and O_2.

17. The species Ne_2 is not known. Show that it is not expected to be a stable molecule.

18. Arrange the following species in order of increasing bond energy: O_2, O_2^+, O_2^-. Which of these is paramagnetic? Explain your answers by drawing an energy level diagram showing the electronic configuration of each of these species.

19. Compare the F—F bond distance in F_2 and in F_2^+. Which is shorter and why? Explain your answer in detail.

Sections 14.7 and 14.8

20. Predict the geometry of each of the following species and account for this geometry by describing the hybrid orbital used by the central atom. (a) NH_4^+ (b) BCl_3 (c) SF_6 (d) BeH_2 (e) SiH_4

21. The nitrate ion is planar and all O—N—O bond angles are 120°. What atomic orbitals are used by N to bond to O in NO_3^-?

22. Describe each of the bonds in formaldehyde in terms of the overlapping of the atomic orbitals used to form the bond.

23. The IF_4^- ion is observed to be square planar. What kind of hybrid atomic orbitals are used by I to bond to F in IF_4^-?

24. Iodoform, CHI_3, is a crystalline yellow solid with a very sharp odor. (When strongly inhaled, the vapor produces a momentary paralysis of the olfactory nerves so that iodoform appears to lose its smell after the first whiff.) Describe the atomic orbitals that overlap to form the molecular orbital used for the (1) C—H bond and (2) C—I bonds in iodoform and predict the I—C—H bond angle.

25. There are three resonance structures that can be written for naphthalene, $C_{10}H_8$, which has the sigma bond skeleton

   Draw the three resonance structures and describe the bonding in naphthalene in terms of the overlap of atomic orbitals used to form both the carbon–carbon bonds and the carbon–hydrogen bonds.

Multiple Choice Questions

1. Which one of the five oxides of chlorine is paramagnetic?
   (a) Cl_2O (b) ClO_2 (c) Cl_2O_4 (d) Cl_2O_6 (e) Cl_2O_7

2. Pi (π) bonding occurs in each of the following EXCEPT
   (a) CO_2 (b) C_2H_4 (c) CN^- (d) C_6H_6 (e) CH_4

3. Each of the following molecules has a nonzero dipole moment EXCEPT
   (a) C_6H_6 (b) CO (c) SO_2 (d) NH_3 (e) LiH

4. Of the elements Sr, Zr, Mo, Cd, and Sb, all of which are in the fifth period of the periodic table, the ones that are paramagnetic are
   (a) Sr, Cd, and Sb only
   (b) Zr, Mo, and Cd only
   (c) Sr, Zr, and Cd only
   (d) Zr, Mo, and Sb only
   (e) Zr and Mo only

5. The geometrical structure of the PF_3 molecule is
   (a) square planar　(b) trigonal bipyramidal　(c) tetrahedral
   (d) trigonal pyramidal　(e) trigonal planar

6. The total number of valence electrons in the phosphonium ion, PH_4^+, is
   (a) 8　(b) 9　(c) 10　(d) 12　(e) 18

7. In the best Lewis structure for ICl_3 the formal charge on I is
   (a) 0　(b) +1　(c) −1　(d) +2　(e) −2

8. In which of the following molecules does the central atom use sp^2 hybrid atomic orbitals in forming bonds?
   (a) H_2S　(b) CS_2　(c) Cl_2O　(d) NH_3　(e) SO_2

9. The geometrical structure of the sulfite ion, SO_3^{2-}, is
   (a) trigonal pyramidal　(b) square planar　(c) trigonal planar　(d) linear
   (e) tetrahedral

10. In the best Lewis structure for perchlorate ion, ClO_4^-, the formal charge on Cl is
    (a) −1　(b) 0　(c) +1　(d) +2　(e) +3

11. Which of the following molecules is nonplanar?
    (a) C_6H_6　(b) SO_3　(c) CF_4　(d) XeF_4　(e) C_2H_4

12. Which hybrids can be used for bonding in a square planar molecule or ion?
    (a) sp^3　(b) dsp^2　(c) sp^2　(d) d^2sp^3　(e) sp^3d

13. In one of the following triatomic molecules the observed bond angle is 116°49′. Which of these molecules would you expect to have a bond angle of about this magnitude?
    (a) H_2O　(b) OF_2　(c) CS_2　(d) O_3　(e) N_2O

14. The total number of valence electrons for the PO_4^{3-} ion is
    (a) 26　(b) 28　(c) 29　(d) 30　(e) 32

15. In which of the following compounds does every atom have eight electrons in its valence shell?
    (a) IF_5　(b) C_2H_4　(c) SF_4　(d) NO_2　(e) KH

16. In which of the following compounds do the bonds have the largest percentage of ionic character?
    (a) N_2O_4　(b) H_2O　(c) HF　(d) CO_2　(e) IBr

17. There are both ionic and covalent bonds in each of the following compounds EXCEPT for
    (a) $CaCO_3$　(b) Sr_3N_2　(c) $NaNO_3$　(d) K_2SO_4　(e) $(NH_4)_3PO_4$

18. Which of the following best describes the hybrids used by S in the sulfite ion, SO_3^{2-}?
    (a) sp　(b) sp^2　(c) sp^3　(d) dsp^2　(e) d^2sp^3

19. Which of the following is an electron deficient compound?
    (a) $NaBII_4$ (b) CO_2 (c) C_2H_6 (d) B_2H_6 (e) C_2H_4

20. Which of the following diatomic species do you expect to have the longest bond length?
    (a) NO^+ (b) O_2^- (c) CO (d) O_2^+ (e) N_2^+

21. Several resonance structures are used to describe the bonding in
    (a) BF_3 (b) NO_3^- (c) $CHCl_3$ (d) NH_4^+ (e) ClO_3^-

22. The geometry of the carbonate ion, CO_3^{2-}, is
    (a) square planar (b) trigonal pyramidal (c) tetrahedral
    (d) trigonal planar (e) T shaped

Questions 23–29. Select that one of the five molecules listed below that fits the description given.

(a) CS_2 (b) $RaCl_2$ (c) HCN (d) F_2 (e) C_2Cl_4

23. Has only one pi (π) bond.

24. Has two double bonds.

25. Has one triple bond.

26. Exhibits ionic bonding.

27. Consists of discrete, polar molecules at room temperature.

28. Has a nonpolar single bond.

29. Is a solid at room temperature.

30. Which of the following is nonpolar, but contains polar bonds?
    (a) HCl (b) H_2O (c) SO_3 (d) NO_2 (e) SO_2

Problems

14.1. Neither CO_2 nor SO_2 dimerizes, but NO_2 dimerizes to N_2O_4. Explain, with the aid of Lewis structures, why NO_2 dimerizes. Draw the Lewis structure for N_2O_4.

14.2. X-ray diffraction studies have established that XeF_4 is square planar and not tetrahedral. State the explanation offered by VSEPR theory to account for the square planar geometry of XeF_4.

14.3. Write Lewis structures for the following species. In all cases give the formal charge on every atom and predict the bond angles. (a) Permanganate ion, MnO_4^- (b) Hydrogen peroxide, $HOOH$ (c) Boric acid, $B(OH)_3$ (d) SF_4 (e) ICl_3 (f) I_2O_5 ($O_2I—O—IO_2$)

14.4. Arrange the following five species in order of *decreasing* N—O bond length. Start with the species with the longest N—O bond distance. Explain the reason for the order you have given.

$$NO, NO^+, NO_2^-, NO_3^-, NO_2^+$$

14.5. Draw the Lewis dot structure for sulfate ion, SO_4^{2-}. Include resonance structures that have more than eight electrons around sulfur, and indicate the formal charge on each atom in each structure. Predict the molecular geometry. What hybrid orbitals are used by the S atom to bond to O in SO_4^{2-}?

14.6. The silver ion, Ag^+, forms a number of complexes such as $AgCl_2^-$ and $Ag(NH_3)_2^+$ in which the X—Ag—X system is linear. Consider the reaction $Ag^+ + 2Cl^- \rightarrow AgCl_2^-$. By drawing dot structures for the reacting species show that this is a Lewis acid–base reaction. Which substance is the Lewis acid? Which the Lewis base? Describe the hybrid orbitals used by Ag^+ to bond to Cl^- in the complex $AgCl_2^-$.

14.7. The element phosphorus exists as P_4 molecules in the liquid phase, in the vapor phase below 800 °C, and in solid white phosphorus (one of several allotropic forms or crystalline modifications of solid phosphorus). In a P_4 molecule each atom is bonded to three others and all bonds are equivalent. Draw the electron dot structure of P_4 What is the P—P—P bond angle? How does this compare with the bond angle in PH_3 that you would predict from VSEPR theory? The P_4 molecule is highly reactive, and it has been said that the structure of the molecule is consistent with its high reactivity. On what basis is that statement made?

14.8. Nitryl chloride, O_2NCl, is a colorless, reactive gas. The molecule is planar. Draw the electron dot structure of O_2NCl, indicating the formal charge on each atom. Predict the bond angles. What is the bond order of the N—O bond in this molecule?

14.9. Explain why xenon difluoride, XeF_2, is linear and not bent like OF_2.

14.10. The molecule BCl_3 is planar with 120° bond angles. The molecule PCl_3 is pyramidal with 100° bond angles. In two or three sentences suggest an explanation for this difference in geometry.

14.11. Antimony forms an oxide with empirical formula Sb_2O_3. At one time it was not known whether the vapor phase of this oxide consisted of Sb_2O_3 units or was Sb_4O_6. If the structure involves alternating O and Sb atoms, show that the only Lewis dot structure for an Sb_2O_3 monomer is inconsistent with the experimental evidence that all Sb—O bond distances in the molecule are equal. (It is now well established that the molecule is Sb_4O_6.)

14.12. Chlorine monoxide, ClO, is a highly reactive molecule believed to be an intermediate in the destruction of ozone by chlorine (generated from freon) in the stratosphere. Draw a molecular orbital energy level diagram indicating the electronic configuration of ClO. Describe the atomic orbitals of Cl and O that overlap to form these molecular orbitals. What is the bond order of the Cl—O bond? How does the electronic configuration of ClO account for its high reactivity?

14.13. The molecule formaldehyde oxime (methanaloxime) has the following approximate geometry

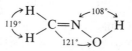

Describe the molecular orbitals for all the bonds in the molecule in terms of the atomic orbitals used by each atom. Draw the structure including lone pairs of electrons and state the orbitals that are occupied by the lone pairs.

14.14. Draw energy level diagrams showing the electronic configurations for the three species NO, NO^+, and NO^-. What is the bond order for each of these? Which of these three molecular species has the largest bond energy? Explain your answers.

14.15. Hydrogen cyanide, HCN, is a poisonous gas. It is considered to be a key precursor molecule in the synthesis of DNA in prebiotic times. The molecule is linear. Describe the molecular orbitals used for the bonds in HCN in terms of the atomic orbitals that are combined to form the MO's. Draw the Lewis dot structure and specify the atomic orbitals occupied by any lone pairs.

14.16. Hydrazoic acid, HN_3, is a dangerously explosive colorless liquid. The molecular geometry of HN_3 is known to be

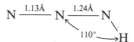

Draw resonance structures that account for this geometry. Include formal charges.

14.17. In BrF_5, four Br—F distances are 1.78 Å, and one Br—F distance is 1.68 Å. Draw the structure of this molecule, indicating the short Br—F bond. The F—Br—F' bond angle is 85°, where F' stands for the fluorine that is closer to Br than the other four. Explain why this bond angle is 85° and not 90°.

14.18. The cyanate ion, NCO^-, is linear. Draw two Lewis resonance structures for this ion. Indicate the formal charges on each. Which of these structures do you expect contributes more importantly to the actual structure of this ion? What experimental measurement would you make to determine which of these structures is closer to the actual structure?

14.19. Consider the three molecular species NO_2, NO_2^+, and NO_2^-.
(a) Draw Lewis dot structures for each of these.
(b) Which of these species is paramagnetic?
(c) Predict the bond angles in NO_2^+ and NO_2^-. The bond angle in NO_2 is 134°. Offer an explanation for the observed value.
(d) Predict relative bond lengths and bond energies in these three species.

14.20. The dipole moments and bond lengths of some diatomic molecules are given below. Calculate an approximate value of the percent ionic character in the bond in each of these molecules.

Molecule	Dipole Moment (D)	Bond Length (Å)
HF	1.82	0.917
HCl	1.08	1.274
BrF	1.29	1.756

14.21. Although the N—F bond is much more polar than the N—H bond, NF_3 has a smaller dipole moment than NH_3. Explain this apparent contradiction.

chapter 15 Oxidation States and Oxidation – Reduction Reactions

Henry Taube, a U.S. inorganic chemist, was born in 1915. He obtained his B.S. and M.S. degrees from the University of Saskatchewan and his Ph.D. from the University of California at Berkeley in 1940. Taube taught at the University of Chicago and is now on the faculty of Stanford. His research concerns the mechanisms of inorganic reactions and the reactivity of inorganic substances. He has investigated electron-transfer reactions in systems of transition metal ions and complexes. He received the Nobel Prize in Chemistry in 1983 for his research on electron-transfer reactions.

We have seen that an important class of reactions that occur in solution are proton-transfer reactions (acid–base reactions), involving the transfer of a proton from a donor (acid) to an acceptor (base). There is another large group of chemical reactions that can be thought of as **electron-transfer reactions.** These reactions may occur in solution, or in the gas phase. They may be **heterogeneous** reactions, involving more than a single phase.

Electron-transfer reactions are also called **oxidation–reduction (redox) reactions.** They are very common, and take place between both inorganic and organic compounds. Oxidation–reduction reactions are important in biological systems; they provide the mechanism for energy transfer in living organisms. They also supply the energy for the batteries we use in our cars, for flashlights, portable radios, flash cameras, and so on. The corrosion of metals occurs as the result of redox reactions.

In this chapter we will study the general features of oxidation–reduction reactions and consider both their similarities to, and differences from, proton-transfer reactions. We will also learn about the oxidation–reduction reactions typical of elements of Groups VA and VIIA of the periodic table, and of the transition metals chromium and manganese.

section 15.1
Definitions of Oxidation and Reduction

Any oxidation–reduction reaction can be considered as the sum of two **half-reactions,** one of which involves a loss of electrons, the other a gain of electrons. A simple example is the reaction of a metal with oxygen, for instance

$$2\text{Li}(s) + \tfrac{1}{2}\text{O}_2(g) \rightleftharpoons \text{Li}_2\text{O}(s) \tag{15-1}$$

In crystalline Li_2O, lithium exists as Li^+ ions and oxygen as oxide ions, O^{2-}, so that the reaction can be thought of as the result of two **half-reactions**

$$2\text{Li}(s) \rightarrow 2\text{Li}^+ + 2e^- \tag{15-1a}$$

and

$$2e^- + \tfrac{1}{2}\text{O}_2 \rightarrow \text{O}^{2-} \tag{15-1b}$$

which sum algebraically to the overall reaction.

In reaction (15-1), metallic lithium loses electrons and becomes lithium ions. This process is known as **oxidation,** and the lithium is said to be **oxidized.** The oxygen gains electrons to become oxide ions; it is said to be **reduced.** The terms **oxidation** and **reduction** are defined as follows:

Oxidation is a loss of electrons.
Reduction is a gain of electrons.

The word *loss* does not imply that the electrons must be removed altogether. Pulling electrons away from an atom oxidizes the atom. Similarly, shifting electrons towards an atom constitutes reduction.

Equation (15-1a) is called an **oxidation half-reaction;** Eq. (15-1b) is called a **reduction half-reaction.** It is particularly important to remember that although we write half-reactions to indicate the electron transfer that has taken place, there are never any free electrons when such an electron transfer between chemical species occurs.

Table 15.1. **Some Common Redox Couples**

$$F_2(g) + 2e^- \rightarrow 2F^-$$
$$Cl_2(g) + 2e^- \rightarrow 2Cl^-$$
$$Ag^+ + e^- \rightarrow Ag(s)$$
$$Fe^{3+} + e^- \rightarrow Fe^{2+}$$
$$I_2(s) + 2e^- \rightarrow 2\,I^-$$
$$Cu^{2+} + 2e^- \rightarrow Cu(s)$$
$$Fe^{2+} + 2e^- \rightarrow Fe(s)$$
$$Zn^{2+} + 2e^- \rightarrow Zn(s)$$
$$Na^+ + e^- \rightarrow Na(s)$$

Oxidants	Reductants
(electron	(electron
acceptors)	donors)

The overall reaction is the one we observe, and no electrons appear in the correctly balanced equation for the overall reaction.

When we discussed acid–base reactions we defined a conjugate pair as an acid (proton donor) and a base (proton acceptor) related by the transfer of a proton. (See Eq. 9-23.) We can similarly define a **redox couple** as a **reducing agent** (electron donor) and an **oxidizing agent** (electron acceptor) related by the transfer of one or more electrons:

$$\text{oxidizing agent} + ne^- \rightleftharpoons \text{reducing agent} \tag{15-2}$$

or simply

$$\text{ox} + ne^- \rightleftharpoons \text{red}$$

In this equation n represents any integer, because redox couples may involve the transfer of one, two, three, or even more electrons. An oxidizing agent is also called an **oxidant,** and a reducing agent is also called a **reductant.** Each half-reaction of an oxidation–reduction reaction involves **one redox couple.** Thus, Li(s) and Li^+ ion are a redox couple, with Li(s) the reducing agent and Li^+ the oxidizing agent.

Table 15.1 lists a number of common redox couples. It is conventional always to write the couple with the oxidizing agent (the oxidized form) on the left and the reducing agent (the reduced form) on the right.

When we discussed acid–base reactions we noted that there are amphiprotic substances (ampholytes), capable of acting either as an acid or as a base. Water is the most prominent ampholyte, but there are a number of others. Similarly, there are substances that can act sometimes as an oxidizing agent and sometimes as a reducing agent. Ferrous ion, Fe^{2+}, is such a species. In Table 15.1 you will note that ferrous ion appears in the column of reducing agents in the half-reaction

$$Fe^{3+} + e^- \rightarrow Fe^{2+}$$

and also in the column of oxidizing agents in the half-reaction

$$Fe^{2+} + 2e^- \rightarrow Fe(s)$$

The reaction between a reducing agent and an oxidizing agent is an **oxidation–reduction reaction.** A reducing *agent* reduces other substances, and when it loses electrons it is thereby oxidized. In reaction (15-1), Li(s) is the reducing agent and $O_2(g)$ is the oxidizing agent. Lithium is oxidized by $O_2(g)$ to Li^+; $O_2(g)$ is reduced by Li(s) to O^{2-}.

section 15.2
The Oxidation State of an Element

To describe the changes that occur in oxidation–reduction reactions and to write correctly balanced equations for such reactions, it is helpful to introduce the concept of the **oxidation state** (or **oxidation number**) of an atom. For monatomic ions, the oxidation state is simply the charge on the ion. For covalently bonded substances, the oxidation state concept is a formalism, a bookkeeping device that makes it easy to keep track of the electrons lost and gained in an oxidation–reduction reaction. The following set of rules is used to assign oxidation states to atoms in ions or molecules:

1. Each pure element has an oxidation state of 0. This is true whether the element is a monatomic gas, a metallic solid, or a polyatomic molecule. Thus $Fe(s)$, N_2, P_4, and S_8 are all in the zero oxidation state.

2. In monatomic ions, the oxidation state of the element is equal to the charge on the ion. Thus the oxidation state of iron is +3 for the ferric ion, Fe^{3+}, and +2 for the ferrous ion, Fe^{2+}. The oxidation state of chlorine in the chloride ion, Cl^-, is −1, and of sulfur in the sulfide ion, S^{2-}, is −2. In discussing an ionic crystalline solid such as K_2S, which consists of K^+ ions and S^{2-} ions, we say that the oxidation states of potassium and sulfur are +1 and −2, respectively. In $AgCl$ the oxidation states of silver and chlorine are +1 and −1, respectively.

3. The oxidation state of hydrogen in any molecule in which it is combined with another element is +1, except in the metallic hydrides such as LiH or CaH_2, where the oxidation state of hydrogen is −1. The metal hydrides are ionic compounds. Crystalline sodium hydride, for instance, consists of Na^+ ions and H^- ions, so the oxidation state of hydrogen is in accord with Rule 2.

4. The oxidation state of oxygen in any molecule or ion in which it is combined with another element is −2, except in the peroxides, the superoxides, and in OF_2. The peroxides are compounds in which there is an O—O covalent bond; the peroxide ion is O_2^{2-}. The oxidation state of oxygen in peroxides is −1. Common peroxides are hydrogen peroxide, H_2O_2, and sodium peroxide, Na_2O_2, but all the alkali and alkaline earth metals form peroxides except for the smallest and least electropositive members of those families, lithium and beryllium. The superoxides of the alkali metals, such as KO_2 and CsO_2, are ionic compounds involving the superoxide ion, O_2^-. The oxidation state of oxygen is $-\frac{1}{2}$ in superoxide ion. Another exception to the rule that oxygen always has a −2 oxidation state is the uncommon gaseous compound OF_2, in which the oxidation state of oxygen is +2 and that of fluorine is −1.

5. In covalent compounds not involving hydrogen or oxygen, the more electronegative element is assigned its common negative oxidation state. Thus chlorine is assigned the −1 oxidation state and sulfur the −2 oxidation state.

6. The algebraic sum of the oxidation numbers of all the atoms combined in a molecule or complex ion must equal the net charge on the molecule or ion. This rule enables us to determine oxidation states not prescribed by the previous rules.

Example 15.1 illustrates the use of these rules to determine the oxidation state of atoms in molecules or ions.

EXAMPLE 15.1. The determination of oxidation states

What is the oxidation state of **(a)** Mn in MnO_4^-, **(b)** C in CCl_4, **(c)** Cr in $Cr_2O_7^{2-}$, **(d)** Cl in $KClO_3$, **(e)** C in CH_3OH, and **(f)** C in CN^-?

Solution

(a) The oxidation state of each of the four oxygen atoms in the deep purple permanganate ion, MnO_4^-, is -2 and the charge on the complex ion is -1. Rule 6 then states:

$$\text{oxidation state of Mn} + 4(-2) = -1$$

The oxidation state of Mn is therefore $+7$ in permanganate ion. To indicate that this is *not* a charge, we write Mn(VII).

(b) According to Rule 5, the oxidation state of Cl in CCl_4 is -1. Since the sum of the oxidation numbers of all the atoms must be zero, the oxidation state of C is $+4$ in CCl_4.

(c) Applying Rules 4 and 6, since $2(+6) + 7(-2) = -2$, the oxidation state of Cr in the bright orange dichromate ion, $Cr_2O_7^{2-}$, is $+6$. We write this as Cr(VI) to emphasize that the $+6$ does not represent a charge.

(d) Potassium chlorate, $KClO_3$, is an ionic crystalline solid containing the ions K^+ and ClO_3^-. The oxidation state of chlorine is $+5$, as $+5 + 3(-2) = -1$, the charge on the chlorate ion.

(e) In methanol, CH_3OH, we assign oxygen the -2 oxidation state and each H is assigned the $+1$ state. Since

$$\text{oxidation state of C} + 4(+1) + (-2) = 0$$

the oxidation state of carbon is -2 in methanol.

(f) In the cyanide ion, CN^-, we utilize Rule 5. Nitrogen is the more electronegative of these two elements, and the common negative oxidation state of N is -3, as in NH_3. This makes the oxidation state of carbon $+2$, since $+2 + (-3) = -1$, the charge on the ion.

Formal Charge versus Oxidation State

It should be clear to you from the examples given that for a covalently bonded compound the oxidation state is *not* the charge on the atom, or anything close to it. Let us consider the four carbon compounds CH_4, CH_3OH, $CHCl_3$, and CCl_4. Using the rules for assigning oxidation states, we find that the oxidation state of carbon is -4 in CH_4, -2 in CH_3OH, $+2$ in $CHCl_3$, and $+4$ in CCl_4. In fact, the electron density around carbon is very similar, although not identical, in all of these compounds. There are four bonded pairs around carbon in each of them and the calculated formal charge (see Section 14.2) on carbon is 0 in all four.

The rules for calculating oxidation states of atoms assign both electrons in a shared pair to the more electronegative atom. The rules for calculating the formal charge assign one electron of a shared pair to each of the bonded atoms. Neither of these assumptions is exactly correct. For covalent bonds, with a small amount of ionic character, the formal charge is closer to the actual charge on the atom. The oxidation state is equal to the actual charge only when the bonding is ionic. In the four carbon compounds we considered above, when more electronegative atoms such as O and Cl are bonded to C, as substitutes for H in CH_4, the electron density around C decreases slightly but the changes are not nearly as large as the changes in the oxidation state.

Because Cl is more electronegative than C, the C—Cl bond is polar covalent, and there is a slight positive charge on C in CCl_4, but it is not anything close to +4. Despite the fact that the oxidation state has no direct physical significance for covalently bonded substances, it is a convenient and useful device for keeping track of electrons lost or gained during an oxidation–reduction reaction, as we shall see in the next section.

EXAMPLE 15.2. Comparison of formal charge and oxidation state

Calculate both the formal charge and the oxidation state of Cr in the chromate ion, CrO_4^{2-}.

Solution. The formal charge on Cr in the Lewis dot structure

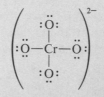

is +2. (See Example 14.2.) As discussed in Example 14.4, experimental evidence on the Cr—O bond length in CrO_4^{2-} indicates that the bond is in between a single and a double bond. In the resonance structures of Example 14.4, in which Cr has an expanded valence shell utilizing d orbitals, the formal charge on Cr is +1. The oxidation state of Cr, however, is +6 in the CrO_4^{2-} ion. Each of the four O atoms is in the −2 oxidation state, and $+6 + 4(-2) = -2$, the charge on the ion. The bond between Cr and O is polar covalent, but the actual charge is close to +1, and the oxidation state of +6 does not represent the actual charge at all.

Intermediate Oxidation States

Many elements exhibit three or more oxidation states. We have already seen one example, that of iron. The oxidation state of iron is 0 for the metallic element, +2 for the ferrous ion, and +3 for the ferric ion. The +2 state of iron is said to be an **intermediate oxidation state** because both a higher and a lower oxidation state exist. Substances in which an element is in an intermediate oxidation state are capable of acting either as an oxidizing agent or as a reducing agent.

Fractional Oxidation States

We have already encountered one example of fractional oxidation states in the superoxides of the alkali metals, KO_2, CsO_2, and RbO_2. Since the oxidation state of an alkali metal in any compound is always +1, the oxidation state of oxygen is $-\frac{1}{2}$ in the superoxides.

In Fe_3O_4, an important iron ore called magnetite, the oxidation state of Fe that is calculated using the rules given above is $\frac{8}{3}$, as

$$3(\text{oxidation state of Fe}) + 4(-2) = 0$$

In fact, there are iron atoms in two different oxidation states, Fe(II) and Fe(III) in Fe_3O_4. Magnetite is a mixed Fe(II)–Fe(III) oxide, and can be formulated as $Fe_2O_3 \cdot FeO$, or as $Fe^{II}Fe_2^{III}O_4$.

In many of the boranes or boron hydrides, the rules for calculating oxidation states yield fractional values for the oxidation state of boron. In tetraborane, B_4H_{10}, boron has an oxidation state of 2.5, and in pentaborane-9, B_5H_9, the oxidation state of boron is 1.8. Such fractional oxidation states occur because of the unusual bonding in these compounds. (Refer to Section 14.9.) While fractional states are not common, boron provides more examples than any other element.

The point to be stressed here is that if you are consistent about following the rules for determining oxidation states you will be able to balance the equations for oxidation–reduction reactions by straightforward methods, even when the oxidation state is fractional. For covalently bonded substances, think of the oxidation state as a useful bookkeeping procedure for counting electrons.

The Oxidation State as a Periodic Property

The oxidation state of an atom depends on the electronic configuration of that atom and is therefore a periodic property. We can summarize information about the oxidation states of various families of elements as follows:

Groups IA and IIA. In addition to the 0 oxidation state, the alkali metals (electronic configuration ns^1) exist only in the $+1$ oxidation state and the alkaline earths (electronic configuration ns^2) only in the $+2$ oxidation state. (Refer to Section 13.10.)

Group IIIA. The outer electronic configuration for members of Group IIIA is ns^2np^1, and the most common oxidation state is $+3$. A $+3$ oxidation state corresponds to forming three bonds to elements more electronegative than the Group IIIA element. The $+3$ state, however, is not the only oxidation state exhibited by members of this family. The $+1$ state, which corresponds to losing or sharing the outer p electron, is also found, particularly for the elements with larger atomic numbers. The most metallic element in Group IIIA, thallium, has more compounds in the $+1$ state than in the $+3$ state. Thallium(I) compounds (thallous compounds) include TlF, TlCl, TlBr, TlI, TlOH, Tl_2CO_3, $TlClO_4$, and Tl_2O. Thallium(III) compounds (thallic compounds) include all the trihalides and Tl_2O_3. There are a few examples of the $+2$ oxidation state for Ga and In. Boron is quite different from other members of this family, and forms a number of electron deficient compounds with fractional oxidation states. Most boron compounds are covalently bonded, and even in those compounds where the oxidation state is $+3$ such as BCl_3, B_2O_3, and $B(OH)_3$, a simple triply charged boron ion does not exist.

Group IVA. The outer electronic configuration for members of this group is ns^2np^2. The highest oxidation state is $+4$, the lowest is -4. Carbon forms only covalent bonds, and exhibits all oxidation states from $+4$ to -4. For the more metallic elements of this group, Sn and Pb, the most common oxidation state is $+2$, but there are several examples of the $+4$ oxidation state, notably PbO_2, the Sn(IV) halides, and SnO_2.

The nonmetals. Nonmetals generally exhibit several oxidation states. For each group of nonmetals in the periodic table there is a maximum and a minimum value of the oxidation state. The maximum value is equal to the group number, that is, $+5$ for the nitrogen family, $+6$ for the chalcogens, and $+7$ for the halogens. The minimum value is the negative number equal in magnitude to the number of electrons that must be added to form an anion with rare gas electronic configuration, that is, -1 for the halogens, -2 for the chalcogens, and -3 for the nitrogen family. The relation between the maximum and minimum values of the oxidation states of a family of nonmetals is

$$(\text{maximum} - \text{minimum}) = 8 \qquad (15\text{-}3)$$

Table 15.2. Common Oxidation States of the First Series of Transition Metals

	Sc	Ti	V	Cr	Mn	Fe	Co	Ni	Cu	Zn
+7					MnO_4^-					
+6				CrO_4^{2-}	MnO_4^{2-}	FeO_4^{2-}				
+5			VO_4^{3-}	$\cdots$	$\cdots$	$\cdots$				
+4		TiO_2	VO^{2+}	$\cdots$	MnO_2	$\cdots$				
+3	Sc^{3+}	Ti^{3+}	VO^+	Cr^{3+}	$MnO(OH)$	Fe^{3+}	Co^{3+}	Ni_2O_3		
+2		TiO	V^{2+}	Cr^{2+}	Mn^{2+}	Fe^{2+}	Co^{2+}	Ni^{2+}	Cu^{2+}	Zn^{2+}
+1									Cu^+	

As an example, consider sulfur, a member of Group VIA. The electronic configuration of the valence electrons of this family is ns^2np^4. The highest oxidation state is +6, which occurs when all six valence electrons are shared with a more electronegative atom. The oxidation state of S is +6 in SO_4^{2-} ion and in SF_6. The lowest oxidation state is −2 (S^{2-}), as two electrons must be gained to achieve the electronic configuration of argon.

Members of these families also exhibit oxidation states with values intermediate between the maximum and minimum. We will discuss the oxidation–reduction chemistry of a few selected nonmetallic elements later on in this chapter.

The transition metals. Most of the transition metals exhibit more than one oxidation state. The +2 and +3 states are the most common. Table 15.2 lists examples of the more commonly encountered oxidation states of the first transition series of metals.

The inner transition metals. All the lanthanides and actinides exhibit the +3 oxidation state, which is the most common by far for these elements. Many, however, form compounds using other oxidation states as well, notably Ce(IV), Eu(II), U(II), U(IV), and U(VI).

section 15.3
Balancing Oxidation–Reduction Equations

We can now expand our previous definitions of oxidation and reduction. Oxidation is the process in which some substance loses one or more electrons and therefore the oxidation state of an atom in that substance increases. The word *loss* does not necessarily mean that the electrons are completely removed. It can correspond merely to a shifting of electron density away from the atom being oxidized. In the half-reaction

$$Zn(s) \rightarrow Zn^{2+} + 2e^-$$

the element zinc has lost two electrons and its oxidation state has increased from 0 to +2. The conversion of CH_4 to CH_3OH is also an oxidation because the oxidation state of carbon increases from −4 to −2. Formally, two electrons must be lost to effect this change in oxidation state. As both CH_4 and CH_3OH are covalently bonded, the two electron loss in this case is merely a formalism. A simple but useful mnemonic is OIL: **O**xidation is an **I**ncrease in the oxidation state and a **L**oss of electrons.

Reduction, conversely, is the process in which some substance gains one or more electrons and the oxidation state of an atom in that substance decreases. In the half-reaction

$$Fe^{2+} + 2e^- \rightarrow Fe(s)$$

iron has gained two electrons and its oxidation state has thereby decreased from $+2$ to 0. The Fe^{2+} ion has been reduced.

Oxidation and reduction always occur together. An oxidation half-reaction and a reduction half-reaction must be combined to produce a correctly balanced oxidation – reduction equation. In doing this, the important thing to remember is that the number of electrons gained by the oxidizing agent must be identical to the number of electrons lost by the reducing agent. While the symbol ne^- for n electrons appears in half-reactions, it never appears in a correctly balanced oxidation – reduction equation.

If we want to write the correctly balanced equation for the reaction between $Zn(s)$ and $Ag^+(aq)$ we start with the half-reactions

$$Zn(s) \rightarrow Zn^{2+}(aq) + 2e^- \tag{15-4a}$$

and

$$Ag^+(aq) + e^- \rightarrow Ag(s) \tag{15-4b}$$

Since the number of electrons lost by the reducing agent (Zn) must be equal to the number of electrons gained by the oxidizing agent (Ag^+), we must multiply the reduction half-reaction (15-4b) by two before adding the two half-reactions together to obtain the correctly balanced equation

$$Zn(s) + 2Ag^+(aq) \rightarrow Zn^{2+}(aq) + 2Ag(s) \tag{15-4}$$

Methods of balancing more complex oxidation – reduction equations will be described with reference to specific examples.

EXAMPLE 15.3. Balancing oxidation – reduction equations for reactions in acidic solution given the skeletal (unbalanced) equation

Metallic cadmium can be dissolved by warm dilute nitric acid. The skeletal equation showing reactants and products is

$$Cd(s) + H^+(aq) + NO_3^- \rightarrow Cd^{2+}(aq) + NO(g) + H_2O$$

Write a correctly balanced equation for this reaction.

Solution. The first thing to do is to determine the oxidation state of all the elements on both sides of the equation so that you can identify the substances that have been oxidized and reduced. You will learn to do this in your head with some experience, but at the beginning it is helpful to write down the oxidation state directly beneath the symbol of the element. Hydrogen in the $+1$ state and oxygen in the -2 state need not be written down unless the reaction involves a change in oxidation state of either or both of these elements. With the oxidation states indicated, the skeletal equation appears as

$$Cd(s) + H^+(aq) + NO_3^- \rightarrow Cd^{2+} + NO(g) + H_2O$$
$$0   +5 +2 \phantom{Cd^{2+}} +2$$

The oxidation state of cadmium has increased from 0 to $+2$, requiring a loss of two electrons. Cadmium is oxidized by nitric acid, and the oxidation half-reaction is

$$Cd(s) \rightarrow Cd^{2+} + 2e^-$$

The oxidation state of nitrogen has decreased from $+5$ to $+2$, requiring a gain of three electrons. Note that the $+5$ state is the maximum oxidation state for nitrogen, a member of Group VA. An element in its maximum oxidation state frequently serves as an oxidizing agent, and is itself reduced to a lower oxidation state. We indicate that this reduction of NO_3^- requires a gain of three electrons by writing the skeletal half-reaction:

$$NO_3^- + 3e^- \rightarrow NO(g)$$

This half-reaction is clearly not balanced because neither the O atoms nor the charge are balanced, although the N atoms are. First balance the oxygen atoms by adding H_2O. If a reaction takes place in aqueous solution so that there is water present, you may always use H_2O for balancing O atoms in oxidation–reduction equations. Since there are three O atoms in the NO_3^- ion but only one in gaseous nitrogen oxide (also called nitric oxide), add two molecules of H_2O on the right-hand side of the equation, yielding

$$NO_3^- + 3e^- \rightarrow NO(g) + 2H_2O$$

Now the N and O atoms are balanced, but the H atoms and the charge are not. Balance the hydrogen atoms next. In acid solution this is done by using $H^+(aq)$ ions. In this case we need $4H^+$ on the left-hand side:

$$4H^+(aq) + NO_3^- + 3e^- \rightarrow NO(g) + 2H_2O$$

If you have done the preceding steps correctly, the charge will now be balanced. Hence be sure to *check the charge balance* at this point in the procedure. You will know that you have a correctly balanced half-reaction if all the atoms and the charge are balanced. In this case the net charge on the left is $4(+1) + (-1) + 3(-1) = 0$, and the net charge on the right is also 0.

If the charge does not balance you have made an error in one of the previous steps and should go back to the very first step of determining the oxidation states of all the atoms and continue through all succeeding steps until you have located your error.

We now have two correctly balanced half-reactions:

Oxidation $\qquad\qquad\qquad Cd(s) \rightarrow Cd^{2+} + 2e^-$

Reduction $\qquad 4H^+(aq) + NO_3^- + 3e^- \rightarrow NO(g) + 2H_2O$

The number of electrons lost in oxidation must be equal to the number of electrons gained in reduction. Therefore multiply the oxidation half-reaction by three and the reduction half-reaction by two, and then add them together. We obtain

$$3Cd(s) + 8H^+(aq) + 2NO_3^- \rightarrow 3Cd^{2+} + 2NO(g) + 4H_2O$$

as the correctly balanced net ionic equation. Check the charge balance of the final equation to insure that you have not made any errors. The net charge on the left is $8(+1) + 2(-1) = +6$, and the net charge on the right is $3(+2) = +6$ also. If you have made any numerical errors the charge will not balance, so you will always be able to know if you have balanced the equation correctly.

The oxidizing agent in this reaction is NO_3^- in acidic solution (nitric acid); the reducing agent is $Cd(s)$. Note that the oxidizing agent oxidizes some other substance and is itself reduced. Similarly, the reducing agent is itself oxidized.

If for some reason you want a total ionic equation rather than a net ionic equation you merely have to add $6NO_3^-$ to each side. The products of this reaction are $Cd(NO_3)_2$, H_2O, and $NO(g)$. In aqueous solution cadmium nitrate exists as Cd^{2+} and NO_3^- ions.

EXAMPLE 15.4. Balancing redox equations for reactions in acidic solution when the skeletal equation is not given

Solid black CuS is an exceedingly insoluble salt. It does not dissolve in HCl, but it reacts with concentrated nitric acid, and the resulting solution contains Cu^{2+} ions. This reaction is the simplest way to "dissolve" CuS. Write a correctly balanced net ionic equation for the reaction between CuS(s) and concentrated nitric acid.

Solution. If a skeletal equation is not given, you must be familiar with the chemistry of the reacting species to be able to write the equation. In this example you must know that nitric acid is frequently used as an oxidizing agent, and that it oxidizes sulfur from the −2 state to the 0 state (elemental sulfur). Concentrated nitric acid is reduced to the red-brown gas, NO_2, when it functions as an oxidizing agent. (You should not expect, right now, that you can predict the products of oxidation–reduction reactions. You will acquire this knowledge by studying the rest of this chapter and working in the laboratory. These worked-out examples will familiarize you with important oxidation–reduction reactions as well as teach you how to balance equations.)

Once you have identified the substances being oxidized and reduced, you can start with half-reactions immediately. A skeletal equation for the reduction of nitric acid to NO_2 is

$$NO_3^- \rightarrow NO_2(g)$$
$$+5 \qquad +4$$

Since nitrogen is being reduced from the +5 state to the +4 state, a gain of one electron is required:

$$NO_3^- + e^- \rightarrow NO_2(g)$$

The N atoms are balanced but the O atoms are not, so we add one H_2O to the right-hand side:

$$NO_3^- + e^- \rightarrow NO_2(g) + H_2O$$

To balance the hydrogen atoms, add $2H^+$ to the left, obtaining

$$2H^+(aq) + NO_3^- + e^- \rightarrow NO_2(g) + H_2O$$

Now check the charge balance. Each half-reaction must have all atoms balanced and have the charge balanced as well. The charge on the left-hand side of this half-reaction is $2(+1) - 1 - 1 = 0$. The charge on the right-hand side is 0 also. Therefore this is the correctly balanced reduction half-reaction.

For the oxidation half-reaction, sulfur is oxidized from the −2 state to the 0 state, requiring a loss of two electrons per S atom:

$$CuS(s) \rightarrow S + 2e^- + Cu^{2+}$$
$$-2 \qquad 0$$

Note that the oxidation state of copper does not change. It is +2 in CuS and +2 in the cupric ion in solution. When CuS reacts with nitric acid the elemental sulfur produced is a pale yellow precipitate that can easily be removed by filtration. The Cu^{2+} ions are free in the solution.

To balance the overall reaction, the reduction half-reaction must be multiplied by two:

$$+ \quad \frac{2 \cdot [2H^+ + NO_3^- + e^- \rightarrow NO_2(g) + H_2O]}{CuS(s) \rightarrow S(s) + 2e^- + Cu^{2+}}$$

Yields $\qquad 4H^+(aq) + 2NO_3^- + CuS(s) \rightarrow 2NO_2(g) + S(s) + 2H_2O + Cu^{2+}$

Check the charge balance to locate any errors. The net charge on the left is

$4(+1) + 2(-1) = +2$. The net charge on the right is $+2$ also, so that the equation is correctly balanced.

In this reaction the oxidizing agent is nitric acid, and the reducing agent is the sulfide ion in CuS.

A consideration of Examples 15.3 and 15.4 provides some useful information about nitric acid, an effective and frequently used oxidizing agent. Nitrogen is reduced from its maximum oxidation state ($+5$) to one of its lower oxidation states when nitric acid serves as an oxidizing agent, but the principal product of the reduction depends on the concentration of the acid. When concentrated ($15\ M$) nitric acid is reduced, the major product is the red-brown gas NO_2. In dilute nitric acid (2 to $6\ M$), the principal reduction product is $NO(g)$. A mixture of oxides of nitrogen is undoubtedly obtained no matter what the concentration of the acid, but we indicate the principal product when writing the net ionic equation. The oxidizing power of nitric acid depends strongly on its concentration; the more concentrated the acid, the more effective an oxidizing agent it is. At concentrations below $2\ M$ the acid has very little oxidizing power.

Example 15.4 illustrates the procedure used to dissolve those sulfides, like CuS, that are insoluble both in water and in HCl. In Section 11.4 we discussed the fact that while many insoluble sulfides dissolve in HCl, there are a sizeable number that do not, because they have very small solubility products. For CuS, as an example

$$CuS(s) \rightleftharpoons Cu^{2+} + S^{2-} \qquad K_{sp} = 8.7 \times 10^{-36} \qquad (15\text{-}5)$$

The nitric acid removes S^{2-} ions from solution by oxidizing S^{2-} to elemental sulfur. This decreases the $[S^{2-}]$ to such a low value that $[Cu^{2+}][S^{2-}] < 8.7 \times 10^{-36}$, which drives reaction (15-5) to the right, thereby dissolving CuS. Other sulfides that are insoluble both in water and in HCl, but dissolve in nitric acid because it is such a good oxidizing agent, are PbS, CdS, and Bi_2S_3.

EXAMPLE 15.5. Balancing redox equations for reactions in acidic solution when the element being oxidized or reduced is not balanced in the skeletal equation

Write a correctly balanced net ionic equation for the reaction between potassium dichromate and oxalic acid in acidic solution. The skeletal equation is

$$Cr_2O_7^{2-} + H^+(aq) + H_2C_2O_4 \rightarrow CO_2(g) + Cr^{3+} + H_2O$$

Solution. We first determine the oxidation state of each element other than hydrogen and oxygen.

$$\begin{array}{ccccccc} Cr_2O_7^{2-} & + & H^+(aq) & + & H_2C_2O_4 & \rightarrow & CO_2(g) & + & Cr^{3+} & + & H_2O \\ +6 & & & & +3 & & +4 & & +3 \end{array}$$

We observe that the oxidation state of chromium has been reduced from $+6$ to $+3$. The maximum oxidation state of Cr, which has six valence electrons [electronic configuration $(Ar)3d^5 4s^1$], is $+6$. In acid solution the bright red-orange dichromate ion is a very powerful oxidizing agent. This is another example of the fact that many substances containing an element in its highest oxidation state are effective oxidizing agents. Dichromate ion is frequently used as an oxidizing agent, and is usually reduced to the $+3$ state (chromic ion).

The skeletal reduction half-reaction is therefore

$$\begin{array}{ccc} Cr_2O_7^{2-} & \rightarrow & Cr^{3+} \\ +6 & & +3 \end{array}$$

The very *first* thing to do is to balance the atoms of the element whose oxidation state is changing, which is Cr in this case. Since there are two chromium atoms in $Cr_2O_7^{2-}$ and only one in Cr^{3+}, we must multiply the Cr^{3+} by two, obtaining as our still unbalanced skeletal half-reaction:

$$Cr_2O_7^{2-} \rightarrow 2Cr^{3+}$$

After you have balanced the atoms of the element being reduced, you can determine the number of electrons that must be gained in order to effect this reduction. A reduction from the +6 state to the +3 state requires a gain of three electrons *for each chromium atom*. Since two chromium atoms are involved, a total of six electrons must be gained. Our unbalanced reduction half-reaction is now

$$Cr_2O_7^{2-} + 6e^- \rightarrow 2Cr^{3+}$$

Add $7H_2O$ on the right to balance the oxygen atoms, and then add $14H^+$ on the left to balance the hydrogen atoms. We obtain

$$14H^+(aq) + Cr_2O_7^{2-} + 6e^- \rightarrow 2Cr^{3+} + 7H_2O$$

Now *check the charge balance*. The net charge on the left-hand side is $14(+1) + (-2) + 6(-1) = +6$. The net charge on the right-hand side is $2(+3) = +6$. Since the charge balances we can be sure that this is the correct reduction half-reaction.

The element being oxidized by the dichromate ion is carbon, in oxalic acid. Oxalic acid is a weak organic acid and is oxidized to gaseous CO_2. The skeletal oxidation half-reaction is

$$\underset{+3}{H_2C_2O_4} \rightarrow \underset{+4}{CO_2(g)}$$

The first step is to multiply the CO_2 by two in order to balance the atom being oxidized, carbon. We then have

$$H_2C_2O_4 \rightarrow 2CO_2(g)$$

The oxidation of carbon from the +3 to the +4 state requires a loss of one electron per carbon atom, so that two electrons must be lost in this oxidation half-reaction:

$$H_2C_2O_4 \rightarrow 2CO_2(g) + 2e^-$$

The oxygen atoms happen to be balanced already, so we only have to add $2H^+$ to the right-hand side to balance the hydrogen atoms. The correctly balanced oxidation half-reaction is

$$H_2C_2O_4 \rightarrow 2CO_2(g) + 2e^- + 2H^+(aq)$$

The net charge is 0 on both sides of this half-reaction.

Because the reduction half-reaction involves a gain of six electrons, the oxidation half-reaction must be multiplied by three, to show a loss of six electrons. The two half-reactions are then

Reduction $\quad 14H^+(aq) + Cr_2O_7^{2-} + 6e^- \rightarrow 2Cr^{3+} + 7H_2O$

Oxidation $\qquad\qquad\qquad 3H_2C_2O_4 \rightarrow 6CO_2(g) + 6e^- + 6H^+(aq)$

In summing these we note that there are $14H^+$ on the left-hand side and $6H^+$ on the right-hand side, leaving a net of $8H^+$ on the left, so that the correctly balanced net ionic equation is

$$8H^+(aq) + Cr_2O_7^{2-} + 3H_2C_2O_4 \rightarrow 2Cr^{3+} + 6CO_2(g) + 7H_2O$$

Note that the charge is balanced in the final equation. In this reaction the oxidizing agent is $Cr_2O_7^{2-}$ in acidic solution, and the reducing agent is oxalic acid, $H_2C_2O_4$.

We can summarize the steps to be followed to balance redox equations in acid or neutral solution as follows:

1. Identify the oxidation state of the elements in both products and reactants that are being oxidized or reduced.
2. In each skeletal half-reaction, balance the atoms of the element whose oxidation state is changing.
3. Calculate the number of electrons lost or gained in the half-reaction by multiplying the number of atoms of the element being oxidized or reduced by the change in oxidation state. Add the correct number of electrons to the proper side of the half-reaction.
4. Balance oxygen atoms by using H_2O.
5. Balance hydrogen atoms by using $H^+(aq)$.
6. Check the charge balance of each half-reaction. If the charge does not balance you have made an error in one of the previous steps. Do *not* go further! Go back to the beginning and try to locate your error.
7. Multiply the half-reactions by appropriate integers to insure that the number of electrons lost in oxidation is equal to the number of electrons gained in reduction.
8. Sum the half-reactions. Simplify the overall equation algebraically so that any H_2O or $H^+(aq)$ appears on only one side of the equation. *Check the charge balance* as well as the balance of all atoms.

The Ion-Electron Method

There is an alternative method of balancing oxidation-reduction reactions that rearranges to some degree the eight steps just outlined. This alternative (and completely equivalent) method is called the **ion-electron method.** The sequence of steps used in balancing a redox equation in acidic solution by the ion-electron method is as follows:

1. Identify the oxidizing agent and its reduced form as well as the reducing agent and its oxidized form, and write two skeletal half-reactions.
2. In each half-reaction, balance the atoms of the elements undergoing changes in oxidation state.
3. Balance oxygen atoms by using H_2O.
4. Balance hydrogen atoms by using $H^+(aq)$.
5. Balance the charge in each half-reaction by adding electrons to equalize the ionic charges. The number of electrons to be added is determined without reference to the oxidation states. Electrons are added simply to balance the charge.
6. To locate any errors in each half-reaction, determine the change in oxidation state, multiply this by the number of atoms undergoing this change and compare the product with the number of electrons lost or gained in your half-reaction. If they are not the same, you have made an error and must go back to the first step.

Steps 7 and 8 are the same for both methods.

Essentially the difference between the two methods is whether you use the change in oxidation state to calculate the number of electrons required and then check your work by verifying that the charge is balanced, or you use the charge balance to calculate the number of electrons required and then check your work by verifying that the number of electrons you have added is the number needed to effect the change in oxidation state.

Balancing Oxidation–Reduction Equations in Basic Solution

Balancing redox equations in basic solution is really not different, in principle, from balancing them in acidic solution. However since $[OH^-] > [H^+]$ in basic solution, the final equation should not contain any H^+ ions, in accord with our general rules that net ionic equations show the principal substances present in solution. Instead of using H_2O and H^+ ions to balance oxygen and hydrogen atoms, H_2O and OH^- should be used. There are several ways to do this, but one simple way is to begin with the first three steps of the "oxidation state method" (that is, determine the number of electrons required by considering the change in oxidation state per atom) and then replace steps 4, 5, and 6 with the following:

4. Balance the charge by using OH^- ions.
5. Balance O atoms by using H_2O. This will simultaneously balance H atoms.
6. Check carefully the balance of all atoms and of the charge.

Steps 7 and 8 are the same as for acidic solutions, except that it is OH^- and H_2O that should appear on only one side of the equation. The following example illustrates this procedure.

EXAMPLE 15.6. **Balancing redox equations in basic solution**

Write a balanced net ionic equation for the reaction between potassium permanganate and potassium nitrite in basic solution. The skeletal (unbalanced) equation is

$$MnO_4^- + NO_2^- \rightarrow MnO_2(s) + NO_3^-$$

Solution. Begin by determining the oxidation state of all atoms:

$$MnO_4^- + NO_2^- \rightarrow MnO_2(s) + NO_3^-$$
$$+7 +3 +4 +5$$

We note that +7 is the highest oxidation state of manganese; MnO_4^- is a powerful oxidizing agent. In acidic solution the manganese in permanganate ion is reduced to the +2 state, manganous ion. In basic solution, permanganate is usually reduced to the +4 state, the insoluble black MnO_2. The skeletal reduction half-reaction in basic solution is therefore

$$MnO_4^- + 3e^- \rightarrow MnO_2(s)$$
$$+7 +4$$

We now add OH^- ions to balance the charge. Since there are four negative charges on the left and none on the right, we add 4 OH^- to the right side. The still unbalanced reduction half-reaction is now

$$MnO_4^- + 3e^- \rightarrow MnO_2(s) + 4\,OH^-$$

Since there are six O atoms on the right-hand side and four on the left, add $2H_2O$ to the left. We obtain

$$2H_2O + MnO_4^- + 3e^- \rightarrow MnO_2(s) + 4\,OH^-$$

Note that the H atoms are also now balanced (four on each side). Since the charge balances (the net charge is -4 on each side), and all atoms are balanced, this is the correct reduction half-reaction.

The nitrite ion is oxidized by the permanganate ion to nitrate ion, which is the $+5$ state of nitrogen. The skeletal oxidation half-reaction is

$$NO_2^- \rightarrow NO_3^- + 2e^-$$

There is a charge of -3 on the right, but only -1 on the left. Therefore add $2\,OH^-$ to the left to balance the charge. We obtain

$$2\,OH^- + NO_2^- \rightarrow NO_3^- + 2e^-$$

To balance the O atoms we must add one H_2O to the right:

$$2\,OH^- + NO_2^- \rightarrow H_2O + NO_3^- + 2e^-$$

This also balances H atoms (two on each side). Since the charge and all atoms are balanced, this is the correct oxidation half-reaction.

To write the correctly balanced overall equation we must multiply the oxidation half-reaction by three and the reduction half-reaction by two before summing them. We obtain

$$6\,OH^- + 3NO_2^- \rightarrow 3H_2O + 3NO_3^- + 6e^-$$
$$4H_2O + 2MnO_4^- + 6e^- \rightarrow 2MnO_2(s) + 8\,OH^-$$

Simple addition results in

$$4H_2O + 2MnO_4^- + 6\,OH^- + 3NO_2^- \rightarrow 3H_2O + 3NO_3^- + 2MnO_2(s) + 8\,OH^-$$

but H_2O and OH^- should not appear on both sides of the equation. Algebraic combination yields

$$H_2O + 2MnO_4^- + 3NO_2^- \rightarrow 3NO_3^- + 2MnO_2(s) + 2\,OH^-$$

as the correctly balanced equation. The net charge on each side of this equation is -5.

Disproportionation

We have already mentioned that an element in an intermediate oxidation state may serve either as an oxidizing agent or as a reducing agent. Sometimes intermediate oxidation states are not stable and spontaneously **disproportionate** to give one substance in which the element is in a higher oxidation state and one in which it is in a lower oxidation state. For instance, copper in the $+1$ oxidation state is not stable as a free ion in solution and spontaneously disproportionates to Cu(II) and Cu(0):

$$2Cu^+ \rightarrow Cu^{2+} + Cu(s) \tag{15-6}$$

Disproportionation equations should be balanced in exactly the same manner as any other oxidation–reduction equation, as the following example demonstrates.

EXAMPLE 15.7. Disproportionation reactions

In basic solution the element phosphorus, P_4, disproportionates to yield the poisonous gas phosphine, PH_3, and phosphite ion, HPO_3^{2-}. Write the correctly balanced net ionic equation for this reaction.

Solution. The oxidation state of phosphorus is -3 in phosphine and $+3$ in phosphite ion. The disproportionation thus involves the reduction of P_4 from the 0 state to the -3 state, and the simultaneous oxidation of P_4 to the $+3$ state. The skeletal reduction half-reaction is

$$P_4 \rightarrow 4PH_3$$

Note that the P atoms must be balanced *first*. Since each P atom undergoes a change in oxidation state from 0 to -3, and there are four P atoms, a total of 12 electrons must be added to the P_4 to effect this change. We now have

$$12e^- + P_4 \rightarrow 4PH_3$$

To balance the charge we must add 12 OH^- ions to the right-hand side, since this is a basic solution. This yields

$$12e^- + P_4 \rightarrow 4PH_3 + 12\,OH^-$$

There are 12 O atoms on the right-hand side, so we must add $12H_2O$ to the left. This balances both the H and O atoms simultaneously:

$$12H_2O + P_4 + 12e^- \rightarrow 4PH_3(g) + 12\,OH^-$$

The oxidation half-reaction involves a loss of 12 electrons as four P atoms are oxidized from the 0 to the $+3$ state. The skeletal half-reaction is

$$P_4 \rightarrow 4HPO_3^{2-} + 12e^-$$
$$0 \qquad +3$$

To balance the charge we add 20 OH^- ions on the left. We need 8 more O atoms and 16 more H atoms on the right-hand side, so we add $8H_2O$. The correctly balanced oxidation half-reaction is

$$20\,OH^- + P_4 \rightarrow 4HPO_3^{2-} + 12e^- + 8H_2O$$

Addition of the reduction half-reaction and the oxidation half-reaction, combining the H_2O and OH^- ions algebraically, yields

$$4H_2O + 8\,OH^- + 2P_4 \rightarrow 4PH_3(g) + 4HPO_3^{2-}$$

Since all coefficients are divisible by two, the simplest correct equation is

$$2H_2O + 4\,OH^- + P_4 \rightarrow 2PH_3(g) + 2HPO_3^{2-}$$

section 15.4
The Redox Chemistry of Group VA Elements

In order to know what the products of a given oxidation–reduction reaction will be, we must become familiar with the various oxidation states of the elements, particularly those elements that are most frequently encountered. Nitrogen is an element with an especially rich oxidation–reduction chemistry, and one that is very important, both in inorganic and organic chemistry.

There are covalently bonded compounds of nitrogen in which every oxidation state from the maximum value of $+5$ to the minimum value of -3 is observed. Table 15.3 summarizes information useful in balancing oxidation–reduction equations involving nitrogen. In the chemistry of nitrogen, the oxidation state concept is only a formalism. There are no grounds whatsoever to consider nitrogen as a cation in its positive oxidation states, or as an anion in its negative oxidation states, except for the ionic nitrides such as Mg_3N_2 and Ca_3N_2, which contain the nitride ion N^{3-}.

Table 15.3. **Summary of the Principal Oxidation – Reduction Chemistry of Nitrogen**

Oxidation State	Species	Remarks
+5	NO_3^-	Nitrate ion, a strong oxidizing agent in acid solution.
+4	NO_2 N_2O_4	Nitrogen dioxide, a red-brown gas; the chief reduction product of concentrated nitric acid. Nitrogen dioxide dimerizes to form dinitrogen tetroxide, N_2O_4.
+3	NO_2^- HNO_2	Nitrite ion, in basic solution. Nitrous acid, in acid solution. Both can act either as an oxidizing agent or as a reducing agent. When acting as a reducing agent, they are oxidized to NO_3^-. When HNO_2 acts as an oxidizing agent in acid solution it is usually reduced to NO(g).
+2	NO	Nitrogen oxide (nitric oxide), a colorless gas; the chief reduction product of dilute nitric acid.
+1	N_2O	Dinitrogen oxide (nitrous oxide), also called laughing gas. Not usually formed in oxidation – reduction reactions in aqueous solution.
0	N_2	Unreactive because of the strong triple bond, $:N\equiv N:$.
−1	NH_2OH	Hydroxylamine, a weak base. Can serve both as an oxidizing agent and as a reducing agent.
−2	N_2H_4	Hydrazine, a colorless liquid that disproportionates to N_2 and NH_3. Hydrazine is a weak base.
−3	NH_3 NH_4^+	In basic solution. In acidic solution. Rarely used as a reducing agent in aqueous solution. When ammonia is burned in O_2 it is oxidized to either N_2 or NO depending on conditions. Oxidation of ammonium salts generally produces N_2.

The use of Table 15.3 is illustrated in the following example.

EXAMPLE 15.8. Predicting products of redox reactions involving nitrogen

In certain laboratory procedures it is necessary to remove NH_4^+ ions from solution. This is accomplished by heating the solution with concentrated nitric acid and evaporating to dryness. Write a balanced net ionic equation for this reaction.

Solution. Table 15.3 informs us that ammonium ions are generally oxidized to N_2. It also states that concentrated nitric acid is a good oxidizing agent, and is reduced to $NO_2(g)$. The two half-reactions are therefore

Oxidation
$$2NH_4^+ \rightarrow N_2(g) + 6e^- + 8H^+(aq)$$
$$\;\;\;\;\;-3 \qquad\qquad 0$$

Reduction
$$2H^+(aq) + NO_3^- + e^- \rightarrow NO_2(g) + H_2O$$
$$\qquad\quad +5 \qquad\qquad\quad +4$$

The correctly balanced overall equation is therefore

$$2NH_4^+ + 6NO_3^- + 4H^+(aq) \rightarrow 6NO_2(g) + N_2(g) + 6H_2O$$

Additional information about the oxidation states of nitrogen is summarized below.

The +5 state. Pure (anhydrous) nitric acid is a colorless liquid that undergoes self-ionization:

$$2HNO_3 \rightleftharpoons NO_2^+ + NO_3^- + H_2O \qquad (15\text{-}7)$$

The nitronium ion, NO_2^+, is involved in the nitration of organic substances. Pure nitric acid is seldom encountered. The laboratory reagent called concentrated nitric acid, frequently used as a strong oxidizing agent, is a 15 M aqueous solution of HNO_3. When freshly prepared, concentrated nitric acid is colorless, but it is decomposed by sunlight and becomes yellow due to dissolved NO_2. If you find dark yellow stains on your fingers after handling nitric acid in the lab, they are due to the reaction between HNO_3 and proteins. Be very careful to wash thoroughly because the damage to skin proteins can cause sores that heal slowly.

In addition to nitric acid and the nitrates, the +5 state is found in dinitrogen pentoxide, N_2O_5. Dinitrogen pentoxide is a volatile white solid at room temperature. Its structure is actually ionic nitronium nitrate, $NO_2^+NO_3^-$. In the gaseous state N_2O_5 is molecular with the structure

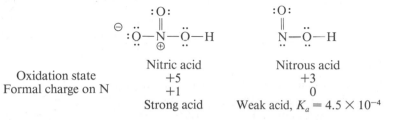

The pentoxide is not too stable; it decomposes to form O_2 and NO_2. This is an auto-oxidation-reduction, with nitrogen being reduced from the +5 to the +4 state, and oxygen being oxidized from the −2 state to the 0 state.

$$N_2O_5(s) \rightarrow 2NO_2(g) + \tfrac{1}{2}O_2(g) \qquad (15\text{-}8)$$

The decomposition of N_2O_5 can be explosive, particularly at elevated temperatures. Dinitrogen pentoxide is an acidic anhydride, yielding nitric acid when it reacts with water:

$$N_2O_5 + H_2O \rightleftharpoons 2NO_3^- + 2H^+(aq) \qquad (15\text{-}9)$$

Note that N_2O_5 is the anhydride of a strong acid, which illustrates the general rule that the higher the oxidation state of an element, the more acidic its oxide. Oxides of nonmetals are usually acidic, as has been discussed in Section 9.5. In general, a higher oxidation state is correlated with a higher formal charge on the central atom in the Lewis dot structure that obeys the octet rule. The two oxyacids of nitrogen are nitric acid, HNO_3, and nitrous acid, HNO_2. Their structures are shown below:

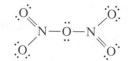

	Nitric acid	Nitrous acid
Oxidation state	+5	+3
Formal charge on N	+1	0
	Strong acid	Weak acid, $K_a = 4.5 \times 10^{-4}$

The higher the positive formal charge on the central atom, the more the central atom will attract electrons from the oxygen atoms to which it is bonded, that is, the more electronegative the central atom will be. Pulling electrons away from the oxygen atoms weakens the O—H bond and makes it easier for the proton to be lost. Therefore the acid strength (the value of the acidity constant, K_a) increases as the formal charge on the central atom and the oxidation state of the central atom increases.

The +4 state. The red-brown gas NO_2 readily dimerizes to form the colorless dinitrogen tetroxide, N_2O_4:

$$2NO_2(g) \rightleftharpoons N_2O_4(g) \qquad (15\text{-}10)$$

Brown   Colorless

Paramagnetic   Diamagnetic

Nitrogen dioxide is paramagnetic because it has an unpaired electron; the dimer has a structure that satisfies the octet rule, and has all electrons paired:

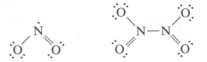

The equilibrium between NO_2 and its dimer is strongly temperature dependent. At 100 °C, the composition is 90% NO_2, 10% N_2O_4. At 60 °C, the equilibrium composition is 50% NO_2 and 50% N_2O_4. As a mixture of the two gases is cooled, one can observe the brown color fading. The mixture of NO_2 and N_2O_4 is poisonous and is one of the major air pollutants. While not commonly used as a laboratory reagent, NO_2 is an oxidizing agent.

The +3 state. Nitrite salts of the alkali metals can be prepared by heating the nitrate with a reducing agent such as C(s), Pb(s), or Fe(s):

$$KNO_3(s) + Pb(s) \xrightarrow{heat} KNO_2(s) + PbO(s) \qquad (15\text{-}11)$$

The products are easy to separate since KNO_2 dissolves in water but PbO does not. Thermal decomposition of the nitrates also produces nitrites:

$$KNO_3(s) \xrightarrow{heat} KNO_2(s) + \tfrac{1}{2}O_2(g) \qquad (15\text{-}12)$$

Acidifying a solution containing a nitrite produces the weak acid HONO, nitrous acid. Solutions of nitrous acid are unstable when heated; disproportionation to NO and NO_3^- occurs.

EXAMPLE 15.9. The disproportionation of nitrous acid

Balance the equation for the disproportionation of nitrous acid.

Solution. The two half-reactions are

Oxidation

$$HNO_2 + H_2O \rightarrow NO_3^- + 2e^- + 3H^+(aq)$$

$$+3 \qquad\qquad\qquad +5$$

Reduction

$$H^+(aq) + HNO_2 + e^- \rightarrow NO(g) + H_2O$$

$$+3 \qquad\qquad +2$$

Overall

$$3HNO_2 \rightarrow H_2O + NO_3^- + 2NO(g) + H^+(aq)$$

Nitrous acid and the nitrites are important in organic chemistry as they are used in the synthesis of the **azo dyes**, which contain the $-N{=}N-$ group. Azo dyes are intensely colored; they constitute the majority of synthetic dyes produced commercially.

The anhydride of nitrous acid is dinitrogen trioxide, N_2O_3. Note that N_2O_3 is the anhydride of a weak acid, whereas N_2O_5, in which nitrogen is in its highest oxidation state, is the anhydride of a strong acid. Pure N_2O_3 exists only in the solid state at low temperatures. At room temperature it is an unstable gas and disproportionates as follows:

$$N_2O_3(g) \rightarrow NO(g) + NO_2(g) \qquad (15\text{-}13)$$

The +2 state. While normal combustion of NH_3 in air produces N_2, if the reaction is carried out at elevated temperatures (750–900 °C) in the presence of a platinum catalyst, NO is produced:

$$4NH_3(g) + 5 O_2(g) \xrightarrow{\text{Pt}} 4NO(g) + 6H_2O(g) \tag{15-14}$$

This reaction is the first step in the commercial **Ostwald process** for making nitric acid, which is used in the production of fertilizers and explosives.

The direct combination of N_2 and O_2 produces NO only if the temperature is very high or large amounts of energy are added, because the reaction $N_2(g) + O_2(g) \rightleftharpoons 2NO(g)$ is extremely endothermic. Nitric oxide, NO, is formed in the exhaust gases of automobile engines because of the high temperatures at which they operate; this contributes to air pollution. In air, NO is rapidly oxidized to the brown NO_2, a major constituent of smog in many cities.

$$2NO(g) + O_2(g) \rightarrow 2NO_2(g) \tag{15-15}$$

The +1 state. Nitrous oxide, or dinitrogen oxide, can be obtained by gently heating ammonium nitrate:

$$NH_4NO_3(s) \xrightarrow{\text{heat}} N_2O(g) + 2H_2O(g) \tag{15-16}$$

Because it is a relatively unreactive substance, N_2O is much less poisonous than other oxides of nitrogen. It is called **laughing gas** because inhaling small doses of N_2O makes people feel giddy. Mixed with oxygen, N_2O is widely used as a general anesthetic for operations of short duration, particularly in dentistry. Its chief commercial use is as the propellant gas in "whipped cream" bombs because N_2O is fairly soluble in cream. The cream is packaged with N_2O under considerable pressure; when the pressure is released the escaping gas forms tiny bubbles through the cream and thereby "whips" it.

The −1 state. Hydroxylamine can be considered to be derived from NH_3 by replacing one hydrogen by a hydroxyl group. Thus hydroxylamine, H_2NOH, is a weak base, just as NH_3 is

$$H_2NOH + H_2O \rightleftharpoons (H_3NOH)^+ + OH^- \tag{15-17}$$

with $K_b(H_2NOH) = 1.2 \times 10^{-8}$ at 25 °C.

Hydroxylammonium chloride, nitrate, and sulfate are white, stable, water-soluble solids. Although hydroxylamine can serve either as an oxidizing agent or as a reducing agent, it is usually used as a reducing agent.

The −2 state. Hydrazine, N_2H_4, can be considered to be derived from NH_3 by replacing one hydrogen by an $-NH_2$ group: H_2N-NH_2. It is a weak base, and its conjugate acid, $N_2H_5^+$, is the **hydrazinium ion:**

$$N_2H_4 + H_2O \rightleftharpoons N_2H_5^+ + OH^- \tag{15-18}$$

Anhydrous hydrazine is a colorless liquid that burns in air to yield N_2:

$$N_2H_4(liq) + O_2(g) \rightarrow N_2(g) + 2H_2O(g) \tag{15-19}$$

This combustion is extremely exothermic, so much so that hydrazine can be used as a rocket fuel. Hydrazine is a powerful reducing agent in basic solution; it is usually oxidized to N_2.

The −3 state. In addition to NH_3 and the ammonium salts nitrogen forms a series of compounds called nitrides in which the −3 oxidation state is utilized. There are ionic nitrides in which the N^{3-} ion exists; Mg_3N_2 and Li_3N are prominent examples.

All the alkaline earths except for Be form ionic nitrides. Beryllium nitride, Be_3N_2, is covalently bonded.

The alkaline earth nitrides react with water to liberate ammonia and form the metal hydroxides. A typical reaction is

$$Ba_3N_2(s) + 6H_2O \rightarrow 2NH_3 + 3Ba^{2+} + 6\,OH^- \qquad (15\text{-}20)$$

The Oxidation States of Phosphorus

Phosphorus, like nitrogen, exhibits all the oxidation states from -3 to $+5$, inclusive. The -3 state is represented by phosphine, PH_3, an extremely poisonous gas. Unlike NH_3, phosphine is not very soluble in water and its aqueous solutions are neither acidic nor basic.

It is the $+5$ oxidation state of phosphorus that is most commonly observed. In nature phosphorus occurs mainly in phosphate rock, which is largely a mixture of $Ca_5(OH)(PO_4)_3$, **hydroxyapatite,** and $Ca_5F(PO_4)_3$, **fluoroapatite.** Animal bones contain about 60% calcium phosphate, and teeth are mostly hydroxyapatite, partly carbonated.

Organic phosphates are of major importance in biochemical processes. Photosynthesis involves the sugar phosphates, and the nucleic acids, DNA and RNA, contain phosphate.

Phosphorus, like other nonmetals, forms only acidic oxides. The oxide of the $+5$ state is called phosphorus pentoxide, but its correct formula is P_4O_{10}. The structure of P_4O_{10} is shown in Fig. 15.1. Phosphorus pentoxide is frequently used as a dehydrating agent because it has such a great affinity for water. It dehydrates many organic compounds and can extract the elements of water from other substances. It converts H_2SO_4 to $SO_3(g)$, for example. The reaction of P_4O_{10} with excess water forms **orthophosphoric acid:**

$$P_4O_{10} + 6H_2O \rightleftharpoons 4H_3PO_4 \qquad (15\text{-}21)$$

which could, alternatively, be written $(HO)_3PO$, as its dot formula is

$$
\begin{array}{c}
:\!O\!: \\
\parallel \\
H\!-\!\ddot{O}\!-\!P\!-\!\ddot{O}\!-\!H \\
\mid \\
:\!O\!: \\
\mid \\
H
\end{array}
$$

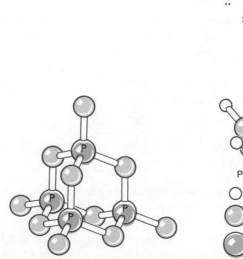

Fig. 15.1. The structure of P_4O_{10}.

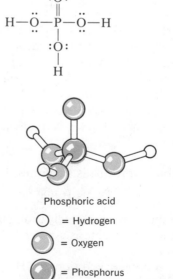

Phosphoric acid

◯ = Hydrogen

◉ = Oxygen

◉ = Phosphorus

Fig. 15.2. The structure of orthophosphoric acid, H_3PO_4 or $(HO)_3PO$.

The structure of orthophosphoric acid (commonly called phosphoric acid) is shown in Fig. 15.2.

Orthophosphoric acid is a moderately weak tribasic acid, with $K_1 = 7.5 \times 10^{-3}$ at 25 °C. The pure acid is a colorless, crystalline solid at room temperature. Unlike nitric acid, it is not a good oxidizing agent. Only at very high temperatures (above 350 °C) does it react with metals.

The +3 oxidation state of phosphorus is represented by phosphorus acid, H_3PO_3, with structure

$$
\begin{array}{c}
\text{H} \\
|\\
\text{H}-\ddot{\text{O}}-\text{P}-\ddot{\text{O}}-\text{H} \\
\|\\
:\text{O}:
\end{array}
$$

by the so-called *trioxide* P_4O_6, and by the trihalides, PF_3, PCl_3, PBr_3, and PI_3.

The Oxidation States of As, Sb, and Bi

Although nitrogen and phosphorus are typical nonmetals, arsenic and antimony are **metalloids** or semimetals, and bismuth is a metal, illustrating the increase in metallic character with increase in atomic number within one family in the periodic table. Only the +5, +3, and −3 oxidation states are important for As, Sb, and Bi, with the +3 state the most common. Since the outer electronic configuration of all Group VA elements is ns^2np^3, the +3 state corresponds to sharing the three p electrons with a more electronegative element, and not sharing the pair of s electrons. A typical compound is $AsCl_3$, with Lewis structure

$$
\begin{array}{c}
:\ddot{\text{Cl}}-\ddot{\text{As}}-\ddot{\text{Cl}}: \\
|\\
:\ddot{\text{Cl}}:
\end{array}
$$

Arsenic and antimony form *pentoxides,* with empirical formulas As_2O_5 and Sb_2O_5, but their true molecular formulas and structures are unknown. They also form *trioxides* with formulas As_4O_6 and Sb_4O_6. The structures of these trioxides is shown in Fig. 15.3.

The +5 oxide of arsenic dissolves in water to form arsenic acid, a moderately weak acid with $K_1 = 5 \times 10^{-3}$:

$$As_2O_5(s) + 3H_2O \rightarrow 2H_3AsO_4 \tag{15-22}$$

The +3 oxide (arsenious oxide) is only slightly soluble in water and the acid formed, H_3AsO_3, is very weak with $K_1 \sim 10^{-10}$. This is another example of the greater acidity of higher oxidation states relative to lower oxidation states.

An acidic solution of arsenious acid is a good reducing agent and can be used to titrate BrO_3^-, MnO_4^-, I_2, and other oxidizing agents. Arsenious acid is oxidized to the

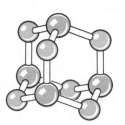

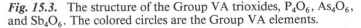

Fig. 15.3. The structure of the Group VA trioxides, P_4O_6, As_4O_6, and Sb_4O_6. The colored circles are the Group VA elements.

+5 state. Because the oxidation is slow at lower pH values, the reaction is usually run at a pH close to 5 or 6, and $HAsO_4^{2-}$ is the principal product.

All arsenic compounds are poisonous. For many years large amounts of As_4O_6 were used in insecticides and weed killers, but as other less dangerous materials have been developed in recent years, the use of arsenic containing insecticides has decreased markedly.

As the atomic number increases in Group VA the +5 oxidation state becomes less stable. The only stable oxide of bismuth is Bi_2O_3, which like other metallic oxides, is a basic oxide.

section 15.5
The Oxidation States of Chlorine and Other Group VIIA Elements

The maximum oxidation state possible for Group VIIA elements is +7, and the minimum is −1. The −1 oxidation state predominates for all these elements. The principal oxidation states of chlorine are −1, +1, +5, and +7, although there are examples of every oxidation state from −1 to +7 with the exception of +2. The higher oxidation states for chlorine, bromine, iodine, and astatine are known principally in compounds with oxygen and in interhalogen compounds. Table 15.4 summarizes information useful in balancing redox equations involving chlorine.

Additional information about the oxidation states of chlorine is given below.

The +7 state. Perchloric acid is an extremely strong acid, the strongest of all oxyacids of chlorine. In general, for any series of oxyacids, the one in which the atom bound to oxygen is in the highest oxidation state is the strongest acid. We have already seen several examples of this behavior. Nitric acid, HNO_3, is stronger than nitrous acid, HNO_2. Sulfuric acid, H_2SO_4, in which sulfur is in its maximum oxidation state of +6, is stronger than sulfurous acid, H_2SO_3, the +4 oxidation state of sulfur. Arsenic acid, H_3AsO_4, the +5 oxidation state of arsenic, is a much stronger acid than arsenious acid, H_3AsO_3. Similarly, the strength of the four oxyacids of chlorine decreases as the oxidation state of Cl decreases, in the series $HClO_4 > HClO_3 > HClO_2 > HOCl$. We can understand this by considering the Lewis structures of these four acids:

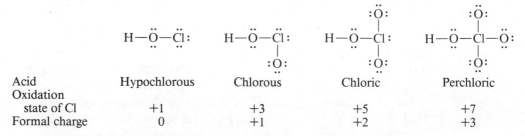

Acid	Hypochlorous	Chlorous	Chloric	Perchloric
Oxidation state of Cl	+1	+3	+5	+7
Formal charge	0	+1	+2	+3

As the oxidation state increases, the formal charge and the actual positive charge on the Cl atom also increases, and the size of the Cl atom decreases as the electron cloud is pulled in more tightly. The electronegativity of Cl increases as the formal charge on Cl increases, and more electron density is pulled away from the O—H bond. This weakens the O—H bond, and makes it easier to lose the proton. In perchlorate ion, ClO_4^-, the Cl atom is small and has the highest positive charge of any oxyanion of chlorine; the O—Cl bond is strong, and $HClO_4$ is 100% dissociated in dilute aqueous solution.

Perchlorate ion shows very little tendency to associate with any cation; it is partic-

Table 15.4. **Summary of the Principal Oxidation – Reduction Chemistry of Chlorine**

Oxidation State	Species	Remarks
+7	ClO_4^-	Perchlorate ion, a powerful oxidizing agent.
+5	ClO_3^-	Chlorate ion, a powerful oxidizing agent.
+4	ClO_2	Chlorine dioxide, a yellow-orange gas. Not a common oxidation state of chlorine, but it is formed when ClO_3^- reacts with a number of reducing agents. Chlorine dioxide can serve as an oxidizing agent also.
+3	ClO_2^-	Chlorite ion, in basic solution. Serves as an oxidizing agent in bleaches. In acid solution $HClO_2$ disproportionates to $ClO_2(g)$ and Cl^-.
+1	ClO^-	Hypochlorite ion, in basic solution.
	HOCl	Hypochlorous acid, in acid solution. Both serve as oxidizing agents. ClO^- disproportionates when warmed, to yield Cl^- and ClO_3^-.
0	Cl_2	Chlorine, a good oxidizing agent.
−1	Cl^-	Chloride ion, a moderately good reducing agent.

ularly useful in studies of complex ions because it does not interfere with the formation of other complexes. Most perchlorates are quite soluble, but those with a large, singly charged cation, such as $KClO_4$, $RbClO_4$, and $CsClO_4$ are only slightly soluble.

Perchlorates and perchloric acid are strong oxidizing agents, but they react slowly at room temperature and below. Hot solutions containing ClO_4^-, however, react vigorously and sometimes violently. Perchloric acid often reacts explosively with organic reducing agents.

The +5 state. The most important chlorate salt is $KClO_3$, which is used as an oxidizing agent in matches and fireworks. The standard laboratory preparation for oxygen involves the decomposition of $KClO_3$ when it is heated in the presence of a very small amount of powdered MnO_2, which serves as a catalyst:

$$KClO_3(s) \xrightarrow{\text{heat, MnO}_2} KCl(s) + \tfrac{3}{2}O_2(g) \tag{15-23}$$

In the absence of a catalyst, if warmed gently, $KClO_3$ disproportionates to yield $KClO_4$ and KCl:

$$4KClO_3(s) \xrightarrow{\text{heat}} 3KClO_4(s) + KCl(s) \tag{15-24}$$

Chlorate ion is the anion of a strong acid, but pure chloric acid, $HClO_3$, has not been prepared because it is unstable and decomposes explosively. The principal reaction is

$$4HClO_3 \rightarrow 4ClO_2(g) + O_2(g) + 2H_2O(g) \tag{15-25}$$

The +4 state. Chlorine dioxide has an odd number of valence electrons and is therefore paramagnetic. It is a highly reactive gas and must be handled with care because it is likely to explode. Chlorine dioxide is an active oxidizing agent and is used commercially for that purpose. It is produced by reducing chlorates such as $NaClO_3$ or $KClO_3$. It is never transported because of its explosive nature. Chlorine dioxide is soluble in water, but neutral or acidic solutions disproportionate to Cl^- and ClO_3^-, whereas alkaline solutions disproportionate to ClO_2^- and ClO_3^-.

The +3 state. Chlorous acid, $HClO_2$, is a moderately weak acid ($K_a \sim 10^{-2}$) that cannot be isolated as a pure substance because it decomposes to form Cl^-, ClO_3^-, and ClO_2. Chlorites, which are used as bleaching agents, are obtained by the disproportionation of ClO_2 in basic solution:

$$2ClO_2 + 2OH^- \rightleftharpoons ClO_2^- + ClO_3^- + H_2O \tag{15-26}$$

The +1 state. Hypochlorous acid, HOCl, is a weak acid, ($K_a = 3 \times 10^{-8}$ at 25 °C) formed when Cl_2 reacts with water:

$$Cl_2(g) + H_2O \rightleftharpoons HOCl + Cl^- + H^+(aq) \tag{15-27}$$

The equilibrium constant for this reaction is small and therefore the yield of HOCl is low. Hypochlorite ion can be produced by the disproportionation of Cl_2 in alkaline solution

$$Cl_2(g) + 2\,OH^- \rightleftharpoons Cl^- + ClO^- + H_2O \tag{15-28}$$

which has a large equilibrium constant. Both HOCl and ClO^- are good oxidizing agents. Solutions of sodium hypochlorite are sold as a laundry bleach. Clorox, for instance, is a 5.25% solution of NaOCl in water. While cold alkaline solutions of ClO^- are fairly stable, the disproportionation reaction

$$3ClO^- \rightarrow 2Cl^- + ClO_3^- \tag{15-29}$$

occurs rapidly when the solution is warmed.

The −1 state. Chlorine is the most abundant of all the halogens in nature, where it exists largely as NaCl in sea water, salt lakes, and deposits of solid NaCl presumably left when prehistoric salt lakes evaporated. Metal halides are predominantly ionic, although for metals that form several oxidation states, the lower oxidation states tend to be ionic while the higher oxidation states are covalent. We have already noted, for example, that $TiCl_2$ is a crystalline solid that decomposes at 475 °C, whereas $TiCl_4$ is a liquid at room temperature and is essentially covalently bonded. Similarly, stannous chloride, $SnCl_2$, is a crystalline solid that melts at 246 °C and the bonding is ionic with partial covalent character, while pure tin(IV) chloride (stannic chloride) is a liquid at room temperature with essentially covalent bonding. The amount of energy required to remove four electrons from a metal is so much larger than the amount of energy required to remove two electrons that the +4 oxidation involves covalent bonding.

Since the −1 state is the lowest oxidation state of chlorine, Cl^- can serve only as a reducing agent. It is seldom used for that purpose, however, since there are many stronger reducing agents available.

Other Halogens

Fluorine differs from chlorine in that only the −1 oxidation state is exhibited (in addition to 0), as fluorine is the most electronegative element. The chemistry of bromine and iodine is similar to that of chlorine.

section 15.6
Oxidation States of the Transition Metals Chromium and Manganese
Redox Chemistry of Chromium

The maximum oxidation state possible for Cr is the +6 state as its electronic configuration is $(Ar)3d^54s^1$, and there are six electrons outside the argon configuration that can be used in bonding. The +6 state corresponds to sharing all six electrons with a more electronegative element. The bright yellow chromate ion, CrO_4^{2-}, is a common anion in neutral or alkaline solution. In acid solution chromate dimerizes to form the red-orange dichromate ion:

$$2CrO_4^{2-} + 2H^+(aq) \rightleftharpoons Cr_2O_7^{2-} + H_2O \tag{15-30}$$

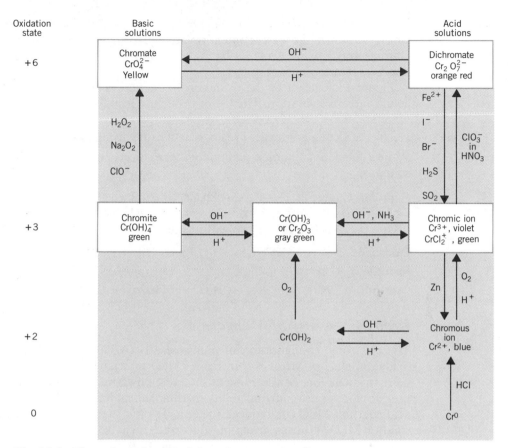

Fig. 15.4. The redox behavior of chromium compounds.

Dichromate ion is a very strong oxidizing agent and is frequently used in the laboratory. It is reduced to chromic ion, Cr^{3+}.

The +3 state is the most prevalent oxidation state of chromium. Chromium(III) forms complexes with a great many anions and neutral molecules; these will be discussed in Chapter 20. In basic solution $Cr(OH)_3$ precipitates. Chromic hydroxide is an **amphoteric** hydroxide; it reacts with both hydronium and hydroxide ions. It dissolves in excess base to form the green chromite ion, $Cr(OH)_4^-$:

$$Cr(OH)_3(s) + OH^- \rightleftharpoons Cr(OH)_4^- \qquad (15\text{-}31)$$

Slowly acidifying a solution containing chromite ion will reprecipitate $Cr(OH)_3$; as more acid is added the $Cr(OH)_3$ dissolves.

With a powerful reducing agent such as Zn, Cr(III) can be reduced to the +2 state, Cr^{2+} ion in acid solution and chromous hydroxide, $Cr(OH)_2$, in base. Chromous ion, Cr^{2+}, is a good reducing agent and is usually oxidized to the +3 state.

The oxides of chromium illustrate the general rule that the higher the oxidation state of an element bonded to oxygen, the more acidic the oxide. Chromium(VI) oxide, CrO_3, is an acidic oxide. It dissolves readily in water to give an acidic solution of dichromate ion:

$$2CrO_3(s) + H_2O \rightarrow Cr_2O_7^{2-} + 2H^+(aq) \qquad (15\text{-}32)$$

It dissolves in base to give a solution of chromate ion. Chromium(III) oxide, Cr_2O_3, is amphoteric, that is, it will dissolve in both acid and base. We have already discussed the amphoteric behavior of chromic hydroxide, $Cr(OH)_3$, which is really hydrated Cr_2O_3. Although the water content is variable, formally $Cr_2O_3 \cdot 3H_2O$ is $2Cr(OH)_3$. Chromium(II) oxide, CrO, is basic, as is chromous hydroxide, $Cr(OH)_2$, which dissolves in acid, but not in base.

Figure 15.4 diagrams the oxidation–reduction behavior of chromium in both acidic and basic solution. The use of this figure is illustrated in Example 15.10.

EXAMPLE 15.10. Predicting products of redox reactions involving chromium

Write a balanced net ionic equation for the reaction between Cr^{3+} and ClO_3^- in acidic solution.

Solution. Using Fig. 15.4 we see that Cr^{3+} is oxidized by chlorate ion to the +6 state, $Cr_2O_7^{2-}$ in acid solution. Table 15.4 states that when ClO_3^- is reduced the product is usually gaseous ClO_2. The two half-reactions are

Oxidation $\qquad 2Cr^{3+} + 7H_2O \rightarrow Cr_2O_7^{2-} + 6e^- + 14H^+(aq)$

Reduction $\quad 2H^+(aq) + ClO_3^- + e^- \rightarrow ClO_2(g) + H_2O$

Overall $\qquad 2Cr^{3+} + 6ClO_3^- + H_2O \rightarrow 6ClO_2(g) + Cr_2O_7^{2-} + 2H^+(aq)$

Redox Chemistry of Manganese

The electronic configuration of manganese is $(Ar)3d^5 4s^2$ so that the maximum oxidation state of Mn is +7. In the +7 state manganese exists as the deep purple permanganate ion, MnO_4^-, one of the most powerful oxidizing agents in frequent use. Because of its oxidizing ability, permanganate ion is used as a disinfectant. In acid solution MnO_4^- is reduced to the pale pink manganous ion, Mn^{2+}. In basic solution MnO_4^- is not as strong an oxidizing agent as it is in acidic solution. In neutral or alkaline solution MnO_4^- is usually reduced to black, insoluble, solid MnO_2, although in strongly alkaline solution it can be reduced to the +6 state, where it exists as the bright green manganate ion, MnO_4^{2-}. Manganate ion disproportionates to MnO_2 and MnO_4^- in acid solution.

Manganese dioxide, called pyrolusite, is the principal ore of manganese. It is an oxidizing agent and its chief use commercially is in the standard dry cell battery used for flashlights, where it serves to oxidize Zn metal to the +2 state.

The +3 state of manganese is rare; in acid solution it exists very briefly as manganic ion, Mn^{3+}. However, Mn^{3+} disproportionates to yield Mn^{2+} and MnO_2:

$$2Mn^{3+} + 2H_2O \rightarrow Mn^{2+} + MnO_2(s) + 4H^+(aq) \qquad (15\text{-}33)$$

Complex ions of Mn(III) such as $Mn(CN)_6^{3-}$ are stable, and in basic solution the insoluble brown manganic hydroxide, usually written as $MnO(OH)$, is formed.

The +2 oxidation state of manganese in acidic solution is the very pale pink Mn^{2+} ion; manganous salts such as $MnSO_4$, $MnCl_2$, and $Mn(NO_3)_2$ are pink or rose-colored soluble solids. In basic solution white, insoluble $Mn(OH)_2$ is formed. The manganous ion is not a good reducing agent and is resistant to oxidizing agents. Acidic solutions containing Mn^{2+} exposed to the air are stable indefinitely. Manganous hydroxide, however, is rapidly oxidized by air to the brown $MnO(OH)$.

Figure 15.5 diagrams the oxidation–reduction behavior of manganese in both acidic and basic solutions.

EXAMPLE 15.11. Predicting the products of redox reactions involving manganese

Write a balanced net ionic equation for the reaction between permanganate ion and hydrogen peroxide, H_2O_2, in acidic solution.

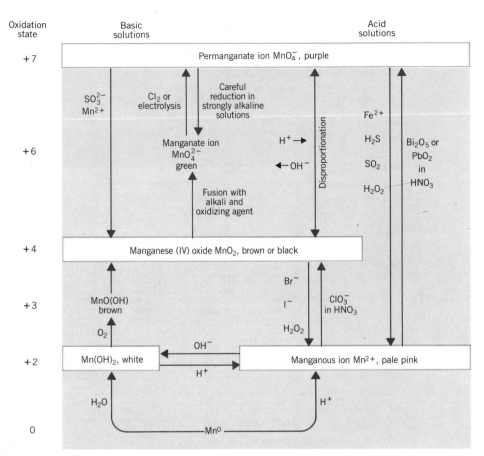

Fig. 15.5. The redox behavior of manganese compounds.

Solution. From Fig. 15.5 we see that permanganate ion is reduced to Mn^{2+} in acidic solution. Oxygen is in its -1 oxidation state in hydrogen peroxide and is oxidized to the 0 state, O_2. The two half-reactions are

Oxidation

$$H_2O_2 \rightarrow O_2(g) + 2H^+(aq) + 2e^-$$
$$ {-1} \qquad\quad 0$$

Reduction

$$8H^+(aq) + MnO_4^- + 5e^- \rightarrow Mn^{2+} + 4H_2O$$
$$ {+7} \qquad\qquad {+2}$$

Overall

$$5H_2O_2 + 2MnO_4^- + 6H^+(aq) \rightarrow 2Mn^{2+} + 5\,O_2(g) + 8H_2O$$

Summary

Oxidation is the process in which electrons are lost by or pulled away from an atom and the **oxidation state** of that atom is thereby increased. **Reduction** is the process in which electrons are gained by or shifted toward some atom and the oxidation state of that atom is thereby decreased. Oxidation and reduction always occur together.

The **oxidation state** of an element is determined by applying six rules set forth in this chapter. For a monatomic ion, the oxidation state is the same as the charge on the ion. For a covalently bonded substance, the oxidation state is a formalism; it is a device that enables us to keep track of electrons lost and gained in oxidation–

reduction reactions. The rules for determining the oxidation state of an element assign both electrons of a shared pair to the more electronegative element. This is in contrast to the rules for calculating the formal charge on an atom in a covalently bonded structure, which assign one electron of a shared pair to each of the bonded atoms. The more ionic character to the bond, the better an approximation the oxidation state is to the actual charge on the atom. The more covalent character the bond has, the better an approximation the formal charge is, and the poorer approximation the oxidation state is, to the actual charge on the atom. If the bonding is polar covalent, the actual charge is generally not equal to either the oxidation state or the formal charge.

A **redox couple** consists of an **oxidizing agent** and a **reducing agent** related by the transfer of one or more electrons. An oxidizing agent oxidizes some other substance and is itself reduced. A reducing agent reduces some other substance and is itself oxidized.

Many elements exhibit three or more oxidation states. An element is in an **intermediate oxidation state** if both a higher and a lower oxidation state exist. Substances in which an element is in an intermediate oxidation state are capable of acting either as an oxidizing agent or as a reducing agent, depending on the species with which they react. Sometimes substances in an intermediate oxidation state spontaneously **disproportionate** to yield one substance in which the element is in a higher and one in which it is in a lower oxidation state.

The oxidation state of an atom depends on the electron configuration of that atom and is therefore a periodic property. Metallic elements generally exhibit only positive oxidation states in addition to the 0 state of the pure element. In all their compounds the alkali metals exhibit only the +1 oxidation state, and the alkaline earths exhibit only the +2 state. For the representative elements, the maximum oxidation state is equal to the group number in the periodic table. For the nonmetals, both negative and positive oxidation states exist, and the minimum oxidation state for a group is related to the maximum by the equation: (maximum − minimum) = 8. Most of the transition metals exhibit several oxidation states; the +2 and +3 are the most common. For the lanthanides and the actinides the +3 state predominates.

There are two alternative but equivalent methods for balancing redox equations, the **oxidation state method** and the **ion–electron method.** Rules for using either method are described in this chapter. To balance oxygen and hydrogen atoms in acidic solution only H_2O and $H^+(aq)$ ions should be used, while in basic solution only H_2O and $OH^-(aq)$ ions should be used.

In order to predict the products of oxidation–reduction reactions, you must become familiar with the various oxidation states of the elements. Nitrogen, chlorine, chromium, and manganese are important elements with an especially rich oxidation–reduction chemistry.

Exercises

Sections 15.1 and 15.2

1. The following pairs of substances constitute a redox couple. Write the half-reaction that should be entered in Table 15.1. (a) Li(s), Li^+ (b) Br_2(liq), Br^- (c) Al(s), Al^{3+} (d) H_2O_2, $O_2(g)$ (e) H_2O, $O_2(g)$ (f) $S_8(s)$, S^{2-} (g) NH_4^+, $N_2(g)$ (h) Ca(s), Ca^{2+}

2. Determine the oxidation state of (a) Mo in MoO_4^{2-} (b) S in SO_3 (c) S in SO_3^{2-} (d) S in $S_2O_3^{2-}$ (e) C in CNO^- (f) In in $In(OH)_3$ (g) In in In_2S_3 (h) I in KIO_3 (i) As in AsO_4^{3-} (j) Fe in $FeAsO_4$ (k) Pd in $PdSO_4$ (l) Pd in PdS_2 (m) B in decaborane, $B_{10}H_{14}$

3. Calculate both the formal charge and the oxidation state of (a) S in SO_4^{2-} and SO_3^{2-} (b) Br in BrO_3^-, BrO_2^-, and BrO^- (c) N in NO_3^- and NO_2^-

4. Which of the following substances is capable of acting both as an oxidizing agent and as a reducing agent? Explain your answers. (a) Cr^{3+} (b) H_2O_2 (c) $NO(g)$ (d) CrO_4^{2-} (e) Ba^{2+} (f) Br^- (g) BrO^- (h) Ti^{3+} (i) $MnO_2(s)$ (j) $F_2(g)$ (k) $HONO$ (l) $SO_2(g)$

5. What is the maximum oxidation state exhibited by (a) C (b) Ba (c) Br (d) Se (e) Sc (f) Mn (g) P

6. What is the minimum oxidation state exhibited by (a) C (b) Ba (c) Br (d) Se (e) Sc (f) Mn (g) P

7. Identify the reducing agent and the oxidizing agent in the following oxidation–reduction reactions:

(a) $Zn(s) + 2H^+(aq) \rightarrow H_2(g) + Zn^{2+}(aq)$
(b) $2Cr^{3+} + H_2O + 6ClO_3^- \rightarrow Cr_2O_7^{2-} + 6ClO_2(g) + 2H^+(aq)$
(c) $Na(s) + Cl_2(g) \rightarrow NaCl(s)$
(d) $8H^+(aq) + MnO_4^- + 5Fe^{2+} \rightarrow Mn^{2+} + 4H_2O + 5Fe^{3+}$

Section 15.3

8. Write correctly balanced half-reactions and the overall equation for the following skeletal equations.

(a) $H_2S + Fe^{3+} \rightarrow Fe^{2+} + S\downarrow$ in acid solution.
(b) $NO_3^- + Bi(s) \rightarrow Bi^{3+} + NO_2\uparrow$ in acid solution.
(c) $Fe(OH)_2(s) + H_2O_2 \rightarrow Fe(OH)_3(s) + H_2O$ in basic solution.
(d) $I^- + IO_3^- \rightarrow I_2$ in acid solution.
(e) $O_2(g) + I^- \rightarrow I_2 + OH^-$ in basic solution.
(f) $Al(s) + NO_3^- \rightarrow Al(OH)_4^- + NH_3$ in basic solution.
(g) $Bi_2S_3(s) + NO_3^- \rightarrow NO(g) + S\downarrow$ in acid solution.
(h) $MnO_4^- + H_2C_2O_4 \rightarrow Mn^{2+} + CO_2\uparrow$ in acid solution.
(i) $ClO^- + Mn(OH)_2(s) \rightarrow MnO_2(s) + Cl^-$ in basic solution.
(j) $N_2H_4 + Cu(OH)_2(s) \rightarrow N_2\uparrow + Cu(s)$ in basic solution.
(k) $Cr_2O_7^{2-} + H_2C=O \rightarrow HCOOH + Cr^{3+}$ in acid solution.
(l) $O_2(g) + SO_3^{2-} \rightarrow SO_4^{2-} + OH^-$ in basic solution.

9. Write correctly balanced half-reactions and an overall net ionic equation for the reaction between
(a) $Bi_2S_3(s)$ and concentrated nitric acid.
(b) $PbS(s)$ and dilute nitric acid.

10. Write a correctly balanced equation for the disproportionation of
(a) Manganate ion, MnO_4^{2-}, to $MnO_2(s)$ and MnO_4^- in acid solution.
(b) Hypochlorite ion, ClO^-, to Cl^- and ClO_3^- in basic solution.

Sections 15.4, 15.5, and 15.6

11. Write correctly balanced half-reactions and the overall net ionic equation for the reaction in which
(a) Nitrous acid reduces MnO_4^- in acid solution.
(b) Nitrous acid oxidizes I^- to I_2 in acid solution.
(c) Nitrous acid and NH_4^+ react to produce $N_2(g)$ in acid solution.
(d) Nitrite ion is reduced to NH_3 by powdered $Zn(s)$ in alkaline solution. The zinc is oxidized to zincate ion, $Zn(OH)_4^{2-}$.
(e) Chloride ion is oxidized to Cl_2 by MnO_2 in acid solution.

(f) Bromide ion is oxidized to Br_2 by MnO_4^- in acid solution.

(g) Iodide ion is oxidized to I_2 by $Cr_2O_7^{2-}$ in acid solution.

(h) Chlorate ion oxidizes Mn^{2+} to $MnO_2(s)$ in acid solution.

(i) Chromite ion is oxidized by H_2O_2 in strongly basic solution.

(j) $Mn(OH)_2(s)$ is oxidized to $MnO_2(s)$ by O_2 in basic solution.

Multiple Choice Questions

1. A sulfur-containing species that cannot be a reducing agent is

   (a) SO_2 (b) SO_3^{2-} (c) H_2SO_4 (d) S^{2-} (e) $S_2O_3^{2-}$

2. A sulfur-containing species that cannot be an oxidizing agent is

   (a) SO_2 (b) SO_3^{2-} (c) H_2SO_4 (d) S^{2-} (e) $S_2O_3^{2-}$

3. Which of the following changes requires a reducing agent?

   (a) $CrO_4^{2-} \rightarrow Cr_2O_7^{2-}$ (b) $BrO_3^- \rightarrow BrO^-$ (c) $H_2O_2 \rightarrow O_2$

   (d) $H_3AsO_3 \rightarrow HAsO_4^{2-}$ (e) $Al(OH)_3 \rightarrow Al(OH)_4^-$

4. In balancing the half-reaction

$$CN^- \rightarrow CNO^- \text{ (skeletal)}$$

   the number of electrons that must be added is

   (a) zero   (b) one on the right   (c) one on the left   (d) two on the right
   (e) two on the left

5. In the following oxidation–reduction reaction

$$8H^+(aq) + 4NO_3^- + 6Cl^- + Sn(s) \rightarrow SnCl_6^{2-} + 4NO_2\uparrow + 4H_2O$$

   the reducing agent is

   (a) $Sn(s)$ (b) Cl^- (c) NO_3^- (d) $H^+(aq)$ (e) $NO_2(g)$

6. In the reaction of question 5 the oxidizing agent is

   (a) $Sn(s)$ (b) Cl^- (c) NO_3^- (d) $H^+(aq)$ (e) $NO_2(g)$

7. For the oxidation of methanol to formic acid

$$CH_3OH \rightarrow HCOOH \text{ (skeletal)}$$

   the number of electrons that must be added to the right side is

   (a) zero   (b) one   (c) two   (d) three   (e) four

8. In the reaction between warm concentrated sulfuric acid and potassium iodide

$$8 I^- + H_2SO_4 + 8H^+(aq) \rightarrow 4 I_2(g) + H_2S(g) + 4H_2O$$

   (a) I^- is reduced   (b) H_2S is the reducing agent   (c) H^+ is reduced
   (d) H_2SO_4 is the oxidizing agent   (e) H^+ is oxidized

9. In balancing the skeletal half-reaction $S_2O_3^{2-} \rightarrow S(s)$ the number of electrons that must be added is

   (a) two on the right   (b) two on the left   (c) three on the right
   (d) four on the left   (e) four on the right

10. When a solution containing CrO_4^{2-} ions is acidified with excess dilute nitric acid

    (a) CrO_4^{2-} is oxidized to the $+7$ state.

    (b) CrO_4^{2-} is reduced to the $+3$ state.

(c) CrO_4^{2-} dimerizes to yield $Cr_2O_7^{2-}$ and H_2O.

(d) CrO_4^{2-} disproportionates to yield Cr^{3+} and $Cr_2O_7^{2-}$.

(e) CrO_4^{2-} disproportionates to yield one product in the $+7$ state and one product in the $+3$ state.

11. For the reaction between permanganate ion and oxalate ion in basic solution the unbalanced equation is

$$MnO_4^- + C_2O_4^{2-} \rightarrow MnO_2(s) + CO_3^{2-}$$

When this equation is balanced the number of OH^- ions is

(a) zero   (b) two on the right   (c) two on the left   (d) four on the right
(e) four on the left

12. Which of the following changes requires an oxidizing agent?

(a) $N_2H_4 \rightarrow N_2(g)$ (b) $MnO_4^- \rightarrow MnO_2(s)$ (c) $H_2SO_3 \rightarrow SO_2(g)$
(d) $Sb(OH)_6^- \rightarrow Sb_4O_6(s)$ (e) $Cu^{2+} \rightarrow Cu(NH_3)_4^{2+}$

13. The equation for the reaction between arsenic(III) sulfide and $KClO_3$ in acid solution is

$$3As_2S_3(s) + 14ClO_3^- + 18H_2O \rightarrow 6H_2AsO_4^- + 9SO_4^{2-} + 14Cl^- + 24H^+(aq)$$

In this reaction the only changes in oxidation state that occur are

(a) Sulfur is oxidized from the -2 to the $+4$ state and chlorine is reduced from the $+5$ to the $+1$ state.

(b) Arsenic is oxidized from the $+3$ to the $+5$ state and chlorine is reduced from the $+5$ to the -1 state.

(c) Sulfur is oxidized from the -2 to the $+6$ state and arsenic is reduced from the $+3$ to the $+1$ state.

(d) Hydrogen is oxidized from the 0 to the $+1$ state and arsenic is reduced from the $+3$ to the $+1$ state.

(e) Arsenic is oxidized from $+3$ to $+5$, sulfur is oxidized from -2 to $+6$, and chlorine is reduced from the $+5$ to the -1 state.

14. When thiosulfate ion, $S_2O_3^{2-}$, reacts with I_2, the products are I^- and tetrathionate ion, $S_4O_6^{2-}$. In a titration of a solution of I_2, 32.78 mL of 0.1000 F $Na_2S_2O_3$ were required to react completely with the I_2. How many millimoles of I_2 were in the solution?

(a) 1.639   (b) 3.278   (c) 4.917   (d) 6.556   (e) 9.834

15. When MnO_4^- and I^- react in a strongly basic solution, the products will most likely be

(a) $Mn(s)$ and I_2 (b) MnO_4^{2-} and IO_3^- (c) MnO_2, O_2, and IO^-
(d) Mn^{2+} and I_2 (e) Mn^{2+} and IO^-

16. Which of the following is a disproportionation reaction?

(a) $CaCO_3(s) + 2H^+(aq) \rightarrow Ca^{2+} + H_2O + CO_2\uparrow$
(b) $Cl_2(g) + 2OH^- \rightarrow Cl^- + ClO^- + H_2O$
(c) $2CrO_4^{2-} + 2H^+(aq) \rightarrow H_2O + Cr_2O_7^{2-}$
(d) $Cu(H_2O)_4^{2+} + 4NH_3 \rightarrow Cu(NH_3)_4^{2+} + 4H_2O$
(e) $Ca(s) + 2H_2O \rightarrow Ca^{2+} + H_2(g) + 2OH^-$

17. When $PbS(s)$ is reacted with warm dilute nitric acid, the products will most likely be

(a) Pb^{2+}, $S(s)$, and $NO_2(g)$
(b) Pb^{2+}, $S(s)$, and $NO(g)$
(c) $PbO(s)$, $S(s)$, and $NO(g)$
(d) $PbO_2(s)$, SO_4^{2-}, and $N_2(g)$
(e) Pb^{2+}, H_2S, and $NO_2(g)$

18. In the reaction

$$2NH_4^+ + 6NO_3^- + 4H^+(aq) \rightarrow 6NO_2(g) + N_2(g) + 6H_2O$$

the reducing agent is

(a) NH_4^+ (b) NO_3^- (c) $H^+(aq)$ (d) NO_2 (e) N_2

Problems

15.1. Write balanced half-reactions and an overall net ionic equation for the following skeletal equations:

(a) $MnO_4^- + C_2O_4^{2-} \rightarrow CO_3^{2-} + MnO_2(s)$ (basic)

(b) $Cr_2O_7^{2-} + CH_3CHO \rightarrow CH_3COOH + Cr^{3+}$ (acid)

(c) $Ag_2S(s) + CN^- + O_2(g) \rightarrow Ag(CN)_2^- + S\downarrow + OH^-$ (basic)

(d) $Pb(s) + PbO_2(s) + SO_4^{2-} \rightarrow PbSO_4(s)$ (acid)

(e) $As_2S_3(s) + NO_3^- \rightarrow H_3AsO_4 + NO_2(g) + S\downarrow$ (acid)

(f) $Cr^{2+} + I^- + Cl_2(g) \rightarrow CrO_4^{2-} + IO_4^- + Cl^-$ (basic)

(g) $V(s) + ClO_3^- \rightarrow HV_2O_7^{3-} + Cl^-$ (basic)

(h) $Hg(liq) + NO_3^- + Cl^- \rightarrow HgCl_4^{2-} + NO_2(g)$ (acid)

15.2. (a) It is not possible to dissolve CdS in dilute hydrochloric acid. Prove that this is so by writing a balanced net ionic equation for the possible dissolution reaction and determining the equilibrium constant for the reaction you have written. The solubility product of CdS is 4×10^{-29} at 25 °C.

(b) It is, however, possible to dissolve CdS in warm dilute nitric acid. Write a balanced net ionic equation for the dissolution of CdS in dilute nitric acid. Then write a *brief* explanation for the difference between the effects of nitric acid and hydrochloric acid on dissolving insoluble sulfides.

(c) How would the dissolution reaction of CdS change if concentrated nitric acid is used instead of dilute? Write the balanced net ionic equation for dissolving CdS in concentrated nitric acid.

15.3. Lead(II) chloride, $PbCl_2$, is a white, crystalline solid, mp 501 °C. Lead(IV) chloride, $PbCl_4$, is a yellow, oily liquid, mp -15 °C. Account for this difference in terms of the electronic structure of the two compounds.

15.4. Permanganate ion oxidizes ferrous ion quantitatively in acid solution at room temperature, and the reaction can be used to determine the amount of iron in a sample.

(a) Write a balanced net ionic equation for the reaction between MnO_4^- and Fe^{2+} in acid solution.

(b) A sample of iron ore weighing 0.7545 g is dissolved in acid and all the iron is completely converted to Fe^{2+} ions. This solution is titrated with 0.0400 F KMnO$_4$, and 31.95 mL of the permanganate solution are required to reach the end point in the titration.

   (1) How many millimoles of Fe^{2+} were titrated?

   (2) What was the percentage of iron in the iron ore?

15.5. Write net ionic equations for the following reactions. Some, but not all, are redox reactions. If the reaction is a redox reaction, balanced half-reactions should be given. Assume dilute aqueous solutions of all soluble substances.

(a) Iron(II) sulfide and hydrochloric acid yields hydrogen sulfide and iron(II) chloride.

(b) Nickel(II) sulfide and dilute nitric acid yields nickel(II) nitrate, nitric oxide gas, and sulfur.

(c) Lead(II) hydroxide and dilute sulfuric acid yields lead(II) sulfate and water.

(d) Sulfurous acid and potassium iodate yields potassium sulfate and iodine.

(e) Potassium chromate, potassium cyanide, and potassium hydroxide yields chromium(III) hydroxide and potassium cyanate.

15.6. Predict the products of the redox reaction between the following pairs of reagents and write correctly balanced net ionic equations for these reactions. Assume aqueous solutions of all soluble reagents.

(a) Hydrazine and I_2 in basic solution.

(b) Potassium dichromate and $SO_2(g)$ in acidic solution.

(c) Potassium chlorate and warm concentrated HCl. [Cl^- is oxidized to $Cl_2(g)$.]

(d) Sodium hypobromite and ammonia in alkaline solution.

(e) Potassium bromate and arsenious acid. (BrO_3^- is reduced to Br^-.)

(f) Potassium permanganate and potassium hexacyanoferrate(II) in acid solution. The hexacyanoferrate(II) ion (also called ferrocyanide ion), is $Fe(CN)_6^{4-}$. It is oxidized to hexacyanoferrate(III) ion.

15.7. Draw the Lewis structures for $N_2O_5(g)$ and $N_2O_3(g)$ and calculate the formal charge on nitrogen in both. Write a balanced equation for the reaction of each of these oxides with water. Explain why N_2O_5 is the anhydride of a strong acid, while N_2O_3 is the anhydride of a weak acid.

15.8. A white solid is known to be *one* of the following compounds: NH_4NO_3, $CaCO_3$, $PbSO_4$, NH_4I, K_2S, or Hg_2Cl_2.

(a) The solid dissolves readily in water, producing a colorless solution that is slightly acidic.

(b) Nitrous acid is added to an aqueous solution of this unknown by first acidifying the solution with H_2SO_4 and then adding $NaNO_2$ crystals. The reaction of nitrous acid with the unknown produces $NO(g)$ and a brownish color in the solution. When the brownish solution is shaken with CCl_4, two liquid layers are visible and the lower layer is deep violet in color.

What is the original solid unknown? Give the reasons that led to your conclusion. Write a correctly balanced net ionic equation for the reaction that produces the acidic solution in part (a) and for the reaction of nitrous acid with the unknown.

15.9. Both Pauling and Ricci proposed the following rule that relates the strength of oxyacids to the number of oxygen atoms bonded to a central atom that do not have hydrogen bonded to them. If an oxyacid can be represented by the formula $XO_m(OH)_n$, the rule states that the larger the value of m, the stronger the acid. If m is 0, the acid is very weak ($K_a \sim 10^{-8}$ or less); if m is 1 the acid is moderately weak ($K_a \sim 10^{-2}$), and if m is 2 or 3 the acid is strong.

Consider the following oxyacids of elements 14 through 17: silicic acid, $Si(OH)_4$; phosphoric acid, $(HO)_3PO$; sulfuric acid, $(HO)_2SO_2$; and perchloric acid, $(HO)ClO_3$. Draw Lewis dot structures of these four oxyacids and under each write the oxidation state and formal charge of the central atom. Predict the order of magnitude of K_a for each of these acids using Ricci's rule. Write a brief explanation for the increasing strength of these acids from silicic to perchloric in terms of the Lewis structure and the electronegativity of the central atom. For an element that can exhibit many oxidation states, how is the electronegativity of the atom related to the oxidation state of that atom?

15.10. Identify the oxidizing agent and the reducing agent in the following oxidation–reduction reactions. Write the two half-reactions that are combined to yield the equation given.

(a) $5ClO_3^- + 9H_2O + 3As_2S_3(s) \rightarrow 6H^+(aq) + 5Cl^- + 6H_2AsO_4^- + 9S\downarrow$

(b) $7CN^- + 2\,OH^- + 2Cu(NH_3)_4^{2+} \rightarrow 2Cu(CN)_3^{2-} + 8NH_3 + CNO^- + H_2O$

chapter 16 Electrochemistry

Walther Hermann Nernst (1864–1941) was a German physical chemist who made fundamental contributions to the fields of electrochemistry, thermodynamics, photochemistry, and the theory of solutions. He studied with Ostwald, van't Hoff, and Arrhenius. His electrochemical research was inspired by the Arrhenius theory of ions and ionic dissociation in aqueous solution. Nernst developed his theory of galvanic cells in 1889. The Nernst Heat Theorem, published in 1906, is one formulation of the third law of thermodynamics. Nernst received the Nobel Prize in chemistry in 1920 for his work in thermodynamics and thermochemistry.

Useful work can be obtained from the movement of electrons through a wire. This electrical flow can be used to drive motors, to turn on light bulbs, to amplify radio signals, and to do a host of other chores for us. In the preceding chapter we learned that electrons are transferred during oxidation–reduction reactions: They are donated by the reducing agent and accepted by the oxidizing agent. A device that causes the transfer of electrons that takes place during a redox reaction to occur through a wire, rather than by direct contact between the reacting species, can serve as a source of electricity. Such a device is called a **galvanic cell,** a **voltaic cell,** or a **battery.**

The various types of dry cell batteries used to power flashlights, cameras, electronic calculators, and children's toys are all galvanic cells, as is the common storage battery used in every automobile. There are many different designs, but all batteries employ a chemical reaction as a source of electrical energy. In recent years fuel cells have been developed, primarily for the space industry, in which the chemical energy released during the combustion of fuels is converted directly into electricity. When a galvanic cell produces electricity, a spontaneous chemical reaction takes place within the cell.

The use of electricity to cause nonspontaneous chemical reactions to occur is known as **electrolysis.** The commercial production of many important substances, such as aluminum, chlorine, and sodium, is carried out by electrolysis.

In this chapter we will study both how we make use of oxidation–reduction reactions to produce electricity and how we use electricity to carry out chemical reactions that do not occur spontaneously.

section 16.1
The Use of Oxidation–Reduction Reactions to Produce Electricity: The Galvanic Cell

The Design of Galvanic Cells

Let us consider a specific oxidation–reduction reaction, the reaction between ferric ions and iodide ions. If we mix solutions containing Fe^{3+} ions and I^- ions a spontaneous reaction in which the I^- ions donate electrons to the Fe^{3+} ions, forming I_2 and Fe^{2+} ions, occurs. Electrons are transferred directly between the I^- and Fe^{3+} ions.

How do we design a galvanic cell that utilizes the redox reaction between Fe^{3+} ions and I^- ions? First of all, we must make sure that the Fe^{3+} ions and I^- ions do not come into contact with one another so that electrons cannot be transferred directly between them. Hence we must put the solution containing Fe^{3+} ions and the solution containing I^- ions into different containers. In the lab we simply use two beakers. In one beaker we might place a solution of KI, in the other a solution of $Fe_2(SO_4)_3$.

In each solution we also need to provide a surface on which electrons can be collected (called an **electrode**), and we need a wire connecting the electrodes through which electrons flow easily. Copper wire is most frequently used. Since we want to utilize the reaction between ferric and iodide ions, we do not want to put anything into these solutions that will react chemically with any constituent of the solution, and therefore we must choose some inert metal for the electrode material. Most often, in the laboratory, a small piece of platinum metal is used when a chemically inert metal is needed as an electrode. (Platinum is not used in commercial batteries as it is too expensive.) The experimental setup as we have described it up till now is shown in Fig. 16.1. Each solution with its electrode is called a **half-cell.** Two half-cells are combined to make one complete galvanic cell.

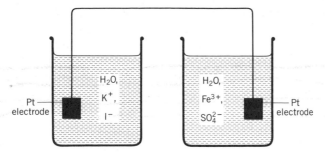

Fig. 16.1. An incomplete galvanic cell. No electrons can flow with the apparatus shown at left. The circuit is not complete and there is no mechanism for maintaining electrical neutrality in each solution if an electron were transferred from an I^- ion to a Fe^{3+} ion through the wire.

The Need for a Salt Bridge

Although I^- ions tend to lose electrons to become I_2, and Fe^{3+} ions tend to gain electrons to become Fe^{2+} ions, no reaction will occur using the apparatus described above and illustrated in Fig. 16.1. If even a single electron left the potassium iodide solution and traveled through the wire to the ferric sulfate solution, the potassium iodide solution would acquire a positive charge. Similarly, the ferric sulfate solution would acquire a net negative charge as soon as a Fe^{3+} ion was reduced to a Fe^{2+} ion. *No solution ever acquires a net macroscopic charge.* Each of these solutions remains electrically neutral at all times, so that we must provide a means of allowing negative charge to move into the KI solution and positive charge to move into the $Fe_2(SO_4)_3$ solution, or no electrons can flow through the wire. A device frequently used in the laboratory to maintain electrical neutrality in galvanic cells is a **salt bridge.**

A salt bridge is typically a U-shaped tube containing a solution of an electrolyte such as NH_4NO_3 or KCl in agar – agar. Agar – agar is a gelatinous substance obtained from certain seaweeds. When it is hot it is a liquid and can easily be mixed with a solution of NH_4NO_3 and poured into the U-tube. At room temperature the agar gels and the tube can be turned upside down without loss of its contents, but the NH_4^+ ions and the NO_3^- ions can still move freely through the agar. One arm of the inverted U-tube is inserted into each beaker, so that the experimental setup is as shown in Fig. 16.2.

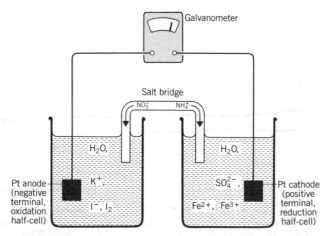

Fig. 16.2. A galvanic cell utilizing the oxidation – reduction reaction $Fe^{3+} + I^- \rightleftharpoons \frac{1}{2}I_2 + Fe^{2+}$. Reduction occurs at the cathode, the right-hand electrode in the figure. The reduction half-reaction is $Fe^{3+} + e^- \rightleftharpoons Fe^{2+}$. Oxidation occurs at the anode, the left-hand electrode in the figure. The oxidation half-reaction is $I^- \rightleftharpoons \frac{1}{2}I_2 + e^-$. The galvanometer measures the flow of current. The salt bridge serves to maintain electrical neutrality in each solution and to complete the circuit so that current can flow.

When an iodide ion in the left-hand beaker moves to the surface of the Pt electrode it gives up an electron, so that the electrode surface acquires a negative charge. Electrons then travel through the wire to the Pt electrode in the right-hand beaker. As electrons leave the left-hand beaker, nitrate ions from the salt bridge move into the solution in that beaker, so that it remains electrically neutral at all times. Similarly, as electrons reach the electrode in the right-hand beaker, Fe^{3+} ions at the electrode surface there are reduced to Fe^{2+}, and ammonium ions from the salt bridge move into that beaker.

Anode and Cathode Reactions

The **anode** is the electrode at which oxidation occurs. It is the negative terminal of a galvanic cell. In the cell being described, the anode acquires a negative charge as I^- ions give up their electrons to the Pt electrode. The oxidation half-reaction is

$$I^- \rightarrow \tfrac{1}{2} I_2 + e^- \tag{16-1a}$$

The **cathode** is the electrode at which reduction occurs. It is the positive terminal of a galvanic cell. The reduction half-reaction in the cell under discussion is

$$Fe^{3+} + e^- \rightarrow Fe^{2+} \tag{16-1b}$$

and the net cell reaction is therefore

$$Fe^{3+} + I^- \rightarrow \tfrac{1}{2} I_2 + Fe^{2+} \tag{16-1}$$

Definition of the Faraday

When 1 mol of Fe^{3+} ions and 1 mol of I^- ions have reacted to form $\tfrac{1}{2}$ mol of I_2 and 1 mol of Fe^{2+} ions, Avogadro's number of electrons (1 mol of electrons) have passed through the wire. The total amount of electrical charge in one mole of electrons is called a **faraday,** denoted $\mathscr{F}$, in honor of the British chemist and physicist Michael Faraday (1791–1867).

$$1 \, \mathscr{F} = \text{charge on 1 mol of electrons}$$

$$= \left(1.60219 \times 10^{-19} \, \frac{C}{\text{electron}}\right)\left(6.022045 \times 10^{23} \, \frac{\text{electron}}{\text{mol}}\right) = 96{,}485 \, \frac{C}{\text{mol}} \tag{16-2}$$

Cells with Metal Electrodes That Take Part in the Reaction

If one of the reacting species in an oxidation–reduction reaction happens to be a metal, the metal itself may be used as the electrode material. A classic example of this is the **Daniell cell**, which utilizes the reaction

$$Zn(s) + Cu^{2+} \rightleftharpoons Cu(s) + Zn^{2+} \tag{16-3}$$

as a source of electricity. Two slightly different designs of a Daniell cell are shown in Fig. 16.3. In Fig. 16.3(b) a porous barrier is used in place of a salt bridge. This barrier has very tiny holes; liquid fills these holes but does not readily flow through them. Ions can pass through to maintain electrical neutrality.

When the half-cell reactions

$$Zn(s) \rightarrow Zn^{2+} + 2e^- \tag{16-3a}$$

$$Cu^{2+} + 2e^- \rightarrow Cu(s) \tag{16-3b}$$

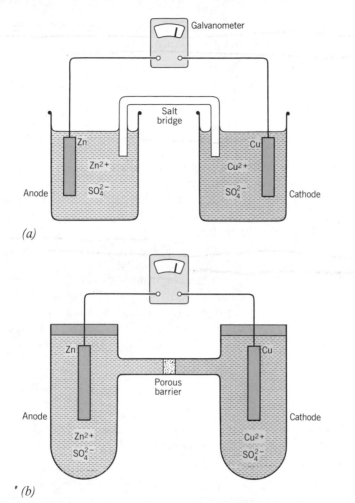

Fig. 16.3. Two ways of designing a Daniell cell, *(a)* using a salt bridge, and *(b)* using a porous barrier to separate the two solutions. The cathode reaction is $Cu^{2+} + 2e^- \rightleftharpoons Cu$, and the anode reaction is $Zn \rightleftharpoons Zn^{2+} + 2e^-$.

take place at the electrode surfaces, sulfate ions pass through the barrier, leaving the $CuSO_4$ solution and entering the $ZnSO_4$ solution. As the reaction proceeds, the Zn electrode is used up; it gets thinner and thinner as Zn goes into solution as Zn^{2+} ions. The copper electrode, on the other hand, gets thicker and heavier as Cu metal is deposited on it by the reaction. We could also use any inert metal for the cathode at the beginning of the reaction; it will soon be plated with copper as the reaction proceeds.

Electrical Current

The flow of electrons through a metal is called an electrical **current**. The SI unit of current is the **ampere** (A), named after the French physicist and mathematician André Marie Ampère (1775–1836). The relation between the current, I, and the magnitude of charge, Q, that flows past a given point in a wire per unit time is

$$I = Q/t \qquad (16\text{-}4a)$$

One ampere (usually denoted one "amp") is defined as the current flow when one coulomb (C) of charge passes any given point in one second:

$$1 \text{ amp} = 1 \text{ coulomb per second (C/s)} \qquad (16\text{-}4b)$$

Equation (16-4a) can be rearranged to solve for the amount of charge used if a known current has passed for a known amount of time:

$$Q \text{ in coulombs} = (I \text{ in amps})(t \text{ in seconds}) \qquad (16\text{-}4c)$$

When we design a galvanic cell, we start with a system that is not at equilibrium. You will recall that in Section 8.2 we discussed the fact that reactions occur to reach a state of equilibrium, and that a net reaction occurs when reagents are mixed only if a system is not at equilibrium. In order for electrons to be transferred during an oxidation–reduction reaction, the initial concentrations of reacting species must be different from equilibrium concentrations. Current will pass as the system moves toward a state of equilibrium. Once equilibrium has been attained, there will be no net reaction and no current will flow.

Many galvanic cells, however, utilize reactions with very large equilibrium constants. For the Daniell cell, the equilibrium constant of reaction (16-3) is 2×10^{37}. With such a large equilibrium constant, the reaction goes to completion, and the cell will function until one of the reagents is depleted. Example 16.1 is a calculation of the time required to deplete a galvanic cell of one of the reacting species.

EXAMPLE 16.1. Calculation of the lifetime of a galvanic cell

A Daniell cell is prepared using a Zn electrode that weighs 23.72 g immersed in a 1.00 F $ZnSO_4$ solution, and a copper electrode immersed in 600 mL of a 1.00 F $CuSO_4$ solution. Assuming that a steady current of 0.10 A is drawn from the cell, how long will the cell continue to supply electricity?

Solution. We must first determine how many moles of Zn metal and of Cu^{2+} ions were in the cell initially. The atomic weight of Zn is 65.38 g · mol^{-1} so that we had 23.72 g/65.38 g · mol^{-1} = 0.3628 mol of Zn to begin with. We also had 0.600 mol of Cu^{2+} ions [(0.600 L)(1.00 M)] so that the Zn will be used up before the Cu^{2+} ions will, as the net cell reaction is $Zn(s) + Cu^{2+} \rightarrow Cu(s) + Zn^{2+}$.

The question can therefore be rephrased as: How long will it take to oxidize 0.3628 mol of Zn to Zn^{2+} ions, if the current being drawn is 0.10 A? Since Eq. (16-3a) tells us that 2 $\mathscr{F}$ of charge are transferred per mole of Zn oxidized, the total amount of charge that must pass through the wire is

$$(0.3628 \text{ mol})(2 \ \mathscr{F}/\text{mol})(96,485 \text{ C}/\mathscr{F}) = 7.001 \times 10^4 \text{ C}$$

To calculate the length of time it will take for this amount of charge to be transferred we utilize Eq. (16-4c):

$$\text{charge in coulombs} = (\text{current in amps})(\text{time in seconds})$$

Solving for t we obtain

$$t = \frac{7.001 \times 10^4 \text{ C}}{0.10 \text{ C} \cdot \text{s}^{-1}} = 7.001 \times 10^5 \text{ s} = 194 \text{ h} = 8.1 \text{ days}$$

The cell will operate continuously for 8.1 days if a steady current of 0.10 A is drawn (an unlikely situation). If smaller currents are drawn the cell will last longer, and if larger currents are drawn the cell will be depleted more quickly.

section 16.2
The EMF of a Galvanic Cell

Defining the Zero of Potential Energy

If you hold a marble of mass m in your hand at a distance above the ground, and then let go of the marble, it will spontaneously fall to the ground. The marble has greater potential energy at a height above the ground than it does at ground level, and spontaneously falls from a point of high potential energy to a point of low potential energy.

The difference between the potential energy that a mass m possesses at a height h_u (the upper height) and the potential energy the same mass possesses at a height h_1 (the lower height) is

$$\text{difference in potential energy} = mg\Delta h = mg(h_u - h_1) \tag{16-5}$$

where g is the acceleration due to gravity.

It is not possible, however, to obtain a value for the absolute potential energy at any height without first defining what is meant by zero height. One cannot say that zero height is ground level because ground level is not the same in Mexico City as it is in Amsterdam. It is a common convention to *define* zero height as sea level. That means, however, that an object at rest on the ground in Denver, Colorado (elevation: 1 mile above sea level) is not at zero potential energy. Someone working exclusively in Denver might find it more convenient to define zero potential energy as the lowest elevation in Denver.

The point is that while the *difference* in potential energy between two positions in the earth's gravitational field is directly measurable, the absolute potential energy at a single position cannot be evaluated until a definition of zero potential energy is agreed upon. The decision about what constitutes zero potential is entirely arbitrary.

Definition of Electromotive Force (EMF)

Exactly the same situation is encountered with electrical potential energy. Current flows from one point to another because there is a difference in electrical potential energy between the two points. In a galvanic cell current flows because there is a difference in the electrical potential energy between the two electrodes.

The difference in electrical potential energy per unit charge is called the electromotive force (EMF), and is measured in volts, so that it is also referred to as the **cell voltage**. Another term for the EMF is the **cell potential**. The definition of EMF with the units of all terms specified is

$$\text{EMF in volts} = \frac{\text{difference in electrical potential energy in joules}}{\text{charge in coulombs}}$$

This equation can be written as

$$\text{PE} = Q \cdot V \tag{16-6a}$$

where PE is the potential energy, Q is the charge in coulombs, and V is the voltage. The amount of energy expended (the work done) to move electrons through a cell is equal to the product of the EMF and the charge on the electrons. One joule of energy is used when 1 C is passed through a cell with an EMF of 1 V. This relation is commonly expressed as

$$\text{volts} \times \text{coulombs} = \text{joules} \tag{16-6b}$$

The equipment used to measure the EMF of a galvanic cell is called a **potentiometer.**

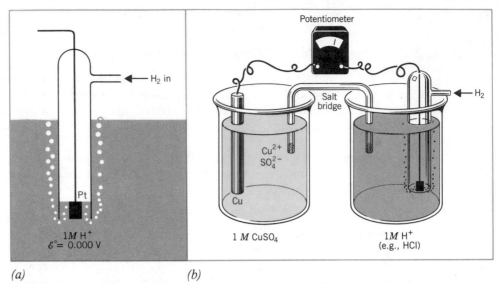

Fig. 16.4. *(a)* The hydrogen electrode. *(b)* The hydrogen electrode used in a galvanic cell.

The Standard Hydrogen Electrode

Using a potentiometer we can measure the EMF of a Daniell cell, that is, we can measure the potential difference between a piece of Zn metal immersed in a zinc sulfate solution of known concentration and a piece of Cu metal immersed in a copper sulfate solution of known concentration. What we cannot measure is the absolute potential of any single electrode (one half-cell). For instance, it is impossible to measure the potential of a piece of Zn metal immersed in a zinc sulfate solution. It has, therefore, been decided to choose one electrode (one half-cell) and arbitrarily define the potential of that electrode as zero. The electrode chosen is called the **standard hydrogen electrode** (SHE), and is shown in Fig. 16.4. It consists of a platinum electrode over which H_2 gas at 1-atm pressure is bubbled, while the electrode is immersed in a solution that is 1.00 *M* in hydronium ion, at 25 °C.*

The half-cell reaction at the standard hydrogen electrode is

$$2H^+(aq, 1.00\ M) + 2e^- \rightleftharpoons H_2(g, 1\ atm) \tag{16-7}$$

The potential of the standard hydrogen electrode at 25 °C is defined to be *exactly* zero volts.

section 16.3
Cell Conventions and Half-Cell Potentials

Suppose we set up the galvanic cell illustrated in Fig. 16.5, which can be described as follows: One half-cell consists of a piece of Zn metal immersed in a 1.00 *M* solution of

---

* The definition of the standard hydrogen electrode given here is not rigorously correct. It is necessary to define a quantity related to the concentration, called the activity. The solution in the SHE contains hydronium ion at unit activity, rather than at 1.0 *M* concentration, but we will neglect the distinction between activity and concentration here, for purposes of simplicity.

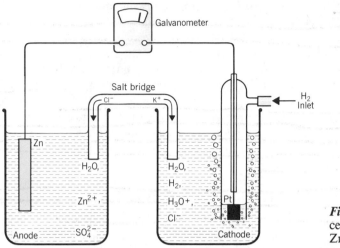

Fig. 16.5. A galvanic cell for measuring the $Zn^{2+}|Zn$ half-cell potential.

some zinc salt, such as $ZnCl_2$, $Zn(NO_3)_2$, or $ZnSO_4$, and the other half-cell consists of a standard hydrogen electrode. A salt bridge completes the circuit.

Both the diagram of the cell (Fig. 16.5) and the written description take a considerable amount of time to put down, and so a shorthand notation has been developed to convey the same information, and even more. The notation for the cell just described is

$$Zn(s) \mid Zn^{2+}(1\ M) \parallel H_3O^+(1\ M) \mid H_2(1\ \text{atm}) \mid Pt(s)$$

A single vertical bar represents a boundary between two phases. A double vertical bar represents the salt bridge, or some other device used to maintain electrical contact between two different liquid solutions that cannot be allowed to mix. Thus the two half-cells are on either side of the double vertical bar. The electrode materials are written at the two extreme ends of this notation. The symbol $Zn(s)|Zn^{2+}(1\ M)$ means: A piece of Zn metal immersed in a solution that is $1\ M$ in Zn^{2+} ions. The symbol $Pt(s)|H_2(g)|H_3O^+(1\ M)$ means: A piece of platinum metal with H_2 gas bubbled over it, immersed in a solution that is $1\ M$ in H_3O^+ ions.

By convention, the half-cell in which oxidation occurs (the anode compartment) is always written on the left, and the half-cell in which reduction occurs (the cathode compartment) is always written on the right. A simple mnemonic to help you remember this convention is that the two words that begin with the letter *"R"* go together:

Reduction occurs in the Right-hand half-cell

Therefore, for the cell being discussed, the two half-cell reactions and the net cell reaction are

Oxidation	$Zn(s) \rightarrow Zn^{2+} + 2e^-$	(16-8a)
Reduction	$2H^+(aq) + 2e^- \rightarrow H_2(g, 1\ \text{atm})$	(16-8b)
Net reaction	$Zn(s) + 2H^+(aq) \rightarrow Zn^{2+} + H_2(g, 1\ \text{atm})$	(16-8)

We can measure the EMF of this galvanic cell using a potentiometer; it is observed to be $+0.763$ V. The anode (Zn electrode) is the negative terminal, and the cathode (hydrogen electrode) is the positive terminal.

The measured cell potential will be denoted here $\Delta\mathscr{E}_{cell}$, with the Δ symbol included to remind us that the measured voltage is a **difference** between two electrode potentials. (It is quite common to omit the Δ sign and simply use the symbol $\mathcal{E}_{cell}$.) When all ions taking part in the reaction are at 1.00 M concentration, and all gases have a partial pressure of 1.00 atm, the species are said to be in their standard states. (As previously noted, the correct definition of the standard state is that the ions are at unit activity rather than unit concentration, but we are ignoring the distinction between activity and concentration for purposes of simplicity.) The voltage of such a standard cell is denoted by using a superscript zero, as $\Delta\mathcal{E}^0_{cell}$. We will see very shortly that the voltage of a galvanic cell depends on the concentrations of the reacting species, and therefore we will need to specify concentrations when giving a numerical value to a cell potential. For the special case of standard conditions, this is done by using the zero superscript.

Another convention is that the term *electrode potential* always means the **reduction potential**. Therefore the measured EMF can be expressed as the difference between two single electrode potentials in the following way:

$$\Delta\mathcal{E}_{cell} = \left(\begin{array}{c}\text{potential at the electrode}\\\text{at which reduction occurs}\end{array}\right) - \left(\begin{array}{c}\text{potential at the electrode}\\\text{at which oxidation occurs}\end{array}\right) \qquad (16\text{-}9a)$$

or

$$\Delta\mathcal{E}_{cell} = \text{cathode potential} - \text{anode potential} \qquad (16\text{-}9b)$$

or

$$\Delta\mathcal{E}_{cell} = \mathcal{E}_{right} - \mathcal{E}_{left} \qquad (16\text{-}9c)$$

where $\mathcal{E}_{right}$ is the potential of the right-hand electrode, the one at which reduction occurs, and $\mathcal{E}_{left}$ is the potential of the left-hand electrode, the one at which oxidation occurs.

If all substances involved in the reaction are in their standard states, the equation becomes

$$\Delta\mathcal{E}^0_{cell} = \mathcal{E}^0_{right} - \mathcal{E}^0_{left} \qquad (16\text{-}9d)$$

For the cell of Fig. 16.5, for which the net cell reaction is given by Eq. (16-8),

$$\Delta\mathcal{E}^0_{cell} = 0.763 \text{ V} = \mathcal{E}^0_{H^+|H_2|Pt} - \mathcal{E}^0_{Zn^{2+}|Zn} \qquad (16\text{-}10)$$

However, the potential of the standard hydrogen electrode is zero volts by definition:

$$\mathcal{E}^0_{H^+|H_2|Pt} = 0.000 \text{ V} \qquad (16\text{-}11)$$

and therefore $\mathcal{E}^0_{Zn^{2+}|Zn} = -0.763$ V. This is the numerical value of the $Zn^{2+}|Zn$ *reduction* potential, the potential for the half-reaction

$$Zn^{2+}(1.00 \text{ } M) + 2e^- \rightarrow Zn(s) \qquad \mathcal{E}^0_{Zn^{2+}|Zn} = -0.763 \text{ V} \qquad (16\text{-}12)$$

Note that in the symbol used to describe a single electrode potential, we put the oxidized form on the left and the reduced form on the right, as $\mathcal{E}^0_{ox|red}$.

For the galvanic cell we have been discussing, the zinc electrode is actually the negative terminal, and the sign of the reduction potential for the $Zn^{2+}|Zn$ electrode is negative (-0.763 V). The sign of the reduction potential for any standard half-cell and the actual sign of the electrode *relative to a standard hydrogen electrode* are always the same.

Suppose we prepare the following galvanic cell

$$Pt(s) \mid H_2(1 \text{ atm}) \mid H_3O^+(1 \text{ } M) \parallel Ag^+(1 \text{ } M) \mid Ag(s)$$

and measure its EMF. We will find that the cell voltage is 0.799 V. For this cell the Ag electrode is the cathode, or positive terminal, and the Pt electrode is the anode, or negative terminal. The cell reactions are

Anode	$H_2(g) \rightarrow 2H^+(aq) + 2e^-$	(16-13a)
Cathode	$Ag^+ + e^- \rightarrow Ag(s)$	(16-13b)
Net cell reaction	$2Ag^+ + H_2(g) \rightarrow 2H^+(aq) + 2Ag(s)$	(16-13)

Since all concentrations were specified to be 1 M and the H_2 gas is at 1-atm pressure, all substances are in their standard states and

$$\Delta\mathscr{E}^0_{cell} = 0.799 \text{ V} = \mathscr{E}^0_{Ag^+|Ag} - \mathscr{E}^0_{H^+|H_2|Pt} \qquad (16\text{-}14a)$$

Since the potential at the standard hydrogen electrode is zero volts by definition,

$$\mathscr{E}^0_{Ag^+|Ag} = 0.799 \text{ V} \qquad (16\text{-}14b)$$

By defining the potential of the standard hydrogen electrode as zero volts, we have made it possible to assign a numerical value to any other single electrode potential. By devising various galvanic cells and measuring their EMF's, we can construct a table of values of standard electrode potentials.

If we set up the following galvanic cell

$$Zn(s) \mid Zn^{2+}(1 \ M) \parallel Ag^+(1 \ M) \mid Ag(s)$$

for which the cell reactions are

Anode	$Zn(s) \rightarrow Zn^{2+} + 2e^-$	(16-15a)
Cathode	$Ag^+ + e^- \rightarrow Ag(s)$	(16-15b)
Net cell reaction	$Zn(s) + 2Ag^+ \rightarrow 2Ag(s) + Zn^{2+}$	(16-15)

then, following the conventions described, we expect the measured cell potential to be

$$\Delta\mathscr{E}^0_{cell} = \mathscr{E}^0_{Ag^+|Ag} - \mathscr{E}^0_{Zn^{2+}|Zn} = 0.799 \text{ V} - (-0.763 \text{ V}) = 1.562 \text{ V} \qquad (16\text{-}16)$$

and this is in agreement with experimental observation.

EXAMPLE 16.2. Cell notation and the standard single electrode potential

For the cell $Zn(s)|Zn^{2+}(1 \ M)\|Cu^{2+}(1 \ M)|Cu(s)$ the measured EMF is 1.100 V. Write the equation for the anode reaction, the cathode reaction and the net cell reaction, and calculate $\mathscr{E}^0_{Cu^{2+}|Cu}$.

Solution. Reduction occurs at the cathode, which is the right-hand electrode of the cell notation. Thus

$$\text{Cathode reaction} \qquad Cu^{2+} + 2e^- \rightarrow Cu(s)$$

Oxidation occurs at the anode, which is the left-hand electrode of the cell notation. Thus

$$\text{Anode reaction} \qquad Zn(s) \rightarrow Zn^{2+} + 2e^-$$

The net cell reaction is $Zn(s) + Cu^{2+} \rightarrow Zn^{2+} + Cu(s)$. This is, of course, a Daniell cell.

$$\Delta\mathscr{E}^0_{cell} = 1.100 \text{ V} = \mathscr{E}^0_{Cu^{2+}|Cu} - \mathscr{E}^0_{Zn^{2+}|Zn}$$
$$1.100 \text{ V} = \mathscr{E}^0_{Cu^{2+}|Cu} - (-0.763 \text{ V}) = \mathscr{E}^0_{Cu^{2+}|Cu} + 0.763 \text{ V}$$

Thus

$$\mathscr{E}^0_{Cu^{2+}|Cu} = 1.100 \text{ V} - 0.763 \text{ V} = 0.337 \text{ V}$$

An **oxidation potential** is the negative of a reduction potential

$$\mathcal{E}_{ox} = -\mathcal{E}_{red}$$ (16-17)

Subtracting the *reduction* potential of the electrode at which oxidation occurs, in Eq. (16-9a), is equivalent to adding the oxidation potential of that electrode. For this reason some people prefer to express Eq. (16-9a) as $\Delta\mathcal{E}_{cell} = \mathcal{E}_{red} + \mathcal{E}_{ox}$, where $\mathcal{E}_{red}$ is the reduction potential at the cathode and $\mathcal{E}_{ox}$ is the oxidation potential at the anode. While this is perfectly correct, it does not emphasize the fact that the cell voltage is the *difference* between the electrode potentials of the two electrodes.

section 16.4
Significance of the Sign of the Cell Potential

If you actually constructed the galvanic cell for which the net cell reaction is Eq. (16-15), and measured the voltage with a potentiometer, you would be able to determine experimentally that the Zn electrode is the negative terminal and the Ag electrode is the positive terminal. But suppose you are not in a lab and you are just thinking about making a galvanic cell from a standard $Ag^+|Ag$ electrode and a standard $Zn^{2+}|Zn$ electrode, and you happen to write the cell notation as $Ag|Ag^+(1\ M)\|Zn^{2+}(1\ M)|Zn$. What will you calculate for the potential of this cell? You must be absolutely consistent about following the conventions defined in the previous section. You should write for the cell reactions

Reduction	$Zn^{2+} + 2e^- \rightarrow Zn(s)$	(16-18a)
Oxidation	$Ag(s) \rightarrow Ag^+ + e^-$	(16-18b)
Net cell reaction	$Zn^{2+} + 2Ag(s) \rightarrow Zn(s) + 2Ag^+$	(16-18)

and calculate the cell potential to be

$$\Delta\mathcal{E}^0_{cell} = \mathcal{E}^0_{Zn^{2+}|Zn} - \mathcal{E}^0_{Ag^+|Ag} = -0.763\ V - 0.799\ V = -1.562\ V \quad (16\text{-}19)$$

The negative sign of the calculated $\Delta\mathcal{E}^0_{cell}$ therefore serves to notify you that you have the cell notation reversed, and that the $Zn^{2+}|Zn$ electrode belongs on the left-hand side, while the $Ag^+|Ag$ electrode belongs on the right-hand side.

The negative sign of $\Delta\mathcal{E}^0_{cell}$ also informs you that when all substances are in their standard states the spontaneous direction for the chemical reaction occurring in the cell is the *reverse* of the one you have written. Thus, if a piece of Zn is inserted into a solution containing Ag^+ ions (for example, a silver nitrate solution), reaction (16-5)

$$Zn(s) + 2Ag^+ \rightarrow Zn^{2+} + 2Ag(s)$$

will spontaneously occur. If, however, a piece of Ag is placed in a solution of zinc nitrate no reaction will occur.

The value of the EMF of a galvanic cell measures the tendency of the net cell reaction to proceed spontaneously from left to right as you have written it. The following statements summarize the information conveyed by the sign of the calculated cell potential.

If $\Delta\mathcal{E}_{cell} > 0$ the net cell reaction proceeds spontaneously from left to right, as written.

If $\Delta\mathcal{E}_{cell} < 0$ the net cell reaction proceeds spontaneously from right to left, that is, the reverse of the reaction you have written is the one that occurs spontaneously.

> If $\Delta \mathcal{E}_{cell} = 0$ no net reaction occurs. The reactants and products are at equilibrium and no current will flow.

Note that it is only possible to obtain electrical work from a system that is *not* at equilibrium. In order for current to flow, there must be a net reaction occurring. As the oxidation–reduction reaction proceeds toward equilibrium, the EMF of the cell decreases to zero. When the system is at equilibrium, the cell potential is zero and we have a dead battery.

section 16.5
The Table of Standard Reduction Potentials

In addition to the potential of the standard hydrogen electrode which is arbitrarily defined to be zero volts, we have already determined the potential of three standard electrodes, namely the $Ag^+|Ag$ electrode, the $Zn^{2+}|Zn$ electrode, and the $Cu^{2+}|Cu$ electrode. By combining any other standard electrode with one of these and measuring the cell potential we could obtain the value of $\mathcal{E}^0$ for any electrode desired. A table that lists the value of the electrode potential for any half-cell in which all concentrations are 1 M and all gases are at 1-atm pressure is a **Table of Standard Electrode Potentials**. By convention, the tabulated values of electrode potentials are standard *reduction* potentials relative to the potential of the standard hydrogen electrode, which is defined to be exactly zero volts. A small sample of such values is given in Table 16.1, and a more extensive table is provided in Appendix F.

Oxidizing agents are written at the left of the Table, with strong oxidizing agents at the top. The reduction potential is positive for strong oxidizing agents, which are species with a great tendency to accept electrons and be reduced. The strongest oxidizing agent known is fluorine gas, F_2, which oxidizes almost every other element in the periodic table, and is reduced to fluoride ion, the -1 oxidation state.

Reducing agents are written at the right of the Table, with strong reducing agents at the bottom. Thus sodium metal is the strongest reducing agent listed in Table 16.1. Sodium is not the strongest reducing agent known; all the other alkali metals are even better reducing agents, as are calcium, strontium, and barium. The more negative the reduction potential, the stronger the reducing agent.

Of all the oxidizing agents in Table 16.1, the weakest is the sodium ion, Na^+. Sodium ion and sodium metal, $Na^+|Na$, constitute a **redox couple**. When comparing redox couples, the stronger the reducing agent, the weaker its corresponding oxidizing agent. Thus F_2 and F^- constitute a redox couple. Fluorine gas is a very strong oxidizing agent, and F^- is a very weak reducing agent. The comparison of relative strengths of a redox couple should remind you of the comparison made for a conju-

Table 16.1. **Table of Standard Reduction Potentials at 25 °C**

	Oxidizing Agent	Reducing Agent		$\mathcal{E}^0$ (V)
Strong	$F_2(g) + 2e^- \rightarrow 2F^-$		Weak	2.87
	$MnO_4^- + 8H^+ + 5e^- \rightarrow Mn^{2+} + 4H_2O$			1.51
	$Br_2(aq) + 2e^- \rightarrow 2Br^-$			1.087
	$Ag^+ + e^- \rightarrow Ag(s)$			0.799
	$Cu^{2+} + 2e^- \rightarrow Cu(s)$			0.337
	$2H^+(aq) + 2e^- \rightarrow H_2(g)$			0.000
	$Pb^{2+} + 2e^- \rightarrow Pb(s)$			−0.126
	$Zn^{2+} + 2e^- \rightarrow Zn(s)$			−0.763
Weak	$Na^+ + e^- \rightarrow Na(s)$		Strong	−2.71

gate acid–base pair in Section 9.3: The weaker an acid, the stronger its conjugate base. Similarly, the weaker an oxidizing agent, the stronger its corresponding reducing agent.

The values of the standard reduction potentials can be used to predict whether or not a given oxidation–reduction reaction will proceed spontaneously when all substances are in their standard states, that is, when all concentrations are 1 M and each gas has a partial pressure of 1 atm. We will do two examples in which we calculate $\Delta \mathcal{E}^0_{cell}$ and use its sign to determine the direction of the spontaneous cell reaction. These examples should convince you that we can predict the direction of the spontaneous reaction from the relative positions of the oxidizing agent and reducing agent in the Table of Standard Electrode Potentials, without doing any numerical calculation, provided that all substances are in their standard states.

EXAMPLE 16.3. Predicting the direction of an oxidation–reduction reaction

Will permanganate ion oxidize bromide ion when all concentrations are 1 M?

Solution. We calculate $\Delta \mathcal{E}^0_{cell}$ for the reaction in which MnO_4^- oxidizes Br^-, to see whether it is negative or positive.

Reduction	$MnO_4^- + 8H^+(aq) + 5e^- \rightarrow Mn^{2+} + 4H_2O$
Oxidation	$2Br^- \rightarrow Br_2(aq) + 2e^-$
Net cell reaction	$2MnO_4^- + 16H^+(aq) + 10Br^- \rightarrow 5Br_2(aq) + 2Mn^{2+} + 8H_2O$

If a galvanic cell is constructed using this reaction with all substances in their standard states, the cell potential will be

$$\Delta \mathcal{E}^0_{cell} = \mathcal{E}^0_{Pt|MnO_4^-,Mn^{2+}} - \mathcal{E}^0_{Pt|Br_2,Br^-} = 1.51 - 1.09 = +0.42 \text{ V}$$

Because the cell potential is positive, $\Delta \mathcal{E}^0_{cell} > 0$, the reaction proceeds spontaneously as written when all substances are in their standard states. Permanganate ion is a very strong oxidizing agent and oxidizes Br^- to Br_2.

Note that in calculating $\Delta \mathcal{E}^0_{cell}$ you use the values of the reduction potentials exactly as they appear in the Table. The cell potential is the difference between the reduction potentials of the two half-cells. The half-cell whose reduction potential is subtracted is always the one in which oxidation actually occurs. *You* never change a sign of any value in the Table. The subtraction carried out in Eq. (16-9) insures that the sign of $\Delta \mathcal{E}_{cell}$ is correct.

Note also that in order to obtain the correct net cell reaction in Example 16.3, we had to multiply the reduction half-reaction by two and the oxidation half-reaction by five, but we did not multiply either $\mathcal{E}^0$ by any number whatsoever. We use the $\mathcal{E}^0$ values *exactly* as they appear in the Table of Standard Reduction Potentials. The cell voltage is a property of the chemical species involved in the oxidation–reduction reaction, but it is independent of the *amount* of any of those species. If we choose to multiply the net cell reaction by some constant, the cell voltage is unaffected. The cell voltage does, however, depend on the concentrations of the reacting species, as we shall see in the following section.

EXAMPLE 16.4. Predicting the spontaneous direction of a redox reaction

Will Pb reduce Zn^{2+} ions to Zn when all concentrations are 1 M?

Solution. Calculate the value of $\Delta \mathcal{E}^0_{cell}$ and use its sign to predict the direction of reaction. If Pb reduces Zn^{2+} the half-reactions are

Reduction	$Zn^{2+} + 2e^- \rightarrow Zn(s)$
Oxidation	$Pb(s) \rightarrow Pb^{2+} + 2e^-$
Net cell reaction	$Pb(s) + Zn^{2+} \rightarrow Zn(s) + Pb^{2+}$

If a galvanic cell were constructed using this reaction with all substances in their standard states, the cell potential would be

$$\Delta\mathcal{E}^0_{cell} = \mathcal{E}^0_{Zn^{2+}|Zn} - \mathcal{E}^0_{Pb^{2+}|Pb} = -0.763 \text{ V} - (-0.126 \text{ V}) = -0.637 \text{ V}$$

Since the calculated standard cell potential is negative, the reaction will *not* occur as written; rather, the reverse reaction will occur. Zinc will reduce Pb^{2+} ions. The spontaneous reaction is

$$Zn(s) + Pb^{2+} \rightarrow Pb(s) + Zn^{2+}$$

Lead will not reduce Zn^{2+} ions. If a piece of Pb metal is immersed in a solution of a zinc salt, nothing at all will happen.

By considering these two examples we learn the following rules about the use of the Table of Standard Reduction Potentials to predict the direction of spontaneous reaction:

1. A reducing agent will react spontaneously with any oxidizing agent stronger than the oxidizing agent with which it is coupled, that is, any oxidizing agent *above* its corresponding oxidizing agent in the Table. Thus Pb will reduce $H^+(aq)$, Cu^{2+}, Ag^+, Br_2, and so on, but will not react with Zn^{2+} or Na^+.

2. An oxidizing agent will react with any reducing agent stronger than the reducing agent with which it is coupled, that is, any reducing agent *below* its corresponding reducing agent in the Table. Thus Br_2 will oxidize Ag, Cu, H_2, and so on, but it will not react with Mn^{2+} or F^- ions.

Once we understand these rules, we do not have to do any calculation to predict the spontaneous direction of a redox reaction for which the half-reactions are listed in the Table. The predictions are rigorously correct only for all substances in their standard states, but the change in voltage with change in concentration is usually small compared to $\Delta\mathcal{E}^0_{cell}$. If $\Delta\mathcal{E}^0_{cell}$ is larger than 0.1 or 0.2 V, the predictions made using the Table will generally be correct. Will Cu reduce Ag^+ ions? Yes, because Ag^+ is a stronger oxidizing agent than Cu^{2+}, since Ag^+ is above Cu^{2+} in Table 16.1. Will Cu reduce Pb^{2+} ions? No, because Pb^{2+} is below Cu^{2+} in Table 16.1.

EXAMPLE 16.5. Using a Table of Standard Reduction Potentials to predict the spontaneous direction of redox reactions

The $\mathcal{E}^0$ values of the following two half-reactions, both of which occur in basic solution, have been obtained from Appendix F.

$$Ni(OH)_2(s) + 2e^- \rightarrow Ni(s) + 2\,OH^- \qquad \mathcal{E}^0 = -0.72 \text{ V}$$
$$CrO_4^{2-} + 4H_2O + 3e^- \rightarrow Cr(OH)_3(s) + 5\,OH^- \qquad \mathcal{E}^0 = -0.12 \text{ V}$$

Write the balanced equation for the oxidation–reduction reaction that occurs spontaneously and involves these half-reactions.

Solution. As written in the Table, both half-reactions are reductions. During reaction, one of them must be reversed to be an oxidation, and that one will have its $\mathcal{E}^0$ value subtracted in determining the cell potential. The only way we can obtain a positive cell potential by subtracting one of these values is for Ni(s) to be oxidized to $Ni(OH)_2$, that is for -0.72 V to be subtracted.

Reduction $CrO_4^{2-} + 4H_2O + 3e^- \rightarrow Cr(OH)_3(s) + 5 OH^-$

Oxidation $Ni(s) + 2 OH^- \rightarrow Ni(OH)_2(s) + 2e^-$

To balance, multiply the oxidation half-reaction by three and the reduction half-reaction by two. The spontaneous redox reaction is

$$3Ni(s) + 2CrO_4^{2-} + 8H_2O \rightarrow 4 OH^- + 3Ni(OH)_2(s) + 2Cr(OH)_3(s)$$

and

$$\Delta \mathcal{E}^0_{cell} = -0.12 \text{ V} - (-0.72 \text{ V}) = +0.60 \text{ V}.$$

EXAMPLE 16.6. Cell conventions and the use of the Table of Standard Electrode Potentials

For each of the following galvanic cells write the anode, cathode, and net cell reactions. Using Appendix F calculate $\Delta \mathcal{E}^0_{cell}$. Does the net cell reaction you have written proceed spontaneously to the right when all substances are in their standard states?

(a) $Pt(s) \mid Cl_2(g) \mid Cl^- \parallel Fe^{2+}, Fe^{3+} \mid Pt(s)$

Solution. Every substance to the right of the double vertical bars is part of the cathode compartment, where reduction occurs. The Pt is an inert metal electrode. Therefore,

Cathode reaction $Fe^{3+} + e^- \rightarrow Fe^{2+}$

Every substance to the left of the double vertical bars is part of the anode compartment, where oxidation occurs. Hence

Anode reaction $2Cl^- \rightarrow Cl_2(g) + 2e^-$

Net cell reaction $2Fe^{3+} + 2Cl^- \rightarrow 2Fe^{2+} + Cl_2(g)$

$$\Delta \mathcal{E}^0_{cell} = \mathcal{E}^0_{Fe^{3+}|Fe^{2+}} - \mathcal{E}^0_{Cl_2|Cl^-} = +0.770 \text{ V} - 1.358 \text{ V} = -0.588 \text{ V}$$

The reaction as written does *not* proceed spontaneously to the right. The reverse reaction, $2Fe^{2+} + Cl_2(g) \rightarrow 2Fe^{3+} + 2Cl^-$, proceeds spontaneously. Chlorine will oxidize ferrous ion to ferric ion.

(b) $Cu(s) \mid CuCl_2(aq) \parallel MnCl_2(aq), HCl(aq) \mid MnO_2(s) \mid Pt(s)$

Solution. The only element with two different oxidation states on the right-hand side of the double vertical bars (the cathode compartment) is manganese, which is in the +4 state in $MnO_2(s)$ and the +2 state in Mn^{2+}. Aqueous $MnCl_2$ consists of Mn^{2+} and Cl^- ions. Thus,

Cathode $4H^+(aq) + MnO_2(s) + 2e^- \rightarrow Mn^{2+} + 2H_2O$

Anode $Cu(s) \rightarrow Cu^{2+} + 2e^-$

Net cell reaction $4H^+(aq) + MnO_2(s) + Cu(s) \rightarrow Cu^{2+} + Mn^{2+} + 2H_2O$

$$\Delta \mathcal{E}^0_{cell} = \mathcal{E}^0_{MnO_2|Mn^{2+}} - \mathcal{E}^0_{Cu^{2+}|Cu} = 1.208 \text{ V} - 0.337 \text{ V} = 0.871 \text{ V}$$

Since the standard cell voltage is positive, the net cell reaction is spontaneous in the direction written when all substances are in their standard states.

section 16.6
Concentration Dependence of the Cell EMF

We have stated earlier that the value of the measured cell potential depends on the concentrations of all the substances involved in the reaction. The simplest experimental device for studying the dependence of $\Delta\mathscr{E}_{cell}$ on concentration is known as a **concentration cell**. A concentration cell is a galvanic cell in which the electrode materials and the solutions in both half-cells are composed of the same substances; only the concentrations of the two solutions differ.

Concentration Cells

An example of a concentration cell is the following:

$$Cu(s) \mid Cu^{2+}(M_1) \parallel Cu^{2+}(M_2) \mid Cu(s)$$

where M_1 and M_2 are the molar concentrations of Cu^{2+} ion in the two half-cells. When we write the cell reaction we must be sure to include the concentrations, because they constitute the only distinction between the two half-cells. We must adhere rigorously to the conventions set forth in Section 16.3. The cell reactions for the concentration cell above are

Reduction	$Cu^{2+}(M_2) + 2e^- \rightarrow Cu(s)$	(16-20a)
Oxidation	$Cu(s) \rightarrow Cu^{2+}(M_1) + 2e^-$	(16-20b)
Net cell reaction	$Cu^{2+}(M_2) \rightarrow Cu^{2+}(M_1)$	(16-20)

Note particularly that the net cell reaction occurring in a concentration cell is simply the change in concentration of some species. Although an oxidation occurs at the anode, the same material is reduced at the cathode, and the net change is merely a change in concentration. Solid copper does not appear in Eq. (16-20) because it is on both the right and the left side of the equation and therefore cancels out.

We find that the net cell reaction is spontaneous as written, and the cell potential is positive, provided that $M_2 > M_1$. When the cell operates, the concentration of Cu^{2+} decreases in the half-cell that was originally more concentrated, while the concentration increases in the solution that was originally more dilute. At equilibrium, the concentration of Cu^{2+} in both solutions is the same, at a value intermediate between M_1 and M_2. At equilibrium, the cell voltage is zero and no current flows. Table 16.2 reports experimental data obtained using the $Cu^{2+}|Cu$ concentration cell at 25 °C.

It is clear from an examination of these data that the measured cell potential depends on the *ratio* of the two concentrations, because when both concentrations

Table 16.2. **Measured EMF at 25 °C of the Concentration Cell[a]**
$$Cu(s) \mid Cu^{2+}(M_1) \parallel Cu^{2+}(M_2) \mid Cu(s)$$

M_2	M_1	$\Delta\mathscr{E}_{cell}$ (V)	$Q = M_1/M_2$	$\log Q = \log(M_1/M_2)$
1.000	0.100	0.0296	1/10	-1
0.100	0.010	0.0296	1/10	-1
0.010	0.001	0.0296	1/10	-1
1.000	0.010	$0.0592 = 2(0.0296)$	1/100	-2
1.000	0.001	$0.0888 = 3(0.0296)$	1/1000	-3

[a] The concentrations M_1 and M_2 are expressed in moles per liter.

Table 16.3. **Measured EMF at 25 °C of the Concentration Cell[a]**
Ag(s) | Ag⁺(M₁) ‖ Ag⁺(M₂) | Ag(s)

M_2	M_1	$\Delta \mathcal{E}'_{cell}$ (V)	$Q = M_1/M_2$	$\log Q = \log(M_1/M_2)$
1.000	0.100	0.0592	1/10	−1
0.100	0.010	0.0592	1/10	−1
0.010	0.001	0.0592	1/10	−1
1.000	0.010	0.1184 = 2(0.0592)	1/100	−2
1.000	0.001	0.1776 = 3(0.0592)	1/1000	−3

[a] The concentrations M_1 and M_2 are expressed in moles per liter.

are changed but the ratio (M_1/M_2) remains the same, the cell voltage remains the same. If the ratio of the concentrations is changed, however, the cell voltage changes.

Suppose we repeat the experiment using an Ag⁺|Ag concentration cell. The data obtained at 25 °C for such a cell are reported in Table 16.3.

You should write the half-reactions for the Ag⁺|Ag concentration cell and show that the net cell reaction is

$$\text{Ag}^+(M_2) \rightarrow \text{Ag}^+(M_1)$$

Examining the data in Table 16.3 we see again that $\Delta \mathcal{E}_{cell}$ depends only on the ratio (M_1/M_2). There is a difference between the two concentration cells, however. When M_1/M_2 is 1/10, $\Delta \mathcal{E}_{cell}$ is 0.0592 V for the Ag⁺|Ag concentration cell, but is only half that value for the Cu²⁺|Cu concentration cell. This is a consequence of the fact that 2 $\mathcal{F}$ of charge must pass through the cell to reduce a mole of Cu²⁺ ions to Cu, whereas only 1 $\mathcal{F}$ of charge must pass to reduce a mole of Ag⁺ ions to Ag.

In Section 8.2 we defined the reaction quotient, Q, for any reaction. Consider Eq. (16-20) carefully. What is the reaction quotient for this reaction? It is simply M_1/M_2. In the numerator we have the concentration of the product, and in the denominator the concentration of the reactant. Similarly, for the net cell reaction of the Ag⁺|Ag concentration cell, the reaction quotient, Q, is also M_1/M_2. Note that in both Tables 16.2 and 16.3 the column giving the ratio of concentrations M_1/M_2 is headed $Q = M_1/M_2$, where Q stands for the reaction quotient for the net cell reaction of the concentration cell.

We can get no electricity from a concentration cell if both solutions are at the same concentration. If $M_1/M_2 = 1$, $\Delta \mathcal{E}_{cell} = 0$. If the two concentrations are the same, the system is at equilibrium and no net change can occur. The further away the system is from equilibrium, that is, the greater the disparity between M_1 and M_2, the larger the cell voltage will be. Examine the data in Tables 16.2 and 16.3 carefully. Note that the smaller the ratio M_1/M_2 (or alternatively, the larger the ratio M_2/M_1), the larger is the value of the cell voltage. For each of these cells, decreasing the ratio from 1/10 to 1/100 doubles the voltage, whereas decreasing the ratio from 1/10 to 1/1000 triples the voltage. The only mathematical function that can account for the observed dependence of the voltage on the ratio M_1/M_2, is the logarithm. You should be able, by substitution, to convince yourself that for concentration cells at 25 °C, the equation

$$\Delta \mathcal{E}_{cell} = -\left(\frac{0.0592}{n}\right) \log Q = -\left(\frac{0.0592}{n}\right) \log \left(\frac{M_1}{M_2}\right) \qquad (16\text{-}21)$$

where n is the number of faradays transferred in each half-reaction, correctly accounts for all the data reported in Tables 16.2 and 16.3. For the Cu²⁺|Cu concentration cell, $n = 2$, while for the $Ag^+|Ag$ concentration cell, $n = 1$. Note that when the ratio M_1/M_2 is less than 1 (that is, $M_2 > M_1$), the $\log(M_1/M_2)$ is negative, and the cell potential is positive. The direction of the spontaneous reaction is from the more concentrated solution to the less concentrated solution. The use of Eq. (16-21) to calculate the EMF of concentration cells at 25 °C is illustrated in Example 16.7.

EXAMPLE 16.7. Calculating the EMF of a concentration cell

(a) What is the EMF of the following concentration cell at 25 °C?

$$Zn(s) \mid Zn^{2+}(0.024\ M) \parallel Zn^{2+}(0.480\ M) \mid Zn(s)$$

Solution. The cell reactions are

Reduction	$Zn^{2+}(0.480\ M) + 2e^- \rightarrow Zn(s)$
Oxidation	$Zn(s) \rightarrow Zn^{2+}(0.024\ M) + 2e^-$
Net cell reaction	$Zn^{2+}(0.480\ M) \rightarrow Zn^{2+}(0.024\ M)$

The reaction quotient, Q, for this reaction is

$$Q = (0.024)/(0.480) = 5.0 \times 10^{-2}$$

and therefore $\log Q = \log(5.0 \times 10^{-2}) = -1.30$. Substitution into Eq. (16-21) yields

$$\Delta \mathscr{E}_{cell} = -\left(\frac{0.0592}{2}\right) \log Q = -\left(\frac{0.0592}{2}\right)(-1.30) = +0.0385\ V$$

(b) If water is added to the more dilute solution so that the $[Zn^{2+}]$ is decreased to $0.012\ M$, will the cell voltage increase, decrease, or remain the same? Explain your answer.

Solution. If water is added to the more dilute solution, the disparity between the two concentrations will become greater, and the voltage will increase. Specifically, the reaction quotient Q will become smaller, and $\log Q$ will become more negative, making $\Delta \mathscr{E}_{cell}$ larger. If the more dilute solution becomes $0.012\ M$, the reaction quotient will become $Q = (0.012)/(0.480) = 2.5 \times 10^{-2}$, and $\log Q = -1.60$. The cell potential after dilution is therefore $\Delta \mathscr{E}_{cell} = -(0.0592/2)(-1.60) = +0.0474\ V$.

Any change that causes the reaction quotient, Q, to be further from unity (that is, causes the concentrations to be further apart) will increase the absolute value of the EMF of a concentration cell.

The Nernst Equation at 25 °C

The equation that relates the measured cell potential, $\Delta \mathscr{E}_{cell}$, of any galvanic cell to the concentrations of the substances involved in the cell reaction, is called the Nernst equation, after the German chemist Walther Nernst (1864–1941). The Nernst equation can be derived using fundamental principles of thermodynamics, and we will discuss it again in Chapter 18, but at this point we will give a preliminary statement of the Nernst equation and consider it as a generalization obtained from experimental observations at 25 °C. The observed relation is

$$\Delta \mathscr{E}_{cell} = \Delta \mathscr{E}^0_{cell} - (0.0592/n)\log Q \qquad \text{at 25 °C} \qquad (16\text{-}22)$$

where n is the number of electrons transferred in the net cell reaction, and Q is the reaction quotient for the net cell reaction. In writing the expression for Q, molar concentrations must be used for substances in solution, and pressure in atmospheres must be used for gases. This is necessary to be consistent with the definition of standard states previously given: 1.00 M for species in solution and 1.00 atm for gases.

You can see the similarity of Eq. (16-22) to Eq. (16-21), which applies only to concentration cells. For concentration cells, the electrode materials of both electrodes are the same, and the substances in both solutions are the same, so the *standard electrode potentials for both electrodes are identical*. Remember that the standard electrode potential refers to cells in which all concentrations are 1 M. If the solutions in both anode and cathode compartments of a concentration cell are 1 M, the cell voltage is zero.

$$\text{for concentration cells } \Delta\mathcal{E}^0_{\text{cell}} = 0 \qquad (16\text{-}23)$$

Therefore the Nernst equation, Eq. (16-22), reduces to Eq. (16-21) for concentration cells. For a galvanic cell that is not a concentration cell, of course, the two standard electrode potentials are different.

You should think of Eq. (16-22) as composed of two terms. The first term, $\Delta\mathcal{E}^0_{\text{cell}}$, gives the cell potential when all concentrations are 1 M, in which case the reaction quotient, Q, must necessarily be 1. The second term, $-(0.0592/n)\log Q$, gives the change in the EMF when the reaction quotient has a value different from 1. The following two examples illustrate the use of the Nernst equation.

EXAMPLE 16.8. Use of the Nernst equation to calculate a cell potential at 25 °C

Calculate the EMF of the galvanic cell

$$\text{Zn(s) | Zn}^{2+}(M_1) \| \text{Ag}^+(M_2) \text{ | Ag(s)} \qquad \text{at 25 °C}$$

when

(a) $M_1 = 1.00\ M$ and $M_2 = 1.00\ M$
(b) $M_1 = 0.100\ M$ and $M_2 = 0.010\ M$
(c) $M_1 = 0.012\ M$ and $M_2 = 0.200\ M$

Solution

(a) When M_1 and M_2 are both 1.00 M, all substances are in their standard states and the cell voltage is equal to the standard cell potential, $\Delta\mathcal{E}^0_{\text{cell}}$, which has already been determined to be 1.562 V, Eq. (16-16). The net cell reaction for this cell, Eq. (16-15), is

$$\text{Zn(s)} + 2\text{Ag}^+ \rightarrow 2\text{Ag(s)} + \text{Zn}^{2+}$$

and therefore Q, the reaction quotient, is given by

$$Q = \frac{[\text{Zn}^{2+}]}{[\text{Ag}^+]^2}$$

(Remember that pure solids do not appear in the expression for either equilibrium constants or reaction quotients. Refer to Sections 8.1 and 8.2.) If both the $[\text{Zn}^{2+}]$ and $[\text{Ag}^+]$ are 1.00 M, $Q = 1$ and $\log Q = 0$, so that the Nernst equation, (16-22), becomes simply

$$\Delta\mathcal{E}_{\text{cell}} = \Delta\mathcal{E}^0_{\text{cell}} = 0.799\ \text{V} - (-0.763\ \text{V}) = 1.562\ \text{V}$$

(b) When $[\text{Zn}^{2+}] = 0.100\ M$ and $[\text{Ag}^+] = 0.010\ M$,

$$Q = \frac{(0.100)}{(0.010)^2} = \frac{1.00 \times 10^{-1}}{1.0 \times 10^{-4}} = 1.0 \times 10^{+3} \quad \text{and} \quad \log Q = 3.0$$

For this reaction $n = 2$, because the two half-reactions that combine to give the net cell reaction are

Reduction	$2Ag^+ + 2e^- \rightarrow 2Ag(s)$
Oxidation	$Zn(s) \rightarrow Zn^{2+} + 2e^-$

Thus Eq. (16-22) becomes

$$\Delta\mathcal{E}_{cell} = 1.562 - (0.0592/2)\log(1.0 \times 10^{+3}) = 1.562 - (0.0296)(3.0)$$
$$= 1.562 - 0.089 = 1.473 \text{ V}$$

Note that when Q is greater than 1 the cell potential is less than the standard cell potential.

(c) When $[Zn^{2+}] = 1.2 \times 10^{-2}$ and $[Ag^+] = 2.00 \times 10^{-1}$

$$Q = \frac{[Zn^{2+}]}{[Ag^+]^2} = \frac{1.2 \times 10^{-2}}{4.00 \times 10^{-2}} = 0.30 = 3.0 \times 10^{-1}$$

Therefore $\log Q = -1 + \log 3 = -1 + 0.48 = -0.52$. The cell potential, calculated using the Nernst equation, is

$$\Delta\mathcal{E}_{cell} = 1.562 - (0.0592/2)(-0.52) = 1.562 + (0.0296)(0.52)$$
$$= 1.562 + 0.015 = 1.577 \text{ V}$$

When Q is less than 1, the cell potential is greater than the standard cell potential.

EXAMPLE 16.9. Use of the Nernst equation to determine the concentration needed to achieve a given cell voltage at 25 °C

If we want the galvanic cell

$$Cu(s) \mid Cu^{2+}(M_1) \parallel Cl^-(0.100 \, M) \mid Cl_2(g, 1 \text{ atm}) \mid Pt(s)$$

to have an EMF of 1.122 V, what value should be used for M_1, the concentration of Cu^{2+} ions?

Solution. Use Appendix F to look up the standard reduction potential of the $Cl_2(g)|Cl^-$ electrode. Its value is 1.358 V. The $\Delta\mathcal{E}^0_{cell}$ for this cell is

$$\Delta\mathcal{E}^0_{cell} = \mathcal{E}^0_{Cl_2|Cl^-} - \mathcal{E}^0_{Cu^{2+}|Cu} = 1.358 \text{ V} - 0.337 \text{ V} = 1.021 \text{ V}$$

The half-cell reactions are

Reduction	$Cl_2(g) + 2e^- \rightarrow 2Cl^-$
Oxidation	$Cu(s) \rightarrow Cu^{2+} + 2e^-$
Net cell reaction	$Cu(s) + Cl_2(g) \rightarrow Cu^{2+} + 2Cl^-$

so that the reaction quotient is

$$Q = \frac{[Cu^{2+}][Cl^-]^2}{P_{Cl_2}}$$

The Nernst equation for this cell is

$$\Delta\mathcal{E}_{cell} = 1.021 \text{ V} - (0.0592/2)\log Q = 1.021 \text{ V} - (0.0296)\log Q$$

We can therefore calculate the value of Q required to make the cell potential 1.122 V. Substitution into the Nernst equation yields $1.122 = 1.021 - (0.0296)\log Q$. Therefore $0.101 = (-0.0296)\log Q$, and

$$\log Q = -(0.101)/(0.0296) = -3.41$$

Hence, $Q = 10^{-3.41} = 3.9 \times 10^{-4}$. Since $[Cl^-] = 0.100$ in this cell, and $P_{Cl_2} = 1.00$ atm,

$$Q = 3.9 \times 10^{-4} = [Cu^{2+}](1.0 \times 10^{-2})/1.00 \quad \text{and} \quad [Cu^{2+}] = 3.9 \times 10^{-2} \, M$$

Thus if we set $[Cu^{2+}] = 3.9 \times 10^{-2} \, M$, the cell voltage will be 1.122 V.

section 16.7
Proof That the Cell Potential Is Independent of the Form of the Net Cell Reaction

Let us consider once again the galvanic cell of Example 16.8, but now let us write the half-reactions as

Reduction	$Ag^+ + e^- \rightarrow Ag(s)$
Oxidation	$\frac{1}{2}Zn(s) \rightarrow \frac{1}{2}Zn^{2+} + e^-$

so that the net cell reaction is

$$\frac{1}{2}Zn(s) + Ag^+ \rightarrow Ag(s) + \frac{1}{2}Zn^{2+}$$

All we have done is to multiply the net cell reaction used in Example 16.8 by $\frac{1}{2}$. Both ways of writing the equation convey exactly the same information. Clearly, the galvanic cell is totally ignorant of the way we choose to write the net cell reaction, and the measured cell potential cannot depend in any way on the form of the equation we write. Let us see what the Nernst equation is when applied to this form of the net cell reaction. The reaction quotient, Q, is now given by

$$Q = \frac{[Zn^{2+}]^{1/2}}{[Ag^+]}$$

and the value of n is now 1. Therefore Eq. (16-22) becomes

$$\Delta \mathcal{E}_{cell} = 1.562 - 0.0592 \log \left(\frac{[Zn^{2+}]^{1/2}}{[Ag^+]} \right)$$

But

$$\frac{[Zn^{2+}]^{1/2}}{[Ag^+]} = \left(\frac{[Zn^{2+}]}{[Ag^+]^2} \right)^{1/2} \quad \text{and} \quad \log \left(\frac{[Zn^{2+}]}{[Ag^+]^2} \right)^{1/2} = \frac{1}{2} \log \frac{[Zn^{2+}]}{[Ag^+]^2}$$

so that the Nernst equation can be rearranged to

$$\Delta \mathcal{E}_{cell} = 1.562 - (0.0592)\left(\frac{1}{2} \log \frac{[Zn^{2+}]}{[Ag^+]^2} \right) = 1.562 - \left(\frac{0.0592}{2} \right) \log \frac{[Zn^{2+}]}{[Ag^+]^2}$$

which is exactly the same expression we had written in Example 16.8.

We see from this example that multiplying the equation for the net cell reaction by any constant whatsoever changes the value of n, and changes the expression for Q, but does not affect the cell voltage, $\Delta \mathcal{E}_{cell}$, in any way. The cell voltage is an intensive property, a property that does not depend on the number of moles of material present.

This discussion should remind you of the fact that the numerical value of, and the expression for, an equilibrium constant both depend on the way the equation is written. What is true of the equilibrium constant is also true of the reaction quotient, Q, of course. Both Q and n, as well as K_{eq}, are defined quantities, and the definitions link them to the form of the equation for the reaction. The cell potential, on the other hand, is an experimentally measured quantity. Whether one constructs an enormous galvanic cell with several liters of each solution, or a small cell with only a few milliliters of each solution, the EMF is the same. The voltage of a cell does not depend on the way anyone chooses to write the spontaneous net cell reaction, and the form of the Nernst equation insures that you will calculate the same numerical value for $\Delta\mathcal{E}_{cell}$ no matter how the equation is written.

section 16.8
Use of the Nernst Equation to Determine the Equilibrium Constant

A galvanic cell can produce electricity only when the cell reaction is not at equilibrium, that is when $Q \neq K_{eq}$. Remember that when the reacting substances in a galvanic cell come to a state of equilibrium, we have a dead battery! When the system is at equilibrium, two conditions must hold, and they are

$$Q = K_{eq} \quad \text{and} \quad \Delta\mathcal{E}_{cell} = 0 \quad \text{for a system at equilibrium} \quad (16\text{-}24)$$

The Nernst equation is valid both for systems *not* at equilibrium and for systems that are at equilibrium. If we write the Nernst equation specifically for a system at equilibrium at 25 °C by substituting the relations of Eq. (16-24) into Eq. (16-22) we obtain

$$0 = \Delta\mathcal{E}^0_{cell} - (0.0592/n)\log K_{eq} \quad (16\text{-}25a)$$

or

$$\Delta\mathcal{E}^0_{cell} = (0.0592/n)\log K_{eq} \quad (16\text{-}25b)$$

or

$$\log K_{eq} = (n\Delta\mathcal{E}^0_{cell})/0.0592 \quad (16\text{-}25c)$$

Thus we can calculate the equilibrium constant at 25 °C from the value of $\Delta\mathcal{E}^0_{cell}$ at 25 °C, as illustrated in Example 16.10.

EXAMPLE 16.10. Calculating K_{eq} from the standard cell potential

Calculate the equilibrium constants for the following reactions at 25 °C.

(a) $Cu(s) + Br_2(aq) = Cu^{2+} + 2Br^-$

Solution. Write the two half-cell reactions and calculate the standard cell potential. For this reaction we have

$$\text{Reduction} \quad Br_2(aq) + 2e^- \rightarrow 2Br^-$$
$$\text{Oxidation} \quad Cu(s) \rightarrow Cu^{2+} + 2e^-$$

Hence $\Delta\mathcal{E}^0_{cell} = \mathcal{E}^0_{Pt|Br_2,Br^-} - \mathcal{E}^0_{Cu^{2+}|Cu} = 1.087 - 0.337 = 0.750$ V. The value of n for this reaction is 2. Equation (16-25c) is then

$$\log K_{eq} = \frac{(2)(0.750)}{0.0592} = 25.3$$
$$K_{eq} = 10^{25.3} = 2 \times 10^{25}$$

Note that $\log K_{eq}$ is known to three significant figures, but two of them define the characteristic of the logarithm (refer to Appendix B), that is, they specify the power of 10 in the antilog. We have only one significant figure in the mantissa (0.3) and therefore we can only give one significant figure in K_{eq}.

Both the large value of K_{eq} and the positive sign of $\Delta\mathcal{E}^0_{cell}$ convey the same information about this reaction: $Cu(s)$ and Br_2 in aqueous solution will spontaneously react to form a solution of cupric bromide. The expression for the equilibrium constant is

$$K_{eq} = \frac{[Cu^{2+}][Br^-]^2}{[Br_2]} = 2 \times 10^{25}$$

(b) $2Ag(s) + 2H^+(aq) = 2Ag^+ + H_2(g)$

Solution. The two half-reactions are

$$\text{Reduction} \qquad 2H^+(aq) + 2e^- \rightarrow H_2(g)$$
$$\text{Oxidation} \qquad\qquad 2Ag(s) \rightarrow 2Ag^+ + 2e^-$$

so that the value of n for this reaction is 2. The expression for the equilibrium constant is

$$K_{eq} = \frac{P_{H_2}[Ag^+]^2}{[H^+]^2}$$

$$\Delta\mathcal{E}^0_{cell} = \mathcal{E}^0_{H^+|H_2|Pt} - \mathcal{E}^0_{Ag^+|Ag} = 0.000 - 0.799 = -0.799 \text{ V}$$

Using Eq. (16-25c) we obtain

$$\log K_{eq} = \frac{(2)(-0.799)}{0.0592} = -27.0$$

and

$$K_{eq} = 1 \times 10^{-27}$$

Again, note that both the negative sign of $\Delta\mathcal{E}^0_{cell}$ and the very small value of K_{eq} tell you the same thing about this reaction: It proceeds to a negligible extent. There will be virtually no reaction if a piece of silver metal is inserted into an aqueous acidic solution. The reverse reaction, however,

$$H_2(g) + 2Ag^+ \rightarrow 2H^+ + 2Ag(s)$$

is spontaneous, with $\Delta\mathcal{E}^0_{cell} = 0 - (-0.799 \text{ V}) = +0.799 \text{ V}$, and

$$K_{eq} = (1 \times 10^{-27})^{-1} = 1 \times 10^{+27}$$

(c) $Ag(s) + H^+(aq) = Ag^+ + \frac{1}{2}H_2(g)$

Solution. The two half-reactions are now

$$\text{Reduction} \qquad H^+(aq) + e^- \rightarrow \tfrac{1}{2}H_2(g)$$
$$\text{Oxidation} \qquad\qquad Ag(s) \rightarrow Ag^+ + e^-$$

so that the value of n is 1. When the equation is written in this way the expression for the equilibrium constant is

$$K_{eq} = \frac{[Ag^+]P_{H_2}^{1/2}}{[H^+]}$$

$\Delta\mathcal{E}^0_{cell} = -0.799 \text{ V}$, just as in part **(b)**. The standard cell potential does not change because we have multiplied the net cell reaction of part **(b)** by $\frac{1}{2}$! Since $n = 1$, Eq. (16-25c) now yields

$$\log K_{eq} = (-0.799)/(0.0592) = -13.5$$

and

$$K_{eq} = 3 \times 10^{-14}$$

Consider the results of parts **(b)** and **(c)**. Note that

$$K_{eq(b)} = \frac{P_{H_2}[Ag^+]^2}{[H^+]^2} = 1._0 \times 10^{-27} \qquad \text{while}$$

$$K_{eq(c)} = \frac{[Ag^+]P_{H_2}^{1/2}}{[H^+]} = 3._2 \times 10^{-14}$$

Therefore $K_{eq(b)} = K_{eq(c)}^2$.

The Use of the Nernst Equation to Determine Solubility Products

Solubility products are difficult to measure directly because the concentrations of the ions in equilibrium with an insoluble solid are so low. Provided that $\mathcal{E}^0$ values are known for two half-reactions that can be combined to yield the solubility product equilibrium (see Table 11.1), we can use the relation between $\Delta\mathcal{E}^0_{cell}$ and an equilibrium constant to determine a solubility product.

As an example, consider the determination of the solubility product of silver bromide, AgBr, which is the equilibrium constant for the reaction

$$AgBr(s) \rightleftharpoons Ag^+ + Br^-$$

If we examine Appendix F, we find a half-reaction involving AgBr(s), namely

$$AgBr(s) + e^- \rightleftharpoons Ag(s) + Br^- \qquad \mathcal{E}^0 = 0.0713 \text{ V}$$

The other half-reaction we will need is

$$Ag^+ + e^- \rightleftharpoons Ag(s) \qquad \mathcal{E}^0 = 0.799 \text{ V}$$

If we combine the oxidation of Ag(s) with the reduction of AgBr(s) we obtain

Oxidation	$Ag(s) \rightleftharpoons Ag^+ + e^-$
Reduction	$AgBr(s) + e^- \rightleftharpoons Ag(s) + Br^-$
Net cell reaction	$AgBr(s) \rightleftharpoons Ag^+ + Br^-$

The galvanic cell for which these are the anode and cathode reactions, respectively, is

$$Ag(s) \mid Ag^+ \parallel Br^- \mid AgBr(s) \mid Ag(s)$$

For this cell $\Delta\mathcal{E}^0_{cell} = 0.0713 - 0.799 = -0.728$ V. Note that $\Delta\mathcal{E}^0_{cell}$ is negative. A solubility product is always a number less than 1. The spontaneous direction of reaction for 1 M Ag$^+$ ions and Br$^-$ ions is the precipitation of AgBr (the *reverse* of the solubility product equilibrium). Using Eq. (16-25c) we obtain

$$\log K_{sp} = -\frac{0.728}{0.0592} = -12.3$$

and

$$K_{sp}(AgBr) = 5 \times 10^{-13}$$

EXAMPLE 16.11. The use of standard reduction potentials to determine a solubility product

Calculate the solubility product for $Cu(OH)_2$ using data from Appendix F.

Solution. The half-reactions that must be combined are

$$Cu(OH)_2(s) + 2e^- \rightleftharpoons Cu(s) + 2\,OH^- \qquad \mathscr{E}^0 = -0.224 \text{ V}$$
$$Cu^{2+} + 2e^- \rightleftharpoons Cu(s) \qquad \mathscr{E}^0 = +0.337 \text{ V}$$

To yield the solubility product equation

$$Cu(OH)_2(s) \rightleftharpoons Cu^{2+} + 2\,OH^-$$

we must make the second of these half-reactions an oxidation. Note that $n = 2$ for the half-reactions that are combined. For the galvanic cell

$$Cu(s)\,|\,Cu^{2+}\,\|\,OH^-\,|\,Cu(OH)_2\,|\,Cu(s)$$
$$\Delta\mathscr{E}^0_{cell} = -0.224 - 0.337 = -0.561 \text{ V}$$

Equation (16-25c) is therefore

$$\log K_{sp} = \frac{(2)(-0.561)}{0.0592} = -18.95$$

and

$$K_{sp} = 1 \times 10^{-19}$$

section 16.9
The Relation between Two Criteria for Predicting the Direction of a Spontaneous Reaction

In Section 8.2 we discussed how we can tell whether or not a reaction will occur when we mix substances together in a known way. The answer to such a question is obtained by comparing the reaction quotient, Q, to the equilibrium constant, K_{eq}, and Eqs. (8-10a), (8-10b), and (8-10c) summarize the relations involved. When we discussed reactions occurring in a galvanic cell, however, we used the sign of the cell potential as a criterion for deciding whether or not a reaction will occur and what the direction of the spontaneous reaction will be. As you might expect, these relations are merely different expressions of the same fundamental concept, and by using the Nernst equation we can show that the two criteria for predicting the direction of a spontaneous reaction are equivalent.

At 25 °C the Nernst equation is

$$\Delta\mathscr{E}_{cell} = \Delta\mathscr{E}^0_{cell} - (0.0592/n)\log Q$$

but we have shown in Eq. (16-25b) that $\Delta\mathscr{E}^0_{cell} = (0.0592/n)\log K_{eq}$. Substitution of this expression for $\Delta\mathscr{E}^0_{cell}$ into the Nernst equation yields

$$\Delta\mathscr{E}_{cell} = (0.0592/n)\log K_{eq} - (0.0592/n)\log Q \qquad (16\text{-}26a)$$

or

$$\Delta\mathscr{E}_{cell} = (0.0592/n)(\log K_{eq} - \log Q) = (0.0592/n)\log (K_{eq}/Q) \qquad (16\text{-}26b)$$

In going from Eq. (16-26a) to Eq. (16-26b) we have used a fundamental property of logarithms: $\log (a/b) = \log a - \log b$.

A consideration of Eq. (16-26b) shows that the sign of the cell potential depends on the ratio K_{eq}/Q. If $K_{eq} > Q$, the reaction will proceed spontaneously to the right in order to increase Q. Then (K_{eq}/Q) is a number greater than 1 and log (K_{eq}/Q) is positive, so that $\Delta\mathcal{E}_{cell}$ is also positive. A positive sign for $\Delta\mathcal{E}_{cell}$ is a necessary consequence of having a reaction quotient smaller than the equilibrium constant. In sum,

If $K_{eq} > Q$, then $\Delta\mathcal{E}_{cell} > 0$ and the reaction is spontaneous from left to right, as written. (16-27a)

If $K_{eq} < Q$, the reverse reaction will occur spontaneously to decrease Q. Then (K_{eq}/Q) is a number less than 1, and log (K_{eq}/Q) is negative, so that $\Delta\mathcal{E}_{cell}$ is also negative. We can summarize this as

If $K_{eq} < Q$, then $\Delta\mathcal{E}_{cell} < 0$ and the *reverse* reaction proceeds spontaneously, that is, the reaction occurs from right to left. (16-27b)

If $K_{eq} = Q$, the system is at equilibrium. The ratio $(K_{eq}/Q) = 1$, and log $(K_{eq}/Q) = 0$, so that $\Delta\mathcal{E}_{cell}$ is also 0. When a system is at equilibrium, no electrical work can be obtained from it. In sum,

If $K_{eq} = Q$, then $\Delta\mathcal{E}_{cell} = 0$ and the system is at equilibrium. No net reaction occurs. (16-27c)

section 16.10
Commercial Batteries

While in principle any oxidation–reduction reaction can be used to produce electricity, in actual practice only a handful of reactions serve as useful sources of an electric current. In commercial use we often want a battery that can be carried around, to power flashlights, cameras, portable radios, or children's toys. Clearly anything that requires a tank of gas, like the hydrogen electrode, is out of the question anywhere but in a laboratory! A practical battery must be relatively inexpensive (that rules out the use of Pt or Ag), able to withstand a lot of moving about, not easily damaged, and capable of delivering a fairly sizeable voltage that does not change appreciably as the cell discharges. A description of a few of the most commonly used commercial batteries is given below.

The Lead Storage Battery

The battery used in all automobiles is a lead storage battery, illustrated in Fig. 16.6. The anode, or negative terminal, consists of a lead electrode. The substance reduced at the cathode is lead dioxide, PbO_2. Since PbO_2 is a powdery solid, the cathode is made of a lead grid with the interstices filled with PbO_2. The anode is also a lead grid, with the interstices filled with spongy lead for greater reactivity. Both electrodes are immersed in the same aqueous solution of sulfuric acid, 35% H_2SO_4 by weight.

At the anode, lead is oxidized from the 0 state to the +2 state, and the insoluble white salt $PbSO_4$ precipitates out. At the cathode, lead is reduced from the +4 state in PbO_2 to the +2 state, and $PbSO_4$ again precipitates out. As the cell is used, therefore, the interstices of both grids fill with $PbSO_4$. The two half-reactions are

Cathode	$PbO_2(s) + HSO_4^- + 3H^+ + 2e^- \rightarrow PbSO_4(s) + 2H_2O$	(16-28a)
Anode	$Pb(s) + HSO_4^- \rightarrow PbSO_4(s) + 2e^- + H^+$	(16-28b)
Net cell	$Pb(s) + PbO_2(s) + 2HSO_4^- + 2H^+ \rightarrow 2PbSO_4(s) + 2H_2O$	(16-28)

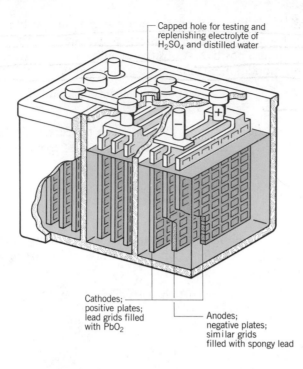

Capped hole for testing and replenishing electrolyte of H_2SO_4 and distilled water

Cathodes; positive plates; lead grids filled with PbO_2

Anodes; negative plates; similar grids filled with spongy lead

Fig. 16.6. Diagram of a lead storage battery. Lead dioxide, PbO_2, is reduced to $PbSO_4$ at the cathode. Lead, Pb, is oxidized to $PbSO_4$ at the anode.

Note that the sulfuric acid solution is quite concentrated so that it is more accurate to represent it as H^+ and HSO_4^- ions, rather than as $2H^+$ and SO_4^{2-} ions.

One of the conveniences of this battery is that no salt bridge or porous barrier is needed; both electrodes are immersed in the same solution. The shorthand notation for this battery therefore does not have a double vertical bar, and appears as follows:

$$Pb(s) \mid PbSO_4(s) \mid H_2SO_4(35\% \text{ aq soln}) \mid PbSO_4(s), PbO_2(s) \mid Pb(s)$$

One cell of a lead storage battery has a potential of about 2.0 V. It is common practice to construct a battery by combining six such cells in series to provide 12.0 V.

As the battery is used, the solution in the cell becomes more dilute, since H_2O is a product of the net cell reaction and both H_3O^+ ions and HSO_4^- ions are used up during the reaction. The density of the electrolytic solution therefore decreases as the battery discharges, and a measurement of the density can be used as a simple way to tell just how far the cell has discharged. This is important because a storage battery has an extremely advantageous feature: It can be recharged. The $PbSO_4$ formed when the cell is discharged remains embedded in the interstices of the grids of the electrodes, so that if electricity is put into the cell by using some external source of energy, the cell reaction can be reversed. The $PbSO_4$ is reconverted into Pb at one electrode and PbO_2 at the other.

The power source for one "all-electric" automobile, which has been suggested to eliminate dependency on gasoline, is a bank of storage batteries. When the car is in use during the day power is drawn from the batteries, which must be recharged for several hours each night. The cell reaction is easily reversed if the $PbSO_4$ is freshly precipitated, but as $PbSO_4$ stands it changes its crystalline structure and ages to a much less reactive form. A battery that has been discharged for a few weeks or more can not be fully recharged. Storage batteries in gasoline powered automobiles are recharged as you drive, so that they never become fully discharged unless there is some malfunction.

The Dry Cell

The most commonly used flashlight battery is the **dry cell,** also called the **Leclanché cell,** after its inventor, Georges Leclanché (1839 – 1882). The anode of a dry cell is a zinc can or cup, which is usually covered with a steel or cardboard jacket to shield it from the atmosphere. The cathode is a graphite rod which serves as an inert electrode. The graphite rod is in the center of the cell (see Fig. 16.7) and is surrounded by a thick paste containing MnO_2 and powdered graphite. It is the MnO_2 that is reduced at the cathode. The electrolyte is a moist paste of a saturated solution of NH_4Cl, $ZnCl_2$, and some inert filler. The cell is not really dry; water is an essential component of the electrolytic paste.

At the anode, Zn is oxidized to the $+2$ state, and at the cathode MnO_2 is reduced to the $+3$ state. There are a number of different chemical substances formed involving the $+3$ oxidation state of manganese, including $Mn_2O_3(s)$, $Mn_2O_4^{2-}$, and $MnO(OH)$. We will write the cell reactions showing the formation of Mn_2O_3 at the cathode.

Anode	$Zn(s) \rightarrow Zn^{2+} + 2e^-$	(16-29a)
Cathode	$2MnO_2(s) + 2NH_4^+ + 2e^- \rightarrow Mn_2O_3(s) + 2NH_3 + H_2O$	(16-29b)
Net cell	$Zn(s) + 2MnO_2(s) + 2NH_4^+ \rightarrow Mn_2O_3(s) + Zn^{2+} + 2NH_3 + H_2O$	(16-29)

The voltage of a dry cell battery is 1.5 V. The dry cell is not rechargeable, as the Zn^{2+} ions formed at the anode migrate through the electrolytic paste and combine with the NH_3 produced at the cathode to form the complex ion $Zn(NH_3)_4^{2+}$.

One of the problems with the dry cell is that the electrolytic paste is acidic, since NH_4Cl is an acidic salt. Thus there is a direct reaction between Zn and NH_4^+ that slowly eats away the zinc can:

$$Zn(s) + 2NH_4^+ \rightarrow Zn^{2+} + H_2(g) + 2NH_3 \qquad (16\text{-}30)$$

A flashlight battery that has never been used but has been sitting on the shelf for more than a year may split and leak as the Zn metal is converted to Zn^{2+} ions by reaction (16-30).

An improved form of the dry cell is the **alkaline dry cell,** in which the NH_4Cl is replaced by KOH. It is more expensive than the acid form, but it lasts longer because there is no corrosion of the Zn by NH_4^+ ions.

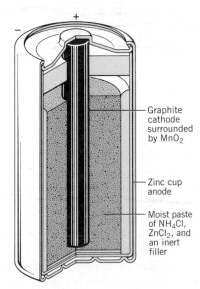

Graphite cathode surrounded by MnO_2

Zinc cup anode

Moist paste of NH_4Cl, $ZnCl_2$, and an inert filler

Fig. 16.7. The dry cell. Manganese dioxide, MnO_2, is reduced to Mn_2O_3 (and other forms of the $+3$ oxidation state of Mn) at the cathode. Zinc, Zn, is oxidized to Zn^{2+} at the anode.

The Mercury Cell

A battery developed during the Second World War is the **mercury cell** which maintains a constant potential of 1.34 V throughout its lifetime. The potential of most cells decreases somewhat as the cell discharges, particularly towards the end of their lifetime. The mercury cell was developed when a steady source of voltage, even while current is being drawn, was needed. The anode material is an **amalgam** of zinc and mercury. Solid mercuric oxide, HgO, is reduced at the cathode. The electrolyte is a paste of $Zn(OH)_2$ in a solution of KOH. It is strongly alkaline and contains the complex zincate ions, $Zn(OH)_4^{2-}$. The cell reactions are

Anode	$Zn(amalgam) + 2\,OH^- \rightarrow ZnO(s) + H_2O + 2e^-$	(16-31a)
Cathode	$HgO(s) + H_2O + 2e^- \rightarrow Hg(liq) + 2\,OH^-$	(16-31b)
Net cell reaction	$HgO(s) + Zn(amalgam) \rightarrow ZnO(s) + Hg(liq)$	(16-31)

Except for the amalgam of Zn and Hg, the substances in the net cell reaction, Eq. (16-31), are pure solids and liquids. It is found experimentally that the variation in the cell voltage with a change in the concentration of zinc in the amalgam is negligible, so that there are no concentration dependent terms in the Nernst equation, and $\Delta\mathcal{E}_{cell} = \Delta\mathcal{E}_{cell}^0$, a constant at constant temperature.

Nickel–Cadmium Batteries

A portable, rechargeable cell has obvious advantages over the dry cell, and the nickel–cadmium battery, recently developed, is becoming increasingly popular. It is used in electronic calculators and other battery powered tools. Although it is considerably more expensive than a dry cell, because it can be recharged the added expense is worthwhile. The cell reactions are

Anode	$Cd(s) + 2\,OH^- \rightarrow Cd(OH)_2(s) + 2e^-$	(16-32a)
Cathode	$NiO_2(s) + 2H_2O + 2e^- \rightarrow Ni(OH)_2 + 2\,OH^-$	(16-32b)
Net reaction	$Cd(s) + NiO_2(s) + 2H_2O \rightarrow Cd(OH)_2(s) + Ni(OH)_2(s)$	(16-32)

The reaction can be readily reversed because the reaction products, $Ni(OH)_2$ and $Cd(OH)_2$, adhere to the electrode surfaces.

Like the mercury cell, this cell also has a voltage that remains constant for the lifetime of the cell, because all the substances in the net cell reaction, Eq. (16-32), are pure solids and liquids, and there are no concentration dependent terms in the Nernst equation.

section 16.11
Electrolysis

The preceding sections have been concerned with the ways in which we convert chemical energy into electrical energy. Now we will examine the chemical reactions that occur when we put electricity into solutions of electrolytes or molten salts. Investigations of the interaction of an electric current with chemical substances were begun early in the nineteenth century. An outstanding study of this subject was carried out by Michael Faraday, and it was his investigations that convinced scientists of the existence of ions and established the charges on many common ions. **Faraday's Laws of Electrolysis** describe accurately the relations between the amount of electricity passed through a cell and the chemical reactions that occur during electrolysis.

Michael Faraday (1791–1867) was an English physicist and chemist. He discovered the laws governing the electrolysis of aqueous solutions, now called Faraday's Laws. He also discovered the law of electromagnetic induction, described the fundamental relationship between light and magnetism, and laid the foundations for the electromagnetic theory developed by James C. Maxwell. In 1826 he began to give Friday evening lectures at the Royal Society in London. He was a superb lecturer, and accompanied his talks with experimental demonstrations. His lectures were immensely popular, and the tradition of Friday evening discourses at the Royal Society continues to this day.

Electricity is obtained from a galvanic cell when an oxidation–reduction reaction proceeds spontaneously toward a state of equilibrium. When electricity is passed through aqueous solutions or molten salts, it is possible to cause nonspontaneous processes to occur, that is, to drive a reaction backwards, away from a position of equilibrium. This process is known as **electrolysis.**

An electrolytic cell looks very much like a galvanic cell, except that a source of direct current electricity is required. It is perfectly possible to use a battery such as a dry cell or a lead storage battery as the source of power. A diagram for a typical electrolytic cell is illustrated in Fig. 16.8. Note the symbol used to represent a battery in electric circuit diagrams: A long skinny line for the positive terminal and a short fatter line for the negative terminal. The battery pumps electrons into the cathode, and substances capable of being reduced obtain electrons at the cathode surface. The battery pumps electrons out of the anode, and substances capable of being oxidized give electrons up at the anode surface. Thus, just as in a galvanic cell*

> Oxidation occurs at the anode.
> Reduction occurs at the cathode.

* A number of my students tell me that they remember this by using the phrase "an ox; red cat." It is also helpful to remember that during electrolysis anions migrate toward the anode (positive terminal), and cations migrate toward the cathode (negative terminal). Similarly, in a galvanic cell, anions from the salt bridge migrate into the anode compartment, and cations from the salt bridge migrate into the cathode compartment.

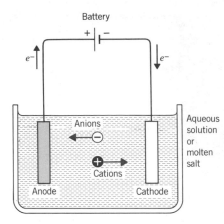

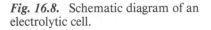

Fig. 16.8. Schematic diagram of an electrolytic cell.

The difference between a galvanic cell and an electrolytic cell is that the negative terminal during electrolysis is the cathode, whereas the cathode is the positive terminal of a galvanic cell that is producing electricity. Similarly, the anode is the positive terminal during electrolysis, and the negative terminal when a galvanic cell produces electricity. As we have previously discussed, some galvanic cells are rechargeable. During the recharging process, the cell is being operated as an electrolytic cell.

Electrolysis of Molten Salts

Although ionic crystalline solids do not conduct electricity, melting such substances permits the ions to move, so that molten salts do conduct electricity. If we pass electricity through molten sodium chloride, for example, we can cause the nonspontaneous reaction

$$NaCl(s) \rightarrow Na(s) + \tfrac{1}{2} Cl_2(g) \tag{16-33}$$

to occur. In fact, this is the method used commercially to prepare sodium metal.

Molten NaCl is electrolyzed in a cell called the **Downs cell.** Care must be taken to prevent Na metal and $Cl_2(g)$ from coming into contact with one another or they will

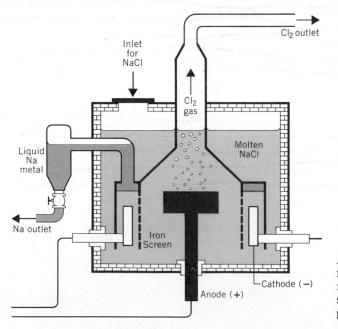

Fig. 16.9. The Downs cell for the electrolysis of molten NaCl to produce sodium metal and chlorine gas.

react spontaneously to form NaCl again. The Downs cell is illustrated in Fig. 16.9. The electrode reactions are

$$\text{Cathode} \qquad Na^+ + e^- \rightarrow Na(liq) \qquad (16\text{-}33a)$$

$$\text{Anode} \qquad 2Cl^- \rightarrow Cl_2(g) + 2e^- \qquad (16\text{-}33b)$$

An iron screen is used to prevent contact between sodium and chlorine gas, but allows ions to pass through. In order to melt NaCl very high temperatures are required. As the melting point of metallic sodium is only 97.8 °C, liquid sodium is the product of the electrolysis. The sodium floats on the molten salt mixture and is drained off into a storage tank. The sodium must also be prevented from coming into contact with air or water.

The Commercial Production of Aluminum

Another metal that is produced commercially by the electrolysis of its molten salts is aluminum. Although aluminum is the most abundant metal to be found in the earth's crust, it never occurs free in nature. Its principal ore is **bauxite,** a hydrated aluminum oxide, Al_2O_3. It is not feasible to use pure Al_2O_3 for electrolysis, as it melts above 2000 °C. The commercial production of aluminum became possible after a discovery by Charles M. Hall (1863–1914) that Al_2O_3 dissolves in molten **cryolite,** Na_3AlF_6, to produce a solution that will conduct electricity. Cryolite is a mineral found mainly in Greenland. Hall devised the method for the electrolytic production of aluminum within eight months after his graduation from Oberlin College, in Ohio. He patented his process in 1889. Hall became vice-president of the company that began commercial production of aluminum, and is now Alcoa, the Aluminum Company of America. At his death, the bulk of the fortune he had acquired went to educational institutions, principally Oberlin.

In the **Hall process,** carbon is used as the anode material. The carbon anodes are consumed during the electrolysis as they are oxidized to $CO_2(g)$. The temperature of the melt is maintained at about 1000 °C. The electrode reactions are

$$\text{Cathode} \qquad\qquad\qquad Al^{3+} + 3e^- \rightarrow Al(liq) \qquad (16\text{-}34a)$$

$$\text{Anode} \qquad\qquad C(s) + 2\,O^{2-} \rightarrow CO_2(g) + 4e^- \qquad (16\text{-}34b)$$

$$\text{Net cell reaction} \quad 3C(s) + 4Al^{3+} + 6\,O^{2-} \rightarrow CO_2(g) + 4Al(liq) \qquad (16\text{-}34)$$

Because liquid aluminum is denser than the molten salt mixture, it collects at the bottom of the cell and can be drawn off there. A diagram of a Hall-process electrolytic cell is shown in Fig. 16.10.

Example 16.12 illustrates calculations of some of the factors involved in the production of aluminum by the electrolysis of molten aluminum salts.

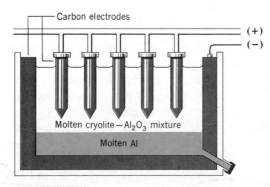

Carbon electrodes
(+)
(−)
Molten cryolite — Al_2O_3 mixture
Molten Al

Fig. 16.10. A Hall-process electrolysis cell for the production of aluminum.

EXAMPLE 16.12. The electrolysis of molten aluminum salts

(a) How many faradays and how many coulombs must be passed through a molten mixture of Al_2O_3 and Na_3AlF_6 to produce 1 kg of aluminum metal?

(b) If 100 A of current are passed through the cell, how long will it take to produce 1 kg of aluminum?

(c) How much current is required if it is desired to produce 1 kg of aluminum in 5 min?

Solution

(a) The atomic weight of aluminum is 26.98, so that 1000 g is $1000/26.98 = 37.06$ mol of Al. Equation (16-34a) tells us that 3 $\mathcal{F}$ of charge are required for each mole of Al produced, so that $3(37.06) = 111.2$ $\mathcal{F}$ are needed per kilogram of Al. As $1 \mathcal{F} = 96{,}485$ C, the number of coulombs required to produce a kilogram of Al is $(111.2 \mathcal{F})(96{,}485 \text{ C}/\mathcal{F}) = 1.073 \times 10^7$ C.

(b) Substitution into Eq. (16-4c), $Q = It$, yields

$$1.073 \times 10^7 \text{ C} = (100 \text{ A})(t)$$

where t is the time in seconds. Therefore the time it will take to produce 1 kg of Al with a current of 100 A is

$$t = 1.073 \times 10^5 \text{ s} = 1788 \text{ min} = 29.8 \text{ h}$$

(c) Five minutes is 300 s. The current required to produce 1 kg of Al in 300 s is

$$I = (1.073 \times 10^7 \text{ C})/(3 \times 10^2 \text{ s}) = 3.58 \times 10^4 \text{ A or } 35{,}800 \text{ A}$$

If you think about the answers calculated in Example 16.12, it should not surprise you to learn that typical currents employed during the commercial production of aluminum are between 20,000 and 50,000 A. The production of aluminum consumes vast amounts of electrical energy in the United States. Just obtaining the high temperature required to melt the aluminum salt mixture requires a great deal of energy. A significant fraction of the aluminum produced goes to make aluminum cans for food and beverages. Recycling aluminum cans can save this country a lot of energy!

Electrolysis of Aqueous Solutions

If an aqueous solution of NaCl is electrolyzed using inert electrodes, chlorine gas is produced at the anode, just as in Eq. (16-33b), but sodium metal is not produced at the cathode. Instead, $H_2(g)$ is formed at the cathode. (See Fig. 16.11.) The reason is simply that water contains hydronium ions, and $H^+(aq)$ ions are a stronger oxidizing agent than Na^+ ions, and are therefore more easily reduced. Electrolysis of aqueous solutions containing any metal ion that is a weaker oxidizing agent than $H^+(aq)$ will produce $H_2(g)$ at the cathode. If we refer to Table 16.1, that means that aqueous solutions containing Pb^{2+}, Zn^{2+}, or Na^+, as well as any of the other alkali metal cations or the alkaline earth cations, will form H_2 on electrolysis. In general, no reducing agent stronger than $H_2(g)$ is produced on electrolysis of an aqueous solu-

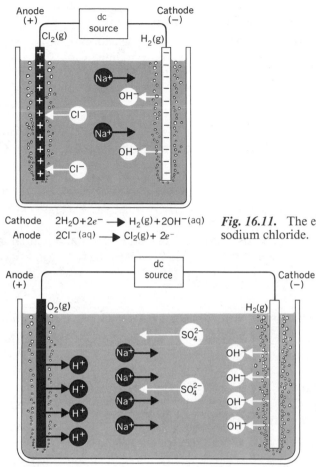

Cathode $2H_2O + 2e^- \longrightarrow H_2(g) + 2OH^-(aq)$
Anode $2Cl^-(aq) \longrightarrow Cl_2(g) + 2e^-$

Fig. 16.11. The electrolysis of aqueous sodium chloride.

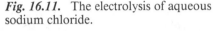

Fig. 16.12. The electrolysis of aqueous Na_2SO_4.

tion.* The cathode reaction for the electrolysis of an aqueous solution of NaCl (or $ZnCl_2$, $CaCl_2$, $ScCl_3$, KCl, and so on) is

$$2H_2O + 2e^- \rightarrow H_2(g) + 2\,OH^- \tag{16-35}$$

so that the products of the electrolysis of aqueous NaCl are $H_2(g)$, $Cl_2(g)$, and a solution of NaOH.

On the other hand, if the metal cation is a stronger oxidizing agent than $H^+(aq)$, the metal will be deposited at the cathode. Thus electrolysis of an aqueous solution of $CuCl_2$ will produce $Cu(s)$ at the cathode and $Cl_2(g)$ at the anode.

Similarly, only anions that are more easily oxidized than OH^- will be oxidized at the anode during electrolysis of an aqueous solution. Solutions of nitrates and sulfates, for instance, will form oxygen gas at the anode, if the electrode is made of an inert metal. The anode reaction in the electrolysis of an aqueous solution of $CuSO_4$, for example, is

$$2H_2O \rightarrow O_2(g) + 4e^- + 4H^+(aq) \tag{16-36}$$

The electrolysis of Na_2SO_4 is depicted in Fig. 16.12.

You should be able to predict the products formed on electrolyzing various aqueous solutions, as is done in Example 16.13.

* Stronger reducing agents than H_2 can be produced if they react very slowly with H^+ ions.

EXAMPLE 16.13. The electrolysis of aqueous solutions

Write the anode and cathode reactions for the electrolysis of aqueous solutions of (a) potassium sulfate, K_2SO_4, (b) calcium iodide, CaI_2, (c) mercuric nitrate, $Hg(NO_3)_2$, and (d) cupric bromide, $CuBr_2$.

Solution

(a) It is more difficult to reduce K^+ than to reduce H_2O or $H^+(aq)$ ions, and it is more difficult to oxidize SO_4^{2-} ions than it is to oxidize H_2O or OH^- ions, so that the electrolysis of aqueous K_2SO_4 is just the electrolysis of water.

Cathode	$2H_2O + 2e^- \rightarrow H_2(g) + 2\,OH^-$
Anode	$2H_2O \rightarrow O_2(g) + 4e^- + 4H^+(aq)$
Overall reaction	$2H_2O \rightarrow 2H_2(g) + O_2(g)$

(b) Calcium is below H_2 in the Table of Standard Reduction Potentials so that H_2 will be formed at the cathode, but I^- is readily oxidized to I_2.

Cathode	$2H_2O + 2e^- \rightarrow H_2(g) + 2\,OH^-$
Anode	$2\,I^- \rightarrow I_2 + 2e^-$
Overall reaction	$2H_2O + 2\,I^- \rightarrow H_2 + 2\,OH^- + I_2$

Thus the products of the electrolysis of an aqueous solution of CaI_2 are $H_2(g)$, I_2, and an aqueous solution of $Ca(OH)_2$.

(c) Mercury is above H_2 in the Table of Standard Reduction Potentials, so that Hg will be formed at the cathode. It is easier to oxidize H_2O than nitrate ions, and therefore O_2 will be formed at the anode.

Cathode	$Hg^{2+} + 2e^- \rightarrow Hg(liq)$
Anode	$2H_2O \rightarrow O_2(g) + 4e^- + 4H^+(aq)$
Overall reaction	$2Hg^{2+} + 2H_2O \rightarrow O_2(g) + 2Hg(liq) + 4H^+(aq)$

(d) Cupric ions will be reduced; bromide ions will be oxidized.

Cathode	$Cu^{2+} + 2e^- \rightarrow Cu(s)$
Anode	$2Br^- \rightarrow Br_2 + 2e^-$
Overall reaction	$Cu^{2+} + 2Br^- \rightarrow Cu(s) + Br_2$

Electroplating

The process of electrolysis is frequently used for electroplating various objects. A fork made of steel, for instance, can be plated with silver by making the fork the cathode of an electrolytic cell, using a piece of silver metal as the anode, and a solution of silver nitrate as the electrolyte. This process is illustrated in Fig. 16.13. At the anode, silver metal is oxidized to Ag^+ ions, which enter the solution to replace the Ag^+ ions that are deposited as silver metal on the surface of the cathode.

Copper is commercially purified by electrolysis, and the process is essentially electroplating also. Impure copper is made the anode and very thin sheets of pure copper serve as the cathode. The impurities in copper are usually other metals, principally silver, gold, zinc, and iron. Silver and gold are not as easily oxidized at the anode as copper is, and therefore as the copper goes into solution as Cu^{2+} ions, bits of silver and gold fall to the bottom of the cell as "anode sludge." These valuable metals

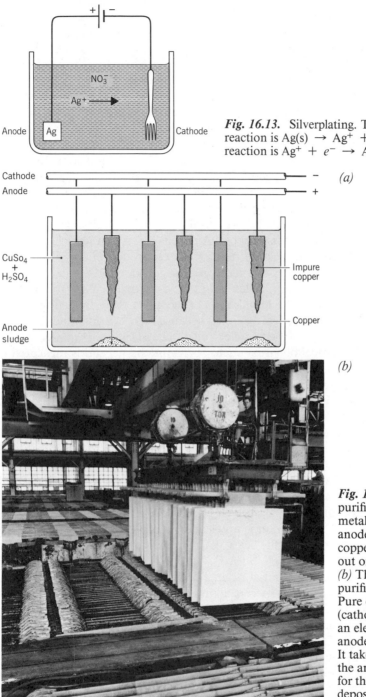

Fig. 16.13. Silverplating. The anode reaction is $Ag(s) \rightarrow Ag^+ + e^-$. The cathode reaction is $Ag^+ + e^- \rightarrow Ag(s)$.

Fig. 16.14. (a) The purification of copper metal by electrolysis. The anodes are made of impure copper. Pure copper plates out on the cathodes. (b) The commercial purification of copper. Pure copper starter plates (cathodes) are lowered into an electrolysis cell between anodes of impure copper. It takes about 28 days for the anodes to dissolve and for the pure copper to be deposited on the cathodes.

are readily recovered from the sludge, which is an important source of gold. Zinc and iron are more easily oxidized at the anode than Cu, and therefore the solution soon contains Zn^{2+} and Fe^{3+} ions. However, these ions are more difficult to reduce at the cathode than Cu^{2+} ions. The voltage is carefully controlled to prevent the deposition of Zn or Fe at the cathode. Figure 16.14 illustrates the purification of copper. Extremely pure copper is needed for the manufacture of electrical wiring because even small amounts of impurities reduce the conductivity of copper wire.

Summary

A **galvanic** or **voltaic cell** is a device for converting chemical energy into electrical energy. In a galvanic cell, the electrons lost by a reducing agent during an oxidation–reduction reaction pass through a wire before reaching the oxidizing agent. The flow of electrons through a wire is an electrical current.

Each galvanic cell must have two surfaces, called **electrodes,** on which electrons can be collected. An electrode must be a good conductor of electricity, and is usually a piece of metal or a graphite rod. Reduction occurs at one electrode, called the **cathode;** oxidation occurs at the other electrode, which is called the **anode.**

Each electrode is immersed in a solution containing ions. In many galvanic cells there are two aqueous solutions that must be kept separate to prevent direct transfer of electrons between the reducing and oxidizing agents. In such a cell some device is necessary for establishing electrical contact between the two solutions, so that each solution remains electrically neutral at all times. In the laboratory, a **salt bridge** is frequently used for this purpose. A porous barrier with holes too tiny to permit mass flow of liquid can also be used. Each electrode, plus the solution in which it is immersed, constitutes one **half-cell** of a galvanic cell.

Electrical **current,** I, is defined as the amount of charge that passes per unit time: $I = Q/t$. A current of 1 coulomb per second (C/s) is, by definition, 1 ampere (1 A).

Current flows in a galvanic cell because there is a difference between the electrical potential energy of the two electrodes. The **electromotive force (EMF)** is the difference in electrical potential energy per unit charge, and is measured in **volts.**

We can measure the EMF or the **cell potential** using an instrument called a **potentiometer.** Although we can measure the potential difference between the two electrodes, we cannot measure the absolute potential of any single electrode. We therefore arbitrarily define the potential of one electrode, called the **standard hydrogen electrode,** to be exactly zero volts. The standard hydrogen electrode consists of a platinum electrode over which H_2 gas at 1-atm pressure is bubbled, while the electrode is immersed in a solution with $[H_3O^+] = 1.00\ M$ at 25 °C.

A galvanic cell in which all ions taking part in the reaction have a 1.00 M concentration and all gases have a partial pressure of 1.00 atm is called a **standard cell,** and the reduction potential at each electrode is then the **standard electrode potential,** denoted $\mathscr{E}°$. By combining any standard half-cell with the standard hydrogen electrode and measuring the cell voltage, we can compile a **Table of Standard Electrode Potentials.**

The electrical potential of an electrode immersed in a solution depends on the concentrations of the species taking part in the reaction at that electrode. The measured cell voltage is the difference between the reduction potential at the cathode and the reduction potential at the anode: $\Delta \mathscr{E}_{cell} = \mathscr{E}_{cathode} - \mathscr{E}_{anode}$.

The Table of Standard Reduction Potentials is arranged with strong oxidizing agents at the top left and strong reducing agents at the bottom right. An oxidizing agent will react spontaneously with any reducing agent stronger than the reducing agent with which it is coupled, that is, any reducing agent *below* its corresponding reducing agent in the Table of Standard Reduction Potentials.

If you want to predict the direction of spontaneous reaction for an oxidation–reduction reaction, calculate $\Delta \mathscr{E}_{cell}$ for either direction. If $\Delta \mathscr{E}_{cell} > 0$, the reaction is spontaneous as you have written it. If $\Delta \mathscr{E}_{cell} < 0$, the reverse reaction is spontaneous. If $\Delta \mathscr{E}_{cell} = 0$, the system is at equilibrium. It is not possible to obtain electrical work from a system at equilibrium.

The relation between the measured cell potential, $\Delta \mathscr{E}_{cell}$, and the concentrations of the substances involved in the cell reaction is given by the **Nernst equation:**

$$\Delta \mathcal{E}_{cell} = \Delta \mathcal{E}^0_{cell} - (0.0592/n)\log Q \quad \text{at } 25 \text{ °C}$$

where n is the number of electrons transferred in the net cell reaction as written, and Q is the reaction quotient for that reaction. If all species are in their standard states, $Q = 1$, and the measured cell voltage is just the standard cell voltage, $\Delta \mathcal{E}^0_{cell}$.

A galvanic cell in which both electrodes are made of the same material and the solutions in which the electrodes are immersed contain the same species but at different concentrations, is known as a **concentration cell**. For a concentration cell, $\Delta \mathcal{E}^0_{cell} = 0$, and the Nernst equation reduces to $\Delta \mathcal{E}_{cell} = -(0.0592/n)\log Q$.

When the reacting species in a galvanic cell have come to equilibrium, the cell potential is zero volts and the reaction quotient, Q, is equal to the equilibrium constant for the net cell reaction, K_{eq}. The Nernst equation can then be solved for K_{eq}. We obtain the relation $\log K_{eq} = (n\Delta \mathcal{E}^0_{cell})/0.0592$ at 25 °C.

A number of galvanic cells are used commercially as a source of electrical energy. Every automobile makes use of the **lead storage battery**, in which lead is oxidized to $PbSO_4$ at the anode, and PbO_2 is reduced to $PbSO_4$ at the cathode. Both electrodes are immersed in the same aqueous solution of sulfuric acid, so there is no need for a salt bridge.

The **dry cell** and the **alkaline dry cell** are used, among other things, for flashlights and children's toys. In a dry cell Zn is oxidized to the +2 state at the anode and MnO_2 is reduced to the +3 state at the cathode. The electrolyte is a moist paste, so the cell is not really "dry."

We obtain electricity from an oxidation–reduction reaction as it moves toward a state of equilibrium. If we pass an electric current through a system already at equilibrium, we can drive the reaction backwards, in the nonspontaneous direction. This process is known as **electrolysis**. Both aqueous solutions of electrolytes and molten salts can be electrolyzed. Some galvanic cells are rechargeable; the recharging process is an electrolysis.

Metallic sodium and chlorine gas are produced by the electrolysis of molten NaCl in a **Downs cell**. Aluminum is produced commercially by the electrolysis of a molten solution of Al_2O_3 in **cryolite, Na_3AlF_6**. This technique is known as the **Hall process**.

Electrolysis of an aqueous solution containing a metal ion that is a weaker oxidizing agent than $H^+(aq)$ produces H_2 at the cathode. Similarly, electrolysis of an aqueous solution containing an anion, such as NO_3^- or SO_4^{2-}, that is harder to oxidize than H_2O or OH^- ions, produces O_2 at the anode.

Electrolysis is used for gold or silverplating, by making the object to be plated the cathode of an electrolytic cell containing an aqueous solution of a gold or silver salt. Electrolysis is also used for purifying impure copper.

Exercises

Section 16.1

1. Write the anode and cathode reactions for a galvanic cell that utilizes each of the following reactions to produce electricity. Specify the metal used as the anode and cathode.

   (a) $Cd(s) + I_2 \rightarrow Cd^{2+} + 2I^-$
   (b) $Ni(s) + 2Fe^{3+} \rightarrow Ni^{2+} + 2Fe^{2+}$
   (c) $2Cr(s) + 3Cu^{2+} \rightarrow 3Cu(s) + 2Cr^{3+}$

2. The reaction $Zn(s) + 2AgCl(s) \rightarrow Zn^{2+} + 2Ag(s) + 2Cl^-$ is used in a galvanic cell. If a steady current of 1.25 A is drawn for 8.00 min, how much Zn has been converted to Zn^{2+} ions?

3. The magnitude of the charge on an electron is 1.60219×10^{-19} C. Calculate the value of the Faraday constant, $\mathcal{F}$.

4. The net cell reaction of a galvanic cell is

$$2H^+(aq) + Pb(s) + SO_4^{2-} \rightarrow PbSO_4(s) + H_2(g)$$

When 21.08 g of $PbSO_4$ have been formed, how many coulombs have passed through the cell?

5. One half-cell of a galvanic cell consists of a piece of platinum dipping into a solution of $FeCl_3$. The other half-cell consists of a piece of cadmium immersed in a solution of $CdCl_2$. The net cell reaction is

$$2Fe^{3+} + Cd(s) \rightarrow Cd^{2+} + 2Fe^{2+}$$

The salt bridge contains a solution of KCl in agar–agar. When the cell is producing electricity, into which half-cell do the K^+ ions move?

Sections 16.2 and 16.3

6. For the following galvanic cells write the anode reaction, the cathode reaction, and the net cell reaction. Specify which electrode is the negative terminal in each cell.
   (a) $Co(s) \mid Co^{2+} \parallel Ag^+ \mid Ag(s)$
   (b) $Zn(s) \mid Zn^{2+} \parallel Br^-, Br_2 \mid Pt(s)$
   (c) $Pb(s) \mid Pb^{2+} \parallel H^+(aq) \mid H_2(g) \mid Pt(s)$
   (d) $Pt(s) \mid H_2(g) \mid H^+(aq) \parallel Cu^{2+} \mid Cu(s)$

7. The measured voltages of the following three standard cells are

$$Zn(s) \mid Zn^{2+}(1\ M) \parallel H_3O^+(1\ M) \mid H_2(g, 1\ atm) \mid Pt(s) \qquad \Delta\mathcal{E}^0_{cell} = 0.763\ V$$
$$Zn(s) \mid Zn^{2+}(1\ M) \parallel Tl^+(1\ M) \mid Tl(s) \qquad \Delta\mathcal{E}^0_{cell} = 0.427\ V$$
$$Sc(s) \mid Sc^{3+}(1\ M) \parallel Zn^{2+}(1\ M) \mid Zn(s) \qquad \Delta\mathcal{E}^0_{cell} = 1.32\ V$$

Using only the data given, calculate $\mathcal{E}^0_{Sc^{3+}|Sc}$ and $\mathcal{E}^0_{Tl^+|Tl}$.

8. A half-cell consisting of a piece of nickel immersed in a 1.00 M $Ni(NO_3)_2$ solution was connected to a standard hydrogen electrode. The measured voltage was 0.25 V, and the nickel electrode was the negative terminal of the cell.

   (a) If this half-cell is connected to a standard $Zn^{2+}|Zn$ half-cell, which electrode will be the negative terminal, zinc or nickel? Explain your answer. Use data from Exercise 7.

   (b) Write the shorthand or line notation for the galvanic cell of part (a) and calculate its cell voltage.

   (c) Which is a stronger oxidizing agent, Zn^{2+} or Ni^{2+}? Explain.

Sections 16.4 and 16.5. **Use Appendix F for $\mathcal{E}^0$ values.**

9. Calculate $\Delta\mathcal{E}^0_{cell}$ for each of the following galvanic cells and state whether the net cell reaction is spontaneous in the direction written when all species are in their standard states.
   (a) $Pt(s) \mid Sn^{4+}, Sn^{2+} \parallel Fe^{2+}, Fe^{3+} \mid Pt(s)$
   (b) $Pt(s) \mid Cl_2(g) \mid Cl^- \parallel H^+(aq) \mid H_2(g) \mid Pt(s)$
   (c) $Pb(s) \mid Pb^{2+} \parallel Cd^{2+} \mid Cd(s)$
   (d) $Ni(s) \mid Ni^{2+} \parallel H^+(aq), Mn^{2+}, MnO_4^- \mid Pt(s)$

10. Calculate the standard cell potential, $\Delta\mathcal{E}^0_{cell}$, for a galvanic cell with each of the following net reactions. Write the shorthand or line notation for each of these cells.

(a) $Cl_2 + 2Br^- \rightarrow Br_2 + 2Cl^-$

(b) $2Al(s) + 3Ni^{2+} \rightarrow 3Ni(s) + 2Al^{3+}$

(c) $Sc(s) + 3H(aq)^+ \rightarrow \frac{3}{2} H_2(g) + Sc^{3+}$

11. If all species are at standard conditions,

(a) Will $Cr_2O_7^{2-}$ oxidize Sn^{2+}?

(b) Will $Ni(s)$ reduce Cr^{2+}?

(c) Will $Cd(s)$ reduce Hg_2^{2+}?

(d) Will dilute nitric acid oxidize Au to Au^{3+}?

Section 16.6

12. Calculate the EMF of the cell

$$Pb(s) \mid Pb(NO_3)_2(M_1) \parallel HCl(M_2) \mid H_2(g) \mid Pt(s)$$

when   (a) $M_1 = 0.100\ M$, $M_2 = 0.300\ M$,   and $P_{H_2} = 1.00$ atm

               (b) $M_1 = 0.040\ M$, $M_2 = 2.000\ M$,   and $P_{H_2} = 0.50$ atm

               (c) $M_1 = 1.00\ M$, $M_2 = 0.050\ M$,   and $P_{H_2} = 2.00$ atm

13. What is the EMF of the following concentration cell at 25 °C?

$$Cd(s) \mid Cd^{2+}(0.0050\ M) \parallel Cd^{2+}(0.600\ M) \mid Cd(s)$$

14. (a) Calculate the EMF of the following concentration cell at 25 °C:

$$Ag(s) \mid AgNO_3(0.018\ M) \parallel AgNO_3(1.20\ M) \mid Ag(s)$$

(b) If a solution of NaCl is added to the 0.018 M AgNO$_3$ solution, will the cell voltage increase, decrease, or remain the same? Explain your answer.

15. For a $Cu^{2+}|Cu$ concentration cell to have an EMF of 0.100 V, what should the ratio of the $[Cu^{2+}]$ in the two solutions be? Is the more dilute solution in the anode or cathode compartment?

Section 16.7

16. Write the Nernst equation for the galvanic cell with the net cell reaction

(a) $Sc(s) + 3H^+(aq) \rightarrow \frac{3}{2} H_2(g) + Sc^{3+}$

(b) $2Sc(s) + 6H^+(aq) \rightarrow 3H_2(g) + 2Sc^{3+}$

and show that the cell voltage is the same regardless of whether we write (a) or (b).

Section 16.8

17. Calculate the equilibrium constants for the following reactions using data from Appendix F:

(a) $Ni(s) + I_2 \rightarrow Ni^{2+} + 2 I^-$

(b) $Zn(s) + 2H^+(aq) \rightarrow H_2(g) + Zn^{2+}$

(c) $2Fe(s) + 3Cd^{2+} \rightarrow 3Cd(s) + 2Fe^{3+}$

(d) $Cl_2(g) + 2Br^- \rightarrow Br_2 + 2Cl^-$

18. For each of the following galvanic cells write the net cell reaction and the expression for the equilibrium constant. Then calculate the numerical value of the equilibrium constant.

(a) $Co(s) \mid Co^{2+} \parallel Br^-, Br_2 \mid Pt(s)$

(b) $Sc(s) \mid Sc^{3+} \parallel H^+(aq) \mid H_2(g) \mid Pt(s)$

(c) $Pb(s) \mid PbSO_4(s) \mid SO_4^{2-} \parallel Zn^{2+} \mid Zn(s)$

Section 16.9

19. For the cell Pb(s) | Pb^{2+}(0.0125 M) ‖ Ag$^+$(0.600 M) | Ag(s), calculate Q, $\Delta\mathscr{E}^0_{cell}$, and $\Delta\mathscr{E}_{cell}$. Write the net cell reaction and calculate K_{eq}. Predict the spontaneous direction of reaction from the sign of $\Delta\mathscr{E}_{cell}$ and also by comparing Q and K_{eq}.

Section 16.10

20. Explain why each of the following galvanic cells is not suitable for use as a flashlight battery.
    (a) Zn(s) | Zn^{2+} ‖ H$^+$(aq) | H$_2$(g) | Pt(s)
    (b) Zn(s) | Zn^{2+} ‖ Ag$^+$ | Ag(s)

21. Why is a lead storage battery rechargeable, but a dry cell not able to be recharged?

22. Write the anode and cathode reactions for an alkaline dry cell. Zinc is oxidized to Zn(OH)$_4^{2-}$ ions and MnO$_2$(s) is reduced to MnO(OH).

Section 16.11

23. During the electrolysis of a solution of Cu(NO$_3$)$_2$, 1.525 g of Cu plated out at the cathode. How many faradays and how many coulombs were passed through the cell?

24. If a solution of Hg(NO$_3$)$_2$ is electrolyzed at a steady current of 1.50 A for 45.0 min, how much Hg will be produced at the cathode?

25. Write the anode, cathode, and net cell reactions for the electrolysis of aqueous solutions of (a) NaClO$_4$ (b) AuCl$_3$ (c) AgNO$_3$ (d) BaBr$_2$.

26. (a) How long will it take to produce 2.00 kg of sodium in a Downs cell if a steady current of 50.0 A is passed through the cell?
    (b) What average current is required if it is desired to produce 1.00 kg of sodium in 10.0 min?

27. Explain why electrolysis of an aqueous solution of Ca(NO$_3$)$_2$ produces H$_2$(g) at the cathode whereas electrolysis of an aqueous solution of Cu(NO$_3$)$_2$ produces Cu(s) at the cathode.

28. If an aqueous solution of HNO$_3$ is electrolyzed for 22.0 min at a steady current of 1.80 A, what volume of H$_2$(g) at 25 °C and 752-mmHg pressure will be collected at the cathode?

Multiple Choice Questions

Use Appendix F for $\mathscr{E}^0$ values.

1. A substance that will reduce Ag$^+$ to Ag but will not reduce Ni^{2+} to Ni is
   (a) Zn   (b) Pb   (c) Mg   (d) Cd   (e) Al

2. The EMF of the cell Zn(s) | Zn^{2+}(1.0 M) ‖ Pb^{2+}(1.0 M) | Pb(s) is
   (a) 0.889 V   (b) 0.637 V   (c) 0.511 V   (d) −0.637 V   (e) −0.889 V

3. The value of the reaction quotient, Q, for the cell

   $$\text{Zn(s) | Zn}^{2+}\text{(0.0100 } M\text{) ‖ Ag}^+\text{(1.25 } M\text{) | Ag(s)}$$

   is   (a) 156   (b) 125   (c) 1.25 × 10^{-2} (d) 8.00 × 10^{-3} (e) 6.40 × 10^{-3}

4. During the electrolysis of aqueous zinc nitrate,
   (a) Zinc plates out at the cathode.
   (b) Zinc plates out at the anode.

(c) Nitrogen gas, N_2, is evolved at the cathode.

(d) Hydrogen gas, H_2, is evolved at the anode.

(e) Oxygen gas, O_2, is evolved at the anode.

5. How many faradays are required to reduce a mole of MnO_4^- to Mn^{2+}?

(a) 1 (b) 2 (c) 3 (d) 4 (e) 5

6. If all species are in their standard states, which of the following is the strongest oxidizing agent?

(a) Br^- (b) Zn^{2+} (c) Fe^{2+} (d) Co^{3+} (e) Mn^{2+}

7. Consider the cell $Cd(s) \mid Cd^{2+}(1.0\ M) \parallel Cu^{2+}(1.0\ M) \mid Cu(s)$. If we wanted to make a cell with a more positive voltage using the same substances, we should

(a) Increase both the $[Cd^{2+}]$ and $[Cu^{2+}]$ to 2.00 M.

(b) Increase only the $[Cd^{2+}]$ to 2.00 M.

(c) Decrease both the $[Cd^{2+}]$ and $[Cu^{2+}]$ to 0.100 M.

(d) Decrease only the $[Cd^{2+}]$ to 0.100 M.

(e) Decrease only the $[Cu^{2+}]$ to 0.100 M.

8. Of the metals listed below,

$$Ag,\ Cd,\ Cu,\ Hg,\ Cr$$

those that plate out at the cathode during an electrolysis of an aqueous solution of one of their salts are

(a) Ag, Cu, and Hg   (b) Ag, Cd, and Cu   (c) Cd, Cu, and Cr

(d) Ag and Cu   (e) Ag, Hg, and Cr

9. When a dilute aqueous solution of Li_2SO_4 is electrolyzed, the products formed at the anode and cathode, respectively, are

(a) S and Li   (b) O_2 and Li   (c) SO_2 and H_2 (d) O_2 and H_2 (e) SO_2 and Li

10. For the cell $Zn(s) \mid Zn^{2+} \parallel Cu^{2+} \mid Cu(s)$, the standard cell voltage, $\Delta\mathscr{E}^0_{cell}$, is 1.10 V. When a cell using these reagents was prepared in the lab the measured cell voltage was 0.98 V. A possible explanation for the observed voltage is

(a) There were 2.00 mol of Zn^{2+} but only 1.00 mol of Cu^{2+}.

(b) The Zn electrode had twice the surface of the Cu electrode.

(c) The $[Zn^{2+}]$ was larger than the $[Cu^{2+}]$.

(d) The volume of the Zn^{2+} solution was larger than the volume of the Cu^{2+} solution.

(e) The $[Zn^{2+}]$ was smaller than the $[Cu^{2+}]$.

11. If a steady current of 15.0 A is passed through an aqueous solution of $CuSO_4$, how long will it take to deposit 0.250 mol of Cu at the cathode, assuming 100% efficiency?

(a) 3.22×10^3 min   (b) 1.61×10^3 min   (c) 53.6 min   (d) 26.8 min

(e) 0.893 min

12. The logarithm of the equilibrium constant, $\log K_{eq}$, of the net cell reaction of the cell $Cd(s) \mid Cd^{2+} \parallel Ag^+ \mid Ag(s)$ is

(a) 67.60   (b) 40.61   (c) 20.30   (d) 13.38   (e) 6.69

13. A steady current of 30.0 A for 70.2 min corresponds to a passage of

(a) 1.26×10^5 electron   (b) 1.31 C   (c) $1.26 \times 10^5\ \mathscr{F}$ (d) $1.31\ \mathscr{F}$ (e) 1.31 V

14. Which of the following changes will increase the EMF of the cell

$$Co(s) \mid CoCl_2(M_1) \parallel HCl(M_2) \mid H_2(g) \mid Pt(s)$$

(a) Increase the volume of the $CoCl_2$ solution from 100 to 200 mL.

(b) Increase M_2 from 0.010 to 0.500 M.

(c) Increase the pressure of the $H_2(g)$ from 1.00 to 2.00 atm.
(d) Increase the weight of the Co electrode from 10.00 to 20.00 g.
(e) Increase M_1 from 0.010 to 0.500 M.

15. A 0.200 F KOH solution was electrolyzed for 1.50 h using a current of 8.00 A. How many moles of O_2 were produced at the anode?
    (a) 0.448   (b) 0.224   (c) 0.112   (d) 2.24×10^{-2} (e) 7.46×10^{-3}

16. What is the value of the reaction quotient, Q, for the cell

$$Ni(s) \mid Ni(NO_3)_2(0.10\ M) \parallel KCl(0.40\ M) \mid Cl_2(g,\ 0.50\ atm) \mid Pt(s)$$

    (a) 3.1×10^{-1} (b) 1.3×10^{-1} (c) 8.0×10^{-2} (d) 3.2×10^{-2}
    (e) 1.6×10^{-2}

17. Warm dilute nitric acid will oxidize
    (a) Pt but not Au   (b) Cu but not Ag   (c) Au but not Ag   (d) Sn but not Cd
    (e) Ag but not Pt

Problems

Use Appendix F for $\mathcal{E}^0$ values.

16.1. In one half-cell of a galvanic cell, a piece of Pt dips into a solution containing 0.850 M Fe^{3+} and 0.010 M Fe^{2+}. In the other half-cell a piece of Cd is immersed in a 0.500 M Cd^{2+} solution.
    (a) When the cell produces electricity, which electrode is the positive terminal?
    (b) Which electrode is the cathode?
    (c) Write the equation for the spontaneous net cell reaction.
    (d) Calculate the equilibrium constant for the net cell reaction.
    (e) Calculate the reaction quotient, Q, for the cell described.
    (f) What is the EMF of this cell?

16.2. Calculate the equilibrium constant for the reaction

$$Hg^{2+} + Hg(liq) \rightleftharpoons Hg_2^{2+}$$

    Hint: Be sure you are not calculating the equilibrium constant for

$$2Hg^{2+} + 2Hg(liq) \rightleftharpoons 2Hg_2^{2+}$$

16.3. The observed EMF of the galvanic cell

$$Tl(s) \mid Tl^+(0.200\ M) \parallel Cl^-(0.500\ M) \mid Cl_2(g,\ 1\ atm) \mid Pt(s)$$

    is 1.755 V.
    (a) Write the equation for the net cell reaction.
    (b) Which electrode is the negative terminal?
    (c) Calculate $\mathcal{E}^0$ for the $Tl^+|Tl$ electrode.
    (d) If the pressure of $Cl_2(g)$ is decreased, will the cell voltage increase, decrease, or remain the same? Explain your answer.
    (e) Calculate the equilibrium constant for the net cell reaction.

16.4. Will acidic solutions of Cr^{2+} be stable if exposed to air, or will O_2 oxidize Cr^{2+} to Cr^{3+}? Show all calculations required to prove your answer.

16.5. For each of the following galvanic cells write the net cell reaction and calculate $\Delta\mathcal{E}^0_{cell}$. Does the reaction you have written proceed spontaneously to the right when all substances are in their standard states?

(a) Zn(s) | Zn²⁺ ‖ H⁺(aq),Cr³⁺,Cr₂O₇²⁻ | Pt(s)

(a) $Zn(s) \mid Zn^{2+} \parallel H^+(aq),Cr^{3+},Cr_2O_7^{2-} \mid Pt(s)$
(b) $Pt(s) \mid I_2,I^- \parallel H^+(aq),Pb^{2+} \mid PbO_2(s) \mid Pt(s)$
(c) $Ag(s) \mid AgCl(s) \mid HCl(aq) \mid H_2(g) \mid Pt(s)$
(d) $Pb(s) \mid Pb(NO_3)_2(aq) \parallel H_2SO_4(aq) \mid PbSO_4(s) \mid Pb(s)$

16.6. (a) From the values of $\mathscr{E}^0_{Cu^{2+}|Cu}$ and $\mathscr{E}^0_{Cu^{2+}|Cu^+}$ in the Table of Standard Reduction Potentials, calculate the equilibrium constant of the reaction

$$Cu(s) + Cu^{2+} \rightleftharpoons 2Cu^+$$

at 25 °C. Does this reaction proceed to the right to an appreciable extent?
(b) Copper(I) bromide, CuBr, is a very slightly soluble salt with $K_{sp} = 4 \times 10^{-8}$ at 25 °C. Calculate the equilibrium constant for the reaction

$$Cu(s) + Cu^{2+} + 2Br^- \rightleftharpoons 2CuBr(s)$$

What do you learn by comparing the equilibrium constants calculated in parts (a) and (b)?

16.7. Calculate $\Delta\mathscr{E}^0_{cell}$ and K_{eq} for the net cell reaction of the cell

$$Ag(s) \mid Ag^+ \parallel Cl^- \mid AgCl(s) \mid Ag(s)$$

Write the expression for this equilibrium constant.

16.8. (a) Calculate $\Delta\mathscr{E}^0_{cell}$ for a galvanic cell that utilizes each of the following reactions:

$$2Ag(s) + 2H^+(aq) + 2Cl^- \rightleftharpoons 2AgCl(s) + H_2(g) \qquad \text{(i)}$$
$$2Ag(s) + 2H^+(aq) + 2I^- \rightleftharpoons 2AgI(s) + H_2(g) \qquad \text{(ii)}$$

(b) Will silver liberate $H_2(g)$ from 1 M HCl? from 1 M HI? Explain.

16.9. The observed EMF of the cell

$$Pt(s) \mid H_2(g, 1 \text{ atm}) \mid H^+(5.0 \times 10^{-4} M) \parallel H^+(M_1) \mid H_2(g, 1 \text{ atm}) \mid Pt(s)$$

is 0.154 V. Calculate the value of M_1.

16.10. The Nernst equation can be used to calculate the single electrode potential of any half-cell. For $ox + ne^- = red$,

$$\mathscr{E} = \mathscr{E}^0 - \left(\frac{0.0592}{n}\right) \log \left(\frac{[red]}{[ox]}\right)$$

For the cell $Zn(s) \mid Zn(NO_3)_2 \parallel Pb(NO_3)_2,HNO_3 \mid PbO_2(s) \mid Pt(s)$ write the expression for the *reduction* potential at both anode and cathode. Using the relation $\Delta\mathscr{E}_{cell} = \mathscr{E}_{cathode} - \mathscr{E}_{anode}$, write the expression for $\Delta\mathscr{E}_{cell}$ and compare it with the expression you would write using Eq. (16-22).

16.11. For the galvanic cell

$$Zn(s) \mid Zn^{2+}(0.0400 M) \parallel Cl^-(5.00 \times 10^{-3} M) \mid Cl_2(g, 1 \text{ atm}) \mid Pt(s)$$

(a) Write equations for the anode, cathode, and net cell reactions.
(b) Calculate the reaction quotient, Q, and the EMF for this cell.
(c) If the [Cl⁻] is changed to 0.500 M, will the EMF increase, decrease, or remain the same? Explain your answer.
(d) If NH_3 is added to the anode compartment, the complex ion $Zn(NH_3)_4^{2+}$ is formed. Will adding NH_3 to the anode compartment increase, decrease, or leave unchanged the cell voltage? Explain your answer.
(e) Calculate the equilibrium constant, K_{eq}, for the net cell reaction written in part (a).

16.12. A piece of Zn metal weighing 3.000 g is placed in 80.00 mL of 0.1000 F $AgNO_3$ at 25 °C. What are the $[Zn^{2+}]$ and $[Ag^+]$ in solution when equilibrium is attained? Assume no change in volume occurs on reaction. What weight of Zn remains unreacted when equilibrium is attained?

16.13. Acidified water was electrolyzed using copper electrodes. A steady current of 1.17 A was passed for 25 min and 15 s. The copper anode lost 0.583 g; no gas evolved at the anode. At the cathode, 232.0 mL of $H_2(g)$ were collected over water at 26.0 °C. The barometric pressure was 762.0 mmHg.
(a) Calculate the number of coulombs and faradays that passed through the cell.
(b) Calculate the number of moles of H_2 produced and the number of moles of Cu that were oxidized.
(c) Show calculations to determine the oxidation state of the copper ion that was formed.
(d) Use data in this problem and the charge on the electron to calculate Avogadro's number.

16.14. Using data from Appendix F calculate the solubility product of silver sulfate, Ag_2SO_4.

16.15. For the cell Pt(s) | H_2(g, 1 atm) | HCl(0.010 F) | AgCl(s) | Ag(s) the measured cell potential is 0.459 V. If we consider the right-hand electrode to be a $Ag^+|Ag$ electrode, what is the $[Ag^+]$ in equilibrium with the AgCl and Cl^-? Calculate $K_{sp}(AgCl)$ from these data.

chapter 17 Energy, Enthalpy, and Thermochemistry

Germain Henri Hess (1802–1850) was born in Geneva, Switzerland, and was professor of chemistry at the University of St. Petersburg (now Leningrad) in Russia. He did research in organic chemistry, but is best known for a series of thermochemical investigations. In 1840 he discovered experimentally that the heat change in a chemical reaction is the same whether the reaction takes place in one step or in several steps. What is now called Hess' Law of Constant Heat Summation is a direct consequence of the law of conservation of energy, the first law of thermodynamics, although it does not appear that Hess recognized this.

The answer to the question "Why do substances A and B react spontaneously to form the products C and D?" can be summed up in the phrase "to reach a state of equilibrium." Whether we are discussing reactions of biological importance, such as the hydrolysis of adenosine triphosphate (ATP) to yield adenosine diphosphate (ADP) and some inorganic phosphate ion, which releases metabolically available energy in living cells, or reactions of interest to inorganic chemists, or reactions involved in some engineering project such as designing an efficient process for refining petroleum, the very same fundamental principles apply when we try to understand the drive to reach a state of equilibrium. These principles are the basics of **thermodynamics,** which is the branch of science concerned with the energy changes accompanying physical and chemical processes.

If we consider a reaction of interest that can be symbolized

$$\alpha A + \beta B \rightleftharpoons \gamma C + \delta D$$

we can enumerate the following important questions:

1. For a given set of concentrations of A, B, C, and D, will the reaction occur spontaneously from left to right, as it is written, at a specified temperature?

2. What is the magnitude of the equilibrium constant

$$K_{eq} = \frac{[C]^\gamma [D]^\delta}{[A]^\alpha [B]^\beta}$$

at the temperature specified? If K_{eq} is large compared to 1, we can expect large yields of the products C and D, but if K_{eq} is small compared to 1, we cannot expect to produce much C and D.

3. If we change the temperature, how is K_{eq} affected? Can we increase the yield by running the reaction at some other temperature?

4. Does this reaction release energy or must we supply energy to make it proceed?

The answers to all of these questions, about any sort of reaction, are to be found in the study of thermodynamics.

section 17.1
State Functions

Whenever you study a new subject, you will find that it has a certain vocabulary you must become familiar with, a "jargon" of its own. Thermodynamics is no exception. In thermodynamics, whatever we are investigating is called the **system.** We must clearly define the boundaries of our system. All the rest of the universe, outside the defined boundaries, is called the **surroundings.** We also define the **state** of the system by specifying a number of properties the system possesses. Each state of each system has a very large number of properties: temperature, chemical composition, pressure, mass, volume, density, color, viscosity, refractive index, and so on. It is not necessary, however, to list the value of every one of these properties in order to define the state of the system completely.

The complete description of a thermodynamic system is accomplished by specifying the values of only a few properties of the system, called **state functions.** A state function is a property that has some definite value for each state of the system; the value of this property is independent of the way in which the state is attained. For instance, we can define a system by specifying that it consists of 1 mol of gaseous hydrogen, H_2, at 20 °C and 1-atm pressure. The volume, density, and viscosity of the H_2 are determined once the temperature, pressure, mass, and chemical composition are specified. All of these properties are state functions. It does not matter whether the $H_2(g)$ was made by electrolyzing water or by reacting zinc with hydrochloric acid. It doesn't matter whether the temperature of 20 °C was attained by cooling the H_2 down from a higher temperature or by warming it up from a lower temperature. The path taken to arrive at the particular state of the system does not affect the value of any state function.

The two most important features of state functions are

1. Assigning values to a few state functions automatically fixes the values of all others.
2. When the state of a system is changed, the value of the *change* in any state function depends only on the initial and final states and not on how the change was accomplished.

We always employ the symbol Δ (the Greek letter *delta*) to mean the change or difference in a state function when the state of the system is changed. To be specific, the symbol for the change in the volume, ΔV, is defined as

$$\Delta V = \left(\begin{array}{c} \text{volume of the} \\ \text{final state} \end{array} \right) - \left(\begin{array}{c} \text{volume of the} \\ \text{initial state} \end{array} \right) = V_2 - V_1 \qquad (17\text{-}1)$$

Note particularly the precise definition of the symbol ΔX. It *always* means the value of the property X in the final state (state 2) minus the value of the property X in the initial state (state 1).

section 17.2
Heat and Work

Work

Work is defined as the product of a force and the distance through which the force is applied. The SI unit of work is the **joule.** One joule is the work done when a force of one newton is applied through a distance of one meter.

Sign Convention for Work

Thermodynamics is concerned with the transfer of energy between the system and its surroundings. At the outset, therefore, we must choose a sign convention that indicates whether energy is leaving the system and entering the surroundings, or being transferred to the system from the surroundings. Since work is a form of energy, this necessitates choosing a sign convention for work. The convention we will use is the following:*

> Work done *on* a system is positive work.
> Work done *by* a system is negative work.

---

* An older sign convention for work, still in use by many scientists, is the opposite of that given here, namely, work done by a system is positive work. If you read other texts always determine the sign convention used by the author to avoid confusion.

When work, w, is positive, it indicates that work has been done on the system and that the process has therefore contributed to an increase in the energy of the system.

When work is negative, it indicates that work has been done by the system on the outside world (the surroundings of the system) and that the process has contributed to a decrease in the energy of the system. If $w = -15$ kJ, the system has done 15 kJ of work on its surroundings.

$$\text{work done by a system} = -(\text{work done on the system}) \tag{17-2}$$

Remember that the sign tells you the *direction* of energy transfer. A positive sign means energy has entered the system; a negative sign means energy has left the system.

The Work of Expansion

One type of work with which we must become familiar is the **work of expansion**, which is the work done when the volume of a system is changed while a confining pressure is being applied. Because we live with the gases of the air exerting a pressure on us and the systems we study, work of expansion is almost always done when chemical or biological reactions are carried out, even though we may not always be aware of it. We will derive the expression for the work of expansion when the confining pressure is constant during the change of volume from an initial V_1 to a final V_2. We restrict the discussion to a constant external pressure because the atmospheric pressure changes very little during the time it takes to carry out most experiments.

Consider a gas enclosed in a cylindrical piston of cross-sectional area A, as illustrated in Fig. 17.1. Let P_{ext} be the constant external pressure applied to the piston. If the gas inside the cylinder has a pressure greater than the external pressure, it will push the piston up, expanding its volume. Let the piston move up a distance x. Since, by definition, pressure is force per unit area, the force against which the expanding gas moves is given by

$$\text{force} = (\text{pressure})(\text{area}) \tag{17-3a}$$

or

$$\text{force} = P_{ext}A \tag{17-3b}$$

The work done *by* the gas when it expands against the external pressure is therefore

$$\text{work} = (\text{force})(\text{distance}) = P_{ext}Ax \tag{17-4}$$

But Ax is just the difference between the final volume of the system and the initial volume. (See Fig. 17.1.)

$$\Delta V = V_{final} - V_{initial} = V_2 - V_1 = Ax \tag{17-5}$$

The work done *by* the gas when it pushes back the piston is therefore $P_{ext}\Delta V$. We have adopted the convention that the work, w, is the work done *on* the system. Hence,

$$w = -P_{ext}(V_2 - V_1) = -P_{ext}\Delta V \tag{17-6}$$

In our example, the system is the gas confined in the cylinder. The surroundings include the source of external pressure on the piston. When the gas expands, $V_2 > V_1$, so that ΔV is positive. The gas does work on the surroundings when it has pushed the piston up. Using Eq. (7-2) we find that the work done on the gas is $-P_{ext}\Delta V$. On the other hand, if the piston moves downward and compresses the gas, V_2 is smaller than

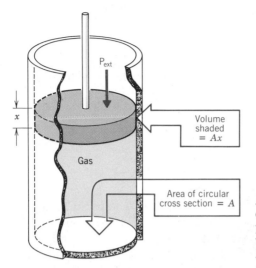

Fig. 17.1. The expansion of a gas against a constant external pressure, P_{ext}. When the piston moves up a distance x, the volume of the gas changes by the amount Ax, so that $\Delta V = Ax$.

V_1, so that ΔV is negative, and the work done on the system, calculated using Eq. (17-6), is positive.

Note that *you* never have to do anything about the sign. Using Eq. (17-6) the work of expansion will be negative if the system expands and does work on its surroundings, and will be positive if the system is compressed and work is done on the system.

A word is in order here about units. It is customary to specify the pressure in atmospheres and the volume change in liters, so that $-P_{ext}\Delta V$ is in liter · atmospheres (L · atm). The liter · atmosphere is not, however, a conventional unit to use for work. In SI units, work should always be expressed in joules. We have already run across this problem, in connection with the ideal gas law. (Refer to Section 4.6.) The conversion factor between liter · atmospheres and joules is given by Eq. (4-26):

$$1 \text{ L} \cdot \text{atm} = 101.32 \text{ J}$$

EXAMPLE 17.1. The work of expansion against a constant external pressure

A sample of an ideal gas, maintained at a constant temperature of 25 °C, is contained in a cylinder sealed by a piston. The gas is initially at a pressure of 6.00 atm and occupies a volume of 400.0 mL. If the external pressure on the piston is constant at 1.50 atm, and the gas expands until its own pressure is equal to the external pressure, how much work does the gas do on expanding?

Solution. We must determine the final volume of the gas in order to calculate the work of expansion. Since the gas is ideal, and the temperature and number of moles of gas are both constant, we can use Boyle's law to calculate the final volume, V_2.

$$P_1 V_1 = P_2 V_2 \qquad \text{so that} \qquad V_2 = P_1 V_1 / P_2$$

Hence

$$V_2 = (6.00 \text{ atm})(0.400 \text{ L})/(1.50 \text{ atm}) = 1.60 \text{ L}$$

$$\text{work of expansion} = -P_{ext}\Delta V = -(1.50)(1.60 - 0.400) \text{ L} \cdot \text{atm}$$

$$= -(1.50)(1.20) = -1.80 \text{ L} \cdot \text{atm}$$

We convert to joules to report the answer.

$$\text{work of expansion} = -(1.80 \text{ L} \cdot \text{atm})(101.32 \text{ J/L} \cdot \text{atm}) = -182 \text{ J}$$

The work done *by* the gas is therefore 182 J.

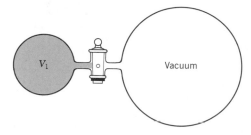

Fig. 17.2. Expansion of a gas into a vacuum. Initially all the gas is in the left-hand bulb. When the stopcock is opened, the gas expands to fill both bulbs. In Example 17.2, $V_1 = 0.400$ L and the evacuated bulb has a volume of 1.20 L.

EXAMPLE 17.2. The work of expansion into a vacuum

Consider the following experiment, which uses two bulbs connected by a stopcock, as in Fig. 17.2. The larger bulb, with a volume of 1.20 L, is evacuated, and an ideal gas is placed in the smaller bulb, which has a volume of 0.400 L. The gas has a pressure of 6.00 atm, and is kept at 25 °C. What is the final volume of the gas when the stopcock is opened, and how much work does the gas do on expanding?

Solution. When the stopcock is opened, the gas expands to fill both bulbs. The final volume is the combined volumes of both bulbs, or 1.60 L. Because the gas expanded into a vacuum, it did not have to push back any external pressure, and so no work was done. When $P_{ext} = 0$, the work of expansion is 0.

Note particularly that the initial and final states of the gas in Examples 17.1 and 17.2 are identical, but the work done by the gas on expanding is quite different for the two methods or paths of expansion.

Heat

Heat is not a material substance, but is energy in transit between system and surroundings as a result of a temperature difference. When the energy of a system changes as the result of a temperature difference between the system and its surroundings we say that there has been a flow of heat.

Temperature is an intrinsic and **intensive property** of the state of a system that is a measure of the average kinetic energy of its molecules. Molecules in a warmer object have greater average kinetic energy than molecules in a cooler object. When a hot object comes in contact with a cold one, the molecules in the hot object transfer some of their kinetic energy during collisions with molecules of the cold object. This transfer of kinetic energy is the flow of heat. Heat is manifested only at the boundary of system and surroundings, and is not a property of the system (that is, heat is not a state function), whereas the temperature is a property of the system.

A sign convention about the direction of heat flow is needed, just as we needed a convention about the sign of work. The following convention for the sign of heat is universally accepted:

Heat flowing into the system (that is, heat absorbed by the system) is positive.

Heat flowing out of the system (that is, heat released by the system) is negative.

Note that this sign convention is consistent with the one adopted for work. When heat, q, is positive it indicates that heat has entered the system and that the process has contributed to an increase in the energy of the system.

When heat is negative, it indicates that heat has left the system and entered the

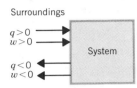

Surroundings

$q > 0$
$w > 0$

System

$q < 0$
$w < 0$

Fig. 17.3. The sign convention used for heat and work. Heat and work are both positive when they enter the system, and are both negative when they leave the system.

surroundings, and that the process has contributed to a decrease in the energy of the system. If a system releases 15 kJ of heat to its surroundings, $q = -15$ kJ.

Figure 17.3 illustrates the sign conventions for heat and work used in this text.

The Mechanical Equivalent of Heat

Imagine that we have a beaker of water at 20 °C and want to raise the temperature of the system to 21 °C. There are several ways we might think of to effect this change, two of which are described below.

Path a. Bring the beaker into contact with a warm object so that heat flows into the beaker and water from the surroundings. (The beaker of water is our system; the warm object is part of the surroundings.)

Path b. Stir the water with an electric motor, so that the work of stirring raises the temperature of the water.

These two paths, which bring about the same change, are illustrated in Fig. 17.4.

In the first method, Path *a*, the temperature change is caused by the absorption of heat. In the second path, Path *b*, work is done to the system in order to raise the temperature. The difference between the two paths is in the surroundings, not in the system. The warm object is cooler in Path *a*; the battery that supplied the energy to run the stirring motor has discharged somewhat in Path *b*. Both *q*, the heat that flowed into the system, and *w*, the work done on the system, are manifested at the boundary between system and surroundings and are not properties of the system. Neither heat nor work are state functions.

Path *b* is, in essence, the classic experiment of James Joule (1818–1889) showing the equivalence of heat and work. Prior to Joule's experiments in the 1840's, work and heat were considered to be different, and different units were used to measure them. Joule demonstrated that expending a given quantity of mechanical energy (the

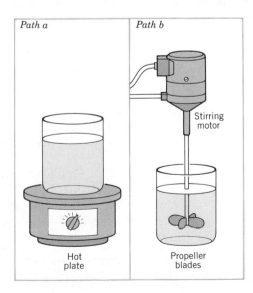

Path a

Path b

Stirring motor

Hot plate

Propeller blades

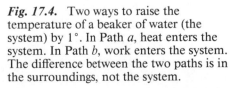

Fig. 17.4. Two ways to raise the temperature of a beaker of water (the system) by 1°. In Path *a*, heat enters the system. In Path *b*, work enters the system. The difference between the two paths is in the surroundings, not the system.

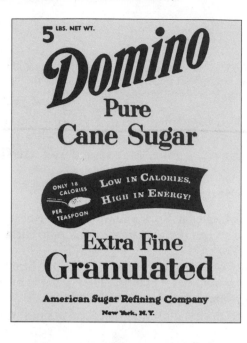

Fig. 17.5. Reproduction of a label on a bag of sugar. What is funny about this label?

motion of the stirring motor) always results in the same increase in temperature as that produced by a fixed quantity of heat flowing into the system. Both heat and work are forms of energy.

The unit originally used to measure a quantity of heat was the calorie, and the calorie continued to be used to measure energy changes accompanying the flow of heat long after it was clear that a separate unit for heat was not necessary. Today the calorie is defined in terms of its mechanical equivalent:

$$1 \text{ cal} = 4.1840 \text{ J} \tag{17-7}$$

Scientists from nations all over the world have decided to avoid the unnecessary duplication of units and use only a single unit, the joule, to measure both heat and work. Many chemists and biologists, however, continue to use the calorie, and although its use has declined in recent years, you will still see it, particularly in older literature. For an amusing misuse of a defined quantity, see Fig. 17.5.

section 17.3
The First Law of Thermodynamics

Let us consider two different processes that can be accomplished in more than one way (in the language of thermodynamics, that can be carried out using several different paths), and examine the heat absorbed by the system, q, and the work done on the system, w, for each of these paths.

Example 1. Let our system be a beaker of water. In the initial state of the system, the water and the beaker are at 20 °C, under 1-atm pressure. In the final state, the water and the beaker are at 21 °C, at the same pressure. This is, of course, the system we discussed in the preceding section. The two paths used to bring about the change from the initial to the final state are illustrated in Fig. 17.4.

Path a. Bring a warm object into contact with the beaker. An amount of heat, q_a, enters the beaker and water in order to raise the temperature from 20 to 21 °C. The magnitude of q_a depends on the size of the beaker and the amount of water in it, but heat must be absorbed by the system, so that q_a is a positive number. No work at all is done, either to the system, or by the system. (The change in volume of a beaker of water when the temperature is increased by $1°$, is negligibly small.) Thus,

$$q_a > 0 \qquad w_a = 0$$

Path b. Put a stirring motor in the beaker, turn on the motor, and let it run until the temperature has risen to 21 °C. No heat enters or leaves the system. Mechanical work is done on the system, and we find that the work done on the system is equal in magnitude to the amount of heat that entered the system during Path *a*. In keeping with the convention that work done on a system is positive work, we obtain

$$q_b = 0 \qquad w_b = q_a$$

We observe that while $q_a \neq q_b$ and $w_a \neq w_b$, $q_a + w_a = q_b + w_b = q_a$.

Example 2. Let our system be the reacting species of an oxidation–reduction reaction. We will use a reaction of importance in biochemistry, the reduction of pyruvate ion to lactate ion by H_2. A knowledge of the structure of these substances is not important for the argument being made here, but we will include the equation for the reaction for the sake of completeness:

$$\underset{\text{Pyruvate ion}}{CH_3-\overset{\overset{O}{\|}}{C}-\overset{\overset{O}{\|}}{C}-O^-} + H_2(g) \;\rightleftharpoons\; \underset{\text{Lactate ion}}{CH_3-\overset{\overset{OH}{|}}{CH}-\overset{\overset{O}{\|}}{C}-O^-} \tag{17-8}$$

State 1 of our system consists of a mole of pyruvate ion in aqueous solution plus a mole of H_2 gas, both at 35 °C. State 2 consists of a mole of lactate ion in aqueous solution at 35 °C.

Path a. Construct a galvanic cell that utilizes this oxidation–reduction reaction. This was actually done by Barron and Hastings in 1934.[*] One half-cell was a hydrogen electrode, and the other was a solution containing both pyruvate and lactate ions, with an inert gold electrode. The solutions were maintained at 35 °C. The two half-reactions for this galvanic cell are

Reduction	pyruvate ion $+ 2H^+(aq) + 2e^- \rightarrow$ lactate ion	(17-8a)
Oxidation	$H_2(g) \rightarrow 2H^+(aq) + 2e^-$	(17-8b)

The amount of electrical work the reaction produces depends on the magnitude of the current drawn. Barrons and Hastings measured the maximum amount of electrical work obtainable from the reaction and found it to be 11,440 cal. When the electrical work done by the cell is the maximum amount, the reaction also releases 10,200 cal of heat to its surroundings. Thus

$$q_a = -10,200 \text{ cal} \qquad w_a = -11,440 \text{ cal}$$

The data are reported in calories because these measurements were made before the introduction of SI units. Note that w_a is a negative quantity because the system is doing work, and q_a is negative because the reaction is exothermic.

Path b. Using a platinum catalyst to speed up the reaction, bubble H_2 gas through an aqueous solution containing pyruvate ion at 35 °C. Since the reacting species are in direct contact, no electrical work is done. The reaction is exothermic and the amount of heat released is 21,640 cal. Thus

$$q_b = -21,640 \text{ cal} \qquad w_b = 0$$

[*] E. S. G. Barron and A. B. Hastings, J. Biol. Chem. **107**, 567 (1934).

We observe that $q_a + w_a = q_b + w_b = -21,640$ cal.

These examples are only a few of the very many reactions scientists have studied carefully. For all the observations made about any reaction, the following statements have been found to be true:

1. If the path of the reaction (that is, the way the reaction is carried out) is changed, the heat absorbed by the system and the work done on the system also change. Thus neither heat nor work is a state function. Heat and work are path-dependent quantities.

2. Although q and w individually depend on the path, their sum, $q + w$, is path independent. The value of $q + w$ is always the same, regardless of how many different paths are used, as long as the initial and final states are the same. Therefore $q + w$ must be the change in some state function.

Statements of the First Law of Thermodynamics

We define the internal energy, E, to be a property of a system, that is, a state function, such that when the state of the system is changed from state 1 to state 2, the change in the internal energy is given by

$$\Delta E = q + w \qquad (17\text{-}9)$$

Equation (17-9) is the **first law of thermodynamics**. The essence of the first law lies in the fact that although the heat absorbed by the system, q, and the work done on the system, w, depend on the path used to carry out a transformation from one state to another, the sum $q + w$ is independent of the path.

The first law is called **the law of conservation of energy**. Because it expresses a very basic relation that has been observed about many different kinds of reactions and systems, the same ideas have been stated in different ways by different people. One of the earliest statements of the first law was the following: Energy can neither be created nor destroyed; the total amount of energy in the universe remains constant.

Think of Example 2 that has just been discussed. The energy of a mole of lactate ion in aqueous solution is 21,640 cal lower than the energy of a mole of pyruvate ion plus a mole of H_2 at the same temperature, 35 °C. When the reaction is carried out, that energy is released to the surroundings. If no electrical work is done, all 21,640 cal are released as heat. If we get electrical work out of the system, less heat is released, but the total amount of energy released to the surroundings remains constant. No matter how we carry out this reaction, the system is lower in energy in the final state and the surroundings are higher in energy by exactly the same amount.

We can change the energy of a system by having either heat or work (or both) enter or leave the system. But if heat or work enters the system, an equivalent amount of energy leaves the surroundings. The total amount of energy of system plus surroundings remains constant.

The first law summarizes observations that have been made about energy changes in physical, chemical, or biological processes. There are three laws of thermodynamics, and none of them can be derived or proved. If the way one of the laws of thermodynamics is stated is not in agreement with an observation that is made, then the law must be amended. This has already happened to the first law, after the creation of the first atomic pile (nuclear reactor) in 1940. Since in an atomic pile energy is created by the process of converting mass to energy, the earlier statement of the first law of thermodynamics (energy can not be created or destroyed) had to be changed to: *In a system of constant mass,* energy can neither be created nor destroyed.

However, mass can be converted into energy and the energy created by a loss of mass of magnitude m is given by the Einstein relation $E = mc^2$. The first law is now called the law of conservation of mass and energy. For ordinary chemical processes not involving nuclear reactions, the mass of the system is essentially constant (mass changes are negligibly small, much too small to be detected), and we can utilize the first law as a law of conservation of energy.

section 17.4
Energy and Enthalpy

When we construct a galvanic cell we utilize a chemical reaction to provide electrical work for us. We burn hydrocarbons to heat our homes, to cook our food, to run automobiles and airplanes, and to generate electricity in power plants. In the laboratory, however, we almost always run a chemical reaction simply to obtain the products of the reaction. We are not usually interested in obtaining work from reactions we run in the lab. We must realize, however, that even when we are not interested in it, if the volume of our system changes, some work of expansion is done. Most of the time chemical reactions are run open to the atmosphere, and therefore an external pressure, equal to atmospheric pressure, is continually exerted by the surroundings on any system of interest. If our system expands, $-P_{ext}\Delta V$ is negative and the system has done work on the surroundings against the confining atmosphere. If our system contracts, $-P_{ext}\Delta V$ is positive and the confining atmosphere has done work on the system. In either case, the work term in the first law, Eq. (17-9), is nonzero whenever the volume of the system changes during a chemical reaction.

If we want to know how much heat is absorbed or released during a chemical reaction, we must first ask about the conditions under which the reaction is carried out, because heat is a path-dependent quantity. The two paths most commonly used for carrying out a process are (1) a constant volume path, and (2) a constant pressure path. The constant pressure path is much more frequently employed than the constant volume path. Let us consider each of these paths separately.

Reactions Carried Out at Constant Volume

In order to run a reaction at constant volume, special equipment is needed. The reaction must take place in a sealed ampule or in a steel "bomb," so-called because the walls are strong enough so that it does *not* explode when subjected to substantial pressure differences inside and out. But if the volume is held constant so that no work of expansion is done, and if no electrical or other mechanical work is done, then $w = 0$, and the first law becomes

$$\Delta E = q_v \tag{17-10}$$

where the subscript "v" on the heat indicates that the reaction is carried out at constant volume. One way, therefore, to measure the value of the change in internal energy when a reaction proceeds is to measure the heat absorbed when the reaction is carried out at constant volume. The device used to make such a measurement is called a bomb calorimeter.

Reactions Carried Out at Constant Pressure

If, as is more common, we carry out a reaction open to the atmosphere, then the pressure of the system is always the same as the external (atmospheric) pressure.

$$P_{sys} = P_{ext} = P \tag{17-11}$$

In that case, work of expansion is done as the reaction proceeds, and provided no electrical work or other mechanical work is done, the first law becomes

$$\Delta E = q_p - P\Delta V \tag{17-12}$$

where the subscript "p" on the heat tells us the reaction is being carried out at constant pressure. The heat absorbed at constant pressure is therefore

$$q_p = \Delta E + P\Delta V = (E_2 - E_1) + P(V_2 - V_1) \tag{17-13}$$

Since the pressure is constant, it is the same in states 1 and 2 and therefore has no subscript.

On examining Eq. (17-13), we find it useful to invent a new function called the enthalpy, H, defined as follows:

$$H = E + PV \tag{17-14}$$

Since E, P, and V are all properties of the system, that is, are all state functions, H must also be a state function. If the state of a system changes while the pressure remains constant

$$\Delta H = H_2 - H_1 = (E_2 + PV_2) - (E_1 + PV_1)$$
$$= E_2 - E_1 + P(V_2 - V_1) \tag{17-15}$$

By substituting Eq. (17-15) into Eq. (17-13) we obtain

$$q_p = \Delta H \tag{17-16}$$

You should consider Eqs. (17-10) and (17-16) and remember that *heat* is a path-dependent term. The heat of reaction at constant volume is ΔE, while the heat of reaction at constant pressure is ΔH. Since chemists are much more likely to run reactions at constant pressure than at constant volume, it is ΔH that is usually meant when the unspecific term "heat of reaction" is used.

section 17.5
Thermochemistry

In 1840 G. H. Hess demonstrated experimentally that the heat absorbed when a given reaction is carried out at constant pressure and temperature is the same whether the reaction occurs in one step or in several steps. This statement is usually referred to as **Hess' Law of Constant Heat Summation.** Carrying out a reaction in several steps is simply carrying it out by another path, and Hess' Law tells us that the heat of reaction at constant pressure, ΔH, is a path-independent quantity. Thus Hess' Law is really another way of stating that the enthalpy, H, is a state function. When the state of a system is changed, the change in any state function depends only on the initial and final states, and not on the path of reaction.

An example of the reactions investigated during the nineteenth century is the following. In aqueous solution, 1 mol of CO_2 reacts with 2 mol of sodium hydroxide to produce a mole of sodium carbonate, and the net ionic equation for the reaction is

$$CO_2(aq) + 2\,OH^-(aq) \rightarrow CO_3^{2-}(aq) + H_2O \qquad \Delta H = -89.36 \text{ kJ} \tag{17-17}$$

The heat absorbed at constant pressure when this reaction is carried out is $q_p = -89.36$ kJ (-21.36 kcal). That means that 89.36 kJ of heat are released to the surroundings per mole of CO_3^{2-} formed. The overall reaction, Eq. (17-17), can be carried out in two steps by adding the 2 mol of sodium hydroxide one at a time, as follows:

$$CO_2(aq) + OH^-(aq) \rightarrow HCO_3^-(aq) \qquad \Delta H = -48.26 \text{ kJ} \quad (17\text{-}17a)$$
$$HCO_3^-(aq) + OH^-(aq) \rightarrow CO_3^{2-}(aq) + H_2O \qquad \Delta H = -41.10 \text{ kJ} \quad (17\text{-}17b)$$

A measurement of the heat absorbed at constant pressure for each of these two steps yields $q_{p_a} = -48.26$ kJ and $q_{p_b} = -41.10$ kJ. We observe that $q_{p_a} + q_{p_b} = -89.36 \text{ kJ} = q_{p_{\text{overall}}}$, just as Eq. (17-17) is the sum of Eq. (17-17a) and Eq. (17-17b).

In general, if there are two different paths for going from state 1 to state 2 of a system, then ΔH for both paths is the same. For instance,

> **Path 1** state 1 $\rightarrow$ state 2   directly
>
> then $\Delta H = H_2 - H_1$
>
> **Path 2** state 1 $\xrightarrow{\Delta H_a}$ state A $\xrightarrow{\Delta H_b}$ state B $\xrightarrow{\Delta H_c}$ state 2
>
> then $\Delta H = \Delta H_a + \Delta H_b + \Delta H_c = H_2 - H_1$

This means that if we can write down a series of equations that, added together, give a desired reaction, we can calculate ΔH for the desired reaction as the sum of the enthalpy changes for the reactions added. Provided that we have a set of reactions for which the ΔH values are known, we do not have to go into the laboratory and measure ΔH for each reaction of interest; we can use Hess' Law to calculate ΔH using pencil and paper. It is not even necessary that the desired equation describes a reaction that we *can* carry out in the laboratory; we can calculate ΔH for a reaction that may be exceedingly difficult or even impossible to perform. The most extensive table of data available for such calculations is a **Table of Standard Enthalpies of Formation,** compiled by Rossini and coworkers at the National Bureau of Standards.

Standard Heats of Formation, ΔH_f^0

The standard heat or enthalpy of formation of a specified compound is defined as ΔH for the reaction in which *one mole* of the compound is formed at 25 °C and 1-atm pressure *from the elements* in their standard states. The standard states of the elements are the most stable (common) forms at 1-atm pressure and any specified temperature. Values of thermodynamic functions of substances in their standard states are usually tabulated at 25 °C (298.15 K). The symbol for a standard enthalpy of formation is ΔH_f^0, where the subscript *"f"* indicates the formation reaction and the superscript zero (^{0}) indicates that all substances are in their standard states, that is, all gases are at 1-atm pressure and all species in solution are at unit activity. As we have done previously, we will neglect the difference between activity and concentration and use 1 M concentration as the standard state for species in solution. Note that the standard states defined here are exactly the same as the standard states used in defining the standard reduction potentials in Chapter 16. A substance can be in its standard state at any temperature, but values of ΔH_f^0 are tabulated at 25 °C, just as $\mathcal{E}^0$ values are.

If the common form of an element is a solid, the crystalline state of the solid must be specified. It may happen that two solid forms can exist at 25 °C, as for instance both graphite and diamond are solid forms of carbon that are found at 25 °C. In that case, the crystalline form that is most stable, that is, lower in energy, is chosen as the standard state. For carbon, the standard state is graphite. The symbol C(gr) is used to denote graphite, although $C(s)$ can be assumed to be graphite if no other information is given.

A major difficulty that students have with thermochemical calculations stems from not learning the precise definition of ΔH_f^0 in every detail. The important things to remember are that you must write an equation that produces exactly 1 mol of the desired compound, and that there must be nothing on the left-hand side of the equation but elements in their standard states. The following examples illustrate the definition of the standard enthalpy of formation.

EXAMPLE 17.3. The definition of ΔH_f^0

Write the equations for which the enthalpy change is the standard heat of formation of **(a)** $HCl(g)$, **(b)** $N_2O(g)$, **(c)** $B_2O_3(s)$, **(d)** $CaCO_3(s)$, and **(e)** glycine, $NH_2CH_2COOH(s)$.

Solution

(a) The stable forms of the elements hydrogen and chlorine at 25 °C are the diatomic gases, so that the correct equation is

$$\tfrac{1}{2} H_2(g) + \tfrac{1}{2} Cl_2(g) \rightarrow HCl(g) \qquad \Delta H_f^0(HCl) = -92.31 \text{ kJ/mol}$$

This means that 92.31 kJ of heat are released for each mole of $HCl(g)$ formed when H_2 and Cl_2 react at 25 °C and 1 atm.

(b) $$N_2(g) + \tfrac{1}{2} O_2(g) \rightarrow N_2O(g) \qquad \Delta H_f^0(N_2O) = +82.05 \text{ kJ/mol}$$

Note that the formation of N_2O from its elements is endothermic.

(c) $$2B(s) + \tfrac{3}{2} O_2(g) \rightarrow B_2O_3(s) \qquad \Delta H_f^0(B_2O_3) = -1272.8 \text{ kJ/mol}$$

(d) $$Ca(s) + C(s) + \tfrac{3}{2} O_2(g) \rightarrow CaCO_3(s) \qquad \Delta H_f^0(CaCO_3) = -1207 \text{ kJ/mol}$$

(e) $$2C(gr) + \tfrac{5}{2} H_2(g) + \tfrac{1}{2} N_2(g) + O_2(g) \rightarrow NH_2CH_2COOH(s)$$
$$\Delta H_f^0(NH_2CH_2COOH) = -537.2 \text{ kJ/mol}$$

A brief table of standard enthalpies of formation is given in Table 17.1. A more extensive table will be found in Appendix G.

By definition, the enthalpy of formation of an element in its standard state is zero, because this means producing one mole of the element in its standard state *from* one mole of the element in its standard state, or no change at all. Since the initial and final states are the same, $\Delta H = 0$. In contrast, the enthalpy of formation of an element in a state different from its standard state is not zero. Thus, while $\Delta H_f^0(\text{graphite}) = 0$, $\Delta H_f^0(\text{diamond}) = +1.897 \text{ kJ/mol}$, and refers to the transformation

Table 17.1. **Standard Enthalpies of Formation,**[a] ΔH_f^0, **at 25 °C (298.15 K), in kJ/mol**

$B_2O_3(s)$	-1272.8	$CHCl_3(\ell)$	-134.5	$Fe_2O_3(s)$	-824.2
$Br(g)$	$+111.8$	$CH_3COOH(\ell)$	-484.5	$NH_2CH_2COOH(s)$	-537.2
$CaCO_3(s)$	-1207.0	$CO_2(g)$	-393.51	$N_2O(g)$	$+82.05$
$CaO(s)$	-635.5	$HCl(g)$	-92.31	$NO_2(g)$	$+33.2$
$CH_4(g)$	-74.81	$H_2O(g)$	-241.83	$SO_2(g)$	-296.8
$C_2H_6(g)$	-84.67	$H_2O(\ell)$	-285.84	$ZnO(s)$	-348.3

[a] The standard state of carbon is graphite.

$$C(s, \text{graphite}) \rightarrow C(s, \text{diamond}) \quad \Delta H^0 = 1.897 \text{ kJ} \quad (17\text{-}18)$$

Similarly, while $\Delta H_f^0(Br_2, \text{liq}) = 0$, $\Delta H_f^0(Br, g) = +111.8$ kJ. This means that 111.8 kJ is needed to convert one-half mole of liquid Br_2 to a mole of gaseous Br atoms. The equation for the formation of $Br(g)$ is

$$\tfrac{1}{2} Br_2(\text{liq}) \rightarrow Br(g) \quad \Delta H_f^0(Br, g) = 111.8 \text{ kJ} \quad (17\text{-}19)$$

Calculations Using Hess' Law

Suppose that we want to determine ΔH at 25 °C and 1 atm for the reaction

$$2Fe_2O_3(s) + 3C(gr) \rightarrow 4Fe(s) + 3CO_2(g) \quad (17\text{-}20)$$

which is utilized in the reduction of iron ore (largely Fe_2O_3) to produce metallic iron. In Table 17.1 we can find the standard heats of formation of $CO_2(g)$ and $Fe_2O_3(s)$. Let us write down the equations that correspond to the tabular values

$$C(gr) + O_2(g) \rightarrow CO_2(g) \quad \Delta H_a = -393.51 \text{ kJ} \quad \text{(a)}$$
$$2Fe(s) + \tfrac{3}{2} O_2(g) \rightarrow Fe_2O_3(s) \quad \Delta H_b = -824.2 \text{ kJ} \quad \text{(b)}$$

We then look for a way to combine these two equations algebraically so that we obtain the desired equation, (17-20). Since Eq. (17-20) has 3 mol of CO_2 on the right-hand side we must multiply Eq. (a) by three. In the desired equation, 4 mol of Fe(s) are on the right-hand side, whereas in Eq. (b) there are 2 mol of Fe(s) on the left. We must therefore multiply Eq. (b) by two and *subtract* it from three times Eq. (a). Subtracting an equation is equivalent to reversing the direction of the arrow because it interchanges the right- and left-hand sides of the equation. Thus

$$\text{Eq. (17-20)} = 3\text{Eq. (a)} - 2\text{Eq. (b)}$$

If it is not clear to you that this is correct, take the trouble to write down each term as follows:

$$
\begin{array}{ll}
3\text{Eq. (a)} & 3C(gr) + 3\,O_2(g) \rightarrow 3CO_2(g) \\
-2\text{Eq. (b)} & 2Fe_2O_3(s) \rightarrow 4Fe(s) + 3\,O_2(g)
\end{array}
$$

and then add the equations written. Applying Hess' Law we therefore obtain for the ΔH^0 of Eq. (17-20)

$$
\begin{aligned}
\Delta H^0 &= 3\Delta H_a - 2\Delta H_b \\
&= 3\Delta H_f^0(CO_2) - 2\Delta H_f^0(Fe_2O_3) \quad (17\text{-}21) \\
&= 3(-393.51) - 2(-824.2) = +467.9 \text{ kJ}
\end{aligned}
$$

We can calculate ΔH^0 for any reaction for which the ΔH_f^0 values of all products and reactants are known. For the general reaction

$$\alpha A + \beta B \rightleftharpoons \gamma C + \delta D$$
$$\Delta H^0 = \gamma \Delta H_f^0(C) + \delta \Delta H_f^0(D) - \alpha \Delta H_f^0(A) - \beta \Delta H_f^0(B) \quad (17\text{-}22a)$$

Since C and D are the products of the reaction, and A and B are the reactants, Eq. (17-22a) is often written in a shorthand form as

$$\Delta H^0 = \sum \Delta H_f^0(\text{prod}) - \sum \Delta H_f^0(\text{react}) \quad (17\text{-}22b)$$

You must remember to use the appropriate coefficients, as in Eq. (17-22a), if you think of the equation in this form. Note that the units of ΔH_f^0 are kilojoules per mole, while the units of ΔH^0 are just kilojoules. The coefficients α, β, γ, and δ have units of moles in Eq. (17-22a).

For example, Eq. (17-22a) applied to reaction (17-20) is

$$\Delta H^0 = 3\Delta H_f^0(CO_2) + 4\Delta H_f^0(Fe) - 2\Delta H_f^0(Fe_2O_3) - 3\Delta H_f^0(C, gr)$$

Since both Fe(s) and C(gr) are elements in their standard states, $\Delta H_f^0(Fe)$ and $\Delta H_f^0(C, gr)$ are 0, so that we obtain

$$\Delta H^0 = 3\Delta H_f^0(CO_2) - 2\Delta H_f^0(Fe_2O_3)$$

which is just what we had obtained as Eq. (17-21).

The following examples further illustrate the use of Hess' Law.

EXAMPLE 17.4. The use of ΔH_f^0 values

Calculate ΔH^0 for the reaction

$$CH_4(g) + 3Cl_2(g) \rightarrow CHCl_3(liq) + 3HCl(g)$$

Utilizing Eq. (17-22a) we obtain

$$\Delta H^0 = 3\Delta H_f^0(HCl) + \Delta H_f^0(CHCl_3) - \Delta H_f^0(CH_4)$$

Note that since $Cl_2(g)$ is an element in its standard state, $\Delta H_f^0(Cl_2)$ is 0. Using the ΔH_f^0 values of Table 17.1,

$$\Delta H^0 = 3(-92.31) + (-134.5) - (-74.81)$$
$$= -276.93 - 134.5 + 74.81 = -336.6 \text{ kJ}$$

EXAMPLE 17.5. The thermite process

The reaction of a mixture of powdered aluminum with iron oxide, Fe_2O_3,

$$2Al + Fe_2O_3 \rightarrow 2Fe + Al_2O_3$$

is so exothermic that it is used for welding iron. The iron produced by the reaction is molten due to the amount of heat released, and can be directed into the space between the pieces that are to be welded. The use of this reaction is called the **thermite process**.

Calculate the amount of heat released when 10.00 g of powdered Al reacts with excess Fe_2O_3 in the thermite process.

Solution. We can calculate ΔH^0 for this reaction using Eq. (17-22a) and data from Appendix G. Since both aluminum and iron are elements and therefore have zero heats of formation, Eq. (17-22a) applied to this reaction yields

$$\Delta H^0 = \Delta H_f^0(Al_2O_3) - \Delta H_f^0(Fe_2O_3)$$
$$= -1676 - (-824.2) = -851 \text{ kJ}$$

When 2 mol of Al react in the thermite process 851 kJ of heat are released. The atomic weight of Al is 26.98 g·mol^{-1}. Hence 10.00 g is 0.3706 mol of Al. The amount of heat released when 10.00 g of powdered Al react in the thermite process is therefore

$$\left(\frac{851 \text{ kJ}}{2 \text{ mol Al}}\right)(0.3706 \text{ mol Al}) = 157.7 \text{ kJ}$$

Heats of Combustion

A Table of Standard Enthalpies of Combustion is often useful, particularly for organic substances. The heat of combustion of a compound containing only C, H, and O is defined as the enthalpy change when *one mole* of that compound combines with

Table 17.2. **Standard Enthalpies of Combustion at 25 °C, in kJ/mol**

C(gr)	−393.51	$C_6H_6(\ell)$	−3302.
$CH_4(g)$	−890.36	$CH_3COOH(\ell)$	−874.2
$C_2H_4(g)$	−1411.	$CH_3OH(\ell)$	−726.1
$C_2H_6(g)$	−1559.9	$C_2H_5OH(\ell)$	−1367.
$C_3H_8(g)$	−2220.1	$H_2(g)$	−285.84

sufficient oxygen to convert all the carbon to $CO_2(g)$ and all the hydrogen to *liquid* water.

Table 17.2 lists some heats of combustion of common substances. Note that enthalpies of combustion of organic compounds are negative, as these combustions are exothermic. Note also that the heat of combustion of graphite is identical to the heat of formation of gaseous CO_2, and the heat of combustion of $H_2(g)$ is identical to the heat of formation of liquid H_2O.

The following example demonstrates how one can use either a Table of Standard Heats of Formation or a Table of Standard Heats of Combustion to calculate a desired ΔH^0 value.

EXAMPLE 17.6. The use of heats of combustion

Calculate ΔH^0 for the reaction

$$2C(gr) + 2H_2O(\ell) \rightarrow CH_3COOH(\ell)$$

using both heats of combustion and heats of formation.

Solution. Using heats of combustion, we write the combustion equations corresponding to the values in Table 17.2 for C(gr) and $CH_3COOH(\ell)$, as follows:

$$CH_3COOH(\ell) + 2\,O_2(g) \rightarrow 2CO_2(g) + 2H_2O(\ell) \tag{a}$$
$$\Delta H_a = \Delta H^0_{comb}(CH_3COOH) = -874.2 \text{ kJ/mol}$$
$$C(gr) + O_2(g) \rightarrow CO_2(g) \quad \Delta H_b = \Delta H^0_{comb}(C,gr) = -393.51 \text{ kJ/mol} \tag{b}$$

The desired equation is 2Eq. (b) − Eq. (a). We have to multiply Eq. (b) by two in order to have 2 mol of C(gr) on the left-hand side, and we must subtract Eq. (a) in order to get CH_3COOH on the right-hand side. Thus, using Hess' Law

$$\Delta H^0 = 2\Delta H_b - \Delta H_a = 2(-393.51) - (-874.2) = +87.2 \text{ kJ}$$

Alternatively, one can use heats of formation. Applying Eq. (17-22a) we obtain

$$\Delta H^0 = \Delta H^0_f(CH_3COOH) - 2\Delta H^0_f(H_2O,\ell)$$
$$= -484.5 - 2(-285.84) = +87.2 \text{ kJ}$$

Clearly, it makes no difference which path is followed to obtain the desired reaction.

Heats of formation of organic compounds are frequently calculated from heats of combustion, because it is easier to obtain experimental values of heats of combustion than of heats of formation. Example 17.7 illustrates this type of calculation.

EXAMPLE 17.7. Calculation of a heat of formation from a heat of combustion

The heat of combustion of liquid ethanol, CH_3CH_2OH, is −1366.8 kJ/mol. Calculate the standard heat of formation of ethanol.

Solution. First write a correctly balanced equation for the combustion of ethanol:

$$CH_3CH_2OH(\ell) + 3\,O_2(g) \rightarrow 2CO_2(g) + 3H_2O(\ell)$$

Applying Eq. (17-22a) to this combustion reaction, we obtain

$$\Delta H^0_{comb} = 2\Delta H^0_f(CO_2) + 3\Delta H^0_f(H_2O,\ell) - \Delta H^0_f(CH_3CH_2OH,\ell)$$

The only unknown in this equation is $\Delta H^0_f(CH_3CH_2OH,\ell)$. Substitution yields

$$-1366.8 = 2(-393.51) + 3(-285.84) - \Delta H^0_f(CH_3CH_2OH,\ell)$$

Hence,

$$\Delta H^0_f(CH_3CH_2OH,\ell) = 2(-393.51) + 3(-285.84) + 1366.8 = -277.7 \text{ kJ/mol}$$

section 17.6
Bond Energies

In order to understand and to predict the reactivity of different molecules, chemists compare the strengths of various chemical bonds. A strong bond is one that requires a great deal of energy to break. It always takes energy to break bonds; bond breaking is an endothermic process. Since we will be considering reactions occurring at constant pressure, we should, speaking strictly, use the term **bond enthalpies,** but most often the phrase that is used is **bond energies.**

For a diatomic molecule, A—B, the bond energy is defined as the amount of energy required to break one mole of isolated, gaseous A—B molecules into isolated gaseous atoms A and B. Note that we must make the products of the bond-breaking individual gaseous atoms, so that no new bonds are formed. Similarly, the reactants must be isolated gaseous molecules so that we are not expending any energy to overcome the forces of attraction between molecules that exist in the liquid and solid phases. Thus the bond energy of the A—B bond is defined to be ΔH_{298} for the reaction

$$AB(g) \rightarrow A(g) + B(g) \tag{17-23}$$

The subscript "298" on ΔH indicates that the value is obtained at 25 °C (298.15 K). Specific examples of Eq. (17-23) are

$$H—\ddot{\underset{..}{C}l: \rightarrow H\cdot + \cdot \ddot{\underset{..}{C}l:} \qquad \Delta H_{298} = 432 \text{ kJ}$$

$$:N{\equiv}N: \rightarrow :\dot{N}\cdot + \cdot\dot{N}: \qquad \Delta H_{298} = 945 \text{ kJ}$$

The C—H Bond Energy

There are, of course, many bonds that never occur in a diatomic molecule. One of the most important of these is the C—H bond, which is found in almost all organic molecules. Chemists have found that a C—H bond has *approximately* the same strength in all molecules. One C—H bond is not identical with another C—H bond in a different molecule or in a different position in the same molecule. A C—H bond formed by the overlap of an sp^3 hybrid atomic orbital on C and the $1s$ atomic orbital of H (sp^3-s) is different from a C—H bond formed by the overlap of an sp^2 hybrid orbital on C and the $1s$ atomic orbital of H (sp^2-s). However, the variation in bond strengths of different C—H bonds is fairly small, and so it is useful to assign a value to the strength of the C—H bond.

It is customary to define the C—H bond energy as one quarter of the enthalpy change for the following dissociation reaction

$$CH_4(g) \rightarrow C(g) + 4H(g) \qquad (17\text{-}24)$$

so that the C—H bond energy is $\frac{1}{4}\Delta H$ for reaction (17-24).

When we dissociate a CH_4 molecule into a gaseous carbon atom and four gaseous hydrogen atoms, we have broken four C—H bonds. We must break all the bonds in the CH_4 molecule and make no new ones. The carbon produced by the dissociation of CH_4 cannot be graphite because there are bonds between the C atoms in graphite, and the products of the dissociation must have no bonds between them.

The dissociation reaction, Eq. (17-24), is difficult to carry out, but we can make use of Hess' Law to calculate its ΔH, using other reactions that we can carry out in the laboratory and for which ΔH values have been measured. The three reactions we can utilize are

$$H_2(g) \rightarrow 2H(g) \qquad (17\text{-}25)$$
$$C(gr) \rightarrow C(g) \qquad (17\text{-}26)$$
$$C(gr) + 2H_2(g) \rightarrow CH_4(g) \qquad (17\text{-}27)$$

Reaction (17-25) is the dissociation of H_2 molecules, and its ΔH is the bond energy of the H—H bond, often written as $D(H_2)$. The symbol D is used to represent a dissociation energy. The bond energy of the H—H bond has been measured very carefully:

$$D(H_2) = 435.9 \text{ kJ/mol} = 104.2 \text{ kcal/mol}$$

Reaction (17-26) is the sublimation of graphite. It is also the standard heat of formation of gaseous carbon. This ΔH value is difficult to measure experimentally, and it has been measured many times by many people because it is used so often in bond energy calculations. You may find different values in different references.

$$\Delta H_f^0(C, g) = \Delta H_{subl}(C, gr) = 716.7 \text{ kJ/mol} = 171.3 \text{ kcal/mol}$$

Reaction (17-27) is, of course, the standard enthalpy of formation of gaseous methane.

$$\Delta H_f^0(CH_4) = -74.81 \text{ kJ/mol} = -17.88 \text{ kcal/mol}$$

To obtain Eq. (17-24) from the three equations given, we must multiply Eq. (17-25) by two, add Eq. (17-26), and subtract Eq. (17-27). Thus

$$\text{Eq. (17-24)} = 2\text{Eq. (17-25)} + \text{Eq. (17-26)} - \text{Eq. (17-27)}$$

Using Hess' Law we obtain the desired ΔH as

$$\Delta H = 2D(H_2) + \Delta H_{subl}(C) - \Delta H_f^0(CH_4)$$
$$= 2(435.9) + 716.7 - (-74.81) = 1663.3 \text{ kJ}$$

The C—H bond energy is $\frac{1}{4}\Delta H = \frac{1}{4}(1663.3) = 415.8$ kJ $= 99.4$ kcal.

How reliable is this value? Will it take 415.8 kJ/mol to break a C—H bond in any organic molecule? The answer is no; the energy required to break a C—H bond in a specified organic molecule is likely to be within ± 10 kJ/mol of this value, but cannot be expected to be exactly 416 kJ/mol. There is clearly no point at all in citing a bond energy to four significant figures.

We must distinguish between a **bond dissociation energy**, the actual value of the energy required to break a specific bond in a polyatomic molecule, and the **bond energy**, which is an average value of the energy required to break similar bonds in a large number of different molecules. For instance, if we break only a single C—H bond in CH_4, obtaining a $\cdot CH_3$ fragment (called a **methyl radical**) plus a single H atom

$$CH_4(g) \rightarrow \cdot CH_3(g) + \cdot H(g) \tag{17-28}$$

it requires 423 kJ and not 416 kJ.

A table of bond energies lists a value obtained from averaging large numbers of calculated values for many molecules. For the C—H bond, we want to average not just the value obtained from methane, but also from ethane, C_2H_6, propane, C_3H_8, and so on. The value of the C—H bond energy usually given is 413 kJ/mol or 99 kcal/mol.

The C—C Bond Energy

A great many organic molecules have C—C single bonds; the simplest of these is ethane, C_2H_6, with structure

$$
\begin{array}{cc}
H & H \\
| & | \\
H-C-C-H \\
| & | \\
H & H
\end{array}
$$

The C—C bond in ethane is formed by the overlap of sp^3 hybrid atomic orbitals on each carbon. If we break all bonds in ethane,

$$C_2H_6(g) \rightarrow 2C(g) + 6H(g) \tag{17-29}$$

we have broken six C—H bonds and one C—C bond. We can calculate the enthalpy change for Eq. (17-29) by using Hess' Law and the three reactions:

$$H_2(g) \rightarrow 2H(g) \tag{17-25}$$
$$C(gr) \rightarrow C(g) \tag{17-26}$$
$$2C(gr) + 3H_2(g) \rightarrow C_2H_6(g) \tag{17-30}$$

The desired reaction, Eq. (17-29), can be obtained algebraically from the three above as

$$\text{Eq. (17-29)} = 3\text{Eq. (17-25)} + 2\text{Eq. (17-26)} - \text{Eq. (17-30)}$$

Therefore the enthalpy change for Eq. (17-29) is

$$\Delta H = 3D(H_2) + 2\Delta H_{subl}(C) - \Delta H_f^0(C_2H_6)$$
$$= 3(435.9) + 2(716.7) - (-84.67) = 2826 \text{ kJ/mol}$$

But $\Delta H = 6(\text{C—H bond energy}) + (\text{C—C bond energy})$. If we use 413 kJ for the C—H bond energy we obtain

$$\text{C—C bond energy} = 2826 - 6(413) = 348 \text{ kJ/mol} \quad \text{or} \quad 83 \text{ kcal/mol}$$

Tabulation of Bond Energies

The following examples illustrate how we build up a table of bond energies.

EXAMPLE 17.8. The O—H bond energy

Calculate the O—H bond energy as one half the enthalpy change for the dissociation reaction

$$H_2O(g) \rightarrow 2H(g) + O(g)$$

Solution. Using Hess' Law, ΔH for the above reaction can be obtained by combining the three reactions

$$H_2(g) \rightarrow 2H(g) \qquad\qquad D(H_2) = 435.9 \text{ kJ/mol}$$
$$O_2(g) \rightarrow 2\,O(g) \qquad\qquad D(O_2) = 495.0 \text{ kJ/mol}$$
$$H_2(g) + \tfrac{1}{2}O_2(g) \rightarrow H_2O(g) \qquad \Delta H_f^0(H_2O, g) = -241.83 \text{ kJ/mol}$$

The enthalpy change for the dissociation of water vapor is

$$\Delta H = D(H_2) + \tfrac{1}{2}D(O_2) - \Delta H_f^0(H_2O, g)$$
$$= 435.9 + \tfrac{1}{2}(495.0) + 241.83 = 925.2 \text{ kJ/mol}$$

The O—H bond energy is therefore $\tfrac{1}{2}(925.2) = 463$ kJ/mol

EXAMPLE 17.9. The C—O bond energy

Calculate the C—O bond energy from the enthalpy change for the dissociation reaction $CH_3CH_2OH(g) \rightarrow 2C(g) + 6H(g) + O(g)$, and the C—H, C—C, and O—H bond energies.

Solution. In order to calculate the ΔH value for the dissociation of ethanol into gaseous atoms we will need to use Hess' Law and the following four reactions:

$$H_2(g) \rightarrow 2H(g) \qquad\qquad D(H_2) = 435.9 \text{ kJ/mol}$$
$$O_2(g) \rightarrow 2\,O(g) \qquad\qquad D(O_2) = 495.0 \text{ kJ/mol}$$
$$C(gr) \rightarrow C(g) \qquad\qquad \Delta H_{subl}(C) = 716.7 \text{ kJ/mol}$$
$$2C(s) + 3H_2(g) + \tfrac{1}{2}O_2(g) \rightarrow CH_3CH_2OH(g) \qquad \Delta H_f^0(C_2H_5OH, g) = -235.4 \text{ kJ/mol}$$

From Hess' Law we obtain ΔH for $CH_3CH_2OH(g) \rightarrow 2C(g) + 6H(g) + O(g)$

$$\Delta H = 2\Delta H_{subl}(C) + 3D(H_2) + \tfrac{1}{2}D(O_2) - \Delta H_f^0(C_2H_5OH, g)$$
$$= 2(716.7) + 3(435.9) + \tfrac{1}{2}(495.0) - (-235.4) = 3224 \text{ kJ}$$

The ethanol molecule, CH_3CH_2OH, has the structure

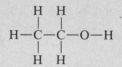

so that if we dissociate ethanol completely into individual atoms we have broken five C—H bonds, one C—C bond, one O—H bond, and one C—O bond. Using 413 kJ for the C—H bond energy, 348 kJ for the C—C bond energy, and 463 kJ for the O—H bond energy, we obtain

$$C\text{—}O \text{ bond energy} = 3224 - 5(413) - 348 - 463 = 348 \text{ kJ}$$

The value usually listed as the C—O single bond energy, obtained by averaging calculated values for many different molecules, is 351 kJ.

By compiling and combining data of this sort, and averaging values for many molecules, we can arrive at a Table of Single Bond Energies, such as Table 17.3. A table of this sort is very useful, but you should always bear in mind that bond energies are not reliable to more than ±5 or ±10 kJ.

Bond energies can be used to obtain approximate values for ΔH for some reactions that may not be measurable. While this is often useful, the uncertainty in bond energy values can cause such calculated values to have significant errors. The following example illustrates the type of calculation that can be made.

Table 17.3. Selected Single Bond Energies in kJ/mol

	H	C	N	O	F	Cl	Br	I	S
H	436								
C	413	348							
N	391	292	161						
O	463	351	175	139					
F	563	441	270	212	158				
Cl	432	328	200	210	251	243			
Br	366	276	...	217	249	218	193		
I	299	240	...	241	281	210	178	151	
S	339	259	...	...	...	277	239	...	266

EXAMPLE 17.10. The use of bond energies

(a) Estimate ΔH_{298} for the isomerization reaction

$$CH_3—O—CH_3(g) \rightarrow CH_3CH_2OH(g)$$

from bond energy values.

Solution. Dimethyl ether, $CH_3—O—CH_3$, has a total of six C—H bonds and two C—O bonds. The amount of energy required to break all bonds in dimethyl ether is therefore $6(413) + 2(351) = 3180$ kJ per mole of $CH_3—O—CH_3$. Ethanol, CH_3CH_2OH, has five C—H bonds, one C—C bond, one O—H bond, and one C—O bond. The amount of energy required to break all bonds is $5(413) + 348 + 463 + 351 = 3227$ kJ/mol.

The isomerization reaction is the sum of the two equations:

$$CH_3—O—CH_3 \rightarrow 2C(g) + 6H(g) + O(g) \qquad \Delta H_1 = 3180 \text{ kJ} \qquad (1)$$
$$2C(g) + 6H(g) + O(g) \rightarrow CH_3CH_2OH(g) \qquad \Delta H_2 = -3227 \text{ kJ} \qquad (2)$$

Therefore the desired $\Delta H = 3180 - 3227 = -47$ kJ.

(b) Calculate ΔH_{298} for the isomerization reaction of part **(a)** using standard heats of formation, and compare it with the estimated value. What is the percentage error of the estimated value?

Solution. Using Eq. (17-22a), the desired ΔH can be written as

$$\Delta H = \Delta H_f^0(CH_3CH_2OH, g) - \Delta H_f^0(CH_3OCH_3, g)$$

The ΔH_f^0 values can be found in Appendix G.

$$\Delta H = -235.1 - (-184.0) = -51.1 \text{ kJ}$$

The estimated value is in error by a little more than 4 kJ. This is certainly as good as can be expected in view of the limits of reliability on bond energy values. The percentage error of the estimate is given by $\left(\dfrac{-47 + 51.1}{51.1} \right) \times 100 = 8.0\%$.

The energies of multiple bonds can be obtained using exactly the same methods we have used for single bonds. Table 17.4 lists some values of multiple bond energies.

Table 17.4. **Multiple Bond Energies in kJ/mol**

C=C	615	C≡C	812
O=O	495	C≡N	891
C=O	728	N≡N	945

section 17.7
Heat Capacity

The **molar heat capacity** of any substance is the amount of heat required to raise the temperature of one mole of that substance one degree, either Celsius or Kelvin, since the size of a degree on these two temperature scales is the same. The units of molar heat capacity are therefore joules per mole per kelvin ($J \cdot mol^{-1}K^{-1}$). [In older literature you will find the molar heat capacity reported in units of calories per mole per kelvin ($cal \cdot mol^{-1}K^{-1}$).]

The **specific heat** of any substance is the amount of heat required to raise the temperature of one gram of that substance one kelvin. The units of specific heat are therefore joules per gram per kelvin ($J \cdot g^{-1}K^{-1}$).

The distinction between the molar heat capacity and the specific heat is illustrated in the following example.

EXAMPLE 17.11. Converting from specific heat to molar heat capacity

Values of the specific heat at 25 °C in calories per gram per kelvin ($cal \cdot g^{-1}K^{-1}$) are given below for several substances.

Al 0.215   Cu 0.0924   Ag 0.0566   Pb 0.0305 H_2O 1.00

For each of these substances, calculate the molar heat capacity in joules per mole per kelvin, and the amount of heat required to raise the temperature of 1 mol from 20.0 to 50.0 °C.

Solution. We will do the calculations in detail for aluminum, and then tabulate the results for all five substances.

To convert from calories to joules, multiply by 4.1840. To convert from specific heat to molar heat capacity, multiply by the molecular weight (atomic weight for elements). The atomic weight of Al is 26.98 g/mol. Hence

$$\frac{\text{molar heat}}{\text{capacity of Al}} = \left(0.215 \frac{cal}{g \cdot K}\right)\left(26.98 \frac{g}{mol}\right)\left(4.1840 \frac{J}{cal}\right) = 24.3 \text{ J} \cdot mol^{-1}K^{-1}$$

For a temperature change of 30°, the amount of heat required is

$$\text{heat} = \left(24.27 \frac{J}{mol \cdot K}\right)(1 \text{ mol})(30°) = 728 \text{ J}$$

	Specific Heat ($cal \cdot g^{-1}K^{-1}$)	*gfw* (g/mol)	*Heat Capacity* ($J \cdot mol^{-1}K^{-1}$)	*Heat Needed to Raise Temperature 30°, (J)*
Al	0.215	26.98	24.3	7.28×10^2
Cu	0.0924	63.546	24.6	7.38×10^2
Ag	0.0566	107.87	25.5	7.66×10^2
Pb	0.0305	207.2	26.4	7.93×10^2
H_2O	1.00	18.02	75.4	2.26×10^3

If we think about the definitions of specific heat and molar heat capacity, and remember that heat is a path-dependent quantity, it is clear that in order to define what is meant by "heat" we must specify the path used. For the two common paths of reaction, there are two kinds of heat capacities. The molar heat capacity at constant pressure, C_p, is the amount of heat required to raise the temperature of one mole of a substance one kelvin when the pressure is maintained constant, and the molar heat capacity at constant volume, C_v, is the amount of heat required to raise the temperature of one mole of a substance one degree when the volume is maintained constant.

For solids and liquids the difference between C_p and C_v is very small because even large pressure changes cause very small changes in volume for solids and liquids. For gases, however, there is a significant difference between C_p and C_v.

If we go into the laboratory and measure the heat capacity of a specific substance, we find that the amount of heat required to raise the temperature of that substance from 20 to 21 °C is not exactly the same as the amount of heat required to raise the temperature some other 1° interval, say from 46 to 47 °C. The two values will be close to one another, but they will not be identical. Experimentally we find that heat capacity is a temperature-dependent function, but it varies slowly with temperature. If we raise the temperature of 1 mol of a substance a few degrees at constant pressure, we would calculate C_p as

$$C_p = \frac{q_p}{\Delta T} = \frac{\Delta H}{\Delta T} \tag{17-31}$$

Similarly, if we raise the temperature of 1 mol of a substance a few degrees at constant volume, we would calculate C_v as

$$C_v = \frac{q_v}{\Delta T} = \frac{\Delta E}{\Delta T} \tag{17-32}$$

Because the heat capacity does vary over the temperature range ΔT, Eqs. (17-31) and (17-32) are not exactly correct,* but unless ΔT is fairly large (usually this means larger than 25°) the error made using these equations is negligible.

The Relation between C_p and C_v

Since the definition of enthalpy, Eq. (17-14), is $H = E + PV$, when the temperature is changed the change in the enthalpy is

$$\Delta H = \Delta E + \Delta(PV) \tag{17-33a}$$

or

$$H_2 - H_1 = E_2 - E_1 + P_2V_2 - P_1V_1 \tag{17-33b}$$

If we divide by the temperature change, ΔT, we obtain

$$\frac{\Delta H}{\Delta T} = \frac{\Delta E}{\Delta T} + \frac{\Delta(PV)}{\Delta T} \tag{17-34a}$$

or

$$C_p = C_v + \frac{\Delta(PV)}{\Delta T} \tag{17-34b}$$

For 1 mol of any gas, assuming the ideal gas approximation is valid, $P_2V_2 = RT_2$

---

* The exact definitions of C_p and C_v necessitate making the temperature interval, ΔT, infinitesimally small. Thus C_p and C_v are defined as first derivatives: $C_p = (\partial H/\partial T)_P$ and $C_v = (\partial E/\partial T)_V$.

and $P_1V_1 = RT_1$, so that $\Delta(PV) = R(T_2 - T_1) = R\Delta T$. Equation (17-34b) therefore becomes

$$\boxed{C_p = C_v + R} \tag{17-35}$$

This equation is valid for real gases to within the ideal gas approximation, that is, to an error of $\pm 1\%$ or less at room temperature and 1-atm pressure.

For solids and liquids the volume change is negligibly small for moderate temperature changes. Therefore $\Delta(PV) \sim 0$ and $C_p = C_v$.

EXAMPLE 17.12. The difference between C_p and C_v

The C_p values at 25 °C in joules per mole per kelvin are given for the following substances: Ar(g) 20.8, N_2(g) 29.3, $H_2O(\ell)$ 75.4, and Pb(s) 26.4. What is C_v for each of these substances?

Solution. For liquids and solids $C_p = C_v$, so that for $H_2O(\ell)$ and Pb(s), C_v is the same as C_p. For the gases, $C_v = C_p - R$, and R is $8.314 \text{ J} \cdot \text{mol}^{-1}\text{K}^{-1}$. Values of C_p and C_v for these four substances are, therefore,

Substance	C_p $(J \cdot mol^{-1}K^{-1})$	C_v $(J \cdot mol^{-1}K^{-1})$
Ar(g)	20.8	12.5
N_2(g)	29.3	21.0
$H_2O(\ell)$	75.4	75.4
Pb(s)	26.4	26.4

Molecular Origins of Heat Capacity

Suppose we have a flask containing 1 mol of gaseous dimethyl ether, CH_3OCH_3. We place the flask in a beaker of hot water and heat is absorbed by the dimethyl ether. What happens to the energy added to the ether? Some of it goes to increase the average translational kinetic energy of the ether molecules; on the average they are moving faster than they were at the lower temperature. This increase in the average translational kinetic energy is manifested in a rise in temperature of the ether. Some of the added energy also goes into increasing the vibrational and rotational kinetic energy of the CH_3—O—CH_3 molecules. Each C—H and C—O bond vibrates, the —CH_3 groups rotate around the C—O bonds, and the molecule bends and is deformed in a number of ways. The more possibilities there are for internal vibration and rotation in a molecule, the more energy is required to raise the temperature of 1 mol of the substance 1 K. Figure 17.6 illustrates rotation and vibration in a diatomic molecule.

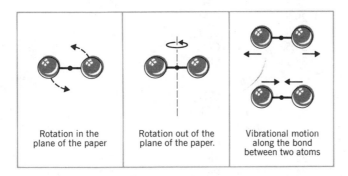

Rotation in the plane of the paper	Rotation out of the plane of the paper.	Vibrational motion along the bond between two atoms

Fig. 17.6. Internal rotations and vibration for a diatomic molecule.

A monatomic gas such as He, Ar, or Ne has no vibrational or rotational kinetic energy. The only kinetic energy possible for a monatomic gas is translational energy. In Chapter 4, when we discussed the kinetic theory of gases, we obtained the expression for the molar translational kinetic energy of a gas:

$$E_{trans} = \tfrac{3}{2} RT \qquad (17\text{-}36)$$

If the temperature of a monatomic gas is raised from T_1 to T_2, the change in its internal energy is

$$\Delta E = E_2 - E_1 = \tfrac{3}{2} RT_2 - \tfrac{3}{2} RT_1 = \tfrac{3}{2} R\Delta T \qquad (17\text{-}37)$$

Therefore, for a monatomic gas

$$C_v = \frac{\Delta E}{\Delta T} = \frac{3}{2} R = 12.5 \text{ J} \cdot \text{mol}^{-1}\text{K}^{-1} \qquad (17\text{-}38)$$

Using Eq. (17-35) we obtain for the heat capacity at constant pressure for a monatomic gas,

$$C_p = \tfrac{3}{2} R + R = \tfrac{5}{2} R = 20.8 \text{ J} \cdot \text{mol}^{-1}\text{K}^{-1} \qquad (17\text{-}39)$$

These values are in accord with experimental measurements.

The heat capacities of polyatomic molecules are larger than those of monatomic gases, and the larger and more complex the molecule, the more possibilities there are for internal vibrations and rotations, therefore the larger the heat capacity. It takes 29.3 J to raise the temperature of 1 mol of N_2 one kelvin at 1-atm pressure and room temperature, whereas for CO_2 it takes 37.2 J and for ethane, C_2H_6, it takes 52.9 J. Table 17.5 lists the values of the molar heat capacities at constant pressure for selected substances.

A calorie is now defined to be exactly 4.1840 J, but was originally defined in terms of the specific heat of liquid water, as the amount of heat required to raise the temperature of 1 g of H_2O from 14.5 to 15.5 °C. The original calorie is very close to, but not exactly, 4.1840 J. If 1 cal is required to raise the temperature of 1 g of liquid water 1 K, then the molar heat capacity of liquid water is 18 cal $\cdot$ mol^{-1}K^{-1} or 75 J $\cdot$ mol^{-1}K^{-1}.

Table 17.5. **Molar Heat Capacities at Constant Pressure, in J $\cdot$ mol^{-1}K^{-1} at Room Temperature**

		Gases		
He	20.8		NH_3	35.6
CO	29.3		CH_4	35.1
N_2	29.3		C_2H_6	52.9
Cl_2	33.9		C_6H_6	82.2
H_2O	33.5		CCl_4	83.3
CO_2	37.2		SF_6	121.
		Liquids		
H_2O	75		C_6H_6	136
CH_3OH	82		CCl_4	133
		Solids		
Ice	37.7		Cu	24.6
Ag	25.5		Fe	24.8
Al	24.3		Pb	26.4

The molar heat capacities of solids listed in Table 17.5 include the values calculated in Example 17.11. Note that the molar heat capacities of the five solid *elements* in Table 17.5 are quite similar. This observation was first made by Pierre Dulong and Alexis Petit in 1819. In SI units the **Law of Dulong and Petit** states that the molar heat capacity of solid elements is 25 ± 1 J $\cdot$ mol^{-1}K^{-1}. While Dulong and Petit made measurements only on metallic elements, experiments conducted since 1819 show that the application of this rule is not limited to elements. At room temperature (~ 300 K) the heat capacity of many (but not all) solids, elements, and compounds, is very close to 25 J $\cdot$ K^{-1} *per mole of atoms*. A theoretical study of the vibrational energy of atoms in solids predicts $C_v = 3R = 25$ J $\cdot$ K^{-1} per mole of atoms provided the temperature is sufficiently high.

section 17.8
Calorimetry

In order to determine ΔE for a reaction, we carry out the reaction in a **bomb calorimeter,** shown in Fig. 17.7. A bomb calorimeter has thick, well-insulated walls to prevent heat exchange with the surroundings. The walls are designed to withstand large differences of pressure inside and out, and therefore to maintain a constant volume.

The thermodynamic system consists of the chemical reagents and products, the calorimeter itself, the thermometer, stirrer, and all contents of the calorimeter. The surroundings consist of the lab bench on which the instrument sits, the air in the laboratory, the person doing the experiment, and so on.

A well designed calorimeter has walls so well insulated that heat leakage to the surroundings is negligibly small. No insulation is perfect, so that heat leakage can never be zero, but if the reaction is complete in a fairly short period of time, heat

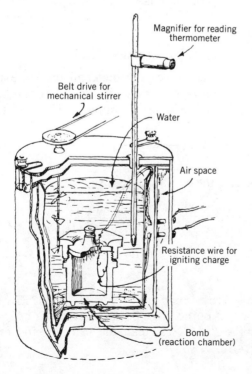

Magnifier for reading thermometer

Belt drive for mechanical stirrer

Water

Air space

Resistance wire for igniting charge

Bomb (reaction chamber)

Fig. 17.7. A bomb calorimeter. [From L. Pauling and P. Pauling, *Chemistry,* W. H. Freeman and Co. (1975).]

leakage can be kept to a value too small to be observed. For best results in calorimetry, the reactions to be measured should be fast, should yield definite products and not involve any side reactions, and should go to completion, that is, should have a very large equilibrium constant. Combustion reactions fit these specifications admirably, and the combustion of organic substances containing only C, H, and O to yield $CO_2(g)$ and liquid H_2O is a reaction well suited for calorimetric measurement.

What we measure in a calorimeter is ΔT, the change in temperature that occurs when the reaction takes place. Combustion of an organic compound is always exothermic, and therefore ΔT is positive; the temperature increases. If n moles of a compound are burned, and ΔE is the molar energy of combustion, the heat released is $-n\Delta E$, because the volume is constant in a bomb calorimeter. (Recall Eq. 17-10.) Remember that ΔE is the heat absorbed at constant volume, so to obtain the heat released we must change the sign.

What happens to the heat released in the combustion reaction? It causes the temperature of the entire calorimeter and its contents to increase. The magnitude of the increase depends on the heat capacity of the calorimeter. Since the volume is constant, we use Eq. (17-32) for the relation between the heat absorbed by the calorimeter and its contents and the observed temperature increase, ΔT:

$$C_v(\text{calorimeter}) = -n\Delta E/\Delta T \qquad (17\text{-}40)$$

Because no two calorimeters are identical, the heat capacity of the calorimeter must be determined by combusting a known quantity of a substance, called the calibration standard, for which the heat of combustion is already known. Thus:

$$C_v(\text{calorimeter}) = -(n_{std}\Delta E_{std})/\Delta T_{std} \qquad (17\text{-}41)$$

Sucrose, $C_{12}H_{22}O_{11}$, and benzoic acid, C_6H_5COOH, are commonly used as standard substances to calibrate a calorimeter.

Once the heat capacity of the calorimeter has been determined, a measured quantity of a substance for which the heat of combustion is unknown can be burned in the same calorimeter. The heat of combustion at constant volume of the unknown is determined using the relation

$$-n_{unk}\Delta E_{unk} = C_v(\text{calorimeter})\Delta T_{unk} \qquad (17\text{-}42)$$

Note that the units of the heat capacity of the calorimeter are just joules per kelvin (a calorimeter is not a compound; we need the heat capacity of the entire calorimeter). The units of each side of Eq. (17-42) are joules. Example 17.13 illustrates the use of Eqs. (17-41) and (17-42).

EXAMPLE 17.13. Calorimetric calculations

(a) Write a balanced equation for the combustion of sucrose.

Solution. Sucrose is $C_{12}H_{22}O_{11}$. The combustion reaction is

$$C_{12}H_{22}O_{11}(s) + 12\,O_2(g) \rightarrow 12CO_2(g) + 11H_2O(\ell)$$

In balancing the oxygen atoms, don't forget the oxygen that is contained in the sucrose.

(b) The heat of combustion of sucrose at constant volume is -5647 kJ/mol at 25 °C. A 2.0026-g sample of sucrose was burned in a bomb calorimeter at 25 °C and the temperature was observed to rise 2.966°. Calculate the heat capacity of the calorimeter.

Solution. The molecular weight of sucrose is 342.30. Thus the sample consists of $(2.0026 \text{ g})/(342.30 \text{ g} \cdot \text{mol}^{-1}) = 5.8504 \times 10^{-3}$ mol of sucrose. The amount of heat released by the combustion of this sample of sucrose is $+(5.8504 \times 10^{-3})(5647) \text{ kJ} = 33.04 \text{ kJ}$. The release of this much heat causes the temperature of the calorimeter and its contents to rise $2.966°$, so that

$$C_v(\text{calorimeter}) = \frac{(5.8504 \times 10^{-3})(5647)}{2.966} = 11.14 \text{ kJ/K}$$

(c) In the same calorimeter, 0.7928 g of benzene, C_6H_6, was burned at 25 °C, and the temperature was observed to rise $3.008°$. Write a balanced equation for the combustion of benzene and calculate ΔE for this combustion.

Solution

$$C_6H_6(\ell) + \tfrac{15}{2} O_2(g) \rightarrow 6CO_2(g) + 3H_2O(\ell)$$

The molecular weight of benzene is 78.11, so that the sample of benzene contains $(0.7928 \text{ g})/(78.11 \text{ g} \cdot \text{mol}^{-1}) = 1.015 \times 10^{-2}$ mol of benzene. Since the same calorimeter is used, $\Delta E(C_6H_6)$ can be calculated using Eq. (17-42):

$$-(1.015 \times 10^{-2} \text{ mol})\Delta E(C_6H_6) = (11.14)(3.008) \text{ kJ}$$

so that $\Delta E(C_6H_6) = -3301 \text{ kJ/mol}$.

The fuel value, or calorie content, of foods is obtained by combusting the food in a bomb calorimeter. What nutritionists call a calorie is the scientific kilocalorie and is usually symbolized with a capital C, as Cal. It is sometimes referred to as a "large calorie." Since foods are not pure substances the heat of combustion (popularly referred to as the calorie content) is reported in Cal/g or Cal/100 g. The following example illustrates how the calorie content of foods is established.

EXAMPLE 17.14. Calorie content of foods

A slice of a banana weighing 2.712 g was burned in a bomb calorimeter and produced a temperature rise of $3.05°$. In the same calorimeter, combustion of a 0.316-g sample of benzoic acid, C_6H_5COOH, produced a temperature rise of $3.24°$. The heat of combustion at constant volume of benzoic acid is -3227 kJ/mol. If an average banana weighs 125 g, how many large calories (kcal or Cal) can be obtained from one (average) banana?

Solution. The molecular weight of benzoic acid is 122.12. Thus the amount of heat released by burning 0.316 g of benzoic acid is

$$\left(\frac{0.316}{122.12} \frac{\text{g}}{\text{g} \cdot \text{mol}^{-1}}\right)\left(3227 \frac{\text{kJ}}{\text{mol}}\right)\left(\frac{1}{4.184} \frac{\text{kcal}}{\text{kJ}}\right) = 1.996 \text{ kcal}$$

Since the same calorimeter is used, the heat capacity of the calorimeter is the same in both experiments, so that

$$C_v(\text{calorimeter}) = \frac{1.996}{3.24} \frac{\text{kcal}}{\text{K}} = -\frac{(2.712 \text{ g})\Delta E(\text{banana})}{3.05 \text{ K}}$$

Hence,

$$\Delta E(\text{banana}) = -\left(\frac{3.05}{3.24}\right)\left(\frac{1.996}{2.712} \frac{\text{kcal}}{\text{g}}\right) = -0.693 \text{ kcal/g}$$

The calorie content of a banana weighing 125 g is therefore

$$(125 \text{ g})(0.693 \text{ Cal/g}) = 87 \text{ Cal}$$

Summary

The system we study is separated from the rest of the universe (the surroundings) by clearly defined boundaries. The state of the system is defined by specifying the value of a number of properties of the system. A property that has a fixed and definite value for each state of the system is called a state function. The value of a state function is independent of the way the state is attained.

When the state of a system is changed, the value of the change in any state function, X, is denoted ΔX, and is defined as the value of the property X in the final state (state 2) minus the value of the property X in the initial state (state 1): $\Delta X = X_2 - X_1$.

For systems open to the pressure of the atmosphere, a change of state is almost always accompanied by a volume change, and therefore work of expansion against the confining atmosphere is done. For a constant external pressure, P_{ext}, the work of expansion is $-P_{ext}\Delta V$.

The temperature of a system is a state function. Temperature is a measure of the average kinetic energy of the molecules. Heat is *not* a state function. It is not a property of the system, but is energy in transit between system and surroundings as a result of a difference in temperature between the system and its surroundings.

Both heat and work are forms of energy that can enter the system from the surroundings, or leave the system and enter the surroundings. The direction of flow into or out of the system is indicated by the sign of the heat, q, or the work, w. The conventions used in this text are the following:

> Heat flowing into a system is positive.
> Heat leaving a system is negative.
> Work entering a system is positive.
> Work leaving a system is negative.

The first law of thermodynamics is also called the law of conservation of mass and energy. It states that in a system of constant mass, energy can neither be created nor destroyed. In reactions that do not involve nuclear changes, any mass change is too small to be detected and the total energy of the universe remains constant. The energy of a given system may change as heat and/or work enter or leave the system. The change in energy of the system is then given by $\Delta E = q + w$.

If the volume of a system is constant, and no electrical or mechanical work is done, $w = 0$, and the first law becomes $\Delta E = q_v$, the heat of reaction at constant volume.

For systems at constant pressure, it is useful to define a new state function, the enthalpy, as $H = E + PV$. The heat of reaction at constant pressure, q_p, is equal to ΔH.

Hess' Law of Constant Heat Summation states that the heat absorbed when a given reaction is carried out at constant pressure and temperature is the same whether the reaction occurs in one step or in several steps. Using a Table of Standard Enthalpies of

Formation and Hess' law, we can calculate the value of ΔH for any reaction. The **standard enthalpy of formation** of a compound is ΔH for the reaction in which one mole of the compound is formed at 25 °C and 1-atm pressure from the elements in their **standard states**. The standard states of the elements are the most stable (common) forms at 25 °C and 1-atm pressure. Provided that the ΔH_f^0 value of substances A, B, C, and D are all known, ΔH^0 for the general reaction

$$\alpha A + \beta B \rightleftharpoons \gamma C + \delta D$$

can be calculated from

$$\Delta H^0 = \gamma \Delta H_f^0(C) + \delta \Delta H_f^0(D) - \alpha \Delta H_f^0(A) - \beta \Delta H_f^0(B)$$

Combustion, particularly of organic molecules, is another important type of reaction for which ΔH^0 values are tabulated. By definition, the **heat of combustion** of a compound containing only C, H, and O is ΔH for the reaction in which one mole of the compound combines with sufficient oxygen to convert all the carbon to $CO_2(g)$ and all the hydrogen to $H_2O(\ell)$.

The **bond energy** of a diatomic molecule, A—B, is defined as ΔH for the dissociation of one mole of gaseous AB molecules into gaseous A and B atoms.

Many bonds occur only in polyatomic molecules. The C—H bond energy is defined as one quarter of the ΔH value for the complete dissociation of methane: $CH_4(g) \rightarrow C(g) + 4H(g)$. The C—C bond energy can then be obtained by assuming the C—H bond energy is the same in ethane and methane, and calculating the amount of energy required to completely dissociate ethane: $C_2H_6(g) \rightarrow 2C(g) + 6H(g)$. By considering a great many reactions of this sort a table of bond energies can be compiled.

The **specific heat** of a substance is the amount of heat required to raise the temperature of one gram of the substance one degree, either Celsius or Kelvin. The **molar heat capacity** is the amount of heat required to raise the temperature of one mole of a substance one kelvin. Because heat is a path-dependent quantity, there are two different heat capacities, the **molar heat capacity at constant pressure**, C_p, and the **molar heat capacity at constant volume**, C_v. For solids and liquids $C_p = C_v$, but for gases $C_p = C_v + R$.

Adding heat to a sample of a compound increases the average translational kinetic energy of the molecules. For substances other than monatomic gases, adding heat also increases the internal vibrational and rotational kinetic energy of the molecules. The larger and more complex the molecule, the more possibilities there are for internal vibrations and rotations, and the more energy is required to raise the temperature of 1 mol of the compound 1°.

The molar heat capacity of solid elements at room temperature and above is usually quite close to $25 \text{ J} \cdot \text{mol}^{-1}\text{K}^{-1}$. This observation was first made in 1819 and is known as the **Law of Dulong and Petit**. The molar heat capacity of solid compounds at room temperature and above is approximately $25 \text{ J} \cdot \text{K}^{-1}$ *per mole of atoms*.

Heats of reaction at constant volume are obtained by using a **bomb calorimeter**. In particular, heats of combustion are determined in a bomb calorimeter. The heat capacity of the calorimeter is first obtained by burning a known amount of some standard substance for which the heat of combustion is known, and observing the rise in temperature. Once the heat capacity of the calorimeter is determined, the heat of combustion of some other compound can be obtained by measuring the temperature rise when a known amount of that compound is burned. The calorie content or fuel value of foods is obtained in this way.

Exercises

Use Appendix G for ΔH_f^0 values.

Sections 17.1 and 17.2

1. A sample of an ideal gas is contained in a cylinder sealed by a piston. The gas is initially at a pressure of 1.00 atm and occupies a volume of 3.00 L. Its temperature is kept constant at 20 °C. If the gas is compressed to a final pressure of 2.50 atm by a constant external pressure on the piston of 2.50 atm, how much work is done on the gas by the surroundings? How much work is done by the gas?

2. Two glass bulbs of equal volume are connected by a stopcock. Initially one bulb is evacuated and the other bulb is filled with an ideal gas at 2.40-atm pressure. Calculate the final pressure of the gas, and the work done by the gas on expanding, when the stopcock is opened.

Section 17.3

3. A gas expands against a constant external pressure of 2.00 atm from an initial volume of 1.40 L to a final volume of 5.90 L. The container is well insulated so that no heat enters or leaves the system. Calculate the change in energy, ΔE, of the gas. Explain why the temperature of the gas falls.

4. Two glass bulbs of equal volume are connected by a stopcock. One bulb is evacuated; the other is filled with an ideal gas at 2.40-atm pressure. The bulbs are insulated so that no heat enters or leaves the system. The stopcock between the two bulbs is opened. What is the change in energy, ΔE, of the gas when it expands into the evacuated bulb? Does the temperature of the gas increase, decrease, or remain the same? Explain briefly.

Sections 17.4 and 17.5

5. Write the equations for the reactions for which the change in enthalpy is ΔH_f^0 of (a) $Cu(OH)_2(s)$, (b) $Na_2SO_3(s)$, (c) $N_2O_5(g)$, (d) $CH_3COOH(\ell)$, and (e) $KMnO_4(s)$.

6. For which of the following is ΔH_f^0 zero? Explain your answers. (a) $Cd(s)$, (b) $Cl(g)$, (c) $Hg(s)$, (d) $Li(s)$, (e) $I_2(g)$, (f) $S(\ell)$, (g) $N_2(g)$, and (h) $H_2O(\ell)$.

7. Calculate ΔH^0 at 25 °C and 1 atm for the following reactions:
   (a) $Ca(s) + C(gr) + \frac{3}{2}O_2(g) \rightleftharpoons CaCO_3(s)$
   (b) $CaO(s) + CO_2(g) \rightleftharpoons CaCO_3(s)$

8. Calculate ΔH^0 at 25 °C and 1 atm for the following reactions:
   (a) $2CH_4(g) \rightleftharpoons C_2H_6(g) + H_2(g)$
   (b) $C_4H_{10}(g) + H_2 \rightleftharpoons 2C_2H_6(g)$

9. Calculate ΔH^0 at 25 °C and 1 atm for the following reactions:
   (a) $2Ag_2S(s) + 2H_2O(\ell) \rightleftharpoons 4Ag(s) + 2H_2S(g) + O_2(g)$
   (b) $Al_2O_3(s) + 3CO(g) \rightleftharpoons 3CO_2(g) + 2Al(s)$

10. Calculate the standard heat of formation of pentane gas, C_5H_{12}, if its standard heat of combustion is -3536 kJ/mol.

11. Calculate ΔH_f^0 of ethyl acetate, $CH_3COOCH_2CH_3(\ell)$, if its standard heat of combustion is -2231 kJ/mol.

Section 17.6

12. Using Table 17.3 calculate $\Delta H_f^0(F, g)$ and $\Delta H_f^0(Cl, g)$.

13. Explain why it is not possible to calculate $\Delta H_f^0(N, g)$ using only data from Table 17.3.

14. Calculate the C—O bond energy in CH_3OH from the enthalpy change for the

dissociation reaction $CH_3OH(g) \rightarrow C(g) + 4H(g) + O(g)$ and the C—H and O—H bond energies.

15. (a) Calculate the C=C bond energy from the enthalpy change for the dissociation of ethylene, $C_2H_4(g)$, into gaseous C and H atoms, and the C—H bond energy.
    (b) Explain why the C=C bond energy is not twice the C—C single bond energy.

16. (a) Using bond energies estimate ΔH^0 for the reaction

$$CH_4(g) + 3Cl_2(g) \rightleftharpoons CHCl_3(g) + 3HCl(g)$$

(b) Calculate ΔH^0 for the reaction of part (a) from a table of ΔH_f^0 values. By what percentage is the estimate of part (a) in error?

Section 17.7

17. The specific heats of the elements at 25 °C in calories per gram per kelvin $(cal \cdot g^{-1}K^{-1})$ are tabulated in the Handbook of Chemistry and Physics. Some of the values listed are Au 0.0308, C(gr) 0.170, Nd 0.0453, and Pd 0.0583. Calculate the molar heat capacities of these elements in joules per mole per kelvin $(J \cdot mol^{-1}K^{-1})$. Do all of these elements obey the Law of Dulong and Petit?

18. How much heat is required to raise the temperature of 1 mol of (a) $C_2H_6(g)$, and (b) $N_2(g)$, from 18 to 40 °C at a constant pressure of 1 atm? Account for the difference between these two values.

19. How much heat is required to raise 10.0 g of $CO_2(g)$ from 20 to 35 °C, (a) at constant pressure and (b) at constant volume? Account for the difference between the two values.

20. How much heat is required to raise the temperature of 0.500 mol of Ar gas from 20.0 to 80.0 °C at constant pressure? Calculate ΔH and ΔE for this process, assuming Ar is an ideal gas.

Section 17.8

21. The heat of combustion of sucrose at constant volume is -5647 kJ/mol at 25 °C. A 1.583-g sample of sucrose was burned in excess oxygen in a bomb calorimeter at 25 °C, and the temperature was observed to rise 4.29°. In the same calorimeter, a 1.420-g sample of glucose, $C_6H_{12}O_6$, was burned in excess oxygen and the temperature was observed to rise 3.64°. Write a balanced equation for the combustion of glucose and calculate the heat of combustion of glucose at constant volume at 25 °C.

22. A sample of ground beef weighing 4.178 g was burned in a bomb calorimeter, producing a temperature rise of 4.77°. In the same calorimeter the combustion of a 2.783-g sample of sucrose produced a temperature rise of 4.63°. The heat of combustion of sucrose at constant volume is -5647 kJ/mol. What is the calorie content of a quarter-pound hamburger made from this ground beef? (1 lb = 453.6 g.)

Multiple Choice Questions

1. For which of the following equations is the enthalpy change at 25 °C and 1 atm equal to $\Delta H_f^0(CH_3OH, \ell)$?
   (a) $CO(g) + 2H_2(g) \rightarrow CH_3OH(\ell)$
   (b) $C(g) + 4H(g) + O(g) \rightarrow CH_3OH(\ell)$
   (c) $C(gr) + H_2O(g) + H_2(g) \rightarrow CH_3OH(\ell)$
   (d) $C(gr) + 2H_2(g) + \frac{1}{2}O_2(g) \rightarrow CH_3OH(\ell)$
   (e) $C(gr) + 2H_2(g) + O(g) \rightarrow CH_3OH(\ell)$

2. For which of the following substances is ΔH_f^0 equal to 0?

   (a) $Br_2(g)$ (b) $N(g)$ (c) $C(g)$ (d) $CO(g)$ (e) $Ne(g)$

3. The molar heat capacity at 25 °C should be close to 25 $J \cdot mol^{-1}K^{-1}$ for all of the following elements EXCEPT

   (a) Pt   (b) Kr   (c) W   (d) K   (e) Sr

4. At 25 °C, ΔH^0 for the reaction $Ca(s) + \frac{1}{2} O_2(g) \rightleftharpoons CaO(s)$ is -635.5 kJ/mol. How many grams of Ca must combine with oxygen to liberate 1000 kJ of heat?

   (a) 65.55   (b) 63.07   (c) 40.08   (d) 25.47   (e) 24.51

5. The standard enthalpy change in kilojoules per mole at 25 °C for the reaction

$$CH_4(g) + 2 O_2(g) \rightleftharpoons CO_2(g) + 2H_2O(g)$$

   is the value of

   (a) $-393.51 + 2(285.84) + 74.81$
   (b) $-393.51 - 2(285.84) - 74.81$
   (c) $-393.51 - 2(285.84) + 74.81$
   (d) $-393.51 + 2(241.83) - 74.81$
   (e) $-393.51 - 2(241.83) + 74.81$

6. The heat of formation of $C_2H_6(g)$ is by definition the enthalpy change for the reaction

   (a) $C_2H_4(g) + H_2(g) \rightarrow C_2H_6(g)$ (b) $2C(g) + 6H(g) \rightarrow C_2H_6(g)$
   (c) $2C(s) + 6H(g) \rightarrow C_2H_6(g)$ (d) $2CH_4(g) \rightarrow C_2H_6(g) + H_2(g)$
   (e) $2C(s) + 3H_2(g) \rightarrow C_2H_6(g)$

7. In the balanced equation for the combustion of 1 mol of butane, $C_4H_{10}(g)$, the coefficient of oxygen is

   (a) $\frac{5}{2}$ (b) $\frac{9}{2}$ (c) 5   (d) $\frac{13}{2}$ (e) 13

8. The standard heat of combustion of ethylene is -1411 kJ/mol. The standard heat of vaporization of liquid water is 44.0 kJ/mol. What is ΔH^0, in kilojoules per mole, for the reaction

$$C_2H_4(g) + 3 O_2(g) \rightarrow 2CO_2(g) + 2H_2O(g)$$

   (a) -1323 (b) -1367 (c) -1411 (d) -1455 (e) -1499

9. If 100.0 J of heat are added to 1.00 mol of $Ne(g)$ at 30.0 °C and constant pressure, how much will its temperature rise?

   (a) 3.3°   (b) 4.8°   (c) 8.0°   (d) 30.0°   (e) 34.8°

10. Given the standard enthalpies at 25 °C, in kilojoules per mole, for the following two reactions

$$Fe_2O_3(s) + \tfrac{3}{2} C(s) \rightleftharpoons \tfrac{3}{2} CO_2(g) + 2Fe(s) \qquad \Delta H^0 = +234.1$$
$$C(s) + O_2(g) \rightleftharpoons CO_2(g) \qquad \Delta H^0 = -393.5$$

    the ΔH^0 value for $4Fe(s) + 3 O_2(g) \rightleftharpoons 2Fe_2O_3(s)$ is calculated as

    (a) $\tfrac{3}{2}(-393.5) - 234.1$ (b) $\tfrac{3}{2}(-393.5) + 234.1$ (c) $-393.5 - 234.1$
    (d) $3(-393.5) - 2(234.1)$ (e) $3(-393.5) + 2(234.1)$

11. A gas absorbs 100 J of heat and is simultaneously compressed by a constant external pressure of 1.50 atm from 8.00 to 2.00 L in volume. What is ΔE in joules for the gas?

    (a) -812 (b) $+812$ (c) -912 (d) $+912$ (e) 1012

12. Given the following data:

$$N_2O(g) + 3H_2(g) \rightarrow N_2H_4(\ell) + H_2O(\ell) \qquad \Delta H^0 = -316.97 \text{ kJ}$$

the standard heat of formation of liquid hydrazine, $\Delta H_f^0(N_2H_4, \ell)$, is

(a) $-316.97 + 82.05 + 285.84$ (b) $+316.97 + 82.05 - 285.84$
(c) $-316.97 - 82.05 + 241.83$ (d) $-316.97 + 82.05 + 241.83$
(e) $+316.97 - 82.05 + 285.84$

13. How much heat, in joules, must be added to 0.250 mol of Ar(g) to raise its temperature from 20.0 to 36.0 °C at constant pressure?

(a) 50.0   (b) 83.2   (c) 187   (d) 200   (e) 333

14. The standard heat of combustion of solid boron is equal to

(a) $\Delta H_f^0(B_2O_3)$ (b) $-\Delta H_f^0(B_2O_3)$ (c) $\frac{1}{2}\Delta H_f^0(B_2O_3)$ (d) $-\frac{1}{2}\Delta H_f^0(B_2O_3)$
(e) $2\Delta H_f^0(B_2O_3)$

15. An adiabatic process is one in which there is no transfer of heat across the boundary between system and surroundings. For such a process

(a) $P_{ext}\Delta V = 0$ (b) $q = w$ (c) $\Delta E = w$ (d) $\Delta H = 0$ (e) $\Delta E = q$

16. A bomb calorimeter has a heat capacity of 2.47 kJ/K. The combustion of a 0.105-g sample of ethylene, $C_2H_4(g)$, in this bomb causes a temperature rise of 2.14°. What is ΔE per mole of ethylene for the combustion reaction?

(a) -5.29 kJ   (b) -50.3 kJ   (c) -572 kJ   (d) -660 kJ   (e) -1.41×10^3 kJ

17. What is ΔH in kilojoules for the vaporization of 10.00 g of liquid bromine at 25 °C and 1 atm?

(a) 1.934   (b) 6.996   (c) 13.99   (d) 30.91   (e) 223.6

18. A gas expands against a constant external pressure of 2.00 atm, increasing its volume by 3.40 L. Simultaneously, the system absorbs 400 J of heat from its surroundings. What is ΔE, in joules, for this gas?

(a) -689 (b) -289 (c) $+400$ (d) $+289$ (e) $+689$

19. The standard heat of combustion of Al(s) is -834.9 kJ per mole of Al. In reacting Al with O_2, which of the following will release 250 kcal of heat?

(a) The reaction of 0.626 mol of Al.
(b) The formation of 0.626 mol of Al_2O_3.
(c) The reaction of 0.299 mol of Al.
(d) The formation of 0.150 mol of Al_2O_3.
(e) The formation of 1.253 mol of Al_2O_3.

20. The C—Cl bond energy can be calculated from

(a) $\Delta H_f^0(CCl_4, \ell)$ only
(b) $\Delta H_f^0(CCl_4, \ell)$ and $D(Cl_2)$
(c) $\Delta H_f^0(CCl_4, \ell)$, $D(Cl_2)$, and $\Delta H_f^0(C, g)$
(d) $\Delta H_f^0(CCl_4, \ell)$, $D(Cl_2)$, $\Delta H_f^0(C, g)$, and $\Delta H_{vap}^0(CCl_4)$
(e) $\Delta H_f^0(CCl_4, \ell)$, $D(Cl_2)$, $\Delta H_f^0(C, g)$, $\Delta H_{vap}^0(CCl_4)$, and $\Delta H_{fus}^0(CCl_4)$

21. Using bond energies (symbolized by ε) an estimated value of ΔH^0 for the reaction
$$H_2C=CH_2(g) + H_2(g) \rightarrow CH_3-CH_3(g)$$ would be

(a) $\varepsilon_{C=C} + \varepsilon_{H-H} - 2\varepsilon_{C-H} - \varepsilon_{C-C}$
(b) $\varepsilon_{C=C} + \varepsilon_{H-H} - 6\varepsilon_{C-H} - \varepsilon_{C-C}$
(c) $\varepsilon_{C=C} - \varepsilon_{H-H} + 4\varepsilon_{C-H} - \varepsilon_{C-C}$
(d) $\varepsilon_{C=C} + \varepsilon_{H-H} - 4\varepsilon_{C-H} + \varepsilon_{C-C}$
(e) $\varepsilon_{C=C} - \varepsilon_{H-H} + 2\varepsilon_{C-H} + \varepsilon_{C-C}$

22. Using only the following data:

$$Fe_2O_3(s) + 3CO(g) \rightarrow 2Fe(s) + 3CO_2(g) \qquad \Delta H^0 = -26.8 \text{ kJ}$$
$$Fe(s) + CO_2(g) \rightarrow FeO(s) + CO(g) \qquad \Delta H^0 = +16.5 \text{ kJ}$$

the ΔH^0 value, in kilojoules, for the reaction

$$Fe_2O_3(s) + CO(g) \rightarrow 2FeO(s) + CO_2(g)$$

is calculated to be

(a) −43.3   (b) −10.3   (c) +6.2   (d) +10.3   (e) +22.7

(*Note:* FeO is a nonstoichiometric compound, so that its composition varies with preparation, and thermochemical data involving FeO have large uncertainties.)

Problems

Use Appendix G for ΔH_f^0 values.

17.1. To calculate $\Delta H_f^0(\text{I, g})$, what information is needed in addition to the data contained in Table 17.3?

17.2. Estimate the ΔH^0 values for the reactions of Exercise 8 from bond energy values. Calculate the percentage errors of these estimates using the actual ΔH^0 values obtained in Exercise 8. Why is the estimate for reaction (b) better than the estimate for reaction (a)?

17.3. (a) The heats of combustion of formaldehyde, $H_2C{=}O(g)$, and formic acid, $HCOOH(\ell)$, are −563 and −270 kJ, respectively. Using only these data, calculate ΔH^0 for

$$H_2C{=}O(g) + \tfrac{1}{2}O_2(g) \rightarrow HCOOH(\ell)$$

(b) Calculate ΔH^0 for the reaction in part (a) using ΔH_f^0 values.

17.4. Given that $\Delta H^0 = -143.0$ kJ for

$$2NH_3(g) + \tfrac{1}{2}O_2(g) \rightarrow N_2H_4(\ell) + H_2O(\ell)$$

calculate $\Delta H_f^0(N_2H_4, \ell)$.

17.5. The combustion of 1.000 g of *n*-propane, $CH_3CH_2CH_3(g)$, releases 50.33 kJ of heat at 25 °C and 1 atm.

(a) Write a balanced equation for the combustion of *n*-propane.

(b) Calculate the standard heat of formation of *n*-propane using the data given above.

17.6. Derive an expression for ΔH^0 for the reaction

$$2XO(s) + CO_2(g) \rightarrow X_2O_3(s) + CO(g)$$

from the following thermochemical data:

$$X(s) + CO_2(g) \rightarrow XO(s) + CO(g) \qquad \Delta H_1^0 \qquad (1)$$
$$CO(g) + \tfrac{1}{2}O_2(g) \rightarrow CO_2(g) \qquad \Delta H_2^0 \qquad (2)$$
$$2X(s) + \tfrac{3}{2}O_2(g) \rightarrow X_2O_3(s) \qquad \Delta H_3^0 \qquad (3)$$

17.7. To 1.000-mol samples of He(g) and $Cl_2(g)$, both at 20.0 °C and 1 atm, 1.000 kJ of heat is added, at constant pressure. Assuming both gases are ideal, calculate the molar translational kinetic energy of each gas after the heat has been added. Account for the difference.

17.8. Calculate the standard enthalpy of formation of $Ca(OH)_2(s)$ from the following data:

$$H_2(g) + \tfrac{1}{2}O_2(g) \rightleftharpoons H_2O(\ell) \qquad \Delta H_a^0 = -285.84 \text{ kJ/mol} \qquad \text{(a)}$$
$$CaO(s) + H_2O(\ell) \rightleftharpoons Ca(OH)_2(s) \qquad \Delta H_b^0 = -65.2 \quad \text{kJ/mol} \qquad \text{(b)}$$
$$Ca(s) + \tfrac{1}{2}O_2(g) \rightleftharpoons CaO(s) \qquad \Delta H_c^0 = -635.5 \quad \text{kJ/mol} \qquad \text{(c)}$$

17.9. The density of liquid water at 100 °C and 1 atm is 0.9584 g/cm³. The density of water vapor at 100 °C and 1 atm is 5.96×10^{-4} g/cm³.
(a) Calculate ΔV (per mole) at 100 °C and 1 atm for the process

$$H_2O(\ell) \rightarrow H_2O(g)$$

(b) Calculate the work of expansion against a constant pressure of 1 atm when the process of part (a) occurs.
(c) If the heat of vaporization of water, ΔH_{vap}, is 40.66 kJ/mol at 100 °C, what is ΔE_{vap}?

17.10. The combustion of cyanamide, $N\equiv CNH_2(s)$, produces $CO_2(g)$, $N_2(g)$, and $H_2O(\ell)$.
(a) Write a balanced equation for the combustion of cyanamide.
(b) Calculate $\Delta H_f^0(N\equiv CNH_2, s)$ from its standard heat of combustion, which is -741 kJ/mol.

17.11. Boron and hydrogen combine to form a large number of compounds called boranes, one of which is B_5H_9, a gas. $\Delta H_f^0(B_5H_9) = 62.76$ kJ/mol. B_5H_9 ignites spontaneously in air with a green flash to produce solid B_2O_3 and liquid H_2O. The molar heat of combustion of $B_5H_9(g)$ is -4507.6 kJ/mol at 25 °C.
(a) Write the correctly balanced equation for which the enthalpy change is $\Delta H_f^0(B_5H_9, g)$.
(b) Write the correctly balanced equation for which the enthalpy change is the standard heat of combustion of $B_5H_9(g)$.
(c) Using the data given above, plus $\Delta H_f^0(H_2O, \ell)$, calculate $\Delta H_f^0(B_2O_3, s)$ in kilojoules per mole.

17.12. The thermite reaction is

$$2Al(s) + Fe_2O_3(s) \rightarrow Al_2O_3(s) + 2Fe(s)$$

This reaction is so exothermic that it is used for welding massive units, such as propellers for large ships.
(a) Calculate ΔH^0 for the thermite reaction at 25 °C.
(b) The heat of fusion of chromium is 14.6 kJ/mol. How many grams of aluminum have to be used up in the thermite reaction in order to melt 100 g of chromium?

17.13. (a) The standard heat of combustion of cyclopropane, $C_3H_6(g)$, has been reported to be -2091 kJ/mol. Calculate $\Delta H_f^0(C_3H_6, g)$.
(b) The structure of cyclopropane is shown below:

Calculate a value for the C—C bond energy in cyclopropane. Compare this with the C—C bond energy listed in Table 17.3. To account for the difference, compare the C—C—C bond angle in cyclopropane with the normal tetrahedral bond angle. What is the effect on the C—C bond of forming this small ring?

17.14. (a) The average specific heat of water between 0 and 100 °C is $1.0 \text{ cal} \cdot \text{g}^{-1}\text{K}^{-1}$. Suppose you want to make a large pot of coffee and you begin by heating 2.00 kg of water from 15 to 95 °C in a kettle open to the atmosphere. Calculate the amount of heat that must be added. Is this ΔH or ΔE?

(b) Between 15 and 95 °C the volume of 1 g of water changes from 1.00087 to 1.03959 mL/g. Calculate $P\Delta V$ for the process of heating 2.00 kg of water from 15 to 95 °C at a constant pressure of 1.00 atm. Note that this is $\Delta H - \Delta E$. For this process is it a good approximation to say $\Delta H = \Delta E$?

17.15. Alanine is an amino acid, synthesized in the human body, and a constituent of proteins. It is commonly abbreviated ala and has the structure

$$\text{CH}_3\text{—CH—COOH}$$
$$|$$
$$\text{NH}_2$$

It is a solid at room temperature.

The following experiment was carried out to determine the heat of combustion of alanine. A 2.840-g sample of urea, $\text{CO(NH}_2)_2$, was combusted in a bomb calorimeter at 25 °C, and the temperature was observed to rise 3.16°. The standard heat of combustion of urea at constant volume is -632.90 kJ/mol. In the same bomb calorimeter, combustion of a 1.818-g sample of alanine produced a temperature rise of 3.49°. The products of both combustions are $\text{CO}_2(g)$, liquid H_2O, and $\text{N}_2(g)$.

Within experimental uncertainty, ΔH and ΔE for the combustion of alanine are the same. Calculate $\Delta H^0_{\text{comb}}(\text{ala})$, and from that value calculate $\Delta H^0_f(\text{ala})$.

17.16. Acetone has the structure

$$\text{O}$$
$$||$$
$$\text{CH}_3\text{—C—CH}_3$$

The heat of vaporization of acetone is 63 kJ/mol. Calculate the bond energy of the $\text{C}{=}\text{O}$ double bond in acetone. Show all your reasoning. Write a correctly balanced equation corresponding to any ΔH you use in the calculation.

17.17. Calculate a value for the bond energy of the O—O bond in hydrogen peroxide, H—O—O—H, if the standard heat of formation of $\text{H}_2\text{O}_2(g)$ is -136.3 kJ/mol. Show all your reasoning. Relative to other single bonds, is the O—O bond in H_2O_2 strong or weak?

17.18. A Handbook reports the following values for the specific heat of some inorganic compounds in calories per gram per kelvin:
(a) Ag_2Se from 37 to 187 °C: 0.0693   (b) AgCl at 50 °C: 0.0906
(c) PbWO_4 at 15 °C: 0.0769   (d) PbBr_2 at 50 °C: 0.0530
Compare the applicability of the Law of Dulong and Petit to these solid compounds with its applicability to the solid elements listed in Table 17.5.

chapter 18 Entropy, Free Energy, and Equilibrium

Josiah Willard Gibbs (1839–1903) was born in New Haven, Connecticut and held the position of Professor of Mathematical Physics at Yale from 1871 until his death in 1903. His work in thermodynamics, and particularly his famous paper "On the Equilibrium of Heterogeneous Substances" led to the development of a new branch of physical chemistry. The free energy function, $H - TS$, is now called the Gibbs Free Energy Function and denoted G, in his honor.

If we want to be able to predict whether or not a particular reaction will occur under a specified set of conditions, the two thermodynamic functions we considered in the previous chapter, ΔE and ΔH, are not sufficient. We will need to introduce two additional state functions, the entropy and the free energy, in order to answer the questions posed at the beginning of Chapter 17.

section 18.1
Energy and Spontaneity: The Need for a Second Law

Many spontaneous processes release heat as they proceed. Indeed so many common spontaneous processes are exothermic that for a time during the nineteenth century it was erroneously believed that all spontaneous processes release heat to their surroundings. Examples of some exothermic spontaneous reactions are

1. The neutralization of a strong acid with a strong base.

$$H_3O^+ + OH^- \rightarrow 2H_2O(\ell) \qquad \Delta H^0_{298} = -55.9 \text{ kJ}$$

2. The hydrolysis of ATP to ADP. This reaction releases metabolically available energy in every cell of every organism.

$$ATP^{3-} + 2H_2O \rightarrow ADP^{2-} + HPO_4^{2-} + H_3O^+$$
$$\Delta H^0_{298} = -24 \text{ kJ} \qquad \text{at pH} = 7$$

3. The rusting of iron.

$$2Fe(s) + \tfrac{3}{2}O_2(g) \rightarrow Fe_2O_3(s) \qquad \Delta H^0_{298} = -824.2 \text{ kJ}$$

Reactions are exothermic if the products are at a lower energy than the reactants; the difference in energy is released as heat.

There are, however, a substantial number of spontaneous processes that are endothermic, that is, spontaneous processes for which the final state is a state of higher energy than the initial state. Examples of endothermic spontaneous processes are

1. The evaporation of water at room temperature and an air pressure of 1 atm:

$$H_2O(\ell) \rightarrow H_2O(g) \qquad \Delta H^0_{298} = +44.0 \text{ kJ}$$

The vaporization of any liquid is spontaneous if the partial pressure of that compound in the gas phase is less than the vapor pressure of the liquid, yet all vaporizations are endothermic. If some alcohol is swabbed on your arm it will spontaneously evaporate. The part of your arm the alcohol touched will feel quite cool as the heat required to vaporize the alcohol comes from your body.

2. The dissolution of solid ammonium nitrate in water.

$$NH_4NO_3(s) + H_2O \rightarrow NH_4^+(aq) + NO_3^-(aq) \qquad \Delta H^0_{298} = +25.8 \text{ kJ}$$

If you add some solid ammonium nitrate to a beaker of water and stir, you can feel the beaker and its contents getting colder as the ammonium nitrate dissolves. The dissolution reaction absorbs heat from the surroundings, but occurs spontaneously.

3. The sublimation of solid naphthalene, $C_{10}H_8$.

$$C_{10}H_8(s) \rightarrow C_{10}H_8(g) \qquad \Delta H^0_{298} = +72.6 \text{ kJ}$$

At one time, naphthalene was used as moth balls. If some solid naphthalene is placed in your clothes closet, it will spontaneously vaporize, yet the sublimation reaction is endothermic.

Since both endothermic and exothermic reactions can be spontaneous, we cannot predict whether or not a given reaction taking place at constant pressure will occur spontaneously simply by determining the sign of ΔE or ΔH. We need to introduce a new function in order to predict spontaneity at constant pressure.

Spontaneity and Molecular Disorder

In order to gain some insight into the factors that cause a reaction to occur spontaneously, let us consider spontaneous processes for which the initial and final states are at the *same* energy. We have already discussed one such process, namely the diffusion of two different gases at the same temperature and pressure. (See Fig. 5.19.) There is no change in energy when two nonreacting gases at the same temperature and pressure diffuse, but the process occurs spontaneously. We associate that spontaneity with an increase in the molecular disorder of the system.

The isothermal expansion of an ideal gas into a vacuum is another example of a spontaneous process for which the initial and final states are at the same energy.

Consider a sample of an ideal gas at a pressure $P_1 = 8.00$ atm in a 1.00-L bulb connected by a stopcock to a 3.00-L evacuated bulb, as in Fig. 18.1. The temperature of the system is 298.2 K (25.0 °C). Since this is an ideal gas, $P_1 V_1 = nRT_1$ and n is 0.327 mol. When the stopcock is opened, the gas spontaneously expands into the evacuated bulb. The final volume of the gas is 4.00 L, the volume of both bulbs, and the pressure drops to 2.00 atm, because $P_1 V_1 = P_2 V_2$ as the temperature is constant. No work is done because there is no external pressure for the gas to push against. We can summarize the initial and final states of this system as follows:

Initial State	*Final State*
$P_1 = 8.00$ atm	$P_2 = 2.00$ atm
$V_1 = 1.00$ L	$V_2 = 4.00$ L
$T_1 = T = 298.2$ K	$T_2 = T = 298.2$ K
$n_1 = n = 0.327$ mol	$n_2 = n = 0.327$ mol

For an ideal gas, the energy is a function of temperature only. As long as the temperature remains constant, $\Delta E = 0$. We can calculate the heat absorbed, q, from the first law of thermodynamics, $\Delta E = q + w$. Since $w = 0$ and $\Delta E = 0$, q must also be 0.

Note that this reaction proceeds spontaneously even though the initial and final states are at the same energy. We associate the spontaneity of this change with an increase in the disorder or freedom of motion of the molecules in the final state relative to the initial state. The molecules in the final state have a larger volume to move around in and are therefore freer from restraints on their motion than the molecules in the initial state.

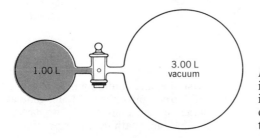

Fig. 18.1. The isothermal expansion of an ideal gas into a vacuum. Initially the gas is in the 1.00-L bulb. When the stopcock is opened, the gas expands to fill both bulbs so that the final volume of the gas is 4.00 L.

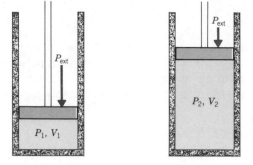

Fig. 18.2. Isothermal expansion of an ideal gas against a constant external pressure equal to the final pressure, P_2, of the gas.

The same spontaneous process described above, from the same initial state to the same final state, can be carried out using a different path of reaction. We can expand the gas against a constant external pressure equal to the final pressure, 2.00 atm, by enclosing the gas in a cylinder fitted with a piston, as shown in Fig. 18.2. We maintain the gas at a constant temperature of 25.0 °C, so that ΔE is still 0, but now the gas has done work against the external pressure. The work, w, done on the gas is

$$w = -P_{ext}\Delta V = -(2.00 \text{ atm})(3.00 \text{ L}) = -6.00 \text{ L} \cdot \text{atm}$$

Using SI units, we must express the work in joules.

$$w = -(6.00 \text{ L} \cdot \text{atm})\left(101.3 \frac{\text{J}}{\text{L} \cdot \text{atm}}\right) = -608 \text{ J} = -0.608 \text{ kJ}$$

To keep the system at the same temperature after the expansion, heat must be absorbed from the surroundings. Since $\Delta E = 0 = q + w$, the first law tells us that $q = -w$. The heat absorbed is therefore 0.608 kJ.

The driving force for this spontaneous process in which the energy remains constant is the increase in freedom of motion of the molecules of the gas, just as when the gas expanded into a vacuum. That increase is exactly the same for this path as for the expansion into the evacuated bulb. The only quantities that change when the path changes are the work done on the gas and the heat absorbed by the gas.

section 18.2
Reversible Processes

For the ideal gas sample described in the preceding section, let us consider an isothermal expansion using a third path of reaction, different from the two just described. Imagine the gas in a cylinder enclosed by a piston on which sits a large pile of very fine grained sand, as shown in Fig. 18.3. The system is at equilibrium in its initial state so that the mass of sand is large enough to make $P_{ext} = P_1 = 8.00$ atm.

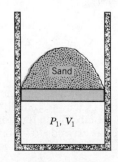

Fig. 18.3. The external pressure confining the gas is due to the mass of sand on the piston. If we remove the sand one grain at a time until the pressure and volume of the gas are P_2, V_2, we have a close approximation to the reversible path for the expansion.

If we remove one of the very fine grains of sand, what will happen? The piston will move up a tiny little bit, as the gas expands a tiny amount. A new position of equilibrium will be attained at which $P_{ext} = P_{gas}$, and this pressure is just a tiny bit less than 8.00 atm. Now remove a second grain of sand. Again, the gas will expand a very small amount, and a new position of equilibrium will be attained. Imagine performing the entire expansion of the gas from $V_1 = 1.00$ L to $V_2 = 4.00$ L by removing the sand one grain at a time. Clearly this will take a very long time, and the smaller the grains of sand the longer it will take. If the grains of sand are infinitesimally small, this method of expanding the gas is called the **reversible path**. If each grain of sand is infinitesimally small, it will take an infinite amount of time to carry out the expansion. Thus the reversible path is one that we can conceive of, describe, and carry out calculations for, but it is not a path that can actually be utilized in the laboratory, although we can approach it closely.

A **reversible process** is one in which at each step the system is at equilibrium with its surroundings. The entire process is a succession of equilibrium states. There is a driving force for the reaction, and an opposing force, and the driving force is only infinitesimally greater than the opposing force at each step of the process. In the example just considered, the driving force is the pressure of the gas, and the opposing force is the external pressure due to the weight of the sand on the piston.

We can summarize the characteristics of a reversible process as follows:

1. The driving force is only infinitesimally greater than the opposing force at any point during the entire process.

2. The process occurs in a series of infinitesimal steps, and at each step the system is at equilibrium with its surroundings.

3. The reversible process would take an infinite amount of time to carry out, in practice.

4. The work done by the system when the process is carried out reversibly is the maximum work possible, because at each step the opposing force has the greatest value possible. If the opposing force (the external pressure) were increased even infinitesimally, the reaction would proceed in the opposite direction.

The essential feature of a reversible process is that by changing some variable in the surroundings by an infinitesimal amount the process can be reversed. In the example given, if one grain of sand were added to the piston, the gas would be compressed.

Processes that are spontaneous are **irreversible.** Be careful to use the word irreversible in its strict thermodynamic meaning. Irreversibility does not mean that the process cannot be reversed; it means only that a reversal cannot be achieved merely by changing some variable by an infinitesimal amount.

section 18.3
Derivation of the Reversible Work of Expansion of an Ideal Gas

During the reversible expansion of an ideal gas the external pressure is equal to the pressure of the gas at each infinitesimal step.

$$P_{ext} = P_{gas} = nRT/V \tag{18-1}$$

For each step, the volume changes by an infinitesimal amount. In order to write equations involving infinitesimally small quantities, the calculus must be employed.

An infinitesimal change in some quantity is denoted mathematically by a differential; thus the change in the volume during each step of the expansion is dV. The work done by the gas during each infinitesimal step is therefore $P_{ext} \, dV$, and

$$P_{ext} \, dV = \frac{(nRT)}{V} \, dV \tag{18-2}$$

The work done by the gas for the entire transformation from state 1 to state 2 is obtained by summing up the work done in all the infinitesimal steps. Such a summation is obtained by integration. For an isothermal reversible expansion we obtain

$$-w_{rev} = \int_{V_1}^{V_2} \frac{(nRT)}{V} \, dV = nRT \int_{V_1}^{V_2} \frac{dV}{V} \tag{18-3}$$

where w_{rev} is the work done on the gas, and $-w_{rev}$ is the work done by the gas, according to Eq. (17-2). Because the process is isothermal, T is constant, and therefore we can take T out of the integral sign.

The integral $\int dV/V$ is equal to the natural logarithm of V. We therefore obtain

$$-w_{rev} = nRT \ln (V_2/V_1) \tag{18-4}$$

section 18.4
The Maximum Work That Can Be Done by
an Isothermal Expansion of an Ideal Gas

We stated in Section 18.2 that the maximum work done by a system during a process is obtained when the process is carried out reversibly. The larger the opposing force, the greater the work done, and the opposing force is at a maximum when the process is reversible. Therefore, for the isothermal expansion of an ideal gas, the maximum work done by the gas is the work done using the reversible path.

$$\text{maximum work done by gas} = -w_{rev} = nRT \ln (V_2/V_1) \tag{18-5}$$

For the expansion we have been discussing in Sections 18.1 and 18.2, $V_2 = 4.00$ L and $V_1 = 1.00$ L so that $V_2/V_1 = 4.00$, $\ln (V_2/V_1) = \ln (4.00) = 1.386$ and Eq. (18-4) yields

$$w_{rev} = -(0.327 \text{ mol}) \left(8.3144 \, \frac{\text{J}}{\text{mol} \cdot \text{K}} \right) (298.2 \text{ K})(1.386) = -1124 \text{ J} = -1.12 \text{ kJ}$$

The value of R used in Eq. (18-4) should be 8.3144 J $\cdot$ mol^{-1}K^{-1}, so that the calculated work will be in joules. If you use R in liter $\cdot$ atmospheres per mol per kelvin (L $\cdot$ atm $\cdot$ mol$^{-1} \cdot$ K^{-1}), you will calculate w_{rev} in liter $\cdot$ atmospheres and will then have to convert to joules. It saves time and effort to use R in joules per mol per kelvin (J $\cdot$ mol^{-1}K^{-1}) directly.

The work done by the gas when the expansion is carried out reversibly, 1.12 kJ, is the maximum amount of work the gas can do in expanding from 1.00 to 4.00 L at a constant temperature of 25 °C. When the expansion occurs into a vacuum no work is done. When the expansion is against a constant external pressure of 2.00 atm, the gas does 0.608 kJ of work. There are other paths in addition to the three discussed here, and the work done by the gas can be any value between 0 and 1.12 kJ, but the maximum amount of work that can be done when 0.327 mol of an ideal gas expands from a volume of 1.00 L to a volume of 4.00 L at a constant temperature of 25 °C, is the reversible work, 1.12 kJ.

section 18.5
The Entropy Change for an Isothermal Process

The direction of a spontaneous process for which the energy is constant is always the one that increases the molecular disorder or randomness. The thermodynamic property that measures the amount of molecular disorder is called the **entropy** and is denoted S. If we want to be able to predict whether or not a given reaction occurs spontaneously, we will need to consider both the change in energy, ΔE (or ΔH if the pressure is constant), and the change in entropy, ΔS, for that reaction.

The entropy is a property of the system, so that the change in entropy, $\Delta S = S_2 - S_1$, is independent of the path used in going from the initial state to the final state. For the isothermal expansion of an ideal gas, ΔS is the same whether the expansion is carried out into a vacuum, against a constant external pressure, or reversibly.

For an isothermal process the change in entropy, ΔS, is defined as

$$\Delta S = q_{rev}/T \tag{18-6}$$

where q_{rev} is the heat absorbed during the reversible path. No matter what path is actually used to change the state of a system from some initial state to a final state, we can imagine the reversible path and calculate the heat absorbed when the transformation is carried out reversibly.

To obtain the change in entropy for the isothermal expansion of an ideal gas, therefore, we must first be able to calculate the heat absorbed during the reversible path, q_{rev}. As long as the temperature remains constant, the energy of an ideal gas is constant, so that $\Delta E = 0$ for any isothermal process, regardless of the path. The first law then tells us that $q = -w$ for any isothermal process for an ideal gas. Since $q = -w$ for any path, it is also true for the reversible path, and therefore

$$q_{rev} = -w_{rev} = nRT \ln (V_2/V_1) \tag{18-7}$$
$$\Delta S = q_{rev}/T = nR \ln (V_2/V_1) \tag{18-8}$$

For the expansion we have been considering,

$$\Delta S = (0.327 \text{ mol}) \left(8.3144 \ \frac{J}{mol \cdot K} \right) \ln (4.00) = 3.77 \text{ J} \cdot K^{-1}$$

We often report an entropy change in units of joules per mole per kelvin, that is, the change in entropy per mole. This is obtained by dividing the total entropy change by the number of moles. For an isothermal expansion of an ideal gas, ΔS per mole is $R \ln (V_2/V_1)$. For the expansion of 1 mol of an ideal gas at constant temperature from 1.00 to 4.00 L, $\Delta S = (8.3144)(1.386) = 11.5 \text{ J} \cdot mol^{-1}K^{-1}$.

Table 18.1. **Summary of Calculations of ΔE, ΔS, q, and w for Three Different Paths Used to Expand 0.327 mol of an Ideal Gas from $V_1 = 1.00$ L to $V_2 = 4.00$ L, at a Constant Temperature of 25 °C**

Path 1	Path 2	Path 3
Expansion into a vacuum	Expansion versus constant external pressure	Reversible expansion
$P_{ext} = 0$	$P_{ext} = 2.00$ atm	$P_{ext} = P_{gas} = nRT/V$
$w = 0$	$w = -0.608$ kJ	$w_{rev} = -1.12$ kJ
$q = 0$	$q = +0.608$ kJ	$q_{rev} = +1.12$ kJ
$\Delta E = 0$	$\Delta E = 0$	$\Delta E = 0$
$\Delta S = 3.77$ J/K	$\Delta S = 3.77$ J/K	$\Delta S = 3.77$ J/K

Table 18.1 summarizes the information we have obtained about the three paths of isothermal expansion of 0.327 mol of an ideal gas from 1.00 to 4.00 L at 25 °C.

Note particularly that while the heat absorbed by the gas and the work done on the gas change when the path is changed, both ΔE and ΔS are independent of the path.

ΔS for Isothermal Phase Changes

The value of ΔS for any process measures the change in the molecular disorder or randomness when that process takes place. If the process is isothermal, Eq. (18-6) defines the change in entropy. Phase changes are isothermal, and we can therefore use Eq. (18-6) to calculate the change in entropy on fusion, vaporization, sublimation, or change in crystal form, at the equilibrium temperature.

The Entropy of Vaporization, ΔS_{vap}

We will consider first vaporization at the normal boiling point, T_{bp}. That is the equilibrium

1 mol of liquid A at temperature T_{bp} $\rightleftharpoons$ 1 mol of vapor A at temperature T_{bp}
and 1-atm pressure   and 1-atm pressure

Since a liquid and its vapor are in equilibrium at the normal boiling point and 1-atm pressure, if we add heat very slowly so that the temperature always remains at T_{bp}, we are carrying out the vaporization reversibly. The amount of heat that must be added at constant temperature and pressure to convert 1 mol of liquid into 1 mol of vapor, keeping the liquid and vapor in equilibrium throughout the process, is called the **heat of vaporization,** and is denoted ΔH_{vap}. Vaporization is an endothermic process and therefore $\Delta H_{vap} > 0$. The entropy change on vaporization, using Eq. (18-6), is

$$\Delta S_{vap} = q_p/T_{bp} = \Delta H_{vap}/T_{bp} \qquad (18-9)$$

As ΔH_{vap} is always positive, ΔS_{vap} is also positive.

$$\Delta S_{vap} = S_{vapor} - S_{liquid} > 0 \qquad (18-10)$$

so that $S_{vapor} > S_{liquid}$. The entropy of 1 mol of a gas at its normal boiling point is always larger than the entropy of 1 mol of the liquid. This is a measure of the large increase in disorder or randomness that occurs on vaporization. Molecules in the vapor phase have more freedom of motion than molecules in the liquid.

Table 18.2 lists the heat of vaporization, the normal boiling point, and the entropy of vaporization of several common liquids.

The entropy of vaporization of most liquids lies between 83 and 93 $J \cdot mol^{-1}K^{-1}$. That generalization is known as **Trouton's Rule,** which is usually stated as follows: The entropy of vaporization of most liquids is 88 ± 5 $J \cdot mol^{-1}K^{-1}$ at the normal boiling point. An examination of Table 18.2 shows that water and ethanol have unusually large entropies of vaporization. That is because these liquids are more ordered than most, due to extensive hydrogen bonding in the liquid phase. Hydrogen bonding orients the molecules and increases the amount of molecular order. (Refer to Sections 5.2 and 7.1 for discussions of hydrogen bonding in water.) As a result, the increase in molecular disorder on converting 1 mol of liquid to the vapor is greater for water and ethanol (and alcohols in general) than for most other liquids, and ΔS_{vap} is larger than usual.

Trouton's Rule can be used to obtain a rough estimate of the heat of vaporization of liquids that neither associate nor dissociate. This type of calculation is illustrated in Example 18.1.

Table 18.2. The Entropy of Vaporization of Some Common Liquids

Compound	$t_{bp}(°C)$	$T_{bp}(K)$	$\Delta H_{vap}(kJ/mol)$	$\Delta S_{vap} = \Delta H_{vap}/T_{bp}$ $(J \cdot mol^{-1}K^{-1})$
Benzene, C_6H_6	80.1	353.3	30.8	87.2
Ether, $C_2H_5OC_2H_5$	34.6	307.8	26.0	84.5
Ethanol, C_2H_5OH	78.5	351.7	39.2	111.
Mercury, Hg	356.6	629.8	59.3	94.1
n-Pentane, C_5H_{12}	36.2	309.4	25.8	83.3
Water, H_2O	100.0	373.2	40.7	109.

EXAMPLE 18.1. The use of Trouton's Rule

The normal boiling point of acetone, CH_3COCH_3, is 56.2 °C, and that of nitrobenzene, $C_6H_5NO_2$, is 210.8. Estimate ΔH_{vap} for these two liquids.

Solution. Neither of these liquids is hydrogen bonded, so that we expect Trouton's Rule to apply to them. We must express the normal boiling point in kelvins, and calculate ΔH_{vap} as $T_{bp}\Delta S_{vap}$. Assuming ΔS_{vap} is $88\ J \cdot mol^{-1}K^{-1}$, we obtain for acetone

$$T_{bp} = 329.4 \quad \text{and} \quad \Delta H_{vap} \sim (329.4\ K)(88\ J \cdot mol^{-1}K^{-1}) = 29\ kJ/mol$$

Similarly, for nitrobenzene,

$$T_{bp} = 484.0 \quad \text{and} \quad \Delta H_{vap} \sim (484.0\ K)(88\ J \cdot mol^{-1}K^{-1}) = 43\ kJ/mol$$

The experimental values of ΔH_{vap} are 30.0 kJ/mol for acetone and 40.4 kJ/mol for nitrobenzene. The estimates, while rough, are within the uncertainties of Trouton's Rule. You should calculate ΔS_{vap} for these two liquids. The values, in joules per mole per kelvin, are 91.1 for acetone, and 83.5 for nitrobenzene.

The Entropy Change on Fusion

A solid and its liquid are in equilibrium at 1 atm at the normal melting point, T_{mp}. The process of fusion is

1 mol of solid A at temperature T_{mp} $\rightleftharpoons$ 1 mol of liquid A at temperature T_{mp}
and 1-atm pressure   and 1-atm pressure

The amount of heat that must be added at constant temperature and pressure to convert one mole of solid into one mole of liquid, keeping the solid and liquid in equilibrium throughout the process, is called the **heat of fusion** and denoted ΔH_{fus}. Fusion is an endothermic process, so ΔH_{fus} is positive for all substances. Because solid and liquid are in equilibrium at all times throughout the fusion, ΔH_{fus} is the reversible heat of the transformation, and the entropy change on fusion, ΔS_{fus}, is given by

$$\Delta S_{fus} = \Delta H_{fus}/T_{mp} \tag{18-11}$$

Because

$$\Delta S_{fus} = S_{liquid} - S_{solid} > 0 \tag{18-12}$$

the entropy of a mole of liquid is always greater than the entropy of a mole of solid of the same substance. This is a measure of the increase in randomness and disorder that

Table 18.3. The Entropy of Fusion of Some Common
Substances at Their Normal Melting Points

Compound	$t_{mp}(°C)$	$T_{mp}(K)$	$\Delta H_{fus}(kJ/mol)$	$\Delta S_{fus} = \Delta H_{fus}/T_{mp}$ $(J \cdot mol^{-1}K^{-1})$
Benzene, C_6H_6	5.5	278.7	9.95	35.7
Ether, $C_2H_5OC_2H_5$	−116.3	156.9	7.27	46.3
Ethanol, C_2H_5OH	−114.5	158.7	5.02	31.6
Mercury, Hg	−38.9	234.3	2.33	9.94
n-Pentane, C_5H_{12}	−129.7	143.5	8.42	58.7
Water, H_2O	0.0	273.2	6.01	22.0

results when the solid phase is converted to a liquid. Table 18.3 lists the heat of fusion, the normal melting point, and the entropy of fusion of several common liquids.

A comparison of the values of ΔS_{fus} and ΔS_{vap} of the same substance (Tables 18.2 and 18.3) shows that ΔS_{fus} is always smaller than ΔS_{vap}. The increase in molecular disorder that occurs when 1 mol of liquid is converted to vapor is considerably larger than the increase that occurs when 1 mol of solid is converted to liquid.

section 18.6
The Second Law of Thermodynamics

The **second law of thermodynamics** is a generalization of all the observations that have been made about spontaneous or naturally occurring processes. Like the first law, it has been stated in many different ways by many different people.

One of the most useful ways to state the second law concerns an **isolated system.** An isolated system is one for which the energy is constant because neither heat nor work cross the boundary between system and surroundings. The second law states that *in an isolated system,* the direction of all spontaneous processes is that which increases the amount of molecular disorder, that is, that which increases the entropy.

The unidirectionality of spontaneous processes is a phenomenon we are all familiar with. If we have two bulbs connected by a stopcock, and one contains $N_2(g)$ while the other contains $CH_4(g)$, within some time after opening the stopcock both bulbs will contain both gases. If we start with both gases in both bulbs, we never expect to observe all the N_2 in one bulb and all the CH_4 in the other.

Similarly, if we have a stack of pennies arranged with all the heads facing up and we let them drop on the floor, we expect to find very close to a 50–50 mixture of heads and tails when they land on the floor. If we drop a stack of pennies which are exactly half heads and half tails, we would be astonished indeed to find all heads facing up when they landed on the floor.

The establishment of a more random distribution of molecules, which occurs spontaneously, is more probable than the establishment of a more ordered state. A disordered state, that is, a state of higher entropy, is a state of higher probability than a state of greater order and lower entropy.

Rudolf Clausius (1882–1888), a German thermodynamicist, succinctly summarized the first and second laws of thermodynamics in two sentences: "The energy of the universe is constant. The entropy of the universe is ever increasing." The second of these statements has, on occasion, been misinterpreted because what is meant by "the universe" is not clearly defined. We can see from the statement of the second law given above that "the universe" must be an isolated system. Our planet, the earth, is *not* an isolated system because we get energy from the sun and other stars.

Defining the boundaries of a truly isolated universe is not trivial, and errors in interpreting the second law have been made by people who apply it to systems that are not isolated.

EXAMPLE 18.2. The second law of thermodynamics

A perplexed friend asks you to explain the following: "Water spontaneously freezes at any temperature below 0 °C, and the final state, ice, has less entropy than the initial state, because there is less molecular disorder in the crystalline state than in the liquid. How is it that this spontaneous process occurs with a decrease in entropy?"

Solution. A spontaneous process must be accompanied by an increase in entropy *only if it occurs within an isolated system.* Water, when it freezes, liberates heat to its surroundings, so it is not an isolated system. (When ice melts, the heat of fusion must be added in order to form liquid water. When water freezes, the heat of fusion is liberated.) The entropy of the water decreases, but the entropy of the surroundings increases because heat has passed into the surroundings. The entropy of the universe, that is, the water plus its surroundings, increases because the increase in entropy of the surroundings is larger in magnitude than the decrease in entropy of the water.

section 18.7
The Gibbs Free Energy Function

The second law tells us that ΔS is positive for spontaneous processes that occur within isolated systems. But we do not usually carry out reactions in isolated systems, and it is useful to have a function of our nonisolated system that can be used as a criterion to predict the direction of a spontaneous reaction that is taking place at constant temperature and pressure, the conditions under which we most frequently run reactions.

In order to see what such a function should be, let us consider a total isolated system consisting of the system we are interested in plus its surroundings. Let a spontaneous process take place within our system. Then, since entropy is an extensive property,

$$\Delta S_{total} = \Delta S_{sys} + \Delta S_{surr} \tag{18-13}$$

where ΔS_{sys} is the change in entropy of our nonisolated system, ΔS_{surr} is the change in entropy of the surroundings, and ΔS_{total} is the change in entropy of the total isolated system. The second law of thermodynamics states that ΔS_{total} must be greater than 0.

$$\Delta S_{total} = \Delta S_{sys} + \Delta S_{surr} > 0 \qquad \text{for a spontaneous process} \tag{18-14}$$

The entropy change in our nonisolated system may be negative (that is, the entropy may decrease) if the entropy change of the surroundings is positive and of greater magnitude than ΔS_{sys}.

Let us consider that the system and surroundings are at constant temperature and pressure. Then as the spontaneous process occurs the heat absorbed by our nonisolated system is ΔH_{sys}. What has happened to the surroundings? They have absorbed an amount of heat equal to $-\Delta H_{sys}$:

$$q_{surr} = -\Delta H_{sys} \tag{18-15}$$

Equation (18-15) is simply a statement of the first law of thermodynamics. Energy must be conserved. No energy enters or leaves the boundaries of the total isolated system, so the heat absorbed by the system must come from the surroundings. If ΔH_{sys} is negative, q_{surr} is positive, that is, the surroundings absorb the heat released by the system.

The entropy of the surroundings changes because they have absorbed heat.

$$\Delta S_{surr} = \frac{q_{surr}}{T} = -\frac{\Delta H_{sys}}{T} \tag{18-16}$$

If we substitute this expression for ΔS_{surr} into Eq. (18-14) we obtain

$$\Delta S_{sys} - \frac{\Delta H_{sys}}{T} > 0 \qquad \text{for a spontaneous process} \tag{18-17a}$$

at constant temperature and pressure as the requirement of the second law of thermodynamics. Multiply this through by the absolute temperature, T, to obtain

$$T\Delta S_{sys} - \Delta H_{sys} > 0 \qquad \text{for a spontaneous process} \tag{18-17b}$$

This is usually rearranged to

$$-(\Delta H_{sys} - T\Delta S_{sys}) > 0 \qquad \text{for a spontaneous process} \tag{18-17c}$$

or

$$(\Delta H_{sys} - T\Delta S_{sys}) < 0 \qquad \text{for a spontaneous process} \tag{18-17d}$$

Equation (18-17d) suggests that the function we require to predict the direction of a spontaneous reaction at constant temperature and pressure is $H - TS$. We define the **Gibbs Free Energy, G,** as

$$G = H - TS \tag{18-18}$$

This function is named in honor of Josiah Willard Gibbs (1839–1903), an American mathematician and physicist who was a professor at Yale University and laid the theoretical foundations for much of that branch of study we now call physical chemistry.

There are two tendencies for spontaneous reactions: (1) the tendency to decrease the energy, and (2) the tendency to increase the entropy. Recall that most spontaneous processes are exothermic, that is, they proceed to a final state of lower energy than the initial state. A function that will serve as a criterion to predict the direction of a spontaneous reaction must combine both the change in energy (or enthalpy, if the pressure is constant) and the change in entropy for the reaction. The Gibbs free energy function does precisely that.

Since H, T, and S are all state functions, G is also a state function. At constant temperature the change in the Gibbs Free Energy is given by

$$\Delta G = \Delta H - T\Delta S \tag{18-19}$$

Note that for an exothermic reaction $\Delta H < 0$, that is, the enthalpy decreases. If the entropy increases, $\Delta S > 0$, and the term $-T\Delta S$ is negative. Thus both a decrease in the enthalpy and an increase in the entropy contribute negative terms to ΔG.

For a reaction taking place at constant temperature and pressure, the sign of ΔG can be used to predict the direction of spontaneous change.

If $\Delta G < 0$ at constant T and P, the reaction is spontaneous.

If $\Delta G > 0$ at constant T and P, the *reverse* reaction is spontaneous.

If $\Delta G = 0$ at constant T and P, the system is at equilibrium and no net reaction will occur.

section 18.8
Calculation of ΔG^0

The standard free energy change, ΔG^0, is the free energy change when all substances are in their **standard states**. The superscript zero (^{0}) denotes "standard state." A gas is in its standard state if its pressure is 1 atm. A species in solution is in its standard state if its concentration is 1.00 M. A pure solid or a pure liquid is, by definition, in its standard state. The usual standard state temperature is 25 °C (298.2 K). It is possible to define the standard state temperature to be any value you are interested in, but it is customary to consider the standard temperature to be 25 °C.

For the general reaction $\alpha A + \beta B \rightleftharpoons \gamma C + \delta D$, ΔG^0 is defined as

$$\Delta G^0 = \gamma G_C^0 + \delta G_D^0 - \alpha G_A^0 - \beta G_B^0 \qquad (18\text{-}20)$$

At constant temperature, Eq. (18-19) is perfectly general and applies whether or not substances are in their standard states. If all species are in their standard states, Eq. (18-19) becomes

$$\Delta G^0 = \Delta H^0 - T\Delta S^0 \qquad (18\text{-}21)$$

We have already learned how to calculate ΔH^0 using tables of heats of formation or combustion, and Hess' law. (See Section 17.5.) In order to calculate ΔG^0 we also need to know how to calculate ΔS^0.

Absolute Entropies

In previous sections we have related the entropy to molecular disorder: The greater the disorder or freedom of motion of the atoms or molecules in a system, the greater the entropy. The most ordered arrangement of any given substance, with the least freedom of motion of the atoms, is a perfect crystal at zero degrees Kelvin (absolute zero). Therefore the lowest entropy any substance can attain is as a perfect crystal at absolute zero.

One formulation of **the third law of thermodynamics** states:[*] The entropy of perfect crystals of all pure elements and compounds is zero at the absolute zero of temperature.

As the temperature of any substance increases, the freedom of motion of the atoms of that substance also increases. Thus the entropy of any compound at a temperature above zero degrees Kelvin is greater than 0. There are straightforward ways of measuring and calculating how the entropy of any substance increases as the temperature increases, but we shall not discuss such methods here. What is important to understand is the following. Starting with the statement that the entropy of a pure substance is 0 at absolute zero, we can calculate what the entropy of that substance will be at 25 °C (298.2 K) and compile a table of values known as **absolute or third-law entropies**, denoted S^0. In contrast, there is no similar table of absolute energies or enthalpies, because the zero of energy is undefined. Absolute entropies can be evaluated because the state of zero entropy is known (from the third law) for all pure substances. Table 18.4 lists the absolute entropies of a few selected substances. A more extensive list will be found in Appendix G.

Examine the values listed in Table 18.4 carefully. Note that as a group, the absolute entropies of solids are lower than the entropies of liquids, which are lower

---

[*] Like the other two laws of thermodynamics, the third law is a generalization of many observations that have been made. It has been stated in several different ways, which can be shown to be equivalent.

Table 18.4. **The Absolute or Third Law Entropies (S^0)**
of Selected Substances in $J \cdot mol^{-1}K^{-1}$ at 25 °C

	Solids		
Ag	42.55	Cu	33.15
C(gr)	5.74	$CuSO_4$	109.
CaO	39.7	Zn	41.6
$CaCO_3$	92.9	ZnO	43.64
	Liquids		
CH_3OH	126.8	Hg	76.02
C_6H_6	172.8	H_2O	69.94
	Gases		
Ar	154.7	N_2	191.5
CO_2	213.6	NO_2	239.9
C_6H_6	269.2	N_2O_4	304.2
He	126.1	O_2	205.03
H_2O	188.7	SO_2	248.1

than the entropies of gases. Also note that the larger and more complex a molecule, the larger is its absolute entropy. Thus the monatomic gas He has lower entropy than the triatomic CO_2, which has lower entropy than gaseous benzene, C_6H_6, which has 12 atoms. Even for single atoms, the absolute entropy increases as the number of electrons and protons in the atom increases, that is, as the mass increases. Thus S^0_{298} for He(g) is 126.1, for Ne(g) is 144.1, and for Kr(g) is 164.0 $J \cdot mol^{-1}K^{-1}$. Among the solid elements the absolute entropy generally increases as the atomic number increases. Thus C ($Z = 6$) has a significantly lower absolute entropy than Cu ($Z = 29$) or Ag ($Z = 47$). All metallic solids have entropies below 85 $J \cdot mol^{-1}K^{-1}$.

The absolute entropy of graphite is higher than that of diamond (5.7 as compared to 2.4 $J \cdot mol^{-1}K^{-1}$ at 25 °C). Compare the crystal structures of diamond and graphite, Figs. 14.35 and 14.42. Note that the delocalized π electrons in graphite have much greater freedom of translational motion than any electrons in diamond. In general, for different crystalline modifications of the same substance, the weaker the bonds, that is, the fewer the restraints on the motions of the atoms, the higher the entropy.

EXAMPLE 18.3. Predicting relative S^0 values

Which member of each of the following pairs has the greater absolute entropy? Assume 1 mol of each substance.

(a) Al(s) and Pb(s), at the same temperature.

(b) Ag(s) at 298 K and Ag(s) at 398 K.

(c) LiCl(s) and $LaCl_3$(s) at the same temperature.

Solution

(a) Pb ($Z = 82$) has greater entropy than Al ($Z = 13$) at the same temperature. For elements, the entropy generally increases with increasing atomic number.

(b) Silver has greater entropy at 398 K than at 298 K. The entropy of any substance increases as its temperature increases.

(c) Lanthanum chloride, $LaCl_3$, has greater entropy than lithium chloride, LiCl, at the same temperature. Entropy increases with increasing chemical complexity.

Note particularly that there is no "Δ" in front of the symbol for absolute entropy, S^0. All absolute entropies are positive quantities since the lowest value of the entropy of any substance is 0. You should be careful to distinguish between the absolute entropy of a substance and the entropy of formation of that substance. For $N_2O_4(g)$, for instance, the entropy of formation refers to the reaction in which $N_2O_4(g)$ is formed from the elements:

$$N_2(g) + 2 O_2(g) \rightarrow N_2O_4(g) \tag{18-22}$$

The entropy change for this reaction is a negative quantity; there is much less freedom of motion for the atoms when they are bound together as a single N_2O_4 molecule than when they exist as separate N_2 and O_2 molecules. The standard absolute entropy of N_2O_4, on the other hand, is a positive quantity. It measures the greater freedom of motion possessed by N_2O_4 at 25 °C and 1 atm as compared to N_2O_4 at absolute zero.

Calculation of ΔS^0

For the general reaction $\alpha A + \beta B \rightleftharpoons \gamma C + \delta D$, the change in entropy when all substances are in their standard states, ΔS^0, is given by

$$\Delta S^0 = \gamma S_C^0 + \delta S_D^0 - \alpha S_A^0 - \beta S_B^0 \tag{18-23}$$

Examples 18.4(a) and 18.4(b) illustrate the use of Eq. (18-23).

EXAMPLE 18.4. Using absolute entropies to calculate ΔS^0 values

(a) Calculate ΔS^0 for $Zn(s) + \frac{1}{2} O_2(g) \rightleftharpoons ZnO(s)$.

Solution. Before you begin the numerical calculation it is helpful to ask yourself "Will this ΔS^0 value be positive or negative?" The left-hand side of this equation contains a solid plus a gas, whereas there is only a solid on the right. There is considerably less molecular disorder on the right-hand side, and therefore the entropy decreases as this reaction proceeds, so that ΔS^0 will be negative. The S^0 values needed are listed in Table 18.4.

$$\Delta S^0 = S_{ZnO}^0 - S_{Zn}^0 - \frac{1}{2} S_{O_2}^0$$
$$= 43.64 - 41.6 - \frac{1}{2}(205.03) = -100.5 \text{ J} \cdot \text{K}^{-1}$$

(b) Calculate ΔS^0 for $2NO_2(g) \rightleftharpoons N_2O_4(g)$.

Solution. The dimerization of NO_2 is also a reaction for which ΔS^0 is negative. There are 2 mol of gas on the left-hand side of this equation, and only 1 mol of gas on the right. In general, gases have so much more disorder than liquids and solids, that the side of an equation with the larger number of moles of gas is the side with greater entropy. Using S^0 values from Table 18.4,

$$\Delta S^0 = S_{N_2O_4}^0 - 2S_{NO_2}^0 = 304.2 - 2(239.9) = -175.6 \text{ J} \cdot \text{K}^{-1}$$

Calculation of ΔG^0 Values

Using Hess' law and a table of heats of formation (Table 17.1 or Appendix G) as well as Eq. (18-23) and a table of absolute entropies (Table 18.4 or Appendix G), we can calculate ΔG^0 for any desired reaction from Eq. (18-21), $\Delta G^0 = \Delta H^0 - T\Delta S^0$. Examples 18.5(a) and 18.5(b) illustrate the method of calculation.

EXAMPLE 18.5. **Calculation of ΔG^0 values**

(a) Calculate ΔG^0 for $2NO_2(g) \rightleftharpoons N_2O_4(g)$.

Solution. First we calculate ΔH^0_{298}.

$$\Delta H^0 = \Delta H^0_f(N_2O_4, g) - 2\Delta H^0_f(NO_2, g) = 9.16 - 2(33.2) = -57.2 \text{ kJ}$$

$\Delta S^0 = -175.6 \text{ J} \cdot \text{K}^{-1}$, as calculated in Example 18.4(b).

It is particularly important to note that ΔH^0 is given in kilojoules while ΔS^0 is in joules per kelvin. In order to use Eq. (18-21) both terms must be in the *same* units. The term $T\Delta S^0$ must be converted to kilojoules before it can be combined with ΔH^0. Probably the most common student error in using Eq. (18-21) is to have different units (kilojoules and joules) for the two terms in this equation. As it is customary to report ΔG^0 in kilojoules multiply the ΔS^0 value by 10^{-3} so that $T\Delta S^0$ is also in kilojoules. For the dimerization of N_2O_4 we obtain

$$\Delta G^0 = -57.2 - (298.15)(-175.6 \times 10^{-3}) = -57.2 + 52.36 = -4.8 \text{ kJ}$$

The negative sign of ΔG^0 indicates that the dimerization reaction $2NO_2(g) \rightleftharpoons N_2O_4(g)$ proceeds spontaneously to the right at 25 °C when both gases are at 1-atm pressure.

(b) Calculate ΔG^0 for $CaCO_3(s) \rightleftharpoons CaO(s) + CO_2(g)$.

Solution. First calculate ΔH^0_{298} using Hess' Law.

$$\Delta H^0 = \Delta H^0_f(CaO) + \Delta H^0_f(CO_2) - \Delta H^0_f(CaCO_3)$$
$$= -635.5 - 393.51 - (-1206.9) = +177.9 \text{ kJ}$$

Note that this is an endothermic reaction; heat must be supplied to decompose calcium carbonate to calcium oxide and carbon dioxide.

Next calculate ΔS^0 using Table 18.4.

$$\Delta S^0 = S^0_{CaO} + S^0_{CO_2} - S^0_{CaCO_3} = 39.7 + 213.6 - 92.9 = 160.4 \text{ J/K}$$

The entropy change is positive for this reaction. There is a large increase in molecular disorder when a solid compound is decomposed into a gas and a simpler solid compound.

$$\Delta G^0 = \Delta H^0 - T\Delta S^0 = 177.9 - (298.15)(160.4 \times 10^{-3}) \text{ kJ}$$
$$= 177.9 - 47.82 = 130.1 \text{ kJ}$$

Since $\Delta G^0 > 0$, this reaction does not proceed spontaneously to the right at 25 °C and 1 atm. Indeed, we have all observed that $CaCO_3$ does not decompose at room temperature, because the shells of many marine animals (clams, oysters, etc.) as well as the large deposits of marble in some mountain ranges, are principally $CaCO_3$.

The function ΔG^0 combines the enthalpy factor (ΔH^0) and the entropy factor ($T\Delta S^0$). A decrease in energy as a reaction proceeds leads to a negative value of ΔH^0, and contributes a negative term to ΔG^0. An increase in entropy as a reaction proceeds leads to a positive value of $T\Delta S^0$, and also contributes a negative term $(-T\Delta S^0)$ to ΔG^0. Thus we say that the enthalpy factor favors the products if $\Delta H^0 < 0$, and the entropy factor favors the products if $\Delta S^0 > 0$.

It is very common (but certainly not invariably the case) that ΔH^0 and ΔS^0 have the same sign, so that the two factors drive the reaction in opposite directions. In Example 18.5(a), for instance, ΔH^0 is negative (-57.2 kJ/mol) so that the enthalpy factor drives the reaction to the right, favoring N_2O_4. On the other hand, ΔS^0 is also negative (-175.6 J $\cdot$ mol^{-1}K^{-1}), so that the entropy factor drives the reaction to the left, favoring the NO_2. In other words, the state of lower energy is N_2O_4, whereas the state of higher entropy is $2NO_2$. It is only by determining the sign of ΔG^0 that we can predict whether or not the reaction does proceed spontaneously to the right. In this case, $\Delta G^0 < 0$ so that the enthalpy factor predominates and the reaction does proceed to the right when both gases are at 1 atm and 25 °C. Note, however, that the two terms in ΔG^0 are of opposite sign and quite close in magnitude, so that ΔG^0 is not very large.

In Example 18.5(b), ΔH^0 and ΔS^0 are both positive, so that the entropy factor favors the products, $CaO + CO_2$, while the enthalpy factor favors the reactant, $CaCO_3$. Again the enthalpy factor predominates. The decomposition of $CaCO_3$ does not proceed spontaneously at 25 °C because ΔG^0 is positive.

At temperatures close to 25 °C it is more common for the enthalpy factor to predominate over the entropy factor in determining the sign of ΔG^0, as in the two examples we have just discussed. However, there are many examples of reactions for which the entropy factor predominates over the enthalpy factor. Furthermore, as the temperature increases, the term $T\Delta S^0$ becomes larger and the entropy factor becomes more important in determining the sign of ΔG^0. Many reactions that do not proceed at 25 °C occur readily at elevated temperatures. The decomposition of $CaCO_3$ is an example of this. Above about 800 K (500 °C), calcium carbonate spontaneously decomposes into CaO(s) and CO_2(g). We will discuss the temperature dependence of ΔH, ΔS, and ΔG in Section 18.11.

Free Energies of Formation, ΔG_f^0

We have just described the calculation of ΔG^0 for any reaction, using Eq. (18-15). Clearly we can use this relation to obtain ΔG^0 for **formation** reactions, that is, reactions in which one mole of compound is formed from the elements in their standard states.

EXAMPLE 18.6. Calculation of ΔG_f^0 values

Calculate $\Delta G_f^0(N_2O_4, g)$.

Solution. The formation reaction is

$$N_2(g) + 2\,O_2(g) \rightleftharpoons N_2O_4(g)$$

$\Delta H_f^0(N_2O_4, g)$ is given in Appendix G. It is 9.16 kJ/mol. $\Delta S_f^0(N_2O_4, g)$ is obtained using Eq. (18-23) and Table 18.4.

$$\Delta S_f^0(N_2O_4) = S_{N_2O_4}^0 - S_{N_2}^0 - 2S_{O_2}^0 = 304.2 - 191.5 - 2(205.03)$$
$$= -297.3 \text{ J} \cdot \text{mol}^{-1}\text{K}^{-1}$$

Therefore

$$\Delta G_f^0(N_2O_4) = 9.16 - (298.15)(-297.3 \times 10^{-3}) = 9.16 + 88.66$$
$$= +97.82 \text{ kJ/mol}$$

Table 18.5. **Standard Free Energies of Formation, ΔG_f^0, in kJ/mol at 25 °C**

CaO(s)	−604.2	$H_2O(g)$	−228.59
$CaCO_3$(s)	−1128.8	H_2O(liq)	−237.18
$CH_4(g)$	−50.75	$NO_2(g)$	+51.30
CH_3OH(liq)	−166.4	$N_2O_4(g)$	+97.82
C_6H_6(liq)	+124.50	$NH_3(g)$	−16.5
$CO_2(g)$	−394.36	$SO_2(g)$	−300.19

In this way we can construct a table of standard free energies of formation at 25 °C. A brief list of ΔG_f^0 values is given in Table 18.5. A more extensive table will be found in Appendix G. Hess' Law can be applied to ΔG values just as well as to ΔH values, because both ΔG and ΔH are path-independent quantities. Thus we can calculate any ΔG^0 value using a table of standard free energies of formation. The method of calculation is illustrated in Example 18.7.

EXAMPLE 18.7. Calculation of ΔG^0 using ΔG_f^0 values

Using data in Table 18.5 calculate ΔG^0 for the following reactions:

(a)
$$2NO_2(g) \rightleftharpoons N_2O_4(g)$$

(b)
$$CaCO_3(s) \rightleftharpoons CaO(s) + CO_2(g)$$

Solution. You will recognize that this is the same problem as Example 18.5. We are just going to use an alternate method of calculation.

(a)
$$\Delta G^0 = \Delta G_f^0(N_2O_4) - 2\Delta G_f^0(NO_2)$$
$$= 97.82 - 2(51.30) = -4.78 \text{ kJ/mol}$$

The value obtained using Eq. (18-21) in Example 18.5(a) is −4.8 kJ/mol. The difference between the two numbers is due to rounding errors and experimental uncertainties in the last significant figure of the tabulated values. In both methods of calculation, ΔG^0 is a relatively small difference between two large values. Such numbers generally have large uncertainties.

(b)
$$\Delta G^0 = \Delta G_f^0(CaO) + \Delta G_f^0(CO_2) - \Delta G_f^0(CaCO_3)$$
$$= -604.2 - 394.4 - (-1128.8) = 130.2 \text{ kJ/mol}$$

As expected, this is identical with the value obtained in Example 18.5(b) within experimental error.

section 18.9
ΔG^0 and the Equilibrium Constant

Since the sign of ΔG for a reaction tells us whether or not the reaction will proceed spontaneously to the right as we have written it, and the ratio of the reaction quotient to the equilibrium constant, Q/K, also tells us whether or not the reaction will proceed spontaneously to the right [see Eqs. (8-14a), (8-14b), and (8-14c) and Section 8.2], there must be a relationship between ΔG and the ratio Q/K. That relationship is one of the most important and fundamental in thermodynamics, and is

$$\Delta G = RT \ln (Q/K) = -RT \ln (K/Q) \qquad (18\text{-}24)$$

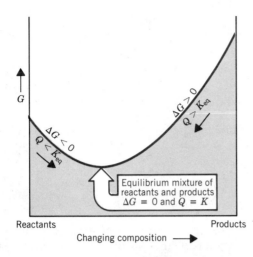

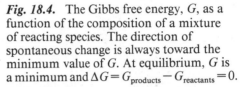

Fig. 18.4. The Gibbs free energy, G, as a function of the composition of a mixture of reacting species. The direction of spontaneous change is always toward the minimum value of G. At equilibrium, G is a minimum and $\Delta G = G_{\text{products}} - G_{\text{reactants}} = 0$.

The derivation of Eq. (18-24) is given in Appendix I.

If a system is not at equilibrium, $Q \neq K$. If $Q < K$, K/Q is greater than 1, and $\ln (K/Q)$ is positive, so that ΔG is negative. Both criteria, $Q < K$ and $\Delta G < 0$ tell us the same thing: The reaction will proceed spontaneously to the right to reach equilibrium.

If, on the other hand, $Q > K$, K/Q is less than 1, and $\ln (K/Q)$ is negative, so that ΔG is positive. Both criteria, $Q > K$ and $\Delta G > 0$ tell us that the reverse reaction will proceed spontaneously to reach equilibrium.

Figure 18.4 illustrates graphically how the free energy changes during a chemical reaction. At equilibrium, the free energy of the system is at a minimum.

From Eq. (18-24) we can obtain an extremely useful relationship by considering a system when all substances are in their standard states. What is the reaction quotient, Q, if all gases are at 1 atm and all solutes have a 1 M concentration? As all terms in the expression for Q would be 1 in this case, the value of Q must be 1 when all substances are in their standard states.

$$Q = 1 \text{ when all substances are in their standard states} \qquad (18\text{-}25)$$

By definition, the value of ΔG is ΔG^0 when all substances are in their standard states. Thus Eq. (18-24) becomes

$$\Delta G^0 = RT \ln (1/K) = -RT \ln K \qquad (18\text{-}26a)$$

If we wish to use base 10 logarithms rather than base e logarithms, Eq. (18-26a) becomes

$$\Delta G^0 = -2.303RT \log K \qquad (18\text{-}26b)$$

The value of R to be used in computations with Eqs. (18-26a) and (18-26b) must make the term RT have the same units as ΔG^0. Since we customarily express ΔG^0 in kilojoules per mole, R must be 8.3144×10^{-3} kJ $\cdot$ mol^{-1}K^{-1}. If ΔG^0 is given in kilocalories per mole and you want to use these equations to calculate the equilibrium constant, use $R = 1.9872 \times 10^{-3}$ kcal $\cdot$ mol^{-1}K^{-1}. Example 18.8 illustrates the use of Eq. (18-26a).

EXAMPLE 18.8. Calculation of K_{eq} from ΔG^0

Calculate the equilibrium constant at 25 °C for the following reactions:

(a) $2NO_2(g) \rightleftharpoons N_2O_4(g)$

(b) $$CaCO_3 \rightleftharpoons CaO(s) + CO_2(g)$$

Solution. These are, of course, the reactions of Example 18.7.

(a) ΔG^0 for the dimerization of NO_2 has been calculated to be -4.78 kJ/mol at 25 °C in Example 18.7**(a)**. Substituting this value into Eq. (18-26a) yields

$$-4.78 \text{ kJ/mol} = -\left(8.3144 \times 10^{-3} \frac{\text{kJ}}{\text{mol} \cdot \text{K}}\right)(298.15 \text{ K})\ln K_{eq}$$

or $$\ln K_{eq} = +1.93$$

Therefore

$$K_{eq} = e^{1.93} = 6.9 = P_{N_2O_4}/P_{NO_2}^2$$

The pressures of both gases must be given in atmospheres in this equilibrium constant, because the standard state is defined as the state in which each gas is at 1-atm pressure.

If you prefer to use base 10 logarithms, the calculation is

$$-4.78 \text{ kJ/mol} = -(8.3144 \times 10^{-3})(298.15 \text{ K})(2.303)\log K_{eq}$$

so that

$$\log K_{eq} = 0.837 \quad \text{and} \quad K_{eq} = 10^{0.837} = 6.9$$

Note that this equilibrium constant is greater than 1. This tells us that if we start with a mixture of the two gases, both at 1-atm pressure at 25 °C, the reaction will proceed to the right to use up some NO_2 and produce more N_2O_4. However, the equilibrium constant is not very much greater than 1, so there will be appreciable quantities of both gases present in an equilibrium mixture at 25 °C.

(b) ΔG^0 for the decomposition of $CaCO_3$ was calculated to be $+130.2$ kJ/mol in Example 18.7**(b)**. Using Eq. (18-26a) we obtain

$$+130.2 \text{ kJ/mol} = -\left(8.3144 \times 10^{-3} \frac{\text{kJ}}{\text{mol} \cdot \text{K}}\right)(298.15 \text{ K})\ln K_{eq}$$

so that $$\ln K_{eq} = -52.52$$

and

$$K_{eq} = e^{-52.52} = 1.6 \times 10^{-23} = P_{CO_2}$$

This equilibrium constant is very small compared to 1. Therefore the position of equilibrium is far to the left. The pressure of CO_2 in the air is greater than the value required to be in equilibrium with solid $CaCO_3$, so there is no decomposition of $CaCO_3$ at 25 °C.

Another useful equation can be obtained by substituting Eq. (18-26a) into Eq. (18-24). If we write Eq. (18-24) as

$$\Delta G = -RT \ln (K/Q) = -RT \ln K + RT \ln Q$$

and substitute $\Delta G^0 = -RT \ln K$, we obtain

$$\Delta G = \Delta G^0 + RT \ln Q \tag{18-27}$$

This equation can be used to determine the direction of reaction when substances are not in their standard states, as shown in Examples 18.9 and 18.10.

EXAMPLE 18.9. Calculation of ΔG from the reaction quotient, Q

If NO_2 and N_2O_4 are mixed in a flask so that $P_{NO_2} = 0.10$ atm and $P_{N_2O_4} = 0.50$ atm at 25 °C, will the reaction $2NO_2(g) \rightleftharpoons N_2O_4(g)$ proceed to the right or to the left?

Solution. The reaction quotient, Q, for these pressures is

$$Q = P_{N_2O_4}/P_{NO_2}^2 = 0.50/(0.10)^2 = 50$$

We have already calculated ΔG^0 for this reaction in Example 18.7(a). We can now calculate ΔG at the pressures given, using Eq. (18-27).

$$\Delta G = \Delta G^0 + RT \ln Q = -4.78 \text{ kJ/mol} + (8.3144 \times 10^{-3})(298.15)\ln 50 \text{ kJ/mol}$$
$$= -4.78 + (2.479)(3.912) = -4.78 + 9.70 = +4.92 \text{ kJ/mol}$$

As ΔG is positive this reaction proceeds to the left, and some of the N_2O_4 decomposes to yield more NO_2. The reaction will continue until $Q = K_{eq} = 6.9$.

EXAMPLE 18.10. The distinction between ΔG and ΔG⁰

A mixture of the three gases NO, Br_2, and NOBr is placed in a container at 25 °C so that at the instant of mixing

$$P_{Br_2} = 0.16 \text{ atm} \qquad P_{NO} = 0.10 \text{ atm} \qquad \text{and} \qquad P_{NOBr} = 0.80 \text{ atm}$$

Will the reaction

$$NOBr(g) \rightleftharpoons NO(g) + \tfrac{1}{2} Br_2(g)$$

proceed spontaneously to the right or to the left? The standard free energies of formation at 25 °C in kilojoules per mole for these three gases are

$$NO(g) \ 86.57 \qquad Br_2(g) \ 3.142 \qquad NOBr(g) \ 82.42$$

Solution. According to Hess' Law

$$\Delta G^0 = \tfrac{1}{2} \Delta G_f^0(Br_2) + \Delta G_f^0(NO) - \Delta G_f^0(NOBr)$$
$$= \tfrac{1}{2} (3.142) + 86.57 - 82.42 = +5.72 \text{ kJ} \cdot \text{mol}^{-1}$$

The reaction quotient, Q, at the instant of mixing is

$$Q = \frac{P_{Br_2}^{1/2} P_{NO}}{P_{NOBr}} = \frac{(0.16)^{1/2}(0.10)}{(0.80)} = 0.050 \text{ atm}^{1/2}$$

Substitution into Eq. (18-27) yields

$$\Delta G = 5.72 + RT \ln (0.050) = 5.72 + (8.3144 \times 10^{-3})(298.2)(-3.0)$$
$$= 5.72 - 7.43 = -1.71 \text{ kJ} \cdot \text{mol}^{-1}$$

Since $\Delta G < 0$, reaction will proceed spontaneously to the right when the three gases are mixed at these pressures. Note that ΔG^0 is positive, but ΔG is negative. If the three gases are each at 1 atm when mixed, the reaction proceeds spontaneously to the left, but at the pressures stipulated at the instant of mixing, the reaction proceeds to the right.

section 18.10
ΔG and the Nernst Equation

In Section 16.4 we used the sign of the cell potential (the EMF) as a criterion for predicting whether or not the net cell reaction occurring in a galvanic cell will proceed spontaneously to the right, as written. In this Chapter, Section 18.7, we have used the sign of ΔG as a criterion for predicting whether or not any reaction will proceed spontaneously to the right. We expect, therefore, that there is a relationship between ΔG for the net cell reaction and the cell potential, $\Delta \mathcal{E}_{cell}$. That relationship is

$$\Delta G = -n\mathcal{F}\Delta\mathcal{E}_{cell} \tag{18-28}$$

We obtain electrical work from a galvanic cell when the net cell reaction proceeds spontaneously toward a state of equilibrium. The maximum amount of work we can obtain is the product of the charge transferred and the cell potential. (Refer to the definition of the EMF in Section 16.2.) For the transfer of n moles of electrons, the amount of charge that passes through the cell is $n\mathcal{F}$, because the **faraday, $\mathcal{F}$,** is the charge on one mole of electrons. The electrical work done is therefore $n\mathcal{F}\Delta\mathcal{E}_{cell}$. The free energy of the system decreases as the reaction proceeds toward equilibrium (see Fig. 18.4) and the system produces electrical work. The decrease in the free energy is equal in magnitude to the maximum electrical work done, so that

$$\Delta G = -w_{max} = -n\mathcal{F}\Delta\mathcal{E}_{cell}$$

For the special situation when all substances are in their standard states, Eq. (18-28) becomes

$$\Delta G^0 = -n\mathcal{F}\Delta\mathcal{E}_{cell}^0 \tag{18-29}$$

Thus we can calculate ΔG^0 for the net cell reaction of a galvanic cell by using a Table of Standard Electrode Potentials, such as Appendix F, or we can calculate $\Delta\mathcal{E}_{cell}^0$ from ΔG^0.

EXAMPLE 18.11. Calculation of $\Delta\mathcal{E}_{cell}^0$ from ΔG^0

From the value of ΔG^0 calculate $\Delta\mathcal{E}_{cell}^0$ for the cell

$$Pt(s) \mid I_2(s) \mid I^- \parallel Fe^{3+}, Fe^{2+} \mid Pt(s)$$

Solution. First write the anode and cathode reactions.

Anode reaction	$2\,I^- \rightarrow I_2 + 2e^-$
Cathode reaction	$Fe^{3+} + e^- \rightarrow Fe^{2+}$
Net cell reaction	$2Fe^{3+} + 2\,I^- \rightarrow I_2 + 2Fe^{2+}$

The values of ΔG_f^0 needed can be found in Appendix G.

$$\Delta G^0 = 2\Delta G_f^0(Fe^{2+}) + \Delta G_f^0(I_2) - 2\Delta G_f^0(I^-) - 2\Delta G_f^0(Fe^{3+})$$
$$= 2(-84.94) + 0 - 2(-51.67) - 2(-10.6) = -45.4 \text{ kJ}$$

To obtain $\Delta\mathcal{E}_{cell}^0$ in volts using Eq. (18-29), ΔG^0 must be in joules, since volts × coulombs = joules. For the net cell reaction, as it is written above, $n = 2$. Substitution into Eq. (18-29) yields

$$\Delta G^0 = -4.54 \times 10^4 \text{ J} = -2(96,485 \text{ C})\Delta\mathcal{E}_{cell}^0, \quad \text{so that } \Delta\mathcal{E}_{cell}^0 = 0.235 \text{ V}$$

We can check this value using Appendix F.

$$\Delta\mathcal{E}_{cell}^0 = \mathcal{E}_{cathode}^0 - \mathcal{E}_{anode}^0 = +0.770 - 0.535 = +0.235 \text{ V}$$

Note that ΔG^0 is negative, and $\Delta\mathcal{E}_{cell}^0$ is positive. Both these criteria predict the reaction written occurs spontaneously.

The Nernst Equation

We have now obtained all the relations needed to derive the Nernst equation. If we substitute the expressions for ΔG and ΔG^0 given in Eqs. (18-28) and (18-29) into Eq. (18-27), $\Delta G = \Delta G^0 + RT \ln Q$ becomes

$$-n\mathcal{F}\Delta\mathcal{E}_{cell} = -n\mathcal{F}\Delta\mathcal{E}^0_{cell} + RT \ln Q$$

Dividing this equation by $-n\mathcal{F}$ yields the **Nernst equation:**

$$\Delta\mathcal{E}_{cell} = \Delta\mathcal{E}^0_{cell} - (RT/n\mathcal{F})\ln Q \tag{18-30}$$

It is customary to express the logarithm of the reaction quotient, Q, as a base 10 logarithm. The Nernst equation is then

$$\Delta\mathcal{E}_{cell} = \Delta\mathcal{E}^0_{cell} - (2.303RT/n\mathcal{F})\log Q \tag{18-31}$$

The value of $2.303RT/n\mathcal{F}$ at 25 °C is

$$\frac{(2.303)(8.3144 \text{ J} \cdot \text{mol}^{-1}\text{K}^{-1})(298.15 \text{ K})}{(n)(96,485 \text{ C/mol})} = \frac{0.05916}{n} \text{ volts}$$

since joules per coulomb = volts. The Nernst equation at 25 °C is therefore $\Delta\mathcal{E}_{cell} = \Delta\mathcal{E}^0_{cell} - (0.05916/n)\log Q$, which is the equation we used in Chapter 16.

By equating the two expressions for ΔG^0, Eq. (18-26a) and Eq. (18-29), we can obtain the general relation between $\Delta\mathcal{E}^0_{cell}$ and the equilibrium constant.

$$\Delta G^0 = -RT \ln K_{eq} = -n\mathcal{F}\Delta\mathcal{E}^0_{cell}$$

so that

$$\Delta\mathcal{E}^0_{cell} = (RT/n\mathcal{F})\ln K_{eq} \tag{18-32}$$

If we use base 10 logarithms and substitute numerical values for R, T, and $\mathcal{F}$, we obtain the relation

$$\log K_{eq} = (n\Delta\mathcal{E}^0_{cell})/0.05916 \quad \text{at 25 °C} \tag{18-33}$$

which we used in Chapter 16.

EXAMPLE 18.12. Calculation of ΔG^0 and K_{eq} from $\Delta\mathcal{E}^0_{cell}$

Is the oxidation of ferrous sulfate in sulfuric acid by the oxygen in the air a spontaneous reaction? Calculate ΔG^0 and the equilibrium constant for the reaction between ferrous ions and O_2 in acidic solution.

Solution. We must first write the two half-reactions.

Reduction	$4H^+(aq) + O_2(g) + 4e^- \rightarrow 2H_2O$
Oxidation	$Fe^{2+} \rightarrow Fe^{3+} + e^-$
Net reaction	$4Fe^{2+} + O_2(g) + 4H^+(aq) \rightarrow 4Fe^{3+} + 2H_2O$

The half-cell potentials are given in Appendix F.

$$\Delta\mathcal{E}^0_{cell} = \mathcal{E}^0_{cathode} - \mathcal{E}^0_{anode} = 1.229 - 0.770 = 0.459 \text{ V}$$

Since $\Delta\mathcal{E}^0_{cell}$ is positive, the reaction will proceed to the right spontaneously if all substances are in their standard states. We can calculate ΔG^0 from Eq. (18-29). The value of n for this reaction is 4. We therefore obtain

$$\Delta G^0 = -4(96,485 \text{ C})(0.459 \text{ V}) = -1.77 \times 10^5 \text{ J} = -177 \text{ kJ}$$

The large negative value of ΔG^0 again indicates that the reaction proceeds spontaneously to the right. We can obtain the equilibrium constant using either Eq. (18-26a) or Eq. (18-33):

$$\Delta G^0 = -177 \text{ kJ} = -RT \ln K_{eq}$$

or

$$\ln K_{eq} = \frac{177}{(8.3144 \times 10^{-3})(298.15)} = 71.4$$

Therefore

$$K_{eq} = e^{71.4} = 1 \times 10^{31}$$

This is a very large equilibrium constant. Ferrous ions in acidic solution are indeed readily air oxidized to ferric ions.

The following example emphasizes the distinction between ΔG and ΔG^0, as well as that between $\Delta \mathcal{E}_{cell}$ and $\Delta \mathcal{E}_{cell}^0$.

EXAMPLE 18.13. Relations between $\Delta \mathcal{E}_{cell}$, $\Delta \mathcal{E}_{cell}^0$, ΔG, and ΔG^0

Consider the cell $Co(s) \mid Co^{2+}(c_1) \parallel Ni^{2+}(c_2) \mid Ni(s)$ at 25 °C.

(a) Write the net cell reaction.

Solution. First write the half-reactions.

Cathode reaction	$Ni^{2+} + 2e^- \rightarrow Ni(s)$
Anode reaction	$Co(s) \rightarrow Co^{2+} + 2e^-$
Net cell reaction	$Ni^{2+} + Co(s) \rightarrow Ni(s) + Co^{2+}$

(b) Calculate $\Delta \mathcal{E}_{cell}^0$, ΔG^0, and K_{eq}.

Solution

$$\Delta \mathcal{E}_{cell}^0 = \mathcal{E}_{cathode}^0 - \mathcal{E}_{anode}^0 = -0.25 - (-0.28) = +0.03 \text{ V}$$
$$\Delta G^0 = -n\mathcal{F}\Delta\mathcal{E}_{cell}^0 = -(2)(96,485 \text{ C})(0.03 \text{ V}) = -5.8 \text{ kJ} = -6 \text{ kJ}$$

We can calculate the equilibrium constant using either Eq. (18-26a) or Eq. (18-33).

$$\log K_{eq} = (n\Delta\mathcal{E}_{cell}^0)/0.0592 = (2)(0.03)/0.0592 = 1.01$$

so that $K_{eq} = 10^{1.01} = 10 = 1 \times 10^1$. We are only entitled to one significant figure, since $\Delta\mathcal{E}_{cell}^0$ is only known to one figure.

(c) Calculate Q, $\Delta\mathcal{E}_{cell}$, and ΔG when $c_1 = 1.00 \ M$ and $c_2 = 0.0010 \ M$.

Solution

$$Q = [Co^{2+}]/[Ni^{2+}] = c_1/c_2 = 1.00/0.0010 = 1.0 \times 10^3$$
$$\Delta\mathcal{E}_{cell} = \Delta\mathcal{E}_{cell}^0 - (0.0592/n) \log Q = +0.03 - (0.0592/2) \log 10^3$$
$$= 0.03 - (3)(0.0592)/2 = 0.03 - 0.089 = -0.06 \text{ V}$$
$$\Delta G = -n\mathcal{F}\Delta\mathcal{E}_{cell} = -2(96,485 \text{ C})(-0.06 \text{ V}) = +11.6 \text{ kJ} = 12 \text{ kJ}$$

(d) Will the reaction proceed spontaneously from left to right when all substances are in their standard states?

Solution. When all substances are in their standard states, $c_1 = c_2 = 1.00\ M$ and $Q = 1$. Since $K_{eq} = 10$, $Q < K_{eq}$, and the reaction will proceed spontaneously to the right. Alternatively, one could say that since $\Delta \mathscr{E}^0_{cell} = +0.03 > 0$, the reaction will proceed spontaneously to the right when all substances are in their standard states. A third way of reaching the same conclusion is to note that when all substances are in their standard states $\Delta G = \Delta G^0 = -6$ kJ. Since $\Delta G < 0$, the reaction will proceed spontaneously to the right.

(e) Will the reaction proceed spontaneously from left to right when $c_1 = 1.00\ M$ and $c_2 = 1.0 \times 10^{-3}\ M$?

Solution. When $[Co^{2+}] = 1.00\ M$ and $[Ni^{2+}] = 1.0 \times 10^{-3}\ M$, $\Delta \mathscr{E}_{cell} = -0.06$ V, $Q = 1.0 \times 10^3$, and $\Delta G = +12$ kJ. Since $\Delta \mathscr{E}_{cell} < 0$, $Q > K_{eq}$, and $\Delta G > 0$, the net cell reaction given in part **(a)** will *not* proceed spontaneously. Instead the reverse reaction, $Ni(s) + Co^{2+} \rightarrow Co(s) + Ni^{2+}$, proceeds spontaneously.

In summary, we can see that the several criteria we have used to predict spontaneity of a reaction are related by Eq. (18-24) and Eq. (18-28):

$$\Delta G = RT \ln (Q/K) \quad \text{and} \quad \Delta G = -n\mathscr{F}\Delta \mathscr{E}_{cell}$$

If $Q < K$, $(Q/K) < 1$, and $\ln (Q/K)$ is negative, so that ΔG is also negative. Similarly, if $\Delta \mathscr{E}_{cell}$ is positive, ΔG is negative.

A reaction will proceed spontaneously from left to right as written, *if*	$\Delta G < 0$	or	$Q < K$	or $\quad \Delta \mathscr{E}_{cell} > 0$	(18-34a)
A reaction will proceed in the *reverse* direction, from right to left, *if*	$\Delta G > 0$	or	$Q > K$	or $\quad \Delta \mathscr{E}_{cell} < 0$	(18-34b)
The system is at equilibrium and no net reaction will occur, *if*	$\Delta G = 0$	or	$Q = K$	or $\quad \Delta \mathscr{E}_{cell} = 0$	(18-34c)

section 18.11
The Temperature Dependence of ΔH, ΔS, ΔG, and K_{eq}

In order to answer the question "Will a given reaction proceed spontaneously to the right as we have written it?" we must determine the sign of ΔG, if the reaction is to occur at a fixed temperature and pressure.

Suppose there is a reaction we want to run in order to obtain certain products. We calculate ΔG for this reaction at 25 °C and find that it is positive. Does that mean that it will be impossible for the reaction to occur? Not necessarily, because it may well be that at a different temperature ΔG will be negative. If we know how ΔG varies with temperature we may be able to find a temperature at which the reaction proceeds and the equilibrium constant is large enough to obtain a reasonable yield of product. Since $\Delta G = \Delta H - T\Delta S$, we must consider the temperature dependence of both ΔH and ΔS in order to determine the temperature dependence of ΔG.

Table 18.6. The Temperature Dependence of ΔH^0 and ΔS^0 for Three Reactions

Reaction	$\Delta H^0_{298}(kJ)$	$\Delta H^0_{600}(kJ)$	$\Delta S^0_{298}(J/K)$	$\Delta S^0_{600}(J/K)$
$NO(g) + \frac{1}{2}O_2(g) \rightleftharpoons NO_2(g)$	-57.0	-59.1	-73.3	-77.9
$SO_3(g) \rightleftharpoons SO_2(g) + \frac{1}{2}O_2(g)$	$+98.9$	$+97.5$	$+94.0$	$+93.4$
$CaCO_3(s) \rightleftharpoons CaO(s) + CO_2(g)$	$+177.9$	$+176.$	$+160.4$	$+156.3$

The Temperature Dependence of ΔH and ΔS

Experimentally, we find that for most reactions ΔH and ΔS change very slowly with temperature. For temperature changes of about 100°, ΔH usually changes by no more than 1% and ΔS by no more than 2–3%. Data for a few reactions are given in Table 18.6. Examine that data and note that for an increase in temperature of 300° both ΔH^0 and ΔS^0 remain approximately constant.

It is not difficult to understand why ΔH and ΔS vary so little with temperature. By definition, $\Delta H = H$ of products $- H$ of reactants. Increasing the temperature will increase the enthalpy of the products and will also increase the enthalpy of the reactants, by about the same amount. Thus the difference between them remains virtually unchanged. A similar argument applies to the change in entropy, ΔS.

The Temperature Dependence of ΔG

It is quite a different story for ΔG, however; ΔG is markedly temperature dependent because of the T in $\Delta G = \Delta H - T\Delta S$. Example 18.14 illustrates the change in ΔG with a change in temperature.

EXAMPLE 18.14. Calculation of ΔG^0 at different temperatures

Using the data of Table 18.6, calculate ΔG^0 at both 298 and 600 K for the reaction $CaCO_3(s) \rightleftharpoons CaO(s) + CO_2(g)$.

Solution. Recall that we have already calculated ΔG^0_{298} in Example 18.5(b).

$$\Delta G^0_{298} = 177.9 - (298.15)(0.1604)$$
$$= 177.9 - 47.8 = 130.1 \text{ kJ}$$
$$\Delta G^0_{600} = 176 - (600)(0.1563) = 176 - 93.8 = 82 \text{ kJ}$$

Note that the entropy factor ($T\Delta S^0$) is much larger at the higher temperature.

Consider two temperatures T_1 and T_2, about 100° apart, so that to a very good degree of approximation ΔH and ΔS are constant between T_1 and T_2. Then,

$$\Delta G_1 = \Delta H - T_1\Delta S \quad \text{and} \quad \Delta G_2 = \Delta H - T_2\Delta S$$

so that ΔG will change significantly as the temperature changes from T_1 to T_2. Note that as the temperature increases the entropy factor becomes increasingly important in determining the sign of ΔG.

The Temperature Dependence of K_{eq}

The equilibrium constant also changes markedly as the temperature changes. In Eq. (18-26a), $\Delta G^0 = -RT \ln K_{eq}$, ΔG^0 is the free energy change when all gases are at 1 atm and all substances in solution are at 1.00 M concentration, at the temperature T. Both ΔG^0 and RT vary as the temperature is changed.

If temperatures T_1 and T_2 are close enough together so that it is a good approximation to assume that ΔH^0 and ΔS^0 are constant, we may combine Eq. (18-26a) and Eq. (18-21) to yield

$$\Delta G_1^0 = -RT_1 \ln K_1 = \Delta H^0 - T_1 \Delta S^0$$

and

$$\Delta G_2^0 = -RT_2 \ln K_2 = \Delta H^0 - T_2 \Delta S^0$$

In these equations K_1 is the equilibrium constant at temperature T_1 and K_2 is the equilibrium constant at temperature T_2.

If we rearrange each of these equations to solve for the logarithm of the equilibrium constant we obtain

$$\ln K_1 = -\frac{\Delta H^0}{RT_1} + \frac{\Delta S^0}{R} \tag{18-30a}$$

and

$$\ln K_2 = -\frac{\Delta H^0}{RT_2} + \frac{\Delta S^0}{R} \tag{18-30b}$$

Note that Eq. (18-30a) and Eq. (18-30b) are correct only if ΔH^0 and ΔS^0 do not vary with temperature. If that assumption is valid, as it usually is, then $\Delta S^0/R$ is a constant, and $-\Delta H^0/R$ is a constant, so that a plot of $\ln K_{eq}$ as a function of $1/T$ will be a straight line with slope $-\Delta H^0/R$.

If we subtract Eq. (18-30b) from Eq. (18-30a) we obtain

$$\ln K_1 - \ln K_2 = -\frac{\Delta H^0}{RT_1} + \frac{\Delta H^0}{RT_2} \tag{18-31a}$$

or

$$\ln (K_1/K_2) = -\frac{\Delta H^0}{R}\left(\frac{1}{T_1} - \frac{1}{T_2}\right) \tag{18-31b}$$

This equation is known as the **van't Hoff equation.** Another useful form for the purposes of computation is

$$\ln (K_1/K_2) = -\frac{\Delta H^0}{R}\left(\frac{T_2 - T_1}{T_1 T_2}\right) \tag{18-31c}$$

The van't Hoff equation may also, of course, be written in terms of base 10 logarithms, as

$$\log (K_1/K_2) = -\frac{\Delta H^0}{2.303R}\left(\frac{1}{T_1} - \frac{1}{T_2}\right) \tag{18-31d}$$

A consideration of this form of the equation shows that a plot of $\log K_{eq}$ versus $1/T$ is a straight line with slope $-\Delta H^0/2.303R$, as long as the temperature range covered is small enough so that the value of ΔH^0 is approximately constant.

Values of the equilibrium constant of the dimerization reaction $2NO_2(g) \rightleftharpoons N_2O_4(g)$ as a function of temperature are listed in Table 18.7. A plot of $\log K$ versus $1/T$ for the dimerization is shown in Fig. 18.5. Since the plot is linear, we can determine ΔH^0 from the slope of the plot. The total change in temperature for the data of Table 18.7 is $61°$, and therefore it is an excellent approximation that ΔH^0 remains constant.

The slope of the plot is determined graphically to be 3.03×10^3. (See Fig. 18.5.) From the slope we can determine ΔH^0 as follows:

$$\text{slope} = -(\Delta H^0)/2.303R = 3.03 \times 10^3$$

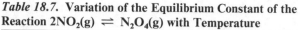

Table 18.7. Variation of the Equilibrium Constant of the Reaction $2NO_2(g) \rightleftharpoons N_2O_4(g)$ with Temperature

$T(K)$	K_{eq}	$\log K$	$(1/T) \times 10^3$
282.2	33.2	1.52	3.54
293.2	13.1	1.12	3.41
298.2	8.79	0.944	3.35
306.2	4.77	0.679	3.27
325.2	1.26	0.100	3.08
333.2	0.751	−0.124	3.00
343.2	0.408	−0.389	2.91

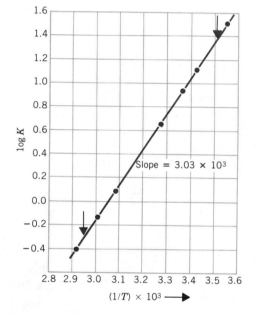

Fig. 18.5. Plot of $\log_{10} K$ versus $1/T$ for the reaction

$$2NO_2(g) \rightleftharpoons N_2O_4(g)$$

ΔH^0 is determined from the slope of this plot. To calculate the slope, choose two widely separated points on the line (not experimental points), read their coordinates, and evaluate the slope, m,

$$m = \frac{y_2 - y_1}{x_2 - x_1}$$

The points used for the slope determination of this plot are marked by colored arrows. Their coordinates are $(2.942 \times 10^{-3}, -0.300)$ and $(3.500 \times 10^{-3}, 1.390)$.

Therefore

$$\Delta H^0 = -(3.03 \times 10^3)(2.303)(8.3144 \times 10^{-3}) = -58.0 \text{ kJ}$$

The dimerization is an exothermic reaction. As the temperature increases, the equilibrium shifts to the left, in accordance with Le Chatelier's principle. Examine the data in Table 18.7 and note that the equilibrium constant gets smaller as the temperature increases.

Calculations using Eq. (18-31) are illustrated in Example 18.15.

EXAMPLE 18.15. The van't Hoff equation

For the reaction $SO_3(g) \rightleftharpoons SO_2(g) + \frac{1}{2}O_2(g)$ the equilibrium constant has been determined to be 1.57×10^{-1} at 900 K and 5.13×10^{-1} at 1000 K.

(a) Calculate ΔG^0_{1000} and ΔH^0 for this reaction between 900 and 1000 K.

Solution

$$\Delta G^0_{1000} = -R(1000)\ln K_{1000} = -(8.314 \times 10^{-3})(1000)\ln 0.513$$
$$= -(8.314)(-0.667) = +5.55 \text{ kJ}$$

Note that to obtain ΔG_{1000}^0 in kilojoules we must use R in kilojoules per mole per kelvin.

We obtain an average value of ΔH^0 between 900 and 1000 K using Eq. (18-31c).

$$\ln\left(\frac{K_{1000}}{K_{900}}\right) = \ln\left(\frac{0.513}{0.157}\right) = -\frac{\Delta H^0}{8.3144 \times 10^{-3}}\left[\frac{9 \times 10^2 - 1 \times 10^3}{(9 \times 10^2)(1 \times 10^3)}\right]$$

$$= \ln(3.27) = +\Delta H^0(100)/(8.3144)(9 \times 10^2)$$

so that

$$\Delta H^0 = (8.3144)(9)(1.184) = 88.6 \text{ kJ}$$

(b) Assuming ΔH^0 remains constant between 900 and 1200 K, calculate K_{1200}, the equilibrium constant at 1200 K. Also calculate ΔG_{1200}^0.

Solution. Again, we use the van't Hoff equation, Eq. (18-31c).

$$\ln\left(\frac{K_{1200}}{K_{1000}}\right) = \frac{88.6}{8.314 \times 10^{-3}}\left[\frac{1000 - 1200}{(1000)(1200)}\right] = \frac{(88.6)(200)}{(8.314 \times 10^{-3})(10^3)(1.2 \times 10^3)}$$

$$= 1.776 = 1.78$$

Therefore

$$K_{1200}/K_{1000} = e^{1.776} = 5.906 = 5.91$$

$$K_{1200} = (5.906)(K_{1000}) = (5.906)(0.513) = 3.03$$

From K_{1200} we determine ΔG_{1200}^0 using Eq. (18-26a)

$$\Delta G_{1200}^0 = -(8.314 \times 10^{-3})(1.2 \times 10^3)(\ln 3.03) = -11.1 \text{ kJ}$$

(c) Using Appendix G, calculate ΔH_{298}^0 and ΔG_{298}^0.

Solution

$$\Delta H_{298}^0 = \Delta H_f^0(SO_2) - \Delta H_f^0(SO_3)$$
$$= -296.8 - (-395.7) = +98.9 \text{ kJ}$$
$$\Delta G_{298}^0 = \Delta G_f^0(SO_2) - \Delta G_f^0(SO_3) = -300.2 - (-371.1) = +70.9 \text{ kJ}$$

It is useful to tabulate ΔH^0 and ΔG^0 at two widely separated temperatures and compare the percentage change in each.

	298 K	1000 K	Change
ΔH^0 (kJ)	98.9	88.6	−9.9%
ΔG^0 (kJ)	70.9	5.5	−92.2%

We observe that over a temperature range of 700°, ΔH^0 has changed relatively little, while ΔG^0 has changed by a large amount.

(d) Does the reaction proceed spontaneously to the right at 25 °C? Which factor, the enthalpy or the entropy, predominates at 25 °C?

Solution. At 25 °C, ΔG_{298}^0 is positive. The reaction does not proceed spontaneously to the right. The entropy factor favors the products, because there are $\frac{3}{2}$ mol of gas on the right and only 1 mol of gas on the left. The enthalpy factor, on the other hand, favors the reactant, SO_3, because the reaction is endothermic. Both ΔH^0 and ΔS^0 are positive. At 25 °C the enthalpy factor predominates and the position of equilibrium is far to the left. Sulfur trioxide does not decompose to SO_2 and O_2 at 25 °C.

(e) Does the reaction proceed spontaneously to the right at 1200 K? Which factor, the entropy or the enthalpy, predominates at 1200 K?

Solution. At 1200 K the reaction proceeds spontaneously to the right because ΔG^0_{1200} is negative. [See part (b).] At this high temperature the equilibrium constant is larger than 1, and SO_3 spontaneously dissociates to form SO_2 and O_2. The enthalpy factor still favors the reactant, SO_3, but at the higher temperature the effect of the increase in disorder becomes more important.

Summary

Spontaneous or naturally occurring processes are frequently exothermic (that is, the final state is lower in energy than the initial state), but there are a number of examples of endothermic spontaneous processes as well.

Processes that occur spontaneously that do not involve any change in energy always occur with an increase in molecular disorder or randomness. The thermodynamic function that measures the amount of molecular disorder is called the **entropy**, S.

The energy of an ideal gas is a function of temperature only. As long as the temperature is constant, $\Delta E = 0$ for an ideal gas. When an ideal gas expands spontaneously into a larger volume, its entropy increases because the molecules have more freedom of motion in a larger volume. The change in entropy depends only on the initial and final state of the system and not on the path used to carry out the expansion. Thus entropy is a state function.

Although there are many paths of reaction, that is, many ways in which a system may change from an initial state to a final state, there is one path of special interest, the **reversible path.** The reversible path is one in which the total change is carried out in a series of infinitesimal steps, and at each step the system is in equilibrium with its surroundings. The work done by the system along the reversible path is the maximum work possible for that process.

For an isothermal process, the change in entropy is defined as $\Delta S = q_{rev}/T$, where q_{rev} is the heat absorbed when the process is carried out reversibly.

The entropy of vaporization, $\Delta S_{vap} = \Delta H_{vap}/T_{bp}$, at the normal boiling point, is approximately the same for most liquids. **Trouton's Rule** states that $\Delta S_{vap} = 88 \pm 5$ $J \cdot mol^{-1}K^{-1}$ for most liquids. Hydrogen bonded liquids are exceptions to Trouton's Rule, as they have unusually large entropies of vaporization due to the greater order that exists in the liquid, relative to nonhydrogen-bonded liquids.

The entropy of fusion, $\Delta S_{fus} = \Delta H_{fus}/T_{mp}$, is always less than the entropy of vaporization of the same substance. The increase in molecular disorder that occurs when a mole of liquid is converted to a mole of vapor is significantly larger than the increase in molecular disorder that occurs when a mole of solid is converted to the liquid phase.

The **second law of thermodynamics** states that spontaneous processes that occur in an isolated system are always accompanied by an increase in entropy. However, we do not usually work with isolated systems, and therefore find it useful to define a thermodynamic function that can be used as a criterion to predict the direction of a spontaneous reaction that occurs at constant temperature and pressure. That function is the **Gibbs free energy**, $G = H - TS$.

The drive to reach a state of equilibrium is composed of two tendencies: (1) the tendency to decrease the energy, and (2) the tendency to increase the entropy. Both a decrease in energy and an increase in entropy lead to a decrease in the Gibbs free energy, G. Any spontaneous process at constant temperature and pressure occurs with a decrease in free energy, that is, $\Delta G < 0$.

The standard free energy change, ΔG^0, is the change in free energy when all substances are in their **standard states.** We can calculate ΔG^0 from $\Delta G^0 = \Delta H^0 - T\Delta S^0$. Values of ΔH^0 are calculated using Hess' law and a table of standard heats of formation or of combustion. Values of ΔS^0 are calculated using a table of **absolute entropies,** S^0.

The **third law of thermodynamics** states that the entropy of perfect crystals of all pure elements and compounds is zero at zero degrees Kelvin. At any temperature above absolute zero, the entropy must be greater than 0, as entropy increases with increasing temperature. The value of the entropy at 25 °C and 1 atm, calculated as the increase in entropy above zero as the temperature is increased from absolute zero to 298.2 K, is tabulated as S^0, the absolute entropy at 25 °C.

In general, the entropies of gases are much larger than the entropies of liquids, which are larger than the entropies of solids. The larger and more complex a molecule, the greater is its absolute entropy. For single atoms, the entropy increases with increasing mass. For a reaction involving a change in the number of moles of gas, the side with the larger number of moles of gas is the side with greater entropy.

ΔS^0 for the general reaction $\alpha A + \beta B \rightleftharpoons \gamma C + \delta D$ is calculated as $\Delta S^0 = \gamma S_C^0 + \delta S_D^0 - \alpha S_A^0 - \beta S_B^0$.

The change in the free energy for the reaction in which one mole of a substance at 25 °C and 1 atm is formed from the elements in their standard states is called the standard **free energy of formation,** ΔG_f^0. Tables of ΔG_f^0 values can be used to calculate ΔG^0 for any reaction, since Hess' law applies to ΔG as well as to ΔH.

Both the ratio of the reaction quotient, Q, to the equilibrium constant, K, and the sign of ΔG can be used to predict the direction of a spontaneous reaction. The relationship between these two quantities is $\Delta G = RT \ln (Q/K) = -RT \ln (K/Q)$. When all substances are in their standard states, $Q = 1$ and $\Delta G = \Delta G^0$. The general relation therefore becomes $\Delta G^0 = -RT \ln K$, which is used to calculate the equilibrium constant from values of ΔG^0. Another useful form of this relation is $\Delta G = \Delta G^0 + RT \ln Q$.

For a reaction that takes place in a galvanic cell, the maximum amount of electrical work that can be done by the system is the product of the charge transferred, $n\mathcal{F}$, and the EMF, $\Delta \mathcal{E}_{cell}$. The decrease in free energy as the reaction proceeds spontaneously toward equilibrium is equal in magnitude to the maximum work done by the system: $\Delta G = -n\mathcal{F} \Delta \mathcal{E}_{cell}$. In this equation, n is the number of moles of electrons transferred when the net cell reaction takes place, and $\mathcal{F}$ is the **faraday,** the charge in coulombs on one mole of electrons. When all substances are in their standard states, this becomes $\Delta G^0 = -n\mathcal{F} \Delta \mathcal{E}_{cell}^0$.

From the relation between ΔG and the cell potential we can derive the **Nernst equation,**

$$\Delta \mathcal{E}_{cell} = \Delta \mathcal{E}_{cell}^0 - (RT/n\mathcal{F}) \ln Q$$

The two equations $\Delta G = RT \ln (Q/K)$ and $\Delta G = -n\mathcal{F} \Delta \mathcal{E}_{cell}$ relate the three criteria we have used to predict the direction of a spontaneous reaction. If $\Delta G < 0$ then $Q < K$ and also $\Delta \mathcal{E}_{cell} > 0$.

Both ΔH and ΔS change slowly with temperature. For most reactions, when the temperature changes by $\sim 100°$, ΔH changes by no more than 1% and ΔS by no more than 2–3%. Both ΔG and the equilibrium constant, however, change markedly with a change in temperature. For changes in temperature small enough so that ΔH^0 is approximately constant, the temperature dependence of the equilibrium constant is given by the van't Hoff equation,

$$\ln (K_1/K_2) = -\frac{\Delta H^0}{R}\left(\frac{T_2 - T_1}{T_1 T_2}\right)$$

Exercises

Section 18.1

1. Consider two bulbs connected by a stopcock. One bulb contains $N_2(g)$, the other contains $Ar(g)$. Both gases are at the same temperature and pressure. Describe what happens when the stopcock is opened. Which state has more molecular disorder, the final state or the initial state? Explain your answer. Assuming the gases are ideal, what is ΔE for this process?

2. Consider two bulbs connected by a stopcock, maintained at a constant temperature of 30.0 °C. One bulb has a volume of 2.00 L and is evacuated. The other bulb has a volume of 3.00 L and contains an ideal gas at 2.50-atm pressure. What are w, the work done on the gas, ΔE, and q, the heat absorbed by the gas, when the stopcock is opened? Which state has more molecular disorder, the final state or the initial state? Explain your answer.

3. An ideal gas at 2.50-atm pressure and 30.0 °C is contained in a cylinder sealed by a piston. The initial volume of the gas is 3.00 L. The external pressure on the piston is changed to 1.50 atm and the gas expands against a constant external pressure of 1.50 atm until its own pressure is 1.50 atm. If the temperature of the gas is constant, calculate w, ΔE, and q for the expansion. Is the change in molecular disorder for this process greater than, less than, or the same as that for the process described in Exercise 2? Explain your answer.

Sections 18.2, 18.3, and 18.4

4. An object I am holding falls to the floor. If I pick it up and hold it at exactly the same height at which I held it originally, is the fall of the object a reversible or irreversible process? Explain your answer.

5. What is the maximum amount of work that can be done by an ideal gas as it expands from an initial state in which $P_1 = 2.50$ atm and $V_1 = 3.00$ L to a final state in which $P_2 = 1.50$ atm and $V_2 = 5.00$ L, at a constant temperature of 30.0 °C?

6. A sample of an ideal gas is expanded reversibly at a constant temperature of 25.0 °C from an initial pressure of 4.00 atm and a volume of 250.0 mL to a final pressure of 1.00 atm. Calculate ΔE, w, and q for this process.

Section 18.5

7. Calculate ΔS for the process described in Exercise 5.

8. Calculate ΔS for the process described in Exercise 6.

9. The melting points, boiling points, heats of fusion, and heats of vaporization of four substances are listed below. Calculate ΔS_{fus} and ΔS_{vap} for each of these substances. Explain why ΔS_{vap} is greater than ΔS_{fus} for all of these compounds.

Compound	$t_{mp}(°C)$	$\Delta H_{fus}(kJ/mol)$	$t_{bp}(°C)$	$\Delta H_{vap}(kJ/mol)$
Arsenic tribromide, $AsBr_3$	+32.8	11.8	+221	41.4
Chloroform, $CHCl_3$	−63.5	9.20	+61.7	29.4
Methyl chloride, CH_3Cl	−97.73	6.45	−24.2	21.6
Methyl mercaptan, CH_3SH	−123.	5.91	+6.2	24.5

10. Calculate ΔS_{vap} for each of the five compounds for which the normal boiling points and heats of vaporization are listed below. Which of these compounds does not obey Trouton's Rule? Account for the exceptions to Trouton's Rule.

Compound	$t_{bp}(°C)$	$\Delta H_{vap}(kJ/mol)$
Boron trichloride, BCl_3	+12.5	23.9
Methanol, CH_3OH	+64.96	37.5
Methyl bromide, CH_3Br	+3.56	24.0
Hydrazine, N_2H_4	+113.5	40.6
Phosphorus trichloride, PCl_3	+76.	30.5

Section 18.6

11. Occasionally one reads in a newspaper or magazine that someone has argued that the theory of evolution is contradictory to the second law of thermodynamics. The argument made is that because higher forms of life are more ordered and structured than lower forms of life, they could not have evolved from the lower forms because that would imply a spontaneous process accompanied by a decrease in entropy. Is this argument a valid criticism of the theory of evolution? Explain your answer.

Sections 18.7 and 18.8

Use Appendix G for the necessary data.

12. Without doing any numerical calculations, predict the sign of ΔS^0 for each of the following reactions. Explain your answers.
   (a) $N_2(g) + 3H_2(g) \rightleftharpoons 2NH_3(g)$
   (b) $(NH_4)_2SO_4(s) \rightleftharpoons 2NH_3(g) + H_2O(g) + SO_3(g)$
   (c) $C_2H_6(g) + \frac{7}{2}O_2(g) \rightleftharpoons 2CO_2(g) + 3H_2O(liq)$

13. Which member of each of the following pairs has the greater absolute entropy? Except where specified to the contrary, assume 1 mol of each substance at the same temperature and pressure.
   (a) He (g, 100 atm) and He (g, 0.200 atm)
   (b) Pt(s) and Fe(s)
   (c) $NH_4F(s)$ and $(NH_4)_2CO_3(s)$
   (d) $C_4H_{10}(g)$ and $CH_4(g)$
   (e) Cu(s, 373 K) and Cu(s, 273 K)

14. From the following data at 25 °C, calculate ΔG^0 for the reaction

$$CH_4(g) + 2O_2(g) \rightleftharpoons CO_2(g) + 2H_2O(liq)$$

	$CH_4(g)$	$O_2(g)$	$CO_2(g)$	$H_2O(liq)$
ΔH_f^0 (kJ/mol)	−74.81	0.00	−393.51	−285.84
S^0 (J · mol⁻¹K⁻¹)	186.2	205.0	213.6	69.9

15. Calculate ΔS^0 for the following spontaneous reactions:

(a) $2Fe(s) + \frac{3}{2}O_2(g) \rightleftharpoons Fe_2O_3(s)$

(b) $2H_2(g) + O_2(g) \rightleftharpoons 2H_2O(liq)$

16. Calculate ΔG^0 for each of the following reactions.

(a) $N_2O_5(s) \rightleftharpoons 2NO_2(g) + \frac{1}{2}O_2(g)$

(b) $N_2(g) + 2O_2(g) \rightleftharpoons 2NO_2(g)$

(c) $2NO(g) + O_2(g) \rightleftharpoons 2NO_2(g)$

17. Predict whether each of the following reactions will go spontaneously to the right when all species are in their standard states at 25 °C.

(a) $3C_2H_2(g) \rightleftharpoons C_6H_6(liq)$

(b) $3C_2H_6(g) \rightleftharpoons C_6H_6(liq) + 6H_2(g)$

18. For each of the reactions in Exercise 17, which factor, the enthalpy or the entropy, predominates in determining the sign of ΔG^0 at 25 °C?

19. Using tables of ΔH_f^0 and S^0 at 25 °C, calculate ΔG_f^0 for (a) $Ni(OH)_2(s)$ and (b) urea, $(NH_2)_2CO(s)$.

20. The following data are given for solid ammonium chloride:

$$\Delta H_f^0(NH_4Cl) = -314.4 \text{ kJ/mol} \qquad \Delta G_f^0(NH_4Cl) = -201.5 \text{ kJ/mol}$$

Explain why $\Delta G_f^0(NH_4Cl)$ is more positive than $\Delta H_f^0(NH_4Cl)$.

Section 18.9

21. Calculate the equilibrium constant at 25 °C for each of the reactions of Exercise 16. Write the expression for each of these equilibrium constants.

22. (a) Calculate ΔG^0 for $PCl_5(g) \rightleftharpoons PCl_3(g) + Cl_2(g)$

(b) In a mixture of these three gases at 25 °C the initial partial pressures are $P_{PCl_5} = 0.25$ atm, $P_{PCl_3} = 2.0 \times 10^{-3}$ atm, and $P_{Cl_2} = 0.50$ atm. In which direction will the reaction proceed?

23. Calculate the equilibrium constant at 25 °C for each of the following reactions. Which factor, the enthalpy or the entropy, predominates in determining the sign of ΔG^0?

(a) $NH_4Cl(s) \rightleftharpoons NH_3(g) + HCl(g)$

(b) $N_2(g) + CO(g) + 2H_2(g) \rightleftharpoons (NH_2)_2CO(s)$

Section 18.10

24. (a) Calculate ΔG^0 for the net cell reaction of the galvanic cell

$$Cu(s) \mid Cu^{2+}(aq) \parallel Ag^+(aq) \mid Ag(s)$$

(b) From ΔG^0 calculate $\Delta\mathcal{E}^0_{cell}$ and the equilibrium constant. Is the net cell reaction spontaneous in the direction written? Explain your answer.

25. (a) Using Appendix F calculate $\Delta\mathcal{E}^0_{cell}$ for the galvanic cell

$$Ni(s) \mid Ni^{2+}(aq) \parallel Br^-(aq) \mid AgBr(s) \mid Ag(s)$$

(b) From $\Delta\mathcal{E}^0_{cell}$ calculate ΔG^0 and the equilibrium constant at 25 °C.

26. For the galvanic cell $Cd(s) \mid Cd^{2+}(0.050\ M) \parallel Co^{2+}(2.00\ M) \mid Co(s)$

(a) Calculate Q, $\Delta\mathcal{E}_{cell}$, and ΔG.

(b) Will the net cell reaction proceed spontaneously from left to right when $[Cd^{2+}] = 0.050\ M$ and $[Co^{2+}] = 2.00\ M$?

Section 18.11

27. (a) Assuming ΔH^0 and ΔS^0 are constant between 25 and 75 °C, calculate ΔG^0_{348} for the reactions of Exercise 23.

28. For the reaction $(NH_4)_2SO_4(s) \rightleftharpoons 2NH_3(g) + H_2O(g) + SO_3(g)$, do you expect ΔG^0_{600} to be more negative, less negative, or the same as ΔG^0_{298}? Explain your answer.

29. Using data from Table 18.7 for the reaction $2NO_2(g) \rightleftharpoons N_2O_4(g)$
    (a) Calculate ΔG^0_{343} and ΔG^0_{298}.
    (b) Calculate K_{eq} at 90 °C from K_{343}, assuming ΔH^0 is constant at -58.0 kJ between 25 and 90 °C.

Multiple Choice Questions

Use Appendices F and G for any data needed.

1. For the reaction $2C(gr) + 3H_2(g) \rightarrow C_2H_6(g)$ the entropy change at 25 °C in joules per mole per kelvin $(J \cdot mol^{-1}K^{-1})$ is
   (a) 229.5   (b) 93.2   (c) -150.9 (d) -162.3 (e) -173.7

2. For the reaction $2SO_2(g) + O_2(g) \rightleftharpoons 2SO_3(g)$ at 25 °C, ΔS^0 is -188.0 $J \cdot mol^{-1}K^{-1}$, and ΔH^0 is -197.7 kJ/mol. What is ΔG^0 for this reaction in kilojoules per mole (kJ/mol)?
   (a) -253.8 (b) -193.0 (c) -141.6 (d) 5.586×10^4 (e) 5.625×10^4

3. If an ideal gas is expanded at constant temperature,
   (a) $\Delta E > 0$ and $\Delta S > 0$ (b) $\Delta E = 0$ and $\Delta S = 0$ (c) $\Delta E = 0$ and $\Delta S < 0$
   (d) $\Delta E < 0$ and $\Delta S > 0$ (e) $\Delta E = 0$ and $\Delta S > 0$

4. If $\Delta G^0_f(HI, g) = +1.7$ kJ, what is the equilibrium constant at 25 °C for $2HI(g) \rightleftharpoons H_2(g) + I_2(s)$?
   (a) 24   (b) 3.9   (c) 2.0   (d) 0.50   (e) 0.25

5. Which of the following is true for the reaction $H_2O(\ell) \rightleftharpoons H_2O(g)$ at 100 °C and 1-atm pressure?
   (a) $\Delta H = 0$ (b) $\Delta S = 0$ (c) $\Delta H = \Delta E$ (d) $\Delta H = T\Delta S$ (e) $\Delta H = \Delta G$

6. Calculate the equilibrium constant at 25 °C for the reaction

$$3C(gr) + 4H_2(g) \rightleftharpoons C_3H_8(g)$$

   (a) 3.02×10^9 (b) 1.31×10^4 (c) 1.00   (d) 7.64×10^{-5} (e) 3.31×10^{-10}

7. For the reaction $A(g) \rightleftharpoons B(g) + C(g)$, the equilibrium constant at a certain temperature is 2.0×10^{-4} atm. A mixture of the three gases is placed in a flask and the initial partial pressures are $P_A = 2.0$ atm, $P_B = 0.50$ atm, and $P_C = 1.0$ atm. Which of the following is true at the instant of mixing?
   (a) $\Delta G^0 = 0$ (b) $\Delta G^0 < 0$ (c) $\Delta G = 0$ (d) $\Delta G < 0$ (e) $\Delta G > 0$

8. For which of the following reactions is $\Delta S^0 < 0$?
   (a) $NH_4Cl(s) \rightleftharpoons NH_3(g) + HCl(g)$
   (b) $2NO_2(g) \rightleftharpoons N_2(g) + 2 O_2(g)$
   (c) $2 IBr(g) \rightleftharpoons I_2(s) + Br_2(liq)$
   (d) $(NH_4)_2CO_3(s) \rightleftharpoons 2NH_3(g) + H_2O(g) + CO_2(g)$
   (e) $C_6H_6(s) \rightleftharpoons C_6H_6(liq)$

9. Which of the following statements is true if the reaction quotient, Q, is equal to 1?
   (a) $\Delta G = 0$ (b) $\Delta G^0 = 0$ (c) $\Delta S^0 = 0$ (d) $\Delta G = \Delta G^0$ (e) $\Delta H^0 = T\Delta S^0$

10. For the reaction $I_2(s) + Cl_2(g) \rightleftharpoons 2\ ICl(g)$, $\Delta G^0_{298} = -10.9$ kJ. If both $Cl_2(g)$ and $ICl(g)$ are put in a flask at 25 °C with some solid I_2, and the initial partial pressures are $P_{ICl} = 0.800$ atm and $P_{Cl_2} = 2.1 \times 10^{-3}$ atm, what is ΔG in kilojoules at the instant of mixing?
    (a) 14.2×10^3 (b) 1.18×10^3 (c) 3.3 (d) -4.7 (e) -25.1

11. For the reaction $\frac{1}{2} N_2(g) + O_2(g) \rightleftharpoons NO_2(g)$, $\Delta H^0_{298} = +33.2$ kJ and $\Delta S^0_{298} = -60.9$ J·mol^{-1}K^{-1}. Using these data the equilibrium constant for the reaction at 25 °C is
    (a) 1×10^{-9} (b) 2×10^{-3} (c) 8×10^{-1} (d) 4×10^2 (e) 1×10^9

12. If a process is both endothermic and spontaneous then
    (a) $\Delta S > 0$ (b) $\Delta S < 0$ (c) $\Delta H < 0$ (d) $\Delta G > 0$ (e) $\Delta E = 0$

13. Assuming HCl obeys Trouton's rule, if $\Delta H^0_{vap} = 16.2$ kJ/mol at the normal boiling point, a good approximation to the normal boiling temperature is
    (a) -114 °C   (b) -89 °C   (c) 0.18 °C   (d) 93 °C   (e) 184 °C

14. For the reaction $HBr(g) = \frac{1}{2} H_2(g) + \frac{1}{2} Br_2(g)$ calculate ΔG^0 in kilojoules.
    (a) -51.86 (b) -50.29 (c) $+53.43$ (d) $+55.00$ (e) $+56.57$

15. Assuming ΔH^0 and ΔS^0 are independent of temperature between 25 and 100 °C for the reaction $N_2(g) + 3H_2(g) \rightleftharpoons 2NH_3(g)$, what is ΔG^0 in kilojoules at 100 °C?
    (a) 28.04 (b) -18.07 (c) -26.24 (d) -32.98 (e) -72.35

16. Which of the following statements about a reaction occurring in a galvanic cell is true?
    (a) If $\Delta \mathcal{E}_{cell} > 0$, then $\Delta G < 0$ (b) If $\Delta \mathcal{E}^0_{cell} < 0$, then $\Delta G < 0$
    (c) If $\Delta \mathcal{E}^0_{cell} < 0$, then $K_{eq} > 1$ (d) If $\Delta \mathcal{E}_{cell} < 0$, then $K_{eq} > 1$
    (e) If $\Delta \mathcal{E}_{cell} > 0$, then $K_{eq} > 1$

17. For the net cell reaction of the cell $Zn(s) \mid Zn^{2+} \parallel Cd^{2+} \mid Cd(s)$, ΔG^0 in kilojoules at 25 °C is
    (a) 112.5 (b) 69.47 (c) -34.73 (d) -69.47 (e) -225.0

18. Assuming ΔH_{vap} and ΔS_{vap} are approximately constant between 25 °C and the normal boiling point of CCl_4, estimate the temperature of the normal boiling point.
    (a) 0.35 °C   (b) 75 °C   (c) 100 °C   (d) 274 °C   (e) 348 °C

19. For a certain reaction $\Delta G^0_{800} = -40.0$ kJ. What is K_{eq} at 800 K for this reaction?
    (a) 1.02×10^6 (b) 409 (c) 6.01 (d) 2.45×10^{-3} (e) 9.77×10^{-7}

20. Which of the following statements about the spontaneous reaction occurring in a galvanic cell is always true?
    (a) $\Delta \mathcal{E}^0_{cell} > 0$, $\Delta G^0 < 0$, and $Q < K$ (b) $\Delta \mathcal{E}^0_{cell} > 0$, $\Delta G^0 > 0$, and $Q < K$
    (c) $\Delta \mathcal{E}^0_{cell} > 0$, $\Delta G > 0$, and $Q > K$ (d) $\Delta \mathcal{E}_{cell} > 0$, $\Delta G < 0$, and $Q > K$
    (e) $\Delta \mathcal{E}_{cell} > 0$, $\Delta G < 0$, and $Q < K$

21. For the gas-phase decomposition $PCl_5 \rightleftharpoons PCl_3(g) + Cl_2(g)$
    (a) $\Delta H < 0$ and $\Delta S < 0$ (b) $\Delta H > 0$ and $\Delta S > 0$ (c) $\Delta H > 0$ and $\Delta S < 0$
    (d) $\Delta H < 0$ and $\Delta S > 0$ (e) $\Delta H = 0$ and $\Delta S > 0$

Problems

Use Appendices F and G for data needed.

18.1. Calculate ΔG^0 and the equilibrium constant at 25 °C for the following reactions:
   (a) $3H_2(g) + SO_2(g) \rightleftharpoons H_2S(g) + 2H_2O(liq)$
   (b) $CO(g) + CaO(s) \rightleftharpoons Ca(s) + CO_2(g)$

18.2. From the following data at 25 °C,

	$NH_4^+(aq)$	$H^+(aq)$	$NH_3(aq)$
ΔH_f^0 (kJ/mol)	−132.8	0.00	−80.83
S^0 (J·mol^{-1}K^{-1})	+112.8	0.00	+110.

   (a) Calculate ΔG^0 and K_{eq} at 25 °C for the reaction

$$NH_4^+(aq) \rightleftharpoons H^+(aq) + NH_3(aq)$$

   (b) Account for the sign of ΔS^0 for this reaction.

18.3. (a) Calculate $\Delta S_f^0(C_2H_6, g)$, that is, ΔS^0 for the formation of ethane from its elements at 25 °C.
   (b) Calculate ΔS^0 for the reaction

$$C_2H_4(g) + H_2(g) \rightleftharpoons C_2H_6(g)$$

   (c) Compare the two ΔS^0 values calculated. One is more negative than the other. Explain the difference in terms of the change in molecular disorder for the two reactions.

18.4. Dissolving KCl(s) in water is an endothermic process; the heat of solution is +17.2 kJ/mol at 25 °C. Explain why KCl spontaneously dissolves in water at 25 °C.

18.5. From data in the table of ΔG_f^0 values, calculate K_{sp} for AgBr(s).

18.6. Although HF is a hydrogen-bonded liquid, its entropy of vaporization is less than 88 J·mol^{-1}K^{-1} at its normal bp, 19.5 °C. Explain why ΔS_{vap} for HF is lower than predicted by Trouton's rule.

18.7. (a) Calculate ΔH^0, ΔS^0, ΔG^0, and K_{eq} at 25 °C for the reaction

$$H_2S(g) + \tfrac{3}{2}O_2(g) \rightleftharpoons H_2O(liq) + SO_2(g)$$

Specify the units of all numerical values.
   (b) Does the tendency for reactions to proceed to a state of minimum energy favor the forward direction (the products) or the backward direction (the reactants) of this reaction? Explain your answer.
   (c) Does the tendency for reactions to proceed to a state of maximum entropy favor the forward or backward direction of this reaction? Explain your answer.
   (d) Which factor, the enthalpy or the entropy, predominates in determining the position of equilibrium for this reaction at 25 °C? Explain your answer.

18.8. At 35 °C and a total pressure of 1 atm, $N_2O_4(g)$ is 27.2% dissociated into $NO_2(g)$. What is ΔG^0 at 35 °C for the reaction

$$N_2O_4(g) \rightleftharpoons 2NO_2(g)$$

18.9. (a) Calculate ΔH_{298}^0, ΔS_{298}^0, ΔG_{298}^0, and K_{298} for the reaction

$$PCl_3(g) + Cl_2(g) \rightleftharpoons PCl_5(g)$$

(b) Which factor, enthalpy or entropy, provides the principal driving force for this reaction at 25.0 °C? Explain your answer.

(c) Calculate K_{400}, the equilibrium constant at 400 K, assuming that ΔH^0 is independent of temperature between 298 and 400 K. Using Le Chatelier's principle, explain the change in the value of the equilibrium constant.

18.10. For the galvanic cell   Ag(s) | Ag$^+$(aq) ‖ Cl$^-$(aq) | AgCl(s) | Ag(s)   calculate ΔG^0, $\Delta \mathcal{E}^0_{cell}$, and K_{eq} at 25 °C. Write the expression for the equilibrium constant.

18.11. Nitric oxide (NO) and carbon monoxide (CO) are air pollutants generated by automobiles. It has been suggested that under suitable conditions these two gases could be made to react to produce N_2 and CO_2, components of unpolluted air.

(a) Write a balanced equation for the proposed reaction. Give the oxidation states of carbon and nitrogen in both reactants and products. Which gas, NO or CO, is being oxidized?

(b) Write the expression for the equilibrium constant for the reaction.

(c) Calculate ΔH^0_{298}, ΔG^0_{298}, and K_{298} for the reaction. If all gases are at 1 atm and 25 °C does the equilibrium favor reactants or products?

(d) In an urban area, typical partial pressures of these gases are $P_{N_2} = 0.781$ atm, $P_{CO_2} = 3.1 \times 10^{-4}$ atm, $P_{NO} = 5.0 \times 10^{-7}$ atm, and $P_{CO} = 5.0 \times 10^{-5}$ atm. In which direction will the reaction proceed with these partial pressures at 25 °C?

(e) Without doing any numerical calculations, will the equilibrium constant for this reaction at the much higher temperatures in the exhaust system of an automobile be greater than, less than, or the same as the equilibrium constant at 25 °C that you calculated in part (c)? Explain your answer.

18.12. From the following data for hydrazine, N_2H_4(g),

$$\Delta H^0_f(N_2H_4) = 95.4 \text{ kJ/mol} \qquad \Delta G^0_f(N_2H_4) = 159.3 \text{ kJ/mol}$$

calculate S^0_{298} for hydrazine. Specify the units of your answer.

18.13. (a) Calculate ΔH^0 and ΔS^0 at 25 °C for the reaction

$$CO_2(g) + 2NH_3(g) \rightleftharpoons CO(NH_2)_2(s) + H_2O(\ell)$$

(b) Will the production of urea, $CO(NH_2)_2$(s), from CO_2 and NH_3 proceed spontaneously at 25 °C and 1 atm? Explain your answer.

18.14. Sulfuryl chloride, SO_2Cl_2(g), is a highly reactive compound used in organic chemistry as a chlorinating agent. When heated, it decomposes as follows: SO_2Cl_2(g) $\rightleftharpoons$ SO_2(g) + Cl_2(g). The decomposition is endothermic.

(a) A 4.386-g sample of SO_2Cl_2 is placed in a 1.000-L bulb and the temperature is raised to 375 K. When the system has come to equilibrium at 375 K, the total pressure in the bulb is found to be 1.76 atm. Calculate the partial pressures of SO_2, Cl_2, and SO_2Cl_2 at equilibrium. *Hint:* First calculate what the pressure in the bulb would be if no dissociation of the SO_2Cl_2 occurred.

(b) Give the expression for, the units of, and the numerical value of the equilibrium constant, K_p, for the decomposition of sulfuryl chloride at 375 K.

(c) Calculate ΔG^0_{375} for this decomposition.

(d) Which factor, the enthalpy or the entropy, predominates in determining the position of equilibrium at 375 K?

18.15. Interest in the bacterial production of protein for synthetic foods has focused on the reaction in which glycine is formed from the simple compounds NH_3, CH_4, and O_2:

$$NH_3(g) + 2CH_4(g) + \tfrac{3}{2} O_2(g) \rightleftharpoons NH_2CH_2COOH(s) + 3H_2O(\ell)$$

(a) Calculate ΔS^0_{298} for this reaction and give a molecular interpretation of the value you obtain.

(b) Will this reaction proceed spontaneously at 25 °C if all species are in their standard states? Explain your answer.

(c) Calculate K_{eq} for this reaction at 25 °C and at normal body temperature, 37.0 °C. Explain the change in the value of K_{eq} with an increase in temperature.

chapter 19 Chemical Kinetics

Henry Eyring (1901–1982), a U.S. theoretical chemist, was a professor at Princeton University and the University of Utah, where he was also Dean of the Graduate School. His interests were in the theory of reaction rates, on the application of quantum mechanics to chemistry, and in the theory of liquids. He developed the concept of the activation energy required for reaction into the theory of the transition state and carried out extensive research in theoretical chemical kinetics. He was coauthor with John Walter and George E. Kimball of a widely used textbook, "Quantum Chemistry," and coauthored "Modern Chemical Kinetics" with his son, Edward.

We have seen that thermodynamic calculations enable us to predict whether or not a reaction will proceed to the right, as written. But there is one very important matter that thermodynamics cannot tell us about, and that is the *rate* at which a reaction will occur. If a reaction takes days or weeks or even longer to occur it may be of little use to know that the desired products are favored thermodynamically. Some reactions occur very quickly, in a fraction of a second or in a few minutes. The neutralization of a solution of a strong acid with a strong base is an example of a reaction that is complete virtually as soon as the reagents are mixed. Other reactions, like the rusting of iron at room temperature, take hours or days or more. Thermodynamics provides no information about this aspect of chemical reactions. Time is not a thermodynamic variable.

If a thermodynamic calculation tells us that a reaction will not proceed as we have written it at a given temperature and pressure, there is no point in going to the lab and trying it out, for it simply will not work. However, if thermodynamics predicts that a reaction will proceed, there is no guarantee that when we go to the lab and try it, we will obtain the desired products in the course of an afternoon's work. The reaction may occur so slowly as to be impractical. We have already discussed one example of this. The formation of ammonia from its elements

$$N_2(g) + 3H_2(g) \rightleftharpoons 2NH_3(g)$$

is thermodynamically favored at 25 °C: $\Delta G_{298}^0 = -33.0$ kJ and $K_{298} = 6.0 \times 10^5$. Nevertheless, the reaction is so slow at 25 °C that the production of ammonia cannot be carried out at that temperature. Commercial production of ammonia occurs at temperatures between 500 and 600 °C, because the rate of reaction is much faster at higher temperatures, even though the equilibrium constant is smaller.

Suppose that when two reagents, A and B, are mixed together, there are two different reactions that might occur:

$$A + B \rightarrow C + D \quad \text{with equilibrium constant } K_1 \tag{1}$$
$$A + B \rightarrow E + F \quad \text{with equilibrium constant } K_2 \tag{2}$$

Thermodynamics enables us to calculate the equilibrium constants K_1 and K_2. Let us assume that $K_1 > K_2$. The products C and D are therefore thermodynamically favored over the products E and F. This does not necessarily mean that when A and B are mixed the products will be C and D, rather than E and F. It is possible that the rate of formation of E and F is very much faster than the rate of formation of C and D. If C and D are the observed products when the reaction is actually carried out, we say that the reaction is **thermodynamically controlled.** On the other hand, if E and F are the observed products we say that the reaction is **kinetically controlled.** If the products of a reaction are the species with lowest Gibbs free energy, the reaction is thermodynamically controlled, whereas if the products formed are not the species with lowest Gibbs free energy, but the species formed fastest, the reaction is kinetically controlled.

Thermodynamics can tell us where the position of equilibrium lies between the reactants and the products, but it provides no information about the **mechanism** of the reaction. The change in any state function is path independent, and the mechanism is basically a description of the path of reaction. The mechanism of a reaction describes, in detail, how the reaction occurs. It is a series of steps that defines the order in which bonds are broken and new ones made, until the final products are obtained.

Understanding the factors that affect the rate of a chemical reaction and elucidating information about the mechanism of a reaction are the principal objectives of that branch of chemistry known as **chemical kinetics.**

section 19.1
Factors Affecting the Rates of Chemical Reactions

(a) **The nature of the reactants.** Each reaction proceeds at its own rate. Changing the nature of any reacting species will, in general, change the rate of reaction. Thus the reaction between gaseous H_2 and I_2 to form $HI(g)$ proceeds at a different rate from the reaction between gaseous H_2 and Br_2 to form $HBr(g)$, even when the temperature and pressures of the gases are the same.

(b) **The concentrations of the reacting species.** In order for a chemical reaction to occur, the reacting species must come close together, that is, they must collide.* Because collisions are more frequent when concentrations are greater, reaction rates increase as the concentrations of the reacting species increase. For most reactions, the rate depends only on the concentrations of the substances that are the reactants in the stoichiometric equation for the reaction. There are, however, many examples of reactions for which the rate depends on the concentration of some substance that does not appear in the correctly balanced stoichiometric equation for the reaction, or for which the rate depends on the concentration of one or more of the products of the reaction. We can learn a great deal about the mechanism of a reaction by investigating how the reaction rate depends on the concentration of various substances.

(c) **The temperature.** In general, increasing the temperature increases the rate of a reaction. As a useful rule of thumb, at temperatures near room temperature, the rate of most chemical reactions increases by a factor of 2 or 3 for a $10°$ rise in temperature. Many biologically important reactions have an even greater temperature dependence and their rates increase by a factor larger than 2 or 3 when the temperature increases by $10°$.

(d) **The presence of a catalyst.** A catalyst is a substance that increases the rate of a chemical reaction but is not used up in the course of the reaction. Thus the amount of catalyst present at the end of the reaction is the same as the amount that was present at the beginning. For a condition of dynamic equilibrium, a catalyst increases the rate of both the forward and backward reactions by the same factor, and therefore does not change the value of the equilibrium constant.

(e) **The nature of the solvent.** Of course not all reactions occur in solution, but a great many reactions are carried out in solution. For these reactions, changing the solvent will generally change the rate of a reaction. Thus if a reaction is carried out in aqueous solution, and some alcohol is added to the reaction mixture, the rate of the reaction will change, even though alcohol takes no part in the reaction.

section 19.2
The Rate Expression

Experiment tells us that the rate of most homogeneous reactions can be written in the form

$$\text{rate} = k[A]^m[B]^n \ldots \tag{19-1}$$

In this expression, A and B are usually the reactants in the stoichiometric equation for the reaction, but, as described above, the rate may depend upon the concentration of some substance that does not appear in the stoichiometric equation, or it may depend

---

* The decomposition of unstable nuclei (radioactive decay) is not considered a chemical reaction in this context. We will discuss the observed rate law governing radioactive decay, but not the mechanism of this process.

on the concentration of one or more of the products of the reaction. In Eq. (19-1), the way in which the rate depends on the concentrations of the chemical species A, B, ... is written explicitly, and the dependence of the rate on all other factors is included in the numerical value of k, which is called the **specific rate constant**. The numerical value of k is independent of the concentration of any species present in the reaction mixture, but it does depend on the temperature, the particular reaction, the solvent (if the reaction occurs in solution), and the catalysts (if any) that are present. The exponents m, n, ... in the rate expression are either small integers (0, 1, 2, 3, ...) or small rational fractions ($\frac{1}{2}$, $\frac{3}{2}$, ...).

Any **rate** is a change in some quantity with time. The speed of an object is the time rate of change of distance. How fast is the car going? A speed of 75 km/h represents the change in distance per unit time, and can be symbolized

$$\frac{\Delta(\text{distance})}{\Delta(\text{time})} = \frac{\Delta s}{\Delta t}$$

where the "Δ" symbol, as always, stands for "the change in," and the letter s is used to represent distance. Because the speed can change from one second to another, we must make the time interval as small as possible in order to describe the speed at a given instant. This means that the speed is a first derivative, ds/dt. Any **rate** is a first derivative with respect to time. If you have never taken any calculus, you will not be familiar with properties of first derivatives, but you do not need to be familiar with derivatives to understand the fundamental ideas of chemical kinetics. You need to understand only that the symbol for a rate, $\Delta s/\Delta t$, becomes ds/dt as the time interval is reduced to an infinitesimally small value. The fact that any rate is a first derivative with respect to time is an important and useful concept and explains the notation we use to represent the rate of a chemical reaction.

What is it that changes with time in a chemical reaction? It is the concentration of any of the reactants or products. As the reaction $A + B \rightarrow C + D$ progresses, the concentrations of C and D increase, and the concentrations of A and B decrease. We can express the rate in terms of the change in concentration of any of these four species, but we must account for the difference in sign for reactants and products. Since [A] is decreasing with time, $\Delta[A]/\Delta t$ is a negative quantity, whereas $\Delta[C]/\Delta t$ is positive. The rate of a reaction is usually defined in terms of the decrease in concentration of one of the reactants, as $-\Delta[A]/\Delta t$.

Consider the hydrolysis of *tert*-butyl bromide to form *tert*-butyl alcohol and HBr

$$(CH_3)_3CBr + H_2O \rightarrow (CH_3)_3COH + HBr \tag{19-2}$$

Data for this reaction at 50 °C in an acetone–H_2O solvent, are shown in Table 19.1. For each time interval we can calculate an average rate. For the initial time interval, for instance, we obtain $-\Delta[(CH_3)_3CBr]/\Delta t$ as $-(0.0951 - 0.1056)/9 = 1.17 \times 10^{-3}$ moles per liter per minute ($mol \cdot L^{-1}min^{-1}$). Since the $[(CH_3)_3CBr]$ is continually decreasing as the reaction proceeds, the rate changes at each instant. To define the rate at any given time, we must make the time interval smaller and smaller until it becomes infinitesimal:

$$\text{rate} = -d[(CH_3)_3CBr]/dt$$

Note that the rate decreases as the reaction proceeds. While the average rate for the initial 9 min is 1.17×10^{-3} $mol \cdot L^{-1}min^{-1}$, between 54.0 and 72.0 min, for instance, the average rate is only

$$-\frac{(0.0432 - 0.0536)\, M}{(72.0 - 54.0)\, \text{min}} = \frac{0.0104\, M}{18\, \text{min}} = 5.78 \times 10^{-4}\, M \cdot \text{min}^{-1}$$

Table 19.1. Rate Data for the Reaction of $(CH_3)_3CBr$ with H_2O in an Acetone–H_2O Mixture at 50°C

Time (min)	$[(CH_3)_3CBr]$ (mol/L)	$\Delta[(CH_3)_3CBr]/\Delta t$ (mol·$L^{-1}min^{-1}$)
0.00	0.1056	
		1.17×10^{-3}
9.00	0.0951	
		1.06×10^{-3}
18.0	0.0856	
		9.89×10^{-4}
27.0	0.0767	
		9.38×10^{-4}
40.0	0.0645	
		7.79×10^{-4}
54.0	0.0536	
		5.78×10^{-4}
72.0	0.0432	
		4.91×10^{-4}
105.0	0.0270	
		3.2×10^{-4}
135.0	0.0174	

As the concentration of *tert*-butyl bromide, $(CH_3)_3CBr$, decreases with time, the concentration of *tert*-butyl alcohol, $(CH_3)_3COH$, increases with time. Figure 19.1 is a plot of the concentration of both of these substances as a function of time. The rate of reaction may be expressed in terms of the change in either of these concentrations with time:

$$\text{rate} = -\frac{d[(CH_3)_3CBr]}{dt} = +\frac{d[(CH_3)_3COH]}{dt} \tag{19-3}$$

The derivative $d[A]/dt$ at any time, t, is the slope of a plot of $[A]$ versus t, that is, it is

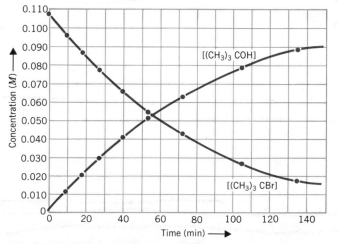

Fig. 19.1. Change in the concentration of a reactant and a product with time. Data of Table 19.1 for the reaction of *tert*-butyl bromide with H_2O to form *tert*-butyl alcohol at 50 °C are plotted. The $[(CH_3)_3CBr]$ decreases with time, and $[(CH_3)_3COH]$ increases with time, at an equal rate.

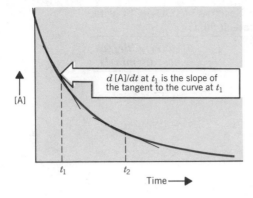

Fig. 19.2. The derivative, $d[A]/dt$, is the slope of the plot of [A] versus time, and changes at each instant.

the slope of the line tangent to the curve at the time t. This is illustrated in Fig. 19.2. If [A] is decreasing with time, the slope of a plot of [A] versus time is negative, and the rate of the reaction $A + B \rightarrow C + D$ is therefore expressed as $-d[A]/dt$.

Consider the reaction between iodine, I_2, and thiosulfate ion, $S_2O_3^{2-}$, in aqueous solution:

$$I_2 + 2S_2O_3^{2-} \rightarrow S_4O_6^{2-} + 2I^- \tag{19-4}$$

The stoichiometry tells us that two $S_2O_3^{2-}$ ions are used for every I_2 that is used. Thus the rate of disappearance of thiosulfate ions is twice the rate of disappearance of I_2. Alternatively, this can be stated as: The rate of disappearance of I_2 is one half the rate of disappearance of $S_2O_3^{2-}$ ions. The change in the concentrations of all the reactants and products of Eq. (19-4) can be expressed as

$$-\frac{d[I_2]}{dt} = -\frac{1}{2}\frac{d[S_2O_3^{2-}]}{dt} = +\frac{d[S_4O_6^{2-}]}{dt} = \frac{1}{2}\frac{d[I^-]}{dt} \tag{19-5}$$

Since all these quantities are equal, any one of them can be considered the rate of the reaction.

For the general reaction $\alpha A + \beta B \rightarrow \gamma C + \delta D$,

$$\text{rate of reaction} = -\frac{1}{\alpha}\frac{d[A]}{dt} = -\frac{1}{\beta}\frac{d[B]}{dt} = \frac{1}{\gamma}\frac{d[C]}{dt} = \frac{1}{\delta}\frac{d[D]}{dt} \tag{19-6}$$

The Order of Reaction

The **order** of a chemical reaction is by definition the sum of the exponents of all the concentration terms in the rate expression. For a rate of the form $-d[A]/dt = k[A]^m[B]^n$, the order of the reaction is $m + n$.

The order of a chemical reaction can be zero, integral, or fractional, although it is most frequently a small integer. There is no way that you can look at the stoichiometric equation for a reaction and guess what the order will be; the order must be determined experimentally. There is no *necessary* connection between the coefficients of the reactants in the stoichiometric equation for the reaction and the kinetic order, although there are examples where they are directly related. For the rate expression $-d[A]/dt = [A]^m[B]^n$, we say that the value of m is the order of the reaction with respect to A, the value of n is the order with respect to B, and the overall order is $m + n$.

Several examples of rate expressions for specific reactions will make these concepts clearer.

1. For the gas-phase decomposition of dinitrogen pentoxide

$$2N_2O_5(g) \rightarrow 4NO_2(g) + O_2(g) \qquad (19\text{-}7)$$

the rate of decomposition of N_2O_5 has been observed experimentally to be

$$\text{rate} = -d[N_2O_5]/dt = k_1[N_2O_5] \qquad (19\text{-}8)$$

When the order of a reaction is 1, we call the reaction **first order**. This reaction is therefore a first-order reaction. It is first order with respect to N_2O_5, and first order overall.

2. For the gas-phase decomposition of the red-brown gas NO_2,

$$2NO_2(g) \rightarrow 2NO(g) + O_2(g) \qquad (19\text{-}9)$$

the rate of decomposition of NO_2 has been found experimentally to be

$$\text{rate} = -d[NO_2]/dt = k_2[NO_2]^2 \qquad (19\text{-}10)$$

The order of this reaction is therefore 2. We say that the reaction is **second order** in NO_2 and second order overall.

Note particularly that it is not possible merely by examining the stoichiometry of these two gas-phase decompositions, Eqs. (19-7) and (19-9), to predict that the decomposition of N_2O_5 is first order while that of NO_2 is second order. The only way to find this out is to go into the laboratory and carry out the measurements necessary to determine the order of these reactions, namely to measure both $[N_2O_5]$ and $[NO_2]$ as a function of time.

3. The reaction between acetic acid, CH_3COOH, and ethanol, CH_3CH_2OH,

$$\underset{\text{Acetic acid}}{CH_3COOH} + \underset{\text{Ethanol}}{CH_3CH_2OH} \rightarrow \underset{\text{Ethyl acetate}}{CH_3COOCH_2CH_3} + H_2O \qquad (19\text{-}11)$$

is called an **esterification**. The product, ethyl acetate, belongs to the class of substances known as **esters**.* Esters often have pleasant, fruit-like odors; ethyl acetate is one of the principal ingredients in the oil extracted from the pineapple. It is also the solvent in various glues, such as Duco cement. The rate of reaction (19-11) can be expressed either as the rate of disappearance of acetic acid or as the rate of disappearance of ethanol; these are equal because they are used up in a 1:1 molar ratio. Thus

$$\text{rate} = -d[CH_3COOH]/dt = -d[C_2H_5OH]/dt = k_2''[CH_3COOH][C_2H_5OH] \quad (19\text{-}12)$$

This is a second-order reaction. It is first order in ethanol and first order in acetic acid, but second order overall.

4. The reaction between iodine and acetone in aqueous solution has the following stoichiometry:

$$\underset{\text{Acetone}}{CH_3-\overset{\overset{\text{O}}{\|}}{C}-CH_3} + I_2 + H_2O \rightarrow \underset{\text{Iodoacetone}}{CH_3-\overset{\overset{\text{O}}{\|}}{C}-CH_2I} + H_3O^+ + I^- \quad (19\text{-}13)$$

The rate of this reaction has been found to be

$$\text{rate} = -d[CH_3COCH_3]/dt = -d[I_2]/dt = k_2'[CH_3COCH_3][H_3O^+] \quad (19\text{-}14)$$

This is a **second-order** reaction; it is first order in acetone, first order in hydronium ion, and second order overall. Note that the rate of this reaction does not

---

* For a fuller discussion of esters, see Section 23.6.

depend on the concentration of iodine. The reaction is zero order in I_2. This could certainly not have been predicted from the stoichiometry of Eq. (19-13). It was determined experimentally by measuring the rate in a series of solutions in which all concentrations were kept constant except for the concentration of iodine. It was then found that the rate of reaction did not change when $[I_2]$ is changed but all other concentrations are held constant. Note also that this is an example of a reaction for which the rate depends on the concentration of a substance that is a product in the stoichiometric equation, H_3O^+. The dependence of the rate on $[H_3O^+]$ was determined experimentally by measuring the rate in a series of solutions of varying pH.

5. The gas-phase decomposition of acetaldehyde

$$CH_3CHO(g) \rightarrow CH_4(g) + CO(g) \tag{19-15}$$

at 720 K is observed to have the rate expression

$$\text{rate} = -d[CH_3CHO]/dt = k[CH_3CHO]^{3/2} \tag{19-16}$$

and is therefore of order 3/2. Again, you could not have predicted this order by considering the stoichiometry of reaction (19-15).

EXAMPLE 19.1. Factors affecting the rate and the specific rate constant

Three aqueous acidic solutions of acetone and iodine at 30 °C are prepared. Descriptions of these solutions are tabulated below.

Solution	Number moles CH_3COCH_3	Number moles I_2	Number moles H_3O^+
I	0.10	0.080	0.020
II	0.30	0.030	0.010
III	0.20	0.010	0.020

Each of these solutions contains enough water to make the final volume, V, the same. Reaction (19-13) occurs, with a rate given by Eq. (19-14).

(a) In which of these solutions is the rate of reaction fastest?

Solution. According to Eq. (19-14), the rate of reaction depends only on the concentrations of acetone and hydronium ion. Thus

$$\text{rate in solution I} = k_2' \left(\frac{0.10}{V}\right)\left(\frac{0.020}{V}\right) = (k_2'/V^2)(2.0 \times 10^{-3})$$

$$\text{rate in solution II} = k_2' \left(\frac{0.30}{V}\right)\left(\frac{0.010}{V}\right) = (k_2'/V^2)(3.0 \times 10^{-3})$$

$$\text{rate in solution III} = k_2' \left(\frac{0.20}{V}\right)\left(\frac{0.020}{V}\right) = (k_2'/V^2)(4.0 \times 10^{-3})$$

We see that the rate of reaction between acetone and I_2 is fastest in solution III and slowest in solution I.

(b) If more water is added to each of these solutions, keeping the temperature constant at 30 °C, will the rate of reaction increase, decrease, or remain the same? Will the specific rate constant, k_2', increase, decrease, or remain the same?

Solution. Adding water to each of these solutions increases the volume of solution and thereby decreases the $[CH_3COCH_3]$, $[H_3O^+]$, and $[I_2]$. Thus the rate of reaction will decrease. The value of the specific rate constant, k'_2, will remain the same because rate constants are constant as long as neither the temperature nor the solvent is changed.

(c) If each of these three solutions is cooled to 10 °C, will the rate of reaction increase, decrease, or remain the same? Will the specific rate constant, k'_2, increase, decrease, or remain the same?

Solution. Decreasing the temperature will cause both the specific rate constant, k'_2, and the rate of reaction, to decrease. The rate of most chemical reactions increases by a factor of 2 or 3 for every 10° rise in temperature. A decrease in temperature of 20° therefore causes a significant decrease in the reaction rate.

Units of Rates and Rate Constants

The units of the rate of any reaction are moles per liter per unit time, which may be $mol \cdot L^{-1}s^{-1}$ ($M \cdot s^{-1}$) or $mol \cdot L^{-1}min^{-1}$ ($M \cdot min^{-1}$), for example. The units of the specific rate constant, k, however, depend on the order of the reaction, because in Eq. (19-1) the units on each side of the equation must, of course, be the same.

For a first-order reaction, for example,

$$\underset{mol \cdot L^{-1}s^{-1}}{\text{rate} = -d[A]/dt} = \underset{(s^{-1})(mol/L)}{k_1[A]} \qquad (19\text{-}17)$$

Since the units of the left-hand side are moles per liter per second, the units of the rate constant, k_1, must be reciprocal seconds. First-order rate constants always have units of reciprocal time.

EXAMPLE 19.2. Determining the units of a rate constant

What are the units of a second-order rate constant?

Solution. A second-order rate expression is of the form

$$\text{rate} = -d[A]/dt = k_2[A]^2$$

In order to make the units of both sides equal moles per liter per second, the units of k_2 must be liters per mole per second, as shown below:

$$mol \cdot L^{-1}s^{-1} = (L \cdot mol^{-1}s^{-1})(mol^2L^{-2})$$

---

section 19.3
Determination of the Order of Reaction by the Method of Initial Rates

There are two principal methods of determining the order of a chemical reaction. One is to measure the concentration of some species as a function of time. For the reaction between acetic acid and ethyl alcohol, for instance, Eq. (19-11), one might determine the concentration of acetic acid in the solution every 15 minutes for several hours, until the concentration of acetic acid is no longer changing with time, and the system has come to equilibrium. We will discuss calculating the rate constant and the order of a reaction from data of this sort in Sections 19.4 and 19.5.

A second method is to perform a series of experiments in which the concentrations of reactants are varied systematically and the effect of these variations on the **initial rate** of reaction is determined. The change in concentration of any species with time is always greatest at the beginning of a reaction; as the system gets closer and closer to equilibrium, the changes in the concentrations of reactants and products get smaller and smaller.

Let us consider a reaction that has been extensively studied, the hydrolysis of sucrose. Ordinary sugar, or sucrose, in aqueous solution breaks down into two simpler sugars, glucose and fructose. This reaction is catalyzed by $H^+(aq)$, and proceeds at a significant rate at room temperature only in the presence of a strong acid.

$$C_{12}H_{22}O_{11}(\text{sucrose}) + H_2O \rightarrow \text{glucose} + \text{fructose} \qquad (19\text{-}18)$$

In a certain experiment, a solution in which the initial concentration of sucrose was 1.0023 M was maintained at 25 °C, and the concentration of sucrose was measured as a function of time. The results are reported in Table 19.2, and a plot of the concentration of sucrose versus time is shown as Fig. 19.3. The **initial rate** of reaction is the slope of the curve at time $t = 0$, that is, the slope of the line tangent to the curve at zero time, the start of the experiment.

During the first 30 minutes the concentration of sucrose changed from 1.0023 to 0.9022 M. Thus,

$$\text{average rate for first 30 min} = -\frac{(0.9022 - 1.0023)\ M}{(30.00 - 0.00)\ \text{min}} = +\frac{0.1001\ M}{30.00\ \text{min}}$$

$$= +3.337 \times 10^{-3}\ \text{mol} \cdot \text{L}^{-1}\text{min}^{-1}$$

This value is very close to, but not exactly, the initial rate of the reaction, which must be determined as the slope of the tangent to the curve at $t = 0$.

The initial rate is the largest rate for any particular experiment. Since it is the largest, it is the most readily measured. In addition, at zero time we know the reactant concentrations more precisely than at any other time, because we choose the initial concentrations and prepare the solutions. Once the reaction is under way we must make measurements while concentrations are changing.

An investigation of how the initial rate changes as the concentrations of reactants are systematically varied enables us to determine the order of reaction. The method is described in Examples 19.3 and 19.4. The initial concentration of a reactant, A, is symbolized $[A]_0$, which is read as "the concentration of A at time zero."

Table 19.2. The Concentration of Sucrose in mol/L, at 25 °C, as a Function of Time in Minutes, for a Given Concentration of Acid

Time (min)	$[C_{12}H_{22}O_{11}]$
0.00	1.0023
30.00	0.9022
60.00	0.8077
90.00	0.7253
130.00	0.6297
160.00	0.5753
200.00	0.5007
240.00	0.4358
280.00	0.3793

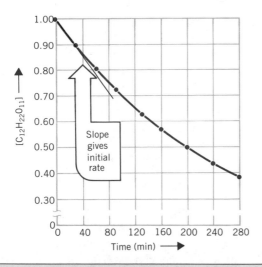

Fig. 19.3. Plot of the concentration of sucrose as a function of time at 25 °C using the data of Table 19.2. The reaction rate at any time is given by the slope of the tangent to the curve at that time.

EXAMPLE 19.3. Determination of the rate expression using the method of initial rates

The initial rate of a reaction

$$A + 2B \rightarrow products$$

is measured for a series of different initial concentrations of A and B. The data obtained are tabulated below:

Experiment Number	Initial Concentrations $[A]_0$ (M)	$[B]_0$ (M)	Initial Rate $(mol \cdot L^{-1}min^{-1})$
I	0.100	0.100	3.20×10^{-3}
II	0.100	0.200	3.20×10^{-3}
III	0.200	0.100	1.28×10^{-2}
IV	0.300	0.200	2.88×10^{-2}

Determine the rate expression for this reaction.

Solution. A comparison of experiments I and II shows us that doubling the initial concentration of B while leaving the initial concentration of A constant has no effect at all on the initial rate of reaction. We can only conclude that the rate does not depend on [B]. A comparison of experiments I and III reveals that doubling the initial concentration of A while keeping the initial concentration of B constant multiplies the initial rate by a factor of 4, for $4(3.20 \times 10^{-3}) = 1.28 \times 10^{-2}$. This indicates that the rate expression is of the form $k[A]^2$. We can check this by comparing experiments II and IV. Tripling the initial concentration of A while holding the initial concentration of B constant increases the rate by a factor of 9 times. We conclude therefore, that the rate is second order in A and zero order in B, and the rate expression is

$$rate = k[A]^2$$

The units of the specific rate constant, k, are liters per mole per minute.

EXAMPLE 19.4. Determination of the order and the rate constant of a reaction by the method of initial rates

A study of the rate of formation of molecular iodine by the reaction

$$2 I(g) \rightarrow I_2(g)$$

in the presence of argon produced the following data:

Experiment Number	Initial Concentrations $[I]_0$ (M)	$[Ar]_0$ (M)	Initial Rate $(mol \cdot L^{-1}s^{-1})$
1	1.0×10^{-5}	1.0×10^{-3}	8.70×10^{-4}
2	2.0×10^{-5}	1.0×10^{-3}	3.48×10^{-3}
3	4.0×10^{-5}	1.0×10^{-3}	1.39×10^{-2}
4	1.0×10^{-5}	5.0×10^{-3}	4.35×10^{-3}
5	2.0×10^{-5}	5.0×10^{-3}	1.74×10^{-2}

Determine the order and the specific rate constant of this reaction.

Solution. A comparison of experiments 1 and 2 shows us that doubling the $[I]_0$ while keeping the $[Ar]_0$ constant multiplies the initial rate by four, since $4(8.70 \times 10^{-4}) = 3.48 \times 10^{-3}$. Comparing experiments 2 and 3 confirms this result, since the $[I]_0$ is doubled again and the rate is increased by a factor of 4, $4(3.48 \times 10^{-3}) = 1.39 \times 10^{-2}$. A comparison of experiments 1 and 4 shows that multiplying the $[Ar]_0$ by five while keeping the $[I]_0$ constant multiplies the initial rate by a factor of 5. This is confirmed by a comparison of experiments 2 and 5. All the data are consistent with the following rate law:　rate $= k_3[I]^2[Ar]$

This is a third-order reaction, second order in I and first order in Ar. The fact that the rate depends on $[Ar]$ indicates that a three-body collision, involving two I atoms and an Ar atom, is necessary for the reaction to occur. The Ar atom absorbs some of the energy of the colliding I atoms and prevents the I_2 molecule from dissociating as soon as it is formed.

The rate constant can be determined by substituting any set of data into the rate expression. For experiment 1 we obtain

$$\text{rate} = 8.70 \times 10^{-4} = k_3(1.0 \times 10^{-5})^2(1.0 \times 10^{-3}) = k_3(1.0 \times 10^{-13})$$

The value of k_3 is therefore 8.7×10^9 mol^{-2}L^{2}s^{-1}. You should verify that this value of k_3 is obtained from any of the experiments.

section 19.4
First-Order Reactions

A reaction of the form A $\rightarrow$ products, with a rate law given by Eq. (19-17)

$$-d[A]/dt = k_1[A]$$

is a **first-order reaction**. An equation having the form of this rate law is called a differential equation. It describes how the rate at any given instant depends on the concentration of A at that instant. If we know the initial concentration of A and want to find out what the concentration of A will be an hour later, or three hours later, or at any time, t, it is necessary to obtain another form of the rate equation, called the integrated form.

To proceed from the differential form of the rate equation to the integrated form requires the use of calculus. Since some of you reading this text are not familiar with the calculus, we will simply write down what the rate equation looks like in integrated form. For those who have some knowledge of calculus, the derivation of the integrated form of the first-order rate equation is given in Appendix J.

The integrated form of Eq. (19-17) is

$$\ln \left(\frac{[A]_t}{[A]_0} \right) = -k_1 t \tag{19-19}$$

where $[A]_t$ is the concentration of A at any time t, $[A]_0$ is the concentration of A at time zero, and ln is the symbol for the natural (base e) logarithm. This equation can be put into exponential form by taking the antilog of both sides:

$$[A]_t/[A]_0 = e^{-k_1 t} \tag{19-20a}$$

or

$$[A]_t = [A]_0 e^{-k_1 t} \tag{19-20b}$$

Equation (19-20b) is the **exponential decay law for first-order reactions. It states that the concentration of A decreases exponentially with time for a first-order reaction.** A plot of [A] as a function of time for a first-order reaction, Fig. 19.4, shows this exponential decline.

Equations (19-19), (19-20a), and (19-20b), are completely equivalent, but the logarithmic form is more convenient for obtaining the numerical value of the specific rate constant, k_1. If we make use of the properties of logarithms we can rewrite Eq. (19-19) as

$$\ln[A]_t - \ln[A]_0 = -k_1 t \tag{19-21a}$$

or

$$\ln[A]_t = -k_1 t + \ln[A]_0 \tag{19-21b}$$

Equation (19-21b) is the equation of a straight line:

$$y = mx + b$$
$$\ln[A]_t = -k_1 t + \ln[A]_0$$

Thus, for a first-order reaction, if we plot $\ln[A]_t$ as ordinate versus t as abscissa, we will obtain a straight line with slope equal to $-k_1$.

If base 10 logarithms are preferred to natural logarithms, we need only utilize the relation $\ln X = 2.303 \log X$ to transform Eq. (19-21b) into

$$\log[A]_t = -(k_1/2.303)t + \log[A]_0 \tag{19-21c}$$

A plot of $\log[A]_t$ versus time has slope $-(k_1/2.303)$. Before the widespread use of electronic calculators, Eq. (19-21c) was almost always used because log tables provide base 10 logarithms. Electronic calculators have made it more convenient to use Eq. (19-21b).

If a reaction is not first order, a plot of either $\ln[A]_t$ or $\log[A]_t$ versus time will not be linear. Therefore, in order to prove that a reaction is first order, the first thing to do is to make such a plot and show that it is indeed a straight line.

For the data given in Table 19.2 for the hydrolysis of sucrose, we can demonstrate that the reaction is first order by tabulating ln[sucrose] as a function of time and making the required plot. A plot of ln[sucrose] versus time is shown in Fig. 19.5, and the data plotted are listed in Table 19.3.

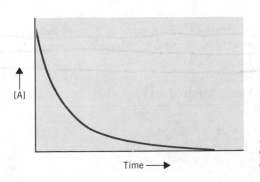

Fig. 19.4. The exponential decay of the concentration of the reactant for a first-order reaction.

Table 19.3. **Data for the Reaction between Sucrose and Water at 25 °C, Used for the Plot of Fig. 19.5**

$[C_{12}H_{22}O_{11}]$	Time(min)	$ln[C_{12}H_{22}O_{11}]$
1.0023	0.00	+0.0023
0.9022	30.00	−0.1029
0.8077	60.00	−0.2136
0.7253	90.00	−0.3212
0.6297	130.00	−0.4625
0.5753	160.00	−0.5529
0.5007	200.00	−0.6916
0.4358	240.00	−0.8306
0.3793	280.00	−0.9694

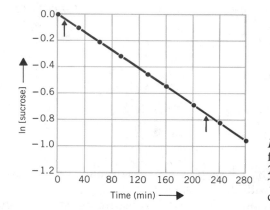

Fig. 19.5. Plot of ln [sucrose] versus time for a particular kinetics experiment at 25 °C. The data of Table 19.3 are plotted. The arrows indicate the points used to calculate the slope.

To obtain the slope of the plot, select two points on the line drawn that are widely separated and read the values of the ordinate and abscissa for both points. Do *not* choose experimental points. On Fig. 19.5 the points selected are (10,−0.039) and (220,−0.768), and are indicated by arrows on the plot. The slope is therefore

$$\text{slope} = -k_1 = \frac{-0.768 - (-0.039)}{(220.0 - 10.0) \text{ min}} = -\frac{0.729}{210 \text{ min}}$$

and the value obtained for the rate constant, k_1, is $3.47 \times 10^{-3} \text{ min}^{-1}$.

The reason we do not choose experimental points in calculating the slope is that each experimental point is subject to experimental uncertainty. Some points may lie above the line and others below it. The best line drawn through *all* the experimental points is more reliable than any particular pair of points used to obtain the line.

The Half-Life

The **half-life** of a reaction is defined as the amount of time it takes for one half the material originally present at time $t = 0$ to decompose or form products. It is the amount of time required for the concentration of A to fall to one half its initial value.

For a first-order reaction we can obtain the expression for the half-life by using Eq. (19-19). Let $\tau_{1/2}$ represent the time required for [A] to decrease to $\frac{1}{2}[A]_0$. Substitution into Eq. (19-19) yields

$$\ln\left(\frac{\frac{1}{2}[A]_0}{[A]_0}\right) = \ln\frac{1}{2} = -\ln 2 = -k_1\tau_{1/2} \tag{19-22a}$$

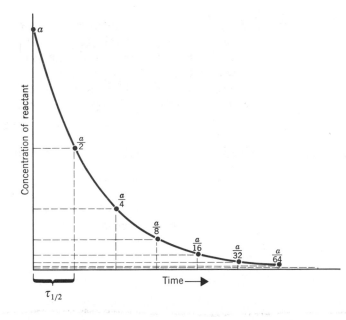

Fig. 19.6. A graph of a first-order reaction showing that the half-life is independent of initial concentration.

which can be rearranged as

$$\tau_{1/2} = (\ln 2)/k_1 = 0.693/k_1 \qquad (19\text{-}22b)$$

The striking thing about this expression is that the half-life of a first-order reaction does not depend on the initial concentration of A. For the experiment described in Table 19.3 on the hydrolysis of sucrose,

$$\tau_{1/2} = \frac{0.693}{3.47 \times 10^{-3} \text{ min}^{-1}} = 200 \text{ min}$$

It takes 200 min (3.33 h) for half of all the sucrose present to be converted to glucose and fructose at 25 °C, whether one starts with 1 g of sucrose per liter or 1 mol (342.3 g) of sucrose per liter, or any other concentration of sucrose. The half-life of a first-order reaction is a characteristic of that reaction. For any other order reaction, the half-life is *not* characteristic of the reaction, but depends on the initial concentration of reactants, as we shall see in the next section. A plot of a first-order exponential decay as a function of time, showing the constancy of the half-life, is depicted in Fig. 19.6.

EXAMPLE 19.5. First-order gas-phase decompositions

At 55 °C the gas-phase decomposition of N_2O_5 is first order in N_2O_5 with specific rate constant 1.42×10^{-3} s^{-1}.

(a) Determine the half-life of this decomposition.

Solution

$$\tau_{1/2} = \frac{\ln 2}{k} = \frac{0.693}{1.42 \times 10^{-3} \text{ s}^{-1}} = 488 \text{ s} = 8.13 \text{ min}$$

(b) If the initial partial pressure of N_2O_5 is 2.84 atm at 55 °C, how long will it take for the partial pressure to fall to 0.355 atm?

Solution. The partial pressure of a gas is directly proportional to its concentration in moles per liter at constant temperature: $P_A = [A]RT$. Rate expressions for gas-phase reactions have the same form as rate expressions for reactions in solution, with partial pressures of gases used rather than concentrations.

If the partial pressure of N_2O_5 falls from 2.84 to 0.355 atm it has decreased to 1/8 its original value. A decrease by a factor of 1/8 takes three half-lives. At the end of the first 8.13 min, the partial pressure falls to 1/2 its initial value (1.42 atm); after 16.26 min (two half-lives) the partial pressure is again cut in half, to 0.710 atm. At the end of three half-lives, or after 24.4 min, the partial pressure is 1/8 its original value, 0.355 atm.

A generalization of the reasoning used in Example 19.5 yields the result that after n half-lives, for a first-order reaction, the fraction of reactant remaining is $(\frac{1}{2})^n$ or $1/2^n$. Remember that this is not true for any other order.

Radioactive Decay

A very important type of reaction that is always first order is the decay of radioactive nuclei. All elements have isotopes that are unstable and decay to other nuclei. For radioactive substances, instead of using the concentration of the radioactive species in the rate laws, Eqs. (19-17), (19-19), (19-20a), and (19-20b), we use the number of radioactive nuclei, N, or the **activity**, usually given in counts per minute (cpm) or disintegrations per minute (dpm). The unit of time may also be seconds, so that the activity may be reported in counts per second (cps) or in dps. The activity of a radioactive sample is directly proportional to the number of radioactive nuclei in the sample. The first-order decay law, Eq. (19-20a), for radioactive substances is therefore usually written as

$$N/N_0 = e^{-\lambda t} \tag{19-23}$$

where λ is the symbol commonly used for the specific rate constant of a radioactive decay, also called the **first-order decay constant**.

The half-life of a radioactive nucleus, $\tau_{1/2} = \ln 2/\lambda$, is a characteristic property of that nucleus and can be used to identify it. Chlorine, for example, has two stable isotopes, ^{35}Cl and ^{37}Cl, and eight radioactive isotopes, each with its own characteristic half-life. The half-life of ^{34}Cl is only 1.56 s, while ^{36}Cl has a half-life of 3×10^5 yr and the half-life of ^{38}Cl is 37.3 min.*

The methods used to solve problems involving radioactive decay are the same as those used for any other first-order decomposition. A typical problem is illustrated in Example 19.6.

EXAMPLE 19.6. Determination of the half-life

The radioactivity of a sample of a transuranium nuclide is counted on a radiation detection device. The following data are obtained:

Time (h)	0.00	2.00	4.00	6.00	8.00	12.00
Activity (cpm)	8780	7527	6452	5531	4741	3484

Determine the half-life of the nuclide.

---

* Half-lives of isotopes can be found in the Table of Isotopes of the *Handbook of Chemistry and Physics*.

Solution. A plot of the natural logarithm of the activity as a function of time should be drawn. (Alternatively, we could plot the base 10 logarithm of the activity versus time.) A table of the values plotted should always appear along with the plot. These values and the plot are shown below.

Time (h)	ln(activity)
0.00	9.080
2.00	8.923
4.00	8.772
6.00	8.618
8.00	8.464
12.00	8.156

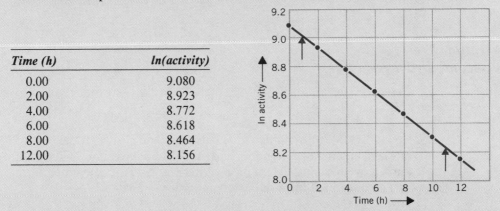

The slope of the plot is determined using two widely separated points (indicated by arrows on the plot above).

$$\text{slope} = -\lambda = \frac{8.23 - 9.00}{11.00 - 1.00} = -0.077$$

The first-order decay constant is therefore $0.077\ \text{h}^{-1}$.

The half-life is obtained from Eq. (19-22b):

$$\tau_{1/2} = (\ln 2)/\lambda = 0.693/0.077 = 9.0\ \text{h}$$

The isotope is $^{234}_{94}\text{Pu}$.

Pseudo-First-Order Reactions

Many reactions in aqueous solution involve water as a reactant and are of the type

$$A + H_2O \rightarrow \text{products} \tag{19-24}$$

with a rate expression

$$\text{rate} = -d[A]/dt = k_1'[A][H_2O] \tag{19-25}$$

For a reaction carried out in dilute aqueous solution, a liter of solution contains approximately 1000 g of water, as the density will be close to that of pure water (1.0 g/mL) if the solution is dilute. Since 1000 g of water is $1000/18.02 = 55.5$ mol of H_2O, $[H_2O] \sim 55.5\ M$ in dilute aqueous solutions, and remains essentially constant at $55.5\ M$ throughout the reaction. If, for instance, the $[A]$ is $0.1\ M$ or less, the largest amount of H_2O that can be used up in a liter of solution is 0.1 mol, and that will change the $[H_2O]$ very little. The argument here is the same as that used in Section 8.1 for not including the $[H_2O]$ in equilibrium constant expressions for reactions occurring in dilute aqueous solution.

For a reaction that occurs in dilute aqueous solution with a rate law of the form of Eq. (19-25), we combine $k_1'[H_2O]$ as a new constant, k_1, so that the rate law becomes

$$\text{rate} = -d[A]/dt = k_1[A]$$

We see that the reaction appears to be first order when in fact it is second order. This kind of reaction is said to be a **pseudo-first-order** reaction. It is first order in A, but only pseudo-first-order overall. The hydrolysis of sucrose, Eq. (19-18), is an example of a pseudo-first-order reaction. Because water is the solvent, we do not observe the dependence of the rate on the $[H_2O]$. It is often difficult to determine how the rate depends on the solvent concentration.

section 19.5
Second-Order Reactions

Consider a second-order reaction of the form $2A \longrightarrow$ products, with a rate expression

$$\text{rate} = -d[A]/dt = k_2[A]^2 \qquad (19\text{-}26)$$

Equation (19-26) is a differential equation, and just as in the case of the first-order rate law, in order to determine what $[A]$ will be at any time t, if its value at time $t = 0$ is known, we need a different form of the rate law, the integrated form.

The integrated form of the second-order rate law is

$$\frac{1}{[A]_t} - \frac{1}{[A]_0} = k_2 t \qquad (19\text{-}27)$$

The derivation of Eq. (19-27) will be found in Appendix J. A simple rearrangement of this equation shows that it is the equation of a straight line:

$$y = mx + b$$

$$\frac{1}{[A]_t} = k_2 t + \frac{1}{[A]_0}$$

A plot of $1/[A]_t$ versus time for a second-order reaction will therefore be linear, and the slope will be the value of k_2, the second-order specific rate constant.

It is easy to show that the half-life of a second-order reaction depends on the initial concentration of A and is therefore *not* a characteristic of the reaction. The time it takes for $[A]$ to fall to $\frac{1}{2}[A]_0$ is obtained by direct substitution into Eq. (19-27):

$$\frac{1}{\frac{1}{2}[A]_0} - \frac{1}{[A]_0} = k_2 \tau_{1/2} \qquad (19\text{-}28a)$$

or

$$\tau_{1/2} = 1/(k_2[A]_0) \qquad (19\text{-}28b)$$

Because the half-life of a second-order reaction changes as the initial concentration changes, it is not a useful quantity and, in general, half-lives are cited only for first-order reactions. It is only for first-order reactions that the half-life is independent of the initial concentration of A.

If the concentration of a substance, A, decreases with time during the course of a reaction, and you do not know whether the reaction is first or second order in A, you should plot *both* $\ln[A]$ versus time and $1/[A]$ versus time. If one of these plots is linear while the other is not, you will know with certainty what the order of the reaction is. Example 19.7 illustrates this procedure.

EXAMPLE 19.7. Determining whether a reaction is first or second order

The rate of dimerization of gaseous butadiene, $CH_2{=}CH{-}CH{=}CH_2$, has been studied at 326 °C. The following data giving the pressure of butadiene as a function of time were obtained.

t (min)	0.00	17.30	42.50	55.08	68.05	90.05	119.00	176.67
P_B (atm)	0.832	0.661	0.509	0.458	0.417	0.361	0.307	0.235

What is the order of this reaction, and what is the rate constant?

Solution. For a gas-phase reaction, the pressure of a gas is used in the rate law instead of the concentration, as they are directly proportional: $P_B = [B]RT$, where $[B] = n_B/V$. If we symbolize butadiene simply as B(g), the dimerization reaction is

$$2B(g) \rightarrow B_2(g)$$

To determine whether the reaction is first or second order, we must plot both $\ln P_B$ versus time and $1/P_B$ versus time. Data for these two plots are tabulated below. The plots are shown in Figs. 19.7 and 19.8.

t (min)	0.00	17.30	42.50	55.08	68.05	90.05	119.00	176.67
P_B (atm)	0.832	0.661	0.509	0.458	0.417	0.361	0.307	0.235
$\ln P_B$	−0.184	−0.414	−0.676	−0.780	−0.874	−1.02	−1.18	−1.45
$1/P_B$	1.20	1.51	1.97	2.18	2.40	2.77	3.25	4.25

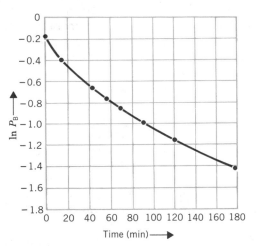

Fig. 19.7. Plot of $\ln P_B$ versus time for the dimerization of butadiene at 326 °C, where P_B is the partial pressure of butadiene. The data plotted are tabulated in Example 19.7. The fact that this plot is not linear proves that the reaction is not first order in butadiene.

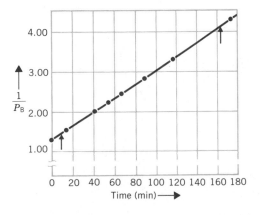

Fig. 19.8. Plot of $1/P_B$ versus time for the dimerization of butadiene at 326 °C, where P_B is the partial pressure of butadiene. The data plotted are tabulated in Example 19.7. As this plot is linear while the plot of $\ln P_B$ versus time is not, the reaction is second order in butadiene. Arrows indicate the points used to calculate the slope, which is 1.73×10^{-2} atm^{-1}min^{-1}.

An examination of Figs. 19.7 and 19.8 shows that the plot of $\ln P_B$ versus time is not linear, but the plot of $1/P_B$ versus time is a straight line. The dimerization of butadiene is therefore second order in butadiene. The rate law is $-dP_B/dt = kP_B^2$. The second-order rate constant is the slope of the plot of $1/P_B$ versus time, and is 1.73×10^{-2} atm^{-1}min^{-1}.

It is possible that neither a plot of $\ln[A]$ nor a plot of $1/[A]$ is a linear function of time. This means that the reaction is not simply either first or second order in A. Procedures to determine other orders of reaction are similar to those we have discussed for first- and second-order reactions, but involve somewhat more complicated integrated rate laws.

section 19.6
The Temperature Dependence of Rate Constants

We have all observed the increase in the rate of reaction with an increase in temperature. It takes more time to cook a hard-boiled egg in Denver than it does in New York, because the temperature of boiling water is lower in Denver due to the lower atmospheric pressure at its mile-high altitude. At the lower temperature, the rate of reaction is less and food cooks more slowly. Anyone who has done photographic developing knows that the warmer the developing mixture, the faster the pictures develop.

The dependence of the rate on temperature is observed in the numerical value of the specific rate constant. Values of the rate constant as a function of temperature for the second-order reaction

$$CH_3I + CH_3CH_2ONa \rightarrow CH_3OCH_2CH_3 + NaI \qquad (19\text{-}29)$$

Methyl   Sodium   Ethyl methyl
iodide   ethoxide   ether

run in ethanol, are listed in Table 19.4 and plotted as Fig. 19.9. The shape of the curve of the rate constant versus temperature for this reaction is typical of that for most chemical reactions, regardless of the order of the reaction. There are other types of temperature dependence (for explosive reactions, or some enzyme-catalyzed reactions) but the majority of chemical reactions have the temperature dependence illustrated in Fig. 19.9.

Table 19.4. The Rate Constant as a Function of Temperature for the Reaction between CH_3I and C_2H_5ONa, Eq. (19-29)

$t(^\circ C)$	$k\ (mol^{-1}L \cdot s^{-1})$
0.0	0.56×10^{-4}
6.0	1.18×10^{-4}
12.0	2.45×10^{-4}
18.0	4.88×10^{-4}
24.0	10.0×10^{-4}
30.0	20.8×10^{-4}

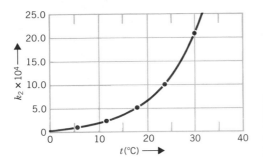

Fig. 19.9. The rate constant for the second-order reaction between methyl iodide and sodium ethoxide [Eq. (19-29)], as a function of temperature. The data plotted are listed in Table 19.4.

Effective Collisions

The rates of gas-phase reactions are of particular theoretical interest because, using the kinetic theory of gases, it is possible to calculate just how many collisions occur between gas molecules in a given time. When the number of collisions per second is compared with the rate of reaction, it is clear that only a small fraction of the collisions that occur actually leads to reaction.

By considering the details of a collision between molecules we can understand why only a small fraction of collisions results in a reaction between the colliding species. A collision between two molecules is quite different from a collision between two golf balls. A molecule has no definite boundary. There is a fairly diffuse electron cloud surrounding all the nuclei. When two molecules approach each other closely, there is a mutual repulsion of their electron clouds. This repulsion causes them to slow down, and the consequent decrease in their kinetic energy is accompanied by an increase in their potential energy. If the molecules were moving slowly before the collision, they would stop, reverse direction, and move apart before any significant interpenetration of their electron clouds could take place. This means that slow-moving molecules simply bounce off each other without any reaction occurring. Figure 19.10 illustrates the different outcomes of collisions between reactant molecules colliding with different relative speeds.

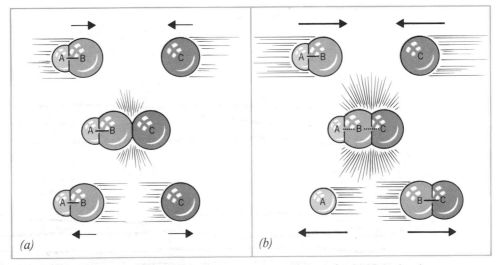

Fig. 19.10. *(a)* When two slow-moving molecules collide, their electron clouds cannot interpenetrate much and they just bounce off each other, chemically unchanged. *(b)* When fast-moving molecules collide, atoms approach each other much more closely as their electron clouds interpenetrate. This can lead to bond making and bond breaking. The net change here is $AB + C \rightarrow A + BC$.

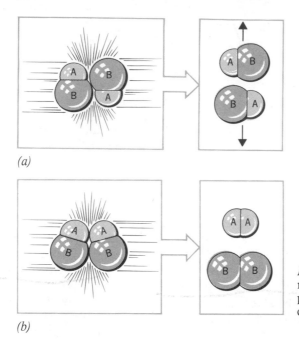

(a)

(b)

Fig. 19.11. Collisions between A—B molecules. *(a)* A collision that cannot produce a net chemical change. *(b)* A collision that can lead to a net reaction.

As an analogy, consider that two people walking slowly can bump into each other without either of them being hurt, but that two people running at each other full tilt can easily break some bones! We conclude, therefore, that only a collision between molecules with sufficiently large speeds can result in reaction.

In addition to the relative speeds with which two molecules approach each other, there is another factor that must be considered in determining whether or not a collision between two molecules is effective at leading to reaction, and that is the **orientation factor.** In order for reaction to occur, specific bonds must be broken, and the colliding molecules must be oriented with respect to one another so that new bonds can be formed.

Consider the reaction $2AB \rightarrow A_2 + B_2$. Figure 19.11*(a)* illustrates a collision between two A—B molecules that is unlikely to be effective in forming A_2 and B_2 because of the way in which the two colliding molecules are oriented relative to one another, while Fig. 19.11*(b)* shows a different orientation of the colliding A—B molecules that is likely to be much more effective at forming the products.

We see, therefore, that only a very small fraction of collisions result in reaction for two reasons: (1) The colliding molecules may not be appropriately oriented to one another, and (2) the collision may not be sufficiently energetic.

The Activation Energy

In 1889 the Swedish chemist Svante Arrhenius suggested that before a reaction can occur there must be some minimum kinetic energy possessed jointly by two colliding molecules. The minimum quantity of energy that must be available for a collision to lead to reaction is called the **activation energy** and denoted E_{act}. If two molecules collide with less than this critical amount of energy, they will recoil without undergoing chemical change.

The kinetic energy of the collision must be transformed into potential energy. Translational energy is converted into both rotational and vibrational energy, and as

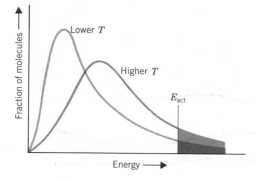

Fig. 19.12. The distribution of kinetic energies in a gas at two different temperatures. The fraction of molecules with an energy greater than the activation energy, E_{act}, is small, but increases markedly with increasing temperature.

atoms within a molecule vibrate with larger amplitude, bonds can be broken. The activation energy is necessary to rupture bonds between atoms in a molecule, or between an ion and its solvation shell if the reacting species are in solution.

The reason why the rates of chemical reactions are so strongly temperature dependent becomes clear once we recognize the need for a minimum kinetic energy for the colliding molecules. As the temperature is increased, the average kinetic energy increases and a greater fraction of molecules have higher kinetic energies. Recall the discussion in Section 4.9 about the temperature dependence of the distribution of molecular speeds for gases. Figure 19.12 shows how the distribution of kinetic energies varies with temperature. Since the average kinetic energy of the reacting species increases as the temperature increases, a greater fraction of collisions leads to reaction at higher temperature. In addition, the frequency of collisions increases as the temperature rises. The net result of both a greater frequency of collision and a larger fraction of collisions having energies above the activation energy leads to a rapid increase in the rate of reaction as the temperature is increased.

The Arrhenius Equation

Arrhenius proposed the following equation to describe the temperature dependence of any specific rate constant:

$$k = Ae^{-E_{act}/RT} \tag{19-30}$$

where T is the absolute temperature and A is a constant (or nearly so), independent of temperature. The constant A is called the **frequency factor**. In using this equation, it is more convenient to take the logarithm of both sides. We obtain

$$\ln k = \ln A - E_{act}/RT \tag{19-31a}$$

which can be rearranged as

$$\ln k = -E_{act}/RT + \text{constant} \tag{19-31b}$$

Since A is a constant, $\ln A$ is also a constant, and we have written the equation in this way to emphasize that it is the equation of a straight line:

$$\ln k = -\left(\frac{E_{act}}{R}\right)\frac{1}{T} + \text{constant}$$
$$y = (m)\,x + b$$

A plot of $\ln k$ versus $1/T$ will be linear, with slope $-E_{act}/R$. Note particularly that Eqs. (19-30) and (19-31a) apply to all rate constants, regardless of the order of reaction.

The value of the activation energy can be obtained as follows. The specific rate constant is measured as a function of temperature. A plot of the natural logarithm of the rate constant versus the reciprocal of the absolute temperature is then drawn. The slope of this plot is $-E_{act}/R$, so that E_{act} is calculated from the slope.

If it is more convenient to use base 10 logarithms, we can employ the relation $\ln X = 2.303 \log X$ to convert Eq. (19-31b) to base 10 logarithms, as follows:

$$\log k = -\left(\frac{E_{act}}{2.303R}\right)\frac{1}{T} + \text{constant}' \qquad (19\text{-}31c)$$

A plot of $\log k$ versus $1/T$ is linear, with slope $-(E_{act}/2.303R)$. For the data of Table 19.4 a plot of $\log k$ versus $1/T$ is shown as Fig. 19.13.

Although the determination of E_{act} is more reliable if we have values of k at several different temperatures, we can obtain a value for E_{act} if we have measurements of k at only two different temperatures. Let k_1 be the specific rate constant at T_1, and k_2 be the specific rate constant at T_2. Equation (19-31a) applied specifically to these two temperatures yields

$$\ln k_1 = \ln A - E_{act}/RT_1$$

and

$$\ln k_2 = \ln A - E_{act}/RT_2$$

Subtraction of the second of these equations from the first yields

$$\ln k_1 - \ln k_2 = -E_{act}/RT_1 + E_{act}/RT_2$$

If we use the property of logarithms that $\ln a - \ln b = \ln (a/b)$, and also factor the common term $-E_{act}/R$ on the right-hand side, we obtain

$$\ln (k_1/k_2) = -\frac{E_{act}}{R}\left(\frac{1}{T_1} - \frac{1}{T_2}\right) = -\frac{E_{act}}{R}\left(\frac{T_2 - T_1}{T_1 T_2}\right) \qquad (19\text{-}32a)$$

If it is more convenient to use base 10 logarithms, this equation can be written as

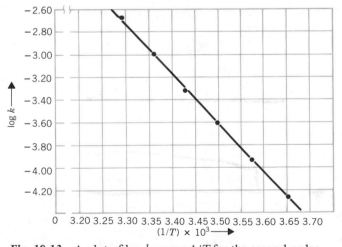

Fig. 19.13. A plot of $\log k$ versus $1/T$ for the second-order rate constant for the reaction between CH_3I and C_2H_5COONa [Eq. (19-29)]. The data plotted are from Table 19.4. The slope of a plot of $\log k$ versus $1/T$ is $-(E_{act}/2.303R)$, so that the activation energy can be calculated from the slope.

$$\log (k_1/k_2) = -\frac{E_{act}}{2.303\ R}\left(\frac{1}{T_1} - \frac{1}{T_2}\right) = -\frac{E_{act}}{2.303R}\left(\frac{T_2 - T_1}{T_1 T_2}\right) \quad (19\text{-}32\text{b})$$

You have probably recognized the similarity between Eq. (19-32a) and the van't Hoff equation, Eq. (18-31b) or Eq. (18-31c), which gives the temperature dependence of the equilibrium constant for a reaction. The similarity is, of course, not coincidental. The equilibrium constant is related to the specific rate constant of the forward and reverse reactions, as we shall see in Section 19.10. In Section 19.7 we shall show the relationship between ΔH and the activation energies of the forward and reverse reactions.

The following example illustrates the use of Eq. (19-32a).

EXAMPLE 19.8. Determination of the activation energy

From the values of the specific rate constant listed in Table 19.4 for the reaction between methyl iodide and sodium ethoxide in ethanol, Eq. (19-29), at 12 and at 30 °C, calculate the activation energy for this reaction.

Solution. The temperatures used in Eq. (19-32a) must, of course, be on the Kelvin scale. At 12 °C, $k_1 = 2.45 \times 10^{-4}$ and $T_1 = 285.2$. At 30 °C, $k_2 = 20.8 \times 10^{-4}$ and $T_2 = 303.2$. Substitution into Eq. (19-32a) yields

$$\ln\left(\frac{2.45 \times 10^{-4}}{2.08 \times 10^{-3}}\right) = -\frac{E_{act}}{R}\left[\frac{18.0}{(285.2)(303.2)}\right]$$

Since $\ln (0.245/2.08) = \ln (0.1178) = -2.139$, the equation becomes

$$-2.139 = -\frac{E_{act}}{R}\left[\frac{18.0}{(285.2)(303.2)}\right]$$

or

$$E_{act}/R = \frac{(2.139)(285.2)(303.2)}{18.0} = 1.028 \times 10^4 = 1.03 \times 10^4$$

The activation energy can then be calculated in any desired units by choosing the appropriate value for R. Thus we obtain

$$E_{act} = (1.028 \times 10^4)(8.314) = 8.55 \times 10^4 \text{ J/mol} = 85.5 \text{ kJ/mol}$$

The answer should only be given to three significant figures because the difference between the two temperatures, 18.0, is only valid to three figures. If we prefer to express E_{act} in calories we obtain

$$E_{act} = (1.028 \times 10^4)(1.987) = 2.04 \times 10^4 \text{ cal/mol} = 20.4 \text{ kcal/mol}$$

If the activation energy for a reaction has been determined from the slope of a plot of $\ln k$ versus $1/T$ for a series of measurements, Eq. (19-32a) can be used to determine the specific rate constant at a temperature for which a measurement of k has not been made. This type of calculation is illustrated in Example 19.9.

EXAMPLE 19.9. Determining a specific rate constant from E_{act}

The activation energy for the reaction between methyl iodide and sodium ethoxide, calculated from the slope of a plot of $\ln k$ versus $1/T$ using the data of Table 19.4, is 82.6 kJ. (Note that this is a better value than the one calculated in Example 19.8.) What is the specific rate constant for reaction (19-29) at 20 °C?

Solution. In addition to the activation energy, we need the specific rate constant at some other temperature. We can use any value listed in Table 19.4; let us choose the value at 24.0 °C, or 297.2 K. Direct substitution into Eq. (19-32a) yields

$$\ln\left(\frac{k_{293}}{k_{297}}\right) = \frac{-82.6 \times 10^3}{8.3144}\left(\frac{1}{293.2} - \frac{1}{297.2}\right)$$
$$= -0.4560$$

Hence,

$$\frac{k_{293}}{k_{297}} = e^{-0.4560} = 6.338$$

and

$$k_{293} = (0.6338)(10.0 \times 10^{-4}) = 6.34 \times 10^{-4}$$

Thus the specific rate constant at 20 °C, or 293.2 K, is 6.34×10^{-4} mol^{-1}L $\cdot$ s^{-1}.

section 19.7
Transition State Theory and the Activated Complex

During the course of a reaction, the potential energy of the reacting species changes as the distances between the atoms change. In principle, we should be able to calculate the potential energy of a collection of atoms as a function of the positions of all the atoms in space. In practice, for all but very small molecules, there are too many variables to make a complete and exact potential energy surface calculation possible. By varying only two distances at a time while holding other distances constant, we can generate many potential energy surfaces and learn a great deal about reaction pathways. On these diagrams, points of equal potential energy are connected, and form what is called a contour. A pathway of minimum potential energy can then be located.

Consider the hypothetical reaction involving only three atoms

$$A{-}B + C \rightleftharpoons A + B{-}C$$

Along the minimum potential energy path by which the reactants, A—B and C, are converted into the products, A and B—C, there is a specific geometric configuration of all three atoms, called the **activated complex**, which we may denote A $\cdots$ B $\cdots$ C, which must be formed before reaction can occur. The state of the system when the activated complex has been formed is called the **activated state** or the **transition state**, as it represents a transition between reactants and products. A study of potential energy contours reveals that, along the minimum potential energy path, the potential energy of all the atoms is at a maximum when the activated complex has been formed.

Potential Energy versus Reaction Coordinate Diagrams

We can plot the potential energy as a function of the **reaction coordinate**, a measure of the extent to which the reaction has proceeded toward completion along the reaction pathway. If we are dealing with an exothermic reaction, the energy of the reactants is higher than that of the products. Energy is released when the products are formed. But before the products can be formed the system must acquire additional energy: it must pass through the **activated state** in which the potential energy is a maximum. The situation is illustrated in Fig. 19.14.

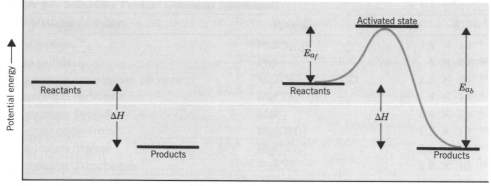

Fig. 19.14. Potential energy profile for an exothermic reaction. The reactants must acquire sufficient energy to reach the activated state before the products can be formed.

The energy that the reactants must acquire to reach the activated state is the activation energy of the forward reaction, denoted E_{a_f}. For the backward reaction to take place the same activated state must be attained. The activation energy of the backward reaction, E_{a_b}, is therefore larger than the activation energy of the forward reaction, for an exothermic reaction. From Fig. 19.14 it is evident that

$$\Delta H = E_{a_f} - E_{a_b} \qquad (19\text{-}33)$$

Note that ΔH is negative, as it must be for an exothermic reaction.

The activated complex, the species at the top of the potential energy barrier is neither the reactants nor the products; it is an unstable combination of all the atoms involved in the reaction for which the potential energy is a maximum along the minimum potential energy pathway that results in reaction.

For an endothermic reaction, the reactants are at a lower potential energy than the products. The activated state is at a higher energy than either the reactants or the products. Figure 19.15 is the potential energy profile for an endothermic reaction. Equation (19-33) is applicable to both endothermic and exothermic reactions.

A reaction with a large activation energy is said to have a high potential energy barrier. Such reactions will have slower rates of reaction than reactions with low potential energy barriers, that is, reactions for which the activation energy is small.

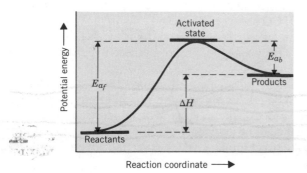

Fig. 19.15. The potential energy profile for an endothermic reaction. The reactants must acquire sufficient energy to reach the activated state before the products can be formed.

section 19.8
The Relation between the Mechanism of a Reaction and Kinetics Data

Consider the reaction

$$NO_2(g) + CO(g) \rightarrow CO_2(g) + NO(g) \tag{19-34}$$

for which, at temperatures below 500 K, the rate expression is

$$\text{rate} = -d[NO_2]/dt = k[NO_2]^2 \tag{19-35}$$

The experimentally observed rate law could not have been predicted on the basis of the correctly balanced equation for the reaction. Indeed, we might say that the observed rate law seems surprising, given the stoichiometry of the reaction. *If* the stoichiometry were an indication of the mechanism, we might expect that a CO molecule and an NO_2 molecule must collide in order for the reaction to occur. Yet the observed rate does not depend on the concentration of CO at all!

Whenever the observed rate cannot be predicted from the stoichiometric equation for the reaction we can be sure that the mechanism of the reaction is not described by the stoichiometry. In general, an overall chemical reaction occurs by a series of simple steps, each of which is called an **elementary process**. The succession of elementary steps is called the **mechanism**, or detailed pathway, for the reaction. The stoichiometry is the sum of all the steps that comprise the mechanism.

The observed rate law provides a clue to the mechanism of the reaction. It tells us that the **rate-determining step** must involve a collision between two NO_2 molecules. A two-step mechanism has been proposed for this reaction, one slow step and one that is rapid.

| (slow) | step 1 | $NO_2 + NO_2 \rightarrow NO + NO_3$ |
| (fast) | step 2 | $NO_3 + CO \rightarrow CO_2 + NO_2$ |

Note that by adding these two equations together we obtain the overall stoichiometry, Eq. (19-34).

The species NO_3 is postulated to be an **intermediate** in this mechanism. It is very short lived, as the second step in the mechanism, which uses up NO_3, occurs very rapidly.

Although the overall reaction occurs in two steps, the rate is determined by step 1 only, which is the slow step relative to step 2. The slowest step in a reaction mechanism is called the **rate-determining step**. Because all other steps must wait for the slowest step to occur, the overall rate is determined by the rate of the slowest step. Not all reaction mechanisms involve a *single* rate-determining step. In some cases two or more steps are considerably slower than the others. For a single step to be rate determining, it must be significantly slower than all others.

The experimental rate expression provides information about the rate-determining step (or steps) of a reaction mechanism. The rate of reaction (19-34) does not depend on the concentration of CO because CO is used only in the fast step. The rate law, Eq. (19-35), tells us that the activated complex contains two N atoms and four O atoms, and may be denoted $N_2O_4\ddagger$, where the symbol $\ddagger$ is used to denote an activated complex.

Each step in a mechanism is called an **elementary process**. For an elementary process the rate expression can be predicted from the stoichiometry, because in the rate law for each elementary process, the exponent of the concentration of each species is the same as the coefficient of that species in the reaction equation for the process. A process for which the stoichiometry and the mechanism are identical is, by definition, an elementary process. Remember, however, that there is no necessary connection between the overall stoichiometry and rate expression.

The observed rate law provides information only about the slowest step or steps in the mechanism. Scientists propose mechanisms that they invent using their imagination and familiarity with other chemical reactions. Sometimes several mechanisms, with different intermediates, can be proposed, all of which are in agreement with the rate law. Often a long series of experiments is required to rule out one or more postulated mechanisms. It can be a very difficult process to establish the correct mechanism.

We have stated that if an observed rate expression could not have been predicted from the overall reaction, the mechanism of the reaction is certainly not described by the stoichiometry. The converse of this statement is not true, however. If an observed rate expression is in agreement with the overall stoichiometry, it does not necessarily follow that the stoichiometric equation represents an elementary process. There may still be a series of steps required for the reaction to occur that are not indicated by the stoichiometry.

The Hydrogen Iodide Reaction

An important example that illustrates many of the statements we have just made is the reaction

$$H_2(g) + I_2(g) \rightarrow 2HI(g) \tag{19-36}$$

for which the observed rate law is

$$\text{rate} = -d[H_2]/dt = k[H_2][I_2] \tag{19-37}$$

This is precisely the rate expression you would expect if the stoichiometry described the mechanism, that is, if the mechanism involved a collision between an H_2 and an I_2 molecule. For many years it was believed that Eq. (19-36) represents an elementary process. However, during the 1960's a two-step mechanism was proposed, both by a Russian chemist, Nikolai N. Semenov, and by an American, Henry Eyring. Their proposed mechanism is

(rapid equilibrium)	step 1	$I_2 \overset{K}{\rightleftharpoons} 2I$
(slow)	step 2	$H_2 + 2I \overset{k'}{\rightarrow} 2HI$

Iodine atoms are an intermediate species in this mechanism. The rate is determined by the slow step, and since this is an elementary process, the rate law is predicted to be

$$\text{rate} = -d[H_2]/dt = k'[H_2][I]^2$$

But the concentration of iodine atoms is determined by the rapid equilibrium: $K = [I]^2/[I_2]$, so that $[I]^2 = K[I_2]$. Substituting this expression for $[I]^2$ into the rate law we obtain

$$\text{rate} = k'K[H_2][I_2] = k[H_2][I_2]$$

which is the observed rate expression. The product of the equilibrium constant, K, for the first step, and the rate constant, k', of the second step, is merely another constant, and is equal to the observed rate constant if this mechanism is correct.

Note that both the single-step mechanism that assumes the stoichiometry is an elementary process, and the proposed two-step mechanism, are in agreement with the observed rate law. How can one decide between them?

In 1967 J. H. Sullivan tested the two theories by irradiating the reaction mixture with ultraviolet light. The ultraviolet light dissociates I_2 and increases the [I]. By studying how the reaction rate depends on the intensity of the ultraviolet light, Sullivan was able to demonstrate that I atoms are an intermediate in this reaction, and that the rate of appearance of HI is proportional to the square of the I atom concentration, as required for the two-step mechanism.

Molecularity versus Order

Elementary processes are classified according to the number of molecules that they involve, and they are either **unimolecular** (involve only one molecule), **bimolecular** (involve two molecules), or more rarely, **termolecular** (involve three molecules). The term **molecularity** of a reaction can be used to describe *only* an elementary process, and it is either 1, 2, or 3. The molecularity of different steps in a complex mechanism may not be the same.

The order of a reaction refers to the overall reaction, and can only be determined experimentally. It can be zero, integral, or fractional. ***For an elementary process the order and the molecularity are identical.*** For an overall reaction, the molecularity has no meaning, and the order cannot be predicted from the balanced stoichiometric equation.

Example 19.10 illustrates how we can eliminate some proposed mechanisms for a reaction on the basis of kinetics data. You should always bear in mind that a mechanism is, originally, the invention of some scientist. It is the product of a creative mind and needs to be proved (or disproved) by experimental evidence.

EXAMPLE 19.10. The relation between kinetics data and mechanism

For the reaction

$$2H_2(g) + 2NO(g) \rightarrow N_2(g) + 2H_2O(g)$$

the observed rate expression is

$$\text{rate} = k[NO]^2[H_2]$$

The following four mechanisms have been proposed for this reaction. Can any of them be ruled out on the basis of the observed rate expression?

mechanism I	$2H_2(g) + 2NO(g) \xrightarrow{k_1} N_2(g) + 2H_2O(g)$	
mechanism II	$H_2(g) + NO(g) \xrightarrow{k_2} H_2O(g) + N(g)$	(slow)
	$N(g) + NO(g) \longrightarrow N_2(g) + O(g)$	(fast)
	$O(g) + H_2(g) \longrightarrow H_2O(g)$	(fast)
mechanism III	$H_2(g) + 2NO(g) \xrightarrow{k_3} N_2O(g) + H_2O(g)$	(slow)
	$N_2O(g) + H_2(g) \longrightarrow N_2(g) + H_2O(g)$	(fast)
mechanism IV	$2NO(g) \underset{}{\overset{K_{eq}}{\rightleftharpoons}} N_2O_2(g)$	(fast equilibrium)
	$N_2O_2(g) + H_2(g) \xrightarrow{k_4} N_2O(g) + H_2O(g)$	(slow)
	$N_2O(g) + H_2(g) \rightarrow N_2(g) + H_2O(g)$	(fast)

Solution. We note first that in all four mechanisms the overall stoichiometry is the algebraic sum of the elementary processes of the mechanism. Note also that in kinetics mechanisms there is usually only a single arrow in the forward direction.

Equilibria must be specifically designated, as in mechanism IV. Only the *reactants* of the elementary process appear in the rate expression.

Mechanism I. When the stoichiometric equation is written as a mechanism, it implies that the reaction actually occurs as indicated by the stoichiometry. In this case it would mean that two H_2 molecules and two NO molecules would all collide simultaneously. If this *were* an elementary process, the rate law would have to be

$$\text{rate for mechanism I} = k_1[H_2]^2[NO]^2$$

that is, the reaction would be fourth order. Since the observed rate is third order, this mechanism can be eliminated. It should be emphasized that the probability of four molecules colliding simultaneously is so small that such a mechanism is highly improbable.

Mechanism II. The overall rate will be the rate of the slow step of the mechanism,

$$H_2(g) + NO(g) \xrightarrow{k_2} H_2O(g) + N(g)$$

Since this is an elementary process, the rate law would be

$$\text{rate for mechanism II} = k_2[H_2][NO]$$

As this is not in agreement with experimental observation, mechanism II can be eliminated as a possibility.

Mechanism III. The overall rate will be the rate of the slow step,

$$H_2(g) + 2NO(g) \xrightarrow{k_3} N_2O(g) + H_2O(g)$$

which would be

$$\text{rate for mechanism III} = k_3[NO]^2[H_2]$$

in agreement with experimental observation. This does not prove that mechanism III is correct. It only means that it is consistent with the observed rate expression and therefore cannot be ruled out as a possibility on this evidence alone. Note that N_2O is postulated to be an intermediate species in this mechanism. In trying to establish the mechanism of this reaction, one would certainly look for evidence that N_2O is actually present in the reaction mixture.

Mechanism IV. The overall rate is the rate of the slow step,

$$N_2O_2(g) + H_2(g) \xrightarrow{k_4} N_2O(g) + H_2O(g)$$

and is

$$\text{rate for mechanism IV} = k_4[N_2O_2][H_2]$$

However, N_2O_2 is an intermediate species, not a reactant or a product. The reaction rate should not be expressed in terms of the concentrations of intermediates. The $[N_2O_2]$ can be related to the [NO] by the equilibrium constant for the first step of this mechanism, $K_{eq} = [N_2O_2]/[NO]^2$, so that $[N_2O_2] = K_{eq}[NO]^2$. Substituting this into the rate of the slow step yields

$$\text{rate for mechanism IV} = k_4 K_{eq}[NO]^2[H_2]$$

Since k_4 and K_{eq} are both constants, $k_4 K_{eq} = k$, a new constant, so that the predicted rate for mechanism IV is in agreement with experimental observation.

We see that the observed rate expression rules out mechanisms I and II, but cannot decide between mechanisms III and IV. It is also possible that neither III nor IV is correct, but that some unpostulated mechanism is the right one.

Note that the rate-determining step in mechanism III is termolecular, while all steps in mechanism IV are bimolecular. That makes mechanism IV more likely than mechanism III, but neither mechanism can be established just from the observed rate expression.

Catalysis

A **catalyst** is a substance that changes the rate of a chemical reaction but does not itself undergo a permanent change when the reaction takes place. Any catalyzed reaction must proceed through a mechanism not apparent in the stoichiometry. A catalyst is used in the rate-determining step of a mechanism, but then is formed again in some subsequent step so that there is no net change in the concentration of the catalyst during the course of the reaction. The catalyst therefore does not appear in the overall stoichiometric equation for the reaction.

Catalyzed reactions are very common. Most reactions occurring in the bodies of humans and other animals are catalyzed, as are the reactions that take place in the atmosphere. Catalysts are added to industrial chemical processes to speed up reactions, and often the major task of developing an efficient industrial process is in finding a suitable catalyst.

There are two categories of catalysts, **homogeneous** and **heterogeneous**. A homogeneous catalyst is present in the same phase as the reactants. We have already discussed one example of homogeneous catalysis. The hydrolysis of sucrose, reaction (19-18), is catalyzed by $H^+(aq)$. A large number of reactions that take place in solution are catalyzed either by acids or by bases.

A **heterogeneous catalyst** is in a different phase from the reacting molecules. Many gas-phase reactions are catalyzed by the presence of a finely divided metal or metal oxide. Without the surface of the solid catalyst, the reaction in the gas phase is very slow. Petroleum refining is an industrial process that relies heavily on heterogeneous catalysis. Silica alumina gel is used as a catalytic surface in cracking heavy petroleum fractions to produce high octane gasoline. Nickel and aluminum oxides, as well as chromic oxide, are used as catalysts in hydrogenating and dehydrogenating hydrocarbons.

Practically all reactions occurring in living systems are catalyzed by species called **enzymes**. Very small quantities of enzymes may be present, but each enzyme is specific for a particular process. There are literally thousands of different enzymes, which are members of a group of molecules called proteins, having molecular weights between 10,000 and 1,000,000. The structure of proteins will be discussed in Chapter 23.

The simplest sort of mechanism for a catalyzed reaction would be

Step 1	A + catalyst → intermediates
Step 2	intermediates → catalyst + B
Overall reaction	A → B

Since the concentration of the catalyst is a constant throughout the reaction, it is included in the value of k, the specific rate constant. Thus the numerical value of k will change if the concentration of catalyst changes. For the hydrolysis of sucrose, Eq.

Table 19.5. **The Observed First-Order Rate Constant for the Hydrolysis of Sucrose in Aqueous Solution at 25 °C as a Function of the [H⁺(aq)]**

$k_1(obs)$ (min^{-1})	$[H^+(aq)]$ (M)	$k_1/[H^+(aq)]$ $(M^{-1}min^{-1})$
2.3×10^{-5}	2.9×10^{-3}	7.9×10^{-3}
4.5×10^{-5}	5.6×10^{-3}	8.0×10^{-3}
7.6×10^{-5}	9.5×10^{-3}	8.0×10^{-3}

(19-18), in any given aqueous acidic solution, the [H⁺] remains constant, and the rate law we observe is

$$-d[\text{sucrose}]/dt = k_1[\text{sucrose}]$$

In fact, the observed k_1 is directly proportional to the [H⁺] as can be seen in the data of Table 19.5. The actual rate expression is

$$-d[\text{sucrose}]/dt = k[\text{H}^+][\text{sucrose}]$$

where $k[\text{H}^+]$ is the observed first-order rate constant.

Another example of a catalyzed reaction is the decomposition of hydrogen peroxide,

$$2\text{H}_2\text{O}_2(\text{aq}) \rightarrow 2\text{H}_2\text{O} + \text{O}_2(\text{g}) \tag{19-38}$$

Without a catalyst this reaction is very slow, but it is homogeneously catalyzed by several species, including iodide ions. A two-step mechanism has been proposed for the catalysis by I⁻, as shown below. Note that I⁻ ions are consumed in step 1 and regenerated in step 2:

$$\text{step 1} \qquad \text{H}_2\text{O}_2 + \text{I}^- \rightarrow \text{H}_2\text{O} + \text{IO}^-$$

$$\text{step 2} \qquad \text{IO}^- + \text{H}_2\text{O}_2 \rightarrow \text{H}_2\text{O} + \text{O}_2 + \text{I}^-$$

These two bimolecular elementary processes sum to give the overall stoichiometry shown in Eq. (19-38).

The role of the catalyst is to provide an alternate activated complex, one with a lower activation energy. The reaction proceeds more quickly because the amount of energy required to reach the activated state is lower. This is illustrated in Fig. 19.16. The two reaction pathways are different; therefore the curves are drawn along different reaction coordinates.

Because a catalyst is regenerated in a step subsequent to the rate-determining step, the same catalyst can be used over and over again. As a result only a small amount of

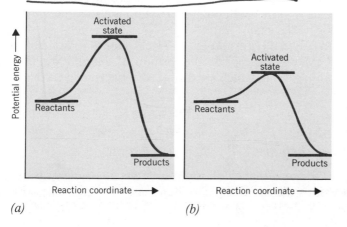

(a)   *(b)*

Fig. 19.16. Potential energy profile for the same exothermic reaction, *(a)* uncatalyzed pathway and *(b)* catalyzed pathway. The role of the catalyst is to provide an alternate mechanism with a lower activation energy.

catalyst need be present in a reaction mixture to enhance the reaction rate by a large factor. This phenomenon is of particular importance in biological systems, where the concentration of enzymes necessary to catalyze the reactions taking place may be very small. Enzymes are the most effective catalysts known. Some enzymes increase reaction rates by factors of 10^{12} or 10^{13}.

section 19.9
Chain Reactions

We have seen in the preceding section that reaction mechanisms involve short-lived intermediate species that do not appear in the stoichiometric equation for the reaction. In certain reactions the intermediate is a **free radical,** a substance containing one or more unpaired electrons. A free radical may be either uncharged or an ion, and it may consist of a single atom such as H· or :Br·, or a group of atoms such as ·CH$_3$, the methyl radical. Free radicals are highly reactive. They can be produced by high temperatures or **photochemically,** by the absorption of either visible or ultraviolet light.

Once free radicals are formed they sometimes react with other molecules to produce a product plus another free radical. If a mechanism involves a sequence of steps in some of which there is a regeneration of reactive intermediates, the reaction is called a **chain reaction.** Once the reactive intermediate is formed the reaction becomes self-propagating. Chain reactions are usually very rapid and can be explosive if there are steps in the chain that increase the number of radicals in the system. The reaction between gaseous H_2 and O_2 to produce H_2O proceeds through a chain mechanism. In the absence of a catalyst this reaction is very slow, but with a suitable catalyst to start the chain mechanism, it proceeds very rapidly and can be explosive.

One of the most famous reactions for which a chain mechanism has been proposed is the gas-phase reaction between hydrogen and bromine to form hydrogen bromide:

$$H_2(g) + Br_2(g) \rightarrow 2HBr(g) \tag{19-39}$$

The mechanism proposed is as follows:

step 1	$Br_2 \rightarrow Br\cdot + Br\cdot$	(chain initiation)	(19-40a)
step 2	$Br\cdot + H_2 \rightarrow HBr + H\cdot$	(chain propagation)	(19-40b)
step 3	$H\cdot + Br_2 \rightarrow HBr + Br\cdot$	(chain propagation)	(19-40c)
step 4	$H\cdot + HBr \rightarrow H_2 + Br\cdot$	(chain inhibition)	(19-40d)
step 5	$Br\cdot + Br\cdot \rightarrow Br_2$	(chain termination)	(19-40e)

Note that step 4 is chain inhibiting because a product molecule is used up and a reactant molecule is formed, although a Br· atom is also formed that can react as in step 2 to propagate the chain.

The reaction between $H_2(g)$ and $Br_2(g)$ is very slow at room temperature. At high temperature, however, thermal decomposition of diatomic bromine takes place, producing bromine atoms, which are the reactive intermediates that begin the chain reaction.

section 19.10
Reaction Rates and Equilibrium

In Section 5.8 we noted that equilibrium in molecular systems is dynamic, that is, opposing reactions are occurring simultaneously at equal rates. By equating the forward and reverse reaction rates we can determine the relationship between the equilibrium constant and the rate constants of the forward and reverse reactions.

Let us consider the equilibrium

$$CH_3COOH + H_2O \rightleftharpoons H_3O^+ + CH_3COO^- \qquad (19\text{-}41)$$

Both the forward and reverse reactions of this equilibrium are elementary processes. The forward reaction is pseudo-first order with a rate law

$$\text{forward rate} = k_f[CH_3COOH] \qquad (19\text{-}42)$$

The numerical value of k_f at 25 °C is $7.9 \times 10^5 \text{ s}^{-1}$. The rate of the backward reaction is

$$\text{reverse rate} = k_r[CH_3COO^-][H_3O^+] \qquad (19\text{-}43)$$

with

$$k_r = 4.5 \times 10^{10} \text{ L} \cdot \text{mol}^{-1}\text{s}^{-1} \qquad \text{at 25 °C}$$

These rate expressions apply whether the system is at equilibrium or not. When the system reaches equilibrium, however, the rates of the forward and reverse reactions must be equal. We obtain, at equilibrium,

$$k_f[CH_3COOH] = k_r[CH_3COO^-][H_3O^+]$$

Rearranging this to solve for the ratio k_f/k_r, we find

$$\frac{k_f}{k_r} = \frac{[CH_3COO^-][H_3O^+]}{[CH_3COOH]} = K_{eq}$$

At 25 °C, $k_f/k_r = (7.9 \times 10^5)/(4.5 \times 10^{10}) = 1.8 \times 10^{-5} = K_a(\text{HOAc})$.

The relation

$$\boxed{K_{eq} = k_f/k_r} \qquad (19\text{-}44)$$

has been obtained for a reaction for which both the forward and reverse directions proceed via a single elementary process. It is, however, a generally valid relationship, regardless of the mechanism. Consider the equilibrium

$$H_3AsO_4 + 3I^- + 2H^+ \rightleftharpoons H_3AsO_3 + I_3^- + H_2O \qquad (19\text{-}45)$$

which was investigated by Wilson and Dickinson in 1937. For this equilibrium

$$\text{forward rate} = k_f[H_3AsO_4][I^-][H^+]$$

and

$$\text{reverse rate} = k_r[H_3AsO_3][I_3^-]/[I^-]^2[H^+]$$

By equating these expressions, you should be able to show that Eq. (19-44) applies to this equilibrium.

If a reaction is catalyzed, the catalyst is a part of the activated complex for the forward reaction. But the activated complex for the forward and reverse directions is exactly the same configuration of atoms. A catalyst decreases the activation energy of both the forward and reverse reactions. (See Fig. 19.16.)

A catalyst for the forward reaction of an equilibrium is also a catalyst for the reverse reaction. The concentration of catalyst therefore cancels out when the forward and reverse rates are equated at equilibrium, and no term in the concentration of catalyst appears in the expression for the equilibrium constant. The catalyst increases the rate of both the forward and the backward reactions by exactly the same factor and therefore does not affect the magnitude of the equilibrium constant.

Summary

Reaction rates depend upon five factors: (1) the nature of the reacting species, (2) the concentrations of the reacting species, (3) the temperature, (4) the presence or absence of a catalyst, and (5) the nature of the solvent, if the reaction occurs in solution.

The dependence of the rate on the concentrations of the reacting species (or the partial pressures if the reactants are gases) is written explicitly in the rate expression for the reaction. The dependence on all other factors is given by the numerical value of the **specific rate constant, k.** The form of the rate law is, therefore, rate $= k[A]^m[B]^n \ldots$, where the exponents m and n may be zero, integral, or fractional. The sum of the exponents, $m + n + \ldots$, is the **order** of the reaction.

The rate of a chemical reaction is the change in concentration of one of the components with time. As each concentration is continually changing as the reaction progresses, the rate is also continually changing and can only be defined at a given instant. It is therefore a first derivative. Since the concentration of a reactant, A, decreases with time, the rate is defined as $-d[A]/dt$ and has units of moles per liter per unit time.

A reaction for which the rate expression is $-d[A]/dt = k_1[A]$ is a **first-order reaction.** There are two types of **second-order** rate laws: Those for which the rate expression is $-d[A]/dt = k_2[A]^2$ and those for which the rate expression is $-d[A]/dt = k_2'[A][B]$. Third and higher-order reactions are less common than first- or second-order reactions. There are also a number of reactions with fractional orders.

The order and the rate constant of a reaction can be determined by performing a series of experiments in which the concentrations of reactants are varied systematically and the effect of those variations on the **initial rate** of the reaction is determined. The initial rate is the slope of a plot of the concentration of reactant versus time, evaluated at time $t = 0$, that is, at the start of the reaction.

The integrated form of the first-order rate equation is an exponential decay: $[A]_t = [A]_0 e^{-k_1 t}$. By taking the logarithm of this equation one can show that a plot of $\ln[A]_t$ versus time is a straight line with slope $-k_1$ for a first-order reaction.

The **half-life** of a reaction is the amount of time it takes for one half the material originally present to be used up. For a first-order reaction, the half-life is independent of the initial concentration of reactant, and is given by $\tau_{1/2} = \ln 2/k_1 = 0.693/k_1$. For all other orders of reaction, the half-life depends on the initial concentration of reactant.

Radioactive nuclei always decay with first-order kinetics, so that each unstable nucleus has a characteristic half-life that can be used to identify it.

The dependence of a rate on the concentration of solvent is difficult to observe because the concentration of solvent is so large compared to other concentrations that it is essentially constant during the course of the reaction. In particular, the dependence of a rate on $[H_2O]$ for reactions in dilute aqueous solution is often not observed. Reactions with a rate law $-d[A]/dt = k_1'[A][H_2O]$ will appear to be first order if they take place in aqueous solution, as $k_1'[H_2O]$ will be essentially constant. Such reactions are said to be **pseudo-first order.**

The integrated form of the second-order rate law $-d[A]/dt = k_2[A]^2$ is

$$\frac{1}{[A]_t} - \frac{1}{[A]_0} = k_2 t$$

A plot of $1/[A]_t$ versus time is therefore linear with slope k_2 if the reaction is second order in A. To determine whether a reaction is first or second order in a reactant A, plot both ln[A] and 1/[A] versus time. The order is determined if one of these plots is linear while the other is not.

Near room temperature, the rate of most chemical reactions increases by a factor of 2 or 3 for a 10° rise in temperature. Many biologically important reactions are even more temperature sensitive. In order for molecules to react when they collide, there must be some minimum kinetic energy, called the **activation energy,** that is possessed jointly by the colliding species. The reason reaction rates increase with temperature is not only that collisions become more frequent as the average kinetic energy of the molecules increases, but also that a larger fraction of molecules has a kinetic energy greater than the activation energy. Therefore a larger fraction of collisions results in reaction at higher temperatures.

If the colliding molecules are not appropriately oriented relative to one another, even sufficiently energetic collisions will not be effective at leading to reaction.

The dependence of a rate constant on temperature is given by the Arrhenius equation: $k = Ae^{-E_{act}/RT}$. In this equation A, the **frequency factor,** and E_{act}, the **activation energy,** are both considered to be constant. When investigations are made over wide ranges of temperature it is found that while both A and E_{act} are nearly constant they do have a small temperature dependence. To obtain the numerical value of the activation energy, the specific rate constant is measured as a function of temperature. A plot of ln k versus $1/T$ is made, and the activation energy is calculated from the slope of this plot, which is $-E_{act}/R$.

Before a reaction can occur, the reacting species must pass through the **activated state,** an unstable combination of all the atoms in which the potential energy is a maximum. The **activation energy** is the difference in energy between the activated state and the average potential energy of the reactant species. For an equilibrium process, the backward reaction must proceed through the same **activated state** or **transition state** as the forward reaction.

Whenever the observed rate of reaction cannot be predicted from the stoichiometric equation for the reaction we can be sure that the **mechanism** is not described by the stoichiometry. The mechanism is a series of **elementary processes** that defines the order in which bonds are broken and new ones made. In the rate law for an elementary process, the exponent of the concentration of each species is the same as the coefficient of that species in the equation for the elementary process.

In a mechanism consisting of several steps, the slowest step is **rate determining.** The observed rate law tells us the species that are involved in the rate-determining step. Mechanisms frequently postulate the formation of **intermediate** species, which are short lived and do not appear in the stoichiometric equation. It is common for several proposed mechanisms to be consistent with the observed rate law and the overall stoichiometry. A great deal of experimental evidence is often necessary to establish the correct mechanism.

It is very important to distinguish between the terms **molecularity** and **order.** The molecularity refers only to an elementary process and must be a small integer, either 1, 2, or 3. The molecularity is the number of molecules that react in the elementary process. In contrast, the **order** is the sum of the exponents in the observed rate law for the overall reaction. It can only be determined by experiment and can be zero, fractional, or integral.

A **catalyst** is a substance that increases the rate of chemical reaction but does not itself undergo a permanent change during the course of the reaction. It is a reactant in the rate-determining step and then a product of some subsequent step. The role of a catalyst is to provide an alternate mechanism with a lower activation energy.

Catalysts are very common. Biological processes are catalyzed by **enzymes,** which are protein molecules. There are thousands of different enzymes, each one specific for a particular process. There are two categories of catalysts, **homogeneous** and **heterogeneous.** A homogeneous catalyst is in the same phase as the reactants, whereas a heterogeneous catalyst is in a different phase.

A number of reactions proceed through a **free-radical mechanism,** that is, a mechanism in which an intermediate species has one or more unpaired electrons. Free radicals are highly reactive and a mechanism involving them sometimes becomes a **chain reaction.**

For a dynamic equilibrium, the rates of the forward and reverse reactions are equal. By equating forward and reverse rates we find that $k_f/k_r = K_{eq}$. A catalyst affects the forward and reverse rates by exactly the same factor and therefore does not change the value of the equilibrium constant.

Exercises

Sections 19.1 and 19.2

1. The rate law for the oxidation of iodide ion by chlorate ion in aqueous acidic solution

$$ClO_3^- + 9\,I^- + 6H^+(aq) \rightarrow 3\,I_3^- + Cl^- + 3H_2O$$

has been found to be

$$\text{rate} = k[ClO_3^-][I^-][H^+]^2$$

An aqueous acidic solution of I^- and ClO_3^- is prepared. What is the effect on (1) the rate of reaction, and (2) the specific rate constant, k, of (a) adding water to this solution, (b) adding ammonia to this solution, and (c) heating the solution from 20 to 35 °C?

2. For the following reactions, state the overall order and the order with respect to each reactant:

(a) $ClO_3^- + 9\,I^- + 6H^+(aq) \rightarrow 3\,I_3^- + Cl^- + 3H_2O$

   $\text{rate} = -d[ClO_3^-]/dt = k[ClO_3^-][I^-][H^+]^2$

(b) $CHCl_3(g) + Cl_2(g) \rightarrow CCl_4(g) + HCl(g)$

   $\text{rate} = -d[CHCl_3]/dt = k[CHCl_3][Cl_2]^{1/2}$

(c) $2NOBr(g) \rightarrow 2NO(g) + Br_2(g)$

   $\text{rate} = -d[NOBr]/dt = k[NOBr]^2$

(d) $Cl_2(g) + CO(g) \rightarrow COCl_2(g)$

   $\text{rate} = -d[CO]/dt = k[CO][Cl_2]^{3/2}$

(e) $S_2O_8^{2-} + 3\,I^- \rightarrow I_3^- + 2SO_4^{2-}$

   $\text{rate} = -d[S_2O_8^{2-}]/dt = k[I^-][S_2O_8^{2-}]$

3. For the reaction $Cl_2(g) + CO(g) \rightarrow COCl_2(g)$

(a) What is the relation between $d[CO]/dt$ and $d[COCl_2]/dt$?

(b) Express the rate of this reaction in terms of $d[Cl_2]/dt$ and $d[COCl_2]/dt$. What are the units of this rate?

4. The rate of decomposition of azomethane, $CH_3N_2CH_3$, at 600 K

$$CH_3N_2CH_3(g) \rightarrow CH_3CH_3(g) + N_2(g)$$

has been studied by measuring the partial pressure of azomethane as a function of time. The following data have been obtained:

t (s)	0	1000.	2000.	3000.	4000.
P_{azo} (mmHg)	8.20×10^{-2}	5.72×10^{-2}	3.99×10^{-2}	2.78×10^{-2}	1.94×10^{-2}

Calculate the average rate of reaction between (a) $t = 0$ and $t = 1000$ s and (b) $t = 3000$ and $t = 4000$ s. Specify the units of the rate of reaction. Explain why the rates for these two equal time intervals are not the same.

5. For the reaction $3H_2(g) + N_2(g) \rightarrow 2NH_3(g)$

   (a) What is the relationship between $\dfrac{d[H_2]}{dt}$ and $\dfrac{d[N_2]}{dt}$?

   (b) What is the relationship between $\dfrac{d[N_2]}{dt}$ and $\dfrac{d[NH_3]}{dt}$?

6. (a) What are the units of the rate of the reaction

$$2H_2(g) + 2NO(g) \rightarrow N_2(g) + 2H_2O(g)$$

   for which the rate law is $-d[NO]/dt = k[NO]^2[H_2]$?

   (b) What are the units of the rate constant for this reaction?

7. The rate of the gas-phase decomposition of azomethane, $CH_3N_2CH_3$, can be expressed as $-dP_{azo}/dt = k_1 P_{azo}$, where P_{azo} is the partial pressure of $CH_3N_2CH_3$. What are the units of this rate if time is measured in minutes and the pressure in atmospheres? What are the units of the specific rate constant, k_1?

Section 19.3

8. For the data of Table 19.1, plot $[(CH_3)_3CBr]$ versus time. Draw the tangent to the curve at $t = 0$, and from the slope of that tangent, determine the initial rate of the reaction at 50 °C.

9. The following data on the initial rate of the reaction

$$NH_4^+ + NO_2^- \rightarrow N_2(g) + 2H_2O$$

in aqueous solution at 25 °C have been obtained:

$[NH_4^+]_0$ (M)	$[NO_2^-]_0$ (M)	Initial Rate (M · s^{-1})
0.300	0.0606	4.86×10^{-6}
0.200	0.0404	2.16×10^{-6}
0.200	0.0101	5.40×10^{-7}
0.100	0.0606	1.62×10^{-6}
0.050	0.0808	1.08×10^{-6}

Deduce the rate law for this reaction and determine the value of the specific rate constant, k. What are the units of k?

10. For the reaction $2A + B \rightarrow C$, the following data were collected:

| Experiment Number | Initial Concentrations | | Initial Rate |
	$[A]_0$ (M)	$[B]_0$ (M)	$-d[A]_0/dt$ (M · s^{-1})
1	0.10	0.20	3.00×10^2
2	0.30	0.40	3.60×10^3
3	0.30	0.80	1.44×10^4

(a) Deduce the rate law for this reaction. Explain your reasoning. What is the order of this reaction?

(b) Calculate the specific rate constant for this reaction and give its units.

Section 19.4

11. For the data of Table 19.1, plot $\ln[(CH_3)_3CBr]$ versus time, calculate the first-order specific rate constant from the plot, and specify its units.

12. For the data of Table 19.1, plot $[(CH_3)_3CBr]$ versus time. Reading your plot, find the time it takes for the concentration of *tert*-butyl bromide to fall to 1/2 its initial value and to 1/4 its initial value. Draw vertical lines to the curve at these two times. From your graph, what is the half-life of this reaction? Calculate $\tau_{1/2}$ from the first-order specific rate constant obtained in Exercise 11 and compare your two values.

13. If a sample containing $^{131}_{53}I$, $\tau_{1/2} = 8.070$ days, has an activity of 14,200 cpm, how long will it take for the activity to fall to 1775 cpm?

14. How long will it take for 90% of a reactant to decompose, if the decomposition follows a first-order rate law? Express your answer in terms of the specific rate constant for the reaction.

15. For a first-order reaction, how long will it take for the concentration of reactant to fall to 1/8 its initial value? Express your answer in terms of the half-life and also in terms of the first-order specific rate constant.

16. The rate law for the hydrolysis of sucrose is $-d[\text{sucrose}]/dt = k_1[\text{sucrose}]$ when the reaction is run in aqueous acidic solution. Describe experiments you could carry out to determine the dependence of this rate on the $[H_2O]$.

Section 19.5

17. The rate of the gas-phase reaction $2A(g) \rightarrow B(g)$ was investigated at 80 °C. The following data on the partial pressure of A as a function of time were obtained:

t (s)	0	100	200	300	400
P_A (mmHg)	400	244	176	136	112

Plot both $\ln P_A$ versus time and $1/P_A$ versus time. Is the reaction first or second order? What is the value of the specific rate constant?

18. For a second-order reaction with rate law $-d[A]/dt = k_2[A]^2$, show that the time it takes for the $[A]$ to fall to 1/4 its initial value is three times larger than the time it takes for the $[A]$ to fall to 1/2 its initial value.

Section 19.6

19. Values of the first-order rate constant for the hydrolysis of an organic chloride as a function of temperature are given below:

$T(K)$	273.2	298.2	308.2	318.2
$k(s^{-1})$	1.06×10^{-5}	3.19×10^{-4}	9.86×10^{-4}	2.92×10^{-3}

Plot $\ln k$ versus $1/T$ and calculate the activation energy.

20. Between 600 and 700 K the activation energy for the gas-phase decomposition of NO_2 is 115 kJ/mol. If the specific rate constant is 4.74 at 656.0 K, calculate its value at 627.0 K.

21. (a) What is the activation energy of a reaction if the rate doubles when the temperature is increased from 20 to 30 °C? (b) For many reactions of biological importance, the rate increases by a factor of 100 or more when the temperature is increased by 10°. Does this imply the activation energy for such reactions is greater than, less than, or the same as activation energies of typical chemical reactions? Explain your answer.

Section 19.8

22. The following mechanism is proposed for the reaction

$$Cl_2(g) + CHCl_3(g) \rightarrow CCl_4(g) + HCl(g)$$

step 1	$Cl_2 \xrightleftharpoons{K_{eq}} 2Cl$	(rapid equilibrium)
step 2	$Cl\cdot + CHCl_3 \rightarrow HCl + \cdot CCl_3$	(slow)
step 3	$Cl\cdot + \cdot CCl_3 \rightarrow CCl_4$	(fast)

Show that this mechanism is consistent with the observed rate law,

$$-d[Cl_2]/dt = +d[HCl]/dt = k[CHCl_3][Cl_2]^{1/2}$$

23. The rate of the reaction between iodide ion and hypochlorite ion

$$I^- + OCl^- \rightarrow OI^- + Cl^-$$

has been found to be   rate $= k[I^-][OCl^-]/[OH^-]$
Two mechanisms have been proposed:

mechanism I	$OCl^- + H_2O \xrightleftharpoons{K_{eq}} HOCl + OH^-$	(rapid equilibrium)
	$I^- + HOCl \xrightarrow{k_2} HOI + Cl^-$	(slow)
	$OH^- + HOI \xrightarrow{k_3} H_2O + OI^-$	(fast)
mechanism II	$OCl^- + H_2O \xrightleftharpoons{K_{eq}} HOCl + OH^-$	(rapid equilibrium)
	$I^- + HOCl \xrightarrow{k_2'} ICl + OH^-$	(slow)
	$ICl + 2\,OH^- \xrightarrow{k_3'} OI^- + Cl^-$	(fast)

Can either of these mechanisms be ruled out on the basis of the observed rate law? Explain your answer.

24. A mechanism proposed for the reaction in which 2-methylpropene, $(CH_3)_2C{=}CH_2$, is converted to *tert*-butyl alcohol, $(CH_3)_3COH$, is

step 1	$(CH_3)_2C{=}CH_2 + H^+ \rightarrow (CH_3)_2\overset{\oplus}{C}{-}CH_3$	(slow)
step 2	$(CH_3)_2\overset{\oplus}{C}{-}CH_3 + H_2O \rightarrow (CH_3)_3COH + H^+$	(fast)

(a) What is the overall stoichiometry of this reaction?
(b) What is the intermediate species in this mechanism?
(c) Is this a catalyzed reaction? If so, what is the catalyst?
(d) What is the rate law for this reaction?
(e) If the reaction is run in aqueous acidic solution, what will the observed rate law be?

Section 19.9

25. The reaction between $H_2(g)$ and $O_2(g)$ to form $H_2O(g)$ has a very complex mechanism that has not yet been fully elucidated. Some of the steps involved are

step 1	$H_2 + O_2 \rightarrow HO_2\cdot + H\cdot$
step 2	$H\cdot + O_2 \rightarrow HO\cdot + O\cdot$
step 3	$HO_2\cdot + H_2 \rightarrow HO\cdot + H_2O$
step 4	$H_2 + HO\cdot \rightarrow H\cdot + H_2O$
step 5	$O\cdot + H_2 \rightarrow HO\cdot + H\cdot$
step 6	$H\cdot + HO\cdot \rightarrow H_2O$

Label each of these steps as either a chain initiation, propagation, inhibition, or termination. Which of these steps increases the number of radicals in the system and therefore can lead to the reaction becoming explosive?

Multiple Choice Questions

1. If concentrations are measured in moles per liter and time in minutes, the units for the rate of a third-order reaction are
   (a) $mol \cdot L^{-1}min^{-1}$ (b) $L^2mol^{-2}min^{-1}$ (c) $L \cdot mol^{-1}min^{-1}$ (d) min^{-1}
   (e) $mol^2L^{-2}min^{-1}$

2. If concentrations are measured in moles per liter and time in minutes, the units for the specific rate constant of a third-order reaction are
   (a) $mol \cdot L^{-1}min^{-1}$ (b) $L^2mol^{-2}min^{-1}$ (c) $L \cdot mol^{-1}min^{-1}$ (d) min^{-1}
   (e) $mol^2L^{-2}min^{-1}$

3. Which of the following statements about the order of a reaction is TRUE?
   (a) The order of a reaction must be a positive integer.
   (b) A second-order reaction is also bimolecular.
   (c) We can determine the order of the reaction from the correctly balanced net ionic equation for the reaction.
   (d) The order of a reaction increases with increasing temperature.
   (e) The order of a reaction can only be determined by experiment.

4. In an experiment to study the reaction $A + 2B \rightarrow C + 2D$, the initial rate, $-d[A]/dt$ at $t = 0$, was found to be $2.6 \times 10^{-2} M \cdot s^{-1}$. What is the value of $-d[B]/dt$ at $t = 0$ in $M \cdot s^{-1}$?
   (a) 2.6×10^{-2} (b) 5.2×10^{-2} (c) 1.3×10^{-2} (d) 1.0×10^{-1}
   (e) 6.5×10^{-3}

5. The reaction $A(g) + 2B(g) \rightarrow C(g) + D(g)$ is an elementary process. In an experiment, the initial partial pressures of A and B are $P_A = 0.60$ atm and $P_B = 0.80$ atm. When $P_C = 0.20$ atm, the rate of the reaction, relative to the initial rate, is
   (a) $\frac{1}{48}$ (b) $\frac{1}{24}$ (c) $\frac{9}{16}$ (d) $\frac{3}{4}$ (e) $\frac{1}{6}$

6. For the reaction between chlorine and nitric oxide,

$$Cl_2(g) + 2NO(g) \rightarrow 2NOCl(g)$$

   it is found that doubling the concentration of both reactants increases the rate by a factor of 8. If only the concentration of Cl_2 is doubled, the rate increases by a factor of 2. The order of this reaction with respect to NO is
   (a) 0   (b) 1   (c) 2   (d) 3   (e) $\frac{1}{2}$

The Following Information Is for Use in Questions 7–9

In an "iodine-clock" experiment, the time, t, it takes for the blue color of the starch–iodine complex to appear is a measure of the initial rate of formation of I_3^-. The rate of the reaction

$$3 I^- + S_2O_8^{2-} \rightarrow I_3^- + 2SO_4^{2-}$$

was studied using the iodine-clock technique, and the following data were obtained:

Run Number	$[I]_0$ (M)	$[S_2O_8^{2-}]_0$ (M)	t (s)
1	0.0800	0.0400	44.0
2	0.0800	0.0800	22.1
3	0.1600	0.0200	43.9
4	0.0400	0.0400	88.0

7. The order of this reaction with respect to $S_2O_8^{2-}$ is
   (a) 0   (b) $\frac{1}{2}$ (c) 1   (d) 2   (e) 3

8. The overall order of this reaction is
   (a) 0   (b) $\frac{1}{2}$ (c) 1   (d) 2   (e) 3

9. If a run is made with $[I]_0 = 0.120\ M$ and $[S_2O_8^{2-}]_0 = 0.0400\ M$, you would expect to see the blue color appear in approximately
   (a) 29 s   (b) 33 s   (c) 44 s   (d) 66 s   (e) 88 s

10. For a hypothetical reaction $A + 2B \rightarrow 3C + D$, $d[C]/dt$ is equal to
    (a) $-d[A]/dt$ (b) $-d[B]/dt$ (c) $+3d[A]/dt$ (d) $-\frac{3}{2}d[B]/dt$ (e) $+d[A]/dt$

11. Which of the following statements is TRUE?
    (a) Endothermic reactions have higher activation energies than exothermic reactions.
    (b) The rate law for a reaction depends on the concentrations of all reactants that appear in the stoichiometric equation.
    (c) The rate of a catalyzed reaction is independent of the concentration of the catalyst.
    (d) The specific rate constant for a reaction is independent of the concentrations of the reacting species.
    (e) In all reaction mechanisms, there is a single rate-determining step.

12. The first-order rate constant for the decomposition of N_2O_5 in CCl_4 solution is 6.2×10^{-4} at 45 °C and 2.1×10^{-3} at 55 °C. What is the value of the activation energy for this reaction in kilojoules per mole?
    (a) 46   (b) 1.1×10^2 (c) 2.5×10^3 (d) 2.5×10^4 (e) 1.1×10^5

13. For the reaction $A + 2B \rightarrow 2C$, the rate law for formation of C is
    (a) rate $= k[A][B]^2$ (b) rate $= k[A][B]$ (c) rate $= [C]^2/[A][B]^2$
    (d) rate $= k[A]^2[B]$ (e) impossible to state from the data given

14. The rate law for the dimerization of NO_2 is $-d[NO_2]/dt = k[NO_2]^2$.
    Which of the following changes will change the value of the specific rate constant, k?
    (a) Doubling the total pressure on the system.
    (b) Doubling the volume of the container in which the reaction is occurring.
    (c) Adding more O_2 to the reaction mixture.
    (d) Adding more NO_2 to the reaction mixture.
    (e) Running the reaction in CCl_4 solution rather than in the gas phase.

15. For a reaction for which the activation energies of the forward and reverse directions are equal in value,
    (a) the stoichiometry is the mechanism   (b) $\Delta H = 0$ (c) $\Delta S = 0$
    (d) the order is 0   (e) there is no catalyst

The Following Information Is for Use in Questions 16–18

The rate law for the hydrolysis of thioacetamide, CH_3CSNH_2,

$$CH_3 \overset{\overset{S}{\|}}{-C} -NH_2 + H_2O \rightarrow H_2S + CH_3 \overset{\overset{O}{\|}}{-C} -NH_2$$

is rate = $k[H^+][TA]$, where TA = thioacetamide.

16. In which of the following solutions, all at 25 °C, will the rate of hydrolysis of thio-acetamide (TA) be LEAST?

(a) 0.10 M in TA, 0.20 F in HNO_3 (b) 0.15 M in TA, 0.15 F in HCl
(c) 0.10 M in TA, 0.080 F in HCl   (d) 0.060 M in TA, 0.10 F in HNO_3
(e) 0.15 M in TA, 0.10 F in CH_3COOH

17. If some sodium acetate is added to a solution that is 0.10 M in both TA and H^+(aq) at 25 °C,

(a) The reaction rate decreases, but k remains the same.
(b) Both the reaction rate and k decrease.
(c) The reaction rate increases, but k remains the same.
(d) Both the reaction rate and k increase.
(e) The reaction rate remains the same, but k decreases.

18. If the temperature of the solution is increased from 25 to 75 °C

(a) The reaction rate decreases, but k remains the same.
(b) Both the reaction rate and k decrease.
(c) The reaction rate increases, but k remains the same.
(d) Both the reaction rate and k increase.
(e) The reaction rate remains the same, but k decreases.

Problems

19.1. The first-order reaction $SO_2Cl_2(g) \rightarrow SO_2(g) + Cl_2(g)$ has a specific rate constant $k_1 = 2.20 \times 10^{-5}$ s^{-1} at 593 K. What percent of a sample of SO_2Cl_2 is decomposed after heating at 593 K for 1.00 h?

19.2. If concentrations are expressed in moles per liter, and time in seconds, what are the units of (a) the rate and (b) the specific rate constant, k, of a reaction that obeys the rate law $k[A][B]^{1/2}$?

19.3. For the hydrolysis of sucrose at 25 °C in an acid solution, using the data of Table 19.3, the first-order rate constant was found to be 3.47×10^{-3} min^{-1}. How long will it take to hydrolyze 80.0% of a kilogram of sucrose? How long will it take to hydrolyze 80.0% of a pound of sucrose?

19.4. In the upper atmosphere (stratosphere) there is a significant concentration of ozone, O_3. Ozone decomposes to O_2. A suggested mechanism for the decomposition of ozone is

$$\text{step 1} \qquad O_3 \overset{K_{eq}}{\rightleftharpoons} O_2 + O \qquad \text{(rapid equilibrium)}$$

$$\text{step 2} \qquad O + O_3 \overset{k_2}{\longrightarrow} 2\,O_2 \qquad \text{(slow)}$$

(a) What is the stoichiometric equation for this reaction?
(b) The mechanism proposed is consistent with the observed rate law. What is the rate law for this reaction?

(c) What is the relation between the observed specific rate constant and the constants K_{eq} and k_2 in the proposed mechanism?

19.5. The rate law for the forward reaction in

$$3HNO_2(aq) \rightleftharpoons H^+(aq) + NO_3^-(aq) + 2NO(g) + H_2O$$

is $k_f[HNO_2]/[NO]^2$. Find the rate law for the reverse reaction.

19.6. The decomposition of dinitrogen pentoxide in CCl_4 at 45 °C

$$N_2O_5(soln) \rightarrow 2NO_2(soln) + \tfrac{1}{2}O_2(g)$$

has been investigated and the following data were obtained:

t (s)	0.00	184	319	526	867	1189	1877	2315	2724
$[N_2O_5]$ (M)	2.33	2.08	1.91	1.67	1.36	1.11	0.72	0.55	0.43

Determine whether this reaction is first or second order in N_2O_5 and compute the value of the specific rate constant.

19.7. The following data give the temperature dependence of the rate constant for the reaction $N_2O_5(g) \rightarrow 2NO_2(g) + \tfrac{1}{2}O_2(g)$.

t (°C)	0.0	25.0	35.0	45.0	55.0
k (s^{-1})	7.87×10^{-7}	3.46×10^{-5}	1.35×10^{-4}	4.98×10^{-4}	1.50×10^{-3}

(a) By means of a suitable plot, determine the activation energy for this reaction.
(b) If you take any two of these five data points and substitute them into Eq. (19-32a), you will obtain a value for the activation energy. Which is more reliable, the value obtained graphically or the value using any pair of data, assuming both calculations are carried out correctly? Explain your answer.

19.8. The uncoiling of DNA (deoxyribonucleic acid) is a first-order process with an activation energy ~420 kJ.
(a) At 50 °C, the half-life for uncoiling is estimated to be 2 min. What is the half-life at normal body temperature, 37 °C?
(b) What do the results of part (a) lead you to conclude about temperature regulation in the human body?

19.9. The following data were obtained by measuring the radioactivity of an unstable nuclide with a radiation detection device.

Time (days)	0.00	5.00	10.00	15.00	20.00	25.00
Activity (cpm)	5640	5300	4981	4681	4398	4133

Compute the half-life of the nuclide.

19.10. The reaction between carbon monoxide and chlorine to form the poisonous gas phosgene, $COCl_2$,

$$Cl_2(g) + CO(g) \rightarrow COCl_2(g)$$

has the experimental rate law $d[COCl_2]/dt = k[CO][Cl_2]^{3/2}$.

Show that the following mechanism is consistent with this rate law. The symbol M represents an inert molecule that serves to exchange kinetic energy with the reactants on collision.

step 1	$Cl_2(g) + M \rightleftharpoons 2Cl(g) + M$	(fast equilibrium, K_1)
step 2	$Cl(g) + CO(g) + M \rightleftharpoons ClCO(g) + M$	(fast equilibrium, K_2)
step 3	$ClCO(g) + Cl_2(g) \rightarrow COCl_2 + Cl(g)$	(slow)

What is the relation between the experimental specific rate constant, k, and the constants K_1, K_2, and k_3 of this mechanism?

19.11. The decomposition of dinitrogen pentoxide

$$N_2O_5(g) \rightarrow 2NO_2(g) + \tfrac{1}{2}O_2(g)$$

is first order in N_2O_5. The following mechanism has been proposed.

step 1	$N_2O_5 \xrightleftharpoons{K_{eq}} NO_2 + NO_3$	(rapid equilibrium)
step 2	$NO_2 + NO_3 \xrightarrow{k_2} NO_2 + O_2 + NO$	(slow)
step 3	$NO + NO_3 \xrightarrow{k_3} 2NO_2$	(fast)

(a) Show that this mechanism is consistent with the rate law. Express the experimental rate constant, k, in terms of the constants for the elementary processes, K_{eq}, k_2, and k_3.
(b) If $k = 5.0 \times 10^{-4}$ s^{-1} at a certain temperature, how long does it take for the concentration of N_2O_5 to fall to one tenth its original value at this temperature?

19.12. The rate law for the reaction $2NO_2(g) + F_2(g) \rightarrow 2NO_2F(g)$ is

$$-d[NO_2]/dt = k[NO_2][F_2]$$

Devise a two-step mechanism with F· as an intermediate species that is consistent with the observed rate law. Indicate the rate-determining step of your mechanism.

19.13. The reaction between aqueous thallous and ceric ions in acidic solution $Tl^+(aq) + 2Ce^{4+}(aq) \rightarrow 2Ce^{3+}(aq) + Tl^{3+}(aq)$, is very slow at room temperature, even though the equilibrium constant is very large ($K_{298} = 1.5 \times 10^{12}$). The stoichiometry represents an elementary process. The reaction is catalyzed by Mn^{2+} ion, with the following mechanism:

$$Ce^{4+} + Mn^{2+} \rightarrow Ce^{3+} + Mn^{3+}$$
$$Ce^{4+} + Mn^{3+} \rightarrow Ce^{3+} + Mn^{4+}$$
$$Mn^{4+} + Tl^+ \rightarrow Tl^{3+} + Mn^{2+}$$

Suggest a reason for the fact that the catalyzed mechanism is faster than the uncatalyzed pathway.

19.14. (a) Determine the rate expression and the specific rate constant for the reaction $2NO(g) + 2H_2(g) \rightarrow N_2(g) + 2H_2O(g)$ from the following data on the initial rate of reaction.

$[H_2]_0$ (M)	$[NO]_0$ (M)	Initial Rate (M·s^{-1})
0.060	0.010	2.6×10^{-7}
0.060	0.0050	6.5×10^{-8}
0.020	0.030	7.8×10^{-7}

Specify the units of the rate constant.

(b) It has been proposed that the initial step in the mechanism is a rapid equilibrium, $2NO \rightleftharpoons N_2O_2$. Propose a subsequent step or steps for the mechanism that would be consistent with the rate law you found. Designate the rate-determining step in your mechanism.

19.15. The kinetics of the reaction $2X + Y \rightarrow Z$ was studied by the method of initial rates, and the following data were obtained at 25 °C:

Run Number	Initial Concentrations (M)		Initial Rate $(mol \cdot L^{-1}s^{-1})$
	$[X]_0$	$[Y]_0$	
1	0.20	0.10	7.0×10^{-4}
2	0.20	0.20	1.4×10^{-3}
3	0.40	0.20	1.4×10^{-3}
4	0.60	0.60	4.2×10^{-3}

(a) Deduce the rate law for this reaction. Calculate the numerical value of the specific rate constant and specify its units.

(b) The following three mechanisms have been proposed for this reaction. In these mechanisms M and N are intermediates. What is the rate law to be expected for each mechanism? Which of these mechanisms is consistent with the rate law you deduced in part (a)?

mechanism I	$X + Y \rightarrow M$	(slow)
	$X + M \rightarrow Z$	(fast)
mechanism II	$X + X \rightarrow M$	(fast)
	$Y + M \rightarrow Z$	(slow)
mechanism III	$Y \rightarrow M$	(slow)
	$M + X \rightarrow N$	(fast)
	$N + X \rightarrow Z$	(fast)

19.16. The reaction

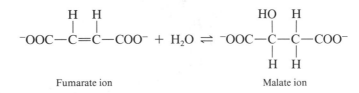

Fumarate ion   Malate ion

is an exothermic, enzyme-catalyzed reaction carried out in aqueous solution. The fumarate ion is the substrate, denoted S, the enzyme is denoted E, and the malate ion is the product, denoted P. The proposed mechanism of this reaction is

$$E + S \rightleftharpoons ES \text{ complex}$$
$$ES \text{ complex} \rightleftharpoons \text{transition state} \rightleftharpoons EP \text{ complex}$$
$$EP \text{ complex} \rightleftharpoons E + P$$

where ES complex is formed when the enzyme is adsorbed by the substrate, fumarate ion. A potential energy profile for the reaction is shown on the next page. Using this diagram, determine the values of

(a) ΔH for E $+$ S $\rightleftharpoons$ ES complex

(b) ΔH for E $+$ P $\rightleftharpoons$ EP complex

(c) ΔH for S $+$ H$_2$O $\rightleftharpoons$ P

(d) The activation energy for the backward reaction.

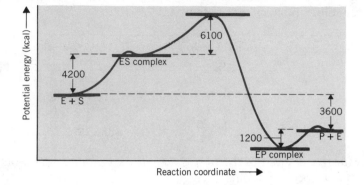

chapter 20 Coordination Compounds

Alfred Werner (1866–1919), a French–Swiss chemist, was born in Alsace and educated in Zurich. He was a professor at the University of Zurich from 1893 until his death. Werner studied the structure of inorganic coordination compounds and developed the coordination theory of valence. He introduced the concept of coordination number. Werner asserted that he awoke one morning at 2 A.M. as the coordination theory flashed into his consciousness, and that by 5 P.M. of the same day he had worked out the essential features of the theory. As a result of his theory, new and unsuspected examples of geometrical and optical isomerism were discovered. More than 200 Ph.D. dissertations were prepared under his direction, and he and his students synthesized many new series of inorganic complex compounds.

Many metals, particularly the transition metals, form coordination compounds in which several chemical groups (most often 4 or 6, but not at all limited to those numbers) are positioned in specific geometric arrangements around the central metal ion or atom. These compounds are frequently brightly colored, and their geometries and magnetic properties are especially interesting. Ions with such structures are called complex ions.

Many coordination compounds have important commercial uses. In film developing, silver salts that have not been exposed to light are removed from the film by complexing Ag^+ ions with thiosulfate ($S_2O_3^{2-}$) ions. Many industrial catalysts are coordination compounds, as are commercial water softeners.

Chlorophyll and hemoglobin are well-known coordination compounds, and a great deal of recent research has focused on the essential role metal ions play in biological processes. Bioinorganic chemistry and organometallic chemistry are important, growing fields of research today. In this chapter we will discuss the properties and structures of coordination compounds.

section 20.1
Ligands and the Coordinate Bond

A **coordination compound** or a **complex ion** consists of a central metal ion or atom and several **ligands.** Ligands are groups that are attached to the central metal ion or atom. They may be anions, such as Cl^-, CN^-, I^-, $C_2O_4^{2-}$, NO_2^-, and SCN^-, or neutral molecules such as NH_3, H_2O, and ethylenediamine, $NH_2CH_2CH_2NH_2$, which is abbreviated en. Much more rarely, ligands may be cations. If the ligands are neutral molecules such as H_2O or NH_3, they are polar molecules and the negative end of the molecule points toward the metal. Examples of coordination complexes are $Cu(NH_3)_4^{2+}$, $Ag(CN)_2^-$, $Fe(H_2O)_6^{3+}$, PtF_4^{2-}, and $Co(NO_2)_6^{3-}$.

The bond between the metal ion and the ligand is a covalent bond with partial ionic character, that is, it consists of a pair of electrons shared unequally between the metal and an atom of the ligand. In all the covalent bonds of the type A:B that we discussed in Chapter 14, one of the electrons of the shared pair was contributed by atom A and the other was contributed by atom B. In coordination compounds, however, *both* electrons of the shared pair are contributed by the ligand. The central metal ion has empty atomic orbitals for use in bonding. A distinctive feature of a ligand, therefore, is that it must possess at least one lone pair of electrons that it can share with the metal.

To point out the fact that both electrons of the shared pair are contributed by the ligand, these bonds in coordination compounds are sometimes referred to as **coordinate bonds.** It should be emphasized, however, that the nature of a coordinate bond is identical with that of all other covalent bonds with partial ionic character. Consider, for instance, the ammonium ion. Ammonia is a base, $:NH_3$, with a lone pair of electrons on the nitrogen atom. When ammonia reacts with a proton, H^+, to form the ammonium ion

$$H^+ \quad + \quad :NH_3 \quad \rightarrow \quad H-\overset{\displaystyle H}{\underset{\displaystyle H}{N}}-H \quad ^{\oplus} \qquad (20\text{-}1)$$

Vacant $1s$ orbital   Lone pair on N

all four bonds between N and H are identical. The coordinate bond to which the N atom contributed both electrons differs in no way from the bonds to the other three H atoms. All four N—H bonds in ammonium ion are equivalent.

Because of the lone pair of electrons on the nitrogen atom, ammonia forms coordinate bonds with a great many metal ions. A typical example is the tetraamminecopper(II) complex ion. The nitrogen atom of each of four ammonia molecules contributes a pair of electrons to form a covalent bond with a Cu^{2+} ion in the reaction

$$Cu^{2+} + 4 : NH_3 \rightarrow Cu(NH_3)_4^{2+} \tag{20-2}$$

Like many other complex ions, the $Cu(NH_3)_4^{2+}$ ion is brightly colored; it is a deep royal blue.

section 20.2
The Coordination Number

The total number of points of attachment by ligands to the metal ion is called the **coordination number** of the cation. Often, but not invariably, the coordination number is a characteristic of the metal ion. For example, the typical coordination number for Ag^+ is 2, and a few common silver complexes are $Ag(NH_3)_2^+$, $AgCl_2^-$, and $Ag(CN)_2^-$. For Zn^{2+} the typical coordination number is 4, as in $Zn(OH)_4^{2-}$, $Zn(NH_3)_4^{2+}$, and $Zn(C_5H_5N)_4^{2+}$, while for Cr^{3+} the coordination number is 6, as in $Cr(CN)_6^{3-}$, $Cr(NH_3)_6^{3+}$, and $Cr(H_2O)_6^{3+}$. Some metal ions exhibit only a single coordination number. The coordination number of Pt^{2+} is always 4, while that of Co^{3+} and Cr^{3+} is invariably 6. Most metal ions, however, exhibit more than one coordination number, and the coordination number depends on the size and charge of the ligand. If the ligand is very large, fewer can coordinate to the metal.

All coordination numbers from 2 up to 12 have been observed, although 4 and 6 are by far the most common. In recent years people have been intrigued with the problem of synthesizing complexes with uncommon coordination numbers and the number of known complexes with coordination numbers of 5, 7, and 8, or even the more unusual 9 and 10, has been increasing.

Ligands that bond to a single metal ion are classified as either **monodentate**, **bidentate**, or **polydentate**, depending on how many points of attachment a given ligand has to the central metal ion. Consider, for example, the complex formed by Cr^{3+} and oxalate ion, $C_2O_4^{2-}$. It has the formula $Cr(C_2O_4)_3^{3-}$, yet the coordination number of Cr^{3+} is 6. This is because oxalate ion is a bidentate ligand. Each oxalate ion uses two lone pairs of electrons, one on each of two oxygen atoms, to bind to the metal ion twice, as pictured below:

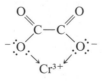

Two other common bidentate ligands are ethylenediamine (en) and dimethylglyoxime anion (DMG)

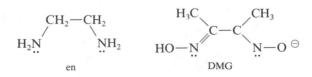

en   DMG

Table 20.1. **Some Common Ligands**

Monodentate Ligands

Ligand	Name
F^-, Cl^-, Br^-, I^-	Fluoro, chloro, bromo, iodo
CN^-	Cyano
$:NO_2^-$	Nitro (bonds to the metal through N)
$:ONO^-$	Nitrito (bonds to the metal through O)
$:OCN^-$	Cyanato (bonds to the metal through O)
$:NCO^-$	Isocyanato (bonds to the metal through N)
$:SCN^-$	Thiocyanato (bonds to the metal through S)
$:NCS^-$	Isothiocyanato (bonds to the metal through N)
CH_3COO^-	Acetato
H_2O	Aquo
NH_3	Ammine
CO	Carbonyl
NO	Nitrosyl
$C_5H_5N:$	Pyridine, symbolized py

Bidentate Ligands

Symbol	Name	Formula
en	Ethylenediamine	$\ddot{N}H_2-CH_2-CH_2-\ddot{N}H_2$
pn	Propylenediamine	$\ddot{N}H_2-CH_2CH(CH_3)-\ddot{N}H_2$
ox	Oxalato	$^-\ddot{O}\!-\!\underset{\displaystyle \overset{\|}{O}}{C}\!-\!\underset{\displaystyle \overset{\|}{O}}{C}\!-\!\ddot{O}^-$
CO_3^{2-}	Carbonato	$O=C{\Big\langle}^{\ddot{O}^-}_{\ddot{O}^-}$
acac	Acetylacetonato	$H_3C-C-CH=C-CH_3$ with $:O:$ and $:O:^-$
bipy	2,2′-Bipyridine	
DMG	Dimethylglyoximato	$H_3C-C-C-CH_3$, $HO-N$, $N-O^-$
gly	Glycinato	$:NH_2-CH_2-C-\ddot{O}:^-$ with O

Hexadentate Ligand

EDTA	Ethylenediaminetetraacetato	$^-:OOCCH_2$ and $CH_2COO:^-$, $^-:OOCCH_2$ $\ddot{N}CH_2CH_2\ddot{N}$ $CH_2COO:^-$

Multidentate ligands form compounds that are called **chelates,** from the Greek word for the claw of a crab *(chela)* since the ligand can be thought of as acting like a pincer and attaching to the metal from two sides.

$$\text{M} \supset$$

Some ligands that have two different atoms with lone pairs of electrons are **bridging ligands,** that is, they are coordinated to two different metal ions.

Table 20.1 lists some common ligands. You should refer to this list frequently enough so that you become familiar with the names and formulas of the more commonly used ligands.

The ethylenediaminetetraacetate ion (EDTA) is a hexadentate ligand, that is, it coordinates with six points of attachment to the metal, four O atoms and two N atoms; it forms complexes with a very large number of metal ions. Many ligands bind only to transition metal ions and do not form complexes with cations having the electronic configuration of one of the rare gases. However, EDTA binds to both Ca^{2+} and Mg^{2+} as well as to transition metal ions. It is so efficient at binding metal ions that it is used to remove traces of metal from distilled water, and is of biological importance because it can remove a metal from an enzyme and destroy the catalytic activity of the enzyme. It has also been used to remove Pb^{2+} from the bloodstream of patients suffering from lead poisoning. This procedure must be monitored carefully, because EDTA will remove essential metal ions as well as the Pb^{2+}.

section 20.3
Lewis Acids and Bases

We have already mentioned that the transition metal ions form a vast number of complexes, whereas cations with the electronic configuration of one of the rare gases (the alkali and alkaline earth cations) form relatively few. In order to understand why this is so, we must consider the **Lewis theory of acids and bases.**

In Section 7.4 we discussed the definitions of acid and base proposed by Brønsted and Lowry: An acid is a proton donor, and a base is a proton acceptor. G. N. Lewis proposed the following, rather different, set of definitions:

> A base is an electron-pair donor.
> An acid is an electron-pair acceptor.

The reaction between NH_3 and H^+ is a good one to use to point out the differences between these two theories of acids and bases. In Brønsted–Lowry theory, reaction (20-1) is a proton-transfer reaction, with NH_3 a proton acceptor, or base, and NH_4^+ ion its conjugate acid. In Lewis theory NH_3 is a base because it has a lone pair of electrons to share, and H^+ is an acid because it has a vacant orbital that can accept (and therefore share) a lone pair of electrons.

The Lewis theory extends the concept of an acid to substances that do not contain a hydrogen atom but have vacant orbitals, such as a great many metal cations. The formation of a coordinate bond between a ligand and a central metal ion with a vacant orbital is therefore an acid–base reaction according to the Lewis definition, but is not considered an acid–base reaction according to Brønsted–Lowry theory.

The molecule BF_3 has no hydrogen atom and therefore cannot donate a proton, but it does have a vacant atomic orbital, and is therefore a Lewis acid. The electronic configuration of boron ($Z = 5$) is $1s^2 2s^2 2p$, and that of fluorine ($Z = 9$) is $1s^2 2s^2 2p^5$. Boron thus has 3 valence electrons and each fluorine has 7, so that BF_3 has $3 + 7(3) = 24$ valence electrons. The Lewis dot formula for BF_3 (see Section 14.2) is

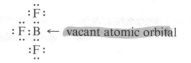

and there is a vacant atomic orbital on boron that can accept a pair of electrons. Indeed, BF_3 is known to react with NH_3 to form a compound with a coordinate bond,

$$BF_3 + :NH_3 \rightarrow F_3B:NH_3 \tag{20-3}$$

According to the Lewis definition, reaction (20-3) is an acid–base reaction.

Examples of reactions that are classified as acid–base reactions by the Lewis theory, but not by Brønsted–Lowry theory are

Acid		*Base*		
Ag^+	$+$	$2NH_3$	$\rightarrow$	$Ag(NH_3)_2^+$
$FeCl_3$	$+$	$:O(CH_2CH_3)_2$	$\rightarrow$	$Cl_3Fe:O(CH_2CH_3)_2$
Co^{2+}	$+$	$6NO_2^-$	$\rightarrow$	$Co(NO_2)_6^{4-}$
Ni^{2+}	$+$	$6H_2O$	$\rightarrow$	$Ni(H_2O)_6^{2+}$

Cation Acids

All ions are hydrated in aqueous solution. Hydrated metal ions that are Lewis acids have aqueous solutions that are, in fact, acidic, as the metal ion binds the OH^- from a water molecule. Some typical reactions are

$$Fe(H_2O)_6^{3+} + H_2O \rightleftharpoons Fe(H_2O)_5OH^{2+} + H_3O^+ \tag{20-4}$$

$$Sn^{2+}(aq) + 2H_2O \rightleftharpoons SnOH^+(aq) + H_3O^+ \tag{20-5}$$

It is common to write $Fe^{3+}(aq)$, $Sn^{2+}(aq)$, etc., to represent the hydrated metal ions. In the Lewis theory, the unhydrated metal ion with a vacant atomic orbital is the acid; in Brønsted–Lowry theory it is the hydrated metal ion that can donate a proton that is the acid.

The acidity of a hydrated metal ion depends on the strength of the bond between the cation and oxygen. If the cation–oxygen bond is very strong, the bond between oxygen and hydrogen is weakened and the proton can be donated to a base. On the other hand, if the cation–oxygen bond is weak, the oxygen–hydrogen bond is difficult to break, and only the entire water molecule can be split from the cation. This is illustrated in Fig. 20.1.

Reasoning solely on the basis of **Coulomb's Law,** and considering nothing but the electrostatic attraction between the cation and oxygen, we expect small, highly charged cations to form the most strongly acidic hydrated cations. Thus trivalent cations, as a group, are more acidic than bivalent cations, and most monovalent

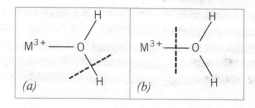

Fig. 20.1. Hydrated metal ion, M^{3+}, showing *(a)* a strong cation–oxygen bond and the splitting off of a proton to yield an acidic solution, and *(b)* a weak cation–oxygen bond that does not result in an acidic solution.

Table 20.2. **Acidity Constants for Hydrated Metal Ions at 25 °C in Perchlorate Solution**

Ion	Acid–Base Equilibrium	K_a
Tl^{3+}	$Tl^{3+}(aq) + 2H_2O \rightleftharpoons TlOH^{2+}(aq) + H_3O^+$	7×10^{-2}
Bi^{3+}	$Bi^{3+}(aq) + 2H_2O \rightleftharpoons BiOH^{2+}(aq) + H_3O^+$	3×10^{-2}
Fe^{3+}	$Fe^{3+}(aq) + 2H_2O \rightleftharpoons FeOH^{2+}(aq) + H_3O^+$	7×10^{-3}
Cr^{3+}	$Cr^{3+}(aq) + 2H_2O \rightleftharpoons CrOH^{2+}(aq) + H_3O^+$	2×10^{-4}
Hg^{2+}	$Hg^{2+}(aq) + 2H_2O \rightleftharpoons HgOH^+(aq) + H_3O^+$	2×10^{-4}
Sn^{2+}	$Sn^{2+}(aq) + 2H_2O \rightleftharpoons SnOH^+(aq) + H_3O^+$	1×10^{-4}
Al^{3+}	$Al^{3+}(aq) + 2H_2O \rightleftharpoons AlOH^{2+}(aq) + H_3O^+$	1×10^{-5}
Sc^{3+}	$Sc^{3+}(aq) + 2H_2O \rightleftharpoons ScOH^{2+}(aq) + H_3O^+$	1×10^{-5}
Fe^{2+}	$Fe^{2+}(aq) + 2H_2O \rightleftharpoons FeOH^+(aq) + H_3O^+$	5×10^{-9}
Cu^{2+}	$Cu^{2+}(aq) + 2H_2O \rightleftharpoons CuOH^+(aq) + H_3O^+$	5×10^{-9}
Ni^{2+}	$Ni^{2+}(aq) + 2H_2O \rightleftharpoons NiOH^+(aq) + H_3O^+$	5×10^{-10}
Zn^{2+}	$Zn^{2+}(aq) + 2H_2O \rightleftharpoons ZnOH^+(aq) + H_3O^+$	3×10^{-10}

cations have negligible acidic character. When the cation is a Lewis acid and has vacant atomic orbitals, the bond between cation and oxygen has covalent character and is considerably strengthened. Cations with rare gas electronic structure have no vacant valence atomic orbitals; the hydration of such cations is essentially due to ion-dipole forces of attraction. (See Section 7.1.)

The larger bivalent alkaline earth cations, Ca^{2+}, Sr^{2+}, and Ba^{2+}, are not acidic and very rarely form complex ions. The smaller bivalent alkaline earth cations, Be^{2+} and Mg^{2+}, are weakly acidic due to the stronger electrostatic attraction between them and the oxygen end of a water molecule. The trivalent Al^{3+}, which has the electronic configuration of neon, has acidic aqueous solutions, but does not form a large number of complex ions.

Table 20.2 lists acidity constants for some cation acids. The experimental uncertainty in these values is larger than that for molecular acids. Because many anions act as ligands, the value of K_a for a cation acid varies with the anion present. Those listed are for perchlorate solutions; ClO_4^- rarely complexes with cations.

The following example illustrates how we can explain the differences in acidity constants observed in Table 20.2.

EXAMPLE 20.1. Relative strengths of hydrated cation acids

Explain the difference in acid strength of (a) Fe^{3+} and Al^{3+} and (b) Fe^{2+} and Fe^{3+}.

Solution

(a) Ferric ion, Fe^{3+}, is a stronger cation acid than Al^{3+}. The acidity constant for the hydrated ferric ion is 7×10^{-3}, while K_a for the hydrated aluminum ion is 700 times smaller, 1×10^{-5}. The Fe^{3+} ion does not have the electronic configuration of a rare gas, while the Al^{3+} ion does. Cations without rare gas electronic configuration are more effective at polarizing the electron cloud of a ligand, that is, at pulling the electron cloud of the ligand towards the cation, than are cations of the same charge with rare gas electronic configuration. Thus the bond between Fe^{3+} ion and the O atom of H_2O is partly covalent, partly ionic. Even though the Al^{3+} ion is a little smaller than Fe^{3+} ion (their ionic radii are 0.50 and 0.67 Å, respectively) so that coulombic attraction is somewhat larger for the Al^{3+} ion, the Fe—O bond is stronger than the Al—O bond because of its covalent character. Alternatively, we might say that the Fe^{3+} ion is more electronegative than the Al^{3+} ion because of the difference in their electronic structures. Ferric ion is more acidic than Al^{3+} ion, and forms many more coordination compounds, because Fe^{3+} is a Lewis acid, and does not have the same electronic configuration as one of the rare gases, whereas Al^{3+} has the same electronic configuration as neon.

(b) Ferric ion is a stronger cation acid than is Fe^{2+}. As a rule, trivalent cations are more acidic than bivalent. The higher positive charge increases the electrostatic attraction between the cation and the oxygen atom of water, and it is easier for the O—H bond to be broken.

section 20.4
Inert versus Labile Complexes

A **ligand substitution reaction** is one in which the ligands attached to the metal are exchanged for others. A typical example is the following:

$$Cu(H_2O)_4^{2+} + 4NH_3 \rightleftharpoons Cu(NH_3)_4^{2+} + 4H_2O \tag{20-6}$$

which shows that the formation of a complex in aqueous solution is, in essence, a competition for the metal ion between H_2O and the other ligand. An aqueous solution of any copper(II) salt contains the *aquo* complex of Cu^{2+} ion, with water molecules as ligands. Almost instantly, on adding excess ammonia, the color changes and the tetraamminecopper(II) complex ion is formed. Both the ammine and aquo complexes of Cu^{2+} are **labile**, that is, they exchange ligands rapidly.

The $Cu(NH_3)_4^{2+}$ complex is only one example of a labile complex; most complexes are labile. A relatively small number of complexes are **inert**. Inert complexes undergo ligand substitution reactions very slowly; they form slowly and exchange ligands slowly.

There is no hard and fast rule as to how slow reactions must be for complexes to be considered inert, but it has been suggested that if a complex does not come to equilibrium with other ligands, including H_2O, within one or two minutes it should be classified as inert.

The difference between a labile complex and an inert one is a matter of kinetics and not of thermodynamic stability. There are thermodynamically stable complexes that are labile, such as $Cu(NH_3)_4^{2+}$, and others that are inert, such as $Cr(NH_3)_6^{3+}$ or $Fe(CN)_6^{4-}$. It is also possible for a complex to be unstable in the thermodynamic sense, but inert, so that the reaction to form a more stable complex is very slow.

The equilibrium constant for Eq. (20-6) is

$$K_{\text{formation}} = K_{\text{stability}} = \frac{[Cu(NH_3)_4^{2+}]}{[Cu^{2+}(aq)][NH_3]^4} = 1.2 \times 10^{12} \tag{20-7}$$

The large equilibrium constant for the formation of this complex from Cu^{2+} ions and the ligand ammonia molecules indicates that the position of equilibrium is far to the right; thermodynamically the $Cu(NH_3)_4^{2+}$ complex ion is very stable. The following series of reactions illustrates the lability of this ion.

If we add a very small amount of an ammonia solution to a solution containing $Cu^{2+}(aq)$ ions, we observe the formation of a pale blue precipitate of $Cu(OH)_2$:

$$Cu^{2+}(aq) + 2NH_3 + 2H_2O \rightleftharpoons Cu(OH)_2(s) + 2NH_4^+ \tag{20-8}$$

If we continue to add ammonia, the pale blue precipitate dissolves and the solution becomes a clear intense royal blue color, the distinctive color of the tetraamminecopper(II) complex ion:

$$Cu(OH)_2(s) + 4NH_3 \rightleftharpoons Cu(NH_3)_4^{2+} + 2\,OH^- \tag{20-9}$$

Adding excess HCl or other strong acid will immediately destroy the complex:

$$Cu(NH_3)_4^{2+} + 4H_3O^+ \rightleftharpoons Cu(H_2O)_4^{2+} + 4NH_4^+ \tag{20-10}$$

and the solution again becomes the pale blue-green color characteristic of cupric ion in aqueous solution, which is, in fact, the $Cu(H_2O)_4^{2+}$ complex ion.

All the equilibria of Eqs. (20-6), (20-8), (20-9), and (20-10) are established very quickly, virtually as soon as the solutions are mixed. The tetraamminecopper(II) ion, a very stable ion, is labile. It is immediately in equilibrium with $Cu^{2+}(aq)$ and NH_3. The addition of H_3O^+ removes NH_3 from solution and drives reaction (20-6) to the left instantaneously.

Complexes of Cr(III), on the other hand, are inert. The inertness of Cr(III) complexes is evident if we try to repeat the series of experiments just described for Cu(II), Eqs. (20-8) and (20-9), using Cr(III). Adding NH_3 to an aqueous solution of Cr^{3+} yields only $Cr(OH)_3$, a gray-green precipitate:

$$Cr^{3+}(aq) + 3NH_3 + 3H_2O \rightleftharpoons Cr(OH)_3(s) + 3NH_4^+ \tag{20-11}$$

No matter how much excess ammonia is added, the $Cr(NH_3)_6^{3+}$ complex ion does not form in any reasonable period of time.

Nevertheless, the $Cr(NH_3)_6^{3+}$ ion does exist, and once formed it is both stable and inert. Adding strong acid to it does *not* form NH_4^+ and destroy the complex, so that a reaction similar to Eq. (20-10) does not occur.

Thermodynamic stability depends upon the difference in energy between the products and the reactants. The rate of reaction depends on the activation energy of the forward reaction, that is, on the difference in energy between the reactants and the activated complex. Thus there is no necessary connection between lability and thermodynamic stability. Tripositive ions having either three or six d electrons, such as Co(III) and Cr(III), form inert complexes. Many complexes of Fe(II), which has six d electrons, are inert. We shall consider the reason that these cations form inert complexes when we discuss theories of bonding in transition metal complexes, in Section 20.11.

section 20.5
Evidence for the Existence of Complex Ions

A great many coordination compounds can be crystallized. For those that can be obtained as crystals, **X-ray diffraction** studies provide direct evidence of the existence of complexes and of the geometric arrangement of the ligands around the central metal ion. Figure 20.2 is an X-ray diffraction photograph of a single crystal of a copper complex. A monochromatic (single wavelength) X-ray beam is diffracted by a crystal in a manner that depends on the distances between atoms and their spatial geometry. From such X-ray data the details of the molecular structure such as bond distances and bond angles can be deduced. X-ray diffraction studies of a very large number of coordination complexes have been carried out so that their geometries have been conclusively established. Figure 20.3 shows the structure of the copper complex of Fig. 20.2, obtained by analyzing the X-ray data. In this particular complex the coordination number of Cu^{2+} is 5; it is bound to five nitrogen atoms.

Long before X-ray diffraction was an available tool, however, chemists had deduced a great deal of information about complexes using many kinds of chemical evidence, less direct than X-ray studies, but nevertheless convincing.

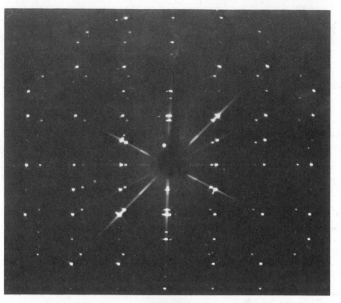

Fig. 20.2. X-ray diffraction photograph of a single crystal of a copper complex.

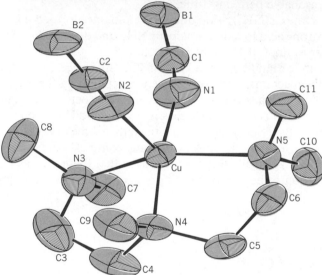

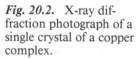

Fig. 20.3. Structure of the copper complex of Fig. 20.2 obtained by analyzing the X-ray data. This is a pentacoordinate copper(II) complex in which Cu^{2+} is bonded to five N atoms. Hydrogen atoms are not shown. Atoms are represented by thermal ellipsoids which show the relative space each atom occupies as a result of its thermal motion.

The scientist responsible for the greatest advances in our knowledge of complex geometries was Alfred Werner, who carried out his investigations between 1890 and 1919, and was awarded the Nobel Prize in 1913. Werner studied the complexes of Co(III) and Cr(III), which are inert. If ligands are tightly bound to a metal in an inert complex, they do not react in the same way as they do when they are free, that is, *not* bound to a metal. For example, free NH_3 reacts with any strong acid to yield NH_4^+, but NH_3 that is tightly bound in an inert complex does not react with H_3O^+ at all. Similarly, free Cl^- ions will precipitate as AgCl when Ag^+ ions are added, but bound Cl^- ions in an inert complex do not react with Ag^+.

An example of the type of evidence and reasoning used by Werner is the following. A yellow-orange compound is known to have the empirical formula $CoCl_3N_6H_{18}$, or $CoCl_3 \cdot 6NH_3$. Remember that the empirical formula can be obtained from an analytical determination of the percent by weight of each element in the compound. The empirical formula does not, however, tell us what ligands are bound to the Co(III). The following experiments are performed on this compound.

1. Excess concentrated H_2SO_4 is added to the solid coordination compound. A test for the presence of NH_4^+ ions in the resulting solution indicates that NH_4^+ ions are absent from this solution. In the vapor above the solution HCl gas is detected. When the solution is gently heated until the water has evaporated, a solid different from the original is left as a residue. When analyzed the residue is found to have the empirical formula $Co_2(SO_4)_3 \cdot 12NH_3$.

2. Excess silver nitrate solution is added to an aqueous solution of the coordination compound. An immediate precipitate of white AgCl is observed. From the mass of the precipitate it is determined that 3 mol of AgCl are formed per mole of $CoCl_3 \cdot 6NH_3$.

The conclusions drawn from these observations are that while all six ammonia molecules are tightly bound to the Co(III), none of the Cl^- ions are. The compound consists of the complex cation $Co(NH_3)_6^{3+}$ and three Cl^- anions, and should be written as $[Co(NH_3)_6]Cl_3$. After the sulfuric acid treatment, the Cl^- ions have been removed by being driven off as gaseous HCl, and the remaining ions are $Co(NH_3)_6^{3+}$ and SO_4^{2-}, so that the residue after evaporation is $[Co(NH_3)_6]_2(SO_4)_3$. We should emphasize that these results are obtained only because $Co(NH_3)_6^{3+}$ is an inert complex. With a labile complex, NH_4^+ ions would be formed when sulfuric acid is added even though NH_3 is tightly bound to the metal ion.

Another piece of chemical evidence used to determine the nature of complex ions is the freezing point depression. You will recall, from Sections 6.5 and 6.6, that the freezing point depression is a **colligative property,** that is, it depends on the number of moles of particles in solution per mole of compound, and not on the nature of the solute particles. For an electrolyte, the number of ions per formula unit has been denoted by the symbol v, and the freezing point depression of an ideal solution is given by Eq. (6-14b),

$$\Delta T_f = vK_f m$$

where K_f is a constant and is a property of the solvent only, and m is the molality of the solute. For H_2O, $K_f = 1.86$. Thus, for an ideal solution, the number of ions per formula unit can be determined by calculating $\Delta T_f/K_f m$. If we examine the value of this quantity for a series of solutions of NaCl, for which v is known to be 2, we obtain data such as that given in Table 20.3. We observe that for an electrolyte in dilute solution, the experimental value of v is very close to, but not identical with, the number of ions per formula unit. The more dilute the solution, the closer v gets to the

Table 20.3. **Freezing Point Depression of Aqueous Solutions of Sodium Chloride**

Molality of NaCl soln	ΔT_f	$\Delta T_f/K_f m = v$
0.100 m	0.348°	0.348/0.186 = 1.87
0.010 m	0.036°	0.036/0.0186 = 1.94

Table 20.4. **The Typical Range of Molar Conductances of 0.001 *M* Aqueous Solutions of Electrolytes**

Type of Electrolyte	Number of Ions	Range of Molar Conductances (ohm⁻¹)
4:1	5	520–570
3:1	4	400–450
2:1	3	210–275
1:1	2	90–135

actual number of ions per formula unit. The reason v is not exactly 2 for NaCl solutions is that these solutions are not ideal due to **interionic forces.**

Ions in solution do not behave independently because of the electrostatic attractions between ions of opposite charge and the electrostatic repulsions between ions of the same charge. The more concentrated the solution, the closer the ions are to one another, and the larger the coulombic interactions are. The coulombic force is inversely proportional to the square of the distance between the charged particles. The more dilute the solution, the smaller the interionic forces and the more nearly ideal the solution.

Even though the value of $\Delta T_f/K_f m$ will not be exactly integral, for dilute solutions it will be slightly less than the true value of v, and we will be able to determine v from the data. For instance, for the yellow-orange compound of empirical formula $CoCl_3 \cdot 6NH_3$ discussed above, the value of $\Delta T_f/K_f m$ for a dilute solution is 3.9. We know, therefore, that there are 4 ions per formula unit, which confirms previous evidence that the compound consists of $Co(NH_3)_6^{3+}$ and $3Cl^-$ ions.

Another property that depends on the number of ions per formula unit is the **molar conductivity.** Ionic solutions conduct electricity, and the molar conductivity is a measure of the ability of the solution to carry current. The greater the number of ions per formula unit, the greater the molar conductivity. If we compare results for a given concentration of electrolyte, we find that we can use their molar conductivities to classify electrolytes according to type. Table 20.4 lists the range of molar conductances observed for 0.001 *M* solutions of electrolytes of various types. The molar conductivity of a 0.001 *M* solution of the yellow-orange compound with empirical formula $CoCl_3 \cdot 6NH_3$ is 432 ohm⁻¹, and using Table 20.4, we see that it is a 3:1 electrolyte with 4 ions per formula unit. Once again, this confirms that the compound is $[Co(NH_3)_6]Cl_3$.

Both the freezing point depression and the molar conductivity provide the same information, and sometimes only one of these measurements is made, but often scientists feel it is important to provide every possible bit of evidence leading to a given conclusion, and will carry out confirmatory experiments. Example 20.2 indicates how we can reason from various types of chemical evidence and obtain information about the nature of complexes.

EXAMPLE 20.2. Chemical evidence for the nature of a complex

Two different compounds of Cr(III) have the same empirical formula, $CrCl_3 \cdot 6H_2O$. One is green and the other is violet. The following information is obtained about each of these salts.

When a solution of the green compound, which we will call salt A, is mixed with excess silver nitrate solution, 1 mol of AgCl is precipitated per mole of salt A. The molar conductivity of a freshly prepared 0.001 M solution of salt A is 132 ohm^{-1}. If salt A is dried in an oven at 110 °C for several hours, it loses weight and the residue has the empirical formula $CrCl_3 \cdot 4H_2O$.

When a solution of the violet compound, which we will call salt B, is mixed with excess silver nitrate solution, 3 mol of AgCl are precipitated per mole of salt B. The molar conductivity of a freshly prepared 0.001 M solution of salt B is 437 ohm^{-1}. If salt B is dried in an oven at 110 °C for several hours, it is unchanged and still has the empirical formula $CrCl_3 \cdot 6H_2O$.

Deduce the formulas of salts A and B.

Solution. Of the three Cl^- ions in salt A, only two can be tightly bound, as 1 mol of AgCl is precipitated per mole of salt A. Two of the six water molecules cannot be tightly bound to the Cr(III) because they are driven off when salt A is heated at 110 °C. We conclude that four H_2O molecules and two Cl^- ions *are* tightly bound to the Cr(III) in an inert complex. The complex cation is therefore $Cr(H_2O)_4Cl_2^+$. The charge on this cation must be +1 because two Cl^- ions are bound to Cr^{3+}, and $+3 + 2(-1) = +1$. We note that the coordination number of chromium in this complex is 6. The green salt A is a 1 : 1 electrolyte, with the ions $Cr(H_2O)_4Cl_2^+$ and Cl^-. The formula of the solid compound is $[Cr(H_2O)_4Cl_2]Cl \cdot 2H_2O$. It is the two waters of hydration in the crystal structure that are removed by heating at 110 °C.

Reasoning in the same way, violet salt B must have all six H_2O molecules tightly bound to Cr(III). None of the three Cl^- ions are tightly bound as they all precipitate as AgCl when excess Ag^+ is added. From the molar conductivity, violet salt B is a 3 : 1 electrolyte, with 4 ions per formula unit. The ions of this salt are therefore $Cr(H_2O)_6^{3+}$ and three Cl^- ions, and the formula of the solid is $[Cr(H_2O)_6]Cl_3$.

It is important to remember that H_2O can substitute for other ligands when coordination compounds are dissolved in water. Example 20.3 describes evidence that such a substitution has occurred.

EXAMPLE 20.3. Evidence for the substitution of a chloro ligand by water in aqueous solution

The molar conductivity of a freshly prepared 0.001 M solution of the green salt $[Cr(H_2O)_4Cl_2]Cl \cdot 2H_2O$ is 132 ohm^{-1}, but after standing for 24 h the molar conductivity has increased to 290 ohm^{-1}, and after 88 h it is 410 ohm^{-1}. Account for these observations.

Solution. The fact that the molar conductance of an aqueous solution of the green salt $[Cr(H_2O)_4Cl_2]Cl \cdot 2H_2O$ increases with time indicates that a reaction is occurring with the water solvent. Water is replacing the two Cl^- ligands, and the cation $Cr(H_2O)_4Cl_2^+$ is being transformed into $Cr(H_2O)_6^{3+}$. We note that after a long period of time (88 h or almost 4 days), the conductivity is that of a 3 : 1 electrolyte. Because Cr(III) complexes are inert, this reaction occurs very slowly. The reaction taking place is

$$Cr(H_2O)_4Cl_2^+ + 2H_2O \rightarrow Cr(H_2O)_6^{3+} + 2Cl^-$$

As the number of ions in solution is increasing with time, the molar conductivity of the solution also increases.

section 20.6
Nomenclature

A system of naming complex ions and coordination compounds has been devised by the Inorganic Nomenclature Committee of the International Union of Pure and Applied Chemistry. There are a number of rules to learn; it takes a little time and effort to master these rules, but with practice you will find that naming these substances becomes straightforward.

Rules for Naming Salts that Contain a Complex Ion or Ions

1. The cation is named first, and then the anion, with a space separating the names of the two ions. This is the same rule for any salt, whether it contains complex ions or not.

2. For a complex cation, the ligands are named first, then the name of the metal that is the central atom or ion is given, followed without a space by the oxidation state of the metal as a Roman numeral placed in parentheses.

3. Anionic ligands that are not hydrocarbons are given names that end in -o. Usually the final -e of an anion is simply changed to an -o, such as

   SO_4^{2-} sulfate ion, becomes sulfato as a ligand
   CO_3^{2-} carbonate ion, becomes carbonato as a ligand

In the case of many anions with names that end in -ide, the entire last syllable is replaced by -o, as you can see by examining Table 20.1. In addition to those listed in Table 20.1, anions in which the -ide is replaced by -o are the following:

   O^{2-} oxide ion, becomes oxo
   OH^- hydroxide ion, becomes hydroxo
   O_2^{2-} peroxide ion, becomes peroxo

H^- as a ligand is called both hydrido and hydro; either of these names is acceptable. When serving as a ligand, S^{2-} is called thio and SH^- is called mercapto.

Hydrocarbon anion ligands are given names that end in -yl, such as $C_5H_5^-$, cyclopentadienyl.

4. When a neutral molecule serves as a ligand the name of the molecule is unchanged. There are several exceptions to this rule: H_2O is called aquo and NH_3 is called ammine. Note particularly that there are two m's in the spelling of ammine when ammonia is meant. Organic amines (compounds with the —NH_2 group) are spelled with only one m. When the gases NO and CO serve as ligands they are referred to as nitrosyl and carbonyl, respectively.

5. The very few ligands that are cations have names that end in -ium, such as $NH_2NH_3^+$, hydrazinium.

6. A Greek prefix indicates the number of ligands of each kind: mono, di, tri, tetra, penta, hexa, and so on. The prefix mono, however, is often omitted. If the name of the ligand itself contains the terms mono, di, or tri, or if there would be any confusion about what the ligand is if these prefixes are used, then bis, tris, and tetrakis are substituted for di, tri, and tetra. The name of the ligand is then enclosed in parentheses.

Examples of the names of complex cations containing only one kind of ligand using the rules above are

$$Cr(NH_3)_6^{3+}$$ hexaamminechromium(III) ion

$$Pt(en)_3^{4+}$$ tris(ethylenediamine)platinum(IV) ion

$$Fe(H_2O)_6^{2+}$$ hexaaquoiron(II) ion

7. If there is more than one type of ligand, anionic ligands are named first, then netural ligands, and lastly positive ligands. If there is more than one ligand in any of these three categories, the ligands are listed alphabetically. Examples are

$$Co(en)_2Cl_2^+$$ dichlorobis(ethylenediamine)cobalt(III) ion

$$Cr(H_2O)_4Br_2^+$$ dibromotetraaquochromium(III) ion

$$Pt(NH_3)_4(NO_2)I^{2+}$$ iodonitrotetraammineplatinum(IV) ion

You deduce the oxidation state of the metal from the charge on the complex and the number of anionic ligands. Note that these names are written with no spaces left between any of the parts or before the parentheses.

8. A complex anion is named in exactly the same way as a complex cation except that the syllable -ate is added to the name of the metal, and the *Latin,* rather than the English, name of the metal is used. Examples are

$$Fe(CN)_6^{4-}$$ hexacyanoferrate(II) ion

$$Pt(NH_3)(NO_2)_5^-$$ pentanitromonoammineplatinate(IV) ion

$$Ag(S_2O_3)_2^{3-}$$ dithiosulfatoargentate(I) ion

$$AlH_4^-$$ tetrahydroaluminate(III) ion or

tetrahydridoaluminate(III) ion

The Rule for Naming Nonionic or Molecular Complexes

9. Molecular complexes are given a one-word name, with no spaces between any part of the name. Rules 2 through 7 above apply. Examples are

$$Cu(acac)_2$$ bis(acetylacetonato)copper(II)

$$Cr(H_2O)_3Cl_3$$ trichlorotriaquochromium(III)

The Rule for Naming Compounds with a Bridging Ligand or Ligands

10. A bridging ligand is indicated by placing μ- (pronounced *mew* and spelled mu) before its name. The syllable di is placed before the name of the metal if the ligand bridges two identical metal atoms. Examples are

$$[(NH_3)_5Co-O_2-Co(NH_3)_5]^{4+}$$ decaammine-μ-peroxodicobalt(III) ion

$$[(H_2O)_4Fe\overset{OH}{\underset{OH}{<>}}Fe(H_2O)_4]^{4+}$$ octaaquo-μ-dihydroxodiiron(III) ion

The following examples illustrate these nomenclature rules.

EXAMPLE 20.4. Naming coordination compounds

Give the names of the following compounds: **(a)** $[Co(NH_3)_4CO_3]NO_3$, **(b)** $Na[Al(OH)_4]$, **(c)** $Cr(NH_3)_3(NO_2)_3$, **(d)** $[Rh(en)_2(H_2O)Cl]Cl_2$, and **(e)** $[Pt(py)_4][PtF_4]$.

Solution

(a) $[Co(NH_3)_4CO_3]NO_3$ carbonatotetraamminecobalt(III) nitrate

(b) $Na[Al(OH)_4]$ sodium tetrahydroxoaluminate(III)

(c) $Cr(NH_3)_3(NO_2)_3$ trinitrotriamminechromium(III)

(d) $[Rh(en)_2(H_2O)Cl]Cl_2$ chloroaquobis(ethylenediamine)rhodium(III) chloride

(e) $[Pt(py)_4][PtF_4]$ tetrapyridineplatinum(II) tetrafluoroplatinate(II)

To deduce the oxidation state of the platinum in this compound, reason as follows: The cation and the anion must have charges of equal magnitude as they are in a 1 : 1 ratio. Since pyridine is uncharged, the charge on the cation is the same as the oxidation state of the platinum. With platinum in the +2 oxidation state, the cation has a charge of +2, and the anion has a charge of −2, as there are four F⁻ ligands. Recall also that Pt(II) is always four coordinate.

EXAMPLE 20.5. Writing formulas of coordination compounds

Write the formula of each of the following compounds: **(a)** tetraiodoethylenediamineplatinum(IV), **(b)** potassium tetrafluoroargentate(III), **(c)** rubidium ethylenediaminetetracetatocobaltate(III) dihydrate, and **(d)** lithium pentahydroxoaquostannate(IV).

Solution

(a) tetraiodoethylenediamineplatinum(IV) $Pt(en)I_4$

(b) potassium tetrafluoroargentate(III) $K[AgF_4]$

(c) rubidium ethylenediaminetetraacetato- $Rb[CoEDTA] \cdot 2H_2O$
      cobaltate(III) dihydrate

Note that the EDTA anion has a charge of −4, so that with cobalt in the +3 oxidation state the complex anion has a charge of −1.

(d) lithium pentahydroxoaquostannate(IV) $Li[Sn(OH)_5(H_2O)]$

section 20.7
The Geometry of Coordination Compounds; Isomers

The geometries of coordination complexes are generally in agreement with the predictions of VSEPR theory, which you may want to review. (See Section 14.3.)

Complexes in which the coordination number of the metal is two are linear, X—M—X, as for example $[H_3N—Ag—NH_3]^+$. If the coordination number is three, the complex is generally trigonal planar, with the metal, M, in the center of an equilateral triangle and the ligands at the corners of the triangle. If all ligands are not identical, the triangle is not equilateral.

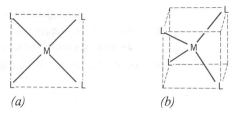

Fig. 20.4. Limiting geometries for coordination number 4. *(a)* Square-planar geometry. All ligands and the metal in one plane. *(b)* Tetrahedral geometry. Solid red lines represent bonds. Dotted lines are used to facilitate visualization of the actual geometry. Distortion from these limiting geometries occurs when all four ligands are not identical.

Geometries for Coordination Number 4

Two limiting geometries are observed for coordination number 4, **square planar** and **tetrahedral**. These two geometries are depicted in Fig. 20.4. There are many more tetrahedral four-coordinate complexes than square planar. Metal ions that commonly form square-planar complexes are those with eight d electrons (d^8 ions), such as Ni(II), Pd(II), Pt(II) and Au(III). Some d^4 and d^9 complexes are also square planar.

Cis–trans Isomerism for Four-Coordinate Complexes

There is a very important distinction between tetrahedral and square-planar geometry for complexes of the type MX_2Z_2, such as $Pd(py)_2Br_2$ or $Pt(NH_3)_2Cl_2$. **Geometric isomers** are possible for square-planar geometry, but not for tetrahedral geometry. Geometric isomers occur when ligands occupy different positions around the central atom. Ligands that are adjacent to one another are said to be in *cis* positions, while ligands that are opposite to one another are in *trans* positions. Figure 20.5 shows the *cis–trans* isomerism possible for a square-planar complex of the type MX_2Z_2.

Since two compounds with the same empirical formula $Pt(NH_3)_2Cl_2$ are known, with different chemical properties and different solubilities, it is clear that the structure of $Pt(NH_3)_2Cl_2$ is square planar. Platinum(II) complexes react slowly and are

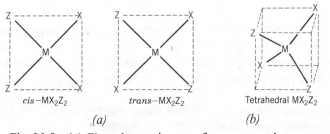

cis–MX_2Z_2 *trans*–MX_2Z_2 Tetrahedral MX_2Z_2

(a)   *(b)*

Fig. 20.5. *(a)* Cis and *trans* isomers for a square-planar complex of type MX_2Z_2. The *cis* compound is not symmetric, and has a dipole moment. The *trans* compound is symmetric, and has zero dipole moment. *(b)* Only one isomer is possible for tetrahedral MX_2Z_2. Examine a three-dimensional model to see this more clearly. The existence of two isomers is conclusive proof that the geometry is not tetrahedral.

very stable; their geometry is characteristically square planar. *Cis* and *trans* isomers are possible for square-planar complexes of the type MX_2YZ, as well as MX_2Z_2, where X, Y, and Z are monodentate ligands. Many bidentate ligands, such as en and oxalate ion, are too small to span the *trans* positions and can only occupy *cis* positions.

Geometric isomerism in square-planar complexes is illustrated in Example 20.6.

EXAMPLE 20.6. *Cis–trans* **isomers of square-planar complexes**

Draw the structures of all the isomers of the square-planar complexes **(a)** $Pd(NH_3)_2BrI$ and **(b)** $Ni(en)Cl_2$.

Solution

(a) Two isomers are possible for $Pd(NH_3)_2BrI$, with the two ammonias occupying *cis* positions in one, and *trans* positions in the other.

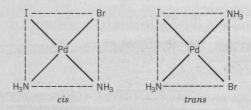

 *cis* *trans*

(b) Only one isomer is possible for $Ni(en)Cl_2$ as ethylenediamine cannot span the trans positions. It is customary to symbolize a chelating ligand as a curved line with the symbol for the ligand in the middle. Thus $\overset{en}{\frown}$ could be used for $\ddot{N}H_2{-}CH_2{-}CH_2{-}\ddot{N}H_2$.

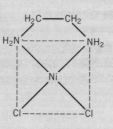

Geometries for Coordination Number 5

Complexes in which the coordination number is 5 have **trigonal bipyramidal** geometry or a **square-based pyramidal** structure, as illustrated in Fig. 20.6. In a square-based pyramid, four ligands are at the corners of a square, the metal ion sits above the center of the square, and the fifth ligand is along the perpendicular to the square. An example of a complex with trigonal bipyramidal geometry is $Fe(CO)_5$. Significant distortions from these idealized geometries occur when there is more than one kind of ligand and the different ligands differ considerably in size.

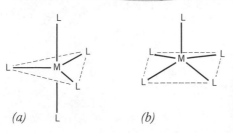

Fig. 20.6. Possible geometries for coordination number 5. *(a)* Trigonal bipyramid. *(b)* Square-based pyramid. Distortions from these limiting geometries occur when the ligands are not identical.

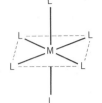

Fig. 20.7. Octahedral geometry for coordination number 6.

Octahedral Geometry for Coordination Number 6

Complexes in which the coordination number of the metal is 6 almost invariably have **octahedral** geometry, illustrated in Fig. 20.7. Geometric isomerism is possible for a six-coordinate complex. Indeed, Werner observed that for six-coordinate complexes of the type MA_4X_2, where A and X are monodentate ligands, he never found more than two geometric isomers, and this observation led him to postulate octahedral geometry long before X-ray diffraction studies had conclusively proved the geometry is octahedral.

Werner considered three possibilities for symmetric six-coordinate structures: (1) planar hexagonal, (2) trigonal prismatic, and (3) octahedral. For complexes of the type MA_4X_2, such as $Co(NH_3)_4Br_2$, only two geometric isomers are observed. For planar hexagonal or trigonal prismatic geometry, three such isomers are possible, while for octahedral geometry only two isomers are possible, as is shown in Fig. 20.8.

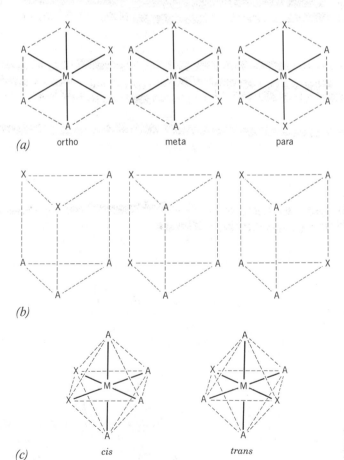

Fig. 20.8. Geometric isomers possible for a six-coordinate complex of formula MA_4X_2 if the geometry were *(a)* planar hexagonal, *(b)* trigonal prismatic, and *(c)* octahedral. Only two isomers of such complexes have been found. This observation led Werner to postulate octahedral geometry for six-coordinate complexes.

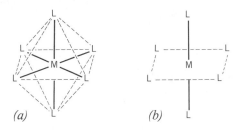

Fig. 20.9. *(a)* An octahedron. *(b)* Two-dimensional representation of an octahedron used for drawing structures of complexes with octahedral geometry.

The six corners of an octahedron are all equivalent, although this may not be apparent if you look only at the two-dimensional representation we use when drawing octahedral structures with paper and pencil. Figure 20.9 compares the conventional two-dimensional drawing with a diagram of an octahedron. It is a great help to use three-dimensional models. Examine such a model until it is clear to you that both

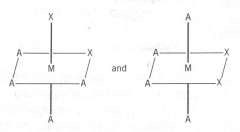

are identical, and represent the *cis* isomer of an MA$_4$X$_2$ complex.

Geometric isomers of the same empirical formula have different properties: different melting points, different solubilities, and often different colors. For instance, *cis*-dichlorotetraamminecobalt(III) ion is violet, while *trans*-dichlorotetraamminecobalt(III) ion is green. (See Fig. 20.10.) In considering geometric isomers it is important to remember that several bidentate ligands are too small to span *trans* positions. For instance, only the *cis* form of Co(en)Cl$_4$ can be made. Unsymmetrical bidentate ligands such as glycinato can also give rise to geometrical isomers. The *cis* and *trans* forms of triglycinatocobalt(III) are depicted in Fig. 20.11.

Octahedral complexes of the type MA$_3$X$_3$ can also have two geometric isomers, as is illustrated in Fig. 20.12 for trichlorotripyridinerhodium(III).

Linkage Isomerism

A special kind of geometric isomerism is called **linkage isomerism**. Certain ligands can bond to the metal in more than one way, as is indicated in Table 20.1. The NO$_2^-$ ion, for example, can bond through N (nitro) or through O (nitrito), and the SCN$^-$ ion can bond through S (thiocyanato) or through N (isothiocyanato).

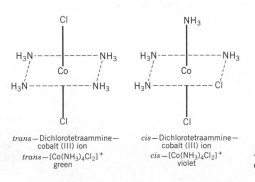

trans—Dichlorotetraammine—
cobalt (III) ion
trans—[Co(NH$_3$)$_4$Cl$_2$]$^+$
green

cis—Dichlorotetraammine—
cobalt (III) ion
cis—[Co(NH$_3$)$_4$Cl$_2$]$^+$
violet

Fig. 20.10. Geometric isomerism in octahedral complexes.

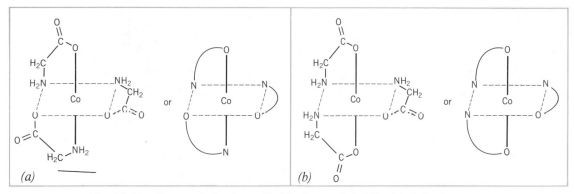

Fig. 20.11. Geometric isomers of triglycinatocobalt(III). *(a) cis* form,
(b) trans form. Note that it is convenient to use abbreviations for chelating ligands in
diagrams. A curved line with the symbol for the ligand written in the middle is frequently used.

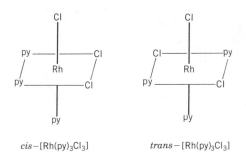

cis−[Rh(py)₃Cl₃]   *trans*−[Rh(py)₃Cl₃]

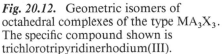

Fig. 20.12. Geometric isomers of
octahedral complexes of the type MA₃X₃.
The specific compound shown is
trichlorotripyridinerhodium(III).

section 20.8
Optical Isomerism

While different geometric isomers have different chemical and physical properties,
another type of isomerism exists in which the two isomers have identical chemical
and physical properties but differ in the direction in which they rotate the plane of
plane-polarized light.

A light ray has an electromagnetic vibration associated with it. (Refer to Section
12.2.) The vibrations occur at right angles to the direction of propagation of the ray, as
shown in Fig. 20.13*(a)*. For a beam of ordinary light, vibrations in all of the perpendicular directions occur simultaneously. **Plane-polarized light,** depicted in Fig.
20.13*(b)*, is light with these vibrations in only one direction, so that there is a single
plane containing the ray and its electromagnetic vibrations. Ordinary light can be
turned into plane-polarized light by passing it through a lens made of a polarizing
medium. Two examples of substances that convert ordinary light into plane-polarized light are **calcite,** a crystalline form of $CaCO_3$, and the material known as
Polaroid.

If you pass polarized light through water, or an aqueous solution of NaCl, it
emerges with the plane of polarization of the light exactly the same as when it entered.
Solutions of certain substances, however, rotate the plane of polarized light, so that
the plane of polarization of the emergent light is at an angle relative to the plane of
polarization of the incident light. Substances that are capable of rotating the plane of
polarized light are said to be **optically active.** An instrument known as a **polarimeter** is
used to measure the angle of rotation of the plane of polarized light after it passes
through a solution of an optically active substance.

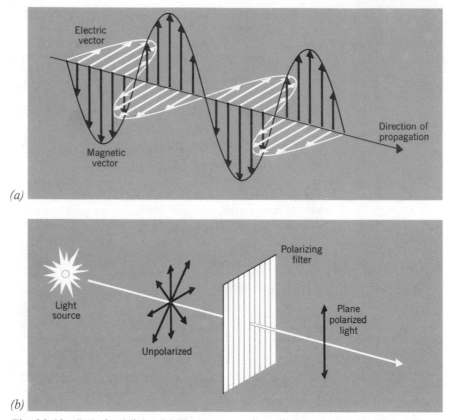

Fig. 20.13. Polarized light. *(a)* Electromagnetic radiation composed of electric and magnetic vectors. *(b)* Orientation of electric vectors in unpolarized (ordinary) light and polarized light.

Two isomers that differ only in the direction in which they rotate the plane of polarized light are called **optical isomers** or **enantiomers**. Optical isomers have the same groups arranged in the same order, but are mirror images of one another, and are not superimposable. They are related to one another in the same way as your left and right hands are related. The extent of rotation of the plane of polarized light by the two isomers is exactly the same; however, one enantiomer rotates the light in a clockwise direction, the other in a counterclockwise direction.

In order for a molecule or ion to be optically active it must not be possible to divide the structure into two identical halves. Another way of describing this requirement for optical activity is to say that there is no plane of symmetry. There are a large number of organic molecules that have optical isomers, and we will discuss optical isomerism again in Chapter 23, with specific examples from organic chemistry.

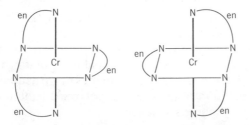

Fig. 20.14. The enantiomers of tris(ethylenediamine)chromium(III) ion.

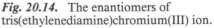

Certain categories of octahedral complexes have optical isomers. Six-coordinate complexes containing three identical bidentate ligands, such as $Cr(en)_3^{3+}$ or $Co(ox)_3^{3-}$ are optically active. The mirror images of tris(ethylenediamine)chromium(III) ion are shown in Fig. 20.14. Octahedral complexes with two identical bidentate ligands plus two identical monodentate ligands also have optical isomers.

EXAMPLE 20.7. Geometric and optical isomers of a complex

How many geometric and optical isomers are there of the complex ion $Co(en)_2Br_2^+$?

Solution. There are a total of three isomers. There are *cis* and *trans* isomers. The *cis* form occurs as a pair of enantiomers, each of which is optically active. The three isomers are depicted below.

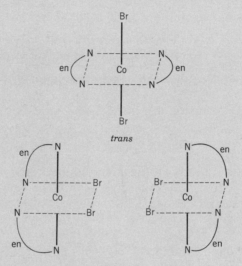

trans

Enantiomers of the *cis* form

For the *trans* form, a plane perpendicular to the square with the four nitrogen atoms at the corners, bisecting a side of that square, and passing through the two Br atoms and the Co atom, is a plane of symmetry.

---

section 20.9
The Effective Atomic Number (EAN) Rule

One of the properties of complexes that we would like to understand is the occurrence of characteristic coordination numbers. Why is it that Co^{3+} invariably has a coordination number of 6, while Cd^{2+} is usually four coordinate? N. V. Sidgwick of Oxford University proposed that characteristic coordination numbers arise because of the stability achieved when the central metal atom is surrounded by the same number of electrons as one of the rare gases. Sidgwick introduced the concept of the **effective atomic number** (EAN), the number of electrons surrounding the coordinated metal atom.

Consider the Cd^{2+} ion, which usually has a coordination number of 4. The atomic number of cadmium is 48, so that the Cd^{2+} ion has 46 electrons. Each ligand contributes a *pair* of electrons, so that four ligands furnish a total of 8 electrons, and the effective atomic number of cadmium in a four-coordinate complex is $46 + 8 = 54$, the same as the atomic number of the rare gas Xenon.

In a great many complexes the metal atom does have an effective atomic number equal to one of the rare gases. There are, however, many exceptions to this rule, as we shall see in Example 20.8. If the bonding between metal and ligand were largely covalent, we would expect the EAN rule to apply. It is not a sure predictor of coordination number because the bonding is partly ionic, and factors such as ligand size and structure play a role as well.

EXAMPLE 20.8. Sidgwick's EAN rule

Calculate the effective atomic number of the metal in the following complexes: (a) $Zn(OH)_4^{2-}$, (b) $Ag(CN)_2^-$, (c) HgI_4^{2-}, and (d) $Cu(NH_3)_4^{2+}$.

Solution

(a) The atomic number of Zn is 30. Therefore,

Zn^{2+} ion contributes 28 electrons
4 OH^- ligands contribute 8 electrons   EAN = 36

The atomic number of the rare gas krypton is 36.

(b) The atomic number of Ag is 47. Therefore,

Ag^+ ion contributes 46 electrons
2 CN^- ligands contribute 4 electrons   EAN = 50

There is no rare gas with an atomic number of 50. The coordination number of Ag^+ is usually 2, so the EAN rule does not hold for Ag^+. There are four-coordinate Ag(III) complexes, such as AgF_4^-, in which the EAN of silver is 52. Again the EAN rule does not apply.

(c) The atomic number of Hg is 80. Therefore,

Hg^{2+} ion contributes 78 electrons
4 I^- ligands contribute 8 electrons   EAN = 86

The atomic number of the rare gas radon is 86.

(d) The atomic number of Cu is 29. Therefore,

Cu^{2+} ion contributes 27 electrons
4 NH_3 ligands contribute 8 electrons   EAN = 35

There is no rare gas with an atomic number of 35. Of course, since a Cu^{2+} ion has an odd number of electrons, there is no coordination number that could achieve the EAN of a rare gas.

One class of compounds that invariably obeys the EAN rule is the metal carbonyls. In the metal carbonyls, CO is bonded to a neutral metal atom. Each CO molecule contributes two electrons. For $Ni(CO)_4$, for example, the neutral Ni atom contributes 28 electrons and the four CO ligands contribute 8 electrons, so the EAN is 36, the atomic number of krypton. We can use the EAN rule to predict the formula of metal carbonyls, as is illustrated in Example 20.9.

EXAMPLE 20.9. Metal carbonyls and the EAN rule

Predict the formulas of the carbonyl complexes of Cr and Ru.

Solution. Let the chromium carbonyl complex be denoted $Cr(CO)_n$, where n is to be determined. In this complex,

Cr metal contributes 24 electrons
n CO ligands contribute $2n$ electrons $EAN = 24 + 2n$

To make the $EAN = 36$, the atomic number of Kr, n must be 6. The chromium carbonyl complex is, in fact, $Cr(CO)_6$.

For $Ru(CO)_n$

Ru metal contributes 44 electrons
n CO ligands contribute $2n$ electrons $EAN = 44 + 2n$

The formula is $Ru(CO)_5$, and the EAN is 54, the atomic number of Xe.

section 20.10
Magnetic Moments of Transition Metal Complexes

Diamagnetism and Paramagnetism

Magnetism in substances other than single atoms is primarily due to the **electron spin** or **intrinsic angular momentum**.* (Refer to Section 13.2.) The spin of electrons is quantized, and only two spin states are possible, $\alpha(\uparrow)$ or $\beta(\downarrow)$. If all the electrons in a molecule or ion have paired spins, the substance is **diamagnetic**. A diamagnetic substance is slightly repelled by an external magnetic field. Since each atomic orbital holds two electrons of opposite spin, substances with filled atomic orbitals are diamagnetic. All ions with rare gas electronic configurations, such as the alkali and alkaline earth ions, are diamagnetic. Because the covalent bond involves a shared pair of electrons with opposite spin, most substances are diamagnetic.

Substances with one or more unpaired electrons are **paramagnetic**. Paramagnetic substances are attracted to an external magnetic field, but lose their magnetism when removed from the field. The greater the number of unpaired electrons, the greater the attraction between the paramagnetic species and the magnetic field. The magnitude of the attraction or repulsion between an external magnetic field and any substance can be measured using a **Gouy balance**, described in Fig. 20.15. The repulsion of diamagnetic substances by a magnetic field is very weak when compared with the attraction experienced by paramagnetic substances. Substances with a strong attraction to an external magnetic field that retain a permanent magnetization when the field is removed are said to be **ferromagnetic**. Metallic iron and Fe_3O_4 are ferromagnetic.

Many of the ions of the transition metals are paramagnetic. Consider the first transition series, the 10 elements from Sc to Zn, atomic numbers 21 through 30. The five $3d$ atomic orbitals are being filled as we proceed through this series. In Section 13.5 we discussed the fact that when ions of the metals of the first transition series are formed, the $4s$ electrons are always lost first so that the valence electrons of these ions are $3d$ electrons.

Titanium, atomic number 22, has electronic configuration $(Ar)^{18}3d^24s^2$. To form the Ti^{3+} ion, the two $4s$ electrons and one of the $3d$ electrons are lost, and the

---

* In some substances the orbital angular momentum of the electrons contributes to the total magnetic moment.

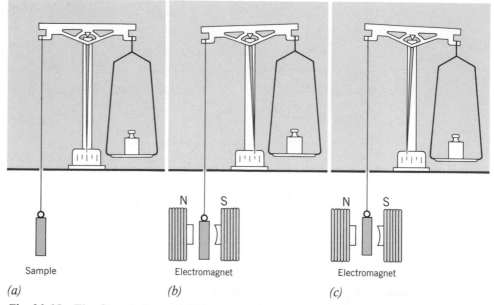

Fig. 20.15. The Gouy balance. *(a)* The sample is weighed in the absence of a magnetic field. *(b)* A paramagnetic sample is attracted by a magnetic field, and appears to weigh more when the field is turned on. *(c)* A diamagnetic sample is slightly repelled by a magnetic field and appears to weigh slightly less when the field is turned on.

electronic configuration of the Ti^{3+} ion is $(Ar)^{18}3d^1$, with one unpaired electron. The Ti^{3+} ion is paramagnetic, and has a **magnetic dipole moment** (like a miniature bar magnet) of 1.73 Bohr magnetons. The **Bohr magneton** (abbreviated BM) is the unit used to measure magnetic moments, and is named in honor of Niels Bohr.

Iron ($Z = 26$), has electronic configuration $(Ar)3d^64s^2$. The ferrous ion has configuration $(Ar)3d^6$, since the two $4s$ electrons are lost to form the Fe^{2+} ion. As there are five $3d$ orbitals, the six $3d$ electrons, in accordance with **Hund's Rule**, have the following configuration: $3d$ ↑↓ ↑ ↑ ↑ ↑. The Fe^{2+} ion therefore has four unpaired electrons, and is observed to be paramagnetic with a magnetic moment of 5.2 BM. If one more electron is lost, forming the ferric ion, Fe^{3+}, with configuration $3d$ ↑ ↑ ↑ ↑ ↑, there are five unpaired electrons and the magnetic moment is observed to be 5.9 BM.

Since the valence electrons of all the first-row transition metal ions are $3d$ electrons, it is common to omit the (Ar) and the 3 and to give the electronic configuration of Fe^{3+}, for example, simply as d^5. The electronic configurations of the more frequently encountered ions of the first transition series are listed in Table 20.5.

Spin-Only Magnetic Moments

For many of the first transition series ions, particularly those having five or fewer $3d$ electrons, the observed magnetic moment is due only to the unpaired electron spins,

Table 20.5. Electronic Configurations of Some First-Row Transition Metal Ions

Configurations	d^1	d^2	d^3	d^4	d^5	d^6	d^7	d^8	d^9	d^{10}
Ion	Ti^{3+}	Ti^{2+}	V^{2+}	Cr^{2+}	Mn^{2+}	Fe^{2+}	Co^{2+}	Ni^{2+}	Cu^{2+}	Zn^{2+}
		V^{3+}	Cr^{3+}	Mn^{3+}	Fe^{3+}	Co^{3+}				

and a theoretical calculation based on the principles of quantum mechanics yields the relation

$$\text{``spin-only'' magnetic moment} = \mu_{\text{spin-only}} = \sqrt{n(n+2)} \qquad (20\text{-}12)$$

where n is the number of unpaired electrons. Since the maximum number of unpaired d electrons is five, the following table can be prepared:

Number of unpaired electrons, n	1	2	3	4	5
$\mu_{\text{spin-only}}$(BM)	1.73	2.83	3.87	4.90	5.92

For those ions with more than five $3d$ electrons, the observed magnetic moment is usually slightly larger than that calculated from the "spin-only" formula, Eq. (20-12). This is due to a small contribution to the magnetic moment from the orbital angular momentum of the electrons (whose magnitude is determined by the value of the quantum number ℓ). For the first transition series ions, the orbital contribution to the magnetic moment is about 0.2–0.4 BM. Thus the observed magnetic moment for Fe^{2+} is 5.2 BM rather than 4.9.

High-Spin and Low-Spin Complexes

When transition metal ions form complexes, it is experimentally observed that some of the complexes have the same magnetic moment as the free (uncomplexed) metal ion, and others have a smaller magnetic moment. Thus the complexes of many, but not all, of these ions can be divided into two groups: (1) **Low-spin complexes,** for which the number of unpaired electrons, and therefore the magnetic moment, is less than that of the free ion, and (2) **high-spin complexes,** for which the magnetic moment is the same as that of the free ion. Table 20.6 summarizes some of these experimental observations for octahedral complexes.

For transition metal ions with configuration $d^1, d^2, d^3, d^8, d^9, d^{10}$, all the octahedral complexes of a given metal ion have approximately the same magnetic moment; there are no "high-spin" versus "low-spin" complexes.

The Colors of Transition Metal Complexes

In the preceding sections we have discussed two striking features of coordination complexes: their geometries and their paramagnetism. A third observation readily made is that transition metal complexes frequently have distinctive colors. The deep royal blue color of $Cu(NH_3)_4^{2+}$ is used as an analytical test to detect the presence of Cu(II), and the blood-red color of $Fe(SCN)_6^{3-}$ is used as a test for the Fe(III) ion. Aqueous solutions of the alkali and alkaline earth metal ions are colorless (unless the anion is colored), but aqueous solutions of most of the transition metal ions are

Table 20.6. **Examples of Octahedral High-Spin and Low-Spin Complexes of d^4, d^5, d^6, and d^7 Metal Ions of the First Transition Series**

Free ion	n	Low-Spin Complex	μ (BM)	High-Spin Complex	μ (BM)
Mn(III) d^4	4	$Mn(CN)_6^{3-}$	3.0	$Mn(C_2O_4)_3^{3-}$	4.9
Fe(III) d^5	5	$Fe(CN)_6^{3-}$	1.9	FeF_6^{3-}	5.9
Co(III) d^6	4	$Co(NO_2)_6^{3-}$	0.0	$CoCl_6^{3-}$	5.3
Co(II) d^7	3	$Co(NO_2)_6^{4-}$	1.9	$Co(NH_3)_6^{2+}$	4.8

colored: Cu(II) is aquamarine, Ni(II) is bright green, Ti(III) is reddish violet, and so on. The colors of these complexes are due to the fact that they absorb light of specific wavelengths in the visible region of the spectrum. Any theory of bonding in coordination complexes must be able to account satisfactorily for the geometry, the magnetic moment, and the **absorption spectrum** (the color) of each complex.

It was not until the 1950's that substantial progress was made in explaining all three of these properties on the basis of a comprehensive theory, and the development of this theory led to a great deal of intensive research in this area during the past twenty-five years. We shall now turn, therefore, to a discussion of current theoretical approaches to bonding in coordination complexes.

section 20.11
Theories of Bonding in Coordination Complexes

The simplest theoretical explanation of the magnetic properties and absorption spectra of transition metal complexes with specific geometries is provided by the **crystal field theory (CFT).** In order to understand crystal field theory, you should have a clear picture of the shape and orientation of the five d atomic orbitals, illustrated in Fig. 13.8. Two of these five orbitals have lobes of electron density with their directions of maximum extent along the coordinate axes; these are $d_{x^2-y^2}$ and d_{z^2}. The other three d orbitals have four lobes of electron density that extend along axes at 45° to each of the coordinate axes; these are called d_{xz}, d_{yz}, and d_{xy}.

Crystal Field Theory for Octahedral Complexes

Crystal field theory focuses on the electrostatic interaction between the ligands and the valence electrons of the central metal ion. There is, of course, an overall attraction between metal and ligands, but there are also repulsions between the negative ligands and the valence electrons of the metal. Let us consider an octahedral complex with six identical ligands located at equal distances along the $\pm x$, $\pm y$, and $\pm z$ axes, as illustrated in Fig. 20.16.

Because ligands are either anions or dipolar molecules like H_2O or NH_3 with their negative ends pointing toward the metal ion, a simple electrostatic picture represents the six ligands as six negative charges located along the coordinate axes, equidistant from the central metal atom. Most of the electron density of an electron in the $d_{x^2-y^2}$ or d_{z^2} atomic orbital is therefore pointing directly at the negative charge of the ligand. There will therefore be a repulsion between the negative ligand and an electron in the $d_{x^2-y^2}$ or d_{z^2} orbital of the metal. The effect of this repulsion is to increase the energy of these two atomic orbitals. An electron in the d_{xy}, d_{xz}, or d_{yz} orbitals, on the other

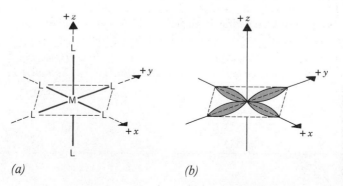

(a)   *(b)*

Fig. 20.16. *(a)* The placement of six ligands along the coordinate axes in an octahedral ligand field. *(b)* The $d_{x^2-y^2}$ atomic orbital drawn on the same axes.

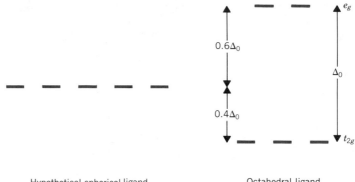

Free metal ion. The five d atomic orbitals are degenerate.

Hypothetical spherical ligand field. All five atomic orbitals are degenerate, but higher in energy than in the free ion.

Octahedral ligand field. Five-fold degeneracy is partially removed.

Fig. 20.17. The effect of an octahedral ligand field: The repulsion between the six negative ligands and the electrons in the d atomic orbitals of the metal raises the energy, and splits the five-fold degenerate energy level into two energy levels, one doubly degenerate and one with three-fold degeneracy.

hand, is directed in between the ligands. There will be a repulsion between electrons in these atomic orbitals and the ligands, but this repulsion will not be nearly as large as the repulsion between the ligands and electrons in the $d_{x^2-y^2}$ or d_{z^2} orbitals.

The result is that the presence of six ligands in an octahedral configuration partially removes the degeneracy of the five d atomic orbitals. We recall from Section 13.1 that for a free atom or ion in the absence of an electric or magnetic field, the five d orbitals are **degenerate,** that is, equal in energy. The six ligands, arranged octahedrally around the central metal ion, produce an electric field that partially removes the five-fold degeneracy.

In the presence of an octahedral ligand field, two of the five d orbitals ($d_{x^2-y^2}$ and d_{z^2}) are higher in energy than the other three (d_{xy}, d_{xz}, and d_{yz}). The $d_{x^2-y^2}$ and d_{z^2} orbitals are referred to collectively as the e_g **orbitals,** and the d_{xy}, d_{xz}, and d_{yz} orbitals are called the t_{2g} **orbitals.***

It is important to remember that all five d orbitals are higher in energy in the presence of the six ligands than when the ion is free. We can imagine a sphere of negative charge with a magnitude equal to that of the six ligands surrounding the metal ion. Such a spherical ligand field would repel the electrons in the five d orbitals equally, and would not disturb the five-fold degeneracy. The octahedral ligand field, however, produces a splitting into two levels. An energy level diagram illustrating these two situations is shown in Fig. 20.17.

The separation between the e_g and t_{2g} levels in an octahedral field is denoted Δ_0, and is called the **crystal field splitting.** The splitting into e_g and t_{2g} levels is not symmetric; the t_{2g} level lies $0.4\Delta_0$ below the average energy of the five d orbitals, and the e_g level lies $0.6\Delta_0$ above the average. When the d orbitals are occupied the gain in energy (compared to the average) that is achieved by preferential filling of the lower lying t_{2g} level is called the **crystal field stabilization energy (CFSE).** Each electron occupying a t_{2g} orbital is stabilized by an amount of energy equal to $0.4\Delta_0$, while an electron in an e_g orbital is destabilized by $0.6\Delta_0$.

---

* The symbols e_g and t_{2g} are derived from a mathematical treatment of symmetry properties of these orbitals called group theory.

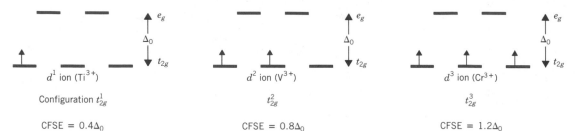

Fig. 20.18. The electronic configuration of d^1, d^2, and d^3 ions in an octahedral ligand field. The magnitude of Δ_0 depends on the ligand and on the metal ion.

High-Spin versus Low-Spin Octahedral Complexes

For ions with configurations d^1, d^2, and d^3, there is only one possible way to fill the orbitals in the presence of an octahedral ligand field. These configurations are depicted in Fig. 20.18, and the crystal field stabilization energy for each is given.

When we consider a d^4 ion such as Cr^{2+} or Mn^{3+}, there are two possible configurations, because there are two factors that tend to stabilize the energy of the four electron system. One of these factors is the stabilization achieved if all four electrons are in different orbitals with parallel spin, because the electrons are then as far apart from one another as possible. If electrons pair up in the same orbital, this stabilization is lost. The energy that is needed to pair the electrons in one orbital is called the **pairing energy.**

The other factor is the crystal field stabilization energy, the lowering of energy that is achieved by having all four electrons in the t_{2g} level. Which of these two factors is more important depends on the magnitude of Δ_0. If Δ_0 is small (**weak ligand field**), the gain in crystal field stabilization energy is not sufficient to overcome the loss in stability due to electron pairing. It is then energetically favorable for the four electrons to occupy different atomic orbitals with parallel spin. The net result is a high-spin complex. On the other hand, if Δ_0 is large (**strong ligand field**), then it is energetically favorable for the four electrons to expend the pairing energy in order to gain the crystal field stabilization energy. These two possibilities are depicted in Fig. 20.19.

In the view of crystal field theory, the distinction between low-spin and high-spin complexes is simply the result of a difference in the magnitude of the crystal field splitting, Δ_0. A large crystal field splitting forces electrons to pair up in the lower lying t_{2g} level, and results in a low-spin complex. A small crystal field splitting, on the other

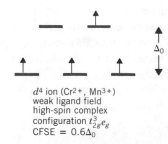

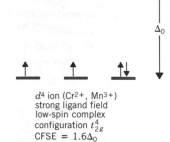

d^4 ion (Cr^{2+}, Mn^{3+})
weak ligand field
high-spin complex
configuration $t_{2g}^3 e_g$
CFSE = $0.6\Delta_0$

d^4 ion (Cr^{2+}, Mn^{3+})
strong ligand field
low-spin complex
configuration t_{2g}^4
CFSE = $1.6\Delta_0$

Fig. 20.19. Low and high-spin complexes of d^4 ions in an octahedral ligand field.

↑   ↑ e_g

↑↓   ↑↓   ↑↓ t_{2g} **Fig. 20.20.** The only configuration possible for an octahedral complex of a d^8 ion.

hand, does not produce sufficient crystal field stabilization energy to cause the electrons to lose the stabilization achieved when all the electrons have parallel spin. A small value of Δ_0, therefore, results in a high-spin complex. The net crystal field stabilization energy of a high-spin d^4 complex is $3(0.4\Delta_0) - 0.6\Delta_0 = 0.6\Delta_0$.

The occurrence of high- and low-spin octahedral complexes of metal ions with d^5, d^6, and d^7 configurations can be similarly explained, as is illustrated by Example 20.10.

EXAMPLE 20.10. Low versus high-spin d^6 octahedral complexes

The $Co(CN)_6^{3-}$ ion is not paramagnetic; its magnetic moment is zero. The CoF_6^{3-} ion, on the other hand, has a magnetic moment of 5.3 BM. Account for the difference between the magnetic properties of these two octahedral complexes using crystal field theory.

Solution. Co(III) is a d^6 ion. In an octahedral field the five d orbitals are split into two levels, the t_{2g} and the e_g. The CN^- ligands produce a large separation between the t_{2g} and e_g levels, and all six d electrons of the metal occupy the lower lying t_{2g} level and are completely paired. Since there are no unpaired electrons, the magnetic moment is zero. The F^- ligands cause only a small separation of the t_{2g} and e_g levels, and the six d electrons occupy both levels. There are four unpaired electrons, as shown in the diagram, and the magnetic moment is slightly larger than 4.9 BM, the value predicted by the "spin-only" formula, Eq. (20-12), because of a small contribution from the orbital angular momentum of the electrons.

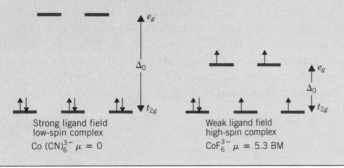

Strong ligand field
low-spin complex
$Co(CN)_6^{3-}$ $\mu = 0$

Weak ligand field
high-spin complex
CoF_6^{3-} $\mu = 5.3$ BM

For d^8, d^9, and d^{10} metal ions there is only one way to fill the t_{2g} and e_g levels, so that there are no "low-spin" versus "high-spin" octahedral complexes. Whether Δ_0 is large or small, there is only one possible configuration. For a d^8 ion, that configuration is depicted in Fig. 20.20.

The Spectrochemical Series

Certain ligands always produce strong ligand fields, while other ligands always cause weak ligand fields. We can arrange ligands in the order of the magnitude of the splitting they produce for a given metal ion and a given geometry. This arrangement is called the **spectrochemical series** and is shown below:

$$I^- < Br^- < Cl^- < F^- < OH^- < C_2O_4^{2-} \sim H_2O < :NCS^- < py \sim NH_3 < en < bipy < NO_2^- < CN^- < CO$$

weak ligand fields ——————————————————————→ strong ligand fields

Thus we note, in Table 20.6, that CN^- and NO_2^- ligands produce low-spin octahedral complexes, whereas F^- and Cl^- result in high-spin complexes.

The value of the crystal field splitting depends not only on the ligand but also on the charge on the central metal ion. The higher the positive charge on the metal ion, the more closely the ligands are pulled in to the metal, the greater the interaction between the ligands and the d electrons of the metal, and the greater the difference in energy between the t_{2g} and the e_g levels. Consider the $Co(NH_3)_6^{3+}$ and $Co(NH_3)_6^{2+}$ ions, for example. Ammonia is a ligand in the middle of the spectrochemical series, and some NH_3 complexes are high spin, while others are low spin. In fact, the Co(II) complex, $Co(NH_3)_6^{2+}$, is a paramagnetic high-spin complex, whereas $Co(NH_3)_6^{3+}$ is a diamagnetic ($\mu = 0$) low-spin complex. The higher charge on the Co^{3+} ion results in a greater value of Δ_0 for $Co(NH_3)_6^{3+}$ than for $Co(NH_3)_6^{2+}$. There are six d electrons for Co^{3+}, and they fill the t_{2g} level of $Co(NH_3)_6^{3+}$, so that there are no unpaired electrons.

You will recall from Section 20.4 that Cr(III) and Co(III) form inert complexes. These are octahedral complexes with the t_{2g} level either half-full (configuration t_{2g}^3 for Cr^{3+}) or completely filled (configuration t_{2g}^6 for Co^{3+}), and these configurations confer exceptional stability. The crystal field stabilization energy for the t_{2g}^6 configuration is $6(0.4\Delta_0) = 2.4\Delta_0$, and is the maximum CFSE possible for octahedral complexes. The activated complex in a ligand substitution reaction is a configuration that includes both the ligand that is leaving and the ligand that will be bound to the metal in the product. There is a large loss in crystal field stabilization energy on formation of the activated complex for octahedral complexes with either the t_{2g}^3 or the t_{2g}^6 configuration, and thus the activation energy is very large, and the reaction occurs very slowly. All complexes with less than three d electrons, as well as all complexes with electrons in the e_g orbitals, are labile.

If we compare the magnitude of the crystal field splitting for complexes of the same ligand with metal ions of the same group and in the same oxidation state, but in different rows of the periodic table, we find that Δ_0 increases by about 25–50% on going from the first transition series to the second, and by another 25–50% on going from the second transition series to the third. This has been attributed to the fact that for the third transition series the valence electrons are $5d$ electrons, which extend farther out from the nucleus than the $4d$ or $3d$, and therefore interact more strongly with the ligands.

Absorption Spectra

One of the most striking features of transition metal complexes is their color. Substances are colored when they absorb light of a particular wavelength (or frequency) in the visible region and transmit other wavelengths. Consider the $Ti(H_2O)_6^{3+}$ ion. It appears reddish violet because it absorbs light of frequency in the yellow-green region and transmits light in the red and blue regions. The visible spectrum is shown in Fig. 20.21.

When a sample absorbs light in the visible region, the color we see is the sum of the transmitted frequencies. If a sample absorbs only green, it appears red to us. Green and red are **complementary colors,** as are orange and blue, and yellow and violet. If a

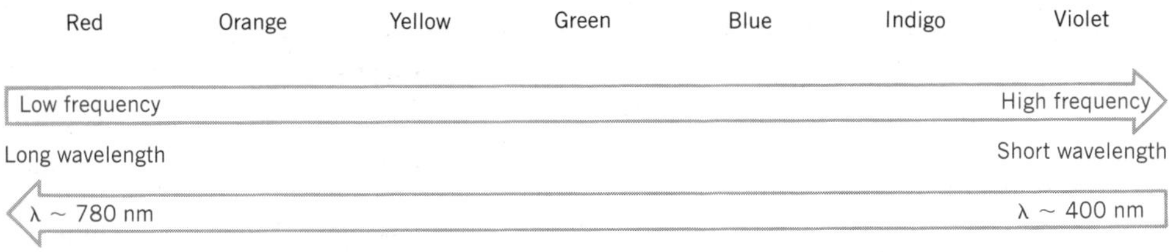

Red	Orange	Yellow	Green	Blue	Indigo	Violet

Low frequency → High frequency

Long wavelength → Short wavelength

← $\lambda \sim 780$ nm → $\lambda \sim 400$ nm

Fig. 20.21. The visible region of the spectrum.

sample absorbs only one color, the color we see will be its complementary color. The $Ti(H_2O)_6^{3+}$ ion absorbs both green and yellow, and so appears red violet.

Since Ti(III) is a d^1 ion, the electronic configuration of any octahedral Ti(III) complex is t_{2g}^1, as shown in Fig. 20.18. The magnitude of Δ_0, however, varies with the ligand. The light that is absorbed is used in the transition of the electron from the t_{2g} to the e_g level, and therefore the frequency of the light absorbed is related to the energy separation Δ_0 by the **Bohr frequency condition**

$$\Delta_0 = \text{energy separation} = h\nu \qquad (20\text{-}13)$$

where h is **Planck's constant**.

The relation between the frequency and the wavelength, $\nu\lambda = c$, where c is the speed of light, enables us to calculate Δ_0 in joules from a measurement of either the frequency or the wavelength of the absorbed light.

The absorption spectrum of the $Ti(H_2O)_6^{3+}$ complex ion is shown in Fig. 20.22. You will notice that the absorption band in the visible region is broad. This is principally because the crystal field splitting depends on the distance between the central metal ion and the ligands, and there is considerable fluctuation in this distance due to vibrational motion of the atoms. The maximum of the absorption band in the visible region is observed to occur at a wavelength $\lambda = 493$ nm, which is in the yellow-green region. Very frequently, instead of reporting the wavelength itself, the reciprocal of the wavelength, $1/\lambda$ (the wavenumber), is given. For the $Ti(H_2O)_6^{3+}$ ion

$$\frac{1}{\lambda} = \frac{1}{4.93 \times 10^{-7} \text{ m}} = 2.03 \times 10^6 \text{ m}^{-1} = 2.03 \times 10^4 \text{ cm}^{-1}$$

The crystal field splitting, Δ_0, for $Ti(H_2O)_6^{3+}$ is commonly stated to be 20,300 cm^{-1}. The splitting, Δ_0, is, of course, an energy difference and should be reported in joules, but since it is straightforward to calculate the energy separation from the wavelength or the frequency of the transition, any one of these quantities may be cited as Δ_0. You can always tell from the units what quantity is actually reported. Thus if Δ_0 is given as 20,300 cm^{-1}, this means the reciprocal wavelength, $1/\lambda$, is 2.03×10^4 cm^{-1}, and therefore the frequency

$$\nu = c/\lambda = (3 \times 10^{10} \text{ cm} \cdot \text{s}^{-1})(2.03 \times 10^4 \text{ cm}^{-1}) = 6.09 \times 10^{14} \text{ s}^{-1}$$

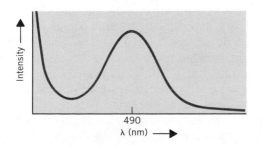

Fig. 20.22. The absorption spectrum of $Ti(H_2O)_6^{3+}$.

Table 20.7. **Absorption Spectra of Complexes of Cr(III)**

Ligand	Cl^-	H_2O	NH_3	CN^-
Δ_0 (kJ/mol)	163	208	258	314
Absorbed light	Near IR	Orange yellow	Blue violet	Violet
Transmitted light (observed color)	Green	Violet	Yellow	Yellow

and the energy separation

$$\Delta_0 = h\nu = (6.626 \times 10^{-34} \text{ J} \cdot \text{s})(6.09 \times 10^{14} \text{ s}^{-1})$$
$$= 4.04 \times 10^{-19} \text{ J}$$

The amount of energy required for 1 mol of $Ti(H_2O)_6^{3+}$ ions to undergo the transition from the t_{2g} level to the e_g level is

$$\Delta_0 = (4.04 \times 10^{-19} \text{ J})(6.022 \times 10^{23} \text{ mol}^{-1}) = 24.3 \times 10^4 \text{ J/mol} = 243 \text{ kJ} \cdot \text{mol}^{-1}$$

The spectrochemical series has been elucidated by examining spectra of various complexes of the same metal. Consider, for example, the data in Table 20.7 on the complexes of Cr(III).

The Cl^- ligands cause the weakest ligand field, that is, the smallest separation between t_{2g} and e_g. The light absorbed is therefore of the lowest frequency, at the very end of the red range, near the infrared region of the spectrum. When red light is absorbed, the color transmitted is green, and $CrCl_6^{3-}$ appears green to us. On the other hand, the CN^- ligands produce the strongest ligand field, that is, the largest value of Δ_0. The light absorbed is therefore of the highest frequency, in the violet region. When violet light is absorbed, the color transmitted is yellow, and the $Cr(CN)_6^{3-}$ ion is yellow. By measuring the frequency or wavelength of the absorbed light for complexes with various ligands, we can arrange the ligands in the proper order in the spectrochemical series.

Ligand Field Theory

Crystal field theory enables us to understand many of the properties of coordination complexes, but because it considers only the electrostatic interaction between the ligands and the valence electrons of the central metal ion, it is too simple a picture to describe the bonding in these complexes adequately. The bond between the central metal ion and its ligands is partly covalent, partly ionic, and a number of properties of complex ions cannot be explained without including the effects of the covalent character of the bond. A description of the bonding in coordination complexes that is better than crystal field theory is provided by **ligand field theory,** which combines the concepts of molecular orbital theory with those of crystal field theory.

Using only crystal field theory it is difficult to understand why CN^- and CO ligands produce a much stronger ligand field than Cl^- ions do. The answer is provided by ligand field theory: Those ligands that produce the strongest ligand fields (CN^-, NO_2^-, and CO) have empty antibonding π-MO's that extend close to the metal and can interact with the metal AO's. This is a metal to ligand(π) interaction, often called π back bonding. The covalent character of the bond between the ligand and the central metal ion is increased by this interaction, which lowers the energy of the t_{2g} orbital and therefore increases the value of Δ_0.

A detailed study of the application of molecular orbital theory to coordination compounds is beyond the scope of this text. Crystal field theory provides an adequate and useful explanation for most of the properties of coordination complexes, but it is important to remember that there is covalent character to the bond between metal and ligands, so that the description given by the crystal field theory is not entirely accurate.

Summary

When a metal is bonded to several surrounding **ligands,** the resulting structure is a **complex ion** or **coordination compound.** Ligands may be either anions or neutral, polar molecules. The bond between the metal and ligand is partly ionic, partly covalent.

The total number of points of attachment by all ligands to the metal ion is the **coordination number** of the cation. The most common coordination numbers are 4 and 6, but there are known examples of complexes with all coordination numbers between 2 and 12. Ligands that bond to a single metal ion may be **monodentate, bidentate,** or **polydentate,** depending on the number of ligand atoms that bond to the metal. A complex in which a multidentate ligand bonds to a metal ion with two or more points of attachment is called a **chelate.** A ligand may also bond to two different metal ions; it is then called a **bridging ligand.**

Ligands are **Lewis bases,** that is, they have a lone pair of electrons they can share with the metal. A metal ion with a vacant atomic orbital that can accept (and share) a pair of electrons is a **Lewis acid.**

Hydrated cations are acidic if the metal – oxygen bond is strong and a proton of the bonded water molecule can be donated to a solvent molecule. Small, highly charged cations that do not have the electronic configuration of one of the rare gases (and therefore have one or more vacant atomic orbitals) form the strongest metal – oxygen bonds. Aqueous solutions of these cations (for example, Fe^{3+}, Cr^{3+}, Tl^{3+}, and Hg^{2+}) are acidic. Such cations also form many coordination compounds.

Complexes may be classified as either **labile** or **inert.** Labile complexes undergo ligand exchange reactions rapidly. Inert complexes undergo ligand exchange reactions slowly. Octahedral complexes of tripositive cations with either three or six d electrons [$Cr(III)$ and $Co(III)$] are inert, as are square-planar $Pt(II)$ complexes. Early studies by Werner of inert complexes led to great advances in our knowledge of the geometry and structure of complexes. The measurement of either the freezing point depression or the molar conductivity of a dilute aqueous solution of an inert complex enables us to determine the number of ions per formula unit. The detailed geometry of coordination compounds that can be crystallized can now be determined by the techniques of **X-ray diffraction,** a method that was not available to Werner.

Isomerism is a common phenomenon for coordination compounds. **Geometric isomers,** with different physical and chemical properties, are formed when the ligands occupy different relative positions around the metal ion. When identical ligands are adjacent to one another, they are said to be in *cis* positions; when they are opposite one another they occupy *trans* positions. *Cis – trans* isomerism is the most common form of geometric isomerism; another type is **linkage isomerism,** in which the same ligand can bond to the metal through two different atoms.

A pair of molecules are **enantiomers** or **optical isomers** if they are mirror images of one another, and are not superimposable. Optical isomers have identical chemical and physical properties; they differ only in the direction in which they rotate the plane of **plane-polarized light. Optically active** molecules lack a plane of symmetry.

In attempting to explain why certain cations have characteristic coordination numbers, N. V. Sidgwick proposed that stability is achieved when the metal is surrounded by the same number of electrons as one of the rare gases, and that the number of ligands necessary to achieve the **effective atomic number (EAN)** of a rare gas determines the coordination number of the metal. While the metal ion in a great many complexes does achieve the EAN of a rare gas, there are also many exceptions to Sidgwick's rule. Factors such as ligand size and structure affect the coordination number of the metal. The EAN rule is less likely to apply if the metal-ligand bond has a large amount of ionic character.

Many transition metal ions and their complexes have one or more unpaired electrons and are therefore **paramagnetic.** They are attracted to an external magnetic field, whereas substances with no unpaired electrons are slightly repelled by a magnetic field and are said to be **diamagnetic.**

Complexes of many transition metal ions may be divided into two categories: **low-spin complexes** and **high-spin complexes.** High-spin complexes have the same number of unpaired electrons as the free ion, and hence the same magnetic moment. Low-spin complexes have fewer unpaired electrons than the free ion, and hence a smaller magnetic moment.

The existence of low- and high-spin complexes is satisfactorily accounted for by **crystal field theory (CFT),** which describes the effect of the electrostatic interaction between the negative ligands and the electrons in the valence d orbitals of transition metal ions. For an octahedral ligand field, the five-fold degeneracy of the d atomic orbitals is partially removed, and there is a splitting into two levels, one doubly degenerate (e_g) of higher energy, and one triply degenerate (t_{2g}) of lower energy. The magnitude of the **crystal field splitting,** Δ_0, determines whether the electrons will occupy only the t_{2g} orbitals (with consequent pairing of some spins that are unpaired in the free ion) or will occupy all five orbitals.

Another feature of complex ions that can be accounted for by crystal field theory is the color of these ions, or more accurately, the **absorption spectrum** in the visible region. For an octahedral complex, the color is due to the frequencies absorbed when an electron makes the transition from the t_{2g} to the e_g level. By measuring the frequency of the light absorbed and using the **Bohr frequency condition,** the energy separation between the t_{2g} and e_g levels can be determined. By comparing Δ_0 for complexes of the same geometry with different ligands bonded to the same cation, the ligands can be arranged in a **spectrochemical series.** The halide ligands, I^-, Br^-, Cl^-, and F^-, produce a small crystal field splitting (a weak ligand field). Complexes with halide ligands are therefore usually high-spin complexes. Ligands that produce a large cyrstal field splitting (a strong ligand field) are CN^-, NO_2^-, and CO. Complexes with these ligands will therefore be low-spin complexes.

Exercises

Section 20.1 and 20.2

1. What is the coordination number of the central metal ion in each of the following compounds?   (a) $Ni(en)_2Cl_2$ (b) $[Co(NH_3)_6]SO_4$ (c) $Fe(CO)_5$ (d) $Cu(acac)_2$
(e) $K_3[Cr(C_2O_4)_3]$ (f) $Na[Co(EDTA)]$

2. What is the oxidation state of the metal ion in each of the following complexes?
(a) $Cr(H_2O)_4Br_2^+$ (b) AgF_4^- (c) $[Ru(bipy)_3]^{2+}$ (d) $Ni(CN)_5^{3-}$ (e) $AuCl_2^-$
(f) $[Co(NH_3)_2(CN)_4]^-$ (g) $[Rh(NH_3)_5Cl]Cl_2$

3. Why does NH_3 form coordination complexes whereas the isoelectronic species CH_4 does not?

Section 20.3

4. Which of the following cations is a Lewis acid? Explain your answers. (a) Ba^{2+} (b) Pd^{2+} (c) Tl^{3+} (d) Al^{3+} (e) Zr^{2+}

5. Which of the following hydrated metal ions is acidic in aqueous solution? (a) $Rh^{3+}(aq)$ (b) $Hg^{2+}(aq)$ (c) $Rb^+(aq)$ (d) $Fe^{3+}(aq)$ (e) $Co^{2+}(aq)$ (f) $Sr^{2+}(aq)$

6. Explain why aqueous solutions of Ba^{2+} are not acidic, but aqueous solutions of Be^{2+} are acidic.

7. Explain why K_a for $Cr^{3+}(aq)$ is larger than K_a for $Cu^{2+}(aq)$.

8. Explain why aqueous solutions of Sr^{2+} are not acidic, but aqueous solutions of Sn^{2+} are acidic.

Section 20.4

9. Write balanced net ionic equations for the reaction that occurs when each of the following pairs of reagents are mixed. If no reaction occurs write NR.
   (a) $Zn(OH)_4^{2-}$ and HCl
   (b) $Cr^{3+}(aq)$ and excess NH_3
   (c) $Co(NH_3)_6^{3+}$ and H_2SO_4
   (d) $Zn^{2+}(aq)$ and excess NH_3
   (e) $Ni(NH_3)_6^{2+}$ and $HClO_4$
   (f) $AgCl(s)$ and excess NH_3
   (g) $Ba(OH)_2(s)$ and excess NaOH
   (h) $Al(OH)_3(s)$ and excess NaOH

10. Which of the following complexes is expected to be inert to ligand substitution?
   (a) $Ni(en)_3^{2+}$ (b) $Mg(EDTA)^{2-}$ (c) $Rh(NH_3)_6^{3+}$ (d) $Sc(H_2O)_6^{3+}$ (e) $Co(NO_2)_6^{3-}$ (f) $Cr(H_2O)_6^{2+}$ (g) $Cr(NH_3)_6^{3+}$ (h) $Al(OH)_4^-$ (i) FeF_6^{3-}

Section 20.5

11. A compound has the empirical formula $CoCl_3 \cdot 5NH_3$. When an aqueous solution of this compound is mixed with excess silver nitrate solution, 2 mol of AgCl precipitate per mole of compound. The molar conductance of a 0.001 M solution of this compound is 245 ohm^{-1}. When excess sulfuric acid is added to this compound, no NH_4^+ ions are detected in the resulting solution. Deduce the correct formula of this compound.

12. A rose-colored compound has the empirical formula $CoCl_3 \cdot 5NH_3 \cdot H_2O$. Two moles of this compound react with concentrated sulfuric acid to yield HCl(g) and 1 mol of a new compound with empirical formula $Co_2(SO_4)_3 \cdot 10NH_3 \cdot 5H_2O$. When this new compound is dried at room temperature, it loses 3 mol of water per mole of $Co_2(SO_4)_3 \cdot 10NH_3 \cdot 5H_2O$. State the significance of each observation and deduce the formula of the complex in the first compound.

13. A 0.0400 m solution of $K_2[Fe(CN)_5NO]$ freezes at -0.206 °C. Calculate the observed value of v in Eq. (6-14b). How many ions are in solution per $K_2[Fe(CN)_5NO]$ unit? What are these ions? Explain why the observed v is not the same as the number of ions per formula unit.

14. Dilute equimolar solutions of each of the following compounds are prepared. Arrange the compounds in order of increasing molar conductivity of these solutions.
   (a) $Na_3Co(NO_2)_6$ (b) $K[Co(EDTA)]$ (c) $Co(py)_3(CN)_3$ (d) $[Cr(NH_3)_5Cl]Cl_2$ (e) $[Pt(NH_3)_6]Br_4$

Section 20.6

15. Name the following compounds: (a) $[Co(NH_3)_4(NO_2)_2]NO_2$ (b) $Pt(en)Cl_4$ (c) $Na_3[Ag(S_2O_3)_2]$ (d) $K_3[Co(NO_2)_6]$ (e) $Na_2[Zn(EDTA)]$ (f) $[Ru(en)_2Br_2]Br$

16. Write the formulas of each of the following compounds: (a) cesium fluorotrichloroiodate(III)   (b) tris(ethylenediamine)nickel(II) nitrate   (c) potassium dibromodioxalatocobaltate(III)   (d) sodium hexacyanoferrate(II)   (e) chloropentaamminechromium(III) sulfate

Section 20.7 and 20.8

17. Draw the structures of the *cis* and *trans* isomers of the following complexes: (a) dichlorobis(ethylenediamine)rhodium(III) ion   (b) dinitrotetraamminecobalt-(III) ion   (c) diglycinatopalladium(II)

18. For which of the following complexes are optical isomers possible? Explain your answers and draw structures of the enantiomers. (a) $[Cr(NH_3)_4(C_2O_4)]^+$
    (b) $[Cr(NH_3)_2(C_2O_4)_2]^-$ (c) $[FeCl_5F]^{3-}$ (d) $[Co(en)_2NH_3Cl]^{2+}$ (e) $Cr(C_2O_4)_3^{3-}$

19. Draw all geometric and optical isomers of the following complexes: (a) dicarbonatodiamminecobalt(III) ion   (b) tribromotripyridineplatinum(IV) ion   (c) dithiocyanatoethylenediamineplatinum(II) ion

Section 20.9

20. Calculate the effective atomic number of the metal in the following complexes, and compare it with the atomic number of the nearest rare gas. What do you conclude about the validity of Sidgwick's EAN rule? (a) $Fe(CN)_6^{4-}$ (b) $AuCl_2^-$ (c) $Al(C_2O_4)_3^{3-}$
    (d) CdI_4^{2-} (e) $[Cr(H_2O)_4(NH_3)_2]^{2+}$ (f) $Co(NO_2)_6^{3-}$ (g) $Fe(CO)_5$

Section 20.10

21. Classify each of the following complexes as either high spin or low spin. Explain your answers.
    (a) $Fe(H_2O)_6^{3+}$, $\mu = 5.94$ BM
    (b) $Co(CN)_6^{4-}$, $\mu = 1.9$ BM
    (c) $Mn(C_2O_4)_3^{3-}$, $\mu = 4.9$ BM
    (d) $Cr(Me_6tren)Cl^+$, $\mu = 4.85$ BM $\{Me_6tren = [(CH_3)_2NCH_2CH_2]_3N\}$
    (e) $Fe(NO_2)_6^{4-}$, $\mu = 0.0$ BM
    (f) $Co(H_2O)_6^{2+}$, $\mu = 4.6$ BM

22. For which of the following transition metal ions are low-spin complexes impossible? Explain your answers. (a) Zn^{2+} (b) Rh^{3+} (c) Zr^{2+} (d) Ag^+ (e) Mn^{3+} (f) Ru^{2+}

23. The observed magnetic moment of $Ni(H_2O)_6^{2+}$ is 3.2 BM. Compare this with the value predicted by Eq. (20-12) and explain why they are different.

Section 20.11

24. The $Fe(CN)_6^{3-}$ ion has a magnetic moment of 1.76 BM, while $Fe(H_2O)_6^{3+}$ has a magnetic moment of 5.94 BM. Explain the difference between the magnetic moments of these two octahedral complexes using crystal field theory.

25. Give the ground state *d*-electron configuration, the number of unpaired electrons, and the predicted spin-only magnetic moment of the following complexes. State whether or not an orbital contribution to the magnetic moment is expected. Give also the magnitude of the CFSE in terms of Δ_0. (a) $Cr(H_2O)_6^{2+}$ (b) $Co(en)_3^{2+}$ (high spin)
    (c) $Rh(C_2O_4)_3^{3-}$ (low spin)   (d) $Mn(CN)_6^{4-}$ (e) $Mn(NH_3)_6^{2+}$ (high spin)

26. Which complex has the larger crystal field splitting in each pair? Give a brief explanation for each answer.
    (a) $Mn(en)_3^{3+}$ or MnF_6^{3-} (b) $Co(H_2O)_6^{3+}$ or $Rh(H_2O)_6^{3+}$
    (c) PtI_4^{2-} or PtF_4^{2-} (d) $Cr(H_2O)_6^{2+}$ or $Cr(H_2O)_6^{3+}$
    (e) $Fe(CN)_6^{3-}$ or $Fe(SCN)_6^{3-}$

27. When excess concentrated ammonia solution is added to an aqueous solution containing $Ni(H_2O)_6^{2+}$, the color of the solution changes from bright green to a beautiful blue violet. Write a balanced equation for the reaction that occurs and explain the observed color change.

28. When excess 6 M HCl is added to a dilute aqueous solution of copper sulfate, the color of the solution changes from blue to green. Write a balanced equation for the reaction that occurs and explain the observed color change.

29. The Δ_0 values for the complexes $[Co(NH_3)_6]^{3+}$, $[Rh(NH_3)_6]^{3+}$, and $[Ir(NH_3)_6]^{3+}$ are, respectively, 23,000, 34,000, and 41,000 cm^{-1}. What are these three Δ_0 values expressed in kilojoules per mole?

30. Explain the difference in the color of aqueous solutions of the following ions:

$[Co(NH_3)_5(NO_2)]^{2+}$ yellow
$[Co(NH_3)_5(H_2O)]^{3+}$ red
$[Co(NH_3)_4Cl_2]^+$ violet

Multiple Choice Questions

1. Which of the following statements about the coordination number of a cation is TRUE?

(a) Most metal ions exhibit only a single, characteristic coordination number.
(b) The coordination number is equal to the number of ligands bonded to the metal atom.
(c) The coordination number is determined solely by the tendency to surround the metal atom with the same number of electrons as one of the rare gases.
(d) The most common coordination numbers are 4, 6, and 8.
(e) For most cations, the coordination number depends on the size, structure, and charge of the ligands.

2. The correctly balanced equation for the reaction of concentrated ammonia solution with a suspension of zinc(II) hydroxide is

(a) $Zn^{2+}(aq) + 4NH_3 \rightarrow Zn(NH_3)_4^{2+}$
(b) $Zn(OH)_2(s) + 4NH_3 \rightarrow Zn(NH_3)_4^{2+} + 2\,OH^-$
(c) $Zn(OH)_2(s) + 2\,OH^- \rightarrow Zn(OH)_4^{2-}$
(d) $Zn(OH)_2(s) + 2NH_4^+ \rightarrow Zn^{2+}(aq) + 2NH_4OH$
(e) $Zn^{2+}(aq) + 2NH_3 \rightarrow Zn(NH_3)_2^{2+}$

3. Which of the following cations is colorless?
(a) $Cu^{2+}(aq)$ (b) $Fe^{3+}(aq)$ (c) $Zn^{2+}(aq)$ (d) $Ni^{2+}(aq)$ (e) $Cr^{2+}(aq)$

4. Potassium hexacyanoferrate(II) is the compound
(a) $K_4[Fe(CN)_6]$ (b) $KFe(SCN)_4$ (c) $K_3[Fe(CN)_6]$ (d) $K_3[Fe(SCN)_6]$
(e) $K_4[Fe(NCO)_6]$

5. Which of the following complexes is diamagnetic?
(a) $Fe(CN)_6^{4-}$ (b) $Cu(NH_3)_4^{2+}$ (c) $Ti(H_2O)_6^{3+}$ (d) $Ni(en)_3^{2+}$ (e) $Co(py)_6^{2+}$

6. Concentrated ammonia solution is added to an aqueous solution of each of the following salts. In which one does a precipitate form?
(a) $Ni(NO_3)_2$ (b) $CuSO_4$ (c) $Mg(NO_3)_2$ (d) $CoSO_4$ (e) $ZnCl_2$

7. The instrument used to measure the optical activity of an enantiomer is a
   (a) potentiometer   (b) refractometer   (c) Gouy balance
   (d) UV spectrometer   (e) polarimeter

8. The crystal field stabilization energy for an octahedral complex with electronic configuration $t_{2g}^3 e_g^2$ is
   (a) 0   (b) $0.4\Delta_0$ (c) $1.2\Delta_0$ (d) $1.6\Delta_0$ (e) $2.0\Delta_0$

9. Which of the following cations is *not* paramagnetic?
   (a) $Sc^{3+}(aq)$ (b) $Ti^{3+}(aq)$ (c) $V^{3+}(aq)$ (d) $Cr^{3+}(aq)$ (e) $Mn^{3+}(aq)$

10. Which of the following complexes is not a chelate?
    (a) bis(dimethylglyoximato)nickel(II)
    (b) potassium tetrathiocyanatoethylenediaminechromate(III)
    (c) carbonatotetraamminecobalt(III) nitrate
    (d) *trans*-diglycinatopalladium(II)
    (e) sodium tetracyanodithiocyanatoferrate(III)

11. Which of the following complexes exhibits optical isomerism?
    (a) *trans*-dithiocyanatotetraamminechromium(III) ion
    (b) *cis*-dicarbonatodiamminecobaltate(III) ion
    (c) *trans*-dicarbonatodiamminecobaltate(III) ion
    (d) *cis*-diglycinatoplatinum(II)
    (e) *trans*-diglycinatoplatinum(II)

12. A 0.020 *m* solution of each of the following compounds is prepared. Which solution would you expect to freeze at $-0.142\ °C$?
    (a) $Na[Co(EDTA)]$ (b) $[Cr(py)_5Cl]Cl_2$ (c) $[Cr(NH_3)_6]Cl_3$ (d) $Co(py)_3Cl_3$
    (e) $[Co(en)_2Cl_2]Cl$

13. Of the following complexes, the one with the largest value of the crystal field splitting, Δ_0, is
    (a) $Fe(H_2O)_6^{2+}$ (b) $Ru(H_2O)_6^{2+}$ (c) $Fe(NH_3)_6^{3+}$ (d) $[Ru(CN)_6]^{3-}$
    (e) $[Fe(CN)_6]^{3-}$

14. If $K_{sp}(AgBr) = 7.7 \times 10^{-13}$ and the formation or stability constant of $Ag(S_2O_3)_2^{3-}$ is 1.0×10^{13}, the equilibrium constant for the reaction

$$AgBr(s) + 2S_2O_3^{2-} \rightleftharpoons Ag(S_2O_3)_2^{3-} + Br^-$$

    is
    (a) 7.7×10^{-26} (b) 1.3×10^{-13} (c) 1.3×10^{-1} (d) 7.7   (e) $7.7 \times 10^{+13}$

15. The crystal field stabilization energy of a low-spin octahedral complex of a d^7 ion is
    (a) $1.6\Delta_0$ (b) $1.8\Delta_0$ (c) $2.0\Delta_0$ (d) $2.2\Delta_0$ (e) $2.4\Delta_0$

16. The crystal field stabilization energy of a high-spin octahedral complex of a d^7 ion is
    (a) $0.4\Delta_0$ (b) $0.8\Delta_0$ (c) $1.2\Delta_0$ (d) $1.6\Delta_0$ (e) $2.0\Delta_0$

17. The effective atomic number of nickel in $Ni(CN)_4^{2-}$ is
    (a) 26   (b) 28   (c) 32   (d) 34   (e) 36

18. What is the ratio of uncomplexed to complexed Zn^{2+} ion in a solution that is 10 *M* in NH_3 if the stability constant of $Zn(NH_3)_4^{2+}$ is 3×10^9?
    (a) 3×10^{-9} (b) 3×10^{-11} (c) 3×10^{-12} (d) 3×10^{-13} (e) 3×10^{-14}

19. The formula of a carbonyl complex of cobalt, $(CO)_n Co—Co(CO)_n$, in which there is a single covalent Co—Co bond is
    (a) $Co_2(CO)_4$ (b) $Co_2(CO)_6$ (c) $Co_2(CO)_8$ (d) $Co_2(CO)_{10}$ (e) $Co_2(CO)_{12}$

20. If excess $AgNO_3$ solution is added to 100.0 mL of a 0.0240 F solution of dichlorobis(ethylenediamine)cobalt(III) chloride, how many moles of $AgCl$ should be precipitated?

(a) 0.00120   (b) 0.00160   (c) 0.00240   (d) 0.00480   (e) 0.00720

Problems

20.1. The simplest empirical formula of a compound is $Pt(NH_3)_2Cl_2$. If excess silver nitrate solution is added to an aqueous solution of this compound, no $AgCl$ precipitates. If sulfuric acid is added to this compound, no HCl is found in the gas phase, and NH_4^+ ions are not detected in the resulting solution. The molar conductivity of a dilute aqueous solution of this compound is about the same as that of a solution of magnesium sulfate of the same concentration. Measurements of the freezing point of dilute aqueous solutions of this compound indicate that its molecular weight is close to 600. Deduce the formula of this compound.

20.2. Manganese carbonyl is a dimer, $(CO)_n Mn-Mn(CO)_n$. If each Mn atom has an EAN of 36, the atomic number of krypton, deduce the value of n. The Mn-Mn single bond is an ordinary covalent bond. Each Mn atom contributes one electron to the pair shared by them both.

20.3. At 25 °C, the stability constant for the formation of $Ag(NH_3)_2^+$

$$Ag^+(aq) + 2NH_3 \rightleftharpoons Ag(NH_3)_2^+(aq)$$

is $K_{stab} = 1.6 \times 10^7$. The solubility product of $AgCl$ is 1.7×10^{-10}.

(a) One drop of a 0.010 F $AgNO_3$ solution is added to a large excess of 2.00 F NH_3. What fraction of the added silver is in the form of $Ag^+(aq)$ when equilibrium is attained?

(b) Write a net ionic equation for the reaction between $AgCl(s)$ and aqueous ammonia, and calculate the equilibrium constant for the reaction you have written.

(c) Calculate the molar solubility of $AgCl$ in 2.00 F NH_3 at 25 °C.

20.4. Analysis of a compound shows that it is 46.2% Pt, 33.6% Cl, 16.6% N, and 3.6% H. The molar conductivity of a freshly prepared 0.001 M solution of this compound is 420 ohm^{-1}. Addition of excess $AgNO_3$ solution to 50.0 mL of a 0.0320 F solution of the compound precipitates 0.6879 g of $AgCl$. State the significance of each observation, write the chemical formula for the compound, give its IUPAC name, and show its probable geometry.

20.5. The relative size of cation and ligand is an important factor for determining the coordination number of the cation. If nothing but the *radius ratio, r_c/r_L*, determines the coordination number, simple geometry (assuming spherical ions) shows that to preclude direct contact between ligands, a linear complex with coordination number 2 will be formed if $r_c/r_L = 0.155$, a tetrahedral complex will be formed if r_c/r_L lies between 0.225 and 0.414, and a square-planar or octahedral complex will be formed if r_c/r_L lies between 0.414 and 0.732. The ionic radius of Zn^{2+} is 0.74 Å, of Cr^{3+} is 0.55 Å, and of both O^{2-} and OH^- is 1.40 Å. If spatial considerations alone determine the coordination number of the zincate ion and the chromite ion, can you predict whether the zincate ion is ZnO_2^{2-}, $Zn(OH)_4^{2-}$ or $Zn(OH)_6^{4-}$? Can you predict whether the chromite ion is CrO_2^-, $Cr(OH)_4^-$, or $Cr(OH)_6^{3-}$? Explain your answers. Does the EAN rule predict the same formula for each of these complexes that is predicted solely on the basis of the radius ratio? Explain.

20.6. Seven coordinate complexes are not common, but recently a number of seven coordinate complexes of Mo(II) have been synthesized. Offer a reason why, if you wanted to make a complex with coordination number 7, you might think of Mo(II) as a cation likely to be useful for achieving your goal.

20.7. Given the following data at 25 °C,

$$\Delta G^0 = -41.4 \text{ kJ} \cdot \text{mol}^{-1} \quad \text{for} \quad Ag^+(aq) + 2NH_3 \rightleftharpoons Ag(NH_3)_2^+(aq)$$
$$\Delta G^0 = -55.6 \text{ kJ} \cdot \text{mol}^{-1} \quad \text{for} \quad Ag^+(aq) + Cl^-(aq) \rightleftharpoons AgCl(s)$$

(a) Calculate ΔG^0 for $AgCl(s) + 2NH_3 \rightleftharpoons Ag(NH_3)_2^+(aq) + Cl^-(aq)$.
(b) Calculate K_{eq} for the reaction of part (a). If all substances are in their standard states, will this reaction proceed from left to right, as written? Explain your answer.
(c) Calculate ΔG for the reaction of part (a) when the ammonia concentration is 6.0 M and the concentrations of Cl^- and $Ag(NH_3)_2^+$ are 0.10 M. Will this reaction proceed spontaneously from left to right at these concentrations? Explain your answer.

20.8. The formation of a coordination complex with many ligands can be considered to occur in steps in which one ligand is added at a time. The equilibrium constant for each step is denoted by K_i. Thus to form a complex with n ligands, we could write n equilibrium constants, $K_1, K_2, \ldots, K_n$. The overall equilibrium constant for adding all n ligands to the cation is denoted β_n.
(a) Write expressions for the following equilibrium constants:
   (i) K_1 for the reaction of Mg^{2+} with ATP^{4-}.
   (ii) β_4 for the complex of Cu^{2+} and pyridine.
   (iii) K_2 for the reaction of Co^{3+} and ethylenediamine.
(b) Prove that for the formation of $Ag(NH_3)_2^+$ from Ag^+ and NH_3

$$\beta_2 = K_1 K_2$$

20.9. Zinc hydroxide is an insoluble white solid, but it is amphoteric and dissolves in both acids and bases. Write net ionic equations for the dissolution of $Zn(OH)_2$ in
(a) excess 6.0 F NaOH (b) concentrated NH_3 (c) excess HCl(aq)
(d) excess 2.0 F CH_3COOH

20.10. (a) There are five complex ions that can be made from Co(III) using ethylenediamine and CN^- as ligands, including geometrical isomers. Give the formulas of each of these complexes and draw structures of all geometrical isomers.
(b) Using NH_4^+ and CN^- to make neutral compounds, give the formula and name of five compounds that can be made from the five complex ions of part (a).

20.11. The Co(II) ion complexes with the anions of several amino acids. Both the glycine anion (gly) and the alanine anion (ala) have a -1 charge. For the formation of $Cogly^+$, $\log K_1 = 4.95$. For the formation of $Coala^+$, $\log K_1 = 4.83$. If a solution contains Co(II) ions and the two amino acid anions in a ratio of 2 gly to 3 ala, how will the cobalt be distributed between the two complexes?

20.12. Which of the following substances would you expect to be colored? Give a reason for each answer.
(a) La^{3+} (b) $Fe(SCN)_2^+$ (c) $MnO_2(s)$ (d) Ti^{4+} (e) $Ni(ClO_4)_2$ (f) $Zn(OH)_4^{2-}$
(g) $Cr(NH_3)_3Br_3$

20.13. Give brief explanations for the following observations.
(a) $Fe(NO_2)_6^{4-}$ is an inert complex while FeF_6^{4-} is labile.

(b) The $Co(H_2O)_6^{2+}$ ion is light red (pink), while the $CoCl_4^{2-}$ ion, which has tetrahedral geometry, is blue.

20.14. Solutions of $Fe(H_2O)_6^{2+}$ are green, while solutions of $[Fe(CN)_6]^{4-}$ are yellow. Reasoning from these observations, explain how you can predict the relative values of Δ_0 for these two complexes.

20.15. (a) Name the salt $K[Rh(py)_2(C_2O_4)_2]$.
(b) Label and draw all possible isomers of this salt.

20.16. Both Ni^{2+} and Mn^{2+} form insoluble hydroxides and it is not possible to separate Ni^{2+} from Mn^{2+} merely by controlling the pH. Both of these ions also form complexes with ethylenediamine. The coordination number of each cation is 4. The formation constant, β_2 (see problem 20.8 for definition) is 6.25×10^4 for $Mn(en)_2^{2+}$ and 4.8×10^{13} for $Ni(en)_2^{2+}$. The solubility product of $Mn(OH)_2$ is 1.6×10^{-13} and of $Ni(OH)_2$ is 2×10^{-15}.
(a) A solution is $0.10\ M$ in both Ni^{2+} and Mn^{2+}. Ethylenediamine is added in excess so that its concentration is $1.0\ M$. A solution of NaOH is now added dropwise. Assuming no volume change, show that $Mn(OH)_2$ will precipitate but $Ni(OH)_2$ will *not* precipitate, when after the addition of two drops, the $[OH^-] = 0.010\ M$.
(b) Calculate the $[Mn(en)_2^{2+}]$ in the solution when the $[OH^-]$ is $0.010\ M$. What important fact do these calculations demonstrate?

20.17. A complex $ML_4L'_2$, in which the L' ligands lie on the z axis and are at a greater distance from the metal, M, than the L ligands, which lie along the x and y axes, is said to have tetragonal geometry. The e_g level of octahedral geometry is split into two levels by a ligand field of tetragonal geometry, as is the t_{2g} level, so that there are four energy levels. The lowest level is doubly degenerate. Draw an energy level diagram showing the splitting of the five d orbitals in a tetragonal ligand field. Label each level with the symbol(s) of the d orbitals. Explain your reasoning in detail.

20.18. At 30 °C the equilibrium constant for the reaction

$$Al(OH)_3(s) + OH^- \rightleftharpoons Al(OH)_4^-$$

has been reported to be 13. If K_{sp} for $Al(OH)_3$ is 1.9×10^{-32} at this temperature, calculate K_{stab} for $Al(OH)_4^-$.

chapter 21 Properties and Structures of Metallic and Ionic Crystalline Solids

Sir William Henry Bragg (1862–1942) and his son Sir William Lawrence Bragg (1890–1971) were British physicists who developed the theory of X-ray diffraction. The fundamental equation used in the elucidation of crystalline structure by X-ray diffraction is known as "Bragg's Law." In 1915 the Braggs were jointly awarded the Nobel Prize for physics "for their services in the analysis of crystal structure by means of X-rays."

Crystalline solids are among nature's great works of art. The symmetry of crystalline structures delights the eye and is a reflection of the repeating, regular arrangements of the atoms, ions, or molecules of which the solid is composed. Because of the regularity of the internal structure of a crystal we can perform a variety of experiments involving interactions between crystals and directed beams of radiation, with electric and magnetic fields in specific directions. Analysis of the results of these experiments has provided much of our knowledge of bond distances and angles, molecular geometry, and of the sizes of atoms and ions.

section 21.1
Crystalline and Amorphous Solids

The great majority of solids are crystalline. The features of crystalline solids that we observe most readily are their characteristic geometries, rigidity, and incompressibility. Crystals are three-dimensional solids bounded by plane surfaces. The **interfacial angles,** that is, the angles of intersections of these plane surfaces, are characteristic of a given substance and are always the same. The sizes and shapes of crystals depend on factors connected with their growth: The degree of supersaturation of the solution in which the crystals form, the temperature, the pressure, the presence of other substances, and so on. But the **constancy of interfacial angles** for a given crystalline substance is always observed, regardless of the actual size or shape of the crystal.

The characteristic geometry of a crystal is associated with the regular arrangement, in a pattern that repeats periodically in three dimensions, of the atoms, ions, or molecules of which the substance is composed.

Pure crystalline solids have sharp melting points. When subjected to high pressures they are highly resistant to change of shape. Another important and striking property of crystalline solids is their **anisotropy.** The mechanical and electrical properties of most crystals depend, in general, on the direction along which they are measured. For instance, it is usually much easier to cleave or shear a crystal in some directions than in others. This is schematically illustrated in Fig. 21.1. You may have observed this type of anisotropy if you have ever examined crystals of mica or asbestos. Mica cleaves into thin sheets, and asbestos cleaves into long, rod-like pieces.

There are also **amorphous solids,** such as the glasses, rubber, and plastics. Many liquids can be **supercooled,** that is, brought to temperatures below their freezing points without crystallizing. As the temperature decreases they become more and more rigid, but their internal structure lacks the definite arrangement of atoms, ions, or molecules that characterizes crystalline solids. The difference between crystalline and amorphous solids is illustrated in Fig. 21.2. Amorphous solids do not have sharp melting points. When heated, they gradually soften over a fairly wide temperature

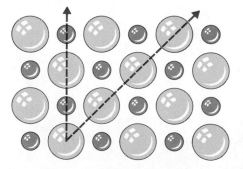

Fig. 21.1. Crystalline anisotropy. A two-dimensional array of two different spherical atoms. Shear stress along the indicated directions is different because the interactions of the constituents will differ as they move along these two directions. Thus resistance to cleavage along these two directions differs, and the crystal is anisotropic.

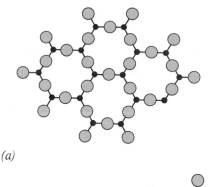

(a)

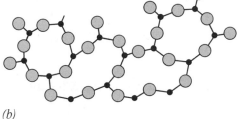

(b)

Fig. 21.2. Comparison of the internal structure of amorphous and crystalline solids. *(a)* Two-dimensional representation of the orderly arrangement of atoms in a crystalline solid. *(b)* Two-dimensional representation of the disorder in an amorphous (glassy) solid of the same substance.

range. Amorphous solids are **isotropic.** Their mechanical strength, electrical, and optical properties are the same in all directions. Liquids and gases are also isotropic; anisotropy is a property unique to crystalline solids. Because of the differences between the properties of amorphous and crystalline solids, many scientists reserve the term "solid" for crystalline solids only. Amorphous solids are then classified as supercooled liquids.

section 21.2
Crystal Defects

While we describe a crystal as a regular arrangement of atoms, ions, or molecules, a real crystal always has some defects. A common type of defect is a **vacancy:** A position in the crystal that should contain an atom or ion is vacant. Another type of defect is a **substitutional impurity:** A different atom or ion, often one of about the same size as the correct one, is incorporated into the crystal. Substances that invariably crystallize with significant numbers of such defects are **nonstoichiometric compounds.** (Refer to Section 2.10.) A third type of defect occurs when some atom or ion is located in between the regular positions, at an **interstitial site.** These defects are illustrated in Fig. 21.3. Sometimes defects are deliberately introduced into crystals by adding very small amounts of impurities, because the defects may change the mechanical or electrical properties of the crystals in desirable ways.

section 21.3
The Structures of Metals

Three quarters of all the elements in the periodic table are metals. The observable properties of metals that distinguish them from nonmetals are their luster, their high thermal and electrical conductivity, their malleability, and ductility. Not all metals possess all of these properties, and certainly not to the same degree. For instance,

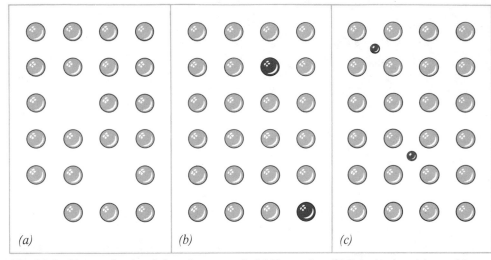

Fig. 21.3. Types of point defects in a crystal. *(a)* Vacancies. *(b)* Substitutional impurities. *(c)* Interstitial impurities.

while lead is soft and easily hammered into any desired shape, it is not a very good conductor of electricity.

A pure metal is a crystalline solid in which the metal atoms are closely packed in a repeating, three-dimensional array.

Hexagonal and Cubic Closest Packing

About 70% of all metals have structures that are described as **close-packed.** In close-packed structures, the metal atoms are considered to be spheres of equal size, with each sphere in contact with six others in a single layer, as shown in Fig. 21.4. If we want to pack spheres of identical diameter in such a way as to leave the minimum possible empty space, we can find two different ways to do it.

The two principal types of stacking of closest packed layers in metallic structures are called **hexagonal close packing** and **cubic close packing.** In both types, a second layer *(B)* is placed on top of the first layer *(A)* in such a way that each sphere in the second layer is in contact with three spheres in the first layer. Spheres of the second layer fit into depressions in the first layer, over the interstices between the spheres of the first layer. The difference in the two types of close-packed structures arises in the positioning of the third layer.

If each sphere in the third layer lies directly above a sphere in the first layer, the stacking arrangement is called **hexagonal close packing,** and can be symbolized as *ABABAB* ⋯ . Figure 21.5 illustrates hexagonal close packing. For an infinite number of layers, each sphere is in close contact with 12 others, 3 in the layer above it, 6 in its own layer, and 3 in the layer below it. Thus the number of closest neighbors of each

Fig. 21.4. Closest packing of spheres in a single layer. Each sphere is in contact with six others in the same layer.

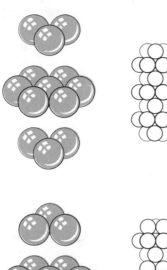

Fig. 21.5. The hexagonal close-packed crystal structure. On the left the structure is expanded to show clearly that the first and third layers are directly above one another and fit into the same interstices of the second layer. On the right is a side view of many layers with a double-headed arrow showing the repeat distance of the stacking pattern. In an infinitely extended structure, each atom has 12 nearest neighbors.

Fig. 21.6. The cubic close-packed crystal structure. On the left the structure is expanded to show that the first and third layers are not directly above one another. The repeat distance of the stacking pattern is longer than for the hexagonal close-packed structure. A cubic close-packed structure has a face-centered cubic unit cell. In an infinitely extended crystal, each atom has 12 nearest neighbors.

sphere, or the coordination number of each atom, is 12. Metals that crystallize in the hexagonal close-packed structure include Be, Mg, Ti, Zn, and Cd.

If the spheres in the first and third layers lie in different interstices of the second layer, rather than in the same interstices, the stacking pattern is called **cubic close packing**, and can by symbolized $ABCABCABC\cdots$. The third layer is not directly above either layer A or layer B. The repeat distance for identical layers is longer for cubic close packing than it is for hexagonal close packing. However, in an infinite number of layers of both hexagonal and cubic close-packed structures, the coordination number of any atom is 12. Cubic close packing is illustrated in Fig. 21.6. Metals that crystallize in the cubic close-packed structure include Al, Ca, Ni, Cu, Ag, Au, and Pt.

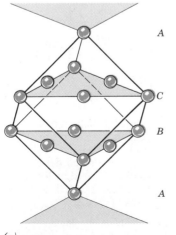

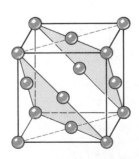

Fig. 21.7. The face-centered cubic (fcc) unit cell. *(a)* A unit cell standing on one corner to show the cubic close packing. *(b)* A unit cell resting on one face to show the face centering.

(a) *(b)*

The cubic close-packed structure is based on a **face-centered cubic (fcc) unit cell,** as shown in Fig. 21.7. Any crystal can be considered to be composed of an infinite number of unit cells stacked in a repeating arrangement in three dimensions. Face-centered cubic unit cells will be discussed in detail in Section 21.5.

Close-packed structures are the most efficient at filling space of any crystalline structures. About 74% of space is filled with atoms in either hexagonal or cubic close-packed structures, with the interstices taking up about 26% of space. In addition to the metals, crystals of the rare gases also contain atoms packed in hexagonal and cubic close-packed structures.

Body-Centered Cubic Structures

While about 70% of all metals crystallize in one of the two close-packed structures, another 25% have a **body-centered cubic (bcc) structure.** In this structure, the metal atoms are located at the corners and the center of a cubic unit cell, as shown in Fig. 21.8. The entire cyrstal is constructed of an infinite number of such unit cells, repeated in three dimensions. The atom at the center of the unit cell belongs entirely to that unit cell, but each atom at a corner is shared by eight unit cells, four in each of two layers, as shown in Fig. 21.9. Each unit cell is therefore considered to have one eighth of the atom at each corner, and as there are eight corners per cube, the total number of atoms per body-centered cubic unit cell is two:

$$(8 \text{ corners})(\tfrac{1}{8} \text{ atom/corner}) + 1 \text{ central atom} = 2 \text{ atoms per unit cell}$$

The packing of atoms in a body-centered cubic structure is somewhat less efficient than in the close-packed structures; about 68% of space is filled with atoms in a body-centered cubic structure. The atoms at the corners of the cubic unit cell are not in contact with one another. Each atom in the crystal touches four atoms in the layer above, and four in the layer below, so that the coordination number, or number of nearest neighbors of each atom, is 8 in a body-centered cubic structure. Metals that crystallize in a body-centered cubic structure include Li, Na, K, Cr, Fe, and Ba.

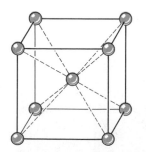

Fig. 21.8. The body-centered cubic unit cell. In a crystal made by repeated stacking of such unit cells in three dimensions, each sphere has eight nearest neighbors. The distance between nearest neighbors is $(\sqrt{3}a)/2$, where a is the axial length of the cubic unit cell.

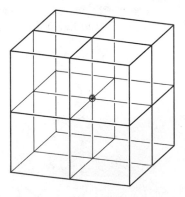

Fig. 21.9. The repeated stacking of unit cells to form a crystal. The corner of any unit cell is shared by eight unit cells. In the diagram, one corner is darkened to show that it is shared by four cells in the bottom layer and four cells in the layer that is stacked on top.

section 21.4
Metallic Bonding

In order to explain the characteristic properties of metals, we must understand the nature of the bonding between metal atoms in the structures just described. There are two important characteristics of all metal atoms. The first is that the number of valence electrons for a metal atom is less than the number of valence orbitals, in contrast to the situation for nonmetals. There are many more metals than nonmetals because in the three series of transition metals and in the lanthanides and actinides, it is inner shells that are being filled with electrons, and the number of valence electrons remains very nearly the same throughout each series. The second characteristic property of metallic atoms is that they have low **ionization energies** relative to the nonmetals. It is easier to remove or ionize a valence electron from a metal atom than from a nonmetal atom.

We have seen that in metallic crystals, each metal atom has either 8 or 12 nearest neighbors, yet the number of valence electrons per metal atom is less than 4, and is frequently only 1 or 2. It is clearly impossible for metal atoms to be linked by typical covalent bonds in which two atoms share a pair of electrons. A relatively small number of electrons must be shared by eight or more other atoms.

Nonmetals typically form diatomic gaseous molecules, such as O_2, N_2, Cl_2, and F_2, in which a covalent bond links the two atoms. In contrast, the gaseous state of most metals (but not all) is monatomic. In the solid state, however, stability is achieved when the valence electrons of one atom are shared by many other nuclei. Because of the relatively low ionization energy of metal atoms, the valence electrons in a solid metallic crystal are free to pass from one atom to another so that they are shared by several. The metallic bond is considered to consist of mobile or **delocalized electrons** shared by virtually all the atoms in a sample of metal. The electronic structure of metallic crystals is commonly described as a collection of metallic ions immersed in a "sea" of mobile valence electrons.

The characteristic properties of solid metals such as high electrical and thermal conductivity and silvery luster, are ascribed to these relatively mobile valence electrons. Such metallic properties are absent in the gaseous state of metals where metallic bonding is not possible. Gaseous sodium and mercury, for instance, are monatomic, and are not good conductors of heat or electricity.

Band Theory of Metals

In order to describe the orbitals occupied by the valence electrons of metals in the solid state, let us imagine constructing a crystal of lithium metal by adding lithium atoms one at a time, forming first Li_2, then Li_3, Li_4, Li_5, and so on, until finally we have Li_N, where N is a very large number, of the order of 10^{20}.

Diatomic Li_2 is a known species in the gaseous state and has been discussed in Section 14.5. Each Li atom has electronic configuration $1s^2 2s^1$, with a single $2s$ valence electron. Two $2s$ atomic orbitals, one from each Li atom, overlap to form two molecular orbitals, $\sigma 2s$ and $\sigma^* 2s$, and both the valence electrons occupy the $\sigma 2s$ molecular orbital. The $\sigma^* 2s$ orbital is vacant.

It is a general rule that n atomic orbitals can be combined to form n molecular orbitals. When three Li atoms form Li_3, the three $2s$ atomic orbitals overlap to form three molecular orbitals. The electronic configuration of Li_3, with three electrons to occupy the three molecular orbitals, is shown in Fig. 21.10(c). Electrons in the

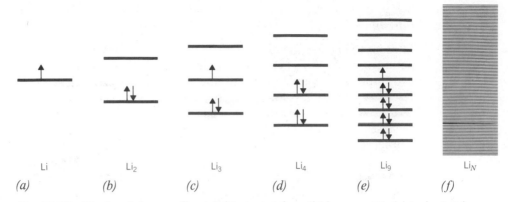

Fig. 21.10. The band theory of metals illustrated for a lithium crystal. *(a)* A single Li atom with one valence $2s$ electron. *(b)* Li_2 with two molecular orbitals, one filled and one vacant. *(c)* Li_3 with three molecular orbitals. *(d)* Li_4 with four molecular orbitals. *(e)* Li_9 with nine molecular orbitals, half-filled. *(f)* The band of N molecular orbitals in a crystal of lithium, where N is a very large number, of the order of 10^{20}. Note that as the number of lithium atoms increases, the spacing between the orbitals decreases. In the crystal the band of N molecular orbitals is half-filled. Electrons in the lower, filled levels of the band can easily be unpaired and excited to vacant levels of the band. Electrons in the band are delocalized and move over the entire crystal.

molecular orbitals are spread out over the three Li nuclei and are the property of the molecule as a whole, not of any one lithium atom.

For N lithium atoms, there are N molecular orbitals that arise from the overlap of the $2s$ valence atomic orbitals of all the atoms. The separation between the molecular

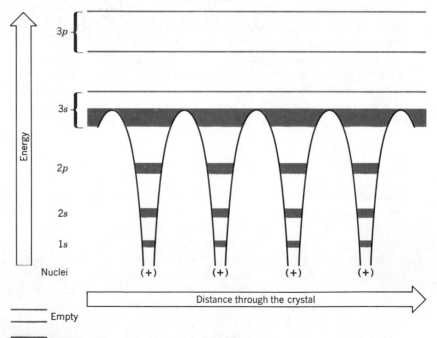

Fig. 21.11. Bands in metallic sodium. The $1s$, $2s$, and $2p$ bands are filled. Electrons in filled bands cannot move through the solid because orbitals in the same band on neighboring atoms are already filled. The $3s$ band is half-filled; it is the valence band. The $3p$ band is empty. The $3s$ and $3p$ bands overlap to form the conduction band.

orbitals decreases as N increases, so that there is now a **band** of N closely spaced molecular orbitals, as shown in Fig. 21.10(f). This band is half-full, as each molecular orbital can hold two electrons, and there are only N valence electrons. The band containing the outer or valence shell electrons is called the **valence band.**

The empty $2p$ atomic orbitals of the Li atoms also overlap to form a wide band of molecular orbitals. Because the $2s$ and $2p$ orbitals do not differ very much in energy, the band of molecular orbitals formed from the $2s$ atomic orbitals and the band of molecular orbitals formed from the $2p$ atomic orbitals join continuously and are called a **conduction band.** A band that is partly or completely vacant and is uninterrupted throughout the crystal is a conduction band. Electrons in the lower, filled levels of the conduction band can easily be unpaired and excited to vacant levels of the band. In these delocalized orbitals they can circulate throughout the crystal. When an electric field is applied, the electrons move in one direction and lithium is therefore a good conductor of electricity. The bands in the next alkali metal, sodium, are shown in Fig. 21.11.

ection 21.5
The Seven Crystal Systems

Because all the atoms in a crystal are arranged in a repeating three-dimensional pattern, if we choose any point at all within the crystal, there will be a very large number of other points with exactly the same surroundings or environment. A set of identical points (that is, points with identical surroundings) within a crystal is called a **lattice.**

If lattice points are connected by straight lines, the space within the crystal is divided into parallelepipeds. One of these parallelepipeds is called a **unit cell.** A crystal can be thought of as an infinite number of unit cells stacked together in three dimensions. A crystal is therefore a repetitive multiple of its unit cell, and has the same symmetry that the unit cell possesses.

All crystals can be classified as belonging to one of only seven crystal systems, defined by the symmetry of the unit cell. A unit cell is described by specifying the lengths along the three coordinate axes (a,b,c) and the angles (α,β,γ) between the axes, as shown in Fig. 21.12.

The unit cell of highest symmetry is the **cubic unit cell,** in which all axial lengths are equal, and the three coordinate axes are perpendicular to one another. The unit cell of lowest symmetry is the **triclinic,** in which all axial lengths are unequal, there are no perpendicular axes, and the three interaxial angles are all different. The characteristics of the unit cells that define the seven crystal systems are listed in Table 21.1.

In a **simple or primitive lattice,** lattice points (points with identical environments) are located only at the corners of each unit cell. In a **body-centered cell,** lattice points are located at the center of the cell as well as at the corners. In a **face-centered unit cell,** lattice points are found at the center of each of the six faces of the cell, as well as at each of the eight corners. Sodium chloride, depicted in Fig. 21.13, crystallizes in a face-centered cubic lattice.

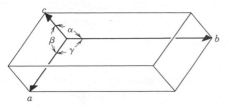

Fig. 21.12. One unit cell. The lengths along the coordinate axes are denoted a,b,c. The angles between the axes are α,β,γ and are defined as shown in the diagram.

Table 21.1. **The Seven Crystal Systems**

System	Axes	Angles	Examples
Cubic	$a = b = c$	$\alpha = \beta = \gamma = 90°$	NaCl, CsCl
Tetragonal	$a = b \neq c$	$\alpha = \beta = \gamma = 90°$	TiO_2 (rutile)
Orthorhombic	$a \neq b \neq c$	$\alpha = \beta = \gamma = 90°$	$CdSO_4$, $HgBr_2$
Rhombohedral	$a = b = c$	$\alpha = \beta = \gamma \neq 90°$	$CaCO_3$ (calcite)
Hexagonal	$a = b \neq c$	$\alpha = \beta = 90°, \gamma = 120°$	SiO_2
Monoclinic	$a \neq b \neq c$	$\alpha = \gamma = 90°, \beta \neq 90°$	KIO_3, $NaHCO_3$
Triclinic	$a \neq b \neq c$	$\alpha \neq \beta \neq \gamma \neq 90°$	$NaHSO_4$, CuF_2

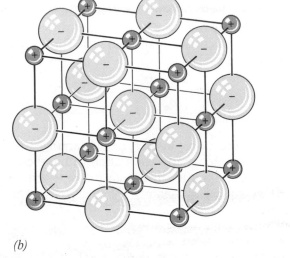

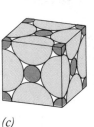

(a)   *(b)*   *(c)*

Fig. 21.13. Various ways of depicting the NaCl crystal. Sodium chloride crystallizes in a face-centered cubic unit cell, with 4 Na^+ ions and 4 Cl^- ions per unit cell. If the unit cell is defined with the Na^+ ions at the eight corners and six face centers, there are 12 Cl^- ions on the edges in between the Na^+ ions, and one in the center of the cell, in contact with the 6 Na^+ ions at the face centers. Because there is anion–cation contact in NaCl, *(a)* and *(c)* are better representations of the three-dimensional structure than *(b)* is.

The Fourteen Bravais Lattices

Figure 21.14 depicts the 14 different lattices (called the **Bravais lattices**) that are observed for the seven crystal systems. The cubic system has 3 different lattices (simple, body-centered, and face-centered), the tetragonal system has only 2 lattices (simple and body-centered), while there are 4 orthorhombic lattices. The monoclinic system has two lattices (simple and end-centered), and the rhombohedral, hexagonal, and triclinic systems have simple lattices only.

In many calculations involving properties of crystals it is important to know how many atoms or molecules are contained in each unit cell. Particles located on edges, faces, or corners of a unit cell are shared by several unit cells and must not be counted more than once. The following rules are used to calculate the number of particles per unit cell, where a "particle" can mean either an atom, an ion, or a molecule.

1. A particle located at the center of a unit cell (or wholly within one unit cell) belongs to that unit cell only.
2. A particle lying on a face-center belongs equally to two unit cells and is therefore counted as one half a particle for each unit cell.

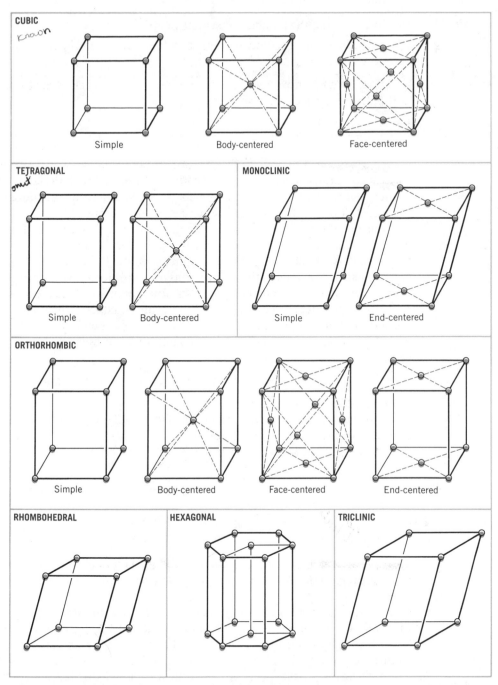

Fig. 21.14. Unit cells of the 14 Bravais lattices of the seven crystal systems. There are 3 lattices in the cubic system, 2 in the tetragonal, 4 in the orthorhombic, 2 in the monoclinic, and 1 each in the hexagonal, rhombohedral, and triclinic systems.

3. A particle occupying a corner is shared by eight unit cells and is counted as one eighth a particle for each unit cell.

4. A particle lying on an edge is shared by four unit cells and is therefore counted as one fourth a particle for each unit cell.

EXAMPLE 21.1. Calculating the number of particles per unit cell

✱ If molecules occupy the lattice points of a face-centered unit cell, how many molecules are there per unit cell?

Solution. In a face-centered cell there are molecules at each of the eight corners and at the center of each of the six faces.

(8 corners)($\frac{1}{8}$ per corner) + (6 face centers)($\frac{1}{2}$ per face center) = 1 + 3 = 4 molecules per unit cell

For ionic crystals there are two or more different kinds of ions in each unit cell. Usually the cations and anions are of different sizes, with anions larger than cations. The crystal is, of course, electrically neutral, and the number of cations and anions per unit cell must reflect the stoichiometry of the compound.

Consider the sodium chloride crystal, for instance, shown in Fig. 21.13. The chloride ions are significantly larger than the sodium ions ($r_{Cl^-} = 1.81$ Å, $r_{Na^+} = 0.95$ Å). Each Na^+ ion is in contact with 6 Cl^- ions, and each Cl^- ion is in contact with 6 Na^+ ions. If the sodium ions are considered to occupy the corners and face centers of the unit cell, there are 4 Na^+ ions per unit cell. There must also then be 4 Cl^- ions per unit cell. There are 12 Cl^- ions on the edges of the unit cell, in between the sodium ions at the corners, and 1 Cl^- ion at the center of the unit cell, in contact with the 6 Na^+ ions at the face centers. Therefore the number of chloride ions per unit cell is

(12 at edges)($\frac{1}{4}$ per edge) + 1 at center = 4 ions per unit cell

as required by the stoichiometry.

An equivalent, alternative description of the NaCl unit cell is to locate the chloride ions at the corners and face centers with the sodium ions on the edges and at the center. In either description, there are 4 Na^+ ions and 4 Cl^- ions per unit cell.

section 21.6
X-Ray Diffraction

The experimental technique widely used today to obtain information about unit cell dimensions and bond distances and angles in crystalline substances is **X-ray diffraction.**

The wave properties of electromagnetic radiation and the regions of the electromagnetic spectrum have been defined in Section 12.2. The observation of the **diffraction** of light was among the evidence that convinced scientists of its wave nature. In order to explain the phenomenon of diffraction we must consider what happens when electromagnetic waves of the same wavelength (or frequency) are superimposed.

Constructive and Destructive Interference

Two waves of the same wavelength are **in-phase** with one another if both of them attain their maximum amplitude at the same time, and at the same point in space. When waves that are in-phase with one another are superimposed, the resultant radiation has a greater intensity than either of the original waves, as is illustrated in Fig. 21.15(a). This increase in intensity is known as **constructive interference.**

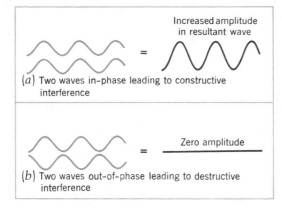

Fig. 21.15. *(a)* Constructive interference when two waves that are in-phase are superimposed. *(b)* Destructive interference when two waves that are out-of-phase are superimposed.

On the other hand, two waves of the same wavelength are **out-of-phase** if one reaches its maximum amplitude at the same time that the other reaches its minimum amplitude. The superposition of two out-of-phase waves results in a wave of zero intensity. This decrease in intensity is called **destructive interference,** and is illustrated in Fig. 21.15*(b)*. A small phase difference between two superimposed waves results in a partial cancellation of the wave intensities.

In a **double-slit diffraction experiment,** illustrated in Fig. 21.16, light from a radiation source impinges on a barrier with two small slits a fixed distance apart. Each slit then serves as a separate radiation source, and the waves passing through the slits are superimposed. What we observe on the detector is a pattern of bright lines on a dark background. Where the waves have interfered constructively, there is a line of high intensity radiation. Where the waves have interfered destructively, there is darkness.

Consider the intensity of radiation at a point P on the detector. (Refer to Fig. 21.16.) Light reaching P after passing through slit S_1 has traveled a distance r_1. Light reaching point P after passing through slit S_2 has traveled a distance r_2. Only if the path-length difference between r_1 and r_2 is an *integral number of wavelengths* of the light will there be constructive interference. In order to observe diffraction patterns, the distance between the two slits must be of the same order of magnitude as the wavelength of the radiation.

X-rays are electromagnetic radiation of very short wavelength, of the order of magnitude of 1 – 10 Å. It is not possible to manufacture diffraction gratings with slits or rulings so closely spaced, but the distance between planes of atoms in crystals is exactly of the order of magnitude of a few angstroms. It was William Bragg, an English physicist, who first showed that a crystal can be considered a natural diffraction grating for X-rays. The scattering or diffraction of X-rays by crystals depends on the exact geometry of the crystal lattice. Interpreting the observed diffraction pattern

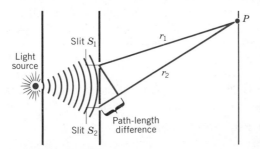

Fig. 21.16. A double-slit diffraction experiment. Only if the difference in path length of the radiation reaching point P from slit S_1 and from slit S_2 is an integral number of wavelengths will there be constructive interference and a spot of maximum intensity at P.

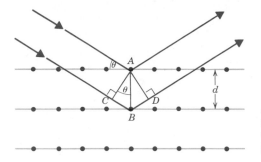

Fig. 21.17. Bragg's Law. Only if the angle θ satisfies the Bragg equation, $n\lambda = 2d \sin \theta$, will the scattered radiation be in-phase and interfere constructively.

provides us with detailed information about interatomic distances and bond angles in the compound composing the crystal.

The basic equation that relates the wavelength, λ, of the X-rays used, the distance, d, between successive planes of atoms in the crystal, and the angle, θ, between the beam of incoming X-rays and the crystal plane, is known as **Bragg's Law.** Figure 21.17 illustrates the scattering of **monochromatic** (single wavelength) X-rays by a crystal.

If θ is the angle of incidence of the incoming X-ray relative to a plane of atoms in the crystal, then the scattered X-ray will be reflected from that plane at the same angle θ. (It is a well-known law for reflected light that the angle of incidence is equal to the angle of reflection.) X-rays that are not scattered by atoms in the first plane pass into the crystal and may be reflected by atoms in the second, third, or any successive plane. All the incident beams of X-rays are in-phase. The reflected waves will, in general, be out-of-phase, and will therefore interfere destructively. Let us see what condition is required for the reflected beam to contain waves that are in-phase, so that the emerging beam will be of sufficient intensity to be observed.

Consider two rays, one that is scattered by an atom in the first plane, and one that is scattered by an atom in the second plane. Let the distance between the two planes be denoted d, as in Fig. 21.17. The emergent ray that has been scattered by the second plane has traveled a greater distance than the ray scattered by the first plane when both rays reach the detector. (See Fig. 21.19 for a schematic diagram of the apparatus.) From Fig. 21.17, the extra distance traveled by the ray scattered from the second plane is $CB + BD$, where AC is perpendicular to the direction of the incident ray, and AD is perpendicular to the direction of the emergent ray.

Because AB is perpendicular to the planes of atoms in the crystal, the angle BAC must also be equal to θ. We can calculate the length of BC using the triangle ABC and the definition of the sine of an angle:

$$\sin BAC = \sin \theta = BC/d \tag{21-1}$$

so that $BC = d \sin \theta$. Note that d is the length of the hypotenuse of the triangle ABC.

Triangles ABD and ABC are congruent, so that angle BAD is also equal to θ. From the definition of the sine of angle BAD, $BD = d \sin \theta$. Thus the extra distance traveled by the ray scattered from the second plane is

$$BC + BD = 2d \sin \theta \tag{21-2}$$

Constructive interference will occur if and only if the extra distance traveled by the second ray is an integral number of wavelengths of the light, that is, if

$$BC + BD = n\lambda \qquad \text{where} \qquad n = 1,2,3,\dots \tag{21-3}$$

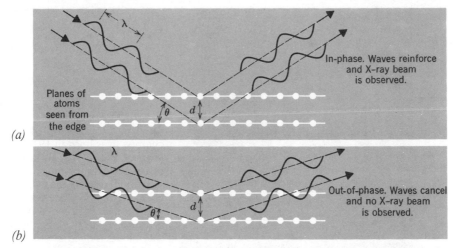

Fig. 21.18. Diffraction from successive planes of atoms in a crystal. *(a)* Scattered waves from different planes of atoms are in-phase. *(b)* Scattered waves from different planes of atoms are out-of-phase.

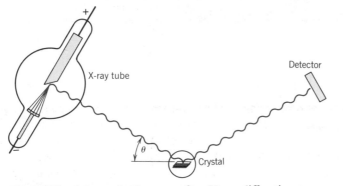

Fig. 21.19. Schematic diagram of an X-ray diffraction experiment.

Bragg's law therefore states that the emergent beam will have an observable intensity only if

$$n\lambda = 2d \sin \theta \qquad (n = 1,2,3, \dots) \qquad (21\text{-}4)$$

If X-rays of wavelength λ strike the crystal planes at angles other than those that satisfy the Bragg equation, Eq. (21-4), they will interfere destructively and the emergent beam will have a zero or nonobservable intensity. (See Fig. 21.18.) In an X-ray diffraction experiment, the crystal is slowly rotated to change the angle between crystal planes and the incident X-ray beam. Since λ is known, the measurement of the angles at which X-ray intensities are observed at the detector (see Fig. 21.19) enables us to calculate the distances between successive planes of atoms in the crystal.

Complete analysis of an X-ray diffraction pattern may take many months of work. The use of a computer has made it possible to undertake X-ray diffraction studies to determine the structures of high molecular weight compounds of biological importance. Because heavier atoms with more electrons scatter X-rays more strongly than light atoms, it is possible to determine both the nature and the positions of the atoms in the crystal.

Dorothy Crowfoot Hodgkin, British chemist, was born in 1910. During the early years of World War II she elucidated the three-dimensional molecular configuration of penicillin using the technique of X-ray crystallography. Once the structure of penicillin was known, it became possible to synthesize it in the laboratory. This led the way to the industrial production of penicillin. In 1964 Dorothy Hodgkin was awarded the Nobel Prize in Chemistry for elucidating the structure of molecules of biochemical importance, notably penicillin and vitamin B_{12}.

section 21.7
Atomic Dimensions and Unit Cell Geometry

Measurements of unit cell dimensions and geometry obtained from X-ray diffraction studies provide information about the magnitudes of atomic and ionic radii.

Consider, for instance, the calculation of the atomic radius of copper. Copper crystallizes in the cubic close-packed structure, which has a face-centered cubic unit cell. There are four Cu atoms per unit cell, and the axial length is 3.615 Å. The face of a unit cell has the appearance shown in Fig. 21.20. The atoms, considered to be spherical, are in contact along the face diagonal, which is 4 atomic radii in length.

Using the Pythagorean theorem, if the side of a square is denoted x and the diagonal d, $x^2 + x^2 = d^2$, so that the face diagonal of a cubic unit cell is $\sqrt{2}x$. The length of a face diagonal of a copper unit cell is therefore $(\sqrt{2})(3.615 \text{ Å})$ or 5.112 Å. The atomic radius of Cu is thus determined to be $5.112/4 = 1.278$ Å.

If the density of the crystal is known, knowledge of the unit cell geometry can be used to calculate Avogadro's number as is illustrated in Example 21.2.

EXAMPLE 21.2. Calculation of Avogadro's number from crystallographic data

Copper crystallizes in a face-centered cubic crystal with a unit cell length of 3.615 Å. The density of metallic copper is 8.935 g·cm^{-3} at 20 °C. Calculate Avogadro's number from this data.

Solution. If the side of the cubic unit cell is 3.615 Å or 3.615×10^{-8} cm, the volume of one unit cell is $(3.615 \times 10^{-8})^3$ cm^3, or 47.24×10^{-24} cm^3. This volume is occupied by four Cu atoms.

The volume of 1 mol of copper can be determined from its density and atomic weight. If 8.935 g occupy 1 cm^3, then 1 mol, or 63.546 g, occupies a volume of

$$\left(63.546 \, \frac{\text{g}}{\text{mol}}\right)\left(\frac{1 \text{ cm}^3}{8.935 \text{ g}}\right) = 7.112 \, \frac{\text{cm}^3}{\text{mol}}$$

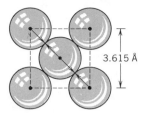

3.615 Å

Fig. 21.20. One face of a unit cell of a copper crystal. There are 4 Cu atoms per unit cell. The atoms are in contact along the face diagonal, which is 4 atomic radii in length.

From the ratio of the volume per mole to the volume per unit cell we can calculate the number of unit cells per mole:

$$\frac{7.112 \text{ cm}^3/\text{mol}}{47.24 \times 10^{-24} \text{ cm}^3/\text{unit cell}} = 1.5055 \times 10^{23} \text{ unit cells/mol}$$

and as there are four Cu atoms per unit cell, the number of Cu atoms per mole is $4(1.5055 \times 10^{23}) = 6.022 \times 10^{23}$, which is, of course, Avogadro's number.

In crystallographic work, using the known value of Avogadro's number, the density of a crystal is often calculated from the dimensions and geometry of the unit cell.

Calculation of Ionic Radii

The size of an ion is not fixed, as the electron cloud surrounding the nucleus has no rigid boundary. If we measure internuclear distances in crystals of differing geometries containing an ion in common, we observe that the size of the ion depends somewhat on the number of nearest neighbors of opposite sign. However, many similar compounds are isomorphous, that is, their crystal structures are the same. In a series of isomorphous compounds with an ion in common, the variation in ionic radius is very small.

By making certain assumptions about the contacts between ions, we can calculate numerical values for ionic radii that are very useful. The following examples illustrate the reasoning used in this type of calculation.

EXAMPLE 21.3. Calculation of ionic radii from unit cell dimensions when the anion is very much larger than the cation

Lithium chloride crystallizes in a face-centered cubic structure. The unit cell length is 5.14 Å. Assuming that Li$^+$ ions are so much smaller than Cl$^-$ ions that there is anion–anion contact, and the Li$^+$ ions fit into the interstices between the Cl$^-$ ions, calculate values for the radii of lithium and chloride ions.

Solution. With the assumptions made about the relative sizes of Li$^+$ and Cl$^-$ ions, a face of the LiCl unit cell has the appearance shown in Fig. 21.21.

The Li$^+$ ions may be so small relative to the Cl$^-$ ions that they are not in actual contact with the anions, as in Fig. 21.21(a), or they may just fit exactly in the space between the Cl$^-$ ions, as in Fig. 21.21(b). In either case, the assumption of anion–anion contact allows us to calculate a value for the radius of the chloride ion.

The length of the unit cell is 5.14 Å. Using the Pythagorean theorem, the length of the diagonal of each face is therefore $(\sqrt{2})(5.14 \text{ Å}) = 7.27 \text{ Å}$, and this must be four times the radius of the chloride ion. Thus $r_{Cl^-} = 7.27/4 = 1.82 \text{ Å}$.

If we assume the Li$^+$ ions are in contact with all the adjacent Cl$^-$ ions, as in Fig. 21.21(b), the length of one side of the unit cell is $2r_{Cl^-} + 2r_{Li^+}$.

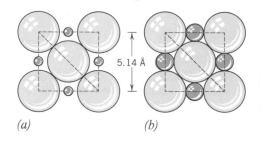

(a)   *(b)*

Fig. 21.21. A face of the LiCl unit cell, assuming anion–anion contact. Small colored circles are Li^+ ions, large open circles are Cl^- ions.

Thus

$$5.14 \text{ Å} = 2r_{Cl^-} + 2r_{Li^+} = 2(1.82 \text{ Å}) + 2r_{Li^+}$$

and

$$r_{Li^+} = \tfrac{1}{2}(5.14 - 3.64) = 0.75 \text{ Å}$$

In fact, by measuring internuclear distances in many lithium compounds, we find that the Li^+ ion is smaller than 0.75 Å. Figure 21.21*(a)* is a better representation of the LiCl crystal than Fig. 21.21*(b)*.

The radius of the Li^+ ion, obtained from a number of measured internuclear distances in lithium compounds, is 0.60 Å. The value of the radius of the Cl^- ion that fits best for a large number of chlorides is 1.81 Å.

EXAMPLE 21.4. Calculation of ionic radii when there is anion–cation contact

Rubidium chloride crystallizes in a face-centered cubic lattice with unit cell length = 6.54 Å. If r_{Cl^-} is 1.81 Å, calculate r_{Rb^+}.

With ions that are roughly equal in size, anions are not in contact with one another, and the face of the unit cell is as shown below.

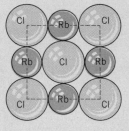

Hence

$$6.54 \text{ Å} = 2r_{Cl^-} + 2r_{Rb^+}$$

and

$$r_{Rb^+} = 3.27 \text{ Å} - 1.81 \text{ Å} = 1.46 \text{ Å}$$

The value of r_{Rb^+} that fits best for many compounds is 1.48 Å.

The radii of the more common monatomic ions are shown in relation to the periodic table in Fig. 21.22.

The following rules relate the ionic radii to the position of the atom in the periodic table.

1. Within one family, ionic radii increase as the atomic number increases, that is, as the number of electron shells increases.

2. The radii of cations with the same number of electrons (that is, with the

| Li⁺ Be²⁺ | | | | | | | | | | | | O²⁻ F⁻ | |

Li⁺	Be²⁺											O²⁻	F⁻
0.60	0.31											1.40	1.36
Na⁺	Mg²⁺											S²⁻	Cl⁻
0.95	0.65											1.84	1.81
K⁺	Ca²⁺	Sc³⁺	...	Cr²⁺	Mn²⁺	Fe²⁺	Co²⁺	Ni²⁺	Cu²⁺	Zn²⁺	...	Se²⁻	Br⁻
1.33	0.99	0.81		0.89	0.80	0.75	0.72	0.70	0.70	0.74		1.98	1.95
Rb⁺	Sr²⁺	Y³⁺	...						Ag⁺	Cd²⁺	...	Te²⁻	I⁻
1.48	1.13	0.93							1.26	0.97		2.21	2.16
Cs⁺	Ba²⁺	La³⁺	...						Au⁺	Hg²⁺	...		
1.69	1.35	1.15							1.37	1.10			

Fig. 21.22. The radii of some monatomic ions, in angstrom units.

configuration of the same rare gas) decrease from left to right across a period in the table. The electronic charge cloud of these isoelectronic cations is pulled in closer to the nucleus as the number of protons in the nucleus increases.

3. For transition metal ions with the same charge, both the nuclear charge and the number of electrons increase from left to right across a period, and the size of the ions does not change by a large amount. There is a slight decrease in size from Cr^{2+} to Cu^{2+}, as the electrons added are all in the $3d$ shell. Outer electrons are shielded from the nuclear charge by the inner shell electrons that penetrate to the nucleus more than the $3d$ electrons do. The inner shell electrons remain the same throughout the series from Cr^{2+} to Cu^{2+}, while the nuclear charge increases and the number of $3d$ electrons increases. Therefore the effective nuclear charge increases, pulling the $3d$ electrons in a little closer to the nucleus, and the ionic radius decreases.

4. Cations are significantly smaller than anions with the same electronic configuration. For example, S^{2-} and Cl^- are larger than the isoelectronic cations K^+, Ca^{2+}, and Sc^{3+}. For ions with the same number of electrons, the larger the nuclear charge, the more closely the electrons are pulled in toward the nucleus, and the smaller the ion.

5. When the same element forms two cations, the one with the higher charge is always smaller in size. The radii of Fe^{2+} and Fe^{3+} ions are 0.75 and 0.64 Å, of Cr^{2+} and Cr^{3+} ions are 0.89 and 0.61 Å, and of Cu^+ and Cu^{2+} ions are 0.96 and 0.70 Å, respectively. The larger positive charge pulls the electrons in closer to the nucleus.

section 21.8
The Lattice Energy of Ionic Crystals

The lattice energy of an ionic crystal is defined as the amount of energy required to separate the ions in one mole of crystal from their positions in the lattice to an infinite separation in the gaseous state. That is, the lattice energy, usually denoted U, is ΔH for the reaction

$$MX(\text{crystal}) \rightarrow M^+(g) + X^-(g) \tag{21-5}$$

Strictly speaking, we should call ΔH for reaction (21-5) the lattice enthalpy, but the term lattice energy is commonly used.

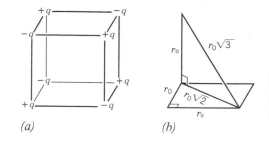

Fig. 21.23. *(a)* Prototype crystal with four ion-pairs. *(b)* Relation between the length of a side of a cube, a face diagonal, and a cube diagonal.

Because the forces holding ions together in an ionic crystal are largely the coulombic interactions, the lattice energy depends on the ionic charges and the distances between the ions in the way one would predict using Coulomb's law. The expression for the lattice energy of a crystal in which the bond between cation and anion is purely ionic, with no covalent character, has been derived by M. Born, A. Landé, and J. Mayer. The principal term in that expression accounts for about 90% of the lattice energy and may be written as

$$U = \frac{N_A A z_+ z_-}{r_0} \tag{21-6}$$

where z_+ and z_- are the magnitudes of the charges on cation and anion, respectively, r_0 is the equilibrium internuclear distance, A is a constant (called the **Madelung constant**) that depends on the crystal structure, and N_A is Avogadro's number. The Madelung constant is determined solely by using Coulomb's law and the distances between ions for the particular geometric arrangement in the crystal.

To understand the calculation of the Madelung constant for a crystal, let us calculate the coulombic interaction energy for a simple prototype crystal, four pairs of ions of charges $+q$ and $-q$ situated on alternate corners of a cube, as in Fig. 21.23(a). The length of the side of the cube is r_0, the equilibrium internuclear distance.

Using the Pythagorean theorem we find [refer to Fig. 21.23(b)]

$$\text{length of face diagonal} = r_0 \sqrt{2} \tag{21-7a}$$
$$\text{length of cube diagonal} = r_0 \sqrt{3} \tag{21-7b}$$

Let us now calculate all the coulombic interactions in this cube.

1. Each ion of charge $+q$ has 3 ions of charge $-q$ a distance r_0 away. As there are 4 ions with charge $+q$, there are 12 terms of magnitude $-q^2/r_0$.
2. Each ion of charge $+q$ has 1 ion of charge $-q$ at the opposite end of a cube diagonal, a distance $r_0 \sqrt{3}$ away. As there are 4 ions with charge $+q$, there are 4 terms of magnitude $-q^2/r_0\sqrt{3}$.
3. There are like charges at the opposite ends of each face diagonal. There are 6 faces and 12 face diagonals, so there are 12 terms of magnitude $q^2/r_0\sqrt{2}$.

The total of all coulombic interactions in this simple prototype crystal is

$$\frac{-12q^2}{r_0} - \frac{4q^2}{r_0\sqrt{3}} + \frac{12q^2}{r_0\sqrt{2}} = \frac{q^2}{r_0}\left(-12 - \frac{4}{\sqrt{3}} + \frac{12}{\sqrt{2}}\right) = -5.825\frac{q^2}{r_0}$$

As there are four pairs of ions, we can write the total of all coulombic interactions as $-4(1.456q^2/r_0)$.

Analogously, for N_A pairs of ions in a sodium chloride crystal, the total of all coulombic interactions is $-N_A(1.748q^2/r_0)$, and the Madelung constant for NaCl is 1.748.

For many ionic crystals, the internuclear distance r_0 is equal to $r_c + r_a$, the sum of the radii of the cation and anion. Even when the cation and anion are not in direct contact, $r_c + r_a$ is a good approximation to r_0.

The expression for the pure ionic lattice energy, Eq. (21-6), therefore tells us that crystals with small, highly charged ions have larger lattice energies than those with larger, singly charged ions. Properties frequently (but *not* invariably) associated with large lattice energies are hardness, high melting point, and low solubility. Solubility, in particular, depends on a number of factors in addition to the lattice energy, and is difficult to predict from purely qualitative considerations. We will return to a discussion of the factors affecting the solubility of ionic crystals at the end of this section.

Let us compare the lattice energies of LiCl and CaO. Referring to Fig. 21.22, we find that for LiCl the sum of the cation and anion radii, $r_c + r_a$, is $0.60 + 1.81 = 2.41$ Å. For CaO the corresponding sum is $0.99 + 1.40 = 2.39$ Å. These sums are essentially equal, so that the denominator of the expression for the lattice energy, Eq. (21-6), can be considered to be the same for the two compounds. Because Ca^{2+} and O^{2-} ions are doubly charged, while Li^+ and Cl^- ions are singly charged, the factor z_+z_- is four times larger for CaO than for LiCl. The lattice energy is therefore close to four times as large for CaO as it is for LiCl. The observed values are 845 kJ/mol for LiCl and 3460 kJ/mol for CaO. Calcium oxide is harder, and has a much higher melting point (2580 °C) than LiCl (614 °C), because of the stronger forces of attraction between the doubly charged ions. Calcium oxide is also very much less soluble than LiCl.

The effect of a change in the interionic distance, $r_c + r_a$, on the lattice energies of crystals with ions of the same charge, is less marked than the effect of doubling the charge of each ion, as is illustrated in Example 21.5.

EXAMPLE 21.5. Trends in lattice energies with a change in internuclear distance

Account for the following variation in lattice energies (in kilojoules per mole) observed for the potassium halides:

KF 817   KCl 718   KBr 688   KI 636

Solution. The only factor in Eq. (21-6) that varies for these four compounds is r_0, which can be considered the sum of the ionic radii, $r_c + r_a$. The cation is the same in all these crystals, but the anion radius increases regularly as the atomic number of the halide increases. Because the lattice energy is inversely proportional to $r_c + r_a$, as that sum increases the lattice energy decreases. As r_a increases regularly for the series F^-, Cl^-, Br^-, and I^-, the lattice energy decreases regularly for KF, KCl, KBr, and KI.

The Effect of Cations without Rare Gas Electronic Configuration

The equation given for the major term in the lattice energy, Eq. (21-6), is correct only if the bond between cation and anion is purely ionic. There are very few pure ionic bonds. The addition of covalent character to the bond strengthens it and increases the lattice energy to a value larger than that calculated using Eq. (21-6). In particular, a cation that does not have the electronic configuration of one of the rare gases has vacant valence atomic orbitals and is much more effective at pulling the electronic charge cloud of the anion into the region between the cation and anion nuclei and therefore of imparting covalent character to the bond, than is a cation that has a rare gas electronic configuration.

As an example, let us compare AgCl and NaCl. The value of $r_c + r_a$ is 2.76 Å for NaCl and 3.07 Å for AgCl. Despite the somewhat larger denominator in Eq. (21-6), the lattice energy of AgCl is larger than that of NaCl by about 15% (904 kJ/mol for AgCl compared to 788 kJ/mol for NaCl). While Na^+ has the electronic configuration of neon, Ag^+ does not have a rare gas electronic configuration. There is significant covalent character to the bond in AgCl, while the bond in NaCl is essentially ionic. Because the bond in AgCl is ionic with partial covalent character, the lattice energy of AgCl is increased above that of NaCl. The greater cohesive forces in the AgCl crystal are largely responsible for the fact that AgCl has a very small solubility in water while NaCl is a soluble salt.

The larger the anion, the farther its electron cloud extends out from its own nucleus, and the easier it is to pull the anionic charge cloud toward the cation. Larger anions are more polarizable than smaller anions. If we compare two crystals with the same cation but different anions, we can expect more covalent character in the cation–anion bond for the compound with the larger anion.

EXAMPLE 21.6. The effect of anion size on the lattice energy when the anion–cation bond has covalent character

(a) The lattice energy of AgBr is 895 kJ/mol. Predict the lattice energy of the isomorphous AgI using Eq. (21-6).

Solution. The value of $r_c + r_a$ is 3.21 Å for AgBr and 3.42 Å for AgI. If the only difference between AgBr and AgI were in the size of the anion, we would expect the lattice energies to be proportional to the inverse ratio of $r_c + r_a$. Hence we expect the lattice energy of AgI to be $(895)(3.21/3.42) = 840$ kJ/mol.

(b) The lattice energy of AgI is 883 kJ/mol. Explain why the actual value is larger than predicted in part (a).

Solution. The Ag^+ ion does not have a rare gas electronic configuration and is a good polarizer of anions. The iodide ion is larger and more polarizable than the bromide ion. Therefore the cation–anion bond has more covalent character in AgI than in AgBr. The lattice energy of AgI is therefore larger than the value calculated assuming the only difference between U for AgBr and for AgI is in the value of $r_c + r_a$ in Eq. (21-6).

Small, highly charged cations are more effective polarizers of anions than larger, singly charged cations. For two cations of the same charge that do not have the electronic configuration of a rare gas, the smaller one has a greater charge density and is the better polarizer of anions. Thus, there is more covalent character to the bond in CuI than in AgI, because Cu^+ is smaller than Ag^+.

As we stated earlier, relative solubilities of different ionic crystals cannot be predicted solely by comparing the lattice energies of the substances. We must also consider the hydration of the gaseous ions. The overall dissolution of an ionic crystal

$$MX(s) \rightarrow M^+(aq) + X^-(aq) \tag{21-8}$$

can be written as the sum of three reactions:

$$MX(s) \rightarrow M^+(g) + X^-(g) \qquad \Delta H = U \tag{21-5}$$

$$M^+(g) \rightarrow M^+(aq) \qquad \Delta H = \Delta H_{hyd}(M^+) \tag{21-9}$$

$$X^-(g) \rightarrow X^-(aq) \qquad \Delta H = \Delta H_{hyd}(X^-) \tag{21-10}$$

The enthalpy changes for reactions (21-9) and (21-10) are called, respectively, the heats of hydration of the M^+ and the X^- ions. Because of the strong ion-dipole forces of attraction between ions and water, the hydration of ions is exothermic and heats of hydration are negative quantities. Small, highly charged ions have more negative heats of hydration than larger, singly charged ions. Thus the same factors that make the lattice energy large and positive and result in strong forces of attraction between ions in the crystal lattice, also make the heats of hydration of the ions large and negative, and result in strong forces of attraction between the ions and water molecules. As a result, it is very difficult to predict relative solubilities from qualitative considerations. We must know ΔH and ΔS for each of the three reactions (21-5), (21-9), and (21-10) in order to calculate ΔG for the dissolution reaction and therefore predict the solubility.

section 21.9
The Born–Haber Cycle   no calc-for BohrHaber

Lattice energies of ionic crystalline solids are calculated using **Hess' Law** (refer to Section 17.5) and a series of six reactions called the **Born–Haber cycle**. The six reactions used in a Born–Haber cycle are described below for the calculation of the lattice energy of NaCl.

1. The vaporization (sublimation) of sodium metal.
$$Na(s) \rightarrow Na(g) \qquad \Delta H_1 = S = 108 \text{ kJ}$$
2. The ionization energy of sodium.
$$Na(g) \rightarrow Na^+(g) + e^- \qquad \Delta H_2 = IE = 496 \text{ kJ}$$
3. The formation of a mole of gaseous Cl atoms from chlorine in its standard state, gaseous Cl_2 molecules.
$$\tfrac{1}{2} Cl_2(g) \rightarrow Cl(g) \qquad \Delta H_3 = \tfrac{1}{2}D = 121 \text{ kJ}$$
4. The electron affinity of chlorine.
$$Cl(g) + e^- \rightarrow Cl^-(g) \qquad \Delta H_4 = -EA = -348 \text{ kJ}$$
5. The formation of 1 mol of solid NaCl from the elements in their standard states.
$$Na(s) + \tfrac{1}{2} Cl_2(g) \rightarrow NaCl(s) \qquad \Delta H_5 = \Delta H_f^0(NaCl) = -411 \text{ kJ}$$
6. The lattice energy of NaCl.
$$NaCl(s) \rightarrow Na^+(g) + Cl^-(g) \qquad \Delta H_6 = U$$

The sum of the first four of these equations is
$$Na(s) + \tfrac{1}{2} Cl_2(g) \rightarrow Na^+(g) + Cl^-(g)$$

If from this we subtract Eq. (5), the formation of NaCl from its elements, we will obtain the equation for the lattice energy of NaCl:
$$Eq.(6) = Eq.(1) + Eq.(2) + Eq.(3) + Eq.(4) - Eq.(5)$$

It is for this reason that the six equations are called a cycle. Applying Hess' Law to this cycle we obtain
$$U = S + IE + \tfrac{1}{2}D - EA - \Delta H_f^0$$

so that $U = 108 + 496 + 121 - 348 + 411 = 788 \text{ kJ}$, the lattice energy of 1 mol of NaCl.

The Born–Haber cycle can be used to obtain any one of the six ΔH values for the six appropriate equations, provided that the other five are known. Of these six, one of the most difficult to obtain experimentally is the electron affinity of the nonmetal. If the cation–anion bond is a pure ionic bond, a theoretical value of the lattice energy can be calculated using the Born–Mayer equation, and the Born–Haber cycle can then be used to obtain a value for the electron affinity of the nonmetal.

Summary

Some observable characteristics of crystalline solids are their distinctive geometries, sharp melting points, incompressibility, and anisotropy in their mechanical, electrical, and optical properties. These characteristics are associated with the regular arrangement of the atoms, ions, or molecules of which the crystal is composed in a repeating three-dimensional pattern. **Amorphous solids,** such as glasses and plastics, have no distinctive geometry, gradually soften over a moderately wide temperature range, and are isotropic. They are supercooled liquids without a regular repeating arrangement of the atoms, molecules, or ions of which they are composed.

All crystals have defects. Sometimes **vacancies** in the crystal structure occur, or there are impurities at atomic positions, or particles at interstitial sites in between atomic positions.

About 70% of all metals have a structure that is described as **close-packed.** In a close-packed structure each atom is in contact with 12 nearest neighbors, six surrounding it in one plane or layer, three in the layer above it, and three in the layer below it. There are two types of close-packed structures, **hexagonal and cubic close packing,** which differ only in the way the layers are stacked on one another. Another 25% of metals crystallize in a **body-centered cubic** structure, with atoms at the eight corners and the center of a **cubic unit cell.** Each atom in a body-centered cubic structure has 8 nearest neighbors.

For any metal atom, the number of valence electrons is less than the number of valence orbitals. With either 8 or 12 nearest neighbors, and a relatively small number of valence electrons, it is impossible for metal atoms to be bonded together in the solid state by typical localized covalent bonds. The valence atomic orbitals of the metal atoms overlap to form **bands** of very closely spaced molecular orbitals that belong to the entire solid. Because of the relatively low ionization energies of the valence electrons of metals, these electrons are easily excited to vacant orbitals in the band, and are **delocalized.** The electronic structure of a metal is described as a collection of metallic ions immersed in a sea of mobile, delocalized valence electrons. It is these mobile electrons that give rise to the high thermal and electrical conductivity of metals, as well as to their typical silvery luster.

All crystals can be classified as belonging to one of seven crystal systems: the cubic, tetragonal, orthorhombic, hexagonal, rhombohedral, monoclinic, or triclinic. These systems are distinguished by the geometry of their **unit cell.** The seven crystal systems can be further divided into 14 different lattice types. A **simple or primitive lattice** has lattice points only at the corners of each unit cell. A **body-centered cell** has lattice points at the corners and at the center, while a **face-centered cell** has lattice points at the corners and the six face centers. In the orthorhombic and monoclinic systems there are also end-centered lattices with lattice points at the centers of only two parallel faces as well as at the corners of the unit cell.

A particle located on an edge, face, or corner of a unit cell is shared by several unit cells and must not be counted more than once in calculating how many particles there are per unit cell.

The experimental technique used to measure interatomic distances and bond angles in the molecules or ions of which crystalline solids are composed is called **X-ray diffraction.** In this technique, monochromatic beams of X-rays impinge on a crystal that is slowly rotated to vary the angle between the incident X-ray beam and planes of atoms in the crystal. Only for those angles specified by **Bragg's Law,** $n\lambda = 2d \sin \theta$, will the beam of X-rays scattered by the crystal interfere constructively and result in an emergent beam of measurable intensity.

By measuring internuclear distances and the geometry of unit cells we obtain information about the sizes of atoms and ions. A summary of values of ionic radii obtained from measurements of ionic crystals is contained in Fig. 21.22. From a knowledge of the dimensions of the unit cell and Avogadro's number we can calculate the crystal density.

The **lattice energy** of an ionic crystal is the amount of energy per mole required to remove the ions from their lattice positions and separate them to an infinite distance in the gaseous state. Properties of solids that are frequently, but not invariably, associated with large values of the lattice energy are hardness, high melting point, and low solubility. The solubility depends on the heats of hydration of the ions as well as on the lattice energy and is difficult to predict from qualitative considerations.

Crystals with small, highly charged ions have larger lattice energies than crystals with larger, singly charged ions. If the bond between cation and anion has a partial covalent character, the lattice energy is increased to a value greater than that for a crystal with ions of the same charge and size but for which the bond is purely ionic. There will be greater covalent character to the bond if the cation does not have the electronic configuration of one of the rare gases. Such cations are effective polarizers of the charge clouds of anions, that is, they pull the anionic charge cloud into the region between anion and cation.

The **Born–Haber cycle** is a series of six reactions to which Hess' Law is applied in order to calculate the lattice energy of an ionic crystal. If the lattice energy can be calculated theoretically, the Born–Haber cycle can be used to determine ΔH for one of the other reactions in the cycle. It has frequently been used to determine the electron affinity of nonmetals.

Multiple Choice Questions

1. Potassium metal crystallizes in the body-centered cubic structure. The number of nearest-neighbor atoms for each potassium atom in the solid is

   (a) 4   (b) 6   (c) 8   (d) 10   (e) 12

2. Which of the following crystals has the largest lattice energy?

   (a) KCl   (b) RbI   (c) LiBr   (d) MgO   (e) NaF

3. The number of molecules per unit cell for a compound that crystallizes in an ortho-rhombic end-centered lattice with a molecule at each lattice site is

   (a) 1   (b) 2   (c) 4   (d) 6   (e) 8

4. A compound contains two types of atoms, X and Y. It crystallizes in a cubic lattice with X atoms at the corners of the unit cells and Y atoms at the body centers. The simplest formula of this compound is

   (a) X_8Y (b) X_2Y (c) XY   (d) XY_2 (e) XY_8

5. Which of the following crystals has the largest lattice energy?

   (a) CaI_2 (b) NiS   (c) KBr   (d) SrO   (e) BaS

6. The melting point of RbBr is 682 °C while that of NaF is 988 °C. The principal reason that the melting point of NaF is much higher than that of RbBr is that
   (a) The two crystals are not isomorphous.
   (b) The gram formula weight of NaF is smaller than that of RbBr.
   (c) The bond in RbBr has more covalent character than the bond in NaF.
   (d) The difference in electronegativity between Rb and Br is smaller than the difference between Na and F.
   (e) The internuclear distance, $r_c + r_a$, is greater for RbBr than for NaF.

7. If the radius of a metal atom is 2.0 Å and its crystal structure is cubic close-packed (face-centered cubic lattice), what is the volume, in cubic centimeters, of one unit cell?
   (a) 8.00×10^{-24} (b) 1.60×10^{-23} (c) 2.26×10^{-23} (d) 3.20×10^{-23}
   (e) 1.81×10^{-22}

8. For which of the following ionic crystalline solids does the cation–anion bond have the largest amount of covalent character?
   (a) CdS   (b) NaBr   (c) SrS   (d) BaO   (e) LiF

9. For a certain crystal the unit cell axial lengths are found to be $a = 5.62$ Å, $b = 7.41$ Å, and $c = 10.13$ Å. The three coordinate axes are mutually perpendicular. The crystal system to which this crystal belongs is the
   (a) tetragonal   (b) orthorhombic   (c) monoclinic   (d) triclinic
   (e) rhombohedral

10. Which of the following ions has the smallest radius?
    (a) Ti^{2+} (b) Pt^{2+} (c) Ni^{2+} (d) Zr^{2+} (e) Cd^{2+}

11. When the following five anions are arranged in order of decreasing ionic radius, the correct sequence is
    (a) $Se^{2-}, I^-, Br^-, O^{2-}, F^-$
    (b) $I^-, Se^{2-}, O^{2-}, Br^-, F^-$
    (c) $Se^{2-}, I^-, Br^-, F^-, O^{2-}$
    (d) $I^-, Br^-, F^-, Se^{2-}, O^{2-}$
    (e) $I^-, Se^{2-}, Br^-, O^{2-}, F^-$

12. For which of the following crystals would you expect the assumption of anion–anion contact to be valid?
    (a) CsBr   (b) NaF   (c) KCl   (d) NaI   (e) SrO

13. Which of the following ions has the largest heat of hydration?
    (a) Na^+ (b) Al^{3+} (c) F^- (d) Sr^{2+} (e) O^{2-}

14. Which of the following anions is most easily polarized?
    (a) Cl^- (b) Se^{2-} (c) Br^- (d) Te^{2-} (e) N^{3-}

15. Which of the following substances has the highest melting point?
    (a) Cs_2O (b) SeO_2 (c) LiF   (d) CaF_2 (e) NiO

16. Of the five ΔH values needed to calculate a lattice energy using the Born–Haber cycle, the one that is most difficult to measure is
    (a) The heat of sublimation of the metal.
    (b) The heat of formation of gaseous atoms of the nonmetal.
    (c) The ionization energy of the metal.
    (d) The electron affinity of the nonmetal.
    (e) The standard heat of formation of the crystal.

17. An ionic crystalline solid, MX_3, has a cubic unit cell. Which of the following arrangements of the ions is consistent with the stoichiometry of the compound?

    (a) M^{3+} ions at the corners, X^- ions at the face centers.
    (b) M^{3+} ions at the corners, X^- ions at the body centers.
    (c) X^- ions at the corners, M^{3+} ions at the face centers.
    (d) X^- ions at the corners, M^{3+} ions at the body centers.
    (e) M^{3+} ions at the corners and the body centers, X^- ions at the face centers.

18. If the three interaxial angles defining the unit cell are all equal in magnitude, the crystal cannot belong to the

    (a) orthorhombic system   (b) hexagonal system   (c) tetragonal system
    (d) cubic system   (e) rhombohedral system

19. Which of the following statements about metallic elements is FALSE?

    (a) These elements form ions larger in size than the neutral atoms.
    (b) These elements have relatively low ionization energies.
    (c) These elements have few outer shell electrons.
    (d) These elements have relatively low electronegativities.
    (e) These elements have relatively high densities.

Problems

21.1. (a) For each of the three cubic lattices, calculate the relationship between the axial length, a, of the unit cell and the atomic radius, r, of the atoms in the cell, assuming all the lattice sites are occupied by identical atoms that are in contact with their nearest neighbors.
    (b) From the number of atoms per unit cell and the volume of a sphere ($V = \frac{4}{3}\pi r^3$), calculate the percentage of the unit cell volume that is filled for each of the three cubic lattices, making the same assumptions as in part (a). Which of the three cubic lattices represents the most efficient packing of spheres into a volume?

21.2. There are several crystalline forms of metallic iron, depending on the pressure and temperature. One form, α-Fe, crystallizes in a body-centered cubic structure with a unit cell axial length of 2.8664 Å.
    (a) Calculate the atomic radius of Fe in α-Fe. Explain why it is not correct to report the atomic radius to five significant figures even though the unit cell length has been measured to five figures.
    (b) Calculate the density of α-Fe.

21.3. A second crystalline form of metallic iron, γ-Fe, crystallizes in the face-centered cubic structure. Assuming the atomic radius of Fe is 1.26 Å in this structure, calculate the length of the unit cell and the density of γ-Fe. Which form has a greater density, α-Fe or γ-Fe? Explain.

21.4. The lattice energy of CaF_2 is 2630 kJ/mol, while that of $CaCl_2$ is 2250 kJ/mol. Account for the difference between them.

21.5. The lattice energy of CuI is 958 kJ/mol, while that of NaI is 690 kJ/mol. Account for the difference between them.

21.6. The following experimental lattice energies are given:

BaO	3100 kJ/mol
CdS	3400 kJ/mol
RbCl	688 kJ/mol

Account for the differences between these three values. Suppose a theoretical calculation of a purely ionic lattice energy was carried out for each of these compounds. Would you expect the calculated value to be significantly larger than, essentially the same as, or significantly smaller than the experimental value given above? Explain.

21.7. Lithium bromide crystallizes in a face-centered cubic lattice, with a unit cell axial length of 5.501 Å. Specify any assumptions you may make about contacts between ions in the unit cell, and calculate values for the radii of Li^+ ions and Br^- ions. Compare your values with those listed in Fig. 21.22 and account for any differences.

21.8. A newly synthesized copper complex with molecular formula $C_{11}H_{29}N_5B_2Cu$ is found to crystallize in the orthorhombic system with $a = 14.749$ Å, $b = 17.604$ Å, and $c = 13.907$ Å. The density of this compound is measured and found to be 1.17 ± 0.01 g/cm³. How many molecules are there per unit cell?

21.9. Calcium crystallizes in a face-centered cubic lattice. The axial length of one unit cell is 5.57 Å. Calculate the radius of a calcium atom and the density of solid calcium.

21.10. The lattice energy of CuI is 958 kJ/mol while that of AgI is 883 kJ/mol. Account for the difference between them.

21.11. Calculate the lattice energy of KF from the following data:
The heat of dissociation of gaseous F_2 is 158.0 kJ/mol. The electron affinity of fluorine is 333 kJ/mol. The heat of sublimation of solid K is 89.91 kJ/mol. The ionization energy of K is 418.6 kJ/mol. The standard heat of formation of KF(crystal) is -562.6 kJ/mol.

21.12. The experimental uncertainties in lattice energies are about ± 5 to ± 10 kJ/mol. Thus the observed lattice energies for NaF (908 kJ/mol) and AgCl (904 kJ/mol) are essentially the same. Account for the similarity in these two values.

21.13. An unknown metal has a density of 21.45 g/cm³. It crystallizes in a face-centered cubic lattice, and the length of the unit cell is 3.9231 Å. Calculate the atomic weight of the metal. What is the metal?

21.14. The X-rays emitted by Mo have a wavelength of 0.7093 Å. When these X-rays impinge upon a platinum crystal, the minimum angle between the beam and a crystal plane for which a diffracted X-ray can be observed at the detector is 9.045°. Calculate the spacing between parallel planes of platinum atoms.

21.15. The "experimental" value of a lattice energy is the value calculated using the Born–Haber cycle, as the five ΔH values summed to calculate the lattice energy have been measured directly. Using the Born–Mayer equation a theoretical "pure-ionic" lattice energy can be calculated. This theoretical value is correct if the cation–anion bond is a pure ionic bond. When a comparison of the experimental lattice energy and the theoretical "pure-ionic" lattice energy is made for the copper(I) halides, the results are shown below.

	CuCl	*CuBr*	*CuI*
U (B–H cycle, kJ/mol)	979	971	958
U (theory, kJ/mol)	904	870	833
Difference (kJ/mol)	75	101	125

Why is there a difference between the experimental and theoretical values? Why does the difference between them increase regularly for the series CuCl, CuBr, and CuI?

21.16. Both magnesium oxide and lithium fluoride crystallize in a face-centered cubic structure. The properties of these two compounds differ markedly, as can be seen from the table below.

Property	*MgO*	*LiF*
Melting point, °C	2800	842
Hardness (Mohs scale)	6.5	3.3
Solubility (g/100 mL H_2O)	0.00062	0.27
Density (g/cm^3)	3.58	2.63
Internuclear distance (Å)	2.05	1.96

Account for the difference in properties between these two compounds.

21.17. The distance between centers of adjacent Na^+ and Cl^- ions in crystalline NaCl has been measured by X-ray diffraction to be 2.820 Å.
(a) What is the volume of one unit cell in the NaCl lattice?
(b) If the density of solid NaCl is $2.163 \text{ g} \cdot \text{cm}^{-3}$, how many unit cells are in 1 mol of NaCl?
(c) From the number of Na^+-Cl^- ion pairs per unit cell, and the data given above, calculate Avogadro's number.

chapter 22 Radioactivity and Nuclear Chemistry

Marie Sklodowska Curie (1867–1934) was born in Warsaw. During her student days she became involved in the students' revolutionary organization, and was forced to leave Poland. She went to Paris and spent the rest of her life there. She was the first person to receive two Nobel Prizes, one in Physics in 1903, for the discovery of radioactivity, which she shared with her husband, Pierre Curie, and with Henri Becquerel, and one in Chemistry in 1911, for the discovery of radium and the study of its properties. She and her husband also discovered the element polonium, which was named in honor of Marie Curie's native land.

In our study of chemistry we have spent very little time discussing properties of the nucleus other than the nuclear charge, and the effects of the nuclear charge on the orbital electrons. There are, however, many chemical uses for isotopes: Tracer studies are used to elucidate reaction mechanisms and isotopic dilution techniques are used to determine yields when the product cannot be completely separated from the reaction mixture. Radioactive isotopes have also provided a means to date material of historical and geological interest. The use of radioisotopes as a diagnostic tool in medicine has been steadily increasing; many hospitals now have a department of nuclear medicine.

The techniques we now have available for releasing the tremendous amounts of energy stored in nuclei have aroused a great deal of interest and controversy, both because we may someday need to rely on nuclear energy to provide us with electricity and because of the possibility of accidental release of harmful radiation. In this chapter we will study the properties of nuclei, emphasizing radioactive isotopes and the ways in which chemists utilize them.

section 22.1
Subatomic Particles and Nuclides

During the last several years a great deal of intensive research has been carried out on the structure of nuclei, and many new particles have been identified and described. Most of these particles, however, are of interest mainly to nuclear physicists. The subatomic particles that are important to chemists and biochemists are relatively few in number; information about these particles is summarized in Table 22.1.

The proton and neutron are collectively referred to as **nucleons.** As we have already mentioned in Section 1.2, **nuclides** are specific nuclear species characterized by a **mass number, A,** which is equal to the number of nucleons, and the atomic number, Z, which is equal to the number of protons in the nucleus. Since A is equal to the sum of the number of protons plus neutrons,

$$A - Z = N = \text{the number of neutrons} \qquad (22\text{-}1)$$

Isotopes are nuclides with the same Z but different A, such as $^{12}_{6}C$, $^{13}_{6}C$, and $^{14}_{6}C$. The symbol for an isotope is $^{A}_{Z}\text{Element}$, but if you read the older literature you may also see $_{Z}\text{Element}^{A}$, a notation that should no longer be used. **Isobars** are nuclides with the same A but different Z, such as $^{19}_{10}Ne$ and $^{19}_{9}F$.

Table 22.1. Subatomic Particles of Primary Interest to Chemists and Biologists

Particle	Symbol	Approximate Atomic Weight[a]	Charge[b]
Proton	p	1	$+1$
Neutron	n	1	0
Electron	e^- or β^-	5.5×10^{-4}	-1
Positron	e^+ or β^+	5.5×10^{-4}	$+1$
Deuteron	D^+ $(n + p)$	2	$+1$
Triton	T^+ $(2n + p)$	3	$+1$
Alpha	α or He^{2+}	4	$+2$

[a] In atomic mass units (amu).

[b] In units equal to the charge on the proton, which is 1.602×10^{-19} C.

section 22.2
Nuclear Binding Energies

The **binding energy** of a nucleus is the energy that would be released if a nucleus were constructed from its individual nucleons. For example,

$$6n + 6p \rightarrow {}^{12}_{6}C(\text{nucleus}) + \text{binding energy} \tag{22-2}$$

Alternatively, the binding energy can be defined as the amount of energy required to break a nucleus into its separate, constituent nucleons.

An atom of a stable isotope always has a mass *less* than the sum of the masses of electrons, protons, and neutrons that comprise the atom. The release of energy when nucleons combine to form a nucleus is accompanied by a decrease in the total mass of the system, in accord with the Einstein relation

$$\Delta E = (\Delta m)c^2 \tag{22-3}$$

where Δm is the loss of mass, ΔE is the energy released, and c is the speed of light. In SI units, $c = 2.997925 \times 10^8$ m $\cdot$ s^{-1}.

The unit of nuclear mass is the **atomic mass unit, amu,** which is defined as exactly $\frac{1}{12}$ of the mass of a ${}^{12}_{6}C$ atom. (See Section 1.3.) The masses of a proton and a neutron are not identical, although they are very close to one another. Values of these masses are listed in Table 22.2.

The mass of a hydrogen atom, ${}^{1}_{1}H$, is the sum of the masses of one proton and one electron, and is very slightly less than the mass of a neutron.

$$\text{mass } {}^{1}_{1}H = 1.00727647 + 0.00054858 = 1.00782505 \tag{22-4}$$

In order to determine the binding energy of the ${}^{12}_{6}C$ nucleus, as defined by Eq. (22-2), we would need to know the mass of the nucleus of the ${}^{12}_{6}C$ atom. It is difficult to determine nuclear masses by direct experimental means, but we do have accurate values of atomic masses. Thus we can compare the mass of six neutrons, six protons, and six electrons with the mass of a ${}^{12}_{6}C$ *atom,* and evaluate the binding energy using the relation

$$6n + 6p + 6e^- \rightarrow {}^{12}_{6}C(\text{atom}) + \text{binding energy} \tag{22-5}$$

instead of Eq. (22-2).

The difference between the binding energy calculated using Eq. (22-5) and that calculated using Eq. (22-2) is insignificant because the energy holding electrons around a nucleus is completely negligible compared to the energy binding together the protons and neutrons within the nucleus.

Because a hydrogen atom consists of one proton and one electron, the mass of six

Table 22.2. **Rest Masses of the Elementary Particles**[a]

Particle	Mass (amu)
Electron	0.0005485803(2)
Proton	1.00727647(1)
Neutron	1.00866501(4)

[a] The numbers in parentheses are the uncertainties in the last digit of the quoted value.

protons and six electrons is six times the mass of a hydrogen atom, so that the mass of the left-hand side of Eq. (22-5) can be calculated as

$$
\begin{aligned}
\text{mass of } 6n &= 6(1.00866501) = 6.05199006 \\
+ \underline{\text{mass of } 6(^1_1H)} &= 6(1.00782505) = \underline{6.04695030} \\
\text{total mass of left-hand side} &= 12.09894036
\end{aligned}
$$

Since the mass of one $^{12}_6C$ atom is exactly 12 amu, there is a mass loss of 0.0989404 amu on forming one $^{12}_6C$ atom from its nucleons, or a mass loss of 0.0989404 g on forming 1 mol of $^{12}_6C$ atoms. The binding energy per mole of $^{12}_6C$ can be calculated using the Einstein relation, Eq. (22-3). We obtain

$$(9.89404 \times 10^{-5} \text{ kg} \cdot \text{mol}^{-1})(2.997925 \times 10^8 \text{ m} \cdot \text{s}^{-1})^2 = 8.892 \times 10^{12} \text{ J} \cdot \text{mol}^{-1}$$

Thus the binding energy of $^{12}_6C$ is 8.892×10^9 kJ · mol⁻¹ or 2.125×10^9 kcal · mol⁻¹. This is an absolutely tremendous amount of energy! Remember that the energy lost or gained in most chemical reactions (typical values of ΔH, for instance) is of the order of 10^2 or 10^3 kJ · mol⁻¹, about *one million times smaller* than the binding energy of a typical nucleus.

Because nuclear binding energies are so large compared to ordinary chemical energies, the units of energy that are used by most chemists, kilojoules or kilocalories, are not sufficiently large to be convenient for nuclear reactions. The unit of energy commonly used by nuclear chemists is the **MeV** or **million electron volts**. One electron volt (eV) is the energy acquired by a single electron when it is accelerated through a potential difference of one volt. As we discussed in Chapter 16, the product of the charge in coulombs and the potential difference in volts is the energy in joules. Thus

$$1 \text{ eV} = (1.6022 \times 10^{-19} \text{ C})(1 \text{ V}) = 1.6022 \times 10^{-19} \text{ J} \qquad (22\text{-}6)$$

For 1 mol of electrons accelerated through a potential difference of 1 V, the energy acquired is $(6.022 \times 10^{23})(1.6022 \times 10^{-19} \text{ J})$, or 9.648×10^4 J. The relation between the energy units is therefore

$$1 \text{ eV/particle} = 96.48 \text{ kJ} \cdot \text{mol}^{-1} = 23.06 \text{ kcal} \cdot \text{mol}^{-1} \qquad (22\text{-}7)$$

One million electron volts is a million times larger, or 9.648×10^7 kJ · mol⁻¹.

The energy released when 1 amu of mass is lost on combining nucleons to form a nucleus is 931.50 MeV. This value is calculated in Example 22.1.

EXAMPLE 22.1. The energy equivalent of one atomic mass unit

Calculate the energy equivalent of 1 amu.

Solution. One amu is $\frac{1}{12}$ the mass of a $^{12}_6C$ atom. Since one $^{12}_6C$ atom has a mass of $12/N_A$ grams, where N_A is Avogadro's number, the mass in grams of 1 amu is

$$1/N_A = 1/(6.022045 \times 10^{23}) = 1.660565 \times 10^{-24} \text{ g} = 1.660565 \times 10^{-27} \text{ kg}$$

Using the Einstein relation, $E = mc^2$,

$$
\begin{aligned}
\text{the energy equivalent of 1 amu} &= (1.660565 \times 10^{-27} \text{ kg})(2.997925 \times 10^8 \text{ m} \cdot \text{s}^{-1})^2 \\
&= 1.49244 \times 10^{-10} \text{ J}
\end{aligned}
$$

Note that the fundamental constants, Avogadro's number and the speed of light, are known to seven significant figures, and to avoid rounding errors and obtain the energy equivalent of 1 amu accurately to five figures, we utilize the most accurate current values of the fundamental constants. The charge on an electron is 1.602189×10^{-19} C, so that there are 1.602189×10^{-19} J/eV, and hence

$$\text{the energy equivalent of 1 amu} = \frac{1.49244 \times 10^{-10} \text{ J}}{1.60219 \times 10^{-19} \text{ J} \cdot \text{eV}^{-1}} = 9.3150 \times 10^8 \text{ eV}$$
$$= 9.3150 \times 10^2 \text{ MeV} = 931.50 \text{ MeV}$$

We have already calculated that the mass loss on forming a $^{12}_6\text{C}$ atom from its nucleons is 0.0989404 amu. Using the energy equivalent of 1 amu we can express the binding energy of $^{12}_6\text{C}$ in million electron volts, the unit usually employed by nuclear chemists.

$$\text{binding energy of } ^{12}_6\text{C} = (0.0989404 \text{ amu})(931.50 \text{ MeV/amu}) = 92.163 \text{ MeV}$$

The binding energy is a measure of the stability of the nucleus. The greater the binding energy *per nucleon,* the more stable the nucleus. Thus to compare relative stabilities of different nuclei, we must calculate the binding energy per nucleon, which is abbreviated *BE/A*. For $^{12}_6\text{C}$, A is 12, so that $BE/A = 92.163/12 = 7.68$ MeV.

EXAMPLE 22.2. Calculating the binding energy per nucleon

Calculate the binding energy per nucleon for ^4_2He.

Solution. The binding energy of ^4_2He is defined by the equation

$$2n + 2p + 2e^- \rightarrow {}^4_2\text{He(atom)} + \text{binding energy}$$

The mass of the left-hand side of this equation is the mass of two neutrons and two hydrogen atoms.

$$\text{mass of } 2n = 2(1.00866501) = 2.01733002$$
$$+ \text{ mass of } 2({}^1_1\text{H}) = 2(1.00782505) = \underline{2.01565010}$$
$$\text{total mass of left-hand side} = 4.03298012$$

Virtually the only isotope of helium found in nature is ^4_2He (there is 0.00013% of ^3He). The mass of ^4_2He is therefore equal to the atomic weight of He. For most elements, several isotopes occur naturally, and the atomic weight is a weighted average of the isotopic masses. The mass of the specific isotope is needed to calculate its binding energy.

The mass of ^4_2He is 4.00260 amu. The loss of mass on forming the atom from its nucleons and electrons is

$$4.032980 - 4.00260 = 0.03038 \text{ amu}$$

Since the energy equivalent of 1 amu is 931.50 MeV,

$$\text{binding energy of } ^4_2\text{He} = (0.03038 \text{ amu})\left(931.50 \frac{\text{MeV}}{\text{amu}}\right) = 28.30 \text{ MeV}$$

As there are four nucleons ($2p$, $2n$), the binding energy per nucleon is

$$BE/A = 28.30/4 = 7.075 \text{ MeV}$$

A plot of the binding energy per nucleon as a function of the number of nucleons is shown in Fig. 22.1. A consideration of this plot provides us with important information. Several features that should be noted are the following:

1. The plot has a maximum at $A \sim 60$ (near the element iron). For mass numbers less than 60, if two light nuclei combine to form a heavier nucleus, the binding energy of the heavier nucleus is larger than the sum of the binding energies of the two lighter nuclei, and energy is released. The process of obtaining energy by fusing or combining two light nuclei to form a heavier one is called **fusion.**

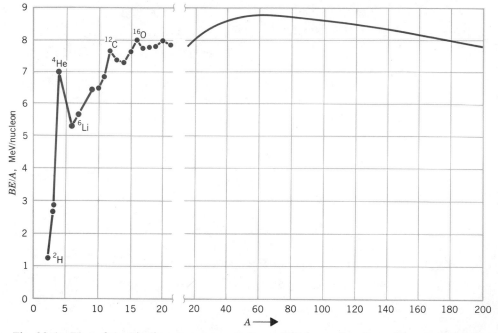

Fig. 22.1. Plot of the binding energy per nucleon, BE/A, as a function of the number of nucleons, A. The scale used for the abscissa is expanded for A less than 20, in order to show details of the curve for the lightest nuclei. The break in the curve indicates where the change in scale occurs.

For mass numbers greater than 60, the fusion of two nuclei to form a heavier nucleus would require energy, because the binding energy per nucleon is decreasing. However, the **fission** of a very heavy nucleus with mass number A greater than 60 into two lighter fragments releases energy. Both fission and fusion processes will be discussed in more detail a little later in this section.

2. The average binding energy per nucleon is remarkably constant for all nuclei except for a few of the lightest ones. For mass numbers greater than $A = 9$, all values of BE/A lie between 6.5 and 8.8 MeV, a rather narrow range. Only H, He, and Li have isotopes for which the binding energy per nucleon is less than 6 MeV. We have already seen that for $^{12}_{6}C$, $BE/A = 7.68$ MeV, and for $^{4}_{2}He$, $BE/A = 7.07$ MeV. For elements with Z greater than 6, if more specific data is not available, as a first approximation we can use the relation

$$\text{binding energy per nucleon} \sim 8 \text{ MeV} \qquad (22\text{-}8)$$

to estimate the binding energy of a given nuclide.

3. For low atomic numbers, the various maxima in the curve correspond to the nuclei $^{4}_{2}He$, $^{12}_{6}C$, $^{16}_{8}O$, $^{20}_{10}Ne$, and $^{24}_{12}Mg$, which are especially stable nuclei. Note that both Z and A are even numbers for these particularly stable nuclei.

Nucleon Pairing

The observation that exceptionally stable nuclei have both an even number of protons and an even number of neutrons suggests that like nucleons pair off in some way. We note, for instance, that ^{13}C is much less stable than ^{12}C; since ^{13}C has six protons and seven neutrons, at least one nucleon must be unpaired. When we examine the

Table 22.3. The Number of Stable Isotopes for Nuclides Characterized by Even or Odd Numbers of Protons and Neutrons

Type	Number
Z even, N even	165
Z even, N odd	56
Z odd, N even	53
Z odd, N odd	9

stable nuclei of all the elements, we find that the majority have both an even number of protons and an even number of neutrons, as shown in Table 22.3. This is explained in terms of the stabilization gained by the combination of like nucleons into pairs. Odd – odd nuclides are usually unstable; only nine stable ones are known. The four light odd – odd nuclei that are stable are ^2_1H, ^6_3Li, $^{10}_5B$, and $^{14}_7N$.

Fusion

When we discussed the information contained in Fig. 22.1, we observed that the fusion of two light nuclei to form a heavier one will release energy, provided that the sum of the mass numbers of the two light nuclei is less than 60. The greatest release of energy is possible when the very lightest nuclei fuse, because the slope of the binding energy per nucleon curve is steepest for low A.

Fusion reactions, in fact, provide the energy radiated by the sun and the stars. The principal reaction in the sun is the fusion of four hydrogen atoms to form a single helium atom. There is spectroscopic evidence that indicates the mass of the sun is 73% hydrogen atoms, 26% helium atoms, and only 1% of all other elements. The fusion of hydrogen atoms into helium takes place through a series of steps involving the production of several intermediate nuclides, including deuterium, ^2_1H, tritium, ^3_1H, and ^3_2He. Among the many nuclear reactions that are believed to occur at the very high temperature of the sun are the following:

$$^1_1H + {}^1_1H \rightarrow {}^2_1H + {}^0_1\beta \tag{22-9}$$

$$^2_1H + {}^1_1H \rightarrow {}^3_2He \tag{22-10}$$

$$^3_2He + {}^3_2He \rightarrow {}^4_2He + 2{}^1_1H \tag{22-11}$$

$$^3_2He + {}^1_1H \rightarrow {}^4_2He + {}^0_1\beta \tag{22-12}$$

$$^{14}_7N + {}^1_1H \rightarrow {}^{15}_7N + {}^0_1\beta \tag{22-13}$$

$$^{15}_7N + {}^1_1H \rightarrow {}^{12}_6C + {}^4_2He \tag{22-14}$$

Note that each of these equations is balanced both with respect to the nuclear charge and to the nuclear mass. For any nuclear reaction, the sum of the Z values on the left- and right-hand sides of the equation must be equal, as must the sum of the A values on each side of the equation. For example, in Eq. (22-14), the Z values balance because $7 + 1 = 6 + 2$, and the A values balance because $15 + 1 = 12 + 4$.

Scientists have been seriously considering how to utilize fusion reactions to provide the energy required to produce electricity. The difficulty is that before nuclei can be fused, the large repulsion between the two positively charged nuclei must be overcome. To overcome the coulombic repulsion of the two nuclei, the collision energy must be very large, and this requires enormous temperatures. Fusion reactions are therefore also called **thermonuclear reactions**. The only fusion reaction that has been carried out here on earth is the one that is used in the hydrogen bomb

$$^2_1H + {}^3_1H \rightarrow {}^4_2He + {}^1_0n + 17.6 \text{ MeV} \tag{22-15}$$

To achieve the high temperature required, 40,000,000 K, an atom bomb is used to ignite the fusion process.

A serious practical problem in utilizing fusion reactions to produce energy is that there is no substance that can remain solid at the very high temperatures necessary to initiate fusion, so there is no material container in which to place the reacting species. At the temperatures required for fusion, all atoms are ionized, so the gaseous matter consists of charged particles, including electrons. In order to confine the species to be fused, matterless "walls" of magnetic fields have been used. These intense magnetic fields keep the hot gases concentrated inside the reaction volume while the density is increased sufficiently to achieve fusion. Recent research has centered on the possibility of using lasers to generate the extremely high temperatures required to initiate fusion. A great deal of research remains to be done before the fusion process becomes a practical means of supplying our energy needs.

Fission

The reactions used in the nuclear power plants operating today are fission reactions, in which a very heavy nucleus, typically ^{235}U, is split into lighter fragments, releasing a portion of the nuclear binding energy of the heavy nucleus as heat, which is then converted into electricity. Fission was also the process used in the atomic bombs the United States dropped on Hiroshima and Nagasaki to end World War II.

Fission is accomplished by bombarding heavy nuclei with other particles such as neutrons, protons, deuterons, and alpha particles. By far the most commonly used projectiles are slow-moving neutrons, also called **thermal neutrons,** meaning a neutron that has the same kinetic energy as a gas molecule at approximately room temperature.

During the 1930's, scientists were trying to produce **transuranium elements,** that is, elements with atomic numbers greater than 92. Attempts to accomplish this involved bombarding uranium with thermal neutrons. Between 1934 and 1938 more than 90 articles were published describing experiments in which uranium was bombarded with neutrons. Among the scientists actively engaged in this type of research were Enrico Fermi in Rome, and Otto Hahn, Fritz Strassman, and Lise Meitner in Berlin. In 1938 Otto Hahn identified barium as one of the products of the reaction between thermal neutrons and uranium. Further work by Hahn and Strassman proved that strontium, yttrium, rubidium, and cesium, in addition to barium, were reaction products. Hahn and Strassman found these results surprising, because they expected to obtain only products with atomic number greater than 92. At this time Lise Meitner, who had been forced to flee from Germany because she was Jewish, was working in Sweden. Hahn wrote to her, describing his experimental results, and she proposed that the uranium nucleus had been split into lighter fragments, a process never observed before then. She named this new process **nuclear fission.**

When a heavy nucleus is struck by a neutron it can split in many different ways, so that large numbers of fragments are obtained. More than 200 different isotopes of 35 different elements have been found among the fission products of ^{235}U. Two of the many ways that the ^{235}U nucleus splits are shown below:

$$_{0}^{1}n + {}_{92}^{235}U \rightarrow {}_{56}^{142}Ba + {}_{36}^{91}Kr + 3\,{}_{0}^{1}n \tag{22-16}$$

$$_{0}^{1}n + {}_{92}^{235}U \rightarrow {}_{52}^{137}Te + {}_{40}^{97}Zr + 2\,{}_{0}^{1}n \tag{22-17}$$

A particularly striking feature of these reactions is that the fission produces more neutrons than are used to cause the fission. These neutrons, in turn, can strike other uranium nuclei and produce additional fission reactions. With each fission reaction,

Lise Meitner (1878–1968) was born in Vienna, Austria, and was one of the first women to receive a doctorate from the University of Vienna. She studied under Max Planck in Berlin, and was associated with Otto Hahn in research on radioactivity for over thirty years. She and Hahn jointly discovered the element protactinium. Meitner did research on the relationship between the beta and gamma radiation of radioactive substances. In the late 1930's she and Hahn and Fritz Strassman carried out the research that led to the discovery of uranium fission. Because she was Jewish, in 1938 Meitner was forced to flee from Nazi-occupied Austria to Sweden. Hahn wrote to her describing the surprising results of experiments they had been carrying out, and it was Meitner who suggested the results indicated that uranium atoms had been split into lighter fragments, and coined the name "fission."

energy is released, and if the number of fissions increases rapidly with the production of more neutrons, a violent explosion can take place. This succession of fission reactions is called a **chain fission reaction**.

If the sample of fissionable material is small, most neutrons escape from the sample before they can strike another uranium nucleus, and a chain reaction does not occur. There is a minimum size necessary for the mass of fissionable material before a chain reaction can take place. A mass too small to permit a chain reaction is said to be **subcritical**. A **critical mass** is just large enough to maintain a chain reaction with a constant rate of fission. The magnitude of the critical mass varies with the nucleus being split. Among the heavy nuclei that have been induced to undergo fission are ^{235}U, ^{233}U, and ^{239}Pu. The critical mass also depends to a certain extent on the shape of the sample. If the mass of the fissionable material is larger than the critical mass, relatively few neutrons escape and the reaction becomes **supercritical**. In the atomic bomb, two subcritical masses are brought together to form a supercritical mass when the bomb is to be exploded.

In nuclear reactors for power generation, the concentration of fissionable material is, of course, kept below supercritical levels. Fuel rods, containing uranium enriched with ^{235}U or some other fissionable nucleus, are used as the reactor core. The most frequently used nuclear fuel in the United States consists of pellets of U_3O_8 in which the abundance of ^{235}U has been increased to about 3% (the natural abundance of ^{235}U is 0.72%). The flux of neutrons is controlled by rods containing cadmium or boron, materials that are good absorbers of neutrons. Boron removes neutrons by the reaction

$$^{10}_{5}B \, + \, ^{1}_{0}n \, \rightarrow \, ^{7}_{3}Li \, + \, ^{4}_{2}\alpha$$

Inserting the **control rods** more deeply into the reactor core can stop the fission reaction. A schematic diagram of a nuclear reactor is shown in Fig. 22.2.

Both the fission products and the used fuel rods are highly radioactive, and the waste products must be stored for many hundreds of years before the radioactivity falls to a safe level. The problems involved with safely storing dangerous materials for such long periods of time are obviously tremendous. To mention only one consideration: If you bury the radioactive waste in the earth, how can you be sure that there will not be an earthquake near that location during the next several hundred years?

Breeder Reactors

In current nuclear reactors in the United States the ^{235}U fuel is used once and all products are discarded. No fuel is reprocessed or reused. It is possible to build reactors in which new fuel is produced as the reactor produces heat. In these reactors, called **breeder reactors,** neutrons are absorbed either by ^{232}Th or ^{238}U, both of which are more plentiful than ^{235}U. The reactions that occur when these two nuclides absorb neutrons

$$^{238}_{92}U \, + \, ^{1}_{0}n \, \rightarrow \, ^{239}_{94}Pu \, + \, 2 \, ^{0}_{-1}\beta$$
$$^{232}_{90}Th \, + \, ^{1}_{0}n \, \rightarrow \, ^{233}_{92}U \, + \, 2 \, ^{0}_{-1}\beta$$

produce fissionable nuclei, $^{239}_{92}U \, + \, ^{233}_{92}U$, that can be used as fuel for the nuclear reactor.

Many questions remain unresolved about the advisability of producing breeder

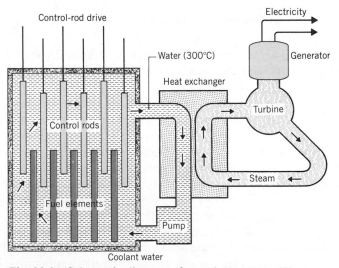

Fig. 22.2. Schematic diagram of a nuclear reactor. The control rods contain some material that is a good absorber of neutrons, such as Cd or B. They can be moved in or out of the reactor core to control the number of nuclei undergoing fission per unit time. The fuel elements contain the fissionable material, usually U_3O_8 pellets enriched with ^{235}U.

reactors. First and foremost, $^{239}_{94}Pu$ is exceedingly toxic. Inhaling even a very small amount can be fatal. The potential dangers of an accident using a breeder reactor are therefore greater than the potential dangers with the nuclear reactors currently in use. Secondly, the major use today of $^{239}_{94}Pu$ is for nuclear weapons. With breeder reactors, there would be a greater supply of $^{239}_{94}Pu$ available, and a consequent risk that some will be stolen and diverted for use in weaponry. There are also difficulties associated with the design of a breeder reactor. Water, which is used as a coolant for current nuclear reactors (see Fig. 22.2) cannot be used for breeder reactors, which operate at higher temperatures. Liquid sodium is used as a coolant instead. Because liquid sodium is highly reactive (see Sections 2.2 and 13.10) it tends to attack the walls of its container, and greater care must be taken in handling and monitoring the cooling system.

The principal reason that scientists are trying to develop nuclear power plants that utilize fusion rather than fission is because fusion reactions produce much less radioactive waste material than fission reactions. At the present time, however, all working nuclear power plants use fission as the source of energy.

section 22.3
Radioactivity

Nuclear Stability

Table 22.3 lists the number of stable isotopes of different types; many more nuclei are unstable and spontaneously emit radiation and are thereby changed into some other nucleus. Often the new nucleus is also unstable and there is a complex decay scheme until a stable nucleus is finally achieved.

The most important factor in determining nuclear stability is the ratio of neutrons to protons in the nucleus. If we plot the number of neutrons, N, as a function of the number of protons, Z, for all the stable nuclei, we find that there is a **belt of stability:** A band of values of the ratio of neutrons to protons (n/p) within which all the stable nuclei lie. This is shown in Fig. 22.3.

For light nuclei, up to $Z = 20$ (calcium), most stable nuclei have equal numbers of protons and neutrons, and the ratio $n/p = 1$ provides the most stability. As Z increases, the coulombic repulsive force between the protons increases, and the number of neutrons in the nucleus must become larger than the number of protons in order for a nucleus to be stable. By increasing the number of neutrons, the size of the nucleus increases, diminishing the coulombic repulsions somewhat, and increasing the total nuclear binding forces. As Z increases above 20, therefore, the value of the n/p ratio necessary for stability increases above unity, and reaches a value of 1.5 about $Z = 80$. Table 22.4 summarizes the values of n/p that lie on the belt of stability at selected values of Z. Note that the belt of stability terminates at $Z = 83$ (bismuth). For higher Z, all isotopes are unstable, although some have very long half-lives. For example, the half-life of ^{238}U is 4.5×10^9 yr.

Table 22.4. **Values of the Ratio of Neutrons to Protons for Nuclei that Lie within the Belt of Stability at Selected Values of the Atomic Number Z**

Up to	$Z = 20$	$n/p \sim 1.0$
At	$Z = 40$	$n/p \sim 1.25$
At	$Z = 60$	$n/p \sim 1.4$
Above	$Z = 80$	$n/p \sim 1.5$

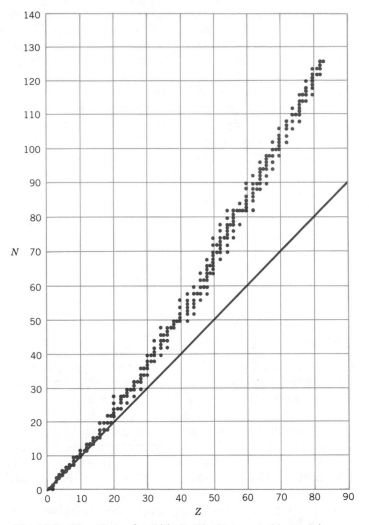

Fig. 22.3. The "belt of stability": The known stable nuclei on a plot of the number of neutrons, N, as a function of the number of protons, Z. Note the gradual increase in the neutron to proton ratio as Z increases. The 45° line corresponds to a neutron to proton ratio of unity.

It is important to remember that the value of n/p that achieves stability is not unique, but covers a small range. Tellurium, for instance, has seven stable isotopes with mass numbers 122, 123, 124, 125, 126, 128, and 130. Since $Z = 52$ for tellurium, the range of values of n/p for stable nuclei varies from $70/52 = 1.35$ to $78/52 = 1.50$ for $Z = 52$. The preceding element, antimony, has only two stable nuclei, $^{121}_{51}\text{Sb}$ and $^{123}_{51}\text{Sb}$, so that the n/p range for $Z = 51$ is $70/51 = 1.37$ to $72/51 = 1.41$.

Shell Model of the Nucleus

One might well inquire why, if adding neutrons tends to diminish the coulombic repulsions between the protons within a nucleus, nuclei are unstable if the value of n/p increases above a certain figure. Why do we observe a belt or band of stability, rather than just a minimum value of n/p required for stability?

This phenomenon has been explained by the **shell model** of the nucleus. Nuclei with 2, 8, 20, 28, 50, 82, or 126 neutrons or protons are particularly stable and abundant in nature, and nuclei that have 2, 8, 20, 28, 50, 82, or 126 neutrons are observed to be unusually inert with respect to capturing neutrons that are projected at them. This suggests that there are shells of nucleons that are filled when either the number of neutrons or protons is 2, 8, 20, 28, 50, 82 or 126. These numbers are referred to as nuclear **magic numbers**. Protons and neutrons exist in separate shells. Thus nuclei with either $Z = 50$ (Sn) or $N = 50$ ($^{88}_{38}$Sr or $^{90}_{40}$Zr) are unusually stable, but nuclei with $N + Z = 50$ are not. One exceptionally stable nucleus, and the end product of many radioactive decay schemes, is $^{208}_{82}$Pb, for which $Z = 82$ and $N = 126$, so that there is a closed (filled) shell of both neutrons and protons.

After a shell of neutrons is filled, any additional neutrons must enter a higher energy shell. This results in decreased stability. Careful mass measurements for nuclides in the neighborhood of $N = 50$, 82, and 126 and $Z = 28$, 50, and 82, have shown that the first nucleon outside one of these closed shells is always weakly bound. The decrease in stability that occurs when neutrons must be added to high energy neutron shells is considered to be responsible for the fact that nuclei with too large an n/p ratio are unstable.

Beta Decay for Neutron-Rich Nuclei

Nuclei lying above the belt of stability are neutron rich, that is, the value of n/p is too high for the nucleus to be stable. Such nuclei achieve stability by emitting beta particles (electrons). This process can be pictured as a transformation of a neutron in the nucleus to a proton and an electron, with the subsequent emission of the electron from the nucleus. The process $^1_0n \rightarrow {}^1_1p + {}^{\ 0}_{-1}\beta$ decreases the number of neutrons by one and increases the number of protons by one, therefore n/p is decreased.

A spontaneous β-decay process releases energy, and although there is no change in the mass number, A, there is a slight decrease in the exact nuclear mass. The heavier nuclide decays to the lighter one, and the decrease in mass appears principally as the kinetic energy of the products. Two examples of β decay are

$$^{14}_{6}C \rightarrow {}^{14}_{7}N^+ + {}^{\ 0}_{-1}\beta^- \tag{22-18}$$

for which the initial value of n/p is $8/6 = 1.33$, which is too large for stability at $Z = 6$, so that decay occurs to produce a nucleus with $n/p = 7/7 = 1.0$, and

$$^{116}_{49}In \rightarrow {}^{116}_{50}Sn^+ + {}^{\ 0}_{-1}\beta^- \tag{22-19}$$

for which the initial value of $n/p = 67/49 = 1.367$ (too large) and the product nucleus has $n/p = 66/50 = 1.320$.

To calculate the energy released in reaction (22-18), we must compare the mass of a $^{14}_{6}C$ atom with the mass of a neutral $^{14}_{7}N$ atom, because the right-hand side consists of a $^{14}_{7}N^+$ ion plus an electron. (Remember that a β^- particle is an electron.) The two species on the right-hand side therefore have the same mass as a neutral $^{14}_{7}N$ atom.

$$\begin{aligned} \text{mass of } {}^{14}C &= 14.0032419 \text{ amu} \\ - \text{ mass of } {}^{14}N &= 14.0030744 \text{ amu} \\ \hline \text{mass loss} &= \ 0.0001675 \text{ amu} \end{aligned}$$

The energy equivalent to this mass loss is

$$\left(931.50 \ \frac{\text{MeV}}{\text{amu}}\right)(1.675 \times 10^{-4} \text{ amu}) = 0.1560 \text{ MeV}$$

We might, therefore, expect that the emitted β^- particles will have a kinetic energy

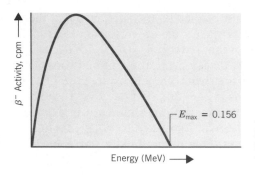

Fig. 22.4. Spectrum of energies of β^- particles emitted when ^{14}C decays. The number of counts per minute (cpm) at each value of the energy is plotted.

equal to 0.1560 MeV. Experiment shows, however, that the emission of β^- particles is more complex than this.

When the energies of β^- particles emitted by ^{14}C are investigated, it is found that every value between 0 and 0.1560 MeV is observed. A plot of the number of electrons emitted per unit time as a function of the energy of the emitted electrons produces the spectrum shown in Fig. 22.4. Very few of the emitted β^- particles are observed to have a kinetic energy equal to 0.1560 MeV; most have kinetic energies roughly one third this value. Conservation of energy requires that the difference between 0.156 MeV and the observed energy of the β rays be accounted for. In 1931, Enrico Fermi, elaborating on a suggestion of Wolfgang Pauli's, proposed that the only way to account for the energy difference is to assume that another particle is emitted simultaneously with the β^- particle.

Experiments to observe such a particle during the 1930's proved that no charged particle or particle of significant mass is emitted. Fermi therefore proposed that the additional particle be called a **neutrino,** indicating a tiny, uncharged particle, of almost zero mass. It was not until 1956, twenty-five years after Fermi's original proposal, that the neutrino was actually detected, and shown to have the properties predicted for it. When any neutron-rich unstable nucleus decays, emitting a β^- particle, a neutrino (denoted v) is emitted simultaneously. Equation (22-18) should therefore be written more exactly as

$$^{14}_{6}C \rightarrow {}^{14}_{7}N^{+} + {}^{0}_{-1}\beta^{-} + v$$

Decay Processes for Neutron-Poor Nuclei

Nuclei that lie below or to the right of the belt of stability have too small an n/p ratio to be stable. They must decrease the number of protons, and/or increase the number of neutrons in order to achieve stability. There are three different decay processes that serve to increase the n/p ratio. These are (1) **electron capture,** (2) **positron emission,** and (3) **alpha (α) decay.** Of these, α decay is observed only for very heavy elements, generally Z greater than 80, and it is only for heavy elements that α decay increases the n/p ratio.

Electron Capture

Electron capture involves the capture by the nucleus of one of the atom's orbital electrons, followed by conversion of a proton to a neutron. The orbital electron captured is usually from the K shell (that is, the shell with the two $1s$ electrons), and the process is then also called K **capture.** Sometimes L shell electrons are captured. Examples of electron capture (EC) are

$$^{205}_{82}\text{Pb} \xrightarrow{\text{EC}} {}^{205}_{81}\text{Tl} \qquad (22\text{-}20)$$

and

$$^{195}_{79}\text{Au} \xrightarrow{\text{EC}} {}^{195}_{78}\text{Pt} \qquad (22\text{-}21)$$

For reaction (22-21) note that the unstable gold nucleus has 116 neutrons and 79 protons, for an n/p ratio of 1.468. The value 1.468 is too low for $Z = 79$, and the capture of an orbital electron that combines with a proton in the nucleus to form a neutron, increases the number of neutrons to 117, while decreasing the number of protons to 78. The ratio n/p becomes $117/78 = 1.50$, and $^{195}_{78}\text{Pt}$ is a stable nuclide.

After electron capture, the new nucleus has the same mass number, A, but one less proton than the original. Electron capture is hard to detect since no particle is emitted. However, the capture of an electron is usually followed by transitions of the orbital electrons. If a K shell electron is captured, there is a vacancy in the K shell. An L shell electron may then drop down into that vacant spot, with emission of radiation in the X-ray region. Sufficient transitions of orbital electrons occur until there are no vacancies in inner electron shells.

Positron Emission

An unstable nucleus with an n/p ratio too low for stability may also emit a **positron,** β^+, a particle with the same mass as an electron but a positive charge equal in magnitude to the electronic charge. Positron emission converts a proton in the nucleus into a neutron: $p^+ \rightarrow n + \beta^+$. An example of positron emission is

$$^{19}_{10}\text{Ne} \rightarrow {}^{19}_{9}\text{F}^- + \beta^+ + \nu \qquad (22\text{-}22)$$

Note that a neutrino is emitted simultaneously with a positron, just as a neutrino is emitted simultaneously with β^- emission. Any β-decay process involves the simultaneous emission of a neutrino.

Since the original ^{19}Ne atom had 10 orbital electrons, there are still 10 electrons around the product F nucleus, and hence it is an F^- ion that is the initial decay product. The right-hand side of Eq. (22-22) therefore has the mass of a neutral $^{19}_{9}\text{F}$ atom plus the mass of one electron plus the mass of one positron. Since an electron and a positron have the same mass, the right-hand side of Eq. (22-22) has the mass of a $^{19}_{9}\text{F}$ atom plus the mass of two electrons. The mass loss when reaction (22-22) occurs is therefore

$$\text{mass of } ^{19}\text{Ne} - \text{mass of } ^{19}\text{F} - \text{mass of } 2\ e^-$$

The mass of an electron is 5.486×10^{-4} amu, while that of ^{19}Ne is 19.001892 amu and of ^{19}F is 18.99840 amu. The energy released when reaction (22-22) occurs is therefore

$$[19.001892 - 18.99840 - 2(5.486 \times 10^{-4})] \left(931.50\ \frac{\text{MeV}}{\text{amu}} \right) = 2.231\ \text{MeV}$$

Positron emission and electron capture are competing processes and it is not possible to predict with certainty which will occur to correct an n/p ratio that is too small. We do observe that electron capture occurs more often when Z is large, and positron emission occurs more often when Z is small.

If $Z < 30$, isotopes that are neutron poor almost always undergo positron emission. There are exceptions to this generalization; $^{7}_{4}\text{Be}$, for example, undergoes electron capture. If $Z > 70$, electron capture is more likely than positron emission. For

Table 22.5. Decay Processes Observed for the Isotopes of Carbon

$_6^{10}C$	$_6^{11}C$	$_6^{12}C$	$_6^{13}C$	$_6^{14}C$	$_6^{15}C$
$\dfrac{n}{p} = \dfrac{4}{6} = 0.67$	$\dfrac{5}{6} = 0.83$	$\dfrac{6}{6} = 1.0$	$\dfrac{7}{6} = 1.17$	$\dfrac{8}{6} = 1.33$	$\dfrac{9}{6} = 1.50$
Unstable, n/p too small positron emitters		Stable		Unstable, n/p too large β^- emitters	

$30 < Z < 70$, both processes are likely to occur and only experiment can answer the question of how the n/p ratio is increased. Frequently both electron capture and positron emission occur for the same unstable nucleus.

Table 22.5 summarizes the decay processes observed for the isotopes of carbon, which illustrate the generalizations made about the decay of neutron-rich and neutron-poor nuclei.

α Decay

For nuclei with very large Z, the third process by which the n/p ratio can be increased is the emission of an α-particle. All the elements above Pb, $Z = 82$, have some isotopes that emit α-particles, and a few elements between $Z = 60$ and $Z = 82$ are also α emitters. An α-particle consists of two protons plus two neutrons, so that α decay results in a decrease of the mass number by 4 and a decrease in the atomic number by 2. Let us consider the following example of an unstable nucleus that emits α-particles:

$$^{208}_{84}Po \rightarrow {}^{204}_{82}Pb^{2-} + {}^4_2He^{2+} \tag{22-23}$$

The initial product is a Pb^{2-} ion because there are still the 84 orbital electrons of the Po atom. As the α-particles collide with molecules in the air or with the product $^{204}Pb^{2-}$ ions, they quickly pick up electrons and become helium gas. The final products of this decay are neutral ^{204}Pb atoms and He gas.

Note that the decay process, Eq. (22-23) has increased the n/p ratio. The original n/p ratio for ^{208}Po was $124/84 = 1.476$, which is too low for $Z = 84$. The product nucleus, ^{204}Pb, has an n/p ratio of $122/82 = 1.488$, and is one of the four stable isotopes of lead.

To calculate the energy released in the decay of ^{208}Po, we note that the right-hand side of Eq. (22-23) has the mass of a neutral ^{204}Pb atom plus a neutral ^4_2He atom. Since the mass of ^{208}Po is 207.98126 amu, while the mass of ^{204}Pb is 203.97307 and the mass of ^4_2He is 4.00260 amu, the mass loss when ^{208}Po decays is

$$207.98126 - 203.97307 - 4.00260 = 0.00559 \text{ amu}$$

The energy released is therefore

$$(0.00559 \text{ amu})\left(931.50 \frac{\text{MeV}}{\text{amu}}\right) = 5.21 \text{ MeV}$$

The 5.21 MeV is shared by the two product nuclei, as the Pb nucleus recoils when the α-particle is emitted. (This is the same mechanical phenomenon as the recoil of a rifle when it shoots forth a bullet.) The law of conservation of momentum requires that each α-particle carry away (204/208) of the kinetic energy, so that the α-particles are emitted with a kinetic energy of

$$\left(\frac{204}{208}\right)(5.21 \text{ MeV}) = 5.11 \text{ MeV}$$

The remainder of the energy, or $(4/208)(5.21) = 0.10$ MeV, goes into the recoil of the Pb nucleus.

It is characteristic of α decay that all the α-particles emitted have a sharply defined energy. This is in contrast to β emission, which consists of a broad spectrum of energies due to the simultaneous emission of a neutrino. Generally the α-particles emitted on nuclear decay have energies between 3 and 9 MeV, but because they are heavy and doubly charged they cannot penetrate very far into matter. A thick sheet of cardboard or a film of water a few millimeters in thickness is sufficient protection against α rays. The penetrating power of β rays is greater than that of α rays.

Example 22.3 illustrates how we can predict the decay products of various unstable nuclei.

EXAMPLE 22.3. Nuclear decay processes

List the most likely modes of decay for the following unstable nuclei (whose half-lives are given for information only) and write a balanced nuclear equation for each decay process.

(a) $^{13}_{7}\text{N}$ ($\tau_{1/2} = 10.0$ min)   (b) $^{25}_{11}\text{Na}$ ($\tau_{1/2} = 60$ s)
(c) $^{149}_{63}\text{Eu}$ ($\tau_{1/2} = 106$ day)   (d) $^{210}_{85}\text{At}$ ($\tau_{1/2} = 8.3$ h)

Solution

(a) For $^{13}_{7}\text{N}$ the n/p ratio is 6/7 or 0.86, which is too low for stability. Because Z is less than 30, the most likely mode of decay is by positron emission:

$$^{13}_{7}\text{N} \rightarrow {}^{13}_{6}\text{C}^{-} + {}^{0}_{1}\beta^{+} + \nu$$

Electron capture is possible, but is not as likely as positron emission. In fact, ^{13}N is observed to be a β^{+} emitter.

(b) For $^{25}_{11}\text{Na}$, the n/p ratio is $14/11 = 1.27$, which is too large for $Z = 11$. The only possible mode of decay is β^{-} emission.

$$^{25}_{11}\text{Na} \rightarrow {}^{25}_{12}\text{Mg}^{+} + {}^{0}_{-1}\beta^{-} + \nu$$

(c) For $^{149}_{63}\text{Eu}$, the n/p ratio is $86/63 = 1.365$, which is too low for stability at $Z = 63$. At this Z both electron capture and positron emission are likely, although electron capture is more likely because Z is fairly high. In fact, $^{149}_{63}\text{Eu}$ does decay by electron capture.

$$^{149}_{63}\text{Eu} \xrightarrow{\text{EC}} {}^{149}_{62}\text{Sm}$$

(d) For $^{210}_{85}\text{At}$, the n/p ratio is $125/85 = 1.47$, which is too low for $Z = 85$. At this value of Z both electron capture and α decay are possible, but α decay is more likely. In fact, both electron capture and α decay are observed for $^{210}_{85}\text{At}$.

$$^{210}_{85}\text{At} \xrightarrow{\text{EC}} {}^{210}_{84}\text{Po}$$
$$^{210}_{85}\text{At} \rightarrow {}^{206}_{83}\text{Bi}^{2-} + {}^{4}_{2}\text{He}^{2+}$$

Gamma-Ray Emission

In addition to α decay and β decay (either β^{-} or β^{+}), nuclei emit a third type of radiation, γ rays. **Gamma rays** are not material particles, but are electromagnetic radiation, or photons, of very short wavelength. Since the relation between wave-

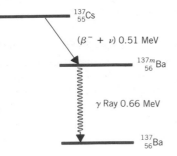

Fig. 22.5. Decay of ^{137}Cs to an isomeric state of ^{137}Ba, followed by γ emission to the nuclear ground state of ^{137}Ba.

length and frequency is $\lambda v = c$ [see Eq. (12-6)], a short wavelength is synonymous with a high frequency. As the energy of a photon is $h v$, high frequency means high energy. Gamma rays are therefore high energy radiation.

The emission of gamma rays is analogous to the emission of light in electronic transitions, as in the emission spectra of atoms, which was discussed in Section 12.2. Nuclei, like electrons, can exist in a discrete set of allowed energy levels, consisting of a nuclear ground state and additional higher energy excited states. An excited state of a nucleus is referred to as an **isomeric** state. A γ ray is emitted during a transition from a higher energy isomeric state to the nuclear ground state or to a lower energy isomeric state.

Consider the decay of the unstable nucleus $^{137}_{55}$Cs, for which the n/p ratio is $82/55 = 1.491$, a value too large for stability at $Z = 55$. The ^{137}Cs nucleus decays by emitting simultaneously a β^- particle and a neutrino, but the product nucleus is an excited or isomeric state of $^{137}_{56}$Ba, denoted $^{137m}_{56}$Ba. The isomeric state further decays to the ground state of ^{137}Ba by emission of a γ ray. The total decay scheme is illustrated in Fig. 22.5.

The γ ray emitted by ^{137m}Ba has a wavelength $\lambda = 0.0188$ Å or 1.88×10^{-3} nm. That is a very short wavelength, and the photons emitted have an energy of 0.66 MeV. Note particularly that the energy differences for nuclear transitions are much greater than for electronic transitions. Recall that transitions in the spectrum of atomic hydrogen (Sections 12.2 and 12.3) are of the order of a few electron volts. Nuclear transitions are about a million times larger, that is, they are of the order of 1 MeV (million electron volts).

An isomeric nucleus that is used extensively in medicine is $^{99m}_{43}$Tc, which decays to the ground state with a 6.0 h half-life, emitting a γ ray of 0.1427 MeV. The relatively short half-life makes this nucleus particularly useful. An injection of ^{99m}Tc bound to other substances that accumulate in specific organs such as the lung or liver, allows for a photographic scan of the organ that is used to diagnose abnormalities such as tumors or damaged tissue.

EXAMPLE 22.4. Calculating the wavelength of γ radiation

What is the wavelength of the 0.1427 MeV γ ray emitted by ^{99m}Tc?

Solution. The relation between the energy of the radiation and its wavelength is $E = h v = hc/\lambda$, so that $\lambda = hc/E$. We must express the energy in joules if we use Planck's constant in joule seconds. The relation between electron volts and joules has been given in Eq. (22-6). The energy of this transition is

$$(0.1427 \times 10^6 \text{ eV})(1.6022 \times 10^{-19} \text{ J/eV}) = 2.286_3 \times 10^{-14} \text{ J}$$

The wavelength of the emitted radiation is

$$\lambda = \frac{(6.6262 \times 10^{-34} \text{ J} \cdot \text{s})(2.9979 \times 10^8 \text{ m} \cdot \text{s}^{-1})}{2.2863 \times 10^{-14} \text{ J}} = 8.689 \times 10^{-12} \text{ m}$$

or,

$$\lambda = 8.689 \times 10^{-3} \text{ nm} = 0.08689 \text{ Å}$$

section 22.4
Naturally Occurring Radioactive Substances

All elements found in nature with atomic numbers greater than 83 (bismuth) are radioactive. They belong to one of three **radioactive decay series,** chains of successive decays that begin with a parent nucleus of very long half-life. All the isotopes produced in one such chain of decays constitute a radioactive series.

The Uranium Series

The parent nucleus of the uranium series is ^{238}U, which has a half-life of 4.5×10^9 yr. After 14 successive decays, 8 by emission of an α-particle and 6 by emission of a β particle, a stable isotope, ^{206}Pb, is produced. Since the atomic mass number, A, decreases by 4 after α emission, and does not change at all following β emission, all isotopes in this series have mass numbers less than 238 by multiples of 4, that is, 238, 234, 230, and so on, down to 206. Any one of these numbers can be written as $4n + 2$, where n is an integer, and so the uranium series is also known as the $4n + 2$ series. (See Fig. 22.6.)

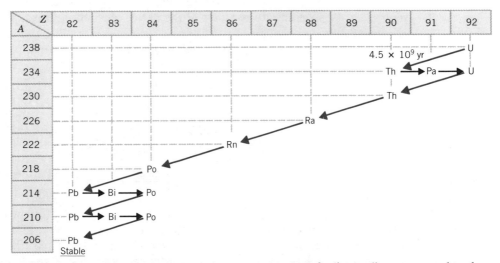

Fig. 22.6. The uranium or $4n + 2$ series. Arrows to the left, diagonally, correspond to the emission of an α-particle. Arrows to the right, horizontally, correspond to the emission of a β particle. A small number of the nuclei shown have alternate modes of decay, but the alternate modes occur much less frequently.

The Thorium Series

The parent nucleus of the thorium series is $^{232}_{90}$Th, which has a half-life of 1.39×10^{10} yr. The stable end product of this series is ^{208}Pb. All isotopes in this series have mass numbers differing from 232 by multiples of 4. Since 232 is an integral multiple of 4, all members of this chain have mass numbers that are integral multiples of 4, and the thorium series is also called the $4n$ series.

The Actinium Series

The parent isotope of this chain is ^{235}U, which has a half-life of 7.1×10^8 yr, and the stable end product is ^{207}Pb. The actinium series is also called the $4n + 3$ series.

section 22.5
The Rate of Radioactive Decay

Let us imagine that we could examine a single unstable nucleus. Will it decay (that is, emit radiation) in the next second, a week from now, or 87 years from now? It is impossible to predict. The moment when α-, β^-, or β^+ particles escape is a matter of chance. However, if we have a sample containing a very large number of unstable nuclei, we can calculate what *fraction* will decay in a given time, because radioactive decay is described by a first-order rate law. (See Section 19.4.)

The rate of decay at any instant (the time rate of change in the number of unstable nuclei) is directly proportional to the number of radioactive nuclei present at that time:

$$\text{rate of decay} = -dN/dt = \lambda N = \text{activity of sample} \tag{22-24}$$

The activity is expressed either as distintegrations per minute or per second (*dpm* or *dps*), or as counts per minute or per second (*cpm* or *cps*). The proportionality constant, λ, is the **specific rate constant** or the **decay constant,** and is a property of the particular radioactive nucleus.

The integrated form of the first-order rate law, Eq. (19-19), can be written either as

$$\ln(N/N_0) = -\lambda t \tag{22-25a}$$

or as

$$N = N_0 e^{-\lambda t} \tag{22-25b}$$

where N_0 is the number of radioactive nuclei present at zero time. Since the activity is directly proportional to N, Eq. (22-25a) can also be written as

$$\ln\left(\frac{\text{activity at time } t}{\text{activity at time zero}}\right) = -\lambda t \tag{22-26}$$

A plot of the natural logarithm of the activity as a function of time will therefore be a straight line with slope $-\lambda$, and it is from the slope of such a plot that λ can be determined. A typical problem has been worked out in Example 19.6.

For radioactive decay, as for all first-order decay processes, the **half-life,**

$$\tau_{1/2} = (\ln 2)/\lambda = 0.693/\lambda$$

is a characteristic property of the decaying nucleus. The half-life is independent of the amount of material present and is used to identify various radioactive species. Both the decay constant λ and the half-life $\tau_{1/2}$ are virtually independent* of the physical or chemical state of the radioactive atom. For instance, ^{14}C is a β^- emitter with a half-life of 5730 yr, and the half-life is the same whether the ^{14}C is in CO_2, or any organic molecule, such as cellulose in plants, or in solid coal.

Because it is usually the half-life, rather than the decay constant, that is reported for radioactive species, it is convenient to write Eq. (22-25a) so that the half-life is specifically indicated. Substituting $\lambda = (\ln 2)/\tau_{1/2}$ into Eq. (22-25a), we obtain

$$\ln(N/N_0) = -(\ln 2)t/\tau_{1/2} = -0.693t/\tau_{1/2} \qquad (22\text{-}27)$$

We can also write this equation using base 10 logarithms:

$$\log(N/N_0) = -(\log 2)t/\tau_{1/2} = -0.301t/\tau_{1/2} \qquad (22\text{-}28)$$

The use of Eqs. (22-27) and (22-28) is illustrated in Examples 22.5 and 22.6.

EXAMPLE 22.5. Definition of the curie†

By definition, one **curie** (Ci) is an activity of 3.700×10^{10} disintegrations per second (dps). The unstable nucleus ^{14}C is a β^- emitter with a half-life of 5730 yr. What mass of ^{14}C is required in order to have an activity of 1 Ci?

Solution. The activity of a sample is given by λN. Since N is the number of radioactive nuclei, λ must be in reciprocal seconds if the activity is expressed in distintegrations per second. To obtain λ in reciprocal seconds we must express the half-life, $\tau_{1/2}$, in seconds.

$$\tau_{1/2} = (5730 \text{ yr}) \left(365 \, \frac{\text{day}}{\text{yr}} \right) \left(24 \, \frac{\text{h}}{\text{day}} \right) \left(60 \, \frac{\text{min}}{\text{h}} \right) \left(60 \, \frac{\text{s}}{\text{min}} \right) = 1.807 \times 10^{11} \text{ s}$$

Therefore

$$\lambda = 0.6931/\tau_{1/2} = 0.6931/1.807 \times 10^{11} = 3.84 \times 10^{-12} \text{ s}^{-1}$$

How many ^{14}C atoms are needed for the activity to be 3.700×10^{10} dps?

$$\lambda N = (3.84 \times 10^{-12} \text{ s}^{-1})(N \text{ atoms}) = 3.700 \times 10^{10} \text{ dps}$$

Thus,

$$N = \frac{3.700 \times 10^{10}}{3.84 \times 10^{-12}} = 0.9645 \times 10^{22} = 9.65 \times 10^{21} \text{ atoms}$$

Since 1 mol of carbon-14 contains 6.022×10^{23} atoms, the number of moles of carbon-14 needed to have an activity of 1 Ci is

$$\frac{9.65 \times 10^{21} \text{ atoms}}{6.022 \times 10^{23} \text{ atoms} \cdot \text{mol}^{-1}} = 1.60 \times 10^{-2} \text{ mol of } ^{14}C$$

A mole of ^{14}C weighs 14.0 g, and therefore we need

$$(14.0 \text{ g} \cdot \text{mol}^{-1})(1.60 \times 10^{-2} \text{ mol}) = 0.224 \text{ g of } ^{14}C$$

to have a sample with an activity of 1 Ci.

---

* Small changes of a few percent in the value of several decay constants have been observed in recent years when atoms were subjected to large changes in pressure or temperature, or to electric and magnetic fields.

† The curie was originally the number of dps emitted by 1 g of pure radium. With that definition, the numerical value of the curie kept changing as more accurate values of the atomic weight and decay constant for radium were determined. In 1950 an international committee adopted the definition now in use.

Remember that the 0.224 g is the mass of the carbon alone. If the ^{14}C were in $^{14}CO_2$, we would need 1.60×10^{-2} mol of $^{14}CO_2$, or

$$(46.0 \text{ g} \cdot \text{mol}^{-1})(1.60 \times 10^{-2} \text{ mol}) = 0.736 \text{ g of } ^{14}CO_2$$

EXAMPLE 22.6. Use of the radioactive decay rate law

A radioactive isotope commonly used in biotracer studies of phosphorus metabolism is ^{32}P, which has a half-life of 14.3 days. Suppose we have been doing experiments with ^{32}P and accumulate waste material that has an activity of 1 **millicurie** (1 mCi = 1×10^{-3} Ci). The radiation safety officer in our laboratories rules that it is not safe to dispose of this waste material until the activity has fallen to 0.010 μCi [1 microcurie $(\mu Ci) = 1 \times 10^{-6}$ Ci]. How long must we store the waste in a lead container before it is safe to dispose of it?

Solution. The activity now is 1×10^{-3} Ci. We want to calculate the length of time it takes for the activity to decrease to 1×10^{-8} Ci, as 0.010 μCi is $10^{-2} \times 10^{-6}$ Ci. Using Eqs. (22-26) and (22-28) we obtain

$$\log\left(\frac{N}{N_0}\right) = \log\left(\frac{10^{-8}}{10^{-3}}\right) = \frac{-0.301t}{\tau_{1/2}} = \frac{-0.301t}{14.3}$$

Simplifying this expression we have

$$\log(10^{-5}) = -5 = -0.301t/14.3$$

which yields

$$t = \frac{(5)(14.3 \text{ days})}{0.301} = 238 \text{ days}$$

Note that there is no need to convert the half-life from 14.3 days to any other unit of time. The units of t and $\tau_{1/2}$ must be identical. If $\tau_{1/2}$ is expressed in days, then t will automatically be in days. Note also that it is more convenient to use base 10 logarithms in this problem because both the initial and final activities are expressed simply as powers of 10.

section 22.6
The Uses of Radioactivity

Isotope Dilution

Isotope dilution is used to determine the yield of a product that is difficult to separate from its reaction mixture. If you can separate some of the product from its mixture, but cannot quantitatively separate all of the product, you can still calculate your yield using the technique of isotope dilution.

As an example of the use of this technique, consider the following problem. Suppose we have hydrolyzed a protein and now have a mixture of the amino acids that constituted the protein. We want to know how much of one specific amino acid, glycine, is present in the mixture, but it is not possible to separate all of the glycine quantitatively from the other amino acids. We therefore add to the reaction mixture a known amount of radioactive glycine. The calculation is simplest if the radioactive isotope used has an exceedingly long half-life so that in the space of time it takes to perform the experiment the activity can be considered to be essentially constant.

Since it will usually take a few hours to perform the experiment, we need an isotope with a half-life so long that the activity does not decrease significantly in a few hours. An ideal isotope for the purpose is ^{14}C, which has a half-life of 5730 yr. Glycine prepared with ^{14}C is commercially available.

Let us assume that the radioactive glycine we have purchased has a **specific activity** of 35×10^3 cpm/g. Note the difference between the total activity, measured in counts per minute, and the **specific activity**, measured in counts per minute per gram. We add to the reaction mixture 0.520 g of the radioactive glycine, and stir until the mixture is homogeneous. We have now diluted the activity, because we have mixed a small amount of radioactive glycine with a much larger amount of nonradioactive glycine. We then extract from the reaction mixture a small amount of glycine; let us say that we can isolate 0.180 g of pure glycine from the mixture. The glycine isolated will contain both radioactive and nonradioactive molecules. The activity of the isolated glycine is measured and is found to be 117 cpm. The specific activity of the isolated glycine is therefore 117 cpm/0.180 g = 650 cpm/g.

If we assume that the total activity has not decreased significantly during the time it took us to add the radioactive glycine, stir, isolate a small amount of glycine from the mixture, and then measure its activity, we can calculate the weight, W, of glycine that was in the original reaction mixture. The total activity of the added glycine was

$$\left(35 \times 10^3 \; \frac{\text{cpm}}{\text{g}}\right)(0.520 \text{ g}) = 18,200 \text{ cpm} = \text{total activity before mixing}$$

The total activity after diluting the radioactive glycine with the nonradioactive glycine is

$$\left(650 \; \frac{\text{cpm}}{\text{g}}\right)(W + 0.520)\text{g} = \text{total activity after mixing}$$

Assuming no measurable decrease in activity during the time of the experiment (because of the very long half-life of ^{14}C), we simply equate the expressions for the total activity before and after mixing:

$$(650)(W + 0.520) = 18,200$$

Therefore $W + 0.520 = 18,200/650 = 28.0$, and $W = 27.5$ g. Thus the hydrolyzed protein mixture contained 27.5 g of glycine.

In Example 22.7 we will derive a formula for use in isotope dilution problems. In solving such problems, however, it is more useful to reason the solution through than to use the derived formula.

EXAMPLE 22.7. Derivation of the isotope dilution formula

Derive an expression that can be used to calculate the weight, W, of product in a mixture, using the technique of isotope dilution. Let

$$w_{\text{add}} = \text{weight of added radioactive material}$$
$$A_i = \text{initial } \textit{specific} \text{ activity of added material}$$
$$A_f = \text{final } \textit{specific} \text{ activity of isolated sample of product}$$

Solution

$$\text{total activity before mixing} = A_i w_{\text{add}}$$
$$\text{total activity after mixing} = A_f(W + w_{\text{add}})$$

If the half-life of the radioactive isotope is large enough so that the decrease in activity during the time of the experiment is negligible, then the total activity before and after mixing are the same. Hence

$$A_i w_{add} = A_f(W + w_{add}) \qquad (22\text{-}29)$$

Solving this equation for W we obtain

$$W = \left(\frac{A_i - A_f}{A_f}\right) w_{add} \qquad (22\text{-}30)$$

If the product is a liquid, the specific activity may be measured in counts per minute per milliliter, and the volume of the product may be used in place of the weight. Example 22.8 illustrates this use of isotope dilution.

EXAMPLE 22.8. The isotope dilution method

In order to determine the volume of blood in an animal without killing it, a 1.00-mL sample of an aqueous solution containing tritium is injected into the animal's bloodstream. The sample injected has an activity of 1.8×10^6 cps. After sufficient time for the sample to be completely mixed with the animal's blood due to normal blood circulation, 2.00 mL of blood are withdrawn from the animal, and the activity of the blood sample withdrawn is found to be 1.2×10^4 cps. Calculate the volume of the animal's blood.

Solution. Let

$$V = \text{volume of blood in this animal}$$
$$\text{total activity before mixing} = 1.8 \times 10^6 \text{ cps}$$
$$\text{total activity after mixing} = (V \text{ mL}) \left(\frac{1.2 \times 10^4 \text{ cps}}{2.00 \text{ mL}}\right)$$

Hence

$$1.8 \times 10^6 = V(0.60 \times 10^4)$$

and solving for V we obtain

$$V = \frac{1.8 \times 10^6}{0.6 \times 10^4} = 3.0 \times 10^2 = 300 \text{ mL}$$

Radiocarbon Dating

Carbon-14 is produced in the upper atmosphere as a result of **cosmic ray** interaction with nitrogen. Cosmic radiation originates in outer space, has extremely high penetrating power, and consists of many elementary particles and high-energy atomic nuclei. Neutrons, produced by the interaction of primary cosmic rays with atoms in the upper atmosphere, react with ^{14}N in the atmosphere to produce ^{14}C:

$$^{14}_{7}N + ^{1}_{0}n \rightarrow ^{14}_{6}C + ^{1}_{1}H \qquad (22\text{-}31)$$

Carbon-14 decays by emitting β^- particles. [See Eq. (22-18).]

The ^{14}C is gradually oxidized to $^{14}CO_2$, which mixes with the ordinary, nonradioactive CO_2 in the atmosphere. Because of winds, the mixing is thorough and the same concentration of ^{14}C is observed in the atmosphere all over the world. This steady state concentration of ^{14}C is about one atom of ^{14}C to 10^{12} atoms of ordinary carbon. The CO_2, both radioactive and nonradioactive, is absorbed by plants during photosynthesis. Animals eat these plants and the radioactive carbon is then fixed in animal tissues. As a result all carbonaceous material from plants or animals has a certain amount of radioactive ^{14}C.

Because there is a continuous intake from the atmosphere and exhalation back into the atmosphere of CO_2 by living plants and animals, the amount of ^{14}C per gram of total carbon remains constant at the steady state value of the atmosphere during the lifetime of any plant or animal. After death, however, the radioactivity decreases as the ^{14}C present decays and there is no longer any intake of CO_2. By determining the radioactivity of a sample of carbon from wood, bone, hide, horn, etc., the number of years that have elapsed since the death of the plant or animal can be calculated. The technique of dating objects by measuring the radioactivity due to ^{14}C (**radiocarbon dating**) was developed by Willard F. Libby, an American physical chemist. Libby won the Nobel Prize in 1960 for his work in this field.

In order to date items on the basis of their ^{14}C content, we must make two important assumptions. First we must assume uniform distribution of ^{14}C over the earth. This has been checked by collecting fresh wood samples from trees in different localities over the globe. The average ^{14}C content of fresh wood is 15.3 ± 0.1 dpm/g of total carbon. Note that the carbon in the wood must be reduced to elementary carbon to determine the specific activity per gram of carbon.

The second assumption is that the steady state concentration of ^{14}C in the atmosphere that we observe today has been the same for many thousands of years. This assumption turns out not to be exactly correct, as we will discuss later, and tables have been prepared to correct for changes in the rate of production of ^{14}C by cosmic radiation thousands of years ago.

Example 22.9 illustrates the calculations used to date an object after its β-ray activity has been measured.

EXAMPLE 22.9. Radiocarbon dating

A specimen of a sequoia tree is observed to emit β^- rays with a specific activity of 10.8 dpm/g of carbon. The radioactivity is due to ^{14}C, which has a half-life of 5730 yr. How long ago did the tree die?

Solution. At the time of death, which is zero time, the specific activity of the wood was 15.3 dpm/g of carbon, the value for living wood all over the earth. Using Eqs. (22-26) and (22-27), we obtain

$$\ln\left(\frac{10.8}{15.3}\right) = -\frac{0.693t}{5730 \text{ yr}} \qquad \text{where } t \text{ is in years}$$

the same time unit used for the half-life. This equation is solved for t as follows:

$$\ln(0.7059) = -0.3483 = -0.693t/5730$$

so that

$$t = (5730 \text{ yr})(0.3483)/0.693 = 2.88 \times 10^3 \text{ yr}$$

This sequoia tree therefore died about 2880 years ago.

How good is the method of radiocarbon dating? Checks have been made by dating objects from Egyptian tombs and other sites that had previously been dated by archaeologists using historical evidence. In general, agreement is very good. Some

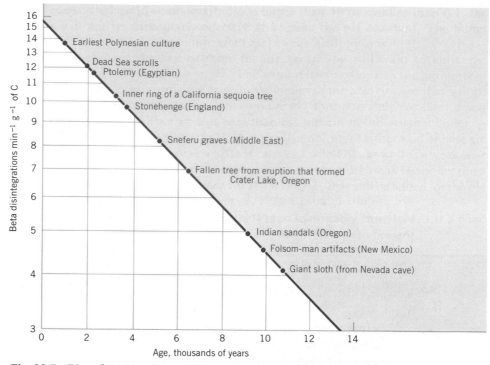

Fig. 22.7. Plot of the logarithm of the activity (the β disintegration rate in counts per minute per gram) as a function of age for some objects that have been dated by the radiocarbon technique.

of the objects that have been dated by this method, and their ages, are shown in Fig. 22.7.

Libby found that for materials more than 7000 years old there is an error of some 600–700 years in the age obtained by radiocarbon dating. Discrepancies between the dates of some Egyptian articles obtained by the ^{14}C method and the dates deduced by archaeologists have been explained as due to a failure in the assumption that the steady state concentration of ^{14}C in the atmosphere has been the same over many thousands of years.

The earth's magnetic field normally deflects about half the cosmic rays from entering the atmosphere. (A beam of moving charged particles is bent or deflected by a magnetic field. See Section 12.1.) If the earth's magnetic field differed from its present value many thousands of years ago, the amount of ^{14}C in the atmosphere would also have been different from the present value. The ages of objects obtained by the radiocarbon dating technique always appear to be several hundred years too young, compared to the dates given by archaeologists. A similar discrepancy is observed when comparing the age of a giant Sequoia by counting the tree rings and the age calculated as in Example 22.9. This means that there is more β-ray activity than expected for an object of that age. The explanation given for this discrepancy is that the earth's magnetic field was somewhat lower 7000 years ago than it is now. A lower magnetic field would have deflected a smaller fraction of the cosmic rays, and therefore the ^{14}C level in the atmosphere was somewhat higher. The present concentration of ^{14}C has been constant for about 5000 years.

Radiocarbon dating cannot be used for objects more than 25,000 years old, because the level of β-ray activity decays during that length of time to a value too small to be measured accurately. Isotopes with much longer half-lives than 5730 years must be used to date objects significantly older than 25,000 years, but the same principles apply.

Geologists use a number of methods involving radioactive decay to determine the time since the formation or solidification of particular minerals and thereby to obtain a value for the age of the earth. One of the most useful techniques is the determination of the uranium to lead ratio in rocks containing ^{238}U, which has a half-life of 4.5×10^9 yr. The lead isotope ^{206}Pb is the stable end product, after a complex series of steps (see Section 22.4), of ^{238}U decay. Provided there is no other source of lead in a mineral that contains uranium, the ratio of uranium-238 to lead-206 can be used to determine the amount of uranium decay that has occurred. In order to find out if the lead in the rock is present only as the result of uranium decay, it is analyzed using a mass spectrometer. Naturally occurring lead in the earth's crust contains a small percentage of ^{204}Pb. The absence of ^{204}Pb in the rock is presumed to indicate the absence of ordinary lead.

The calculations required to determine the age of rock samples containing uranium-238 by measuring the relative amounts of ^{206}Pb and ^{238}U are illustrated in Example 22.10.

EXAMPLE 22.10. Determining the age of rocks containing uranium

A sample of a rock is analyzed and found to contain 0.232 g of ^{206}Pb and 1.605 g of ^{238}U. Assuming all the ^{206}Pb now present came from the decay of ^{238}U, calculate the time since the formation of this rock.

Solution. The first thing to calculate is the amount of ^{238}U that decayed to yield the 0.232 g of ^{206}Pb. Since 238 g of uranium will decay to yield only 206 g of lead, the mass of uranium that decayed to lead-206 was more than 0.232 g.

$$\text{mass of } ^{238}U \text{ that decayed} = \left(\frac{238}{206}\right)(0.232 \text{ g}) = 0.268 \text{ g}$$

Thus the rock originally contained $1.605 + 0.268 = 1.873$ g of uranium-238. Over the years, 0.268 g of ^{238}U decayed, leaving 1.605 g of ^{238}U and producing 0.232 g of ^{206}Pb.

If zero time is the time when the rock was formed and there was no ^{206}Pb present, the time, t, it took for the present composition to be attained is calculated by utilizing Eq. (22-27) and the known half-life of ^{238}U, 4.5×10^9 yr. We obtain

$$\ln(1.605/1.873) = -(0.693t)/4.5 \times 10^9$$

Note that the ratio N/N_0 is the same as the ratio of the weight of ^{238}U at time t to the weight of ^{238}U at time zero, because the number of ^{238}U atoms is directly proportional to the weight of ^{238}U. Solving for t we obtain

$$\ln(0.8569) = -0.1544 = -(0.693t)/4.5 \times 10^9$$

and therefore

$$t = (0.1544)(4.5 \times 10^9)/0.693 = 1.00 \times 10^9$$

This rock was formed one billion years ago.

Note particularly that this calculation is correct only if all the ^{206}Pb now found in the rock is the product of ^{238}U decay.

Radioactive Tracer Techniques

Radioactive isotopes are widely used in chemistry, biochemistry, and medicine in **tracer studies,** to follow the course of molecules as they take part in a series of chemical reactions.

The elucidation of mechanisms of complex organic reactions has often been carried out with the help of radioisotopes. For instance, by feeding CO_2 labeled with ^{14}C to living plants, it has been possible to determine the molecules in which the radioactivity appears as time passes, and thus to figure out the course of the photosynthetic process. A simple example will illustrate the concepts utilized in this type of reasoning.

Consider the reaction between any carboxylic acid and an alcohol, known as **esterification.** To be specific, we will use methanol, CH_3OH, as the alcohol, but will describe the carboxylic acid only as RCOOH, where R stands for some organic group of atoms. The esterification reaction is

$$\underset{\text{Carboxylic acid}}{R-\overset{\overset{\displaystyle O}{\|}}{C}-OH} + \underset{\text{Methyl alcohol}}{CH_3OH} \rightarrow \underset{\text{Methyl ester}}{R-\overset{\overset{\displaystyle O}{\|}}{C}-OCH_3} + H_2O \qquad (22\text{-}32)$$

Two possible mechanisms can be formulated for the elimination of the water molecule when the alcohol and acid react:

Mechanism 1. The bond broken in the acid is the O—H bond, and the bond broken in the alcohol is the C—O bond, as follows:

$$R-\overset{\overset{\displaystyle O}{\|}}{C}-O\!\!\dashv\!\!H + HO\!\!\dashv\!\!CH_3 \rightarrow R-\overset{\overset{\displaystyle O}{\|}}{C}-OCH_3 + H_2O$$

Mechanism 2. The bond broken in the acid is the C—O bond, and the bond broken in the alcohol is the O—H bond, as follows:

$$R-\overset{\overset{\displaystyle O}{\|}}{C}\!\!\dashv\!\!OH + H\!\!\dashv\!\!OCH_3 \rightarrow R-\overset{\overset{\displaystyle O}{\|}}{C}-OCH_3 + H_2O$$

The problem of determining which of these two mechanisms is correct was solved by using CH_3OH labeled with ^{18}O. If the methyl ester and water are completely separated after the reaction is complete, we need only find out where the ^{18}O appears in the products. If mechanism 1 is correct, all the ^{18}O will be in the H_2O formed, whereas if mechanism 2 is correct, all the ^{18}O will be in the methyl ester, in the —OCH_3 group.

Using several different acids, as well as different alcohols, it has been conclusively established by this technique that mechanism 2 is correct, since the ^{18}O always appears in the ester formed, and not in the water, if the alcohol has the labeled oxygen.

Summary

When **nucleons** (protons and neutrons) combine to form a nucleus, energy is released and the mass of the system decreases. The energy released is the **binding energy** of the nucleus, and the decrease in mass is given by the Einstein relation $E = mc^2$, where E is the binding energy and m is the mass loss.

Nuclear binding energies are very large compared to ordinary chemical energies. The unit of energy used to report nuclear binding energies is the million electron volt, **MeV**, 1×10^6 eV. One MeV per particle is 9.648×10^7 kJ/mol. A mass loss of 1 amu releases 931.50 MeV.

The larger the binding energy per nucleon, the more stable the nucleus. A plot of binding energy per nucleon (BE/A) versus the mass number, A, has a maximum at

$A \sim 60$ (near iron). The average binding energy per nucleon is remarkably constant for all nuclei except for a few of the lightest ones. Only H, He, and Li have isotopes for which BE/A is less than 6 MeV. For $A > 9$, all values of BE/A lie between 6.5 and 8.8 MeV. The increase in BE/A with increase in A is steepest for the very lightest nuclei.

Nuclei with even Z and even N are particularly stable, and are more abundant than all other stable nuclei. Stable nuclei with odd Z and odd N are very uncommon. From these facts it is concluded that like nucleons pair off in some way.

If two light nuclei combine to form a heavier nucleus of mass number $A < 60$, energy is released. This process is called **fusion.** Fusion reactions provide the energy radiated by the sun and the stars. Because fusion reactions can only occur at enormously high temperatures, they are also called **thermonuclear reactions.**

The splitting of a very heavy nucleus into two lighter fragments releases energy, and is called **nuclear fission.** Uranium-235 and plutonium-239 undergo fission when bombarded with **thermal neutrons,** that is, neutrons with a kinetic energy equal to that of a typical gas molecule at room temperature.

A characteristic of fission reactions is that they produce more neutrons than are used to cause the fission. If the sample of fissionable material has a sufficiently large mass, called the **critical mass,** the neutrons produced cause additional fission reactions and a **chain fission reaction** ensues. For masses larger than the critical mass, the number of fissions increases very rapidly and a violent explosion takes place. A chain fission reaction causes the explosion produced by the atom bomb.

Nuclear fission, usually of ^{235}U, is the source of energy in nuclear power reactors. The heat produced by the fission reaction is converted into electricity. The flux of neutrons is kept at a safe level by inserting **control rods** containing cadmium or boron into the reactor core. Cadmium and boron are good absorbers of neutrons.

If we plot the number of neutrons, N, as a function of the number of protons, Z, for all the stable nuclei, we find that there is a band of values of the ratio of neutrons to protons within which all stable nuclei lie. This band is called the **belt of stability.** Up to $Z = 20$, most stable nuclei have equal numbers of protons and neutrons, but as Z increases, a nucleus must have more neutrons than protons to be stable, and the value of the neutron to proton ratio required for stability rises to 1.50 by $Z = 80$. The belt of stability terminates at $Z = 83$ (bismuth). All nuclei with $Z > 83$ are radioactive.

The existence of the belt of stability can be explained by the **shell model** of the nucleus. According to this model, protons and neutrons occupy separate shells, analogous to the shells of the orbital electrons. Successive nuclear shells are filled when occupied by 2, 8, 20, 28, 50, 82, or 126 neutrons or protons. These numbers are called the nuclear **magic numbers.**

Nuclei for which the value of n/p is too large for stability decay by emitting electrons (β^- particles). The emission of an electron is accompanied by the simultaneous emission of a **neutrino,** an uncharged particle of very small mass.

Nuclei for which the value of n/p is too small for stability decay by one of three processes: **electron capture, positron (β^+) emission,** or **alpha (α) emission.** Alpha decay is observed only for very heavy nuclei. Positron emission is the most likely mode of decay for neutron-poor nuclei with $Z < 30$. Electron capture is more likely than positron emission for $Z > 70$, but positron emission and electron capture are competing processes. Frequently the same unstable nucleus can decay by both electron capture and positron emission. For the very heavy nuclei, $Z > 80$, both electron capture and α decay are likely, and both may occur for the same nucleus.

Gamma rays are emitted by nuclei undergoing transitions from an excited (isomeric) state to the ground state. Gamma rays are very high energy electromagnetic radiation, with wavelengths less than 0.1 Å.

All radioactive decay follows a first-order rate law. Because it is usually the **half-life**, $\tau_{1/2}$, rather than the **specific rate constant** or **decay constant**, λ, that is reported for radioactive nuclei, the rate law is most conveniently written in terms of the half-life as

$$\ln\left(\frac{\text{activity at time } t}{\text{activity at time zero}}\right) = -\frac{0.693t}{\tau_{1/2}}$$

The unit of activity is the **curie** (Ci), which is equal to 3.700×10^{10} dps. Samples used in the lab more frequently have activities measured in **millicuries** (mCi) or **microcuries** (μCi).

Isotope dilution is a technique used to determine the yield of a product that cannot be separated quantitatively from its reaction mixture. A known amount of the product labeled with a radioactive isotope is added to the reaction mixture. From the decrease in the **specific activity** (activity per gram) due to the dilution of the radioactive product by the nonradioactive product, the amount of nonradioactive product can be determined.

The activity of ^{14}C (a β^- emitter with a half-life of 5730 yr) has been used to date carbon-containing objects that are several thousand years old. All living plants and animals have an approximately constant ^{14}C concentration due to the intake of ^{14}CO$_2$ from the atmosphere. After death the concentration decreases due to radioactive decay.

The time of formation of uranium containing minerals can be calculated using the observed ratio of ^{238}U to ^{206}Pb. Lead-206 is the stable end product of the decay of ^{238}U, which has a half-life of 4.5×10^9 yr.

Tracer studies, in which a reactant is labeled with a radioactive isotope, are used to elucidate the mechanism by which a reaction occurs.

Exercises

Sections 22.1 and 22.2

1. Calculate the binding energy per nucleon of $^{60}_{28}$Ni, which has an isotopic mass of 59.9332 amu.

2. Balance the following nuclear reactions by filling in the blanks in the equations:

   (a) $^{14}_{7}$N + $^{4}_{2}$He $\rightarrow$ $^{17}_{8}$O + ____

   (b) $^{10}_{5}$B + ____ $\rightarrow$ $^{3}_{1}$H + 2($^{4}_{2}$He)

   (c) $^{235}_{92}$U + $^{1}_{0}$n $\rightarrow$ $^{90}_{38}$Sr + $^{143}_{54}$Xe + ____

   (d) $^{238}_{92}$U + $^{4}_{2}$He $\rightarrow$ ____ + 3($^{1}_{0}$n)

   (e) $^{232}_{90}$Th + ____ $\rightarrow$ $^{240}_{96}$Cm + 4($^{1}_{0}$n)

3. The isotopic mass of $^{6}_{3}$Li is 6.015126. If the following fusion reaction could be made to occur, how much energy in million electron volts would be released? How many kilojoules of energy would be released per mole of $^{12}_{6}$C formed?

$$^{6}_{3}\text{Li} + ^{6}_{3}\text{Li} \rightarrow ^{12}_{6}\text{C}$$

4. The binding energy per nucleon is 7.075 MeV for $^{4}_{2}$He and 6.46 MeV for $^{9}_{4}$Be. If the process

$$2(^{4}_{2}\text{He}) + ^{1}_{0}n \rightarrow ^{9}_{4}\text{Be}$$

could be carried out, would it release energy or require energy? Explain your answer.

5. In the *Handbook of Chemistry and Physics* published by the Chemical Rubber Co., there is a Table of Isotopes. Using this Table, write the formulas of all nuclides that

constitute more than 20% of the naturally occurring element, for the first 20 elements of the periodic table. Classify these nuclides into four categories: (a) even Z, even N; (b) odd Z, even N; (c) even Z, odd N; (d) odd Z, odd N. For what conclusion(s) do you think these results provide evidence?

Sections 22.3 and 22.4

6. List the most likely modes of decay for the following unstable nuclei and write a balanced nuclear equation for each decay process:

   (a) $^{206}_{86}Rn$ (b) $^{120}_{49}In$ (c) $^{20}_{12}Mg$ (d) $^{84}_{39}Y$

7. An isomeric state of zinc, $^{69m}_{30}Zn$, emits a 0.4387 MeV γ ray and has a half-life of 13.9 h. Calculate the frequency and wavelength of the emitted γ radiation.

8. Naturally occurring fluorine consists only of a single isotope, $^{19}_{9}F$. Predict the decay modes of the radioactive fluorine isotopes ^{17}F, ^{18}F, ^{20}F, and ^{21}F.

9. The isotopic mass of $^{214}_{85}At$ is 213.9963 amu. It decays by emitting an α-particle, to $^{210}_{83}Bi$, which has a mass of 209.9841 amu. The mass of $^{4}_{2}He$ is 4.00260 amu. Calculate the energy released when astatine-214 decays. What is the kinetic energy of the emitted α-particle?

Section 22.5

10. Radon-222 has a half-life of 3.82 days. What weight of ^{222}Rn is required to have an activity of 1 Ci?

11. The activity of a 0.100-mg sample of pure $^{239}_{94}Pu$ (an α emitter) is 1.36×10^7 dpm. Calculate the half-life of $^{239}_{94}Pu$ in years.

12. A sample of a radioactive material injected into a patient for diagnostic tests has an activity of 5.8×10^4 cpm. What is this activity expressed in microcuries?

13. A lab purchases some ^{60}Co (a β^- emitter) with an activity of 5.00 mCi to serve as a radioactive source. The half-life of ^{60}Co is 5.26 yr. When the activity falls to 3.00 mCi, the sample is no longer useful for the experiments being carried out. How long will the source last?

14. Iodine-131 has a half-life of 8.06 days. What fraction of a sample of ^{131}I will be left after (a) 3.00 days and (b) 30.0 days?

Section 22.6

15. A mixture is to be assayed for penicillin. You add 10.0 mg of penicillin labeled with ^{14}C that has a specific activity of 0.785 $\mu Ci \cdot mg^{-1}$. From this mixture you are able to isolate only 0.42 mg of pure penicillin. The specific activity of the isolated penicillin is 0.102 $\mu Ci \cdot mg^{-1}$. How much penicillin was in the original mixture?

16. A wooden bowl is unearthed in the remains of an ancient Indian village in Central America. The β^- activity of the bowl due to ^{14}C is determined to be 11.9 dpm/g of total carbon. How long ago was the log from which the bowl was made cut from its tree?

17. The cleavage of ATP (adenosine triphosphate) to ADP (adenosine diphosphate) and phosphoric acid may be written as follows:

It was of interest to determine whether the P—O bond marked "*a*" or the one marked "*b*" is cleaved by the hydrolysis.

Outline an experiment that can be used to determine the answer to this question. Describe the results that would lead you to conclude that cleavage was at "*a,*" and the results that would lead you to conclude that cleavage was at "*b.*" (This experiment has been carried out and cleavage is at "*b.*")

Multiple Choice Questions

1. The half-life of ^{99m}Tc is 6.0 h. The delivery of a sample of ^{99m}Tc from the reactor to the nuclear medicine lab of a certain hospital takes 3.0 h. What is the minimum amount of ^{99m}Tc that must be shipped in order for the lab to receive 10.0 mg?

   (a) 20.0 g   (b) 18.6 mg   (c) 15.0 mg   (d) 14.1 mg   (e) 12.5 mg

2. Which of the following pairs are isobars?

   (a) $^{138}_{58}$Ce, $^{140}_{58}$Ce   (b) $^{106}_{48}$Cd, $^{106}_{46}$Pd   (c) $^{83}_{36}$Kr, $^{84}_{36}$Kr   (d) $^{74}_{32}$Ge, $^{75}_{33}$As
   (e) $^{235}_{92}$U, $^{239}_{94}$Pu

3. Of the following particles, the one with the greatest mass is the

   (a) triton   (b) proton   (c) deuteron   (d) neutron   (e) alpha particle

4. Aluminum-25 decays by emitting a positron. The species immediately produced has

   (a) 12 protons, 13 neutrons, 13 electrons
   (b) 13 protons, 12 neutrons, 13 electrons
   (c) 12 protons, 13 neutrons, 12 electrons
   (d) 14 protons, 11 neutrons, 14 electrons
   (e) 13 protons, 13 neutrons, 13 electrons

5. The most likely mode of decay for $^{13}_{7}$N is

   (a) α emission   (b) β^- emission   (c) β^+ emission   (d) γ emission
   (e) electron capture

6. The most likely mode of decay for $^{25}_{11}$Na is

   (a) α emission   (b) β^- emission   (c) β^+ emission   (d) γ emission
   (e) electron capture

7. Of the following nuclei, the one that has the greatest binding energy per nucleon is

   (a) $^{56}_{26}$Fe   (b) $^{2}_{1}$H   (c) $^{238}_{92}$U   (d) $^{239}_{94}$Pu   (e) $^{6}_{3}$Li

8. The half-life of ^{99m}Tc is 6.0 h. If the total residual activity in a patient 36.0 h after receiving an injection containing ^{99m}Tc must be no more than 0.01 μCi, what is the maximum activity the sample injected can have?

   (a) 0.06 μCi   (b) 0.16 μCi   (c) 0.32 μCi   (d) 0.36 μCi   (e) 0.64 μCi

9. The name thermonuclear reaction is applied to fusion reactions because

   (a) They are initiated by thermal neutrons.
   (b) They are exothermic.
   (c) They produce more neutrons than the number used to initiate the reaction.
   (d) They occur only at very high temperatures.
   (e) They are endothermic.

10. Of the following particles, the one with the smallest mass is the

    (a) electron   (b) proton   (c) positron   (d) H atom   (e) neutrino

11. One atomic mass unit in kilograms is

    (a) $1/N_A$ (b) $12/N_A$ (c) $1 \times 10^{-3}/N_A$ (d) $12 \times 10^{-3}/N_A$ (e) $1/12N_A$

12. The missing fission product in the reaction

$$^{235}_{92}U + ^1_0n \rightarrow ^{146}_{57}La + \underline{\quad} + 3(^1_0n)$$

is (a) $^{86}_{35}Br$ (b) $^{87}_{35}Br$ (c) $^{87}_{32}Ge$ (d) $^{89}_{35}Br$ (e) $^{89}_{32}Ge$

13. The transuranium element nobelium was made by bombarding another trans-uranium element with ^{12}C. The reaction is

$$\underline{\quad} + ^{12}_6C \rightarrow ^{254}_{102}No + 4(^1_0n)$$

The target atom of the bombardment is
(a) $^{246}_{96}Cm$ (b) $^{246}_{100}Fm$ (c) $^{243}_{96}Cm$ (d) $^{243}_{100}Fm$ (e) $^{242}_{98}Cf$

14. Which of the following atoms has only a single stable nuclide?
(a) $_{20}Ca$ (b) $_{28}Ni$ (c) $_{36}Kr$ (d) $_{45}Rh$ (e) $_{50}Sn$

15. When $^{118}_{51}Sb$ emits a positron, the product (daughter) nuclide is
(a) $^{117}_{50}Sn$ (b) $^{118}_{50}Sn$ (c) $^{119}_{50}Sn$ (d) $^{118}_{52}Te$ (e) $^{119}_{52}Te$

16. Tritium, ^3_1H, has a half-life of 12.26 yr. A 5.00-mL sample of tritiated water has an activity of 2.40×10^9 cpm. How many years will it take for the activity to fall to 3.00×10^8 cpm?
(a) 6.13 (b) 12.26 (c) 24.52 (d) 36.78 (e) 49.04

17. The activity of ^{14}C in living bones is 15.3 dpm/g of carbon. The half-life of ^{14}C is 5730 yr. A fossil animal bone found in the American southwest has an activity of 3.83 dpm/g of carbon. How many years ago did the animal die?
(a) 5730 (b) 8600 (c) 11,400 (d) 14,300 (e) 17,200

18. When $^{226}_{88}Rn$ emits an alpha particle, the nuclide formed is
(a) $^{222}_{87}Fr$ (b) $^{222}_{86}Rn$ (c) $^{224}_{86}Rn$ (d) $^{224}_{84}Po$ (e) $^{222}_{84}Po$

19. A radioisotope decays at such a rate that after 96.0 min only $\frac{1}{8}$ of the original amount remains. The half-life of this nuclide, in minutes, is
(a) 12.0 (b) 24.0 (c) 32.0 (d) 48.0 (e) 64.0

20. In a series of three steps in a radioactive disintegration sequence starting with $^{228}_{88}Ra$, the particles emitted are, successively, a β^- particle, a β^- particle, and an α-particle. The resulting product is an isotope of
(a) $_{92}U$ (b) $_{90}Th$ (c) $_{88}Ra$ (d) $_{86}Rn$ (e) $_{84}Po$

Problems

22.1. Naturally occurring silver is 51.82% $^{107}_{47}Ag$ and 48.18% $^{109}_{47}Ag$. The masses of these two isotopes are 106.90509 and 108.9047 amu, respectively.
(a) Calculate the atomic weight of silver from these data.
(b) Silver-111 is a β^- emitter, with a half-life of 7.5 days. The β^- particles emitted have a continuous spectrum of energies, with a maximum energy of 1.05 MeV. Explain why the emitted β^- particles have a continuous spectrum of energies, rather than a single sharply defined energy.
(c) The product of this decay is $^{111}_{48}Cd$, which has a mass of 110.9042 amu. Calculate the mass of $^{111}_{47}Ag$ in amu.

22.2. The element potassium has three naturally occurring isotopes, ^{39}K, ^{40}K, and ^{41}K. Which of these isotopes would you expect to have the lowest natural abundance? Explain the reason for your choice.

22.3. In the $^{238}_{92}U$ decay scheme, the first step is the emission of an α-particle. The product (daughter nuclide) emits a β^- particle, the next product emits a β^- particle, and the next emits an α-particle. Write balanced equations for these first four steps in the decay scheme of $^{238}_{92}U$.

22.4. Calculate the binding energy per nucleon of $^{16}_{8}O$ (natural abundance 99.759%) and $^{17}_{8}O$ (natural abundance 0.0374%), for which the isotopic masses are 15.9949149 and 16.999133 amu, respectively. Which of these two isotopes has a larger binding energy per nucleon? For what conclusion(s) does this provide evidence?

22.5. Tritium, $^{3}_{1}H$, is a β^- emitter with a half-life of 12.26 yr. What is the activity, in millicuries, of a 1.00-mL sample of tritium at 25 °C and 0.0100-atm pressure?

22.6. Calculate the energy, in million electron volts, released on binding one additional proton to $^{52}_{25}Mn$. The isotopic masses of ^{52}Mn and ^{53}Fe are, respectively, 51.94556 and 52.94558 amu.

22.7. The masses of ^{38}Ar, ^{34}S, and ^{4}He are, respectively, 37.96272, 33.96786, and 4.00260. Can the reaction below occur? Explain your answer.

$$^{38}_{18}Ar \rightarrow \; ^{34}_{16}S + \; ^{4}_{2}He$$

22.8. Assuming that the practical lower limit, because of counting uncertainties, on the usefulness of the radiocarbon dating technique is an activity of 1.0 cpm/g of carbon, what is the oldest object that can be reliably dated by this method? The half-life of ^{14}C is 5730 yr, and living plants have an activity of 15.3 cpm/g of carbon.

22.9. The chemical oxidation of graphite releases 393.51 kJ/mol.

$$C(gr) + O_2(g) \rightarrow CO_2(g) \qquad \Delta H = -393.51 \text{ kJ}$$

Calculate the difference in mass between products and reactants for this reaction. What important fact does this illustrate?

22.10. The isotopic masses of ^{239}Pu and ^{4}He are 239.05216 and 4.00260 amu, respectively. If the decay energy for α decay of ^{239}Pu is 5.243 MeV, what is the mass of ^{235}U?

22.11. The half-life for the decay of ^{238}U is 4.5×10^9 yr. The end product of the decay is ^{206}Pb. Estimate the age of a mineral in which the $^{206}Pb/^{238}U$ ratio (corrected for any Pb present in the mineral at the time of its formation) is found to be 0.297.

22.12. Radiometric techniques can be used to determine concentrations too low to be measured by the usual analytical techniques. A 20.00-mL sample of a solution containing Ba^{2+} ions is titrated by adding successive 0.100-mL portions of a 0.0100 M SO_4^{2-} solution labeled with the radioactive isotope ^{35}S. After each addition, a 2.00-mL aliquot of solution is withdrawn, filtered, and its activity is measured. The aliquot is then returned to the main solution before the next addition of SO_4^{2-}. No activity is observed in the solution after the first six additions of SO_4^{2-}. The activity of each aliquot after subsequent additions is as follows:

mL SO_4^{2-} added	0.700	0.800	0.900	1.00	1.10	1.20
Activity (cpm)	245	1270	2310	3360	4420	5470

(a) Why was there no activity in the solution following the first six additions of the radioactive SO_4^{2-} solution?
(b) Plot the activity of the aliquots versus the volume of SO_4^{2-} added, and then determine the $[Ba^{2+}]$ in the original solution.

22.13. The half-life of $^{131}_{53}$I (a β^- emitter) is 8.06 days.

(a) If HI is made with ^{131}I, when the iodine decays gaseous H_2 plus the product of the ^{131}I decay is formed. Write a balanced equation for the decomposition of gaseous HI made with ^{131}I.

(b) A sample of 0.200 mol of HI made entirely with ^{131}I is placed in an evacuated flask of volume 4.00 L at 25.0 °C. The temperature of the flask is kept constant for 16.12 days. At the end of that time, what is the pressure in the flask?

22.14. Calculate the maximum energy in million electron volts of an emitted positron when ^{21}Na (mass = 20.99764 amu) decays to ^{21}Ne (mass = 20.99385 amu).

22.15. The radioactivity of a sample of ^{209}Pb is counted on a radiation detection device. The following data are obtained:

Time (min)	0.00	30.00	60.00	90.00	120.00	150.00
Activity (cpm)	39,400	35,500	31,900	28,800	25,900	23,300

(a) Using just the activities at 0.00 and at 60.00 min, obtain a value for the half-life of this nuclide.

(b) By means of a suitable plot, use all six data points and calculate a value for the half-life of this nuclide.

(c) Assuming all calculations are correct, which value is more reliable, the value calculated in part (a) or the one calculated in part (b)? Explain your answer.

22.16. The overall reaction in the sun responsible for the energy it radiates is

$$4(^1_1H) \rightarrow {}^4_2He^{2-} + 2(^0_1\beta^+)$$

It has been estimated that the sun produces 3×10^{22} kJ/s. How many moles of H atoms are converted to He in one second to produce this energy?

chapter 23 Introduction to Organic Chemistry*

Friedrich August Kekulé (1829 – 1896) was a German chemist who laid the foundations for the modern theories of structure in organic chemistry. In 1858 Kekulé showed that carbon is tetrahedral and is able to link with itself to form long chains. This theory was proposed at the same time, completely independently, by the Scottish chemist A. S. Couper. The concept of open chains of carbon atoms led the way to an understanding of aliphatic compounds. The structure of benzene and related compounds, however, could not be described using open chains and remained a great puzzle. In 1865 Kekulé proposed that the structure of benzene is a closed ring of six carbon atoms, and from his structural work, the field of aromatic chemistry was developed. Kekulé reported that the structure of benzene came to him in a dream:

I was sitting writing at my textbook but the work did not progress; my thoughts were elsewhere. I turned my chair to the fire and dozed.... The atoms were gamboling before my eyes.... My mental eye ... could not distinguish larger structures of manifold conformation: long rows, sometimes more closely fitted together, all twining and twisting in snakelike motion. But look! What was that? One of the snakes had seized hold of its own tail, and the form whirled mockingly before my eyes. As if by a flash of lightning I awoke.... I spent the rest of the night in working out the consequences of the hypothesis.

* This chapter was coauthored by Professor L. G. Wade, Jr.

Organic chemistry is the study of the compounds of the element carbon, atomic number 6. Many such compounds are found in living plants and animals, and that is the origin of the term organic. Many more compounds of carbon have been synthesized by chemists in the laboratory. More than a million carbon compounds are now known, and in all but a very small number of them, the carbon atoms are **tetravalent,** that is, the carbon atom forms four bonds to other atoms. Carbon atoms bond to one another, and it is because of this that such a multitude of organic compounds are possible. An understanding of the types of bonds used by carbon is the foundation of the study of organic chemistry.

section 23.1
Bonding in Organic Compounds

The electronic configuration of an isolated gaseous carbon atom is $1s^2 2s^2 2p^2$, but in forming bonds to other atoms a carbon atom combines its outer $2s$ and $2p$ orbitals to form **hybrid atomic orbitals.** We have already discussed hybrid atomic orbitals in Section 14.7 and therefore will only briefly review the ideas presented there. Three types of hybrid orbitals are utilized by carbon atoms, namely the sp^3, sp^2, and sp hybrids.

Methane and Ethane

Most of the organic molecules whose bonding we will discuss here are **hydrocarbons,** compounds that contain only carbon and hydrogen. The simplest hydrocarbon is methane, CH_4, which has tetrahedral geometry. Bonding in methane has already been described in Section 14.7. Carbon uses four sp^3 hybrid orbitals, formed by combining the four atomic orbitals $2s$, $2p_x$, $2p_y$, and $2p_z$, to bond to the four hydrogen atoms in methane.

The simplest compound involving more than a single carbon atom in which carbon uses only sp^3 hybrid orbitals is ethane, C_2H_6. In ethane, each carbon atom uses four tetrahedral sp^3 hybrid orbitals to bond to three hydrogen atoms and to the other carbon atom. The ethane molecule is therefore not planar, but can be described as two tetrahedra joined at a vertex. The structure of ethane is shown in Fig. 23.1. The conventional way of writing ethane to indicate its structure is

The two carbon atoms are joined by a σ bond, formed by the overlap of the two sp^3 hybrid orbitals. The distance between the two carbon atoms is 1.54 Å (0.154 nm). Ethane can be considered to be two **methyl** ($-CH_3$) groups joined by a single bond, as H_3C-CH_3.

The two methyl groups in ethane can rotate with respect to one another without breaking the σ bond, and indeed we find that such rotations occur constantly. Because of this rotation there are essentially an infinite number of **conformations** of ethane, that is, different orientations of the two methyl groups relative to one another. The potential energies of these different conformations are not the same, but they do not differ greatly.

Two of the infinite number of conformations of ethane have been given names.

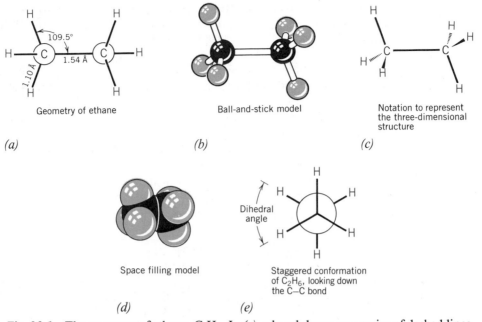

Fig. 23.1. The structure of ethane, C_2H_6. In *(c)*, a bond drawn as a series of dashed lines represents a bond directed away from the viewer, that is, behind the plane of the paper. A bond drawn as a solid wedge represents a bond coming toward the viewer, that is, in front of the plane of the paper. The pointed edge of the wedge indicates the atom further away from the viewer. In *(b)* and *(c)* the C—H bond lengths appear to be of different lengths because a three-dimensional molecule has been projected onto two-dimensional space. The six C—H bond lengths in ethane are, of course, all equal in length, and are 1.10 Å, as shown in *(a)*.

The **eclipsed conformation** is the one in which the three C—H bonds on the two ends of the molecule are aligned so that if you look down the C—C bond you can see only three hydrogen atoms. (See Fig. 23.2.) If we start with the two methyl groups lined up, and then rotate one —CH₃ group relative to the other, the angle between the two

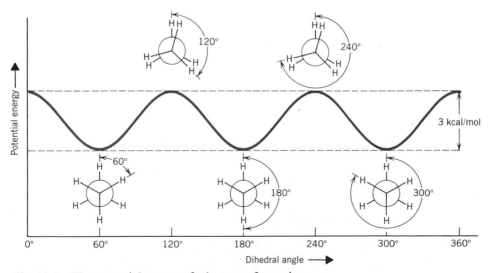

Fig. 23.2. The potential energy of ethane conformations.

groups (the angle we have rotated one group through) is called the **dihedral angle.** The dihedral angle is 0° for the eclipsed conformation. The conformation of lowest potential energy is the **staggered conformation,** in which the dihedral angle is 60°, and the C—H bonds are as far apart from one another as possible. The difference in potential energy between the eclipsed and the staggered forms is about 12.5 kJ/mol (3 kcal/mol). Because the electrons in the C—H bonds repel each other, an ethane molecule spends most of its time in or near the staggered conformation, but rotations of the two methyl groups about the C—C σ bond continually take place.

Figure 23.2 is a plot of the potential energy of the ethane molecule as it rotates through all of the possible conformations. Notice that the threefold symmetry of the ethane molecule implies that any two conformations whose dihedral angles differ by 120° (360°/3) have the same energy. For instance, the 0° (eclipsed) conformation has the same energy as the 120 and 240° conformations. Similarly, the staggered conformations at 60, 180, and 300° have the same energy.

Ethylene

There are three sp^2 hybrid orbitals, formed by combining the $2s$ and two of the three $2p$ atomic orbitals. The third $2p$ atomic orbital remains unhybridized, and is used in forming a π bond. (Refer to Sections 14.5 and 14.8.) The simplest hydrocarbon involving sp^2 hybrid orbitals is ethylene, C_2H_4.

The geometry of ethylene, which is also called **ethene,** is depicted in Fig. 23.3. In striking contrast to the free rotation about the C—C bond that is observed for ethane, the two **methylene** ($>CH_2$) groups joined by a double bond in ethylene, $H_2C=CH_2$, cannot rotate with respect to one another. If they were to rotate, the two $2p$ atomic orbitals would no longer overlap, and the π bond would be broken. The ethylene molecule, pictured also in Fig. 14.38, is therefore planar. The carbon–carbon distance is considerably shorter in ethylene (1.34 Å) than it is in ethane (1.54 Å). The presence of the π bond, in addition to the σ bond, pulls the carbon nuclei closer together. Because the energy of a π bond is less than that of a σ bond, the amount of energy required to break ethylene into two $:CH_2$ fragments is somewhat less than twice the energy required to split ethane into two $·CH_3$ groups. (See Table 23.1.)

Acetylene

There are also a number of hydrocarbons in which a carbon atom is bonded to only two other atoms, using sp hybrid orbitals. The simplest of these is **acetylene,** C_2H_2. In this compound each carbon is bonded to one C and one H atom, and the molecule is linear, as the angle between two sp hybrids is 180°. Since a carbon atom has four valence electrons, if two sp hybrids are formed there is still an electron in each of the two remaining (that is, unused in hybrid formation) $2p$ atomic orbitals. These $2p$

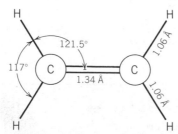

Fig. 23.3. The geometry of ethylene.

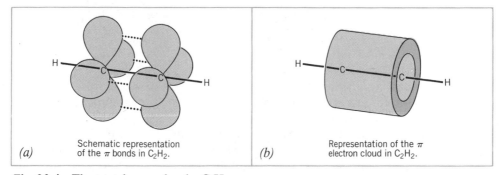

(a) Schematic representation of the π bonds in C_2H_2.

(b) Representation of the π electron cloud in C_2H_2.

Fig. 23.4. The acetylene molecule, C_2H_2.

Table 23.1. The Effects of the Hybridization of the Carbon Atomic Orbitals on the Carbon–Carbon Bond Order, Bond Length, and Bond Energy, and on the C—H Bond Length and Bond Energy

Molecule	Hybridization Used by Carbon; Geometry about Each C Atom	C—C Bond Length (Å)	C—C Bond Order	C—C Bond Energy[a] (kJ/mol)	C—H Bond Length (Å)	C—H Bond Energy[a] (kJ/mol)
C_2H_6	sp^3, tetrahedral	1.54	1	347	1.10	413
C_2H_4	sp^2, trigonal planar	1.34	2	615	1.08	440
C_2H_2	sp, linear	1.20	3	812	1.06	500

[a] The uncertainty in bond energies is about ±10 kJ/mol.

atomic orbitals on the two carbon atoms can overlap to form two π-molecular orbitals. The Lewis formula for acetylene is H:C:::C:H, or H—C≡C—H. The electron clouds of the two π-MO's merge, forming a cylindrical charge distribution about the two carbon atoms.

Representations of the bonding in acetylene are shown in Fig. 23.4. The bond between the two carbon atoms in C_2H_2 is a triple bond, consisting of one σ and two π bonds. As a result of the extra negative charge density between the two carbon nuclei, the carbon–carbon distance is only 1.20 Å (0.120 nm) in acetylene.

Comparing sp^3, sp^2, and sp Hybridization

The hybridization used by carbon affects not only the carbon–carbon bond distances and energies, but also the carbon–hydrogen bond distances and energies. In general, C—H bond distances are shortened and C—H bond energies increased somewhat as the multiple bond character of the carbon–carbon bond increases. Table 23.1 summarizes the effects of different hybridizations on bond lengths and energies in ethane, ethylene, and acetylene.

Using ethane, ethylene, and acetylene as models, we can predict the geometry of larger hydrocarbons, as illustrated in Example 23.1.

EXAMPLE 23.1. Expected geometries of hydrocarbons

Predict the geometry of 2-methylpropene, $(CH_3)_2C=CH_2$, which is also called isobutylene.

Solution. To answer this question, it is helpful to number the carbon atoms. The standard system of numbering is to begin with the extreme carbon at the end of the double bond, as

$$\overset{③}{H_3C}-\overset{②}{C}=\overset{①}{CH_2} \quad \text{or the equivalent} \quad \overset{①}{H_2C}=\overset{②}{C}-\overset{③}{CH_3}$$
$$\underset{CH_3}{|} \qquad\qquad\qquad\qquad \underset{CH_3}{|}$$

The numbering clarifies the name 2-methylpropene, which signifies that there is a methyl group on carbon atom number 2 of a chain of three carbon atoms, with one double bond between carbon atoms number 1 and number 2.

In this molecule, carbon atoms number 3 and number 4 employ tetrahedral sp^3 orbitals to bond to the three H atoms and to carbon atom number 2. The two methyl groups are tetrahedral in shape. There is essentially free rotation about the C—C bond between C number 2 and C number 3, as well as between C number 2 and C number 4. Since there is a double bond between C number 1 and C number 2, we expect bond angles close to 120° for all the C—C—C bond angles, and the four C atoms plus the two methylene hydrogens should lie in one plane. The atomic orbitals that overlap to form each bond are indicated in the diagram below.

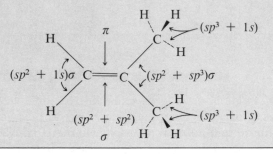

section 23.2
Geometric Isomerism

The great resistance to rotation about a double bond, as just described for ethylene, leads to an interesting phenomenon known as **geometric isomerism.** We have already defined geometric isomerism in coordination compounds in Section 20.7. The concepts discussed there apply also to organic molecules that are geometric isomers.

Consider, for example, the molecule 1,2-dibromoethene, in which a bromine atom has replaced a hydrogen atom on each carbon of ethylene. Because of the rigidity of the double bond, it is possible for both bromine atoms to be on the same side of the double bond (*cis*-dibromoethene), or for the two bromine atoms to be on opposite sides of the double bond (*trans*-dibromoethene), as shown in Fig. 23.5. The only difference between these two isomers is their geometry.

Cis- and *trans*-dibromoethene have different physical properties. Their densities differ slightly, and their melting points differ considerably. The *trans* isomer has a higher melting point (−6.5 °C) than does the *cis* isomer (−53 °C). Because the *trans* isomer has a symmetric structure, the dipole moments of the individual bonds cancel

Br   Br
 \\   /
  C = C
 /   \\
H   H

The *cis* isomer

Br   H
 \\   /
  C = C
 /   \\
H   Br

The *trans* isomer

Fig. 23.5. Geometric isomers of 1,2-dibromoethene.

each other out, and the molecule has a zero dipole moment. In *cis*-dibromoethene the dipole moments of the two C—Br bonds and the two C—H bonds do not cancel, and there is a net resultant dipole moment for the molecule. There is more repulsion between the electron charge clouds of the two large bromine atoms in the *cis* configuration than there is in the *trans* configuration, and therefore the *trans* isomer is more stable than the *cis* isomer, by a small amount.

EXAMPLE 23.2. Geometric isomerism

For which of the following compounds is geometric isomerism possible? Explain your answer and draw the structures of any geometric isomers. (a) $H_3C—CH=CH_2$ and (b) $H_3C—CH=CHBr$.

Solution

(a) No geometric isomers are possible for $CH_3—CH=CH_2$. There are two H atoms bonded to carbon atom number 1. In order for *cis* and *trans* isomers to exist, each of the two carbon atoms joined by a double bond must have two different groups bonded to it.

(b) The *cis* and *trans* isomers of $H_3C—CH=CHBr$ are shown below.

cis-1-Bromopropene   trans-1-Bromopropene

---

section 23.3
Optical Isomerism

Certain molecules have mirror images that are not superimposable on one another, in the same way that your left and right hands are not superimposable. The essential feature of a molecule that can exist in two nonequivalent mirror-image forms is a certain amount of dissymmetry; the molecule must not possess a plane or a point of symmetry.

Any molecule containing one tetrahedrally substituted carbon atom bonded to four different atoms or groups of atoms will have two isomers that are mirror images of one another. This is illustrated in Fig. 23.6 for chlorofluoroiodomethane. Such isomers are called **optical isomers** or **enantiomers**. The property of not being superimposable on one's mirror image is termed **chirality**.

From the discussion in Section 20.8 of optical isomerism in coordination compounds, you will recall that the name optical isomer arises from the ability of solutions of such molecules to rotate the plane of **plane-polarized light.** In ordinary light, there are transverse vibrations in all directions perpendicular to the direction of propagation of the light ray. In plane-polarized light, all vibrations are in a single direction, so that the wave motion of the light lies in a plane. (See Fig. 20.13.)

Ordinary light is polarized into two plane-polarized rays when it passes through a crystal of **calcite** (a form of $CaCO_3$). A device for separating the two plane-polarized rays so that one obtains a single ray of plane-polarized light is known as a **Nicol prism,** invented in 1828 by the Scottish physicist William Nicol. If you pass plane-polarized

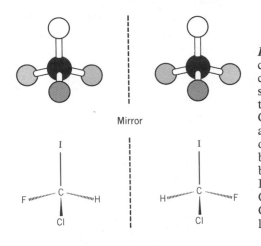

Fig. 23.6. Chirality at an asymmetric carbon atom. The mirror image of chlorofluoroiodomethane is not superimposable on the molecule. Examine the figure until it is clear to you that if the I, Cl, and C atoms are superimposed, the F atom of one isomer will fall on the H atom of the other isomer. A dashed bond is behind the plane of the paper; a solid wedge bond is in front of the plane of the paper. Because iodine, I, is such a large atom, the C—I bond is significantly longer than the C—H or C—F bonds, and somewhat longer than the C—Cl bond.

light through a solution of most compounds, the plane of polarization of the light that emerges from the solution is identical with the plane of polarization of the incident light. When plane-polarized light passes through a solution of a chiral compound, however, it emerges with the plane of polarization rotated with respect to the plane of polarization of the incident light.

A molecule that is an enantiomer is said to be **optically active.** A pair of enantiomers or optical isomers have identical physical properties, except that one molecule rotates the plane of polarized light in the clockwise direction, and the other rotates the plane in the counterclockwise direction. The instrument used to measure the angle through which the plane of polarization is rotated is called a **polarimeter.** Figure 23.7 schematically illustrates the essential elements of a polarimeter.

A solution that contains equal concentrations of two enantiomers will not rotate plane-polarized light because the rotations of the two isomers cancel each other out. Such a solution is known as a **racemic mixture.**

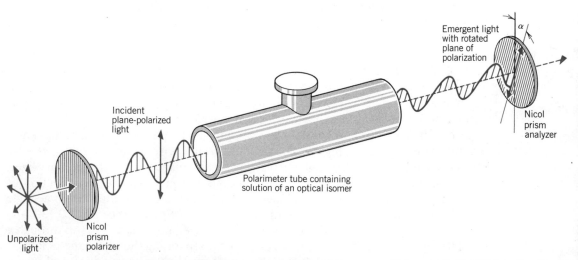

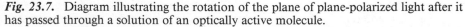

Fig. 23.7. Diagram illustrating the rotation of the plane of plane-polarized light after it has passed through a solution of an optically active molecule.

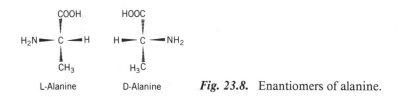

Fig. 23.8. Enantiomers of alanine.

A very large number of the organic compounds in living organisms are optically active, and this has important biological consequences. Although the physical properties of enantiomers are the same, if the molecule is involved in some biological process, the physiological activity is often profoundly different. The reaction sites in the biological system are sensitive to the asymmetry of the optically active species, and it is quite common for one enantiomer to be essential for some biological function while the other enantiomer is of no use.

Amino acids, the building blocks of proteins, are examples of optically active molecules necessary for life. Amino acids are molecules with the general formula

The different amino acids differ in the nature of the R group. Twenty amino acids are found in naturally occurring proteins. One of these is alanine, in which R is a methyl group. The two enantiomers of alanine are shown in Fig. 23.8.

To distinguish one enantiomer from its mirror image, the symbols D- and L- are used. These letters derive from the words *dextro* (right handed) and *levo* (left handed), and can be related to the configuration of the molecule. It is a striking fact that only the L-amino acids are used in building proteins.

The enzymes that catalyze metabolic reactions are also optically active molecules. One enantiomer catalyzes a specific reaction; the other has little or no catalytic activity. Similarly, if an optically active substance is used as a drug, one of the two isomers will be effective and the other will not.

section 23.4
Aromaticity

In Section 14.8 we discussed the concept of resonance and its application to the electronic structure of benzene, C_6H_6. In benzene the six carbon atoms lie in a plane and form a regular hexagon. Each carbon atom uses sp^2 hybrid orbitals. The σ-bonded structure of the molecule consists of the overlap of sp^2 hybrids on adjacent carbons, and the overlap of carbon sp^2 with hydrogen $1s$ atomic orbitals to form the six C—H bonds. Each carbon has one electron in the $2p$ atomic orbital not used in forming the sp^2 hybrids. The essential feature of the bonding in benzene is that these six electrons are **delocalized** into a ring of π-electron density above and below the plane of the molecule. The bonding in benzene is depicted in Fig. 14.41. There are a great many organic compounds with structures that are derived from benzene, and they are referred to as **aromatic compounds.**

Delocalized orbitals or electron clouds are **multicentered,** that is, they are spread out over more than two nuclei. We can compare the energy of the actual molecule, in which the π electrons are delocalized, with the calculated energy of a hypothetical resonance structure that we draw using only Lewis localized electron-pair bonds. In the case of benzene, this means comparing the energy of benzene,

with the calculated energy of one of the Kekulé structures,

In all aromatic compounds, multicenter bonding leads to a lowering of the energy. That is, the real molecule has an energy lower than any hypothetical resonance structure with only localized bonds.

For benzene, the energy lowering produced by multicenter bonding is demonstrated by considering the heat of hydrogenation of cyclohexene,

which has a single (localized) π bond, compared with the heat of hydrogenation of benzene. The enthalpy change of the reaction

$$+ \text{ H}_2 \rightarrow \qquad\qquad (23\text{-}1)$$

C_6H_{10}
Cyclohexene

C_6H_{12}
Cyclohexane

is -120 kJ/mol (-28.6 kcal/mol). For a hypothetical Kekulé structure with three localized π bonds, we might expect the heat of hydrogenation to be three times as large. Thus we would predict for

$$+ 3 \text{ H}_2 \rightarrow \qquad\qquad (23\text{-}2)$$

Kekulé
structure

C_6H_{12}
Cyclohexane

that ΔH would be $3(-120)$ kJ/mol or -360 kJ/mol. We find, however, that the heat of hydrogenation of benzene is considerably *smaller* in magnitude. The value of ΔH for the reaction

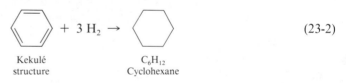

$$+ 3 \text{ H}_2 \rightarrow \qquad\qquad (23\text{-}3)$$

C_6H_6
Benzene

C_6H_{12}
Cyclohexane

is -208 kJ/mol (-49.8 kcal/mol), so that much less heat is released when benzene is hydrogenated than we would expect for the hydrogenation of three localized π bonds.

Figure 23.9 shows the relative energies of cyclohexene, benzene, the hypothetical Kekulé structure, and cyclohexane. The difference between 360 and 208 kJ/mol, or 152 kJ/mol (36 kcal/mol) is termed the **resonance energy of benzene.** It is important to remember that the resonance energy is the difference between the energy of a structure that has no physical reality and the energy of the real molecule. Nevertheless, it is a very useful concept, and enables us to appreciate the effects of delocalization of electrons.

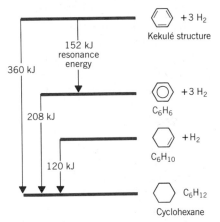

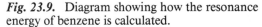

Fig. 23.9. Diagram showing how the resonance energy of benzene is calculated.

All the carbon–carbon bonds in benzene are of equal length, and the carbon–carbon distance is 1.39 Å, intermediate between the value for a single bond (1.54 Å) and that for a double bond (1.34 Å). Because the six π electrons are delocalized over the entire ring, each of the six identical bonds has a **bond order** of $\frac{3}{2}$ (one full σ bond and half a π bond, as there are six π electrons delocalized over six C atoms).

The chemical reactions of benzene demonstrate the stability achieved by the delocalization of the six π electrons. While cyclohexene, with its single localized π bond, rapidly reacts with many species to add atoms to the ring, benzene either reacts very slowly or not at all. For instance, cyclohexene adds HI

$$
\langle \text{C}_6\text{H}_{10} \rangle + \text{HI} \rightarrow \langle \text{C}_6\text{H}_{11}\text{I} \rangle{-}\text{I} \tag{23-4}
$$

but benzene does not. Furthermore, the hydrogenation of cyclohexene is rapid at room temperature while the hydrogenation of benzene is slow even at elevated temperatures. Benzene does not undergo addition reactions readily, because addition would destroy the "resonance stabilized" ring of π electrons.

In recent years, a great deal of attention has been devoted to the **polynuclear aromatics.** These are compounds with several benzene rings sharing one or more

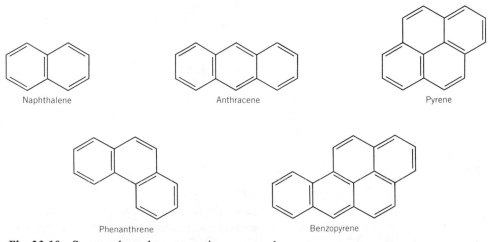

Fig. 23.10. Some polynuclear aromatic compounds.

sides of their rings. Produced primarily in combustion processes, many of these polynuclear aromatics are known to be **carcinogens** (cancer-causing substances). Perhaps the best-known carcinogenic polynuclear aromatic compound is benzopyrene. Benzopyrene is found in tobacco smoke, diesel exhaust, and the smoke from coal-burning and wood-burning stoves and power plants. Figure 23.10 shows the structure of some common polynuclear aromatic compounds.

Heterocyclic Aromatic Compounds

Although most aromatic compounds are either substituted benzenes or chains of fused benzene rings, there are also aromatic compounds that are **heterocyclic.** In heterocyclic compounds the ring contains at least one atom that is not carbon. A common example is **pyridine,** C_5H_5N.

Pyridine can be considered a resonance hybrid of two Kekulé structures:

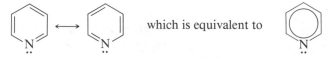

The nitrogen atom in pyridine, like each of the carbon atoms, uses sp^2 hybrid orbitals to bond to other members of the ring. The σ bonds to the ring thus use two of the five valence electrons of nitrogen. The electron in the $2p$ atomic orbital unused in hybrid formation becomes part of the π-electron cloud, and the remaining two electrons are a lone pair occupying the third sp^2 hybrid orbital. Note that the nitrogen atom in pyridine is not bonded to a hydrogen atom.

Pyridine is aromatic because of the six delocalized π electrons; it has many properties similar to those of benzene. It is a planar molecule with bond angles of 120°. The four C—C bonds are all of the same length, and the two C—N bonds are equal in length. Like benzene, pyridine resists addition reactions. The resonance energy of pyridine has been calculated to be 96 kJ/mol (23 kcal/mol).

The lone pair of electrons occupying the third sp^2 hybrid orbital makes pyridine a base, like ammonia. Pyridine is a base both according to the Brønsted–Lowry definition, and the Lewis definition. The nitrogen atom can share its lone pair of electrons with an acid, such as H^+, that has a vacant orbital:

$$\bigcirc N: + H^+ \rightarrow \bigcirc NH^+ \qquad (23\text{-}5)$$

Pyridine   Pyridinium ion

The basicity constant of pyridine, which is the equilibrium constant for its reaction with water,

$$\bigcirc N: + H_2O \rightarrow \bigcirc NH^+ + OH^- \qquad (23\text{-}6)$$

is $K_b = 2.9 \times 10^{-9}$.

section 23.5
The Alkanes

The **alkanes** are compounds containing only carbon and hydrogen (hydrocarbons) in which each carbon atom uses sp^3 hybrid orbitals to bond to four other atoms. The alkanes are relatively unreactive molecules; because they engage in so few chemical

Table 23.2. The First Ten Straight-Chain Alkanes

Name	Formula	mp (°C)	bp (°C)
Methane	CH_4	−182	−164
Ethane	C_2H_6	−183	− 89
Propane	C_3H_8	−190	− 42
n-Butane	$n\text{-}C_4H_{10}$	−138	− 0.5
n-Pentane	$n\text{-}C_5H_{12}$	−130	+ 36
n-Hexane	$n\text{-}C_6H_{14}$	− 95	+ 69
n-Heptane	$n\text{-}C_7H_{16}$	− 91	+ 98
n-Octane	$n\text{-}C_8H_{18}$	− 57	+126
n-Nonane	$n\text{-}C_9H_{20}$	− 51	+151
n-Decane	$n\text{-}C_{10}H_{22}$	− 30	+174

reactions with other substances they have also been called **paraffins** (from the Latin, meaning "little affinity"). We have already discussed bonding in the two simplest alkanes, methane, CH_4, and ethane, C_2H_6, both of which are gases at room temperature.

Straight-Chain or Normal Alkanes

By continually replacing a terminal hydrogen atom with a methyl group, we can build the series of **straight-chain alkanes** (also called **normal** or *n*-alkanes). The names, boiling points, and melting points of the first ten *n*-alkanes are given in Table 23.2.

Any noncyclic alkane can be described by the formula C_nH_{2n+2}, where n represents some integer. For an alkane with five or more carbon atoms, the name consists of either the Latin or Greek root for the number of carbon atoms in the molecule, followed by the suffix -ane.

Notice that the boiling points of the normal alkanes increase regularly with an increasing number of carbon atoms. Thus the first four *n*-alkanes are gases, and the remainder of those listed in Table 23.2 are liquids at room temperature. Still higher members of the series are solids at room temperature. A plot of the boiling point as a function of the number of carbon atoms is a smooth curve, as shown in Fig. 23.11. This effect is not due to the increasing molecular weight of these molecules, but rather to the increasing molecular surface area, which results in an increase in the intermolecular van der Waals attractive forces.

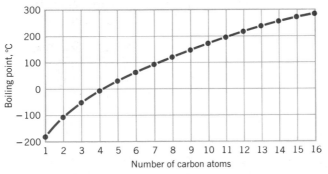
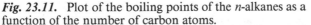

Fig. 23.11. Plot of the boiling points of the *n*-alkanes as a function of the number of carbon atoms.

Branched-Chain Alkanes

For an alkane with four or more carbon atoms, **branched-chain** as well as **straight-chain** molecules are possible. For instance, there are two isomers of butane, C_4H_{10}, *n*-butane and isobutane,

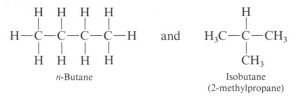

n-Butane	Isobutane (2-methylpropane)

and there are three isomers of pentane, C_5H_{12}, with the common names *n*-pentane, isopentane, and neopentane:

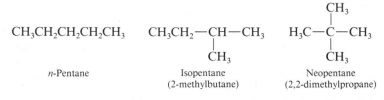

n-Pentane	Isopentane (2-methylbutane)	Neopentane (2,2-dimethylpropane)

The larger the number of carbon atoms, the larger the number of possible isomers.

Do not be misled by the term "straight chain" into thinking that the carbon atoms in the normal alkanes lie along a straight line. On the contrary, alkane chains tend to have a zig–zag geometry, because all bond angles in the alkanes are very close to the tetrahedral angle, 109.5°. Figure 23.12 is a schematic diagram showing the zig–zag geometry of *n*-pentane.

Branching of the chain always results in a lowering of the boiling point relative to the straight-chain isomer. The branched-chain alkanes are more compact than the straight-chain species, and the molecular surface area is decreased, which decreases the intermolecular van der Waals attractive forces. Thus while *n*-pentane boils at 36 °C, isopentane (2-methylbutane) boils at 28 °C, and neopentane (2,2-dimethylpropane) boils at 9.5 °C.

Nomenclature of Alkanes

As you can see from the names specified above, many of these hydrocarbons have common names, but there is also a systematic way of naming organic molecules. To name a branched-chain alkane, find the longest continuous carbon chain and use its name as the root name for the compound. Any other groups that are attached to the continuous chain are then called **alkyl groups,** and are named by changing the -ane suffix of the alkane name to -yl. Table 23.3 lists some common alkyl groups.

Table 23.3. **Some Common Alkyl Groups**

$-CH_3$	Methyl
$-CH_2CH_3$	Ethyl
$-CH_2CH_2CH_3$	*n*-Propyl
$-CH \begin{smallmatrix} CH_3 \\ \\ CH_3 \end{smallmatrix}$	Isopropyl
$-CH_2CH_2CH_2CH_3$	*n*-Butyl

Fig. 23.12. The zig–zag geometry of *n*-pentane.

If a particular kind of alkyl group appears in a molecule more than once, we use the prefixes di-, tri-, tetra-, penta-, and so on, to denote two, three, four, or five of the groups. The continuous carbon chain is numbered from one end to the other in whichever direction allows the substituent groups to be on the carbon atoms with the lowest possible numbers. If there are two possible longest continuous carbon chains of the same length, choose the one with the simplest alkyl groups. Example 23.3 describes how to name some larger branched-chain alkanes.

EXAMPLE 23.3. Naming branched-chain alkanes

Give the systematic names of the following molecules:

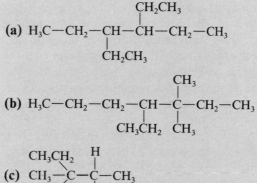

Solution

(a) The longest chain of carbon atoms contains six carbons, so that this compound is a hexane. It does not matter at which end we begin numbering, since the branches are on carbon atoms number 3 and number 4, whether we number from left to right or from right to left. The correct name is 3,4-diethylhexane.

$$①\quad②\quad③\quad \overset{\displaystyle CH_2CH_3}{\underset{\displaystyle CH_3CH_2}{H_3C—CH_2—CH—CH—CH_2—CH_3}}\quad④\quad⑤\quad⑥$$

(b) The longest chain contains seven carbon atoms, so this is a heptane. We must begin numbering on the right-hand side, because the substituent groups are then on carbon atoms number 3 and number 4. If we begin numbering from left to right, the branches would be on carbon atoms with larger numbers (number 4 and number 5). The correct name for this molecule is 3,3-dimethyl-4-ethylheptane.

$$⑦\,⑥\qquad⑤\qquad④\quad \overset{\displaystyle CH_3}{\underset{\displaystyle CH_3CH_2\;\;CH_3}{H_3C—CH_2—CH_2—CH—C—CH_2—CH_3}}\quad②\quad①$$

(c) The longest continuous chain contains five carbon atoms.

$$\underset{\displaystyle CH_3CH_2\;\;CH_3}{\overset{\displaystyle CH_3CH_2}{H_3C—C——CH—CH_3}}\quad\text{is the same as}\quad⑤\,④\;\underset{\displaystyle H_3C\;\;\;\;CH_3}{\overset{\displaystyle CH_3CH_2}{H_3C—CH_2—C——CH—CH_3}}\;②\;①$$

Note that the longest chain need not be written in a straight line. The correct name of this compound is 2,3-dimethyl-3-ethylpentane.

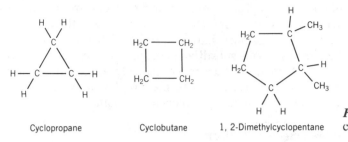

Fig. 23.13. Some simple cycloalkanes.

Cyclopropane   Cyclobutane   1, 2-Dimethylcyclopentane

Alkanes can also form rings, which are denoted by the prefix cyclo-. The first two cyclic alkanes, cyclopropane and cyclobutane, are relatively strained compared with other alkanes, because the ring angles are 60 and 90°, respectively. These values are quite far from the normal tetrahedral angle of 109.5°. Both cyclopropane and cyclobutane are more reactive than other alkanes, and when they react the ring structure is likely to be broken.

The structures of a few simple cycloalkanes are shown in Fig. 23.13. For cyclic alkanes with five or more carbon atoms, the rings are not planar, and all bond angles are close to 109.5°.

Alkanes in general are unreactive, and with the exception of combustion, there are relatively few reactions that they undergo readily. Alkanes are widely used as fuels. Gasoline and kerosene are mixtures containing several alkanes, and **natural gas,** which is used for heating and cooking, is principally methane. In the laboratory, alkanes are primarily used as nonpolar solvents for other organic compounds, because they do not react with other substances.

section 23.6
Functional Groups

Although the alkane portions of molecules are normally unreactive, most organic compounds have groups of atoms that can be induced to undergo desired reactions under the proper conditions. These reactive groups of atoms are called **functional groups.** As a rule of thumb, you can assume that any part of a molecule that doesn't look like an alkane is a functional group. In general, the functional groups a molecule possesses determine the kinds of reactions the molecule undergoes. It is important, therefore, to be familiar with the functional groups most commonly found in organic compounds.

Alkenes

Hydrocarbons containing at least one carbon–carbon double bond,

$$\begin{matrix} \diagdown & & \diagup \\ & C=C & \\ \diagup & & \diagdown \end{matrix}$$

are called **alkenes.** An older name for these compounds that is still in use is **olefins.** A double bond consists of a σ bond plus a π bond, and as the π bond is not as strong as the σ bond, it is more readily broken. Alkenes are therefore much more reactive than alkanes, although they are nonpolar molecules and have physical properties similar to the alkanes. The π bond of alkenes is relatively electron rich, and is **nucleophilic,** meaning attracted to nuclei.

Alkenes readily undergo addition reactions, as illustrated by the addition of HBr to ethylene, Eq. (23-7). The π-electron cloud attracts the proton of HBr, and both π electrons are used to form a bond between the hydrogen and one of the carbon atoms, leaving the other carbon atom with a vacant $2p$ atomic orbital, and therefore a positive charge. This intermediate species is called a **carbonium ion** or a **carbocation,** the name given to any ionic species in which a carbon atom has a positive charge. Once the proton has been pulled off the HBr, a bromide ion remains. The carbocation and the bromide ion react quickly to form bromoethane, the final product. The mechanism of the addition reaction is therefore

$$\underset{\text{Ethylene}}{H_2C\!=\!CH_2} + HBr \rightarrow \underset{\text{Carbocation}}{H_2\overset{+}{C}\!-\!CH_3} + Br^- \rightarrow \underset{\text{Bromoethane}}{\overset{\displaystyle Br}{\overset{|}{H_2C}}\!-\!CH_3} \qquad (23\text{-}7)$$

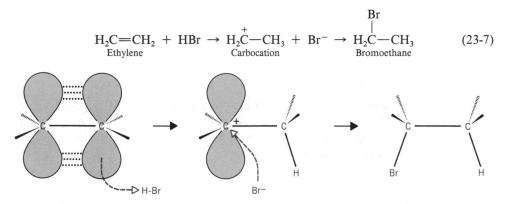

Alkenes are named in the same way as alkanes, except that the -ane ending is changed to -ene. The carbon chain is numbered so that the double-bonded carbons receive the lowest possible numbers, and the lower of the two numbers of the double-bonded carbons is used in the name. Example 23.3 illustrates the way alkenes are named.

EXAMPLE 23.4. Nomenclature of alkenes

Give the systematic names of the following molecules:

(a) $H_2C\!=\!CHCH_2CH_3$ (b) $CH_3CH\!=\!CHCH_3$

(c) $H_2C\!=\!C\!\!\begin{smallmatrix}\nearrow CH_3 \\ \searrow CH_2CH_3\end{smallmatrix}$ (d) $H_3C\!-\!\underset{\underset{CH_3}{|}}{CH}\!-\!CH\!=\!CH_2$

Solution

(a) $\overset{①\;②\;③\;④}{H_2C\!=\!CHCH_2CH_3}$ (b) $\overset{①\;②\quad③\;④}{CH_3CH\!=\!CHCH_3}$

      1-Butene   2-Butene

(c) $\underset{①\;②\quad③\;④}{H_2C\!=\!\overset{\overset{\displaystyle CH_3}{|}}{C}\!-\!CH_2CH_3}$ (d) $\overset{④\;③\quad②\quad①}{H_3C\!-\!\underset{\underset{CH_3}{|}}{CH}\!-\!CH\!=\!CH_2}$

   2-Methyl-1-butene   3-Methyl-1-butene

As we have already discussed in Sections 23.1 and 23.2, rotation about a double bond is restricted due to the fact that the $2p$ orbitals of the π bond would no longer be aligned for maximum overlap if one end of the molecule were to be rotated with respect to the other end. The rigidity of the double bond leads to the existence of geometric isomers, and therefore many alkenes occur in *cis* and *trans* forms. The geometric isomers of 2-butene are shown below:

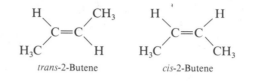

trans-2-Butene   *cis*-2-Butene

Alcohols

Organic compounds containing the hydroxyl (—OH) group are called **alcohols.** Alcohols can be named by changing the -ane ending of the alkane name to -anol, or by adding the word alcohol to the alkyl group to which the hydroxyl group is attached. Several examples are given below:

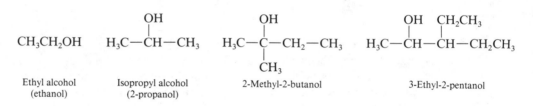

Because of the highly polar nature of the —OH group, alcohols are much more polar compounds than are the alkanes and the alkenes. In fact, methanol, ethanol, the propanols, and the butanols are very soluble in water. As the number of carbon atoms increases, however, the solubility of the alcohol in water decreases. If the number of carbon atoms is small, the —OH group is a predominant feature of the molecule and the alcohol is water soluble. Alcohols with larger numbers of carbon atoms are more alkane-like, and are therefore more soluble in nonpolar solvents and less soluble in water.

The reason alcohols are water soluble is that they can form **hydrogen bonds** with water, as has already been discussed in Sections 7.2 and 7.3. The hydrogen-bonded structure of an aqueous solution of ethanol has been depicted in Fig. 7.5. Alcohols also hydrogen bond with themselves, so that there are strong intermolecular attractions in pure liquid alcohols. As a result, the boiling points of alcohols are relatively high, much higher than would be predicted on the basis of their molecular weights alone. In the solid or liquid phases, alcohols are associated through chains of hydrogen bonds. Hydrogen bonding in liquid ethanol is depicted in Fig. 23.14.

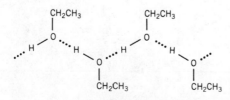

Fig. 23.14. Hydrogen bonding in pure liquid ethanol.

Ethers

Ethers are compounds that contain two alkyl groups bonded to an oxygen atom. They are generally named according to their alkyl groups, as in the following examples:

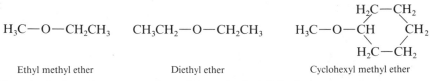

$$H_3C—O—CH_2CH_3 \qquad CH_3CH_2—O—CH_2CH_3$$

Ethyl methyl ether · · · · · · · · · · · · Diethyl ether · · · · · · · · · · · · Cyclohexyl methyl ether

Like alcohols, ethers are polar, although their dipole moments are somewhat smaller than the dipole moments of alcohols of the same molecular weight. Ethers differ markedly from the alcohols in their physical properties, such as solubility in water and boiling point. Ethers are relatively insoluble in water, and an ether boils at a temperature significantly lower than an alcohol of the same molecular weight. Consider, for example, diethyl ether and *n*-butanol, both of which have the same formula, $C_4H_{10}O$.

$$CH_3CH_2—O—CH_2CH_3 \qquad CH_3CH_2CH_2CH_2OH$$

Diethyl ether · *n*-Butanol
sparingly soluble in water · · · · · · · · · · very soluble in water
bp 35 °C · bp 118 °C

n-Butanol is 10 times as soluble in water as is diethyl ether, and diethyl ether boils at 35 °C, just a little above room temperature, whereas *n*-butanol boils at a much higher temperature, 118 °C.

The reason for these differences is that while alcohols can engage in hydrogen bonding, ethers cannot. Pure water is extensively hydrogen bonded, and if an ether molecule is forced into the midst of a water solution, a number of hydrogen bonds between the water molecules must be broken. The net result is that ethers are very nearly immiscible with water, and two separate layers are formed when one tries to mix water with diethyl ether. An alcohol, on the other hand, can engage in hydrogen bonding with water molecules, and the free energy of a mixture of water and alcohol is less than the free energy of the separate liquids.

Ether molecules cannot hydrogen bond intermolecularly as they lack a hydrogen atom bonded to an electronegative atom. Intermolecular forces between ether molecules in the liquid phase are considerably smaller than intermolecular forces between hydrogen-bonded alcohol molecules, and the boiling point of an alcohol is significantly higher than that of an ether of the same molecular weight.

EXAMPLE 23.5. Comparative boiling points

Arrange the following three molecules in order of increasing boiling point: *n*-pentanol (molecular weight 88.2); 2,2-dimethylbutane (molecular weight 86.2); ethyl *n*-propyl ether (molecular weight 88.2).

Solution. The branched-chain alkane, 2,2-dimethylbutane, has the lowest boiling point of these three compounds. It is nonpolar and has a smaller molecular surface area than the ether, so that intermolecular van der Waals attractions are smallest for this compound. The alcohol has the highest boiling point because it can engage in hydrogen bonding, while the ether cannot.

$$\overset{\displaystyle CH_3}{\underset{\displaystyle CH_3}{H_3C—CH_2—\overset{|}{\underset{|}{C}}—CH_3}} \qquad CH_3CH_2—O—CH_2CH_2CH_3 \qquad CH_3(CH_2)_3CH_2OH$$

bp 49.7 °C · · · · · · · · · · · · · · · · · · bp 63.6 °C · · · · · · · · · · · · · · · · · · bp 137 °C
2,2-Dimethylbutane · · · · · · · · · · Ethyl *n*-propyl ether · · · · · · · · · · *n*-Pentanol

Note: The boiling points of *n*-hexane and ethyl *n*-propyl ether are similar. Presumably, *n*-hexane has a larger molecular surface area, as it is nonpolar while the ether is polar.

Amines

A compound in which one or more of the hydrogen atoms of ammonia, NH_3, has been replaced by an organic group is called an **amine**. If the organic groups are alkyl groups, the amine is said to be **aliphatic**. There are also aromatic amines, in which the organic group bonded to the nitrogen atom is derived from benzene.

Primary, secondary, and **tertiary amines** have one, two, and three hydrogens replaced by organic substituents, respectively. Aliphatic amines are commonly named in the same way that ethers are named: We use the names of the alkyl groups followed by the suffix -amine. Some common aliphatic amines are listed in Table 23.4. The simplest aromatic amine is called **aniline,** and has the formula

$$C_6H_5NH_2 \quad \text{or} \quad \text{\raisebox{-0.5ex}{⬡}}-NH_2$$

The most striking characteristic of amines is their basicity. Because amines are derived from ammonia, they are weak bases, just as ammonia is. Amines are proton acceptors, and are therefore Brønsted–Lowry bases. (See Sections 7.4 and 9.2.) Trimethylamine, for instance, accepts a proton from water to form trimethylammonium ion and hydroxide ion:

$$(CH_3)_3N \colon + H_2O \rightleftharpoons (CH_3)_3NH^+ + OH^- \tag{23-8}$$

In organic chemistry amines most commonly function as **Lewis bases** or **nucleophiles.** A Lewis base is a substance that has a pair of electrons available for forming a coordinate covalent bond with an electron-pair acceptor, or **Lewis acid.** (Refer to Section 20.3.) The nitrogen atom of any amine has a lone pair of electrons that it can share with a Lewis acid.

Amines react with acids to form ionic salts. A typical reaction is the following:

$$\underset{\text{Triethylamine}}{(CH_3CH_2)_3N \colon} + HCl \rightarrow \underset{\text{Triethylammonium chloride}}{(CH_3CH_2)_3\overset{+}{N}H \ Cl^-} \tag{23-9}$$

These ionic salts are usually water soluble, even when the amine itself is not soluble in water.

Table 23.4. Some Common Aliphatic Amines

$$\underset{\text{Methylamine}}{CH_3NH_2} \qquad \underset{\text{Ethylamine}}{CH_3CH_2NH_2}$$

(a) Primary Amines

$$\underset{\text{Dimethylamine}}{\overset{H_3C}{\underset{H_3C}{>}}NH} \qquad \underset{\text{Ethylisopropylamine}}{CH_3CH_2-N-\overset{CH_3}{\underset{CH_3}{CH}}}$$

(b) Secondary Amines

$$\underset{\text{Triethylamine}}{(CH_3CH_2)_3N} \qquad \underset{\text{Ethyldimethylamine}}{CH_3CH_2-N(CH_3)_2}$$

(c) Tertiary Amines

The **nucleophilic** property of amines is demonstrated by their reactions with methyl iodide, CH_3I, to form alkylammonium iodides. Triethylamine, for example, reacts vigorously with methyl iodide at room temperature, as follows:

$$(CH_3CH_2)_3N\colon + \quad CH_3I \quad \rightarrow \quad (CH_3CH_2)_3\overset{+}{N}CH_3 \; I^- \qquad (23\text{-}10)$$

Nucleophile   Electrophile   Methyltriethylammonium iodide

Compounds in which four alkyl groups replace the four hydrogens of the ammonium ion are called **quaternary ammonium compounds.** Quaternary ammonium salts are ionic substances and therefore generally are water soluble and have fairly high melting points.

Aldehydes and Ketones

Aldehydes and ketones are characterized by the presence of a **carbonyl group,** or

If one of the groups bonded to the carbon is a hydrogen atom, the compound is an **aldehyde.** The general formula for an aldehyde is therefore

$$\overset{\displaystyle H}{\underset{\displaystyle |}{R-C}}=O$$

where R is some organic substituent. If R is an alkyl group, the aldehyde is aliphatic. There are also aromatic aldehydes, where R is a substance derived from benzene. Aliphatic aldehydes are named by changing the alkane ending to -anal. No number is necessary, since the

group must come at the end of a carbon chain, and the aldehydic carbon is therefore always number 1. The simpler aldehydes have common names that are almost universally used. A list of some common aldehydes is given in Table 23.5.

Table 23.5. **Some Common Aldehydes**

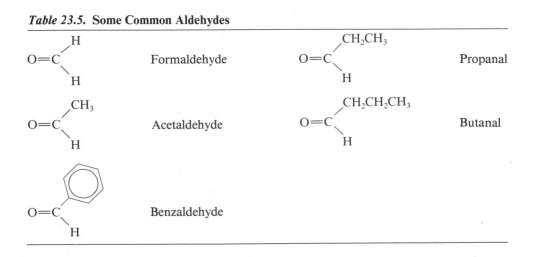

Ketones are compounds in which both groups bonded to the carbon atom of a carbonyl group are organic substituents. The generic formula for a ketone is therefore

The R groups may be either aliphatic or aromatic. Ketones may be named by changing the alkane ending to -anone, and using a number to denote the carbonyl carbon. Alternatively, aliphatic ketones can be named by naming the two alkyl groups attached to the carbonyl carbon. The simplest ketone, $(CH_3)_2C=O$, is universally called **acetone** rather than dimethyl ketone. Table 23.6 gives the names and formulas of some common ketones.

The most important characteristic of the carbonyl group is its bond polarization. Because oxygen is considerably more electronegative than carbon, the more loosely held (and therefore polarizable) π electrons of the carbon–oxygen double bond are pulled toward the oxygen atom, so that the oxygen end of the bond acquires a small net negative charge, and the carbon atom acquires a small net positive charge.

One result of this bond polarity is that aldehydes and ketones boil at higher temperatures than do alkanes of comparable molecular weights. Dipole-dipole interactions lead to stronger intermolecular attractive forces for aldehydes and ketones than for alkanes, and therefore the boiling point is raised.

Another result of the polarity of the carbonyl bond is that the carbon atom, with its positive charge, is electrophilic and is subject to nucleophilic attack. Thus aldehydes and ketones react with ammonia and primary amines to form compounds called

Table 23.6. **Some Common Ketones**

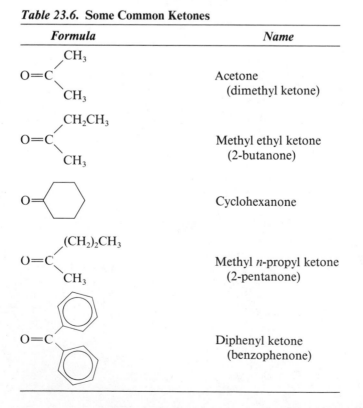

Formula	Name
	Acetone (dimethyl ketone)
	Methyl ethyl ketone (2-butanone)
	Cyclohexanone
	Methyl *n*-propyl ketone (2-pentanone)
	Diphenyl ketone (benzophenone)

imines. An imine contains the nitrogen analog of a carbonyl group. A typical reaction is

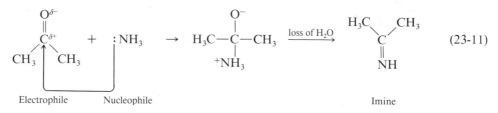

Imines react rapidly with water to regenerate the carbonyl compound in the presence of acids or bases. They are important intermediates in certain biochemical processes.

Carboxylic Acids and Their Derivatives

Carboxylic acids are characterized by the presence of the carboxyl group,

$$\begin{matrix} \text{O} \\ \parallel \\ -\text{C}-\text{OH} \end{matrix}$$

which is often represented as —COOH. The representation —COOH is not strictly correct as it is written, and you must remember that one of the two oxygen atoms is part of a carbonyl group.

A carboxylic acid is a Brønsted–Lowry acid, because the proton of the

$$\begin{matrix} \text{O} \\ \parallel \\ -\text{C}-\text{OH} \end{matrix}$$

group can be donated to a base, leaving a **carboxylate ion.** Carboxylic acids are weak acids, and their conjugate bases, the carboxylate ions, are weak bases. We have already discussed acetic acid, CH_3COOH, the most common carboxylic acid, in Sections 7.3 and 7.4. The other carboxylic acids have similar acidic properties. The acidity constants, K_a, for several carboxylic acids are listed in Table 23.7, and it is readily seen that these values are remarkably similar. Aliphatic carboxylic acids are named by changing the -ane ending of the corresponding alkane to -anoic acid. The first few carboxylic acids have common names that are widely used.

Consideration of the proton-transfer reaction between a carboxylic acid and water, such as

$$\begin{matrix} & \text{O} & & & & \text{H} & & & \text{O} \\ & \parallel & & & & \diagup & & & \parallel \\ CH_3CH_2 & -C-OH & + & :\text{O} & & & \rightleftharpoons & CH_3CH_2C-O^- & + & H_3O^+ \\ & & & & \diagdown & & & & \\ & & & & \text{H} & & & & \end{matrix} \qquad (23\text{-}12)$$

Propanoic acid   Propanoate ion

might lead us to inquire whether or not alcohols, which also contain the —OH group, are acidic. In fact, alcohols are extremely weak acids, so weak that for practical purposes they are not considered acids. The acidity constant of an alcohol is about 10^{13} times smaller than the acidity constant of the corresponding carboxylic acid. For instance, K_a for the reaction

$$CH_3CH_2CH_2OH + H_2O \rightleftharpoons CH_3CH_2CH_2O^- + H_3O^+ \qquad (23\text{-}13)$$

is about 10^{-18}, whereas K_a for propanoic acid [Eq. (23-12)], is about 10^{-5}.

Table 23.7. Some Common Carboxylic Acids

Formula	Name	K_a
$\text{H}-\overset{\displaystyle \text{O}}{\overset{\|}{\text{C}}}-\text{OH}$	Formic acid (methanoic acid)	1.8×10^{-4}
$\text{CH}_3\overset{\displaystyle \text{O}}{\overset{\|}{\text{C}}}-\text{OH}$	Acetic acid (ethanoic acid)	1.8×10^{-5}
$\text{CH}_3\text{CH}_2\overset{\displaystyle \text{O}}{\overset{\|}{\text{C}}}-\text{OH}$	Propionic acid (propanoic acid)	1.3×10^{-5}
$\text{CH}_3\text{CH}_2\text{CH}_2\overset{\displaystyle \text{O}}{\overset{\|}{\text{C}}}-\text{OH}$	Butyric acid (butanoic acid)	1.5×10^{-5}
$\text{C}_6\text{H}_5-\overset{\displaystyle \text{O}}{\overset{\|}{\text{C}}}-\text{OH}$	Benzoic acid	6.3×10^{-5}
$\text{CH}_3(\text{CH}_2)_6\overset{\displaystyle \text{O}}{\overset{\|}{\text{C}}}-\text{OH}$	Caprylic acid (octanoic acid)	1.3×10^{-5}

The reason it is so much easier to remove the —OH proton from a carboxylic acid than from an alcohol, is that the carboxylate anion is much more stable (lower in energy) than the anion formed when the proton is removed from the alcohol. The carboxylate ion does not have one carbon–oxygen double bond and one carbon–oxygen single bond. The two carbon–oxygen bonds are identical, and the negative charge on the ion is spread equally over the two oxygen atoms. The actual ion is therefore described as a resonance hybrid of two structures, as shown below for propanoate ion:

$$\text{CH}_3\text{CH}_2-\overset{\displaystyle \text{O}}{\underset{\displaystyle \text{O}^-}{\text{C}}} \quad \longleftrightarrow \quad \text{CH}_3\text{CH}_2-\overset{\displaystyle \text{O}^-}{\underset{\displaystyle \text{O}}{\text{C}}}$$

In the case of the alcohol anion, however, one oxygen atom must bear the entire negative charge, and there is no resonance stabilization.

Compared with compounds of similar molecular weights, carboxylic acids have unusually high boiling points. For instance, formic acid, dimethyl ether, and ethanol all have molecular weights of 46. The boiling point of dimethyl ether is −25 °C, of hydrogen-bonded ethanol is 79 °C, and of formic acid, 101 °C. The high boiling points of carboxylic acids can be explained by their tendency to form hydrogen-bonded dimers, like that shown below for formic acid.

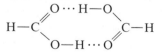

Esters

Carboxylic acids react with alcohols to form compounds called **esters.** The esterification reaction also produces water. The general reaction may be formulated as

$$R-\overset{\overset{\displaystyle O}{\|}}{C}-OH + R'OH \rightleftharpoons R-\overset{\overset{\displaystyle O}{\|}}{C}-OR' + H_2O \qquad (23\text{-}14)$$

Carboxylic   Alcohol   Ester
acid

A specific example of an esterification reaction is the following:

$$H_3C-\overset{\overset{\displaystyle O}{\|}}{C}-OH + CH_3CH_2OH \rightleftharpoons H_3C-\overset{\overset{\displaystyle O}{\|}}{C}-OCH_2CH_3 + H_2O \qquad (23\text{-}15)$$

Acetic acid   Ethanol   Ethyl acetate
(ethanoic acid)   (ethyl ethanoate)

The mechanism for the elimination of water when the alcohol and acid react has been investigated using ^{18}O as a tracer, and is discussed in Section 22.6.

The name of the ester is derived from the alkyl group of the alcohol, followed by the name of the carboxylic acid, in which the -oic acid ending is changed to -oate. Another example of an esterification, to illustrate the way the ester is named, is given below:

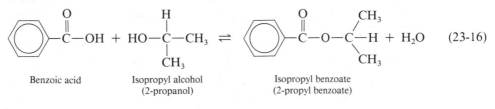

Benzoic acid   Isopropyl alcohol   Isopropyl benzoate
                 (2-propanol)   (2-propyl benzoate)

Amides

Because amines are bases, a proton-transfer reaction always occurs between a carboxylic acid and an amine. The application of heat to the salt that is formed drives off water, and results in the formation of an **amide.** A typical reaction is shown below:

$$\begin{array}{l} H_3C \\ \diagdown \\ \overset{..}{N}H \\ \diagup \\ H_3C \end{array} + CH_3COOH \rightarrow (CH_3)_2\overset{+}{N}H_2\ CH_3COO^- \xrightarrow{\text{heat}} (CH_3)_2N-\overset{\overset{\displaystyle O}{\|}}{C}-CH_3 \qquad (23\text{-}17)$$

Dimethyl-   Acetic acid   Dimethylammonium   *N,N*-Dimethyl acetamide
amine   acetate   (*N,N*-dimethyl ethanamide)

Names of amides are formed by using the names of the alkyl groups on nitrogen (preceded by *N*- to indicate they are on nitrogen), followed by the common name of the acid with the suffix -amide instead of -oic acid.

An interesting characteristic of amides is that the amide nitrogen is no longer nucleophilic or basic. This property can be attributed to the fact that the bonding between nitrogen and carbon is not correctly represented as a single bond. The actual electronic structure is best described as a resonance hybrid of two structures, shown below:

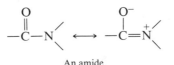

An amide
(not basic, not nucleophilic)

Thus the amide bond has a partial double bond character, and there is a partial positive charge on the nitrogen atom. The lone pair of electrons on nitrogen in one of the resonance structures is used in bonding in the other resonance structure. In the amide, therefore, these electrons are much less available for use in nucleophilic bond-forming reactions than are the lone pair of electrons on the nitrogen atom in amines.

As we will see later, a number of interesting substances, including nylon and proteins, are held together by amide bonds between acids and amines.

section 23.7
Oxidation States of Carbon

In Section 15.2 we defined the oxidation state of an element, and stressed that for covalently bonded compounds the oxidation state or number is a convenient formalism, but is *not* the same as the actual charge on the atom. For organic compounds we can define the oxidation state of carbon using the rules outlined in Section 15.2, and it is then immediately apparent that the oxidation number of carbon is not the charge on the carbon atom. The oxidation state of carbon in CH_4, for example, is -4, but the C—H bonds are covalent, and the actual charge on carbon is only slightly negative, much smaller than -1.

As oxygen atoms are bonded to a carbon atom, electron density is pulled away from the carbon towards the more electronegative oxygen, and the carbon atom acquires a somewhat more positive charge. It is therefore oxidized. Bonding hydrogen atoms to a carbon atom tends to increase the electron density on carbon, since carbon is slightly more electronegative than hydrogen. Thus a carbon atom is reduced if the number of H atoms bonded to it increases, and is oxidized if the number of O atoms bonded to it increases.

A consistent set of oxidation numbers for carbon can be obtained by assigning a 0 oxidation state to any undefined "R" group, and then using the standard values of $+1$ for hydrogen and -2 for oxygen. Oxidation states for common organic groups are listed in Table 23.8. Complete oxidation of any organic compound yields CO_2, in which the oxidation state of carbon is $+4$.

Table 23.8. The Oxidation State of Carbon in Common Organic Groups

Organic Group		Oxidation State of Carbon	
Methyl	$R-CH_3$	-3	Least oxidized
Methylene	$\begin{array}{c} R' \\ \diagdown \\ CH_2 \\ \diagup \\ R \end{array}$	-2	
Alcohol	RCH_2OH	-1	
Aldehyde	$R-CH{=}O$	$+1$	
Ketone	$\begin{array}{c} R' \\ \diagdown \\ C{=}O \\ \diagup \\ R \end{array}$	$+2$	
Carboxylic acid	$R-COOH$	$+3$	Most oxidized

section 23.8
Brief Survey of Biologically Important Compounds

Living tissue is a complex mixture of organic compounds. If we were to put a piece of steak into a blender, homogenize it, and then shake it up with some hexane and some water, we would have separated the tissue into three kinds of compounds. There would be some material insoluble in both hexane and water that would remain as a solid precipitate at the bottom of this mixture. Proteins, DNA, cellulose, and other polymers are examples of substances found in tissue that will not dissolve either in water or in common organic solvents such as hexane.

Above this solid material would be the water layer containing sugars, amino acids, and other water soluble substances. Organic compounds that dissolve in water are classified as **hydrophilic,** or "water loving." These are polar compounds and substances that can form hydrogen bonds to water.

In the uppermost hexane layer would be found those natural substances that are soluble in hydrocarbons, but insoluble in water. These are nonpolar, **hydrophobic** ("water hating") substances, and include fats, waxes, natural hydrocarbons, and phosphoglycerides. The term **lipid** is used to describe all the naturally occurring organic species that are insoluble in water but soluble in hydrocarbons and other nonpolar solvents such as benzene and ethers.

Fats

The fats are perhaps the most obvious examples of lipids. **Fats** are naturally occurring esters of **glycerol** (1,2,3-propanetriol) with long-chain carboxylic acids. Glycerol is a **triol,** that is, a substance with three —OH groups. These —OH groups can react with three molecules of a carboxylic acid to produce a **triester.** The long-chain carboxylic acids that combine with glycerol to produce fats are called **fatty acids.** A typical reaction that produces a fat is the following:

$$
\begin{array}{cccc}
\text{CH}_2\text{—OH} & & & \text{CH}_2\text{—O—}\overset{\displaystyle O}{\overset{\|}{\text{C}}}\text{—(CH}_2)_{16}\text{CH}_3 \\
| & & & | \quad\quad \overset{\displaystyle O}{} \\
\text{CH—OH} & + & 3\ \text{CH}_3(\text{CH}_2)_{16}\text{—}\overset{\displaystyle O}{\overset{\|}{\text{C}}}\text{—OH} \rightarrow & \text{CH—O—}\overset{\|}{\text{C}}\text{—(CH}_2)_{16}\text{CH}_3 \quad + \quad 3\ \text{H}_2\text{O} \quad (23\text{-}18) \\
| & & & | \quad\quad \overset{\displaystyle O}{} \\
\text{CH}_2\text{—OH} & & & \text{CH}_2\text{—O—}\overset{\|}{\text{C}}\text{—(CH}_2)_{16}\text{CH}_3
\end{array}
$$

Glycerol   Stearic acid   Stearin, a fat

Because of the long-chain hydrocarbon portions of fats, and their lack of hydrogen-bonding groups, fats are relatively insoluble in water but are soluble in nonpolar organic substances.

Phospholipids

Phosphoglycerides are fats in which one of the three —OH groups of glycerol is esterified to phosphoric acid, and the other two to fatty acids. A typical phosphoglyceride therefore has the formula

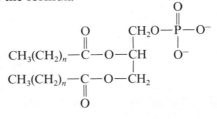

where n is some integer, usually 10, 12, 14, or 16. Monophosphate esters are called **phosphatidic acids,** and are rare in nature. Usually the phosphoric acid part of the molecule is further esterified to an alcohol, as in the biologically important compound phosphatidyl choline, shown below:

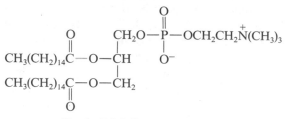

Phosphatidyl choline

Phosphoglycerides, collectively referred to as phospholipids, occur widely in the cell membranes of both plants and animals. The phosphate ester group can serve as a polar, hydrophilic region of the molecule, which can be described as having long nonpolar "tails" and a small, highly polar "head." Thus a phosphoglyceride is often schematically represented as shown in Fig. 23.15(a).

In aqueous solution phospholipids form **micelles.** A micelle is a spherical structure of many phospholipid molecules, with the hydrophobic long-chain fatty acid "tails" clustering together and pointing in toward the center of the sphere, while the surface of the spherical cluster consists of the polar ends of the molecules. Thus it is only the

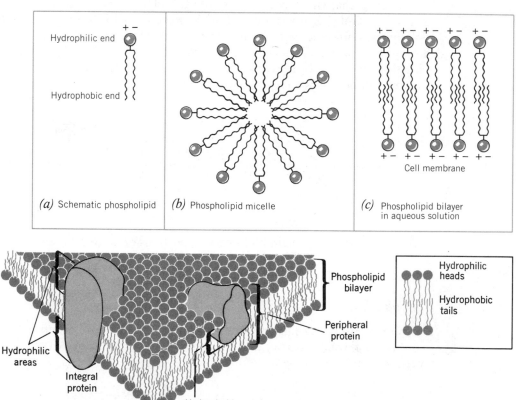

(a) Schematic phospholipid   *(b)* Phospholipid micelle   *(c)* Phospholipid bilayer in aqueous solution

(d) Bilayer structure of a cell membrane. Lipid mosaic model: Irregularly shaped proteins float randomly in a lipid sea. From S. J. Singer, *Annals of the New York Academy of Sciences,* Vol. 195, p. 21, 1962. Used by permission of the publisher and author.

Fig. 23.15.   Schematic diagram showing the long hydrophobic tails and the small hydrophilic head of a phospholipid, and the clusters that form in aqueous solution.

polar groups that come into contact with the water. A schematic diagram of a micelle is shown in Fig. 23.15*(b)*.

Phospholipids also form bilayers, depicted in Fig. 23.15*(c)*. Bilayers are particularly likely to form at the interface between two aqueous surfaces. Just as in the case of a micelle, this bilayer structure allows only the polar, hydrophilic heads of the phospholipid molecules to come into contact with the water. The tendency of phospholipids to form bilayers explains their usefulness as principal components of cell membranes in living tissues.

Proteins

In the discussion of carboxylic acids in Section 23.6, we described the reaction between a carboxylic acid and an amine, which produces an amide if the reaction mixture is heated to drive off water. This same general process takes place inside living organisms, catalyzed by specific enzymes, so that the reaction occurs readily at body temperature. Furthermore, the molecules used to form the amides are the **amino acids,** substances with both an amino group, $-NH_2$, and a carboxyl group, $-COOH$, on the same molecule.

The general formula for an amino acid is

$$\overset{\displaystyle NH_2}{\underset{\displaystyle H}{R-C-COOH}}$$

There are 20 different amino acids commonly found in nature. The names, formulas, and conventionally used abbreviations for some of the amino acids are given in Table 23.9. Most of the amino acids exist primarily in the form of a **zwitterion,** or inner salt, in which the carboxylic acid proton is donated to the amino group.

$$\overset{\displaystyle CH_3}{\underset{}{\overset{+}{H_3N}-CH-\overset{\displaystyle O}{\overset{\|}{C}}-O^-}}$$

Zwitterion of alanine

Amino acids can form amide linkages on both ends, allowing a long chain, or **polymer,** to be formed. A polymer of amino acids is called a **peptide,** and the linkage between amino acid molecules is called a **peptide bond.** If only two amino acid molecules are linked, the resulting molecule is a **dipeptide.** For instance, alanine and serine react to form the dipeptide alanylserine, as shown below:

$$\overset{\displaystyle CH_3}{\overset{|}{H_2N-CH-\overset{\displaystyle O}{\overset{\|}{C}}-OH}} \ + \ H_2N-\underset{\displaystyle CH_2OH}{\overset{}{CH}}-\overset{\displaystyle O}{\overset{\|}{C}}-OH \ \rightarrow$$

Alanine   Serine

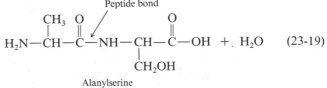

Peptide bond

Alanylserine

$$\overset{\displaystyle CH_3}{H_2N-CH-\overset{\displaystyle O}{\overset{\|}{C}}-NH-\underset{\displaystyle CH_2OH}{CH}-\overset{\displaystyle O}{\overset{\|}{C}}-OH \ + \ H_2O} \qquad (23\text{-}19)$$

Alanylserine also exists primarily in the **zwitterion** form.

If three amino acid molecules combine, a tripeptide is formed, and as more amino

Table 23.9. A Few Naturally Occurring Amino Acids

Amino Acid	Name	Abbreviation
H—CHCOOH with NH$_2$	Glycine	gly
H$_3$C—CHCOOH with NH$_2$	Alanine	ala
(H$_3$C)$_2$CH—CHCOOH with NH$_2$	Valine	val
HO—CH$_2$—CHCOOH with NH$_2$	Serine	ser
HS—CH$_2$—CHCOOH with NH$_2$	Cysteine	cys
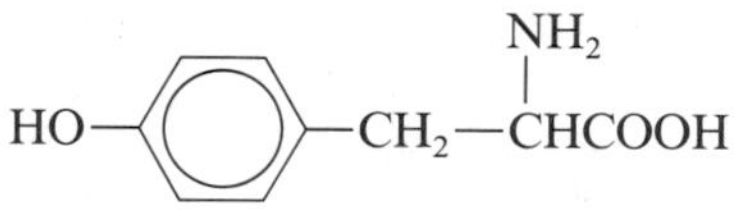 HO—⬡—CH$_2$—CHCOOH with NH$_2$	Tyrosine	tyr
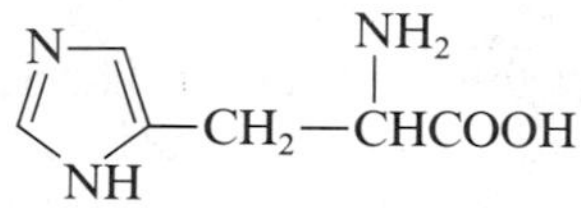 imidazole—CH$_2$—CHCOOH with NH$_2$	Histidine	his

acids are added to the chain, a polymer of any length can be formed. The general term **polypeptide** is used to indicate a polymer of many amino acids.

Proteins are very large molecules containing anywhere from 50 to more than 8000 amino acids per molecule. The formation of a typical polypeptide is shown below:

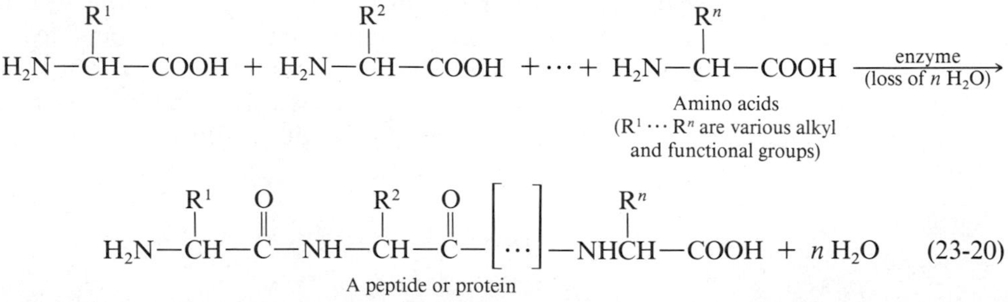

$$\underset{\substack{\text{Amino acids} \\ (R^1 \cdots R^n \text{ are various alkyl} \\ \text{and functional groups})}}{H_2N-\underset{R^1}{CH}-COOH \ + \ H_2N-\underset{R^2}{CH}-COOH \ + \cdots + \ H_2N-\underset{R^n}{CH}-COOH} \xrightarrow[\text{(loss of } n \text{ H}_2\text{O)}]{\text{enzyme}}$$

$$\underset{\text{A peptide or protein}}{H_2N-\underset{R^1}{CH}-\overset{O}{\underset{\|}{C}}-NH-\underset{R^2}{CH}-\overset{O}{\underset{\|}{C}}-\left[\cdots\right]-NH\underset{R^n}{CH}-COOH} \ + \ n \ H_2O \qquad (23\text{-}20)$$

The different R groups used in a protein give the final molecule the kinds of structural, hormonal, or catalytic properties needed by a particular organism.

A few examples of common types of proteins are the structural proteins, as in fingernails and hair; the peptide hormones such as insulin, oxytocin, and vasopressin; and the catalytic enzymes such as pepsin and trypsin (digestive enzymes). When a protein is being assembled, the sequence of amino acids is directed by the cell's ribonucleic acids (RNA), whose synthesis is in turn directed by the deoxyribonucleic acid in the cell's chromosomes. Just how these biopolymers direct the synthesis of proteins will be covered in detail in your organic chemistry and biochemistry courses.

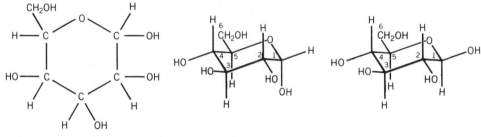

α-Glucose   β-Glucose

Fig. 23.16. The two stereoisomers of glucose.

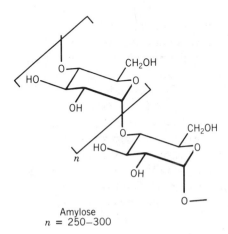

Amylose
n = 250–300

Fig. 23.17. The structure of amylose.

Carbohydrates

The sugars and their derivatives came to be called **carbohydrates** because their molecular formulas often are of the form $C_n(H_2O)_n$, where n is some integer. For instance, glucose has the formula $C_6H_{12}O_6$. Glucose is the most important fundamental food for many types of cells. It is a cyclic molecule, with five carbons and one oxygen forming a six-membered ring. The ring is not planar. Two stereoisomers are possible, called α- and β-glucose. The structure of glucose is shown in Fig. 23.16.

Glucose is stored in the liver in the form of **glycogen,** for use when quick energy is needed. Glycogen is a polymer of glucose containing thousands of glucose units per molecule. **Amylose,** a digestible form of starch, is another polymer of glucose, containing from 250–300 glucose units. Its structure is shown in Fig. 23.17.

Cellulose is also a glucose polymer, and occurs naturally in most plants. Cellulose has a very high molecular weight, as it contains about three thousand glucose units. It is insoluble in water, and is the chief structural component of plant cells. It is exceptionally abundant; about 50% of the weight of wood and bark, and about 90% of the weight of cotton fibers is cellulose. The structure of cellulose is shown in Fig. 23.18.

Polymers of many glucose units are collectively termed **polysaccharides.** Glycogen and amylose can be broken down into glucose by enzymes that occur in animals. Cellulose, on the other hand, can only be broken up into its constituent glucose units by certain species of bacteria. It is for this reason that wood is not a food for humans or other animals.

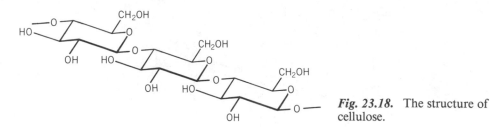

Fig. 23.18. The structure of cellulose.

section 23.9
Synthetic Polymers

Many of the solid objects we encounter are **polymers** of one sort or another. **Polymerization** is the bonding together of many small molecules (monomers) to make much larger molecules. We have already seen that wood, cotton, paper, and other cellulose products are all composed of polymeric glucose. Hair, leather, and aardvarks are largely held together by proteins, which are polymers of amino acids.

One of the most useful achievements of modern chemistry has been the development of techniques for synthesizing polymers from nonbiological sources, such as petroleum derivatives. This development not only allowed industry to avoid the limitations imposed by the numbers of rubber trees, cattle, and sheep in the world, but also allowed the design of specifically tailored polymeric materials with exactly the desired degrees of hardness and elasticity. The two major types of polymers are the **condensation polymers** and the **addition polymers**. We shall briefly consider one example of each type.

Nylon: A Condensation Polymer

We saw in the preceding section that proteins consist of amino acids held together by the formation of amide bonds:

$$\underset{\text{R}-\overset{\displaystyle O}{\overset{\displaystyle \|}{\text{C}}}-\text{OH}}{} + \text{H}_2\text{N}-\text{R}' \xrightarrow{\text{heat}} \text{R}-\overset{\displaystyle O}{\overset{\displaystyle \|}{\text{C}}}-\overset{\displaystyle H}{\overset{\displaystyle |}{\text{N}}}-\text{R}' + \text{H}_2\text{O} \qquad (23\text{-}21)$$

A reaction in which a new bond is formed with the extrusion of a small molecule (usually water, as in the formation of an amide), is called a **condensation reaction.** Polymeric materials that are held together by bonds formed in condensation reactions are called **condensation polymers.**

Nylon is a synthetic condensation polymer that is held together by amide bonds, and is therefore analogous to a protein. Because of the high strength and stability of amide bonds, nylon is one of the strongest and most durable of the synthetic polymers. The carboxylic acid used in making nylon is adipic acid (hexanedioic acid), a molecule with a carboxyl group on each end. The amine used is hexamethylenediamine, a molecule with an amino group on each end. The synthesis of nylon can be represented as:

$$n\ \text{HO}-\overset{\displaystyle O}{\overset{\displaystyle \|}{\text{C}}}-(\text{CH}_2)_4-\overset{\displaystyle O}{\overset{\displaystyle \|}{\text{C}}}-\text{OH} + n\ \text{H}_2\text{N}-(\text{CH}_2)_6-\text{NH}_2 \rightarrow \left[\begin{array}{c} {}^-\text{O}-\overset{\displaystyle O}{\overset{\displaystyle \|}{\text{C}}}-(\text{CH}_2)_4-\overset{\displaystyle O}{\overset{\displaystyle \|}{\text{C}}}-\text{O}^- \\ \overset{+}{\text{H}_3\text{N}}-(\text{CH}_2)_6-\overset{+}{\text{N}}\text{H}_3 \end{array} \right]_n$$

Adipic acid
(hexanedioic acid)

Hexamethylenediamine
(1,6-diaminohexane)

Nylon salt

$$\xrightarrow[\text{loss of } n \text{ H}_2\text{O}]{\text{heat}} \quad -(\text{CH}_2)_6-\underset{\underset{H}{|}}{N}-\left[\underset{}{\overset{\overset{O}{\|}}{C}}-(\text{CH}_2)_4-\overset{\overset{O}{\|}}{C}-\underset{\underset{H}{|}}{N}-(\text{CH}_2)_6-\underset{\underset{H}{|}}{N}-\overset{\overset{O}{\|}}{C}-(\text{CH}_2)_4-\right]_n \quad (23\text{-}22)$$

Amide bonds
nylon

Polyvinyl Chloride: An Addition Polymer

When a polymer is formed simply by adding the monomer molecules together, that polymer is called an **addition polymer.** Polyethylene, polypropylene, polystyrene, butyl rubber, and polyvinyl chloride (often called vinyl) are examples of addition polymers.

The polymerization of vinyl chloride occurs via a **free-radical chain reaction.** A free radical is a substance containing an odd number of electrons. We symbolize the unpaired electron by a dot, so that $\cdot\text{CH}_3$ is the methyl radical, and $\text{Br}\cdot$ is the bromine atom, which is also a free radical. In order to initiate polymer formation in the synthesis of polyethylene and polyvinyl chloride, a free radical is necessary. Radical-initiated polymerization is usually carried out by adding a small quantity of an organic peroxide, ROOR, which is readily cleaved by light to form radicals:

$$\text{ROOR} \xrightarrow{h\nu} 2\,\text{RO}\cdot \qquad\qquad (23\text{-}23)$$

Using the symbol $\text{R}\cdot$ to represent any free radical that may be employed to initiate polymerization, the series of reactions leading to polyvinyl chloride can be described as:

$$\text{R}\cdot + \text{H}_2\text{C}=\underset{\underset{\text{Cl}}{|}}{\text{CH}} \rightarrow \text{R}-\text{CH}_2-\underset{\underset{\text{Cl}}{|}}{\overset{\overset{\text{H}}{|}}{\text{C}}}\cdot \qquad (23\text{-}24a)$$

Vinyl chloride

$$\text{R}-\text{CH}_2-\underset{\underset{\text{Cl}}{|}}{\overset{\overset{\text{H}}{|}}{\text{C}}}\cdot + \text{H}_2\text{C}=\underset{\underset{\text{Cl}}{|}}{\text{CH}} \rightarrow \text{RCH}_2-\underset{\underset{\text{Cl}}{|}}{\overset{\overset{\text{H}}{|}}{\text{C}}}-\text{CH}_2-\underset{\underset{\text{Cl}}{|}}{\overset{\overset{\text{H}}{|}}{\text{C}}}\cdot \qquad (23\text{-}24b)$$

and so on, with the radical produced in one step reacting with more vinyl chloride in a succeeding step.

In its pure form polyvinyl chloride is a brittle substance that is not suitable for most of its common uses. It is most often treated with liquid plasticizers, however, that allow the polymer to stay supple and flexible. Many plasticizers gradually evaporate from the polymer, causing it to grow harder and more brittle as it ages.

section 23.10
Drugs

One of the most important applications of organic chemistry in recent years has been in the development of new drugs. When one speaks of the "wonders of modern medicine," one is most often praising the achievements of modern chemistry that happen to be prescribed by medical practitioners. Two major classes of drugs will be considered here: The hormones, which occur naturally in the body but often need to be supplemented for various reasons; and the antibiotics, which are used to help the body overcome bacterial infections.

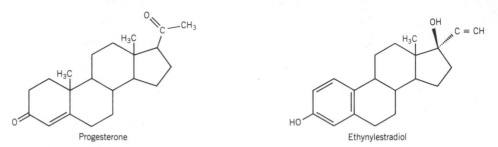

Fig. 23.19. The structures of the two hormones contained in the most widely used birth control pill. The natural hormone, estradiol, lacks the —C≡CH group of the synthetic ethynylestradiol.

Synthetic Hormones

Perhaps the most serious long-term problem facing mankind is the prospect of overpopulation and its accompanying food shortages and starvation. The most convenient and effective birth control method yet discovered involves the use of a daily dose of an estrogen and a progestin (female hormones) combined into "the Pill," which makes a woman's body assume certain characteristics of pregnancy, such as the cessation of ovulation.

One of the widely used birth control pills is a combination of ethynylestradiol and progesterone. Both of these compounds are synthesized in the laboratory. The synthetic progesterone is identical to that found naturally in the body. Ethynylestradiol differs from the natural hormone, estradiol, by the presence of the **ethynyl** (—C≡CH) **group.** The structures of progesterone and ethynylestradiol are shown in Fig. 23.19.

Dopamine

Dopamine is one of several substances used to transmit messages to the brain. An overabundance of this compound has been observed in cases of schizophrenia, and an insufficient supply of dopamine results in the condition known as Parkinson's disease, which causes muscular rigidity and tremors. Oral administration or injections of dopamine are ineffective in treating Parkinson's disease, however, because dopamine cannot cross the "blood-brain barrier," that is, it cannot get into the cerebrospinal fluid from the bloodstream.

Chemical synthesis has provided a way to solve this problem. A compound related to dopamine, called **dopa** (L-dopa or levodopa) has been synthesized. Dopa is metabolized to dopamine by brain cells. Dopa is an amino acid, related to tyrosine, and can therefore cross the blood-brain barrier. It is then converted to dopamine in the brain. The discovery of dopa has allowed many victims of Parkinson's disease to resume normal lives. The structure of dopa and dopamine are shown in Fig. 23.20.

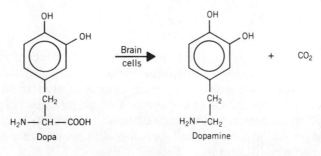

Fig. 23.20. The structures of dopa and dopamine. Dopa can be synthesized. It crosses the "blood-brain barrier" and is converted to dopamine by brain cells.

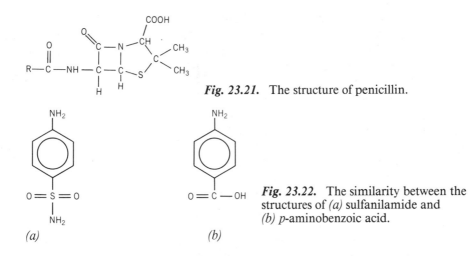

Fig. 23.21. The structure of penicillin.

Fig. 23.22. The similarity between the structures of *(a)* sulfanilamide and *(b)* p-aminobenzoic acid.

Antibiotics

Many wounds and infections considered minor today were frequently fatal before the discovery of penicillin. Penicillin inhibits the action of an enzyme that is necessary for the synthesis of the bacterial cell wall. Different forms of penicillin are made by changing the alkyl group, R, in the structure shown in Fig. 23.21.

Another antibiotic, sulfanilamide, stops bacteria from growing by acting as an ineffective imitation of p-aminobenzoic acid, a necessary nutrient for the production of folic acid. Without folic acid, bacteria cannot grow. Sulfanilamide fits into the active site of the enzyme that synthesizes folic acid, but cannot be made into active folic acid. In this way, the organism loses an essential nutrient and fails to grow, giving the body's natural defenses an opportunity to eliminate the invading bacteria. The similarity of sulfanilamide and p-aminobenzoic acid can be seen by comparing the two structures, shown in Fig. 23.22.

Summary

Organic chemistry is the study of the compounds of carbon. More than a million carbon compounds are known. In bonding to other atoms, including other carbon atoms, a carbon atom combines its valence $2s$ and $2p$ AO's to form either sp, sp^2, or sp^3 hybrid AO's. The effects of the type of hybridization used by carbon on the C—C bond length and bond energy, and on the C—H bond length and bond energy are described by comparing ethane, ethene, and acetylene. An understanding of the geometry of these three molecules enables us to predict the geometry of larger hydro-carbons.

There is free rotation about a C—C single bond, as in ethane, but a larger barrier to rotation about a C=C double bond, because rotation of the two groups connected by a double bond would break the π bond. As a result, if each of the two carbon atoms joined by a double bond has two different groups attached to it, *cis* and *trans* isomers will exist. A great many organic compounds containing one or more double bonds form geometric isomers.

A molecule containing at least one tetrahedrally substituted carbon atom bonded to four different atoms or groups will have two isomers that are nonsuperimposable mirror images of one another. Such a molecule is said to be **chiral,** and the pair of isomers are called **enantiomers.** Each enantiomer is **optically active,** that is, one enantiomer rotates the plane of plane-polarized light in the clockwise direction, the other rotates it in the counterclockwise direction. Many organic molecules in living organisms, including the amino acids, are optically active. In biological processes, it is common for one enantiomer to be essential for a particular reaction to occur, and the other enantiomer to be of little or no use.

Compounds with structures derived from benzene are called **aromatic** compounds; they contain multicentered, delocalized molecular orbitals. The difference in energy between the actual molecule and a hypothetical Lewis structure with localized bonds is the **resonance energy** of the molecule. Molecules with delocalized MO's are said to be **resonance stabilized.**

The **alkanes** are hydrocarbons of the general formula C_nH_{2n+2}, where n is an integer. They are relatively unreactive molecules. They do undergo combustion and are widely used for fuels. In the laboratory, they are used principally as solvents for other organic molecules.

Organic compounds frequently have **functional groups** that are reactive under appropriate conditions. Compounds are then classified by the functional groups they contain. Examples are the **alkenes** (which contain at least one carbon–carbon double bond), the **alcohols** (which contain at least one —OH group), the **ethers** (which contain the R—O—R′ structure, where R and R′ are organic groups), the **amines** (in which one or more hydrogen atoms of NH_3 is replaced by an organic group), the **aldehydes** (which contain the —CH=O group), the **ketones** (which contain the $>$C=O group), the **carboxylic acids** (which contain the —COOH group), the **esters** (which contain the —COOR group, where R is not hydrogen, but an organic group), and the **amides** (which contain the O=C—N group).

Organic compounds found in living organisms can be classified into three categories: (a) **hydrophilic compounds** that dissolve in water because they are polar or can form hydrogen bonds to water,   (b) **hydrophobic compounds** that dissolve in common organic solvents such as the alkanes, benzene, or the ethers, but do not dissolve in water, and   (c) compounds that do not dissolve either in water or in common organic solvents. Proteins, cellulose, and other polymeric molecules fall into this category.

The **fats** are naturally occurring esters of glycerol and long-chain carboxylic acids called **fatty acids.** If one or more of the three —OH groups of glycerol is esterified to phosphoric acid, a **phosphoglyceride** is formed. Phosphoglycerides are major components of cell membranes in living tissues.

Because **amino acids** have both a carboxylic acid group (—COOH) and a primary amine group (—NH_2), they can polymerize by forming amide bonds with the elimination of H_2O. A polymer of amino acids is called a **peptide. Proteins** are very large peptides containing anywhere from 50 to more than 8000 amino acids per molecule.

The sugars and their derivatives constitute the **carbohydrates. Glucose,** a simple sugar, is the most important food for many types of cells in living organisms. Glucose is a cyclic molecule, with five carbon atoms and one oxygen atom in a nonplanar six-membered ring. Many polymers of glucose, including glycogen, amylose, and cellulose, are of great biological importance.

One of the most useful achievements of modern chemistry has been the development of techniques for synthesizing polymers. Nylon, orlon, dacron, polyvinyl chloride, and polystyrene are only a few of the many synthetic polymers widely used today. The two major types of polymers are **condensation polymers** and **addition polymers.**

Synthetic organic chemistry has also produced a large number of drugs that have virtually revolutionized the field of medicine. A few of major importance are the penicillins, sulfa drugs, synthetic hormones, and levodopa.

Multiple Choice Questions

1. The $-\overset{\overset{\displaystyle O}{\|}}{C}-\overset{\overset{\displaystyle H}{|}}{N}-$ linkage occurs in all of the following substances EXCEPT

   (a) nylon (b) glycogen (c) protein (d) valylglycine (e) *N*-ethylacetamide

2. Which of the following can exist as *cis* and *trans* isomers?

   (a) $H_2C{=}CHCl$ (b) $H_2C{=}CH-CH{=}CH_2$ (c) $CH_3CH_2CH{=}CHCl$
   (d) $ClCH{=}CCl_2$ (e) $(CH_3)_2C{=}O$

3. A racemic mixture contains

   (a) Equal amounts of *cis* and *trans* isomers.
   (b) Both straight-chain and branched-chain alkanes.
   (c) A catalyst to increase the rate of reaction.
   (d) Equal amounts of a primary and a secondary amine.
   (e) Equal amounts of a pair of enantiomers.

4. Of the following substances the least soluble in water is

   (a) $CH_3CH_2OCH_3$ (b) $CH_3CH_2CH_2OH$ (c) $(CH_3)_2C{=}O$

   (d) $CH_3CH_2\overset{\overset{\displaystyle O}{\|}}{C}H$ (e) CH_3CH_2COOH

5. The correct systematic name for

   $$\underset{\underset{\displaystyle CH_3}{\diagdown}}{H_3C-CH-CH}\overset{\overset{\displaystyle CH_3CH_2}{|}}{}\overset{\diagup CH_3}{}$$

   is

   (a) 2-methyl-3-ethylbutane (b) 1,1-dimethyl-2-ethylpropane
   (c) 2-ethyl-3-methylhexane (d) 1,1,2-trimethylbutane
   (e) 2,3-dimethylpentane

6. A heterocyclic compound

   (a) Contains two or more fused benzene rings.
   (b) Contains both polar and nonpolar groups.
   (c) Is a reactive compound due to ring strain.
   (d) Is a cyclic compound soluble in both water and organic solvents.
   (e) Has a ring of atoms, at least one of which is not carbon.

7.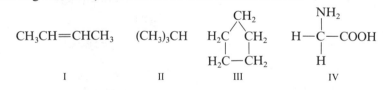

   Of the compounds listed above, which can exist as optical isomers?
   (a) none (b) IV (c) III and IV (d) I, III, and IV (e) I and II

8. Ordinary light is polarized into two plane-polarized rays when it passes through a crystal of

   (a) nickel (b) calcium (c) aragonite (d) alanine (e) calcite

9. The correct systematic name of

   $$H_2C{=}\overset{\overset{\displaystyle CH_3}{|}}{\underset{\underset{\displaystyle CH_2CH_3}{|}}{C}}-CH_3$$

   is

   (a) 2-ethyl-1-propene (b) 1-methyl-1-ethyl-ethylene (c) 2-ethyl-2-propene
   (d) 2-methyl-1-butene (e) 3-methyl-3-butene

10. The monomer from which the polymer $-(CH_2CCl_2CH_2CCl_2)_n-$ (Saran) is made is
    (a) $H_2C=CCl_2$ (b) $ClCH=CHCl$ (c) $Cl_2C=CCl_2$ (d) $H_2C=CHCl$
    (e) $ClCH=CCl_2$

11.

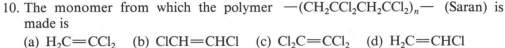

    I   II   III   IV

Of the compounds listed above, which would react readily with HBr?
    (a) I   (b) I and II   (c) II   (d) II and IV   (e) I and III

12. The correct systematic name for

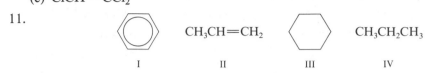

is
    (a) 3-ethylbutane   (b) 2-ethylbutane   (c) 3-methylhexane
    (d) 3-methylpentane   (e) 1,1-diethylethane

13. $CH_3CH_2CH_2CH_2OH$ $CH_3CH_2-O-CH_2CH_2CH_3$
                            I   II

$$\begin{matrix} CH_2-CH_2 \\ | \quad\quad\; \searrow \\ \quad\quad\quad CHOH \qquad CH_3CH=CHCH_2CH_2OH \\ | \quad\quad\; \nearrow \\ CH_2-CH_2 \end{matrix}$$

                    III   IV

Which of the compounds listed above are isomers?
    (a) III and IV   (b) I and IV   (c) II and III   (d) I and II   (e) all four

14. Which of the following compounds is a tertiary amine?
    (a) $CH_3CH_2NH_2$ (b) $(CH_3)_2NH$ (c) $(CH_3)_3N$ (d) $(CH_3CH_2)_4NCl$
    (e) $H_3C-\underset{\underset{NH_2}{|}}{CH}-COOH$

15. Which of the following compounds can exist as a pair of enantiomers?
    (a) $H_3C-CHCl_2$ (b) $H_2N-\underset{\underset{H}{|}}{\overset{\overset{CH_2OH}{|}}{C}}-CH_3$ (c) FCH_2-COOH

    (d) $H_3C-\underset{\underset{CH_3}{|}}{\overset{\overset{NH_2}{|}}{C}}-COOH$ (e) Cl_2CH-CH_2Br

16. In which of the following compounds is there a carbon atom in the +3 oxidation state?
    (a) CH_3CH_2OH (b) $H_3C-O-CH_3$ (c) CH_3COOH
    (d) $H_3C-\overset{\overset{O}{||}}{C}-CH_3$ (e) $H_3C-\underset{\underset{H}{|}}{C}=O$

17.

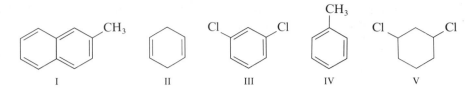

I   II   III   IV   V

Which of the compounds listed above are aromatic?

(a) III and V   (b) III, IV, and V   (c) I, II, and IV   (d) I   (e) I, III, and IV

18. A secondary amine is

(a) An organic compound with two —NH₂ groups.
(b) A compound with two carbon atoms and an —NH₂ group.
(c) A compound with an —NH₂ group on the carbon atom in the number 2 position.
(d) A compound in which two of the hydrogens of NH₃ have been replaced by organic groups.
(e) A compound with both an —NH₂ and a —COOH group.

19. The product of the reaction of 2-butene with HBr is

(a) 1-bromobutane   (b) 2-bromobutane   (c) 3-bromobutane
(d) 1,3-dibromobutane   (e) 2,3-dibromobutane

20. Which of the following is a polysaccharide?

(a) nylon   (b) amylose   (c) polyvinyl chloride   (d) penicillin
(e) polyethylene

Problems

23.1. Draw a diagram of the conformation of lowest potential energy of 1,2-dichloroethane, looking down the C—C bond. Also draw a diagram of the conformation of highest potential energy. What is the dihedral angle for these two conformations?

23.2. (a) If dichloromethane were square planar with the C atom in the center of the square, how many isomers would be possible?
(b) Dichloromethane is actually tetrahedral. How many isomers of dichloromethane are possible?

23.3. (a) Draw a Lewis electron dot diagram for H_3C—CH=N—CH_3.
(b) For each of the bonds in this molecule, specify the atomic orbitals on each atom that overlap to form the bond.
(c) Are geometric isomers possible for this compound?

23.4. Which of the compounds below can exhibit geometric isomerism?

(a) H_3C—CH=$C(CH_3)_2$ (b) H_3C—CH=$CHCl$ (c) H_3C—CH=$CBrCl$

(d)   (e)   (f)

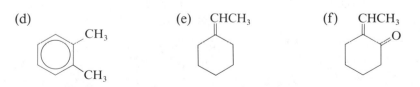

23.5. Although cyclohexene is an inexpensive, common organic reagent, cyclohexyne cannot be purchased. Explain the unavailability of cyclohexyne.

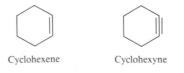

Cyclohexene   Cyclohexyne

23.6. Describe the orbitals used in bonding for allene, $H_2C=C=CH_2$. Predict the geometry, including bond angles, for this molecule.

23.7. (a) How many isomers of dibromobenzene ($C_6H_4Br_2$) are possible? Draw them.
(b) If the Kekulé structure for benzene were correct and benzene had three localized double bonds (1,3,5-cyclohexatriene), how many isomers of dibromobenzene would there be?

23.8. How many isomers of dibromopyridine ($C_5H_3Br_2N$) are possible? Draw them.

23.9. Name the following compounds:

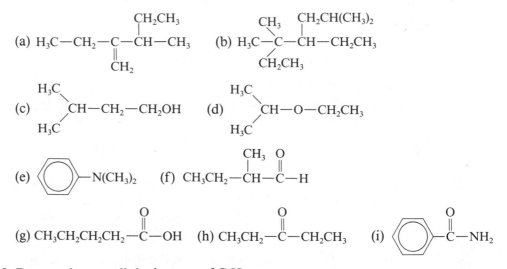

23.10. Draw and name all the isomers of C_6H_{14}.

23.11. Which of the following compounds exist as pairs of enantiomers? For the ones that do, draw the enantiomers.

(a) $H_3C-\underset{\underset{Cl}{|}}{CH}-CH_3$ (b) $H_3C-\underset{\underset{Cl}{|}}{CH}-CH_2Cl$ (c) H_2N-CH_2-COOH

(d) $H_2N-\underset{\overset{|}{CH}}{\overset{CH_2OH}{|}}-COOH$ (e) $H_3C-CH_2-\underset{\underset{I}{|}}{\overset{\overset{Br}{|}}{C}}-CH_2Br$

23.12. Show the individual steps in the reaction of 2-pentene with HI. Would you expect to get a single product or a mixture of products? Explain your answer.

23.13. The following five compounds have similar molecular weights, but a wide range of boiling points. Arrange them in order of increasing boiling point, and explain your reasoning.
(a) dimethyl ether   (b) formic acid   (c) propane   (d) ethanol
(e) acetaldehyde

23.14. Draw the structures and name the acid derivatives that would be formed when the following compounds combine:
(a) benzoic acid and ethanol   (b) benzoic acid and dimethylamine
(c) acetic acid and ammonia   (d) acetic acid and butanol

23.15. Each of the following reactions modifies the substituents on one or more carbon atoms. For the carbon atoms involved, indicate the oxidation state in the starting material and in the product. Has the carbon atom been oxidized or reduced?

(a) $H_3C—CH=CH—CH_3 \xrightarrow{KMnO_4} H_3C—CH—CH—CH_3$
$\qquad\qquad\qquad\qquad\qquad\qquad\quad$ OH   OH

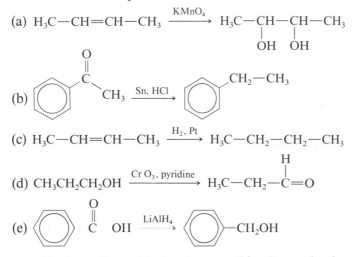

(c) $H_3C—CH=CH—CH_3 \xrightarrow{H_2,\ Pt} H_3C—CH_2—CH_2—CH_3$

(d) $CH_3CH_2CH_2OH \xrightarrow{Cr\,O_3,\ pyridine} H_3C—CH_2—\overset{\displaystyle H}{\underset{\displaystyle }{C}}=O$

23.16. Identify the amino acids that have combined to make the piece of a peptide shown below. Classify each of the six side chains that extend from the peptide linkages as hydrophilic, hydrophobic, or in neither category.

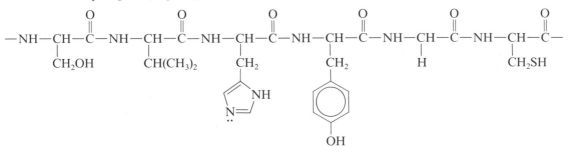

23.17. State whether each of the polymers shown below is an addition polymer or a condensation polymer. Explain your answer.

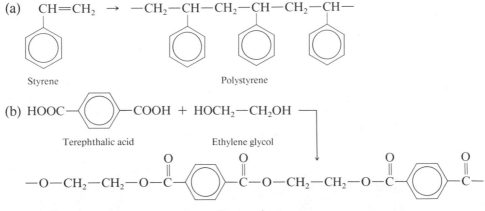

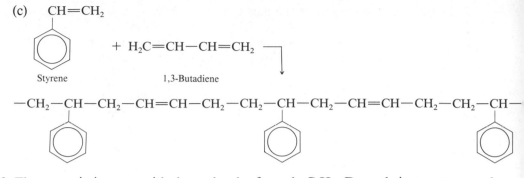

23.18. There are six isomers with the molecular formula C_4H_8. Draw their structures and name each compound.

23.19. There are six isomeric alkenes with the formula C_5H_{10}. Draw their structures and name each compound.

23.20. There are four cyclic alkane isomers with the formula C_5H_{10}. Draw their structures and name each compound.

appendix A Units and Conversion Factors

Base Units of the International System (SI Units)

Physical Quantity	Name of Unit	Symbol
Length	Meter	m
Mass	Kilogram	kg
Time	Second	s
Electric current	Ampere	A
Temperature	Kelvin	K
Luminous intensity	Candela	cd
Amount of substance	Mole	mol

Derived SI Units

Physical Quantity	Name of Unit	Symbol	Basic Units
Area	Square meter	m^2	m^2
Volume	Cubic meter	m^3	m^3
Density	Kilogram per cubic meter	$kg \cdot m^{-3}$	$kg \cdot m^{-3}$
Velocity	Meter per second	$m \cdot s^{-1}$	$m \cdot s^{-1}$
Acceleration	Meter per second per second	$m \cdot s^{-2}$	$m \cdot s^{-2}$
Force	Newton	N	$kg \cdot m \cdot s^{-2}$
Pressure	Pascal	Pa	$kg \cdot m^{-1}s^{-2}$ or N/m^2
Energy	Joule	J	$kg \cdot m^2s^{-2}$ or $N \cdot m$
Electric charge	Coulomb	C	$A \cdot s$
Electric potential difference	Volt	V	$kg \cdot m^2s^{-3}A^{-1}$ or J/C

Conversion Factors for Other Units of Length

Metric System	English System
1 kilometer (km) $= 1 \times 10^3$ m	1 inch(in.) $= 2.54$ cm
1 centimeter (cm) $= 1 \times 10^{-2}$ m	1 yard(yd) $= 0.9144$ m
1 millimeter (mm) $= 1 \times 10^{-3}$ m	1 mile(mi) $= 1.6093$ km
1 nanometer (nm) $= 1 \times 10^{-9}$ m	1 m $= 1.0936$ yd
1 angstrom (Å) $= 1 \times 10^{-10}$ m	$= 39.370$ in.
$= 1 \times 10^{-8}$ cm	$= 3.2808$ ft
$= 1 \times 10^{-1}$ nm	1 km $= 0.6214$ mi

Conversion Factors for Other Units of Mass and Weight

Metric System	English System
1 gram(g) $= 1 \times 10^{-3}$ kg	1 pound(lb) $= 453.59$ g
1 milligram(mg) $= 1 \times 10^{-3}$ g	1 ounce(oz) $= 28.35$ g
$= 1 \times 10^{-6}$ kg	1 (short) ton $= 907.18$ kg
1 metric ton $= 1 \times 10^3$ kg	1 kg $= 2.2046$ lb

1 atomic mass unit (amu) $= 1.6606 \times 10^{-27}$ kg $= 1.6606 \times 10^{-24}$ g

Conversion Factors for Other Units of Volume

Metric System	*English System*
1 liter(L) = 1×10^{-3} m³	1 U.S. quart(qt) $= 9.4635 \times 10^{-4}$ m³
$= 1$ dm³	$= 0.94635$ L
$= 1 \times 10^3$ cm³	1 cubic foot(ft³) $= 2.8317 \times 10^{-2}$ m³
1 milliliter(mL) $= 1 \times 10^{-6}$ m³	1 cubic inch(in.³) $= 16.387$ cm³
$= 1$ cm³	1 L $= 1.0567$ qt

Conversion Factors for Other Units of Force

$$1 \text{ dyne (dyn)} = 1 \times 10^{-5} \text{ N} \qquad 1 \text{ newton (N)} = 1 \times 10^{5} \text{ dyn}$$

Definitions

One newton (N) is the force that imparts an acceleration of one meter per second per second to a mass of one kilogram.

One dyne (dyn) is the force that imparts an acceleration of one centimeter per second per second to a mass of one gram.

Conversion Factors for Other Units of Pressure

1 atmosphere (atm) $= 760$ mmHg $= 1.01325 \times 10^5$ Pa $= 14.696$ lb/in.²
1 bar $= 1 \times 10^5$ Pa   1 torr $= 1$ mmHg   1 lb/in.² $= 6.895 \times 10^3$ Pa
1 dyne per square centimeter (dyn/cm²) $= 1 \times 10^{-1}$ Pa

Definition

Pressure is force per unit area. One pascal (Pa) is the pressure exerted when a force of one newton is applied per square meter.

$$1 \text{ Pa} = 1 \text{ N/m}^2$$

Conversion Factors for Other Units of Energy

1 thermochemical calorie (cal) $= 4.184$ J (exactly)
$$1 \text{ erg} = 1 \times 10^{-7} \text{ J} \qquad 1 \text{ J} = 1 \times 10^{7} \text{ erg}$$

1 liter · atmosphere(L · atm) $= 101.325$ J $= 24.217$ cal
1 electron volt (eV) $= 1.6022 \times 10^{-19}$ J $= 1.6022 \times 10^{-12}$ erg
1 eV/particle is equivalent to 96.485 kJ/mol or 23.06 kcal/mol
1 million electron volts (MeV) $= 1.602189 \times 10^{-13}$ J

Definitions

One joule of energy is expended when a force of one newton is applied through a distance of one meter.

$$1 \text{ J} = 1 \text{ N} \cdot \text{m}$$

One calorie was originally defined as the amount of heat required to raise the temperature of one gram of water from 14.5 to 15.5 °C. The thermochemical calorie is very close to, but not exactly the same as, the original calorie.

appendix B Mathematical Operations and Techniques

B1. Exponents

Exponential (Scientific) Notation

To express a number in exponential notation (also called scientific notation), the following form is used:

$$N \times 10^n$$

where N is a number between 1 and 10, and n is an integer, called the power of 10, or the exponent.

In order to determine the value of the exponent, n, count the number of places the decimal point must be moved in order to obtain N, the number between 1 and 10. *If the decimal point must be moved n places to the left, the exponent is +n. If the decimal point must be moved n places to the right, the exponent is −n.*

Examples
(a) $97,652,000 = 9.7652 \times 10^7$ (b) $57,830 = 5.783 \times 10^4$
(c) $0.000072065 = 7.2065 \times 10^{-5}$ (d) $0.00206 = 2.06 \times 10^{-3}$

It is important to remember that $10^0 = 1$. Indeed, $a^0 = 1$, where a (called the base), is any number at all.

Addition and Subtraction of Exponential Numbers

In order to add or subtract numbers written in exponential form, each number must be expressed to the same power of 10. The various N values are then added or subtracted in the usual manner. If necessary, the power of 10 of the final answer is changed to express the result in conventional scientific notation.

Examples
(a) $5.76 \times 10^4 - 2.1 \times 10^3 = (5.76 - 0.21) \times 10^4 = 5.55 \times 10^4$
(b) $9.85 \times 10^{-3} + 7.42 \times 10^{-4} = (9.85 + 0.742) \times 10^{-3} = 10.59 \times 10^{-3}$
$$= 1.059 \times 10^{-2}$$
(c) $8.73 \times 10^4 + 4.29 \times 10^2 = (8.73 + 0.0429) \times 10^4 = 8.77 \times 10^4$
(d) $1.26 \times 10^{-2} - 9.48 \times 10^{-3} - 6.91 \times 10^{-4} = (12.6 - 9.48 - 0.691) \times 10^{-3}$
$$= (12.6 - 10.17) \times 10^{-3}$$
$$= 2.4 \times 10^{-3}$$
(e) $5.407 \times 10^7 - 638 = (5.407 - 0.0000638) \times 10^7 = 5.407 \times 10^7$

The number 638 is insignificant with respect to 5.407×10^7.

Note that the rules concerning significant figures, discussed in the Introduction, are followed in all of these examples.

Multiplication of Exponential Numbers

The rule for multiplying powers of 10 is that *the exponents must be added algebraically* to give the correct power of 10:

$$10^a \cdot 10^b = 10^{(a+b)}$$

Thus,

$$(N_1 \times 10^{n_1})(N_2 \times 10^{n_2}) = N_1 N_2 \times 10^{(n_1+n_2)}$$

The product should then be converted to conventional scientific notation.

Examples

(a) $(5.3 \times 10^6)(8.468 \times 10^5) = 44.9 \times 10^{11} = 4.5 \times 10^{12}$

The product should only be expressed to two significant figures as 5.3×10^6 is only known to two figures. In changing 44.9 to 4.49 we divided by 10; therefore the 10^{11} factor must be multiplied by 10 to keep the product constant.

(b) $(7.684 \times 10^{-8})(8.61 \times 10^3) = 66.16 \times 10^{-5} = 6.62 \times 10^{-4}$

Division of Exponential Numbers

The rule for dividing powers of 10 is that the exponent of the denominator is subtracted from the exponent of the numerator:

$$10^a/10^b = 10^{(a-b)}$$

Thus,

$$\frac{N_1 \times 10^{n_1}}{N_2 \times 10^{n_2}} = \left(\frac{N_1}{N_2}\right) \times 10^{(n_1-n_2)}$$

The result should then be converted to conventional exponential notation.

Examples

(a) $\dfrac{5.3 \times 10^{-8}}{6.294 \times 10^4} = 0.842 \times 10^{-12} = 8.4 \times 10^{-13}$

(b) $\dfrac{4.712 \times 10^7}{6.581 \times 10^{-3}} = 0.7160 \times 10^{[7-(-3)]} = 0.7160 \times 10^{10} = 7.160 \times 10^9$

Of course, a computation may involve both multiplication and division of exponential numbers, as in the following example:

$$\frac{(6.52 \times 10^{-4})(9.467 \times 10^2)}{(1.873 \times 10^3)(8.51 \times 10^5)} = \frac{(6.52)(9.467)}{(1.873)(8.51)} \times 10^{(-4+2-3-5)} = 3.87 \times 10^{-10}$$

Powers of Exponential Numbers

When a power of 10 is raised to a power, the exponent is multiplied by the power:

$$(10^a)^b = 10^{ab}$$

Thus,

$$(N \times 10^n)^b = N^b \times 10^{bn}$$

Examples

(a) $(5.7 \times 10^6)^2 = (5.7)^2 \times 10^{12} = 32.49 \times 10^{12} = 3.2 \times 10^{13}$
(b) $(6.413 \times 10^{-4})^3 = (6.413)^3 \times 10^{-12} = 263.7 \times 10^{-12} = 2.637 \times 10^{-10}$
(c) $(4.91 \times 10^{-2})^5 = (4.91)^5 \times 10^{-10} = 2854 \times 10^{-10} = 2.85 \times 10^{-7}$

Roots of Exponential Numbers

In taking the root of a power of 10, the exponent is divided by the value of the desired root. To ensure that the result is expressed with an integral power of 10, make the original exponent divisible by the desired root before dividing. (A calculator does this automatically, but there are times when it is useful to compute the exponent in your head.)

Examples

(a) $\sqrt{5.79 \times 10^7} = (5.79 \times 10^7)^{1/2} = (57.9 \times 10^6)^{1/2} = (57.9)^{1/2} \times 10^3$
$$= 7.61 \times 10^3$$

(b) $\sqrt[3]{8.61 \times 10^{-8}} = (86.1 \times 10^{-9})^{1/3} = (86.1)^{1/3} \times 10^{-3} = 4.42 \times 10^{-3}$

The following example illustrates a computation involving several different operations:

$$\frac{(3.49 \times 10^{-2})(6.073 \times 10^{-1})^3}{(0.08206)(8.152 \times 10^{-7})^{1/2}} = ?$$

Note at the start that the answer is only good to three significant figures. First evaluate the terms raised to powers, and then collect all powers of 10 together.

$$(6.073 \times 10^{-1})^3 = (6.073)^3 \times 10^{-3}$$
$$(8.152 \times 10^{-7})^{1/2} = (81.52 \times 10^{-8})^{1/2} = (81.52)^{1/2} \times 10^{-4}$$

The calculation is then rearranged as

$$\frac{(3.49)(6.073)^3 \times 10^{-5}}{(8.206)(81.52)^{1/2} \times 10^{-6}} = \frac{(3.49)(223.98) \times 10}{(8.206)(9.029)} = 10.55 \times 10 = 1.06 \times 10^2$$

Exercises Involving Exponential Numbers

Calculate the following quantities. Express all answers to the correct number of significant figures.

B1. $1.72 \times 10^5 - 9.47 \times 10^4$

B2. $4.51 \times 10^{-3} + 6.40 \times 10^{-5} - 8.27 \times 10^{-4}$

B3. $2.50 \times 10^{-6} + 7.03 \times 10^{-8}$

B4. $3.92 \times 10^4 - 2.5 \times 10^3 + 1.63$

B5. $(1.643 \times 10^{-3})(7.2 \times 10^2)(6.03 \times 10^{-5})$

B6. $(2.10 \times 10^{-4})(6.591 \times 10^{-8})(4.20 \times 10^3)(1.093 \times 10^5)$

B7. $\dfrac{4.637 \times 10^{-8}}{7.20 \times 10^{-5}}$

B8. $\dfrac{(3.8037 \times 10^{-6})(4.19 \times 10^4)}{(2.068 \times 10^3)(9.108 \times 10^{-3})}$

B9. $\dfrac{(3.70 \times 10^{-3})^2(8.503 \times 10^8)^{1/3}}{(6.421 \times 10^{-4})(9.56 \times 10^3)^{1/2}}$

B10. $\dfrac{8.536 \times 10^{-2} - 4.91 \times 10^{-1}}{6.3704 \times 10^{-4}}$

B11. $\dfrac{(5.63 \times 10^3 - 8.72 \times 10^2)^{1/2}}{9.686 \times 10^{-4} + 2.058 \times 10^{-2}}$

B2. *Logarithms*

Base 10 Logarithms

The common, or base 10, logarithm of any number is the power to which 10 must be raised to obtain that number. A logarithm is therefore an exponent. Base 10 logarithms are abbreviated log. The logarithms of integral powers of 10 can be obtained by inspection. Examples are (a) $\log(10{,}000) = \log(10^4) = 4$, (b) $\log(10^{-15}) = -15$, and (c) $\log 0.0000010 = \log(10^{-6}) = -6$.

In order to obtain the logarithm of a number that is not an integral power of 10, we use a log table. Most calculators have log tables in their memory banks, so you can obtain a base 10 logarithm simply by pushing the log button. However, there are times when it is important to know how to read a log table.

A logarithm consists of two parts: A decimal part, called the mantissa, and an integer, called the characteristic. A log table lists *only* the decimal portions of the logarithm, that is, only the mantissas. A table of four-place logarithms lists four decimal places for each mantissa. Appendix B3 is a four-place table to use with the following examples. Note that it is customary not to print the decimal point in the table.

Suppose we want the log of 37,800. First express the number in exponential notation, as 3.78×10^4. The characteristic, or integral part of the logarithm, is 4 (the integral power of 10). The log of 3.78 is obtained by looking in the table. The first two digits (3,7) are found in the first column at the left of the table, and the third digit is located across the top of the table. The mantissa of 3.78 is 0.5775. You will find it in the row beginning 37, under the column headed 8.

$$\log(37{,}800) = \log(3.78 \times 10^4) = 4.5775$$

Since the characteristic is always given by the integral power of 10, the value of the mantissa is independent of the position of the decimal point, and depends only on the digits in the number between 1 and 10. For example, all the following numbers have the same mantissa.

Number	Characteristic	Mantissa	Log
6.54	0	0.8156	0.8156
6540	3	0.8156	3.8156
65,400,000	7	0.8156	7.8156

The mantissas listed in a log table are positive numbers. If the characteristic is negative, the terms should be combined algebraically.

Example   What is $\log(0.003710)$?

Solution

$$\log(0.003710) = \log(3.710 \times 10^{-3}) = -3 + \log 3.710$$
$$= -3 + 0.5694 = -2.4306$$

It is also common to write this logarithm as $\bar{3}.5694$, indicating that the characteristic is negative, but the mantissa is positive.

Example   Find the log of 0.000284, 0.284, 28.4, and 284,000.

Number	Characteristic	Mantissa	Logarithm
0.000284	−4	0.4533	−3.5467
0.284	−1	0.4533	−0.5467
28.4	1	0.4533	1.4533
284,000	5	0.4533	5.4533

Base 10 Antilogarithms

In doing problems, we will sometimes have to solve the following type of equation: $\log K = 3.7135$. The number K is called the antilogarithm (abbreviated antilog) of 3.7135. An equivalent way of writing this equation is $K = 10^{3.7135}$.

If your calculator has a 10^x button, you can use it to obtain $K = 10^{3.7135} = 5.17 \times 10^3$. You can also find the antilog by locating the mantissa 0.7135 in a table of logarithms. It is found in the row beginning 51, under the column headed 7. The characteristic, 3, gives the power of 10, so that antilog$(3.7135) = 5.17 \times 10^3$.

Example Find the antilog of each of the following numbers: 6.5740, 0.5740, $\bar{1}$.5740, −1.5740, and −8.5740.

Logarithm	Characteristic	Mantissa	Antilog
6.5740	6	0.5740	3.75×10^6
0.5740	0	0.5740	3.75
$\bar{1}$.5740	−1	0.5740	3.75×10^{-1}
−1.5740	−2	0.4260	2.67×10^{-2}
−8.5740	−9	0.4260	2.67×10^{-9}

Note that if the logarithm is given with a negative decimal portion, you must rewrite it with a positive mantissa, in order to use a log table. Thus −1.5740 is written as $-2 + 0.4260$ before using a log table. A calculator does this automatically. Simply use the 10^x button to find $10^{-1.5740} = 0.0267 = 2.67 \times 10^{-2}$.

There are occasions when you will not be able to use your calculator to obtain an antilog, because most calculators cannot handle numbers smaller than 10^{-99} or larger than 10^{99}. (There may be different limits on some models.) The antilog of −108 is 10^{-108}, but many calculators will display zero, or indicate that an error has been made, if you try to enter 10^{-108}. Similarly, if $\log x = 132$, $x = 10^{132}$, but most calculators will not display this number. How do you solve the equation $\log y = 104.32$? Determine the power of 10 in your head (10^{104}), and find the antilog of 0.32, either using your calculator, or reading a log table. If $\log y = 104.32$, $y = 2.1 \times 10^{104}$.

Example Find K if $\log K = -99.7$.

Write

$$\log K = -100 + 0.3 \qquad K = (\text{antilog } 0.3) \times 10^{-100} = 2 \times 10^{-100}$$

Exercises Involving Logarithms

B12. Find the base 10 logarithm of (a) 6.022×10^{23}, (b) 4.87×10^{-11}, (c) 0.00519, and (d) 8×10^{124}.

B13. Evaluate x if (a) $\log x = 15.314$, (b) $\log x = -12.410$, (c) $\log x = -0.3307$, (d) $\log x = 99.84$, and (e) $\log x = -115.7$.

Fundamental Properties of Logarithms

1. The log of a product is the sum of the logs of the factors:

$$\log (ab) = \log a + \log b$$

We have already used this property. For example: Find log (0.00764).

$$\log (7.64 \times 10^{-3}) = \log 7.64 + \log 10^{-3} = 0.8831 - 3 = -2.1169$$

2. The logarithm of a quotient is the logarithm of the numerator minus the logarithm of the denominator:

$$\log (a/b) = \log a - \log b$$

As a special case of this general rule, $\log (1/b) = -\log b$. Since 10^0 is 1, $\log (1) = 0$.

3. The logarithm of the nth power of a number is n times the logarithm of the number:

$$\log a^n = n \log a$$

The relation stated above is true for all values of n, fractions as well as integers. It is therefore possible to extract higher-order roots using logarithms, if your calculator can readily extract only square roots and cube roots.

Example   Evaluate $(9.63 \times 10^{-8})^{1/5}$.

Solution   First make the exponent divisible by five. (This is not necessary, but it simplifies the calculation.)

$$(9.63 \times 10^{-8})^{1/5} = (963 \times 10^{-10})^{1/5} = (963)^{1/5} \times 10^{-2}$$

Now we must evaluate the fifth root of 963. Using logarithms, this is done as follows:

$$\text{let} \quad x = (963)^{1/5}$$

Then,

$$\log x = (1/5)\log 963 = (1/5)(2.9836) = 0.5967$$

The antilog of 0.5967 is 3.95, so $x = 3.95$. Therefore,

$$(9.63 \times 10^{-8})^{1/5} = 3.95 \times 10^{-2}$$

Natural Logarithms

In addition to base 10 logarithms, there are logarithms that are the exponents of the base e, where

$$e = 2.718281828 \cdots$$

These logarithms are called natural logarithms and are commonly abbreviated ln. (Many people pronounce ln as if it were *lin*.) Thus,

$$e^1 = 2.7183 \quad \text{therefore} \quad \ln 2.7183 = 1$$
$$e^2 = 7.3891 \quad \text{therefore} \quad \ln 7.3891 = 2$$

Many calculators have ln buttons as well as log buttons. To take the antilog of a natural logarithm, use the e^x button on your calculator. Check the following examples on your own calculator:

(a) $\ln(x) = 3.4691$ therefore $x = e^{3.4691} = 32.11$
(b) $\ln(x) = -24.05$ therefore $x = e^{-24.05} = 3.59 \times 10^{-11}$

B3. *Four-Place Table of Logarithms*

N	0	1	2	3	4	5	6	7	8	9	1	2	3	4	5	6	7	8	9
															Proportional Parts				
10	0000	0043	0086	0128	0170	0212	0253	0294	0334	0374	ª4	8	12	17	21	25	29	33	37
11	0414	0453	0492	0531	0569	0607	0645	0682	0719	0755	4	8	11	15	19	23	26	30	34
12	0792	0828	0864	0899	0934	0969	1004	1038	1072	1106	3	7	10	14	17	21	24	28	31
13	1139	1173	1206	1239	1271	1303	1335	1367	1399	1430	3	6	10	13	16	19	23	26	29
14	1461	1492	1523	1553	1584	1614	1644	1673	1703	1732	3	6	9	12	15	18	21	24	27
15	1761	1790	1818	1847	1875	1903	1931	1959	1987	2014	ª3	6	8	11	14	17	20	22	25
16	2041	2068	2095	2122	2148	2175	2201	2227	2253	2279	3	5	8	11	13	16	18	21	24
17	2304	2330	2355	2380	2405	2430	2455	2480	2504	2529	2	5	7	10	12	15	17	20	22
18	2553	2577	2601	2625	2648	2672	2695	2718	2742	2765	2	5	7	9	12	14	16	19	21
19	2788	2810	2833	2856	2878	2900	2923	2945	2967	2989	2	4	7	9	11	13	16	18	20
20	3010	3032	3054	3075	3096	3118	3139	3160	3181	3201	2	4	6	8	11	13	15	17	19
21	3222	3243	3263	3284	3304	3324	3345	3365	3385	3404	2	4	6	8	10	12	14	16	18
22	3424	3444	3464	3483	3502	3522	3541	3560	3579	3598	2	4	6	8	10	12	14	15	17
23	3617	3636	3655	3674	3692	3711	3729	3747	3766	3784	2	4	6	7	9	11	13	15	17
24	3802	3820	3838	3856	3874	3892	3909	3927	3945	3962	2	4	5	7	9	11	12	14	16
25	3979	3997	4014	4031	4048	4065	4082	4099	4116	4133	2	3	5	7	9	10	12	14	15
26	4150	4166	4183	4200	4216	4232	4249	4265	4281	4298	2	3	5	7	8	10	11	13	15
27	4314	4330	4346	4362	4378	4393	4409	4425	4440	4456	2	3	5	6	8	9	11	13	14
28	4472	4487	4502	4518	4533	4548	4564	4579	4594	4609	2	3	5	6	8	9	11	12	14
29	4624	4639	4654	4669	4683	4698	4713	4728	4742	4757	1	3	4	6	7	9	10	12	13
30	4771	4786	4800	4814	4829	4843	4857	4871	4886	4900	1	3	4	6	7	9	10	11	13
31	4914	4928	4942	4955	4969	4983	4997	5011	5024	5038	1	3	4	6	7	8	10	11	12
32	5051	5065	5079	5092	5105	5119	5132	5145	5159	5172	1	3	4	5	7	8	9	11	12
33	5185	5198	5211	5224	5237	5250	5263	5276	5289	5302	1	3	4	5	6	8	9	10	12
34	5315	5328	5340	5353	5366	5378	5391	5403	5416	5428	1	3	4	5	6	8	9	10	11
35	5441	5453	5465	5478	5490	5502	5514	5527	5539	5551	1	2	4	5	6	7	9	10	11
36	5563	5575	5587	5599	5611	5623	5635	5647	5658	5670	1	2	4	5	6	7	8	10	11
37	5682	5694	5705	5717	5729	5740	5752	5763	5775	5786	1	2	3	5	6	7	8	9	10
38	5798	5809	5821	5832	5843	5855	5866	5877	5888	5899	1	2	3	5	6	7	8	9	10
39	5911	5922	5933	5944	5955	5966	5977	5988	5999	6010	1	2	3	4	5	7	8	9	10
40	6021	6031	6042	6053	6064	6075	6085	6096	6107	6117	1	2	3	4	5	6	8	9	10
41	6128	6138	6149	6160	6170	6180	6191	6201	6212	6222	1	2	3	4	5	6	7	8	9
42	6232	6243	6253	6263	6274	6284	6294	6304	6314	6325	1	2	3	4	5	6	7	8	9
43	6335	6345	6355	6365	6375	6385	6395	6405	6415	6425	1	2	3	4	5	6	7	8	9
44	6435	6444	6454	6464	6474	6484	6493	6503	6513	6522	1	2	3	4	5	6	7	8	9
45	6532	6542	6551	6561	6571	6580	6590	6599	6609	6618	1	2	3	4	5	6	7	8	9
46	6628	6637	6646	6656	6665	6675	6684	6693	6702	6712	1	2	3	4	5	6	7	7	8
47	6721	6730	6739	6749	6758	6767	6776	6785	6794	6803	1	2	3	4	5	5	6	7	8
48	6812	6821	6830	6839	6848	6857	6866	6875	6884	6893	1	2	3	4	4	5	6	7	8
49	6902	6911	6920	6928	6937	6946	6955	6964	6972	6981	1	2	3	4	4	5	6	7	8
50	6990	6998	7007	7016	7024	7033	7042	7050	7059	7067	1	2	3	3	4	5	6	7	8
51	7076	7084	7093	7101	7110	7118	7126	7135	7143	7152	1	2	3	3	4	5	6	7	8
52	7160	7168	7177	7185	7193	7202	7210	7218	7226	7235	1	2	2	3	4	5	6	7	7
53	7243	7251	7259	7267	7275	7284	7292	7300	7308	7316	1	2	2	3	4	5	6	6	7
54	7324	7332	7340	7348	7356	7364	7372	7380	7388	7396	1	2	2	3	4	5	6	6	7
N	0	1	2	3	4	5	6	7	8	9	1	2	3	4	5	6	7	8	9

ª Interpolation in this section of the table is inaccurate.

(continued)

B3. *Four-Place Table of Logarithms (continued)*

N	0	1	2	3	4	5	6	7	8	9	1	2	3	4	5	6	7	8	9
55	7404	7412	7419	7427	7435	7443	7451	7459	7466	7474	1	2	2	3	4	5	5	6	7
56	7482	7490	7497	7505	7513	7520	7258	7536	7543	7551	1	2	2	3	4	5	5	6	7
57	7559	7566	7574	7582	7589	7597	7604	7612	7619	7627	1	2	2	3	4	5	5	6	7
58	7634	7642	7649	7657	7664	7672	7679	7686	7694	7701	1	1	2	3	4	4	5	6	7
59	7709	7716	7723	7731	7738	7745	7752	7760	7767	7774	1	1	2	3	4	4	5	6	7
60	7782	7789	7796	7803	7810	7818	7825	7832	7839	7846	1	1	2	3	4	4	5	6	6
61	7853	7860	7868	7875	7882	7889	7896	7903	7910	7917	1	1	2	3	4	4	5	6	6
62	7924	7931	7938	7945	7952	7959	7966	7973	7980	7987	1	1	2	3	3	4	5	6	6
63	7993	8000	8007	8014	8021	8028	8035	8041	8048	8055	1	1	2	3	3	4	5	5	6
64	8062	8069	8075	8082	8089	8096	8102	8109	8116	8122	1	1	2	3	3	4	5	5	6
65	8129	8136	8142	8149	8156	8162	8169	8176	8182	8189	1	1	2	3	3	4	5	5	6
66	8195	8202	8209	8215	8222	8228	8235	8241	8248	8254	1	1	2	3	3	4	5	5	6
67	8261	8267	8274	8280	8287	8293	8299	8306	8312	8319	1	1	2	3	3	4	5	5	6
68	8325	8331	8338	8344	8351	8357	8363	8370	8376	8382	1	1	2	3	3	4	4	5	6
69	8388	8395	8401	8407	8414	8420	8426	8432	8439	8445	1	1	2	2	3	4	4	5	6
70	8451	8457	8463	8470	8476	8482	8488	8494	8500	8506	1	1	2	2	3	4	4	5	6
71	8513	8519	8525	8531	8537	8543	8549	8555	8561	8567	1	1	2	2	3	4	4	5	5
72	8573	8579	8585	8591	8597	8603	8609	8615	8621	8627	1	1	2	2	3	4	4	5	5
73	8633	8639	8645	8651	8657	8663	8669	8675	8681	8686	1	1	2	2	3	4	4	5	5
74	8692	8698	8704	8710	8716	8722	8727	8733	8739	8745	1	1	2	2	3	4	4	5	5
75	8751	8756	8762	8768	8774	8779	8785	8791	8797	8802	1	1	2	2	3	3	4	5	5
76	8808	8814	8820	8825	8831	8837	8842	8848	8854	8859	1	1	2	2	3	3	4	5	5
77	8865	8871	8876	8882	8887	8893	8899	8904	8910	8915	1	1	2	2	3	3	4	4	5
78	8921	8927	8932	8938	8943	8949	8954	8960	8965	8971	1	1	2	2	3	3	4	4	5
79	8976	8982	8987	8993	8998	9004	9009	9015	9020	9025	1	1	2	2	3	3	4	4	5
80	9031	9036	9042	9047	9053	9058	9063	9069	9074	9079	1	1	2	2	3	3	4	4	5
81	9085	9090	9096	9101	9106	9112	9117	9122	9128	9133	1	1	2	2	3	3	4	4	5
82	9138	9143	9149	9154	9159	9165	9170	9175	9180	9186	1	1	2	2	3	3	4	4	5
83	9191	9196	9201	9206	9212	9217	9222	9227	9232	9238	1	1	2	2	3	3	4	4	5
84	9243	9248	9253	9258	9263	9269	9274	9279	9284	9289	1	1	2	2	3	3	4	4	5
85	9294	9299	9304	9309	9315	9320	9325	9330	9335	9340	1	1	2	2	3	3	4	4	5
86	9345	9350	9355	9360	9365	9370	9375	9380	9385	9390	1	1	2	2	3	3	4	4	5
87	9395	9400	9405	9410	9415	9420	9425	9430	9435	9440	0	1	1	2	2	3	3	4	4
88	9445	9450	9455	9460	9465	9469	9474	9479	9484	9489	0	1	1	2	2	3	3	4	4
89	9494	9499	9504	9509	9513	9518	9523	9528	9533	9538	0	1	1	2	2	3	3	4	4
90	9542	9547	9552	9557	9562	9566	9571	9576	9581	9586	0	1	1	2	2	3	3	4	4
91	9590	9595	9600	9605	9609	9614	9619	9624	9628	9633	0	1	1	2	2	3	3	4	4
92	9638	9643	9647	9652	9657	9661	9666	9671	9675	9680	0	1	1	2	2	3	3	4	4
93	9685	9689	9694	9699	9703	9708	9713	9717	9722	9727	0	1	1	2	2	3	3	4	4
94	9731	9736	9741	9745	9750	9754	9759	9763	9768	9773	0	1	1	2	2	3	3	4	4
95	9777	9782	9786	9791	9795	9800	9805	9809	9814	9818	0	1	1	2	2	3	3	4	4
96	9823	9827	9832	9836	9841	9845	9850	9854	9859	9863	0	1	1	2	2	3	3	4	4
97	9868	9872	9877	9881	9886	9890	9894	9899	9903	9908	0	1	1	2	2	3	3	4	4
98	9912	9917	9921	9926	9930	9934	9939	9943	9948	9952	0	1	1	2	2	3	3	4	4
99	9956	9961	9965	9969	9974	9978	9983	9987	9991	9996	0	1	1	2	2	3	3	3	4
N	0	1	2	3	4	5	6	7	8	9	1	2	3	4	5	6	7	8	9

Proportional Parts

B4. Graphs

It is quite common in a laboratory experiment that we vary some property (called the independent variable) and observe how a different property (called the dependent variable) changes. For instance, we may have observed how the volume of a sample of gas changes as we vary the temperature at atmospheric pressure, and obtained the following data:

t (°C)	6.0	10.0	18.0	25.0	32.0
V (L)	1.237	1.255	1.290	1.321	1.352

The clearest way to represent the relationship between the dependent and independent variables is to graph them. It is customary to plot the independent variable along the horizontal axis (also called the abscissa or x axis), and the dependent variable along the vertical axis (also called the ordinate or y axis).

If the data given above are plotted with temperature as the abscissa, and volume as the ordinate, the graph shown below is obtained. We see immediately that there is a linear relationship between the temperature and the volume.

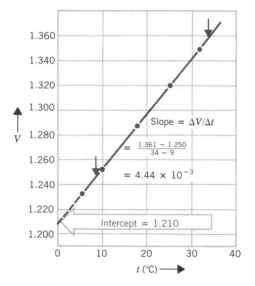

The equation for any straight line is

$$y = mx + b$$

where m is the slope of the line, and b is the intercept on the y axis. Since we have plotted V as a function of t, the equation relating volume in liters (L) and temperature in degrees Celsius (°C) for this sample of gas, at 1 atm, is $V = mt + b$.

We can read the value of b directly on our graph, by extending (extrapolating) the line back so that it crosses the V axis. We obtain $b = 1.210$ L.

The slope is defined as $\Delta V/\Delta t$, and is determined by choosing two widely separated points (not experimental points), and evaluating ΔV, the difference in V, and Δt, the difference in t, for these two points. The points used for the calculation of the slope are indicated on the graph by red arrows. The calculation of the slope is shown on the graph. The equation for the line representing the experimental data is therefore

$$V = 4.44 \times 10^{-3}t + 1.210$$

The units of the slope are liters per degree Celsius, and the units of the intercept are liters.

B5. *Quadratic Equations*

An algebraic equation of the form $ax^2 + bx + c = 0$ is called a quadratic equation, or a second-order equation. Every quadratic equation has two solutions (two roots), given by the quadratic formula

$$x = \frac{-b \pm (b^2 - 4ac)^{1/2}}{2a}$$

Example   Solve the equation $3x^2 - 5x = 4$.

Solution   First rearrange the equation to the form

$$3x^2 - 5x - 4 = 0$$

Inspection then yields $a = 3$, $b = -5$, and $c = -4$. Evaluate $b^2 - 4ac$:

$$b^2 - 4ac = 25 - (4)(3)(-4) = 25 + 48 = 73$$

Then

$$x = \frac{5 \pm (73)^{1/2}}{6} = \frac{5 \pm 8.544}{6}$$

One root is therefore

$$\frac{5 + 8.544}{6} = 2.257$$

and the second root is

$$\frac{5 - 8.544}{6} = \frac{-1.2867}{6} = -0.2144$$

While there are always two roots to every quadratic equation, only one of the roots may be physically meaningful. In that case, the other root can be discarded. If, in the equation above, for example, x represented a volume, the negative root would be meaningless physically.

Exercises in the Use of the Quadratic Formula

B14.   (a) Solve the following equation for y:

$$4y^2 - 7y = 3$$

(b) Solve the following equation for x:

$$2x^2 + 5x = -1$$

B15.   Find the positive root of the following equation:

$$4.0 \times 10^{-3} = (z + 5.0 \times 10^{-2})(z)$$

B16.   If $k = 6.3 \times 10^{-4}$ and $a = 1.0 \times 10^{-2}$, solve the following equation for x:

$$k = \frac{x^2}{a - x}$$

Calculate only the positive root.

appendix C Experimental Uncertainties

Uncertainties that are expressed in the same units as the quantities themselves are called **absolute uncertainties** or **absolute errors**. For instance, the mass of an object may be reported as 2.036 ± 0.002 g. The absolute uncertainty in the mass is ± 0.002 g. The absolute uncertainty, which is also referred to as the **precision** of an experimental value, is the range of reliability of that value. The smaller the absolute uncertainty, the better the precision of the value.

A clear distinction should be made between the terms **precision** and **accuracy**. The precision of a numerical result is the degree of agreement between that number and other values obtained by repeating the experiment. It is a measure of the degree to which results check. The accuracy of a numerical result is the degree of agreement between that number and the true value. In many cases the true value is not known and the accuracy of an experimental result cannot be determined. It is quite possible to repeat an experiment several times and obtain results that agree closely with one another, but are all quite far from the true value, if some systematic error is being made. In that case, the experimental result has a high precision and a low accuracy.

The **relative uncertainty** (or **relative error**) in a quantity is obtained by dividing the absolute uncertainty by the value of the quantity. If the relative uncertainty is multiplied by 100, we obtain the **percent uncertainty** or **percent error**. For example, for the mass cited above, 2.036 ± 0.002 g, the relative uncertainty is

$$\frac{\pm 0.002 \text{ g}}{2.036 \text{ g}} = \pm 0.001$$

and the percent uncertainty is $\pm 0.1\%$. If the volume of a liquid, measured in a graduated cylinder, is 6.5 ± 0.2 mL, the relative uncertainty is

$$\frac{\pm 0.2 \text{ mL}}{6.5 \text{ mL}} = \pm 0.03$$

and the percent uncertainty is $\pm 3\%$. The relative and percent uncertainties are dimensionless (unitless) quantities. Since the absolute uncertainty always has the same units as the quantity itself, when we divide the absolute uncertainty by the quantity, the result has no units.

As you will have noticed in the examples discussed so far, both absolute and relative uncertainties are usually known to only one significant figure. If you are calculating a relative uncertainty from an absolute uncertainty that is only known to one significant figure, the relative uncertainty should be reported only to one significant figure.

Suppose we have two objects weighed on two different balances. One object has mass 47.2 ± 0.1 g, and the second object has mass 4.72 ± 0.01 g. The absolute uncertainties of these two objects are quite different, but each has the same relative uncertainty, since

$$\frac{\pm 0.1 \text{ g}}{47.2 \text{ g}} = \frac{\pm 0.01 \text{ g}}{4.72 \text{ g}} = \pm 0.002$$

Each of the numbers 47.2 g and 4.72 g has three significant figures, and the same relative uncertainty, but different absolute uncertainties. If the two masses are 51.8

and 4.36 g, the absolute uncertainties are still ±0.1 and ±0.01 g, respectively, and the relative uncertainties of both are still ±0.002.

As a quick rule of thumb, it is approximately true that values with the same relative uncertainty have the same number of significant figures. This is not true, however, if one value is slightly less than a power of 10, and the other is slightly more than the same power of 10. For instance, compare masses of 94.6 ± 0.1 g and 102.7 ± 0.1 g. Both these numbers have the same relative uncertainty:

$$\frac{\pm 0.1 \text{ g}}{94.6 \text{ g}} = \pm 0.001 \quad \text{and} \quad \frac{\pm 0.1 \text{ g}}{102.7 \text{ g}} = \pm 0.001$$

but 94.6 has three significant figures, while 102.7 has four significant figures. It is the relative uncertainty, and not the number of significant figures, that gives the correct measure of the reliability of the value.

The rules discussed in the Introduction for determining the number of significant figures in a calculated answer are based on the approximation that the number of significant figures is a measure of the relative uncertainty in a value. The distinction between the rule to apply for multiplication and/or division, as opposed to the rule to apply for addition and/or subtraction, merits emphasizing. *For a value calculated by multiplying or dividing factors, the answer cannot have a relative uncertainty smaller than the factor with the largest relative uncertainty.* Because the larger the relative uncertainty, the smaller the number of significant figures, the answer cannot have more significant figures than the factor with the smallest number of significant figures. *For a value calculated by addition and/or subtraction, the answer cannot have an absolute uncertainty smaller than the term with the largest absolute uncertainty.* The absolute uncertainty is *not* directly related to the number of significant figures.

Problems

C1. The National Bureau of Standards establishes the limits of uncertainty allowable for volumetric equipment. A 25-mL volumetric pipet rated "Class *A*" by the National Bureau of Standards must deliver a volume within the limits 25.00 ± 0.05 mL.
(a) What is the absolute uncertainty of a 25-mL Class *A* volumetric pipet?
(b) What is the relative uncertainty of a 25-mL Class *A* volumetric pipet?

Answer (a) ±0.05 mL (b) ±0.2%

C2. A 25-mL graduated cylinder can be used to measure the volume of a liquid to the nearest 0.2 mL. Twenty milliliters of a dilute nitric acid solution are measured with a graduated cylinder.
(a) Express the volume of the solution to the correct number of significant figures and give the absolute uncertainty of the volume.
(b) Calculate the relative uncertainty of the volume of solution.

Answer (a) 20.0 ± 0.2 mL (b) ±1%

C3. The masses of two objects, weighed on the same balance, are reported as 29.042 and 0.286 g. What are the absolute and relative uncertainties in each of these values?

C4. The mass of a crucible, weighed on a triple beam balance, is reported as 13.3 g. When the same crucible is weighed on a single-pan Mettler balance, its mass is reported as 13.284 g. What are the absolute and relative uncertainties in each of these masses?

appendix D Derivation of the Conversion Factor between Two Units of Pressure, the Atmosphere and the Pascal

Pressure is defined as force per unit area. The gases in the atmosphere above the surface of the earth are attracted to the earth by the force of gravity. The instrument used to measure the pressure exerted by the atmosphere is a barometer, which contains a column of mercury supported by the atmospheric pressure. The mercury in the barometer is attracted to the earth by the force of gravity.

Force is defined by Newton's second law of motion:

$$\text{force} = (\text{mass}) \cdot (\text{acceleration}) \tag{D-1}$$

Acceleration is the time rate of change of velocity. An object dropped from a height accelerates as it falls to the earth, that is, its speed increases each second that it falls. This acceleration is a result of the force of gravity on the object. The acceleration due to gravity varies somewhat from place to place on earth, and the International Committee on Weights and Measures has adopted as the **standard acceleration due to gravity,** g, a value of $9.80665 \text{ m} \cdot \text{s}^{-2}$. At standard gravity, the speed of a freely falling body increases by 9.80665 meters per second each second that it falls.

By definition, one atmosphere is the pressure that supports a column of mercury exactly 76-cm high at 0 °C and standard gravity. The density of mercury at 0 °C is $13.595 \text{ g} \cdot \text{cm}^{-3}$.

Let us calculate the force of gravity on a column of mercury exactly 76-cm high at 0 °C and standard gravity. Since force is (mass) · (acceleration), we must know the mass of the mercury in the column. We can calculate the mass of the mercury if we know its density and its volume, since

$$\text{mass in g} = (\text{density in g} \cdot \text{cm}^{-3})(\text{volume in cm}^3) \tag{D-2}$$

The volume of the mercury column is given by

$$\text{volume in cm}^3 = (\text{height in cm})(\text{cross-sectional area in cm}^2)$$

Let A equal the cross-sectional area of the column of mercury in centimeters squared. Then the volume of mercury in the column is $76A \text{ cm}^3$. By substitution into Eq. (D-2),

$$\begin{aligned}\text{mass of mercury} &= (13.595 \text{ g} \cdot \text{cm}^{-3})(76A \text{ cm}^3) \\ &= 1033.22A \text{ g} = 1.03322A \text{ kg}\end{aligned}$$

Using Eq. (D-1) we find

$$\begin{aligned}\text{force of gravity} &= (\text{mass}) \cdot (\text{acceleration due to gravity}) \\ &= (1.03322A \text{ kg})(9.80665 \text{ m} \cdot \text{s}^{-2}) \\ &= 10.1324A \text{ kg} \cdot \text{m} \cdot \text{s}^{-2} = 10.1324A \text{ N}\end{aligned}$$

See Appendix A for the definition of a newton (N).

Pressure is force per unit area, so to obtain the pressure we must divide the force by the area in square meters. Note that we have expressed the force of gravity in SI units (newtons), and therefore must express the area in square meters. As the area is A cm^2, and 1 cm $= 1 \times 10^{-2}$ m, the area is $A \times 10^{-4}$ m^2.

$$\text{pressure} = \frac{10.1324A \text{ N}}{A \times 10^{-4} \text{ m}^2} = 1.0132 \times 10^5 \text{ Pa}$$

By definition, one pascal is one newton per square meter. As the density of mercury at 0 °C is only given to five figures, the pressure of the column of mercury can only be expressed to five figures.

Since 1 atm is the pressure that supports a column of mercury exactly 76-cm high at 0 °C at standard gravity,

$$1 \text{ atm} = 1.0132 \times 10^5 \text{ Pa}$$

appendix E Equilibrium Constants at 25 °C

Table E1. Acidity Constants of Weak Acids

Name of Acid	Formula	K_a
Acetic acid	CH_3COOH	1.8×10^{-5}
Ammonium ion	NH_4^+	5.7×10^{-10}
Anilinium ion	$C_6H_5NH_3^+$	2.4×10^{-5}
Arsenic acid	H_3AsO_4	6.0×10^{-3}
Dihydrogen arsenate ion	$H_2AsO_4^-$	1×10^{-7}
Monohydrogen arsenate ion	$HAsO_4^{2-}$	4×10^{-12}
Benzoic acid	C_6H_5COOH	6.3×10^{-5}
Boric acid	H_3BO_3	5.8×10^{-10}
Carbonic acid	$H_2CO_3 (H_2O + CO_2)$	4.3×10^{-7}
Hydrogen carbonate ion	HCO_3^-	4.7×10^{-11}
Dimethylammonium ion	$(CH_3)_2NH_2^+$	1.4×10^{-11}
Formic acid	$HCOOH$	1.8×10^{-4}
Hydrocyanic acid	HCN	4×10^{-10}
Hydrofluoric acid	HF	7.2×10^{-4}
Hydrogen chromate ion	$HCrO_4^-$	3.0×10^{-7}
Hydrogen sulfate ion	HSO_4^-	1.2×10^{-2}
Hydrosulfuric acid	H_2S	1.0×10^{-7}
Hydrogen sulfide ion	HS^-	1.3×10^{-13}
Hypobromous acid	$HOBr$	2.5×10^{-9}
Hypochlorous acid	$HOCl$	3.5×10^{-8}
Methylammonium ion	$CH_3NH_3^+$	2.4×10^{-11}
Monochloroacetic acid	$CH_2ClCOOH$	1.4×10^{-3}
Nitrous acid	$HONO$	4.5×10^{-4}
Oxalic acid	$H_2C_2O_4$	5.9×10^{-2}
Hydrogen oxalate ion	$HC_2O_4^-$	6.4×10^{-5}
Phenol	C_6H_5OH	1.3×10^{-10}
Phosphoric acid	H_3PO_4	7.5×10^{-3}
Dihydrogen phosphate ion	$H_2PO_4^-$	6.2×10^{-8}
Monohydrogen phosphate ion	HPO_4^{2-}	4×10^{-13}
Propanoic acid	CH_3CH_2COOH	1.3×10^{-5}
Pyridinium ion	$C_5H_5NH^+$	5.0×10^{-6}
Sulfurous acid	$H_2SO_3 (H_2O + SO_2)$	1.2×10^{-2}
Hydrogen sulfite ion	HSO_3^-	6.2×10^{-8}
Trimethylammonium ion	$(CH_3)_3NH^+$	1.6×10^{-10}

(continued)

Table E2. **Solubility Product Constants**

Name of Solid	Formula	K_{sp}
Aluminum hydroxide	$Al(OH)_3$	1.9×10^{-33}
Aluminum phosphate	$AlPO_4$	1.3×10^{-20}
Antimony sulfide	Sb_2S_3	2×10^{-93}
Barium carbonate	$BaCO_3$	8.1×10^{-9}
Barium chromate	$BaCrO_4$	2.0×10^{-10}
Barium fluoride	BaF_2	1.7×10^{-6}
Barium iodate	$Ba(IO_3)_2$	6×10^{-10}
Barium oxalate	$BaC_2O_4 \cdot 2H_2O$	1.1×10^{-7}
Barium sulfate	$BaSO_4$	1.1×10^{-10}
Barium sulfite	$BaSO_3$	8.0×10^{-7}
Bismuth oxychloride	$BiOCl$	7.0×10^{-9}
Bismuth sulfide	Bi_2S_3	2×10^{-72}
Cadmium cyanide	$Cd(CN)_2$	1×10^{-8}
Cadmium hydroxide	$Cd(OH)_2$	1.2×10^{-14}
Cadmium sulfide	CdS	4×10^{-29}
Calcium carbonate	$CaCO_3$	4.8×10^{-9}
Calcium chromate	$CaCrO_4$	7.1×10^{-4}
Calcium fluoride	CaF_2	3.9×10^{-11}
Calcium hydroxide	$Ca(OH)_2$	7.9×10^{-6}
Calcium oxalate	$CaC_2O_4 \cdot H_2O$	2.3×10^{-9}
Calcium sulfate	$CaSO_4 \cdot 2H_2O$	2.4×10^{-5}
Chromium(III) hydroxide	$Cr(OH)_3$	7×10^{-31}
Cobalt(II) hydroxide	$Co(OH)_2$	2×10^{-16}
Cobalt(III) hydroxide	$Co(OH)_3$	4×10^{-45}
Cobalt sulfide (α)[a]	$CoS(\alpha)$	6×10^{-21}
Cobalt sulfide (β)[a]	$CoS(\beta)$	8×10^{-23}
Copper(I) chloride	$CuCl$	2×10^{-7}
Copper(II) chromate	$CuCrO_4$	3.6×10^{-6}
Copper(II) hydroxide	$Cu(OH)_2$	2×10^{-19}
Copper(I) iodide	CuI	5×10^{-12}
Copper(I) sulfide	Cu_2S	1.6×10^{-48}
Copper(II) sulfide	CuS	8.7×10^{-36}
Iron(II) hydroxide	$Fe(OH)_2$	8×10^{-15}
Iron(III) hydroxide	$Fe(OH)_3$	6.3×10^{-38}
Iron(II) sulfide	FeS	5×10^{-18}
Iron(III) sulfide	Fe_2S_3	1×10^{-88}
Lead bromide	$PbBr_2$	6.3×10^{-6}
Lead carbonate	$PbCO_3$	1.5×10^{-13}
Lead chloride	$PbCl_2$	1.6×10^{-5}
Lead chromate	$PbCrO_4$	1.8×10^{-14}
Lead fluoride	PbF_2	3.7×10^{-8}
Lead hydroxide	$Pb(OH)_2$	2.8×10^{-16}
Lead iodide	PbI_2	8.7×10^{-9}
Lead oxalate	PbC_2O_4	8×10^{-12}

Table E2. **Solubility Product Constants** *(continued)*

Name of Solid	Formula	K_{sp}
Lead sulfate	$PbSO_4$	1.8×10^{-8}
Lead sulfide	PbS	8×10^{-28}
Magnesium ammonium phosphate[b]	$MgNH_4PO_4 \cdot 6H_2O$	2.5×10^{-13}
Magnesium carbonate	$MgCO_3$	4×10^{-5}
Magnesium fluoride	MgF_2	6.4×10^{-9}
Magnesium hydroxide	$Mg(OH)_2$	1.5×10^{-11}
Magnesium oxalate	MgC_2O_4	8.6×10^{-5}
Manganese(II) carbonate	$MnCO_3$	8.8×10^{-11}
Manganese(II) hydroxide	$Mn(OH)_2$	4.5×10^{-14}
Manganese(II) sulfide	MnS	5×10^{-15}
Mercury(I) bromide	Hg_2Br_2	1.3×10^{-22}
Mercury(I) chloride	Hg_2Cl_2	1.1×10^{-18}
Mercury(I) chromate	Hg_2CrO_4	2×10^{-9}
Mercury(I) cyanide	$Hg_2(CN)_2$	5×10^{-40}
Mercury(I) iodide	Hg_2I_2	4.5×10^{-29}
Mercury(I) sulfate	Hg_2SO_4	6.8×10^{-7}
Mercury(II) sulfide	HgS	3×10^{-53}
Nickel hydroxide	$Ni(OH)_2$	2×10^{-15}
Nickel sulfide (α)[a]	$NiS(\alpha)$	3×10^{-21}
Silver acetate	CH_3COOAg	2.3×10^{-3}
Silver arsenate	Ag_3AsO_4	1×10^{-20}
Silver bromide	$AgBr$	5.0×10^{-13}
Silver carbonate	Ag_2CO_3	8.2×10^{-12}
Silver chloride	$AgCl$	1.8×10^{-10}
Silver chromate	Ag_2CrO_4	9×10^{-12}
Silver cyanide	$AgCN$	1.2×10^{-16}
Silver iodide	AgI	1.5×10^{-16}
Silver phosphate	Ag_3PO_4	1×10^{-20}
Silver sulfate	Ag_2SO_4	1.7×10^{-5}
Silver sulfide	Ag_2S	7×10^{-50}
Silver thiocyanate	$AgSCN$	1.0×10^{-12}
Strontium carbonate	$SrCO_3$	9.4×10^{-10}
Strontium chromate	$SrCrO_4$	3.6×10^{-5}
Strontium fluoride	SrF_2	7.9×10^{-10}
Strontium oxalate	$SrC_2O_4 \cdot H_2O$	5.6×10^{-8}
Strontium sulfate	$SrSO_4$	2.8×10^{-7}
Thallium(I) sulfide	Tl_2S	1×10^{-22}
Tin(II) hydroxide	$Sn(OH)_2$	2×10^{-26}
Tin(II) sulfide	SnS	8×10^{-29}
Zinc carbonate	$ZnCO_3$	2×10^{-11}
Zinc hydroxide	$Zn(OH)_2$	4.5×10^{-17}
Zinc sulfide	ZnS	1.1×10^{-21}

[a] When freshly precipitated from basic solution, the more soluble alpha (α) forms of CoS and NiS are formed. On standing, the crystal structures change to a less soluble form.

[b] $K_{sp} = [Mg^{2+}][NH_4^+][PO_4^{3-}]$.

appendix F Standard Electrode Potentials at 25 °C

Half-Reaction	$\mathcal{E}^0$ (volts)
$F_2(g) + 2e^- \rightarrow F^-$	+2.87
$H_2O_2 + 2H^+ + 2e^- \rightarrow 2H_2O$	+1.776
$PbO_2(s) + SO_4^{2-} + 4H^+ + 2e^- \rightarrow PbSO_4(s) + H_2O$	+1.685
$MnO_4^- + 4H^+ + 3e^- \rightarrow MnO_2(s) + 2H_2O$	+1.679
$HOCl + H^+ + e^- \rightarrow \frac{1}{2} Cl_2(g) + H_2O$	+1.63
$BrO_3^- + 6H^+ + 5e^- \rightarrow \frac{1}{2} Br_2 + 3H_2O$	+1.52
$MnO_4^- + 8H^+ + 5e^- \rightarrow Mn^{2+} + 4H_2O$	+1.51
$ClO_3^- + 6H^+ + 5e^- \rightarrow \frac{1}{2} Cl_2(g) + 3H_2O$	+1.47
$Ce^{4+} + e^- \rightarrow Ce^{3+}$	+1.443
$Au^{3+} + 3e^- \rightarrow Au(s)$	+1.42
$Cl_2(g) + 2e^- \rightarrow 2Cl^-$	+1.36
$Cr_2O_7^{2-} + 14H^+ + 6e^- \rightarrow 2Cr^{3+} + 7H_2O$	+1.33
$O_2(g) + 4H^+ + 4e^- \rightarrow 2H_2O$	+1.229
$MnO_2(s) + 4H^+ + 2e^- \rightarrow Mn^{2+} + 2H_2O$	+1.23
$2 IO_3^- + 12H^+ + 10e^- \rightarrow I_2 + 6H_2O$	+1.195
$ClO_4^- + 2H^+ + 2e^- \rightarrow ClO_3^- + H_2O$	+1.19
$Br_2(aq) + 2e^- \rightarrow 2Br^-$	+1.087
$Br_2(\ell) + 2e^- \rightarrow 2Br^-$	+1.0652
$NO_3^- + 4H^+ + 3e^- \rightarrow NO(g) + 2H_2O$	+0.96
$NO_3^- + 3H^+ + 2e^- \rightarrow HONO + H_2O$	+0.94
$2Hg^{2+} + 2e^- \rightarrow Hg_2^{2+}$	+0.920
$Hg^{2+} + 2e^- \rightarrow Hg(\ell)$	+0.855
$Ag^+ + e^- \rightarrow Ag(s)$	+0.799
$Hg_2^{2+} + 2e^- \rightarrow 2Hg(\ell)$	+0.789
$Fe^{3+} + e^- \rightarrow Fe^{2+}$	+0.771
$O_2(g) + 2H^+ + 2e^- \rightarrow H_2O_2$	+0.682
$Ag_2SO_4(s) + 2e^- \rightarrow 2Ag(s) + SO_4^{2-}$	+0.653
$MnO_4^- + 2H_2O + 3e^- \rightarrow MnO_2(s) + 4 OH^-$	+0.588
$MnO_4^- + e^- \rightarrow MnO_4^{2-}$	+0.564
$I_2(s) + 2e^- \rightarrow 2 I^-$	+0.535
$Cu^+ + e^- \rightarrow Cu(s)$	+0.521
$Ag_2CrO_4(s) + 2e^- \rightarrow 2Ag(s) + CrO_4^{2-}$	+0.446
$O_2(g) + 2H_2O + 4e^- \rightarrow 4 OH^-$	+0.401
$Cu^{2+} + 2e^- \rightarrow Cu(s)$	+0.337
$Hg_2Cl_2(s) + 2e^- \rightarrow 2Hg(\ell) + 2Cl^-$	+0.268
$AgCl(s) + e^- \rightarrow Ag(s) + Cl^-$	+0.2223
$Co(OH)_3(s) + e^- \rightarrow Co(OH)_2(s) + OH^-$	+0.17
$Cu^{2+} + e^- \rightarrow Cu^+$	+0.153

(continued)

Half-Reaction	$\mathcal{E}^0$ *(volts)*
$Sn^{4+} + 2e^- \rightarrow Sn^{2+}$	+0.15
$S + 2H^+ + 2e^- \rightarrow H_2S(aq)$	+0.14
$S_4O_6^{2-} + 2e^- \rightarrow 2S_2O_3^{2-}$	+0.09
$AgBr(s) + e^- \rightarrow Ag(s) + Br^-$	+0.0713
$NO_3^- + H_2O + 2e^- \rightarrow NO_2^- + 2\,OH^-$	+0.01
$2H^+(aq) + 2e^- \rightarrow H_2(g)$	0.0000
$Fe^{3+} + 3e^- \rightarrow Fe(s)$	−0.036
$CrO_4^{2-} + 4H_2O + 3e^- \rightarrow Cr(OH)_3(s) + 5\,OH^-$	−0.12
$Pb^{2+} + 2e^- \rightarrow Pb(s)$	−0.126
$Sn^{2+} + 2e^- \rightarrow Sn(s)$	−0.136
$O_2(g) + 2H_2O + 2e^- \rightarrow H_2O_2 + 2\,OH^-$	−0.146
$AgI(s) + e^- \rightarrow Ag(s) + I^-$	−0.152
$Cu(OH)_2(s) + 2e^- \rightarrow Cu(s) + 2\,OH^-$	−0.224
$Ni^{2+} + 2e^- \rightarrow Ni(s)$	−0.25
$Co^{2+} + 2e^- \rightarrow Co(s)$	−0.28
$PbSO_4(s) + 2e^- \rightarrow Pb(s) + SO_4^{2-}$	−0.356
$Cd^{2+} + 2e^- \rightarrow Cd(s)$	−0.403
$Cr^{3+} + e^- \rightarrow Cr^{2+}$	−0.41
$Fe^{2+} + 2e^- \rightarrow Fe(s)$	−0.44
$2CO_2(g) + 2H^+ + 2e^- \rightarrow H_2C_2O_4$	−0.49
$S(s) + 2e^- \rightarrow S^{2-}$	−0.508
$Cr^{2+} + 2e^- \rightarrow Cr(s)$	−0.557
$Ag_2S(s) + 2e^- \rightarrow 2Ag(s) + S^{2-}$	−0.705
$Ni(OH)_2(s) + 2e^- \rightarrow Ni(s) + 2\,OH^-$	−0.72
$Cr^{3+} + 3e^- \rightarrow Cr(s)$	−0.74
$Zn^{2+} + 2e^- \rightarrow Zn(s)$	−0.763
$2H_2O + 2e^- \rightarrow H_2(g) + 2\,OH^-$	−0.8277
$SO_4^{2-} + H_2O + 2e^- \rightarrow SO_3^{2-} + 2\,OH^-$	−0.93
$Mn^{2+} + 2e^- \rightarrow Mn(s)$	−1.18
$Cr(OH)_3(s) + 3e^- \rightarrow Cr(s) + 3\,OH^-$	−1.3
$Ti^{2+} + 2e^- \rightarrow Ti(s)$	−1.63
$Al^{3+} + 3e^- \rightarrow Al(s)$	−1.66
$Be^{2+} + 2e^- \rightarrow Be(s)$	−1.85
$Ti^{3+} + e^- \rightarrow Ti^{2+}$	−2.0
$Sc^{3+} + 3e^- \rightarrow Sc(s)$	−2.08
$Mg^{2+} + 2e^- \rightarrow Mg(s)$	−2.363
$Na^+ + e^- \rightarrow Na(s)$	−2.714
$Ca^{2+} + 2e^- \rightarrow Ca(s)$	−2.866
$Sr^{2+} + 2e^- \rightarrow Sr(s)$	−2.89
$Ba^{2+} + 2e^- \rightarrow Ba(s)$	−2.90
$Cs^+ + e^- \rightarrow Cs(s)$	−2.92
$K^+ + e^- \rightarrow K(s)$	−2.924
$Li^+ + e^- \rightarrow Li(s)$	−3.045

appendix G Standard Molar Enthalpies of Formation, ΔH_f^0, Standard Molar Free Energies of Formation, ΔG_f^0, and Standard Absolute Entropies, S^0, at 25 °C

Substance	$\Delta H_f^0(kJ/mol)$	$\Delta G_f^0(kJ/mol)$	$S_{298}^0(J \cdot mol^{-1}K^{-1})$
Aluminum			
Al(s)	0.00	0.00	28.32
Al(g)	326.	286.	164.4
Al^{3+}(aq)	−524.7	−481.2	−313.
Al_2O_3(s)	−1676.	−1582.	50.92
AlF_3(s)	−1504.	−1425.	66.44
$AlCl_3$(s)	−704.2	−628.9	110.7
$AlBr_3$(s)	−526.3	−505.0	184.
$Al_2(SO_4)_3$(s)	−3440.8	−3100.1	239.
Antimony			
Sb(s)	0.00	0.00	45.69
Sb_4O_6(s)	−1441.	−1268.	221.
$SbCl_3$(g)	−314.	−301.	337.7
$SbCl_3$(s)	−382.2	−323.7	184.
Arsenic			
As(grey metal)	0.00	0.00	35.
As_4(g)	144.	92.5	314.
$AsCl_3$(g)	−258.6	−245.9	327.1
$AsCl_3$(liq)	−336.	−295.	233.
Barium			
Ba(s)	0.00	0.00	66.9
Ba^{2+}(aq)	−538.36	−560.7	13.
BaO(s)	−558.1	−528.4	70.3
BaF_2(s)	−1200.	−1149.	96.2
$BaCl_2$(s)	−860.06	−810.0	126.
$BaCl_2 \cdot 2H_2O$(s)	−1461.7	−1296.	203.
Beryllium			
Be(s)	0.00	0.00	9.54
Be(g)	320.6	282.8	136.17
BeO(s)	−610.9	−581.6	14.1
Boron			
B(s)	0.00	0.00	5.86
B(g)	563.	519.	153.3
B_2O_3(s)	−1272.8	−1193.7	53.97
B_2H_6(g)	36.	86.6	232.0
H_3BO_3(s)	−1094.3	−969.0	88.83
BF_3(g)	−1137.3	−1120.3	254.0

(Continued)

Substance	$\Delta H_f^0 (kJ/mol)$	$\Delta G_f^0 (kJ/mol)$	$S_{298}^0 (J \cdot mol^{-1}K^{-1})$
Bromine			
$Br_2(liq)$	0.00	0.00	152.23
$Br_2(g)$	30.91	3.142	245.35
$Br(g)$	111.8	82.43	174.91
$HBr(g)$	−36.4	−53.43	198.59
$Br^-(aq)$	−120.9	−102.8	80.71
Cadmium			
$Cd(s)$	0.00	0.00	51.76
$Cd(g)$	112.0	77.45	167.64
$Cd^{2+}(aq)$	−72.38	−77.74	−61.09
$CdO(s)$	−258.	−228.	54.8
$Cd(OH)_2(s)$	−557.56	−470.53	95.4
$CdCl_2(s)$	−391.5	−344.0	115.3
$CdS(s)$	−162.	−156.	64.9
$CdSO_4(s)$	−933.28	−822.78	123.04
Calcium			
$Ca(s)$	0.00	0.00	41.6
$Ca(g)$	192.6	158.9	154.8
$Ca^{2+}(aq)$	−542.96	−553.04	−55.2
$CaO(s)$	−635.5	−604.2	39.7
$Ca(OH)_2(s)$	−986.6	−896.8	76.1
$CaF_2(s)$	−1215.	−1162.	68.87
$CaBr_2(s)$	−674.9	−656.1	130.
$CaCO_3(calcite)$	−1206.9	−1128.8	92.9
$CaSO_4(s,anh)$	−1432.7	−1320.3	107.
$CaSO_4 \cdot 2H_2O(s)$	−2021.1	−1795.7	194.0
$Ca(NO_3)_2(s)$	−937.22	−741.99	193.
Carbon			
$C(graphite)$	0.00	0.00	5.740
$C(diamond)$	1.897	2.900	2.38
$C(g)$	716.7	671.3	157.99
$CO(g)$	−110.52	−137.15	197.56
$CO_2(g)$	−393.51	−394.36	213.64
$CH_4(g)$	−74.81	−50.75	186.15
$C_2H_2(g)$	226.75	209.2	200.8
$C_2H_4(g)$	52.27	68.12	219.5
$C_2H_6(g)$	−84.67	−32.89	229.5
$C_3H_8(g)$	−103.8	−23.5	269.9
$n\text{-}C_4H_{10}(g)$	−126.1	−17.15	310.1
$C_6H_6(g)$	82.927	129.66	269.2
$C_6H_6(liq)$	49.028	124.5	172.8
$CH_3OH(g)$	−200.7	−162.0	239.7
$CH_3OH(liq)$	−238.7	−166.4	127.
$CH_3CH_2OH(g)$	−235.1	−168.6	282.6
$CH_3CH_2OH(liq)$	−277.7	−174.9	161.
$HCHO(g)$	−116.	−110.	218.7
$CH_3CHO(g)$	−166.4	−133.7	266.
$HCOOH(g)$	−362.6	−335.7	251.
$HCOOH(liq)$	−409.	−346.	129.0
$HCOOH(aq)$	−410.	−356.	164.

(Continued)

Substance	$\Delta H_f^0(kJ/mol)$	$\Delta G_f^0(kJ/mol)$	$S_{298}^0(J \cdot mol^{-1}K^{-1})$
$CH_3COOH(liq)$	−484.5	−390.	160.
$CH_3COOH(aq)$	−488.45	−399.6	...
$CH_3COO^-(aq)$	−488.87	−372.5	...
$(CH_3)_2CO(liq)$	−248.	−155.4	200.
$CH_3OCH_3(g)$	−184.0	−112.9	267.1
$H_2C_2O_4(s)$	−826.8	−697.9	120.
$C_2O_4^{2-}(aq)$	−824.2	−674.9	51.0
$CHCl_3(g)$	−103.1	−70.37	296.5
$CHCl_3(liq)$	−134.5	−73.72	202.
$CCl_4(g)$	−102.9	−60.63	309.7
$CCl_4(liq)$	−135.4	−65.27	216.4
$CO(NH_2)_2(s)$	−333.2	−197.15	104.6
$NH_2CH_2COOH(s)$	−537.2	−377.7	103.5
Chlorine			
$Cl_2(g)$	0.00	0.00	222.96
$Cl(g)$	121.68	105.70	165.09
$Cl^-(aq)$	−167.46	−131.17	55.2
$ClO_2(g)$	103.	123.	249.
$HCl(g)$	−92.307	−95.299	186.80
Chromium			
$Cr(s)$	0.00	0.00	23.8
$Cr_2O_3(s)$	−1140.	−1058.	81.2
$CrCl_3(s)$	−563.2	−493.7	126.
Cobalt			
$Co(s)$	0.00	0.00	30.0
$CoCl_2(s)$	−317.	−274.	106.
$CoSO_4(s)$	−859.8	−753.5	113.
Copper			
$Cu(s)$	0.00	0.00	33.15
$Cu^{2+}(aq)$	64.39	64.98	−98.7
$CuO(s)$	−157.	−130.	42.63
$CuS(s)$	−53.1	−53.6	66.5
$CuSO_4(s,anh)$	−771.4	−661.9	109.
$CuSO_4 \cdot 5H_2O(s)$	−2278.0	−1880.	305.
Fluorine			
$F_2(g)$	0.00	0.00	202.7
$F(g)$	78.99	61.91	158.754
$F^-(aq)$	−332.6	−278.8	−13.8
$HF(g)$	−271.1	−273.2	173.78
Hydrogen			
$H_2(g)$	0.00	0.00	130.59
$H(g)$	217.965	203.247	114.713
$H^+(aq)$	0.00	0.00	0.00
$H_3O^+(aq)$	−285.84	−237.19	69.94

(Continued)

Substance	$\Delta H_f^0(kJ/mol)$	$\Delta G_f^0(kJ/mol)$	$S_{298}^0(J \cdot mol^{-1}K^{-1})$
Iodine			
$I_2(s)$	0.00	0.00	116.14
$I_2(g)$	62.438	19.36	260.6
$I(g)$	106.84	70.283	180.68
$I^-(aq)$	−55.94	−51.67	109.4
$HI(g)$	26.5	1.7	206.48
$ICl(g)$	17.8	−5.44	247.44
$IBr(g)$	40.8	3.7	258.66
Iron			
$Fe(s)$	0.00	0.00	27.2
$Fe^{2+}(aq)$	−87.9	−84.94	−113.
$Fe^{3+}(aq)$	−47.7	−10.6	−293.
$Fe_2O_3(s)$	−824.2	−742.2	87.40
$Fe_3O_4(s)$	−1118.	−1015.	146.
Lead			
$Pb(s)$	0.00	0.00	64.81
$Pb^{2+}(aq)$	1.6	−24.3	21.
$PbO(s,red)$	−219.0	−188.9	66.5
$PbO(s,yellow)$	−217.3	−187.9	68.70
$PbO_2(s)$	−277.	−217.4	68.6
$PbF_2(s)$	−663.2	−619.7	121.3
$PbCl_2(s)$	−359.4	−314.1	136.
$PbBr_2(s)$	−277.0	−260.4	162.
$PbI_2(s)$	−175.1	−173.8	177.
$PbS(s)$	−100.	−98.7	91.2
$PbSO_4(s)$	−918.4	−811.24	147.
Mercury			
$Hg(liq)$	0.00	0.00	76.02
$Hg(g)$	61.317	31.85	174.8
$HgO(s,red)$	−90.83	−58.555	70.29
$HgS(s,black)$	−53.6	−47.7	88.3
$HgCl_2(s)$	−224.	−179.	146.
$Hg_2Cl_2(s)$	−265.2	−210.78	192.
Nitrogen			
$N_2(g)$	0.00	0.00	191.5
$N(g)$	472.704	455.579	153.19
$NO(g)$	90.25	86.57	210.65
$NO_2(g)$	33.2	51.30	239.9
$N_2O(g)$	82.05	104.2	219.7
$N_2O_4(g)$	9.16	97.82	304.2
$N_2O_5(s)$	−41.8	134.	113.
$NH_3(g)$	−46.11	−16.5	192.3
$NH_4^+(aq)$	−132.8	−79.50	112.8
$NOCl(g)$	52.59	66.36	264.
$NH_4Cl(s)$	−314.4	−201.5	94.6
$NH_4Br(s)$	−270.8	−175.	113.
$(NH_4)_2SO_4(s)$	−1179.3	−900.36	220.3
$NH_4NO_3(s)$	−365.6	−184.0	151.1

(Continued)

Substance	$\Delta H_f^0(kJ/mol)$	$\Delta G_f^0(kJ/mol)$	$S_{298}^0(J \cdot mol^{-1}K^{-1})$
Oxygen			
$O_2(g)$	0.00	0.00	205.14
$O(g)$	249.17	231.73	161.055
$O_3(g)$	142.7	163.2	238.9
$OH^-(aq)$	−229.94	−157.30	−10.5
$H_2O(g)$	−241.83	−228.59	188.72
$H_2O(liq)$	−285.84	−237.18	69.94
$H_2O_2(liq)$	−187.8	−120.4	110.
Phosphorus			
$P(s,white)$	0.00	0.00	41.1
$P_4(g)$	58.91	24.5	280.
$PH_3(g)$	5.4	13.	210.1
$PCl_3(g)$	−287.	−268.	311.7
$PCl_5(g)$	−375.	−305.	364.5
$P_4O_{10}(s)$	−2984.	−2698.	228.9
Potassium			
$K(s)$	0.00	0.00	63.6
$K(g)$	90.00	61.17	160.23
$KF(s)$	−562.58	−533.12	66.57
$KCl(s)$	−435.87	−408.32	82.68
$KBr(s)$	−392.2	−379.2	96.44
$KI(s)$	−327.6	−322.3	104.3
Silver			
$Ag(s)$	0.00	0.00	42.55
$Ag^+(aq)$	105.9	77.11	73.93
$Ag_2O(s)$	−31.0	−11.2	121.
$AgCl(s)$	−127.1	−109.8	96.2
$AgBr(s)$	−99.50	−95.94	107.1
$AgI(s)$	−62.38	−66.32	114.
$Ag_2S(rhombic)$	−31.8	−40.3	146.
$Ag_2SO_4(s)$	−713.37	−615.76	200.
Sodium			
$Na(s)$	0.00	0.00	51.0
$Na(g)$	108.7	78.11	153.62
$Na_2O(s)$	−415.9	−377.	72.8
$NaCl(s)$	−411.00	−384.03	72.38
Sulfur			
$S(rhombic)$	0.00	0.00	31.8
$S(monoclinic)$	0.30	0.096	32.6
$S(g)$	278.80	238.27	167.75
$SO_2(g)$	−296.83	−300.19	248.1
$SO_3(g)$	−395.7	−371.1	256.6
$SO_4^{2-}(aq)$	−907.51	−741.99	17.
$H_2S(g)$	−20.6	−33.6	205.7
$H_2SO_4(liq)$	−813.99	−690.10	156.90
$SF_6(g)$	−1210.	−1105.	291.5

(Continued)

Substance	$\Delta H_f^0 (kJ/mol)$	$\Delta G_f^0 (kJ/mol)$	$S_{298}^0 (J \cdot mol^{-1}K^{-1})$
Zinc			
Zn(s)	0.00	0.00	41.6
Zn^{2+}(aq)	−152.4	−147.2	106.5
ZnO(s)	−348.3	−318.3	43.64
ZnS(sphalerite)	−206.0	−201.3	57.7
$ZnCl_2$(s)	−415.1	−369.43	111.5
$ZnSO_4$(s,anh)	−982.8	−874.5	120.
$ZnSO_4 \cdot 7H_2O$(s)	−3076.	−2563.	387.

appendix H Bohr Theory of the H Atom

Bohr's theory of the hydrogen atom begins with the following four postulates:

1. ***The energy of an electron in a hydrogen atom is quantized.*** The electron has certain stationary states of motion allowed, and in each of these stationary states it has a fixed and definite energy. All other values of the energy are forbidden.

2. ***A hydrogen atom radiates or absorbs energy only when the electron makes a transition from one allowed stationary state to another.*** When changing from a higher energy state to a lower energy state, a quantum of radiation is emitted with frequency v, given by $\Delta E = hv$, where ΔE is the difference in energy between the two allowed states.

3. In each allowed state the electron moves around the nucleus in a circular orbit of fixed radius.

4. In each allowed state the angular momentum of the electron is quantized. The allowed values of the angular momentum are given by $nh/2\pi$, where $n = 1,2,3, \ldots, \infty$.

Of these four postulates, only the first two are still considered to be correct and are retained in quantum mechanics. The third is entirely incorrect; it violates the Uncertainty Principle. According to modern quantum theory, electrons do not travel in any fixed orbit. The fourth postulate is partially correct. The angular momentum of the electron in the hydrogen atom is quantized, but the allowed values are given by $\sqrt{\ell(\ell + 1)}(h/2\pi)$, where $\ell = 0,1,2, \ldots, (n - 1)$, for a state with energy E_n, $n = 1,2,3, \ldots, \infty$.

Starting with his four postulates, Bohr was able to derive the correct expression for the allowed energies of the electron in a hydrogen atom or H-like ion. The argument combines well-known concepts of Newtonian mechanics with Bohr's new postulates.

The force of attraction between the electron and the nucleus is given by Coulomb's law:

$$\text{coulombic attraction} = Ze^2/r^2$$

where Z is the atomic number, so that Ze is the magnitude of positive charge in the nucleus, and r is the distance of the electron from the nucleus. The coulombic attraction pulls the electron in toward the nucleus.

When a body travels in a circular orbit there is a centrifugal force outward of magnitude mv^2/r, where m is the mass of the body and v is its speed. The electron remains in a stable circular orbit because the coulombic force pulling it in to the nucleus is exactly balanced by the outward centrifugal force.

$$\text{coulombic force inward} = \text{centrifugal force outward}$$
$$Ze^2/r^2 = mv^2/r \tag{H-1}$$

Newtonian mechanics defines the angular momentum of a body moving in a circular orbit of radius r as mvr. Bohr's fourth postulate can therefore be written as

$$mvr = nh/2\pi \tag{H-2}$$

where $n = 1,2,3, \ldots, \infty$.

The total energy of the electron is the sum of its kinetic energy, $\frac{1}{2}mv^2$, and its coulombic potential energy, $-Ze^2/r$.

$$E = \tfrac{1}{2}mv^2 - Ze^2/r \tag{H-3}$$

If Eq. (H-1) is rearranged to solve for mv^2 we obtain

$$mv^2 = Ze^2/r$$

Substituting this into Eq. (H-3) yields the following expression for the energy:

$$E = \tfrac{1}{2}Ze^2/r - Ze^2/r = -\tfrac{1}{2}Ze^2/r \tag{H-4}$$

An expression for r can be obtained by eliminating v between Eqs. (H-1) and (H-2).

$$v^2 = Ze^2/mr \qquad \text{from Eq. (H-1)}$$
$$v^2 = n^2h^2/4\pi^2m^2r^2 \qquad \text{from Eq. (H-2)}$$

Equating the two expressions for v^2 and solving for r, we obtain

$$r = n^2h^2/4\pi^2mZe^2 \tag{H-5}$$

Substitution of this expression for r into the expression for E given in Eq. (H-4) yields

$$E = -\frac{\tfrac{1}{2}Ze^2(4\pi^2mZe^2)}{n^2h^2} = -\frac{2\pi^2me^4Z^2}{n^2h^2} \tag{H-6}$$

This expression gives the values of the allowed energies for an electron in a hydrogen atom or H-like ion. The frequency emitted during a transition from a higher energy state with quantum number n_H to a lower energy state with quantum number n_L is obtained by using the Bohr frequency condition.

$$E_{n_H} - E_{n_L} = h\nu = \frac{2\pi^2me^4Z^2}{h^2}\left(\frac{1}{n_L^2} - \frac{1}{n_H^2}\right)$$

Hence $\tag{H-7}$

$$\nu = \frac{2\pi^2me^4Z^2}{h^3}\left(\frac{1}{n_L^2} - \frac{1}{n_H^2}\right)$$

The wavenumber, $\tilde{\nu}$, is $1/\lambda = \nu/c$, so that

$$\tilde{\nu} = \frac{1}{\lambda} = \frac{2\pi^2me^4Z^2}{ch^3}\left(\frac{1}{n_L^2} - \frac{1}{n_H^2}\right) \tag{H-8}$$

If the equation for the allowed energies, Eq. (H-6), is correct then Eq. (H-8) must be identical with the Balmer formula for the hydrogen atom, Eq. (12-11):

$$\frac{1}{\lambda} = \mathcal{R}\left(\frac{1}{n_L^2} - \frac{1}{n_H^2}\right) \text{ cm}^{-1}$$

The two equations are identical only if

$$\mathcal{R} = \frac{2\pi^2me^4}{h^3c} \tag{H-9}$$

since $Z = 1$ for hydrogen. It was a great triumph for Bohr that substitution of the fundamental constants into the right-hand side of Eq. (H-9) yields 109,737 cm^{-1}, in very close agreement with the experimental value for the Rydberg constant, 109,678 cm^{-1}.

The radius of the first Bohr orbit of hydrogen is denoted a_0, and is the atomic unit of length. By setting $n = 1$ and $Z = 1$ in Eq. (H-5) we obtain

$$a_0 = \frac{h^2}{4\pi^2 me^2} = 0.529 \text{ Å} = 0.0529 \text{ nm} \tag{H-10}$$

Recall that for the $1s$ electron of hydrogen the maximum value of the radial probability distribution function, $4\pi^2 r^2 \psi^2$ (see Fig. 13.3), occurs at $r = a_0$.

Note that the Bohr theory predicts that the speed of the electron is inversely proportional to the radius of the orbit, so that the electron in a hydrogen atom is moving at maximum speed when it is in the first Bohr orbit, $n = 1$, $r = 0.529$ Å, $v = 2.188 \times 10^6$ m/s.

appendix I Derivation of the Relations $\Delta G = \Delta G^0 + RT \ln Q$ and $\Delta G = RT\ln(Q/K_{eq})$

Let us consider the general reaction

$$\alpha A + \beta B \rightleftharpoons \gamma C + \delta D \tag{I-1}$$

The free energy change for this reaction, ΔG, is given by

$$\Delta G = \gamma G_C + \delta G_D - \alpha G_A - \beta G_B \tag{I-2}$$

where G_A is the molar free energy of substance A in the reaction mixture, G_B is the molar free energy of substance B, and so on.

The free energy of a substance in a mixture is a function of the concentration of that substance in the mixture. If the substance is a gas, we express the free energy of the substance as a function of its partial pressure in the gas phase, because the concentration is directly proportional to the partial pressure.

When a substance is in its standard state, its molar free energy is denoted by using the superscript zero (^{0}). Thus the molar free energy of substance A in its standard state is G_A^0.

The molar free energy of substance A in a state other than its standard state is related to its standard state molar free energy by the relation

$$G_A = G_A^0 + RT \ln P_A \qquad \text{if A is a gas} \tag{I-3}$$

or by

$$G_A = G_A^0 + RT \ln [A] \qquad \text{if A is in solution} \tag{I-4}$$

Equation (I-3) is correct only if substance A can be treated as an ideal gas. An ideal gas is in its standard state at temperature T if its pressure is 1 atm; therefore the partial pressure P_A must be given in atmospheres. Equation (I-4) is also an approximation, consistent with the approximation made throughout that a solute is in its standard state if its concentration is $1.00\ M$. The exact expression is

$$G_A = G_A^0 + RT \ln (a_A) \tag{I-5}$$

where a_A is the activity of substance A, and the standard state is one in which the substance is at unit activity. The activity is directly proportional to the concentration of the solute. If we use Eq. (I-4) rather than Eq. (I-5), we will derive the ideal law of chemical equilibrium that has been used throughout this text.

To be specific, let us consider a gas-phase reaction in which we can treat all the reacting species as ideal gases. If we mulitply Eq. (I-3) by the constant α, we obtain

$$\alpha G_A = \alpha G_A^0 + \alpha RT \ln P_A = \alpha G_A^0 + RT \ln P_A^{\alpha} \tag{I-6}$$

In writing the last term of Eq. (I-6) we have made use of a fundamental property of logarithms:

$$y \ln x = \ln x^y$$

937

For each of the four gases in our reaction mixture we can write an equation analogous to Eq. (I-6), namely,

$$\alpha G_A = \alpha G_A^0 + RT\ln P_A^\alpha$$
$$\beta G_B = \beta G_B^0 + RT\ln P_B^\beta$$
$$\gamma G_C = \gamma G_C^0 + RT\ln P_C^\gamma$$
$$\delta G_D = \delta G_D^0 + RT\ln P_D^\delta$$

If we substitute each of these expressions into Eq. (I-2), grouping all the logarithmic terms together and factoring the term RT, we obtain

$$\Delta G = \gamma G_C^0 + \delta G_D^0 - \alpha G_A^0 - \beta G_B^0 + RT(\ln P_C^\gamma + \ln P_D^\delta - \ln P_A^\alpha - \ln P_B^\beta) \qquad (I\text{-}7)$$

But by definition,

$$\Delta G^0 = \gamma G_C^0 + \delta G_D^0 - \alpha G_A^0 - \beta G_B^0 \qquad (I\text{-}8)$$

We can also simplify Eq. (I-7) by making use of another property of logarithms:

$$\ln x + \ln y = \ln(xy) \qquad \text{and} \qquad \ln x - \ln y = \ln(x/y)$$

By combining the logarithmic terms and using the definition of ΔG^0, Eq. (I-8), we can write Eq. (I-7) in the following simpler form:

$$\Delta G = \Delta G^0 + RT\ln \frac{P_C^\gamma P_D^\delta}{P_A^\alpha P_B^\beta} \qquad (I\text{-}9)$$

We now recognize the quantity whose logarithm is needed in Eq. (I-9). It is the reaction quotient, Q:

$$Q = \frac{P_C^\gamma P_D^\delta}{P_A^\alpha P_B^\beta} \qquad (I\text{-}10)$$

Equation (I-9) is therefore

$$\Delta G = \Delta G^0 + RT\ln Q \qquad (I\text{-}11)$$

which we set out to derive. This equation is valid for any and all values of the pressures of the four gases. In particular, it is valid whether the reaction mixture is at equilibrium or not. But if the reaction is at equilibrium, then

$$\Delta G = 0 \qquad \text{and} \qquad Q = K_{eq} \qquad (I\text{-}12)$$

When the reaction mixture is at equilibrium, therefore, Eq. (I-11) becomes

$$0 = \Delta G^0 + RT\ln K_{eq}$$

or

$$\Delta G^0 = -RT\ln K_{eq} \qquad (I\text{-}13)$$

If we substitute this expression for ΔG^0 back into Eq. (I-11) we obtain

$$\Delta G = -RT\ln K_{eq} + RT\ln Q = RT\ln(Q/K_{eq}) \qquad (I\text{-}14)$$

appendix J Derivation of the Integrated Forms of the First- and Second-Order Rate Laws

For a first-order reaction, $-d[A]/dt = k_1[A]$. If we rearrange this equation so that all terms in [A] are on one side of the equation, and terms in t are on the other side, we obtain

$$\frac{d[A]}{[A]} = -k_1 dt \tag{J-1}$$

We will integrate this equation between two limits (a definite integral). Let the lower limit be the start of the experiment, when the concentration of A is $[A]_0$ and $t = 0$. Let the upper limit be any later time, t, when the concentration of A is $[A]_t$. We then obtain

$$\int_{[A]_0}^{[A]_t} \frac{d[A]}{[A]} = -k_1 \int_{t=0}^{t} dt \tag{J-2}$$

Since

$$\int \frac{dx}{x} = \ln x$$

the result of the integration is

$$\ln[A]_t - \ln[A]_0 = -k_1 t \tag{J-3}$$

which is Eq. (19-19), or its equivalent, Eq. (19-21a).

To obtain the integrated form of a second-order rate law of the type

$$-d[A]/dt = k_2[A]^2$$

we first rearrange this equation to

$$-\frac{d[A]}{[A]^2} = k_2 \, dt \tag{J-4}$$

Taking the definite integral between the lower limit at $t = 0$ and the upper limit at any later time, t, we obtain

$$-\int_{[A]_0}^{[A]_t} \frac{d[A]}{[A]^2} = k_2 \int_{t=0}^{t} dt \tag{J-5}$$

Since

$$\int \frac{dx}{x^2} = -\frac{1}{x}$$

this integrates to

$$\frac{1}{[A]_t} - \frac{1}{[A]_0} = k_2 t \tag{J-6}$$

which is Eq. (19-27).

appendix K Methods of Solving Problems in Acid–Base Equilibria when the Self-Ionization of Water Cannot Be Neglected

K1. Solutions of Strong Acids

In Example 9.1 and the discussion thereafter we presented the argument that validates the statement: In an aqueous solution of a strong acid, if the concentration of strong acid is greater than or equal to $1 \times 10^{-6} M$, then virtually all the H_3O^+ ions in solution come from the strong acid and we can neglect the contribution of the self-ionization of water to the total $[H_3O^+]$.

In most experiments involving strong acids the concentration of strong acid is greater than $1 \times 10^{-6} M$, but there are occasions when we use extremely dilute solutions. How do we calculate the pH of a solution of a strong acid with a concentration less than $1 \times 10^{-6} M$? For $1 \times 10^{-6} F$ HCl, pH = 6, but for $1 \times 10^{-8} F$ HCl the pH is clearly *not* 8, because the solution does not become basic when the acid is more dilute. The general method of approach to such a problem is to utilize the electroneutrality equation, that is, to equate the total concentrations of positive and negative charge. The following example illustrates the method to be used.

Example What is the pH of $3.0 \times 10^{-7} F$ HCl?

Solution There are two sources of H_3O^+ in this solution: (1) the strong acid (HCl), and (2) the self-ionization of water. The ions present in solution are H_3O^+, Cl^-, and OH^- ions. The electroneutrality equation is

$$[H_3O^+] = [Cl^-] + [OH^-]$$

Because HCl is 100% dissociated, $[Cl^-] = 3.0 \times 10^{-7} M$. The $[OH^-]$ is related to the $[H_3O^+]$ by the expression for K_w, Eq. (9-6a). The electroneutrality equation is therefore

$$[H_3O^+] = 3.0 \times 10^{-7} + \frac{K_w}{[H_3O^+]}$$

which, at 25 °C, is

$$[H_3O^+] = 3.0 \times 10^{-7} + \frac{1.0 \times 10^{-14}}{[H_3O^+]}$$

By combining terms we obtain a quadratic equation in $[H_3O^+]$:

$$[H_3O^+]^2 - 3.0 \times 10^{-7}[H_3O^+] - 1.0 \times 10^{-14} = 0$$

We can solve this by using the quadratric formula (see Appendix B5), discarding the negative root, as only a positive value of $[H_3O^+]$ is physically meaningful. We obtain

$$[H_3O^+] = \frac{3.0 \times 10^{-7} + (9 \times 10^{-14} + 4 \times 10^{-14})^{1/2}}{2} = \frac{3.0 \times 10^{-7} + (13 \times 10^{-14})^{1/2}}{2}$$

$$= \frac{6.6 \times 10^{-7}}{2} = 3.3 \times 10^{-7}$$

The pH of this solution is therefore $7 - 0.52 = 6.48$.

As an exercise, calculate the pH of $1.0 \times 10^{-8}\ F$ HCl. The correct answer is 6.98.

K2. *Solutions of Weak Acids*

For solutions of weak acids, the contribution of the self-ionization of water to the total $[H_3O^+]$ may be significant if either the acid is quite dilute or if the acidity constant, K_a, is quite small. Instead of the approximate solution described in Chapter 9, an exact solution that includes the contribution of the self-ionization of water must be used if $CK_a < K_w$, where C is the stoichiometric concentration of the weak acid.

The determination of the $[H_3O^+]$ or pH of a dilute solution of a very weak acid requires solving a cubic equation, obtained as follows:

Consider a C formal solution of the weak acid HA. At equilibrium the solution contains H_3O^+, OH^-, and A^- ions, and HA molecules. Conservation of mass for A requires that

$$C = [HA] + [A^-] \tag{K-1}$$

which is called the mass balance equation. The electroneutrality equation for this solution is

$$[H_3O^+] = [A^-] + [OH^-] \tag{K-2}$$

In addition to these two equations, we have the equilibrium constant expressions

$$K_a = \frac{[H_3O^+][A^-]}{[HA]} \tag{K-3}$$

and

$$K_w = [H_3O^+][OH^-] \tag{K-4}$$

Combining these four equations and eliminating terms in $[A^-]$, $[HA]$, and $[OH^-]$ leads to a cubic equation in $[H_3O^+]$.

Solve Eq. (K-4) for $[OH^-]$ and substitute that expression into Eq. (K-2). We obtain

$$[H_3O^+] = [A^-] + \frac{K_w}{[H_3O^+]}$$

or

$$[A^-] = \frac{[H_3O^+]^2 - K_w}{[H_3O^+]} \tag{K-5}$$

Substitute this expression for $[A^-]$ into Eq. (K-1) and solve for $[HA]$.

$$[HA] = \frac{C[H_3O^+] - [H_3O^+]^2 + K_w}{[H_3O^+]} \tag{K-6}$$

Now substitute both Eqs. (K-5) and (K-6) into Eq. (K-3) to obtain

$$K_a = \frac{[H_3O^+]([H_3O^+]^2 - K_w)}{C[H_3O^+] - [H_3O^+]^2 + K_w}$$

This can be algebraically rearranged to yield

$$[H_3O^+]^3 - K_a[H_3O^+]^2 - (CK_a + K_w)[H_3O^+] - K_wK_a = 0 \tag{K-7}$$

There are straightforward methods of solving cubic equations, but a method of successive approximations is probably the best way to go about solving Eq. (K-7) for the $[H_3O^+]$.

glossary

Absolute Entropy (Third Law Entropy) The entropy of any substance at a temperature above absolute zero. The third law of thermodynamics states that the entropy of perfect crystals of all pure elements and compounds is zero at absolute zero. At any temperature above zero Kelvin, the entropy is greater than 0. The value of the entropy of any substance at 25 °C and 1 atm is the standard absolute entropy, S^0.

Absolute Temperature The two fixed points that define the absolute temperature scale are **absolute zero** (the theoretical lower limit of temperature), and the **triple point of water,** the temperature at which air-free water freezes at the pressure of its own vapor, which is assigned the value 273.16 K. The zero on the **Celsius scale,** 0 °C, is 273.15 K. The **absolute temperature,** T, is $273.15 + t$, where t is the temperature in degrees Celsius. A degree of temperature on the absolute scale is called a **kelvin,** and has the same magnitude as a degree of temperature on the Celsius scale.

Absolute Uncertainty The range of reliability of an experimental value. Thus, if a volume is reported as (18.94 ± 0.02) mL, the absolute uncertainty in the volume is ± 0.02 mL. See Appendix C.

Absolute Zero The zero of temperature on the Kelvin scale or -273.15 °C.

Absorption Spectrum A plot of the intensity of radiation absorbed by a substance as a function of the frequency or wavelength of the radiation.

Acceleration The change in velocity per unit time. The SI units of acceleration are meters per second per second.

Acceleration of Gravity, g The proportionality constant between the weight of a body (the force with which it is attracted to the earth) and its mass: $W = mg$. A freely falling body near the earth's surface falls to the earth with a constant acceleration of approximately 980 cm $\cdot$ s^{-2}. The acceleration due to gravity varies somewhat from place to place on the earth.

Acetone, ($CH_3 COCH_3$) 2-Propanone or dimethyl ketone. Acetone is widely used as an organic solvent.

Acetylene, (H—$C\equiv C$—H) A linear molecule in which the carbon atoms are *sp* hybridized. Acetylene is a gas at room temperature.

Acid A substance with a sharp, sour taste that turns litmus red. **Arrhenius definition:** A compound whose cation is H$^+$. **Brønsted–Lowry definition:** A proton donor, a substance able to donate an H$^+$ to some other species. **Lewis definition:** An electron-pair acceptor.

Acid–Base Indicator A substance, usually an organic weak acid, HIn, that changes color within a fairly narrow range of pH values. The weak acid, HIn, and its conjugate weak base, In$^-$, have two different colors

Acid–Base Titration A laboratory procedure for determining the concentration of an unknown acidic or basic solution. If the unknown solution is acidic, a solution of strong base of known concentration is added from a buret until the number of moles of added OH$^-$ is exactly equal to the number required to react completely with the amount of acid originally present.

Acidic Anhydride A substance that reacts with water to form an acidic solution.

Acidic Proton The nucleus of a hydrogen atom that can be donated to some other species. Hydrogen atoms whose nuclei cannot be transferred to other species are not acidic. Acidic hydrogen atoms are frequently bonded to oxygen atoms.

Acidic Salt A salt whose aqueous solution has a pH < 7 at 25 °C.

Acidity Constant, K_a The equilibrium constant for the proton-transfer reaction between a weak acid, HA, and water. An acidity constant has the form $K_a(HA) = [A^-][H_3O^+]/[HA]$.

Actinide Elements A series of 14 radioactive elements from thorium $Z = 90$, to lawrencium, $Z = 103$. The seven inner $5f$ AO's are being filled in this series. As their outer electronic configurations are the same, these elements are very similar chemically.

Activated Complex The chemical species that exists in the **activated state (transition state).** The activated complex is a specific geometric configuration of all the atoms involved in the reaction. Along the minimum potential energy pathway that leads to reaction, the potential energy of the activated state is a maximum.

Activation Energy The difference in energy between the activated state and the average potential energy of the reacting molecules. The reacting species must acquire the activation energy before reaction can occur.

Active Metals The **alkali metals,** Group IA (Li, Na, K, Rb, Cs, and Fr), and the **alkaline earth metals,** Group IIA (Be, Mg, Ca, Sr, Ba, and Ra). These elements are very reactive and tend to transfer their valence electrons to other substances during chemical reactions.

Activity (Radioactivity) The number of counts or disintegrations per unit time, as measured on a counting device such as a Geiger counter. The activity of a radioactive sample is directly proportional to the number of radioactive nuclei in the sample.

Activity A quantity proportional to the concentration of a species in solution, which must be used in the law of mass action to make the law exact. The proportionality constant between activity and concentration, the **activity coefficient,** varies with the concentration of the species and of all other species in solution.

Addition Compound The combination of two molecules to form a single larger unit. If one of the molecules can accept a pair of electrons and the other has a lone pair of electrons to donate, the product is called a **Lewis acid–base adduct.**

Addition Polymerization The process of forming a polymer or macromolecule simply by adding together monomer molecules. Most addition polymers are formed via a chain reaction.

Aldehyde A compound with general form

$$\overset{\displaystyle O}{\overset{\displaystyle \|}{R-C-H}}$$

Aliphatic Compound An organic compound that does not contain any benzene rings. The carbon–carbon bonds in aliphatic compounds are single, double, or triple bonds; none has a fractional bond order.

Alkali Metals Group IA: Li, Na, K, Rb, Cs, and Fr.

Alkaline Dry Cell A commercial galvanic cell with a zinc anode and a graphite cathode surrounded by MnO_2. The electrolyte is a moist paste of KOH and $ZnCl_2$.

Alkaline Earth Metals Group IIA: Be, Mg, Ca, Sr, Ba, and Ra.

Alkane A hydrocarbon in which all carbon–carbon bonds are single bonds.

Alkene A hydrocarbon containing at least one carbon–carbon double bond.

Alkyl Group The portion of an alkane that remains when one hydrogen is removed. An alkyl group has the general formula C_nH_{2n+1}.

Allotropes Two or more forms in which the same element exists. Ozone, O_3, and dioxygen, O_2, are allotropes, as are orthorhombic and monoclinic sulfur.

Alpha (α) Particle The nucleus of a helium atom, $^4_2He^{2+}$, with two neutrons, two protons, and a charge of $+2$. Unstable nuclei that emit α-particles are said to undergo α **decay.**

Amalgam An alloy of mercury with another metal or metals.

Amide A derivative of a carboxylic acid in which the —OH of the —COOH group is replaced by —NH_2 or —NRR′. The **amide linkage** is .

Amide Ion, NH_2^- The anion formed when NH_3 loses a proton. It is a very strong base and does not exist in aqueous solution.

Amine A substance derived from ammonia, NH_3, by substituting some other group (usually an organic group) for one or more of the H atoms. Amines, like ammonia, are weak bases.

Amino Acid A compound containing both an amino group, —NH_2, and a carboxyl group, —COOH. The general formula of an amino acid is

$$\underset{\displaystyle R}{\overset{}{H_2N-\underset{\displaystyle |}{CH}-COOH}}$$

Amorphous Solids Solids without distinctive geometries or sharp melting points. Their internal structure lacks the definite arrangement of atoms, molecules, or ions, that characterizes crystalline solids.

Ampere (A) The SI unit of electrical current. One ampere (1 amp) is one coulomb per second ($C \cdot s^{-1}$).

Amphiprotic Able to act either as an acid (to donate a proton) or as a base (to accept a proton). An amphiprotic substance is called an **ampholyte.**

Amphoteric Having both acidic and basic properties. Amphoteric oxides and hydroxides react with both H_3O^+ and OH^- ions.

Amplitude The amplitude of a wave is the maximum value of the wave height above the average value.

Amylose A polymer of glucose containing from 250–300 glucose units. It is a digestible form of starch.

Analytical Separation A laboratory procedure to remove more than 99.9% of one substance from solution while leaving 100% of a second species in solution.

Angstrom (Å) A unit of length equal to 1×10^{-10} m or 1×10^{-8} cm. It is not an SI unit, but is convenient because atomic and molecular diameters are typically a few angstroms in length.

Angular Motion Rotation of a body in which all points in the body describe concentric circles about a fixed axis. The number of radians per second through which the body rotates is its **angular speed.** The product of the moment of inertia of the body and its angular velocity is its **angular momentum.**

Anhydrous Having no water.

Aniline The simplest aromatic amine, $C_6H_5NH_2$.

Anion A negatively charged ion. An atom becomes an anion by gaining one or more electrons. A group of atoms bearing a negative charge is a polyatomic anion.

Anisotropic Exhibiting different values of a property when measured along different axes. Anisotropy is characteristic of the crystalline state.

Anode The electrode at which oxidation occurs. During electrolysis the anode is the positive terminal. In a galvanic cell the anode is the negative terminal.

Antibonding Molecular Orbital An orbital centered about two or more nuclei, with an energy higher than any of the atomic orbitals that combined to form the molecular orbital.

Aromatic Compounds Benzene and its derivatives. In aromatic molecules electrons are delocalized over one or more rings of atoms, and the carbon–carbon bond order is not integral.

Arrhenius Definitions of Acid, Base, and Salt An acid is an electrolyte with cation H^+, a base is an electrolyte with anion OH^-, a salt is an electrolyte with a cation other than H^+ and an anion other than OH^-.

Arsenious Oxide (Arsenic Trioxide) The oxide of the +3 oxidation state of arsenic. Its correct molecular formula is As_4O_6.

Atmosphere A unit of pressure. One atmosphere is the pressure that supports a column of mercury exactly 760-mm high at 0 °C, under standard gravity.

Atom A unit of matter, the smallest unit of an element that can combine chemically with other elements. An atom consists of a positively charged nucleus and a sufficient number of negatively charged electrons so that the atom is electrically neutral.

Atomic Emission Spectra The frequencies of electromagnetic radiation emitted by atoms when they are heated or subjected to an electrical discharge or arc.

Atomic Mass Unit (amu) A unit of mass equal to one twelfth the mass of a single atom of the isotope carbon-12 (^{12}C). In SI units, 1 amu = 1.660565×10^{-27} kg.

Atomic Number The number of protons in the nucleus of an atom. It is usually denoted by Z. The number of electrons in an electrically neutral atom is also equal to the atomic number.

Atomic Orbital (AO) A one-electron wave function; a solution of the Schrödinger equation for the hydrogen atom or an H-like ion. For a many-electron atom an AO is a subunit of the wave function for all the electrons. Each orbital describes how a single electron behaves in the field of a nucleus shielded by all the other electrons and is defined by three quantum numbers, n, ℓ, and m_ℓ.

Atomic Theory of Matter The theory that all material substances are composed of atoms of a relatively small number of substances called elements.

Atomic Weight The average mass, in amu, of an atom in a naturally occurring sample of an element, with each isotope weighted according to its natural abundance.

Aufbau Process The building-up of the periodic table by describing the ground state electronic configuration of each element in order of increasing atomic number, Z.

Average Translational Kinetic Energy per Molecule, $\langle \varepsilon_k \rangle$ The total translational kinetic energy of a sample of gas divided by the number of molecules in the sample. It is independent of the nature of the gas, the amount of gas, and the pressure of the gas. The value of $\langle \varepsilon_k \rangle$ depends only on the absolute temperature, and is $\frac{3}{2} kT$, where k is Boltzmann's constant.

Avogadro's Law or Hypothesis Equal volumes of different gases at the same temperature and pressure contain equal numbers of molecules, and therefore equal numbers of moles of gas. An alternative statement of Avogadro's law is the following: The volume of a gas is directly proportional to the number of moles of gas at constant temperature and pressure, with the same proportionality constant for all gases.

Avogadro's Number, N_A The number of atoms contained in exactly 12 g of the isotope carbon-12 (^{12}C). Avogadro's number has been determined by experiment to be 6.022045×10^{23}.

Axial Positions In a trigonal bipyramidal structure, five atoms are bonded to a central atom. Three of the bonded atoms lie in a plane, and the other two occupy axial positions along the perpendicular to that plane.

Azimuthal (Angular) Quantum Number, ℓ The quantum number that determines the shape or angular distribution of an atomic orbital. The allowed values of ℓ are $0, 1, \ldots, (n-1)$. An orbital with $\ell = 0$ is an s orbital, and is spherical. An orbital with $\ell = 1$ is a p orbital, has two lobes, and is dumbbell shaped. For all atoms other than hydrogen, the energy of an orbital depends on both n and ℓ.

Azo Dyes Organic compounds containing the $-N{=}N-$ group. They are intensely colored and are produced from nitrites and nitrous acid.

Balance A device for measuring the mass of an object

by comparing its weight with the weight of an object whose mass is known. An **analytical balance** can measure a mass to ± 0.0001 g.

Balanced Equation An equation that describes a physical or chemical change and indicates that in any such process both mass and charge are conserved.

Balmer Formula A formula for calculating the wavenumbers or wavelengths of the emitted radiation in the spectrum of atomic hydrogen.

Balmer Series The frequencies in the visible and near ultraviolet region in the emission spectrum of atomic hydrogen. These frequencies are emitted when the electron makes a transition from any higher energy state to the state with quantum number $n = 2$.

Band Theory of Metals A description of the bonding in metallic solids. The valence electrons occupy bands of closely spaced orbitals that are delocalized over the entire solid.

Barometer A device for measuring atmospheric pressure in terms of the height of a column of liquid supported by that pressure. The liquid normally used in a barometer is mercury, because of its high density, but other liquids may also be employed.

Base A substance with a bitter taste that turns litmus blue. **Arrhenius definition:** A compound whose anion is OH^-. **Brønsted–Lowry definition:** A proton acceptor. **Lewis definition:** An electron-pair donor.

Basicity Constant, K_b The equilibrium constant for the proton-transfer reaction between a weak base and water.

Basic Salt A salt whose aqueous solution has a $pH > 7$ at 25 °C.

Battery A galvanic cell.

Bauxite The principal ore of aluminum. A hydrated aluminum oxide, $Al_2O_3 \cdot xH_2O$, where x is variable.

Belt of Stability A narrow band of values of the neutron to proton ratio (n/p) within which all stable nuclei lie. For $Z < 20$, n/p must be quite close to 1.00 for a nucleus to be stable. As Z increases, the value of n/p required for stability increases, and reaches a value of 1.5 at $Z = 80$.

Bent, V-shaped, or Angular Molecule A triatomic molecule in which the three atoms do not lie in a line.

Berthollides (Nonstoichiometric Compounds) Compounds whose composition varies within a certain range, depending on the conditions under which they are prepared. Transition metal oxides, sulfides, and hydrides are frequently berthollides.

Beta (β) Decay The emission by an unstable nucleus of electrons (β^- particles) or of positrons (β^+ particles). Emission of a neutrino always accompanies

β emission. Neutron-rich nuclei emit β^- particles. **Positron emission** is a decay process of neutron-poor nuclei that is more common for light nuclei than for heavy ones.

Bidentate Ligand A ligand that has two atoms with lone pairs of electrons, both of which bind to the same metal ion to form a coordination complex.

Bimolecular Elementary Process A single step in a mechanism in which two molecules react to form the activated complex.

Binary Liquid Solution A mixture of two liquids, having various possible proportions of the two constituents.

Binding Energy (Nuclear) The amount of energy that must be expended to break a nucleus apart into its separate neutrons and protons. Alternatively, the amount of energy that would be released if the separate nucleons combined to form a nucleus.

Body-Centered Cubic Lattice A crystal structure with lattice points at the eight corners and the center of a cubic unit cell.

Bohr Frequency Condition The equation $\Delta E = h\nu$, relating the frequency, ν, of the radiation emitted or absorbed when a transition occurs between two energy levels, and the difference in energy, ΔE, between the two levels. The proportionality constant, h, is **Planck's constant.**

Boiling Chips Glass beads, small pieces of porous ceramic material, or carborundum bits that are used to promote the formation of gaseous bubbles in liquids that are being heated to the boiling point. Boiling chips are used to prevent **bumping,** which is the rapid expansion of bubbles in a superheated liquid.

Boiling Point Elevation, ΔT_b The difference between the boiling point of a solution and the boiling point of the pure solvent used to prepare the solution. It is a positive quantity, as a solution of a nonvolatile solute in a volatile solvent always boils at a *higher* temperature than does the pure solvent.

Boltzmann's Constant, k The gas constant per molecule. The ratio of the gas constant, R, to Avogadro's number: $k = R/N_A$. The numerical value of k is 1.38066×10^{-23} J $\cdot$ K^{-1}.

Bomb Calorimeter A device for measuring the heat of reaction at constant volume, ΔE. It has thick, well-insulated walls to prevent heat exchange with the surroundings, and to withstand large differences of pressure inside and out.

Bond-Dipole Vector Addition The process of summing the vectors repesenting the dipoles of all the bonds within a molecule to obtain the resultant vector, which is the net dipole of the entire molecule.

Bond Dissociation Energy The amount of energy required per mole to break a specific bond in a specific molecule. It should be distinguished from the **bond energy,** which is an average value of the energy per mole required to break similar bonds in a large number of different molecules.

Bonded Pair A pair of electrons shared by two atoms, constituting the chemical bond between them.

Bonding Molecular Orbital An orbital centered about two or more nuclei, lower in energy than any of the atomic orbitals that combined to form the molecular orbital.

Bond Length The distance between two nuclei of atoms that are bonded together in a molecule.

Bond Order One half the difference between the number of electrons in bonding MO's and the number in antibonding MO's. A stable bond results only when the bond order is greater than 0.

Boranes (Boron Hydrides) A series of more than 15 different compounds of boron and hydrogen. The boranes are electron deficient compounds employing B—H—B bridge bonds, or three-center bonds, in which a single pair of electrons binds three atoms.

Born–Haber Cycle A series of six reactions to which Hess' Law is applied in order to calculate the lattice energy of an ionic crystal from experimental data.

Boyle's Law The pressure and volume of a fixed quantity of gas are inversely proportional, provided the temperature is constant.

Bragg's Law The equation specifying the conditions required for constructive interference in X-ray diffraction by a crystal. If λ is the wavelength of the incident monochromatic X-rays and d is the distance between successive planes of atoms in the crystal, constructive interference occurs only if the angle θ between the X-rays and the crystal planes satisfies the relation $n\lambda = 2d \sin \theta$, where n is an integer.

Branched-Chain Alkane An alkane in which at least one carbon is bonded to three or four other carbon atoms.

Bravais Lattices The 14 lattice types observed in crystals. The cubic system has 3 lattices, the tetragonal 2, the orthorhombic 4, the monoclinic 2, and the hexagonal, rhombohedral, and triclinic systems have only a single primitive lattice each.

Bridging Ligand A ligand that is simultaneously bonded to two different metal ions in a coordination complex.

Brønsted–Lowry Definitions of Acid and Base An acid is a proton donor; a base is a proton acceptor.

Buffer A solution that is able to maintain an approximately constant pH, even when moderate amounts of either a strong acid or a strong base are added. A solution containing both a weak acid and its conjugate weak base functions as a buffer solution.

Buret A piece of equipment used for measuring the volume (usually up to 50 mL) of solution added to another solution. It consists of a long, narrow, graduated glass tube and a stopcock to control the flow of liquid.

Bystander Ion Another term for **spectator ion.**

Calcite A crystalline form of calcium carbonate, $CaCO_3$. If ordinary light is passed through a crystal of calcite it emerges as two rays of plane-polarized light.

Calorie A unit of energy, originally defined as the amount of heat required to raise the temperature of one gram of water from 14.5 to 15.5 °C. The modern definition of a calorie is simply 4.1840 J. The amount of energy specified by the two definitions is almost, but not exactly, the same.

Carbohydrates The sugars and their derivatives, such as starch and cellulose. Many (but not all) of the sugars have molecular formulas that can be expressed as $C_n(H_2O)_n$, which gave rise to the name carbohydrate.

Carbocation (Carbonium Ion) An ion in which there is a positive charge on a carbon atom.

Carbonyl Group The $>C=O$ group.

Carboxylate Ion The conjugate base of a carboxylic acid. All carboxylate ions contain the

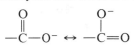

group, in which the negative charge is shared equally by the two oxygen atoms. A **carboxylic acid** is a weak acid containing the **carboxyl group,**

$$\overset{\displaystyle O}{\overset{\displaystyle \|}{-C}}-OH$$

Carcinogen A substance that causes cancer.

Catalyst A substance that increases the rate of a reaction without itself undergoing permanent change during the reaction. It is used in the rate-determining step, and formed as a product in some subsequent step of the reaction.

Cathode The electrode at which reduction occurs. The cathode is the negative terminal during electrolysis, and the positive terminal of a galvanic cell.

Cathode Rays Streams of electrons emitted by the negatively charged terminal, or cathode, when a potential difference of several thousand volts is applied between two metal plates in an evacuated tube.

Cation A positively charged ion. An atom becomes a cation by losing one or more electrons.

Cell Potential or Cell Voltage The electromotive force (EMF) of a galvanic cell.

Cellulose A naturally occurring polymer of glucose containing about 3000 glucose units. It is the chief structural component of plant cells.

Celsius Temperature Scale A temperature scale numerically identical to the **centigrade** temperature scale. On this scale the melting point of pure ice in equilibrium with air-saturated water at 1-atm pressure is zero degrees (0 °C). The boiling point of water at 1 atm is assigned the value 100 °C.

Chain Fission Reaction The explosive reaction that occurs when a single fission reaction produces more neutrons than the number of neutrons that caused the original fission, and the mass of the sample of fissionable material is large enough so that the neutrons produced strike other fissionable nuclei and cause additional fission reactions.

Chain Reaction A mechanism of several steps, in some of which there is a regeneration of reactive intermediates. Free radicals are frequently the reactive intermediates in chain reactions.

Chalcogens The oxygen family of elements, Group VIA of the periodic table: O, S, Se, Te, and Po.

Charge Cloud Density The plot of the square of the wave function, $|\psi|^2$, for an electron in an atom or molecule. It is the probability distribution function for an electron.

Charles' Law or Gay-Lussac's Law The volume of a fixed quantity of gas is directly proportional to its absolute temperature, provided the pressure of the gas remains constant.

Chelate A coordination compound in which two or more different atoms of a polydentate ligand are bonded to a single metal ion.

Chemical Kinetics The study of the factors affecting the rate of a chemical reaction and the elucidation of the mechanism by which a reaction occurs.

Chemical Properties The characteristic transformations observed when a substance reacts with other species.

Chiral Molecule A molecule that is not superimposable on its mirror image. A compound whose molecules are chiral can exist as a pair of **enantiomers** or **optical isomers.**

Chlorites Salts of **chlorous acid,** $HClO_2$. The chlorite ion, ClO_2^-, is an oxidizing agent. Chlorites are used in bleaches.

Cis–trans Isomers Two molecules that differ only in the geometric arrangement of like atoms or groups. In the *cis* form the identical groups are adjacent to one another, or on the same side of a double bond. In the *trans* form, the identical groups are opposite one another, or an opposite sides of a double bond.

Classical (Newtonian) Mechanics The set of physical laws describing the motions of macroscopic bodies.

Classical Electromagnetic Theory The set of physical laws deduced prior to 1900 to describe the electrical and magnetic properties of radiation.

Close-Packed Structure A crystal structure in which each atom has 12 nearest neighbors, 6 in the plane of the atom, 3 in the plane above, and 3 in the plane below. About 74% of space is filled with atoms in a close-packed structure.

Colligative Properties Quantities that are directly proportional to the total concentration of all solute particles and independent of the nature of the solute particles. There are four colligative properties: the **vapor pressure lowering,** the **boiling point elevation,** the **freezing point depression,** and **osmotic pressure.**

Common Ion Effect The solubility of an insoluble or slightly soluble electrolyte is less in a solution containing a soluble electrolyte having an ion in common with the insoluble one than it is in pure water.

Complementary Colors Red and green, blue and orange, and violet and yellow. If a solution absorbs radiation of frequency corresponding only to one of these colors, the color of the solution (the transmitted radiation) will be the complementary color.

Complex Ion An ion with a central metal atom bonded to two or more ligands.

Compound A substance that contains two or more elements in a definite proportion by weight. The composition of a pure compound is invariant, and independent of the method of preparation of the compound.

Concentration Cell A galvanic cell in which both electrodes are made of the same substance, and the two solutions in which the electrodes are immersed differ only in the concentration of one of the reacting species.

Condensation The conversion of a gas to its liquid. All condensations are exothermic.

Condensation Polymerization The process in which monomer molecules combine with loss of some simple molecule like water.

Condensed Phases The liquid and solid phases, as contrasted with the gaseous phase.

Conductance Titration An experimental procedure in which the electrical conductivity of a solution is measured after each addition of a small volume of a second solution. A reaction, frequently a precipitation reaction, occurs when the two solutions are mixed. The electrical conductivity changes as the reaction proceeds.

Conduction Band in a Metal A band of very closely spaced energy levels that can be occupied by the valence electrons of the metal. Many of the levels in the band are vacant.

Conformations Different geometric arrangements of

atoms of a molecule that are interconvertible by rotation about single bonds. Usually conformations interconvert at room temperature, so individual conformations cannot be isolated.

Conjugate Acid–Base Pair An acid and a base related by the equation   acid = base + H^+.

Conservation of Mass In an ordinary chemical reaction the number of moles of atoms of each element is the same after reaction as it is before the reaction.

Constancy of Interfacial Angles The angle of intersection of corresponding plane surfaces in crystals of the same substance is always the same, regardless of size or shape of the crystal.

Constructive Interference The superposition of two waves with the same wavelength that are in phase with one another. The resultant wave has a greater intensity than that of either of the superposed waves.

Control Rods Rods used in a nuclear reactor containing material that is a good absorber of neutrons (such as cadmium or boron). These rods can be inserted into the core of the reactor to slow down or stop the fission reactions that are the energy source of the reactor.

Coordinate Bond A covalent bond in which both of the electrons of the shared pair were originally on one of the two bonded atoms, and the other atom had a vacant AO to accept the electron pair.

Coordination Compound A neutral molecule in which a central metal atom or ion is bonded to two or more ligands.

Coordination Number The number of donor atoms (atoms with a lone pair of electrons to share) to which a given cation or metal atom is bonded. A ligand may have two or more donor atoms.

Core The nucleus and the inner electrons of an atom, that is, all the electrons except the valence electrons. The core is also referred to as the **kernel** of the atom.

Corpuscular Composed of particles or corpuscles.

Cosmic Radiation Highly penetrating radiation from outer space. Primary cosmic radiation is mostly high energy protons, but the protons interact with other atoms in space and in the upper atmosphere to produce secondary cosmic rays that consist of many elementary particles and high energy atomic nuclei.

Coulomb's Law $F = kq_1q_2/Dr^2$, where F is the force of attraction or repulsion between two particles of charge q_1 and q_2, separated by a distance r, and immersed in a medium of dielectric constant D. The value of the constant k depends on the units used for the force, the charges, and the distance between the charges.

Covalent Bond A shared pair of electrons between two atoms. If the electron cloud of the shared pair spends more time around one of the atoms than the other, the bond is described as **covalent with partial ionic character,** or **polar covalent.**

C_p and C_v The molar heat capacities at constant pressure and at constant volume, respectively.

Criterion of Maximum Overlap The greater the amount of overlap between two atomic orbitals in the region between two nuclei, the stronger the bond between the two atoms.

Critical Mass The mass of a sample containing fissionable nuclei that is just large enough to maintain a chain fission reaction with a constant rate of fission.

Critical Temperature That temperature above which it is impossible to liquefy a specific gas, no matter how great the applied pressure. The minimum pressure required to liquefy a gas at its critical temperature is the **critical pressure.** The point on a phase diagram whose coordinates are the critical temperature and pressure is known as the **critical point.**

Crookes Tube A vacuum tube containing a gas at very low pressure, and an anode and a cathode that can be subjected to an electrical potential difference of several thousand volts.

Cryolite An aluminum ore, Na_3AlF_6, that melts at a temperature significantly below the melting point of Al_2O_3.

Cryoscopic Constant, K_f The proportionality constant between the **freezing point depression, ΔT_f,** and the **molality,** m, of a dilute solution. The cryoscopic constant (also called the **molal freezing point depression constant**) is a property of the solvent only.

Crystal Field Splitting The separation between the energies of the d atomic orbitals produced by the electric field of the ligands surrounding a transition metal atom or ion.

Crystal Field Stabilization Energy (CFSE) For a coordination complex with an octahedral ligand field, the gain in energy as a result of the preferential filling of the lower lying t_{2g} energy level.

Crystal Field Theory A theory used to explain the absorption spectra and magnetic properties of coordination complexes with specific geometries by considering the effect of the electrostatic interaction between the negative ligands and the electrons in the valence orbitals of the metal to which the ligands are bonded.

Cubic Centimeter (cc or cm^3) A unit of volume. A cube with dimensions 1 cm $\times$ 1 cm $\times$ 1 cm has a volume of 1 cm^3.

Cubic Close-Packed Structure A crystal structure in which the stacking of close-packed layers has an

$ABCABC \cdots$ pattern. The third layer is not directly above the first, but fits into different interstices of the second layer than those occupied by the first.

Cubic Unit Cell A three-dimensional repeating unit in a crystal in which the three coordinate axes are of equal length and are mutually perpendicular.

Curie, Ci A unit of radioactivity. One curie is 3.700×10^{10} disintegrations per second.

Current (Electrical) The amount of charge that passes a given point per unit time. $I = Q/t$, where I is current, Q is charge, and t is time.

Cycle A complete repeating unit of a wave.

Dalton Another term for the **atomic mass unit.**

Daltonides (Stoichiometric Compounds) Compounds with a fixed and definite atomic composition, independent of the method of preparation.

Dalton's Law of Partial Pressures The total pressure of a mixture of gases is the sum of the partial pressures of all the components of the mixture. Alternatively, the partial pressure of the ith gas in a mixture is equal to the product of the total pressure and the mole fraction of the ith gas.

Daniell Cell A galvanic cell utilizing the reduction of Cu^{2+} ions by zinc.

De Broglie Relation, $p = h/\lambda$ The relation between the wave properties (the wavelength, λ) and the corpuscular properties (the linear momentum, p) of both matter and electromagnetic radiation. For a particle of mass m moving with speed v, $p = mv$, and $\lambda = h/mv$.

Debye (D) The unit of dipole moment, named after Nobel laureate Peter Debye.

Decay Constant The first-order specific rate constant for an unstable (radioactive) nucleus.

Defect Any deviation from the regular, ordered, geometric array of ions in an ionic crystalline solid.

Degeneracy of an Energy Level The number of atomic orbitals having the same energy. **Degenerate atomic orbitals** have the same energy. In the absence of an external magnetic field, p orbitals are three-fold degenerate, and d orbitals are five-fold degenerate.

Degree of Dissociation, α For a solution of a weak acid with formality C, $\alpha = [H_3O^+]/C$. For a solution of a weak base with formality C, $\alpha = [OH^-]/C$.

Delocalized Molecular Orbital An MO that extends over more than two bonded atoms, and is therefore multicentered. Electrons in a delocalized MO move freely around all the bonded atoms, and are referred to as **delocalized electrons.**

Delta (Δ) X The value of property X in the final state minus the value of property X in the initial state.

Density Mass per unit volume.

Depth of the Potential Well The difference between zero potential energy (infinite separation of two particles) and the minimum value of potential energy that can be attained by varying the distance between two particles.

Destructive Interference The decrease in intensity of a resultant wave on superposition of two or more waves of the same wavelength that are out-of-phase.

Deuteron The nucleus of deuterium, an isotope of hydrogen with one neutron and one proton. It is denoted either $^2_1H^+$ or D^+.

Diamagnetic Having all electron spins paired. Diamagnetic substances are very slightly repelled by an external magnetic field.

Dielectric Constant of a Liquid The ratio of the work required to separate two oppositely charged particles a given distance in a vacuum, to the work required to separate them to that same distance when they are immersed in the liquid.

Diffraction A property of waves. A modification that light undergoes on being reflected from ruled surfaces, such as a grating. A **diffraction pattern** of alternate light and dark bands is produced.

Diffusion The gradual mixing of two or more substances as the result of random motion of the molecules of the substances.

Digonal (sp) Hybrids Hybrid atomic orbitals formed by combining the s and one of the p valence AO's of the same atom. There are two sp hybrid AO's, directed at $180°$ to one another.

Dihedral Angle in Ethane The angle of rotation between the two $—CH_3$ groups. A zero dihedral angle means that the three $C—H$ bonds on the two ends of the molecule are aligned so that when you look down the $C—C$ bond you see only three hydrogen atoms.

Dimensional Analysis A technique for solving numerical problems. The appropriate mathematical process to use is determined by performing the same operations on the units of all quantities that you perform on the numbers, and insuring that all units cancel except the correct units of the desired answer.

Dimerization A reaction in which two identical molecules (monomers) combine to form a single larger molecule, the **dimer.**

Dipole or Dipolar Molecule A molecule for which the centers of positive and negative charge do not coincide. One end of a dipole has a small net positive charge; the other end has a small net negative charge. The **dipole moment** is the product of the magnitude of the charge at each center and the distance separating the centers of positive and negative charge.

Dipole-Dipole Interaction The force of attraction between two polar molecules when they are aligned so that the negative end of one molecule is adjacent to the positive end of the other.

Diprotic Acid A substance with two acidic protons, that is, two protons that can be donated to a base.

Discrete Consisting of discontinuous parts. A **discrete spectrum** is one that contains only certain frequencies, as opposed to a continuous spectrum that contains every frequency.

Dispersion The separation of polychromatic light (light consisting of many frequencies) into rays of different frequency. Dispersion occurs when light passes through a prism because different frequencies are refracted by different amounts.

Dispersion Force The weak force of attraction between two atoms or molecules that are very close to one another, due to the correlation of their electronic motions to reduce the mutual repulsion of the two electron clouds. A fluctuating dipole in one molecule induces a fluctuating dipole in the second.

Disproportionation The process in which a substance containing an element in an intermediate oxidation state spontaneously reacts to produce one substance in which the element is in a higher oxidation state and one in which it is in a lower oxidation state.

Dissociation Constant For a molecular weak acid, the acidity constant of that weak acid. The dissociation constant of a molecular weak base is the basicity constant of that weak base.

Dissolution The process of dissolving a solute in a solution. It is most commonly used with regard to solid solutes in liquid solutions.

Distillate The substance that comes out of the top of a **distillation column,** and is then condensed, during a **fractional distillation** to separate a mixture of two liquids into its pure components. The **distillate** is the more volatile of the two liquids.

Distribution Constant An equilibrium constant for the equilibrium of a single component distributed between two different phases.

Distribution Function, $D(u)$, for Gas Velocities A function of the speed, u, such that the area under a plot of $D(u)$ versus u between any two values u_1 and u_2 is equal to the fraction of molecules with speeds between u_1 and u_2. This function is also called the **Maxwell–Boltzmann distribution function.**

Dopa (Levodopa or 1-Dopa) An amino acid that is converted by the brain to **dopamine,** one of the substances that transmits messages to the brain. A person with an insufficient supply of dopamine suffers from Parkinson's disease.

Dot Formula A symbol for an atom or molecule in which dots are used to represent the valence elec-

trons. Four pairs of dots around the symbol for an element represent a complete octet, the electronic configuration of a rare gas.

Double Bond The sharing of two pairs of electrons by two atoms.

Downs Cell A cell used for the commercial preparation of metallic sodium and $Cl_2(g)$ by the electrolysis of molten $NaCl$.

Dry Cell A commercial galvanic cell with a zinc anode and a graphite cathode surrounded by solid MnO_2. The electrolyte is a moist paste of NH_4Cl, $ZnCl_2$, and some inert filler.

Dynamic Equilibrium The condition that exists when two opposing reactions are occurring at the same rate. In molecular systems, all equilibria are dynamic.

Ebullioscopic Constant, K_b The proportionality constant between the **boiling point elevation, ΔT_b,** and the **molality,** m, of a dilute solution. The ebullioscopic constant (also called the **molal boiling point elevation constant**) is a property of the solvent only.

Eclipsed Conformation of Ethane A conformation in which the two $—CH_3$ groups are aligned so that if you look down the $C—C$ bond you see only three hydrogen atoms.

Effective Atomic Number (EAN) The effective atomic number of a metal ion in a coordination complex is the total number of electrons surrounding the metal, including all pairs shared with ligands. The EAN rule proposed by Sidgwick states that the coordination number of a cation is determined by the tendency to make the EAN equal to the atomic number of one of the rare gases.

Effusion The passage of gas molecules through a very small orifice or a porous membrane. The hole through which the molecules pass must be so small that only one molecule at a time can pass through the hole.

e_g Level The higher energy, doubly degenerate energy level formed when an octahedral ligand field splits the five d AO's into two levels.

Eigenfunction A solution of the Schrödinger wave equation.

Elastic Collisions Collisions for which the total kinetic energy is conserved.

Electrode A conducting surface on which electrons enter or leave an electrolytic or a galvanic cell. There are two electrodes per cell, an anode and a cathode.

Electrolysis The process in which electricity is passed through a molten salt or an aqueous solution of an electrolyte. Electrolysis is used to recharge a rechargeable battery.

Electrolyte A compound whose aqueous solution contains ions.

Electromagnet A device consisting of a very large number of coils of wire wound around a metallic core. When electric current is passed through the coils, a magnetic field is produced.

Electromagnetic Radiation The radiation produced by an oscillating electric charge. It consists of an oscillating electric field and an oscillating magnetic field, perpendicular to one another, and propagated through space by wave motion.

Electromotive Force (EMF) The difference in electrical potential energy per unit charge between the two electrodes of a galvanic cell. The EMF is measured in volts.

Electron An extremely light, negatively charged sub-atomic particle found outside the nucleus of every atom. The magnitude of the electronic charge is the fundamental unit of electrical charge in chemistry.

Electron Affinity The energy released when an electron is added to an isolated gaseous atom of an element, to form a gaseous uninegative ion. The electron affinity is equal to $-\Delta H$ for the reaction $X(g) + e^- \rightarrow X^-(g)$.

Electron Capture A process used by neutron-poor nuclei to achieve stability. The nucleus captures one of the orbital electrons (usually from the K shell), and a proton in the nucleus is converted to a neutron, increasing the n/p ratio. The nucleus formed has the same value of A as the original, but Z has decreased by 1.

Electron Cloud The term used to describe the negative charge density surrounding the nucleus of any atom. Because electrons travel very rapidly, and follow no simple path, we observe negative charge density throughout the entire volume of the atom.

Electron Deficient Compounds Those in which one atom of the compound has less than eight electrons in its valence shell. Boron forms a substantial number of electron deficient compounds.

Electronegativity A measure of the ability of a bonded atom to attract electrons from other atoms to which it is bonded. The atoms with the largest electronegativities are F, O, N, and Cl.

Electroneutrality Equation An equation that relates the concentrations of all positive ions to the concentrations of all negative ions in a solution. It expresses the fact that the total amount of positive charge must equal the total amount of negative charge in any solution.

Electron Spin An intrinsic angular momentum possessed by every electron. The electron spin produces a magnetic moment not associated with the orbital motion of the electron. There are only two possible spin states, α and β, $m_s = +\frac{1}{2}$ and $-\frac{1}{2}$.

Electron-Transfer Reactions Oxidation–reduction reactions.

Electron Volt (eV) A unit of energy. The amount of energy acquired by a single electron accelerated through a potential difference of one volt, which is equal to 1.6022×10^{-19} J.

Electroplating A process for covering an object made of some less expensive metal with a more precious metal, usually silver or gold. The object to be plated is made the cathode of an electrolytic cell. The anode is made of pure silver or gold, and the electrolyte is an aqueous solution of a silver or gold salt.

Element A substance that cannot be separated into two or more substances different from itself by ordinary chemical means.

Elementary Process A single step in a reaction mechanism. It is only for an elementary process that the stoichiometry is identical with the mechanism.

Empirical Constants Values that are determined by experiment, rather than being calculated theoretically.

Empirical Formula A chemical formula that identifies the atoms of which a compound is composed and describes the *ratios* in which these atoms are combined. It is to be distinguished from the molecular formula, which gives the exact number of each kind of atom in one molecule of compound.

Enantiomers Isomers that are nonsuperimposable mirror images of one another. Enantiomers are optically active molecules that differ only in the direction in which they rotate the plane of **plane-polarized light.**

Endothermic Process One that requires the absorption of heat in order for the reaction to proceed. If heat is released when a reaction occurs, the process is **exothermic.**

Energy The product of a force and the distance through which the force is exerted. The energy expended when a force of one newton is exerted through a distance of one meter is a joule.

Energy Level Diagram A set of horizontal lines representing the allowed energies of an atomic or molecular system.

Enthalpy, H A state function, defined as

$$H = E + PV.$$

Enthalpy Change, ΔH The amount of heat absorbed when a process takes place at constant pressure. If the process is endothermic, ΔH is positive; if the process is exothermic, ΔH is negative. The release of heat is equivalent to the absorption of a negative quantity of heat.

Enthalpy of Formation ΔH for the reaction when one mole of a compound is formed from the elements in their standard states. See also **standard enthalpy of formation.**

Entropy, S A property of a system that measures the amount of molecular disorder or randomness.

Entropy of Vaporization, ΔS_{vap} The difference between the entropy of one mole of vapor and one mole of liquid at the same temperature and pressure. It is always positive, as the vapor is more disordered than the liquid.

Equation of State A single equation that relates the four variables n, V, T, and P of a sample of gas. The ideal gas law is an equation of state, and there are several others as well.

Equatorial Positions In a trigonal bipyramidal structure, five atoms are bonded to a central atom. Three of the bonded atoms lie in a plane, at the corners of an equilateral triangle. The locations of these three atoms are called the **equatorial positions.**

Equilibrium Constant For the reaction

$$\alpha A + \beta B \rightleftharpoons \gamma C + \delta D$$

the function

$$\frac{[C]^\gamma [D]^\delta}{[A]^\alpha [B]^\beta}$$

which has the same numerical value for any equilibrium mixture of all the substances involved in the reaction at a fixed temperature.

Equilibrium Vapor Pressure of a Liquid At a given temperature, the pressure at which the rate of condensation of the vapor is equal to the rate of evaporation of the liquid. Since the rate of evaporation is a constant at constant temperature, the equilibrium vapor pressure is also a constant at constant temperature.

Equivalence Point The point in a titration at which the number of moles of the reagent being added is exactly the amount needed to react completely with the original reagent, so that neither reagent is in excess.

Erg A unit of energy in the cgs (centimeter-gram-second) system of units. It is defined as the energy expended when a force of one dyne is exerted through a distance of one centimeter. An erg is a very small unit of energy: 10^7 erg = 1 J.

Ester A derivative of a carboxylic acid in which the —OH group is replaced by —OR, where R is an organic substituent. Many esters have pleasant, fruity odors.

Esterification The reaction between a carboxylic acid and an alcohol to produce water and an ester.

Ethene Another name for ethylene, C_2H_4 or $H_2C{=}CH_2$.

Ether A compound in which two organic groups are bonded to an oxygen atom. The general formula of an ether is R—O—R'.

Ethoxide Ion, $CH_3CH_2O^-$ The anion formed when ethanol loses a proton. It is a very strong base and does not exist in aqueous solution as it reacts with water to yield OH^- and ethanol.

Ethynyl Group —C≡C—H.

Exact Number One that has no experimental uncertainty. Conversion factors within one system of units are exact numbers.

Excited State An allowed energy state of an atom or a molecule with an energy higher than the ground state.

Exothermic Reaction A reaction that releases heat as it proceeds.

Expanded Valence Shell An outer shell containing more than eight electrons.

Experimental Error or Uncertainty The range of reliability of an experimentally measured value.

Exponential Decay Law The integrated form of a first-order rate of reaction: $[A]_t = [A]_0 e^{-k_1 t}$.

Exponential (Scientific) Notation The expression of a number as the product of a number between 1 and 10 and some integral power of 10.

Extensive Property One that depends on the amount of material present. Examples are mass and volume.

Extrapolation The process of extending a plot of experimental data as a function of some variable back to values of the variable that cannot be measured.

Face-Centered Lattice A crystal lattice in which lattice points are at the eight corners and six face centers of the unit cell.

Fahrenheit Temperature Scale A temperatue scale on which the melting point of pure ice at 1 atm is 32 °F, and the boiling point of water at 1 atm is 212 °F. A Fahrenheit degree has $\frac{5}{9}$ the magnitude of a Celsius degree.

Family (Group) of Elements The elements in one vertical column of the periodic table. The members of a family have similar physical and chemical properties, and the same number of valence electrons.

Faraday, $\mathscr{F}$ The charge on one mole of electrons. The numerical value of the faraday is 96,485 C.

Faraday's Laws of Electrolysis The relationships between the amount of electricity passed through a cell and the chemical changes that occur during electrolysis.

Fat A naturally occurring triester of glycerol and three long-chain carboxylic acids called **fatty acids.**

Ferromagnetic Having a magnetic moment larger than paramagnetic substances. Ferromagnetic substances are attracted to a magnetic field, but do not lose their magnetism when removed from an external field, as paramagnetic substances do.

First Ionization Energy The energy required to remove one electron from an isolated gaseous atom to form a gaseous unipositive ion. It is ΔH for the reaction $X(g) \rightarrow X^+(g) + e^-$.

First Law of Thermodynamics (The law of conservation of mass and energy.) In a system of constant mass, energy can neither be created nor destroyed. The energy of a system can change if heat and/or work enter or leave the system. The change in energy is then $\Delta E = q + w$, where q is the heat absorbed by the system and w is the work done on the system.

First-Order Decay Constant The specific rate constant for a first-order reaction.

First Transition Series The 10 elements from scandium through zinc ($Z = 21$ to 30), in which the $3d$ atomic orbitals are being filled.

Fission The splitting of a heavy nucleus into lighter fragments with the release of energy. Most fission processes are initiated by bombarding the heavy nucleus with **thermal neutrons.**

Flame Tests Tests used in analytical chemistry to detect the presence of certain metallic elements. The color imparted to a flame is characteristic of the element because it is the visible portion of the atomic spectrum of that element.

Fluoroapatite, $Ca_5F(PO_4)_3$ A major constituent of phosphate rock, the principal source of phosphorus in nature.

Formal Charge The charge on an atom in a molecule or ion that is calculated by assuming that all the lone pair electrons and one half of the electrons in pairs shared by an atom, belong to that atom. It is not the actual charge on an atom unless the bonding is pure covalent.

Formality The concentration of an electrolyte in moles per liter. The term **formality** (rather than **molarity**) is used to emphasize that solute particles are ions rather than molecules. A solution containing one mole per liter of an electrolyte such as NaCl is said to be "one **formal**" (1 F).

Formula A symbol or group of symbols that specifies the atomic composition of a substance.

Formula Weight The mass of one mole of a substance. For a substance that consists of discrete molecules, the formula weight and the molecular weight are identical. The term formula weight is used for an ionic crystalline solid to emphasize that no discrete molecules exist.

Fractional Distillation The process of separating a liquid mixture into its pure components. The apparatus used is called a **still,** and its principal component is a **distillation column.**

Fraction Ionized A synonym for the degree of dissociation.

Free Energy of Formation The change in Gibbs free energy for the reaction in which one mole of a substance is formed from the elements in their standard states. See also **standard free energy of formation.**

Free Radical A substance with one or more unpaired electrons. Many chemical reactions proceed via a mechanism that involves free-radical intermediate species.

Free-Radical Chain Reaction A reaction that proceeds through a series of steps in which one free radical is consumed but another is formed.

Freezing Point Depression, ΔT_f The difference between the freezing point of a pure solvent and the freezing point of a solution prepared using that solvent. It is a positive quantity, as the freezing point of a solution of a nonvolatile solute in a volatile solvent is always *lower* than the freezing point of the pure solvent.

Frequency For electromagnetic radiation, the number of cycles passing any given point in space in one second. Frequency is denoted by the symbol v, and has units of reciprocal seconds, s^{-1}.

Frequency Factor, A A term in the Arrhenius equation for the temperature dependence of specific rate constants, $k = Ae^{-E_{act}/RT}$. The frequency factor is approximately constant, but has a small temperature dependence.

Functional Group A reactive portion of an organic molecule that, for the most part, determines the chemical reactions the molecule will undergo.

Fusion The conversion of a solid to its liquid. Fusion and melting are synonymous.

Fusion (Nuclear) The process in which two light nuclei combine to produce a heavier nucleus with mass number A less than 60, with the release of energy. The fusion of four hydrogen nuclei to form a helium nucleus (a multistep process) is the principal source of energy radiated by the sun.

Galvanic Cell A device for converting chemical energy into electrical energy. It consists of an anode and a cathode, each immersed in a solution containing ions. If the solutions are different, they must be kept separated. Electrical contact between them is then made by using a **salt bridge** or some porous barrier.

Gamma Rays Electromagnetic radiation of very high energy (very short wavelength) that is emitted when nuclei in excited **(isomeric)** states make the transition to the nuclear ground state.

Gay-Lussac's Law of Combining Volumes The volumes of gases that react with one another or are produced in a chemical reaction, are in the ratios of small integers, provided that all the gases are at the same temperature and pressure.

Geometric Isomers Molecules with the same empirical formula but different chemical and physical properties. They differ in the positions occupied by the atoms or groups of atoms within the molecule.

Gibbs Free Energy, *G* A state function, a thermodynamic property of a system, defined as

$$G = H - TS$$

where *H* is enthalpy, *S* is entropy, and *T* is the absolute temperature.

Glycerol, CH_2OH—$CHOH$—CH_2OH 1, 2, 3-Propanetriol. Triesters of glycerol and long-chain carboxylic acids constitute that class of compounds known as **fats.**

Glycogen A polymer of glucose stored in the liver of animals.

Goes to Completion A term used to describe a reaction for which essentially 100% of one or more of the reacting species is used up during the reaction.

Gouy Balance An instrument used to measure the magnetic moment of a molecule. The experimental procedure consists of measuring the apparent weight of a sample in the presence and absence of a magnetic field of known magnitude.

Graham's Law of Gaseous Effusion The relative rates of effusion of two different gases at the same temperature are inversely proportional to the square root of the molecular weights of the gases.

Gram-Atom The mass, in grams, of Avogadro's number of atoms of an element. It is equivalent to a mole of atoms.

Ground State The state of lowest energy.

Half-Full Effect The extra stability of an electronic configuration in which a set of degenerate atomic orbitals is half-full with all electrons having parallel spin. This configuration minimizes electron–electron repulsion, as the electrons are as far apart as possible.

Half-Life The time required for the concentration or amount of a reacting substance to fall to one half its original value. For first-order reactions *only,* the half-life is a characteristic of the reaction and is independent of the initial concentration.

Half-Reactions Equations used to balance oxidation–reduction equations. The oxidation half-reaction shows the species being oxidized, the product of the oxidation, and the electrons lost. The reduction half-reaction shows the species being reduced, the product of the reduction, and the electrons gained.

Halide Ions Uninegative ions formed when a halogen atom adds an electron to achieve the electronic configuration of the succeeding rare gas.

Hall Process A technique for producing aluminum by the electrolysis of a molten solution of Al_2O_3 in cryolite, Na_3AlF_6.

Halogens Group VIIA of the periodic table: F, Cl, Br, I, and At. These are reactive, nonmetallic elements. The electronic configuration of the valence electrons is $(ns)^2(np)^5$.

Heat of Formation Synonymous with **enthalpy of formation.**

Heat of Fusion, ΔH_{fus} The amount of heat required to convert one mole of solid to one mole of liquid at constant temperature and pressure, when solid and liquid are in equilibrium throughout the process. All heats of fusion are positive quantities.

Heat of Hydration of an Ion ΔH for the reaction in which a mole of gaseous ions combines with water to produce hydrated ions.

Heat of Mixing, ΔH_{mix} The heat absorbed when two substances are mixed to form a solution. If the solution is ideal, the heat of mixing is zero.

Heat of Reaction at Constant Pressure, q_p The heat absorbed when a reaction occurs at constant pressure and temperature. It is equal to ΔH for the reaction.

Heat of Reaction at Constant Volume, q_v The heat absorbed when a reaction occurs at constant volume and temperature. It is equal to ΔE for the reaction.

Heat of Vaporization, ΔH_{vap} The amount of heat required to convert one mole of liquid to one mole of vapor at constant temperature and pressure, keeping the liquid and vapor in equilibrium throughout the process. The heat of vaporization is a temperature dependent quantity, and is positive for all substances.

Helium-Molecule Ion The species He_2^+.

Henderson–Hasselbalch Equation An equation used to calculate the pH of a buffer solution. It is

$$pH = pK_a + \log([\text{weak base}]/[\text{conjugate weak acid}])$$

Henry's Law The pressure of a substance in the gas phase in equilibrium with a solution containing that substance as a solute, is directly proportional to the concentration of the solute:

$$P_A = K'[A]$$

where K' is Henry's Law constant.

Hess' Law of Constant Heat Summation The heat absorbed when a given reaction is carried out at constant pressure and temperature is the same whether the reaction occurs in one step or in several steps.

Heterocyclic Compound An organic compound containing a ring of atoms, at least one of which is not carbon.

Heterogeneous Mixture A mixture whose properties are not uniform throughout the sample. The properties of **homogeneous mixtures,** in contrast, are uniform throughout the sample.

Heterogeneous Reaction A reaction that occurs in more than a single phase, such as a reaction between a gas and a solid, or between a solid and a liquid phase.

Heteronuclear Diatomic Molecule A molecule in which two atoms of different elements are bonded, in contrast to a **homonuclear diatomic molecule,** in which two atoms of the same element are bonded.

Hexagonal Close-Packed Structure A crystal structure in which the layers are stacked so that the third layer lies directly above the first. The pattern of stacking is described as $ABABAB\cdots$.

High-Spin Complex of a Metal Ion One that has the same number of unpaired electrons, and hence the same magnetic moment, as the free ion.

H-Like Ions Ions consisting of only a single electron and a nucleus of charge $+Ze$. Examples are He^+, Li^{2+}, and Be^{3+}.

Homogeneous Reaction A reaction that takes place in a single phase. All reactants and products may be gases, or all may be in the same liquid solution.

Homonuclear Diatomic Molecule A molecule in which two atoms of the same element are bonded.

Hund's Rule When two or more atomic orbitals of equal energy are available, the state of lowest energy for a many-electron system is the state in which electrons occupy different orbitals with parallel spin.

Hybrid Atomic Orbital An atomic orbital formed by combining valence atomic orbitals of a single atom.

Hydrated Ion An ion surrounded by closely associated water molecules that are attracted to it by ion-dipole forces.

Hydration Number of an Ion The number of H_2O molecules that, on the average, are closely associated with the ion in aqueous solution.

Hydrazine, N_2H_4 A weak base. It combines with H^+ to form the **hydrazinium ion, $N_2H_5^+$.** These substances are the principal species in which nitrogen exhibits the -2 oxidation state. In aqueous basic solution, hydrazine is a good reducing agent.

Hydrides Compounds containing the hydride ion, H^-.

Hydrocarbon A molecule that is composed entirely of carbon and hydrogen atoms.

Hydrogen Bond or Bridge An attractive interaction between two molecules or ions due to having a hydrogen atom situated between two strongly electronegative atoms. Species in which there is a polar $F-H$, $O-H$, or $N-H$ bond orient themselves to form hydrogen bonds.

Hydrogen-Molecule Ion The species H_2^+.

Hydronium Ion The name of the ion formed when an H^+ ion bonds to one or more water molecules. It is commonly written either as $H_3O^+(aq)$ or $H^+(aq)$.

Hydrophilic Compound A water soluble organic compound. A **hydrophobic** compound dissolves only in nonpolar organic solvents.

Hydroxyapatite, $Ca_5(OH)(PO_4)_3$ One of the chief sources of phosphorus in nature. Hydroxyapatite is a major constituent of the teeth of animals.

Hydroxylamine, H_2NOH A weak base derived from ammonia by replacing one $-H$ with $-OH$. Hydroxylamine combines with H^+ to yield **hydroxylammonium ion, $(H_3NOH)^+$.** Hydroxylamine can serve either as an oxidizing agent or a reducing agent, but is usually used as a reducing agent.

Hypochlorous Acid, $HOCl$ A weak acid in which chlorine is in the $+1$ oxidation state. Its conjugate base, OCl^-, hypochlorite ion, is chiefly used as an oxidizing agent in bleaches. Both $HOCl$ and OCl^- are usually reduced to Cl^- when they serve as oxidizing agents.

Hypothetical Valence State of an Atom A state we imagine to exist immediately prior to the formation of bonds using hybrid atomic orbitals to overlap the atomic orbitals of another atom.

Ideal Gas A gas that obeys the **ideal gas law, $PV = nRT$,** where P is the pressure of a gas, V is its volume, n is the number of moles, R is a universal constant, and T is the absolute temperature. No real gas is ideal.

Ideal Law of Chemical Equilibrium The equilibrium constant for the reaction

$$\alpha A + \beta B \rightleftharpoons \gamma C + \delta D$$

is

$$[C]^\gamma [D]^\delta / [A]^\alpha [B]^\beta$$

and is a constant at constant temperature for any equilibrium mixture of the substances A, B, C, and D.

Ideal Solution A solution for which all volatile components obey **Raoult's law** over the entire range of concentrations. Most real solutions are not ideal.

Imine A compound containing a carbon–nitrogen double bond, $>C=NH$. It is the product of the reaction between an aldehyde or ketone with ammonia, in which the carbonyl oxygen is replaced by NH.

Indicator A substance added to a titration mixture to signal the equivalence point. Only a very small amount of indicator is added. Usually the indicator changes color at the equivalence point.

Induced Dipole Moment A transitory dipole moment that is formed by the temporary **polarization** of the electron cloud of one atom or molecule when another approaches very closely.

Inert Complex A coordination complex that undergoes ligand substitution reactions slowly. An inert complex in aqueous solution is not in equilibrium with its dissociation products in less than 1 or 2 min.

Inert Gases An older name for the **rare** or **noble gases.**

Initial Rate of a Reaction The slope of a curve of the concentration of reactant versus time, evaluated at time $t = 0$.

Inner Transition Elements The **lanthanides** ($Z = 58$ to 71) and the **actinides** ($Z = 90$ to 103), in which, respectively, the $4f$ and $5f$ orbitals are being filled.

Insulator A substance that does not conduct electricity.

Intensive Property A property that does not depend on the number of moles of material present. Pressure, temperature, density, and the cell voltage, or EMF, are intensive properties.

Interatomic Bond A chemical bond between two atoms.

Interfacial Angle in a Crystal The angle of intersection of two plane surfaces that bound the crystalline solid.

Interionic Forces The coulombic interactions between ions.

Intermediate Oxidation State An oxidation state that is neither the highest nor the lowest oxidation state of a given element.

Intermediate Species A substance that does not appear in the stoichiometric equation for a reaction, but is formed in one step of the mechanism of the reaction, and used up in some subsequent step. Intermediates are generally short-lived species.

Intermolecular Forces Forces between two nonbonded atoms or molecules. They are contrasted with **intramolecular forces,** which are the forces between bonded atoms in a molecule or complex ion.

Interstitial Site A location in between regular atomic positions in the crystal structure.

Intrinsic Magnetic Moment A magnetism associated with the spin of an electron, and not due to the orbital angular momentum of the electron.

Ion A charged particle, formed when an atom or a group of atoms loses or gains one or more electrons.

Ion-Dipole Force of Attraction The force of attraction between an ion and the oppositely charged end of a dipolar molecule.

Ion–Electron Method A procedure for balancing oxidation–reduction equations. In this method, the number of electrons lost or gained is determined after all atoms have been balanced, by requiring that the charge be balanced.

Ionic Compound A substance built of positive and negative ions in a ratio such that the compound is electrically neutral, that is, has no net charge. Ionic compounds are crystalline solids at room temperature. They have high melting points relative to molecular compounds, and conduct electricity when molten.

Ionization Constant For a molecular weak acid or base, the ionization constant is identical with the dissociation constant of that acid or base.

Ionization Energy (IE) The amount of energy required per mole to remove a valence electron from an isolated, gaseous atom and produce a gaseous ion. Metallic elements have lower ionization energies than nonmetallic elements.

Ion-Pair A positive and a negative ion so close together that they move as a unit for some period of time.

Ion Product of H_2S The quantity $[H_3O^+]^2[S^{2-}]$ for a saturated aqueous solution of H_2S. The ion product of H_2S is 1.3×10^{-20} at 25 °C.

Ion Product of Water, K_w The quantity $[H_3O^+][OH^-]$. At 25 °C, the ion product of water is 1×10^{-14}.

Irreversible Process One that cannot be reversed merely by changing some variable by an infinitesimal amount. All spontaneous processes are irreversible.

Isobars Nuclei with the same mass number but different atomic numbers.

Isoelectronic Having the same number of electrons and identical electronic configurations.

Isolated System One for which the energy is constant because neither heat nor work cross the boundary between system and surroundings.

Isomeric State of a Nucleus An excited state, that is, one higher in energy than the nuclear ground state.

Isomorphous Having the same crystal structure as a different substance.

Isotope Dilution A technique for determining the amount of a substance in a mixture when the substance cannot be quantitatively separated from the mixture. A known amount of the substance, labeled with a radioactive isotope, is added to the mixture. The amount of nonradioactive material can be determined from the decrease in the **specific activity** of the substance.

Isotopes Nuclei with the same atomic number but different mass numbers.

Isotropic Substance One whose properties are the same in all directions. Gases, liquids, and amorphous solids are isotropic.

Joule The SI unit of energy. The amount of work expended when a force of one newton is exerted through a distance of one meter.

K Capture The capture of an electron from the K shell by an unstable (radioactive) nucleus.

Kekulé Structure of Benzene A structure with alternating carbon–carbon single and double bonds. In the actual structure of benzene all carbon–carbon bonds are equivalent and have bond order $\frac{3}{2}$. The real molecule is considered a resonance hybrid of two Kekulé structures.

Kelvin The fundamental SI unit of temperature. A kelvin is equal in magnitude to a degree on the **Celsius scale.** The **Kelvin temperature scale** is the same as the **absolute temperature scale.**

Kernel Another term for the **core** of an atom.

Ketone An organic molecule in which both groups bonded to the carbon atom of a carbonyl group, $>C=O$, are organic substituents (not hydrogen).

Kilocalorie (kcal) One thousand calories.

Kilogram The fundamental SI unit of mass. One thousand grams.

Kilojoule (kJ) One thousand joules.

Kinetically Controlled Reaction A reaction for which the products formed are not thermodynamically favored (those with lowest Gibbs free energy), but are the products formed fastest.

Kinetic Energy The energy a body possesses due to its motion. If a body is moving with speed u, the magnitude of its kinetic energy is $\frac{1}{2} mu^2$, where m is the mass of the body.

Kinetic-Molecular Theory of Gases A set of postulates that describes the nature of a gas. These postulates describe an ideal gas, and are an excellent description of real gases at low gas densities, that is, at low pressures and high temperatures.

Labile Complex A complex that undergoes ligand substitution reactions rapidly. A complex that is not labile is **inert.**

Lanthanides (Rare Earths) The 14 elements from Ce to Lu ($Z = 58$ to 71). These elements are very similar chemically as their outer electronic configurations are essentially the same. The seven inner $4f$ AO's are being filled as we progress from Ce to Lu.

Lattice or Space Lattice A set of all points with identical environments within the crystal.

Lattice Energy of an Ionic Crystal The amount of energy required, per mole, to separate the ions from their lattice positions to an infinite distance in the gas phase.

Laughing Gas Dinitrogen oxide (nitrous oxide), N_2O. It is used as an anesthetic, particularly in dentistry.

Law of Conservation of Energy Energy can neither be created nor destroyed. An early form of the **first law of thermodynamics,** now more properly called the **law of conservation of mass and energy.**

Law of Dulong and Petit The molar heat capacity of any solid element is 25 ± 1 J·mol^{-1}K^{-1} at room temperature.

Law of Mass Action The ideal law of chemical equilibrium.

Lead Storage Battery The battery used in automobiles. At the anode lead is oxidized to $PbSO_4$. At

the cathode, PbO_2 is reduced to $PbSO_4$. Both electrodes are immersed in aqueous H_2SO_4.

Le Chatelier's Principle If a system in a state of dynamic equilibrium is subjected to a stress that changes any of the factors that determine the state of equilibrium, the reaction will shift in the direction that offsets or minimizes the stress.

Leclanché Cell The common dry cell.

Leveling Effect of Water on Acids and Bases Any acid stronger than H_3O^+ reacts with water to form H_3O^+. Any base stronger than OH^- reacts with water to form OH^-.

Lewis Acid An electron-pair acceptor, a substance having at least one atom with a vacant atomic orbital. A **Lewis base** is an electron-pair donor. In Lewis theory, the reaction between a metal ion with a vacant atomic orbital and a ligand with a lone pair of electrons to form a coordination compound is an acid–base reaction.

Lewis Dot Formula A formula for an atom, molecule, or ion in which electrons are represented by dots and four pairs of dots around the symbol for an element represent a complete octet.

Ligand An anion or a neutral, polar molecule, with one or more lone pairs of electrons it can share with a metal ion to form a coordination complex.

Ligand Field Theory A combination of crystal field theory with molecular orbital theory to provide a more accurate description of the bonding in coordination compounds than is provided by crystal field theory alone.

Lime (Quicklime) Calcium oxide, CaO, a white solid commercially used in the preparation of cements and mortars.

Limiting Law (Applied to Gases) A law that is approximately correct at measurable gas pressures, but becomes exact in the limit of zero pressure. The error made in using a limiting law to predict the behavior of gases may be significant at very high pressure, but becomes smaller and smaller as the pressure decreases.

Limiting Reagent A reagent in a reaction mixture that is completely consumed during reaction, whereas other reagents are present in excess.

Linear Combination of Atomic Orbitals (LCAO–MO) A molecular orbital constructed by adding and/or subtracting the wave functions of the valence electrons of the atoms bonded, with specific numerical coefficients.

Linear Momentum, p For a particle of mass m moving with speed v, p is the product mv.

Linkage Isomerism A type of geometric isomerism that occurs when a ligand can bond to a metal through two different atoms. An example is the NO_2^- ligand, which can bond through either N (nitro) or O (nitrito).

Lipid A naturally occurring compound insoluble in water but soluble in hydrocarbons and other non-polar organic solvents.

Liquid Junction Potential The difference in electrical potential energy at the boundary between two different liquid solutions.

Liter (L) A unit of volume. The conversion factor between the liter and the SI unit of volume, the cubic meter (m^3), is 10^3 L/m^3.

Litmus A substance, extracted from lichens, that turns red in acidic solution and blue in basic solution.

London or Dispersion Forces The weak attraction between two atoms or molecules whose electronic motions are correlated to reduce the mutual repulsion of their electron charge clouds. London forces are named after Fritz London.

Lone Pair A pair of electrons with opposite spin in an atomic orbital that is not being used for bonding.

Low-Spin Complex of a Metal Ion A complex that has fewer unpaired electrons, and hence a smaller magnetic moment, than the free ion.

Madelung Constant A constant in the theoretical expression for the lattice energy of an ionic crystalline solid. The numerical value of the Madelung constant is determined solely by the geometry of the crystal.

Magic Numbers (Nuclear) The numbers of protons or neutrons necessary to fill successive nuclear energy shells: 2, 8, 20, 28, 50, 82, and 126.

Magnetic Dipole Moment For a bar magnet, the product of the strength of a magnetic pole and the distance between the north and south poles. Atoms and molecules behave like tiny bar magnets and have magnetic moments measured in units of the **Bohr magneton.**

Magnetic Quantum Number, m_ℓ A number that determines the direction or orientation of the magnetic field associated with the orbital angular momentum of the electron. For a given value of ℓ, the allowed values of m_ℓ are $0, \pm 1, \ldots, \pm \ell$. There are $(2\ell + 1)$ values of m_ℓ for each value of ℓ.

Mass A measure of the quantity of matter in a sample of any substance. The SI unit of mass is the **kilogram.**

Mass Number, A The sum of the number of protons and neutrons in the nucleus of an atom.

Material Balance Equation An equation that expresses the conservation of mass before and after a chemical reaction by relating the concentration of several species in a mixture.

Maxwell–Boltzmann Distribution The distribution of speeds in a sample of gas. See **distribution function, $D(u)$.**

Mean or Average Speed, $\langle u \rangle$ For a sample of gas, the sum of the speeds of all the molecules at any given instant, divided by the number of molecules. Although the individual speeds continually change, the mean speed is a constant at constant temperature. The mean speed is proportional to $(T/M)^{1/2}$, where T is the absolute temperature and M is the molecular weight of the gas.

Mean-Square Speed, $\langle u^2 \rangle$ For a sample of gas, the sum of the squares of the speeds of all the molecules at any given instant, divided by the number of molecules. It is equal to $3RT/M$.

Mechanical Equivalent of Heat The amount of work equal in energy to a given amount of heat. Because heat and work are both forms of energy, only one unit is needed. We now define the calorie in terms of its mechanical equivalent, 1 cal = 4.1840 J.

Mechanism of a Reaction A series of elementary processes that define the order in which bonds are broken and new ones made.

Mercury Cell A commercial galvanic cell that provides a constant 1.34 V throughout its lifetime. The anode is an amalgam of zinc and mercury. Solid HgO is reduced at the cathode, and the electrolyte is a paste of $Zn(OH)_2$ in KOH.

Metalloid An element that possesses both metallic and nonmetallic properties. The elements usually classified as metalloids are B, Si, Ge, As, Sb, Te, and Po.

Metals Elements that tend to lose electrons and become cations when they take part in chemical reactions. Most metals are solid at room temperature, and are malleable, ductile, and good conductors of heat and electricity. The metallic elements are on the left side and the bottom of the periodic table.

Meter The fundamental SI unit of length.

Methyl Group, $-CH_3$ A carbon atom bonded to three H atoms.

Methylene Group, $>CH_2$ A carbon atom bonded to two H atoms.

Methyl Radical, $\cdot CH_3$ The fragment left after removing one H atom from methane.

Metric System A decimal system of measures and weights in which the meter is the fundamental unit of length. SI units are an expansion of the original metric system.

MeV A unit of energy, one million electron volts, 1×10^6 eV.

Micelle A spherical structure of a great many phospholipid molecules in which the long-chain fatty acid hydrophobic "tails" of the phospholipid cluster together, pointing in toward the center of the sphere, and the small hydrophilic phosphate "heads" of the phospholipid form the surface of the sphere.

Microcurie A unit of radioactivity, 1×10^{-6} Ci. A **millicurie** is 1×10^{-3} Ci.

Millikan's Oil Drop Experiment The first accurate measurement of the charge on an electron.

Milliliter (mL) A unit of volume equal to one cubic centimeter (cc or cm^3) or 10^{-3} L. There are 10^6 mL in 1 m^3.

Mixture Two or more substances, combined in varying proportions, and each retaining its own specific properties. The components of a mixture can be separated by physical methods.

Molal Boiling Point Elevation Constant, K_b Another term for the **ebullioscopic constant**. The **molal freezing point depression constant, K_f,** is another term for the **cryoscopic constant**.

Molality, m The number of moles of solute per kilogram of solvent.

Molar Conductance A measure of the current carrying ability of a solution per mole of electrolyte. For solutions of different electrolytes of the same concentration, it increases as the number of ions per formula unit increases.

Molar Heat Capacity The amount of heat required to raise the temperature of one mole of a substance one degree, either Kelvin or Celsius. Because heat is a path-dependent quantity, we define both a **molar heat capacity at constant volume, C_v,** and a **molar heat capacity at constant pressure, C_p.**

Molarity, M The number of moles of solute per liter of *solution*. A solution containing one mole of solute per liter of solution is said to be "one **molar**" (1 M).

Molar Ratios The ratios of different kinds of atoms that are combined in a compound, or the ratios of moles of different substances taking part in a chemical reaction.

Molar Solubility The number of moles of solid that dissolves per liter of solution. If the composition of the solution is not specified, the solvent is pure water.

Molar Translational Kinetic Energy The translational kinetic energy of one mole of gas. It is equal to $\frac{3}{2} RT$.

Mole The amount of any substance containing Avogadro's number of particles of that substance. One mole of any element contains 6.022×10^{23} atoms of that element, and has a mass, in grams, equal to the atomic weight of the element.

Molecular Compound A compound for which a discrete, individual molecule is a recognizable entity. Solid molecular compounds are soft and low melting compared to ionic crystalline solids.

Molecularity of an Elementary Process The number (either 1, 2, or 3) of molecules involved in the process. The term molecularity can be applied *only* to an elementary process.

Molecular Orbital (MO) Theory A model for describing the bonding in molecules. Electrons occupy orbitals centered around two or more nuclei. Molecular orbitals are constructed by making linear combinations of the atomic orbitals of the valence electrons of the bonded atoms in the molecule.

Molecular Weight The mass, in grams, of one mole of a substance that consists of discrete molecules, or the mass of one molecule of that substance in amu. The molecular weight is the sum of the atomic weights of all the atoms in the molecule.

Molecule A combination of atoms bound together so tightly that the molecule behaves as a single particle. A molecule is the smallest unit of matter of a molecular compound. For ionic crystalline solids no true molecule exists, and the smallest unit is termed simply a formula unit.

Mole Fraction The number of moles of one component in a mixture, divided by the total number of moles of all substances present in the mixture. The mole fraction of the ith component in a mixture is denoted, X_i, and is equal to n_i/n_{total}.

Mole Percent The mole fraction expressed as a percentage.

Monochromatic Radiation Radiation of a single wavelength.

Monodentate Ligand A ligand that binds to a metal through a single donor atom.

Monomer A small molecule that can combine with identical molecules to form a polymer.

Monoprotic Acid A substance with only one acidic proton.

Most Probable Speed of a Gas The speed at which the Maxwell–Boltzmann distribution function, $D(u)$, is a maximum.

Multicentered Electron Cloud One in which the electrons occupy a delocalized molecular orbital, extending over more than two bonded atoms.

Natural Gas The gas found with petroleum deposits. It consists largely of methane, but also contains ethane, propane, and butanes. It is used for heating our homes and cooking our food.

Negative Deviation from Raoult's Law A term describing a solution for which the vapor pressure of a component is *less* than predicted by Raoult's law. For a mixture of two liquids, A and B, a negative deviation from Raoult's law is due to a stronger force of attraction between the unlike molecules, A and B, than between the like molecules, A and A or B and B.

Nernst Equation The relation between the cell voltage and the concentrations of the species taking part in the net cell reaction. The Nernst equation is $\Delta \mathscr{E}_{\text{cell}} = \Delta \mathscr{E}^0_{\text{cell}} - (RT/n\mathscr{F})\ln Q$, where n is the number of moles of electrons transferred, Q is the reaction quotient for the net cell reaction, $\mathscr{F}$ is the faraday, and $\Delta\mathscr{E}^0_{cell}$ is the cell potential (EMF) when all substances are in their standard states.

Net Ionic Equation A balanced chemical equation in which only the species actually taking part in the reaction are included. Substances that exist in aqueous solutions as ions are written in ionic form, and ions that are present but take no part in the reaction **(spectator ions)** are omitted.

Neutralization The reaction between an acid and a base to produce H_2O. If the acid and base are both compounds, neutralization produces a salt plus water.

Neutral Salt A salt whose aqueous solution has a $pH = 7$ at 25 °C.

Neutrino An elementary particle that has no charge and a very small mass, much smaller than the mass of an electron.

Neutron A subatomic particle contained in the nucleus of an atom. A neutron is an uncharged particle with a mass almost (but not exactly) identical to the mass of a proton. The mass of a neutron is approximately one atomic mass unit.

Nicol Prism A device for obtaining a single ray of plane-polarized light by separating the two plane-polarized rays that emerge when ordinary light is passed through a crystal of calcite.

Nitrides Compounds containing the **nitride ion, N^{3-}.**

Noble or Rare Gases Group 0 of the periodic table: He, Ne, Ar, Kr, Xe, and Rn. These elements are chemically unreactive and are the only monatomic gases at room temperature and 1-atm pressure. All except He have valence shell electronic configurations $(ns)^2(np)^6$.

Node A point or a region in space where a wave or a wave function has zero amplitude. There is zero probability of locating an electron at a node in the electronic wave function.

Nonelectrolyte A substance that exists entirely as molecules when dissolved in water.

Nonmetals Elements that tend to accept electrons in their reactions with other substances. They are located in the upper right-hand corner of the periodic table.

Nonstoichiometric Compounds (Berthollides) Solid compounds whose atomic composition varies within a certain range, depending on the conditions of preparation.

Nonvolatile Solute A substance that, dissolved in a volatile solvent, has a vapor pressure so small that it cannot be measured.

Normal Alkane A **straight-chain alkane,** in which each carbon atom is bonded to only one or two other carbon atoms.

Normal Boiling Point The temperature at which a liquid is in equilibrium with its vapor at a pressure of 1 atm.

Normal Melting Point The temperature at which the solid and liquid phases of the same substance are in equilibrium at a pressure of 1 atm.

Nucleon Either a proton or a neutron. The protons and neutrons in a nucleus are referred to collectively as **nucleons.**

Nucleophile An atom or group of atoms that can donate electron density during chemical reactions.

Nucleus The part of an atom that contains all the protons and neutrons, and therefore nearly all the mass of the atom. It has a positive charge equal to the atomic number, Z.

Nuclide A specific nuclear species, characterized by its mass number, A, and its atomic number, Z.

Octahedral Geometry The structure of a molecule or complex ion with a central metal atom bonded to six other atoms situated at the corners of a regular octahedron. All bond angles are 90° and the six bond lengths are equal if the six bonded atoms are identical.

Octet Rule The statement that except for H, Li, Be, and B, an atom combines with other atoms to form bonds in order to achieve the electronic configuration of a rare gas and have eight electrons in its valence shell. Most compounds obey the octet rule, but there are exceptions.

Oil Drop Experiment The experiment carried out by Robert Millikan in 1910 that provided the first accurate measurement of the charge on an electron.

Olefin Another name for an **alkene.**

One-Component Phase Diagram A plot that gives the relation between pressure and temperature for the equilibria between the gaseous, liquid, and solid phases of a single pure substance.

Optical Isomers (Enantiomers) A pair of molecules that are nonsuperimposable mirror images of one another. Each is **optically active,** and rotates the plane of polarization of polarized light. The two isomers rotate polarized light in opposite directions.

Order of a Chemical Reaction The sum of the exponents of the concentrations of the reacting species in the observed rate law. It cannot be predicted from the overall stoichiometry and must be determined experimentally.

Orthophosphoric Acid, H_3PO_4 Commonly called phosphoric acid. Its structure is $(HO)_3P{=}O$.

Osmosis The flow of solvent through a **semipermeable membrane** from a solution of lower solute concentration into a solution of greater solute concentration. The amount of pressure that must be applied to the more concentrated solution just to prevent solvent flow through the membrane is called the **osmotic pressure.**

Ostwald Dilution Law The expression for the dissociation constant of a weak molecular acid or base in terms of the degree of dissociation of that species, α, and the formality of the solution, C. The expression is

$$K_a = \alpha^2 C/(1 - \alpha) \text{ or } K_b = \alpha^2 C/(1 - \alpha)$$

Ostwald Process A commercial process for the manufacture of nitric acid. The first step in the Ostwald process is the oxidation of ammonia by O_2 at high temperature, using a platinum catalyst, to produce $NO(g)$.

Oxidant or Oxidizing Agent A substance that oxidizes some other species and is itself reduced.

Oxidation The process in which a substance loses electrons and the oxidation state of some element in the substance increases.

Oxidation Potential The negative of a reduction potential.

Oxidation–Reduction Reaction An electron-transfer reaction, in which one substance loses electrons and is oxidized, while some other substance gains electrons and is reduced.

Oxidation State or Oxidation Number For a monatomic ion, the oxidation state is the charge on the ion. For a covalently bonded atom, the oxidation state is the charge on an atom calculated by assigning both electrons of a shared pair to the more electronegative atom. The oxidation state is a formalism; a useful device for counting electrons lost or gained in an oxidation–reduction reaction.

Oxidation State Method A method of balancing redox equations. The number of electrons lost or gained in a half-reaction is determined by multiplying the change in oxidation state per atom by the number of atoms undergoing that change. Errors are found by checking the charge balance.

Oxides Compounds containing the **oxide ion,** O^{2-}.

Oxyanions Polyatomic anions containing oxygen in addition to at least one other element. Among the most common oxyanions are nitrate (NO_3^-), sulfate (SO_4^{2-}), and carbonate (CO_3^{2-}) ions.

Oxygen Family The chalcogens: O, S, Se, Te, and Po. Group VIA of the periodic table.

Ozone, O_3 An allotropic form of oxygen found in the upper atmosphere.

Packing Material The material placed inside a **distillation column** to provide many surfaces on which the vapor can condense as it rises through the column. The packing material may be glass beads or coils, steel wool, or a twisted steel gauze.

Paired Electrons Two electrons with opposite spin occupying a single atomic orbital.

Paraffins Another term for the **alkanes.**

Parallel Spin Electrons having the same value of the

spin quantum number m_s are said to have parallel spin.

Paramagnetic Having one or more unpaired electrons. A paramagnetic substance is attracted to, and magnetized by, an external magnetic field, but loses its magnetism when removed from the field.

Partial Pressure For a component in a mixture of gases, the pressure that component would exert if it alone occupied the volume of the container at the same temperature. To the ideal gas approximation, the partial pressure of the ith component, P_i, is given by $n_i(RT/V)$.

Particulate Consisting of particles; **corpuscular.**

Pascal (Pa) The SI unit of pressure. One pascal is one newton per square meter. One atmosphere is equal to 1.0132×10^5 Pa.

Pauli Exclusion Principle No two electrons in the same atom can have all four quantum numbers n, ℓ, m_ℓ, and m_s alike.

Penetration of an Atomic Orbital The fraction of time an electron in the orbital spends near the nucleus. The greater the penetration, the lower the energy of the orbital. For orbitals of the same n but different ℓ, the lower the value of ℓ the greater the penetration.

Peptide A molecule formed when amino acid molecules combine via peptide bonds. If only two amino acids combine, the compound is a **dipeptide.** A peptide bond is

$$\begin{array}{cc} O & H \\ \parallel & \mid \\ -C\!-\!N\!- \end{array}$$

Percentage by Weight For a component of a mixture or compound,

$$\left(\frac{\text{mass of component in sample}}{\text{mass of total sample}}\right) \times 100$$

Percentage Yield The percentage yield of the product of a chemical reaction is

$$\left(\frac{\text{actual yield}}{\text{theoretical maximum yield}}\right) \times 100$$

Period A horizontal row of the periodic table that terminates in one of the rare gases. There are seven periods, consisting of 2, 8, 8, 18, 18, 32, and 32 elements, respectively, but not all the elements of the seventh period are known.

Periodic Properties Physical or chemical properties of the elements that vary with atomic number in a regular, recurring pattern.

Periodic Table An arrangement of the elements in order of increasing atomic number. Elements with similar physical and chemical properties are listed in a vertical column.

Peroxides Compounds containing the **peroxide ion,** O_2^{2-}.

Phosphoglycerides or Phospholipids Fats in which one of the three —OH groups of glycerol is esterified to phosphoric acid or a partially esterified phosphoric acid, and the other two —OH groups are esterified to long-chain fatty acids. If the phosphate group is esterified only to glycerol, the compound is a **phosphatidic acid.**

Photochemical Reaction A reaction that is initiated by shining light, either visible or ultraviolet, on the reactants.

Photon A quantum of energy. Electromagnetic radiation of frequency v consists of a stream of packets of energy, or photons, of amount hv.

pH Scale A logarithmic scale for the $[H_3O^+]$. $pH = -\log[H_3O^+]$

Physical Properties Properties of a substance that can be observed without changing the identity of the substance, and that do not depend on its reaction with other substances.

Pi (π) Bond A bond formed by the sideways overlap of two or more p orbitals on different atoms.

Planck's Constant, h The proportionality constant between the energy of a photon and the frequency of the radiation. Its value in SI units is 6.6262×10^{-34} Joule · seconds (J · s).

Plane-Polarized Light Light for which the electromagnetic vibrations perpendicular to the direction of propagation of the light are in a single direction, so that there is only a single plane containing the ray and its electromagnetic vibrations. For ordinary light the electromagnetic vibrations are in all directions perpendicular to the direction of propagation of the light ray.

Polar Covalent Bond A bond in which the shared pair of electrons is shared unequally by the bonded atoms. Also called a covalent bond with partial ionic character.

Polarimeter An instrument for measuring the angle through which the plane of polarization of polarized light has been rotated after the light has passed through an optically active substance.

Polarization The deformation or distortion of the electron charge cloud of an atom or molecule that results in the separation of the centers of positive and negative charge.

Polarizer A substance that can pull or distort the electron cloud of a neighboring atom or ion towards itself. In general, cations polarize anions, and cations without rare gas electronic configuration are more effective polarizers than cations with the electronic configuration of one of the rare gases.

Polar Molecule A molecule that possesses a permanent **dipole moment.**

Polaroid A material capable of converting ordinary light into **plane-polarized light.**

Polydentate Ligand A ligand that has more than two donor atoms that bind to a metal ion, forming a **chelate.**

Polymerization The process in which many small molecules (the **monomers**) are joined to form a very large molecule, the **polymer.**

Polynuclear Aromatics Compounds composed of two or more fused benzene rings, sharing one or more sides.

Polypeptide A compound containing four or more amino acid molecules linked by the **peptide bond.**

Polyprotic Acid A substance with more than one acidic proton.

Polysaccharide A polymer of many glucose units.

Positive Deviation from Raoult's Law A term describing a solution for which the vapor pressure of a component is **greater** than predicted by Raoult's law. For a mixture of two liquids, A and B, a positive deviation from Raoult's law is due to a weaker force of attraction between the unlike molecules, A and B, than between the like molecules, A and A or B and B.

Positron An elementary particle, β^+, that has the same mass as an electron and a positive charge equal in magnitude to the electronic charge.

Positron Emission A process by which neutron-poor nuclei increase their neutron to proton ratio. A proton in the nucleus is converted to a neutron with emission of a positron and a neutrino.

Potential Energy For a particle or a system of particles, the potential energy is due to the position of the particle or to the relative distances between particles. **Potential energy** is to be contrasted with **kinetic energy,** which is the energy a body possesses due to its motion.

Potentiometer An instrument for measuring cell potentials.

Precipitate A solid substance that forms when two or more solutions are combined.

Precipitation The process in which a solid solute comes out of solution and settles to the bottom of the container as a separate solid phase.

Precipitation Reaction One in which an insoluble solid is formed when two or more solutions are mixed together.

Pressure The force exerted per unit area: $P = f/A$.

Primary Amine, R—NH$_2$ A molecule in which an organic group has been substituted for one of the hydrogens of ammonia.

Primitive or Simple Lattice One in which lattice points are located only at the eight corners of the unit cell.

Principal Quantum Number, n The quantum number that governs the overall size and the energy of an atomic orbital. Allowed values of n are 1, 2, 3, ..., ∞.

Probability Density or Probability Distribution The square of the wave function, $|\psi|^2$, plotted in three dimensions. Since it is impossible to know simultaneously both the position and the velocity of a particle in an allowed energy state, the probability distribution provides as much information as we can obtain about the position of the particle in space.

Promotion Energy The amount of energy required to bring (or **promote**) the outer electrons of an atom from the ground state of the isolated gaseous atom to a hypothetical valence state in which hybrid orbitals are occupied. If strong covalent bonds are formed using the hybrid orbitals, the resultant molecule is lower in energy than the isolated atoms by an amount greater than the promotion energy.

Proton A subatomic particle contained in the nucleus of an atom. A proton has a positive charge equal in magnitude to the charge on the electron. The mass of a proton is approximately one atomic mass unit, and is about 1836 times the mass of an electron.

Proton-Transfer Reaction (Acid–Base Reaction) A reaction in which one substance (the acid) donates a proton to a second substance (the base).

Pseudo-First-Order Reaction One that appears to be first order in reactant but actually depends on the concentration of the solvent as well. The dependence of the rate on the solvent concentration is not observed because the solvent concentration is so large compared to the concentrations of other reactants that it remains essentially constant throughout the reaction. Reactions in aqueous solution with rates that depend on $[H_2O]$ are often pseudo-first-order reactions.

$P–V$ Isotherm A plot of the pressure of a fixed quantity of a gas as a function of its volume, at constant temperature.

Pyridine A heterocyclic aromatic base, C_5H_5N.

Qualitative Analysis A branch of chemistry. It consists of a series of laboratory procedures used to identify the ions present in an unknown mixture.

Quantitative Reaction One that goes to completion, that is, virtually 100% of one or more of the reacting species is used up during the reaction.

Quantized Existing as multiples of a definite quantity. It is the opposite of being continuous.

Quantum Mechanics or Wave Mechanics The set of physical laws used to describe the behavior of atomic, subatomic, and molecular systems.

Quantum Number A number associated with some property of a system that is restricted to a set of discrete values, usually integers or half-integers.

Quaternary Ammonium Compound A compound in which each of the four hydrogen atoms of the ammonium ion has been replaced by an organic group.

Quicklime, CaO Calcium oxide.

Racemic Mixture A mixture containing equal amounts of a pair of enantiomers. A solution of a racemic mixture will not rotate the plane of plane-polarized light.

Radial Probability Distribution The probability of finding an electron at a given distance from the nucleus after averaging over all angular variables. It is the probability of finding the electron in an infinitesimally thin spherical shell of radius r around the nucleus, and is $4\pi r^2|\psi|^2$.

Radioactive Decay Series Chains of successive decays of radioactive species that begin with a parent nucleus of very long half-life. There are three naturally occurring decay series: the uranium, thorium, and actinium series.

Radioactive Isotope An isotope that spontaneously decays to become an isotope of a different element.

Radiocarbon Dating A technique for dating carbon-containing objects up to about 25,000 years of age, by measuring the β^- activity of ^{14}C.

Raoult's Law The vapor pressure of a volatile component of a solution is directly proportional to the mole fraction of that component, and the proportionality constant is the vapor pressure of the pure component at the same temperature. It is symbolized as $P_A = P_A^0 X_A^{liq}$.

Rare Earths The **lanthanides,** elements 58 to 71.

Rare Gases The noble gases: He, Ne, Ar, Kr, Xe, and Rn.

Rate of a Reaction The change in the concentration of one of the components with time. Because the rate is continually changing, it must be expressed as a first derivative with respect to time. For a reaction in which one of the reacting species, A, decreases with time, the rate is $-d[A]/dt$.

Rate of Condensation The number of molecules that leave the vapor phase and enter the liquid phase per unit time per unit surface area. The **rate of evaporation** is the number of molecules that enter the vapor phase and leave the liquid phase per unit time per unit surface area. When a liquid is in equilibrium with its vapor, the rate of evaporation is equal to the rate of condensation.

Rate-Determining Step In a complex mechanism, the slowest step. Sometimes two or more steps are much slower than all others and there is more than one rate-determining step.

Reaction Coordinate A measure of how far a reaction has proceeded toward completion, that is, toward the equilibrium state.

Reaction Quotient, Q The same function as the equilibrium constant, but evaluated using any given set of concentrations of the reactants, and not necessarily equilibrium concentrations.

Redox Couple An oxidizing agent and a reducing agent related by the transfer of one or more electrons.

Reductant or Reducing Agent A substance that reduces some other species and is itself oxidized.

Reduction The process in which a substance gains electrons and the oxidation state of some element in the substance decreases.

Reduction Potential A value, in volts, that compares the tendency of any half-reaction,

$$ox + ne^- \rightarrow red$$

to proceed to the right with the tendency of the reaction

$$2H^+(aq) + 2e^- \rightarrow H_2(g)$$

to proceed to the right. The larger the value of the reduction potential, the greater the tendency of the reduction half-reaction to proceed. A positive value of a reduction potential means that the oxidized form is a stronger oxidizing agent than $H^+(aq)$. A negative value of the reduction potential means that the reduced form is a stronger reducing agent than $H_2(g)$.

Refraction The bending of a ray of light when it passes from one medium (such as air) into another (such as a prism). The amount refracted depends on the frequency of the radiation.

Relative Uncertainty The ratio of the absolute uncertainty of a measured quantity to its numerical value. See Appendix C.

Representative Elements Those elements in which the *s* and *p* orbitals are being filled: Groups IA through VIIA and the rare gases.

Representative Metals The alkali and alkaline earth metals, plus the metallic elements of Groups IIIA, IVA, and VA.

Residue in a Fractional Distillation The material left behind in the pot. The residue consists of the less volatile of two liquids in the distillation of a binary liquid mixture.

Resonance Energy of Benzene The difference between the (theoretical) energy of a hypothetical Kekulé structure and the energy of an actual benzene molecule.

Resonance Hybrid A description of a molecular structure for which there is no single Lewis structure that is a correct description of the bonding but there are two or more Lewis structures that satisfy the octet rule.

Reversible Process One that occurs in a series of infinitesimal steps. At each step the system is at equilibrium with its surroundings. By changing some variable by an infinitesimal amount, the process can be reversed.

Root-Mean-Square Speed, u_{rms} The square root of the mean-square speed in a sample of a gas. Although the velocity of each molecule is constantly changing, the root-mean-square speed is a constant at constant temperature, and is $(3RT/M)^{1/2}$.

Rounding The process of discarding digits that are not significant in a number. If the last retained digit is increased by 1, the number has been **rounded up.** If the last digit retained is left unchanged, the number has been **rounded down.**

Rydberg Constant, $\mathcal{R}$ The constant 109,678 cm^{-1} in the Balmer formula for the wavelengths or wavenumbers of all the radiation in the emission spectrum of atomic hydrogen.

Salt An electrolyte whose cation is anything other than H^+ ion, and whose anion is anything other than OH^- ion.

Salt Bridge A device for maintaining electrical contact between two separate electrolytic solutions that are components of a galvanic cell. It consists of a U-tube containing a solution of an electrolyte dissolved in agar–agar. One end of the inverted U-tube is immersed in each solution.

Saturated Solution A solution in equilibrium with excess undissolved solute. The equilibrium is **dynamic,** and the two opposing reactions are (1) dissolution of solute into the solution, and (2) precipitation of solute out of solution. The concentration of solute in a saturated solution is a constant at constant temperature.

Schrödinger Wave Equation The fundamental physical law describing the behavior of atomic, subatomic, and molecular systems. It is a differential equation and its solutions are called **wave functions.**

Scientific Notation Another name for **exponential notation.**

Secondary Amine, $\overset{R'}{\underset{R}{>}}NH$ A molecule in which two organic groups have been substituted for two of the three hydrogens of ammonia.

Second Ionization Energy, $(IE)_2$ The amount of energy required to remove an electron from a unipositive gaseous ion to produce a dipositive gaseous ion. It is ΔH for $X^+(g) \rightarrow X^{2+}(g) + e^-$.

Second Law of Thermodynamics In an isolated system, the direction of all spontaneous processes is that which increases the entropy or molecular disorder of the system.

Second-Order Reaction One in which the sum of the exponents of all concentration terms in the rate expression is 2. There are two types of second-order reactions: (1) $-d[A]/dt = k_2[A]^2$ and (2) $-d[A]/dt = k'_2[A][B]$.

Second Series of Transition Metals The 10 elements yttrium through cadmium ($Z = 39$ to 48) in which the 5 inner $4d$ atomic orbitals are being filled.

Self-Ionization A reaction that occurs in hydrogen bonded liquids in which a proton shifts from one molecule to an identical molecule along a hydrogen bond.

Semimetal A metalloid.

Semipermeable Membrane One with pores of such a size that small molecules, like H_2O, pass through the membrane, but larger molecules cannot pass through.

Shared Pair Two electrons of opposite spin located in the region between two nuclei and constituting a covalent bond between the two atoms.

Shell Model of the Nucleus Neutrons and protons exist in separate nuclear energy levels that are filled when occupied by either 2, 8, 20, 28, 50, 82, or 126 nucleons.

Shells (Electronic) Atomic orbitals are divided into shells according to the principal quantum number, n. For the K shell, $n = 1$; for the L shell $n = 2$, and so on, in alphabetical order.

Shielding The screening of the nuclear charge for an electron in a many-electron atom by other electrons that have greater penetration to the nucleus. A valence electron is shielded by inner electrons and experiences an effective nuclear charge much lower than the actual nuclear charge.

Short-Range Forces Forces that become significant only when the distance between two particles is very small, that is, when the particles are almost in contact.

Sigma (σ) Bond An electron charge cloud centered around two nuclei that is completely symmetric with respect to rotation by any angle about the internuclear axis.

Sigma 1s ($\sigma 1s$) Molecular Orbital The bonding MO formed by combining the $1s$ atomic orbitals of two different atoms.

Significant Figures Those that are known with certainty, plus one additional digit that has experimental uncertainty.

Simplest Formula of a Compound A formula that indicates only the relative number of atoms of each type in the compound, and specifies only the simplest molar ratios in which the atoms combine.

Single Covalent Bond A single pair of electrons shared by two atoms.

SI Units An expanded and modernized form of the metric system, adopted at an international conference in 1960.

Slaked Lime $Ca(OH)_2$ Calcium hydroxide. So-called because it is the product of adding water to lime: $CaO(s) + H_2O \rightarrow Ca(OH)_2(s)$.

Solubility Product An equilibrium constant for the equilibrium between an insoluble or slightly soluble electrolyte and a saturated aqueous solution containing the ions of that electrolyte. The reaction is always written with the insoluble or slightly soluble solid on the left-hand side.

Solubility Product Principle The solubility product of an insoluble or slightly soluble electrolyte is a constant at constant temperature.

Solute A substance in relatively small concentration in a solution; the **solvent** is the substance in largest concentration. A **solution** is a homogeneous mixture of two or more substances, having various possible proportions of the constituents, which may be solids, liquids, or gases.

Solvated Ion An ion surrounded by closely associated solvent molecules. If the solvent is water, a solvated ion is a hydrated ion. Other dipolar solvents are also attracted to ions by ion-dipole forces.

Specific Activity The activity per gram of a radioactive substance.

Specific Heat The amount of heat required to raise the temperature of one gram of a substance one kelvin, or one degree Celsius.

Specific Rate Constant, k The number in the rate expression: rate = $k[A]^m[B]^n$. The value of the specific rate constant depends on the particular reaction, the temperature, the concentration of catalyst, and the solvent if the reaction occurs in solution. For a radioactive nucleus, the specific rate constant or decay constant is the number, λ, in the rate expression: $-dN/dt = \lambda N$, where N is the number of radioactive nuclei.

Spectator Ion An ion that is present in a solution in which a chemical reaction is occurring, but takes no part in the reaction. Spectator ions are also called **bystander ions.**

Spectrochemical Series A listing of ligands in the order of the magnitude of the crystal field splitting they produce, for complexes with a given geometry and a given metal ion.

Spectrum The series of images formed when a beam of radiation is passed through a slit and then dispersed so that the component waves are separated and arranged in order of their frequencies.

Speed of Light, c The magnitude of the velocity with which all electromagnetic radiation travels through a vacuum. Its value is 2.997925×10^8 m/s.

sp (Diagonal) Hybrids Atomic orbitals formed by combining the s and one of the p valence atomic orbitals of the same atom. The two sp hybrids are directed at 180° to one another.

sp^2 (Trigonal) Hybrids Atomic orbitals formed by combining the s and two p valence atomic orbitals of the same atom. The three sp^2 hybrids lie in a plane and are directed at 120° to one another.

sp^3 (Tetrahedral) Hybrids Atomic orbitals formed by combining the s and three p valence atomic orbitals of the same atom. The four sp^3 hybrids make angles of 109°28′ with one another.

Spin-Only Magnetic Moment The magnetic moment due only to unpaired electron spins, and not to the orbital angular momentum of the electrons. For n unpaired electrons, its value is $\sqrt{n(n+2)}$.

Spin Quantum Number, m_s The quantum number that determines the direction or orientation of the intrinsic magnetic moment of the electron. The value of m_s can be either $+\frac{1}{2}$ or $-\frac{1}{2}$.

Spin State α The state in which the electron spin has spin quantum number $m_s - \frac{1}{2}$. An electron in **spin state β** has spin quantum number $m_s = -\frac{1}{2}$.

Square-Based Pyramidal Geometry One of the limiting geometries for coordination number 5. Four ligand donor atoms are at the corners of a square, the metal ion is located slightly above the center of the square, and the fifth ligand is on the perpendicular to the square, directly above the metal.

Square-Planar Geometry One of the limiting geometries for coordination number 4. The four ligand donor atoms are at the corners of a square, with the metal atom at the center.

Staggered Conformation of Ethane The one in which the dihedral angle is 60°, and the C—H bonds are as far apart as possible. It is the conformation of lowest potential energy.

Standard Cell A galvanic cell in which all substances are in their **standard states,** that is, all species in solution are at 1 M concentration and all gases are at a partial pressure of 1 atm. The temperature must be specified but is usually 25 °C.

Standard Enthalpy of Formation, ΔH_f^0 The amount of heat absorbed at constant pressure when 1 mol of a compound is formed from its elements in their standard states at 1 atm and 25 °C.

Standard Free Energy of Formation, ΔG_f^0 The change in the Gibbs free energy when 1 mol of a compound is formed from its elements in their standard states at 1 atm and 25 °C.

Standard Hydrogen Electrode (SHE) A platinum electrode over which H_2 gas at 1-atm pressure is bubbled, while the electrode is immersed in a solution that is 1.00 M in hydronium ion, at 25 °C.

Standardized Solution One for which the concentration of solute has been accurately and quantitatively determined, usually to four significant figures.

Standard State of an Element The most stable form of the element at 25 °C and 1-atm pressure. A pure solid or a pure liquid is, by definition, in its standard state. A gas is in its standard state if its pressure is 1 atm. A species in solution is in its standard state if its concentration is 1.00 M. Any temperature can be chosen to be the standard state temperature, but in tables of thermodynamic functions the standard state temperature is 25 °C.

Standing Waves The modes of vibration of a plucked string that is tied down at both ends.

State Function A property of a system that has a fixed and definite value for each state of a system. When the state of a system is changed, the value of the change in any state function depends only on the initial and final states and not on the path.

State of a System A term describing a system for which the values of a few state functions are specified. For instance, the state of a system may be defined by specifying the temperature, pressure, mass, and chemical composition.

Stationary State An allowed state of a system in which the energy has a fixed and definite value.

Steady State Distribution of Gas Velocities A distribution in which the fraction of molecules with speeds between any two specified values remains constant with time, even though the velocity of each gas molecule is continually changing due to collisions with other molecules and the walls of the container.

Stoichiometric Compounds (Daltonides) Compounds with a fixed and definite atomic composition.

Stoichiometry The study of the quantitative relationships between the reactants and products of a chemical reaction.

Straight-Chain (Normal) Alkane A molecule of general formula C_nH_{2n+2} in which each carbon atom is bonded to no more than two other carbon atoms.

Strong Electrolyte One that exists as virtually 100% ions in dilute aqueous solution. Most acids and bases are weak electrolytes; the few exceptions are called **strong acids** and **strong bases.**

Strong Ligand Field One that produces a large crystal field splitting, and therefore results in a **low-spin complex.**

Subatomic Particles The particles of which all atoms are composed. There are a great many subatomic particles, but the three of greatest importance in chemistry are the **electron,** the **proton,** and the **neutron.**

Subcritical Mass of Fissionable Material A mass small enough so that most neurons produced by a fission process escape from the sample without striking other fissionable nuclei.

Sublimation The conversion of a solid directly to its vapor, without the formation of liquid.

Subshells In a shell of atomic orbitals, orbitals with the same principal quantum number n, but different values of ℓ. For example, the L shell has two subshells, the $2s$ and the $2p$.

Substitutional Impurity A defect in a crystal in which an atom or ion other than the constituents of the substance composing the crystal occupies an atomic site.

Supercooled Liquid One that has been cooled to

below the freezing point of the substance but has not crystallized. Amorphous solids are supercooled liquids that appear rigid but actually continue to flow extremely slowly.

Supercritical Mass of Fissionable Material A mass large enough so that most neutrons produced by a fission process strike other fissionable nuclei, causing additional fissions, before they can escape from the sample. As the number of fissions increases rapidly in a supercritical mass, a violent explosion occurs.

Superheated Liquid One that has been heated to a temperature above its boiling point, but has not begun to boil because gaseous bubbles of sufficient size have not formed in the body of the liquid.

Superoxides Compounds containing the **superoxide ion, O_2^-.**

System In thermodynamics, a portion of space with clearly defined boundaries containing the substances being studied or investigated. Everything outside those boundaries is called the **surroundings.**

Table of Standard Electrode Potentials A list of reduction potentials of standard half-cells relative to the reduction potential of the standard hydrogen electrode, which is, by definition, exactly zero volts.

Temperature An intrinsic property of a system. It is a measure of the average kinetic energy of the molecules of the system.

Termolecular Describing an elementary process that involves the reaction of three molecules.

Tertiary Amine A molecule in which three organic groups have been substituted for the three hydrogens of ammonia.

Tetrahedral Geometry A molecular geometry in which four atoms or groups are bonded to a central atom. The bonded atoms are at the corners of a regular tetrahedron. A majority of coordination complexes with four ligands have tetrahedral geometry.

Tetrahedral (sp^3) Hybrids Atomic orbitals formed by combining the s and three p valence orbitals of the same atom.

Tetravalent Forming four bonds to other atoms. Carbon is almost always tetravalent.

Thermal Neutrons Slow-moving neutrons with a kinetic energy of about the same magnitude as a typical gas molecule at room temperature.

Thermodynamically Controlled Reaction One for which the products are those with lowest Gibbs free energy. This is to be contrasted with a **kinetically controlled reaction.**

Thermonuclear Reaction A nuclear fusion reaction. The term thermonuclear is used because fusion reactions require extremely high temperatures.

Third Law Entropy See **absolute entropy.**

Third Law of Thermodynamics The entropy of a perfect crystal of all pure elements and compounds is zero at absolute zero.

Three-Center Bond (Bridge Bond) A bond in which a single pair of electrons bonds three atoms.

Thorium Series A radioactive decay series with parent nucleus thorium-232. The stable end product is lead-208. All isotopes in the series have mass numbers that are integral multiples of 4, so it is also called the $4n$ series.

Titrant A solution of known concentration, added from a buret during a titration, to a solution containing a substance of unknown concentration that reacts with the titrant.

Titration Curve (Acid–Base) A plot of the pH of the solution versus the volume of added titrant. Because there is a sharp change in pH at the equivalence point, the volume of titrant required to reach the equivalence point can be determined precisely.

t_{2g} Level Of the two levels into which the five d orbitals are split by an octahedral ligand field, the lower energy, triply degenerate level consisting of the d_{xy}, d_{yz}, and d_{xz} orbitals, is the t_{2g} level.

Torr A unit of pressure, named to honor Torricelli, equal to the pressure that can support a column of mercury exactly 1-mm high. A torr is therefore 1/760 of an atmosphere.

Tracer Studies Experiments carried out to elucidate the mechanism of a reaction by labeling one of the reactants with a radioactive isotope and observing in which product molecules the radioactivity appears.

***Trans* Geometric Arrangement** A structure in which identical atoms or groups are opposite one another.

Transition Elements Three groups of 10 elements each, which correspond to the filling of the 3d, 4d, and 5d atomic orbitals. The first series comprises elements 21 through 30 (Sc to Zn); the second series comprises elements 39 through 48 (Y to Cd); and the third series comprises elements 71 through 80 (Lu to Hg).

Transition State Another term for the **activated state.**

Translational Motion Motion in which the center-of-mass of a body changes its position in space. Gas molecules are translating through the space of their container. Translational motion is to be contrasted with vibrational and rotational motion, in which the center-of-mass of a body does not change its position in space.

Transuranium Elements Those with $Z > 92$. All transuranium elements are man-made and radioactive.

Triclinic Unit Cell A unit cell of a crystal in which the three axial lengths are unequal and none of the axes is perpendicular to any other.

Triester A molecule in which all three —OH groups on a **triol** have been esterified, that is, replaced by —OR groups, where R is an organic substituent.

Trigonal Bipyramidal Geometry A structure in which a central atom is bonded to five other atoms. Three of the bonded atoms lie in a plane (the equatorial plane) at the corners of an equilateral triangle surrounding the central atom. The other two bonded atoms lie on an axis perpendicular to the equatorial plane. Trigonal bipyramidal geometry is one of the limiting geometries for coordination number 5.

Trigonal Hybrids sp^2 Hybrid atomic orbitals.

Trigonal Planar Geometry A structure in which a central atom is bonded to three other atoms located at the corners of an equilateral triangle surrounding the central atom. All four atoms lie in a single plane.

Triol An organic molecule having three —OH groups.

Triple Bond The bond between two atoms sharing three pairs of electrons.

Triple Point A unique pressure and temperature at which the solid, liquid, and vapor phases of a single pure substance are all in equilibrium.

Triprotic Acid A substance with three acidic protons. The most common triprotic acid is orthophosphoric acid, H_3PO_4.

Triton The nucleus of **tritium**, an isotope of hydrogen with two neutrons and one proton, ${}^3_1H^+$ or T^+.

Trouton's Rule A generalization based on the fact that the increase in molecular disorder when a mole of liquid vaporizes at its normal boiling point is about the same for all substances. Trouton's rule states that the entropy of vaporization of most liquids is 88 ± 5 J · mol^{-1}K^{-1} at the normal boiling point. Hydrogen bonded liquids have significantly larger entropies of vaporization than predicted by Trouton's rule.

Uncertainly Principle It is impossible to determine simultaneously exact values of both the position and the linear momentum (or velocity) of a particle. The uncertainty principle, first put forth by Werner Heisenberg, is one of the fundamental concepts of quantum mechanics.

Unimolecular Describing an elementary process that involves only a single molecule. The decay of a radioactive nucleus is unimolecular.

Unit Cell The smallest parallelepiped from which a crystal can be constructed by translational repetition along the crystal axes. Any crystal can be considered to be constructed by stacking together an infinite number of unit cells in three dimensions, in a regular, repeating pattern.

Universal Gas Constant, R The proportionality constant in the ideal gas law, $PV = nRT$. The numerical value of R depends on the units chosen to measure P and V. If P is in atmospheres and V in liters, $R = 0.082057$ L · atm · mol^{-1}K^{-1}.

Unsaturated Solution A solution able to dissolve more solute than is at present in solution, at the same temperature.

Uranium Series A naturally occurring radioactive decay series with parent nucleus uranium-238. The stable end product is lead-206. All isotopes in this series have mass numbers that can be described as $4n + 2$, where n is integral, so this is also called the $4n + 2$ series.

Vacancy A site in a crystal that should be occupied by an atom or ion, but is empty. A vacancy is a common type of defect in crystals.

Valence Electrons The outermost electrons of an atom, those that, on the average, are farthest from the nucleus. Atoms of different elements that have the same number of valence electrons have similar chemical properties.

Valence-Shell Electron-Pair Repulsion (VSEPR) Theory A theory that the geometry of a molecule or ion is the structure that minimizes the repulsion between pairs of electrons in the valence shell of the central atom.

van der Waals' Equation An equation of state relating the pressure, volume, temperature, and number of moles of a gas. It involves two empirical constants, a and b, and is designed to fit experimental values at high pressures and low temperatures better than the ideal gas law does. It is

$$\left(P + \frac{na}{V^2}\right)(V - nb) = nRT$$

van der Waals Forces Weak attractive forces between uncharged atoms or molecules. Both **dispersion forces** and **dipole-dipole interactions** are van der Waals forces.

van't Hoff Equation The relation between temperature and the numerical value of an equilibrium constant for a reaction, for a temperature range sufficiently small so that ΔH for the reaction is approximately constant. The equation is

$$\ln\left(\frac{K_1}{K_2}\right) = -\frac{\Delta H^0}{R}\left(\frac{T_2 - T_1}{T_1 T_2}\right)$$

van't Hoff Equation for Osmotic Pressure The osmotic pressure is directly proportional to the concentration of solute in moles per liter, and the proportionality constant is RT. Thus the equation is $\pi = cRT$.

van't Hoff Mole Number The ratio $\Delta T_f/K_f m$ or $\Delta T_b/K_b m$. For an electrolytic solute, the van't Hoff mole number is somewhat less than the number of ions per formula unit, but approaches that value more and more closely the more dilute the solution.

Vaporization The conversion of a liquid to its vapor.

Vapor Pressure Lowering, ΔP The difference between the vapor pressure of a pure solvent and the vapor pressure of a solution prepared using that solvent. It is a positive quantity, as the vapor pressure of a solution of a nonvolatile solute is always *lower* than the vapor pressure of the pure solvent.

Vapor Pressure of a Solution For a nonvolatile solute in a volatile solvent, the vapor pressure of the solution is the pressure of solvent in equilibrium with the solution. The vapor pressure of a solution of two volatile liquids is the sum of the vapor pressures of the two components in the vapor phase in equilibrium with the solution.

Vector A quantity having both magnitude and direction. Velocity is a vector quantity.

Volatile Having a measurable vapor pressure. Liquid A is more volatile than liquid B if the vapor pressure of liquid A at a given temperature is higher than the vapor pressure of liquid B at that same temperature.

Volt A unit of electrical potential. One volt corresponds to a potential difference of one joule per coulomb.

Voltaic Cell Another term for a **galvanic cell.**

Volumetric Flask A piece of glassware designed to contain a specified volume of liquid at a specified temperature, usually 20 °C. A **volumetric pipet** is a piece of glassware designed to deliver a specified volume of liquid at a specified temperature, usually 20 °C. A volumetric pipet is used to transfer a specific volume of liquid from one container to another, and is also called a **transfer pipet.**

Wave Equation A differential equation describing the motion of any kind of wave: sound wave, light wave, ocean wave, etc.

Wave Function A solution of the Schrödinger wave equation.

Wavelength, λ The distance between successive maxima (or successive minima) of a wave.

Wave Mechanics Another name for **quantum mechanics.**

Wavenumber, $\tilde{\nu}$ The reciprocal of the wavelength: $\tilde{\nu} = 1/\lambda$.

Weak Electrolyte An electrolyte for which the extent of dissociation into ions is significantly less than 100%, even in dilute aqueous solution. Most acids are **weak acids,** and most bases are **weak bases.**

Weak Ligand Field One that produces a small crystal field splitting and therefore results in a **high-spin complex.**

Weight The force of attraction exerted on a body by gravity. It is equal to mg, where m is mass and g is the acceleration of gravity.

Weight Percentage For a component in a mixture, the weight of that component divided by the total weight of the mixture, multiplied by 100 to make it a percentage.

Work The product of a force and the distance through which the force operates.

Work of Expansion The work done when the volume of a system changes while a confining pressure is being applied. For a constant external pressure, the work of expansion is $P_{\text{ext}}\Delta V$.

X-ray Diffraction An experimental technique for determining interatomic distances and bond angles in molecules and ions in the crystalline state. In this technique, X-rays impinge on a crystal that is slowly rotated to change the angle of incidence between the X-ray beam and planes of atoms in the crystal. By measuring those angles for which constructive interference occurs and the emerging beam has a measurable intensity, both the molecular and crystal geometries can be determined.

Zeeman Effect The splitting of lines in an atomic spectrum into a multiplet of lines, when an external magnetic field is applied.

Zwitterion (Inner Salt) A form of an amino acid molecule in which the carboxyl proton has been donated to the amino group. The general form of a zwitterion is $\overset{+}{N}H_3-\underset{\underset{R}{|}}{C}H-COO^-$.

answers to even numbered questions

Problems, introduction

I.2. (a) 16.00, 16.0, 16; (b) 1.007, 1.01, 1.0; (c) 2.876×10^4, 2.88×10^4, 2.9×10^4; (d) 2.604×10^6, 2.60×10^6, 2.6×10^6; (e) 2.045×10^{-3}, 2.05×10^{-3}, 2.0×10^{-3}. **I.4.** (a) 2.008×10^{-4}, 4 sig. figures; (b) 2.0772×10^7, 5 sig. figures; (c) 1.0570×10^{-2}, 5 sig. figures; (d) 7.03×10^3, 3 sig. figures. **I.6.** (a) 297 K; (b) 298.2 K; (c) 300.0 K; (d) 297.50 K; (e) 350 K. **I.8.** $1 \, g \cdot L^{-1} = 1 \, kg \cdot m^{-3}$. **I.10.** 235.6 mL or 0.2356 L. **I.12.** 505 g. **I.14.** $3.0 \, m \cdot s^{-1}$. **I.16.** 25.055 mL. **I.18.** (a) 2.61×10^{-3} and (b) 5.788×10^{-3}. **I.20.** (a) 351.6 K; (b) 388 K; (c) 398.81 K. **I.22.** $0.8871 \, g \cdot mL^{-1}$. **I.24.** 9.74 g. **I.26.** $-263 \, °F$, 109 K. **I.28.** $-309.44 \, °F$, 83.46 K.

Exercises, chapter 1

2. 80 protons and 80 electrons in all isotopes. The number of neutrons in the seven isotopes is 116, 118, 119, 120, 121, 122, 124. **4.** Be 2, Mg 10, Ca 18, Sr 36, Ba 54. **6.** $5.8872 \times 10^{-23} \, g \cdot atom^{-1}$. There are two chlorine isotopes, ^{35}Cl and ^{37}Cl. **8.** 1.82600. **10.** 2.546×10^{-2} mol. **12.** 0.4891 g. **14.** Li^+, Na^+, K^+, Rb^+, Cs^+. **16.** 1.182 mol SO_4^{2-} ions, 3.546 mol H_2O. **18.** C_6H_6. **20.** $(C_3H_8)_n$, C_3H_8. **22.** 4.489 g. **24.** 0.3262 mol, 51.55 g. **26.** 33.601 g. **28.** C_3H_8, 8.264 g.

Multiple Choice Questions, chapter 1

2. (a) **4.** (c) **6.** (d) **8.** (d) **10.** (c) **12.** (d) **14.** (e) **16.** (d) **18.** (e) **20.** (a)

Problems, chapter 1

1.2. $24.31 \, g \cdot mol^{-1}$. **1.4.** $(Gd_2O_3)_n$. **1.6.** (a) 0.4560 mol; (b) 1.848×10^{24} atoms; (c) 641 mL; (d) 135.21-g amphetamine. **1.8.** $173.05 \, g \cdot mol^{-1}$. **1.10.** 15.999. **1.12.** (a) 315 C; (b) 3.26×10^{-3} mol; (c) 9.66×10^4 C; (d) 6.03×10^{23}. **1.14.** 94.79%. **1.16.** (a) 0.18868 g H, 3.748 g C, 0.874 g N; (b) $(C_5H_3N)_n$; (c) $C_{10}H_6N_2$, $154.171 \, g \cdot mol^{-1}$. **1.18.** 47.327%, 7.6332 g.

Multiple Choice Questions, chapter 2

2. (c) **4.** (a) **6.** (e) **8.** (a) **10.** (c) **12.** (b) **14.** (b) **16.** (a)

Problems, chapter 2

2.2. (a) strontium nitrate; (b) silver carbonate; (c) cobalt(II) chloride or cobaltous chloride; (d) cobalt(III) chloride or cobaltic chloride; (e) gold(I) cyanide or aurous cyanide; (f) ammonium phosphate; (g) potassium hydrogen sulfate or potassium bisulfate; (h) sodium sulfide; (i) calcium hydrogen sulfite or calcium bisulfite; (j) copper(I) bromide or cuprous bromide; (k) iron(III) sulfate or ferric sulfate; (l) iron(II) sulfate or ferrous sulfate; (m) sodium sulfite; (n) potassium permanganate; (o) copper(II) iodide or cupric iodide. **2.4.** (a) CaC_2O_4; (b) $KHCO_3$; (c) $Ba(NO_2)_2$; (d) NH_4MnO_4; (e) $LiNO_3$; (f) $Al(ClO_4)_3 \cdot 6H_2O$; (g) $Na_2S_2O_3$; (h) Rb_2O_2; (i) Cu_2O; (j) $Co_2(SO_4)_3$; (k) Fe_2O_3; (l) $Sr(OH)_2$; (m) CrF_2; (n) $CuSeO_4 \cdot 5H_2O$.
2.6. (a) $2K(s) + 2H_2O \rightarrow 2K^+ + 2 OH^- + H_2(g)$
   (b) $2K(s) + O_2(g) \rightarrow K_2O_2(s)$ and $K(s) + O_2(g) \rightarrow KO_2(s)$
   (c) $Sr(s) + 2H_2O \rightarrow Sr^{2+} + 2 OH^- + H_2(g)$
   (d) $Sr(s) + \frac{1}{2} O_2(g) \rightarrow SrO(s)$ and $Sr(s) + O_2(g) \rightarrow SrO_2(s)$
2.8. (a) 118; (b) 111; (c) thallium.

2.10. (a) $Cl_2(g) + H_2O \rightarrow H^+(aq) + Cl^-(aq) + HOCl(aq)$
(b) $Cl_2(g) + H_2(g) \rightarrow 2HCl(g)$
(c) $Cl_2(g) + Sr(s) \rightarrow SrCl_2(s)$
(d) $H_2(g) + \frac{1}{2}O_2(g) \rightarrow H_2O(\ell)$
2.12. (a) Bi, Sb, As, P, N and (b) Li, B, N, O, F.
2.14. $Mg(s) + \frac{1}{2}O_2(g) \rightarrow MgO(s)$
$3Mg(s) + N_2(g) \rightarrow Mg_3N_2(s)$
$Mg(s) + H_2(g) \rightarrow MgH_2(s)$
$Mg(s) + Cl_2(g) \rightarrow MgCl_2(s)$
2.16. $NH_4NO_3(s) \rightarrow 2H_2O(g) + N_2O(g)$

Exercises, chapter 3

2. (a) false and (b) false. **4.** 6.00 L. **6.** 4/1. **8.** 0.065 mmHg or 8.5×10^{-5} atm.
10. 5.5 °C. **12.** 3.32×10^{-3} mol. **14.** 24.5 L/mol. **16.** (a) 1.02 g/L and (b) 0.16 g.
18. 174 °C. **20.** 750 mmHg. **22.** (a) 0.495 and (b) 0.722 atm.

Multiple Choice Questions, chapter 3

2. (2) **4.** (a) **6.** (d) **8.** (b) **10.** (b) **12.** (c) **14.** (e) **16.** (c) **18.** (a)

Problems, chapter 3

3.2. (a) false and (b) false. **3.4.** 354.0 mL. **3.6.** (a) 227 g/mol; (b) 1; (c) $SbCl_3$;
(d) 228.11 g/mol. **3.8.** 73.6% Al. **3.10.** 6.06×10^{23}.
3.12. (a) $T = (742.0)(0.900)/(0.0375)(62.36)$ (2 errors).
(b) $M = (2.08)(0.082057)(423)(760)/(756)(0.350)$ (4 errors).
(c) Caspar omitted the phrase "provided that the gases are at the same temperature and pressure."
(d) Boyle's law does not apply here. $P_2 = 2.00$ atm.
(e) $n = (300/400)(0.0600)$ as $n_1T_1 = n_2T_2$.
(f) $(3)(745)(0.420)/(2)(62.36)(303)$ mol (3 errors).
3.14. (a) 2.500×10^{-3}; (b) 7.500×10^{-3}; (c) 3.750×10^{-3}; (d) 172 mL; (e) 272 mmHg.
3.16. (a) 1.101×10^{-2} mol and (b) 19.01%.

Exercises, chapter 4

2. The gas will be liquefied. **4.** The volume occupied by the molecules themselves is a significant fraction of the volume of the container. **6.** All collisions the molecules make are elastic collisions. **8.** 250 K. **10.** 821 J; 4.54×10^{-21} J. **14.** No, 1.66 times as fast as CO_2. **16.** SO_2.

Multiple Choice Questions, chapter 4

2. (c) **4.** (b) **6.** (e) **8.** (e) **10.** (a) **12.** (b) **14.** (b)

Problems, chapter 4

4.2. 515.3 m/s or 1153 mile/h. **4.4.** expand. **4.6.** (a) false; (b) false; (c) true; (d) true;
(e) true. **4.8.** (a) true; (b) false; (c) false. **4.10.** (a) 2.00 L; (b) 96.0 J or 22.9 cal;
(c) 681 m/s; (d) 1:1. **4.12.** 378.6 °C. **4.14.** N_2.

Exercises, chapter 5

2. The four bond-dipole vectors sum to zero; SiF_4 is symmetric. **4.** Dispersion forces in $SiCl_4$ are greater than in SiH_3Cl. **6.** Water expands on freezing. **8.** Ethanol is hydrogen bonded, dimethyl ether is not. **12.** 50.87 kJ/mol. **14.** Intermolecular forces are smaller at 100 °C than at 20 °C. **18.** exothermic. **20.** 99.6 mL. **22.** (a) $\Delta S_{fus} > 0$ and (b) $\Delta S_{vap} > \Delta S_{fus}$. **24.** 600 mmHg. **28.** Less than. **30.** (a) Point A rhombic, monoclinic, and gaseous S; Point B liquid, monoclinic, and gaseous S; point C rhombic, monoclinic, and liquid S and (b) rhombic S is denser than monoclinic S.

Multiple Choice Questions, chapter 5
2. (b) **4.** (a) **6.** (a) **8.** (e) **10.** (d) **12.** (c) **14.** (a) **16.** (e) **18.** (d) **20.** (d)

Exercises, chapter 6
2. (a) 0.18500 M and (b) 3.29×10^{-2} M. **4.** 0.200 F. **6.** (a) $[Cu^{2+}] = 0.200$ M, $[SO_4^{2-}] = 0.0520$ M, $[K^+] = 0.064$ M and (b) $[Ba^{2+}] = 0.160$ M, $[NO_3^-] = 0.420$ M, $[NH_4^+] = 0.1000$ M. **10.** 0.185 g. **14.** (a) $X_B = 0.982$, $X_Q = 0.018$ and (b) 1.4×10^2. **16.** No, it is not ideal. A fraction of the phenol has dissociated. **18.** 1.1×10^2 g/mol. **20.** -0.430 °C. **22.** $C_6H_{12}Cl_2$, 155.068 g · mol^{-1}. **24.** (b), (a), (d), and (c). **26.** 0.263 atm or 200 mmHg. **28.** 82.0 mmHg. **30.** (a) $\underline{D}$ obeys Raoult's Law, $\underline{C}$ does not and (b) positive deviations.

Multiple Choice Questions, chapter 6
2. (b) **4.** (c) **6.** (e) **8.** (d) **10.** (d) **12.** (d) **14.** (a) **16.** (c) **18.** (d) **20.** (c)

Problems, chapter 6
6.2. $C_{10}H_{16}$, 136.236 g/mol. **6.4.** $[NH_4^+] = 0.106$ M, $[SO_4^{2-}] = 0.1080$ M, $[Cl^-] = 0.15$ M, $[Zn^{2+}] = 0.040$ M. **6.6.** (b) Negative deviations and (d) $\Delta H_{mix} < 0$. **6.8.** $[Pb^{2+}] = 1.52 \times 10^{-3}$ M, $[I^-] = 3.04 \times 10^{-3}$ M. **6.10.** 776 g · mol^{-1}. **6.12.** (a) 0.2000 m; (b) 0.01538; (c) 80.7 °C; (d) 4.5 °C; (e) 98.46 mmHg. **6.14.** (a) $P_{\underline{Y}} = 195$ mmHg, $P_{\underline{Z}} = 105$ mmHg and (b) $P_{\underline{Y}}^0 = 650$ mmHg, $P_{\underline{Z}}^0 = 150$ mmHg. **6.16.** S_8, 256.48 g/mol.

Exercises, chapter 7
2. Ethanol is a hydrogen-bonded liquid, acetone is not. **4.** (b) Strong and dilute and (c) concentrated and weak. **6.** Test solution with litmus paper. **8.** one, $CH_3CH_2COO\textcircled{H}$. **12.** 5.4%.
14. (a) $Ca^{2+}(aq) + CO_3^{2-}(aq) \rightarrow CaCO_3\downarrow$
(b) $NH_4^+(aq) + OH^-(aq) \rightarrow NH_3 + H_2O$
(c) NR
(d) $Zn^{2+}(aq) + S^{2-}(aq) \rightarrow ZnS\downarrow$
(e) $OH^-(aq) + H^+(aq) \rightarrow H_2O$
(f) $2Ag^+(aq) + CrO_4^{2-}(aq) \rightarrow Ag_2CrO_4\downarrow$
(g) $CH_3COOH + NH_3 \rightarrow CH_3COO^-(aq) + NH_4^+(aq)$
(h) $Sr^{2+}(aq) + SO_4^{2-}(aq) \rightarrow SrSO_4\downarrow$
(i) NR
(j) $Ba^{2+}(aq) + SO_3^{2-}(aq) \rightarrow BaSO_3\downarrow$
(k) $H_2S + Cu^{2+}(aq) \rightarrow CuS\downarrow + 2H^+(aq)$
(l) $Ni^{2+}(aq) + 2 OH^- \rightarrow Ni(OH)_2\downarrow$
16. (a) $MnS(s) + 2H^+(aq) \rightarrow H_2S + Mn^{2+}(aq)$
(b) $AgCl(s) + CO_3^{2-}(aq) \rightarrow Ag_2CO_3\downarrow + Cl^-(aq)$
(c) $BaCO_3(s) + 2H^+(aq) \rightarrow Ba^{2+}(aq) + CO_2\uparrow + H_2O$
(d) $Sr^{2+}(aq) + H_2SO_3 \rightleftharpoons SrSO_3(s) + 2H^+(aq)$

(e) $Zn(OH)_2(s) + 2CH_3COOH \rightarrow Zn^{2+} + 2CH_3COO^-(aq) + 2H_2O$
(f) $2Fe^{3+} + 3SO_4^{2-} + 3Ba^{2+} + 6OH^- \rightarrow 3BaSO_4\downarrow + 2Fe(OH)_3\downarrow$
18. $[K^+] = 0.0567\ M$, $[NO_3^-] = 0.0367\ M$, $[Cl^-] = 0.0200\ M$. **20.** 0.292 g.

Multiple Choice Questions, chapter 7

2. (d) **4.** (d) **6.** (d) **8.** (c) **10.** (d) **12.** (e) **14.** (c) **16.** (a) **18.** (b)

Problems, chapter 7

7.2. (a) $Zn(OH)_2(s) + 2CH_3COOH \rightarrow Zn^{2+}(aq) + 2H_2O + 2CH_3COO^-$
(b) $MgCO_3(s) + 2H^+(aq) \rightarrow Mg^{2+}(aq) + H_2O + CO_2\uparrow$
(c) $Fe(OH)_3(s) + 3H^+(aq) \rightarrow Fe^{3+}(aq) + 3H_2O$
(d) $BaSO_3(s) + 2H^+(aq) \rightarrow Ba^{2+}(aq) + H_2O + SO_2\uparrow$
(e) $CuCO_3(s) + 2CH_3COOH \rightarrow Cu^{2+}(aq) + 2H_2O + CO_2\uparrow + 2CH_3COO^-(aq)$
7.4. 0.340 g, 0.0445 M. **7.6.** 0.0642 M. **7.8.** 0.2037 F. **7.10.** 0.713 g,
$[K^+] = 0.05120\ M$, $[NO_3^-] = 0.812\ M$, $[Pb^{2+}] = 0.0150\ M$. **7.12.** 15.12% Fe.
7.14. 31.9%.

Exercises, chapter 8

2. (a) M^4; (b) M^3; (c) M; (d) $atm^{-1/2}$; (e) atm; (f) atm^{-1}; (g) $atm \cdot M^{-1}$; (h) atm.
4. (c) and (e). **6.** (a) Not at equilibrium, $Q > K_{eq}$, reaction to the left. (b) At equilibrium.
(c) Not at equilibrium, $Q < K_{eq}$ reaction to the right. **8.** (a) yes and (b) no. **10.** There is
more SO_3 at equilibrium than at the instant of mixing. **12.** The weaker the acid formed
when an insoluble electrolyte reacts with $H^+(aq)$, the more the dissolution reaction will
proceed to the right. H_2SO_3 is a weak acid, H_2SO_4 is a strong acid. **14.** (a) K_p remains the
same. (b) The amount of HCl increases. **16.** (a) right-hand side; (b) left-hand side;
(c) to the left. **18.** (a) Increase and (b) remain the same. **20.** $3.8 \times 10^{-6}\ M^2$.
22. $1.27 \times 10^6\ atm^{-1/2}$ and (b) $7.86 \times 10^{-7}\ atm^{1/2}$. **26.** $K_p = K_c = 7.04 \times 10^{-2}$.
28. (a) 0.800 atm; (b) $P_{SO_2} = P_{Cl_2} = 0.63$ atm, $P_{SO_2Cl_2} = 0.17$ atm; (c) $K_p = 2.3$ atm.

Multiple Choice Questions, chapter 8

2. (c) **4.** (d) **6.** (e) **8.** (c) **10.** (a) **12.** (d) **14.** (b)

Problems, chapter 8

8.2. (a) $K_{eq} = [CO_2]/P_{CO_2}$; (b) An increase in P_{CO_2} causes a shift to the right, and therefore
an increase in $[CO_2]$. **8.4.** $K_p = 1.49 \times 10^{-5}\ atm^{-2}$, $K_c = 6.00 \times 10^{-2}\ M^{-2}$ or $(L/mol)^2$.
8.6. (a) Less than and (b) $K_p = 4.9 \times 10^2\ atm^{-1/2}$. **8.8.** (a) increase; (b) decrease;
(c) remain the same; (d) increase; (e) If $P_{NH_3} = P_{H_2S}$ before the volume is doubled, then
P_{NH_3} remains the same. If $P_{NH_3} > P_{H_2S}$ before the volume is doubled, then P_{NH_3} will
decrease. If $P_{NH_3} < P_{H_2S}$ before the volume is doubled, then P_{NH_3} will increase.
8.10. (a) 1.54×10^{15}, no units; (b) 1.15×10^{-4}; (c) to the left; (d) greater at 1800 K than
at 298 K. **8.12.** $P_{I_2} = P_{Br_2} = 2.7 \times 10^{-2}$ atm. **8.14.** (a) $P_{HCl} = 0.055$ atm,
$P_{NH_3} = 1.031$ atm. (b) 1.24 g. **8.16.** $P_{HI} = 0.729$ atm, $P_{H_2} = P_{I_2} = 0.0988$ atm.

Exercises, chapter 9

2. (a) -0.15; (b) 0.52; (c) 2.34. **4.** (a) pH $= -0.05$, pOH $= 14.05$; (b) pH $= 11.68$,
pOH $= 2.32$; (c) pH $= 13.80$, pOH $= 0.20$. **6.** $[H_3O^+] = 2.9 \times 10^{-2}\ M$,
$[OH^-] = 3.5 \times 10^{-3}\ M$.
8. (a) $C_6H_5COOH + H_2O \rightleftharpoons C_6H_5COO^- + H_3O^+$
(b) $CH_3NH_2 + H_2O \rightleftharpoons CH_3NH_3^+ + OH^-$

(c) $H_2C_2O_4 + H_2O \rightleftharpoons HC_2O_4^- + H_3O^+$
(d) $HSO_3^- + H_2O \rightleftharpoons SO_3^{2-} + H_3O^+$
(e) $OCl^- + H_2O \rightleftharpoons HOCl + OH^-$

10. $[H_3O^+] = 3.9 \times 10^{-12} M$, $[OH^-] = 2.6 \times 10^{-3} M$. **12.** $[H_3O^+] = 1.4 \times 10^{-5} M$, $[OH^-] = 6.9 \times 10^{-10} M$. **14.** (a) H_2O, water; (b) SO_4^{2-}, sulfate ion; (c) NH_3, ammonia; (d) F^-, fluoride ion; (e) $H_2PO_4^-$, dihydrogen phosphate ion. **16.** (a) 2.5×10^{-5}; (b) 5.6×10^{-10}; (c) 4.2×10^{-4}; (d) 2.5×10^{-2}. **18.** S^{2-}, CN^-, NH_3, CH_3COO^-, F^-. **20.** 1.3×10^{-19}. **22.** (a) acidic; (b) basic; (c) basic; (d) neutral; (e) basic; (f) acidic; (g) neutral; (h) basic; (i) neutral; (j) acidic. **24.** (a) basic; (b) acidic; (c) basic; (d) acidic. **26.** acidic.

Multiple Choice Questions, chapter 9

2. (e) **4.** (e) **6.** (b) **8.** (d) **10.** (c) **12.** (a) **14.** (a) **16.** (b) **18.** (e) **20.** (c) **22.** (b) **24.** (b)

Problems, chapter 9

9.2. $[CH_3COO^-] = 0.2000 M$, $[CH_3COOH] = [OH^-] = 1.1 \times 10^{-5} M$, $[H_3O^+] = 9.5 \times 10^{-10} M$. **9.4.** $[H_3O^+] = 0.080 M$, $[CH_3COOH] = 0.150 M$, $[CH_3COO^-] = 3.3 \times 10^{-5} M$, $[OH^-] = 1.3 \times 10^{-13} M$, pH = 1.1. **9.6.** (a) 3.0×10^{-6}; (b) $0.017 F$. **9.8.** (a) 9.1×10^{-5} and (b) $0.12 F$.
9.10. (a) $OBr^- + H_2O \rightleftharpoons HOBr + OH^-$; (b) 5.0×10^{-6}; (c) 2.0×10^{-9}.
9.12. $HClO_4$, $KHSO_4$, CH_3NH_3Br, $NaHCO_3$, Na_2O. **9.14.** (a) $[(CH_3)_3N] = 0.117 M$, $[OH^-] = [(CH_3)_3NH^+] = 2.75 \times 10^{-3}$, $[H_3O^+] = 3.64 \times 10^{-12}$, pH = 11.44; (b) 6.48×10^{-5}; (c) 2.60×10^{-2}, increase. **9.16.** (a) pH = 4.34, $\alpha = 7.0 \times 10^{-4}$; (b) 3.1×10^{-7}; (c) 10.05; (d) 1.19. **9.18.** $[OH^-] = 0.12 M$, $[NH_3] = 0.090 M$, $[H_3O^+] = 8.3 \times 10^{-14} M$, $[NH_4^+] = 1.35 \times 10^{-5} M$, pH = 13.08.

Exercises, chapter 10

2. 4.50. **4.** 4.27. **6.** 3.57. **8.** 9.24. **10.** The pK_a of $NH_4^+ = 9.24$, so the pH range for efficient NH_4^+/NH_3 buffers is 8.24 to 10.24. **12.** $0.139 F$. **14.** $0.1273 F$. **16.** (a) yellow; (b) yellow; (c) green; (d) blue.
18. (a) $C_6H_5COOH + OH^- \rightleftharpoons C_6H_5COO^- + H_2O$
(b) $NH_3 + H_3O^+ \rightleftharpoons NH_4^+ + H_2O$
(c) $HCOOH + OH^- \rightleftharpoons HCOO^- + H_2O$
(d) $C_6H_5NH_2 + H_3O^+ \rightleftharpoons C_6H_5NH_3^+ + H_2O$
20. 8.37. **22.** phenolphthalein. **24.** $K_a = 1.35 \times 10^{-5}$, $0.151 F$. **26.** (a) $3.3 \times 10^{-2} M$; (b) 0.978 atm; (c) 3.92. **28.** (a) 7.87, phenol red; (b) 30.0 mL; (c) 60.0 mL.
30. H_2O, Na^+, S^{2-}, HS^-, OH^-, H_2S, H_3O^+.
electroneutrality: $[Na^+] + [H_3O^+] = [OH^-] + [HS^-] + 2[S^{2-}]$
material balance: $C = 0.100 M = [S^{2-}] + [HS^-] + [H_2S]$
32. 8.60. **34.** $[K^+] = 0.0800 M$, $[HCO_3^-] = 0.0784 M$, $[CO_3^{2-}] = [CO_2] = 8.2 \times 10^{-4} M$, $[OH^-] = 2.2 \times 10^{-6} M$, $[H_3O^+] = 4.5 \times 10^{-9} M$.

Multiple Choice Questions, chapter 10

2. (a) **4.** (b) **6.** (e) **8.** (b) **10.** (b) **12.** (d) **14.** (d) **16.** (e) **18.** (c) **20.** (e) **22.** (c) **24.** (d) **26.** (a) **28.** (b) **30.** (c)

Problems, chapter 10

10.2. $[C_5H_5NH^+] = 2.61 \times 10^{-2} M$, $[C_5H_5N] = [H_3O^+] = 4.21 \times 10^{-4} M$, $[OH^-] = 2.4 \times 10^{-11} M$. **10.4.** (a) $[NH_4^+] \approx 0.300 M$, $[NH_3] = 0.200 M$,

$[H_3O^+] = 8.6 \times 10^{-10}$ M, $[OH^-] = 1.2 \times 10^{-5}$ M; (b) Yes. pH = 9.07, within one pH unit of pK_a. **10.6.** Mix 1.00 mol NH_3 with 106 mL of 6.00 F strong acid, add water to make volume 1.00 L. **10.8.** (a) 4.52 and (b) 4.48. **10.10.** (a) $[HCOO^-] = 0.500$ M, $[HCOOH] = 0.250$ M, $[H_3O^+] = 9.0 \times 10^{-5}$ M, $[OH^-] = 1.1 \times 10^{-10}$ M; (b) Yes. pH = 4.05, within one pH unit of pK_a.

10.12. (a) $CH_3NH_2 + HCOOH \rightleftharpoons CH_3NH_3^+ + HCOO^-$

(b) $2NH_4^+ + SO_4^{2-} + Ba^{2+} + 2 OH^- \rightleftharpoons 2NH_3 + 2H_2O + BaSO_4\downarrow$

(c) $Ag_2CO_3(s) + 2H^+(aq) \rightleftharpoons 2Ag^+ + H_2O + CO_2\uparrow$

(d) $NH_4^+ + OH^- \rightleftharpoons NH_3 + H_2O$

(e) $HCO_3^- + H_3O^+ \rightleftharpoons 2H_2O + CO_2\uparrow$

(f) $HCO_3^- + OH^- \rightleftharpoons H_2O + CO_3^{2-}$

10.14. (a) 4.62 and (b) 4.62. **10.16.** (a) 6.4×10^{-4} and (b) 10.7. **10.18.** Riva.

Exercises, chapter 11

2. $Hg_2SO_4(s) \rightleftharpoons Hg_2^{2+} + SO_4^{2-}$

$PbCO_3(s) \rightleftharpoons Pb^{2+} + CO_3^{2-}$

$AgIO_3(s) \rightleftharpoons Ag^+ + IO_3^-$

$Cr(OH)_3(s) \rightleftharpoons Cr^{3+} + 3 OH^-$

$La(IO_3)_3(s) \rightleftharpoons La^{3+} + 3 IO_3^-$

4. (a) 7.1×10^{-7}; (b) 5.3×10^{-4}; (c) 5.3×10^{-4}; (d) 1.3×10^{-3}. **6.** (a) $s = (K_{sp})^{1/2}$; (b) $s = (K_{sp}/27)^{1/4}$; (c) $s = (K_{sp}/4)^{1/3}$; (d) $s = (K_{sp})^{1/2}$; (e) $s = (K_{sp}/4)^{1/3}$. **8.** (a) $(0.086 + z)(4z^2) = K_{sp}$; (b) $(0.086 + z)(z) = K_{sp}$; (c) $(0.086 + z)(z) = K_{sp}$; (d) $(0.086 + z)(4z^2) = K_{sp}$. **10.** 2.2×10^{-14}. **12.** 4×10^{-8}. **14.** $s = 2.2 \times 10^{-5}$ in water and 3.4×10^{-7} in 0.10 F $Mn(NO_3)_2$. Smaller in solution with common ion. **16.** Yes, AgSCN will precipitate. $Q = 2.5 \times 10^{-3} > K_{sp}$. **18.** Yes, $Ba(IO_3)_2$ will precipitate. $Q = 2.9 \times 10^{-4} > K_{sp}$. **20.** $[Ca^{2+}] = 4.00 \times 10^{-2}$ M, $[CO_3^{2-}] = 1.2 \times 10^{-7}$ M. **22.** 0.097 g. **24.** (a) 3.3×10^8; (b) 7.8×10^{-3}; (c) 2.2×10^9. **26.** (a) 1.1×10^{-9} M; (b) 1.1×10^{-6} M; (c) no.

Multiple Choice Questions, chapter 11

2. (a) **4.** (d) **6.** (a) **8.** (b) **10.** (a) **12.** (c) **14.** (c) **16.** (e) **18.** (e) **20.** (e) **22.** (d) **24.** (b) **26.** (b)

Problems, chapter 11

11.2. (a) 1.3×10^{-3}; (b) 1.5×10^{-4}; (c) 8.7×10^{-7}. **11.4.** (a) CaF_2 and (b) $La(OH)_3$. **11.6.** $Fe(OH)_3$ will precipitate. **11.8.** Calculated value = 7.4×10^{-3} g per 100 mL. Actual value larger because $C_2O_4^{2-}$ reacts with H_2O. **11.10.** Yes, $PbSO_4$ forms. $K_{eq} = 8.9 \times 10^2$. **11.12.** No, Ag_2CO_3 will form. **11.14.** $[Pb^{2+}] = 0.012$ M, $[K^+] = 0.060$ M, $[NO_3^-] = 0.0840$ M, $[IO_3^{2-}] = 4.7 \times 10^{-6}$ M. **11.16.** (a) $K_{sp} = [Ag^+]^2[SO_4^{2-}] = 2.4 \times 10^{-5}$ M^3 and (b) $[Ag^+] = 8.9 \times 10^{-3}$ M, $s = 4.5 \times 10^{-3}$ M. **11.18.** (a) 2×10^{-14} M; (b) 0.002%; (c) 3.6. **11.20.** Yes, $Mn(OH)_2$ will precipitate. **11.22.** 99.996%. **11.24.** CuS will precipitate, MnS will not.

Exercises, chapter 12

4. $\lambda = 2000$ nm $= 2.000 \times 10^{-4}$ cm, $v = 1.499 \times 10^{14}$ s^{-1}, $\tilde{v} = 5.000 \times 10^5$ m^{-1}, near IR. **8.** $\tilde{v} = 5331.57$ cm^{-1}, $\lambda = 1875.62$ nm, $v = 1.59836 \times 10^{14}$ s^{-1}, near IR. **10.** Only the four frequencies listed in Table 12.2. **12.** $v = 6.906 \times 10^{14}$ s^{-1}, $\lambda = 434.1$ nm, visible. **14.** The violet ray. **16.** 7.3×10^{-18} J. **18.** $\lambda = 4.5 \times 10^{-7}$ m, $E = 4.4 \times 10^{-19}$ J. **20.** For the He$^+$ ion, the energy is four times lower than for the H atom, or -6.04 eV. **22.** $E = 91.80$ eV, $\lambda = 13.50$ nm, far UV.

Multiple Choice Questions, chapter 12
2. (d) 4. (c) 6. (d) 8. (e) 10. (d) 12. (e) 14. (a)

Problems, chapter 12
12.2. Lyman, $n_H = 3 \rightarrow n_L = 1$. 12.4. (a) 0.12 Å and (b) 5.5×10^{-22} Å.
12.6. (a) (1) -2.175 eV, (2) -13.60 eV; (b) 11.42 eV, 1.830×10^{-18} J;
(c) $v = 2.762 \times 10^{15}$ s^{-1}, $\lambda = 108.6$ nm, far UV. 12.8. v_5 from $E_3 \rightarrow E_1$, v_6 from
$E_4 \rightarrow E_1$, v_2 from $E_3 \rightarrow E_2$, v_3 from $E_4 \rightarrow E_2$, v_1 from $E_4 \rightarrow E_3$.

Exercises, chapter 13
2. (a) unacceptable; (b) acceptable; (c) unacceptable; (d) acceptable. 4. K 1, L 4, M 9,
N 16, O 25, number of orbitals $= n^2$. 6. zero. 8. 32. 10. (a), (d), (e) excited; (b),
(f) ground; (c) impossible. 12. (a) $(Ar)^{18}$; (b) $(Ar)^{18}(3d)^1$; (c) $(Ar)^{18}(3d)^2$;
(d) $(Ar)^{18}(3d)^3$. 14. (a) 0; (b) 3; (c) 0; (d) 1. 16. VIIA. 22. (a) HI, HBr, HCl, HF and
(b) BrCl, NO, HCl, ClF, LiI. 24. BrCl, AsCl$_3$, GeCl$_4$, GaCl$_3$, CaCl$_2$, KCl. 28. IVA.
30. Number unpaired electrons (a) 2; (b) 1; (c) 0; (d) 1.
32. (a) $2K(s) + Cl_2(g) \rightarrow 2KCl(s)$
(b) $Sr(s) + Cl_2(g) \rightarrow SrCl_2(s)$
Reaction (a) is more exothermic.

Multiple Choice Questions, chapter 13
2. (e) 4. (d) 6. (c) 8. (d) 10. (e) 12. (d) 14. (a) 16. (b) 18. (e) 20. (e) 22. (b)
24. (b) 26. (b)

Problems, chapter 13
13.2. (a) Ca^{2+}, S^{2-}, Sc^{3+} and (b) Ag$^+$, Cd^{2+}, Fe^{3+}, Ti^{2+}, Zn^{2+}.
13.4. Ca^{2+}(aq) + CO$_2$(g) + 2 OH$^-$(aq) $\rightarrow$ CaCO$_3$(s) + H$_2$O. 13.12. (a) Cl$^-$ largest,
Sc^{3+} smallest; (b) Ar largest, Al smallest; (c) Na largest, Mg, smallest. 13.16. (b) between
41.1 and 75.6 eV.

Exercises, chapter 14
2. Li$^+$(g) + Br$^-$(g) is higher in energy by 196 kJ/mol. 8. (a) tetrahedral; (b) bent;
(c) trigonal pyramidal; (d) trigonal bipyramidal; (e) trigonal planar. 10. Bond angles
$\sim 120°$. 12. Axial bonds longer than equatorial. 14. IF$_4^-$ is an AX$_4$E$_2$ molecule with lone
pairs in the axial positions. 18. O$_2^-$, O$_2$, O$_2^+$, all are paramagnetic. 20. (a) tetrahedral,
sp^3; (b) trigonal planar, sp^2; (c) octahedral, sp^3d^2; (d) linear, sp; (e) tetrahedral, sp^3.
24. I—C—H bond angle slightly greater than 109.5°.

Multiple Choice Questions, chapter 14
2. (e) 4. (d) 6. (a) 8. (e) 10. (e) 12. (b) 14. (e) 16. (c) 18. (c) 20. (b) 22. (d)
24. (a) 26. (b) 28. (d) 30. (c)

Problems, chapter 14
14.2. XeF$_4$ is an AX$_4$E$_2$ molecule with lone pairs in the axial positions. 14.4. NO$_3^-$, NO$_2$,
NO, NO$_2^+$, NO$^+$. 14.6. Ag$^+$ Lewis acid, Cl$^-$ Lewis base. 14.8. Formal charge: 0 on Cl,
$+1$ on N, $-\frac{1}{2}$ on each O, bond angles $\sim 120°$, bond order 1.5. 14.12. Bond order 1.5.
14.14. Bond order NO 2.5, NO$^+$ 3.0, NO$^-$ 2.0. NO$^+$ has largest bond energy.
14.20. HF 41.3%, HCl 17.7%, BrF 15.3%.

Exercises, chapter 15

2. (a) +6; (b) +6; (c) +4; (d) +2; (e) +4; (f) +3; (g) +3; (h) +5; (i) +5; (j) +3; (k) +2; (l) +4; (m) +1.4. **4.** (a), (b), (c), (g), (h), (i), (k), (l). **6.** (a) −4; (b) 0; (c) −1; (d) −2; (e) 0; (f) 0; (g) −3. **10.** (a) $4H^+(aq) + 3MnO_4^{2-} \rightarrow 2MnO_4^- + MnO_2(s) + 2H_2O$; (b) $3ClO^-(aq) \rightarrow 2Cl^-(aq) + ClO_3^-(aq)$.

Multiple Choice Questions, chapter 15

2. (d) **4.** (d) **6.** (c) **8.** (d) **10.** (c) **12.** (a) **14.** (a) **16.** (b) **18.** (a)

Problems, chapter 15

15.2. (a) $CdS(s) + 2H^+(aq) \rightarrow Cd^{2+}(aq) + H_2S \qquad K_{eq} = 3 \times 10^{-9}$
(b) $3CdS(s) + 8H^+(aq) + 2NO_3^-(aq) \rightarrow 3Cd^{2+}(aq) + 3S\downarrow + 2NO(g) + 4H_2O$
(c) $CdS(s) + 4H^+(aq) + 2NO_3^-(aq) \rightarrow Cd^{2+}(aq) + S\downarrow + 2NO_2(g) + 2H_2O$
15.4. (a) $8H^+(aq) + MnO_4^- + 5Fe^{2+} \rightarrow Mn^{2+} + 5Fe^{3+} + 4H_2O$. (b) (1) 6.39 mmol and (2) 47.3%. **15.8.** NH_4I. **15.10.** (a) Oxidizing agent ClO_3^-, reducing agent As_2S_3 and (b) oxidizing agent CN^-, reducing agent $Cu(NH_3)_4^{2+}$.

Exercises, chapter 16

2. 0.203 g. **4.** 1.341×10^4 C.
6. (a) $Co(s) + 2Ag^+ \rightarrow 2Ag(s) + Co^{2+}$
(b) $Zn(s) + Br_2 \rightarrow Zn^{2+} + 2Br^-$
(c) $Pb(s) + 2H^+(aq) \rightarrow H_2(g) + Pb^{2+}$
(d) $H_2(g) + Cu^{2+} \rightarrow Cu(s) + 2H^+(aq)$
8. (a) zinc; (b) $\Delta\mathcal{E}^0_{cell} = 0.51$ V; (c) Ni^{2+}. **10.** (a) 0.27 V; (b) 1.41 V; (c) 2.08 V. **12.** (a) 0.125 V; (b) 0.194 V; (c) 0.040 V. **14.** (a) 0.108 V and (b) increase.
18. (a) $Co(s) + Br_2 \rightarrow Co^{2+} + 2Br^- \qquad K_{eq} = 2 \times 10^{46}$.
(b) $Sc(s) + 3H^+(aq) \rightarrow Sc^{3+} + \frac{3}{2}H_2(g) \qquad K_{eq} = 10^{105}$.
(c) $Pb(s) + Zn^{2+} + SO_4^{2-} \rightarrow PbSO_4(s) + Zn(s) \qquad K_{eq} = 2 \times 10^{-14}$.
20. (a) A gas occupies a large volume and (b) silver is too expensive. **24.** 4.21 g.
26. (a) 46.6 h and (b) 6.99×10^3 A. **28.** 0.304 L.

Multiple Choice Questions, chapter 16

2. (b) **4.** (e) **6.** (d) **8.** (a) **10.** (c) **12.** (b) **14.** (b) **16.** (d)

Problems, chapter 16

16.2. 1.6×10^2. **16.4.** Air oxidizes Cr^{2+} to Cr^{3+}. **16.6.** (a) 6.1×10^{-7}, no and (b) 4×10^8. **16.8.** (a) (i) −0.2223 V, (ii) +0.152 V; (b) from HCl, no; from HI, yes. **16.12.** $[Zn^{2+}] = 5.000 \times 10^{-2}$ M, $[Ag^+] = 9 \times 10^{-28}$ M, 2.738 g Zn unreacted. **16.14.** 1.2×10^{-5}.

Exercises, chapter 17

2. $P_2 = 1.20$ atm, $w = 0$. **4.** $\Delta E = 0$, T remains constant. **6.** (a), (d), and (g). **8.** (a) 64.95 kJ and (b) −43.2 kJ. **10.** −147 kJ/mol. **12.** ΔH^0_f for F is 79 kJ/mol, for Cl is 122 kJ/mol. **14.** 335 kJ/mol. **16.** (a) −312 kJ/mol and (b) −305.2 kJ/mol, −2%. **18.** (a) 1.16 kJ and (b) 645 J. **20.** 624 J, $\Delta H = 624$ J, $\Delta E = 374$ J. **22.** 307 cal.

Multiple Choice Questions, chapter 17

2. (e) **4.** (b) **6.** (e) **8.** (a) **10.** (d) **12.** (a) **14.** (c) **16.** (e) **18.** (b) **20.** (d) **22.** (c)

Problems, chapter 17

17.2. (a) 42 kJ, -35% error and (b) -42 kJ, $+2.8\%$ error. **17.4.** 50.6 kJ/mol.
17.6. $\Delta H^0 = \Delta H_3^0 - 2\Delta H_1^0 - 3\Delta H_2^0$. **17.8.** -986.5 kJ/mol.
17.10. (a) $NCNH_2(s) + \frac{3}{2}O_2(g) \rightarrow CO_2(g) + N_2(g) + H_2O(\ell)$; (b) 62 kJ/mol.
17.12. (a) -852 kJ/mol and (b) 1.78 g. **17.14.** (a) ΔH and (b) 1.88 cal, excellent
approximation. **17.16.** 718 kJ/mol. **17.18.** (a) 28.5 J/K per mol atoms;
(b) 27.2 J/K per mol atoms; (c) 24.4 J/K per mol atoms; (d) 27.1 J/K per mol atoms.

Exercises, chapter 18

2. $w = q = \Delta E = 0$, final state has more molecular disorder. **4.** irreversible. **6.** $\Delta E = 0$,
$w_{rev} = -140$ J, $q_{rev} = 140$ J. **8.** 0.471 J/K. **10.** In $J \cdot mol^{-1}K^{-1}$, BCl_3 83.7, CH_3OH 110.9,
CH_3Br 86.7, N_2H_4 105, PCl_3 87.4, CH_3OH and N_2H_4 are hydrogen bonded and do not
obey Trouton's Rule. **12.** (a) negative; (b) positive; and (c) negative. **14.** -818 kJ.
16. (a) -31.4 kJ; (b) 102.6 kJ; and (c) -70.54 kJ. **18.** enthalpy for both (a) and (b).
20. ΔS_f^0 is negative. **22.** (a) $+37$ kJ and (b) to the left. **24.** (a) -89.1 kJ and
(b) 0.462 V, 4.3×10^{15}, yes. **26.** (a) $Q = 0.025$, $\Delta \mathscr{E}_{cell} = 0.17$ V, $\Delta G = -32.8$ kJ and
(b) yes. **28.** More negative.

Multiple Choice Questions, chapter 18

2. (c) **4.** (b) **6.** (b) **8.** (c) **10.** (c) **12.** (a) **14.** (d) **16.** (a) **18.** (b) **20.** (e)

Problems, chapter 18

18.2. (a) $\Delta G^0 = 52.8$ kJ, $K_{eq} = 5.6 \times 10^{-10}$. **18.6.** Hydrogen bonding persists in the
vapor phase. **18.8.** 2.92 kJ. **18.10.** $\Delta G^0 = 55.7$ kJ, $\Delta \mathscr{E}_{cell}^0 = -0.577$ V,
$K_{eq} = [Ag^+][Cl^-] = 1.8 \times 10^{-10}$. **18.12.** 238 $J \cdot mol^{-1}K^{-1}$.
18.14. (a) $P_{SO_2} = P_{Cl_2} = 0.76$ atm, $P_{SO_2Cl_2} = 0.24$ atm; (b) $K_p = 2.41$ atm; (c) -2.74 kJ;
and (d) entropy.

Exercises, chapter 19

2. (a) 4th order; 1st order in ClO_3^- and I^-, 2nd order in H^+ (aq); (b) $\frac{3}{2}$ order;
1st order in $CHCl_3$, $\frac{1}{2}$ order in Cl_2; (c) 2nd order; 2nd order in $NOBr$; (d) $\frac{3}{2}$ order;
1st order in CO, $\frac{3}{2}$ order in Cl_2; and (e) 2nd order; 1st order in I^- and $S_2O_8^-$.
4. (a) 2.48×10^{-5} $mmHg \cdot s^{-1}$ and (b) 8.4×10^{-6} $mmHg \cdot s^{-1}$ rate decreases with time as
the concentration of reactants (pressure of gaseous reactants) decreases with time.
6. (a) M/s or $mol \cdot L^{-1}s^{-1}$ and (b) $M^{-2}s^{-1}$ or $L^2mol^{-2}s^{-1}$.
8. 1.28×10^{-3} $mol \cdot L^{-1}min^{-1}$. **10.** $k[A][B]^2$. **12.** From a well-drawn plot,
half-life $= 53 \pm 1$ min. **14.** $2.303/k$. **16.** Run the reaction in a different solvent, varying
the $[H_2O]$. **20.** 1.79. **24.** (a) $(CH_3)_2C{=}CH_2 + H_2O \rightarrow (CH_3)_3COH$; (b) $(CH_3)_3C^+$;
(c) yes, H^+; (d) rate $= k[(CH_3)_2C{=}CH_2][H^+(aq)]$; and (e) rate $= k'[(CH_3)_2C{=}CH_2]$.

Multiple Choice Questions, chapter 19

2. (b) **4.** (b) **6.** (c) **8.** (d) **10.** (d) **12.** (b) **14.** (e) **16.** (e) **18.** (d)

Problems, chapter 19

19.2. (a) $mol \cdot L^{-1}s^{-1}$ and (b) $L^{1/2}mol^{-1/2}s^{-1}$. **19.4.** (a) $2 O_3 \rightarrow 3 O_2$; (b) $k[O_3]^2[O_2]^{-1}$;
and (c) $k = k_2K_{eq}$. **19.6.** first order, 6.22×10^{-3} s^{-1} **19.8.** (a) 1.4×10^3 min or 23 h.
19.14. (a) rate $= k[H_2][NO]^2$, $k = 4.33 \times 10^{-2}$ $L^2mol^{-2}s^{-1}$ and
(b) $N_2O_2 + H_2 \rightarrow N_2 + 2H_2O$, slow step. **19.16.** (a) 4200 kcal; (b) -1200 kcal;
(c) -3600 kcal; and (d) 15,100 kcal

Exercises, chapter 20

2. (a) +3; (b) +3; (c) +2; (d) +2; (e) +1; (f) +3; and (g) +3.   **4.** (b), (c), and (e).
6. Due to the extremely small size of the Be^{2+} ion.   **8.** The Sr^{2+} ion has rare gas electronic configuration, the Sn^{2+} ion does not.   **10.** (c), (e), and (g).   **12.** $[Co(NH_3)_5(H_2O)]Cl_3$.
14. (c), (b), (d), (a), and (e).   **16.** (a) $Cs[IFCl_3]$; (b) $[Ni(en)_3](NO_3)_2$; (c) $K_3[Co(C_2O_4)_2Br_2]$; (d) $Na_4[Fe(CN)_6]$; and (e) $[Cr(NH_3)_5Cl]SO_4$.   **18.** (b) and (e).   **20.** (a), (d), (f), and (g) obey Sidgwick's EAN rule.   **22.** (a), (c), and (d).   **24.** The CFS, Δ_0, is much smaller for H_2O than for CN^-.   **26.** (a) $Mn(en)_3^{3+}$; (b) $Rh(H_2O)_6^{3+}$; (c) PtF_4^{2-}; (d) $Cr(H_2O)_6^{3+}$; and (e) $Fe(CN)_6^{3-}$.   **28.** $CuCl_4^{2-}$ absorbs in the red and transmits green, $Cu(H_2O)_4^{2+}$ absorbs in the orange and transmits blue, $Cu(H_2O)_4^{2+} + 4Cl^- \rightleftharpoons CuCl_4^{2-} + 4H_2O$.

Multiple Choice Questions, chapter 20

2. (b)   **4.** (a)   **6.** (c)   **8.** (a)   **10.** (e)   **12.** (c)   **14.** (d)   **16.** (b)   **18.** (e)   **20.** (c)

Problems, chapter 20

20.2. $(CO)_5Mn—Mn(CO)_5$.
20.4. $[Pt(NH_3)_5Cl]Cl_3$ chloropentaammineplatinum(IV) chloride, octahedral geometry.
20.6. Mo(II) has $40e^-$. With seven ligands the EAN of Mo(II) is 54 (Kr).
20.10. (a) $Co(CN)_6^{3-}$, $Co(en)(CN)_4^-$, $Co(en)_2(CN)_2^+$ *cis* and *trans*, $Co(en)_3^{3+}$; and (b) ammonium hexacyanocobaltate(III) $(NH_4)_3[Co(CN)_6]$, ammonium tetracyanoethylenediaminecobaltate(III) $(NH_4)[Co(en)(CN)_4]$, *cis* and *trans* dicyanobis(ethylenediamine)cobalt(III) cyanide $[Co(en)_2(CN)_2]CN$, tris(ethylenediamine)cobalt(III) cyanide $[Co(en)_3](CN)_3$.   **20.12.** (b), (c), (e), and (g).
20.16. (b) $[Mn(en)_2^{2+}] = 1 \times 10^{-4}\ M$. It is possible to separate Mn^{2+} and Ni^{2+} using en to tie Ni^{2+} up as $Ni(en)_2^{2+}$ while Mn^{2+} is precipitated as $Mn(OH)_2$.   **20.18.** 6.8×10^{32}.

Multiple Choice Questions, chapter 21

2. (d)   **4.** (c)   **6.** (e)   **8.** (a)   **10.** (c)   **12.** (d)   **14.** (d)   **16.** (d)   **18.** (b)

Problems, chapter 21

21.2. (a) 1.24 Å and (b) $7.875\ g \cdot cm^{-3}$.   **21.6.** Theoretical pure ionic value should be fairly close to the experimental value for RbCl and BaO, but significantly smaller than the experimental value for CdS.   **21.8.** 8.   **21.14.** 2.256 Å.   **21.16.** Both Mg^{2+} and O^{2-} are doubly charged ions, while Li^+ and F^- are singly charged.

Exercises, chapter 22

2. (a) $_1^1H$; (b) $_0^1n$; (c) $3(_0^1n)$; (d) $_{94}^{239}Pu$; and (e) $_6^{12}C$.   **4.** release energy.
6. (a) electron capture and α emission; (b) β^- emission; (c) β^+ emission; and (d) electron capture and β^+ emission.   **8.** For ^{17}F and ^{19}F, β^+ emission, for ^{20}F and ^{21}F, β^- emission.   **10.** $6.50\ \mu g$.   **12.** $0.026\ \mu Ci$.   **14.** (a) 77.3% and (b) 7.58%.
16. 2.08×10^3 yr.

Multiple Choice Questions, chapter 22

2. (b)   **4.** (a)   **6.** (b)   **8.** (e)   **10.** (e)   **12.** (b)   **14.** (d)   **16.** (d)   **18.** (b)   **20.** (c)

Problems, chapter 22

22.2. $_{19}^{40}K$, an odd–odd nuclide.   **22.4.** For ^{16}O, $BE/A = 7.976$ MeV. For ^{17}O, $BE/A = 7.751$ MeV.   **22.6.** 7.27 MeV.   **22.8.** 22,600 yr.

22.10. 235.04393 amu. **22.12.** (a) The added SO_4^{2-} ions are precipitated as $BaSO_4$ and (b) $[Ba^{2+}] = 3.4 \times 10^{-4}$ M. **22.14.** 2.51 MeV. **22.16.** 5×10^{13} mol H atoms fused per second.

Multiple Choice Questions, chapter 23
> **2.** (c) **4.** (a) **6.** (e) **8.** (e) **10.** (a) **12.** (d) **14.** (c) **16.** (c) **18.** (d) **20.** (b)

Problems, chapter 23
> **23.2.** (a) Two isomers (*cis* and *trans*) would be possible if dichloromethane were square planar and (b) only one isomer is possible for tetrahedral dichloromethane.
> **23.4.** (b), (c), and (f). **23.8.** Six. **23.10.** There are five isomeric hexanes: hexane, 2-methylpentane, 3-methylpentane, 2,2-dimethylbutane, and 2,2-dimethylbutane.
> **23.12.** A mixture of products. **23.14.** (a) ethylbenzoate; (b) *N,N*-dimethylbenzamide; (c) ammonium acetate; and (d) butyl acetate. **23.16.** From left to right: serine (hydrophilic), valine (hydrophobic), histidine (hydrophilic), tyrosine (hydrophilic), glycine (neither category), and cysteine (hydrophilic). **23.18.** 1-butene, *cis*- and *trans*-2-butene, 2-methylpropene, methylcyclopropane, and cyclobutane. **23.20.** cyclopentane, methylcyclobutane, *cis*-dimethylcyclopropane, and *trans*-dimethylcyclopropane.

Exercises, appendix B
> **B2.** 3.75×10^{-3}. **B4.** 3.67×10^4. **B6.** 6.35×10^{-3}. **B8.** 8.46×10^{-3}. **B10.** -637.
> **B12.** (a) 23.7797; (b) -10.312; (c) -2.28; and (d) 124.9 **B14.** (a) 2.106 and -0.4561 and (b) -2.281 and -0.219. **B16.** 2.2×10^{-3}.

credits

introduction

Figure I.3.: (*a*) Mettler Instrument Corporation: (*b,c*) Ohaus Scale Corporation.

chapter 1

Opener: Edgar Fahs Smith Collection, University of Pennsylvania. Page 27: New York Public Library. Page 31: Peter M. Lerman. Page 38: Peter M. Lerman. Page 41: Peter M. Lerman. Figure 1.2: Peter Lerman. Figure 1.4: Peter Lerman.

chapter 2

Opener: Edgar Fahs Smith Collection, University of Pennsylvania. Page 69: Elyse Rieder. Page 73: Peter M. Lerman. Page 75: Runk/Schoenberger/Grant Heilman. Page 75: Mario Fantin/Photo Researchers. Figure 2.7: (*d*) Ruck/Schoenberger/Grant Heilman. (*c*) Mario Fantain/Photo Researchers.

chapter 3

Opener: New York Public Library, Picture Collection.

chapter 4

Opener: Library of Congress.

chapter 5

Opener: Mary Evans Picture Library/Photo Researchers. Page 163: National Oceanic and Atmospheric Administration.

chapter 6

Opener: Edgar Fahs Smith Collection, University of Pennsylvania. Page 195: (*a–d*) Peter M. Lerman. Figure 6.4: (*a–d*) Peter Lerman.

chapter 7

Opener: New York Public Library Picture Collection.

chapter 8

Opener: Edgar Fahs Smith Collection, University of Pennsylvania.

chapter 9

Opener: Edgar Fahs Smith Collection, University of Pennsylvania. Page 305: Corning.

chapter 10

Opener: Edgar Fahs Smith Collection, University of Pennsylvania.

chapter 11

Opener: Smithsonian Institution. Page 397: Schmidt-Thomsen, Landesden Kmalamt, Westfalen-Lippe, Muenster, Germany.

chapter 12

Opener: AP/Wide World Photos. Page 432: AIP Niels Bohr Library. Page 434: (*a,b*) Pssc film, *Mother Waves,* Education Development Center, Inc., Newton, Massachusetts. Figure 12.9: From *Qualitative Analysis and Electrolytic Solutions* by Edward J. King. Copyright © 1959 by Harcourt Brace Jovanovich, Inc. Reproduced by permission of the publisher. Figure 12.14: Courtesy of Education Development Center, Inc., Newton, Massachusetts.

chapter 13

Opener: Ullstein Bilderdienst, Berlin, West Germany. Page 455: *X-Rays to Quarks,* E. Segle, W. H. Freeman, San Francisco. Page 477: J. Allan Cash/Photo Researchers. Figure 13.19: Adapted from Linus Pauling, *The Nature of the Chemical Bond.* Third edition copyright © 1960 by Cornell University. Used by permission of the publisher, Cornell University Press.

chapter 14

Opener: Edgar Fahs Smith Collection, University of Pennsylvania. Page 528: Linus Pauling Institute of Science and Medicine.

chapter 15

Opener: AP/Wide World Photos.

index

$Q < K_{sp}$ unsaturated

$Q > K_{sp}$ supersaturated

$Q = K_{sp}$ saturated

frequency $\nu = \dfrac{c}{\lambda}$ speed of light / wavelength

$\bar{\nu}$ wavenumber $= \dfrac{1}{\lambda} = \dfrac{k}{c}$

λ

Microwave
far Infrared
near Infrared
Visible
Near Ultraviolet
Far Ultraviolet
X-ray

2×10^8

2×10^6

2×10^4

780

380

200

10

Derived SI Units

Physical Quantity	Name of Unit	Symbol	Basic Units
Area	Square meter	m^2	m^2
Volume	Cubic meter	m^3	m^3
Density	Kilogram per cubic meter	$kg \cdot m^{-3}$	$kg \cdot m^{-3}$
Velocity	Meter per second	$m \cdot s^{-1}$	$m \cdot s^{-1}$
Acceleration	Meter per second per second	$m \cdot s^{-2}$	$m \cdot s^{-2}$
Force	Newton	N	$kg \cdot m \cdot s^{-2}$
Pressure	Pascal	Pa	$kg \cdot m^{-1}s^{-2}$ or N/m^2
Energy	Joule	J	$kg \cdot m^2s^{-2}$ or $N \cdot m$
Electric charge	Coulomb	C	$A \cdot s$
Electric potential difference	Volt	V	$kg \cdot m^2s^{-3}A^{-1}$ or J/C

Conversion Factors for Other Units of Length

Metric System	English System
1 kilometer (km) $= 1 \times 10^3$ m	1 inch(in.) $= 2.54$ cm
1 centimeter (cm) $= 1 \times 10^{-2}$ m	1 yard(yd) $= 0.9144$ m
1 millimeter (mm) $= 1 \times 10^{-3}$ m	1 mile(mi) $= 1.6093$ km
1 nanometer (nm) $= 1 \times 10^{-9}$ m	1 m $= 1.0936$ yd
1 angstrom (Å) $= 1 \times 10^{-10}$ m	$= 39.370$ in.
$= 1 \times 10^{-8}$ cm	$= 3.2808$ ft
$= 1 \times 10^{-1}$ nm	1 km $= 0.6214$ mi

Conversion Factors for Other Units of Mass and Weight

Metric System	English System
1 gram(g) $= 1 \times 10^{-3}$ kg	1 pound(lb) $= 453.59$ g
1 milligram(mg) $= 1 \times 10^{-3}$ g	1 ounce(oz) $= 28.35$ g
$= 1 \times 10^{-6}$ kg	1 (short) ton $= 907.18$ kg
1 metric ton $= 1 \times 10^3$ kg	1 kg $= 2.2046$ lb

1 atomic mass unit (amu) $= 1.6606 \times 10^{-27}$ kg $= 1.6606 \times 10^{-24}$ g

---

Conversion Factors for Other Units of Energy

1 thermochemical calorie (cal) $= 4.184$ J (exactly)

1 erg $= 1 \times 10^{-7}$ J   1 J $= 1 \times 10^7$ erg

1 liter $\cdot$ atmosphere(L $\cdot$ atm) $= 101.325$ J $= 24.217$ cal

1 electron volt (eV) $= 1.6022 \times 10^{-19}$ J $= 1.6022 \times 10^{-12}$ erg

1 eV/particle is equivalent to 96.485 kJ/mol or 23.06 kcal/mol

1 million electron volts (MeV) $= 1.602189 \times 10^{-13}$ J

Definitions

One joule of energy is expended when a force of one newton is applied through a distance of one meter.

$$1 \, J = 1 \, N \cdot m$$

One calorie was originally defined as the amount of heat required to raise the temperature of one gram of water from 14.5 to 15.5 °C. The thermochemical calorie is very close to, but not exactly the same as, the original calorie.